M000266412

# Practical Business Statistics
## Sixth Edition

**Andrew F. Siegel**
Department of Information Systems
and Operations Management
Department of Finance and Business Economics
Department of Statistics
Michael G. Foster School of Business
University of Washington

AMSTERDAM • BOSTON • HEIDELBERG • LONDON
NEW YORK • OXFORD • PARIS • SAN DIEGO
SAN FRANCISCO • SINGAPORE • SYDNEY • TOKYO
Academic Press is an imprint of Elsevier

Academic Press is an imprint of Elsevier
30 Corporate Drive, Suite 400, Burlington, MA 01803, USA
The Boulevard, Langford Lane, Kidlington, Oxford, OX5 1GB, UK

© 2012 Andrew F. Siegel. Published by Elsevier Inc. All rights reserved.

No part of this publication may be reproduced or transmitted in any form or by any means, electronic or mechanical, including photocopying, recording, or any information storage and retrieval system, without permission in writing from the publisher. Details on how to seek permission, further information about the Publisher's permissions policies and our arrangements with organizations such as the Copyright Clearance Center and the Copyright Licensing Agency, can be found at our website: *www.elsevier.com/permissions*.

This book and the individual contributions contained in it are protected under copyright by the Publisher (other than as may be noted herein).

**Notices**
Knowledge and best practice in this field are constantly changing. As new research and experience broaden our understanding, changes in research methods, professional practices, or medical treatment may become necessary.

Practitioners and researchers must always rely on their own experience and knowledge in evaluating and using any information, methods, compounds, or experiments described herein. In using such information or methods they should be mindful of their own safety and the safety of others, including parties for whom they have a professional responsibility.

To the fullest extent of the law, neither the Publisher nor the authors, contributors, or editors assume any liability for any injury and/or damage to persons or property as a matter of products liability, negligence or otherwise, or from any use or operation of any methods, products, instructions, or ideas contained in the material herein.

**Library of Congress Cataloging-in-Publication Data**
Siegel, Andrew F.
  Practical business statistics / Andrew F. Siegel. – 6th ed.
    p. cm.
  Includes bibliographical references and index.
  ISBN 978-0-12-385208-3 (alk. paper)
  1. Industrial management–Statistical methods. I. Title.
  HD30.215.S57 2012
  519.5024'65–dc22                                    2010041182

**British Library Cataloguing-in-Publication Data**
A catalogue record for this book is available from the British Library.

For information on all Academic Press publications
visit our Web site at *www.elsevierdirect.com*

*Typeset by*: diacriTech, Chennai, India

Printed in the United States of America
11 12 13 14  9 8 7 6 5 4 3 2 1

Working together to grow
libraries in developing countries

www.elsevier.com | www.bookaid.org | www.sabre.org

ELSEVIER    BOOK AID International    Sabre Foundation

*To Ann, Bonnie, Clara, Michael, and Mildred*

# Contents in Brief

# Contents

## 4. Landmark Summaries: Interpreting Typical Values and Percentiles

## 5. Variability: Dealing with Diversity

## Part II
# Probability

## 6. Probability: Understanding Random Situations

# 12. Multiple Regression: Predicting One Variable from Several Others

# 13. Report Writing: Communicating the Results of a Multiple Regression

# 14. Time Series: Understanding Changes over Time

## Part V
## Methods and Applications

Statistical literacy has become a necessity for anyone in business, simply because your competition has already learned how to interpret numbers and how to measure many of the risks involved in this uncertain world. Can you afford to ignore the tons of data now available (to anyone) online when you are searching for a competitive, strategic advantage? We are not born with an intuitive ability to assess randomness or process massive data sets, but fortunately there are fundamental basic principles that let us compute, for example, the risk of a future payoff, the way in which the chances for success change as we continually receive new information, and the best information summaries from a data warehouse. This book will guide you through foundational activities, including how to collect data so that the results are useful, how to explore data to efficiently visualize its basic features, how to use mathematical models to help separate meaningful characteristics from noise, how to determine the *quality* of your summaries so that you are in a position to make judgments, and how to know when it would be better to ignore the set of data because it is indistinguishable from random noise.

## EXAMPLES

Examples bring statistics to life, making each topic relevant and useful. There are many real-world examples used throughout *Practical Business Statistics*, chosen from a wide variety of business sources, and many of them of current interest as of 2010 (take a look at the status of Facebook relative to other top websites in Chapter 11). The donations database, which gives characteristics of 20,000 individuals together with the amount that they contributed in response to a mailing, is introduced in Chapter 1 and used in many chapters to illustrate how statistical methods can be used for data mining. The stock market is used in Chapter 5 to illustrate volatility, risk, and diversification as measured by the standard deviation, while the *systematic* component of market risk is summarized by the regression coefficient in Chapter 11. Because we are all curious about the salaries of others, I have used top executive compensation in several examples and, yes, Enron was an outlier even before the company filed for bankruptcy and the CEO resigned. Quality control is used throughout the book to illustrate individual topics and is

also covered in its own chapter (18). Opinion surveys and election polls are used throughout the book (and especially in Chapter 9) because they represent a very pure kind of real-life statistical inference that we are all familiar with and use frequently in business. Using the Internet to locate data is featured in Chapter 2. Prices of magazine advertisements are used in Chapter 12 to show how multiple regression can uncover relationships in complex data sets, and we learn the value of a larger audience with a higher income simply by crunching the numbers. Microsoft's revenues and U.S. unemployment rates are used in Chapter 14 to demonstrate what goes on behind the scenes in time-series forecasting. Students learn better through the use of motivating examples and applications. All numerical examples are included in the Excel® files on the companion website, with ranges named appropriately for easy analysis.

## STATISTICAL GRAPHICS

To help show what is going on in the data sets, *Practical Business Statistics* includes over 200 figures to illustrate important features and relationships. The graphs are exact because they were initially drawn with the help of a computer. For example, the bell-shaped normal curves here are accurate, unlike those in many books, which are distorted because they appear to be an artist's enhancement of a casual, hand-drawn sketch. There is no substitute for accuracy!

## EXTENSIVE DEVELOPMENT: REVIEWS AND CLASS TESTING

This book began as a collection of readings I handed out to my students as a supplement to the assigned textbook. All of the available books seemed to make statistics seem unnecessarily difficult, and I wanted to develop and present straightforward ways to think about the subject. I also wanted to add more of a real-world business flavor to the topic. All of the helpful feedback I have received from students over the years has been acted upon and has improved the book. *Practical Business Statistics* has been through several stages of reviewing and classroom testing. Now that five editions have been used in colleges and universities across the country and around the world, preparing the sixth edition has given me the chance to fine-tune the book, based on the additional reviews and all the helpful, encouraging comments that I have received.

## WRITING STYLE

I enjoy writing. I have presented the "inside scoop" wherever possible, explaining how we statisticians *really* think about a topic, what it implies, and how it is useful. This approach helps bring some sorely needed life to a subject that unfortunately suffers from dreadful public relations. Of course, the traditional explanations are also given here so that you can see it both ways: here is what we say, and here is what it means, all the while maintaining technical rigor.

It thrilled me to hear even some of my more quantitative-phobic students tell me that the text is actually *enjoyable to read!* And this was *after* the final grades were in!

## CASES

To show how statistical thinking can be useful as an integrated part of a larger business activity, cases are included at the end of each of Chapters 3–12. These cases provide extended and open-ended situations as an opportunity for thought and discussion, often with no single correct answer.

## ORGANIZATION

The reader should always know *why* the current material is important. For this reason, each part begins with a brief look at the subject of that part and the chapters to come. Each chapter begins with an overview of its topic, showing why the subject is important to business, before proceeding to the details and examples.

Key words, the most important terms and phrases, are presented in bold in the sentence of the text where they are defined. They are collected in the Key Words list at the end of each chapter and also included in the glossary at the back of the book (hint! this could be very useful!). This makes it easy to study by focusing attention on the main ideas. An extensive index helps you find main topics as well as small details. Try looking up "examples," "correlation," "unpaired *t* test," or even "mortgage."

Extensive end-of-chapter materials are included, beginning with a *summary* of the important material covered. Next is the list of *key words*. The *questions* provide a review of the main topics, indicating why they are important. The *problems* give the student a chance to apply statistics to new situations. The *database exercises* (included in most chapters) give further practice problems based on the employee database in Appendix A. The *projects* bring statistics closer to the students' needs and interests by allowing them to help define the problem and choose the data set

from their work experience or interests from sources including the Internet, current publications, or their company. Finally, the *cases* (one each for Chapters 3–12) provide extended and open-ended situations as an opportunity for thought and discussion, often with no single correct answer.

Several special topics are covered in addition to the foundations of statistics and their applications to business. Data mining is introduced in Chapter 1 and carried throughout the book. Because communication is so important in the business world, Chapter 13 shows how to gather and present statistical material in a report. Chapter 14 includes an intuitive discussion of the Box–Jenkins forecasting approach to time series using ARIMA models. Chapter 18 shows how statistical methods can help you achieve and improve quality; discussion of quality control techniques is also interspersed throughout the text.

*Practical Business Statistics* is organized into five parts, plus appendices, as follows:

- Part I, Chapters 1 through 5, is "Introduction and Descriptive Statistics." Chapter 1 motivates by showing how the use of statistics provides a competitive edge in business and then outlines the basic activities of statistics and offers varied examples including data mining with large databases. Chapter 2 surveys the various types of data sets (quantitative, qualitative, ordinal, nominal, bivariate, time-series, etc.), the distinction between primary and secondary data, and use of the Internet. Chapter 3 shows how the histogram lets you see what's in the data set, which would otherwise be difficult to determine just from staring at a list of numbers. Chapter 4 covers the basic landmark summaries, including the average, median, mode, and percentiles, which are displayed in the box plot and the cumulative distribution function. Chapter 5 discusses variability, which often translates to *risk* in business terms, featuring the standard deviation as well as the range and coefficient of variation.

- Part II, including Chapters 6 and 7, is "Probability." Chapter 6 covers probabilities of events and their combinations, using probability trees both as a way of visualizing the situation and as an efficient method for computing probabilities. Conditional probabilities are interpreted as a way of making the best use of the information you have. Chapter 7 covers random variables (numerical outcomes), which often represent those numbers that are important to your business but are not yet available. Details are provided concerning general discrete distributions, the binomial distribution, the normal distribution, the Poisson distribution, and the exponential distribution.

- Part III, Chapters 8 through 10, is "Statistical Inference." These chapters pull together the descriptive summaries of Part I and the formal probability assessments of Part II, allowing you to reach probability conclusions about an unknown population based on a sample. Chapter 8 covers random sampling, which forms the basis for the exact probability statements of statistical inference and introduces the central limit theorem and the all-important notion of the standard error of a statistic. Chapter 9 shows how confidence intervals lead to an exact probability statement about an unknown quantity based on statistical data. Both two-sided and one-sided confidence intervals for a population mean are covered, in addition to prediction intervals for a new observation. Chapter 10 covers hypothesis testing, often from the point of view of distinguishing the presence of a real pattern from mere random coincidence. By building on the intuitive process of constructing confidence intervals from Chapter 9, hypothesis testing can be performed in a relatively painless intuitive manner while ensuring strict statistical correctness (I learned about this in graduate school and was surprised to learn that it was not yet routinely taught in introductory courses—why throw away the intuitive confidence interval just as we are starting to test hypotheses?)

- Part IV, Chapters 11 through 14, is "Regression and Time Series." These chapters apply the concepts and methods of the previous parts to more complex and more realistic situations. Chapter 11 shows how relationships can be studied and predictions can be made using correlation and regression methods on bivariate data. Chapter 12 extends these ideas to multiple regression, perhaps the most important method in statistics, with careful attention to interpretation, diagnostics, and the idea of "controlling for" or "adjusting for" some factors while measuring the effects of other factors. Chapter 13 provides a guide to report writing (with a sample report) to help the student communicate the results of a multiple regression analysis to other business people. Chapter 14 introduces two of the most important methods that are needed for time-series analysis. The trend-seasonal approach is used to give an intuitive feeling for the basic features of a time series, while Box–Jenkins models are covered to show how these complex and powerful methods can handle more difficult situations.
- Part V, Chapters 15 through 18, is "Methods and Applications," a grab bag of optional, special topics that extend the basic material covered so far. Chapter 15 shows how the analysis of variance allows you to use hypothesis testing in more complex situations, especially involving categories along with numeric data. Chapter 16 covers nonparametric methods, which can be used when the basic assumptions for statistical inference are not satisfied, that is, for cases where the distributions might not be normal or the data set might be merely ordinal. Chapter 17 shows how chi-squared analysis can be used to test relationships among the

categories of nominal data. Finally, Chapter 18 shows how quality control relies heavily on statistical methods such as Pareto diagrams and control charts.

- Appendix A is the "Employee Database," consisting of information on salary, experience, age, gender, and training level for a number of administrative employees. This data set is used in the *database exercises* section at the end of most chapters. Appendix B describes the donations database on the companion website (giving characteristics of 20,000 individuals together with the amount that they contributed in response to a mailing) that is introduced in Chapter 1 and used in many chapters to illustrate how statistical methods can be used for data mining. Appendix C gives detailed solutions to selected parts of problems and database exercises (marked with an asterisk in the text). Appendix D collects all of the statistical tables used throughout the text.

## POWERPOINT SLIDES

A complete set of PowerPoint slides, that I developed for my own classes, is available on the companion website.

## EXCEL® GUIDE

The Excel® Guide, prepared by me (and I have enjoyed spreadsheet computing since its early days) provides examples of statistical analysis using Excel® using data taken chapter-by-chapter from *Practical Business Statistics*. It's a convenient way for students to learn how to use computers if your class is using Excel®.

## COMPANION WEBSITE

The companion website http://www.elsevierdirect.com includes the PowerPoint presentation slides, the Excel Guide, and Excel files with all quantitative examples and problem data.

## INSTRUCTOR'S MANUAL

The instructor's manual is designed to help save time in preparing lectures. A brief discussion of teaching objectives and how to motivate students is provided for each chapter. Also included are detailed solutions to questions, problems, and database exercises, as well as analysis and discussion material for each case. The instructor's manual is available at the companion website.

## ACKNOWLEDGMENTS

Many thanks to all of the reviewers and students who have read and commented on drafts and previous editions of *Practical Business Statistics* over the years. I have been lucky to have dedicated, careful readers at a variety of institutions who were not afraid to say what it would take to meet their needs.

I am fortunate to have been able to work with my parents, Mildred and Armand Siegel, who provided many careful and detailed suggestions for the text.

Very special thanks go to Lauren Schultz Yuhasz, Lisa Lamenzo, Gavin Becker, and Jeff Freeland, who have been very helpful and encouraging with the development and production of this edition. Warm thanks go to Michael Antonucci, who started this whole thing when he stopped by my office to talk about computers and see what I was up to and encourage me to write it all down. I am also grateful to those who were involved with previous editions, including Scott Isenberg, Christina Sanders, Catherine Schultz, Richard T. Hercher, Carol Rose, Gail Korosa, Ann Granacki, Colleen Tuscher, Adam Rooke, Ted Tsukahara, and Margaret Haywood. It's a big job producing a work like this, and I was lucky to have people with so much knowledge, dedication, and organizational skill.

Thanks also go out to David Auer, Eric Russell, Dayton Robinson, Eric J. Bean, Michael R. Fancher, Susan Stapleton, Sara S. Hemphill, Nancy J. Silberg, A. Ronald Hauver, Hirokuni Tamura, John Chiu, June Morita, Brian McMullen, David B. Foster, Pablo Ferrero, Rolf R. Anderson, Gordon Klug, Reed Hunt, E. N. Funk, Rob Gullette, David Hartnett, Mickey Lass, Judyann Morgan, Kimberly V. Orchard, Richard Richings, Mark Roellig, Scott H. Pattison, Thomas J. Virgin, Carl Stork, Gerald Bernstein, and Jeremiah J. Sullivan.

A special mention is given to a distinguished group of colleagues who have provided helpful guidance, including Bruce Barrett, University of Alabama; Brian Goff, Western Kentucky University; Anthony Seraphin, University of Delaware; Abbott Packard, Hawkeye Community College; William Seaver, University of Tennessee–Knoxville; Nicholas Jewell, University of California–Berkeley; Howard Clayton, Auburn University; Giorgio Canarello, California State University–Los Angeles; Lyle Brenner, University of Florida–Gainesville; P. S. Sundararaghavan, University of Toledo; Julien Bramel, Columbia University, Ronald Bremer, Texas Tech University; Stergios Fotopoulos, Washington State University; Michael Ghanen, Webster University; Phillip Musa, Texas Tech University; Thomas Obremski, University of Denver; Darrell Radson, University of Wisconsin, Milwaukee; Terrence Reilly, Babson College; Peter Schuhmann, University of Richmond; Bala Shetty, Texas A&M University; L. Dwight Sneathen Jr., University of Arizona; Ted Tsukahara, St. Mary's College; Edward A. Wasil, American University; Michael Wegmann, Keller Graduate School of Management; Mustafa Yilmaz,

Northeastern University; Gary Yoshimoto, St. Cloud State University; Sangit Chatterjee, Northeastern University; Jay Devore, California Polytechnic State University; Burt Holland, Temple University; Winston Lin, State University of New York at Buffalo; Herbert Spirer, University of Connecticut; Donald Westerfield; Webster University; Wayne Winston, Indiana University; Jack Yurkiewicz, Pace University; Betty Thorne, Stetson University; Dennis Petruska, Youngstown State University; H. Karim, West Coast University; Martin Young, University of Michigan; Richard Spinetto, University of Colorado at Boulder; Paul Paschke, Oregon State University; Larry Ammann, University of Texas at Dallas; Donald Marx, University of Alaska; Kevin Ng, University of Ottawa; Rahmat Tavallali, Walsh University; David Auer, Western Washington University; Murray Cote, Texas A&M University; Peter Lakner, New York University; Donald Adolphson, Brigham Young University; and A. Rahulji Parsa, Drake University.

## TO THE STUDENT

As you begin this course, you may have some preconceived notions of what statistics is all about. If you have positive notions, please keep them and share them with your classmates. But if you have negative notions, please set them aside and remain open-minded until you've given statistics another chance to prove its value in analyzing business risk and providing insight into piles of numbers.

In some ways, statistics is easier for your generation than for those of the past. Now that computers can do the messy numerical work, you are free to develop a deeper understanding of the concepts and how they can help you compete over the course of your business career.

Make good use of the introductory material so that you will always know why statistics is worth the effort. Focus on examples to help with understanding and motivation. Take advantage of the summary, key words, and other materials at the ends of the chapters. Don't forget about the detailed problem solutions and the glossary at the back when you need a quick reminder! And don't worry. Once you realize how much statistics can help you in business, the things you need to learn will fall into place much more easily.

Why not keep this book as a reference? You'll be glad you did when the boss needs you to draft a memo immediately that requires a quick look at some data or a response to an adversary's analysis. With the help of *Practical Business Statistics* on your bookshelf, you'll be able to finish early and still go out to dinner. *Bon appétit!*

ANDREW F. SIEGEL

Andrew F. Siegel is Professor, Departments of ISOM (Information Systems and Operations Management) and Finance, at the Michael G. Foster School of Business, University of Washington, Seattle. He is also Adjunct Professor in the Department of Statistics. He has a Ph.D. in statistics from Stanford University (1977), an M.S. in mathematics from Stanford University (1975), and a B.A. in mathematics and physics summa cum laude with distinction from Boston University (1973). Before settling in Seattle, he held teaching and/or research positions at Harvard University, the University of Wisconsin, the RAND Corporation, the Smithsonian Institution, and Princeton University. He has also been a visiting professor at the University of Burgundy at Dijon, France, at the Sorbonne in Paris, and at HEC Business School near Paris. The very first time he taught statistics in a business school (University of Washington, 1983) he was granted the Professor of the Quarter award by the MBA students. He was named the Grant I. Butterbaugh Professor beginning in 1993; this endowed professorship was created by a highly successful executive in honor of Professor Butterbaugh, a business statistics teacher. (Students: Perhaps you will feel this way about your teacher 20 years from now.) Other honors and awards include Burlington Northern Foundation Faculty Achievement Awards, 1986 and 1992; Research Associate, Center for the Study of Futures Markets, Columbia University, 1988; Excellence in Teaching Awards, Executive MBA Program, University of Washington, 1986 and 1988; Research Opportunities in Auditing Award, Peat Marwick Foundation, 1987; and Phi Beta Kappa, 1973.

He belongs to the American Statistical Association, where he has served as Secretary–Treasurer of the Section on Business and Economic Statistics. He has written three other books: *Statistics and Data Analysis: An Introduction* (Second Edition, Wiley, 1996, with Charles J. Morgan), *Counterexamples in Probability and Statistics* (Wadsworth, 1986, with Joseph P. Romano), and *Modern Data Analysis* (Academic Press, 1982, co-edited with Robert L. Launer). His articles have appeared in many publications, including the *Journal of the American Statistical Association*, the *Journal of Business, Management Science*, the *Journal of Finance*, the *Encyclopedia of Statistical Sciences*, the *American Statistician*, the *Review of Financial Studies, Proceedings of the National Academy of Sciences of the United States of America*, the *Journal of Financial and Quantitative Analysis, Nature*, the *Journal of Portfolio Management*, the *American Mathematical Monthly*, the *Journal of the Royal Statistical Society*, the *Annals of Statistics*, the *Annals of Probability*, the *Society for Industrial and Applied Mathematics Journal on Scientific and Statistical Computing, Statistics in Medicine, Genomics*, the *Journal of Computational Biology, Genome Research, Biometrika, Journal of Bacteriology, Statistical Applications in Genetics and Molecular Biology, Discourse Processes, Auditing: A Journal of Practice and Theory, Contemporary Accounting Research*, the *Journal of Futures Markets*, and the *Journal of Applied Probability*. His work has been translated into Chinese and Russian. He has consulted in a variety of business areas, including election predictions for a major television network, statistical algorithms in speech recognition for a prominent research laboratory, television advertisement testing for an active marketing firm, quality control techniques for a supplier to a large manufacturing company, biotechnology process feasibility and efficiency for a large-scale laboratory, electronics design automation for a Silicon Valley startup, and portfolio diversification analysis for a fund management company.

# Introduction and Descriptive Statistics

Welcome to the world of statistics. This is a world you will want to get comfortable with because you will make better management decisions when you know how to assess the available information and how to ask for additional facts as needed. How else can you expect to manage 12 divisions, 683 products, and 5,809 employees? And even for a small business, you will need to understand the larger business environment of potential customers and competitors it operates within. These first five chapters will introduce you to the role of statistics and data mining in business management (Chapter 1) and to the various types of data sets (Chapter 2). Summaries help you see the "big picture" that might otherwise remain obscured in a collection of data. Chapter 3 will show you a good way to see the basic facts about a list of numbers—by looking at a *histogram*. Fundamental summary numbers (such as the average, median, percentiles, etc.) will be explained in Chapter 4. One reason statistical methods are so important is that there is so much *variability* out there that gets in the way of the message in the data. Chapter 5 will show you how to measure the extent of this diversity problem.

## Chapter 1

# Introduction

## Defining the Role of Statistics in Business

A business executive must constantly make decisions under pressure, often with only incomplete and imperfect information available. Naturally, whatever information is available must be utilized to the fullest extent possible. *Statistical analysis* helps extract information from data and provides an indication of the quality of that information. *Data mining* combines statistical methods with computer science and optimization in order to help businesses make the best use of the information contained in large data sets. *Probability* helps you understand risky and random events and provides a way of evaluating the likelihood of various potential outcomes.

Even those who would argue that business decision making should be based on expert intuition and experience (and therefore should not be overly quantified) must admit that all available relevant information should be considered. Thus, statistical techniques should be viewed as an important part of the decision process, allowing informed strategic decisions to be made that combine executive intuition with a thorough understanding of the facts available. This is a powerful combination.

We will begin with an overview of the competitive advantage provided by a knowledge of statistical methods, followed by some basic facts about statistics and probability and their role in business.

## 1.1 WHY STATISTICS?

Is knowledge of statistics really necessary to be successful in business? Or is it enough to rely on intuition, experience, and hunches? Let's put it another way: Do you really want to ignore much of the vast potentially useful information out there that comes in the form of data?

## Why Should You Learn Statistics?

By learning statistics, you acquire the competitive advantage of being comfortable and competent around data and uncertainty. A vast amount of information is contained in data, but this information is often not immediately accessible—statistics helps you extract and understand this information. A great deal of skill goes into creating strategy from knowledge, experience, and intuition. Statistics helps you deal with the knowledge component, especially when this knowledge is in the form of numbers, by answering questions such as, To what extent should you really believe these figures and their implications? and, How should we summarize this mountain of data? By using statistics to acquire knowledge, you will add to the value of your experience and intuition, ultimately resulting in better decision making.

You won't be able to avoid statistics. These methods are already used routinely throughout the corporate world, and the lower cost of computers is increasing your need to be able to make decisions based on quantitative information.

## Is Statistics Difficult?

Statistics is no more difficult than any other field of study. Naturally, some hard work is needed to achieve understanding of the general ideas and concepts. Although some attention to details and computations is necessary, it is much easier to become an expert *user* of statistics than it is to become an expert statistician trained in all of the fine details. Statistics is easier than it used to be now that personal computers can do the repetitive number-crunching tasks, allowing you to concentrate on interpreting the results and their meaning. Although a few die-hard purists may bemoan the decline of technical detail in statistics teaching, it is good to see that these details are now in their proper place; life is too short for all human beings to work out the intricate details of techniques such as long division and matrix inversion.

## Does Learning Statistics Decrease Your Decision-Making Flexibility?

Knowledge of statistics *enhances* your ability to make good decisions. Statistics is not a rigid, exact science and should not get in the way of your experience and intuition. By learning about data and the basic properties of uncertain events, you will help solidify the information on which your decisions are based, and you will add a new dimension to your intuition. Think of statistical methods as a component of decision making, but not the whole story. You want to supplement—not replace—business experience, common sense, and intuition.

## 1.2 WHAT IS STATISTICS?

**Statistics** is the art and science of collecting and understanding data. Since *data* refers to any kind of recorded information, statistics plays an important role in many human endeavors.

## Statistics Looks at the Big Picture

When you have a large, complex assemblage of many small pieces of information, statistics can help you classify and analyze the situation, providing a useful overview and summary of the fundamental features in the data. If you don't yet have the data, then statistics can help you collect them, ensuring that your questions can be answered and that you spend enough (but not too much) effort in the process.

## Statistics Doesn't Ignore the Individual

If used carefully, statistics pays appropriate attention to all individuals. A complete and careful statistical analysis will summarize the general facts that apply to everyone and *will also alert you to any exceptions*. If there are special cases in the data that are not adequately summarized in the "big picture," the statistician's job is not yet complete. For example, you may read that in 2008 the average U.S. household size was 2.56 people.[1] Although this is a useful statistic, it doesn't come close to giving a complete picture of the size of all households in the United States. As you will see, statistical methods can easily be used to describe the entire distribution of household sizes.

### Example
#### Data in Management

Data sets are very common in management. Here is a short list of kinds of everyday managerial information that are, in fact, data:

1. Financial statements (and other accounting numbers).
2. Security prices and volumes and interest rates (and other investment information).
3. Money supply figures (and other government announcements).
4. Sales reports (and other internal records).
5. Market survey results (and other marketing data).
6. Production quality measures (and other manufacturing records).
7. Human resource productivity records (and other internal databases).
8. Product price and quantity sold (and other sales data).
9. Publicity expenditures and results (and other advertising information).

Think about it. Probably much of what you do depends at least indirectly on data. Perhaps someone works for you and advises you on these matters, but you rarely see the actual data. From time to time, you might ask to see the "raw data" in order to keep some perspective. Looking at data and asking some questions about them may reveal surprises: You may find out that the quality of the data is not as high as you had thought (you mean that's what we base our forecasts on?), or you may find out the opposite and be reassured. Either way, it's worthwhile.

## Looking at Data

What do you see when you look hard at tables of data (for example, the back pages of the *Wall Street Journal*)? What does a professional statistician see? The surprising answer

---

1. U.S. Census Bureau, Statistical Abstract of the United States: 2010 (129th Edition) Washington, DC, 2009; http://www.census.gov/statab/www/, Table 59, accessed June 29, 2010.

to both of these questions often is, Not much. You've got to go to work on the numbers—draw pictures of them, compute summaries from them, and so on—before their messages will come through. This is what professional statisticians do; they find this much easier and more rewarding than staring at large lists of numbers for long periods of time. So don't be discouraged if a list of numbers looks to you like, well, a list of numbers.

## Statistics in Management

What should a manager know about statistics? Your knowledge should include a broad overview of the basic concepts of statistics, with some (but not necessarily all) details. You should be aware that the world is random and uncertain in many aspects. Furthermore, you should be able to effectively perform two important activities:

1. Understand and use the results of statistical analysis as background information in your work.
2. Play the appropriate leadership role during the course of a statistical study if you are responsible for the actual data collection and/or analysis.

To fulfill these roles, you do not need to be able to perform a complex statistical analysis by yourself. However, some experience with actual statistical analysis is essential for you to obtain the perspective that leads to effective interpretation. Experience with actual analysis will also help you to lead others to sound results and to understand what they are going through. Moreover, there may be times when it will be most convenient for you to do some analysis on your own. Thus, we will concentrate on the ideas and concepts of statistics, reinforcing these with practical examples.

## 1.3 THE FIVE BASIC ACTIVITIES OF STATISTICS

In the beginning stages of a statistical study, either there are not yet any data or else it has not yet been decided what data to look closely at. The *design* phase will resolve these issues so that useful data will result. Once data are available, an initial inspection is called for, provided by the *exploratory* phase. In the *modeling phase*, a system of assumptions and equations is selected in order to provide a framework for further analysis. A numerical summary of an unknown quantity, based on data, is the result of the *estimation* process. The last of these basic activities is *hypothesis testing*, which uses the data to help you decide what the world is really like in some respect. We will now consider these five activities in turn.

## Designing a Plan for Data Collection

Designing a plan for data collection might be called *sample survey design* for a marketing study or *experimental design*

for a chemical manufacturing process optimization study. This phase of **designing the study** involves planning the details of data gathering. A careful design can avoid the costs and disappointment of finding out—too late—that the data collected are not adequate to answer the important questions. A good design will also collect just the right amount of data: enough to be useful, but not so much as to be wasteful. Thus, by planning ahead, you can help ensure that the analysis phase will go smoothly and hold down the cost of the project.

Statistics is particularly useful when you have a large group of people, firms, or other items (the *population*) that you would like to know about but can't reasonably afford to investigate completely. Instead, to achieve a useful but imperfect understanding of this population, you select a smaller group (the *sample*) consisting of some—but not all—of the items in the population. The process of generalizing from the observed sample to the larger population is known as *statistical inference*. The *random sample* is one of the best ways to select a practical sample, to be studied in detail, from a population that is too large to be examined in its entirety.[2] By selecting randomly, you accomplish two goals:

1. You are guaranteed that the selection process is fair and proceeds without bias; that is, all items have an equal chance of being selected. This assures you that, on average, samples will be representative of the population (although each particular random sample is usually only approximately, and not perfectly, representative).
2. The randomness, introduced in a controlled way during the design phase of the project, will help ensure validity of the statistical inferences drawn later.

## Exploring the Data

As soon as you have a set of data, you will want to check it out. **Exploring the data** involves looking at your data set from many angles, describing it, and summarizing it. In this way you will be able to make sure that the data are really what they are claimed to be and that there are no obvious problems.[3] But good exploration also prepares you for the formal analysis in either of two ways:

1. By verifying that the expected relationships actually exist in the data, thereby validating the planned techniques of analysis.
2. By finding some unexpected structure in the data that must be taken into account, thereby suggesting some changes in the planned analysis.

---

2. Details of random sampling will be presented in Chapter 8.
3. Data exploration is used throughout the book, where appropriate, and especially in Chapters 3, 4, 11, 12, and 14.

Exploration is the first phase once you have data to look at. It is often not enough to rely on a formal, automated analysis, which can be only as good as the data that go into the computer and which assumes that the data set is "well behaved." Whenever possible, examine the data directly to make sure they look OK; that is, there are no large errors, and the relationships observable in the data are appropriate to the kind of analysis to be performed. This phase can help in (1) editing the data for errors, (2) selecting an appropriate analysis, and (3) validating the statistical techniques that are to be used in further analysis.

## Modeling the Data

In statistics, a **model** is a system of assumptions and equations that can generate artificial data similar to the data you are interested in, so that you can work with a few numbers (called *parameters*) that represent the important aspects of the data. A model can be a very effective system within which questions about large-scale properties of the data can be answered.

Having the additional structure of a statistical model can be important for the next two activities of estimation and hypothesis testing. We often try to explore the data before deciding on the model, so that you can discover whatever structure—whether expected or unexpected—is actually in the data. In this way, data exploration can help you with modeling. Often, a model says that

data equals structure plus random noise

For example, with a data set of 3,258 numbers, a model with a single parameter representing "average additional sales dollars generated per dollar of advertising expense" could help you study advertising effectiveness by adjusting this parameter until the model produces artificial data similar to the real data. Figure 1.3.1 illustrates how a model, with useful parameters, can be made to match a real data set.

Here are some models that can be useful in analyzing data. Notice that each model generates data with the general approach "data equals structure plus noise," specifying the structure in different ways. In selecting a model, it can be very useful to consider what you have learned by exploring the data.

1. Consider a simple model that generates artificial data consisting of a *single number* plus noise. Chapter 4 (landmark summaries) shows how to extract information about the single number, while Chapter 5 (variability) shows how to describe the noise.
2. Consider a model that generates *pairs* of artificial noisy data values that are related to each other. Chapters 11 and 12 (correlation, regression, and multiple regression) show some useful models for describing the nature and extent of the relationship and the noise.
3. Consider a model that generates a *series* of noisy data values where the next one is related to the previous one. Chapter 14 (time series) presents two systems of models that have been useful in working with business time series data.

## Estimating an Unknown Quantity

**Estimating an unknown quantity** produces the best educated guess possible based on the available data. We all want (and often need) estimates of things that are just plain impossible to know exactly. Here are some examples of unknowns to be estimated:

1. Next quarter's sales.
2. What the government will do next to our tax rates.
3. How the population of Seattle will react to a new product.
4. How your portfolio of investments will fare next year.
5. The productivity gains of a change in strategy.
6. The defect rate in a manufacturing process.

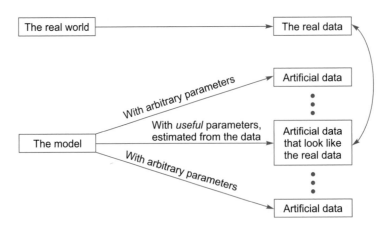

**FIGURE 1.3.1**   A model is a system of assumptions and equations that can generate artificial data. When you carefully choose the parameters of the model, the artificial data (from the model) can be made to match the real data, and these useful parameters help you understand the real situation.

7. The winners in the next election.
8. The long-term health effects of computer screens.

Statistics can shed light on some of these situations by producing a good, educated guess when reliable data are available. Keep in mind that all statistical estimates are just guesses and are, consequently, often wrong. However, they will serve their purpose when they are close enough to the unknown truth to be useful. If you knew how accurate these estimates were (approximately), you could decide how much attention to give them.

Statistical estimation also provides an indication of the amount of uncertainty or error involved in the guess, accounting for the consequences of random selection of a sample from a large population. The *confidence interval* gives probable upper and lower bounds on the unknown quantity being estimated, as if to say, I'm not sure exactly what the answer is, but I'm quite confident it's between these two numbers.

You should routinely expect to see confidence intervals (and ask for them if you don't) because they show you how reliable an estimated value actually is. For example, there is certainly some information in the statement that sales next quarter are expected to be

$11.3 million

However, additional and deeper understanding comes from also being told that you are 95% confident that next quarter's sales will be

between $5.9 million and $16.7 million

The confidence interval puts the estimate in perspective and helps you avoid the tendency to treat a single number as very precise when, in fact, it might not be precise at all.[4]

## Hypothesis Testing

Statistical **hypothesis testing** is the use of data in deciding between two (or more) different possibilities in order to resolve an issue in an ambiguous situation. Hypothesis testing produces a definite decision about which of the possibilities is correct, based on data. The procedure is to collect data that will help decide among the possibilities and to use careful statistical analysis for extra power when the answer is not obvious from just glancing at the data.[5]

Here are some examples of hypotheses that might be tested using data:

1. The average New Yorker plans to spend at least $10 on your product next month.
2. You will win tomorrow's election.

3. A new medical treatment is safe and effective.
4. Brand X produces a whiter, brighter wash.
5. The error in a financial statement is smaller than some material amount.
6. It is possible to predict the stock market based on careful analysis of the past.
7. The manufacturing defect rate is below that expected by customers.

Note that each hypothesis makes a definite statement, and it may be either true or false. The result of a statistical hypothesis test is the conclusion that either the data support the hypothesis or they don't.

Often, statistical methods are used to decide whether you can rule out "pure randomness" as a possibility. For example, if a poll of 300 people shows that 53% plan to vote for you tomorrow, can you conclude that the election will go in your favor? Although many issues are involved here, we will (for the moment) ignore details, such as the (real) possibility that some people will change their minds between now and tomorrow, and instead concentrate only on the element of randomness (due to the fact that you can't call and ask every voter's preference). In this example, a careful analysis would reveal that it is a real possibility that less than 50% of voters prefer you and that the 53% observed is within the range of the expected random variation.

### Example
#### Statistical Quality Control

Your manufacturing processes are not perfect (nobody's are), and every now and then a product has to be reworked or tossed out. Thank goodness for your inspection team, which keeps these bad pieces from reaching the public. Meanwhile, however, you're losing lots of money manufacturing, inspecting, fixing, and disposing of these problems. This is why so many firms have begun using statistical quality control.

To simplify the situation, consider your assembly line to be *in control* if it produces similar results over time that are within the required specifications. Otherwise, your line will be considered to be *out of control*. Statistical methods help you monitor the production process so that you can save money in three ways: (1) keep the monitoring costs down, (2) detect problems quickly so that waste is minimized, and (3) whenever possible, don't spend time fixing it if it's not broken. Following is an outline of how the five basic activities of statistics apply to this situation.

During the design phase, you have to decide *what* to measure and *how often* to measure it. You might decide to select a random sample of 5 products to represent every batch of 500 produced. For each one sampled, you might have someone (or something) measure its length and width

*(Continued)*

---

4. Details of confidence intervals will be presented in Chapter 9 and used in Chapters 9–15.
5. Details of hypothesis testing will be presented in Chapter 10 and used in Chapters 10–18.

**Example—cont'd**

as well as inspect it visually for any obvious flaws. The result of the design phase is a plan for the early detection of problems. The plan must work in *real time* so that problems are discovered immediately, not next week.

Data exploration is accomplished by plotting the measured data on *quality control charts* and looking for patterns that suggest trouble. By spotting trends in the data, you may even be able to anticipate and fix a problem before any production is lost!

In the modeling phase, you might choose a standard statistical model, asserting that the observed measurements fluctuate randomly about a long-term average. Such a model then allows you to estimate both the long-term average and the amount of randomness, and then to test whether these values are acceptable.

Statistical estimation can provide management with useful answers to questions about how the production process is going. You might assign a higher grade of quality to the production when it is well controlled within precise limits; such high-grade items command a higher price. Estimates of the quality grade of the current production will be needed to meet current orders, and forecasting of future quality grades will help with strategic planning and pricing decisions.

Statistical hypothesis testing can be used to answer the important question: Is this process in control, or has it gone out of control? Because a production process can be large, long, and complicated, you can't always tell just by looking at a few machines. By making the best use of the statistical information in your data, you hope to achieve two goals. First, you want to detect when the system has gone out of control even before the quality has become unacceptable. Second, you want to minimize the "false alarm" rate so that you're not always spending time and money trying to fix a process that is really still in control.

**Example**
*A New Product Launch*

Deciding whether or not to launch a new product is one of the most important decisions a company makes, and many different kinds of information can be helpful along the way. Much of this information comes from statistical studies. For example, a marketing study of the target consumer group could be used to estimate how many people would buy the product at each of several different prices. Historical production-cost data for similar items could be used to assess how much it would cost to manufacture. Analysis of past product launches, both successful and unsuccessful, could provide guidance by indicating what has worked (and failed) in the past. A look at statistical profiles of national and international firms with similar products will help you size up the nature of possible competition. Individual advertisements could be tested on a sample of viewers to assess consumer

reaction before spending large amounts on a few selected advertisements.

The five basic activities of statistics show up in many ways. Because the population of consumers is too large to be examined completely, you could *design* a study, choosing a sample to represent the population (e.g., to look at consumer product purchase decisions, or for reactions to specific advertisements). Data *exploration* could be used throughout, wherever there are data to be explored, in order to learn about the situation (for example, are there separate groups of customers, suggesting market segmentation?) and as a routine check before other statistical procedures are used. A variety of statistical *models* could be chosen, adapted to specific tasks. One model might include parameters that relate consumer characteristics to their likelihood of purchase, while another model might help in forecasting future economic conditions at the projected time of the launch. Many *estimates* would be computed, for example, indicating the potential size of the market, the likely initial purchase rate, and the cost of production. Finally, various *hypothesis tests* could be used, for example, to tell whether there is sufficient consumer interest to justify going ahead with the project or to decide whether one ad is measurably better (instead of just randomly better) than another in terms of consumer reaction.

## 1.4 DATA MINING

Most companies routinely collect data—at the cash register for each purchase, on the factory floor from each step of production, or on the Internet from each visit to its website—resulting in huge databases containing potentially useful information about how to increase sales, how to improve production, or how to turn mouse clicks into purchases. **Data mining** is a collection of methods for obtaining useful knowledge by analyzing large amounts of data, often by searching for hidden patterns. Once a business has collected information for some purpose, it would be wasteful to leave it unexplored when it might be useful in many other ways. The goal of data mining is to obtain value from these vast stores of data, in order to improve the company with higher sales, lower costs, and better products. Here are just a few of the many areas of business in which data mining can be helpful:

1. *Marketing and sales:* Companies have lots of information about past contacts with potential customers and their results. These data can be mined for guidance on how (and when) to better reach customers in the future. One example is the difficult decision of when a store should reduce prices: reduce too soon and you lose money (on items that might have been sold for more); reduce too late and you may be stuck (with

items no longer in season). As reported in the *Wall Street Journal:*[6]

*A big challenge: trying to outfox customers who have been more willing to wait and wait for a bargain.... The stores analyze historical sales data to pinpoint just how long to hold out before they need to cut a price—and by just how much.... The technology, still fairly new and untested, requires detailed and accurate sales data to work well.*

Another example is the supermarket affinity card, allowing the company to collect data on every purchase, while knowing your mailing address! This could allow personalized coupon books to be sent, for example, if no peanut butter had been purchased for two months by a customer who usually buys some each month.

2. *Finance:* Mining of financial data can be useful in forming and evaluating investment strategies and in hedging (or reducing) risk. In the stock markets alone, there are many companies: about 3,298 listed on the New York Stock Exchange and about 2,942 companies listed on the NASDAQ Stock Market.[7] Historical information on price and volume (number of shares traded) is easily available (for example, at http://finance.yahoo.com) to anyone interested in exploring investment strategies. Statistical methods, such as hypothesis testing, are helpful as part of data mining to distinguish random from systematic behavior because stocks that performed well last year will not necessarily perform well next year. Imagine that you toss 100 coins six times each and then carefully choose the one that came up "heads" all six times—this coin is not as special as it might seem!

3. *Product design:* What particular combinations of features are customers ordering in larger-than-expected quantities? The answers could help you create products to appeal to a group of potential customers who would not take the trouble to place special orders.

4. *Production:* Imagine a factory running 24/7 with thousands of partially completed units, each with its bar code, being carefully tracked by the computer system, with efficiency and quality being recorded as well. This is a tremendous source of information that can tell you about the kinds of situations that cause trouble (such as finding a machine that needs adjustment by noticing clusters of units that don't work) or the kinds of situations that lead to extra-fast production of the highest quality.

5. *Fraud detection:* Fraud can affect many areas of business, including consumer finance, insurance, and networks (including telephone and the Internet). One of the best methods of protection involves mining data to distinguish between ordinary and fraudulent patterns of usage, then using the results to classify new transactions, and looking carefully at suspicious new occurrences to decide whether or not fraud is actually involved. I once received a telephone call from my credit card company asking me to verify recent transactions—identified by its statistical analysis—that departed from my typical pattern of spending. One fraud risk identification system that helps detect fraudulent use of credit cards is Falcon Fraud Manager from Fair Isaac, which uses the flexible "neural network" data-mining technique, related to the methods of Chapter 12. Consider:[8]

*Falcon manages 65 percent of card accounts worldwide, including 90 percent of credit cards in the U.S. Falcon reviews card transactions and "scores" them based on their likelihood of being fraudulent, enabling card issuers to stop losses faster and to react dynamically to changing fraud activity in real time. Falcon's fraud detection is based on innovative neural network models that are "trained" on large sets of consortium data.... The neural network models search through masses of data to identify very subtle signs of fraud. The size and diversity of the data are critical factors in the power of the models. We have created a fraud consortium that includes information on 1.8 billion card accounts, contributed by lenders that subscribe to the Falcon product.*

Data mining is a large task that involves combining resources from many fields. Here is how statistics, computer science, and optimization are used in data mining:

- *Statistics:* All of the basic activities of statistics are involved: a design for collecting the data, exploring for patterns, a modeling framework, estimation of features, and hypothesis testing to assess significance of patterns as a "reality check" on the results. Nearly every method in the rest of this book has the potential to be useful in data mining, depending on the database and the needs of the company. Some specialized statistical methods are particularly useful, including *classification analysis* (also called *discriminant analysis*) to assign a new case to a category (such as "likely purchaser" or "fraudulent"), *cluster analysis* to identify homogeneous groups of individuals, and *prediction analysis* (also called *regression analysis*).

6. A. Merrick, "Priced to Move: Retailers Try to Get Leg Up on Markdowns with New Software," *The Wall Street Journal*, August 7, 2001, p. A1.

7. Information accessed at http://www.nasdaq.com/screening/company-list.aspx on June 29, 2010.

8. M. N. Greene, "Divided We Fall: Fighting Payments Fraud Together," *Economic Perspectives*, 2009, Vol. 33, pp. 37–42.

- *Computer science:* Efficient algorithms (computer instructions) are needed for collecting, maintaining, organizing, and analyzing data. Creative methods involving *artificial intelligence* are useful, including *machine learning* techniques for prediction analysis such as *neural networks* and *boosting*, to learn from the data by identifying useful patterns automatically. Some of these methods from computer science are closely related to statistical prediction analysis.

- *Optimization:* These methods help you achieve a goal, which might be very specific such as maximizing profits, lowering production cost, finding new customers, developing profitable new product models, or increasing sales volume. Alternatively, the goal might be more vague such as obtaining a better understanding of the different types of customers you serve, characterizing the differences in production quality that occur under different circumstances, or identifying relationships that occur more or less consistently throughout the data. Optimization is often accomplished by *adjusting the parameters of a model* until the objective is achieved.

### Example

*Mining U.S. Neighborhood Data for Potential Customers*

Ideally, when deciding where to locate a new store, restaurant, or factory, or where your company should send its catalog, you would want to look everywhere in the whole country (perhaps even beyond) before deciding. A tremendous amount of information is collected, both by the government and by private companies, on the characteristics of neighborhoods across the United States. The U.S. Census

Bureau, a government agency that does much more than just count people every 10 years, conducts an ongoing detailed survey of over 1,000 counties as shown in Figure 1.4.1, collecting information on variables as diverse as income, household size, mortgage status, and travel time to work. The survey is described as follows:[9]

*The American Community Survey (ACS), part of the 2010 Decennial Census Program, gathers demographic, social, economic, and housing information about the nation's people and communities on a continuous basis. As the largest survey in the United States, it is the only source of small-area data on a wide range of important social and economic characteristics for all communities in the country.... The ACS is sent each month to a sample of roughly 250,000 addresses in the United States and Puerto Rico, or 3 million a year (about 2.5 percent of all residential addresses).... Over a 5-year period, the ACS will sample about 15 million addresses and complete interviews for about 11 million. This sample is sufficient to produce estimates for small geographic areas, such as neighborhoods and sparsely-populated rural counties.*

Private companies also collect and analyze detailed information on the characteristics of U.S. neighborhoods. One such company is Bamberg-Handley, Inc., which maintains the system "A Classification of Residential Neighborhoods" originally developed by CACI Marketing. Considerable data-mining and statistical methods went into classifying consumers by developing 43 clusters within nine summary groups, and much more statistical work goes into providing detailed current information on specific neighborhoods to businesses to help them find customers. Figure 1.4.2 shows the system's clusters, which were developed as follows:[10]

*... using multivariate statistical methods.... The CACI statisticians carefully analyzed and sorted each U.S. neighborhood by 61 lifestyle characteristics such as income, age, household type,*

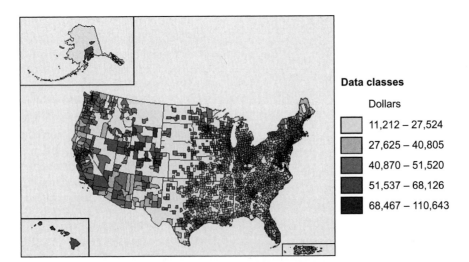

**Data classes**

Dollars

| | |
|---|---|
| | 11,212 – 27,524 |
| | 27,625 – 40,805 |
| | 40,870 – 51,520 |
| | 51,537 – 68,126 |
| | 68,467 – 110,643 |

**FIGURE 1.4.1**  Median income displayed county-by-county as sampled by the U.S. Census Bureau as part of its American Community Survey that provides demographic, social, economic, and housing data. Availability of free government data from large-scale surveys like this provides opportunities for data mining to help businesses better understand their customers and where they live. (*Source: Accessed at http://factfinder.census.gov on June 29, 2010.*)

Summary Groups                                    Segments

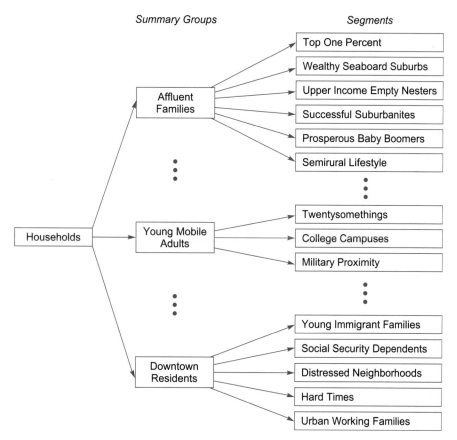

**FIGURE 1.4.2**   Some results of data mining to identify clusters of households that tend to be similar across many observable characteristics, as determined by the system "A Classification of Residential Neighborhoods" for understanding customers and markets. There are two levels of clusters. Based on data, each household sampled from a neighborhood can be classified into one of nine summary groups (the top level of clusters) of which three are shown, and further classified into one of the detailed segments (the next level of clusters) for its summary group. There are many business uses for systems that can help find customers, including where to locate a store and where to send mailings. *(Source: Based on information accessed at http://www.bhimarketing. com/acorn_summary.htm on June 30, 2010.)*

*home value, occupation, education and other key determinants of consumer behavior. These market segments were created using a combination of cluster analysis techniques.... The statistical staff checked the stability of the segments by replicating the clusters with independent samples.*

9. Accessed at http://www.census.gov/acs/www/Downloads/ACSCongressHandbook.pdf on June 29, 2010.
10. Accessed at http://www.demographics.caci.com/products/life_seg.htm on October 30, 2001.

### Example
#### Mining Data to Identify People Who Will Donate to a Good Cause

Many people send money to charity in response to requests received in the mail, but many more do not respond—and sending letters to these nonresponders is costly. If you worked for a charitable organization, you would want to be able to predict the likelihood of donation and the likely amount of the donation ahead of time—before sending a letter—to help you decide where and when to send a request for money. Managers of nonprofit companies (such as charities) need to use many of the same techniques as those of for-profit companies, and data-mining methods can be very helpful to a manager of any company hoping to make better use of data collected on the results of past mailings in order to help plan for the future.

A difficult decision is how often to keep sending requests to people who have responded in the past, but not recently. Some of them will become active donors again—but which ones? Table 1.4.1 shows part of a database that gives information on 20,000 such individuals at the time of a mailing, together with the amount (if any, in the first column) that each one gave as a result of that mailing.[11] The columns in

*(Continued)*

## TABLE 1.4.1 Charitable Donations Mailing Database*

| Donation | Life-time | Gifts | Years Since First | Years Since Last | Average Gift | Major Donor | Promos | Recent Gifts | Age | Home Phone | PC Owner | Catalog Shopper | Per Capita Income | Median Household Income | Professional | Technical | Sales | Clerical | Farmers | Self-Employed | Cars | Owner Occupied | Age 55-59 | Age 60-64 | School |
|---|---|---|---|---|---|---|---|---|---|---|---|---|---|---|---|---|---|---|---|---|---|---|---|---|---|
| $ 0.00 | $ 81.00 | 15 | 6.4 | 1.2 | $ 5.40 | 0 | 58 | 3 |  | 0 | 0 | 0 | 16,838 | 30,500 | 12% | 7% | 17% | 22% | 1% | 2% | 16% | 41% | 4% | 5% | 14.0 |
| 15.00 | 15.00 | 1 | 1.2 | 1.2 | 15.00 | 0 | 13 | 1 | 33 | 1 | 0 | 1 | 17,728 | 33,000 | 11% | 1% | 14% | 16% | 1% | 6% | 8% | 90% | 7% | 11% | 12.0 |
| 0.00 | 15.00 | 1 | 1.8 | 1.8 | 15.00 | 0 | 16 | 1 |  | 1 | 0 | 0 | 6,094 | 9,300 | 3% | 0% | 5% | 32% | 0% | 0% | 3% | 12% | 6% | 3% | 12.0 |
| 0.00 | 25.00 | 2 | 3.5 | 1.3 | 12.50 | 0 | 26 | 1 | 55 | 0 | 0 | 0 | 16,119 | 50,200 | 4% | 7% | 16% | 19% | 6% | 21% | 52% | 79% | 3% | 2% | 12.3 |
| 0.00 | 20.00 | 1 | 1.3 | 1.3 | 20.00 | 0 | 12 | 1 | 71 | 1 | 0 | 0 | 11,236 | 24,700 | 7% | 3% | 7% | 15% | 2% | 5% | 22% | 78% | 6% | 6% | 12.0 |
| 0.00 | 68.00 | 6 | 7.0 | 1.6 | 11.33 | 0 | 38 | 2 | 42 | 0 | 0 | 0 | 13,454 | 40,400 | 15% | 2% | 7% | 4% | 14% | 17% | 26% | 67% | 6% | 5% | 12.0 |
| 0.00 | 110.00 | 11 | 10.2 | 1.4 | 10.00 | 0 | 38 | 2 | 75 | 1 | 0 | 0 | 8,655 | 17,000 | 8% | 3% | 5% | 12% | 15% | 15% | 21% | 82% | 8% | 5% | 12.0 |
| 0.00 | 174.00 | 26 | 10.4 | 1.5 | 6.69 | 0 | 72 | 3 |  | 0 | 0 | 0 | 6,461 | 13,800 | 7% | 4% | 9% | 12% | 1% | 4% | 12% | 57% | 6% | 6% | 12.0 |
| 0.00 | 20.00 | 1 | 1.8 | 1.8 | 20.00 | 0 | 15 | 1 | 67 | 1 | 0 | 0 | 12,338 | 37,400 | 11% | 2% | 16% | 18% | 3% | 3% | 22% | 90% | 10% | 9% | 12.0 |
| 14.00 | 95.00 | 7 | 6.1 | 1.3 | 13.57 | 0 | 56 | 2 | 61 | 0 | 0 | 0 | 10,766 | 20,300 | 13% | 4% | 11% | 8% | 2% | 7% | 20% | 67% | 7% | 7% | 12.0 |
| . | . | . | . | . | . | . | . | . | . | . | . | . | . | . | . | . | . | . | . | . | . | . | . | . | . |
| 0.00 | 25.00 | 2 | 1.5 | 1.1 | 12.50 | 0 | 18 | 2 |  | 0 | 0 | 1 | 9,989 | 23,400 | 14% | 2% | 9% | 10% | 0% | 7% | 20% | 73% | 7% | 6% | 12.0 |
| 0.00 | 30.00 | 2 | 2.2 | 1.4 | 15.00 | 0 | 19 | 1 | 74 | 1 | 0 | 0 | 11,691 | 27,800 | 4% | 1% | 8% | 14% | 0% | 2% | 10% | 65% | 6% | 8% | 12.0 |
| 0.00 | 471.00 | 22 | 10.6 | 1.5 | 21.41 | 0 | 83 | 1 | 87 | 0 | 0 | 0 | 20,648 | 34,000 | 13% | 4% | 20% | 20% | 0% | 2% | 5% | 46% | 8% | 9% | 12.4 |
| 0.00 | 33.00 | 3 | 6.1 | 1.2 | 11.00 | 0 | 31 | 1 | 42 | 1 | 0 | 0 | 12,410 | 21,900 | 9% | 3% | 12% | 20% | 0% | 9% | 13% | 49% | 5% | 8% | 12.0 |
| 0.00 | 94.00 | 10 | 1.1 | 0.3 | 9.40 | 0 | 42 | 1 | 51 | 0 | 0 | 0 | 14,436 | 41,300 | 15% | 7% | 9% | 15% | 1% | 9% | 29% | 85% | 6% | 5% | 13.2 |
| 0.00 | 47.00 | 8 | 3.4 | 1.0 | 5.88 | 0 | 24 | 4 | 38 | 0 | 1 | 0 | 17,689 | 31,800 | 11% | 3% | 17% | 21% | 0% | 6% | 12% | 16% | 2% | 3% | 14.0 |
| 0.00 | 125.00 | 7 | 5.2 | 1.2 | 17.86 | 0 | 49 | 3 | 58 | 0 | 1 | 0 | 26,435 | 43,300 | 15% | 1% | 5% | 9% | 0% | 3% | 16% | 89% | 5% | 24% | 14.0 |
| 0.00 | 109.50 | 16 | 10.6 | 1.3 | 6.84 | 0 | 68 | 4 | 67 | 0 | 0 | 0 | 17,904 | 44,800 | 8% | 3% | 1% | 20% | 4% | 15% | 26% | 88% | 6% | 5% | 12.0 |
| 0.00 | 112.00 | 11 | 10.2 | 1.6 | 10.18 | 0 | 66 | 2 | 82 | 0 | 0 | 0 | 11,840 | 28,200 | 13% | 4% | 12% | 14% | 2% | 6% | 13% | 77% | 5% | 5% | 12.0 |
| 0.00 | 243.00 | 15 | 10.1 | 1.2 | 16.20 | 0 | 67 | 2 | 67 | 0 | 0 | 0 | 17,755 | 40,100 | 10% | 3% | 13% | 24% | 2% | 7% | 24% | 41% | 2% | 4% | 14.0 |

*The first column shows how much each person gave as a result of this mailing, while the other columns show information that was available before the mailing was sent. Data mining can use this information to statistically predict the mailing result, giving us useful information about characteristics that are linked to the likelihood and amount of donations.

**TABLE 1.4.2 Definitions for the Variables in the Donations Database***

| Name of Variable | Description |
| --- | --- |
| Donation | Donation amount in dollars in response to this mailing |
| Lifetime | Donation lifetime total before this mailing |
| Gifts | Number of lifetime gifts before this mailing |
| Years Since First | Years since first gift |
| Years Since Last | Years since most recent gift before this mailing |
| Average Gift | Average of gifts before this mailing |
| Major Donor | Major donor indicator |
| Promos | Number of promotions received before this mailing |
| Recent Gifts | Number of gifts in past two years |
| Age | Age in years |
| Home Phone | Published home phone number indicator |
| PC Owner | Home PC owner indicator |
| Catalog Shopper | Shop by catalog indicator |
| Per Capita Income | Per capita neighborhood income |
| Median Household Income | Median household neighborhood income |
| Professional | Percent professional in neighborhood |
| Technical | Percent technical in neighborhood |
| Sales | Percent sales in neighborhood |
| Clerical | Percent clerical in neighborhood |
| Farmers | Percent farmers in neighborhood |
| Self-Employed | Percent self-employed in neighborhood |
| Cars | Percent households with 3+ vehicles |
| Owner Occupied | Percent owner-occupied housing units in neighborhood |
| Age 55–59 | Percent adults age 55–59 in neighborhood |
| Age 60–64 | Percent adults age 60–64 in neighborhood |
| School | Median years in school completed by adults in neighborhood |

*The first group of variables represents information about the person who received the mailing. For example, the second variable, headed "Lifetime," shows the total dollar amount of all previous gifts by this person, and variable 12 headed "PC Owner" is 1 if he or she owns a PC and is 0 otherwise. The remaining variables represent information about the person's neighborhood, beginning with column 14 headed "Per Capita Income" and continuing through all of the percentages to the last column.

**Example—cont'd**

the database are defined in Table 1.4.2. We will revisit this database in future chapters—from description, through summaries, statistical inference, and prediction—to show how many of the various statistical techniques can be used to help with data mining. One quick discovery is shown in Figure 1.4.3: apparently the more gifts given over the previous two years (from the column headed "Recent Gifts"), the greater the chances that the person gave a gift in response to this mailing.

11. This database was adapted from a large data set originally used in The Second International Knowledge Discovery and Data Mining Tools Competition and is available as part of the UCI Knowledge Discovery in Databases Archive; Hettich, S. and Bay, S. D., 1999, The UCI KDD Archive http://kdd.ics.uci.edu, Irvine, CA, University of California, Department of Information and Computer Science, now maintained as part of the UCI Machine Learning Archive at http://archive.ics.uci.edu/ml/.

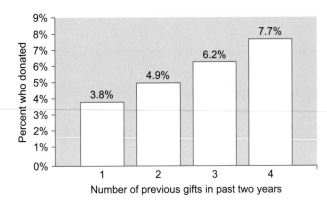

**FIGURE 1.4.3**  A result of data mining the donations database of 20,000 people. The more gifts given over the previous two years (from the database column headed "Recent Gifts"), the greater the chances that the person gave a gift in response to this mailing. For example, out of the 9,924 who gave just one previous gift, 381 (or 3.8%) gave a gift. Out of the 2,486 who gave four previous gifts, 192 (for a larger percentage of 7.7%) donated.

## 1.5  WHAT IS PROBABILITY?

Probability is a *what if* tool for understanding risk and uncertainty. **Probability** shows you the likelihood, or chances, for each of the various potential future events, based on a set of assumptions about how the world works. For example, you might assume that you know basically how the world works (i.e., all of the details of the process that will produce success or failure or payoffs in between). Probabilities of various outcomes would then be computed for each of several strategies to indicate how successful each strategy would be.

You might learn, for example, that an international project has only an 8% chance of success (i.e., the probability of success is 0.08), but if you assume that the government can keep inflation low, then the chance of success rises to 35%—still very risky, but a much better situation than the 8% chance. Probability will not tell you whether to invest in the project, but it will help you keep your eyes open to the realities of the situation.

Here are additional examples of situations where finding the appropriate answer requires computing or estimating a probability number:

1. Given the nature of an investment portfolio and a set of assumptions that describe how financial markets work, what are the chances that you will profit over a one-year horizon? Over a 10-year horizon?
2. What are the chances of rain tomorrow? What are the chances that next winter will be cold enough so that your heating-oil business will make a profit?
3. What are the chances that a foreign country (where you have a manufacturing plant) will become involved in civil war over the next two years?

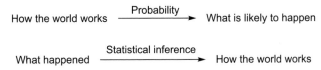

**FIGURE 1.5.1**  Probability and statistics take you in opposite directions. If you make assumptions about how the world works, then probability can help you figure out how likely various outcomes are and thus help you understand what is likely to happen. If you have data that tell you something about what has happened, then statistics can help you move from this particular data set to a more general understanding of how things work.

4. What are the chances that the college student you just interviewed for a job will become a valued employee over the coming months?

Probability is the inverse of statistics. Whereas statistics helps you go from observed data to generalizations about how the world works, probability goes the other direction: if you assume you know how the world works, then you can figure out what kinds of data you are likely to see and the likelihood for each. Figure 1.5.1 shows this inverse relation.

Probability also works together with statistics by providing a solid foundation for statistical inference. When there is uncertainty, you cannot know exactly what will happen, and there is some chance of error. Using probability, you will learn ways to control the error rate so that it is, say, less than 5% or less than 1% of the time.

## 1.6  GENERAL ADVICE

Statistical results should be explainable in a straightforward way (even though their theory may be much more complicated). Here are some general words of advice:

1. Trust your judgment; common sense counts.
2. Maintain a healthy skepticism.
3. Don't be snowed by a seemingly ingenious statistical analysis; it may well rely on unrealistic and inappropriate assumptions.

Because of the vast flexibility available to the analyst in each phase of a study, one of the most important factors to consider in evaluating the results of a statistical study is: *Who funded it?* Remember that the analyst made many choices along the way—in defining the problems, designing the plan to select the data, choosing a framework or model for analysis, and interpreting the results.

## 1.7  END-OF-CHAPTER MATERIALS

### Summary

**Statistics** is the art and science of collecting and understanding data. Statistical techniques should be viewed as an important part of the decision process, allowing informed strategic decisions to be made that combine intuition and expertise with a thorough (statistical) understanding of the facts available. Use of statistics is becoming increasingly important in maintaining a competitive edge.

The basic activities of statistics are as follows:

1.  The phase of **designing the study** involves planning the details of data gathering, perhaps using a random sample from a larger population.
2.  **Exploring the data** involves looking at your data set from many angles, describing it, and summarizing it. This helps you make sure that the planned analysis is appropriate and allows you to modify the analysis if necessary.
3.  **Modeling the data** involves choosing a system of assumptions and equations that behaves like the data you are interested in, so that you can work with a few numbers (called *parameters*) that represent the important aspects of the data. A model can be a very effective system within which questions about large-scale properties of the data can be answered. Often, a model has the form "data equals structure plus noise."
4.  **Estimating an unknown quantity** produces the best educated guess possible based on the available data. You will also want to have some indication of the size of the error involved by using this estimated value in place of the actual unknown value.
5.  **Statistical hypothesis testing** uses data to decide between two (or more) different possibilities in order to resolve an issue in an ambiguous situation. This is often done to see if some apparently interesting feature of the data is really there as opposed to being an artifact of "pure randomness," which is basically uninteresting.

**Data mining** is a collection of methods for obtaining useful knowledge by analyzing large amounts of data, often by searching for hidden patterns. It would be wasteful to leave this information unexplored, after having been collected for some purpose, when it could be useful in many other ways. The goal of data mining is to obtain value from these vast stores of data, in order to improve the company with higher sales, lower costs, and better products.

**Probability** shows you the likelihood, or chances, for each of the various potential future events, based on a set of assumptions about how the world works. Probability is the inverse of statistics: probability tells you what the data will be like when you know how the world is, whereas statistics helps you figure out what the world is like after you have seen some data that it generated.

Statistics works best when you combine it with your own expert judgment and common sense. When statistical results go against your intuition, be prepared to work hard to find the reason why. The statistical analysis may well be incorrect due to wrong assumptions, or your intuition may be wrong because it was not based on facts.

## Key Words

**data mining**, *8*
**designing the study**, *5*
**estimating an unknown quantity**, *6*
**exploring the data**, *5*
**hypothesis testing**, *7*
**model**, *6*
**probability**, *14*
**statistics**, *4*

## Questions

1.  Why is it worth the effort to learn about statistics?
    a.  Answer for management in general.
    b.  Answer for one particular area of business of special interest to you.
2.  Choose a business firm, and list the ways in which statistical analysis could be used in decision-making activities within that firm.
3.  How should statistical analysis and business experience interact with each other?
4.  What is statistics?
5.  What is the design phase of a statistical study?
6.  Why is random sampling a good method to use for selecting items for study?
7.  What can you gain by exploring data in addition to looking at summary results from an automated analysis?
8.  What can a statistical model help you accomplish? Which basic activity of statistics can help you choose an appropriate model for your data?
9.  Are statistical estimates always correct? If not, what else will you need (in addition to the estimated values) in order to use them effectively?
10. Why is a confidence interval more useful than an estimated value?
11. Give two examples of hypothesis testing situations that a business firm would be interested in.
12. What distinguishes data mining from other statistical methods? What methods, in addition to those of statistics, are often used in data mining?
13. Differentiate between probability and statistics.
14. A consultant has just presented a very complicated statistical analysis, complete with lots of mathematical symbols and equations. The results of this impressive analysis go against your intuition and experience. What should you do?
15. Why is it important to identify the source of funding when evaluating the results of a statistical study?

## Problems

*Problems marked with an asterisk (\*) are solved in the Self Test in Appendix C.*

1. Describe a recent decision you made that depended, in part, on information that came from data. Identify the underlying data set and tell how a deeper understanding of statistics might have helped you use these data more effectively.

2. Name three numerical quantities a firm might be concerned with for which exact values are unavailable. For each quantity, describe an estimate that might be useful. In general terms, how reliable would you expect these estimates to be?

3. Reconsider the three estimates from the previous problem. Are confidence intervals readily available? If so, how would they be useful?

4. List two kinds of routine decisions that you make in which statistical hypothesis testing could play a helpful role.

5. Look through a recent copy of the *Wall Street Journal*. Identify an article that relies directly or indirectly on statistics. Briefly describe the article (also be sure to give the title, date, and page number), and attach a copy. Which of the five activities of statistics is represented here?

6.\* Which of the five basic activities of statistics is represented by each of the following situations?

   a. A factory's quality control division is examining detailed quantitative information about recent productivity in order to identify possible trouble spots.

   b. A focus group is discussing the audience that would best be targeted by advertising, with the goal of drawing up and administering a questionnaire to this group.

   c. In order to get the most out of your firm's Internet activity data, it would help to have a framework or structure of equations to allow you to identify and work with the relationships in the data.

   d. A firm is being sued for gender discrimination. Data that show salaries for men and women are presented to the jury to convince them that there is a consistent pattern of discrimination and that such a disparity could not be due to randomness alone.

   e. The size of next quarter's gross national product must be known so that a firm's sales can be forecast. Since it is unavailable at this time, an educated guess is used.

7. Overseas sales dropped sharply last month, and you don't know why. Moreover, you realize that you don't even have the numbers needed in order to tell what the problem is. You call a meeting to discuss how to solve the problem. Which statistical activity is involved at this stage?

8. If your factories produce too much, then you will have to pay to store the extra inventory. If you produce too little, then customers will be turned away and profits will be lost. Therefore, you would like to produce exactly the right amount to avoid these costs to your company. Unfortunately, however, you don't know the correct production level. Which is the main statistical activity required to solve this problem?

9. Before you proceed with the analysis of a large accounting data set that has just been collected, your boss has asked you to take a close look at the data to check for problems and surprises and ensure its basic integrity. Identify the basic statistical activity you are performing.

10. Your company has been collecting detailed data for years on customer contacts, including store purchases, telephone inquiries, and Internet orders, and you would like to systematically use this resource to learn more about your customers and, ultimately, to improve sales. What is the name of the collection of methods that will be most useful to you in this project?

11. Your work group would like to estimate the size of the market for high-quality stereo components in New Orleans but cannot find any reliable data that are readily available. Which basic activity of statistics is involved initially in proceeding with this project?

12. You are wondering whom to interview, how many to interview, and how to process the results so that your questions can be answered at the lowest possible cost. Identify the basic activity of statistics involved here.

13. You have collected and explored the data on Internet information requests. Before continuing on to use the data for estimation and hypothesis testing, you want to develop a framework that identifies meaningful parameters to describe relationships in the data. What basic activity of statistics is involved here?

14. Your firm has been accused of discrimination. Your defense will argue in part that the imbalance is so small that it could have happened at random and that, in fact, no discrimination exists. Which basic activity of statistics is involved?

15. By looking carefully at graphs of data, your marketing department has identified three distinct segments of the marketplace with different needs and price levels. This helpful information was obtained through which basic activity of statistics?

16. You are trying to determine the quality of the latest shipment of construction materials based on careful observation of a sample. Which basic activity of statistics will help you reach your goal?

17. You think that one of the machines may be broken, but you are not sure because even when it is working properly there are a few badly manufactured parts. When you analyze the rate at which defective parts are being produced to decide whether or not there has been an increase in the defect rate, which basic activity of statistics is involved?

18. Your boss has asked you to take a close look at the marketing data that just came in and would like you to report back with your overall impressions of its quality and usefulness. Which main activity of statistics will you be performing?

19. Using data on the characteristics of houses that sold recently in a city of interest, you would like to specify

the way in which features such as the size (in square feet) and number of bedrooms relate to the sale price. You are working out an equation that asserts that the sales price is given by a formula that involves the house's characteristics and parameters (such as the dollar value of an additional bedroom) that are estimated from the data. What main activity of statistics are you involved with?

## Project

Find the results of an opinion poll in a newspaper or magazine or on the Internet. Discuss how each of the five basic activities of statistics was applied (if it is clear from the article) or might have been applied to the problem of understanding what people are thinking. Attach a copy of the article to your discussion.

## Chapter 2

# Data Structures

## Classifying the Various Types of Data Sets

Data can come to you in several different forms. It will be useful to have a basic catalog of the different kinds of data so that you can recognize them and use appropriate techniques for each. A **data set** consists of observations on items, typically with the same information being recorded for each item. We define the **elementary units** as the items themselves (for example, companies, people, households, cities, TV sets) in order to distinguish them from the measurement or observation (for example, sales, weight, income, population, size). Data sets can be classified in four basic ways:

**One:** By the number of pieces of information (variables) there are for each elementary unit

**Two:** By the kind of measurement (numbers or categories) recorded in each case

**Three:** By whether or not the time sequence of recording is relevant

**Four:** By whether or not the information was newly created or had previously been created by others for their own purposes

## 2.1 HOW MANY VARIABLES?

A piece of information recorded for every item (its cost, for example) is called a **variable**. The number of variables, or pieces of information, recorded for each item indicates the complexity of the data set and will guide you toward the proper kinds of analyses. Depending on whether one,

two, or many variables are present, you have *univariate, bivariate,* or *multivariate* data, respectively.

## Univariate Data

**Univariate** (one-variable) data sets have just one piece of information recorded for each item. Statistical methods are used to summarize the basic properties of this single piece of information, answering such questions as:

1. What is a typical (summary) value?
2. How diverse are these items?
3. Do any individuals or groups require special attention?

Here is a table of univariate data, showing the profits of the 10 food services companies in the Fortune 500.

| Company | 2008 Profits ($ millions) |
|---|---|
| McDonald's | $4,313.2 |
| Yum Brands | 964.0 |
| Starbucks | 315.5 |
| Darden Restaurants | 377.2 |
| Brinker International | 51.7 |
| Jack in the Box | 119.3 |
| Burger King Holdings | 190.0 |
| Cracker Barrel Old Country Store | 65.6 |
| Wendy's/Arby's Group | −479.7 |
| Bob Evans Farms | 64.9 |

**Source:** Data are from http://money.cnn.com/magazines/fortune/fortune500/2009/industries/147/index.html, accessed on July 1, 2010.

Here are some additional examples of univariate data sets:

1. The incomes of subjects in a marketing survey. Statistical analysis would reveal the profile (or distribution) of incomes, indicating a typical income level, the extent of variation in incomes, and the percentage of people within any given income range.
2. The number of defects in each TV set in a sample of 50 manufactured this morning. Statistical analysis could be used to keep tabs on quality (estimate) and to see if things are getting out of control (hypothesis testing).
3. The interest rate forecasts of 25 experts. Analysis would reveal, as you might suspect, that the experts don't all agree and (if you check up on them later) that they can all be wrong. Although statistics can't combine these 25 forecasts into an exact, accurate prediction, it at least enables you to explore the data for the extent of consensus.
4. The colors chosen by members of a focus group. Analysis could be used to help in choosing an agreeable selection for a product line.
5. The bond ratings of the firms in an investment portfolio. Analysis would indicate the risk of the portfolio.

## Bivariate Data

**Bivariate** (two-variable) data sets have exactly two pieces of information recorded for each item. In addition to summarizing each of these two variables separately as univariate data sets, statistical methods would also be used to explore the relationship between the two factors being measured in the following ways:

1. Is there a simple relationship between the two?
2. How strongly are they related?
3. Can you predict one from the other? If so, with what degree of reliability?
4. Do any individuals or groups require special attention?

Here is a table of bivariate data, showing the profits of the 10 food services companies in the Fortune 500 along with the total return to investors.

| Company | 2008 Profits ($ millions) | Total Return to Investors 2008 (%) |
|---|---|---|
| McDonald's | $4,313.2 | 8.5% |
| Yum Brands | 964.0 | −16.0% |
| Starbucks | 315.5 | −53.8% |
| Darden Restaurants | 377.2 | 4.4% |
| Brinker International | 51.7 | −44.5% |
| Jack in the Box | 119.3 | −14.3% |
| Burger King Holdings | 190.0 | −15.4% |
| Cracker Barrel Old Country Store | 65.6 | −34.6% |
| Wendy's/Arby's Group | −479.7 | −39.8% |
| Bob Evans Farms | 64.9 | −22.3% |

Source: Data are from http://money.cnn.com/magazines/fortune/fortune500/2009/industries/147/index.html, accessed on July 1, 2010.

Here are some additional examples of bivariate data sets:

1. The cost of production (first variable) and the number produced (second variable) for each of seven factories (items, or elementary units) producing integrated circuits, for the past quarter. A bivariate statistical analysis would indicate the basic relationship between cost and number produced. In particular, the analysis might identify a *fixed cost* of setting up production facilities and a *variable cost* of producing one extra circuit.[1] An analyst might then look at individual factories to see how efficient each is compared with the others.
2. The price of one share of your firm's common stock (first variable) and the date (second variable), recorded every day for the past six months. The relationship between price and time would show you any recent trends in the value of your investment. Whether or not you could then forecast future value is a subject of some controversy (is it an unpredictable "random walk," or are those apparent patterns real?).
3. The purchase or nonpurchase of an item (first variable, recorded as yes/no or as 1/0) and whether an advertisement for the item is recalled (second variable, recorded similarly) by each of 100 people in a shopping mall. Such data (as well as data from more careful studies) help shed light on the effectiveness of advertising: What is the relationship between advertising recall and purchase?

## Multivariate Data

**Multivariate** (many-variable) data sets have three or more pieces of information recorded for each item. In addition to summarizing each of these variables separately (as a univariate data set), and in addition to looking at the relationship between any two variables (as a bivariate data set), statistical methods would also be used to look at the interrelationships among all the items, addressing the following questions:

1. Is there a simple relationship among them?
2. How strongly are they related?
3. Can you predict one (a "special variable") from the others? With what degree of reliability?
4. Do any individuals or groups require special attention?

Here is a table of multivariate data, showing the profits of the 10 food services companies in the Fortune 500 along with the total return to investors, number of employees, and revenues.

---

1. *Variable cost* refers to the cost that varies according to the number of units produced; it is not related to the concept of a *statistical* variable.

| Company | 2008 Profits ($ millions) | Total Return to Investors 2008 (%) | Employees | Revenues ($ millions) |
|---|---|---|---|---|
| McDonald's | $4,313.2 | 8.5% | 400,000 | $23,522.4 |
| Yum Brands | 964.0 | −16.0% | 193,200 | 11,279.0 |
| Starbucks | 315.5 | −53.8% | 176,000 | 10,383.0 |
| Darden Restaurants | 377.2 | 4.4% | 179,000 | 6,747.2 |
| Brinker International | 51.7 | −44.5% | 100,400 | 4,235.2 |
| Jack in the Box | 119.3 | −14.3% | 42,700 | 3,001.4 |
| Burger King Holdings | 190.0 | −15.4% | 41,000 | 2,455.0 |
| Cracker Barrel Old Country Store | 65.6 | −34.6% | 65,000 | 2,384.5 |
| Wendy's/Arby's Group | −479.7 | −39.8% | 70,000 | 1,822.8 |
| Bob Evans Farms | 64.9 | −22.3% | 49,149 | 1,737.0 |

**Source:** Data are from http://money.cnn.com/magazines/fortune/fortune500/2009/industries/147/index.html, accessed on July 1, 2010.

Here are some additional examples of multivariate data sets:

1. The growth rate (special variable) and a collection of measures of strategy (the other variables), such as type of equipment, extent of investment, and management style, for each of a number of new entrepreneurial firms. The analysis would give an indication of which combinations have been successful and which have not.
2. Salary (special variable) and gender (recorded as male/female or as 0/1), number of years of experience, job category, and performance record, for each employee. Situations such as this come up in litigation about whether women are discriminated against by being paid less than men on the average. A key question, which a multivariate analysis can help answer, is, Can this discrepancy be explained by factors other than gender? Statistical methods can remove the effects of these other factors and then measure the average salary differential between a man and a woman who are equal in all other respects.
3. The price of a house (special variable) and a collection of variables that contribute to the value of real estate, such as lot size, square footage, number of rooms, presence or absence of swimming pool, and age of house, for each of a collection of houses in a neighborhood. Results of the analysis would give a picture of how real estate is valued in this neighborhood. The analysis might be used as part of an appraisal to determine fair market value of a house in that neighborhood, or it might be used by builders to decide which combination of features will best serve to enhance the value of a new home.

## 2.2 QUANTITATIVE DATA: NUMBERS

Meaningful numbers are numbers that directly represent the measured or observed *amount* of some characteristic or quality of the elementary units. Meaningful numbers include, for example, dollar amounts, counts, sizes, numbers of employees, and miles per gallon. They *exclude* numbers that are merely used to code for or keep track of something else, such as football uniform numbers or transaction codings like 1 = buy stock, 2 = sell stock, 3 = buy bond, 4 = sell bond. If the data come to you as meaningful numbers, you have **quantitative** data (i.e., they represent quantities). With quantitative data, you can do all of the usual number-crunching tasks, such as finding the average (see Chapter 4) and measuring the variability (see Chapter 5). It is straightforward to compute directly with numerical data. There are two kinds of quantitative data, *discrete* and *continuous*, depending on the values potentially observable.

### Discrete Quantitative Data

A **discrete** variable can assume values only from a list of specific numbers.[2] For example, the number of children in a household is a discrete variable. Since the possible values can be listed, it is relatively simple to work with discrete data sets. Here are some examples of discrete variables:

1. The number of computer shutdowns in a factory in the past 24 hours.
2. The number of contracts, out of the 18 for which you submitted bids, that were awarded.
3. The number of foreign tankers docked at a certain port today.
4. The gender of an employee, if this is recorded using the number 0 or 1.

### Continuous Quantitative Data

We will consider any numerical variable that is not discrete to be **continuous**.[3] This word is used because the possible values form a "continuum," such as the set of all positive numbers, all numbers, or all values between 0 and 100%. For example, the actual weight of a candy bar marked "net weight 1.7 oz." is a continuous random variable; the actual weight might be 1.70235 or 1.69481 ounces instead of exactly 1.7. If you're not yet thinking statistically, you might have assumed that the actual weight was 1.7 ounces exactly; in fact, when you measure in the real world, there are invariably small (and sometimes large) deviations from expected values.

2. Note the difference between a *discrete* variable (as defined here) and a *discreet* variable, which would be much more careful and quiet about its activities.

3. Although this definition is suitable for many business applications, the mathematical theory is more complex and requires a more elaborate definition involving integral calculus (not presented here). We will also refrain from discussing *hybrid* variables, which are neither discrete nor continuous.

Here are some examples of continuous variables:

1. The price of an ounce of gold, in dollars, right now. You might think that this value is discrete (and you would be technically correct, since a number such as $1,206.90 is part of a list of discrete numbers of pennies: 0.00, 0.01, 0.02,…). However, try to view such cases as examples of continuous data because the discrete divisions are so small as to be unimportant to the analysis. If gold ever began trading at a few cents per ounce, it would become important to view it as a case of discrete data; however, it is more likely that the price would be quoted in thousandths of a cent at that point, again essentially a continuous quantity.

2. Investment and accounting ratios such as earnings per share, rate of return on investment, current ratio, and beta.

3. The amount of energy used per item in a production process.

## Watch Out for Meaningless Numbers

One warning is necessary before you begin analyzing quantitative data: Make sure that the numbers are meaningful! Unfortunately, numbers can be used to record anything at all. If the coding is arbitrary, the results of analysis will be meaningless.

### Example
#### Alphabetical Order of States

Suppose the states of the United States are listed in alphabetical order and coded as 1, 2, 3,…, as follows:

| | |
|---|---|
| 1 | Alabama |
| 2 | Alaska |
| 3 | Arizona |
| 4 | Arkansas |
| : | : |

Now suppose you ask for the average of the states of residence for all employees in your firm's database. The answer is certainly computable. However, the result would be absurd because the numbers assigned to states are not numerically meaningful (although they are convenient for other purposes). To know that the average is 28.35, or somewhere between Nevada and New Hampshire, is not likely to be of use to anybody. The moral is: Be sure that your numbers have meaningful magnitudes before computing with them.

## 2.3 QUALITATIVE DATA: CATEGORIES

If the data set tells you which one of several nonnumerical categories each item falls into, then the data are **qualitative** (because they record some quality that the item possesses). As you have just seen, care must be taken to avoid the temptation to assign numbers to the categories and then compute

with them. If there are just a few categories, then you might work with the percentage of cases in each category (effectively creating something numerical from categorical data). If there are exactly two categories, you can assign the number 0 or 1 to each item and then (for many purposes) continue as if you had quantitative data. But let us first consider the general case in which there are three or more categories.

There are two kinds of qualitative data: *ordinal* (for which there is a meaningful ordering but no meaningful numerical assignment) and *nominal* (for which there is no meaningful order).

## Ordinal Qualitative Data

A data set is **ordinal** if there is a meaningful ordering: You can speak of the first (perhaps the "best"), the second, the third, and so on. You can rank the data according to this ordering, and this ranking will probably play a role in the analysis, particularly if it is relevant to the questions being addressed. The *median* value (the middle one, once the data are put in order) will be defined in Chapter 4 as an example of a statistical summary.

Here are some examples of ordinal data:

1. Job classifications such as president, vice president, department head, and associate department head, recorded for each of a group of executives. Although there are no numbers involved here, and no clear way to assign them, there is certainly a standard way of ordering items by rank.

2. Bond ratings such as AA+, AA, AA−, A+, A, A−, B+, B, and B−, recorded for a collection of debt issues. These are clearly ordinal categorical data because the ordering is meaningful in terms of the risk involved in the investment, which is certainly relevant to investment analysis.

3. Questionnaire answers to a question such as "Please rate your feelings about working with your firm on a scale from 1 to 5, where 1 represents 'can hardly wait to get home each day' and 5 represents 'all my dreams have been fulfilled through my work.'" Although the answers are numbers, these results are better thought of as ordinal qualitative data because the scale is so subjective. It is really not clear if the difference between answers of 5 and 4 is of the same magnitude as the difference between a 2 and a 1. Nor do we assume that 2 is twice as good as 1. But at least the notion of ranking and ordering is present here.[4]

---

4. A very careful reader might validly question whether a 4 for one person is necessarily higher than a 3 for another. Thus, due to the lack of standardization among human perceptions, we may not even have a truly ordinal scale here. This is just one of the many complications involved in statistical analysis for which we cannot give a complete account.

## Nominal Qualitative Data

For **nominal** data there are only categories, with no meaningful ordering. There are no meaningful numbers to compute with, and no basis to use for ranking. About all that can be done is to count and work with the percentage of cases falling into each category, using the *mode* (the category occurring the most often, to be formally defined in Chapter 4) as a summary measure.

Here are some examples of nominal data:

1. The states of residence for all employees in a database. As observed earlier, these are really just categories. Any ordering of states that might be done would actually involve some other variable (such as population or per capita income), which might better be used directly.
2. The primary product of each of several manufacturing plants owned by a diversified business, such as plastics, electronics, and lumber. These are really unordered categories. Although they might be put into a sensible ordering, this would require some other factor (such as growth potential in the industry) not intrinsic to these categories themselves.
3. The names of all firms mentioned on the front page of today's issue of the *Wall Street Journal*.

## 2.4 TIME-SERIES AND CROSS-SECTIONAL DATA

If the data values are recorded in a meaningful sequence, such as daily stock market prices, then you have **time-series** data. If the sequence in which the data are recorded is irrelevant, such as the first-quarter 2011 earnings of eight aerospace firms, you have **cross-sectional** data. *Cross-sectional* is just a fancy way of saying that no time sequence is involved; you simply have a cross-section, or snapshot, of how things are at one particular time.

Analysis of time-series data is generally more complex than cross-sectional data analysis because the ordering of the observations must be carefully taken into account. For this reason, in coming chapters we will initially concentrate on cross-sectional data. Time-series analysis will be covered in Chapter 14.

### Example
#### *The Stock Market*

Figure 2.4.1 shows a chart of the Dow Jones Industrial Average stock market index, monthly closing value, starting in October, 1928. This time-series data set indicates how the value of a portfolio of stocks has changed through time. Note how the stock market value has risen impressively through much of its history (at least until the fall of 2007)

although not entirely smoothly. Note the occasional downward bumps (such as the crash of October 1987, the "dot-com bust" of 2000, and the more recent financial difficulties) that represent the risk that you take by holding a portfolio of stocks that often (but not always) increases in value.

Here are some additional examples of time-series data:

1. The price of wheat each year for the past 50 years, adjusted for inflation. These time trends might be useful for long-range planning, to the extent that the variation in future events follows the patterns of the past.
2. Retail sales, recorded monthly for the past 20 years. This data set has a structure showing generally increasing activity over time as well as a distinct seasonal pattern, with peaks around the December holiday season.
3. The thickness of paper as it emerges from a rolling machine, measured once each minute. This kind of data might be important to quality control. The time sequence is important because small variations in thickness may either "drift" steadily toward an unacceptable level, or "oscillate," becoming wider and narrower within fairly stable limits.

Following are some examples of cross-sectional data:

1. The number of hours of sleep last night, measured for 30 people being examined to test the effectiveness of a new over-the-counter medication.
2. Today's book values of a random sample of a bank's savings certificates.
3. The number of phone calls processed yesterday by each of a firm's order-taking employees.

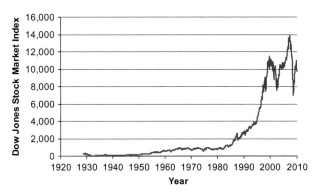

**FIGURE 2.4.1** The Dow Jones Industrial Average stock market index, monthly since 1928, is a time-series data set that provides an overview of the history of the stock market.

## 2.5 SOURCES OF DATA, INCLUDING THE INTERNET

Where do data sets come from? There are many sources of data, varying according to cost, convenience, and how well they will satisfy your business needs. When you control the design of the data-collection plan (even if the work is done by others) you obtain **primary data**. When you use data previously collected by others for their own purposes, you are using **secondary data**.

The main advantage of primary data is that you are more likely to be able to get exactly the information you want because *you* control the data-generating process by designing the types of questions or measurements as well as specifying the sample of elementary units to be measured. Unfortunately, primary data sets are often expensive and time-consuming to obtain. On the other hand, secondary data sets are often inexpensive (or even free) and you might find exactly (or nearly) what you need. This suggests the following strategy for obtaining data: First look for secondary data that will quickly satisfy your needs at a reasonable cost. If this cannot be done, then look also at the cost of designing a plan to collect primary data and use your judgment to decide which source (primary or secondary) to use based on the costs and benefits of each approach.

Here are some examples of primary sources of data:

1. Production data from your manufacturing facility, which may be collected automatically by your company's information systems, including how many units of each type were made each day, together with quality control information such as defect rates.
2. Questionnaire data collected by a marketing company you hired to examine the effects of potential advertising campaigns on customer purchasing behavior.
3. Polling data collected by a political campaign in order to assess the issues that are most important to registered voters who are most likely to vote in an upcoming election.

Here are some examples of secondary sources of data:

1. Economic and demographic data collected and tabulated by the U.S. government and freely available over the Internet.
2. Data reported in specialized trade journals (e.g., advertising, manufacturing, finance) helping those in that industry group remain up-to-date regarding market share and the degree of success of various products.
3. Data collected by companies specializing in data collection and sold to other companies in need of that information. For example, Nielsen Media Research sells television ratings (based on observing the shows watched by a sample of people) to television networks, independent stations, advertisers, advertising agencies, and others. Many library reference collections include volumes of specialized data produced by such companies.

To look for data on the Internet, most people use a search engine (such as Google.com or Yahoo!, at yahoo.com) and specify some key words (such as "government statistics" or "financial data"). The search engine will return a list of links to sites throughout the world that have content relating to your search. Click on a link (or right-click to open in a new browser tab) to check out a site that seems interesting to you (note that Internet addresses that end with ".com" are commercial sites, while ".gov" is government and ".edu" is educational). The Internet has changed considerably in just the past few years, and the trend seems to lead to more and more useful information being available. In particular, nearly all major firms now have their own website. Unfortunately, however, it is still too common for a search to fail to find the information you really want.

### Example

#### Searching the Internet for Government Data on Consumer Prices

The Internet has become a vast resource for data on many topics. Let's see how you might go about searching for government data to be copied to a spreadsheet such as Microsoft Excel® for further analysis.

Begin at your favorite search engine (e.g., Google) and type the key words ("government statistics" in this case) into the search box; then press Enter to see a long list of (possibly) relevant websites. In this case, you are shown the first 10 of over a million web pages that were found. Your screen will look something like Figure 2.5.1.[5]

Selecting FedStats at the top of the list, you find access to a wide collection of statistics from U.S. Federal agencies, as shown in Figure 2.5.2.

Next, clicking on Topic links—A to Z near the top left and then clicking on the letter C from the list of letters across the top (remember that we are looking for information on "Consumer Price Index"), we find ourselves in the index and can scroll down, as indicated in Figure 2.5.3.

Selecting the link Consumer Price Indexes, we find ourselves at the CPI site maintained by the U.S. Bureau of Labor Statistics, as shown in Figure 2.5.4.

Clicking on Historical Data under the heading Latest Numbers at the right, we find a chart and the data for consumer price index for all urban consumers as monthly percentage changes, as shown in Figure 2.5.5.

There are several methods for putting your data into Excel® for analysis. While this particular web page includes a link that will do this task for you (see Figure 2.5.5), other websites might not. If there is no link for this purpose, you could copy and paste the data, or you could right-click in the data table in the web page and choose Export to Microsoft Excel, or you could use Excel's Data Ribbon by choosing Get External Data and then From Web and copying the URL web address from your browser window. Figure 2.5.6 shows how you

*(Continued)*

**FIGURE 2.5.1**   The results of an Internet search for "Government Statistics" as accessed at http://www.google.com/ on July 1, 2010.

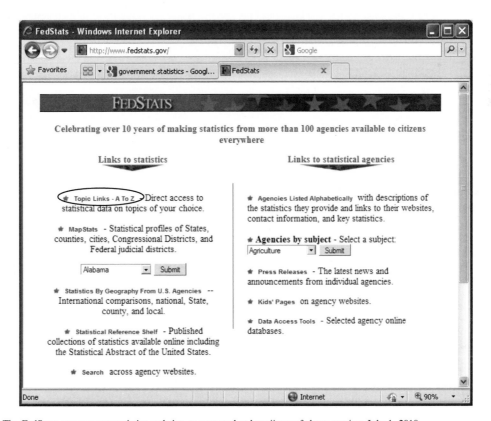

**FIGURE 2.5.2**   The FedStats government statistics website, as accessed at http://www.fedstats.gov/ on July 1, 2010.

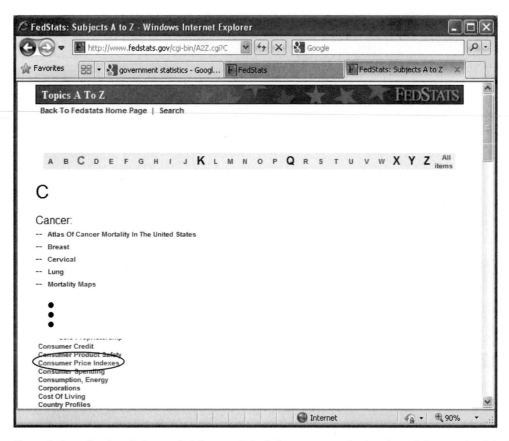

**FIGURE 2.5.3**   Choose the letter C and scroll down to find Consumer Price Indexes, as accessed at http://www.fedstats.gov/ on July 1, 2010.

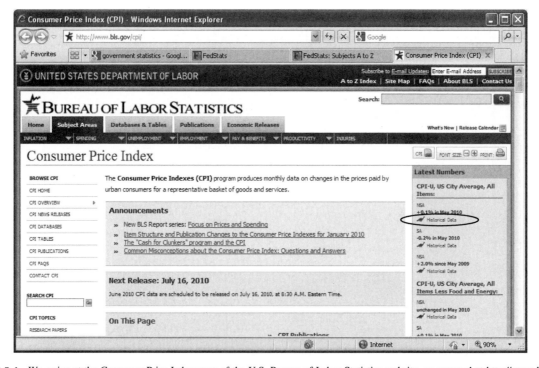

**FIGURE 2.5.4**   We arrive at the Consumer Price Index page of the U.S. Bureau of Labor Statistics website, as accessed at http://www.bls.gov/cpi/ on July 1, 2010.

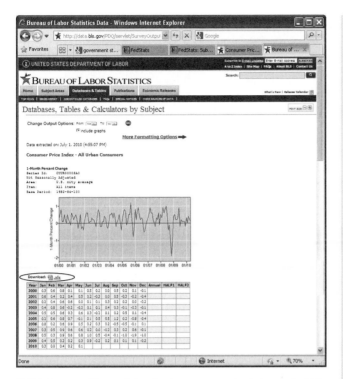

**FIGURE 2.5.5** Chart and data for the consumer price index on the U.S. Bureau of Labor Statistics website, as accessed at http://www.bls.gov/cpi/ on July 1, 2010. Note that this website gives you a link to download the data to Excel®.

**Example—cont'd**

might select and copy the data by dragging from the Year at the top left of the data set, down to the lower right of the data table, and then choosing Edit Copy from the menu.

The data have now been placed on the clipboard, and you can paste them into a worksheet by switching to Excel® and choosing Edit Paste from the menu system, as shown in Figure 2.5.7. Your data set is now in Excel®, and ready for further analysis![6]

Congratulations! The data were located and successfully transferred to a spreadsheet program. Note that many steps were needed along the way, following one link to another in hopes of finding interesting data. Good researchers are persistent, changing the key words and changing to a different search engine as needed to help the process along, as well as often branching to a link only to return and check out a different link.

---

5. Your screens may look different from these for a number of reasons, including the changing nature of the Internet.
6. Note that in some circumstances you may find all your data pasted into a single column. To fix this, first select the data in the column, then choose Text-to-Columns from the Data Tools group of Excel's Data Ribbon, choose "Delimited" from the Convert-Text-To-Columns Wizard, and then experiment to see if selecting "Space" and/or "Tab" will correctly arrange the data across the columns for you. Note also that you may sometimes prefer to use "Paste Special" (instead of "Paste") from the Edit menu in order to control the details of the data transfer.

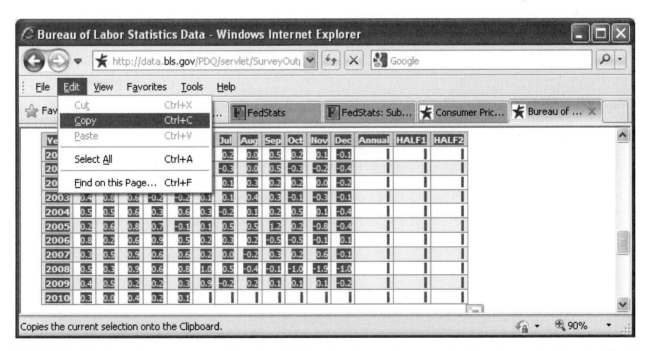

**FIGURE 2.5.6** Select the data by dragging across it; then choose Edit Copy from the menu system (press the Alt key if you don't see the menu). The data are now in the clipboard, ready to be pasted into a different application such as Excel®.

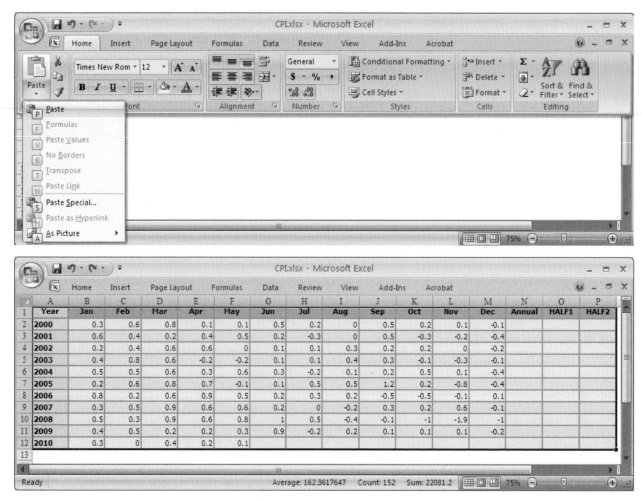

**FIGURE 2.5.7**   Paste the data from the clipboard into Excel® by choosing Paste as shown at top. The consumer price index data have been transferred as shown below.

### Example
#### Finding Home Depot Stock Market Data on Yahoo!

Recent trading data showing daily, weekly, or monthly stock market prices for individual companies (as well as for indexes such as the Dow Jones, Standard & Poor's, and Nasdaq) have been available at Yahoo![7] All it takes is a little searching, and you can often manage to find the data and arrange it in your spreadsheet. Begin at http://finance.yahoo.com where you can enter the stock market symbol ("HD" for Home Depot), as shown in Figure 2.5.8. If you do not know the symbol for a company's stock, you may type the company's name to obtain a list of potential matches. Clicking on GET QUOTES, we find current information about Home Depot stock, as shown in Figure 2.5.9. In order to find detailed data, we will need to choose Historical Prices at the left. The result, shown in Figure 2.5.10, is that we have found our data! To transfer it to a spreadsheet, we scroll down and choose Download to Spreadsheet, as shown in Figure 2.5.11. If a File Download
*(Continued)*

**FIGURE 2.5.8**   Financial data are available for download at many sites, including http://finance.yahoo.com, where we have entered "HD" for Home Depot (accessed July 2, 2010) and will then choose Get Quotes.

FIGURE 2.5.9 The Home Depot stock information page, from which we will choose Historical Prices to obtain data, as accessed at http://finance.yahoo.com on July 2, 2010.

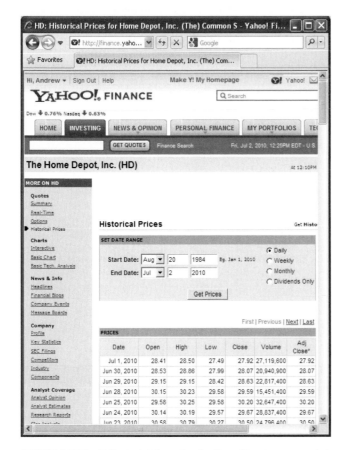

FIGURE 2.5.10 Daily stock information for Home Depot, as accessed at http://finance.yahoo.com on July 2, 2010. Note that we can also choose weekly or monthly, and can change the time span.

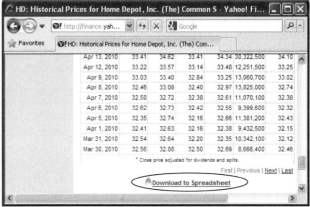

FIGURE 2.5.11 Scrolling down the page, we find the link to allow us to download our data to a spreadsheet for analysis, as accessed at http://finance.yahoo.com on July 2, 2010.

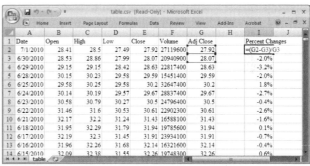

FIGURE 2.5.12 After the data have been downloaded to a worksheet, you can compute percentage changes by copying and pasting the formula shown at the top of Column I.

**Example—cont'd**

box opens, you choose Open. The result is that the data are transferred to a worksheet, where we have also computed the percentage changes (which are often used in financial analysis) using the formula as shown in Figure 2.5.12, and the data are ready for further analysis.

7. Please keep in mind that the Internet changes over time, that web pages may change their format, that new sources of information may appear, and that some information may no longer be available.

## 2.6 END-OF-CHAPTER MATERIALS

### Summary

A **data set** consists of some basic measurement or measurements of individual items or things called **elementary units,** which may be people, households, firms, cities, TV sets, or just about anything of interest. The same piece or pieces of information are recorded for each one. A piece of information recorded for every item (its cost, for example) is called a **variable**.

There are three basic ways of classifying a data set: (1) by the number of variables (univariate, bivariate, or multivariate), (2) by the kind of information (numbers or categories) represented by each variable, and (3) by whether the data set is a time sequence or comprises cross-sectional data.

**Univariate** (one-variable) data sets have just one piece of information recorded for each item. For univariate data, you can identify a typical summary value and get an indication of diversity, as well as note any special features or problems with the data.

**Bivariate** (two-variable) data sets have exactly two pieces of information recorded for each item. For bivariate data, in addition to looking at each variable as a univariate data set, you can study the relationship between the two variables and predict one variable from the other.

**Multivariate** (many-variable) data sets have three or more pieces of information recorded for each item. Also with multivariate data, you can look at each variable individually, as well as examine the relationship among the variables and predict one variable from the others.

Values of a variable that are recorded as meaningful numbers are called **quantitative** data. A **discrete** quantitative variable can assume values only from a list of specific numbers (such as 0 or 1, or the list 0, 1, 2, 3,…, for example). Any quantitative variable that is not discrete is, for our purposes, **continuous**. A continuous quantity is not restricted to a simple list of possible values.

If a variable indicates which of several nonnumerical categories an item falls into, it is a **qualitative** variable. If the categories have a natural, meaningful order, then it is an **ordinal** qualitative variable. If there is no such order, it is a **nominal** qualitative variable. Although it is often possible to record a qualitative variable using numbers, the variable remains qualitative; it is not quantitative because the numbers are not inherently meaningful.

With quantitative data, you have available all of the operations that can be used for numbers: counting, ranking, and arithmetic. With ordinal data, you have counting and ranking only. With nominal data, you have only counting.

If the data values are recorded in a meaningful sequence, then the data set is a **time series**. If the order of recording is not relevant, then the data set is **cross-sectional**. Time series analysis is more complex than that of cross-sectional data.

When you control the design of the data-collection plan (even if the work is done by others), you obtain **primary data**. When you use data previously collected by others for their own purposes, you are using **secondary data**. Primary data sets are often extensive and time-consuming to obtain, but can target exactly what you need. Secondary data sets are often inexpensive (or even free), but you might or might not find what you need.

## Key Words

bivariate, *20*
continuous, *21*
cross-sectional, *23*
data set, *19*
discrete, *21*
elementary units, *19*
multivariate, *20*
nominal, *23*
ordinal, *22*
primary data, *24*
qualitative, *22*
quantitative, *21*
secondary data, *24*
time series, *30*
univariate, *19*
variable, *19*

## Questions

1. What is a data set?
2. What is a variable?
3. What is an elementary unit?
4. What are the three basic ways in which data sets can be classified? (*Hint:* The answer is not "univariate, bivariate, and multivariate" but is at a higher level.)
5. What general questions can be answered by analysis of
   a.  Univariate data?
   b.  Bivariate data?
   c.  Multivariate data?
6. In what way do bivariate data represent more than just two separate univariate data sets?
7. What can be done with multivariate data?
8. What is the difference between quantitative and qualitative data?
9. What is the difference between discrete and continuous quantitative variables?
10. What are qualitative data?
11. What is the difference between ordinal and nominal qualitative data?
12. Differentiate between time-series data and cross-sectional data.
13. Which are simpler to analyze, time-series or cross-sectional data?
14. Distinguish between primary and secondary data.

## Problems

*Problems marked with an asterisk (\*) are solved in the Self Test in Appendix C.*

1. Name two different bivariate data sets that relate directly or indirectly to your responsibilities. In each case, identify the meaning of the relationship between the two factors, and indicate whether or not it would be useful to be able to predict one from the other.

2. Repeat the previous problem, but for multivariate data.

3. Choose a firm and name two quantitative variables that might be important to that firm. For each variable, indicate whether it is discrete or continuous.

4. Choose a firm and name two qualitative variables that might be important to that firm. For each variable, indicate whether it is nominal or ordinal.

5. Identify three time-series data sets of interest to you. For each one,
   a. Is there a definite time trend?
   b. Are there seasonal effects?

6. Choose a firm and identify a database (in general terms) that would be important to it. Identify three different kinds of data sets contained within this database. For each of these three data sets, identify the elementary unit and indicate what might be learned from an appropriate analysis.

7. Your firm has decided to sue an unreliable supplier. What kind of analysis would be used to estimate the forgone profit opportunities based on the performance of competitors, the overall state of the economy, and the time of year?

8. For each of the following data sets, say whether it is primary or secondary data.
   a. U.S. government data on recent economic activity, by state, being used by a company planning to expand.
   b. Production-cost data on recent items produced at your firm's factory, collected as part of a cost-reduction effort.
   c. Industry survey data purchased by your company in order to see how it compares to its competitors.

9. Classify a data set found on the Internet consisting of sales, profits, and number of employees for 100 banking firms. Be sure to provide standard information about the number of variables, who controlled the design of the data-gathering plan, and whether or not the data were recorded in a meaningful sequence.

10. If you think about a telephone directory as a large data set, what are the elementary units?

11.* Table 2.6.1 shows some items from a human resources database, representing the status of five people on May 3, 2010.
   a. What is an elementary unit for this data set?
   b. What kind of data set is this: univariate, bivariate, or multivariate?
   c. Which of these four variables are quantitative? Which are qualitative?
   d. Which variables, if any, are ordinal qualitative? Why?
   e. Is this a time series, or are these cross-sectional data?

12. Consider the data set in Table 2.6.2, which consists of observations on five production facilities (identified by their group ID).
   a. What is an elementary unit for this data set?
   b. What kind of data set is this: univariate, bivariate, or multivariate?
   c. Identify the qualitative variables, if any.

   d. Is there an ordinal variable here? If so, please identify it.
   e. Is this a time series, or are these cross-sectional data?

13. Table 2.6.3 consists of sales and income, both in hundred thousands of dollars, for a six-month period.
   a. What is an elementary unit for this data set?
   b. What kind of data set is this: univariate, bivariate, or multivariate?
   c. Which of these two variables are quantitative? Which are qualitative?
   d. Are these time-series or cross-sectional data?

**TABLE 2.6.1 Employment/History Status of Five People**

| Gender | Salary | Education* | Years of Experience |
|--------|--------|-----------|---------------------|
| M | $42,300 | HS | 8 |
| F | 42,400 | BA | 5 |
| M | 69,800 | MBA | 3 |
| F | 88,500 | MBA | 15 |
| F | 46,700 | BA | 6 |

*HS = High school diploma, BA = College degree, MBA = Master's degree in business administration.

**TABLE 2.6.2 Selected Product Output of Five Production Facilities**

| Group ID | Part | Quality | Employees |
|----------|------|---------|-----------|
| A-235-86 | Brakes | Good | 53 |
| W-186-74 | Fuel line | Better | 37 |
| X-937-85 | Radio | Fair | 26 |
| C-447-91 | Chassis | Excellent | 85 |
| F-258-89 | Wire | Good | 16 |

**TABLE 2.6.3 Sales and Income January through June**

| Sales | Income (loss) |
|-------|---------------|
| 350 | 30 |
| 270 | 23 |
| 140 | (2) |
| 280 | 14 |
| 410 | 53 |
| 390 | 47 |

14. Table 2.6.4 is an excerpt from a salesperson's database of customers.
    a. What is an elementary unit for this data set?
    b. What kind of data set is this: univariate, bivariate, or multivariate?
    c. Which of these variables are quantitative? Which are qualitative?
    d. Which of these variables are nominal? Which are ordinal?
    e. Are these time-series or cross-sectional data?

15. In order to figure out how much of the advertising budget to spend on various types of media (TV, radio, newspapers, etc.), you are looking at a data set that lists how much each of your competitors spent last year for TV, how much they spent for radio, and how much for newspaper advertising. Give a complete description of the type of data set you are looking at.

16. Your firm's sales, listed each quarter for the past five years, should be helpful for strategic planning.
    a. Is this data set cross-sectional or time-series?
    b. Is this univariate, bivariate, or multivariate data?

17. Consider a listing of the bid price and the ask price for 18 different U.S. Treasury bonds at the close of trading on a particular day.
    a. Is this univariate, bivariate, or multivariate data?
    b. Is this cross-sectional or time-series data?

18. You are looking at the sales figures for 35 companies.
    a. Is this data set univariate, bivariate, or multivariate?
    b. Is this variable qualitative or quantitative?
    c. Is this variable ordinal, nominal, or neither?

19. A quality control inspector has rated each batch produced today on a scale from A through E, where A represents the best quality and E is the worst.
    a. Is this variable quantitative or qualitative?
    b. Is this variable ordinal, nominal, or neither?

20. One column of a large inventory spreadsheet shows the name of the company that sold you each part.
    a. Is this variable quantitative or qualitative?
    b. Is this variable ordinal, nominal, or neither?

21. Consider the information about selected cell phones shown in Table 2.6.5.
    a. What is an elementary unit for this data set?
    b. Is this a univariate, bivariate, or multivariate data set?
    c. Is this a cross-sectional or time-series data set?
    d. Is "Service Provider" quantitative, ordinal, or nominal?
    e. Is "Price" quantitative, ordinal, or nominal?
    f. Is "Screen Size" quantitative, ordinal, or nominal?

22. Suppose a database includes the variable "security type" for which 1 = common stock, 2 = preferred stock, 3 = bond, 4 = futures contract, and 5 = option. Is this a quantitative or qualitative variable?

23. The ease of assembling products is recorded using the scale 1 = very easy, 2 = easy, 3 = moderate, 4 = difficult, 5 = very difficult. Is this a quantitative, ordinal, or nominal variable?

24. Suppose a data set includes the variable "business organization" recorded as 1 = sole proprietor, 2 = partnership, 3 = S corporation, 4 = C corporation. Is this a quantitative or qualitative variable?

25. Consider the information recorded in Table 2.6.6 for a selection of household upright vacuum cleaners.
    a. What is an elementary unit for this data set?
    b. What kind of a data set is this: univariate, bivariate, or multivariate?
    c. Which of these variables are quantitative? Which are qualitative?
    d. For each qualitative variable in this data set (if any), determine if it is nominal or ordinal.
    e. Is this cross-sectional or time-series data?

**TABLE 2.6.4 Selected Customers and Purchases**

| Level of Interest in New Products | Last Year's Total Purchases | Geographical Region |
|---|---|---|
| Weak | $88,906 | West |
| Moderate | 396,808 | South |
| Very strong | 438,442 | South |
| Weak | 2,486 | Midwest |
| Weak | 37,375 | West |
| Very strong | 2,314 | Northeast |
| Moderate | 1,244,096 | Midwest |
| Weak | 857,248 | South |
| Strong | 119,650 | Northeast |
| Moderate | 711,514 | West |
| Weak | 22,616 | West |

**TABLE 2.6.5 Information about Cell Phones**

| Model | Service Provider | Price | Screen Size |
|---|---|---|---|
| T-Mobile G2 | T-Mobile | $229 | Large |
| LG Encore | AT&T | 49 | Medium |
| Motorola Rambler | Boost Mobile | 99 | Small |
| BlackBerry Curve 8530 | Virgin Mobile | 249 | Small |
| Samsung Craft SCH-R900 | MetroPCS | 299 | Large |

Source: Data were adapted from information accessed at http://www.pcmag.com on October 2, 2010.

26. The Dow Jones company calculates a number of stock market index numbers that are used as indicators of the performance of the New York Stock Exchange. The best known of these is the Dow Jones Industrial Average (DJIA), which is calculated based on the performance of 30 stocks from companies categorized as general industry. Observations for each of the 30 companies in the DJIA are shown in Table 2.6.7.
   a. What is an elementary unit for this data set?
   b. What kind of a data set is this: univariate, bivariate, or multivariate?
   c. Which of these variables are quantitative? Which are qualitative?
   d. If there are any qualitative variables in this data set, are they nominal or ordinal?
   e. Is this cross-sectional or time-series data?

27. Let's continue to look at the DJIA discussed in problem 26. Table 2.6.8 shows 23 daily observations of the value of the DJIA, with 22 observations of the net change from one observation to the next, and the percent change in the DJIA from one observation to the next.
   a. What is an elementary unit for this data set?
   b. What kind of a data set is this: univariate, bivariate, or multivariate?

   c. Which of these variables are quantitative? Which are qualitative?
   d. If there are any qualitative variables in this data set, are they nominal or ordinal?
   e. Is this cross-sectional or time-series data?

TABLE 2.6.6 Comparison of Upright Vacuum Cleaners

| Price | Weight (lbs) | Quality | Type |
| --- | --- | --- | --- |
| $170 | 17 | Good | Hard-body |
| 260 | 17 | Fair | Soft-body, self-propelled |
| 100 | 21 | Good | Soft-body |
| 90 | 14 | Good | Hard-body |
| 340 | 13 | Excellent | Soft-body |
| 120 | 24 | Good | Soft-body, self-propelled |
| 130 | 17 | Fair | Soft-body, self-propelled |

TABLE 2.6.7  Closing Price and Monthly Percent Change for the Companies in the DJIA

| Company Name | Closing Price February 1, 2002 | Percent Change from January 2, 2002 | Company Name | Closing Price February 1, 2002 | Percent Change from January 2, 2002 |
| --- | --- | --- | --- | --- | --- |
| 3M | 113.27 | −3.30% | Honeywell | 33.68 | 0.84% |
| Alcoa | 35.15 | −1.49 | IBM | 108.00 | −11.11 |
| American Express | 35.10 | −2.17 | Intel Corp. | 34.67 | 5.06 |
| AT&T | 17.33 | −7.33 | International Paper | 41.77 | 4.03 |
| Boeing | 41.46 | 8.82 | J. P. Morgan Chase & Co. | 32.16 | −11.53 |
| Caterpillar | 50.52 | −2.19 | Johnson & Johnson | 57.60 | −1.87 |
| Citigroup | 46.49 | −8.86 | McDonald's | 26.63 | 0.53 |
| Coca-Cola | 44.68 | −5.88 | Merck | 59.43 | −0.55 |
| DuPont | 43.58 | 1.54 | Microsoft Corp. | 62.66 | −6.53 |
| Eastman Kodak | 28.25 | −3.62 | Philip Morris | 49.73 | 6.63 |
| Exxon Mobil | 39.00 | −1.52 | Procter & Gamble | 82.57 | 3.21 |
| General Electric | 36.85 | −10.01 | SBC Communications | 39.96 | 0.15 |
| General Motors | 51.11 | 5.08 | United Technologies | 70.05 | 8.87 |
| Hewlett-Packard | 22.00 | 1.62 | Wal-Mart Stores | 59.26 | 2.08 |
| Home Depot | 49.40 | −1.96 | Walt Disney | 22.45 | 4.66 |

**Source:** Data accessed at http://finance.yahoo.com/?u on February 4, 2002.

**TABLE 2.6.8 Daily Values and Changes of the DJIA during January 2002**

| Date | DJIA | Net Change | Percent Change |
|------|------|-----------|----------------|
| 31-Jan-02 | 9,920.00 | 157.14 | 1.58% |
| 30-Jan-02 | 9,762.86 | 144.62 | 1.48 |
| 29-Jan-02 | 9,618.24 | –247.51 | –2.57 |
| 28-Jan-02 | 9,865.75 | 25.67 | 0.26 |
| 25-Jan-02 | 9,840.08 | 44.01 | 0.45 |
| 24-Jan-02 | 9,796.07 | 65.11 | 0.66 |
| 23-Jan-02 | 9,730.96 | 17.16 | 0.18 |
| 22-Jan-02 | 9,713.80 | –58.05 | –0.60 |
| 18-Jan-02 | 9,771.85 | –78.19 | –0.80 |
| 17-Jan-02 | 9,850.04 | 137.77 | 1.40 |
| 16-Jan-02 | 9,712.27 | –211.88 | –2.18 |
| 15-Jan-02 | 9,924.15 | 32.73 | 0.33 |
| 14-Jan-02 | 9,891.42 | –96.11 | –0.97 |
| 11-Jan-02 | 9,987.53 | –80.37 | –0.80 |
| 10-Jan-02 | 10,067.90 | –26.20 | –0.26 |
| 9-Jan-02 | 10,094.10 | –56.40 | –0.56 |
| 8-Jan-02 | 10,150.50 | –46.50 | –0.46 |
| 7-Jan-02 | 10,197.00 | –62.70 | –0.61 |
| 4-Jan-02 | 10,259.70 | 87.60 | 0.85 |
| 3-Jan-02 | 10,172.10 | 98.70 | 0.97 |
| 2-Jan-02 | 10,073.40 | 51.90 | 0.52 |
| 31-Dec-01 | 10,021.50 | – | – |

**Source:** Data accessed at http://finance.yahoo.com/?u on February 4, 2002.

## Database Exercises

*Problems marked with an asterisk (\*) are solved in the Self Test in Appendix C.*

Refer to the employee database in Appendix A.

1. Describe and classify this database and its parts:
   a.\* Is this a univariate, bivariate, or multivariate data set?
   b. What are the elementary units?
   c. Which variables are qualitative and which are quantitative?
   d.\* Is "training level" ordinal or nominal? Why?
   e. Would you ever want to do arithmetic on "employee number"? What does this tell you about whether this is truly a quantitative variable?
   f. Is this a time series, or are these cross-sectional data?
2.\* For each variable in this database, tell which of the following operations would be appropriate:
   a. Arithmetic (adding, subtracting, etc.).
   b. Counting the number of employees in each category.
   c. Rank ordering.
   d. Finding the percentage of employees in each category.

## Projects

1. Find a table of data on the Internet or used in an article in a business magazine or newspaper and copy the article and table.
   a. Classify the data set according to the number of variables.
   b. What are the elementary units?
   c. Are the data time-series or cross-sectional?
   d. Classify each variable according to its type.
   e. For each variable, tell which operations are appropriate.
   f. Discuss (in general terms) some questions of interest in business that might be answered by close examination of this kind of data set.
2. Find a table of data in one of your firm's internal reports or record one from your own experience. Answer each part of project 1 for this data table.
3. Search the Internet for investment data on a company of interest to you. Write one page reporting on the various kinds of information available, and attach a printout with some numerical data.

## Chapter 3

# Histograms

## Looking at the Distribution of Data

Your partner has been staring at that huge table of customer expenditures on competitors' products for half an hour now, hoping for enlightenment, trying to learn as much as possible from the numbers in the column, and even making some progress (as you can tell from occasional exclamations of "They're mostly spending $10 to $15!" "Hardly anybody is spending over $35!" and "Ooh—here's one at $58!"). You know you really should tell your partner to use a chart instead, such as a histogram, because it would save time and give a more complete picture. The only problem here is the psychology of how to bring up the subject without bruising your partner's ego.

In this chapter, you will learn how to make sense of a list of numbers. A *histogram* is a picture that gives you a visual impression of many of the basic properties of the data set as a whole, answering the following kinds of questions:

**One:** What values are typical in this data set?
**Two:** How different are the numbers from one another?
**Three:** Are the data values strongly concentrated near some typical value?
**Four:** What is the pattern of concentration? In particular, do data values "trail off" at the same rate at lower values as they do at higher values?

**Five:** Are there any special data values, very different from the rest, that might require special treatment?
**Six:** Do you basically have a single, homogeneous collection, or are there distinct groupings within the data that might require separate analysis?

Many standard methods of statistical analysis require that the data be approximately *normally distributed*. You will learn how to recognize this basic bell-shaped pattern and see how to transform the data if they do not already satisfy this assumption.

## 3.1 A LIST OF DATA

The simplest kind of data set is a **list of numbers** representing some kind of information (a single statistical variable) measured on each item of interest (each elementary unit). A list of numbers can show up in several forms that may look very different at first. It may help you to ask yourself, What are the elementary units being measured here? to distinguish the actual measurements from their frequencies.

**Practical Business Statistics, Sixth Edition.**
© 2012 Andrew F. Siegel. Published by Elsevier, Inc. All rights reserved.

## Example
### *Performance of Regional Sales Managers*

Here is an example of a very short list (only three observations), for which the variable is "last quarter sales" and the elementary units are "regional sales managers":

| Name | Sales (ten thousands) |
| --- | --- |
| Bill | 28 |
| Jennifer | 32 |
| Henry | 18 |

This data set contains information for interpretation (i.e., the first name of the sales manager responsible, indicating the elementary unit in each case) in addition to the list of three numbers. In other cases, the column of elementary units may be omitted; the first column would then be a variable instead.

## Example
### *Household Size*

Sometimes a list of numbers is given as a table of frequencies, as in this example of family sizes from a sample of 17 households:

| Household Size (number of people) | Number of Households (frequency) |
| --- | --- |
| 1 | 3 |
| 2 | 5 |
| 3 | 6 |
| 4 | 2 |
| 5 | 0 |
| 6 | 1 |

The key to interpreting a table like this is to observe that it represents a list of numbers in which each number on the left (household size) is repeated according to the number to its right (the frequency—in this case, the number of households). The resulting list of numbers represents the number of people in each household:

**1, 1, 1**, 2, 2, 2, 2, 2, 3, 3, 3, 3, 3, 3, 4, 4, 6

Note that 1 is repeated three times (as indicated by the first row in the data table), 2 is repeated five times (as indicated by the second row), and so on.

The frequency table is especially useful for representing a very long list of numbers with relatively few values. Thus, for a large sample, you might summarize household size as follows:

| Household Size (number of people) | Number of Households (frequency) |
| --- | --- |
| 1 | 342 |
| 2 | 581 |
| 3 | 847 |
| 4 | 265 |
| 5 | 23 |
| 6 | 11 |
| 7 | 2 |

This table represents a lot of data! The corresponding list of numbers would begin by listing 1 a total of 342 times, 2 a total of 581 times, and so on. The table represents the sizes of all 2,071 households in this large sample.[1]

---

1. The number 2,071 is the total frequency, the sum of the right-hand column.

## The Number Line

In order to visualize the relative magnitudes of a list of numbers, we will use locations along a line to represent numbers. The **number line** is a straight line with the scale indicated by numbers:

It is important that the numbers be regularly spaced on a number line so that there is no distortion.[2] You can show the location of each number in the list by placing a mark at its location on the number line. For example, the list of sales figures

28, 32, 18

could be displayed on the number line as follows:

This diagram gives you a very clear impression of how these numbers relate to one another. In particular, you immediately see that the top two are relatively close to one another and are a good deal larger than the smallest number.

Using graphs such as the number line and others that you will study is more informative than looking at lists of numbers. Although numbers do a good job of recording information, they do not provide you with an appropriate visual hint as to their magnitudes. For example, the sequence

0 1 2 3 4 5 6 7 8 9

gives no particular *visual* indication of progressively larger magnitudes; the numerals do not get larger in size or darker as you move through the list. The number line, in contrast, does a nice job of showing you these important magnitudes.

---

2. When it is necessary to distort the line, for example, by skipping over some uninteresting intermediate values, you should show a break in the line. In this way, you won't give the misleading impression of a regular, continuous line.

## 3.2 USING A HISTOGRAM TO DISPLAY THE FREQUENCIES

The **histogram** displays the frequencies as a bar chart rising above the number line, indicating how often the various values occur in the data set. The horizontal axis represents the measurements of the data set (in dollars, number of people, miles per gallon, or whatever), and the vertical axis represents how often these values occur. An especially high bar indicates that many cases had data values at this position on the horizontal number line, while a shorter bar indicates a less common value.

### Example

#### Mortgage Interest Rates

Consider the interest rate for 30-year fixed-rate home mortgages charged by mortgage companies in Seattle, shown in Table 3.2.1. The histogram is shown in Figure 3.2.1. We will now describe how to interpret a histogram in general and at the same time will explain what this particular picture tells you about interest rates.

The horizontal number line at the bottom of the figure indicates mortgage rates, in percentage points, while the vertical line at the left indicates the frequency of occurrence of a mortgage rate. For example, the next-to-last bar at the right (extending horizontally from a mortgage rate of 4.6% to 4.8%) has a frequency (height) of 5, indicating that there are five financial institutions offering a mortgage rate between 4.6% and 4.8%.[3] Thus, you have a picture of the pattern of interest rates, indicating which values are most common, which are less common, and which are not offered at all.

What can you learn about interest rates from this histogram?

1. The range of values. Interest rates range over slightly more than a percentage point, from a low of about 4.0% to a high of about 5.4% (these are the left and right boundaries of the histogram; while the exact highest and lowest can be found by sorting the data, we are interested here in reading the histogram, which gives us a good overall impression).
2. The typical values. Rates from about 4.2% to 4.8% are the most common (note the taller bars in this region).
3. The diversity. It is not unusual for institutions to differ from one another by about 0.5% (there are moderately high bars separated by about half of a percentage point).
4. The overall pattern. Most institutions are concentrated slightly to the left of the middle of the range of values (tall bars here), with some institutions offering higher rates (the bar at the right), and one institution at the far left daring to offer an attractive lower rate (final bar with frequency of one at the left side).
5. Any special features. Perhaps you noticed that the histogram for this example appears to be missing

two bars—from 4.8% to 5.2%. Apparently, no institution offered a rate of 4.8% or more but less than 5.2%.

---

3. It is conventional to count all data values that fall exactly on the boundary between two bars of a histogram as belonging to the bar on the right. In this particular case, the bar from 4.6% to 4.8% along the number line includes all companies whose mortgage rate is equal to or greater than the left endpoint (4.6%) but less than the right endpoint (4.8%). An institution offering 4.8% (if there were one) would be in the next bar, to the right of 4.8 and extending to 5.

### TABLE 3.2.1 Home Mortgage Rates

| Lender | Interest Rate |
|---|---|
| AimLoan.com | 4.125% |
| America Funding, Inc | 4.250% |
| Bank of America | 4.625% |
| CapWest Mortgage Corp | 4.500% |
| Cascade Pacific Mortgage | 4.500% |
| CenturyPoint Mortgage | 4.250% |
| CloseYourOwnLoan.com | 4.625% |
| Envoy Mortgage | 4.375% |
| First Savings Bank Northwest | 5.375% |
| Guild Mortgage Co | 5.250% |
| Habitat Financial | 4.375% |
| Hart West Financial Inc | 4.250% |
| LendingTree Loans | 4.750% |
| Loan Network LLC | 4.250% |
| National Bank of Kansas City | 4.250% |
| National Mortgage Alliance | 4.250% |
| Nationwide Bank | 4.250% |
| Pentagon Federal Credit Union Mtg | 4.250% |
| Quicken Loans | 4.500% |
| RMC Vanguard Mortgage Corp | 4.250% |
| SurePoint Lending | 4.750% |
| The Lending Company | 4.250% |
| The Money Store | 4.500% |
| Washington Trust Bank | 4.750% |
| Your Equity Services | 4.250% |

Source: Data are from http://realestate.yahoo.com, http://www.zillow.com/, http://www.bankrate.com, and https://www.google.com on July 2, 2010.

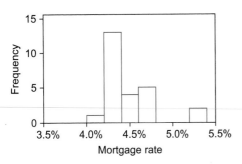

**FIGURE 3.2.1**   A histogram of mortgage interest rates.

While Microsoft® Excel® comes with an add-in that can be used to draw a histogram, it is often preferable to use either a different add-in or to use stand-alone statistical software. To use Excel® to construct a histogram, you can use the Data Analysis choice in the Analysis category under the Data Ribbon[4] and select Histogram from the options presented:

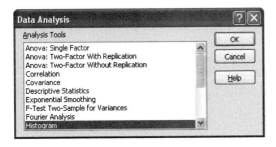

Next, in the dialog box that appears, select your data (by dragging across it or, if it has been named, by typing the name), place a checkmark for Chart Output, and specify a location for the output:

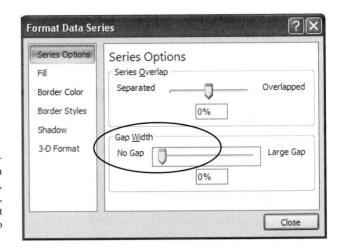

After you choose OK, the result appears as follows:

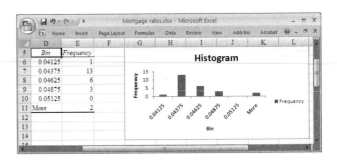

Here, the bars are too skinny for this to be a true histogram because they do not fully cover the part of the (horizontal) number line that they represent. This can be fixed by right-clicking on a bar and choosing Format Data Series:

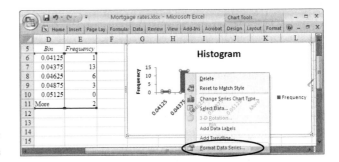

Next, select the Series Options tab in the dialog box and use the slider to set the Gap Width to zero, as follows:

Finally after clicking Close, we obtain an actual histogram where the gaps would not be confused with a lack of data:

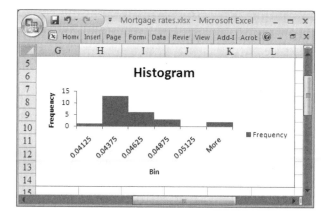

As you can see, creating a histogram in Excel® is not a simple process, especially if you choose to customize your histogram by specifying the bar width (by specifying the Bin Range in the dialog box). As an alternative, you might choose to use StatPad (an Excel add-in) or another software product to correct these problems.

## Histograms and Bar Charts

*A histogram is a bar chart of the frequencies, not of the data.* The height of each bar in the histogram indicates how frequently the values on the horizontal axis occur in the data set. This gives you a visual indication of where data values are concentrated and where they are scarce. Each bar of the histogram may represent many data values (in fact, the height of the bar shows you exactly how many data values are included in the corresponding range). This is different from a bar chart of the actual data, where there is one bar for each data value. Also note that the horizontal axis is always meaningful for a histogram but not necessarily so for a bar chart.

### Example
#### Starting Salaries for Business Graduates

Consider the typical starting salaries for graduating business students in various fields, as shown in Table 3.2.2. Compare the histogram of these data values in Figure 3.2.2 to the bar chart shown in Figure 3.2.3. Note that the bars in the histogram show the number of fields in each salary range, while the bars in the bar chart show the actual salary for that field of business.

Both graphs are useful. The bar chart is most helpful when you want to see all of the details including the identification of each individual data value, when the data set is small enough to allow you to see each one. However, the histogram is far superior for visualizing the data set as a whole, especially for a large data set representing many numbers.

**TABLE 3.2.2 Starting Salaries for Business Graduates**

| Field | Salary |
|---|---|
| Accounting | $67,250 |
| Administrative Services Manager | 70,720 |
| Advertising and Promotions Manager | 57,130 |
| Economics | 77,657 |
| Health Care Management | 56,000 |
| Hotel Administration | 44,638 |
| Human Resources | 69,500 |
| Management Information Systems | 105,980 |
| Marketing Manager | 84,000 |
| Nonprofit Organization Manager | 42,772 |
| Sales Manager | 75,040 |
| Sports Administrator | 49,637 |

**Source:** Accessed at http://www.allbusinessschools.com/faqs/salaries on July 2, 2010.

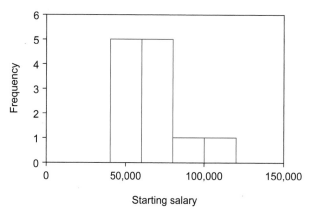

**FIGURE 3.2.2** A histogram of the starting salaries. Note that each bar may represent more than one field of business (read the number on the left). The bars show which salary ranges are most and least typical in this data set. In particular, note that most salaries fall within the range from $40,000 to $80,000 as represented by the tallest two bars representing five fields each.

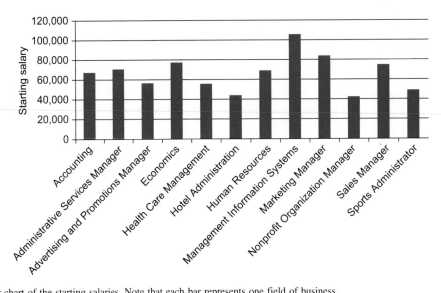

**FIGURE 3.2.3**   A bar chart of the starting salaries. Note that each bar represents one field of business.

## 3.3 NORMAL DISTRIBUTIONS

A **normal distribution** is an idealized, smooth, bell-shaped histogram with all of the randomness removed. It represents an ideal data set that has lots of numbers concentrated in the middle of the range, with the remaining numbers trailing off symmetrically on both sides. This degree of smoothness is not attainable by real data. Figure 3.3.1 is a picture of a normal distribution.[5]

There are actually many different normal distributions, all symmetrically bell-shaped. They differ in that the center can be anywhere, and the scale (the width of the bell) can have any size.[6] Think of these operations as taking the basic bell shape and sliding it horizontally to wherever you'd like the center to be and then stretching it out (or compressing it) so that it extends outward just the right amount. Figure 3.3.2 shows a few normal distributions.

Why is the normal distribution so important? It is common for statistical procedures to assume that the data set is reasonably approximated by a normal distribution.[7] Statisticians know a lot about properties of normal distributions; this knowledge can be exploited whenever the histogram resembles a normal distribution.

How do you tell if a data set is normally distributed? One good way is to look at the histogram. Figure 3.3.3

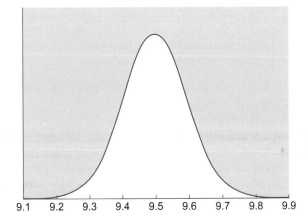

**FIGURE 3.3.1**   A normal distribution, in its idealized form. Actual data sets that follow a normal distribution will show some random variations from this perfectly smooth curve.

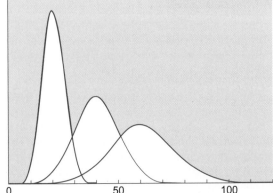

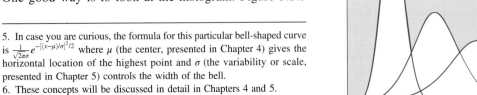

**FIGURE 3.3.2**   Some normal distributions with various centers and scales.

---

5. In case you are curious, the formula for this particular bell-shaped curve is $\frac{1}{\sqrt{2\pi}\sigma}e^{-[(x-\mu)/\sigma]^2/2}$ where $\mu$ (the center, presented in Chapter 4) gives the horizontal location of the highest point and $\sigma$ (the variability or scale, presented in Chapter 5) controls the width of the bell.

6. These concepts will be discussed in detail in Chapters 4 and 5.

7. In particular, many standard methods for computing confidence intervals and hypothesis tests (which you will learn later on) require a normal distribution, at least approximately, for the data.

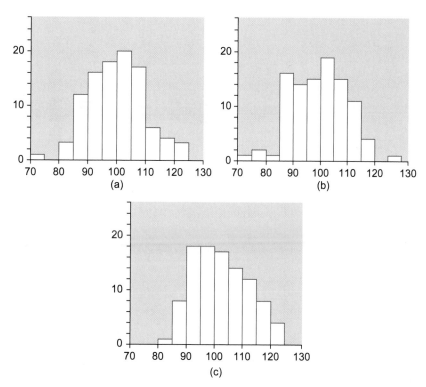

**FIGURE 3.3.3**  Histograms of data drawn from an ideal normal distribution. In each case, there are 100 data values. Comparing the three histograms, you can see how much randomness to expect.

shows different histograms for samples of 100 data values from a normal distribution. From these, you can see how random the shape of the distribution can be when you have only a finite amount of data. Fewer data values imply more randomness because there is less information available to show you the big picture. This is shown in Figure 3.3.4, which displays histograms of 20 data values from a normal distribution.

**Example**
*Stock Price Gains*

Consider the percentage gain in stock price for a collection of northwest firms, as shown in Table 3.3.1. These stock price gains appear to be approximately normally distributed, with a symmetric bell shape, even though we also can see from the histogram that 2008 was not a good year for these companies (nor for the economy in general) because the typical firm's stock lost about 50% of its value. See Figure 3.3.5.

In real life, are all data sets normally distributed? No. It is important to explore the data, by looking at a histogram, to determine whether or not it is normally distributed. This is especially important if, later in the analysis, a standard statistical calculation will be used that requires a normal distribution. The next section shows

**TABLE 3.3.1 Stock Price Percentage Gains for Northwest Companies in 2008**

| Company | Stock Price Gain |
|---|---|
| Alaska Air Group | 17.0% |
| Amazon.com | −44.6% |
| Ambassadors Group | −49.8% |
| American Ecology | −13.8% |
| Avista | −10.0% |
| Banner | −67.2% |
| Barrett Business Services | −39.5% |
| Blue Nile | −64.0% |
| Cardiac Science | −7.3% |
| Cascade Bancorp | −51.5% |
| Cascade Corp | −35.7% |
| Cascade Financial | −60.0% |
| Cascade Microtech | −80.9% |
| City Bank | −76.8% |
| Coeur d'Alene Mines | −82.2% |

*(Continued)*

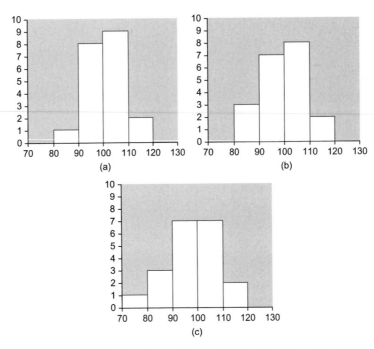

**FIGURE 3.3.4**  Data drawn from a normal distribution. In each case, there are 20 data values. Comparing the histograms, you can see how much randomness to expect.

| TABLE 3.3.1 Stock Price Percentage Gains for Northwest Companies in 2008—cont'd | |
| --- | --- |
| **Company** | **Stock Price Gain** |
| Coinstar | −30.7% |
| Coldwater Creek | −57.4% |
| Columbia Bancorp | −87.8% |
| Columbia Banking System | −59.9% |
| Columbia Sportswear | −19.8% |
| Concur Technologies | −9.4% |
| Costco Wholesale | −24.7% |
| Cowlitz Bancorporation | −49.9% |
| Data I/O | −63.4% |
| Esterline Technologies | −26.8% |
| Expedia | −73.9% |
| Expeditors International | −25.5% |
| F5 Networks | −19.8% |
| FEI | −24.0% |
| Fisher Communications | −40.3% |
| Flir Systems | −2.0% |
| Flow International | −74.0% |
| Frontier Financial | −76.5% |

| | |
| --- | --- |
| Greenbrier | −69.1% |
| Hecla Mining | −70.1% |
| Heritage Financial | −38.4% |
| Home Federal Bancorp | 6.8% |
| Horizon Financial | −72.8% |
| Idacorp | −16.4% |
| InfoSpace | −19.2% |
| Intermec | −34.6% |
| Itron | −33.6% |
| Jones Soda | −95.7% |
| Key Technology | −45.3% |
| Key Tronic | −76.8% |
| LaCrosse Footwear | −24.9% |
| Lattice Semiconductor | −53.5% |
| Lithia Motors | −76.3% |
| Marchex | −46.3% |
| McCormick & Schmick's | −66.3% |
| Merix | −94.0% |
| Micron Technology | −63.6% |
| Microsoft | −45.4% |

**TABLE 3.3.1 Stock Price Percentage Gains for Northwest Companies in 2008—cont'd**

| Company | Stock Price Gain |
|---|---|
| MWI Veterinary Supply | –32.6% |
| Nautilus Group | –54.4% |
| Nike | –20.6% |
| Nordstrom | –63.8% |
| Northwest Natural Gas | –9.1% |
| Northwest Pipe | 8.9% |
| Paccar | –47.3% |
| Pacific Continental | 19.6% |
| Planar Systems | –90.5% |
| Plum Creek Timber | –24.5% |
| Pope Resources | –53.2% |
| Portland General Electric | –29.9% |
| Precision Castparts | –57.1% |
| PremierWest Bancorp | –41.5% |
| Puget Energy | –0.6% |
| RadiSys | –58.7% |
| Rainier Pacific Financial Group | –90.5% |
| RealNetworks | –42.0% |
| Red Lion Hotels | –76.1% |
| Rentrak | –18.5% |
| Riverview Bancorp | –80.5% |
| Schmitt Industries | –37.8% |
| Schnitzer Steel Industries | –45.5% |
| SeaBright Insurance Holdings | –22.1% |
| SonoSite | –43.3% |
| StanCorp Financial Group | –17.1% |
| Starbucks | –53.8% |
| Sterling Financial | –47.6% |
| Timberland Bancorp | –38.8% |
| Todd Shipyards | –36.9% |
| TriQuint Semiconductor | –48.1% |
| Umpqua Holdings | –5.7% |
| Washington Banking | –44.9% |
| Washington Federal | –29.1% |

| West Coast Bancorp | –64.4% |
|---|---|
| Weyerhaeuser | –58.5% |
| Zumiez | –69.4% |

**Source:** Accessed at http://seattletimes.nwsource.com/flatpages/business technology/2009northwestcompaniesdatabase.html on March 27, 2010.

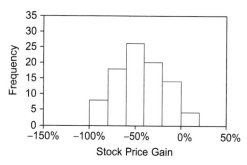

**FIGURE 3.3.5**  A histogram of the stock percentage price gains for these companies shows that the distribution is approximately normal for this economically difficult time period.

one way in which many data sets in business deviate from a normal distribution and suggests a way to deal with the problem.

## 3.4 SKEWED DISTRIBUTIONS AND DATA TRANSFORMATION

A **skewed distribution** is neither symmetric nor normal because the data values trail off more sharply on one side than on the other. In business, you often find skewness in data sets that represent sizes using positive numbers (for example, sales or assets). The reason is that data values cannot be less than zero (imposing a boundary on one side) but are not restricted by a definite upper boundary. The result is that there are lots of data values concentrated near zero, and they become systematically fewer and fewer as you move to the right in the histogram. Figure 3.4.1 gives some examples of idealized shapes of skewed distributions.

### Example
#### Deposits of Banks and Savings Institutions

An example of a highly skewed distribution is provided by the deposits of large banks and savings institutions, shown in Table 3.4.1. A histogram of this data set is shown in Figure 3.4.2. This is not at all like a normal distribution because of the lack of symmetry. The very high bar at the left represents the majority of these banks, which have less than $50 billion in deposits. The bars to the right represent the (relatively few) banks that are larger. Each of the six very short bars at far right represents a single bank, with the very largest being Bank of America with $818 billion.

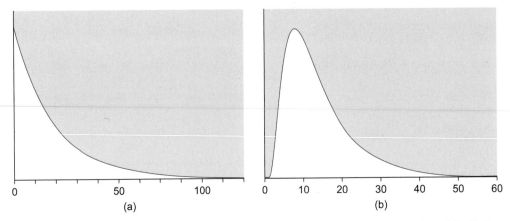

**FIGURE 3.4.1** Some examples of skewed distributions, in smooth, idealized form. Actual data sets that follow skewed distributions will show some random differences from this kind of perfectly smooth curve.

### Example
#### Populations of States

Another example of a skewed distribution is the populations of the states of the United States, viewed as a list of numbers.[8] The skewness reflects the fact that there are many states with small or medium populations and a few states with very large populations (the three largest are California, Texas, and New York). A histogram is shown in Figure 3.4.3.

8. U.S. Census Bureau, *Statistical Abstract of the United States: 2010* (129th Edition), Washington, DC, 2009, accessed at http://www.census .gov/compendia/statab/rankings.html on July 3, 2010.

**TABLE 3.4.1 Deposits of Large Banks and Savings Institutions**

| Bank | Deposits ($ billions) |
|---|---|
| Bank of America | 818 |
| JPMorgan Chase Bank | 618 |
| Wachovia Bank | 394 |
| Wells Fargo Bank | 325 |
| Citibank | 266 |
| U.S. Bank | 152 |
| SunTrust Bank | 119 |
| National City Bank | 101 |
| Branch Banking and Trust Company | 94 |
| Regions Bank | 94 |
| PNC Bank | 84 |
| HSBC Bank USA | 84 |
| TD Bank | 79 |
| RBS Citizens | 78 |
| ING Bank, fsb | 75 |
| Capital One | 73 |
| Keybank | 67 |
| Merrill Lynch Bank USA | 58 |
| The Bank of New York Mellon | 57 |
| Morgan Stanley Bank | 56 |
| Union Bank | 56 |
| Sovereign Bank | 49 |
| Citibank (South Dakota) N.A. | 47 |
| Manufacturers and Traders Trust Company | 45 |
| Fifth Third Bank | 41 |
| Comerica Bank | 40 |
| The Huntington National Bank | 39 |
| Compass Bank | 37 |
| Goldman Sachs Bank | 36 |
| Bank of the West | 34 |
| Marshall and Ilsley Bank | 33 |
| Charles Schwab Bank | 32 |
| Fifth Third Bank | 32 |
| USAA Federal Savings Bank | 32 |
| E-Trade Bank | 30 |
| UBS Bank | 30 |
| Discover Bank | 29 |
| Merrill Lynch Bank and Trust Co | 29 |

**TABLE 3.4.1** Deposits of Large Banks and Savings Institutions—cont'd

| Bank | Deposits ($ billions) |
|---|---|
| Capital One Bank (USA) | 27 |
| Harris National Association | 27 |
| TD Bank USA, National Association | 26 |
| Ally Bank | 25 |
| Citizens Bank of Pennsylvania | 25 |
| Hudson City Savings Bank | 22 |
| Chase Bank USA | 21 |
| State Street Bank and Trust Co | 21 |
| Colonial Bank | 20 |
| RBC Bank (USA) | 19 |
| Banco Popular de Puerto Rico | 18 |
| Associated Bank | 16 |

Source: Accessed at http://nyjobsource.com/banks.html on July 2, 2010.

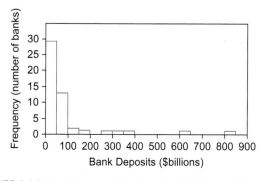

**FIGURE 3.4.2**  A histogram of the deposits (in billions of dollars) of large banks and savings institutions. This is a skewed distribution, not a normal distribution, and has a long tail toward high values (to the right).

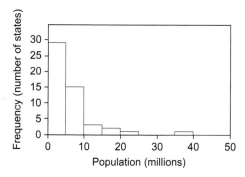

**FIGURE 3.4.3**  A histogram of the 2009 populations of the states of the United States: a skewed distribution.

## The Trouble with Skewness

One of the problems with skewness in data is that, as mentioned earlier, many of the most common statistical methods (which you will learn more about in future chapters) require at least an approximately normal distribution. When these methods are used on skewed data, the answers may well be misleading or just plain wrong. Even when the answers are basically correct, there is often some efficiency lost; essentially, the analysis has not made the best use of all of the information in the data set.

## Transformation to the Rescue

One solution to this dilemma of skewness is to use *transformation* to make a skewed distribution more symmetric. **Transformation** is replacing each data value by a different number (such as its logarithm) to facilitate statistical analysis. The most common transformation in business and economics is the logarithm, which can be used only on positive numbers (i.e., if your data include negative numbers or zero, this technique cannot be used). Using the **logarithm** often transforms skewness into symmetry because it stretches the scale near zero, spreading out all of the small values, which had been bunched together. It also pulls together the very large data values, which had been thinly spread out at the high end. Both types of logarithms (base 10 "common logs" and base *e* "natural logs") work equally well for this purpose. In this section, base 10 logs will be used.

### Example

*Transforming State Populations*

Comparing the histogram of state populations in Figure 3.4.3 to the histogram of the logarithms (base 10) of these numbers in Figure 3.4.4, you can see that the skewness vanishes when these numbers are viewed on the logarithmic scale. Although there is some randomness here, and the result is not perfectly symmetric, there is no longer the combination of a sharp drop on one side and a slow decline on the other, as there was in Figure 3.4.3.

The logarithmic scale may be interpreted as a *multiplicative* or *percentage* scale rather than an additive one. On the logarithmic scale, as displayed in Figure 3.4.4, the distance of 0.2 across each bar corresponds to a 58% increase in population from the left to the right side of the bar.[9] A span of five bars—for example, from points 6 to 7 on the horizontal axis—indicates a ten-fold increase in state population.[10] On the original scale (i.e., displaying actual numbers of people instead of logarithms), it is difficult to make a percentage comparison. Instead, in Figure 3.4.3, you see a difference of 5 million people as you move from left to right across one bar, and a difference of 5 million people is a much larger percentage on the left side than on the right side of the figure.

9. The reaspm is that $10^{0.2}$ is 1.58, which is 58% larger than 1.
10. The reason is that $10^1$ is 10.

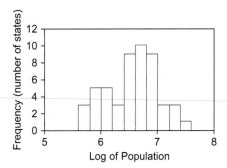

**FIGURE 3.4.4** Transformation can turn skewness into symmetry. A histogram of the logarithms (base 10) of the 2009 populations of the states of the United States is basically symmetric, except for randomness. Essentially no systematic skewness remains.

## Interpreting and Computing the Logarithm

A difference of 1 in the logarithm (to the base 10) corresponds to a factor of 10 in the original data. For example, the data values 392.1 and 3,921 (a ratio of 1 to 10) have logarithms of 2.59 and 3.59 (a difference of 1), respectively. Table 3.4.2 gives some examples of numbers and their logarithms.

From this, you can see how the logarithm pulls in the very large numbers, minimizing their difference from other values in the set (changing 100 million to 8, for example). Also note how the logarithm shows roughly how many digits are in the nondecimal part of a number. California's population of 31,878,234, for example, has a logarithm of 7.5035 (corresponding to the bar on the far right side of Figures 3.4.3 and 3.4.4).

There are two kinds of logarithms. We have looked at the base 10 logarithms. The other kind is the *natural logarithm*,

**TABLE 3.4.2 Some Examples of Logarithms to the Base 10**

| Number | Logarithm |
|---|---|
| 0.001 | −3 |
| 0.01 | −2 |
| 0.1 | −1 |
| 1 | 0 |
| 2 | 0.301 |
| 5 | 0.699 |
| 9 | 0.954 |
| 10 | 1 |
| 100 | 2 |
| 10,000 | 4 |
| 20,000 | 4.301 |
| 100,000,000 | 8 |

abbreviated ln, which uses base $e$ (=2.71828 ...) and is important in computing compound interest, growth rates, economic elasticity, and other applications. For the purpose of transforming data, both kinds of logarithms have the same effect, pulling in high values and stretching out the low values.

Your calculator may have a logarithm key, denoted LOG.[11] Simply key in the number and press the LOG key. Many spreadsheets, such as Microsoft® Excel®, have built-in functions for logarithms. You might enter =LOG(5) to a cell to find the (base 10) logarithm of 5, which is 0.69897. Alternatively, entering =LN(5) would give you the base $e$ value, 1.60944, instead. To find the logarithms of a data set in a column, you can use the Copy and Paste commands to copy the logarithm formula from the first cell down the entire column, greatly shortening the task of finding the logs of a list of numbers. An even faster way to create a column of transformed values, shown below, is to double-click the "fill handle" (the little square at the lower right of the selected cell) after entering the transformation formula (alternatively, you may drag the fill handle).

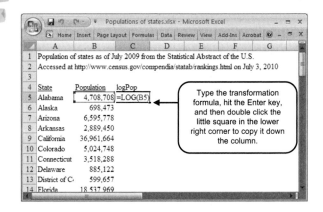

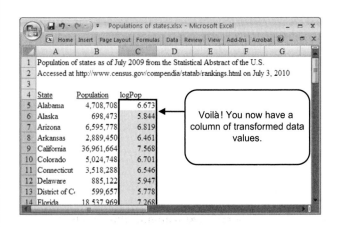

11. Some calculators do not have a LOG key to compute the base 10 logarithm but instead have only an LN key to compute the natural logarithm (base $e$). To find the common logarithm on such a calculator, divide the result of LN by 2.302585, the natural log of 10.

## 3.5 BIMODAL DISTRIBUTIONS

It is important to be able to recognize when a data set consists of two or more distinct groups so that they may be analyzed separately, if appropriate. This can be seen in a histogram as a distinct gap between two cohesive groups of bars. When two clearly separate groups are visible in a histogram, you have a **bimodal distribution**. Literally, a bimodal distribution has *two modes*, or two distinct clusters of data.[12]

A bimodal distribution may be an indication that the situation is more complex than you had thought, and that extra care is required. At the very least, you should find out the reason for the two groups. Perhaps only one group is of interest to you, and you should exclude the other as irrelevant to the situation you are studying. Or perhaps both groups are needed, but some adjustment has to be done to account for the fact that they are so different.

### Example

#### Corporate Bond Yields

Consider yields of bonds expressed as an interest rate representing the annualized percentage return on investment as promised by the bond's future payments, as shown in Table 3.5.1. A histogram of the complete data set, as shown in Figure 3.5.1, looks like two separate histograms. One group indicates yields from about 2% to 6%, and the other extends from about 7% to 10%. This kind of separation is unlikely to be due to pure randomness from a single cohesive data set. There must be some other reason (perhaps you'd like to try to guess the reason before consulting the footnote below for the answer).[13]

---

13. There are two different risk classes of bonds listed here, and, naturally, investors require a higher rate of return to entice them to invest. The B rated bonds are riskier and correspond to the right-hand group of the histogram, while the AA rated bonds are less risky on the left. In addition to the risk differences between the groups, there is also a maturity difference, with the B rated bonds lasting somewhat longer before they mature.

### Is It Really Bimodal?

Don't get carried away and start seeing bimodal distributions when they aren't there. The two groups must be large enough, be individually cohesive, and either have a fair gap between them or else represent a large enough sample to be sure that the lower frequencies between the groups are not just random fluctuations. It may take judgment to distinguish a "random" gap within a single group from a true gap separating two distinct groups.

---

12. The *mode* as a summary measure will be presented in Chapter 4.

**TABLE 3.5.1** Yields of Corporate Bonds

| Issue | Yield | Maturity | Rating |
|---|---|---|---|
| Abbott Labs | 3.314% | 1-Apr-19 | AA |
| African Dev Bk | 3.566% | 1-Sep-19 | AA |
| Bank New York Mtn Bk Ent | 3.623% | 15-May-19 | AA |
| Bank New York Mtn Bk Ent | 3.288% | 15-Jan-20 | AA |
| Barclays Bank Plc | 4.759% | 8-Jan-20 | AA |
| Barclays Bk Plc | 4.703% | 22-May-19 | AA |
| Becton Dickinson & Co | 3.234% | 15-May-19 | AA |
| Chevron Corporation | 3.123% | 3-Mar-19 | AA |
| Coca Cola Co | 3.153% | 15-Mar-19 | AA |
| Columbia Healthcare Corp | 8.117% | 15-Dec-23 | B |
| Credit Suisse New York Branch | 4.185% | 13-Aug-19 | AA |
| Credit Suisse New York Branch | 5.126% | 14-Jan-20 | AA |
| Federal Home Ln Mtg Corp | 3.978% | 14-Dec-18 | AA |
| Ford Mtr Co Del | 8.268% | 15-Sep-21 | B |
| Ford Mtr Co Del | 8.081% | 15-Jan-22 | B |
| Fort James Corp | 7.403% | 15-Nov-23 | B |
| GE Capital Internotes | 5.448% | 15-Sep-19 | AA |
| GE Capital Internotes | 5.111% | 15-Nov-19 | AA |
| General Elec Cap Corp Mtn Be | 4.544% | 7-Aug-19 | AA |
| General Elec Cap Corp Mtn Be | 4.473% | 8-Jan-20 | AA |
| General Mtrs Accep Corp | 8.598% | 15-Jul-20 | B |
| General Mtrs Accep Corp | 8.696% | 15-Nov-24 | B |
| General Mtrs Accep Corp | 8.724% | 15-Mar-25 | B |
| General Mtrs Accep Cpsmartnbe | 8.771% | 15-Jun-22 | B |
| Goodyear Tire & Rubr Co | 7.703% | 15-Aug-20 | B |
| Iron Mtn Inc Del | 7.468% | 15-Aug-21 | B |
| JPMorgan Chase & Co | 4.270% | 23-Apr-19 | AA |
| Medtronic Inc | 3.088% | 15-Mar-19 | AA |
| Merck & Co Inc | 3.232% | 30-Jun-19 | AA |
| Northern Trust Co Mtns Bk Ent | 3.439% | 15-Aug-18 | AA |
| Novartis Securities Investment | 3.179% | 10-Feb-19 | AA |
| Pepsico Inc | 3.489% | 1-Nov-18 | AA |
| Pfizer Inc | 3.432% | 15-Mar-19 | AA |
| Pharmacia Corp | 3.386% | 1-Dec-18 | AA |

(Continued)

**TABLE 3.5.1 Yields of Corporate Bonds—cont'd**

| Issue | Yield | Maturity | Rating |
|---|---|---|---|
| Procter & Gamble Co | 3.126% | 15-Feb-19 | AA |
| Rinker Matls Corp | 9.457% | 21-Jul-25 | B |
| Roche Hldgs Inc | 3.385% | 1-Mar-19 | AA |
| Shell International Fin Bv | 3.551% | 22-Sep-19 | AA |
| United Parcel Service Inc | 2.990% | 1-Apr-19 | AA |
| Wal Mart Stores Inc | 2.973% | 1-Feb-19 | AA |
| Westpac Bkg Corp | 4.128% | 19-Nov-19 | AA |
| : | : | : | : |

**Source:** Corporate bond data accessed at http://screen.yahoo.com/bonds.html on July 3, 2010. Two searches were combined: AA rated bonds with 8- to 10-year maturities, and the B rated bonds with 10- to 15-year maturities.

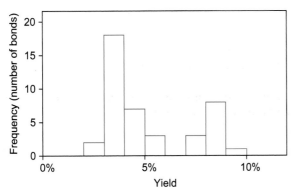

**FIGURE 3.5.1**  Yields of corporate bonds. This is a highly bimodal distribution, with two clear and separate groups, probably not due to chance alone.

**Example**
*Rates of Computer Ownership*

Consider the extent of computer ownership by state as presented in Table 3.5.2. It is interesting to reflect on the large variability from one state to another: Computer ownership is nearly double in Utah (66.1%) what it is in Mississippi (37.2%). To see the big picture among all the states, look at the histogram of this data set shown in Figure 3.5.2. This is a fairly symmetric distribution ("fairly symmetric" implies that it may not be perfectly symmetric, but at least it's not very skewed). The distribution is basically normal, and you see one single group.

However, if you display the histogram on a finer scale, with smaller bars (width 0.2 instead of 5 percentage points), as in Figure 3.5.3, the extra detail suggests that there might be two groups: the two states with lowest computer ownership (on the left) and all other states (on the right) with a gap in between. However, this is not really bimodal, for two reasons. First, the gap is a small one compared to the diversity

among computer ownership rates. Second, and more important, the histogram bars are really too small because many represent just one state. Remember that one of the main goals of statistical techniques (such as the histogram) is to see the big picture and not get lost by reading too much into the details.

**TABLE 3.5.2 Rates of Computer Ownership**

| State | Percent of Households |
|---|---|
| Alabama | 44.2% |
| Alaska | 64.8 |
| Arizona | 53.5 |
| Arkansas | 37.3 |
| California | 56.6 |
| Colorado | 62.6 |
| Connecticut | 60.4 |
| Delaware | 58.6 |
| District of Columbia | 48.8 |
| Florida | 50.1 |
| Georgia | 47.1 |
| Hawaii | 52.4 |
| Idaho | 54.5 |
| Illinois | 50.2 |
| Indiana | 48.8 |
| Iowa | 53.6 |
| Kansas | 55.8 |
| Kentucky | 46.2 |
| Louisiana | 41.2 |
| Maine | 54.7 |
| Maryland | 53.7 |
| Massachusetts | 53.0 |
| Michigan | 51.5 |
| Minnesota | 57.0 |
| Mississippi | 37.2 |
| Missouri | 52.6 |
| Montana | 51.5 |
| Nebraska | 48.5 |
| Nevada | 48.8 |
| New Hampshire | 63.7 |

**TABLE 3.5.2 Rates of Computer Ownership—cont'd**

| State | Percent of Households |
|---|---|
| New Jersey | 54.3 |
| New Mexico | 47.6 |
| New York | 48.7 |
| North Carolina | 45.3 |
| North Dakota | 47.5 |
| Ohio | 49.5 |
| Oklahoma | 41.5 |
| Oregon | 61.1 |
| Pennsylvania | 48.4 |
| Rhode Island | 47.9 |
| South Carolina | 43.3 |
| South Dakota | 50.4 |
| Tennessee | 45.7 |
| Texas | 47.9 |
| Utah | 66.1 |
| Vermont | 53.7 |
| Virginia | 53.9 |
| Washington | 60.7 |
| West Virginia | 42.8 |
| Wisconsin | 50.9 |
| Wyoming | 58.2 |

**Source:** Data are from U.S. Bureau of the Census, *Statistical Abstract of the United States: 2000* on CD-ROM (Washington, DC, 2000), Table 915, and represent 2000 ownership. Their source is the U.S. Department of Commerce, National Telecommunications and Information Administration.

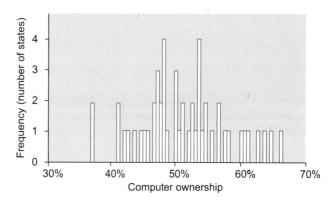

**FIGURE 3.5.3** Computer ownership rates (same data as in previous figure, but displayed with smaller bars). Since too much detail is shown here, it appears (probably wrongly) that there might be two groups. The two states in the first bar at the left with the lowest ownership rates are slightly separated from the others. This is probably just randomness and not true bimodality.

## 3.6 OUTLIERS

Sometimes you will find **outliers**, which are data values that don't seem to belong with the others because they are either far too big or far too small. How you deal with outliers depends on what caused them. There are two main kinds of outliers: (1) mistakes and (2) correct but "different" data values. Outliers are discussed here because they are often noticed when the histogram is examined; a formal calculation to determine outliers (to construct a detailed box plot) will be covered in the next chapter.

### Dealing with Outliers

Mistakes are easy to deal with: Simply change the data value to the number it should have been in the first place. For example, if a sales figure of $1,597.00 was wrongly recorded as $159,700 because of a misplaced decimal point, it might show up as being far too big compared to other sales figures in a histogram. Having been alerted to the existence of this strange data value, you should investigate and find the error. The situation would be resolved by correcting the figure to $1,597, the value it should have been originally.

Unfortunately, the correct outliers are more difficult to deal with. If it can be argued convincingly that the outliers do not belong to the general case under study, they may then be set aside so that the analysis can proceed with only the coherent data. For example, a few tax-free money market funds may appear as outliers in a data set of yields. If the purpose of the study is to summarize the marketplace for general-purpose funds, it may be appropriate to leave these special tax-free funds out of the picture. For another example, suppose your company is evaluating a new pharmaceutical product. In one of the trials, the laboratory technician sneezed into the sample before it

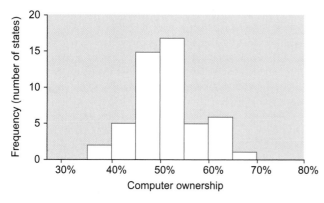

**FIGURE 3.5.2** The rate of computer ownership by state. This is a fairly normal distribution, forming just one cohesive group.

was analyzed. If you are not studying laboratory accidents, it might be appropriate to omit this outlier.

If you wish to set aside some outliers in this way, you must be prepared to convince not just yourself that it is appropriate, but any person (possibly hostile) for whom your report is intended. Thus, the issue of exactly when it is or isn't OK to omit outliers may not have a single, objective answer. For an internal initial feasibility study, for example, it may be appropriate to delete some outliers. However, if the study were intended for public release or for governmental scrutiny, then you would want to be much more careful about omitting outliers.

One compromise solution, which can be used even when you don't have a strong argument for omitting the outlier, is to perform *two different analyses:* one with the outlier included and one with it omitted. By reporting the results of both analyses, you have not unfairly slanted the results. In the happiest case, should it turn out that the conclusions are identical for both analyses, you may conclude that the outlier makes no difference. In the more problematic case, where the two analyses produce different results, your interpretation and recommendations are more difficult. Unfortunately, there is no complete solution to this subtle problem.[14]

There is an important rule to be followed whenever any outlier is omitted, in order to inform others and protect yourself from any possible accusations:

**Whenever an Outlier is Omitted:**

Explain what you did and why!

That is, explain clearly somewhere in your report (perhaps a footnote would suffice) that there is an outlier problem with the data. Describe the outlier, and tell what you did about it. Be sure to justify your actions.

Why should you deal with outliers at all? There are two main ways in which they cause trouble. First, it is difficult to interpret the detailed structure in a data set when one value dominates the scene and calls too much attention to itself. Second, as also occurs with skewness, many of the most common statistical methods can fail when used on a data set that doesn't appear to have a normal distribution. Normal distributions aren't skewed and don't usually produce outliers. Consequently, you will have to deal with any outliers in your data before relying heavily on statistical inference.

---

14. There is a branch of statistics called *robustness* that seeks to use computing power to adjust for the presence of outliers, and robust methods are available for many (but not all) kinds of data sets. For more detail, see D. C. Hoaglin, F. Mosteller, and J. W. Tukey, *Understanding Robust and Exploratory Data Analysis* (New York: Wiley, 1983); and V. Barnett and T. Lewis, *Outliers in Statistical Data* (New York: Wiley, 1978).

**Example**

*Did Net Earnings Increase or Decrease?*

As reported in the *Wall Street Journal*,[15] second-quarter net income of major U.S. companies increased by 27%, a strong increase based on analysis of 677 publicly traded companies. However, there is an outlier in the data: MediaOne had a $24.5 billion gain in the quarter, due to its separation from U.S. West. When this outlier is omitted, net income actually *fell*, by 1.5%.

Much the same situation apparently happened the quarter before, when net income rose 20% due to Ford Motor's sale of a financing unit. If this outlier is omitted, the strong increase fades to an increase of merely 2.5% for that quarter.

As you can see from these two examples, statistical summaries can be misleading when an outlier is present. If you read only that net income was up 27% (or 20%) for large companies, you might (wrongly) conclude that most of the companies enjoyed strong earnings. By omitting the outlier and reanalyzing the data, we obtain a better impression of what actually happened to these companies as a group.

---

15. M. M. Phillips, "MediaOne Item Pushes Earnings of U.S. Firms to Gain, but Asia and Competition Hurt Results," *The Wall Street Journal*, August 3, 1998, pp. A1 and C15.

**Example**

*CEO Compensation by Prepackaged Software Companies*

Compensation for chief executive officers (CEOs) of companies varies from one company to another, and here we focus on prepackaged software companies (see Table 3.6.1). In the histogram shown in Figure 3.6.1, the presence of an outlier (Lawrence J. Ellison of Oracle Corp, with compensation of $56.81 million) seems to have forced nearly all the other companies into just one bar (actually two bars, since Robert E. Beauchamp of Bmc Software with compensation of $10.90 million is represented by the very short bar from 10 to 20 million), showing us that these companies tend to pay their CEOs somewhere between $0 and $10 million. This obscures much of the detail in the distribution of the compensation figures (e.g., just by looking at the numbers you can see that most are under $5 million). Even with the smaller bar width used in the histogram in Figure 3.6.2, details are still obscured. Making the bar width smaller still, as in the histogram of Figure 3.6.3, we find that we now have enough detail, but the interesting part of the distribution occupies just a small part of the figure. Unfortunately, these histograms of the full data set are not as helpful as we would like.

Omitting L. J. Ellison of Oracle Corporation, the largest value and clearly an outlier at over $50 million (but not forgetting this special value), we find a histogram in Figure 3.6.4 that gracefully shows us the skewed distribution generally followed by these compensation numbers, on a scale that reveals the details and, in particular, that most earn less than $5 million and follow a fairly smooth skewed pattern.

## TABLE 3.6.1 CEO Compensation by Prepackaged Software Companies ($ millions)

| Company | CEO Name | Compensation |
| --- | --- | --- |
| Accelrys Inc | Mark J. Emkjer | 2.70 |
| Aci Worldwide Inc | Philip G. Heasley | 2.37 |
| Activision Blizzard Inc | Robert A. Kotick | 3.15 |
| Actuate Corp | Peter I. Cittadini | 2.12 |
| Adobe Systems Inc | Shantanu Narayen | 6.66 |
| Advent Software Inc | Stephanie G. DiMarco | 0.78 |
| American Software -Cl A | James C. Edenfield | 0.67 |
| Amicas Inc | Stephen N. Kahane | 0.85 |
| Ansys Inc | James E. Cashman III | 2.34 |
| Arcsight Inc | Thomas Reilly | 2.11 |
| Ariba Inc | Robert M. Calderoni | 6.27 |
| Art Technology Group Inc | Robert D. Burke | 1.61 |
| Asiainfo Holdings Inc | Steve Zhang | 0.87 |
| Autodesk Inc | Carl Bass | 6.23 |
| Blackbaud Inc | Marc E. Chardon | 2.55 |
| Blackboard Inc | Michael L. Chasen | 8.42 |
| Bmc Software Inc | Robert E. Beauchamp | 10.90 |
| Bottomline Technologies Inc | Robert A. Eberle | 1.77 |
| Ca Inc | John A. Swainson | 8.80 |
| Cadence Design Systems Inc | Lip-Bu Tan | 6.28 |
| Callidus Software Inc | Leslie J. Stretch | 0.87 |
| Chordiant Software Inc | Steven R. Springsteel | 1.82 |
| Citrix Systems Inc | Mark B. Templeton | 5.17 |
| Commvault Systems Inc | N. Robert Hammer | 1.68 |
| Compuware Corp | Peter Karmanos Jr. | 2.81 |
| Concur Technologies Inc | S. Steven Singh | 2.22 |
| Dealertrack Holdings Inc | Mark F. O'Neil | 2.70 |
| Deltek Inc | Kevin T. Parker | 1.58 |
| Demandtec Inc | Daniel R. Fishback | 1.97 |
| Double-Take Software Inc | Dean Goodermote | 0.89 |

| | | |
| --- | --- | --- |
| Ebix Inc | Robin Raina | 2.78 |
| Electronic Arts Inc | John S. Riccitiello | 6.37 |
| Entrust Inc | F. William Conner | 1.56 |
| Epicor Software Corp | L. George Klaus | 3.91 |
| Epiq Systems Inc | Tom W. Olofson | 3.07 |
| Eresearchtechnology Inc | Michael J. McKelvey | 1.15 |
| Gse Systems Inc | John V. Moran | 0.34 |
| I2 Technologies Inc | Pallab K. Chatterjee | 4.86 |
| Informatica Corp | Sohaib Abbasi | 2.78 |
| Interactive Intelligence Inc | Donald E. Brown | 1.03 |
| Intuit Inc | Brad D. Smith | 4.81 |
| Jda Software Group Inc | Hamish N. Brewer | 2.38 |
| Kenexa Corp | Nooruddin (Rudy) S. Karsan | 0.81 |
| Lawson Software Inc | Harry Debes | 3.76 |
| Lionbridge Technologies Inc | Rory J. Cowan | 1.50 |
| Liveperson Inc | Robert P. LoCascio | 0.63 |
| Logility Inc | J. Michael Edenfield | 0.43 |
| Mcafee Inc | David G. DeWalt | 7.53 |
| Medassets Inc | John A. Bardis | 4.45 |
| Microsoft Corp | Steven A. Ballmer | 1.28 |
| Microstrategy Inc | Michael J. Saylor | 4.71 |
| Monotype Imaging Holdings | Douglas J. Shaw | 0.81 |
| Msc Software Corp | William J. Weyand | 1.96 |
| National Instruments Corp | James J. Truchard | 0.19 |
| Nuance Communications Inc | Paul A. Ricci | 9.91 |
| Omniture Inc | Joshua G. James | 3.11 |
| Opentv Corp | Nigel W. Bennett | 1.30 |
| Openwave Systems Inc | Kenneth D. Denman | 0.59 |
| Opnet Technologies Inc | Marc A. Cohen | 0.39 |
| Oracle Corp | Lawrence J. Ellison | 56.81 |
| Parametric Technology Corp | C. Richard Harrison | 5.15 |
| Pegasystems Inc | Alan Trefler | 0.53 |
| Pervasive Software Inc | John Farr | 0.75 |
| Phase Forward Inc | Robert K. Weiler | 7.07 |

*(Continued)*

**TABLE 3.6.1 CEO Compensation by Prepackaged Software Companies ($ millions)—cont'd**

| Company | CEO Name | Compensation |
|---|---|---|
| Phoenix Technologies Ltd | Woodson Hobbs | 3.85 |
| Progress Software Corp | Joseph W. Alsop | 5.71 |
| Pros Holdings Inc | Albert E. Winemiller | 1.56 |
| Qad Inc | Karl F. Lopker | 1.17 |
| Quest Software Inc | Vincent C. Smith | 3.72 |
| Realnetworks Inc | Robert Glaser | 0.74 |
| Red Hat Inc | James M. Whitehurst | 5.00 |
| Renaissance Learning Inc | Terrance D. Paul | 0.59 |
| Rightnow Technologies Inc | Greg R. Gianforte | 1.16 |
| Rosetta Stone Inc | Tom P. H. Adams | 9.51 |
| Saba Software Inc | Bobby Yazdani | 0.99 |
| Salesforce.Com Inc | Marc Benioff | 0.34 |
| Sapient Corp | Alan J. Herrick | 2.01 |
| Seachange International Inc | William C. Styslinger III | 1.33 |
| Solarwinds Inc | Kevin B. Thompson | 2.47 |
| Solera Holdings Inc | Tony Aquila | 3.23 |
| Spss Inc | Jack Noonan | 4.19 |
| Successfactors Inc | Lars Dalgaard | 2.92 |
| Support.Com Inc | Joshua Pickus | 2.39 |
| Sybase Inc | John S. Chen | 9.29 |
| Symantec Corp | John W. Thompson | 7.03 |
| Symyx Technologies Inc | Isy Goldwasser | 1.04 |
| Synopsys Inc | Aart J. de Geus | 4.54 |
| Take-Two Interactive Sftwr | Benjamin Feder | 0.01 |
| Taleo Corp | Michael Gregoire | 2.39 |
| Thq Inc | Brian J. Farrell | 2.28 |
| Tibco Software Inc | Vivek Y. Ranadivé | 4.10 |
| Ultimate Software Group Inc | Scott Scherr | 2.12 |
| Unica Corp | Yuchun Lee | 0.56 |
| Vignette Corp | Michael A. Aviles | 2.55 |
| Vital Images Inc | Michael H. Carrel | 0.60 |
| Vocus Inc | Richard Rudman | 3.70 |
| Websense Inc | Gene Hodges | 2.55 |

**Source:** Executive PayWatch Database of the AFL-CIO, accessed at http://www.aflcio.org/corporatewatch/paywatch/ceou/industry.cfm on July 4, 2010.

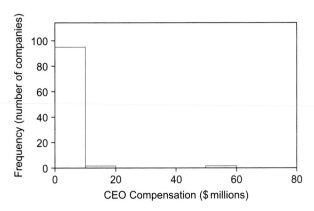

**FIGURE 3.6.1** Histogram of CEO compensation by prepackaged software companies. Note the presence of an outlier at the far right (L. J. Ellison of Oracle Corp, at $56.81 million) that obscures the details of the majority of the companies, forcing nearly all of them into a single bar from 0 to $10 million.

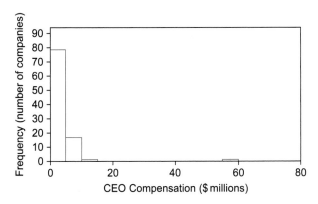

**FIGURE 3.6.2** Another histogram of all 97 companies, but with a smaller bar width. The outlier at the far right still obscures the details of most of the data, although we now see clearly that most are paid less than $5 million.

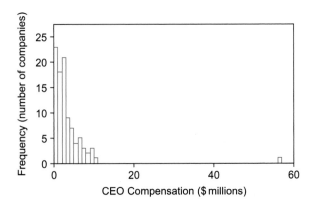

**FIGURE 3.6.3** Another histogram of all 97 companies, but with an even smaller bar width. While the details of the distribution are now available, they are jumbled together at the left.

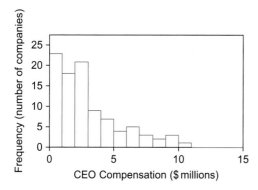

FIGURE 3.6.4    Histogram of CEO compensation for 96 companies, after omitting the largest outlier (Oracle Corp, at $56.81 million) and expanding the scale. Now you have an informative picture of the details of the distribution of CEO compensation across companies in this industry group. We do not forget this outlier: We remember it while expanding the scale to see the details of the rest of the data.

## 3.7  DATA MINING WITH HISTOGRAMS

The histogram is a particularly useful tool for large data sets because you can see the entire data set at a glance. It is not practical to examine each data value individually—and even if you could, would you really want to spend 6 hours of your time giving one second to each of 20,000 numbers? As always, the histogram gives you a visual impression of the data set, and with large data sets you will be able to see more of the detailed structure.

Consider the donations database with 20,000 entries available on the companion site (as introduced in Chapter 1). Figure 3.7.1 shows a histogram of the number of promotions (asking for a donation) that each person had previously received. Along with noting that each person received,

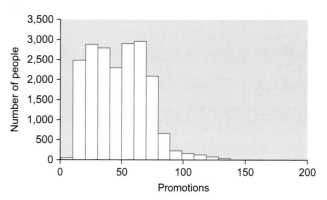

FIGURE 3.7.1    A histogram of the number of promotions received by the 20,000 people in the donations database.

typically, somewhere from about 10 to 100 promotions, we also notice that the distribution is too flat on top to be approximately normal (with such a large sample size—the tall bars represent over 2,000 people each—this is not just randomly different from a normal distribution).

One advantage of data mining with a large data set is that we can ask for more detail. Figure 3.7.2 shows more histogram bars by reducing the width of the bar from 10 promotions to 1 promotion. Even though there are many thin bars, we clearly have enough data here to interpret the result because most of the bars represent over 100 people. In particular, note the relatively large group of people who received about 15 promotions (tall bars at the left). This could be the result of a past campaign to reach new potential donors.

When we look at a histogram of the dollar amounts of the donations people gave in response to the mailing (Figure 3.7.3), the initial impression is that the vast majority

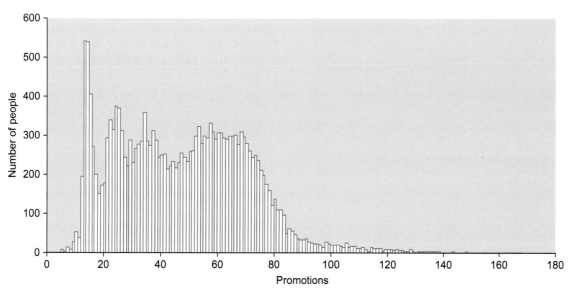

FIGURE 3.7.2    Greater detail is available when more histogram bars are used (with bar width reduced from 10 to 1 promotion) in data mining the donations database. Note the relatively large group of people at the left who received about 15 promotions.

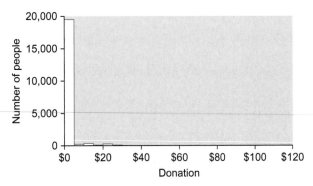

FIGURE 3.7.3   The initial histogram of the 20,000 donation amounts is dominated by the 19,011 people who did not make a donation (and were counted as zero). The 6 people who donated $100 do not even show up on this scale!

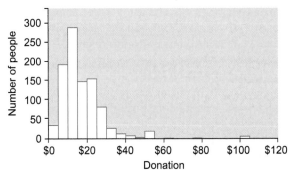

FIGURE 3.7.4   A histogram of the donations of the 989 people who actually made a (nonzero) donation.

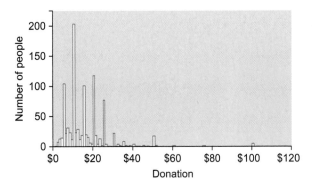

FIGURE 3.7.5   A histogram showing more detail of the sizes of the donations. Note the tendency for people to give "round" amounts such as $5, $10, or $20 instead of, say, $17.

gave little or nothing (the tall bar at the left). Due to this tall bar (19,048 people who donated less than $5), it is difficult to see any detail at all in the remaining fairly large group of 952 people who gave $5 or more (or the 989 people who gave at least something). In particular, we can't even see the 6 people who donated $100.

By setting aside the 19,011 people who did not make a donation, the histogram in Figure 3.7.4 lets you see some details of 989 people who actually donated something. Because we have so much data, we can see even more detail in Figure 3.7.5 using more, but smaller, bins. Note the tall thin spikes at $5 intervals apart representing the tendency for people to prefer donation amounts that are evenly divisible by $5.

## 3.8 HISTOGRAMS BY HAND: STEM-AND-LEAF

These days, the most effective way to construct a histogram is probably to use a statistical software package on a computer. However, there are times when you might want to construct a histogram by hand. For example, there might not be a computer available before the deadline on the project, and you might want to check just one more possibility by examining a histogram. Also, for a small list of numbers it is actually faster to scratch down a histogram on paper than it would be to operate the computer and type the data into it. Finally, by drawing a histogram yourself, you end up "closer to the data," with an intuitive feel for the numbers that cannot be obtained by letting the computer do the work.

The easiest way to construct a histogram by hand is to use the **stem-and-leaf** style, in which the histogram bars are constructed by stacking numbers one on top of the other (or side-by-side). Doing it this way has the advantage of letting the histogram grow before your eyes, showing you useful information from the very start and making your efforts worthwhile.

Begin by identifying the initial digits to be included in the scale underneath the figure, ignoring the fine details. For example, you might include the millions and hundred thousands digits, but leave the ten thousands (and smaller digits) out of the scale. Then use the *next-level digit* (ten thousands) to record each data value, building columns upward (or sideways) to be interpreted as the columns of a histogram.

### Example
#### Employees in Food Services

Consider the number of employees for each Fortune 1000 food-services firm, as shown in Table 3.8.1. Using hundreds of thousands to construct the scale, you would begin with the horizontal scale indicating the range 0 to 4.

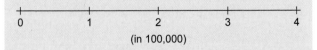

How should you record the first data value, representing the 471,000 employees of PepsiCo? This number has 4 in the hundred thousands place, so put the next digit, 7, above the 4 along the horizontal scale. Since the next digit (1, in the thousands place) is much less important, you need not trouble yourself with it for now. The result so far is as follows:

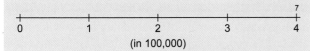

The next data value, 183,000 for McDonald's, will be recorded as an 8 (the ten thousands digit) over the 1 (the hundred thousands digit). Next is Aramark with 133,000, which will be a 3 also over the 1, so stack them up. Recording these two values produces the following:

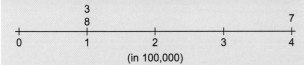

As you record more and more numbers, place them in columns that grow upward. In this way, you get to watch the histogram grow before your eyes. This is much more satisfying (and informative) than having to calculate and count for a while before getting to see any results. The finished histogram in the stem-and-leaf style looks like this:

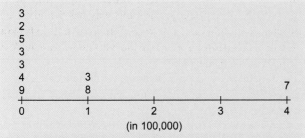

To represent this stem-and-leaf histogram in the more traditional style, simply replace the columns of numbers with bars of the same height, as shown in Figure 3.8.1.

Many computer programs produce a stem-and-leaf histogram that is sideways compared to the traditional histogram, with the bars growing to the right instead of upward. In the following computer stem-and-leaf, note also that each of our groupings has been divided into two groups: for example, the top row would contain 10,000 to 49,999, while the second row contains 50,000 to 99,999 employees.

Looking at either display (stem-and-leaf or traditional histogram), you can see the basic information in the data set. The numbers of employees for major food-services firms ranged from about a few tens of thousands to nearly half a million. Although it's difficult to say with precision, due to the relatively small amount of data, it certainly appears that the distribution is somewhat skewed.

| Leaf Unit = 10000 | Stem-and-leaf of employee N = 10 |
|---|---|
| 5 | 0 23334 |
| 5 | 0 59 |
| 3 | 1 3 |
| 2 | 1 8 |
| 1 | 2 |
| 1 | 2 |
| 1 | 3 |
| 1 | 3 |
| 1 | 4 |
| 1 | 4 7 |

**TABLE 3.8.1  The Number of Employees for Food-Services Firms**

| Firm | Employees | Firm | Employees |
|---|---|---|---|
| PepsiCo | 471,000 | Morrison Restaurants | 33,000 |
| McDonald's | 183,000 | Shoney's | 30,000 |
| Aramark | 133,000 | Family Restaurants | 51,700 |
| Flagstar | 90,000 | Foodmaker | 26,170 |
| Wendy's International | 44,000 | Brinker International | 38,000 |

Source: Data are from *Fortune*, May 15, 1995, p. F-52.

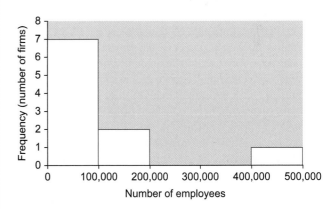

FIGURE 3.8.1   A traditional histogram, using bars with the same heights as the columns of numbers in the stem-and-leaf histogram.

## 3.9  END-OF-CHAPTER MATERIALS

### Summary

The simplest kind of data set is a **list of numbers** representing some kind of information (a single statistical variable) measured on each item of interest (each elementary unit).

A list of numbers may come to you either as a list or as a table showing how many times each number should be repeated to form a list.

The first step toward understanding a list of numbers is to view its histogram in order to see its basic properties, such as typical values, special values, concentration, spread, the general pattern, and any separate groupings. The **histogram** displays the frequencies as a bar chart rising above the number line, indicating how often the various values occur in the data set. The **number line** is a straight line, usually horizontal, with the scale indicated by numbers below it.

A **normal distribution** is a particular idealized, smooth, bell-shaped histogram with all of the randomness removed. It represents an ideal data set that has lots of numbers concentrated in the middle of the range and trails off symmetrically on both sides. A data set follows a normal distribution if it resembles the smooth, symmetric, bell-shaped normal curve, except for some randomness. The normal distribution plays an important role in statistical theory and practice.

A **skewed distribution** is neither symmetric nor normal because the data values trail off more sharply on one side than on the other. Skewed distributions are very common in business. Unfortunately, many standard statistical methods do not work properly if your data set is very skewed.

**Transformation** is replacing each data value by a different number (such as its logarithm) to facilitate statistical analysis. The **logarithm** often transforms skewness into symmetry because it stretches the scale near zero, spreading out all of the small values that had been bunched together. The logarithm also pulls together the very large data values, which had been thinly scattered at the high end of the scale. The logarithm can only be computed for positive numbers. To interpret the logarithm, note that equal distances on the logarithmic scale correspond to equal percent increases instead of equal value increases (dollar amounts, for example).

When two clear and separate groups are visible in a histogram, you have a **bimodal distribution.** It is important to recognize when you have a bimodal distribution so that you can take appropriate action. You might find that only one of the groups is actually of interest to you, and that the other should be omitted. Or you might decide to make some changes in the analysis in order to cope with this more complex situation.

Sometimes you will find **outliers,** which are one or more data values that just don't seem to belong with the others because they are either far too big or far too small. Outliers can cause trouble with statistical analysis, so they should be identified and acted on. If the outlier is a mistake, correct it and continue with the analysis. If it is correct but different, you might or might not omit it from the analysis. If you can convince yourself and others that the outlier is not part of the system you wish to study, you may continue without the outlier. If you cannot justify omitting the outlier, you may proceed with two projects: analyze the data

with and without the outlier. In any case, be sure to state clearly somewhere in your report the existence of an outlier and the action taken.

The most efficient way to draw a histogram is with a computer, using a statistical software package. However, there are times when it is necessary (and even desirable) to construct a histogram by hand. The **stem-and-leaf** histogram constructs the bars of a histogram by stacking numbers one on top of another (or side-by-side). Because data values are recorded successively, you can build intuition about the data set by watching the histogram grow.

## Key Words

**bimodal distribution,** *47*
**histogram,** *37*
**list of numbers,** *35*
**logarithm,** *45*
**normal distribution,** *40*
**number line,** *36*
**outliers,** *49*
**skewed distribution,** *43*
**stem-and-leaf,** *54*
**transformation,** *45*

## Questions

1. What is a list of numbers?
2. Name six properties of a data set that are displayed by a histogram.
3. What is a number line?
4. What is the difference between a histogram and a bar chart?
5. What is a normal distribution?
6. Why is the normal distribution important in statistics?
7. When a real data set is normally distributed, should you expect the histogram to be a perfectly smooth bell-shaped curve? Why or why not?
8. Are all data sets normally distributed?
9. What is a skewed distribution?
10. What is the main problem with skewness? How can it be solved in many cases?
11. How can you interpret the logarithm of a number?
12. What is a bimodal distribution? What should you do if you find one?
13. What is an outlier?
14. Why is it important in a report to explain how you dealt with an outlier?
15. What kinds of trouble do outliers cause?
16. When is it appropriate to set aside an outlier and analyze only the rest of the data?
17. Suppose there is an outlier in your data. You plan to analyze the data twice: once with and once without the outlier. What result would you be most pleased with? Why?
18. What is a stem-and-leaf histogram?
19. What are the advantages of a stem-and-leaf histogram?

## Problems

*Problems marked with an asterisk (\*) are solved in the Self Text in Appendix C.*

1. What distribution shape is represented by the histogram in Figure 3.9.1 of voltages measured for incoming components as part of a quality control program?
2. What distribution shape is represented by the histogram in Figure 3.9.2 of profit margins for consumer products?
3. What distribution shape is represented by the histogram in Figure 3.9.3 of volume (in thousands of units) by sales region?
4. What distribution shape is represented by the histogram in Figure 3.9.4 of hospital length of stay (in days)?
5. Consider the histogram in Figure 3.9.5, which indicates performance of recent on-site service contracts as a rate of return.

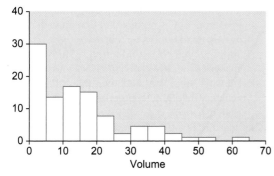

FIGURE 3.9.1

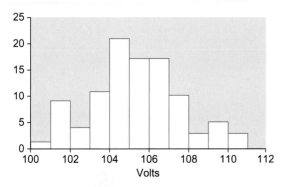

FIGURE 3.9.2

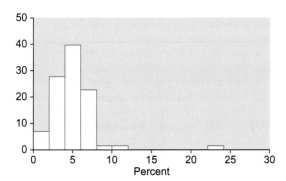

FIGURE 3.9.3

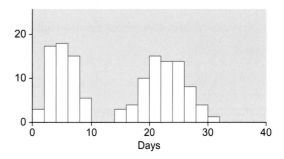

FIGURE 3.9.4

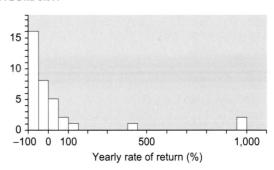

FIGURE 3.9.5

a. At the very high end, how many contracts were extreme outliers that earned over 900% per year?
b. How many contracts are outliers, earning 400% or more?
c. One contract, with a real estate firm that went bankrupt, lost all of its initial investment a few years after work began (hence, the −100% rate of return). Can you tell from the histogram that a contract lost all of its value? If not, what can you say about the worst-performing contracts?
d. How many contracts lost money (i.e., had negative rates of return)?
e. Describe the shape of this distribution.
6.\* Consider the yields (as an interest rate, in percent per year) of municipal bonds, as shown in Table 3.9.1.
a. Construct a histogram of this data set.
b. Based on the histogram, what values appear to be typical for this group of tax-exempt bonds?
c. Describe the shape of the distribution.

### TABLE 3.9.1 Yields of Municipal Bonds

| Issue | Yield |
|---|---|
| CA EdFcsAthRefRev | 4.91% |
| CapProjectsFinAuthFL | 5.18 |
| Chcg ILarptRvSr2001Mdwy | 5.34 |
| ChcgILGOSr2001A | 5.27 |
| CleveOH arptRev200 | 5.21 |
| ClrdoSprgsCO UtilSysSub | 5.16 |

*(Continued)*

**TABLE 3.9.1 Yields of Municipal Bonds—cont'd**

| Issue | Yield |
| --- | --- |
| ClrkCoNVarptSysRev200 | 5.33 |
| Detroit MI wtr sply Sys | 5.26 |
| DL Ar Rpd Trnst TX | 5.19 |
| DL Ar Rpds Trnst TX | 5.23 |
| Est Bay Mud CA wtr Sub | 5.01 |
| HghlndCoHlthFcs FL hospRV | 5.83 |
| Hnlu(cty&cny)HI Wstwr Sy | 5.20 |
| LA Comm Coll Dist CA | 5.03 |
| Lr CoOHhospfcsRvRf & Imp | 5.46 |
| MA Pt AthSpclcRV bds Sr | 5.29 |
| MD Hlt&Ed Fc At Rf Rv S | 5.18 |
| Metro WAS arpt Auth Sys | 5.16 |
| MI St ste trklne fund | 5.16 |
| MI StrgcFnd ltd Rf Rv | 5.44 |
| MI StrgcFnd ltd Rf Rv | 5.64 |
| MO Hlth & Ed FacAuth | 5.26 |
| MO Hlth & Ed FacAuth | 5.17 |
| NH Hlth & Educ Fac | 5.24 |
| NYC Mn Wtr Fin Auth Rf Sr | 5.23 |
| NYC Mn Wtr Fin Auth Rf Sr | 5.25 |
| NYS Drmtry AthRv Sr2001 | 5.36 |
| PA Tpke Comm Rgr fee | 5.23 |
| Phil Ind Dev Ath Pa arptRV | 5.32 |
| Phil PAgas worksRv 3rd Sr | 5.27 |
| Plm Bch Co Schl Bd FL | 5.12 |
| PR Publ Fn Corp 2001Sr | 5.03 |
| PrtoRico Elec Pwr Auth Rv | 5.06 |
| PrtoRico Pub Fnn Corp | 5.03 |
| PrtoRico pub imprvmt | 5.06 |
| Rnco Ccmg Rdv Agri CA tax | 5.04 |
| Seattle WA muni Lt pwr imp | 5.20 |
| SnJose CA aprt Rvbd Sr | 5.00 |
| TmpByWtr FL util SysRf Impr | 5.13 |
| VA Clg Bldg Ath ed fcs Rv | 5.22 |

**Source:** Data are from *Barrons*, October 1, 2001, p. MW43. Their source is The Bond Buyer.

**TABLE 3.9.2 Market Response to Stock Buyback Announcements**

| Company | Three-Month Price Change | Company | Three-Month Price Change |
| --- | --- | --- | --- |
| Tektronix | 17.0% | ITT Corp | −7.5% |
| General Motors | 12.7 | Ohio Casualty | 13.9 |
| Firestone | 26.2 | Kimberly-Clark | 14.0 |
| GAF Corp | 14.3 | Anheuser-Busch | 19.2 |
| Rockwell Intl. | −1.1 | Hewlett-Packard | 10.2 |

**Source:** Data are from the *Wall Street Journal,* September 18, 1987, p. 17. Their source is Salomon Brothers.

7. Business firms occasionally buy back their own stock for various reasons, sometimes when they view the market price as a bargain compared to their view of its true worth. It has been observed that the market price of stock often increases around the time of the announcement of such a buyback. Consider the data on actual percent changes over three months in stock prices for firms announcing stock buybacks shown in Table 3.9.2.
   a. Construct a histogram of this data set.
   b. Construct a stem-and-leaf style histogram of this data set using pen and paper. Your horizontal axis might include four columns of numbers (–0, 0, 1, and 2) representing the tens place.[16]
   c. Based on these histograms, what can you say to summarize typical behavior of these stock prices following a buyback announcement?

8. Consider the percentage change in stock price of the most active issues traded on the NASDAQ stock exchange, as shown in Table 3.9.3.
   a. Construct a histogram of this data set.
   b. Describe the distribution shape.
   c. Identify the outlier.
   d. Interpret the outlier. In particular, what is it telling you about UAL Corporation as compared to other heavily traded stocks on this day?
   e. Suppose you are conducting a study of price changes of heavily traded stocks. Discuss the different ways you might deal with this outlier. In particular, would it be appropriate to omit it from the analysis?

9. Consider CREF, the College Retirement Equities Fund, which manages retirement accounts for employees of nonprofit educational and research organizations. CREF manages a large and diversified portfolio in its stock account, somewhere around $121 billion.

$-38870, 31379$
$111,838, 1138$

### TABLE 3.9.3 Active NASDAQ Stock Market Issues

| Firm | Change |
|---|---|
| PowerShares QQQ Trust Series 1 (QQQQ) | –0.28% |
| Microsoft (MSFT) | 0.47% |
| Intel (INTC) | –0.26% |
| Cisco Systems (CSCO) | –0.61% |
| Sirius XM Radio (SIRI) | 3.14% |
| Oracle (ORCL) | 1.30% |
| Apple (AAPL) | –0.62% |
| YRC Worldwide (YRCW) | –2.72% |
| Micron Technology (MU) | –1.91% |
| Applied Materials (AMAT) | 0.00% |
| Comcast Cl A (CMCSA) | –1.05% |
| Popular (BPOP) | –2.34% |
| Yahoo! (YHOO) | –0.14% |
| NVIDIA (NVDA) | –1.25% |
| Qualcomm (QCOM) | 1.28% |
| eBay (EBAY) | –1.93% |
| Dell (DELL) | 0.00% |
| News Corp. Cl A (NWSA) | –0.76% |
| UAL (UAUA) | –10.28% |
| Huntington Bancshares (HBAN) | –1.66% |

**Source:** Data are from the *Wall Street Journal*, accessed at http://online.wsj.com/ on July 3, 2010.

### TABLE 3.9.4 CREF's Investments

| Company | Portfolio Value ($ Thousands) |
|---|---|
| AAR Corp | $2,035 |
| Alliant Techsystems, Inc | 5,133 |
| Armor Holdings, Inc | 1,758 |
| BAE Systems PLC | 31,984 |
| Boeing Co | 364,299 |
| Echostar Communications Corp | 14,464 |
| Empresa Brasileira de Aeronautica S.A. | 317 |
| General Dynamics Corp | 150,671 |
| General Motors Corp | 183,967 |
| Heico Corp | 740 |
| Hexcel Corp | 1,162 |
| Kaman Corp | 2,141 |
| Lockheed Martin Corp | 81,234 |
| Moog, Inc | 745 |
| Motient Corp | 784 |
| Northrop Grumman Corp | 29,878 |
| Orbital Sciences Corp | 770 |
| Panamsat Corp | 4,861 |
| Pegasus Communications Corp | 4,640 |
| Perkinelmer, Inc | 28,371 |
| Precision Cast Parts Corp | 9,822 |
| Raytheon Co A | 31,952 |
| Raytheon Co B | 25,787 |
| Remec, Inc | 2,147 |
| Rolls-Royce PLC | 40,110 |
| Smith Group PLC | 9,263 |
| Teledyne Technologies, Inc | 4,009 |
| Thales (Ex Thomson CFS) | 45,169 |
| Triumph Group, Inc | 2,875 |
| Zodiac S.A. | 13,429 |

**Source:** Data are from CREF 2000 Annual Report, p. 11.

Investment in aerospace and defense represents 0.91% of this portfolio. Data on the market value of these investments are shown in Table 3.9.4.

a. Construct a histogram of this data set.

b. Based on this histogram, describe the distribution of CREF's investment in aerospace and defense.

c. Describe the shape of the distribution. In particular, is it skewed or symmetric?

d. Find the logarithm of each data value.

e. Construct a histogram of these logarithms.

f. Describe the distribution shape of the logarithms. In particular, is it skewed or symmetric?

10. Consider the 20,000 median household income values in the donations database (available at the companion site). These represent the median household income for the neighborhood of each potential donor in the database.

a. Construct a histogram.

b. Describe the distribution shape.

11. Consider the number of gifts previously given by the 20,000 donors in the donations database (available at the companion site).

a. Construct a histogram.

b. Describe the distribution shape.

**TABLE 3.9.5** Percent Change in Revenues, 2008 to 2009, for Food-Related Companies in the Fortune 500

| Company | Revenue Change |
|---|---|
| Campbell Soup | –9.6% |
| ConAgra Foods | –6.0% |
| CVS Caremark | 12.9% |
| Dean Foods | –10.4% |
| Dole Food | –12.3% |
| General Mills | 7.6% |
| Great Atlantic & Pacific Tea | 36.7% |
| H.J. Heinz | 0.8% |
| Hershey | 3.2% |
| Hormel Foods | –3.3% |
| Kellogg | –1.9% |
| Kraft Foods | –5.8% |
| Kroger | 1.0% |
| Land O'Lakes | –13.5% |
| PepsiCo | 0.0% |
| Publix Super Markets | 1.7% |
| Rite Aid | 7.7% |
| Safeway | –7.4% |
| Sara Lee | –4.2% |
| Supervalu | 1.2% |
| Walgreen | 7.3% |
| Whole Foods Market | 1.0% |
| Winn-Dixie Stores | 1.2% |

**Source:** Data for Food Consumer Products accessed at http://money.cnn.com/magazines/fortune/fortune500/2010/industries/198/index.html; data for Food and Drug Stores accessed at http://money.cnn.com/magazines/fortune/fortune500/2010/industries/148/index.html on July 4, 2010.

12. Consider the percent change in revenues for food-related companies in the Fortune 500, in Table 3.9.5.
    a. Construct a histogram for this data set.
    b. Describe the distribution shape.
    c. Land O'Lakes had the largest decrease, falling by 13.5% and appears at first glance to be somewhat different from the others. Based on the perspective given by your histogram from part a, is Land O'Lakes an outlier? Why or why not?

13. Draw a stem-and-leaf histogram of the average hospital charge in $thousands for treating a patient who had the diagnosis group "Inguinal & femoral hernia procedures w MCC" for a group of hospitals in Washington State

(data accessed at http://www.doh.wa.gov/EHSPHL/hospdata/CHARS/2007FYHospitalCensusandChargesby DRG.xls on July 4, 2010).

29, 37, 57, 71, 38, 44, 36, 13, 42, 19, 16, 53, 37, 18, 54, 71, 10, 38, 43, 42, 58, 15, 31, 25, 47

14. Consider the costs charged for treatment of heart failure and shock by hospitals in the Puget Sound area, as shown in Table 3.9.6.
    a. Construct a histogram.
    b. Describe the distribution shape.

15. Consider the compensation paid to chief executive officers of food processing firms, as shown in Table 3.9.7.
    a. Construct a histogram.
    b. Describe the distribution shape.

16. There are many different and varied formats and strategies for radio stations, but one thing they all have in common is the need for an audience in order to attract advertisers. Table 3.9.8 shows the percent of listeners for radio stations in the Seattle–Tacoma area (averages for ages 12 and older, 6 a.m. to midnight all week).
    a. Construct a histogram.
    b. Describe the distribution shape.

17. Consider the net income as reported by selected firms in Table 3.9.9.
    a. Construct a histogram.
    b. Describe the distribution shape.

18. Many people do not realize how much a funeral costs and how much these costs can vary from one provider to another. Consider the price of a traditional funeral service with visitation (excluding casket and grave liner) as shown in Table 3.9.10 for the Puget Sound Region of Washington State.
    a. Construct a histogram for this data set.
    b. Describe the distribution shape.

19.* When the IRS tax code was revised in 1986, Congress granted some special exemptions to specific corporations. The U.S. government's revenue losses due to some of these special transition rules for corporate provisions are shown in Table 3.9.11.
    a. Construct a histogram for this data set.
    b. Describe the distribution shape.

20. Continuing with the revenue loss data of Table 3.9.11:
    a. Find the logarithm for each data value. Omit the two firms with zero revenue loss from your answers to this problem.
    b. Construct a histogram for this data set.
    c. Describe the distribution shape.
    d. Compare this analysis of the transformed data to your analysis of the original data in problem 19.

21. The number of small electric motors rejected for poor quality, per batch of 250, were recorded for recent batches. The results were as follows:

3, 2, 7, 5, 1, 3, 1, 7, 0, 6, 2, 3, 4, 1, 2, 25, 2, 4, 5, 0, 5, 3, 5, 3, 1, 2, 3, 1, 3, 0, 1, 6, 3, 5, 41, 1, 0, 6, 4, 1, 3
    a. Construct a histogram for this data set.
    b. Describe the distribution shape.
    c. Identify the outlier(s).

**TABLE 3.9.6 Hospital Charges for Heart Failure and Shock at Puget Sound Area Hospitals (Not Including Doctor Fees)**

| Hospital | Charges | Hospital | Charges |
|---|---|---|---|
| Affiliated Health Services | $6,415 | Overlake Hospital Medical Center | $6,364 |
| Allenmore Community Hospital | 5,355 | Providence General Medical Center | 5,235 |
| Auburn Regional Medical Center | 7,189 | Providence Saint Peter Hospital | 5,527 |
| Cascade Valley Hospital | 4,690 | Providence Seattle Medical Center | 7,222 |
| Children's Hospital & Medical Center | 8,585 | Puget Sound Hospital | 9,351 |
| Columbia Capital Medical Center | 6,739 | Saint Clare Hospital | 6,628 |
| Community Memorial Hospital | 4,906 | Saint Francis Community Hospital | 6,235 |
| Evergreen Hospital Medical Center | 5,805 | Saint Joseph Hospital | 7,110 |
| Good Samaritan Hospital | 4,762 | Saint Joseph Medical Center | 6,893 |
| Group Health Central Hospital | 3,289 | Stevens Memorial Hospital | 5,730 |
| Group Health Eastside Hospital | 2,324 | Swedish Medical Center | 7,661 |
| Harborview Medical Center | 7,107 | Tacoma General Hospital | 5,835 |
| Harrison Memorial Hospital | 5,617 | University of Washington Medical Center | 7,893 |
| Highline Community Hospital | 6,269 | Valley General Hospital | 4,279 |
| Island Hospital | 4,811 | Valley Medical Center | 4,863 |
| Mary Bridge Children's Health Center | 5,582 | Virginia Mason Medical Center | 5,773 |
| Northwest Hospital | 4,759 | Whidbey General Hospital | 4,142 |

Source: *Book of Lists 1998,* Puget Sound Business Journal, Vol. 18, Number 33. Their source is the Washington State Department of Health.

**TABLE 3.9.7 CEO Compensation for Food Processing Firms**

| Firm | CEO Compensation | Firm | CEO Compensation |
|---|---|---|---|
| Archer-Daniels-Midland | $3,171,000 | Kellogg | $1,489,000 |
| Campbell Soup | 1,810,000 | Pet | 1,023,000 |
| ConAgra | 1,600,000 | Quaker Oats | 1,398,000 |
| CPC International | 1,202,000 | Ralston Purina Group | 1,363,000 |
| General Mills | 850,000 | Sara Lee | 1,736,000 |
| Heinz | 895,000 | Sysco | 1,015,000 |
| Hershey Foods | 897,000 | Tyson Foods | 1,174,000 |
| Hormel Foods | 985,000 | Wrigley | 475,000 |

Source: Data are from "Executive Compensation Scoreboard," *Business Week,* April 24, 1995, p. 102.

 d. Remove the outlier(s), and construct a histogram for the remaining batches.

 e. Summarize this firm's recent experience with quality of production.

22. Consider the price of renting a car for a week, with manual transmission but declining the collision damage waiver, in 13 European countries (Table 3.9.12).

 a. Draw a histogram of this data set.

 b. Describe the distribution shape.

23. Draw a histogram of interest rates offered by banks on certificates of deposit and describe the distribution shape:

 9.9%, 9.5%, 10.3%, 9.3%, 10.4%, 10.7%, 9.1%, 10.0%, 8.8%, 9.7%, 9.9%, 10.3%, 9.8%, 9.1%, 9.8%

### TABLE 3.9.8 Market Share for Seattle Radio Stations

| Station | Format | Percent of Listeners 12 and Older |
|---|---|---|
| KIXI-AM | '50s–'60s hits | 4.5% |
| KBSG-FM-AM | '60s–'70s hits | 5.5 |
| KJR-FM | '70s hits | 3.8 |
| KLSY-FM | adult-contemporary | 4.2 |
| KPLZ-FM | adult-contemporary | 4.0 |
| KRWM-FM | adult-contemporary | 3.1 |
| KMTT-FM-AM | adult alternative | 3.5 |
| KNWX-AM | all news | 1.7 |
| KCMS-FM | Christian music | 1.6 |
| KCIS-AM | Christian news, info | 0.4 |
| KZOK-FM | classic rock | 5.4 |
| KING-FM | classical | 3.7 |
| KMPS-FM-AM | country | 5.0 |
| KRPM-FM-AM | country | 3.2 |
| KYCW-FM | country | 3.2 |
| KWJZ-FM | modern jazz | 2.7 |
| KIRO-AM | news-talk | 6.3 |
| KOMO-AM | news-talk-music | 2.6 |
| KISW-FM | rock | 4.0 |
| KNDD-FM | rock | 4.6 |
| KJR-AM | sports-talk | 1.5 |
| KIRO-FM | talk-news | 2.3 |
| KVI-AM | talk-news | 4.9 |
| KUBE-FM | Top 40/rhythm | 6.0 |

**Source:** Data are from the *Seattle Times,* October 20, 1995, p. F3. Their source is The Arbitron Co., copyright.

### TABLE 3.9.9 Net Income of Selected Firms

| Firm | Net Income ($ thousands) |
|---|---|
| Bay State Bancorp | $1,423 |
| Bedford Bancshrs | 677 |
| CGI Group Inc | 30,612 |
| CNB Finl-PA | 1,890 |
| Camco Financial | 2,522 |
| Comm Bancorp Inc | 1,340 |
| Concord Communctn | 28 |
| East Penn Bank | 479 |
| Eastern VA Bkshrs | 1,104 |
| FFLC Bancorp Inc | 1,818 |
| FPL Group Inc | 118,000 |
| Fauquier Bankshrs | 620 |
| First Banks Amer | 15,965 |
| First Busey Corp | 3,667 |
| First Finl Bcp-OH | 7,353 |
| First Finl Holdings | 6,804 |
| Firstbank Corp-MI | 2,588 |
| Frankfort First | 354 |

**Source:** Data are selected from Digest of Earnings, *Wall Street Journal,* accessed at http://interactive.wsj.com/public/resources/documents/digest_earnings.htm on January 18, 2002.

### TABLE 3.9.10 Cost of Traditional Funeral Service

| Funeral Home | Cost |
|---|---|
| Bleitz | $2,180 |
| Bonney-Watson | 2,250 |
| Butterworth's Arthur A. Wright | 2,265 |
| Dayspring & Fitch | 1,795 |
| Evergreen-Washelli | 1,895 |
| Faull-Stokes | 2,660 |
| Flintoft's | 2,280 |
| Green | 3,195 |
| Price-Helton | 2,995 |
| Purdy & Walters at Floral Hills | 2,665 |
| Southwest Mortuary | 2,360 |
| Yahn & Son | 2,210 |

**Source:** *Seattle Times,* December 11, 1996, p. D5.

24. Draw a histogram of the market values of your main competitors (in millions of dollars) and describe the distribution shape:

    3.7, 28.3, 10.6, 0.1, 9.8, 6.2, 19.7, 23.8, 17.8, 7.8, 10.8, 10.9, 5.1, 4.1, 2.0, 24.2, 9.0, 3.1, 1.6, 3.7, 27.0, 1.2, 45.1, 20.4, 2.3

25. Consider the salaries (in thousands of dollars) of a group of business executives:

    177, 54, 98, 57, 209, 56, 45, 98, 58, 90, 116, 42, 142, 152, 85, 53, 52, 85, 72, 45, 168, 47, 93, 49, 79, 145, 149, 60, 58

TABLE 3.9.11 Special Exemptions to the 1986 Revision of the IRS Tax Code

| Firm | Estimated Government Revenue Loss ($ millions) | Firm | Estimated Government Revenue Loss ($ millions) |
|---|---|---|---|
| Paramount Cards | $7 | New England Patriots | $6 |
| Banks of Iowa | 7 | Ireton Coal | 18 |
| Ideal Basic Industries | 0 | Ala-Tenn Resources | 0 |
| Goldrus Drilling | 13 | Metropolitan-First Minnesota Merger | 9 |
| Original Appalachian Artworks | 6 | Texas Air/ Eastern Merger | 47 |
| Candle Corp | 13 | Brunswick | 61 |
| S.A. Horvitz Testamentary Trust | 1 | Liberty Bell Park | 5 |
| Green Bay Packaging | 2 | Beneficial Corp | 67 |

**Source:** Data are from "Special Exemptions in the Tax Bill, as Disclosed by the Senate," *New York Times,* September 27, 1986, p. 33. These particular firms are grouped under the heading "Transition Rules for Corporate Provisions." Don't you wish you could have qualified for some of these?

a. Construct a histogram of this data set.
b. Describe the distribution shape.
c. Based on the histogram, what values appear to have been typical for this group of salaries?

26. Consider the order size of recent customers (in thousands of dollars):

    31, 14, 10, 3, 17, 5, 1, 17, 1, 2, 7, 12, 28, 4, 4, 10, 4, 3, 9, 28, 4, 3

a. Construct a histogram for this data set.
b. Describe the distribution shape.

27. Draw a histogram for the following list of prices charged by different stores for a box of envelopes (in dollars) and describe the distribution shape:

    4.40, 4.20, 4.55, 4.45, 4.40, 4.10, 4.10, 3.80, 3.80, 4.30, 4.90, 4.20, 4.05

28. Consider the following list of your product's market share of 20 major metropolitan areas:

    0.7%, 20.8%, 2.3%, 7.7%, 5.6%, 4.2%, 0.8%, 8.4%, 5.2%, 17.2%, 2.7%, 1.4%, 1.7%, 26.7%, 4.6%, 15.6%, 2.8%, 21.6%, 13.3%, 0.5%

a. Construct an appropriate histogram of this data set.
b. Describe the distribution shape.

TABLE 3.9.12 Cost to Rent a Car

| Country | Rental Price (U.S. dollars) | Country | Rental Price (U.S. dollars) |
|---|---|---|---|
| Austria | $239 | Netherlands | $194 |
| Belgium | 179 | Norway | 241 |
| Britain | 229 | Spain | 154 |
| Denmark | 181 | Sweden | 280 |
| France | 237 | Switzerland | 254 |
| Ireland | 216 | West Germany | 192 |
| Italy | 236 | | |

TABLE 3.9.13 Percentage Change in Dollar Value

| Foreign Currency | Change in Dollar Value | Foreign Currency | Change in Dollar Value |
|---|---|---|---|
| Belgium | −5.3% | Singapore | −1.5 |
| Japan | −6.7 | France | −4.9 |
| Brazil | 26.0 | South Korea | −1.0 |
| Mexico | −1.2 | Hong Kong | 0.0 |
| Britain | −3.7 | Taiwan | −0.1 |
| Netherlands | −5.1 | Italy | −4.7 |
| Canada | −1.9 | West Germany | −5.1 |

29. Consider the percentage change in the value of the dollar with respect to other currencies over a four-week period (Table 3.9.13).
a. Construct an appropriate histogram of this data set.
b. Describe the distribution shape.

30. Consider the following list of prices (in dollars) charged by different pharmacies for twelve 60-mg tablets of the prescription drug Tylenol No. 4 with Codeine:[17]

    6.75, 12.19, 9.09, 9.09, 13.09, 13.45, 7.89, 12.00, 10.49, 15.30, 13.29

a. Construct a histogram of these prices.
b. Describe the distribution shape.
c. Comment on the following statement: It really doesn't matter very much where you have a prescription filled.

31. Using the data in problem 26 of Chapter 2 on the 30 Dow Jones Industrials:
a. Construct a stem-and-leaf diagram for percent change during January 2002.
b. Construct a histogram for percent change during January 2002.
c. Describe the shape of the distribution.

**32.** Using the data in problem 27 of Chapter 2 on the Dow Jones Industrial Average:
  a. Construct a stem-and-leaf diagram for net change during January 2002.
  b. Construct a histogram for net change.
  c. Describe the shape of the distribution.
  d. Construct a stem-and-leaf diagram for percent change during January 2002.
  e. Construct a histogram for percent change.
  f. Describe the shape of the distribution.

16. The data value –7.5 would be recorded as a 7 placed over the 20 column (the number 27.5 has a negative 0 for its tens place, and the next digit, 7, is placed above it).
17. Data are from S. Gilje, "What Health-Care Revision Means to Prescription Drug Sales," *Seattle Times*, February 28, 1993, p. K1, and were compiled by C. Morningstar and M. Hendrickson.

## Database Exercises

*Problems marked with an asterisk (*) are solved in the Self Text in Appendix C.*

Refer to the employee database in Appendix A.

**1.** For the salary numbers:
  a. Construct a histogram.
  b. Describe the shape of the distribution.
  c. Summarize the distribution in general terms by giving the smallest salary and the largest salary.
**2.*** For the age numbers:
  a. Construct a histogram.
  b. Describe the shape of the distribution.
  c. Summarize the distribution in general terms.
**3.** For the experience numbers:
  a. Construct a histogram.
  b. Describe the shape of the distribution.
  c. Summarize the distribution in general terms.
**4.** For the salary numbers, separated according to gender:
  a. Construct a histogram for just the males.
  b. Construct a histogram for just the females using the same scale as in part a to facilitate comparison of male and female salaries.
  c. Compare these two salary distributions, and write a paragraph describing any gender differences in salary that you see from comparing these two histograms.[18]

18. Statistical methods for comparing two groups such as these will be presented in Chapter 10.

## Project

Draw a histogram for each of three data sets related to your business interests. Choose your own business data from sources such as the Internet, *Wall Street Journal,* or your firm. Each data set should contain at least 15 numbers. Write a page (including the histogram) for each data set, commenting on the histogram as follows:

  a. What is the distribution shape?
  b. Are there any outliers? What might you do if there are?
  c. Summarize the distribution in general terms.
  d. What have you learned from examining the histogram?

## Case

*Let's Control Waste in Production*

"That Owen is costing us money!" stated Billings in a clear, loud voice at the meeting. "Look, I have proof. Here's a histogram of the materials used in production. You can clearly see two groups here, and it looks as though Owen uses up a few hundred dollars more in materials each and every shift than does Purcell."

You're in charge of the meeting and this is more emotion than you'd like to see. To calm things down, you try to gracefully tone down the discussion and move toward a more deliberate resolution. You're not the only one; a suggestion is made to look into the matter and put it on the agenda for the next meeting.

You know, as do most of the others, that Owen has a reputation for carelessness. However, you've never seen it firsthand, and you'd like to reserve judgment just in case others have jealously planted that suggestion and because Owen is well respected for expertise and productivity. You also know that Billings and Purcell are good friends. Nothing wrong there, but it's worth a careful look at all available information before jumping to conclusions.

After the meeting, you ask Billings to e-mail you a copy of the data. He sends you just the first two columns you see below, and it looks familiar. In fact, there is already a report in your computer that includes all three of the columns below, with one row per shift supervised. Now you are ready to spend some time getting ready for the meeting next week.

| Materials Used | Manager in Charge | Inventory Produced | Materials Used | Manager in Charge | Inventory Produced |
|---|---|---|---|---|---|
| $1,459 | Owen | $4,669 | $1,434 | Owen | $4,589 |
| 1,502 | Owen | 4,806 | 1,127 | Purcell | 3,606 |
| 1,492 | Owen | 4,774 | 1,457 | Owen | 4,662 |
| 1,120 | Purcell | 3,584 | 1,109 | Purcell | 3,549 |
| 1,483 | Owen | 4,746 | 1,236 | Purcell | 3,955 |
| 1,136 | Purcell | 3,635 | 1,188 | Purcell | 3,802 |
| 1,123 | Purcell | 3,594 | 1,512 | Owen | 4,838 |
| 1,542 | Owen | 4,934 | 1,131 | Purcell | 3,619 |
| 1,484 | Owen | 4,749 | 1,108 | Purcell | 3,546 |
| 1,379 | Owen | 4,413 | 1,135 | Purcell | 3,632 |
| 1,406 | Owen | 4,499 | 1,416 | Owen | 4,531 |
| 1,487 | Owen | 4,758 | 1,170 | Purcell | 3,744 |
| 1,138 | Purcell | 3,642 | 1,417 | Owen | 4,534 |
| 1,529 | Owen | 4,893 | 1,381 | Owen | 4,419 |
| 1,142 | Purcell | 3,654 | 1,248 | Purcell | 3,994 |
| 1,127 | Purcell | 3,606 | 1,171 | Purcell | 3,747 |
| 1,457 | Owen | 4,662 | 1,471 | Owen | 4,707 |
| 1,479 | Owen | 4,733 | 1,142 | Purcell | 3,654 |
| 1,407 | Owen | 4,502 | 1,161 | Purcell | 3,715 |
| 1,105 | Purcell | 3,536 | 1,135 | Purcell | 3,632 |
| 1,126 | Purcell | 3,603 | 1,500 | Owen | 4,800 |

### Discussion Questions

1. Does the distribution of Materials Used look truly bimodal? Or could it reasonably be normally distributed with just a single group?
2. Do separate histograms for Owen and Purcell agree with the contention by Billings that Owen spends more?
3. Should we agree with Billings at the next meeting? Justify your answer by careful analysis of the available data.

## Chapter 4

# Landmark Summaries

## Interpreting Typical Values and Percentiles

One of the most effective ways to "see the big picture" in complex situations is through **summarization**, that is, using one or more selected or computed values to represent the data set. Studying each individual case in detail is not, in itself, a statistical activity,[1] but discovering and identifying the features that the cases have in common are statistical activities because they treat the information as a whole.

In statistics, one goal is to condense a data set down to one number (or two or a few numbers) that expresses the most fundamental characteristics of the data. The methods most appropriate for a single list of numbers (i.e., univariate data) include the following:

**One:** The *average, median,* and *mode* are different ways of selecting a single number that closely describes all the numbers in a data set. Such a single-number summary is referred to as a *typical value, center,* or *location.*

**Two:** A *percentile* summarizes information about *ranks,* characterizing the value attained by a given percentage of the data after they have been ordered from smallest to largest.

**Three:** The *standard deviation* is an indication of how different the numbers in the data set are from one another. This concept is also referred to as *diversity* or *variability* and will be deferred to Chapter 5.

---

1. However, this activity may be worthwhile if there is time enough to study every one!

What if there are individuals not adequately described by these summaries? Such outliers may simply be described separately. Thus, you can summarize a large group of data by (1) summarizing the basic structure of most of its elements and then (2) making a list of any special exceptions. In this way, you can achieve the statistical goal of efficiently describing a large data set and still take account of the special nature of the individual.

## 4.1 WHAT IS THE MOST TYPICAL VALUE?

The ultimate summary of any data set is a single number that best represents all of the data values. You might call such a number *a typical value* for the data set. Unless all numbers in the data set are the same, you should expect some differences of opinion regarding exactly which number is "most typical" of the entire data set. There are three different ways to obtain such a summary measure:

1. The *average* or *mean*, which can be computed only for meaningful numbers (quantitative data).
2. The *median*, or halfway point, which can be computed either for ordered categories (ordinal data) or for numbers.
3. The *mode*, or most common category, which can be computed for unordered categories (nominal data), ordered categories, or numbers.

**Practical Business Statistics, Sixth Edition.**
© 2012 Andrew F. Siegel. Published by Elsevier, Inc. All rights reserved.

## The Average: A Typical Value for Quantitative Data

The **average** (also called the **mean**) is the most common method for finding a typical value for a list of numbers, found by adding up all the values and then dividing by the number of items. The sample average expressed as a formula is:

**The Sample Average**

$$\text{Sample Average} = \frac{\text{Sum of data items}}{\text{Number of data items}}$$

$$\overline{X} = \frac{X_1 + X_2 + \cdots + X_n}{n}$$

$$= \frac{1}{n} \sum_{i=1}^{n} X_i$$

where $n$ is the number of items in the list of data and $X_1$, $X_2, \ldots, X_n$ stand for the values themselves. The Greek capital letter *sigma*, $\Sigma$, tells you to add up the symbols that follow it, substituting the values 1 through $n$ in turn for $i$. The symbol for the average, $\overline{X}$, is read aloud as "$X$ bar."

For example, the average of the three-number data set 4, 9, 8 is

$$\overline{X} = \frac{4 + 9 + 8}{3} = \frac{21}{3} = 7$$

Here is how Excel's Average function can be used to find the average of a list of numbers, with the heading "Salary" just above the list:

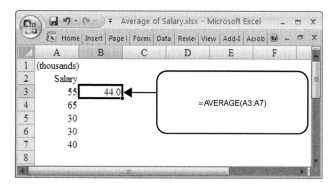

Excel's menu system can help guide you through the process of finding an average (or other statistical function). Begin by selecting the cell where you want the average to go. Then choose Insert Function from the Formulas Ribbon, select Statistical as the category, and choose AVERAGE as the function name. A dialog box will then pop up, allowing you to drag down your list of numbers

and choose OK to complete the process. Here's how it looks:

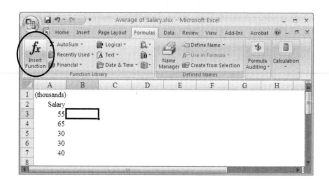

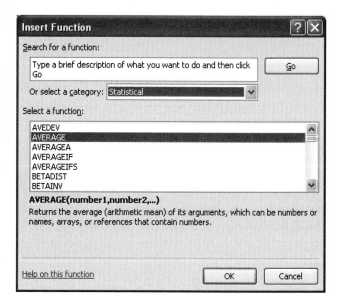

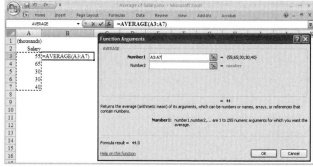

The idea of an average is the same whether you view your list of numbers as a complete population or as a representative sample from a larger population.[2] However, the notation differs slightly. For an entire population, the

---

2. The concept of sampling from a population is crucial to statistical inference and will be covered in detail in Chapters 8–10.

convention is to use $N$ to represent the number of items and to let $\mu$ (the Greek letter *mu*) represent the population mean value. The calculation of the mean remains the same whether you have a population or a sample.

Since the data values must be added together as part of the process of finding the average, it is clear that this method cannot apply to qualitative data (how would you add colors or bond ratings together?).

The average may be interpreted as spreading the total evenly among the elementary units. That is, if you replace each data value by the average, then the total remains unchanged. For example, from an employee database, you could compute the average salary for all employees in Houston. The resulting average would have the following interpretation: If we were to pay all Houston employees the same salary, without changing the total salary for all of Houston, it would be this average amount. Note that you don't have to be contemplating the institution of such a level salary structure to use the average as an indication of typical salary (particularly when you are concerned with the total payroll amount as a budget item).

Since the average preserves the total while spreading amounts out evenly, it is most useful as a summary when there are no extreme values (outliers) present and the data set is a more-or-less homogeneous group with randomness. If one employee earns vastly more than the others, then the average will not be useful as a summary measure. Although it will still preserve the total salary amount, it will not give as useful an indication of the salaries of individuals since the average will be too high for most employees and too low for the exceptional worker.

The average is the only summary measure capable of preserving the total. This makes it particularly useful in situations where a total amount is to be projected for a large group. First, the average would be found for a smaller sample of data representing the larger group. Then this average would be scaled up by multiplying it by the number of individuals in the larger group. The result gives an estimate, or forecast, of the total for the larger population. Generally, when a total is to be determined, the average should be used.

### Example
#### How Much Will Consumers Spend?

A firm is interested in total consumer expenditures on personal health items in the city of Cleveland. It has obtained the results of a random sample of 300 individuals in the area, showing that an average of $6.58 was spent last month by each one.

Naturally, some spent more and others spent less than this average amount. Rather than work with all 300 numbers, we use the average to summarize the typical amount each spent. More importantly, when we multiply it by the population of Cleveland, we obtain a reasonable

estimate of the total personal health expenditures for the entire city:[3]

> Estimated Cleveland personal health expenditures
> = (Average per person from sample)
>   × (Population of Cleveland)
> = ($6.58) × (433,748)
> = $2,854,062

This projection of total sales of $2.9 million is reasonable and probably useful. However, it is almost certainly wrong (in the sense that it does not represent the exact amount spent). Later, when you study confidence intervals (in Chapter 9), you will see how to account for the statistical error arising from the projection from a sample of 300 to a population of 433,748 individuals.

---

3. This is the 2008 population estimate from Table 27 of the U.S. Census Bureau, *Statistical Abstract of the United States: 2010* (129th edition), Washington, DC, 2009, accessed at http://www.census.gov/compendia/statab/cats/population.html on July 5, 2010.

### Example
#### How Many Defective Parts?

The Globular Ball Bearing Company manufactures 1,000 items per lot. From the day's production of 253 lots, a sample of 10 lots was chosen randomly for quality control inspection. The number of defective items in each of these lots was

3, 8, 2, 5, 0, 7, 14, 7, 4, 1

The average of this data set is

$$\frac{3+8+2+5+0+7+14+7+4+1}{10} = \frac{51}{10} = 5.1$$

*(Continued)*

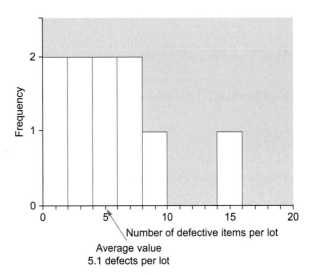

FIGURE 4.1.1  A histogram of the number of defective items in each of 10 lots of 1,000 items, with the average (5.1) indicated.

**Example—cont'd**

which tells you that there were 5.1 defects per lot, on the average. This represents a defect rate of 5.1 per 1,000, or 0.51% (about half of a percent). Scaling up to the day's production of 253 lots, you would expect there to be about

$$5.1 \times 253 = 1,290.3$$

defective items in the entire day's production, out of the total of 253,000 items produced.

To show how the average is indeed a reasonable summary of a list of numbers, Figure 4.1.1 shows a histogram of the 10 numbers in this data set with the average value indicated. Note how the average is nicely in the middle, reasonably close to all of the data values.

## The Weighted Average: Adjusting for Importance

The **weighted average** is like the average, except that it allows you to give a different importance, or "weight," to each data item. The weighted average gives you the flexibility to define your own system of importance when it is not appropriate to treat each item equally.

If a firm has three plants and the employee pension expenses for each plant are being analyzed, it may not make sense to take a simple average of these three numbers as a summary of typical pension expense, especially if the plants are very different in size. If one plant has twice as many employees as another, it seems reasonable that its pension expense should count double in the summarization process. The weighted average allows you to do this by using weights defined according to the size of each plant.

The weights are usually positive numbers that sum to 1, expressing the relative importance of each data item. Don't worry if your initial weights don't add up to 1. You can always force them to do this by dividing each weight by the sum of all of the weights. Your initial weights might be determined by number of employees, market value, or some other objective measure, or by a subjective method (i.e., using someone's personal or expert opinion). Sometimes it's easier to set initial weights without worrying about whether they add up to 1, and then convert later by dividing by the total.

Suppose you decide to compute a weighted average of pension expenses for the three plants using weights determined by the number of employees at each plant. If the three plants have 182, 386, and 697 employees, you would use respective weights of

$$182/1,265 = 0.144$$
$$386/1,265 = 0.305$$
$$697/1,265 = 0.551$$

Note that in each case, weights were obtained by dividing by the sum $182 + 386 + 697 = 1,265$. The resulting weights add to 1, as they must:[4] $0.144 + 0.305 + 0.551 = 1$.

To compute the weighted average, multiply each data item by its weight and sum these results. That's all there is to it. The formula looks like this:

**The Weighted Average**

$$\text{Weighted Average} = \text{Sum of (weight times data item)}$$
$$= w_1 X_1 + w_2 X_2 + \cdots + w_n X_n$$
$$= \sum_{i=1}^{n} w_i X_i$$

where $w_1, w_2, \ldots, w_n$ stand for the respective weights, which add up to 1. You may think of the regular (unweighted) average as a weighted average in which all data items have the same weight, namely $1/n$.

The weighted average of 63, 47, and 98, with respective weights of 0.144, 0.305, and 0.551, is

$$(0.144 \times 63) + (0.305 \times 47) + (0.551 \times 98)$$
$$= 9.072 + 14.335 + 53.998 = 77.405$$

Note that this differs from the regular (unweighted) average value of $(63 + 47 + 98)/3 = 69.333$, as you might expect. The weighted average gives the largest number extra emphasis, namely, weight 0.551, which is more than one-third of the weight. It is therefore reasonable that the weighted average should be larger than the unweighted average in this case.

The weighted average may best be interpreted as an average to be used when some items have more importance than others; the items with greater importance have more of a say in the value of the weighted average.

**Example**
*Your Grade Point Average*

A grade point average, or GPA, computed for your university program is a weighted average. This is necessary because some courses have more credit hours, and therefore more importance, than others. It seems reasonable that a course that meets twice as often as another should have twice the impact, and the GPA reflects this.

Different schools have different systems. Suppose your system involves grades from 0.0 (flunking) to 4.0 (perfect), and your report card for the term looks like this:

| Course | Credits | Grade |
|---|---|---|
| Statistics | 5 | 3.7 |
| Economics | 5 | 3.3 |
| Marketing | 4 | 3.5 |
| Track | 1 | 2.8 |
| Total | 15 | |

---

4. The actual sum might be 0.999 or 1.001 due to round-off error. You need not be concerned when this happens.

The weights will be computed by dividing each of the credits by 15, the total number of credits. Your GPA will then be computed as a weighted average of your grades, weighted according to the number of credits:

$$\left(\frac{5}{15} \times 3.7\right) + \left(\frac{5}{15} \times 3.3\right) + \left(\frac{4}{15} \times 3.5\right) + \left(\frac{1}{15} \times 2.8\right) = 3.45$$

To find the weighted average in Excel®, first assign a name to the numbers in each column. The easiest way is to highlight all columns, including the titles at the top, and then use Excel's Create from Selection in the Defined Names section of the Formulas Ribbon, and click OK, as follows:

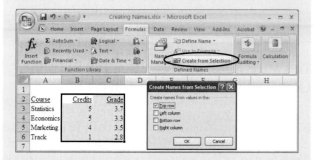

Now you are ready to use Excel's SumProduct function (which will multiply each credit value by the corresponding grade value and then add up the results) divided by the sum of the credits (so that the weights sum to 1). The result is the weighted average, 3.45, as follows:

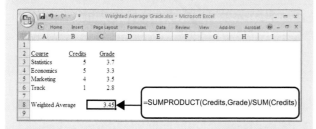

Your GPA, 3.45, fortunately was not greatly affected by the low grade (for track) due to its small weight (just 1 credit). Had these four grades simply been averaged, the result would have been lower (3.33). Thank goodness you didn't hurt your grades very much by concentrating on your business courses when the end-of-term crunch hit and letting the other one slide!

## Example
### The Firm's Cost of Capital

The firm's cost of capital, a concept from corporate finance, is computed as a weighted average. The idea is that a firm has raised money by selling a variety of financial instruments: stocks, bonds, commercial paper, and so forth. Since each of these securities has its own rate of return (a cost of capital to the firm), it would be useful to combine and summarize the

different rates as a single number representing the aggregate cost to the firm of raising money with this selection of securities.

The firm's cost of capital is a simple weighted average of the cost of capital of each security (this is its rate of return, or interest rate), where the weights are determined according to the total market value of that security. For example, if preferred stock represents only 3% of the market value of a firm's outstanding securities, then its cost should be given that low weight.

Consider the situation for Leveraged Industries, Inc., a hypothetical firm with lots of debt resulting from recent merger and acquisition activity:[5]

| Security | Market Value | Rate of Return |
|---|---|---|
| Common stock | $100,000 | 18.5% |
| Preferred stock | 15,000 | 14.9 |
| Bonds (9% coupon) | 225,000 | 11.2 |
| Bonds (8.5% coupon) | 115,000 | 11.2 |
| Total | $455,000 | |

Divide each market value by the total to find the weights, which express the market's assessment of the proportion of each type of security:[6]

| Security | Weight |
|---|---|
| Common stock | 0.220 |
| Preferred stock | 0.033 |
| Bonds (9% coupon) | 0.495 |
| Bonds (8.5% coupon) | 0.253 |

Since the weight of common stock is 0.220, it follows that 22% of the firm is financed with common stock, in terms of market value. The cost of capital is then computed by multiplying market rates of return by these weights and summing:

$$(0.220 \times 18.5) + (0.033 \times 14.9) + (0.495 \times 11.2)$$
$$+ (0.253 \times 11.2) = 12.94$$

The cost of capital for Leveraged Industries, Inc., is therefore 12.9%. This weighted average has combined the individual costs of capital (18.5%, 14.9%, and 11.2%) into a single number.

There would have been no change in this result (12.9%) if you had combined the two bond issues into one line item, with a combined market value of $340,000 and 11.2% rate of return, as may be verified by performing the calculation. This is as it should be, since such a distinction should have no practical consequences: The two bond issues have different coupons because they were issued at different times, and since that time, their market prices have adjusted so that their yields (rates of return) are identical.

This weighted average cost of capital may be interpreted as follows. If Leveraged Industries decided to raise additional capital without changing its basic business strategy (i.e., types of projects, risk of projects) and keeping the same relative mix of securities, then it will have to pay a return of

(Continued)

## Example—cont'd

12.9% or $129 per $1,000 raised per year. This $129 will be paid to the different types of securities according to the weights given.

5. In cost of capital computations, we always use current market values (rather than book values) due to the fact that market values indicate the likely cost to the firm of raising additional capital. Thus, in the case of bonds, you would use the current market yield (interest rate) rather than the coupon rate based on the face value because the bond is probably not trading at its face value. You would also use the market value per bond times the number of bonds outstanding to derive the total market value of outstanding bonds. Estimating the return demanded by the market for common stock is a challenging exercise you are likely to learn about in your finance course.

6. For example, the first weight is 100,000/455,000 = 0.220. The weights do not sum to 1 here due to round-off error. This is not a concern. If greater accuracy were needed or desired, you could work with four, five, or more decimal places for the weight and thereby increase the accuracy of the result.

## Example

### Adjusting for Misrepresentation

Another use of the weighted average is to correct for known misrepresentation in a sample as compared to the population you wish to know about. Since the average of the sample treats all individuals the same, but you know that (compared to the population) some groups of individuals are overrepresented and others underrepresented, reweighting the average will give you a better result. A weighted average is better because it combines the known information about each group (from the sample) with better information about each group's representation (from the population rather than the sample). Since the best information of each type is used, the result is improved.

Reconsider the sample of 300 individuals from Cleveland analyzed in the earlier example of consumer personal health expenditures. Suppose that the percentage of young people (under 18 years old) in the sample (21.7%) did not match the known percentage for the entire population (25.8%) and that the average expenditures were computed separately for each group:

Average expenditure for people under 18 = $4.86
Average expenditure for people 18 or over = $7.06

The weighted average of these expenditures will use the population (rather than the sample) weights, namely, 25.8% younger and 74.2% (which is 100% − 25.8%) older people since you know that these are the correct percentages to use. Of course, if you had expenditure information for the entire city, you would use that too, but you don't; you have it only for the 300 people in the sample. Converting the percentages to weights, the weighted average may be computed as follows:

Weighted average expenditure = $(0.258 \times \$4.86)$
$+ (0.742 \times \$7.06) = \$6.49$

This weighted average, $6.49, gives a better estimate of average personal health expenditures in Cleveland than the regular (unweighted) average value ($6.58). The weighted average is better because it has *corrected for* the fact that there were too many older people in the sample of 300.[7]

Since these older people tend to spend more, without this adjustment, the estimate would have been too large ($6.58 compared to $6.49).

Of course, even this new weighted estimate is probably wrong. But it is based on better information and has a smaller expected error, as can be proven using mathematical models. The new estimate is not necessarily better in every case (that is, the truth may actually be closer to the regular, unweighted average in this example), but the weighted procedure will give an answer that is *more likely* to be closer to the truth.

7. Statisticians often speak of "correcting for" or "adjusting for" one factor or another. This is one way to do it; later you may learn about multiple regression as another powerful way to correct for the influences of factors not under your control.

## The Median: A Typical Value for Quantitative and Ordinal Data

The **median** is the middle value; half of the items in the set are larger and half are smaller. Thus, it must be in the *center* of the data and provide an effective summary of the list of data. You find it by first putting the data in order and then locating the middle value. To be precise, you have to pay attention to some details; for example, you might have to average the two middle values if there is no single value in the middle.

One way to define the median is in terms of *ranks*.[8] **Ranks** associate the numbers 1, 2, 3, . . . , $n$ with the data values so that the smallest has rank 1, the next smallest has rank 2, and so forth up to the largest, which has rank $n$. One basic principle involved in finding the median is as follows:

### The Rank of the Median

The median has rank $(1 + n)/2$.

Paying attention to all of the special cases, you find the median for a list of $n$ items as follows:

1. Put the data items in order from smallest to largest (or largest to smallest; it doesn't matter).
2. Find the middle value. There are two cases to consider:
   a. If $n$ is an odd number, the median is the middle data value, which is located $(1 + n)/2$ steps in from either end of the ordered data list. For example, the median of the list 15, 27, 14, 18, 21, with $n = 5$ items, is

$$\text{Median}\,(15, 27, 14, 18, 21) = \text{Median}\,(14, 15, 18, 21, 27)$$
$$= 18$$

Note that to find the median, 18, you counted three steps into the ordered list, which is as the formula suggested, since $(1 + n)/2 = (1 + 5)/2 = 3$.

8. The ranks form the basis for *nonparametric methods*, which will be presented in Chapter 16.

For an ordinal data example, consider the list of bond ratings AAA, A, B, AA, A. The median is

$$\text{Median (AAA, A, B, AA, A)} =$$
$$\text{Median (B, A, A, AA, AAA)} = A$$

**b.** If $n$ is an even number, there are two middle values instead of just one. They are located $(1 + n)/2$ steps in from either end of the ordered data list.

**i.** If the data set is *quantitative* (i.e., consists of numbers), the median is the average of these two middle values. For example, the median of the list 15, 27, 18, 14, with $n = 4$ items, is

$$\text{Median (15, 27, 18, 14)} = \text{Median (14, 15, 18, 27)}$$
$$= (15 + 18)/2$$
$$= 16.5$$

The formula $(1 + n)/2$ gives $(1 + 4)/2 = 2.5$ in this case, which tells you to go halfway between the second and third number in the ordered list by averaging them.

**ii.** If the data set is *ordinal* (i.e., contains ordered categories) and if the two middle values represent the same category, this category is the median. If they represent different categories, you would have to report them both as defining the median. For example, the median of bond ratings A, B, AA, A is

$$\text{Median (A, B, AA, A)} = \text{Median (B, A, A, AA)} = A$$

since both middle values are rated A.

For another example, the median of bond ratings A, AAA, B, AA, AAA, B is

$$\text{Median (A, AAA, B, AA, AAA, B)}$$
$$= \text{Median (B, B, A, AA, AAA, AAA)} = A \text{ and } AA$$

This is the best you can do because you cannot find the average of two values with ordinal data.

To find the median in Excel®, you would use the Median function, as follows:

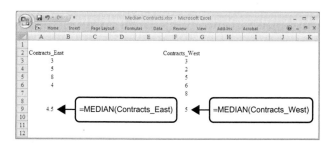

How does the median compare to the average? When the data set is normally distributed, they will be close to one another since the normal distribution is so symmetric and has such a clear middle point. However, the average and median will usually be a little different even for a normal

distribution because each summarizes in a different way, and there is nearly always some randomness in real data. When the data set is not normally distributed, the median and average can be very different because a skewed distribution does not have a well-defined center point. Typically, the average is more in the direction of the longer tail or of the outlier than the median is because the average "knows" the actual values of these extreme observations, whereas the median knows only that each value is either on one side or on the other.

### Example
#### The Crash of October 19, 1987: Stocks Drop at Opening
The stock market crash of 1987 was an extraordinary event in which the market lost about 20% of its value in one day. In this example we will examine how much value was lost as stocks first began trading that day.

Consider the percentage of value lost by 29 of the Dow Industrial stocks between the close of trading on Friday, October 16, and the opening of trading on Monday, October 19, 1987, the day of the crash. Even at the opening, these stocks had lost substantial value, as is shown in Table 4.1.1.

*(Continued)*

**TABLE 4.1.1 Loss at Opening on the Day of the Stock Market Crash of 1987**

| Firm | Change in Value | Firm | Change in Value |
|------|------|------|------|
| Union Carbide | −4.1% | Primerica | −6.8% |
| USX | −5.1 | Navistar | −2.1 |
| Bethlehem Steel | −4.5 | General Electric | −17.2 |
| AT&T | −5.4 | Westinghouse | −15.7 |
| Boeing | −4.0 | Alcoa | −8.9 |
| International Paper | −11.6 | Kodak | −15.7 |
| Chevron | −4.0 | Texaco | −12.3 |
| Woolworth | −3.0 | IBM | −9.6 |
| United Technologies | −4.4 | Merck | −12.0 |
| Allied-Signal | −9.3 | Philip Morris | −12.4 |
| General Motors | −0.9 | Du Pont | −8.6 |
| Procter & Gamble | −3.5 | Sears Roebuck | −11.4 |
| Coca-Cola | −10.5 | Goodyear Tire | −10.9 |
| McDonald's | −7.2 | Exxon | −8.6 |
| Minnesota Mining | −8.9 | | |

**Source:** Data are from "Trading in the 30 Dow Industrials Shows Wide Damage of Oct. 19 Crash," *Wall Street Journal*, December 16, 1987, p. 20. This source included only the 29 securities listed here. Negative numbers indicate a loss in value. All are negative, indicating that none of these 29 industrial stocks opened "up" from its previous close.

**Example—cont'd**

The histogram in Figure 4.1.2 shows that the distribution is fairly normal. There is a hint of skewness toward low values (that is, the tail is slightly longer on the left than on the right), but the distribution is still essentially normal except for randomness. The average percentage change, −8.2%, and the median percentage change, −8.6%, are close to one another. Indeed, this histogram has a clear central

region, and any reasonable summary measure must be near this middle region.

The average percentage change, −8.2%, may be interpreted as follows. If you had invested in a portfolio at the close of trading on Friday, with the same dollar amount invested in each of these stock securities (in terms of their value at the close of trading on Friday), then your portfolio would have lost 8.2% of its worth when it was sold at the start of trading on Monday. Different stocks clearly lost different percentages of their worth, but an equally weighted portfolio would have lost this average amount. What if you had invested different amounts in the different stocks? Then the portfolio loss could be computed as a weighted average using the initial investment values to define the weights.

The median percentage change, −8.6%, may be interpreted as follows. If you arrange these percentages in order, then about half of the securities fell by 8.6% or more and about half fell by 8.6% or less. Thus, a drop of 8.6% in value represents the middle experience of this group of stocks. Table 4.1.2 shows the ordered list; note that Exxon is at the middle, at 15th, since (29 + 1)/2 = 15.

Whenever stocks lose more than a few percentage points of their value, it is noticed. For them to lose as much as 8% at the beginning of the day, just at the start of trading, was clearly an ominous signal. The Dow Jones Industrial Average lost 508 points that day—a record—representing a loss for the day of 22.6%, a tragedy for many people and institutions.[9]

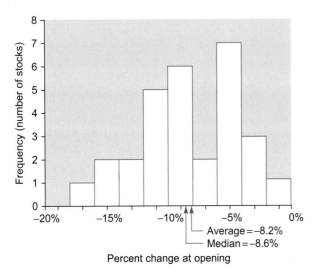

**FIGURE 4.1.2**　The distribution of the percentage of value lost by 29 industrial stocks at the opening of trading on the day of the crash of October 19, 1987.

9. The figure 22.6% was reported in *Wall Street Journal*, October 20, 1987, p. 1.

**TABLE 4.1.2 Stocks Ranked by Loss at Opening, Stock Market Crash of 1987**

| Firm | Change in Value (sorted) | Rank | Firm | Change in Value (sorted) | Rank |
|---|---|---|---|---|---|
| General Motors | −0.9% | 1 | Minnesota Mining | −8.9% | 16 |
| Navistar | −2.1 | 2 | Alcoa | −8.9 | 17 |
| Woolworth | −3.0 | 3 | Allied-Signal | −9.3 | 18 |
| Procter & Gamble | −3.5 | 4 | IBM | −9.6 | 19 |
| Boeing | −4.0 | 5 | Coca-Cola | −10.5 | 20 |
| Chevron | −4.0 | 6 | Goodyear Tire | −10.9 | 21 |
| Union Carbide | −4.1 | 7 | Sears Roebuck | −11.4 | 22 |
| United Technologies | −4.4 | 8 | International Paper | −11.6 | 23 |
| Bethlehem Steel | −4.5 | 9 | Merck | −12.0 | 24 |
| USX | −5.1 | 10 | Texaco | −12.3 | 25 |
| AT&T | −5.4 | 11 | Philip Morris | −12.4 | 26 |
| Primerica | −6.8 | 12 | Westinghouse | −15.7 | 27 |
| McDonald's | −7.2 | 13 | Kodak | −15.7 | 28 |
| Du Pont | −8.6 | 14 | General Electric | −17.2 | 29 |
| Exxon | −8.6 | 15 | | | |

## Example
### Personal Incomes

The distribution of amounts such as incomes of individuals and families (as well as the distribution of sales, expenses, prices, etc.) is often skewed toward high values. The reason is that such data sets often contain many small values, some moderate values, and a few large and very large values. The usual result is that the average will be larger than the median. The reason is that the average, by summing all data items, pays more attention to the large values. Consider the incomes of all households in the United States in 2007:[10]

Average household income: $67,609
Median household income: $50,233

The average income is higher than the median because the average has paid more attention to the relatively few very well-off households. Remember that these high incomes are added in when the average is found, but they are merely "high incomes" to the median (which allows these very high incomes to offset the low incomes on a household-by-household basis).

The histogram in Figure 4.1.3 shows how the distribution of incomes might look for a sample of 100 people from a town. It is strongly skewed toward high values, since there are many low-income people (indicated by the high bars at the left) together with some moderate-income and high-income people (the shorter bars in the middle and on the right). The average income of $38,710 is higher than the median income of $27,216. Evidently, the median (the half-way point) is low because most people are lower income here, and the existence of some higher incomes has boosted the average substantially.

10. From Table 676 of the U.S. Census Bureau, *Statistical Abstract of the United States: 2010* (129th edition), Washington, DC, 2009, accessed at http://www.census.gov/compendia/statab/cats/income_expenditures_poverty_wealth.html on July 5, 2010.

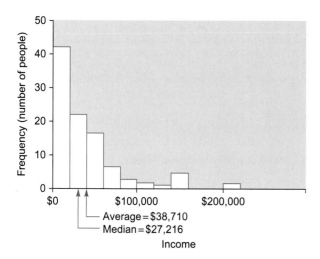

FIGURE 4.1.3   A histogram showing the distribution of incomes for 100 people. This is a skewed distribution, and the average is substantially larger than the median.

## Example
### Stages of Completion of Inventory

Consider a computer manufacturer's inventory of work in progress, consisting of the following stages of production for each unit:

**A.** The basic motherboard (the main circuitboard) is produced.
**B.** Sockets are installed on the motherboard.
**C.** Chips are installed in the sockets.
**D.** The resulting populated motherboard is tested.
**E.** The populated motherboard is installed in the system unit.
**F.** The completed system unit is tested.

If you are given a data set consisting of the production stage for each unit in the factory, the univariate ordinal data set might look like this:

A, C, E, F, C, C, D, C, A, E, E, … ,

This data set is ordinal since there is a natural ordering for each category, namely, the order in which a unit proceeds through production from beginning to end. You might use a frequency listing for such a data set, which would look something like this:

| Stage of Production | Number of Units |
|---|---|
| A | 57 |
| B | 38 |
| C | 86 |
| D | 45 |
| E | 119 |
| F | 42 |
| Total | 387 |

Since these are ordinal data, you can compute the median but not the average. The median will be found as unit (1 + 387)/2 = 194 after units have been placed in order by stage of production. Here's how you might find the median here:

The units at ranks 1 through 57 are in stage A. Thus, the median (at rank 194) is beyond stage A.
The units at ranks 58 (= 57 + 1) through 95 (= 57 + 38) are in stage B. Thus, the median is beyond stage B.
The units at ranks 96 (= 95 + 1) through 181 (= 95 + 86) are in stage C. Thus, the median is beyond stage C.
The units at ranks 182 (= 181 + 1) through 226 (= 181 + 45) are in stage D. Thus, the median is stage D because the median's rank (194) is between ranks 182 and 226.

Thus, about half of the units are less finished and half are more finished than units in stage D. Thus, stage D summarizes the middle point, in terms of completion, of all of the units currently in production.

## The Mode: A Typical Value Even for Nominal Data

The **mode** is the most common category, the one listed most often in the data set. It is the only summary measure available for nominal qualitative data because unordered categories cannot be summed (as for the average) and cannot be ranked (as for the median). The mode is easily found for ordinal data by ignoring the ordering of the categories and proceeding as if you had a nominal data set with unordered categories.

The mode is also defined for quantitative data (numbers), although it can get a little ambiguous in this case. For quantitative data, the mode may be defined as the value at the highest point of the histogram, perhaps at the midpoint of the tallest bar. The ambiguity enters in several ways. There may be two "tallest" bars. Or, even worse, the definition of the mode will depend on how the histogram was constructed; changing the bar width and location will make small (or medium) changes in the shape of the distribution, and the mode can change as a result. The mode is a slightly imprecise general concept when used with quantitative data.

It's easy to find the mode. Looking at either the number of items or at the percentage of items in each category, select the category with the largest number or percentage. If two or more categories tie for first place, you would report all of them as sharing the title of "mode" for that data set.

### Example
#### Voting

As votes come in to be counted during an election, they may be thought of as a nominal qualitative data set. Although you may have your own personal preference for the ordering of the candidates, since this is not universally agreed on, you will regard the data set as unordered. The list of data might begin as follows:

Smith, Jones, Buttersworth, Smith, Smith, Buttersworth, Smith, ...

The results of the election could be summarized as follows:

| Name | Votes | Percent |
|------|-------|---------|
| Buttersworth | 7,175 | 15.1% |
| Jones | 18,956 | 39.9 |
| Harvey | 502 | 1.1 |
| Smith | 20,817 | 43.9 |
| Total | 47,450 | 100.0 |

The mode is clearly Smith, with the most votes (20,817) as well as the largest percentage (43.9%). Note that the mode need not represent more than half (a majority) of the items, although it certainly could in some cases. It just needs to represent more items than any of the other categories.

### Example
#### Quality Control: Controlling Variation in Manufacturing

One of the important activities involved in creating quality products is the understanding of *variation* in manufacturing processes. Some variations are unavoidable and small enough to be tolerable, and other variations result from a process being "out of control" and producing inferior products. The subject of quality control is covered in more detail in Chapter 18.

W. Edwards Deming brought quality control to the Japanese in the 1950s. Some of his methods may be summarized as follows:

*The heart of Deming's method for achieving high quality is statistical. Every process, whether it be on the factory floor or in the office, has variations from the ideal. Deming shows clients a systematic method for measuring these variations, finding out what causes them, reducing them, and so steadily improving the process and thereby the product.[11]*

Gathering data and then analyzing them are key components of good quality control. Consider a factory that has recorded the cause of failure each time an item emerges that is not of acceptable quality:

| Cause of Problem | Number of Cases |
|------------------|-----------------|
| Solder joint | 37 |
| Plastic case | 86 |
| Power supply | 194 |
| Dirt | 8 |
| Shock (was dropped) | 1 |

The mode for this data set is clearly "power supply," since this cause accounted for more quality problems than any other.

The mode helps you focus on the most important category (in terms of its rate of occurrence). There would be little need to create a campaign to educate workers in cleanliness or the need to avoid dropping the boxes, since these causes were responsible for only a few of the problems. The mode, on the other hand, is the best candidate for immediate attention.

In this case, the firm would try to identify the problem with the power supplies and then take appropriate action. Perhaps it is the wrong power supply for the product, and one with larger capacity is needed. Or perhaps a more reliable supplier should be found. In any case, the mode has helped to pinpoint the trouble.

11. "The Curmudgeon Who Talks Tough on Quality," *Fortune,* June 25, 1984, p. 119.

### Example
#### Inventory Completion Stages Revisited

Reconsider the earlier example of a computer manufacturer's inventory of work in progress. The data set is as follows:

| Stage of Production | Number of Units |
|---------------------|-----------------|
| A | 57 |
| B | 38 |
| C | 86 |
| D | 45 |
| E | 119 |
| F | 42 |
| Total | 387 |

The median was found to be at stage D of production, since this stage divides the more primitive half of the items in production from the more advanced half. However, the median is not the mode in this case (although the median and the mode might well be the same in other examples).

The mode here is clearly stage E, with 119 units, more than any other stage. Management may well wish to be aware of the mode in a case like this, because if a troublesome "bottleneck" were to develop in the production process, it would likely show up as the mode.

In this example, stage E represents the installation of the motherboard in the system unit. It could be the natural result of a large order that has now reached this stage. On the other hand, those responsible for installation may be having trouble (perhaps they are understaffed or have had excessive absences), causing items to pile up at stage E. In a case like this, management attention and action may be warranted.

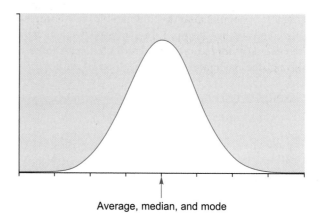

Average, median, and mode

**FIGURE 4.1.4**  The average, median, and mode are identical in the case of a perfect normal distribution. With the randomness of real data, they would be approximately but not exactly equal.

## Which Summary Should You Use?

Given these three summaries (average, median, and mode), which one should be used in a given circumstance? There are two kinds of answers. The first depends on which ones can be computed, and the second depends on which ones are most useful.

The mode can be computed for any univariate data set (although it does suffer from some ambiguity with quantitative data). However, the average can be computed only from quantitative data (meaningful numbers), and the median can be computed for anything except nominal data (unordered categories). Thus, sometimes your choices are restricted, and in the case of nominal data, you have no choice at all and can only use the mode. Here is a guide to which summaries may be used with each type of data:

|  | Quantitative | Ordinal | Nominal |
|---|---|---|---|
| Average | Yes | | |
| Median | Yes | Yes | |
| Mode | Yes | Yes | Yes |

In the case of quantitative data, where all three summaries can be computed, how are they different? For a normal distribution, there is very little difference among the measures since each is trying to find the well-defined middle of that bell-shaped distribution, as illustrated in Figure 4.1.4. However, with skewed data, there can be noticeable differences among them (as we noted earlier for the average and median). Figure 4.1.5 contrasts these summaries for skewed data.

The average should be used when the data set is normally distributed (at least approximately) since it is known to be the most efficient in this case. The average should also be seriously considered in other cases where

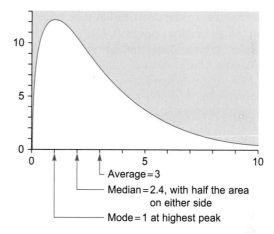

Average = 3

Median = 2.4, with half the area on either side

Mode = 1 at highest peak

**FIGURE 4.1.5**  The average, median, and mode are different in the case of a skewed distribution. The mode corresponds to the highest part of the distribution. The median has half of the area on each side. The average is where the distribution would balance, as if on a seesaw.

the need to preserve or forecast total amounts is important, since the other summaries do not do this as well.

The median can be a good summary for skewed distributions since it is not "distracted" by a few very large data items. It therefore summarizes most of the data better than the average does in cases of extreme skewness. The median is also useful when outliers are present because of its ability to resist their effects. The median is useful with ordinal data (ordered categories) although the mode should be considered also, depending on the questions to be addressed.

The mode must be used with nominal data (unordered categories) since the others cannot be computed. It is also useful with ordinal data (ordered categories) when the most represented category is important.

There are many more summaries than these three. One promising kind of estimator is the *biweight*, a "robust" estimator, which manages to combine the best features of the average and the median.[12] It is a fairly efficient choice when the data set is normally distributed but shares the ability of the median to resist the effects of outliers.

## 4.2  WHAT PERCENTILE IS IT?

**Percentiles** are summary measures expressing ranks as percentages from 0% to 100% rather than from 1 to $n$ so that the 0th percentile is the smallest number, the 100th percentile is the largest, the 50th percentile is the median, and so on. Percentiles may be thought of as indicating landmarks within the data set and are available for quantitative and ordinal data.

Note that a percentile is a number, in the same units as the data set, at a given rank. For example, the 60th percentile of sales performance might be $385,062 (a dollar amount, like the items in the data set, *not* a percentage). This 60th percentile of $385,062 might represent Mary's performance, with about 60% of sales representatives having lower sales and about 40% having higher sales.

Percentiles are used in two ways:

1. To indicate the data value at a given percentage (as in "the 10th percentile is $156,293").
2. To indicate the percentage ranking of a given data value (as in "John's performance, $296,994, was in the 55th percentile").

### Extremes, Quartiles, and Box Plots

One important use of percentiles is as landmark summary values. You can use a few percentiles to summarize important features of the entire distribution. You have already seen the median, which is the 50th percentile since it is ranked halfway between the smallest and largest. The **extremes**, the *smallest* and *largest* values, are often interesting. These are the 0th and 100th percentiles, respectively. To complete a small set of landmark summaries, you could also use the **quartiles**, defined as the 25th and 75th percentiles.

It may come as a surprise to learn that statisticians cannot agree on exactly what a quartile is and that there are many different ways to compute a quartile. The idea is clear: Quartiles are the data values ranked one-fourth of the way in from the smallest and largest values; however, there is ambiguity as to exactly how to find them. John

Tukey, who created exploratory data analysis, defines quartiles as follows:[13]

1. Find the median rank, $(1 + n)/2$, and discard any fraction. For example, with $n = 13$, use $(1 + 13)/2 = 7$. However, with $n = 24$, you would drop the decimal part of $(1 + 24)/2 = 12.5$ and use 12.
2. Add 1 to this and divide by 2. This gives the *rank of the lower quartile*. For example, with $n = 13$, you find $(1 + 7)/2 = 4$. With $n = 24$, you find $(1 + 12)/2 = 6.5$, which tells you to average the data values with ranks 6 and 7.
3. Subtract this rank from $(n + 1)$. This gives the *rank of the upper quartile*. For example, with $n = 13$, you have $(13 + 1) - 4 = 10$. With $n = 24$, you have $(1 + 24) - 6.5 = 18.5$, which tells you to average the data values with ranks 18 and 19.

The quartiles themselves may then be found based on these ranks. A general formula for the ranks of the quartiles, expressing the steps just given, may be written as follows:

**Ranks for the Quartiles**

$$\text{Rank of Lower Quartile} = \frac{1 + \text{int}[(1 + n)/2]}{2}$$

$$\text{Rank of Upper Quartile} = n + 1 - \text{Rank of Lower Quartile}$$

where *int* refers to the integer part function, which discards any decimal portion.

The **five-number summary** is defined as the following set of five landmark summaries: smallest, lower quartile, median, upper quartile, and largest.

**The Five-Number Summary**

- The *smallest data value* (the 0th percentile).
- The *lower quartile* (the 25th percentile, 1/4 of the way in from the smallest).
- The *median* (the 50th percentile, in the middle).
- The *upper quartile* (the 75th percentile, 3/4 of the way in from the smallest and 1/4 of the way in from the largest).
- The *largest data value* (the 100th percentile).

These five numbers taken together give a clear look at many features of the unprocessed data. The two extremes indicate the range spanned by the data, the median indicates the center, the two quartiles indicate the edges of the "middle half of the data," and the position of the median between

---

12. Further details about robust estimators may be found in D. C. Hoaglin, F. Mosteller, and J. W. Tukey, *Understanding Robust and Exploratory Data Analysis* (New York: Wiley, 1983).

13. J. W. Tukey, *Exploratory Data Analysis* (Reading, MA: Addison-Wesley, 1977). Tukey refers to quartiles as *hinges* and defines them on page 33. Excel's quartile function can give slightly different values than these because a weighted average is sometimes used.

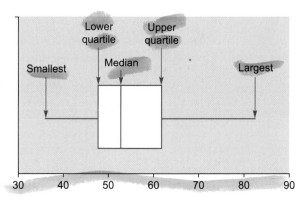

**FIGURE 4.2.1** A box plot displays the five-number summary for a univariate data set, giving a quick impression of the distribution.

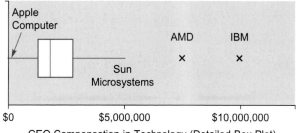

**CEO Compensation in Technology (Detailed Box Plot)**

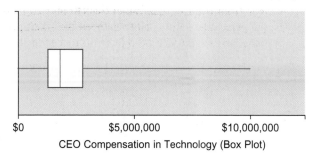

**CEO Compensation in Technology (Box Plot)**

**FIGURE 4.2.2** Box plot (bottom) and detailed box plot (top) for CEO compensation in the technology industry. Both plots show the five-number summary, but the detailed plot provides further important information about outliers (and the largest and smallest values that are not outliers) by identifying the companies. In this example, outliers represent firms with exceptionally high CEO compensation.

the quartiles gives a rough indication of skewness or symmetry.

The **box plot** is a picture of the five-number summary, as shown in Figure 4.2.1. The box plot serves the same purpose as a histogram—namely, to provide a visual impression of the distribution—but it does this in a different way. Box plots show less detail and are therefore more useful for seeing the big picture and comparing several groups of numbers without the distraction of every detail of each group. The histogram is still preferable for a more detailed look at the shape of the distribution.

The **detailed box plot** is a box plot, modified to display the outliers, which are identified by labels (which are also used for the most extreme observations that are not outliers). These labels can be very useful in calling attention to cases that may deserve special attention. For the purpose of creating a detailed box plot, **outliers** are defined as those data points (if any) that are far from the middle of the data set. Specifically, a large data value will be declared to be an outlier if it is bigger than

$$\text{Upper quartile} + 1.5 \times (\text{Upper quartile} - \text{Lower quartile})$$

A small data value will be declared to be an outlier if it is smaller than

$$\text{Lower quartile} - 1.5 \times (\text{Upper quartile} - \text{Lower quartile})$$

This definition of outliers is due to Tukey.[14] In addition to displaying and labeling outliers, you may also label the most extreme cases that are *not* outliers (one on each side) since these are often worth special attention. See Figure 4.2.2 for a comparison of a box plot and a detailed box plot.

---

14. Ibid., p. 44. Also, see Chapter 3 of Hoaglin et al., *Understanding Robust and Exploratory Data Analysis.*

**Example**
*Executive Compensation*

How much money do chief executive officers make? Table 4.2.1 shows the 2000 compensation (salary and bonus) received by CEOs of major technology companies. The data have been sorted and ranked, with the five-number summary values indicated in the table. There are $n = 23$ firms listed; hence, the median ($1,723,600) has rank $(1 + 23)/2 = 12$, which is the rank of Irwin Jacobs, then CEO of Qualcomm. The lower quartile ($1,211,650) has rank $(1 + 12)/2 = 6.5$ and is the average of CEO compensation at Lucent Technologies (rank 6) with Cisco Systems (rank 7). The upper quartile ($2,792,350) has rank $23 + 1 - 6.5 = 17.5$ and is the average of compensation for Micron Technology (rank 17) with EMC (rank 18). The five-number summary of CEO compensation for these 23 technology companies is therefore

| | |
|---|---|
| Smallest | $0 |
| Lower quartile | 1,211,650 |
| Median | 1,723,600 |
| Upper quartile | 2,792,350 |
| Largest | 10,000,000 |

Are there any outliers? If we compute using the quartiles, at the high end, any compensation larger than $2,792,350 + 1.5 \times (2,792,350 - 1,211,650) = \$5,163,400$ will be an outlier.

*(Continued)*

### TABLE 4.2.1 CEO Compensation in Technology

| Company | Executive | Salary and Bonus | Rank | Five-Number Summary |
|---|---|---|---|---|
| IBM* | Louis V. Gerstner Jr. | $10,000,000 | 23 | Largest is $10,000,000 |
| Advanced Micro Devices* | W. J. Sanders III | 7,328,600 | 22 | |
| Sun Microsystems | Scott G. McNealy | 4,871,300 | 21 | |
| Compaq Computer | Michael D. Capellas | 3,891,000 | 20 | |
| Applied Materials | James C. Morgan | 3,835,800 | 19 | |
| EMC | Michael C. Ruettgers | 2,809,900 | 18 | |
| | | | | Upper quartile is $2,792,350 |
| Micron Technology | Steven R. Appleton | 2,774,800 | 17 | |
| Hewlett-Packard | Carleton S. Fiorina | 2,766,300 | 16 | |
| Motorola | Christopher B. Galvin | 2,525,000 | 15 | |
| National Semiconductor | Brian L. Halla | 2,369,800 | 14 | |
| Texas Instruments | Thomas J. Engibous | 2,096,200 | 13 | |
| Qualcomm | Irwin Mark Jacobs | 1,723,600 | 12 | Median is $1,723,600 |
| Unisys | Lawrence A. Weinbach | 1,716,000 | 11 | |
| Pitney Bowes | Michael J. Critelli | 1,519,000 | 10 | |
| NCR | Lars Nyberg | 1,452,100 | 9 | |
| Harris | Phillip W. Farmer | 1,450,000 | 8 | |
| Cisco Systems | John T. Chambers | 1,323,300 | 7 | |
| | | | | Lower quartile is $1,211,650 |
| Lucent Technologies | Richard A. McGinn | 1,100,000 | 6 | |
| Silicon Graphics | Robert R. Bishop | 692,300 | 5 | |
| Microsoft | Steven A. Ballmer | 628,400 | 4 | |
| Western Digital | Matthew E. Massengill | 580,500 | 3 | |
| Oracle | Lawrence J. Ellison | 208,000 | 2 | |
| Apple Computer | Steven P. Jobs | 0 | 1 | Smallest is $0 |

*These values are outliers.

**Source:** Data are from the *Wall Street Journal*, April 12, 2001, pp. R12–R15. Their source is William M. Mercer Inc., New York.

**Example—cont'd**

Thus, the two largest data values, IBM and Advanced Micro Devices (AMD), are outliers. At the low end, any compensation smaller than $1,211,650 - 1.5 \times (2,792,350 - 1,211,650) = -1,159,400$, a negative number, would be an outlier. Since the smallest compensation is $0 (for Steve Jobs at Apple Computer), there are no outliers at the low end of the distribution.

Box plots (in two styles) for these 23 technology companies are displayed in Figure 4.2.2. The detailed box plot conveys more information by identifying the outlying firms (and the most extreme firms that are not outliers). Although ordinarily you would use only one of the two styles, we show both here for comparison.

One of the strengths of box plots is their ability to help you concentrate on the important overall features of several data sets at once, without being overwhelmed by the details. Consider the 2000 CEO compensation for major companies in utilities, financial, and energy as well as technology.[15] This consists of four individual data sets: a univariate data set (a group of numbers) for each of these four industry groups. Thus, there is a five-number summary and a box plot for each group.

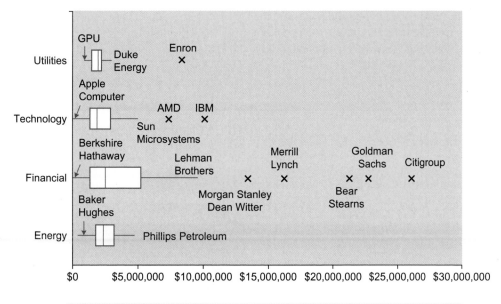

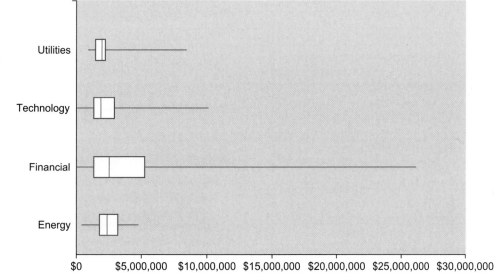

**FIGURE 4.2.3** Box plots for CEO compensation in major firms in selected industry groups, arranged on the same scale so that you may easily compare one group to another. The top figure gives details about the outliers (and most extreme nonoutliers) while the bottom figure shows only the five-number summary.

By placing box plots near each other and on the same scale in Figure 4.2.3, we facilitate the comparison of typical CEO compensation from one industry group to another. Note, for the detailed box plots, how helpful it is to have exceptional CEO firms labeled, compared to the box plots that display only the five-number summaries. Although the highest-paid CEOs come from financial companies, this industry is similar to the others near the lower end (look at the lower quartiles, for example). While risks come along with the big bucks (for example, Enron, the outlier in the Utilities group, filed for bankruptcy protection in December 2001 and its CEO resigned in January 2002), wouldn't it be nice to be in a job category where the lower quartile pays over a million dollars a year? Another

way to look at this situation is to recognize that statistical methods have highlighted Enron as an unusual case based on data available well before the difficulties of this company became famous.

15. Data are from the *Wall Street Journal*, April 12, 2001, pp. R12–R15. Their source is William M. Mercer Inc., New York.

Which kind of box plot is the best? A cost-benefit analysis suggests balancing the extra time and energy involved in making the detailed box plots (showing the individual outliers) against their clear benefits in terms of added information. A good strategy might be to begin by quickly completing displays of the five-number summaries and then to decide whether the additional detail is worth the time and trouble. Naturally, when the computer is doing the work, you would (nearly) always prefer the detailed display.

### Example
#### Data Mining the Donations Database

Consider the donations database of information on 20,000 people available at the companion site. In a data-mining example in Chapter 1, we found a greater percentage of donations among people who had given more often over the previous two years. But what about the size of the donations given? Do those who donated more frequently tend to give larger or smaller contributions than the others? Box plots can help us see what is going on in this large database.

We first focus attention on just the 989 donors out of the 20,000 people (eliminating, for now, the 19,011 who did not donate in response to the mailing). Next, using the ninth column of the database, we separate these 989 donors into four groups: 381 made just one previous gift over the past two years, 215 made two, 201 made three, and 192 made four or more. Taking the current donation amount (from the first column), we now have four univariate data sets.

One of the advantages of data mining is that when we break up the database into interesting parts like these, we have enough data in each part to work with. In this case, although the smallest of the four pieces is less than 1% of the database, it is still enough data (192 people) to see the distribution.

Box plots of the current donation amount for these four groups are shown in Figure 4.2.4. Note the tendency for larger donations, typically, to come from those who have given fewer gifts! You can see this by noticing that the central box moves to the left (toward smaller donation amounts) as you

go up in the figure toward those with more previous gifts. It seems that those who give more often tend to give *less*, not more each time! This reminds us of the importance of those who donate less frequently.

## The Cumulative Distribution Function Displays the Percentiles

The **cumulative distribution function** is a plot of the data specifically designed to display the percentiles by plotting the percentages against the data values. With percentages from 0% to 100% on the vertical axis and percentiles (i.e., data values) along the horizontal axis, it is easy to find either (1) the percentile value for a given percentage or (2) the percentage corresponding to a given data value.

The cumulative distribution function has a vertical jump of height $1/n$ at each of the $n$ data values and continues horizontally between data points. Figure 4.2.5 shows the cumulative distribution function for a small data set consisting of $n = 5$ data values (1, 4, 3, 7, 3), with one of them (3) occurring twice.

If you are given a number and wish to find its percentile ranking, proceed as follows:

### Finding the Percentile Ranking for a Given Number

1. Find the data value along the horizontal axis in the cumulative distribution function.
2. Move vertically up to the cumulative distribution function. If you hit a vertical portion, move halfway up.
3. Move horizontally to the left and read the percentile ranking.

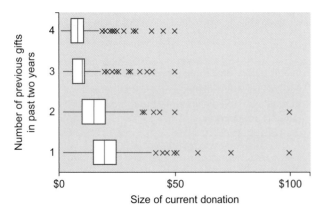

**FIGURE 4.2.4** Box plots showing that those who donated more frequently (the number of previous gifts, increasing as you move upward) tend to give *smaller* current donation amounts (as you can see from the generally leftward shift in the box plots as you move upward from one box plot to the next).

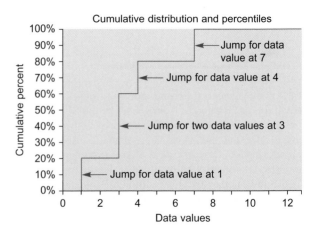

**FIGURE 4.2.5** The cumulative distribution function for the data set 1, 4, 3, 7, 3. Note the jump of size $1/n = 20\%$ at each data value, and the double jump at 3 (since there are two such data values).

In this example, the number 4 is the 70th percentile, since its percentile ranking is halfway between 60% and 80%, as shown in Figure 4.2.6.

If you are given a percentage and wish to find the corresponding percentile, proceed as follows:

### Finding the Percentile for a Given Percentage

1. Find the percentage along the vertical axis in the cumulative distribution function.
2. Move right horizontally to the cumulative distribution function. If you hit a horizontal portion, move halfway across it.
3. Move down and read the percentile on the data axis.

In this example, the 44th percentile is 3, as shown in Figure 4.2.7.

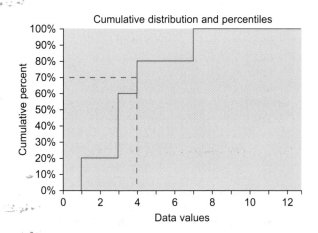

**FIGURE 4.2.6** The data value 4 represents the 70th percentile. Move vertically up from 4. Since you hit a vertical region, move halfway up. Then move across to read the answer, 70%.

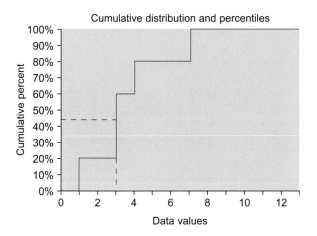

**FIGURE 4.2.7** To find the 44th percentile, move across to the right from 44% and then down to read the answer, 3.

### Example
#### Business Bankruptcies

Consider business bankruptcies per million people, by state, for the United States, sorted in order from least to most bankruptcies per capita and shown in Table 4.2.2. The cumulative distribution function for this data set is displayed in Figure 4.2.8. You can see that the experience of most states (say, from 10% through 90%) was around 75 to 150 business bankruptcies per million population.

*(Continued)*

**TABLE 4.2.2 Business Bankruptcy Rate by State, Sorted (per million people)**

| State | Business Bankruptcy Rate |
|---|---|
| Hawaii | 39.6 |
| South Carolina | 43.8 |
| Massachusetts | 51.4 |
| District of Columbia | 71.0 |
| Maryland | 71.9 |
| Montana | 73.4 |
| Oregon | 74.7 |
| New York | 78.7 |
| Wyoming | 78.8 |
| North Carolina | 80.0 |
| Vermont | 83.7 |
| North Dakota | 84.2 |
| Alabama | 84.7 |
| Washington | 86.3 |
| Pennsylvania | 86.6 |
| Kansas | 87.1 |
| Iowa | 89.9 |
| Missouri | 90.3 |
| Illinois | 91.3 |
| Kentucky | 91.4 |
| West Virginia | 93.1 |
| New Mexico | 93.2 |
| Wisconsin | 93.3 |
| Mississippi | 101.4 |
| Virginia | 102.7 |
| Connecticut | 104.5 |

*(Continued)*

**TABLE 4.2.2  Business Bankruptcy Rate by State, Sorted (per million people)—cont'd**

| | |
|---|---|
| Tennessee | 105.1 |
| Arizona | 106.3 |
| New Jersey | 106.5 |
| Oklahoma | 107.6 |
| Indiana | 108.5 |
| Idaho | 109.6 |
| Utah | 110.4 |
| Texas | 112.1 |
| Alaska | 113.7 |
| Rhode Island | 116.1 |
| Maine | 117.0 |
| South Dakota | 118.1 |
| Nebraska | 123.9 |
| Ohio | 125.0 |
| Minnesota | 125.7 |
| California | 127.8 |
| Louisiana | 129.5 |
| Michigan | 139.4 |
| Arkansas | 150.2 |
| Florida | 150.5 |
| Nevada | 151.9 |
| Colorado | 153.1 |
| Georgia | 177.0 |
| New Hampshire | 241.7 |
| Delaware | 413.5 |

**Source:** Data are calculated from state populations in 2008 (Table 12) and business bankruptcies in 2008 (Table 748) of U.S. Census Bureau, *Statistical Abstract of the United States: 2010* (129th edition), Washington, DC, 2009, accessed from http://www.census.gov/compendia/statab/cats/population.html and http://www.census.gov/compendia/statab/cats/business_enterprise/establishments_employees_payroll.html on July 5, 2010.

### Example—cont'd

Figure 4.2.9 shows how to find percentiles from the cumulative distribution function. The 50th percentile is 104.5 (for Connecticut, as may be seen from the data table), corresponding to the median value of 104.5 business bankruptcies per million people. The 90th percentile is 150.5 (for Florida) and the 95th percentile is 177.0 (for Georgia).

You actually have the choice of three different graphs to display a group of numbers: the histogram, the box plot,

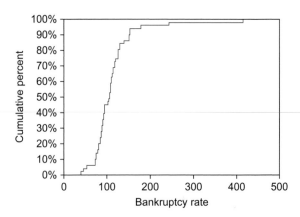

**FIGURE 4.2.8**  Cumulative distribution function for business bankruptcies per million people, by state.

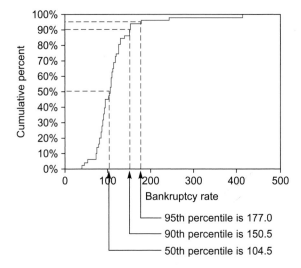

95th percentile is 177.0
90th percentile is 150.5
50th percentile is 104.5

**FIGURE 4.2.9**  Cumulative distribution function for business bankruptcy rates, with 50th, 90th, and 95th percentiles indicated.

and the cumulative distribution function. Each technique displays the same information (the data values) in a different way. For this example of business bankruptcy rates, all three representations are displayed in Figure 4.2.10, arranged vertically so that the relationships are clear.

Regions of high data concentration (i.e., lots of data) correspond to high peaks of the histogram and to *steepness* in the cumulative distribution. Usually, a high concentration of data is found near the middle, as is the case here. Regions of low data concentration (i.e., just a few data values) correspond to low histogram bars and to *flatness* in the cumulative distribution.

The box plot shows you the five-number summary, which may be read from the cumulative distribution as the smallest (at 0%), lower quartile (at 25%), median (at 50%), upper quartile (75%), and largest (at 100%).

Note that the cumulative distribution function is the only display here that actually shows all of the data. The histogram has lost information because it displays only the number of states within each group (for example,

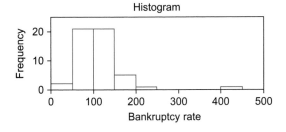

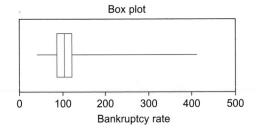

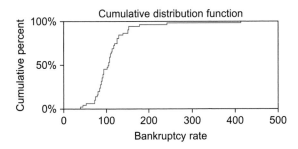

FIGURE 4.2.10   Three graphs of the business bankruptcies data: histogram, box plot, and cumulative distribution, respectively. Note that the cumulative distribution function is steepest in regions of high concentration of data.

from 50 to 100 bankruptcies). The box plot also has lost information because it preserves only the five landmark summaries. Only the cumulative distribution shows enough information to let you reconstruct each number from the original data set.

## 4.3 END-OF-CHAPTER MATERIALS

### Summary

**Summarization** is using one or more selected or computed values to represent the data set. To completely summarize, you would first describe the basic structure of most of the data and then list any exceptions or outliers.

The **average** (also called the **mean**) is the most common method for finding a typical value for a list of numbers, found by adding all the values and then dividing by the number of items. The formula is as follows:

$$\text{Sample Average} = \frac{\text{Sum of data items}}{\text{Number of data items}}$$
$$\overline{X} = \frac{X_1 + X_2 + \cdots + X_n}{n}$$
$$= \frac{1}{n}\sum_{i=1}^{n}X_i$$

For an entire population, the convention is to use $N$ to represent the number of items and to let $\mu$ (the Greek letter *mu*) represent the population mean value. The average spreads the total equally among all of the cases and is most useful when there are no extreme outliers and total amounts are relevant. The data set must be quantitative in order for the average to be computed.

The **weighted average** is like the average, except that it allows you to give a different importance, or weight, to each data item. This extends the average to the case in which some data values are more important than others and need to be counted more heavily. The formula is as follows:

$$\text{Weighted Average} = \text{Sum of (weight times data item)}$$
$$= w_1X_1 + w_2X_2 + \cdots + w_nX_n$$
$$= \sum_{i=1}^{n}w_iX_i$$

The weights usually add up to 1 (you might divide them by their sum if they do not). The weighted average can be computed only for quantitative data.

The **median** is the *middle value*; half of the data items are larger, and half are smaller. **Ranks** associate the numbers 1, 2, 3, ..., $n$ with the data values so that the smallest has rank 1, the next to smallest has rank 2, and so forth up to the largest, which has rank $n$. The rank of the median is $(1 + n)/2$, indicating how many cases to count in from the smallest (or largest) in order to find the median data value; if this rank contains a decimal portion (for example, 13.5 for $n = 26$), then average the two data values at either side of this value (for example, the values at ranks 13 and 14). The median may be computed for either quantitative data or ordinal data (ordered categories).

The median summarizes the "typical" value in a different way than the average; however, the two values will be similar when the data distribution is symmetric, as is the normal distribution. For a skewed distribution or in the presence of outliers, the average and median can give very different summaries.

The **mode** is the most common category, the one listed most often in the data set. It may be computed for any data set: quantitative, ordinal, or nominal (unordered categories), and it is the only summary available for nominal data. In the case of quantitative data, the mode is often defined as the midpoint of the tallest histogram bar; however, this can be ambiguous since it can vary depending on the choice of scale for the histogram.

To decide which of these summaries to use, proceed as follows. If the data set is *nominal*, then only the mode is available. If the data set is *ordinal*, then either the mode or the median can be used; the mode expresses the most popular category, and the median finds the central category with respect to the ordering. For *quantitative* data, all three methods may be used. For a normal distribution, all three methods give similar answers, and the average is best.

For a skewed distribution, all three methods usually give different answers; the median is generally a good summary in this case since it is less sensitive to the extreme values in the longer tail of the distribution. However, if totals are important, the average would be preferred.

**Percentiles** are summary measures expressing ranks as percentages from 0% to 100% rather than from 1 to *n;* the 0th percentile is the smallest number, the 100th percentile is the largest, the 50th percentile is the median, and so on. Note that a percentile is a number in the same units as the data set (such as dollars, gallons, etc.). Percentiles can be used either to find the data value at a given percentage ranking or to find the percentage ranking of a given value. The **extremes**, the *smallest* and *largest* values, are often interesting. The **quartiles** are the 25th and 75th percentiles, whose ranks are determined as follows:

$$\text{Rank of Lower Quartile} = \frac{1 + \text{int}[(1 + n)/2]}{2}$$

$$\text{Rank of Upper Quartile} = n + 1 - \text{Rank of Lower Quartile}$$

where *int* refers to the integer part function, which discards any decimal portion.

The **five-number summary** gives special landmark summaries of a data set: the smallest, lower quartile, median, upper quartile, and largest. The **box plot** displays the five-number summary as a graph. **Outliers** are defined as those data points (if any) that are far out with respect to the data values near the middle of the data set. A **detailed box plot** displays the outliers and labels them separately along with the most extreme observations that are not outliers. Several data sets measured in the same units may be compared by placing their box plots alongside one another on the same scale.

The **cumulative distribution function** is a plot of the data specifically designed to display the percentiles by plotting the percentages against the data values. This graph has a jump of height $1/n$ at each data value. Given a percentage, the percentile may be found by reading across and then down. Given a number, the percentile ranking (percentage) may be found by reading up and then across. Thus, the cumulative distribution function displays the percentiles and helps you compute them. It is the only display that "archives" the data by preserving enough information for you to reconstruct the data values. The cumulative distribution function is steep in regions of high data concentration (i.e., where histogram bars are tall).

## Key Words

**average**, *66*
**box plot**, *77*
**cumulative distribution function**, *80*
**detailed box plot**, *77*
**extremes**, *76*
**five-number summary**, *76*
**mean**, *66*
**median**, *70*
**mode**, *73*
**outlier**, *77*
**percentile**, *76*
**quartiles**, *76*
**rank**, *70*
**summarization**, *65*
**weighted average**, *68*

## Questions

1. What is summarization of a data set? Why is it important?
2. List and briefly describe the different methods for summarizing a data set.
3. How should you deal with exceptions when summarizing a set of data?
4. What is meant by a typical value for a list of numbers? Name three different ways of finding one.
5. What is the average? Interpret it in terms of the total of all values in the data set.
6. What is a weighted average? When should it be used instead of a simple average?
7. What is the median? How can it be found from its rank?
8. How do you find the median for a data set:
    a. With an odd number of values?
    b. With an even number of values?
9. What is the mode?
10. How do you usually define the mode for a quantitative data set? Why is this definition ambiguous?
11. Which summary measure(s) may be used on
    a. Nominal data?
    b. Ordinal data?
    c. Quantitative data?
12. Which summary measure is best for
    a. A normal distribution?
    b. Projecting total amounts?
    c. A skewed distribution when totals are not important?
13. What is a percentile? In particular, is it a percentage (e.g., 23%), or is it specified in the same units as the data (e.g., $35.62)?
14. Name two ways in which percentiles are used.
15. What are the quartiles?
16. What is the five-number summary?
17. What is a box plot? What additional detail is often included in a box plot?
18. What is an outlier? How do you decide whether a data point is an outlier or not?
19. Consider the cumulative distribution function:
    a. What is it?
    b. How is it drawn?
    c. What is it used for?
    d. How is it related to the histogram and box plot?

## Problems

*Problems marked with an asterisk (\*) are solved in the Self Test in Appendix C.*

1.\* Consider the quality of cars, as measured by the number of cars requiring extra work after assembly, in each day's production for 15 days:

30, 34, 9, 14, 28, 9, 23, 0, 5, 23, 25, 7, 0, 3, 24

a. Find the average number of defects per day.
b. Find the median number of defects per day.
c. Draw a histogram of the data.
d. Find the mode number of defects per day for your histogram in part c.
e. Find the quartiles.
f. Find the extremes (the smallest and largest).
g. Draw a box plot of the data.
h. Draw a cumulative distribution function of the data.
i. Find the 90th percentile for this data set.
j. Find the percentile ranking for the next day's value of 29 defects.

2. Table 4.3.1 provides a list of the amounts that your regular customers spent on your products last month:

a. Find the average sales per regular customer.
b. Find the median and quartiles.
c. Draw the box plot.
d. Find the outliers, if any.
e. Draw the detailed box plot.
f. Comment briefly on the differences between these two box plots.
g. If you could expand your list of regular customers to include 3 more, and if their purchasing patterns were like those of these firms, what would you expect total monthly sales for all 13 regular customers to be?
h. Write a paragraph telling what you have learned about these customers using these statistical methods.

3. Your company is trying to estimate the total size of its potential market. A survey has been designed, and data have been collected. A histogram of the data shows a small amount of skewness. Which summary measure would you recommend to the company for this purpose and why?

4. Some people who work at your company would like to visually compare the income distributions of people who buy various products in order to better understand customer selections. For each of 16 products, a list of incomes of representative customers (who bought that product) has been obtained. What method would you recommend?

5. Many countries (but not the United States) have a "value-added tax" that is paid by businesses based on how much value they add to a product (e.g., the difference between sales revenues and the cost of materials). This is different from a sales tax because the consumer does not see it added on at the cash register. Consider the VAT (value-added tax) percentages for various countries, as shown in Table 4.3.2.

a. Draw a histogram of this data set and briefly describe the shape of the distribution.
b. Find the VAT tax level of the average country.
c. Find the median VAT tax level.
d. Compare the average and median. Is this what you expect for a distribution with this shape?
e. Draw the cumulative distribution function.
f. What VAT tax level is at the 20th percentile? The 80th percentile?
g. What percentile is a VAT tax of 16%?

6. Consider the profits of health care companies in the Fortune 500, as shown in Table 4.3.3.

a. Draw a histogram of this data set, and briefly describe the shape of the distribution.
b. Find the profit of the average firm.
c. Find the median profit level.
d. Compare the average and median; in particular, which is larger? Is this what you would expect for a distribution with this shape?
e. Draw the cumulative distribution function.
f. Your firm has a strategic plan that would increase profits to $200 million. What percentile would this profit level represent, with respect to this data set?
g. Your firm's strategic plan indicates that the profit level might actually reach the 60th percentile. What value of profits does this represent?

7. The beta of a firm's stock indicates the degree to which changes in stock price track changes in the stock market as a whole and is interpreted as the market risk of the portfolio. A beta of 1.0 indicates that, on average, the stock rises (or falls) the same percentage

### TABLE 4.3.1 Last Month's Sales

| Customer | Sales ($000) | Customer | Sales ($000) |
|---|---|---|---|
| Consolidated, Inc | $142 | Associated, Inc | $93 |
| International, Ltd | 23 | Structural, Inc | 17 |
| Business Corp | 41 | Communications Co | 174 |
| Computer Corp | 10 | Technologies, Inc | 420 |
| Information Corp | 7 | Complexity, Ltd | 13 |

## TABLE 4.3.2 Value-Added Tax Rates by Country

| Country | Standard VAT Rate |
|---|---|
| Australia | 10.0% |
| Austria | 20.0% |
| Belarus | 20.0% |
| Belgium | 21.0% |
| Canada | 7.0% |
| Czech Republic | 22.0% |
| Denmark | 25.0% |
| Estonia | 18.0% |
| Finland | 22.0% |
| France | 19.6% |
| Georgia | 20.0% |
| Germany | 16.0% |
| Greece | 18.0% |
| Hungary | 25.0% |
| Iceland | 24.5% |
| Ireland | 21.0% |
| Italy | 20.0% |
| Japan | 5.0% |
| Kazakhstan | 15.0% |
| Korea | 10.0% |
| Kyrgyzstan | 20.0% |
| Latvia | 18.0% |
| Lithuania | 18.0% |
| Luxembourg | 15.0% |
| Netherlands | 19.0% |
| New Zealand | 12.5% |
| Norway | 24.0% |
| Poland | 22.0% |
| Portugal | 19.0% |
| Russia | 20.0% |
| Slovakia | 23.0% |
| Spain | 16.0% |
| Sweden | 25.0% |
| Switzerland | 7.5% |
| Turkey | 17.0% |

| | |
|---|---|
| Ukraine | 20.0% |
| United Kingdom | 17.5% |

**Source:** World Taxpayers Associations, accessed at http://www.worldtaxpayers.org/stat_vat.htm on July 5, 2010.

## TABLE 4.3.3 Profits for Health Care Companies in the Fortune 500

| Firm | Profits ($ millions) |
|---|---|
| Aetna | $1,276.5 |
| Amerigroup | 149.3 |
| Centene | 83.7 |
| Cigna | 1,302.0 |
| Community Health Systems | 243.2 |
| Coventry Health Care | 242.3 |
| DaVita | 422.7 |
| Express Scripts | 827.6 |
| HCA | 1,054.0 |
| Health Management Associates | 138.2 |
| Health Net | −49.0 |
| Humana | 1,039.7 |
| Kindred Healthcare | 40.1 |
| Laboratory Corp of America | 543.3 |
| Medco Health Solutions | 1,280.3 |
| Omnicare | 211.9 |
| Quest Diagnostics | 729.1 |
| Tenet Healthcare | 187.0 |
| UnitedHealth Group | 3,822.0 |
| Universal American | 140.3 |
| Universal Health Services | 260.4 |
| WellCare Health Plans | 39.9 |
| WellPoint | 4,745.9 |

**Source:** Data are for 2009, accessed at http://money.cnn.com/magazines/fortune/fortune500/2010/industries/223/index.html on July 5, 2010.

as does the market. A beta of 2.0 indicates a stock that rises or falls at twice the percentage of the market. The beta of a stock portfolio is the weighted average of the betas of the individual stock securities, weighted by the current market values (market value is share price times number of shares). Consider the following portfolio:

100 shares Speculative Computer at $35 per share, beta = 2.4

200 shares Conservative Industries at $88 per share, beta = 0.6

150 shares Dependable Conglomerate at $53 per share, beta = 1.2

a.  Find the beta of this portfolio.

b.  To decrease the risk of the portfolio, you have decided to sell all shares of Speculative Computer and use the money to buy as many shares of Dependable Conglomerate as you can.[16] Describe the new portfolio, find its beta, and verify that the market risk has indeed decreased.

8.* Your firm has the following securities outstanding: common stock (market value $4,500,000; investors demand 17% annual rate of return), preferred stock (market value $1,700,000; current annual yield is 13%), and 20-year bonds (market value $2,200,000; current annual yield is 11%). Find your cost of capital.

9.  Active consumers make up 13.6% of the market and spend an average of $16.23 per month on your product. Passive consumers make up 23.8% of the market and spend $9.85. The remaining consumers have average spending of $14.77. Find the average spending for all consumers.

10.  Consider the 20,000 median household income values in the donations database (available on the companion site). These represent the median household income for the neighborhood of each potential donor in the database.

a.  Construct a cumulative distribution.

b.  Find the 60th and 90th percentile.

11.  Consider the 20,000 people in the donations database (on the companion site).

a.  Construct box plots to compare median household income and per capita income (these specify two columns in the database) by putting the two box plots on the same scale.

b.  Describe what you find in your box plots.

12.  A survey of 613 representative people within your current area indicates that they plan to spend a total of $2,135 on your products next year. You are considering expansion to a new city with 2.1 million people.

a.  Find the average expenditure per person based on the survey of your current area.

b.  If you can gain the same market presence in the new city that you have in your current area, what annual sales level do you expect?

c.  If you expect to gain in the new city only 60% of the market presence in your current area, what annual sales level do you expect?

13.  Your marketing research team has identified four distinct groups of people (Type A, B, C, and D, where D represents "all others") according to personality traits. You figure that market penetration for a new product will be highest among Type A personalities, with 38% purchasing the product within two years. Similarly,

**TABLE 4.3.4  State Population and Taxes**

| State | Population (thousands) | State taxes (per capita) |
|---|---|---|
| Ohio | 11,486 | $2,085 |
| Indiana | 6,377 | 2,337 |
| Illinois | 12,902 | 2,269 |
| Michigan | 10,003 | 2,355 |
| Wisconsin | 5,628 | 2,575 |

for Types B, C, and D, the market penetration will be 23%, 8%, and 3%, respectively. Assume that the personality types represent 18%, 46%, 25%, and 11%, respectively, of your target population. What overall market penetration should you expect among your target population?

14.  A large outdoor recreational facility has three entrances. According to automatic vehicle counters, last year 11,976 vehicles entered at the first entrance, 24,205 at the second, and 7,474 at the third. A survey done at each entrance showed that the average planned length of stay was 3.5 days at the first location, 1.3 days at the second, and 6.0 days at the third. Estimate the typical planned length of stay for the entire facility on a per-vehicle basis.

15.  Given the state taxes and populations for the East North Central States, as shown in Table 4.3.4, compute the per-capita state tax burden for this entire region.[17]

16.  You have begun a quality improvement campaign in your paper mill, and, as a result, lots of pieces of paper come to your desk. Each one describes a recent problem with customers according to the following codes: A = paper unavailable, B = paper too thick, C = paper too thin, D = paper width too uneven, E = paper color not correct, F = paper edges too rough. Here are the results:

A, A, E, A, A, A, B, A, A, A, B, A, B, F, F, A, A, A, A, A, B, A, A, A, A, C, D, F, A, A, E, A, C, A, A, A, F, F

a.  Summarize this data set by finding the percentage that each problem represents out of all problems.

b.  Summarize this data set by finding the mode.

c.  Write a brief (one-paragraph) memo to management recommending the most effective action to improve the situation.

d.  Could the median or average be computed here? Why or why not?

17.  Find the upper quartile for the following box plot.

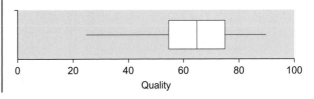

Quality

18. Consider the percent change in housing values over a five-year period for regions of the United States, as shown in Table 4.3.5.
    a. Find the mean and median percent change in housing values.
    b. Find the five-number summary for this data set.
    c. Draw a box plot.
19. Consider the revenues (in $ millions) for the top 12 companies in the Fortune 500 (from www.fortune.com/fortune/fortune500/ on January 30, 2001), as shown in Table 4.3.6.

a. Find the five-number summary.
b. Draw a box plot.
20. Table 4.3.7 shows percent increases from the offer price of initial public stock offerings, as most of these newly traded companies increased in value, whereas some of them lost money.
    a. Draw a cumulative distribution function for this data set.
    b. Find the 35th percentile.
21. Consider the loan fees charged for granting home mortgages, as shown in Table 4.3.8. These are given as a percentage of the loan amount and are one-time fees paid when the loan is closed.
    a. Find the average loan fee.
    b. Find the median loan fee.
    c. Find the mode.
    d. Which summary is most useful as a description of the "typical" loan fee, the average, median, or mode? Why?
22. A mail-order sales company sent its new catalog initially to a representative sample of 10,000 people from its mailing list and received orders totaling $36,851.
    a. Find the average dollar amount ordered per person in this initial mailing.
    b. What total dollar amount in orders should the company expect when the catalog is sent to everyone on the mailing list of 563,000 people?
    c. From the sample of 10,000 people generating $36,851 in orders, only 973 people actually placed

**TABLE 4.3.5 Percent Change in Housing Values over Five Years for U.S. Regions**

| Region | Percent Change | Region | Percent Change |
|---|---|---|---|
| New England | 54.5% | West North Central | 38.3% |
| Pacific | 48.9 | West South Central | 29.5 |
| Middle Atlantic | 35.3 | East North Central | 32.0 |
| South Atlantic | 33.6 | East South Central | 26.0 |
| Mountain | 34.2 | | |

Source: Office of Federal Housing Enterprise Oversight, House Price Index, Third Quarter 2001, November 30, 2001, p. 13, accessed at http://www.ofheo.gov/house/3q01hpi.pdf on February 7, 2002.

**TABLE 4.3.6 Revenues for Selected Fortune 500 Companies (in $ millions)**

| | | | |
|---|---|---|---|
| General Motors | $189,058 | Citigroup | $82,005 |
| Wal-Mart Stores | 166,809 | AT&T | 62,391 |
| Exxon Mobil | 163,881 | Philip Morris | 61,751 |
| Ford Motor | 162,558 | Boeing | 57,993 |
| General Electric | 111,630 | Bank of America Corp | 51,392 |
| Intl. Business Machines | 87,548 | SBC Communications | 49,489 |

**TABLE 4.3.7 Percent Increases from the Offer Price of Initial Public Stock Offerings**

| | | | |
|---|---|---|---|
| American Pharmaceutical | −16% | Northwest Biotherapeutics | −11% |
| Bruker AXS | 2 | Prudential Financial | 13 |
| Carolina Group | 5 | Sunoco Logistics | 12 |
| Centene | 44 | Synaptics | 23 |
| Nassda | 36 | United Defense Industries | 35 |
| Netscreen Technologies | 14 | ZymoGenetics | −16 |

Source: Wall Street Journal, February 7, 2002, p. C13; their sources are WSJ Market Data Group and Dow Jones Newswires.

### TABLE 4.3.8 Home Mortgage Loan Fees

| Institution | Loan Fee | Institution | Loan Fee |
|---|---|---|---|
| Allied Pacific Mortgage | 1.25% | Mortgage Associates | 2% |
| Alternative Mortgage | 2 | Normandy Mortgage | 1.25 |
| Bankplus Mortgage | 1 | Performance Mortgage | 2 |
| Bay Mortgage | 2 | PNC Mortgage | 1 |
| CTX Mortgage | 1.5 | Qpoint Home Mortgage | 1.5 |
| First Mark Mortgage | 2 | Sammamish Mortgage | 2 |
| Mariner Mortgage | 1 | U.S. Discount Mortgage | 2 |

**Source:** From "Summer Mortgage Rates," *Seattle Times,* July 16, 1995, p. G1.

an order. Find the average dollar amount ordered per person who placed an order.

d. Given the information in part c, how many orders should the company expect when the catalog is sent to everyone on the mailing list of 563,000 people?

23. Consider the strength of cotton yarn used in a weaving factory, in pounds of force at breakage, measured from a sample of yarn from the supplies room:

    117, 135, 94, 79, 90, 85, 173, 102, 78, 85, 100, 205, 93, 93, 177, 148, 107

    a. Find the average breaking strength.
    b. Find the median breaking strength.
    c. Draw a histogram indicating the average and median values, and briefly comment on their relationship. Are they the same? Why or why not?
    d. Draw a cumulative distribution.
    e. Find the 10th and the 90th percentiles.
    f. Management would like its supplies to provide a breakage value of 100 pounds or more at least 90% of the time. Based on this data set, do these supplies qualify? In particular, which percentile will you compare to?

24. Your factory's inventory level was measured 12 times last year, with the results shown below. Find the average inventory level during the year.

    313, 891, 153, 387, 584, 162, 742, 684, 277, 271, 285, 845

25. Consider the following list of your products' share of 20 major metropolitan areas:

    0.7%, 20.8%, 2.3%, 7.7%, 5.6%, 4.2%, 0.8%, 8.4%, 5.2%, 17.2%, 2.7%, 1.4%, 1.7%, 26.7%, 4.6%, 15.6%, 2.8%, 21.6%, 13.3%, 0.5%

    a. Find the average and the median.
    b. Draw a cumulative distribution function for this data set.
    c. Find the 80th percentile.

26. Consider the monthly sales of 17 selected sales representatives (in thousands of dollars):

    23, 14, 26, 22, 28, 21, 34, 25, 32, 32, 24, 34, 22, 25, 22, 17, 20

### TABLE 4.3.9 Changing Value of the Dollar

| Country | % Change | Country | % Change |
|---|---|---|---|
| Belgium | –5.3% | Singapore | –1.5% |
| Japan | –6.7 | France | –4.9 |
| Brazil | 26.0 | South Korea | –1.0 |
| Mexico | –1.2 | Hong Kong | 0.0 |
| Britain | –3.7 | Taiwan | –0.1 |
| Netherlands | –5.1 | Italy | –4.7 |
| Canada | –1.9 | West Germany | –5.1 |

a. Find the average and median.
b. Draw the box plot.

27. Consider the percentage change in the value of the dollar with respect to other currencies over a four-week period (Table 4.3.9).

    a. Find the average percentage change in the value of the dollar, averaging over all of these countries.
    b. On average during this time period, did the dollar strengthen or weaken against these currencies?
    c. Find the median. Why is it so different from the average in this case?
    d. Draw a box plot.

28. If you had a list of the miles per gallon for various cars, which of the following is the only possibility for the 65th percentile: 65 cars, 65%, $13,860, or 27 miles per gallon?

29. For the yields of municipal bonds (Table 3.9.1 in Chapter 3):

    a. Find the average yield.
    b. Find the median yield.
    c. Find the quartiles.
    d. Find the five-number summary.
    e. Draw a box plot of these yields.
    f. Identify the outliers, if any, and draw a detailed box plot.

g.  Draw the cumulative distribution function for the data set.

h.  Find the percentile value of 5.40%.

i.  Find the value of the 60th percentile.

30. Using the data from Table 3.9.2 in Chapter 3, find the average and median to summarize the typical size of the market response to stock buyback announcements.

31. Using the data from Table 3.9.4 in Chapter 3, for the portfolio investments of College Retirement Equities Fund in aerospace and defense:

a.  Find the average market value for each firm's stock in CREF's portfolio.

b.  Find the median of these market values.

c.  Compare the average to the median.

d.  Find the five-number summary.

e.  Draw a box plot, and comment on the distribution shape. In particular, are there any signs of skewness?

f.  Is the relationship between the average and median consistent with the distribution shape? Why or why not?

32. Consider the running times of selected films from a video library as shown in Table 4.3.10.

a.  Find the average running time.

b.  Find the median running time.

c.  Which is larger, the average or the median? Based on your answer, do you expect to find strong skewness toward high values?

d.  Draw a histogram and comment on its relationship to your answer to part c.

33. A social group shows only movies of 100 minutes or less at its meetings. Consider the running times of selected films from a video library as shown in Table 4.3.10.

**TABLE 4.3.10 Length in Minutes for Selected Films from a Video Library**

| Time | Film | Time | Film |
|---|---|---|---|
| 133 | *Flower Drum Song* | 84 | *Origins of American Animation* |
| 111 | *Woman of Paris, A* | 109 | *Dust in the Wind* (Chinese) |
| 88 | *Dim Sum: A Little Bit of Heart* | 57 | *Blood of Jesus, The* |
| 120 | *Do the Right Thing* | 60 | *Media: Zbig Rybczynski Collection* |
| 87 | *Modern Times* | 106 | *Life* (Tape 2) (Chinese) |
| 100 | *Law of Desire* (Spanish) | 101 | *Dodsworth* |
| 104 | *Crowd, The* | 123 | *Rickshaw Boy* (Chinese) |
| 112 | *Native Son* | 91 | *Gulliver's Travels* |
| 134 | *Red River* | 136 | *Henry V* (Olivier) |
| 99 | *Top Hat* | | |

a.  What percentage of these movies can the group show?

b.  What is the name of the longest of these movies that could be shown?

c.  Comment on the relationship between your answer to part a and the percentile ranking of your answer to part b.

34. A wine store carries 86 types of wine produced in 2007, 125 types from 2008, 73 from 2009, and 22 from 2010. Identify the types of wine as the elementary units for analysis.

a.  Find the mode of the year of production. What does this tell you?

b.  Find the average year of production and compare it to the mode.

c.  Draw a histogram of year of production.

d.  If the average selling price is $17.99 for 2007 wine, $17.74 for 2008, $18.57 for 2009, and $16.99 for 2010, find the average selling price for all of these types of wine together. (*Hint:* Be careful!)

35. Recall in the example on CEO compensation by prepackaged software companies from Chapter 3 (Table 3.6.1) that we identified an outlier (Lawrence J. Ellison of Oracle Corp, with compensation of $56.81 million).

a.  Draw a detailed box plot for this data set. How many outliers are there?

b.  Omit the largest data value and draw a detailed box plot for the remaining data values, identifying and labeling the outliers (if any) with the company name.

36. Consider the costs charged for treatment of heart failure and shock by hospitals in the Puget Sound area, using the data from Table 3.9.6 of Chapter 3.

a.  Summarize the costs.

b.  Draw a box plot.

c.  Draw a cumulative distribution function.

d.  If your hospital wanted to place itself in the 65th percentile relative to the costs of this area, how much should be charged for this procedure?

37. Consider the data on CEO compensation in food processing firms from Table 3.9.7 of Chapter 3.

a.  Draw a detailed box plot.

b.  Find the 10th percentile of compensation.

38. Summarize prices of funeral services using the average and median, based on the data in Table 3.9.10 of Chapter 3.

39. Use the data set from problem 21 of Chapter 3 on poor quality in the production of electric motors.

a.  Find the average and median to summarize the typical level of problems with quality of production.

b.  Remove the two outliers, and recompute the average and median.

c.  Compare the average and median values with and without the outliers. In particular, how sensitive is each of these summary measures to the presence of outliers?

### TABLE 4.3.11 Sales of Some "Light" Foods

| "Light" Food | Sales ($ millions) |
| --- | --- |
| Entenmann's Fat Free baked goods | $125.5 |
| Healthy Request soup | 123.0 |
| Kraft Free processed cheese | 83.4 |
| Aunt Jemima Lite and Butter Lite pancake syrup | 58.0 |
| Fat Free Fig Newtons | 44.4 |
| Hellmann's Light mayonnaise | 38.0 |
| Louis Rich turkey bacon | 32.1 |
| Kraft Miracle Whip free | 30.3 |
| Ben & Jerry's frozen yogurt | 24.4 |
| Hostess Lights snack cakes | 19.3 |
| Perdue chicken/turkey franks | 3.8 |
| Milky Way II candy bar | 1.1 |

**Source:** Data are from "'Light' Foods Are Having Heavy Going," *Wall Street Journal,* March 4, 1993, p. B1. Their source is Information Resources Inc.

40. Many marketers assumed that consumers would go for reduced-calorie foods in a big way. While these "light" foods caught on to some extent, they hadn't yet sold in the large quantities their producers would have liked (with some exceptions). Table 4.3.11 shows the sales levels of some brands of "light" foods.
    a. Find the size of the total market for these brands.
    b. Find the average sales for these brands.
    c. Draw the cumulative distribution function.
    d. Your company is planning to launch a new brand of light food. The goal is to reach at least the 20th percentile of current brands. Find the yearly sales goal in dollars.
41. For the percent changes in January 2002 for the 30 companies in the Dow Jones Industrial Average (Table 2.6.7 in Chapter 2):
    a. Find the mean percent change.
    b. Find the median percent change.
    c. Find the five-number summary for percent change.
    d. Draw the box plot for percent change.
    e. Draw the cumulative distribution function for percent change.
    f. Find the percentile of a data value of 5% and the data value of the 70th percentile.
42. For the January 2002 daily values of the Dow Jones Industrial Average (Table 2.6.8 in Chapter 2):
    a. Find the mean net change.
    b. Find the median net change.
    c. Find the five-number summary for net change.
    d. Draw the box plot for net change.

e. Find the mean percent change.
f. Find the median percent change.
g. Find the five-number summary for percent change.
h. Draw the box plot for percent change.

16. You may assume that it's OK to trade any number of shares. For now, please ignore real-life problems with "odd lots" of fewer than 100 shares.
17. Populations for 2008 and state taxes for 2009 are data from U.S. Census Bureau, *Statistical Abstract of the United States: 2010* (129th edition), Washington, DC, 2009, accessed from http://www.census.gov/govs/statetax/09staxrank.html and http://www.census.gov/compendia/statab/cats/population.html on July 5, 2010.

### Database Exercises

*Problems marked with an asterisk (\*) are solved in the Self Test in Appendix C.*

Please refer to the employee database in Appendix A.
1.\* For the annual salary levels:
   a. Find the average.
   b. Find the median.
   c. Construct a histogram, and give an approximate value for the mode.
   d. Compare these three summary measures. What do they tell you about typical salaries in this administrative division?
2. For the annual salary levels:
   a. Construct a cumulative distribution function.
   b. Find the median, quartiles, and extremes.
   c. Construct a box plot, and comment on its appearance.
   d. Find the 10th percentile and the 90th percentile.
   e. What is the percentile ranking for employee number 6?
3. For the genders:
   a. Summarize by finding the percent of each category.
   b. Find the mode. What does this tell you?
4. For the ages: Answer the parts of exercise 1.
5. For the ages: Answer the parts of exercise 2.
6. For the experience variable: Answer the parts of exercise 1.
7. For the experience variable: Answer the parts of exercise 2.
8. For the training level: Answer the parts of exercise 3.

### Projects

1. Find a data set consisting of at least 25 numbers relating to a firm or an industry group of interest to you, from sources such as the Internet or trade journals in your library. Summarize the data using all of the techniques you have learned so far that are appropriate to your data. Be sure to use both numerical and graphical methods. Present your results as a two-page report to management, with an executive summary as the first paragraph. (You may find it helpful to keep graphs small to fit in the report.)
2. Find statistical summaries for two quantitative univariate data sets of your own choosing, related to your work, firm, or industry group. For each data set:
   a. Compute the average, the median, and a mode.
   b. For each of these summaries, explain what it tells you about the data set and the business situation.

c. Construct a histogram, and indicate these three summary measures on the horizontal axis. Comment on the distribution shape and the relationship between the histogram and the summaries.

d. Construct the box plot and comment on the costs and benefits of having details (the histogram) as compared to having a bigger picture (the box plot).

## Case

*Managerial Projections for Production and Marketing, or "The Case of the Suspicious Customer"*

B. R. Harris arrived at work and found, as expected, the recommendations of H. E. McRorie waiting on the desk. These recommendations would form the basis for a quarterly presentation Harris would give this afternoon to top management regarding production levels for the next three months. The projections would serve as a planning guide, ideally indicating appropriate levels for purchasing, inventory, and human resources in the immediate future. However, customers have a habit of not always behaving as expected, and so these forecasts were always difficult to prepare with considerable judgment (guesswork?) traditionally used in their preparation.

Harris and McRorie wanted to change this and create a more objective foundation for these necessary projections. McRorie had worked late analyzing the customer survey (a new procedure they were experimenting with, based on responses of 30 representative customers) and had produced a draft report that read, in part:

*We anticipate quarterly sales of $1,478,958, with projected sales by region given in the accompanying table. We recommend that production be increased from current levels in anticipation of these increased sales …*

| | 2010 Quarter II Projections | 2010 Quarter I Actual | 2009 Quarter II Actual |
|---|---|---|---|
| **Sales:** | | | |
| Northeast | $441,067 | $331,309 | $306,718 |
| Northwest | 292,589 | 222,185 | 200,201 |
| South | 149,934 | 118,151 | 101,721 |
| Midwest | 371,195 | 277,952 | 254,315 |
| Southwest | 224,173 | 165,332 | 157,843 |
| Total | 1,478,958 | 1,114,929 | 1,020,798 |
| **Production (Valued at Wholesale):** | | | |
| Chairs | $515,112 | $425,925 | $389,115 |
| Tables | 228,600 | 201,125 | 197,250 |
| Bookshelves | 272,966 | 209,105 | 189,475 |
| Cabinets | 462,280 | 276,500 | 295,400 |
| Total | 1,478,958 | 1,112,655 | 1,071,240 |
| **Production (Units):** | | | |
| Chairs | 11,446.9 | 9,465 | 8,647 |
| Tables | 1,828.8 | 1,609 | 1,578 |
| Bookshelves | 4,199.5 | 3,217 | 2,915 |
| Cabinets | 1,320.8 | 790 | 844 |

Harris was uneasy. They were projecting a large increase both from the previous quarter (32.7%) and from the same quarter last year (44.9%). Historically, the firm has not been growing at near these rates in recent years. Along with that came a recommendation for increased production in order to be prepared for the increased sales. Why the hesitation? Because if these projections turned out to be wrong, and sales did not increase, the firm would be left with expensive inventory (produced at a higher cost than usual due to overtime, hiring of temporary help, and leasing of additional equipment) with its usual carrying costs (including the time value of money: the interest that could have been earned by waiting to spend on the additional production).

Harris asked about this, and McRorie also seemed hesitant. Yet it seemed simple enough: Take the average anticipated spending by customers as reported in the survey and then multiply by the total number of customers in that region. What could be wrong with that? They decided to take a closer look at the data. Here is their spreadsheet, including background information (the wholesale price the firm receives for each item and the number of active customers by region) and the sampling results. Each of the 30 selected customers reported the number of each item they plan to order during the coming quarter. The Value column indicates the cash to be received by the firm (e.g., Customer 1 plans to buy 3 chairs at $45 and 4 bookshelves at $65 for a total value of $395).

| Wholesale | Price |
|---|---|
| Chairs | $45 |
| Tables | 125 |
| Bookshelves | 65 |
| Cabinets | 350 |

| Active Customers | No. of Customers |
|---|---|
| Northeast | 303 |
| Northwest | 201 |
| South | 103 |
| Midwest | 255 |
| Southwest | 154 |
| TOTAL | 1,016 |

### Sample Results

| Customer # | Chairs | Tables | Bookshelves | Cabinets | Value |
|---|---|---|---|---|---|
| 1 | 3 | 0 | 4 | 0 | $395 |
| 2 | 9 | 1 | 6 | 1 | 1,270 |
| 3 | 23 | 2 | 1 | 2 | 2,050 |
| 4 | 7 | 0 | 3 | 0 | 510 |
| 5 | 4 | 0 | 0 | 0 | 180 |
| 6 | 14 | 1 | 5 | 0 | 1,080 |
| 7 | 6 | 0 | 5 | 0 | 595 |
| 8 | 14 | 1 | 0 | 0 | 755 |
| 9 | 1 | 5 | 17 | 3 | 2,825 |
| 10 | 2 | 0 | 4 | 1 | 700 |
| 11 | 16 | 1 | 1 | 1 | 1,260 |
| 12 | 4 | 0 | 4 | 0 | 440 |
| 13 | 6 | 0 | 4 | 1 | 880 |
| 14 | 2 | 1 | 8 | 2 | 1,435 |
| 15 | 42 | 15 | 21 | 18 | 11,430 |

## Sample Results—cont'd

| Customer # | Chairs | Tables | Bookshelves | Cabinets | Value |
|---|---|---|---|---|---|
| 16 | 3 | 0 | 0 | 2 | 835 |
| 17 | 7 | 3 | 0 | 0 | 690 |
| 18 | 1 | 4 | 2 | 0 | 675 |
| 19 | 43 | 0 | 4 | 0 | 2,195 |
| 20 | 6 | 2 | 4 | 2 | 1,480 |
| 21 | 3 | 1 | 1 | 0 | 325 |
| 22 | 45 | 6 | 1 | 0 | 2,840 |
| 23 | 0 | 2 | 7 | 1 | 1,055 |
| 24 | 13 | 6 | 3 | 0 | 1,530 |
| 25 | 19 | 0 | 2 | 2 | 1,685 |
| 26 | 0 | 0 | 0 | 0 | 0 |
| 27 | 8 | 0 | 7 | 0 | 815 |
| 28 | 14 | 3 | 3 | 1 | 1,550 |
| 29 | 6 | 0 | 1 | 2 | 1,035 |
| 30 | 17 | 0 | 6 | 0 | 1,155 |

| Customer # | Chairs | Tables | Bookshelves | Cabinets | Value |
|---|---|---|---|---|---|
| Total (sample) | 338 | 54 | 124 | 39 | 43,670 |
| Average | 11.267 | 1.8 | 4.133 | 1.3 | 1,455.667 |
| Avg Value | $507 | $225 | $268.667 | $455 | $1,455.667 |
| Total Projections (multiplied by 1,016 customers): | | | | | |
| Value | $515,112 | $228,600 | $272,966 | $462,280 | $1,478,958 |
| Units | 11,446.9 | 1,828.8 | 4,199.5 | 1,320.8 | |

## Discussion Questions

1. Would the average-based procedure they are currently using ordinarily be a good method? Or is it fundamentally flawed? Justify your answers.

2. Take a close look at the data using summaries and graphs. What do you find?

3. What would you recommend that Harris and McRorie do to prepare for their presentation this afternoon?

# Variability

## Dealing with Diversity

One reason we need statistical analysis is that there is variability in data. If there were no variability, many answers would be obvious, and there would be no need for statistical methods.[1] A situation with variability often has risk because, even using all available information, you still may not know exactly what will happen next. To manage risk well, you certainly need to understand its nature and how to measure the variability of outcomes it produces. Following are some situations in which variability is important:

**One:** Consider the variability of worker productivity. Certainly, the average productivity of workers summarizes a department's overall performance. However, any efforts to improve that productivity would probably have to take into account individual differences among workers. For example, some programs may be aimed at improving all workers, whereas others may specifically target the quickest or slowest people. A measure of variability in productivity would summarize the extent of these individual differences and provide helpful information in your quest for improved performance.

**Two:** The stock market provides a higher return on your money, on average, than do safer securities such as money market funds. However, the stock market is riskier, and you can actually lose money by investing in stocks. Thus, the average or "expected" return does not tell the whole story. Variability in returns could be summarized for each investment and would indicate the level of risk you would be taking on with any particular investment.

**Three:** You compare your firm's marketing expenditures to those of similar firms in your industry group, and you find that your firm spends less than is typical for this industry. To put your number in perspective, you might wish to take into account the extent of diversity within your industry group. Taking the difference between your firm and the group average and comparing it to a measure of variability for the industry group will indicate whether you are slightly low or are a special exception compared to these other firms. This information would help with strategic planning in setting next year's marketing budget.

**Variability** may be defined as the extent to which the data values differ from each other. Other terms that have a similar meaning include **diversity, uncertainty, dispersion,** and **spread.** You will see three different ways of summarizing the amount of the variability in a data set, all of which require numerical data:

**One:** The *standard deviation* is the traditional choice and is the most widely used. It summarizes how far

---

1. Some statisticians have commented informally that it is variability that keeps them in business!

an observation typically is from the average. If you multiply the standard deviation by itself, you find the *variance*.

**Two:** The *range* is quick and superficial and is of limited use. It summarizes the extent of the entire data set, using the distance from the smallest to the largest data value.

**Three:** The *coefficient of variation* is the traditional choice for a *relative* (as opposed to an *absolute*) variability measure and is used moderately often. It summarizes how far an observation typically is from the average as a percentage of the average value, using the ratio of standard deviation to average.

Finally, you will learn how rescaling the data (for example, converting from Japanese yen to U.S. dollars or from units produced to monetary cost) changes the variability.

## 5.1  THE STANDARD DEVIATION: THE TRADITIONAL CHOICE

The **standard deviation** is a number that summarizes *how far away from the average* the data values typically are. The standard deviation is a very important concept in statistics since it is the basic tool for summarizing the amount of randomness in a situation. Specifically, it measures the extent of randomness of individuals about their average.

If all numbers are the same, such as the simple data set

$$5.5, \ 5.5, \ 5.5, \ 5.5$$

the average will be $\overline{X} = 5.5$ and the standard deviation will be $S = 0$, expressing the fact that this trivial data set shows no variability whatsoever.

In reality, most data sets do have some variability. Each data value will be some distance away from the average, and the standard deviation will summarize the extent of this variability. Consider another simple data set, but with some variability:

$$43.0, \ 17.7, \ 8.7, \ -47.4$$

These numbers represent the rates of return (e.g., 43.0%) for four stocks (Maytag, Boston Scientific, Catalytica, and Mitcham Industries) selected at random by throwing darts at a newspaper page of stock listings.[2] The average value again is $\overline{X} = 5.5$, telling you that these stocks had a 5.5% average rate of return (in fact, a portfolio with equal amounts of money invested in each stock would have had this 5.5% average performance). Although the average is the same as before, the data values are considerably different from one another. The first data value, 43.0, is at a distance $X_1 - \overline{X} = 43.0 - 5.5 = 37.5$ from average, telling you that Maytag's rate of return was 37.5 percentage points above average. The last data value, -47.4, is at a distance

**TABLE 5.1.1  Finding the Deviation from the Average**

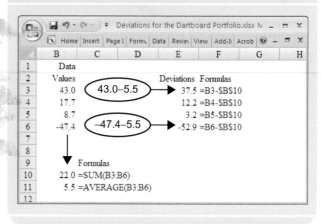

$X_4 - \overline{X} = -47.4 - 5.5 = -52.9$ from average, telling you that Mitcham Industries' rate of return was 52.9 percentage points *below* average (below because it is negative). Table 5.1.1 shows how far each data value is from the average.[3]

These distances from the average are called **deviations** or residuals, and they indicate how far above the average (if positive) or below the average (if negative) each data value is. The deviations form a data set centered around zero that is much like the original data set, which is centered around the average.

The standard deviation summarizes the deviations. Unfortunately, you can't just average them since some are negative and some are positive, and the end result is always an unhelpful zero.[4] Instead, the standard method will first square each number (that is, multiply it by itself) to eliminate the minus sign, sum, divide by $n - 1$, and finally take the square root (which undoes the squaring you did earlier).[5]

---

2. From Georgette Jasen, "Your Money Matters: Winds of Chance Blow Cold on the Pros," *Wall Street Journal*, April 9, 1998, p. C1.

3. There are dollar signs in the formulas to tell Excel to use the same mean value ($B$10) for successive data values, making it easy to copy a formula down a column after you enter it (in cell D3 here). To copy, first select the cell, then either use Edit Copy and Edit Paste, or drag the lower right-hand corner of the selected cell down the column.

4. In fact, when you use algebra, it is possible to prove that the sum of the deviations from average will *always* be zero for any data set. You might suspect that you could simply discard the minus signs and then average, but it can be shown that this simple method does not efficiently use all the information in the data if it follows a normal distribution.

5. Dividing by $n - 1$ instead of $n$ (as you would usually do to find an average) is a way of adjusting for the technical fact that when you have a *sample*, you do not know the true population mean. It may also be thought of as an adjustment for the fact that you have lost one piece of information (*a degree of freedom*, in statistics jargon) in computing the deviations. This lost piece of information may be identified as an indication of the true sizes of the data values (since they are now centered around zero instead of around the average).

**TABLE 5.1.2 Finding the Sum of Squared Deviations, the Variance, and the Standard Deviation**

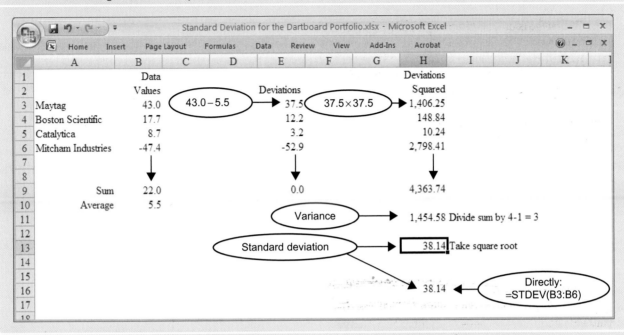

## Definition and Formula for the Standard Deviation and the Variance

The standard deviation is defined as the result of the following procedure. Note that, along the way, the **variance** (the square of the standard deviation) is computed. The variance is sometimes used as a variability measure in statistics, especially by those who work directly with the formulas (as is often done with the *analysis of variance* or *ANOVA*, in Chapter 15), but the standard deviation is often a better choice. The variance contains no extra information and is more difficult to interpret than the standard deviation in practice. For example, for a data set consisting of dollars spent, the variance would be in units of "squared dollars," a measure that is difficult to relate to; however, the standard deviation would be a number measured in good old familiar dollars.

**Finding the Sample Standard Deviation**

1. Find the deviations by subtracting the average from each data value.
2. Square these deviations, add them up, and divide the resulting sum by $n-1$. This is the variance.
3. Take the square root. This is the standard deviation.

Table 5.1.2 shows how this procedure works on our randomly chosen companies. Dividing the sum of squared deviations, 4,363.74, by $n-1 = 4-1 = 3$, you get the variance $4{,}363.74/3 = 1{,}454.58$. Taking the square root, we find the standard deviation, 38.14, which does indeed

appear to be a reasonable summary of the deviations themselves (ignoring the minus signs to concentrate on the *size* of the deviations). The last formula, in the lower-right corner, shows how to compute the standard deviation in Excel® directly in one step.

The formula for the standard deviation puts the preceding procedure in mathematical shorthand. The standard deviation for a sample of data is denoted by the letter $S$, and the formulas for the standard deviation and the variance are as follows:[6]

**The Standard Deviation for a Sample**

$$S = \sqrt{\frac{\text{Sum of squared deviations}}{\text{Number of data items} - 1}}$$

$$= \sqrt{\frac{(X_1 - \bar{X})^2 + (X_2 - \bar{X})^2 + \cdots + (X_n - \bar{X})^2}{n-1}}$$

$$= \sqrt{\frac{1}{n-1}\sum_{i=1}^{n}(X_i - \bar{X})^2}$$

**The Variance for a Sample**

$$\text{Variance} = S^2 = \frac{1}{n-1}\sum_{i=1}^{n}(X_i - \bar{X})^2$$

---

6. There is also a computational formula for the variance, $\frac{1}{n-1}\left(\sum_{i=1}^{n} X_i^2 - n\bar{X}^2\right)$, which gives the same answer in theory, but can be badly behaved for a data set consisting of large numbers that are close together.

Computing the standard deviation for our simple example using the formula produces the same result, 38.14:

$$S = \sqrt{\frac{1}{n-1}\sum_{i=1}^{n}(X_i - \overline{X})^2} = \sqrt{\frac{4,363.74}{4-1}} = \sqrt{1,454.58} = 38.14$$

## Using a Calculator or a Computer

Of course, there is another, much simpler way to find the standard deviation: Use a calculator or a computer. This is how people *really* compute a standard deviation, by delegating the task of actual calculation to an electronic device. The steps and the formula you just learned were not a waste of time, however, since you now understand the basis for the number that will be computed for you automatically. That is, in interpreting a standard deviation, it is important to recognize it as a measure of typical (or *standard*) deviation size.

If your calculator has a $\sum$ or a $\sum +$ key (a summation key), it can probably be used to find a standard deviation. Consult the calculator's instruction manual for details, which will likely proceed as follows: First, clear the calculator's storage registers to get it ready for a new problem. Then enter the data into the calculator, following each value by the summation key. You may now compute the standard deviation by pressing the appropriate key, probably labeled with an $S$ or a sigma $(\sigma)$.[7]

There are many different ways in which computers calculate a standard deviation, depending on which of the many available products you are using: spreadsheets, database programs, programming languages, or stand-alone statistical analysis programs.

## Interpreting the Standard Deviation

The standard deviation has a simple, direct interpretation: It summarizes the *typical distance from average* for the individual data values. The result is a measure of the variability of these individuals. Because the standard deviation represents the typical deviation size, we expect some individuals to be closer to the average than this standard, while others will be farther away. That is, you expect some data values to be less than one standard deviation from the average, while others will be more than one standard deviation away from the average.

Figure 5.1.1 shows how to picture the standard deviation in terms of distance from the average value. Since the average indicates the center of the data, you would expect individuals to deviate to both sides of the average.

---

7. If your calculator has two standard deviation keys, one marked $n$ and the other marked $n-1$, choose the $n-1$ key for now. The other key computes the *population* instead of the *sample* standard deviation, a distinction you will learn about later.

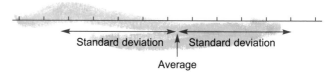

**FIGURE 5.1.1** The number line with an average and a standard deviation indicated. Note that the average is a position on the number line (indicating an absolute number), whereas the standard deviation indicates a distance along the number line, namely, the typical distance from the average.

### Example
#### The Advertising Budget

Your firm spends $19 million per year on advertising, and top management is wondering if that figure is appropriate. Although there are many ways to decide this strategic number, it will always be helpful to understand your competitive position. Other firms in your industry, of size similar to yours, spend an average of $22.3 million annually. You can use the standard deviation as a way to place your difference (22.3 – 19 = $3.3 million) into perspective to see just how low your advertising budget is relative to these other firms.

Here are the budgets, in millions of dollars, for the group of $n = 17$ similar firms:

8, 19, 22, 20, 27, 37, 38, 23, 23,
12, 11, 32, 20, 18, 23, 35, 11

You may verify that the average is $22.3 million (rounding $22.29411) and that the standard deviation is $9.18 million (rounding $9.177177 to three significant digits) for your peer group.

Since your difference from the peer group average ($3.3 million) is even smaller than the standard deviation ($9.18 million), you may conclude that your advertising budget is quite typical. Although your budget is smaller than the average, it is closer to the average than the typical firm in your peer group is.

To visualize your position with respect to the reference group, Figure 5.1.2 shows a histogram of your peer group, with the average and standard deviation indicated (note how effectively the standard deviation shows the typical extent of the data on either side of the average). Your firm, with a $19 million advertising budget, is indeed fairly typical compared to your peers. Although the difference of $3.3 million (between your budget and the peer group average) seems like a lot of money, it is small compared to the individual differences among your peers. Your firm is only slightly below average.

### Example
#### Customer Diversity

Your customers are not all alike; there are individual differences in size of order, product preference, yearly cycle, demands for information, loyalty, and so forth. Nonetheless, you probably have a "typical customer" profile in mind, together with some feeling for the extent of the diversity.

You can summarize customer orders by reporting the average and the standard deviation:

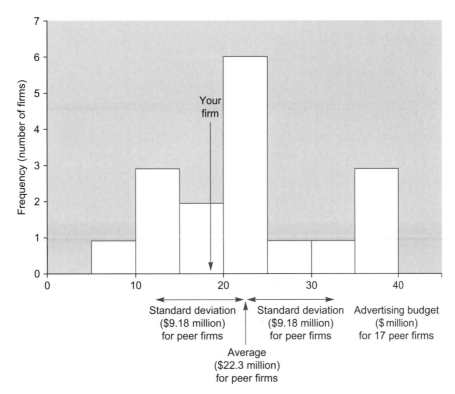

**FIGURE 5.1.2** A histogram of the advertising budgets of your peer group of 17 firms, with the average and standard deviation. Your firm's budget of $19 million is quite typical relative to your peers. In fact, you are not even one standard deviation away from the average.

**Yearly total order, per customer:**

| | |
|---|---|
| Average | $85,600 |
| Standard deviation | $28,300 |

Thus, on average, each customer placed $85,600 worth of orders last year. To indicate the diversity among customers, the standard deviation of $28,300 shows you that, *typically,* customers ordered approximately $28,300 more or less than the average value of $85,600. The *approximately* carries a lot of weight here: Some customers may have been quite close to the average, whereas others were much more than $28,300 away from the average. The average indicates the typical size of yearly orders per customer, and the standard deviation indicates the typical deviation from average.

Note also that the standard deviation is in the same units of measurement as the average; in this example, both are measured in dollars. More precisely, the units of measurement are "dollars per year per customer." This matches the units of the original data set, which would be a list of dollars per year, with one number for each of your customers.

## Interpreting the Standard Deviation for a Normal Distribution

When a data set is approximately normally distributed, the standard deviation has a special interpretation.

Approximately two-thirds of the data values will be *within one standard deviation of the average,* on either side of the average, as shown in Figure 5.1.3.

For example, if your employees' abilities are approximately normally distributed, then you may expect to find about two-thirds of them to be within one standard deviation either above or below the average. In fact, about one-third of them will be within one standard deviation above average, and about one-third of them will be within one standard deviation below average. The remaining (approximately) one-third of your employees would also divide up equally: About one-sixth (half of this one-third) will be more than one standard deviation above the average, and about one-sixth of your employees will (unfortunately!) be more than one standard deviation below the average.

Figure 5.1.3 also shows that, for a normal distribution, we expect to find about 95% of the data *within two standard deviations from the average.*[8] This fact will play a key role later in the study of statistical inference, since error rates are often limited to 5%.

Finally, we expect nearly all of the data (99.7%) to be *within three standard deviations from the average.* This

---

8. Exactly 95% of the data values in a perfect normal distribution are actually within 1.960 standard deviations from the average. Since 1.96 is close to 2, we use "two standard deviations" as a convenient and close approximation.

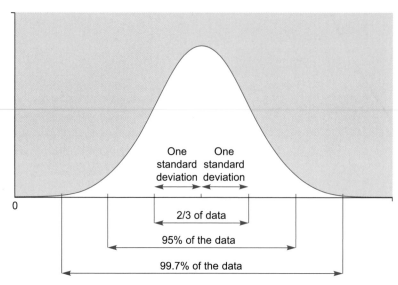

**FIGURE 5.1.3**   When you have a normal distribution, your data set is conveniently divided up according to number of standard deviations from the average. About two-thirds of the data will be within one standard deviation from the average. About 95% of the data will be within two standard deviations from the average. Finally, nearly all (99.7%) of the data are expected to be within three standard deviations from the average.

leaves only about 0.3% of the data more extreme than this. In Figure 5.1.3 you can see how the normal distribution is nearly zero when you reach three standard deviations from the average. The limits of control charts, used extensively for quality control, are often set up so that any observation that is more than three standard deviations away from the average will be brought to your attention as a problem to be fixed.

What happens if your data set is not normally distributed? Then these percentages do not apply. Unfortunately, since there are so many different kinds of skewed (or other nonnormal) distributions, there is no single exact rule that gives percentages for any distribution.[9] Figure 5.1.4 shows an example of a skewed distribution. Instead of two-thirds of the data being within one standard deviation from the average, you actually find about three-fourths of the data values here. Furthermore, most of these data values are to the left of the average (since the distribution is higher here).

**Example**
*A Quality Control Chart for Picture-Scanning Quality*

A factory produces monitor screens and uses control charts to help maintain and improve quality. In particular, the size of an individual dot on the screen (the "dot pitch") must be small so that details will be visible to the user. The control chart contains the individual measurements (which change

somewhat from one monitor to the next) with their average (which you see going through the middle of the data) and the control limits (which are set at three standard deviations above and below the average; more details will be presented in Chapter 18). Figure 5.1.5 shows a control chart with a system that is "in control" with all measurements within the control limits. Figure 5.1.6 shows a control chart with an "out of control" point at monitor 22. The control chart has helped you identify a problem; it is up to you (the manager) to investigate and correct the situation.

**Example**
*Stock Market Returns Vary from Day to Day*

In this example, we examine stock-market volatility (as measured by the appropriate standard deviation) during the time period leading up to the crash of 1987. Consider daily stock market prices, as measured by the Dow Jones Industrial Average at the close of each trading day from July 31 through October 9, 1987, and shown in Table 5.1.3. The Dow Jones Average is a scaled and weighted average of the stock prices of 30 specially selected large industrial firms. One of the usual ways investors look at these data is as a graph of the index plotted against time, as shown in Figure 5.1.7.

Financial analysts and researchers often look instead at the *daily return,* which is the interest rate earned by investing in stocks for just one day. This is computed by taking the change in the index and dividing it by its value the previous day. For example, the first daily return, for August 3, is

$$\frac{2557.08 - 2572.07}{2572.07} = -0.006$$

*(Continued)*

---

9. However, there is a bound called *Chebyshev's rule* which assures that you will find at least $1 - 1/a^2$ of the data within $a$ standard deviations of the average. For example, with $a = 2$, at least 75% of the data (computed as $1 - 1/2^2$) must be within two standard deviations of the average *even if the distribution is not normal* (compare to approximately 95% if the data are normal). With $a = 3$, we see that at least 88.9% of the data fall within three standard deviations of the average.

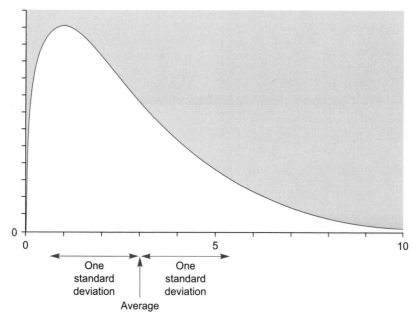

FIGURE 5.1.4   When your distribution is skewed, there are no simple rules for finding the proportion of the data within one (or two or three) standard deviation from the average.

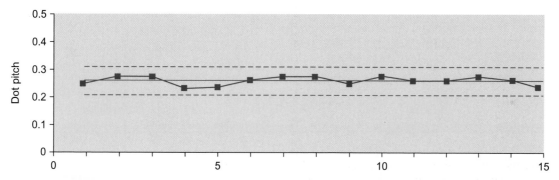

FIGURE 5.1.5   A control chart with measurements for monitor screens, with upper and lower control limits defined using three standard deviations (the average is also shown, going through the middle of the data). The system is in control, with only random deviations, because there are no strong patterns and no observations extend beyond the control limits.

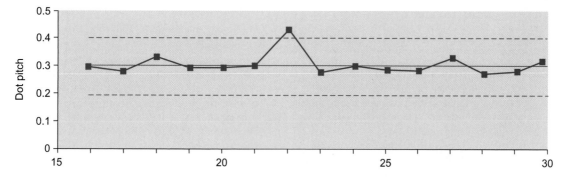

FIGURE 5.1.6   The system is now out of control. Note that screen number 22 is more than three standard deviations above the average. This is not within ordinary system variation, and you would want to investigate to avoid similar problems in the future.

**TABLE 5.1.3** Closing Stock Prices

| Dow Jones Industrial Average | Date | Dow Jones Industrial Average | Date |
|---|---|---|---|
| 2,572.07 | July 31, 1987 | 2,561.38 | |
| 2,557.08 | | 2,545.12 | |
| 2,546.72 | | 2,549.27 | |
| 2,566.65 | | 2,576.05 | |
| 2,594.23 | | 2,608.74 | |
| 2,592.00 | | 2,613.04 | |
| 2,635.84 | | 2,566.58 | |
| 2,680.48 | | 2,530.19 | |
| 2,669.32 | | 2,527.90 | |
| 2,691.49 | | 2,524.64 | |
| 2,685.43 | | 2,492.82 | |
| 2,700.57 | | 2,568.05 | |
| 2,654.66 | | 2,585.67 | |
| 2,665.82 | | 2,566.42 | |
| 2,706.79 | | 2,570.17 | |
| 2,709.50 | | 2,601.50 | |
| 2,697.07 | | 2,590.57 | |
| 2,722.42 | | 2,596.28 | |
| 2,701.85 | | 2,639.20 | |
| 2,675.06 | | 2,640.99 | |
| 2,639.35 | | 2,640.18 | |
| 2,662.95 | | 2,548.63 | |
| 2,610.97 | | 2,551.08 | |
| 2,602.04 | | 2,516.64 | |
| 2,599.49 | | 2,482.21 | October 9, 1987 |

**Source:** This data set is from the *Daily Stock Price Record, New York Stock Exchange*, Standard & Poor's Corporation, 1987.

## Example—cont'd

representing a downturn somewhat *under* 1%.[10] These daily returns represent, in a more direct way than the average itself, what is really happening in the market from day to day in a dynamic sense. We will focus our attention on these daily returns as a data set, shown in Table 5.1.4.

Figure 5.1.8 shows a histogram of these daily returns, indicating a normal distribution. The average daily return during this time was –0.0007, or approximately zero (an average downturn of seven hundredths of a percent). Thus,

the market was heading, on average, neither higher nor lower during this time. The standard deviation is 0.0117, indicating that the value of $1 invested in the market would change, on the average, by approximately $0.0117 each day in the sense that the value might go up or down by somewhat less or more than $0.0117.

The extreme values at either end of Figure 5.1.8 represent the largest daily up and down movements. On September 22, the market went up from 2492.82 to 2568.05, an upswing of 75.23 points, for a daily return of 0.030 (a gain of $0.030 per dollar invested the day before). On October 6, the market went down from 2640.18 to 2548.63, or 91.55 points, for a daily return of –0.035 (a loss of $0.035 per dollar invested the day before).

To be within one standard deviation (0.0117) of the average (–0.0007), a daily return would have to be between –0.0007 – 0.0117 = –0.0124 and –0.0007 + 0.0117 = 0.0110. Of the 49 daily returns, 32 fit this description. Thus, we have found that 32/49, or 65.3%, of daily returns are within one standard deviation of the average. This percentage is fairly close to the approximately two-thirds (66.7%) you would expect for a perfect normal distribution. The two-thirds rule is working.

To be within two standard deviations of the average, a daily return would have to be between –0.0007 – (2 × 0.0117) = –0.0241 and –0.0007 + (2 × 0.0117) = 0.0227. Of the 49 daily returns, 47 fit this description (all except the two extreme observations we noted earlier). Thus, 47/49, or 95.9%, of daily returns are within two standard deviations of the average. This percentage is quite close to the 95% you would expect for a perfectly normal distribution.

This example conforms to the normal distribution rules fairly closely. With other approximately normally distributed examples, you should not be surprised to find a larger difference from the two-thirds or the 95% you expect for a perfect normal distribution.

---

10. This is the return from July 31 through August 3. We will consider it to be a daily return, since the intervening two days were a weekend, with no trading.

## Example
### The Stock Market Crash of 1987: 19 Standard Deviations!

On Monday, October 19, 1987, the Dow Jones Industrial Average fell 508 points from 2246.74 (the previous Friday) to 1738.74. This represents a daily return of –0.2261; that is, the stock market lost 22.61% of its value. This unexpected loss in value, shown in Figure 5.1.9, was the worst since the "Great Crash" of 1929.

To get an idea of just how extreme this crash was in statistical terms, let's compare it to what you would have expected based on previous market behavior. For the baseline period, let's use the previous example, with its July 31 to October 9 time period, extending up to Friday, one week before the crash.

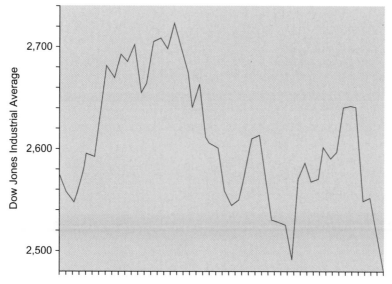

**FIGURE 5.1.7**   The Dow Jones Industrial Average closing stock price, daily from July 31 to October 9, 1987.

For the baseline period, we found an average of –0.0007 and a standard deviation of 0.0117 for daily returns. How many of these standard deviations below this average does the loss of October 19 represent? The answer is

$$\frac{-0.2261 - (-0.0007)}{0.0117} = -19.26 \text{ standard deviations}$$
$$\text{(below the average)}$$

This shows how incredibly extreme the crash was. If daily stock returns were truly normally distributed (and if the distribution didn't change quickly over time), you would essentially *never* expect to see such an extreme result. We would expect to see daily returns more than one standard deviation away from the average fairly often (about one-third of the time). We would see two standard deviations or more from time to time (about 5% of the time). We would see three standard deviations or more only very rarely—about 0.3% of the time or, roughly speaking, about once a year.[11] Even five standard deviations would be pretty much out of the question for a perfect normal distribution. To see 19.26 standard deviations is quite incredible indeed.

But we did see a daily return of 19.26 standard deviations below the average. This should remind you that stock market returns do *not* follow a perfect normal distribution. There is nothing wrong with the theory; it's just that the theory doesn't apply in this case. Although the normal distribution appears to apply most of the time to daily returns, the crash of 1987 should remind you of the need to check the validity of assumptions to protect yourself from special cases.

11. Something that happens only 0.3% of all days will happen about once a year for the following reasons. First, 0.3% expressed as a proportion is 0.003. Second, its reciprocal is 1/0.003 = 333 (approximately), which means that it happens about every 333 days, or (very roughly) about once a year.

**Example**
*Market Volatility before and after the Crash*

In the period following the crash of October 19, 1987, the market was generally believed to be in a volatile state. You can measure the extent of this volatility by using the standard deviation of daily returns, as defined in an earlier example. Here is a table of these standard deviations:

| Standard Deviation | Time Period |
| --- | --- |
| 1.17% | August 1 to October 9 |
| 8.36% | October 12 (1 week before) to October 26 (1 week after) |
| 2.09% | October 27 to December 31, 1987 |

The standard deviation was about seven times higher during the period surrounding the crash (from one week before to one week after) than before this period. After the crash, the standard deviation was lower but remained at nearly double its earlier value (2.09% compared to 1.17%). Apparently, the market got "back to business" to a large degree following the crash but still remained "nervous," as indicated by the volatility, which is measured by standard deviation.

You can see the heightened volatility in Figure 5.1.10. Aside from the wild gyrations of the market around October 19, the vertical swings of the graph are roughly double on the right as compared to the left. These volatilities were summarized using the standard deviations (roughly corresponding to vertical distances in the figure) in the preceding table.

**Example**
*Recent Market Volatility*

More recently, stock market volatility has settled back down somewhat, although it rose during the market turmoil of

*(Continued)*

**TABLE 5.1.4 Daily Stock Market Returns**

| Dow Jones Industrial Average | Daily Return | Dow Jones Industrial Average | Daily Return |
|---|---|---|---|
| 2572.07 (7/31/87) | | 2561.38 | −0.015 |
| 2557.08 (8/3/87) | −0.006 = (2557.08 22572.07)/2572.07 | 2545.12 | −0.006 |
| 2546.72 (8/4/87) | −0.004 | 2549.27 | 0.002 |
| 2566.65 (8/5/87) | 0.008 | 2576.05 | 0.011 |
| 2594.23 | 0.011 | 2608.74 | 0.013 |
| 2592.00 | −0.001 | 2613.04 | 0.002 |
| 2635.84 | 0.017 | 2566.58 | −0.018 |
| 2680.48 | 0.017 | 2530.19 | −0.014 |
| 2669.32 | −0.004 | 2527.90 | −0.001 |
| 2691.49 | 0.008 | 2524.64 | −0.001 |
| 2685.43 | −0.002 | 2492.82 | −0.013 |
| 2700.57 | 0.006 | 2568.05 | 0.030 |
| 2654.66 | −0.017 | 2585.67 | 0.007 |
| 2665.82 | 0.004 | 2566.42 | −0.007 |
| 2706.79 | 0.015 | 2570.17 | 0.001 |
| 2709.50 | 0.001 | 2601.50 | 0.012 |
| 2697.07 | −0.005 | 2590.57 | −0.004 |
| 2722.42 | 0.009 | 2596.28 | 0.002 |
| 2701.85 | −0.008 | 2639.20 | 0.017 |
| 2675.06 | −0.010 | 2640.99 | 0.001 |
| 2639.35 | −0.013 | 2640.18 | −0.000 |
| 2662.95 | 0.009 | 2548.63 | −0.035 |
| 2610.97 | −0.020 | 2551.08 | 0.001 |
| 2602.04 | −0.003 | 2516.64 | −0.014 |
| 2599.49 | −0.001 | 2482.21 (10/9/87) | −0.014 |

**Example—cont'd**

2008. Following is a table of standard deviations of daily returns (measuring volatility) for each year from 2000 through 2010 for the Dow Jones Industrial Average stock market index. Note that a typical price movement from 2004 to 2006 was just over a half of one percent (of the portfolio value) per day, although market volatility has risen since then, more than tripling (rising to 2.39%) during the turmoil year of 2008 before settling down again.

| Year | Standard Deviation |
|---|---|
| 2010 | 1.14% (through July 6) |
| 2009 | 1.53% |
| 2008 | 2.39% |
| 2007 | 0.92% |
| 2006 | 0.62% |
| 2005 | 0.65% |
| 2004 | 0.68% |
| 2003 | 1.04% |
| 2002 | 1.61% |
| 2001 | 1.35% |
| 2000 | 1.31% |

We have continued to see volatility rising just after unusual market events. For example, when the financial services firm Lehman Brothers declared bankruptcy on September 15, 2008, the standard deviation of daily volatility

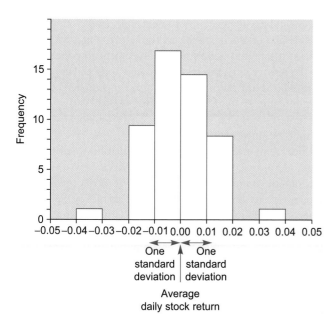

**FIGURE 5.1.8** A histogram of daily returns in stock prices. The average daily return is nearly zero, suggesting that ups and downs were about equally likely in the short term. The standard deviation, 0.0117, represents the size of typical day-to-day fluctuations. A dollar investment in the market would change in value by about a penny per day during this time.

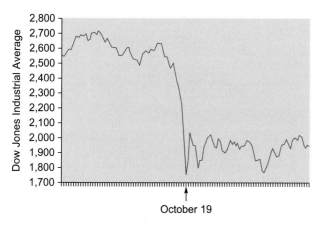

**FIGURE 5.1.9** The Dow Jones Industrial Average closing stock price, daily from July 31 to December 31, 1987.

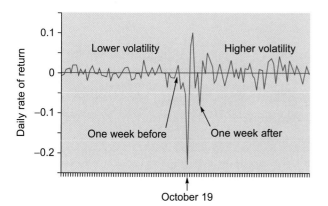

**FIGURE 5.1.10** Daily returns from August 1 to December 31, 1987. Note how the market's volatility was larger after the crash than before.

rose from 1.51% (for the two months just before) to 4.13% (for the two months just after). Similarly, when the "Flash Crash" occurred on May 6, 2010, daily volatility rose from 0.71% (for the two months just before) to 1.53% (for the two months just after).

### Example

*Diversification in the Stock Market*

When you buy stock, you are taking a risk because the price can go up or down as time goes by. One advantage of holding more than just one stock is called diversification. This is the reduction in risk due to the fact that your exposure to possible extreme movements of one of the stocks is limited. Following are measures of risk for three situations: (1) hold Corning stock only, (2) hold JPMorgan Chase stock only, and (3) hold equal amounts of both in a portfolio. Standard deviations of daily rates of return for each case (for the first two quarters of 2010) were as follows:

| Portfolio | Standard Deviation, Daily Return, First Half of 2010 |
|---|---|
| Corning | 2.12% |
| JPMorgan Chase | 2.07% |
| Both together | 1.87% |

Note how the risk is reduced by holding more than one stock (from about 2.1% each day down to about 1.9% per day). If you hold even more stocks, you can reduce the risk even more. The risk of the Dow Jones Industrial Stock Market Index (holding the stock of 30 different companies) during this same time period was even less, at 1.15% per day.

### Example

*Data Mining to Understand Variability in the Donations Database*

Consider the donations database of information on 20,000 people available on the companion site. While it is useful to learn that among those who made a current donation (989 out of 20,000) the average amount was $15.77, it is also important to recognize the variability in the size of the donations. After all, each person who gave did not give exactly $15.77 (although the total amount would have been unchanged if the donors had).

The standard deviation of $11.68 measures the size of the variability or diversity among these donation amounts: A typical donation differed from the average ($15.77) by about one standard deviation ($11.68). While many donation amounts were closer than one standard deviation away from

*(Continued)*

**Example—cont'd**

the average (for example, 101 people donated exactly $15, and 118 people gave $20), there were also donations made that were much more than one standard deviation away from the average (for example, 18 people donated exactly $50, and 6 people gave $100). Figure 5.1.11 shows a histogram of these donation amounts with the average and standard deviation indicated.

## The Sample and the Population Standard Deviations

There are actually two different (but related) kinds of standard deviation: the **sample standard deviation** (for a sample from a larger population, denoted $S$) and the **population standard deviation** (for an entire population, denoted $\sigma$, the lowercase Greek sigma).

Their names suggest their uses. If you have a sample of data selected at random from a larger population, then the sample standard deviation is appropriate. If, on the other hand, you have an entire population, then the population standard deviation should be used. The sample standard deviation is slightly larger in order to adjust for the randomness of sampling.

Some situations are ambiguous. For example, the salaries of all people who work for you might be viewed either as population data (based on the population of all people who work for you) or as sample data (viewing those who work for you, in effect, as a sample from all similar people in the population at large). Some of this ambiguity depends on how you view the situation, rather than on the data themselves. If you view the data as the entire universe of interest, then you are clearly dealing with a population. However, if

you would like to generalize (for example, from your workers to similar workers in similar industries), you may view your data as a sample from a (perhaps hypothetical) population.

To resolve any remaining ambiguity, proceed as follows: *If in doubt, use the sample standard deviation.* Using the larger value is usually the careful, conservative choice since it ensures that you will not be systematically understating the uncertainty.

For computation, the only difference between the two methods is that you subtract 1 (i.e., divide by $n - 1$) for the sample standard deviation, but you do not subtract 1 (i.e., divide by $N$) for the population. This makes the sample standard deviation calculation slightly larger when the sample size is small, reflecting the added uncertainties of having a sample instead of the entire population.[12] There are also some conventional changes in notation: The sample average of the $n$ items is denoted $\overline{X}$, whereas the population mean of the $N$ items is denoted by the Greek letter $\mu$ (mu). The formulas are as follows:

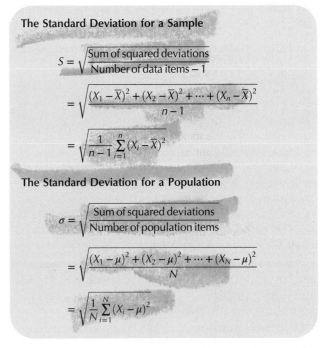

**The Standard Deviation for a Sample**

$$S = \sqrt{\frac{\text{Sum of squared deviations}}{\text{Number of data items} - 1}}$$

$$= \sqrt{\frac{(X_1 - \overline{X})^2 + (X_2 - \overline{X})^2 + \cdots + (X_n - \overline{X})^2}{n - 1}}$$

$$= \sqrt{\frac{1}{n - 1}\sum_{i=1}^{n}(X_i - \overline{X})^2}$$

**The Standard Deviation for a Population**

$$\sigma = \sqrt{\frac{\text{Sum of squared deviations}}{\text{Number of population items}}}$$

$$= \sqrt{\frac{(X_1 - \mu)^2 + (X_2 - \mu)^2 + \cdots + (X_N - \mu)^2}{N}}$$

$$= \sqrt{\frac{1}{N}\sum_{i=1}^{N}(X_i - \mu)^2}$$

The smaller the number of items ($N$ or $n$), the larger the difference between these two formulas. With 10 items, the sample standard deviation is 5.4% larger than the population standard deviation. With 25 items, there is a 2.1% difference, which narrows to 1.0% for 50 items

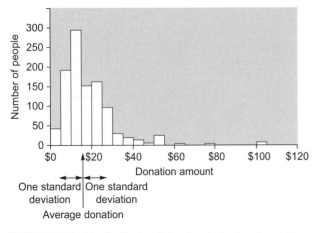

FIGURE 5.1.11 The distribution of donations in the donations database of 20,000 people, showing the amounts for the 989 people who donated something in response to the current mailing. The average donation for this group is $15.77, and the standard deviation is $11.68.

---

12. It is also true that by dividing by $n - 1$ instead of $n$, the sample variance (the square of the standard deviation) is made "unbiased" (i.e., correct for the population, on average). However, the sample standard deviation is still a "biased" estimator of the population standard deviation. Details of sampling from populations will be presented in Chapter 8.

and 0.5% for 100 items. Thus, with reasonably large amounts of data, there is little difference between the two methods.

## 5.2 THE RANGE: QUICK AND SUPERFICIAL

The **range** is the largest minus the smallest data value and represents the size or extent of the data. Here is the range of a small data set representing the number of orders taken recently for each of five product lines.[13]

Range of data set (185, 246, 92, 508, 153)
= Largest – Smallest
= 508 – 92
= 416

| | A | B | C | D | E | F | G | H |
|---|---|---|---|---|---|---|---|---|
| | The Range is the Maximum Less the Minimum.xlsx - Microsoft Excel | | | | | | | |
| 1 | Orders | | Range | | 416 | =MAX(Orders)-MIN(Orders) | | |
| 2 | 185 | | | | | | | |
| 3 | 246 | | Standard deviation | | 161.48 | =STDEV(Orders) | | |
| 4 | 92 | | | | | | | |
| 5 | 508 | | Variance | | 26,076.70 | =VAR(Orders) | | |
| 6 | 153 | | | | | | | |
| 7 | | | | | | | | |

Note that the range is very quickly computed by scanning a list of numbers to pick out the smallest and largest and then subtracting. Way back in the olden days, before we had electronic calculators and computers, ease of computation led many people to use the range as a variability measure. Now that the standard deviation is more easily calculated, the range is not used as often.

When the extremes of the data (i.e., the largest and smallest values) are important, the range is a sensible measure of diversity. This might be the case when you are seeking to describe the extent of the data. This can be useful for two purposes: (1) to *describe* the total extent of the data or (2) to *search for errors* in the data. Since an extreme error made in recording the data will tend to turn up as an especially large (or small) value, the range will immediately seem too large, judging by common sense. This makes the range useful for *editing* the data, that is, for error checking.

On the other hand, because of its sensitivity to the extremes, the range is not very useful as a statistical measure of diversity in the sense of summarizing the data set as a whole. The range does not summarize the typical variability in the data but rather focuses too much attention on just two data values. The standard deviation is more sensitive to all of the data and therefore provides a better look at the big picture. In fact, the range will *always* be larger than the standard deviation.

---

13. For the Excel® formulas to work as shown, you first need to give the name "Orders" to the five numbers. This is done by highlighting the five numbers, then choosing Define Name from the Defined Names group of the Formulas Ribbon, typing the name ("Orders"), and choosing OK.

### Example
#### Employee Salaries

Consider the salaries of employees working in an engineering department of a consumer electronics firm, as shown in Table 5.2.1. We will ignore the ID numbers and concentrate on the salary figures. The highest-paid individual earns $138,000 per year (an engineering management position, head of the department), and the lowest-paid individual earns just $51,000 (a very junior person with a promising future who has not yet completed basic engineering education). The range is $87,000 (= 138,000 – 51,000), representing the dollar amount separating the lowest- and the highest-paid people, as shown in Figure 5.2.1.

Note that the range was computed for the two extremes: those with the least and the most pay. The range does not pretend to indicate the typical variation in salary within the department; the standard deviation would be used to do that.

For a more complete analysis (and to satisfy your curiosity or to help you check your answers), the average salary within the department is $86,750 and the standard deviation (which more reliably indicates the *typical* variability in salary) is $26,634. This is the sample standard deviation, where these engineers are viewed as a sample of typical engineers doing this kind of work.

In summary, $87,000 (the range) separates the lowest and highest amounts. However, $26,634 (the standard deviation) indicates approximately how far individual people are from $86,750 (the average salary for this group).

### Example
#### Duration of Hospital Stays

Hospitals are now being run more like businesses than they were in the past. Part of this is due to the more competitive atmosphere in medical care, with more health maintenance organizations (HMOs), who hire doctors as employees, supplementing traditional hospitals, whose doctors act more independently. Another reason is that the Medicare program currently pays a fixed amount depending on the diagnosis, rather than a flexible amount based on the extent of treatment. This produces a strong incentive to limit, rather than expand, the amount of treatment necessary for a given person's illness.

One measure of the intensity of medical care is the number of days spent in the hospital. Table 5.2.2 shows a list of data representing the number of days spent last year by a sample of patients.[14] The range of this data set is 385, which is 386 – 1 and is impossible since there are only 365 (or 366) days in a year, and this data set is (supposed to be) for one year only. This example illustrates the use of the range in editing a data set as a way of identifying errors before proceeding with the analysis. Examining the smallest and largest values is also useful for this purpose.

*(Continued)*

## 5.3 THE COEFFICIENT OF VARIATION: A *RELATIVE* VARIABILITY MEASURE

**Example—cont'd**

A careful examination of the original records indicated a typing error. The actual value, 286, was mistakenly transcribed as 386. The range for the corrected data set is 285 (that is, 286 – 1).

14. This hypothetical data set is based on the experience and problems of a friend of mine at an economics research center, who spent weeks trying to get the computer to understand a large data tape of health care statistics as part of a study of the efficiency of health care delivery systems. Successful researchers, both in academia and business, often have to overcome many petty problems along the way toward a deeper understanding.

### TABLE 5.2.1  Employee Salaries

| Employee ID Number | Salary | Employee ID Number | Salary |
|---|---|---|---|
| 918886653 | $105,500 | 743594601 | $102,500 |
| 771631111 | 81,000 | 731866668 | 51,000 |
| 148609612 | 84,000 | 490731488 | 138,000 |
| 742149808 | 70,000 | 733401899 | 108,500 |
| 968454888 | 65,000 | 589246387 | 62,000 |

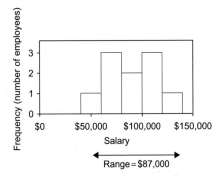

**FIGURE 5.2.1**   The range in salaries is $87,000 for the salary data (from $51,000 to $138,000); it indicates the width of the entire histogram.

### TABLE 5.2.2  Hospital Length of Stay for a Sample of Patients (Patient Days Last Year)

| | | |
|---|---|---|
| 17 | 33 | 5 |
| 16 | 5 | 6 |
| 1 | 1 | 12 |
| 1 | 7 | 16 |
| 7 | 4 | 386 |
| 74 | 13 | 2 |
| 2 | 6 | 7 |
| 163 | 33 | 28 |
| 51 | | |

The **coefficient of variation**, defined as the standard deviation divided by the average, is a relative measure of variability as a percentage or proportion of the average. In general, this is most useful when there are no negative numbers possible in your data set. The formula is as follows:

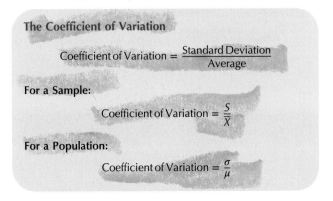

Note that the standard deviation is the numerator, as is appropriate because the result is primarily an indication of variability.

For example, if the average spent by a customer per trip to the supermarket is $35.26 and the standard deviation is $14.08, then the coefficient of variation is 14.08/35.26 = 0.399, or 39.9%. This means that, typically, the amount spent per shopping trip differs by about 39.9% from the average value. In absolute terms, this typical difference is $14.08 (the standard deviation), but it amounts to 39.9% (the coefficient of variation) relative to the average.

The coefficient of variation has *no measurement units*. It is a pure number, a proportion or percentage, whose measurement units have canceled each other in the process of dividing standard deviation by average. This makes the coefficient of variation useful in those situations where you don't care about the actual (absolute) size of the differences, and only the relative size is important.

Using the coefficient of variation, you may reasonably compare a large to a small firm to see which one has more variation on a "size-adjusted basis." Ordinarily, a firm that deals in hundreds of millions of dollars will have its sales vary on a similarly large scale, say, tens of millions. Another firm, with sales in the millions, might have its sales vary by hundreds of thousands. In each case, the variation is about 10% of the size of the average sales. The larger firm has a larger absolute variation (larger standard deviation), but both firms have the same relative or size-adjusted amount of variation (the coefficient of variation).

Note that the coefficient of variation can be larger than 100% even with positive numbers. This could happen with a very skewed distribution or with extreme outliers. It would indicate that the situation is *very* variable with respect to the average value.

### Example

#### Uncertainty in Portfolio Performance

Suppose you have invested $10,000 in 200 shares of XYZ Corporation stock selling for $50 per share. Your friend has purchased 100 shares of XYZ for $5,000. You and your friend expect the per-share price to grow to $60 next year, representing a 20% rate of return, (60 − 50)/50. You both agree that there is considerable risk in XYZ's marketing strategy, represented by a standard deviation of $9 in share price. This says that, although you expect the per-share price to be $60 next year, you wouldn't be surprised if it were approximately $9 larger or smaller than this.

You expect the value of your investment to grow to $12,000 next year ($60 × 200) with a standard deviation of $1,800 ($9 × 200; see Section 5.4 for further details). Your friend's investment is expected to grow to $6,000, with a standard deviation of $900 next year.

It looks as if your risk (standard deviation of $1,800) is double your friend's risk ($900). This makes sense since your investment is twice as large in absolute terms. However, you are both investing in the same security, namely, XYZ stock. Thus, except for the size of your investments, your experiences will be identical. In a relative sense (relative to the size of the initial investment), your risks are identical. This is indeed the case if you compute the coefficient of variation (standard deviation of next year's value divided by its average or expected value). Your coefficient of variation is $1,800/$12,000 = 0.15, which matches your friend's coefficient of variation of $900/$6,000 = 0.15. Both of you agree that the uncertainty (or risk) is about 15% of the expected portfolio value next year.

### Example

#### Employee Productivity in Telemarketing

Consider a telemarketing operation with 19 employees making phone calls to sell symphony tickets. Employees sell 23 tickets per hour, on the average, with a standard deviation of 6 tickets per hour. That is, you should not be at all surprised to hear of an employee selling approximately 6 tickets more or less than the average value (23).

Expressing the employee variability in relative terms using the coefficient of variation, you find that it is 6/23 = 0.261, or 26.1%. This says that the variation of employee sales productivity is about 26.1% of the average sales level.

For the high-level analysis and strategy used by top management, the figure of 26.1% (coefficient of variation) may well be more useful than the figure of 6 tickets per hour (standard deviation). Top management can look separately at the level of productivity (23 tickets per employee per hour) and the variation in productivity (employees may typically be 26.1% above or below the average level).

The coefficient of variation is especially useful in making comparisons between situations of different sizes. Consider another telemarketing operation selling theater tickets, with an average of 35 tickets per hour and a standard deviation of 7. Since the theater ticket productivity is higher overall than the symphony ticket productivity (35 compared to 23,

on average), you should not be surprised to see more variation (7 compared to 6). However, the coefficient of variation for the theater operation is 7/35 = 0.200, or 20.0%. Compared to the 26.1% figure for the symphony, management can quickly conclude that the theater marketing group is actually more homogeneous, relatively speaking, than the symphony group.

## 5.4 EFFECTS OF ADDING TO OR RESCALING THE DATA

When a situation is changed in a systematic way, there is no need to recompute summaries such as typical value (average, median, mode), percentiles, or variability measures (standard deviation, range, coefficient of variation). A few basic rules show how to quickly find the summaries for a new situation.

If a fixed number is *added* to each data value, then this same number is added to the average, median, mode, and percentiles to obtain the corresponding summaries for the new data set. For example, adding a new access fee of $5 to accounts formerly worth $38, $93, $25, and $89 says that the accounts are now worth $43, $98, $30, and $94. The average value per account has jumped exactly $5, from $61.25 to $66.25. Rather than recompute the average for the new account values, you can simply add $5 to the old average. This rule applies to other measures; for example, the median rises by $5, from $63.50 to $68.50. However, the standard deviation and range are unchanged, since the data values are shifted but maintain the same distance from each other. The coefficient of variation does change, and may be easily computed from the standard deviation and the new average.

If each data value is *multiplied by* a fixed number, the average, median, mode, percentiles, standard deviation, and range are each multiplied by this same number to obtain the corresponding summaries for the new data set. The coefficient of variation is unaffected.[15]

These two effects act in combination if the data values are multiplied by a factor $c$ and an amount $d$ is then added; $X$ becomes $cX + d$. The new average is $c \times$ (Old average) $+ d$; likewise for the median, mode, and percentiles. The new standard deviation is $|c| \times$ (Old standard deviation), and the range is adjusted similarly (note that the added number, $d$, plays no role here).[16] The new coefficient of variation is easily computed from the new average and standard deviation.

---

15. This assumes that the fixed number is positive. If it is negative, then make it positive before multiplying for the standard deviation and the range.

16. Note that the standard deviation is multiplied by the *absolute value* of this factor so that it remains a positive number. For example, if $c = -3$, the standard deviation would be multiplied by 3.

**TABLE 5.4.1 Effects of Adding to or Rescaling the Data**

|  | Original Data | Add $d$ | Multiply by $c$ | Multiply Then Add |
|---|---|---|---|---|
| Data | $X$ | $X + d$ | $cX$ | $cX + d$ |
| Average (similarly for median, mode, and percentiles) | $\overline{X}$ | $\overline{X} + d$ | $c\overline{X}$ | $c\overline{X} + d$ |
| Standard deviation (similarly for range) | $S$ | $S$ | $|c|S$ | $|c|S$ |

Table 5.4.1 is a summary chart of these rules. The new coefficient of variation may be easily computed from the new standard deviation and average.

### Example
#### Uncertainty of Costs in Japanese Yen and in U.S. Dollars

Your firm's overseas production division has projected its costs for next year as follows:

| Expected costs | 325,700,000 Japanese yen |
|---|---|
| Standard deviation | 50,000,000 Japanese yen |

To complete your budget, you will need to convert these into U.S. dollars. For simplicity, we will consider only business risk (represented by the standard deviation). A more comprehensive analysis would also consider exchange rate risk, which would allow for the risk of movements in the currency conversion factor.

Japanese yen are easily converted to U.S. dollars using the current exchange rate.[17] To convert yen to dollars, you multiply by 0.0114, which represents the number of dollars per yen. Multiplying both the average (expected) amount and the standard deviation by this conversion factor, you find the expected amount and risk in dollars (rounded to the nearest thousand):

| Expected costs | $3,725,266 |
|---|---|
| Standard deviation | $571,886 |

By using the basic rules, you are able to convert the summaries from yen to dollars without going through the entire budgeting process all over again in dollars instead of yen!

---

17. The exchange rate may be found, for example, by starting at http://www.google.com/finance, choosing "USD/JPY" under Currencies at the right, and then selecting " View JPY in USD" at the middle near the top. This value was accessed on July 7, 2010.

### Example
#### Total Cost and Units Produced

In cost accounting and in finance, a production facility often views costs as either *fixed* or *variable*. The fixed costs will apply regardless of how many units are produced, whereas the variable costs are charged on a per-unit basis. Fixed costs might represent rent and investment in production machinery, and variable costs might represent the cost of the production materials actually used.

Consider a shampoo manufacturing facility, with fixed costs of $1,000,000 per month and variable costs of $0.50 per bottle of shampoo produced. Based on a careful analysis of market demand, management has forecast next month's production at 1,200,000 bottles. Based on past experience with these forecasts, the firm recognizes an uncertainty of about 250,000 bottles. Thus, you expect to produce an average of 1,200,000 bottles of shampoo, with a standard deviation of 250,000 bottles.

Given this forecast of units to be produced, how do you forecast costs? Note that units are converted to costs by multiplying by $0.50 (the variable cost) and then adding $1,000,000 (the fixed cost). That is,

$$\text{Total cost} = \$0.50 \times \text{Units produced} + \$1,000,000$$

Using the appropriate rule, you find that the average (expected) cost and the standard deviation are

$$\text{Average cost} = \$0.50 \times 1,200,000 + \$1,000,000$$
$$= \$1,600,000$$
$$\text{Standard deviation of cost} = \$0.50 \times 250,000$$
$$= \$125,000$$

Your budget of costs is complete. You expect $1,600,000, with a standard deviation (uncertainty) of $125,000.

The coefficient of variation of units produced is 250,000/1,200,000 = 20.8%. The coefficient of variation for costs is quickly computed as $125,000/$1,600,000 = 7.8%. Note that the relative variation in costs is much smaller due to the fact that the large fixed costs make the absolute level of variation seem smaller when compared to the larger cost base.

## 5.5 END-OF-CHAPTER MATERIALS

### Summary

**Variability** (also called **diversity, uncertainty, dispersion**, and **spread**) is the extent to which data values differ from one another. Although measures of center (such as the average, median, or mode) indicate the typical *size* of the data values, a measure of variability will indicate *how close* to this central size measure the data values typically are. If all data values are identical, then the variability is zero. The more spread out things are, the larger the variability.

The **standard deviation** is the traditional choice for measuring variability, summarizing the typical distance from the average to a data value. The standard deviation indicates the extent of randomness of individuals about their common average. The **deviations** are the distances from each data value to the average. Positive deviations represent above-average individuals, and negative deviations indicate below-average individuals. The average of these deviations is always zero. The standard deviation indicates the typical size of these deviations (ignoring the minus signs) and is a number in the same unit of measurement as the original data (such as dollars, miles per gallon, or kilograms).

To find the sample standard deviation:

1. Find the deviations by subtracting the average from each data value.
2. Square these deviations, add them up, and divide the resulting sum by $n-1$. This is the *variance*.
3. Take the square root. You now have the standard deviation.

When you have data for the entire population, you may use the **population standard deviation** (denoted by $\sigma$). Whenever you wish to generalize beyond the immediate data set to some larger population (either real or hypothetical), be sure to use the **sample standard deviation** (denoted by $S$). When in doubt, use the sample standard deviation. Here are their formulas:

$$S = \sqrt{\frac{\text{Sum of squared deviations}}{\text{Number of data items} - 1}}$$

$$= \sqrt{\frac{(X_1 - \overline{X})^2 + (X_2 - \overline{X})^2 + \cdots + (X_n - \overline{X})^2}{n-1}}$$

$$= \sqrt{\frac{1}{n-1}\sum_{i=1}^{n}(X_i - \overline{X})^2}$$

$$\sigma = \sqrt{\frac{\text{Sum of squared deviations}}{\text{Number of population items}}}$$

$$= \sqrt{\frac{(X_1 - \mu)^2 + (X_2 - \mu)^2 + \cdots + (X_N - \mu)^2}{N}}$$

$$= \sqrt{\frac{1}{N}\sum_{i=1}^{N}(X_i - \mu)^2}$$

Both formulas add up the squared deviations, divide, and then take the square root to undo the initial squaring of the individual deviations. For the sample standard deviation, you divide by $n-1$ because the deviations are computed using the somewhat uncertain sample average instead of the exact population mean.

The **variance** is the square of the standard deviation. It provides the same information as the standard deviation but is more difficult to interpret since its unit of measurement is the square of the original data units (such as dollars squared, squared miles per squared gallon, or squared kilograms, whatever these things are). We therefore prefer the standard deviation for reporting variability measurement.

*If your data follow a normal distribution,* the standard deviation represents the approximate width of the middle two-thirds of your cases. That is, about two-thirds of your data values will be within one standard deviation from the average (either above or below). Approximately 95% will be within two standard deviations from the average, and about 99.7% will be within three standard deviations from average. However, don't expect these rules to hold for other (nonnormal) distributions.

The **range** is equal to the largest data value minus the smallest data value and represents the size or extent of the entire data set. The range may be used to describe the data and to help identify problems. However, the range is not very useful as a statistical measure of variability because it concentrates too much attention on the extremes rather than on the more typical data values. For most statistical purposes, the standard deviation is a better measure of variability.

The **coefficient of variation** is the standard deviation divided by the average and summarizes the *relative variability* in the data as a percentage of the average. The coefficient of variation has no measurement units and thus may be useful in comparing the variability of different situations on a size-adjusted basis.

When a fixed number is added to each data value, the average, median, percentiles, and mode all increase by this same amount; the standard deviation and range remain unchanged. When each data value is multiplied by a fixed number, the average, median, percentiles, mode, standard deviation, and range all change by this same factor, and the coefficient of variation remains unchanged.[18] When each data value is multiplied by a fixed number and then another fixed number is added, these two effects act in combination. The coefficient of variation can easily be computed after using these rules to find the average and standard deviation.

---

18. The standard deviation and range are multiplied by the *absolute value* of this fixed number so that they remain positive numbers.

## Key Words

### Questions

1. What is variability?
2. a. What is the traditional measure of variability?
   b. What other measures are also used?
3. a. What is a deviation from the average?
   b. What is the average of all of the deviations?
4. a. What is the standard deviation?
   b. What does the standard deviation tell you about the relationship between individual data values and the average?
   c. What are the measurement units of the standard deviation?
   d. What is the difference between the sample standard deviation and the population standard deviation?
5. a. What is the variance?
   b. What are the measurement units of the variance?
   c. Which is the more easily interpreted variability measure, the standard deviation or the variance? Why?
   d. Once you know the standard deviation, does the variance provide any additional real information about the variability?
6. If your data set is normally distributed, what proportion of the individuals do you expect to find:
   a. Within one standard deviation from the average?
   b. Within two standard deviations from the average?
   c. Within three standard deviations from the average?
   d. More than one standard deviation from the average?
   e. More than one standard deviation *above* the average? (Be careful!)
7. How would your answers to question 6 change if the data were not normally distributed?
8. a. What is the range?
   b. What are the measurement units of the range?
   c. For what purposes is the range useful?
   d. Is the range a very useful statistical measure of variability? Why or why not?
9. a. What is the coefficient of variation?
   b. What are the measurement units of the coefficient of variation?

10. Which variability measure is most useful for comparing variability in two different situations, adjusting for the fact that the situations have very different average sizes? Justify your choice.
11. When a fixed number is added to each data value, what happens to
    a. The average, median, and mode?
    b. The standard deviation and range?
    c. The coefficient of variation?
12. When each data value is multiplied by a fixed number, what happens to
    a. The average, median, and mode?
    b. The standard deviation and range?
    c. The coefficient of variation?

### Problems

*Problems marked with an asterisk (*) are solved in the Self Test in Appendix C.*

1.* Planning to start an advertising agency? Table 5.5.1 reports the size of account budgets for selected firms in the top 100 for spending on advertising.
   a. Find the average budget size.
   b. Find the standard deviation of budget sizes, viewing these firms as a sample of companies with large advertising accounts. What are the units of measurement?
   c. Briefly summarize the interpretation of the standard deviation (from part b) in terms of the differences among these firms.
   d. Find the range. What are the units of measurement?
   e. Briefly summarize the interpretation of the range (from part d) in terms of the differences among these firms.

**TABLE 5.5.1 Advertising Budgets**

| Firm | Budget ($ millions) |
| --- | --- |
| Estee Lauder Cos. | $707.1 |
| Sanofi-Aventis | 487.1 |
| Macy's | 1,087.0 |
| Nissan Motor Co. | 690.9 |
| Honda Motor Co. | 935.9 |
| AstraZeneca | 760.4 |
| Mattel | 356.6 |
| Target | 1,167.0 |
| Campbell Soup Co. | 699.3 |
| Nike | 592.8 |

**Source:** Advertising Age 2009 U.S. ad spending accessed at http://adage.com/marketertrees2010/ on July 7, 2010.

f. Find the coefficient of variation. What are the units of measurement?

g. Briefly summarize the interpretation of the coefficient of variation (from part f) in terms of the differences among these firms.

h. Find the variance. What are the units of measurement?

i. Briefly summarize the interpretation of the variance (from part h) or indicate why there is no simple interpretation.

j. Draw a histogram of this data set. Indicate the average, standard deviation, and range on your graph.

2. Consider the annualized stock return over the decade from 2000 to 2010, July to July, expressed as an annual interest rate in percentage points per year, for major pharmaceutical companies as shown in Table 5.5.2. For these top firms in this industry group, there was definitely a variety of experiences. You may view these firms as a sample, indicating the performance of large pharmaceutical firms in general during this time period.

a. Find the standard deviation of the stock return. What are the units of measurement?

b. Briefly summarize the interpretation of the standard deviation (from part a) in terms of the differences among these firms.

c. Find the range. What are the units of measurement?

d. Briefly summarize the interpretation of the range (from part c) in terms of the differences among these firms.

e. Find the coefficient of variation. What are the units of measurement?

f. Briefly summarize the interpretation of the coefficient of variation (from part e) in terms of the differences among these firms.

g. Find the variance. What are the units of measurement?

h. Briefly summarize the interpretation of the variance (from part g) or indicate why there is no simple interpretation.

i. Draw a histogram of this data set. Indicate the average, standard deviation, and range on your graph.

3. Consider the advertising budgets from problem 1, but expressed in European euros instead of U.S. dollars. Use the exchange rate listed in a recent issue of the *Wall Street Journal* or another source. Based on your answers to problem 1 (that is, without recalculating from the data), compute

a. The average budget in euros.

b. The standard deviation.

c. The range.

d. The coefficient of variation.

4. Mutual funds that specialize in the stock of natural resources companies showed considerable variation in their performance during the 12-month period ending July 2010. Consider the Rate of Return column in Table 5.5.3.

### TABLE 5.5.2 Performance of Pharmaceutical Firms

| Firm | Annualized Stock Return |
|------|------------------------|
| Abbott | 4.54% |
| Amgen | −2.31% |
| Bristol-Myers | −2.12% |
| Eli Lilly | −7.69% |
| Genzyme | 4.15% |
| Gilead Sciences | 22.40% |
| J&J | 4.93% |
| Merck | −2.86% |
| Mylan | 6.89% |
| Pfizer | −7.42% |

**Source:** These are the top 10 pharmaceutical firms in the Fortune 500, accessed at http://money.cnn.com/magazines/fortune/fortune500/2010/industries/21/index.html on July 8, 2010. Annualized stock return was computed from stock prices accessed at Yahoo.com on July 8, 2010, for the decade from July 1, 2000, to July 1, 2010.

### TABLE 5.5.3 Natural Resources Mutual Funds: Rates of Return (12 months ending July 2010) and Assets (as of July 2010)

| Fund | Rate of Return | Assets (millions) |
|------|----------------|-------------------|
| Columbia:Eng&Nat Rs;A | 1.35% | 45.5 |
| Columbia:Eng&Nat Rs;C | 5.83% | 14.6 |
| Columbia:Eng&Nat Rs;Z | 7.90% | 573.9 |
| Dreyfus Natural Res;A | −0.97% | 16.4 |
| Dreyfus Natural Res;B | 0.20% | 1.4 |
| Dreyfus Natural Res;C | 3.28% | 4.3 |
| Dreyfus Natural Res;I | 5.39% | 1.8 |
| Fidelity Adv Energy;A | −4.01% | 202.2 |
| Fidelity Adv Energy;B | −2.93% | 41.8 |
| Fidelity Adv Energy;C | 1.10% | 84.6 |
| Fidelity Adv Energy;I | 2.17% | 23.8 |
| Fidelity Adv Energy;T | −1.91% | 205.0 |
| Fidelity Sel Energy | 2.08% | 1,746.5 |
| Fidelity Sel Nat Gas | −2.10% | 854.9 |

(Continued)

**TABLE 5.5.3** Natural Resources Mutual Funds: Rates of Return (12 months ending July 2010) and Assets (as of July 2010)—cont'd

| Fund | Rate of Return | Assets (millions) |
|---|---|---|
| ICON:Energy | –0.84% | 484.3 |
| Invesco Energy;A | –5.56% | 597.8 |
| Invesco Energy;B | –5.75% | 84.8 |
| Invesco Energy;C | –1.77% | 167.9 |
| Invesco Energy;Inst | 0.35% | 6.9 |
| Invesco Energy;Inv | –0.06% | 384.1 |
| Invesco Energy;Y | 0.20% | 39.3 |
| Ivy:Energy;A | –1.45% | 56.8 |
| Ivy:Energy;B | –1.39% | 2.8 |
| Ivy:Energy;C | 2.81% | 12.0 |
| Ivy:Energy;E | –1.27% | 0.1 |
| Ivy:Energy;I | 4.95% | 2.9 |
| Ivy:Energy;Y | 4.65% | 5.7 |
| Rydex:Energy Fund;A | –1.53% | 3.0 |
| Rydex:Energy Fund;Adv | 3.14% | 5.2 |
| Rydex:Energy Fund;C | 1.65% | 11.2 |
| Rydex:Energy Fund;Inv | 3.68% | 31.7 |
| Rydex:Energy Svcs;A | –3.00% | 7.3 |
| Rydex:Energy Svcs;Adv | 1.60% | 5.0 |
| Rydex:Energy Svcs;C | 0.08% | 8.5 |
| Rydex:Energy Svcs;Inv | 2.10% | 27.3 |
| SAM Sust Water;Inst | 19.91% | 4.0 |
| Saratoga:Energy&BM;A | 5.01% | 2.0 |
| Saratoga:Energy&BM;B | 6.80% | 0.1 |
| Saratoga:Energy&BM;C | 9.88% | 0.2 |
| Saratoga:Energy&BM;I | 11.96% | 3.0 |
| Vanguard Energy Ix;Adm | 3.00% | 129.6 |
| W&R Adv:Energy;A | –0.78% | 178.0 |
| W&R Adv:Energy;B | –0.94% | 4.6 |
| W&R Adv:Energy;C | 3.34% | 4.8 |
| W&R Adv:Energy;Y | 5.83% | 1.7 |
| Ivy:Energy;C | 2.81% | 12.0 |

**Source:** Data are from the *Wall Street Journal* Mutual Fund Screener accessed at http://online.wsj.com/public/quotes/mutualfund_screener.html on July 8, 2010. Their source is Lipper, Inc.

a. Find the average rate of return for these funds.

b. Find the standard deviation and briefly interpret this value.

c. How many of these mutual funds are within one standard deviation from the average? How does this compare to what you expect for a normal distribution?

d. How many of these mutual funds are within two standard deviations from the average? How does this compare to what you expect for a normal distribution?

e. How many of these mutual funds are within three standard deviations from the average? How does this compare to what you expect for a normal distribution?

f. Draw a histogram of this data set and indicate limits of one, two, and three standard deviations on the graph. Interpret your answers to parts c, d, and e in light of the shape of the distribution.

5. Consider the assets of stock mutual funds, as shown in Table 5.5.3. Answer the parts of the previous problem using the column of assets instead of the rates of return.

6.* Consider the number of executives for all Seattle corporations with 500 or more employees:[19]

   12, 15, 5, 16, 7, 18, 15, 12, 4, 3, 22, 4, 12, 4, 6, 8, 4, 5, 6, 4, 22, 10, 11, 4, 7, 6, 10, 10, 7, 8, 26, 9, 11, 41, 4, 16, 10, 11, 12, 8, 5, 9, 18, 6, 5

a. Find the average number of executives per firm.

b. Find the (sample) standard deviation, and briefly interpret this value.

c. How many corporations are within one standard deviation from the average? How does this compare to what you expect for a normal distribution?

d. How many corporations are within two standard deviations from the average? How does this compare to what you expect for a normal distribution?

e. How many corporations are within three standard deviations from the average? How does this compare to what you expect for a normal distribution?

f. Draw a histogram of this data set and indicate limits of one, two, and three standard deviations on the graph. Interpret your answers to parts c, d, and e in light of the shape of the distribution.

7. Repeat problem 6 with the extreme outlier omitted, and write a paragraph comparing the results with and without the outlier.

8. All 18 people in a department have just received across-the-board pay raises of 3%. What has happened to

a. The average salary for the department?

b. The standard deviation of salaries?

c. The range in salaries?

d. The coefficient of variation of salaries?

9. Based on a demand analysis forecast, a factory plans to produce 80,000 video game cartridges this quarter, on average, with an estimated uncertainty of 25,000 cartridges as the standard deviation. The fixed costs for this equipment are $72,000 per quarter, and the variable cost is $1.43 per cartridge produced.

a. What is the forecast expected total cost of the cartridges produced?

b. What is the uncertainty involved in this forecast of total cost, expressed as a standard deviation?

c. Find the coefficient of variation for the number of cartridges produced and for the total cost. Write a paragraph interpreting and comparing these coefficients of variation.

d. After the quarter is over, you find that the factory actually produced 100,000 cartridges. How many standard deviations above or below the average is this figure?

e. Suppose the firm actually produces 200,000 cartridges. How many standard deviations above or below the average is this figure? Would this be a surprise in light of the earlier forecast? Why or why not?

10. Consider the number of gifts (lifetime gifts, previous to this mailing) given by the 20,000 people represented in the donations database (on the companion site).

a. Find the average and standard deviation.

b. Draw a histogram for this data set. Indicate the average and standard deviation on this graph.

11. Let's compare the distribution of the number of lifetime gifts of those who made a current donation to that of those who did not. We will use two data sets, each indicating the number of gifts (lifetime gifts, previous to this mailing) given by the 20,000 people represented in the donations database (on the companion site). One data set is for those who did not make a donation in response to this mailing (named "gifts_D0" with 19,011 people), while the other is for those who did (named "gifts_D1" with 989 people).

a. Find the average and standard deviation for each data set. Compare and interpret these values.

b. Compare these averages and standard deviations to the average and standard deviation of the lifetime gifts for the full database of 20,000 people.

12. Your firm's total advertising budget has been set for the year. You (as marketing manager) expect to spend about $1,500,000 on TV commercials, with an uncertainty of $200,000 as the standard deviation. Your advertising agency collects a fee of 15% of this budget. Find the expected size of your agency's fee and its level of uncertainty.

13. You have been trying to control the weight of a chocolate and peanut butter candy bar by intervening in the production process. Table 5.5.4 shows the weights of two representative samples of candy bars from the day's production, one taken before and the other taken after your intervention.

a. Find the average weight of the candy bars before intervention.

b. Find the standard deviation of the weights before intervention.

c. Find the average weight of the candy bars after intervention.

d. Find the standard deviation of the weights after intervention.

**TABLE 5.5.4 Weight (in Ounces) for Two Samples of Candy Bars**

| Before Intervention | Before Intervention | After Intervention | After Intervention |
|---|---|---|---|
| 1.62 | 1.68 | 1.60 | 1.69 |
| 1.71 | 1.66 | 1.71 | 1.59 |
| 1.63 | 1.64 | 1.65 | 1.66 |
| 1.62 | 1.70 | 1.64 | 1.68 |
| 1.63 | 1.66 | 1.63 | 1.59 |
| 1.69 | 1.71 | 1.65 | 1.57 |
| 1.64 | 1.63 | 1.74 | 1.62 |
| 1.63 | 1.65 | 1.75 | 1.75 |
| 1.62 | 1.70 | 1.66 | 1.72 |
| 1.70 | 1.64 | 1.73 | 1.63 |

e. Compare the standard deviations before and after your intervention, and write a paragraph summarizing what you find. In particular, have you been successful in reducing the variability in this production process?

14. We've all been stopped by traffic at times and have had to sit there while freeway traffic has slowed to a crawl. If you have someone with you (or some good music), the experience may be easier to put up with, but what does traffic congestion cost society? Consider the data presented in Table 5.5.5, grouping all data together as a single univariate data set.

a. Summarize the congestion costs for all these cities by finding the average.

b. Summarize the variation in congestion costs from one city to another using the standard deviation. You may view this data set as a population consisting of all available information.

c. Draw a histogram of this data set. Indicate the average and standard deviation on this graph.

d. Write a paragraph summarizing what you have learned from examining and summarizing this data set.

15. Consider the variability in traffic congestion in Table 5.5.5 for northeastern and for southwestern cities.

a. Compare population variability of these two groups of cities. In particular, which group shows more variability in congestion from city to city?

b. Compare the relative variability of these two groups of cities. In particular, which is more similar from group to group: the (ordinary) variability or the relative variability?

16. Summarize the variability in admission prices for the theme parks shown in Table 5.5.6 by reporting the standard deviation, the range, and the coefficient of variation.

**TABLE 5.5.5** Cost Due to Traffic Congestion, per Registered Vehicle

| Northeastern Cities | Congestion Cost |
|---|---|
| Baltimore, MD | $550 |
| Boston, MA | 475 |
| Hartford, CT | 227 |
| New York, NY | 449 |
| Philadelphia, PA | 436 |
| Pittsburgh, PA | 168 |
| Washington, DC | 638 |

| Midwestern Cities | Congestion Cost |
|---|---|
| Chicago, IL | 498 |
| Cincinnati, OH | 304 |
| Cleveland, OH | 134 |
| Columbus, OH | 346 |
| Detroit, MI | 610 |
| Indianapolis, IN | 488 |
| Kansas City, MO | 175 |
| Louisville, KY | 447 |
| Milwaukee, WI | 210 |
| Minneapolis-St. Paul, MN | 455 |
| Oklahoma City, OK | 294 |
| St. Louis, MO | 315 |

| Southern Cities | Congestion Cost |
|---|---|
| Atlanta, GA | 671 |
| Charlotte, NC | 491 |
| Jacksonville, FL | 439 |
| Memphis, TN | 301 |
| Miami, FL | 545 |
| Nashville, TN | 428 |
| New Orleans, LA | 222 |
| Orlando, FL | 605 |
| Tampa, FL | 519 |

| Southwestern Cities | Congestion Cost |
|---|---|
| Albuquerque, NM | 416 |
| Austin, TX | 455 |
| Corpus Christi, TX | 99 |
| Dallas, TX | 641 |
| Denver, CO | 569 |
| El Paso, TX | 210 |
| Houston, TX | 651 |
| Phoenix, AZ | 552 |
| Salt Lake City, UT | 294 |
| San Antonio, TX | 428 |

| Western Cities | Congestion Cost |
|---|---|
| Honolulu, HI | 283 |
| Los Angeles, CA | 807 |
| Portland, OR | 395 |
| Sacramento, CA | 433 |
| San Diego, CA | 605 |
| San Francisco-Oakland, CA | 597 |
| San Jose, CA | 594 |
| Seattle, WA | 513 |

**Source:** U.S. Census Bureau, *Statistical Abstract of the United States: 2010* (129th edition), Washington, DC, 2009, accessed at http://www.census.gov/compendia/statab/cats/transportation.html on July 8, 2010. Their source is Texas Transportation Institute, College Station, Texas, 2009 Urban Mobility Study.

**TABLE 5.5.6** Theme Park Admission Prices

| Theme Park | Admission Price |
|---|---|
| Adventuredome NV | 25 |
| Luna Park NY | 34 |
| Beach Boardwalk CA | 30 |
| Busch Gardens FL | 65 |
| Cedar Point OH | 46 |
| Disney World FL | 46 |
| Disneyland CA | 97 |
| Dollywood TN | 56 |
| Kings Dominion VA | 47 |
| Knott's Berry Farm CA | 38 |
| Legoland CA | 67 |
| Six Flags Great Adventure NJ | 55 |
| Six Flags Over Georgia GA | 45 |
| Universal Studios Orlando FL | 109 |

**Source:** Theme park websites, accessed on July 8, 2010. Prices are for one adult.

17. Here are rates of return for a sample of recent on-site service contracts:

    78.9%, 22.5%, –5.2%, 997.3%, –20.7%, –13.5%, 429.7%, 88.4%, –52.1%, 960.1%, –38.8%, –70.9%, –73.3%, 47.0%, –1.5%, 23.9%, –35.6%, –62.0%, –75.7%, –14.0%, –81.2%, 46.9%, 135.1%, –34.6%, –85.3%, –73.6%, –9.0%, 19.6%, –86.7%, –87.6%, –88.7%, –75.5%, –91.0%, –97.9%, –100.0%

    a. Find the average and standard deviation of these rates of return.
    b. Write a paragraph discussing the level of risk (a cost) and average return (a benefit) involved in this area of business.

18. Consider interest rates on accounts at a sample of local banks:

    3.00%, 4.50%, 4.90%, 3.50%, 4.75%, 3.50%, 3.50%, 4.25%, 3.75%, 4.00%

    a. Find the standard deviation of these interest rates.
    b. What does this standard deviation tell you about banks in this area?

19. Consider the percentage change in the value of the dollar with respect to other currencies over a four-week period (Table 5.5.7).

    a. Find the standard deviation of these percentages.
    b. Interpret this standard deviation. In particular, what does it measure about the foreign exchange markets?

20. Here are weights of recently produced sinks:

    20.8, 20.9, 19.5, 20.8, 20.0, 19.8, 20.1, 20.5, 19.8, 20.3, 20.0, 19.7, 20.3, 19.5, 20.2, 20.2, 19.5, 20.5

    Find the usual summary measure that tells approximately how far from average these weights are.

21. Consider the price of a hotel room in 20 U.S. cities (Table 5.5.8).

    a. Find the average price of a major-city hotel room in the United States, based on this data set.
    b. Find the sample standard deviation of these prices.
    c. What does the standard deviation tell you about hotel prices in major cities in the United States?

### TABLE 5.5.7 Changing Value of the Dollar

| Country | % Change | Country | % Change |
|---|---|---|---|
| Belgium | –5.3% | Singapore | –1.5% |
| Japan | –6.7 | France | –4.9 |
| Brazil | 26.0 | South Korea | –1.0 |
| Mexico | –1.2 | Hong Kong | 0.0 |
| Britain | –3.7 | Taiwan | –0.1 |
| Netherlands | –5.1 | Italy | –4.7 |
| Canada | –1.9 | West Germany | –5.1 |

### TABLE 5.5.8 Hotel Room Prices

| City | Hotel Price |
|---|---|
| Atlanta | $176 |
| Boston | 244 |
| Chicago | 251 |
| Cleveland | 127 |
| Dallas | 152 |
| Denver | 145 |
| Detroit | 158 |
| Houston | 179 |
| Los Angeles | 174 |
| Miami | 136 |
| Minneapolis | 155 |
| New Orleans | 169 |
| New York | 296 |
| Orlando | 129 |
| Phoenix | 191 |
| Pittsburgh | 137 |
| St. Louis | 141 |
| San Francisco | 227 |
| Seattle | 174 |
| Washington, DC | 224 |

Source: Data are from the *Wall Street Journal*, November 23, 2001, p. W11C.

22. Consider the dollar value (in thousands) of gifts returned to each of your department stores after the holiday season (Table 5.5.9).

    a. Compute the sample standard deviation.
    b. Interpret the standard deviation in a paragraph discussing the variation from one store to another.

23. Airline ticket prices are generally optimized for the airline's benefit, not for the consumer. On July 8, 2010, at the online travel agency http://www.expedia.com, the following airfares were proposed for roundtrip travel from Seattle to Boston leaving one week later: AirTran: $721; Alaska: $787; American: $1,319; Continental: $729; Delta: $520; Frontier: $661; Jet Blue: $657; U.S. Airways: $676; and United: $510.

    a. Compute the standard deviation, viewing these airfares as a sample of fares that might be obtained under similar circumstances.
    b. Write a paragraph interpreting the standard deviation and discussing variation in airline fares.

**TABLE 5.5.9 Gifts Returned ($ thousands)**

| Store | Returned |
|-------|----------|
| A | 13 |
| B | 8 |
| C | 36 |
| D | 18 |
| E | 6 |
| F | 21 |

24. Consider the following productivity measures for a population of employees:

    85.7, 78.1, 69.1, 73.3, 86.8, 72.4,
    67.5, 76.8, 80.2, 70.0

    a. Find the average productivity.
    b. Find and interpret the standard deviation of productivity.
    c. Find and interpret the coefficient of variation of productivity.
    d. Find and interpret the range of productivity.

25. Here are first-year sales (in thousands) for some recent new product introductions that are similar to one you are considering.

    10, 12, 16, 47, 39, 22, 10, 29

    a. Find the average and standard deviation. Interpret the standard deviation.
    b. After you went ahead with the product introduction you were considering, it turned out to have first-year sales of 38 (thousand). How far from average is this? Compare this answer to the standard deviation.
    c. The next new product introduction rang up 92 (thousand) the first year. How many standard deviations (from part a) is this from the average (also from part a)?
    d. For the new product introductions of parts b and c, say whether each is typical for your firm and indicate how you know.

26. Samples from the mine show the following percentages of gold:

    1.1, 0.3, 1.5, 0.4, 0.8, 2.2, 0.7, 1.4, 0.2, 4.5, 0.2, 0.8

    a. Compute and interpret the sample standard deviation.
    b. Compute and interpret the coefficient of variation.
    c. Which data value has the largest positive deviation? Why is this location special to you?
    d. How many standard deviations above the mean is the data value with the largest positive deviation?

27. Consider the return on equity, expressed like an interest rate in percentage points per year, for a sample of companies:

    5.5, 10.6, 19.0, 24.5, 6.6, 26.8, 6.2, −2.4, −28.3, 2.3

    a. Find the average and standard deviation of the return on equity.
    b. Interpret the standard deviation.

    c. How many standard deviations below average is the worst performance?
    d. Draw a histogram and indicate the average, the standard deviation, and the deviation of the worst performance.

28. Your costs had been forecast as having an average of $138,000 with a standard deviation of $35,000. You have just learned that your suppliers are raising prices by 4% across the board. Now what are the average and standard deviation of your costs?

29. For the previous problem, compare the coefficient of variation before and after the price increase. Why does it change (or not change) in this way?

30. You are sales manager for a regional division of a beverage company. The sales goals for your representatives have an average of $768,000 with a standard deviation of $240,000. You have been instructed to raise the sales goal of each representative by $85,000. What happens to the standard deviation?

31. For the preceding problem, compare the coefficient of variation before and after the sales goal adjustment. Why does it change (or not change) in this way?

32. Find the standard deviation of the VAT taxes from Table 4.3.2 in Chapter 4. What does this tell you about international taxation practices from one country to another?

33. Consider the running time of movies from Table 4.3.10.
    a. Find the standard deviation. What does this tell you about these movie times?
    b. Find the range. What does this tell you about these movie times?
    c. How many standard deviations from the average is the longest movie?

34. Different countries have different taxation strategies: Some tax income more heavily than others, while others concentrate on goods and services taxes. Consider the relative size of goods and services taxes for selected countries' international tax rates, as presented in Table 5.5.10. This relative size is measured in two ways: total goods and services taxes as a percentage of overall economic activity (gross domestic product) and as a percentage of all taxes collected by that country.
    a. Find the standard deviation of each variable.
    b. Find the range of each variable.
    c. Find the coefficient of variation of each variable.
    d. Compare each of these variability measures. Which variability measure is the most similar from one variable to the other? Why?

35. For the goods and services tax data of Table 5.5.10:
    a. Draw a box plot for each variable using the same scale.
    b. Based on these box plots, comment on the distribution of each variable.

36. For the municipal bond yields of Table 3.9.1 in Chapter 3:
    a. Find the standard deviation of the yield.
    b. Find the range.
    c. Find the coefficient of variation.
    d. Use these summaries to describe the extent of variability among these yields.

**TABLE 5.5.10 International Taxation: Goods and Services Taxes as a Percent of Gross Domestic Product (GDP) and of Taxes**

| Country | Goods and Services Taxes as % of GDP | Taxes | Country | Goods and Services Taxes as % of GDP | Taxes |
|---------|------|-------|---------|------|-------|
| Belgium | 7.2% | 16.0% | Netherlands | 7.3% | 15.6% |
| Canada | 5.3 | 14.1 | New Zealand | 8.6 | 23.8 |
| Denmark | 9.9 | 20.6 | Norway | 8.2 | 17.4 |
| France | 7.9 | 17.8 | Portugal | 6.8 | 19.0 |
| Germany | 6.4 | 16.4 | Spain | 5.5 | 15.9 |
| Greece | 9.9 | 25.9 | Switzerland | 3.0 | 9.7 |
| Italy | 5.7 | 14.3 | Turkey | 6.5 | 22.2 |
| Japan | 1.4 | 4.4 | United Kingdom | 6.7 | 18.5 |
| Luxembourg | 7.2 | 14.9 | | | |

**Source:** Data are from Gilbert E. Metcalf, "Value-Added Taxation: A Tax Whose Time Has Come?" *Journal of Economic Perspectives* 9, No. 1 (Winter 1995), p. 129. Their source is "Price Waterhouse Guide to Doing Business in …," various countries, Eurostat (1993), and OECD Revenue Statistics.

37. Using the data from Table 3.9.2, find the standard deviation and range to summarize the typical variability (or uncertainty) of the market response to stock buy-back announcements.

38. Using the data from Table 3.9.4 for the portfolio investments of CREF in aerospace and defense:
    a. Find the standard deviation of portfolio value for these firms' stock in CREF's portfolio.
    b. Find the range of these portfolio values.
    c. Find the coefficient of variation.
    d. Interpret the variability based on these three summaries.
    e. How many portfolio values are within one standard deviation of the average? How does this compare to what you would expect for a normal distribution?
    f. How many portfolio values are within two standard deviations of the average? How does this compare to what you would expect for a normal distribution?

39. Summarize the variability in the cost of a traditional funeral service using the standard deviation, based on the data in Table 3.9.10.

40. Use the data set from problem 21 of Chapter 3 on poor quality in the production of electric motors.
    a. Find the standard deviation and range to summarize the typical batch-to-batch variability in quality of production.
    b. Remove the two outliers and recompute the standard deviation and range.
    c. Compare the standard deviation and range values with and without the outliers. In particular, how sensitive is each of these summary measures to the presence of outliers?

41. Compute the standard deviation of the data from Table 4.3.1 to find the variability in spending levels from one regular customer to another for last month. Write a paragraph summarizing these differences.

42. Find the amount of variability in the five-year percent change in housing prices for U.S. regions using the data from Table 4.3.5.

43. How much variability is there in loan fees for home mortgages? Find and interpret the standard deviation, range, and coefficient of variation for the data in Table 4.3.8.

44. The performance claimed by mutual funds is often considerably better than what you would experience if you actually put your money on the line. Table 5.5.11 shows the annual return for internationally diversified bond funds both before adjustment and after subtracting the various expenses, brokerage costs, sales loads, and taxes you might have to pay.
    a. Find the standard deviation of these returns both before and after adjustment.
    b. After adjustment, are these funds (taken as a group) more homogeneous or less homogeneous? Explain how you reached your conclusion.

45. Consider the ages (in years) and maintenance costs (in thousands of dollars per year) for five similar printing presses (Table 5.5.12).
    a. Calculate the average age of the presses.
    b. Calculate the standard deviation of the ages of the presses.
    c. Calculate the range of the ages of the presses.
    d. Calculate the coefficient of variation of the ages of the presses.

### TABLE 5.5.11 International Bond Mutual Fund Performance

| Fund | Annual Return | |
|------|---------------|--|
| | Before Adjustment | After Loads and Taxes |
| T. Rowe Price International Bond | 6.3% | 3.3% |
| Merrill Lynch Global Bond B | 9.5 | 2.6 |
| Merrill Lynch Global Bond A | 10.4 | 2.1 |
| IDS Global Bond | 13.1 | 4.3 |
| Merrill Lynch World Income A | 6.5 | –0.6 |
| Fidelity Global Bond | 4.6 | 2.3 |
| Putnam Global Governmental Income | 10.3 | 0.5 |
| Shearson Global Bond B | 7.5 | 1.2 |
| Paine Webber Global Income B | 3.4 | –2.7 |
| MFS Worldwide Governments | 5.4 | –2.5 |

**Source:** Data are from *Fortune*, March 22, 1993, p.156.

### TABLE 5.5.12 Age versus Maintenance Costs for Similar Presses

| Age | Maintenance Cost |
|-----|------------------|
| 2 | 6 |
| 5 | 13 |
| 9 | 23 |
| 3 | 5 |
| 8 | 22 |

46. Using the data set from the previous problem concerning the ages and maintenance costs of five similar printing presses:
   a. Calculate the average maintenance cost of the presses.
   b. Calculate the standard deviation of the maintenance costs of the presses.
   c. Calculate the range of the maintenance costs of the presses.
   d. Calculate the coefficient of variation of the maintenance costs of the presses.

47. For the percent changes in January 2002 for the 30 companies in the Dow Jones Industrial Average (Table 2.6.7):
   a. Find the standard deviation of the percent change.
   b. Find the range of the percent change.

48. For the January 2002 daily values of the Dow Jones Industrial Average (Table 2.6.8):
   a. Find the standard deviation of the net change.
   b. Find the range of net change.
   c. Find the standard deviation of the percent change.
   d. Find the range of the percent change.

19. Data are from *Pacific Northwest Executive*, April 1988, p. 20.

### Database Exercises

*Problems marked with an asterisk (*) are solved in the Self Test in Appendix C.*

Please refer to the employee database in Appendix A.

1.* For the annual salary levels:
   a. Find the range.
   b. Find the standard deviation.
   c. Find the coefficient of variation.
   d. Compare these three summary measures. What do they tell you about typical salaries in this administrative division?

2. For the annual salary levels:
   a. Construct a histogram and indicate the average and standard deviation.
   b. How many employees are within one standard deviation from the average? How does this compare to what you would expect for a normal distribution?
   c. How many employees are within two standard deviations from the average? How does this compare to what you would expect for a normal distribution?
   d. How many employees are within three standard deviations from the average? How does this compare to what you would expect for a normal distribution?

3. For the ages: Answer the parts of exercise 1.
4. For the ages: Answer the parts of exercise 2.
5. For the experience variable: Answer the parts of exercise 1.
6. For the experience variable: Answer the parts of exercise 2.

### Projects

1. Find a data set of interest to you consisting of some quantity (for you to choose) measured for each firm in two different industry groups (with at least 15 firms in each group).
   a. For each group:
   - Summarize the variability using all of the techniques presented in this chapter that are appropriate to your data.
   - Indicate these variability measures on a histogram and/or box plot of each data set.
   - Write a paragraph summarizing what you have learned about this industry group by examining its variability.

b. For both groups, perform the following comparisons:
- Compare their standard deviations.
- Compare their coefficients of variation.
- Compare their ranges.
- Summarize what you have learned about how these industry groups relate to one another by comparing their variabilities. In particular, which variability measure is most helpful for your particular situation?

2. Find a data set consisting of at least 25 numbers relating to a firm or industry group of interest to you. Summarize the data using all of the techniques you have learned so far that are appropriate to your data. Use both numerical and graphical methods, and address both the typical value and the variability. Present your results as a two-page background report to management, beginning with an executive summary as the first paragraph.

## Case

### Should We Keep or Get Rid of This Supplier?

You and your co-worker B. W. Kellerman have been assigned the task of evaluating a new supplier of parts that your firm uses to manufacture home and garden equipment. One particular part is supposed to measure 8.5 centimeters, but in fact any measurement between 8.4 and 8.6 cm is considered acceptable. Kellerman has recently presented analysis of measurements of 99 recently delivered parts. The executive summary of Kellerman's rough draft of your report reads as follows:

*The quality of parts delivered by HypoTech does not meet our needs. Although their prices are attractively low and their deliveries meet our scheduling needs, the quality of their production is not high enough. We recommend serious consideration of alternative sources.*

Now it's your turn. In addition to reviewing Kellerman's figures and rough draft, you know you are expected to confirm (or reject) these findings by your own independent analysis.

It certainly looks as though the conclusions are reasonable. The main argument is that while the mean is 8.494, very close to the 8.5-cm standard, the standard deviation is so large, at 0.103, that defective parts occur about a third of the time. In fact, Kellerman was obviously proud of having remembered a fact from statistics class long ago, something about being within a standard deviation from the mean about a third of the time. And defective parts might be tolerated 10% or even 20% of the time for this particular application at these prices, but 30% or 33% is beyond reasonable possibility.

It looks so clear, and yet, just to be sure, you decide to take a quick look at the data. Naturally, you expect it to confirm all this. Here is the data set:

| | | | | | | | | | |
|---|---|---|---|---|---|---|---|---|---|
| 8.503 | 8.503 | 8.500 | 8.496 | 8.500 | 8.503 | 8.497 | 8.504 | 8.503 | 8.506 |
| 8.502 | 8.501 | 8.489 | 8.499 | 8.492 | 8.497 | 8.506 | 8.502 | 8.505 | 8.489 |
| 8.505 | 8.499 | 8.489 | 8.505 | 8.504 | 8.499 | 8.499 | 8.506 | 8.493 | 8.494 |
| 8.510 | 8.310 | 8.804 | 8.503 | 8.782 | 8.502 | 8.509 | 8.499 | 8.498 | 8.493 |
| 8.346 | 8.499 | 8.505 | 8.509 | 8.499 | 8.503 | 8.494 | 8.511 | 8.501 | 8.497 |
| 8.501 | 8.502 | 7.780 | 8.494 | 8.500 | 8.498 | 8.500 | 8.502 | 8.501 | 8.491 |
| 8.511 | 8.494 | 8.374 | 8.492 | 8.497 | 8.150 | 8.496 | 8.501 | 8.489 | 8.506 |
| 8.493 | 8.498 | 8.505 | 8.490 | 8.493 | 8.501 | 8.497 | 8.501 | 8.498 | 8.503 |
| 8.508 | 8.501 | 8.499 | 8.504 | 8.505 | 8.461 | 8.497 | 8.495 | 8.504 | 8.501 |
| 8.493 | 8.504 | 8.897 | 8.505 | 8.490 | 8.492 | 8.503 | 8.507 | 8.497 | |

### Discussion Questions

1. Are Kellerman's calculations correct? These are the first items to verify.
2. Take a close look at the data using appropriate statistical methods.
3. Are Kellerman's conclusions correct? If so, why do you think so? If not, why not and what should be done instead?

# Probability

How do you deal with uncertainty? By understanding how it works. *Probability* starts out by encouraging you to clarify your thinking in order to separate the truly uncertain things from the hard facts you are sure of. What exactly *is* the uncertain situation you're interested in? By what exact procedure is it determined? How likely are the various possibilities? Often there is some event that either will happen or won't happen: Will you get the contract? Will the customer send in the order form? Will they fix the machinery on time? In Chapter 6, you will learn about these kinds of situations, their combinations, and how uncertainty is reduced when you learn new information. In other situations, you are interested in an uncertain *number* that has not yet been revealed: What will the announced earnings be? How high will the quarterly sales be? How much time will be lost due to computer trouble? In Chapter 7, you will see how to find a typical summary number, how to assess the risk (or variability) of the situation, and how to find the likelihoods of various scenarios based on an uncertain number.

## Chapter 6

# Probability

## Understanding Random Situations

## Chapter Outline

Our goal is to understand uncertain situations, at least to the greatest extent possible. Unfortunately, we will probably never be able to say "for sure" exactly what will happen in the future. However, by recognizing that some possibilities are more likely than others, and by quantifying (using numbers to describe) these relationships, you will find yourself in a much more competitive position than either those who have no idea of what will happen or those who proceed by a "gut feeling" that has no real basis. The best policy is to combine an understanding of the probabilities with whatever wisdom and experience are available.

Here are some examples of uncertain situations:

**One:** Your division is just about to decide whether or not to introduce a new digital audio tape player into the consumer market. Although your marketing studies show that typical consumers like the product and feel that its price is reasonable, its success is hardly assured. Uncertainty comes from many factors. For example, what will the competition do? Will your suppliers provide quality components in a timely manner? Is there a flaw that hasn't been noticed yet? Will there be a recession or expansion in the economy? Will consumers spend real money on it, or do they just say that they will? Your firm will have to make the best decision it can based on the information available.

**Two:** Your uncle, a farmer in Minnesota, writes to say that he thinks a drought is likely this year. Since he correctly predicted similar problems in 2007, you immediately purchase call options on corn and soybean futures on the Chicago Board of Trade. What will happen to your investment? It is quite speculative. If farmers have a good year, prices could fall and you could lose everything. On the other hand, if there is a serious drought, prices will rise and you will profit handsomely.

**Three:** As manager of operations at a large chemical plant, you have many responsibilities. You must keep costs low while producing large quantities of products. Because some of these chemicals are poisonous, you have put a safety system into place. Still, despite your best efforts, you wonder: How likely is a major disaster during the next year? This is a very uncertain situation, and we do hear about such disasters in the media. To properly understand the costs of safety and the potential benefits, you have decided to study the probability of failure in your plant.

**Four:** Did you ever wonder about your chances of winning a sweepstakes? You would probably guess that is a very improbable event. But how unlikely is it? An answer is provided by a newspaper article, which reports:

*The odds of [not] winning the $10 million grand prize in the current Publishers Clearing House magazine sweepstakes mailing are 427,600,000 to 1. That is about 300 times more unlikely than getting struck by lightning, which is a mere 1.5 million-to-1 shot.[1]*

How can you understand uncertain, random situations without being troubled by their imprecise nature? The best approach is to begin with solid, exact concepts. This can help you know "for sure" what is going on (to the greatest extent possible) even when you are analyzing a situation in which nothing is "for sure."

Such an approach establishes clear boundaries, starting with an exact description of the process of interest (the *random experiment*) together with a list of every outcome that could possibly result (the *sample space*). You may have a number of special cases to study that either happen or don't (e.g., "the project succeeded"); these are called *events*. The likelihood of an event happening is indicated by an exact number called its *probability*.

You may want to combine information about more than one event, for example, to see how likely it is that the reactor *and* the safety system will both fail. Or you may wish to keep your probabilities up-to-date by using *conditional probabilities* that reflect all available information. The easiest way to solve such probability problems is to first build a *probability tree* to organize your knowledge about a situation; this is much easier than trying to choose just the right combination of formulas. Some people also find

*joint probability tables* and *Venn diagrams* useful in building intuition and solving problems.

## 6.1 AN EXAMPLE: IS IT BEHIND DOOR NUMBER 1, DOOR NUMBER 2, OR DOOR NUMBER 3?

You are a contestant on a television game show, and the prize of your dreams (a trip to Hawaii or perhaps a perfect grade on the midterm?) is behind either door number 1, door number 2, or door number 3. Behind the other two doors, there is nothing. You can see all three closed doors, and you have no hints. While the crowd shouts its encouragement, you think over your options, make up your mind, and tell everyone which door you choose. But before opening your choice of door, the show's hosts tell you that they will first open a different door that does not have the prize behind it. After they do this, they offer you a chance to switch. There are now two doors left unopened: your choice and one other. Should you switch? The crowd goes wild, shouting encouragement and suggestions. Would you stay with your original choice or switch to the other door? Please think about it before looking at the answer in the following paragraphs.[2]

\* \* \* \* \*

If you decided to stay with your original choice, you are in good company because nearly all students answer this way. Unfortunately, however, you are wrong, and your decision-making abilities will definitely benefit from the study of probability.

If you decided to switch, congratulations.[3] You have *doubled* your chances of winning, from 1/3 to 2/3.

The principle at work here is that the switchers have made use of new information in an effective way, whereas those who didn't switch have not changed their chances at all. What is the new information? In order to open a door that is not yours and that does not have the prize behind it, the people running the contest have to know something about which door the prize is really behind. In their opening such a door, partial information is revealed to you.

Here's an informal explanation. Imagine having a twin who switches while you do not. Since there are just two doors left and the prize must be behind one of them, your twin will win every time that you do not. Since your overall chances of winning are unchanged at 1/3, it follows that your switching twin will win the remaining 2/3 of the time. For

1. William P. Barrett, "Bank on Lightning, Not Mail Contests," *Seattle Times*, January 14, 1986, p. D1.

2. A similar situation was described by B. Nalebuff, "Puzzles: Choose a Curtain, Duel-ity, Two Point Conversions, and More," *Journal of Economic Perspectives* 1 (1987), pp. 157–63. More recently, this problem received considerable attention after appearing in Marilyn Vos Savant's column in the *Parade* magazine supplement to many Sunday newspapers.

3. Also, please write me a brief note describing why you decided to switch. I am curious about the different intuitive ways people use to get to the correct answer. Please send your note to Andy Siegel, Foster School of Business, University of Washington, Seattle, Washington 98195. Thank you!

those of you who are not convinced, a detailed, formal solution will be presented as an example in Section 6.5.

Don't be discouraged if you didn't make the best choice. Instead, think about the powers you will develop by learning something about probability. In fact, you should even feel optimistic, since your decisions will then be better than those of many others who continue to "stand pat" in the face of new, important information.

## 6.2 HOW CAN YOU ANALYZE UNCERTAINTY?

Our precise framework for studying random situations will begin by carefully identifying and limiting the situation. The result is a *random experiment* that produces one *outcome* from a list of possibilities (called the *sample space*) each time the experiment is run. There will usually also be a number of *events,* each of which either happens or doesn't happen, depending on the outcome.

### The Random Experiment: A Precise Definition of a Random Situation

A **random experiment** is any well-defined procedure that produces an observable outcome that could not be perfectly predicted in advance. A random experiment must be well defined to eliminate any vagueness or surprise. It must produce a definite, observable outcome so that you know what happened after the random experiment is run. Finally, the outcome must not be perfectly predictable in advance since, if it were known with certainty, you would not really have a random situation.[4] Here are some examples of random experiments we will continue to discuss throughout this section:

1. A survey is designed to study family income in the area around a proposed new restaurant. A family is telephoned through random dialing, and its income is recorded to the nearest dollar. In order to be completely precise, you must specify what the outcome is in case nobody answers the telephone or those who do answer refuse to tell you their income. In this case, suppose you select a new number to dial and repeat the process until you get an income number. In this way, each time the random experiment is run, you obtain an income figure as the outcome. Although this solves the immediate problem for probability, nonresponse remains a problem with statistical analysis because there is a group (the nonrespondents) for which you have no income information.

2. The marketing department plans to select 10 representative consumers to form a focus group to discuss and vote on the design of a new reading lamp, choosing one of seven proposed designs. (The outcome is the selected design.)

3. From tomorrow's production, choose five frozen gourmet dinners according to a carefully specified random selection process, cook them, and record their quality as a whole number on a scale from 1 to 4. (The careful selection process ensures that the experiment is well defined, and the outcome is the list of recorded measures of quality. Finally, the outcome is not known with certainty since, occasionally, problems do occur in any manufacturing process.)

A complex situation usually contains many different random experiments; you are free to define the particular one (or ones) that will be most useful. For example, the selection of one dinner for quality inspection is a random experiment all by itself, producing just one quality number. The selection of all five dinners is also a random experiment—a larger one—producing a list of five quality numbers as its outcome.

Think of a random experiment as the arena in which things happen. By narrowing things down to a relatively small situation, you can think more clearly about the possible results.

### The Sample Space: A List of What Might Happen

Each random experiment has a **sample space**, which is a list of *all possible outcomes* of the random experiment, prepared in advance without knowledge of what will happen when the experiment is run. Note that there is nothing random about the sample space. It is the (definite) list of things that might happen. This is an effective way to make a random situation more definite, and it will often help to clarify your thinking as well. Here are the sample spaces corresponding to the previous random experiments:

1. For the family income survey, the sample space is a list of all possible income values. Let's assume that income must be zero or a positive number and think of the sample space as a list of all nonnegative dollar amounts:

   $0
   $1
   $2
   .
   .
   .
   $34,999
   $35,000
   $35,001
   .
   .
   .

---

4. In some cases, the outcome may seem perfectly predictable, such as whether class will meet tomorrow or whether Colgate will still be selling toothpaste in the year 2018. Feel free to still call these "random experiments" so long as there is any conceivable doubt about the outcome, even if it is *nearly* certain.

**2.** For the focus group choosing the design of a new reading lamp, the sample space is much smaller, consisting of the seven proposed designs:

| | |
|---|---|
| Design A | Design E |
| Design B | Design F |
| Design C | Design G |
| Design D | |

**3.** For the quality testing of frozen dinners, the sample space is the collection of all possible lists of quality numbers, one for each dinner tested. This is a collection of lists of five numbers (one for each dinner tested) where each number is from 1 to 4 (representing the quality of that dinner). This sample space is too large to be presented here in its entirety but might begin with a list of all 1's (i.e., "1, 1, 1, 1, 1" indicates horrible quality for all five dinners) and end with a list of all 4's (i.e., "4, 4, 4, 4, 4" indicates marvelous gourmet quality for all five dinners tested). Somewhere within this collection would be included all possible quality score lists (including, e.g., "3, 2, 3, 3, 4"):[5]

| | | | | | |
|---|---|---|---|---|---|
| 1 | 1 | 1 | 1 | 1 | (First list) |
| 1 | 1 | 1 | 1 | 2 | |
| 1 | 1 | 1 | 1 | 3 | |
| 1 | 1 | 1 | 1 | 4 | |
| 1 | 1 | 1 | 2 | 1 | |
| 1 | 1 | 1 | 2 | 2 | |
| 1 | 1 | 1 | 2 | 3 | |
| 1 | 1 | 1 | 2 | 4 | |
| 1 | 1 | 1 | 3 | 1 | |
| 1 | 1 | 1 | 3 | 2 | |
| 1 | 1 | 1 | 3 | 3 | |
| 1 | 1 | 1 | 3 | 4 | |
| 1 | 1 | 1 | 4 | 1 | |
| 1 | 1 | 1 | 4 | 2 | |
| 1 | 1 | 1 | 4 | 3 | |
| 1 | 1 | 1 | 4 | 4 | |
| 1 | 1 | 2 | 1 | 1 | |
| 1 | 1 | 2 | 1 | 2 | |
| 1 | 1 | 2 | 1 | 3 | |
| 1 | 1 | 2 | 1 | 4 | |
| 1 | 1 | 2 | 2 | 1 | |
| 1 | 1 | 2 | 2 | 2 | |
| 1 | 1 | 2 | 2 | 3 | |
| 1 | 1 | 2 | 2 | 4 | |
| . | . | . | . | . | |
| . | . | . | . | . | |
| . | . | . | . | . | |
| 4 | 4 | 4 | 3 | 1 | |
| 4 | 4 | 4 | 3 | 2 | |
| 4 | 4 | 4 | 3 | 3 | |
| 4 | 4 | 4 | 3 | 4 | |

| | | | | | |
|---|---|---|---|---|---|
| 4 | 4 | 4 | 4 | 1 | |
| 4 | 4 | 4 | 4 | 2 | |
| 4 | 4 | 4 | 4 | 3 | |
| 4 | 4 | 4 | 4 | 4 | (Last list) |

## The Outcome: What Actually Happens

Each time a random experiment is run, it produces exactly one **outcome**. Since the sample space contains all possible outcomes, there cannot be any surprises: The outcome of a random experiment must be contained in the sample space.

Here are outcomes from one run of each of our random experiments:

**1.** For the family income survey, after one "nobody home" and one "none of your business," you reached an individual who reported the family income (the outcome):

$$\$36,500$$

**2.** For the focus group, after much discussion, the outcome is the selected design:

Design D

**3.** For the quality testing of frozen dinners, the outcome is a list of the quality measures for each of the five dinners tested:

3 3 1 2 2

## Events: Either They Happen or They Don't

The formal definition of an **event** is any collection of outcomes specified in advance, before the random experiment is run. In practical terms, all that matters is that you can tell whether the event happened or didn't each time the random experiment is run. You may have many events (or just one) for a random experiment, each event corresponding to some feature of interest to you.

Here are some events for each of our random experiments:

**1.** For the family income survey, consider three different events:

| First event: | Low income:* | $10,000 to $24,999 |
|---|---|---|
| Second event: | Middle income: | $25,000 to $44,999 |
| Third event: | Qualifying income: | $20,000 or over |

*The list of outcomes for the even "Low income" might be written in more detail as $10,000, $10,001, $10,002, . . . , $24,997, $24,998, $24,999.

For the observed outcome, $36,500, you see that the first event did not happen, and the other two did happen. Thus, the selected family is "middle income" and has a "qualifying income" but is not "low income."

---

5. In fact, there are 1,024 lists in this sample space. This is $4 \times 4 \times 4 \times 4 \times 4 = 4^5$, since there are 4 possible quality numbers for the first dinner, times 4 for the second, and so on through the 5th dinner tested.

2. For the focus group, consider the event "the chosen design is easy to produce," which includes only the following designs:

Design A
Design B
Design C
Design F

Since the observed outcome, Design D, is not in this list, this event "did not happen." Unfortunately, the focus group selected a design that is not easily produced.

3. For the quality testing of frozen dinners, consider the event "all good or excellent quality," meaning that all five dinners had a score of 3 or 4. For example, the observed outcome

3 3 1 2 2

does not qualify as "all good or excellent quality" (because the last three dinners scored low), so this event *did not happen*. However, had the outcome been

3 4 3 4 4

instead, the event would have happened (because dinners were all of quality 3 or 4).

To view this event in terms of the formal definition, that is, as the collection of outcomes for which the event happens, you would have the following:

| | | | | | | | | | | |
|---|---|---|---|---|---|---|---|---|---|---|
| 3 | 3 | 3 | 3 | 3 | (First list) | 4 | 3 | 3 | 3 | 3 |
| 3 | 3 | 3 | 3 | 4 | | 4 | 3 | 3 | 3 | 4 |
| 3 | 3 | 3 | 4 | 3 | | 4 | 3 | 3 | 4 | 3 |
| 3 | 3 | 3 | 4 | 4 | | 4 | 3 | 3 | 4 | 4 |
| 3 | 3 | 4 | 3 | 3 | | 4 | 3 | 4 | 3 | 3 |
| 3 | 3 | 4 | 3 | 4 | | 4 | 3 | 4 | 3 | 4 |
| 3 | 3 | 4 | 4 | 3 | | 4 | 3 | 4 | 4 | 3 |
| 3 | 3 | 4 | 4 | 4 | | 4 | 3 | 4 | 4 | 4 |
| 3 | 4 | 3 | 3 | 3 | | 4 | 4 | 3 | 3 | 3 |
| 3 | 4 | 3 | 3 | 4 | | 4 | 4 | 3 | 3 | 4 |
| 3 | 4 | 3 | 4 | 3 | | 4 | 4 | 3 | 4 | 3 |
| 3 | 4 | 3 | 4 | 4 | | 4 | 4 | 3 | 4 | 4 |
| 3 | 4 | 4 | 3 | 3 | | 4 | 4 | 4 | 3 | 3 |
| 3 | 4 | 4 | 3 | 4 | | 4 | 4 | 4 | 3 | 4 |
| 3 | 4 | 4 | 4 | 3 | | 4 | 4 | 4 | 4 | 3 |
| 3 | 4 | 4 | 4 | 4 | | 4 | 4 | 4 | 4 | 4 | (Last list) |

Thus, the event "all good or excellent quality" happens if and only if the actual outcome is in this list. Can you find the outcome "3, 4, 3, 4, 4" in this list?

As you can see, there is nothing mysterious going on here. You have to be very careful in your definitions of the random experiment, the sample space, outcomes, and events in order to impose order upon an uncertain situation. However, in the end, events lead to reasonable, ordinary statements such as "Today's production passed quality inspection" or "Hey, Marge, I found another qualifying family!"

## 6.3  HOW LIKELY IS AN EVENT?

You know that each event either happens or doesn't happen every time the random experiment is run. This really doesn't say much. You want to know how *likely* the event is. This is provided by a number called the *probability of the event*. We will define this concept, indicate where the probability numbers come from, and show how the probability indicates the approximate proportion of time the event will occur when the random experiment is run many times.

## Every Event Has a Probability

Every event has a number between 0 and 1, called its **probability**, which expresses how likely it is that the event will happen each time the random experiment is run. A probability of 0 indicates that the event essentially never happens, and a probability of 1 indicates that the event essentially always happens.[6] In general, the probability indicates approximately what proportion of the time the event is expected to happen:

| Probability | Interpretation |
|---|---|
| 1.00 | Always happens (essentially) |
| .95 | Happens about 95% of the time (very likely) |
| .50 | Happens about half of the time |
| .37 | Happens about 37% of the time |
| .02 | Happens about 2% of the time (unlikely, but possible) |
| .00 | Never happens (essentially) |

Some care is needed in interpreting these numbers. If you have a one-shot situation that cannot be repeated (such as the probability that an old building will be successfully dynamited), a probability of 0.97 expresses a high likelihood of success with a small possibility of failure. However, it would be inappropriate to say, "we expect success about 97% of the time" unless you plan to repeat this exact procedure on many similar buildings in the future.

Another way to express likelihood is with *odds*, a positive number defined as the probability that the event happens divided by the probability that it does not happen:

**The Odds**

$$\text{Odds} = \frac{\text{Probability that the event happens}}{\text{Probability that the event does not happen}}$$

$$= \frac{\text{Probability}}{1 - \text{Probability}}$$

$$\text{Probability} = \frac{\text{Odds}}{1 + \text{Odds}}$$

---

6. Due to mathematical technicalities, an event with probability 0 actually can happen. However, it cannot happen very often, and indeed, in technical language, we say that it "almost never happens." If you like to wonder about these things, consider, for example, the probability of pumping exactly 254,896.3542082947839478 . . . barrels of oil tomorrow.

For example, a probability of 0.5 corresponds to odds of 0.5/(1 − 0.5) = 1, sometimes stated as "odds of 1 to 1." A probability of 0.8 has odds of 0.8/(1 − 0.8) = 4, or 4 to 1. If the odds are 2 (or 2-to-1), then the probability is 2/(1 + 2) = 2/3. Higher odds indicate higher probabilities and greater likelihood. Note that although the probability cannot be outside the range from 0 to 1, the odds can be any nonnegative number.

## Where Do Probabilities Come From?

In textbooks, numbers to use as probabilities are often given along with other information about a problem. In real life, however, there are no tags attached saying, "By the way, the probability that the bolt will be defective and cause a problem is 0.003." How do we get the probability numbers to use in real life? There are three main ways: *relative frequency* (by experiment), *theoretical probability* (by formula), and *subjective probability* (by opinion).

## Relative Frequency and the Law of Large Numbers

Suppose you are free to run a random experiment as many times as you wish under the same exact circumstances each time, except for the randomness. The **relative frequency** of an event is the proportion of times the event occurs out of the number of times the experiment is run. It may be given either as a proportion (such as 0.148) or as a percentage (such as 14.8%). The formula for the relative frequency of an event is given by

Relative Frequency of an Event
$$= \frac{\text{Number of times the event occurs}}{\text{Number of times the random experiment is run}}$$

For example, if you interview 536 willing people and find 212 with qualifying incomes of $20,000 or over, the relative frequency is

$$\frac{212}{536} = 0.396 \text{ or } 39.6\%$$

For this relative frequency of 39.6% to make sense, the random experiment must be "find the income of a willing person," and the event of interest must be "income is $20,000 or over." With these definitions, it is clear that you have run the random experiment n = 536 times. There is another, different random experiment of interest here—namely, "find the incomes of 536 willing people"; however, the notion of relative frequency does not make sense for this larger random experiment because it has been run only once.

These are not trivial differences; managers often find this kind of careful "mental discipline" useful in dealing with more complex problems. The relative frequency and the probability of an event are similar but different concepts. The important difference is that the probability is an *exact number*, whereas the relative frequency is a *random number*. The reason is that the probability of an event is a property of the basic situation (random experiment, sample space, event), whereas the relative frequency depends on the (random) outcomes that were produced when the random experiment was run n times.

Relative frequencies may be used as estimates (best guesses) of probability numbers when you have information from previous experience. For example, you might use this relative frequency (39.6% for qualified income) as an approximation to the true probability that a randomly chosen, willing person will have a qualifying income. You could then use this number, 0.396, as if it were a probability. Keep in mind, however, that there is a difference between the true (unknown) probability number and your best guess using the relative frequency approach.

The **law of large numbers** says that the relative frequency will be close to the probability if the experiment is run many times, that is, when n is large. How close, typically, will the (random) relative frequency be to the (fixed) probability? The answer, which depends on how likely the event is and how many runs (n) are made, is provided in terms of a standard deviation in Table 6.3.1. For example, if an event has probability 0.75 and you run the random experiment n = 100 times, then you can expect the relative frequency to be approximately 0.04 above or below the true probability (0.75) on average.

Figures 6.3.1 and 6.3.2 show an example of how the relative frequency (the random line in each graph) approaches the probability (the fixed horizontal line at 0.25 in each graph). Note that the vertical scale has been enlarged in Figure 6.3.2; this can be done because the relative frequency is closer to the probability for larger n.

Since the relative frequency is fairly close to the probability (at least for larger values of n) due to the law of large numbers, it is appropriate to use a relative frequency as a "best guess" of the probability based on available data.

**TABLE 6.3.1 Standard Deviation of the Relative Frequency\***

|  | Probability of 0.50 | Probability of 0.25 or 0.75 | Probability of 0.10 or 0.90 |
|---|---|---|---|
| n = 10 | 0.16 | 0.14 | 0.09 |
| 25 | 0.10 | 0.09 | 0.06 |
| 50 | 0.07 | 0.06 | 0.04 |
| 100 | 0.05 | 0.04 | 0.03 |
| 1,000 | 0.02 | 0.01 | 0.01 |

*\* These have been calculated using the formula for the standard deviation of a binomial percentage, to be covered in Chapter 8.*

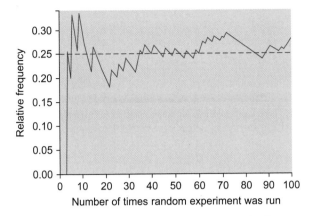

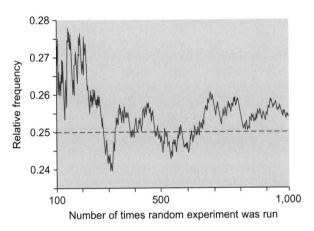

**FIGURE 6.3.1** The relative frequency of an event approximates its probability (0.25 in this case) and generally gets closer as *n* grows. With *n* = 100, you expect to be within approximately 0.04 of the probability.

**FIGURE 6.3.2** The relative frequency of an event, shown for *n* = 100 to 1,000 runs of a random experiment. Most of the time, the relative frequency will be within about 0.01 of the probability (0.25 in this example) when *n* = 1,000. Note the change in scale from the previous graph.

### Example
#### How Variable Is Today's Extra-High-Quality Production?

You have scheduled 50 items for production today, and from past experience, you know that over the long term 25% of items produced fall into the "extra-high-quality" category and are suitable for export to your most demanding customers overseas. What should you expect today? Will you get exactly 25% of 50, or 12.5 extra-high-quality items? Obviously not; but how far below or above this number should you expect to be?

The standard deviation of the relative frequency (0.06 or 6%) from Table 6.3.1 will help answer the question. Think of it this way: Today you have scheduled 50 repetitions of the random experiment "produce an item and measure its quality."[7] The relative frequency of the event "extra-high quality" will be the percent of such items you produce today. This relative frequency will be close to the probability (25% or 0.25) and will (according to the table) be approximately 0.06 or 6 percentage points away from 0.25. Translating

from percentages to items, this variability number is 0.06 × 50 = 3 items. Thus, you expect to see "12.5 plus or minus 3" items of extra-high quality produced today.

The variability, 3 items, is interpreted in the usual way for standard deviation. You would not be surprised to see 10 or 14 such items produced, or even 7 or 18 (around two standard deviations away from 12.5). But you would be surprised and disappointed if only a few were produced (it would signal a need for fixing the process). And you would be surprised and very pleased if, say, 23 were produced (break out the champagne and help employees remember what they did right!).

---

7. This analysis assumes that items are produced independently of one another.

## Theoretical Probability

A **theoretical probability** is a probability number computed using an exact formula based on a mathematical theory or model. This approach can be used only with systems that can be described in mathematical terms. Most theoretical probability methods are too involved to be presented here. We will cover only the special case of the *equally likely* rule and, in Chapter 7, special probability distributions such as the normal and the binomial.

## The Equally Likely Rule

If all outcomes are equally likely—and this is a big "if"—it is easy to find the probability of any event using theoretical probability. The probability is proportional to the number of outcomes represented by the event, according to the following formula:

### If All Outcomes Are Equally Likely, Then

$$\text{Probability of an Event} = \frac{\text{Number of outcomes in the event}}{\text{Total number of possible outcomes}}$$

### Example
#### Coin Tossing and Cards

Since a flipped coin is as likely to land "heads" as "tails," the probability of each of these events is 1/2. What about the three outcomes "heads," "tails," and "on edge"? Since there are three outcomes, does it follow that each of these has probability 1/3? Of course not, since the requirement that "all outcomes be equally likely" is violated here; the rule does not apply since the probability of a flipped coin landing on its edge is minuscule compared to the other two possibilities.

Since cards are generally shuffled before playing, which helps ensure randomness, the *equally likely* rule should apply to any situation in which you specify a property of

*(Continued)*

**Example—cont'd**

one random card. For example, since 13 out of the 52 cards are hearts, the probability of receiving a heart is 0.25. Similarly, the probability of receiving an ace is 4/52 = 7.7%, and the probability of receiving a one-eyed jack is 2/52 = 3.8%.

**Example**
*Gender and Hiring*

For another example, suppose that 15 equally qualified people have applied for a job and 6 are women. If the hiring choice is made from this group randomly (and, thus, without regard to gender), the probability that a woman will be chosen is 6/15 = 0.40 or 40%. This situation fits the equally likely rule if you view the random experiment as choosing one of these people and the event as "a woman is chosen." This event consists of 6 outcomes (the 6 women), and the probability is, then, 6 out of 15, the total number of possible job candidates (outcomes).

**Example**
*Defective Raw Materials*

Suppose your supplier has a warehouse containing 83 automatic transmissions, of which 2 are defective. If one is chosen at random (without knowledge of which are defective), what is the probability that you will receive a defective part? The answer, by the equally likely rule, is 2 out of 83, or 2.4%.

## Subjective Probability

A **subjective probability** is anyone's opinion of what the probability is for an event. While this may not seem very scientific, it is often the best you can do when you have no past experience (so you can't use relative frequency) and no theory (so you can't use theoretical probability). One way to improve the quality of a subjective probability is to use the opinion of an expert in that field, for example, an investment banker's opinion of the probability that a hostile takeover will succeed or an engineer's opinion of the feasibility of a new energy technology.

**Example**
*Settling a Lawsuit*

Your firm has just been sued. Your first reaction is "Oh no!" but then you realize that firms get sued for various reasons all the time. So you sit down to review the details, which include claimed consequential damages of $5 million. Next, you call a meeting of your firm's executives and lawyers to discuss strategy. So that you can choose the most effective strategy, it will help you to know how likely the various possible consequences are.

This is, of course, a probability situation. You are sitting right in the middle of a huge random experiment: Observe the progress of the lawsuit, and record the outcome in terms of (1) dollars spent (for fees and damages, if any) and (2) the resolution (dismissal, settlement without trial, judge trial, or jury trial).

Since so many lawsuits reach settlement without going to trial, you decide initially to consider the various possible costs of settlement. Following are three events of interest here:

1. Inexpensive settlement: less than $100,000.
2. Moderate settlement: $100,000 to $1,000,000.
3. Expensive settlement: over $1,000,000.

How will you find probabilities for these events? There have not been enough similar cases to rely on the relative frequency approach, even after consulting computerized legal databases to identify related lawsuits. There is no scientific theory that would give a formula for these probabilities, which rules out the theoretical probability approach. This leaves only the avenue of subjective probability.

To find subjective probability values for these events, at the meeting it is decided to use the opinions of a legal expert who is familiar with this kind of litigation. After studying the particular details of the case and evaluating it with respect to a few previous similar cases and knowledge about the current "mood" in the legal system, your expert reports the following subjective probabilities:

1. Probability of inexpensive settlement: 0.10.
2. Probability of moderate settlement: 0.65.
3. Probability of expensive settlement: 0.15.

Note that these probabilities add up to 0.90, leaving a 10% chance of going to trial (i.e., not settling).

These subjective probabilities, representing your best assessment of the various likelihoods, tell you that there will probably be a moderate settlement. You are relieved that the likely dollar figure is substantially less than the $5 million mentioned in the initial papers. You can then use these probability numbers to help guide you in making the difficult decisions about the type and quantity of resources to bring forward in your defense.

## Bayesian and Non-Bayesian Analysis

The methods of **Bayesian analysis** in statistics involve the use of subjective probabilities in a formal, mathematical way. Figure 6.3.3 (top) shows how a Bayesian analysis puts the observed data together with prior probabilities and a model (a mathematical description of the situation) to compute the results.

A non-Bayesian analysis is called a frequentist analysis and appears initially to be more objective since its calculations depend only on the observed data and the model. However, a more careful examination, shown in Figure 6.3.3 (bottom), reveals the hidden role that subjective opinions

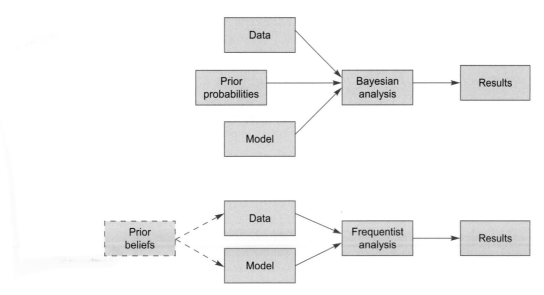

**FIGURE 6.3.3** The general approaches of the Bayesian and the frequentist (non-Bayesian) statisticians. Although both use prior subjective information, the Bayesian uses it in a direct way in computing the results. The frequentist, on the other hand, uses prior information in an informal way to set the stage for a seemingly more "objective" calculation.

play by guiding the selection of the design (which produces the data) and the selection of a model.

Bayesian analysis can be useful in complex statistical studies in business, especially when there is an expert available who has fairly reliable information that you want to include in the analysis along with the data. If this expert opinion can be used effectively in the design of the study and the choice of model, then there would be no need to use complex Bayesian calculations. However, if the situation is important enough and the expert opinion is precise enough to warrant special attention and care, then a Bayesian analysis should be considered.

## 6.4 HOW CAN YOU COMBINE INFORMATION ABOUT MORE THAN ONE EVENT?

There are two "big wins" you can get from understanding probability. The first one, which we have already covered, is appreciation of the concept of the likelihood of an event as a probability number and the corresponding mind-focusing activity of identifying the random experiment. This is a big step toward understanding a vague, uncertain situation. The second "big win" is the ability to consider *combinations of events* and to learn how to take information about the probabilities of some events and use it to find the probabilities of other events that are perhaps more interesting or important.

Keep in mind the objective: to combine the information you have with the basic rules of probability to obtain new, more useful information. Strictly speaking, this derived information is not really "new" since it logically follows from the old information. Rather, it represents a better presentation of the information you already have. The ability to

quickly find these logical consequences will help you explore *what if* scenarios to guide your business decisions.

## Venn Diagrams Help You See All the Possibilities

A **Venn diagram** is a picture that represents the universe of all possible outcomes (the sample space) as a rectangle with events indicated inside, often as circles or ovals, as in Figure 6.4.1. Each point inside the rectangle represents a possible outcome. Each time the random experiment is run, a point (outcome) is selected at random; if this point falls within an event's circle, then the event "happens"; otherwise the event does not happen.

### Not an Event

The **complement** of an event is another event that happens only when the first event does *not* happen. Every event has a complement.[8] Here are some examples of complements of events:

| Event | Complement of the Event |
| --- | --- |
| Successful product launch | Unsuccessful launch |
| Stock price went up | Stock price held steady or dropped |
| Production had acceptable quality | Production quality was not acceptable |

When you think of an event as a collection of outcomes, the complementary event consists of all outcomes in the

---

8. Please note the spelling. While every event has a *complement*, as defined here, only those events that are especially nice and well behaved will deserve a "*compliment*" (spelled with an "i").

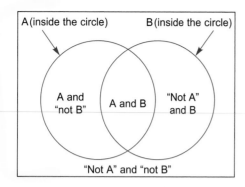

FIGURE 6.4.1  A Venn diagram with two events, each indicated by a circle. Note that some points (outcomes) fall inside both circles, some fall outside both, and some fall within one but not the other. This is a convenient visual representation of all possibilities with two events.

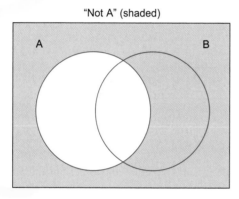

FIGURE 6.4.2  A Venn diagram with the complement "not A" indicated. The complement of A is an event that happens when (and only when) A does not happen.

sample space that are *not* represented by the first event. This is illustrated by the Venn diagram in Figure 6.4.2.

When you are constructing the complement of an event, be sure that you consider all possibilities. For example, the complement of "price went up" is *not* "price went down" because you must also include the possibility that the price remained steady.

## The Complement (Not) Rule

Since an event and its complement together represent all possible outcomes, with no duplication, the sum of their probabilities is 1. This leads to the following rule for an event we'll call A and its complement, called "not A":

Probability of A + Probability of "not A" = 1

This formula can be rewritten to determine the probability of the complement:

**The Complement Rule**

Probability of "not A" = 1 − Probability of A

For example, if the probability of a successful product launch is known to be 0.4, the probability of the complementary event "unsuccessful launch" is 1 − 0.4 = 0.6.

The complement rule is the first illustration here of a method for getting "new" probability information (the probability of the complementary event) from facts that are already known (the probability of the event itself). If you're not impressed yet, please read on.

## One Event *and* Another

Any two events can be combined to produce another event, called their **intersection**, which occurs whenever one event *and* the other event both happen as a result of a single run of the random experiment.

When you think of two events as collections of outcomes, their intersection is a new collection consisting of all outcomes contained in both collections, as illustrated in Figure 6.4.3.

For example, as manager of a business supplies firm, you may want to consider what to do if a recession occurs next year and your competitors respond by lowering prices. The random experiment here is "at the end of next year, observe and record the state of the economy and the pricing policies of your competitors." The two events are "recession" (which will either occur or not occur) and "competitors lowered prices." Your current concern is with a new event, "recession *and* lower competitor prices." In order to formulate a policy, it will be helpful to consider the likelihood of this new possibility.

## What If Both Events Can't Happen at Once?

Two events that cannot both happen at once are said to be **mutually exclusive events**. For example, you could not end the year with both "very good" and "very bad" profits. It is

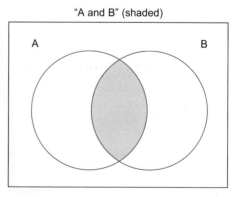

FIGURE 6.4.3  A Venn diagram with the intersection "A and B" indicated. This event happens when (and only when) A happens *and* B happens in a single run of the random experiment.

Two mutually exclusive events

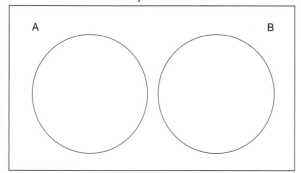

**FIGURE 6.4.4** A Venn diagram for two mutually exclusive events, which cannot both happen on a single run of the random experiment. Because the circles do not overlap, there are no points common to both events.

"A or B" (shaded)

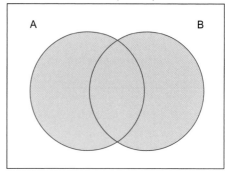

**FIGURE 6.4.5** A Venn diagram with the union "A or B" indicated. This event happens whenever either A happens, B happens, or both events happen in a single run of the random experiment.

also impossible for your next recruit to be both a "male minority Ph.D." and a "female engineer with an MBA." Figure 6.4.4 shows a Venn diagram for two mutually exclusive events as two circles that do not overlap.

## The Intersection (*and*) Rule for Mutually Exclusive Events

Since two mutually exclusive events cannot both happen at once, the probability of their intersection is 0. There is no problem with defining the intersection of two mutually exclusive events; this intersection is a perfectly valid event, but it can never happen. This implies that its probability is 0.

> **The Intersection (*and*) Rule for Two Mutually Exclusive Events**
>
> Probability of "A and B" = 0

## One Event *or* Another

Any two events can be combined to produce another event, called their **union**, which happens whenever either one event *or* the other event (or both events) happen as a result of a single run of the random experiment.[9]

When you think of two events as collections of outcomes, their union is a new collection consisting of all outcomes contained in either (or both) collections, as illustrated in Figure 6.4.5.

---

9. By convention, we will use the word *or* to signify "one or the other or both." Thus, you could say that the event "inflation or recession" happened not only during purely inflationary times or purely recessionary times, but also during those relatively rare time periods when both economic conditions prevailed.

For example, suppose your labor contract is coming up for renegotiation soon, and you expect your workers to demand large increases in pay and benefits. With respect to management's most likely contract proposal, you want to know how likely the various consequences are so that you can plan ahead for contingencies. In particular, your workers might go on strike. Another possibility is a work slowdown. Either one would be a problem for you. The random experiment here is "wait and record labor's reaction to management's contract proposal." The two events under consideration are "strike" and "slowdown." The union of these two events, "strike or slowdown," is clearly an important possibility since it represents a broad class of problem situations that will require your attention.

## The Union (*or*) Rule for Mutually Exclusive Events

If two events are mutually exclusive (i.e., they cannot both happen), you can find the exact probability of their union by adding their individual probabilities together.

Since two mutually exclusive events have no outcomes in common, when you add their probabilities, you count each outcome exactly once. Thus, the probability of their union is the sum of their individual probabilities:

> **The Union (*or*) Rule for Two Mutually Exclusive Events**
>
> Probability of "A or B" = Probability of A + Probability of B

## Finding *or* from *and* and Vice Versa

If you know the probabilities of the three events A, B, and "A and B," the probability of "A or B" can be found. This probability is determined by summing the two probabilities of the basic events and subtracting the probability of their

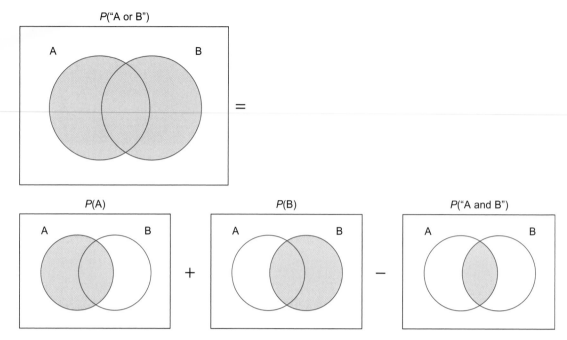

**FIGURE 6.4.6**    A Venn diagram showing how the probability of "A or B" may be found. First, add the probability of A and the probability of B. Then subtract the amount that has been counted twice, namely, the probability of the event "A and B."

intersection. The subtraction represents those outcomes that would be counted *twice* otherwise, as illustrated in Figure 6.4.6. The probability of the event "A or B" is given by the following formula:

> **Finding *or* from *and***
>
> Probability of "A or B" = Probability of A
> + Probability of B − Probability of "A and B"

For example, based on the past experience in your repair shop, suppose the probability of a blown fuse is 6% and the probability of a broken wire is 4%. Suppose also that you know that 1% of appliances to be repaired come in with both a blown fuse *and* a broken wire. With these three pieces of information, you can easily find the probability that an appliance has one problem or the other (or both):

Probability of "blown fuse or broken wire"
= 0.06 + 0.04 − 0.01 = 0.09

Thus, 9% of these appliances have one or the other problem (or both).

Using algebra, we can find a formula for the probability of "A and B" in terms of the probabilities of A, B, and "A or B":

> **Finding *and* from *or***
>
> Probability of "A and B" = Probability of A
> + Probability of B − Probability of "A or B"

In fact, based on knowledge of any three of the four probabilities (for A, B, "A and B," and "A or B"), the remaining probability can be found.

When are these formulas useful? One application is to take known information about probabilities and use a formula to find the probability for another event, perhaps a more meaningful or useful one. Another application is to ensure that the information you are basing decisions on is logically consistent. Suppose, for example, that you have probabilities for A and for B from the relative frequency approach using past data, and you plan to use subjective probabilities for the events "A and B" and "A or B." You will want to make sure that the four resulting probabilities taken together satisfy the preceding formulas.

## One Event *Given* Another: Reflecting Current Information

When you revise the probability of an event to reflect information that another event has occurred, the result is the **conditional probability** of the first event *given* the other event. (All of the ordinary probabilities you have learned about so far can be called **unconditional probabilities** if necessary to avoid confusion.) Here are some examples of conditional probabilities:

1. Suppose the home team has a 70% chance of winning the big game. Now introduce new information in terms of the event "the team is ahead at halftime." Depending on how this event turns out, the probability of winning should be revised. The probability of

winning given that we are ahead at halftime would be larger—say, 85%. This 85% is the conditional probability of the event "win" given the event "ahead at halftime." The probability of winning given that we are *behind* at halftime would be less than the 70% overall chance of winning—say, 35%. This 35% is the conditional probability of "win" given "behind at halftime."

2. Success of a new business project is influenced by many factors. To describe their effects, you could discuss the conditional probability of success given various factors, such as a favorable or unfavorable economic climate and actions by competitors. An expanding economy would increase the chances of success; that is, the conditional probability of success given an expanding economy would be larger than the (unconditional) probability of success.

## The Rule for Finding a Conditional Probability Given Certain Information

To find the conditional probability of the event "success" given "expanding economy," you would proceed by computing, out of all "expanding economy" scenarios, the proportion that corresponds to "success." This is the probability of "success and expanding economy" divided by the probability of "expanding economy."

This is the general rule for finding conditional probabilities. The conditional probability of A given B, provided the probability of B is positive, is[10]

> **Conditional Probability**
>
> Conditional Probability of A Given B
> $$= \frac{\text{Probability of "A and B"}}{\text{Probability of B}}$$

It makes a difference whether you are determining the conditional probability of A given B (which is the probability for A updated to reflect B) or the conditional probability of B given A (which is the quite different probability of B updated to reflect A). For completeness, here is the formula for the other conditional probability:

$$\text{Conditional Probability of B Given A} = \frac{\text{Probability of "A and B"}}{\text{Probability of A}}$$

For example, if the probability of a blown fuse is 6%, the probability of a broken wire is 4%, and the probability of "blown fuse and broken wire" is 1%, you can compute the conditional probability of a broken wire given a blown fuse:

$$\left( \begin{array}{c} \text{Conditional probability} \\ \text{of broken wire} \\ \text{given blown fuse} \end{array} \right)$$

$$= \frac{\text{Probability of "broken wire and blown fuse"}}{\text{Probability of blown fuse}}$$

$$= \frac{0.01}{0.06} = 0.167$$

In this case, having a blown fuse implies a greater likelihood that the wire is broken.

This conditional probability tells you that, out of all appliances you typically see that have a blown fuse, 16.7% also have a broken wire. Note how much greater this conditional probability is than the unconditional probability for a broken wire (4%). This happens because when you are given the fact that the fuse is blown, you are no longer talking about "all appliances" but are considering relatively few (6%) of them. The probability of "broken wire" is then revised from 4% to 16.7% to reflect this new information.

Figure 6.4.7 shows a Venn diagram for this situation. Note that the unconditional probabilities within each circle

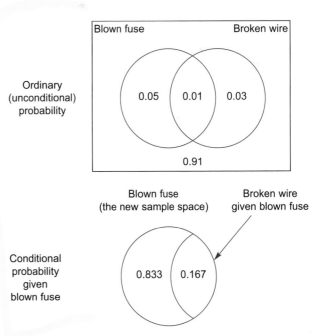

**FIGURE 6.4.7** A Venn diagram for unconditional probabilities (top) and conditional probabilities (bottom) given a blown fuse. With the given information, only the "blown fuse" circle is relevant; thus, it becomes the entire sample space. Dividing the old probabilities by 0.06 (the probability of a blown fuse) gives you their conditional probabilities within the new sample space.

---

10. Since you divide by the probability of B, the formula will not work if the event B has probability 0. In this special case, the conditional probability would be undefined. This is not a problem in practice since an event with probability 0 (essentially) never happens; hence, it doesn't really matter what happens "conditionally" on B.

add up to the correct numbers (0.06 for blown fuse and 0.04 for broken wire). For the conditional probabilities, since you are assured (by the given information) that there is a blown fuse, this circle becomes the new sample space because no other outcomes are possible. Within this new sample space, the old probabilities must be divided by 0.06 (the unconditional probability of a blown fuse) so that they represent 100% of the new situation.

## Conditional Probabilities for Mutually Exclusive Events

Since two mutually exclusive events cannot both happen, if you are given information that one has happened, it follows that the other did *not* happen. That is, the conditional probability of one event given the other is 0, provided the probability of the other event is not 0.

> **Conditional Probabilities for Two Mutually Exclusive Events**
>
> Conditional probability of "A given B" = 0 provided the probability of B is not 0.

## Independent Events

Two events are said to be **independent events** if information about one does not change your assessment of the probability of the other. If information about one event *does* change your assessment of the probability of the other, the events are **dependent events**. For example, "being a smoker" and "getting cancer" are dependent events because we know that smokers are more likely to get cancer than nonsmokers. On the other hand, the events "your stock market portfolio will go up tomorrow" and "you will oversleep tomorrow morning" are independent events since your oversleeping will not affect the behavior of stock market prices.[11]

Formally, events A and B are independent if the probability of A is the same as the conditional probability of A given B.

> **Events A and B are independent if**
>
> Probability of A = Conditional Probability of A given B
>
> **Events A and B are dependent if**
>
> Probability of A ≠ Conditional Probability of A given B

There are several ways to find out whether two events are independent or dependent. You will usually need to use a formula; just "thinking hard" about whether they *should* be independent or not should be used only as a last resort when there is not enough information available to use the formulas. Following are three formulas. Use the formula that is easiest in terms of the information at hand, since all three methods must (by algebra) always give the same answer.[12]

> **Events A and B are independent if any one of these formulas is true:**
>
> Probability of A = Conditional probability of A given B
> Probability of B = Conditional probability of B given A
> Probability of A and B = Probability of A × Probability of B

The third formula may be used to find the probability of "A and B" for two events that are known to be independent. However, this formula gives the wrong answer for dependent events.

> **Example**
> *Women in Executive Positions*
>
> Of the top executives of the 100 fastest-growing public companies headquartered in Washington state, just one out of 100 was a woman.[13] Using the relative frequency approach, you could say that the conditional probability of a (randomly selected) person being a woman given that the person was such an executive is 1/100 = 0.01 (i.e., 1% of such executives are women). In the population at large, the (unconditional) probability of being a woman is 51.1%.[14] Since the probability of a person being a woman changes from 51.1% to only 1% when the extra information "was such an executive" is given, these are not independent events. That is, "being a woman" and "being a top executive" were dependent events. Note that this conclusion follows from the rules (given by the previous equations) and the numbers, and not from pure thought about the situation.
>
> The fact that these are dependent events expresses the historical fact that there have been gender differences in this time and place: Men were more likely to be top executives than are women (different probabilities of being an executive), and executives were more likely to be men than are people in the population at large (different probabilities of being male).
>
> This probability analysis shows that there were gender differences, but it does not explain why. When you look at the percentages, a dependence, indicating gender differences, has clearly been found. These differences might be due to discrimination in hiring, a restricted supply of qualified job

---

11. Unless, of course, you are a very powerful player in the financial markets and miss an important "power breakfast" meeting.

12. There is just one technical difficulty to be considered: If one of the events has probability 0, it is not possible to compute the conditional probability of the other. Whenever one (or both) events has probability 0, we will declare them (automatically) independent of each other.

candidates, or some other reason; probability analysis will not (by itself) tell you which explanation is correct.

---

13. Based on information in *The Puget Sound Business Journal 2001 Book of Lists*, pp. 140–46.
14. This is based on the 2000 total U.S. population of males (134,554,000) and of females (140,752,000), as reported in U.S. Bureau of the Census, *Statistical Abstract of the United States: 2000* on CD-ROM (Washington, DC, 2000), Table 10.

### Example
#### Market Efficiency

Financial markets are said to be *efficient* if current prices reflect all available information. According to the theory of market efficiency, it is not possible to make special profits based on analyzing past price information, since this information is already reflected in the current price. Another conclusion is that prices must change randomly, since any systematic changes would be anticipated by the market.

One way to test market efficiency is to see whether there is any connection between yesterday's price change and today's price change. If the two events "price went up yesterday" and "price went up today" are independent, this would support the theory of market efficiency. In this case, knowing yesterday's price behavior would not help you predict the course of the market today.

On the other hand, if these events are dependent, then markets are not efficient. For example, if markets have "momentum" and tend to continue their upward or downward swings, then knowledge that the market went up yesterday would increase the likelihood of a rise today. This is incompatible with market efficiency, which holds that the markets would anticipate this further rise, and there would be no difference between the unconditional and the conditional probabilities.

## The Intersection (*and*) Rule for Independent Events

As mentioned earlier, for independent events (and *only* for independent events), you can find the probability of "A and B" by simply multiplying the probabilities of the two events.

---

**The Intersection (*and*) Rule for Independent Events**

Probability of "A and B"

= Probability of A × Probability of B

---

### Example
#### Risk Assessment for a Large Power Plant

Large power plants present some risk due to the potential for disaster. Although this possibility is small, the news media remind you from time to time that disasters do happen. Suppose that the probability of overheating at your plant is 0.001 (one in a thousand) per day and the probability of failure of the backup cooling system is 0.000001 (one in a million). If you assume that these events are independent, the probability of "big trouble" (that is, the event "overheating and failure of backup cooling system") would be 0.001 × 0.000001 = 0.000000001 (one in a billion), which is a tolerably small likelihood for some people.

However, it may not be appropriate to assume that these events are independent. It may appear that there is no direct connection between failure of one system (causing overheating) and failure of the other (leaving you without backup cooling). However, independence is not determined by subjective thoughts but by looking at the probabilities themselves. It may well be that these are dependent events; for example, there might be a natural disaster (flood or earthquake) that causes *both* systems to fail. If the events are not independent, the "one in a billion" probability of "big trouble" would be wrong, and the actual probability could be much larger.

## The Relationship between Independent and Mutually Exclusive Events

There is a big difference between *independent* and *mutually exclusive* events. In fact, two events that are independent *cannot* be mutually exclusive (unless one event has probability 0). Also, two events that are mutually exclusive cannot be independent (again, unless one event has probability 0). If one event (or both events) has probability 0, the events are independent *and* mutually exclusive.

## 6.5 WHAT'S THE BEST WAY TO SOLVE PROBABILITY PROBLEMS?

There are two ways to solve probability problems: the hard way and the easy way. The hard way involves creative application of the right combination of the rules you have just learned. The easy way involves seeing the bigger picture by constructing a *probability tree* and reading the answers directly from the tree. Another useful method that helps you see what is going on is the construction of a *joint probability table*. Yet another method, already discussed, is the Venn diagram. Whether you use the easy way or the hard way, the answer will be the same.

## Probability Trees

A **probability tree** is a picture indicating probabilities and conditional probabilities for combinations of two or more events. We will begin with an example of a completed tree and follow up with the details of how to construct the tree. Probability trees are closely related to *decision trees*, which are used in finance and other fields in business.

**Example**
*Managing Software Support*

Software support is a demanding task. Some customers call to ask for advice on how to use the software. Others need help with problems they are having with it. Based on past experience as manager of software support, you have figured some probabilities for a typical support call and summarized your results in the probability tree shown in Figure 6.5.1.

Figure 6.5.1 includes a lot of information. Starting from the left, note that the probability of the event "irate caller" is 0.20 (that is, 20% of callers were irate, which says that 80% were not irate).

Some conditional probabilities are shown in the figure in the set of four branches directly under the heading "was caller helped?" Note that 15% of the irate callers were helped (this is the conditional probability of "helped" given "irate caller") and 85% of irate callers were not helped. Continuing down the remaining branches, we see that 70% of "not irate" callers were helped and 30% of these were not helped. We certainly seem to be doing a better job of helping nonirate callers (70% versus 15% help rates).

The circled numbers at the right in Figure 6.5.1 indicate the probabilities of various events formed using combinations of *and* and *not*. The probability that the caller is irate and is helped is 0.03. That is, 3% of all callers were both irate and helped. Next, 17% of callers were both irate and not helped; 56% of callers were not irate and were helped; and, finally, 24% of callers were both not irate and not helped.

From this tree, you can find any probability of interest. For the event "irate caller," the probability is listed in the first column of circled numbers (0.20), with the probability of its complement, "not irate," just below. For the event "caller helped," the probability is found by adding the two circled probabilities at the right representing help: 0.03 + 0.56 = 0.59. The conditional probability for "caller helped" given "irate caller" is listed along that branch as 0.15. A little more work is required to find the conditional probability of "irate caller" given "caller helped." One solution is to use the definition of conditional probability:[15]

$$
\left( \begin{array}{c} \text{Conditional probability} \\ \text{of "irate caller"} \\ \text{given "caller helped"} \end{array} \right)
$$

$$
= \frac{\text{Probability of "irate caller and caller helped"}}{\text{Probability of "caller helped"}}
$$

$$
= \frac{0.03}{0.03 + 0.56} = 0.051
$$

Thus, of all callers we helped, 5.1% were irate. The other way to find this conditional probability is to construct a new tree that begins with "was caller helped?" instead of "irate caller," since the *given* information must come first in order for the conditional probability to be listed in a probability tree.

---

15. This is the one formula you may still need from the previous section. Probability trees do everything else for you!

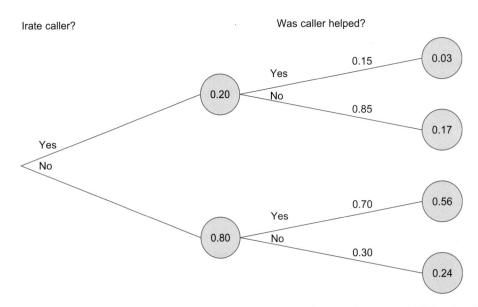

**FIGURE 6.5.1**   Probability tree for the events "irate caller" and "was caller helped." The circled numbers are probabilities; the others are conditional probabilities.

## Rules for Probability Trees

Constructing a probability tree consists of first drawing the basic tree, next recording all information given to you in the problem, and finally applying the fundamental rules to find the remaining numbers and complete the tree. Here are the rules for constructing a probability tree:

1. Probabilities are listed at each endpoint and circled. These add up to 1 (or 100%) at each level of the tree. For example, in Figure 6.5.1, at the first level is 0.20 + 0.80 = 1.00, and at the second level is 0.03 + 0.17 + 0.56 + 0.24 = 1.00. Use this rule if you have all but one circled probability in a column.
2. Conditional probabilities are listed along each branch (except perhaps for the first level) and add up to 1 (or 100%) for each group of branches radiating from a single point. For example, in the first branch of Figure 6.5.1 we have 0.15 + 0.85 = 1.00, and at the second branch we have 0.70 + 0.30 = 1.00. Use this rule if you have all but one conditional probability coming out from a single point.
3. The circled probability at a branch point times the conditional probability along a branch gives the circled probability at the end of the branch. For example, in Figure 6.5.1, for the upper-right branch line we have 0.20 × 0.15 = 0.03, for the next branch line we have 0.20 × 0.85 = 0.17, and similarly for the last two branch lines. Use this rule if you have all but one of the three numbers for a single branch line.
4. The circled probability at a branch point is the sum of the circled probabilities at the ends of all branches extending from it to the right. For example, in Figure 6.5.1, extending from the branch point labeled 0.20 are two branches ending with circled probabilities adding up to this number: 0.03 + 0.17 = 0.20. Use this rule if you have all but one of the circled probabilities in a group consisting of one together with all of its branches.

Using these rules, if you know all but one of the probabilities for a particular branch or level, you can find the remaining probability. For reference, a generic probability tree is shown in Figure 6.5.2a, indicating the meanings of all numbers in the tree. The generic Venn diagram in Figure 6.5.2b is included for comparison.

### Example
#### Drug Testing of Employees

Your firm is considering mandatory drug testing of all employees. To assess the costs (resources for testing and possible morale problems) and benefits (a more productive workforce), you have decided to study the probabilities of the various outcomes on a per-worker basis. The laboratory has provided you with some information, but not everything

you need to know. Using a probability tree, you hope to compute the missing, helpful information.

The testing procedure is not perfect. The laboratory tells you that if an employee uses drugs, the test will be "positive" with probability 0.90. Also, for employees who do not use drugs, the test will be "negative" (that is, "not positive") 95% of the time. Based on an informal poll of selected workers, you estimate that 8% of your employees use drugs.

Your basic probability tree for this situation is shown in Figure 6.5.3. The event "drug user" has been placed first since some of the initial information comes as a conditional probability given this information.

After recording the initial information, you have the tree in Figure 6.5.4. Note that 90% and 95% are conditional probabilities along the branches; the value 8% for drug users is an unconditional probability.

Although other problems may also provide three pieces of information from which to complete the tree, these items may well go in different places on the tree. For example, had you been given a probability (unconditional) for "test positive," you would not have been able to record it directly; you would have had to make a note that the first and third circles at the far right add up to this number.

Next, apply the fundamental rules to complete the probability tree. This is like solving a puzzle, and there are many ways to do it that all lead to the right answer. For example, you might apply rule 1 to find that 0.92 goes below 0.08. Rule 2 also gives the conditional probabilities 0.10 and 0.05. Finally, rule 3 gives all of the probabilities for the ending circles. These results, which compose the completed tree, are shown in Figure 6.5.5.

It is now easy to make a Venn diagram, like that shown in Figure 6.5.6, using the results at the far right of the probability tree in Figure 6.5.5. Although you don't have to draw the Venn diagram to get answers, some people find the Venn diagram helpful.

You can now find any probability or conditional probability quickly and easily from the probability tree (Figure 6.5.5) or the Venn diagram (Figure 6.5.6). Here are some examples:

Probability of "drug user and test positive" = 0.072
Probability of "'not drug user' and test positive" = 0.046
Probability of "test positive" = 0.072 + 0.046 = 0.118
Conditional probability of "test positive" given "not drug user" = 0.05

Other conditional probabilities can be found through application of formulas, for example:

$$\left(\begin{array}{c}\text{Conditional probability}\\\text{of "drug user" given}\\\text{"test positive"}\end{array}\right)$$
$$=\frac{\text{Probability of "drug user and test positive"}}{\text{Probability of "test positive"}}$$
$$=\frac{0.072}{0.072+0.046}=0.610$$

*(Continued)*

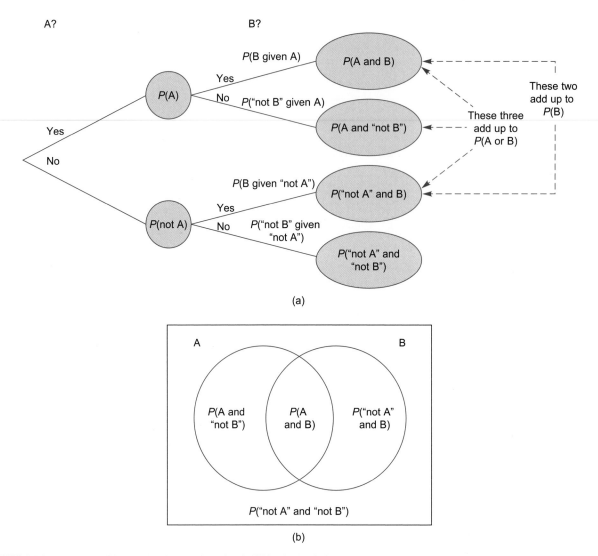

(a)

(b)

**FIGURE 6.5.2**   (a) A probability tree showing a variety of probabilities (in the circles) and conditional probabilities given A (along the branches). (b) The generic Venn diagram, with the four basic probabilities indicated. These four probabilities are the same as the ending probabilities at the far right of a probability tree.

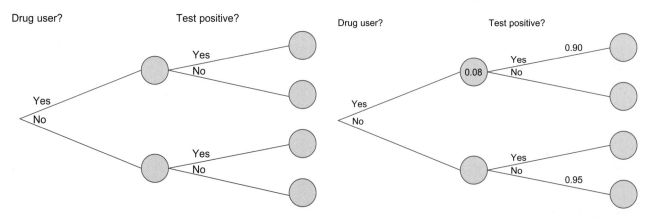

**FIGURE 6.5.3**   Probability tree for the events "drug user" and "test positive" before any information is recorded.

**FIGURE 6.5.4**   The probability tree after the initial information has been recorded.

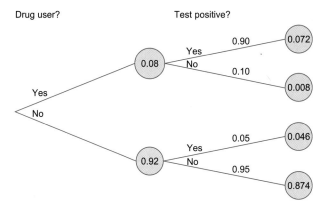

**FIGURE 6.5.5**   The completed probability tree, after the fundamental rules have been applied.

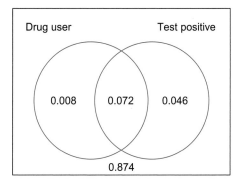

**FIGURE 6.5.6**   The Venn diagram for the employee drug testing example, with the four basic probabilities indicated.

**Example—cont'd**

This conditional probability is especially interesting. Despite the apparent reliability of the drug-testing procedure (90% positive for drug users and 95% negative for nonusers), there is a conditional probability of only 61% for "drug user" given "test positive." This means that, among all of your workers who test positive, only 61% will be drug users, and the remaining 39% will not be drug users.

This is a sobering thought indeed! Are you really ready to implement a procedure that includes 39% innocent people among those who test positive for drugs? Probability analysis has played an important role here in changing the initial information into other, more useful probabilities.

**Example**

*A Pilot Project Helps Predict Success of a Product Launch*

Your firm is considering the introduction of a new product, and you have the responsibility of presenting the case to upper-level management. Two key issues are (1) whether or not to proceed with the project at all, and (2) whether or not it is worthwhile to invest initially in a *pilot project* in a test market, which would cost less and provide some information as to whether the product is likely to succeed.

In getting your thoughts together, you have decided to assume the *what if* scenario that includes both the pilot project and the product launch. You also believe that the following probabilities are reasonable:

1. The probability that the product launch will succeed is 0.60.
2. The probability that the pilot project will succeed is 0.70. (This probability is slightly higher because the pilot project involves a more receptive market.)
3. The probability that either the pilot project or the product launch (or both) will succeed is 0.75.

To help you decide if the pilot project is worthwhile, you want to find out (1) the conditional probability that the product launch is successful given that the pilot project is successful, and (2) the conditional probability that the product launch is successful given that the pilot project fails. In addition, for general background you also need (3) the probability that both the pilot project and the product launch succeed, and (4) the probability that they both fail.

All of these probabilities can be found by creative application of the basic formulas from Section 6.4. However, it can take a person a long time to determine the appropriate combination to use. It's easier to construct the probability tree and read off the answers.

Figure 6.5.7 shows the basic probability tree with the initial information recorded. Note that two of the three numbers have no immediate place in the tree, but you may record them on the side, as shown.

What to do next? Rule 1 for probability trees (or the complement rule) gives 1.00 – 0.70 = 0.30 for the lower circle on the left. However, the circles on the right require more thought. If the top three numbers add up to 0.75 and two of them add up to 0.60, then the third number must be the difference, 0.75 – 0.60 = 0.15, which goes in the second circle from the top (the probability of "successful pilot and 'not successful product launch'"). Now you can use rule 4 to find the probability for the top circle: 0.70 – 0.15 = 0.55. You now have two of the top three circles on the right; since all three add up to 0.75, the remaining one must be 0.75 – 0.55 – 0.15 = 0.05. At this point you can use the basic rules to complete the tree, as shown in Figure 6.5.8.

From this completed tree, you can read off the required probabilities (and any other probabilities of interest to you). Here they are, with some brief discussion of the interpretation of the conditional probabilities:

1. The conditional probability that the product launch is successful given that the pilot project is successful is 0.786. If the pilot project were perfectly predictive of success of the product launch, this number would be 1.00. With real-world imperfections, there is only a 78.6% chance of eventual success with the product following a successful pilot project.
2. The conditional probability that the product launch is successful given that the pilot project fails is 0.167. If the pilot project were perfectly predictive of later success,

*(Continued)*

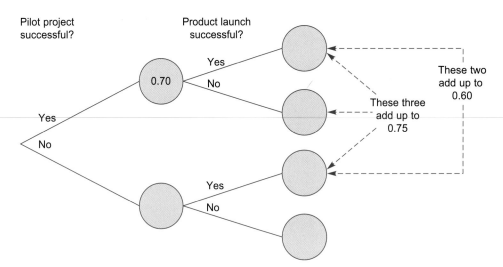

**FIGURE 6.5.7**   The initial probability tree, before the fundamental rules are applied.

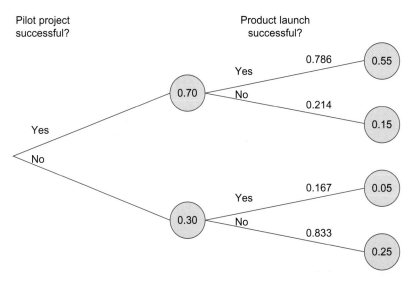

**FIGURE 6.5.8**   The completed probability tree, after the fundamental rules have been applied.

**Example—cont'd**

then this number would be 0. However, here you find a 16.7% chance for success even if the pilot project fails.

3. The probability that both the pilot project and the product launch succeed is 0.55.
4. The probability that they both fail is 0.25.

**Example**

*Solution to "Is It behind Door Number 1, Door Number 2, or Door Number 3?"*

We can now construct a probability tree for the example that was presented in Section 6.1. It is reasonable to assume that the prize is placed behind a random door (i.e., you have no clue as to which door it is) and that your guess is also

random.[16] The probability tree for this situation is shown in Figure 6.5.9. This tree is a little different from those you have seen before, since there are *three* possibilities stemming from each branch. However, the fundamental rules for probability trees still hold true.

The basic numbers used to construct the tree are the probabilities of the prize being behind a given door (1/3) and the conditional probabilities of your guess (also 1/3—always the same conditional probability, since you don't know which door the prize is behind). Then, following the details of completing the tree, the answer is clear: Switching doubles your chances of winning from 1/3 to 2/3.

16. This assumption of random guessing is not necessary; in fact, you could always guess door 1, but the assumption simplifies the analysis.

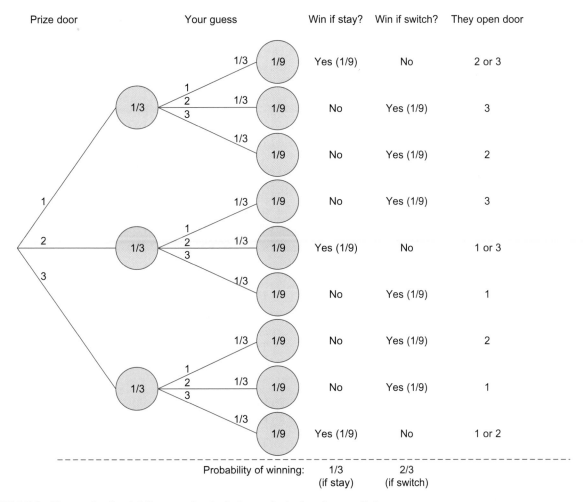

FIGURE 6.5.9 The completed probability tree, after the fundamental rules have been applied.

## Joint Probability Tables

The **joint probability table** for two events gives you probabilities for the events, their complements, and combinations using *and*. Here is the joint probability table for the previous example of employee drug testing:

|  |  | Test positive? | | |
|---|---|---|---|---|
|  |  | Yes | No | |
|  | Yes | 0.072 | 0.008 | 0.08 |
| Drug user? | No | 0.046 | 0.874 | 0.92 |
|  |  | 0.118 | 0.882 | 1 |

The numbers inside the table are the four circled numbers at the right of the probability tree. The numbers outside the table are totals, called *marginal probabilities*, and represent the probabilities of each event and its complement.

More generally, Figure 6.5.10 shows how to interpret the numbers in a generic joint probability table. Note that no conditional probabilities are given; they are easily computed using the basic formula.

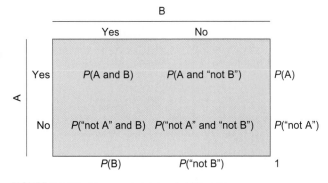

FIGURE 6.5.10 The generic joint probability table.

## 6.6 END-OF-CHAPTER MATERIALS

### Summary

To understand random, unpredictable real-world situations, we begin by narrowing down the possibilities and carefully setting up an exact framework for the study of probability. A **random experiment** is any well-defined procedure that produces an observable outcome that could not be perfectly predicted in advance. Each random experiment has a **sample space**, which is a list of *all possible outcomes* of the random experiment, prepared in advance without knowing what will happen when the experiment is run. Each time a random experiment is run, it produces exactly one **outcome**, the result of the random experiment, describing and summarizing the observable consequences. An **event** either happens or does not happen each time the random experiment is run; formally, an event is any collection of outcomes specified in advance before the random experiment is run. There may be more than one event of interest for a given situation.

Associated with every event is a number between 0 and 1, called its **probability**, which expresses how likely it is that the event will happen each time the random experiment is run. If you run a random experiment many times, the **relative frequency** of an event is the proportion of times the event occurs out of the number of times the experiment is run. The **law of large numbers** says that the relative frequency (a random number) will be close to the probability (an exact, fixed number) if the experiment is run many times. Thus, a relative frequency based on past data may be used as an approximation to a probability value. A **theoretical probability** is computed using an exact formula based on a mathematical theory or model such as the *equally likely* rule; that is, if all outcomes are equally likely, then

$$\text{Probability of an Event} = \frac{\text{Number of outcomes in the event}}{\text{Total number of possible outcomes}}$$

A **subjective probability** is anyone's opinion (use an expert, if possible) of what the probability is for an event. The methods of **Bayesian analysis** in statistics involve the use of subjective probabilities in a formal, mathematical way. A non-Bayesian analysis is called a **frequentist analysis** and does not use subjective probabilities in its computations, although it is not totally objective, since opinions will have some effect on the choice of data and model (the mathematical framework).

A **Venn diagram** is a picture that represents the universe of all possible outcomes (the sample space) as a rectangle with events indicated inside, often as circles or ovals, as in Figure 6.6.1. The **complement** of an event is another event that happens only when the first event does *not* happen. The complement rule is

Probability of "not A" = 1 − Probability of A

The **intersection** of two events is an event that occurs whenever one event *and* the other event both happen as a result of a single run of the random experiment.

Two events that cannot both happen at once are said to be **mutually exclusive events**. The rules for two mutually exclusive events are

Probability of "A and B" = 0
Probability of "A or B" = Probability of A + Probability of B

The **union** of two events is an event that occurs whenever one event *or* the other event happens (or both events happen) as a result of a single run of the random experiment. Based on knowledge of any three of the four probabilities (for A, B,

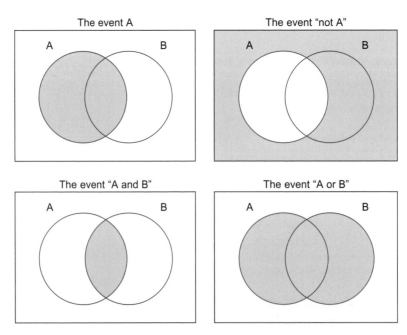

**FIGURE 6.6.1**  Venn diagrams provide an effective display of the meanings of the operations *not, and,* and *or,* which define events in terms of other events.

"A and B," and "A or B"), the remaining probability can be found using one of the following formulas:

Probability of "A or B" = Probability of A
  + Probability of B − Probability of "A and B"

Probability of "A and B" = Probability of A
  + Probability of B − Probability of "A or B"

When you revise the probability of an event to reflect information that another event has occurred, the result is the **conditional probability** of that event *given* the other event. Ordinary (unrevised) probabilities are called **unconditional probabilities**. Conditional probabilities are found as follows (and are left undefined if the given information has probability 0):

Conditional Probability of A Given B

$$= \frac{\text{Probability of ``A and B''}}{\text{Probability of B}}$$

For two mutually exclusive events, the conditional probability is always 0 (unless it is undefined for technical reasons).

Two events are said to be **independent events** if information about one event does not change your assessment of the probability of the other event. If information about one event *does* change your assessment of the probability of the other, we say that the events are **dependent events**. Use any one of the following formulas to find out if two events are independent. Events A and B are independent if any one of these formulas is true:

Probability of A = Conditional probability of A given B
Probability of B = Conditional probability of B given A
Probability of "A and B" = Probability of A
                                × Probability of B

The third formula may be used to find the probability of "A and B" for two events that are known to be independent but gives wrong answers for dependent events. Two independent events cannot be mutually exclusive unless one of the events has probability 0.

A **probability tree** is a picture indicating probabilities and some conditional probabilities for combinations of two or more events. One of the easiest ways to solve a probability problem is to construct the probability tree and then read the answer. All probabilities and conditional probabilities may be found by adding up numbers from the tree or by applying the formula for a conditional probability. Here are the four rules used to construct and validate probability trees:

1. Probabilities are listed at each endpoint and circled. These add up to 1 (or 100%) at each level of the tree.
2. Conditional probabilities are listed along each branch (except perhaps for the first level) and add up to 1 (or 100%) for each group of branches coming out from a single point.
3. The circled probability at a branch point times the conditional probability along a branch gives the circled probability at the end of the branch (on the right).
4. The circled probability at a branch point is the sum of the circled probabilities at the ends of all branches coming out from it to the right.

The **joint probability table** for two events lists probabilities for the events, their complements, and combinations using *and*. Conditional probabilities can then be found by using the formula for a conditional probability.

## Key Words

**Bayesian analysis**, *132*
**complement** *(not)*, *133*
**conditional probability**, *136*
**dependent events**, *138*
**event**, *128*
**frequentist (non-Bayesian) analysis**, *146*
**independent events**, *138*
**intersection** *(and)*, *134*
**joint probability table**, *145*
**law of large numbers**, *130*
**mutually exclusive events**, *134*
**outcome**, *128*
**probability**, *129*
**probability tree**, *139*
**random experiment**, *127*
**relative frequency**, *130*
**sample space**, *127*
**subjective probability**, *132*
**theoretical probability**, *131*
**unconditional probability**, *136*
**union** *(or)*, *135*
**Venn diagram**, *133*

### Questions

1. **a.** What is a random experiment?
   **b.** Why does defining a random experiment help to focus your thoughts about an uncertain situation?
2. **a.** What is a sample space?
   **b.** Is there anything random or uncertain about a sample space?
3. **a.** What is an outcome?
   **b.** Must the outcome be a number?
4. **a.** What is an event?
   **b.** Can a random experiment have more than one event of interest?
5. **a.** What is a probability?
   **b.** Which of the following has a probability number: a random experiment, a sample space, or an event?
   **c.** If a random experiment is to be run just once, how can you interpret an event with a probability of 0.06?

6. **a.** What is the relative frequency of an event?
   **b.** How is the relative frequency different from the probability of an event?
   **c.** What is the law of large numbers?
7. **a.** Name the three main sources of probability numbers.
   **b.** What is the *equally likely* rule?
   **c.** Are you allowed to use someone's guess as a probability number?
   **d.** What is the difference between a Bayesian and a frequentist analysis?
8. What are mutually exclusive events?
9. **a.** What is the complement of an event?
   **b.** What is the probability of the complement of an event?
10. **a.** What is the intersection of two events?
    **b.** What is the probability of "one event and another" if you know
       **(1)** Their probabilities and the probability of "one event or the other"?
       **(2)** Their probabilities and that they are independent?
       **(3)** That they are mutually exclusive?
11. **a.** What is the union of two events?
    **b.** What is the probability of "one event or another" if you know
       **(1)** Their probabilities and the probability of "one event and the other"?
       **(2)** That they are mutually exclusive?
12. **a.** What is the interpretation of conditional probability in terms of new information?
    **b.** Is the conditional probability of A given B always the same number as the conditional probability of B given A?
    **c.** How can you find the conditional probability from the probabilities of two events and the probability of their intersection?
    **d.** What are the conditional probabilities for two independent events?
    **e.** Is the conditional probability of A given B a probability about A or a probability about B?
13. **a.** What is the interpretation of independence of two events?
    **b.** How can you tell whether two events are independent or not?
    **c.** Under what conditions can two mutually exclusive events be independent?
14. **a.** What is a probability tree?
    **b.** What are the four rules for probability trees?
15. What is a joint probability table?
16. What is a Venn diagram?

## Problems

*Problems marked with an asterisk (\*) are solved in the Self Test in Appendix C.*

1.\* As a stock market analyst for a major brokerage house, you are responsible for knowing everything about the automotive companies. In particular, Ford is scheduled to release its net earnings for the past quarter soon, and you do not know what that number will be.
   **a.** Describe the random experiment identified here.
   **b.** What is the sample space?
   **c.** What will the outcome tell you?
   **d.** Based on everything you know currently, you have computed a dollar figure for net earnings that you feel is likely to be close to the actual figure to be announced. Identify precisely the event "the announced earnings are higher than expected" by listing the qualifying outcomes from the sample space.
   **e.** You have an idea of the probability that the announced earnings will be higher than expected. What kind of probability is this if it is based not just on past experience but also on your opinion as to the current situation?

2. You are operations manager for a plant that produces copy machines. At the end of the day tomorrow, you will find out how many machines were produced and, of these, how many are defective.
   **a.** Describe the random experiment identified here.
   **b.** What is the sample space? (*Hint:* Note that each outcome consists of two pieces of information.)
   **c.** What will the outcome tell you?
   **d.** Identify precisely the event "met the goal of at least 500 working (nondefective) machines with 2 or fewer defective machines" by listing the qualifying outcomes from the sample space.
   **e.** In 22 of the past 25 days, this goal has been met. Find the appropriate relative frequency.
   **f.** About how far away from the true, unknown probability of meeting the goal is the relative frequency from part e? (For simplicity, you may assume that these were independent runs of the same random experiment.)

3. As production manager, you are responsible for the scheduling of workers and machines in the manufacturing process. At the end of the day, you will learn how many seat covers were produced.
   **a.** Describe the random experiment identified here.
   **b.** What is the sample space?
   **c.** What will the outcome tell you?
   **d.** Identify precisely the event "produced within plus or minus 5 of the daily goal of 750" by listing the qualifying outcomes from the sample space.
   **e.** In 8 of the past 15 days, this event (from part d) has occurred. Find the appropriate relative frequency.

4. Of the 925 tires your factory just produced, 13 are defective.
   **a.** Find the probability that a randomly selected tire is defective.
   **b.** Find the probability that a randomly selected tire is not defective.
   **c.** What kind of probability numbers are these?

5.\* You are responsible for a staff of 35 out of the 118 workers in a rug-weaving factory. Next Monday a representative will be chosen from these 118 workers. Assume that the representative is chosen at random, without regard to whether he or she works for you.

a. What is the probability that one of your workers will be chosen?

b. What is the probability that the representative will not be one of yours?

6. Suppose two events are independent. One event has probability 0.25, while the other has probability 0.59.

a. Find the probability that both events happen.

b. Find the probability of the union of these events.

c. Find the probability that neither event happens.

7. As part of a project to determine the reliability of construction materials, 100 samples were subject to a test simulating five years of constant use. Of these, 11 samples showed unacceptable breakage. Find the relative frequency of the event "unacceptable breakage" and tell approximately how far this number is from the probability that a sample will show unacceptable breakage.

8.* You are responsible for scheduling a construction project to build a convention center. In order to avoid big trouble, you will need the concrete to be placed before July 27 and for the financing to be arranged before August 6. Based on your experience and that of others, using subjective probability, you have fixed probabilities of 0.83 and 0.91 for these two events, respectively. Assume also that you have a 96% chance of meeting one deadline or the other (or both).

a. Find the probability of "big trouble."

b. Are these events mutually exclusive? How do you know?

c. Are these events independent? How do you know?

9. Two divisions are cooperating in the production of a communications satellite. In order for the launch to be on time, both divisions must meet the deadline. You believe that each has an 83% chance of meeting the deadline. Assuming the two divisions work independently (so that you have independent events), what is the probability that the launch will have to be delayed due to a missed deadline?

10. The probability of getting a big order currently under negotiation is 0.4. The probability of losing money this quarter is 0.5.

a. Assume that these are mutually exclusive events. Find the probability of getting the order or losing money.

b. Again assume that these are mutually exclusive events. Does this rule out the possibility that you fail to get the order and make money this quarter?

c. Now assume instead that the events "get the order" and "lose money" are independent (since the order will not show up in this quarter's financial statements). Find the probability of getting the order and losing money.

d. Now assume instead that the probability of getting the order and losing money is 0.1. Are the events "get the order" and "lose money" independent? How do you know?

11. Your firm has classified orders as either large or small in dollar amount and as either light or heavy in shipping weight. In the recent past, 28% of orders have been large dollar amounts, 13% of orders have been heavy, and 10% of orders have been large in dollar amounts and heavy in weight.

a. Complete a probability tree for this situation, with the first branch being the event "large dollar amount."

b. Construct the joint probability table for this situation.

c. Draw a Venn diagram for this situation.

d. Find the probability that an order is large in dollar amount or heavy (or both).

e. Find the probability that an order is for a large dollar amount and is not heavy.

f. Of the orders with large dollar amounts, what percentage are heavy? What conditional probability does this represent?

g. Of the heavy orders, what percentage are for large dollar amounts? What conditional probability does this represent?

h. Are the events "large dollar amount" and "heavy" mutually exclusive? How do you know?

i. Are the events "large dollar amount" and "heavy" independent? How do you know?

12. Two events are mutually exclusive, one with probability 0.38 and the other with probability 0.54.

a. Find the conditional probability that the first event happens given that the second event happens.

b. Find the probability of the union of these two events.

13. Your company maintains a database with information on your customers, and you are interested in analyzing patterns observed over the past quarter. In particular, 23% of customers in the database placed new orders within this period. However, for those customers who had a salesperson assigned to them, the new order rate was 58%. Overall, 14% of customers within the database had salespeople assigned to them.

a. Draw a probability tree for this situation.

b. What percentage of customers in the database placed a new order but did not have a salesperson assigned to them?

c. Given that a customer did not place a new order, what is the probability that the customer had a salesperson assigned to him or her?

d. If a customer did not have a salesperson assigned to him or her, what is the probability that the customer placed a new order?

14. The human resources department of a company is considering using a screening test as part of the hiring process for new employees and is analyzing the results of a recent study. It was found that 60% of applicants score high on the test, but only 83% of those who score high on the test perform well on the job. Moreover, among those who did not score high on the test, 37.5% perform well on the job.

a. Draw a probability tree for this situation.

b. What percentage of applicants perform well on the job?

c. Of those who did not score high, what percentage did not perform well?

d. Given that an applicant performed well on the job, what is the probability that the applicant did not score high on the test?

**15.** A repair shop has two technicians with different levels of training. The technician with advanced training is able to fix problems 92% of the time, while the other has a success rate of 80%. Suppose you have a 30% chance of obtaining the technician with advanced training.

a. Draw a probability tree for this situation.

b. Find the probability that your problem is fixed.

c. Given that your problem is fixed, find the probability that you did not obtain the technician with advanced training.

**16.** It is currently difficult to hire in the technology sector. Your company believes that the chances of successfully hiring this year are 0.13. Given that your company is successful in hiring, the chances of finishing the project on time are 0.78, but if hiring is not successful, the chances are reduced to 0.53.

a. Draw a probability tree for this situation.

b. Find the probability of finishing the project on time.

**17.** Your firm is planning a new style of advertising and figures that the probability of increasing the number of customers is 0.63, while the probability of increasing sales is 0.55. The probability of increasing sales given an increase in the number of customers is 0.651.

a. Draw a probability tree for this situation.

b. Find the probability of increasing both sales and customers.

c. If sales increase, what is the probability that the number of customers increases?

d. If sales do not increase, what is the probability that the number of customers increases?

e. Find the probability that neither customers nor sales increase.

**18.*** There are two locations in town (north and south) under consideration for a new restaurant, but only one location will actually become available. If it is built in the north, the restaurant stands a 90% chance of successfully surviving its first year. However, if it is built in the south, its chances of survival are only 65%. It is estimated that the chances of the northern location being available are 40%.

a. Draw a probability tree for this situation, with the first branch being "location."

b. Find the probability that the restaurant will survive its first year.

c. Find the probability that the restaurant is built in the south and is successful.

d. Find the probability that the restaurant is in the south given that it is successful.

e. Find the probability of failure given that it is in the north.

**19.** The coming year is expected to be a good one with probability 0.70. Given that it is a good year, you expect that a dividend will be declared with probability 0.90. However, if it is not a good year, then a dividend will occur with probability 0.20.

a. Draw a probability tree for this situation, with the most appropriate choice for the first branch.

b. Find the probability that it is a good year and a dividend is issued.

c. Find the probability that a dividend is issued.

d. Find the conditional probability that it is a good year, given that a dividend is issued.

**20.** Your firm is considering the introduction of a new toothpaste. At a strategy session, it is agreed that a marketing study will be successful with probability 0.65. It is also agreed that the probability of a successful product launch is 0.40. However, given that the marketing study is successful, the probability of a successful product launch is 0.55.

a. Draw the appropriate probability tree for this situation.

b. Find the probability that the marketing study is successful and the product launch is successful.

c. Given that the product launch succeeds, find the conditional probability that the marketing study was favorable.

d. Find the conditional probability that the product launch is successful, given that the marketing study was not successful.

e. Are the two events "marketing study successful" and "product launch successful" independent? How do you know?

**21.** Your store is interested in learning more about customers' purchasing patterns and how they relate to the frequency of store visits. The probability that a customer's visit will result in a purchase is 0.35. The probability that a customer has been to the store within the past month is 0.20. Of those who did not buy anything, 12% had been there within the past month.

a. Draw a probability tree for this situation.

b. Find the conditional probability of making a purchase given that the customer had been to the store within the past month.

c. What percent of customers are frequent shoppers, defined as making a purchase and having been to the store within the past month?

**22.** Your group has been analyzing quality control problems. Suppose that the probability of a defective shape is 0.03, the probability of a defective paint job is 0.06, and that these events are independent.

a. Find the probability of defective shape and defective paint job.

b. Find the probability of defective shape or defective paint job.

c. Find the probability of a nondefective item (i.e., with neither of these defects).

**23.** With your typical convenience store customer, there is a 0.23 probability of buying gasoline. The probability of buying groceries is 0.76 and the conditional probability of buying groceries given that they buy gasoline is 0.85.

   a. Find the probability that a typical customer buys both gasoline and groceries.

   b. Find the probability that a typical customer buys gasoline or groceries.

   c. Find the conditional probability of buying gasoline given that the customer buys groceries.

   d. Find the conditional probability of buying groceries given that the customer did not buy gasoline.

   e. Are these two events (groceries, gasoline) mutually exclusive?

   f. Are these two events independent?

24. You just learned good news: A prototype of the new product was completed ahead of schedule and it works better than expected. Would you expect "the conditional probability that this product will be successful given the good news" to be larger than, smaller than, or equal to the (unconditional) probability of success?

25. Your company sends out bids on a variety of projects. Some (actually 30% of all bids) involve a lot of work in preparing bids for projects you are likely to win, while the others are quick calculations sent in even though you feel it is unlikely that your company will win. Given that you put a lot of work into the bid, there is an 80% chance you will win the contract to do the project. Given that you submit a quick calculation, the conditional probability is only 10% that you will win.

   a. Draw a probability tree for this situation.

   b. What is the probability that you will win a contract?

   c. Given that you win a contract, what is the conditional probability that you put a lot of work into the bid?

   d. Given that you do not win a contract, what is the conditional probability that you put a lot of work into the bid?

26. Suppose 35.0% of employees are staff scientists, 26.0% are senior employees, and 9.1% are both. Are "staff scientist" and "senior employee" independent events?

27. Your marketing department has surveyed potential customers and found that (1) 27% read the trade publication *Industrial Chemistry*, (2) 18% have bought your products, and (3) of those who read *Industrial Chemistry*, 63% have never bought your products.

   a. Draw a probability tree for this situation.

   b. What percentage of the potential customers neither read *Industrial Chemistry* nor have bought your products? (This group represents future expansion potential for your business.)

   c. Find the conditional probability of reading *Industrial Chemistry* given that they have bought your products. (This indicates the presence of the publication among your current customers.)

28. Based on analysis of data from last year, you have found that 40% of the people who came to your store had not been there before. While some just came to browse, 30% of people who came to your store actually bought something. However, of the people who had not been there before, only 20% bought something. You are thinking about these percentages as probabilities representing what will happen each time a person comes to your store.

   a. What kind of probability numbers are these, in terms of where they come from?

   b. Draw a probability tree for this situation.

   c. Find the probability that a person coming to your store will have been there before and will make a purchase.

   d. What is the probability that a customer had been there before given that the customer did not purchase anything during his or her visit?

29. Your telephone operators receive many different types of calls. Requests for information account for 75% of all calls, while 15% of calls result in an actual order. Also, 10% of calls involve both information requests and order placement.

   a. What is the conditional probability that a call generated an order, given that it requested information? (This tells you something about the immediate value to your business of handling information requests.)

   b. What is the conditional probability that a call did not request information, given that it generated an order? (This represents the fraction of your orders that were "easy.")

   c. What is the probability that a call generated an order and did not request information? Interpret this number.

   d. Why are the answers to parts b and c different?

   e. Are the two events "requested information" and "generated an order" independent? How do you know?

30. You've just put in a bid for a large communications network. According to your best information you figure there is a 35% chance that your competitors will outbid you. If they do outbid you, you figure you still have a 10% chance of getting the contract by successfully suing them. However, if you outbid them, there is a 5% chance you will lose the contract as a result of their suing you.

   a. Complete a probability tree for this situation.

   b. Find the probability that you will be awarded the contract.

   c. Find the probability that you will outbid your competitors and be awarded the contract.

   d. Find the conditional probability that you will outbid your competitors given that you are awarded the contract.

   e. Are the events "you were not awarded the contract" and "you outbid the competition" mutually exclusive? Why or why not?

31. The probability that the project succeeds in New York is 0.6, the probability that it succeeds in Chicago is 0.7, and the probability that it succeeds in both markets is 0.55. Find the conditional probability that it succeeds in Chicago given that it succeeds in New York.

32. The espresso project will do well with probability 0.80. Given that it does well, you believe the herbal tea project is likely to do well with probability 0.70. However, if the espresso project does not do well, then the herbal tea project has only a 25% chance of doing well.
    a. Complete a probability tree for this situation.
    b. Find the probability that the herbal tea project does well.
    c. Find the probability that both projects do well.
    d. Find the conditional probability that the espresso project does well given that the herbal tea project does well. Compare it to the appropriate unconditional probability and interpret the result.

33. You have followed up on people who received your catalog mailing. You found that 4% ordered the hat and 6% ordered the mittens. Given that they ordered the hat, 55% also ordered the mittens.
    a. What percentage ordered both items?
    b. What percentage ordered neither?
    c. Given that they did not buy the hat, what percentage did order the mittens nonetheless?

34. Of your customers, 24% have high income, 17% are well educated. Furthermore, 12% both have high income and are well educated. What percentage of the well-educated customers have high income? What does this tell you about a marketing effort that is currently reaching well-educated individuals although you would really prefer to target high-income people?

35. Your production line has an automatic scanner to detect defects. In recent production, 2% of items have been defective. Given that an item is defective, the scanner has a 90% chance of identifying it as defective. Of the nondefective items, the scanner has a 90% chance of identifying it correctly as nondefective. Given that the scanner identifies a part as defective, find the conditional probability that the part is truly defective.

36. You have determined that 2.1% of the CDs that your factory manufactures are defective due to a problem with materials and that 1.3% are defective due to human error. Assuming that these are independent events, find the probability that a CD will have at least one of these defects.

37. You feel that the schedule is reasonable provided the new manager can be hired in time, but the situation is risky regardless. You figure there's a 70% chance of hiring the new manager in time. If the new manager is hired in time, the chances for success are 80%; however, if the new manager cannot be hired in time, the chances for success are only 40%. Find the probability of success.

38. There are 5% defective parts manufactured by your production line, and you would like to find these before they are shipped. A quick and inexpensive inspection method has been devised that identifies 8% of parts as defective. Of parts identified as defective, 50% are truly defective.
    a. Complete a probability tree for this situation.
    b. Find the probability that a defective part will be identified (i.e., the conditional probability of being identified given that the part was defective).
    c. Find the probability that a part is defective or is identified as being defective.

d. Are the events "identified" and "defective" independent? How do you know?
    e. Could an inspection method be useful if the events "identified" and "defective" were independent? Please explain.

39. There is a saying about initial public offerings (IPOs) of stock: "If you want it, you can't get it; if you can get it, you don't want it." The reason is that it is often difficult for the general public to obtain shares initially when a "hot" new company first goes on sale. Instead, most of us have to wait until it starts trading on the open market, often at a substantially higher price. Suppose that, given that you can obtain shares at the initial offering, the probability of the stock performing well is 0.35. However, given that you are unable to initially purchase shares, the conditional probability is 0.8 of performing well. Overall, assume that you can obtain shares in about 15% of IPOs.
    a. Draw a probability tree for this situation.
    b. Find the probability of both (1) your being able to purchase the stock at the initial offering and (2) the stock performing well.
    c. How much access to successful IPOs do you have? Answer this by finding the conditional probability that you are able to purchase stock initially, given that the stock performs well.
    d. What percentage of the time, over the long run, will you be pleased with the outcome? That is, either you were able to initially obtain shares that performed well, or else you were unable to obtain shares that turned out not to perform well.

40. The probability of getting the patent is 0.6. If you get the patent, the conditional probability of being profitable is 0.9. However, given that you do not get the patent, the conditional probability of being profitable is only 0.3. Find the probability of being profitable.

41. You are a contestant on a TV game show with five doors. There is just one prize behind one door, randomly selected. After you choose, the hosts deliberately open three doors (other than your choice) that do not have the prize. You have the opportunity to switch doors from your original choice to the other unopened door.
    a. What is the probability of getting the prize if you switch?
    b. What is the probability of getting the prize if you do not switch?

42. Consider a game in a gambling casino that pays off with probability 0.40. Yesterday 42,652 people played, and 17,122 won.
    a. Find the relative frequency of winning, and compare it to the probability.
    b. As the owner of a casino where many people play this game, how does the law of large numbers help you eliminate much of the uncertainty of gambling?
    c. As an individual who plays once or twice, does the law of large numbers help you limit the uncertainty? Why or why not?

43. Your new firm is introducing two products, a bicycle trailer and a baby carriage for jogging. Your subjective

probabilities of success for these products are 0.85 and 0.70, respectively. If the trailer is successful, you will be able to market the carriage to these customers; therefore, you feel that if the trailer is successful, the carriage will succeed with probability 0.80.

a. Draw a probability tree for this situation.

b. Find the probability that both products succeed.

c. Find the probability that neither product succeeds.

d. Find the probability that the trailer succeeds but the carriage does not.

e. In order for your firm to survive, at least one of the products must succeed. Find the probability that your firm will survive.

44. As the House of Representatives prepared to release the videotape of his grand jury testimony in 1998, President Clinton's approval ratings were as follows: 36% approved of him as a person, 63% approved of him as a president, and 30% approved of him as a president but not as a person.[17] Find the percentage of people who approved of him as a person but not as a president and draw a Venn diagram for this situation.

17. Based upon R. Mishra, "'Swing' Group Holds the Key," in *Seattle Times*, September 19, 1998, p. A2. Their sources are Knight Ridder newspapers and CNN/USA Today Gallup Poll with research by J. Treible.

## Database Exercises

*Problems marked with an asterisk (*) are solved in the Self Test in Appendix C.*

1. View this database as the sample space of a random experiment in which an employee is selected at random. That is, each employee represents one outcome, and all possible outcomes are equally likely.

a.* Find the probability of selecting a woman.

b.* Find the probability that the salary is over $35,000.

c. Find the probability that the employee is at training level B.

d. Find the probability that the salary is over $35,000 and the employee is at training level B.

e.* Find the probability that the salary is over $35,000 given that the employee is at training level B.

f. Is the event "salary over $35,000" independent of being at training level B? How do you know?

g. Find the probability that the salary is over $35,000 given that the employee is at training level C.

2. Continue to view this database as the sample space as in database exercise 1. Consider the two events "high experience (six years or more)" and "female."

a. Find the probabilities of these two events.

b. Find the probability of their intersection. What does this represent?

c. Draw a probability tree for these two events, where the first branch is for the event "female."

d. Find the conditional probability of high experience given female.

e. Find the conditional probability of being female given high experience.

f. Find the probability of being male without high experience.

g. Are the two events "female" and "high experience" independent? How do you know?

h. Are the two events "female" and "high experience" mutually exclusive? How do you know?

3. Continue to view this database as the sample space as in database exercise 1.

a. Are the two events "training level A" and "training level B" independent? How do you know?

b. Are the two events "training level A" and "training level B" mutually exclusive? How do you know?

## Projects

Choose a specific decision problem related to your business interests that depends on two uncertain events.

a. Select reasonable initial values for three probabilities.

b. Complete a probability tree.

c. Report two relevant probabilities and two relevant conditional probabilities, and interpret each one.

d. Write a paragraph discussing what you have learned about your decision problem from this project.

## Case

*Whodunit? Who, If Anyone, Is Responsible for the Recent Rise in the Defect Rate?*

Uh-oh. The defect rate has risen recently and the responsibility has fallen on your shoulders to identify the problem so that it can be fixed. Two of the three managers (Jones, Wallace, and Lundvall) who supervise the production line have already been in to see you (as have some of the workers), and their stories are fascinating.

Some accuse Jones of being the problem, using words such as "careless" and "still learning the ropes" based on anecdotal evidence of performance. Some of this is ordinary office politics to be discounted, of course, but you feel that the possibility should certainly be investigated nonetheless. Jones has countered by telling you that defects are actually produced at a higher rate when others are in charge and that, in fact, Wallace has a much higher error rate. Here are figures from Jones:

**Percent Defective**

| | |
|---|---|
| Wallace | 14.35% |
| Jones | 7.84% |

Soon after, Wallace (who is not exactly known for tact) comes into your office, yelling that Jones is an (unprintable) ... and is not to be believed. After calming down somewhat, he begins mumbling—something about being given difficult assignments by the upper-level management. However, even when he is asked directly, the high error rate is not denied. You are suspicious: It certainly looks as though you've found the problem. However, you are also aware that Wallace (although clearly no relation to Miss Manners) has a good reputation among technical experts and should not be accused without your first considering possible explanations and alternatives.

While you're at it, you decide that it would be prudent to also look at Lundvall's error rates, as well as the two different types of production: one for domestic clients and one for overseas clients (who are much more demanding as to the specifications). Here is the more complete data set you assemble, consisting of counts of items produced recently:

|  | Defective | Nondefective |
|---|---|---|
| **Domestic clients:** | | |
| Wallace | 3 | 293 |
| Lundvall | 12 | 307 |
| Jones | 131 | 2,368 |
| **Overseas clients:** | | |
| Wallace | 255 | 1,247 |
| Lundvall | 75 | 359 |
| Jones | 81 | 123 |

**Discussion Questions**

1. Is Jones correct? That is, using the more complete data set, is it true that Jones has the lowest defect rate overall? Are Jones's percentages correct overall (i.e., combining domestic and overseas production)?

2. Is Wallace correct? That is, what percent of Wallace's production was the more demanding? How does this compare to the other two managers? (*Note:* You may wish to compare conditional probabilities, given the manager, combining defective and nondefective production.)

3. Look carefully at conditional defect rates given various combinations of manager and production client. What do you find?

4. Should you recommend that Wallace start looking for another job? If not, what do you suggest?

## Chapter 7

# Random Variables

## Working with Uncertain Numbers

Many business situations involve random variables, such as waiting to find out your investment portfolio performance or asking customers in a marketing survey how much they would spend. Whenever a random experiment produces a number (or several numbers) as part of its outcome, you can be sure that random variables are involved. Naturally, you will want to be able to compute and interpret summary measures (such as typical value and risk) as well as probabilities of events that depend on the observed random quantity—for example, the probability that your portfolio grows by 10% or more.

You can also think about random variables as being where data sets come from. That is, many of the data sets you worked with in Chapters 2–5 were obtained as observations of random variables. In this sense, the random variable itself represents the population (or the process of sampling from the population), whereas the observed values of the random variable represent the sample data. Much more on population and samples is coming in Chapter 8 and beyond, but the fundamentals of random numbers are covered here in this chapter.

Here are some examples of random variables. Note that each one is random until its value is observed:

**One:** Next quarter's sales—a number that is currently unknown and that can take on one of a number of different values.
**Two:** The number of defective machines produced next week.

**Three:** The number of qualified people who will respond to your "help wanted" advertisement for a new full-time employee.
**Four:** The price per barrel of oil next year.
**Five:** The reported income of the next family to respond to your information poll.

A **random variable** may be defined as a specification or description of a numerical result from a random experiment. The value itself is called an **observation**. For example, "next quarter's sales" is a random variable because it specifies and describes the number that will be produced by the random experiment of waiting until next quarter's numbers are in and computing the sales. The actual future value, $3,955,846, is an observation of this random variable. Note the distinction between a random variable (which refers to the random process involved) and an observation (which is a fixed number, once it has been observed). The pattern of probabilities for a random variable is called its **probability distribution**.

Many random variables have a mean and a standard deviation.[1] In addition, there is a probability for each event based on a random variable. We will consider two

---

1. All of the random variables considered in this chapter have a mean and a standard deviation, although in theory there do exist random variables that have neither a mean nor a standard deviation.

types of random variables: *discrete* and *continuous*. It is easier to work with a discrete random variable because you can make a list of all of its possible values. You will learn about two particular distributions that are especially useful: the *binomial distribution* (which is discrete) and the *normal distribution* (which is continuous). Furthermore, in many cases, you may use a (much simpler) normal probability calculation as a close approximation to a binomial probability.

Since there are so many different types of situations in which data values can arise, there are many types of random variables. The *exponential* and the *Poisson* distributions provide a look at the tip of this iceberg.

A random variable is **discrete** if you can list all possible values it can take on when it is observed. A random variable is **continuous** if it can take on any value in a range (for example, any positive number). For some random variables, it is unclear whether they are discrete or continuous. For example, next quarter's sales might be $385,298.61, or $385,298.62, or $385,298.63, or a similar amount up to some very large number such as $4,000,000.00. Literally speaking, this is discrete (since you can list all the possible outcomes). However, from a practical viewpoint, since the dividing points are so close together and no single outcome is very likely, you may work with it as if it were continuous.

## 7.1 DISCRETE RANDOM VARIABLES

When you have the list of values and probabilities (which defines the *probability distribution*) for a discrete random variable, you know everything possible about the process that will produce a random and uncertain number. Using this list, you will be able to calculate any summary measure (e.g., of typical value or of risk) or probability (of any event determined by the observed value) that might be of interest to you.

Here are some examples of discrete random variables:

1. The number of defective machines produced next week. The list of possible values is 0, 1, 2,....
2. The number of qualified people who will respond to your "help wanted" advertisement for a new full-time employee. Again, the list of possible values is 0, 1, 2,....
3. The resulting budget when a project is selected from four possibilities with costs of $26,000, $43,000, $54,000, and $83,000. The list of possible values is (in thousands of dollars) 26, 43, 54, and 83.

Such a list of possible values, together with the probability of each happening, is the probability distribution of the discrete random variable. These probabilities must be positive numbers (or 0) and must add up to 1. From this distribution, you can find the mean and standard deviation of the random variable. You can also find the probability of any event, simply by adding up the probabilities in the table that correspond to the event.

### Example
#### Profit under Various Economic Scenarios

During a brainstorming session devoted to evaluation of your firm's future prospects, there was a general discussion of what might happen in the future. It was agreed to simplify the situation by considering a best-case scenario, a worst-case scenario, and two intermediate possibilities. For each of these four scenarios, after considerable discussion, there was general agreement on the approximate profit that might occur and its likelihood. Note that this defines the *probability distribution* for the random variable "profits" because we have a list of values and probabilities: one column shows the values (in this case, profit) and another column shows the probabilities.

| Economic Scenario | Profit ($ millions) | Probability |
|---|---|---|
| Great | 10 | 0.20 |
| Good | 5 | 0.40 |
| OK | 1 | 0.25 |
| Lousy | –4 | 0.15 |

This probability distribution can be easily used to find probabilities of all events concerning profit. The probability that the profit is $10 million, for example, is 0.20. The probability of making $3 million or more is found as follows: 0.20 + 0.40 = 0.60—because there are two outcomes ("Great" and "Good") that correspond to this event by having profit of $3 million or more.

## Finding the Mean and Standard Deviation

The **mean or expected value** of a discrete random variable is an exact number that summarizes it in terms of a typical value, in much the same way that the average summarizes a list of data.[2] The mean is denoted by the lowercase Greek letter $\mu$ (*mu*) or by $E(X)$ (read as "expected value of $X$") for a random variable $X$. The formula is

**Mean or Expected Value of a Discrete Random Variable $X$**

$$\mu = E(X) = \text{Sum of (value times probability)}$$
$$= \sum XP(X)$$

If the probabilities were all equal, this would be the average of the values. In general, the mean of a random variable is a weighted average of the values using the probabilities as weights.

This mean profit in the preceding example is

$$\text{Expected profit} = (10 \times 0.20) + (5 \times 0.40) + (1 \times 0.25)$$
$$+ (-4 \times 0.15) = 3.65$$

---

2. In fact, the mean of a random variable is also called its average; however, we will often use *mean* for random variables and *average* for data.

Thus, the expected profit is $3.65 million. This number summarizes the various possible outcomes (10, 5, 1, –4) using a single number that reflects their likelihoods.

The **standard deviation** of a discrete random variable indicates approximately how far you expect it to be from its mean. In many business situations, the standard deviation indicates the *risk* by showing just how uncertain the situation is. The standard deviation is denoted by $\sigma$, which matches our use of $\sigma$ as the population standard deviation. The formula is

**Standard Deviation of a Discrete Random Variable $X$**

$$\sigma = \sqrt{\text{Sum of (squared deviation times probability)}}$$

$$= \sqrt{\sum (X - \mu)^2 P(X)}$$

Note that you would *not* get the correct answer by simply using the $\Sigma$ key on your calculator to accumulate only the single column of values, since this would not make proper use of the probabilities.

The standard deviation of profit for our example is

$$\begin{aligned}
\sigma &= \sqrt{\{[(10 - 3.65)^2 0.20] + [(5 - 3.65)^2 0.40]} \\
&\quad \overline{+ [(1 - 3.65)^2 0.25] + [(-4 - 3.65)^2 0.15]\}} \\
&= \sqrt{8.064500 + 0.729000 + 1.755625 + 8.778375} \\
&= \sqrt{19.3275} = 4.40
\end{aligned}$$

The standard deviation of $4.40 million shows that there is considerable risk involved here. Profit might reasonably be about $4.40 million above or below its mean value of $3.65 million. Table 7.1.1 shows the details of the computations involved in finding the standard deviation.

To use Excel® to compute the mean and standard deviation of a discrete random variable, you might proceed as follows. Using Excel's menu commands, give names to these columns by selecting the numbers with the titles, then choosing Excel's Create from Selection in the Defined names section of the Formulas Ribbon. The mean (3.65) is the sum of the products of value times probability; hence, the formula is "=SUMPRODUCT (Profit,Probability)." Give this cell (which now contains the mean) the name "Mean." The standard deviation (4.40) is the square root (SQRT) of the sum of the products of the square of value minus mean times probability. Hence, the formula is

$$= \text{SQRT(SUMPRODUCT((Profit} - \text{Mean)}\verb|^|2,$$

$$\text{Probability))}$$

These formulas give us 3.65 for the mean and 4.40 for the standard deviation, as before.

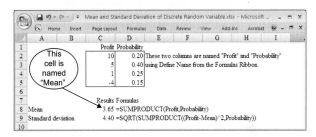

Figure 7.1.1 shows the probability distribution, with the heights of the lines indicating the probability and the location of the lines indicating the amount of profit in each case. Also indicated is the expected value, $3.65 million, and the standard deviation, $4.40 million.

**Example**

*Evaluating Risk and Return*

Your job is to evaluate three different projects ($X$, $Y$, and $Z$) and make a recommendation to upper management. Each project requires an investment of $12,000 and pays off next year. Project $X$ pays $14,000 for sure. Project $Y$ pays either $10,000 or $20,000 with probability 0.5 in each case. Project $Z$ pays nothing with probability 0.98 and $1,000,000 with probability 0.02. A summary is shown in Table 7.1.2.

*(Continued)*

**TABLE 7.1.1 Finding the Standard Deviation for a Discrete Random Variable**

| Profit | Probability | Deviation from Mean | Squared Deviation | Squared Deviation Times Probability |
|---|---|---|---|---|
| 10 | 0.20 | 6.35 | 40.3225 | 8.064500 |
| 5 | 0.40 | 1.35 | 1.8225 | 0.729000 |
| 1 | 0.25 | −2.65 | 7.0225 | 1.755625 |
| −4 | 0.15 | −7.65 | 58.5225 | 8.778375 |
| | | | | Sum: 19.3275 |
| | | | | Square root: 4.40 |

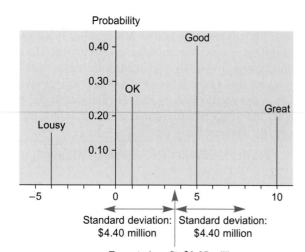

**FIGURE 7.1.1**  The probability distribution of future profits, with the mean (expected profits) and standard deviation (risk) indicated.

**TABLE 7.1.2  Payoffs and Probabilities for Three Projects**

| Project | Payoff | Probability |
|---------|--------|-------------|
| X | 14,000 | 1.00 |
| Y | 10,000 | 0.50 |
|  | 20,000 | 0.50 |
| Z | 0 | 0.98 |
|  | 1,000,000 | 0.02 |

**Example—cont'd**

The means are easily found: $14,000 for X, 10,000 × 0.50 + 20,000 × 0.50 = $15,000 for Y, and 0 × 0.98 + 1,000,000 × 0.02 = $20,000 for Z. We could write these as follows:

$$E(X) = \mu_X = \$14,000$$
$$E(Y) = \mu_Y = \$15,000$$
$$E(Z) = \mu_Z = \$20,000$$

Based only on these expected values, it would appear that Z is best and X is worst. However, these mean values don't tell the whole story. For example, although project Z has the highest expected payoff, it also involves considerable risk: 98% of the time there would be no payoff at all! The risks involved here are summarized by the standard deviations:

$$\sigma_X = \sqrt{(14,000 - 14,000)^2 \times 1.00} = \$0$$

$$\sigma_Y = \sqrt{(10,000 - 15,000)^2 \times 0.50 + (20,000 - 15,000)^2 \times 0.50}$$
$$= \$5,000$$

$$\sigma_Z = \sqrt{(0 - 20,000)^2 \times 0.98 + (1,000,000 - 20,000)^2 \times 0.02}$$
$$= \$140,000$$

These standard deviations confirm your suspicions. Project Z is indeed the riskiest—far more so than either of the others. Project X is the safest—a sure thing with no risk at all. Project Y involves a risk of $5,000.

Which project should be chosen? This question cannot be answered by statistical analysis alone. Although the expected value and the standard deviation provide helpful summaries to guide you in choosing a project, they do not finish the task. Generally, people prefer larger expected payoffs and lower risk. However, with the choices presented here, to achieve a larger expected payoff, you must take a greater risk. The ultimate choice of project will involve your (and your firm's) "risk versus return" preference to determine whether or not the increased expected payoff justifies the increased risk.[3]

What if you measure projects in terms of profit instead of payoff? Since each project involves an initial investment of $12,000, you can convert from payoff to profit by subtracting $12,000 from each payoff value in the probability distribution table:

$$\text{Profit} = \text{Payoff} - \$12,000$$

Using the rules from Section 5.4, which apply to summaries of random variables as well as to data, subtract $12,000 from each mean value and leave the standard deviation alone. Thus, without doing any detailed calculations, you come up with the following expected profits:

| | |
|---|---|
| X: | $2,000 |
| Y: | $3,000 |
| Z: | $8,000 |

The standard deviations of profits are the same as for payoffs, namely:

| | |
|---|---|
| X: | $0 |
| Y: | $5,000 |
| Z: | $140,000 |

---

3. In your finance courses, you may learn about another factor that is often used in valuing projects, namely, the correlation (if any) between the random payoffs and the payoffs of a market portfolio. This helps measure the *diversifiable* and *nondiversifiable risk* of a project. Correlation (a statistical measure of association) will be presented in Chapter 11. The nondiversifiable component of risk is also known as systematic or systemic risk because it is part of the entire economic system and cannot be diversified away.

## 7.2  THE BINOMIAL DISTRIBUTION

Percentages play a key role in business. When a percentage is arrived at by counting the number of times something happens out of the total number of possibilities, the number of occurrences might follow a *binomial distribution*. If so, there are a number of time-saving shortcuts available for finding the expected value, standard deviation, and probabilities of various events. Sometimes you will be interested in the percentage; at other times the number of occurrences will be more relevant. The binomial distribution can give answers in either case. Here are some examples of random variables that follow a binomial distribution:

1. The number of orders placed, out of the next three telephone calls to your catalog order desk.
2. The number of defective products out of 10 items produced.
3. The number of people who said they would buy your product, out of 200 interviewed.
4. The number of stocks that went up yesterday, out of all issues traded on major exchanges.
5. The number of female employees in a division of 75 people.
6. The number of Republican (or Democratic) votes cast in the next election.

## Definition of Binomial Distribution and Proportion

Focus attention on a particular event. Each time the random experiment is run, either the event happens or it doesn't. These *two* possible outcomes give us the *bi* in *binomial*. A random variable $X$, defined as the *number of occurrences* of a particular event out of $n$ trials, has a **binomial distribution** if

1. For each of the $n$ trials, the event always has the same probability $\pi$ of happening.
2. The trials are independent of one another.

The independence requirement rules out "peeking," as in the case of the distribution of people who order the special at a restaurant. If some people order the special because they see other customers obviously enjoying the rich, delicious combination of special aromatic ingredients, and say, "WOW! I'll have that too!" the number who order the special would *not* follow a binomial distribution. Choices have to be made independently in order to get a binomial distribution.

The **binomial proportion** $p$ is the binomial random variable $X$ expressed as a fraction of $n$:

**Binomial Proportion**

$$p = \frac{X}{n} = \frac{\text{Number of occurrences}}{\text{Number of trials}}$$

(Note that $\pi$ is a fixed number, the probability of occurrence, whereas $p$ is a random quantity based on the data.) For example, if you interviewed $n = 600$ shoppers and found that $X = 38$ plan to buy your product, then the binomial proportion would be

$$p = \frac{X}{n} = \frac{38}{600} = 0.063, \text{ or } 6.3\%$$

The binomial proportion $p$ is also called a *binomial fraction*. You may have recognized it as a relative frequency, which was defined in Chapter 6.

---

**Example**
*How Many Orders Are Placed? The Hard Way to Compute*

This example shows the hard way to analyze a binomial random variable. Although it is rarely necessary to draw the probability tree, since it is usually quite large, seeing it once will help you understand what is really going on with the binomial distribution. Furthermore, when the shortcut computations are presented (the easy way) you will appreciate the time they save!

Suppose you are interested in the next $n = 3$ telephone calls to the catalog order desk, and you know from experience (or are willing to assume[4]) that $\pi = 0.6$, so that 60% of calls will result in an order (the others are primarily calls for information, or misdirected). What can we say about the number of calls that will result in an order? Certainly, this number will be either 0, 1, 2, or 3 calls. Since a call is more likely to result in an order than not, we should probably expect the probability of getting three orders to be larger than the probability of getting none at all. But how can we find these probabilities? The probability tree provides a complete analysis, as shown in Figure 7.2.1a, indicating the result of each of the three phone calls.

Note that the conditional probabilities along the branches are always 0.60 and 0.40 (the individual probabilities for each call) since we assume orders occur independently and do not influence each other. The number of orders is listed at the far right in Figure 7.2.1a; for example, the second number from the top, 2, reports the fact that the first and second (but not the third) callers placed an order, resulting in two orders placed. Note that there are three ways in which two orders could be placed. To construct the probability distribution of the number of orders placed, you could add up the probabilities for the different ways that each number could happen:

| Number of Callers Who Ordered, $X$ | Percentage Who Ordered, $p = X/n$ | Probability |
|---|---|---|
| 0 | 0.0% | 0.064 |
| 1 | 33.3% | 0.288 (= 0.096 + 0.096 + 0.096) |
| 2 | 66.7% | 0.432 (= 0.144 + 0.144 + 0.144) |
| 3 | 100.0% | 0.216 |

This probability distribution is displayed in Figure 7.2.1b.

Now that you have the probability distribution, you can find all of the probabilities by adding the appropriate ones. For example, the probability of at least two orders is 0.432 + 0.216 = 0.648. You can also use the formulas for the mean and standard deviation from Section 7.1 to find the mean value (1.80 orders) and the standard deviation (0.849 orders). However, *this would be too much work!* There is a much quicker formula for finding the mean, standard deviation,

*(Continued)*

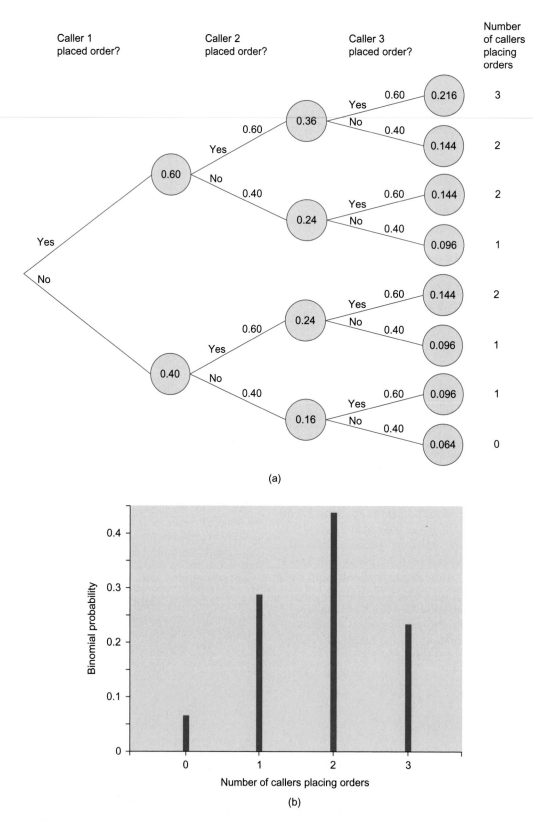

(a)

(b)

**FIGURE 7.2.1** a. The probability tree for three successive telephone calls, each of which either does or does not result in an order being placed. There are eight combinations (the circles at the far right). In particular, there are three ways in which exactly two calls could result in an order: the second, third, and fifth circles from the top, giving a probability of 3 × 0.144 = 0.432. b. The binomial probability distribution of the number of calls that result in an order being placed.

**Example—cont'd**

and probabilities. Although it was possible to compute directly in this small example, you will not usually be so lucky. For example, had you considered 10 successive calls instead of 3, there would have been 1,024 probabilities at the right of the probability tree instead of the 8 in Figure 7.2.1.

4. The probability $\pi$ is usually given in textbook problems involving a binomial distribution. In real life, they arise just as other probabilities do: from relative frequency, theoretical probability, or subjective probability.

Think of this example as a way of seeing the underlying situation and all combinations and then simplifying to a probability distribution of the *number of occurrences*. Conceptually, this is the right way to view the situation. Now let's learn the easy way to compute the answers.

## Finding the Mean and Standard Deviation the Easy Way

The mean number of occurrences in a binomial situation is $E(X) = n\pi$, the number of possibilities times the probability of occurrence. The mean proportion is

$$E\left(\frac{X}{n}\right) = E(p) = \pi$$

which is the same as the individual probability of occurrence.[5]

This is what you would expect. For example, in a poll of a sample of 200 voters, if each has a 58% chance of being in favor of your candidate, on average, you would expect that

$$E\left(\frac{X}{n}\right) = E(p) = \pi = 0.58$$

or 58% of the sample will be in your favor. In terms of the number of people, you would expect $E(X) = n\pi = 200 \times 0.58 = 116$ people out of the 200 in the sample to be in your favor. Of course, the actually observed number and percentage will probably randomly differ from these expected values.

There are formulas for the standard deviation of the binomial number and percentage, summarized along with the expected values in the following table:

---

5. You might have recognized $X/n$ as the relative frequency of the event. The fact that $E(X/n)$ is equal to $\pi$ says that, on average, the relative frequency of an event is equal to its probability. In Chapter 8 we will learn that this property says that $p$ is an unbiased estimator of $\pi$.

**Mean and Standard Deviation for a Binomial Distribution**

|  | Number of Occurrences, $X$ | Proportion or Percentage, $p = X/n$ |
|---|---|---|
| Mean | $E(X) = \mu_X = n\pi$ | $E(p) = \mu_p = \pi$ |
| Standard deviation | $\sigma_X = \sqrt{n\pi(1-\pi)}$ | $\sigma_p = \sqrt{\dfrac{\pi(1-\pi)}{n}}$ |

For the "telephone orders" example, we have $n = 3$ and $p = 0.60$. Using the formulas, the mean and standard deviation are

|  | Number of Occurrences, $X$ | Proportion or Percentage, $p = X/n$ |
|---|---|---|
| Mean | $E(X) = n\pi$ $= 3 \times 0.60$ $= 1.80\,\text{calls}$ | $E(X) = \pi$ $= 0.60 \text{ or } 60\%$ |
| Standard deviation | $\sigma_X = \sqrt{n\pi(1-\pi)}$ $= \sqrt{3 \times 0.60(1-0.60)}$ $= 0.849\,\text{calls}$ | $\sigma_p = \sqrt{\dfrac{\pi(1-\pi)}{n}}$ $= \sqrt{\dfrac{0.60(1-0.60)}{3}}$ $= 0.283 \text{ or } 28.3\%$ |

Thus, we expect 1.80 of these 3 telephone calls to result in an order. Sometimes more (i.e., 2 or 3) and sometimes fewer (i.e., 0 or 1) calls will result in an order. The extent of this uncertainty is measured (as usual) by the standard deviation, 0.849 calls. Similarly, we expect 60% of these 3 calls to result in an order. The last number, 28.3%, representing the standard deviation of the percentage, is interpreted as *percentage points* rather than as a percentage of some number. That is, while the expected percentage is 60%, the actual observed percentage is typically about 28.3 percentage points above this value (at $60 + 28.3 = 88.3\%$) or below (at $60 - 28.3 = 31.7\%$). This is natural if you remember that a standard deviation is stated in the same units as the data, which are percentage points in the case of $p$.

**Example**
*Recalling Advertisements*

Your company is negotiating with a marketing research firm to provide information on how your advertisements are doing with the American consumer. Selected people are to come in one day to watch TV programs and ads (for many products from many companies) and return the next day to answer questions. In particular, you plan to measure the *rate of recall*, which is the percentage of people who remember your ad the day after seeing it.

Before you contract with the firm to do the work, you are curious about how reliable and accurate the results

(Continued)

**Example—cont'd**

are likely to be. Your budget allows 50 people to be tested. From your discussions with the research firm, it seems reasonable initially to assume that 35% of people will recall the ad, although you really don't know the exact proportion. Based on the assumption that it really is 35%, how accurate will the results be? That is, about how far will the measured recall percentage be from the assumed value $\pi = 0.35$ with $n = 50$ for a binomial distribution? The answer is

$$\sigma_p = \sqrt{\frac{\pi(1-\pi)}{n}}$$

$$= \sqrt{\frac{0.35(1-0.35)}{50}}$$

$$= 0.0675 \text{ or } 6.75\%$$

This says that the standard deviation of the result of the recall test (namely, the percentage of people tested who remembered the ad) is likely to differ from the true percentage for the entire population typically by about 7 percentage points in either direction (above or below).

You decide that the results need to be more precise than that. The way to improve the precision of the results is to gather more information by increasing the sample size, $n$. Checking the budget and negotiating over the rates, you find that $n = 150$ is a possibility. With this larger sample, the standard deviation decreases to reflect the extra information:

$$\sigma_p = \sqrt{\frac{\pi(1-\pi)}{n}}$$

$$= \sqrt{\frac{0.35(1-0.35)}{150}}$$

$$= 0.0389 \text{ or } 3.89\%$$

You are disappointed that the extra cost didn't bring a greater improvement in the results. When the size of the study was tripled, the precision didn't even double! This is due, technically, to the fact that it is the *square root of n*, rather than $n$ itself, that is involved. Nevertheless, you decide that the extra accuracy is worth the cost.

## Finding the Probabilities

Suppose you have a binomial distribution, you know the values of $n$ and $\pi$, and you want to know the probability that $X$ will be exactly equal to some number $a$. There is a formula for this probability that is useful for small to moderate $n$. (When $n$ is large, an approximation based on the normal distribution, to be covered in Section 7.4, will be much easier than the exact method presented here.) In addition, Table D–3 in Appendix D gives exact binomial probabilities and cumulative probabilities for $n = 1$ to 20

and $\pi = 0.05, 0.1, 0.2, 0.3, 0.4, 0.5, 0.6, 0.7, 0.8, 0.9,$ and $0.95$. Here is the exact formula:[6]

**Binomial Probability That X Equals a**

$$P(X=a) = \binom{n}{a}\pi^a(1-\pi)^{n-a}$$

$$= \frac{n!}{a!(n-a)!}\pi^a(1-\pi)^{n-a}$$

$$= \frac{1\times2\times3\times\cdots\times n}{(1\times2\times3\times\cdots\times a)[1\times2\times3\times\cdots\times(n-a)]}\pi^a(1-\pi)^{n-a}$$

By using this formula with each value of $a$ from 0 to $n$ (sometimes a lot of work), you (or a computer) can generate the entire probability distribution. From these values, you can find any probability you want involving $X$ by adding together the appropriate probabilities from this formula.

To see how to use the formula, suppose there are $n = 5$ possibilities with a success probability $p = 0.8$ for each one, and you want to find the probability of exactly $a = 3$ successes. The answer is

$$P(X = 3) = \binom{5}{3}0.8^3(1-0.8)^{5-3}$$

$$= \frac{5!}{3!(5-3)!}0.8^3 \times 0.2^2$$

$$= \frac{1\times2\times3\times4\times5}{(1\times2\times3)(1\times2)}0.512 \times 0.040$$

$$= 10 \times 0.02048 = 0.2048$$

---

6. The notation $n!$ is read as "$n$ factorial" and is the product of the numbers from 1 to $n$. For example, $4! = 1 \times 2 \times 3 \times 4 = 24$. (By convention, to get the correct answers, we define $0!$ to be 1.) Many calculators have a factorial key that works for values of $n$ from 0 through 69. The notation

$$\binom{n}{a} = \frac{n!}{a!(n-a)!}$$

is the *binomial coefficient*, read aloud as "$n$ choose $a$," and also represents the number of *combinations* you can make by choosing $a$ items from $n$ items (where the order of selection does not matter). Thus, it represents the number of different ways in which you could assign exactly $a$ occurrences to the $n$ possibilities. For example, with $n = 5$ and $a = 3$, the binomial coefficient is

$$\binom{5}{3} = \frac{5!}{3!(5-3)!} = \frac{120}{6\times2} = 10$$

Thus, there are 10 different ways (combinations) in which three out of five people could buy our product: the first three people could, or the first two and the fourth might, and so forth. The full list of the 10 combinations is (1,2,3), (1,2,4), (1,2,5), (1,3,4), (1,3,5), (1,4,5), (2,3,4), (2,3,5), (2,4,5), and (3,4,5).

This is the probability of *exactly* three successes. If you want the probability of *three or more* successes, you could compute the formula twice more: once for $a = 4$ and once for $a = 5$; the probability of three or more successes would be the total of these numbers. Alternatively, you could use a computer to obtain the probabilities, for example:

**Probability Density Function and Cumulative Distribution Function**

**Binomial with $n = 5$ and $p = 0.800000$**

| $a$ | $P(X = a)$ | $P(X <= a)$ |
|---|---|---|
| 0 | 0.0003 | 0.0003 |
| 1 | 0.0064 | 0.0067 |
| 2 | 0.0512 | 0.0579 |
| 3 | 0.2048 | 0.2627 |
| 4 | 0.4096 | 0.6723 |
| 5 | 0.3277 | 1.0000 |

In either case, once you have the individual probabilities (for 3, 4, and 5 successes), the answer is

$$P(X \geq 3) = P(X = 3) + P(X = 4) + P(X = 5)$$
$$= 0.2048 + 0.4096 + 0.3277$$
$$= 0.9421$$

Thus, you have a 94.2% chance of achieving three or more successes out of these five. Alternatively, using the complement rule, the probability of *three or more* must be one minus the probability of *two or less*, which is listed as 0.0579 in the computer output. The answer would then be found as $1 - 0.0579 = 0.9421$.

To use Excel® to compute binomial probabilities, use the formula "=BINOMDIST($a,n,\pi$, FALSE)" to find the probability $P(X = a)$ of being equal to $a$, and use the formula "=BINOMDIST($a,n,\pi$,TRUE)" to find the probability $P(X \leq a)$ of being *less than or equal* to $a$, as follows:[7]

---

7. The "FALSE" and "TRUE" in Excel's binomial distribution formula refer to whether or not the probability distribution is cumulative, i.e., whether or not it accumulates probabilities for all of the previous (smaller) values of $a$ as well.

**Example**
*How Many Major Customers Will Call Tomorrow?*

How many of your $n = 6$ major customers will call tomorrow? You are willing to assume that each one has a probability $\pi = 0.25$ of calling and that they call independently of one another. Thus, the number of major customers that will call tomorrow, $X$, follows a binomial distribution.

How many do you expect will call? That is, what is the expected value of $X$? The answer is $E(X) = n \times \pi = 1.5$ major customers. The standard deviation is $\sigma_X = \sqrt{6 \times 0.25 \times (1 - 0.25)} = 1.060660$, indicating that you can reasonably anticipate 1 or 2 more or less than the 1.5 you expect. Although this gives you an idea of what to expect, it doesn't tell you the chances that a given number will call. Let's compute the probabilities for this.

What is the probability that exactly $a = 2$ out of your $n = 6$ major customers will call? The answer is

$$P(X = 2) = \binom{6}{2} 0.25^2 (1 - 0.25)^{(6-2)}$$
$$= 15 \times 0.0625 \times 0.316406 = 0.297$$

Here is the entire probability distribution of the number of major customers who will call you tomorrow, including all possibilities for the number $a$ from 0 through $n = 6$:

**Probability Density Function and Cumulative Distribution Function**
**Binomial with $n = 6$ and $p = 0.250000$**

| $a$ | $P(X = a)$ | $P(X <= a)$ |
|---|---|---|
| 0 | 0.1780 | 0.1780 |
| 1 | 0.3560 | 0.5339 |
| 2 | 0.2966 | 0.8306 |
| 3 | 0.1318 | 0.9624 |
| 4 | 0.0330 | 0.9954 |
| 5 | 0.0044 | 0.9998 |
| 6 | 0.0002 | 1.0000 |

Note that the most likely outcomes are 1 or 2 calls, just as you suspected based on the mean value of 1.5 calls.

From this probability distribution, you can compute any probability about the number of major customers who will call you tomorrow. It is highly unlikely that all 6 will call (0.0002 or 0.02%, much less than a 1% chance). The probability that 4 or more will call is $0.0330 + 0.0044 + 0.0002 = 0.0376$. From the second column, you can see that the probability that 3 or fewer will call is 0.9624. Your chances of spending a quiet day with no calls is 0.178. This probability distribution is shown in Figure 7.2.2.

**Example**
*How Many Logic Analyzers to Schedule for Manufacturing?*

You pay close attention to quality in your production facilities, but the logic analyzers you make are so complex that
*(Continued)*

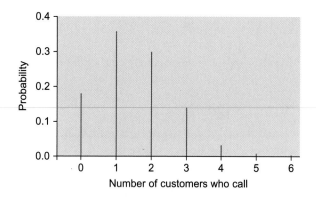

FIGURE 7.2.2　The probability distribution of the number of major customers who will call you tomorrow. These are binomial probabilities, with each vertical bar found using the formula based on $n = 6$ and $p = 0.25$. The number $a$ is found along the horizontal axis.

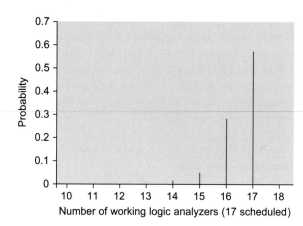

FIGURE 7.2.3　The probability distribution of the number of working logic analyzers produced if you plan to produce only 17. This is binomial, with $n = 17$ and $\pi = 0.97$.

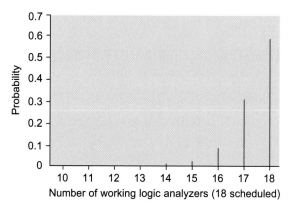

FIGURE 7.2.4　The probability distribution of the number of working logic analyzers produced *if you plan to produce 18*. This is binomial, with $n = 18$ and $\pi = 0.97$.

**Example—cont'd**

there are still some failures. In fact, based on past experience, about 97% of the finished products are in good working order. Today you will have to ship 17 of these machines. The question is: How many should you schedule for production to be reasonably certain that 17 working logic analyzers will be shipped?

It is reasonable to assume a binomial distribution for the number of working machines produced, with $n$ being the number that you schedule and $\pi$ being each one's probability (0.97) of working. Then you can compute the probability that 17 or more of the scheduled machines will work.

What happens if you schedule 17 machines, with no margin for error? You might think that the high (97%) rate would help you, but, in fact, the probability that all 17 machines will work (using $n = 17$ and $a = 17$) is just 0.596:

$$P(X = 17 \text{ working machines}) = \binom{17}{17} 0.97^{17} 0.03^0$$

$$= 1 \times 0.595826 \times 1 = 0.596$$

Thus, if you schedule the same number, 17, that you need to ship, you will be taking a big chance! There is only a 59.6% chance that you will meet the order, and a 40.4% chance that you will fail to ship the entire order in working condition. The probability distribution is shown in Figure 7.2.3.

It looks as though you'd better schedule more than 17. What if you schedule $n = 18$ units for production? To find the probability that *at least* 17 working analyzers will be shipped, you'll need to find the probabilities for $a = 17$ and $a = 18$ and add them up:

$$P(X \geq 17) = P(X = 17) + P(X = 18)$$

$$= \binom{18}{17} 0.97^{17} 0.03^1 + \binom{18}{18} 0.97^{18} 0.03^0$$

$$= 18 \times 0.595826 \times 0.03 + 1 \times 0.577951 \times 1$$

$$= 0.322 + 0.578 = 0.900$$

So if you schedule 18 for production, you have a 90% chance of shipping 17 good machines. It looks likely, but you would still be taking a 10% chance of failure. This probability distribution is shown in Figure 7.2.4.

Similar tedious calculations reveal that if you schedule 19 machines for production, you have a 98.2% chance of shipping 17 good machines (9.2% + 32.9% + 56.1%). So, to be reasonably sure of success, you'd better schedule at least 19 machines to get 17 good ones!

## 7.3 THE NORMAL DISTRIBUTION

You already know from Chapter 3 how to tell if a data set is approximately normally distributed. Now it's time to learn how to compute probabilities for this familiar bell-shaped distribution. One reason the normal distribution is particularly useful is the fact that, given only a mean and a standard deviation, you can compute any

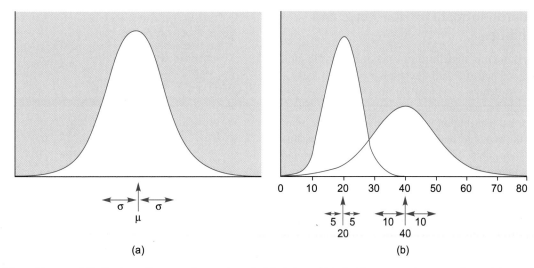

FIGURE 7.3.1   a. The normal distribution, with mean value $\mu$ and standard deviation $\sigma$. Note that the mean can be any number, and the standard deviation can be any positive number. b. Two different normal distributions. The one on the left has a smaller mean value (20) and a smaller standard deviation (5) than the other. The one on the right has mean 40 and standard deviation 10.

probability of interest (provided that the distribution really is normal).

The **normal distribution**, a continuous distribution, is represented by the familiar bell-shaped curve shown in Figure 7.3.1a. Note that there is a normal distribution for each combination of a mean value and a positive standard deviation value.[8] Just slide the curve to the right or left until the peak is centered above the mean value; then stretch it wider or narrower until the scale matches the standard deviation. Two different normal distributions are shown in Figure 7.3.1b.

## Visualize Probabilities as the Area under the Curve

The bell-shaped curve gives you a guide for visualizing the probabilities for a normal distribution. You are more likely to see values occurring near the middle, where the curve is high. At the edges, where the curve is lower, values are not as likely to occur. Formally, it is the *area under the curve* that gives you the probability of being within a region, as illustrated in Figure 7.3.2.

Note that a shaded strip near the middle of the curve will have a larger area than a strip of the same width located nearer to the edge. Compare Figure 7.3.2 to Figure 7.3.3 to see this.

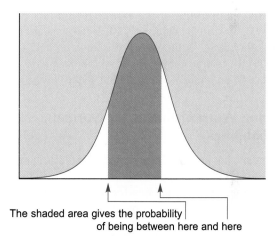

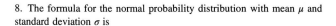

FIGURE 7.3.2   The probability that a normally distributed random variable is between any two values is equal to the area under the normal curve between these two values. You are more likely to see values in regions close to the mean.

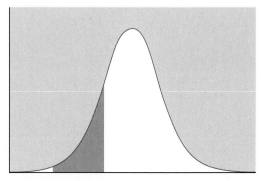

FIGURE 7.3.3   The probability of falling within a region that is farther from the middle of the curve. Since the normal curve is lower here, the probability is smaller than that shown in Figure 7.3.2.

---

8. The formula for the normal probability distribution with mean $\mu$ and standard deviation $\sigma$ is

$$\frac{1}{\sqrt{2\pi}\sigma} e^{-[(x-\mu)/\sigma]^2/2}.$$

## The Standard Normal Distribution *Z* and Its Probabilities

The **standard normal distribution** is a normal distribution with mean $\mu = 0$ and standard deviation $\sigma = 1$. The letter *Z* is often used to denote a random variable that follows this standard normal distribution. One way to compute probabilities for a normal distribution is to use tables that give probabilities for the standard one, since it would be impossible to keep different tables for each combination of mean and standard deviation. The standard normal distribution can represent any normal distribution, provided you think in terms of the *number of standard deviations above or below the mean* instead of the actual units (e.g., dollars) of the situation. The standard normal distribution is shown in Figure 7.3.4.

The **standard normal probability table**, shown in Table 7.3.1, gives the probability that a standard normal random variable *Z* is *less than* any given number *z*. For example, the probability of being less than 1.38 is 0.9162, illustrated as an area in Figure 7.3.5. Doesn't it look like about 90% of the area? To find this number (0.9162), look up the value $z = 1.38$ in the standard normal probability table. While you're at it, look up 2.35 (to find 0.9906), 0 (to find 0.5000), and –0.82 (to find 0.2061). What is the probability corresponding to the value $z = 0.36$?

## Solving Word Problems for Normal Probabilities

A typical word problem involving a normal distribution is a story involving some application to business that gives you a value for the mean and one for the standard deviation. Then you are asked to find one or more probabilities of interest. Here is an example of such a word problem:

The upper-management people at Simplified Technologies, Inc., have finally noticed that sales forecasts are

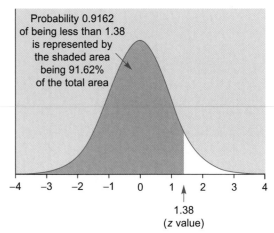

Using the standard normal probability table

**FIGURE 7.3.5**   The probability that a standard normal random variable is *less than* $z = 1.38$ is 0.9162, as found in the standard normal probability table. This corresponds to the shaded region to the left of 1.38, which is 91.62% of the total area under the curve.

usually wrong. Last quarter's sales were forecast as $18 million but came in at $21.3 million. Sales for the next quarter are forecast as $20 million, with a standard deviation (based on previous experience) of $3 million. Assuming a normal distribution centered at the forecast value, find the probability of a "really bad quarter," which is defined as sales lower than $15 million.

The beginning part sets the stage. The first numbers (18 and 21.3) describe past events but play no further role here. Instead, you should focus on the following facts:

- There is a normal distribution involved here.
- Its mean is $\mu = \$20$ million.
- Its standard deviation is $\sigma = \$3$ million.
- You are asked to find the probability that sales will be lower than $15 million.

The next step is to convert all of these numbers (except for the mean and standard deviation) into standardized numbers; this has to be done before you can look up the answer in the standard normal probability table. A **standardized number** (often written as *z*) is the number of standard deviations above the mean (or below the mean, if the standardized number is negative). This conversion is done as follows:

$$z = \text{Standardized number} = \frac{\text{Number} - \text{Mean}}{\text{Standard deviation}}$$
$$= \frac{\text{Number} - \mu}{\sigma}$$

In this example, the number $15 million is standardized as follows:

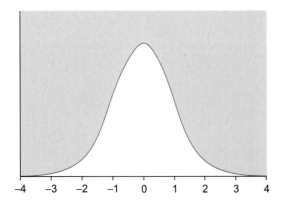

**FIGURE 7.3.4**   The standard normal distribution *Z* with mean value $\mu = 0$ and standard deviation $\sigma = 1$. The standard normal distribution may be used to represent any normal distribution, provided you think in terms of the number of standard deviations above or below the mean.

$$z = \frac{15 - \mu}{\sigma} = \frac{15 - 20}{3} = -1.67$$

**TABLE 7.3.1 Standard Normal Probability Table (See Figure 7.3.5)**

| z Value | Probability | z Value | Probability | z Value | Probability | z Value | Probability | z Value | Probability | z Value | Probability |
|---|---|---|---|---|---|---|---|---|---|---|---|
| -2.00 | 0.0228 | -1.00 | 0.1587 | 0.00 | 0.5000 | 0.00 | 0.5000 | 1.00 | 0.8413 | 2.00 | 0.9772 |
| -2.01 | 0.0222 | -1.01 | 0.1562 | -0.01 | 0.4960 | 0.01 | 0.5040 | 1.01 | 0.8438 | 2.01 | 0.9778 |
| -2.02 | 0.0217 | -1.02 | 0.1539 | -0.02 | 0.4920 | 0.02 | 0.5080 | 1.02 | 0.8461 | 2.02 | 0.9783 |
| -2.03 | 0.0212 | -1.03 | 0.1515 | -0.03 | 0.4880 | 0.03 | 0.5120 | 1.03 | 0.8485 | 2.03 | 0.9788 |
| -2.04 | 0.0207 | -1.04 | 0.1492 | -0.04 | 0.4840 | 0.04 | 0.5160 | 1.04 | 0.8508 | 2.04 | 0.9793 |
| -2.05 | 0.0202 | -1.05 | 0.1469 | -0.05 | 0.4801 | 0.05 | 0.5199 | 1.05 | 0.8531 | 2.05 | 0.9798 |
| -2.06 | 0.0197 | -1.06 | 0.1446 | -0.06 | 0.4761 | 0.06 | 0.5239 | 1.06 | 0.8554 | 2.06 | 0.9803 |
| -2.07 | 0.0192 | -1.07 | 0.1423 | -0.07 | 0.4721 | 0.07 | 0.5279 | 1.07 | 0.8577 | 2.07 | 0.9808 |
| -2.08 | 0.0188 | -1.08 | 0.1401 | -0.08 | 0.4681 | 0.08 | 0.5319 | 1.08 | 0.8599 | 2.08 | 0.9812 |
| -2.09 | 0.0183 | -1.09 | 0.1379 | -0.09 | 0.4641 | 0.09 | 0.5359 | 1.09 | 0.8621 | 2.09 | 0.9817 |
| -2.10 | 0.0179 | -1.10 | 0.1357 | -0.10 | 0.4602 | 0.10 | 0.5398 | 1.10 | 0.8643 | 2.10 | 0.9821 |
| -2.11 | 0.0174 | -1.11 | 0.1335 | -0.11 | 0.4562 | 0.11 | 0.5438 | 1.11 | 0.8665 | 2.11 | 0.9826 |
| -2.12 | 0.0170 | -1.12 | 0.1314 | -0.12 | 0.4522 | 0.12 | 0.5478 | 1.12 | 0.8686 | 2.12 | 0.9830 |
| -2.13 | 0.0166 | -1.13 | 0.1292 | -0.13 | 0.4483 | 0.13 | 0.5517 | 1.13 | 0.8708 | 2.13 | 0.9834 |
| -2.14 | 0.0162 | -1.14 | 0.1271 | -0.14 | 0.4443 | 0.14 | 0.5557 | 1.14 | 0.8729 | 2.14 | 0.9838 |
| -2.15 | 0.0158 | -1.15 | 0.1251 | -0.15 | 0.4404 | 0.15 | 0.5596 | 1.15 | 0.8749 | 2.15 | 0.9842 |
| -2.16 | 0.0154 | -1.16 | 0.1230 | -0.16 | 0.4364 | 0.16 | 0.5636 | 1.16 | 0.8770 | 2.16 | 0.9846 |
| -2.17 | 0.0150 | -1.17 | 0.1210 | -0.17 | 0.4325 | 0.17 | 0.5675 | 1.17 | 0.8790 | 2.17 | 0.9850 |
| -2.18 | 0.0146 | -1.18 | 0.1190 | -0.18 | 0.4286 | 0.18 | 0.5714 | 1.18 | 0.8810 | 2.18 | 0.9854 |
| -2.19 | 0.0143 | -1.19 | 0.1170 | -0.19 | 0.4247 | 0.19 | 0.5753 | 1.19 | 0.8830 | 2.19 | 0.9857 |
| -2.20 | 0.0139 | -1.20 | 0.1151 | -0.20 | 0.4207 | 0.20 | 0.5793 | 1.20 | 0.8849 | 2.20 | 0.9861 |
| -2.21 | 0.0136 | -1.21 | 0.1131 | -0.21 | 0.4168 | 0.21 | 0.5832 | 1.21 | 0.8869 | 2.21 | 0.9864 |
| -2.22 | 0.0132 | -1.22 | 0.1112 | -0.22 | 0.4129 | 0.22 | 0.5871 | 1.22 | 0.8888 | 2.22 | 0.9868 |
| -2.23 | 0.0129 | -1.23 | 0.1093 | -0.23 | 0.4090 | 0.23 | 0.5910 | 1.23 | 0.8907 | 2.23 | 0.9871 |
| -2.24 | 0.0125 | -1.24 | 0.1075 | -0.24 | 0.4052 | 0.24 | 0.5948 | 1.24 | 0.8925 | 2.24 | 0.9875 |
| -2.25 | 0.0122 | -1.25 | 0.1056 | -0.25 | 0.4013 | 0.25 | 0.5987 | 1.25 | 0.8944 | 2.25 | 0.9878 |

**TABLE 7.3.1 Standard Normal Probability Table (See Figure 7.3.5)—cont'd**

| z Value | Probability | z Value | Probability | z Value | Probability | z Value | Probability | z Value | Probability | z Value | Probability |
|---|---|---|---|---|---|---|---|---|---|---|---|
| -2.26 | 0.0119 | -1.26 | 0.1038 | -0.26 | 0.3974 | 0.26 | 0.6026 | 1.26 | 0.8962 | 2.26 | 0.9881 |
| -2.27 | 0.0116 | -1.27 | 0.1020 | -0.27 | 0.3936 | 0.27 | 0.6064 | 1.27 | 0.8980 | 2.27 | 0.9884 |
| -2.28 | 0.0113 | -1.28 | 0.1003 | -0.28 | 0.3897 | 0.28 | 0.6103 | 1.28 | 0.8997 | 2.28 | 0.9887 |
| -2.29 | 0.0110 | -1.29 | 0.0985 | -0.29 | 0.3859 | 0.29 | 0.6141 | 1.29 | 0.9015 | 2.29 | 0.9890 |
| -2.30 | 0.0107 | -1.30 | 0.0968 | -0.30 | 0.3821 | 0.30 | 0.6179 | 1.30 | 0.9032 | 2.30 | 0.9893 |
| -2.31 | 0.0104 | -1.31 | 0.0951 | -0.31 | 0.3783 | 0.31 | 0.6217 | 1.31 | 0.9049 | 2.31 | 0.9896 |
| -2.32 | 0.0102 | -1.32 | 0.0934 | -0.32 | 0.3745 | 0.32 | 0.6255 | 1.32 | 0.9066 | 2.32 | 0.9898 |
| -2.33 | 0.0099 | -1.33 | 0.0918 | -0.33 | 0.3707 | 0.33 | 0.6293 | 1.33 | 0.9082 | 2.33 | 0.9901 |
| -2.34 | 0.0096 | -1.34 | 0.0901 | -0.34 | 0.3669 | 0.34 | 0.6331 | 1.34 | 0.9099 | 2.34 | 0.9904 |
| -2.35 | 0.0094 | -1.35 | 0.0885 | -0.35 | 0.3632 | 0.35 | 0.6368 | 1.35 | 0.9115 | 2.35 | 0.9906 |
| -2.36 | 0.0091 | -1.36 | 0.0869 | -0.36 | 0.3594 | 0.36 | 0.6406 | 1.36 | 0.9131 | 2.36 | 0.9909 |
| -2.37 | 0.0089 | -1.37 | 0.0853 | -0.37 | 0.3557 | 0.37 | 0.6443 | 1.37 | 0.9147 | 2.37 | 0.9911 |
| -2.38 | 0.0087 | -1.38 | 0.0838 | -0.38 | 0.3520 | 0.38 | 0.6480 | 1.38 | 0.9162 | 2.38 | 0.9913 |
| -2.39 | 0.0084 | -1.39 | 0.0823 | -0.39 | 0.3483 | 0.39 | 0.6517 | 1.39 | 0.9177 | 2.39 | 0.9916 |
| -2.40 | 0.0082 | -1.40 | 0.0808 | -0.40 | 0.3446 | 0.40 | 0.6554 | 1.40 | 0.9192 | 2.40 | 0.9918 |
| -2.41 | 0.0080 | -1.41 | 0.0793 | -0.41 | 0.3409 | 0.41 | 0.6591 | 1.41 | 0.9207 | 2.41 | 0.9920 |
| -2.42 | 0.0078 | -1.42 | 0.0778 | -0.42 | 0.3372 | 0.42 | 0.6628 | 1.42 | 0.9222 | 2.42 | 0.9922 |
| -2.43 | 0.0075 | -1.43 | 0.0764 | -0.43 | 0.3336 | 0.43 | 0.6664 | 1.43 | 0.9236 | 2.43 | 0.9925 |
| -2.44 | 0.0073 | -1.44 | 0.0749 | -0.44 | 0.3300 | 0.44 | 0.6700 | 1.44 | 0.9251 | 2.44 | 0.9927 |
| -2.45 | 0.0071 | -1.45 | 0.0735 | -0.45 | 0.3264 | 0.45 | 0.6736 | 1.45 | 0.9265 | 2.45 | 0.9929 |
| -2.46 | 0.0069 | -1.46 | 0.0721 | -0.46 | 0.3228 | 0.46 | 0.6772 | 1.46 | 0.9279 | 2.46 | 0.9931 |
| -2.47 | 0.0068 | -1.47 | 0.0708 | -0.47 | 0.3192 | 0.47 | 0.6808 | 1.47 | 0.9292 | 2.47 | 0.9932 |
| -2.48 | 0.0066 | -1.48 | 0.0694 | -0.48 | 0.3156 | 0.48 | 0.6844 | 1.48 | 0.9306 | 2.48 | 0.9934 |
| -2.49 | 0.0064 | -1.49 | 0.0681 | -0.49 | 0.3121 | 0.49 | 0.6879 | 1.49 | 0.9319 | 2.49 | 0.9936 |
| -2.50 | 0.0062 | -1.50 | 0.0668 | -0.50 | 0.3085 | 0.50 | 0.6915 | 1.50 | 0.9332 | 2.50 | 0.9938 |
| -2.51 | 0.0060 | -1.51 | 0.0655 | -0.51 | 0.3050 | 0.51 | 0.6950 | 1.51 | 0.9345 | 2.51 | 0.9940 |

**TABLE 7.3.1** Standard Normal Probability Table (See Figure 7.3.5)—cont'd

| z Value | Probability | z Value | Probability | z Value | Probability | z Value | Probability | z Value | Probability | z Value | Probability |
|---------|-------------|---------|-------------|---------|-------------|---------|-------------|---------|-------------|---------|-------------|
| -2.52 | 0.0059 | -1.52 | 0.0643 | -0.52 | 0.3015 | 0.52 | 0.6985 | 1.52 | 0.9357 | 2.52 | 0.9941 |
| -2.53 | 0.0057 | -1.53 | 0.0630 | -0.53 | 0.2981 | 0.53 | 0.7019 | 1.53 | 0.9370 | 2.53 | 0.9943 |
| -2.54 | 0.0055 | -1.54 | 0.0618 | -0.54 | 0.2946 | 0.54 | 0.7054 | 1.54 | 0.9382 | 2.54 | 0.9945 |
| -2.55 | 0.0054 | -1.55 | 0.0606 | -0.55 | 0.2912 | 0.55 | 0.7088 | 1.55 | 0.9394 | 2.55 | 0.9946 |
| -2.56 | 0.0052 | -1.56 | 0.0594 | -0.56 | 0.2877 | 0.56 | 0.7123 | 1.56 | 0.9406 | 2.56 | 0.9948 |
| -2.57 | 0.0051 | -1.57 | 0.0582 | -0.57 | 0.2843 | 0.57 | 0.7157 | 1.57 | 0.9418 | 2.57 | 0.9949 |
| -2.58 | 0.0049 | -1.58 | 0.0571 | -0.58 | 0.2810 | 0.58 | 0.7190 | 1.58 | 0.9429 | 2.58 | 0.9951 |
| -2.59 | 0.0048 | -1.59 | 0.0559 | -0.59 | 0.2776 | 0.59 | 0.7224 | 1.59 | 0.9441 | 2.59 | 0.9952 |
| -2.60 | 0.0047 | -1.60 | 0.0548 | -0.60 | 0.2743 | 0.60 | 0.7257 | 1.60 | 0.9452 | 2.60 | 0.9953 |
| -2.61 | 0.0045 | -1.61 | 0.0537 | -0.61 | 0.2709 | 0.61 | 0.7291 | 1.61 | 0.9463 | 2.61 | 0.9955 |
| -2.62 | 0.0044 | -1.62 | 0.0526 | -0.62 | 0.2676 | 0.62 | 0.7324 | 1.62 | 0.9474 | 2.62 | 0.9956 |
| -2.63 | 0.0043 | -1.63 | 0.0516 | -0.63 | 0.2643 | 0.63 | 0.7357 | 1.63 | 0.9484 | 2.63 | 0.9957 |
| -2.64 | 0.0041 | -1.64 | 0.0505 | -0.64 | 0.2611 | 0.64 | 0.7389 | 1.64 | 0.9495 | 2.64 | 0.9959 |
| -2.65 | 0.0040 | -1.65 | 0.0495 | -0.65 | 0.2578 | 0.65 | 0.7422 | 1.65 | 0.9505 | 2.65 | 0.9960 |
| -2.66 | 0.0039 | -1.66 | 0.0485 | -0.66 | 0.2546 | 0.66 | 0.7454 | 1.66 | 0.9515 | 2.66 | 0.9961 |
| -2.67 | 0.0038 | -1.67 | 0.0475 | -0.67 | 0.2514 | 0.67 | 0.7486 | 1.67 | 0.9525 | 2.67 | 0.9962 |
| -2.68 | 0.0037 | -1.68 | 0.0465 | -0.68 | 0.2483 | 0.68 | 0.7517 | 1.68 | 0.9535 | 2.68 | 0.9963 |
| -2.69 | 0.0036 | -1.69 | 0.0455 | -0.69 | 0.2451 | 0.69 | 0.7549 | 1.69 | 0.9545 | 2.69 | 0.9964 |
| -2.70 | 0.0035 | -1.70 | 0.0446 | -0.70 | 0.2420 | 0.70 | 0.7580 | 1.70 | 0.9554 | 2.70 | 0.9965 |
| -2.71 | 0.0034 | -1.71 | 0.0436 | -0.71 | 0.2389 | 0.71 | 0.7611 | 1.71 | 0.9564 | 2.71 | 0.9966 |
| -2.72 | 0.0033 | -1.72 | 0.0427 | -0.72 | 0.2358 | 0.72 | 0.7642 | 1.72 | 0.9573 | 2.72 | 0.9967 |
| -2.73 | 0.0032 | -1.73 | 0.0418 | -0.73 | 0.2327 | 0.73 | 0.7673 | 1.73 | 0.9582 | 2.73 | 0.9968 |
| -2.74 | 0.0031 | -1.74 | 0.0409 | -0.74 | 0.2296 | 0.74 | 0.7704 | 1.74 | 0.9591 | 2.74 | 0.9969 |
| -2.75 | 0.0030 | -1.75 | 0.0401 | -0.75 | 0.2266 | 0.75 | 0.7734 | 1.75 | 0.9599 | 2.75 | 0.9970 |
| -2.76 | 0.0029 | -1.76 | 0.0392 | -0.76 | 0.2236 | 0.76 | 0.7764 | 1.76 | 0.9608 | 2.76 | 0.9971 |
| -2.77 | 0.0028 | -1.77 | 0.0384 | -0.77 | 0.2206 | 0.77 | 0.7794 | 1.77 | 0.9616 | 2.77 | 0.9972 |

**TABLE 7.3.1** Standard Normal Probability Table (See Figure 7.3.5)—cont'd

| z Value | Probability | z Value | Probability | z Value | Probability | z Value | Probability | z Value | Probability | z Value | Probability |
|---|---|---|---|---|---|---|---|---|---|---|---|
| -2.78 | 0.0027 | -1.78 | 0.0375 | -0.78 | 0.2177 | 0.78 | 0.7823 | 1.78 | 0.9625 | 2.78 | 0.9973 |
| -2.79 | 0.0026 | -1.79 | 0.0367 | -0.79 | 0.2148 | 0.79 | 0.7852 | 1.79 | 0.9633 | 2.79 | 0.9974 |
| -2.80 | 0.0026 | -1.80 | 0.0359 | -0.80 | 0.2119 | 0.80 | 0.7881 | 1.80 | 0.9641 | 2.80 | 0.9974 |
| -2.81 | 0.0025 | -1.81 | 0.0351 | -0.81 | 0.2090 | 0.81 | 0.7910 | 1.81 | 0.9649 | 2.81 | 0.9975 |
| -2.82 | 0.0024 | -1.82 | 0.0344 | -0.82 | 0.2061 | 0.82 | 0.7939 | 1.82 | 0.9656 | 2.82 | 0.9976 |
| -2.83 | 0.0023 | -1.83 | 0.0336 | -0.83 | 0.2033 | 0.83 | 0.7967 | 1.83 | 0.9664 | 2.83 | 0.9977 |
| -2.84 | 0.0023 | -1.84 | 0.0329 | -0.84 | 0.2005 | 0.84 | 0.7995 | 1.84 | 0.9671 | 2.84 | 0.9977 |
| -2.85 | 0.0022 | -1.85 | 0.0322 | -0.85 | 0.1977 | 0.85 | 0.8023 | 1.85 | 0.9678 | 2.85 | 0.9978 |
| -2.86 | 0.0021 | -1.86 | 0.0314 | -0.86 | 0.1949 | 0.86 | 0.8051 | 1.86 | 0.9686 | 2.86 | 0.9979 |
| -2.87 | 0.0021 | -1.87 | 0.0307 | -0.87 | 0.1922 | 0.87 | 0.8078 | 1.87 | 0.9693 | 2.87 | 0.9979 |
| -2.88 | 0.0020 | -1.88 | 0.0301 | -0.88 | 0.1894 | 0.88 | 0.8106 | 1.88 | 0.9699 | 2.88 | 0.9980 |
| -2.89 | 0.0019 | -1.89 | 0.0294 | -0.89 | 0.1867 | 0.89 | 0.8133 | 1.89 | 0.9706 | 2.89 | 0.9981 |
| -2.90 | 0.0019 | -1.90 | 0.0287 | -0.90 | 0.1841 | 0.90 | 0.8159 | 1.90 | 0.9713 | 2.90 | 0.9981 |
| -2.91 | 0.0018 | -1.91 | 0.0281 | -0.91 | 0.1814 | 0.91 | 0.8186 | 1.91 | 0.9719 | 2.91 | 0.9982 |
| -2.92 | 0.0018 | -1.92 | 0.0274 | -0.92 | 0.1788 | 0.92 | 0.8212 | 1.92 | 0.9726 | 2.92 | 0.9982 |
| -2.93 | 0.0017 | -1.93 | 0.0268 | -0.93 | 0.1762 | 0.93 | 0.8238 | 1.93 | 0.9732 | 2.93 | 0.9983 |
| -2.94 | 0.0016 | -1.94 | 0.0262 | -0.94 | 0.1736 | 0.94 | 0.8264 | 1.94 | 0.9738 | 2.94 | 0.9984 |
| -2.95 | 0.0016 | -1.95 | 0.0256 | -0.95 | 0.1711 | 0.95 | 0.8289 | 1.95 | 0.9744 | 2.95 | 0.9984 |
| -2.96 | 0.0015 | -1.96 | 0.0250 | -0.96 | 0.1685 | 0.96 | 0.8315 | 1.96 | 0.9750 | 2.96 | 0.9985 |
| -2.97 | 0.0015 | -1.97 | 0.0244 | -0.97 | 0.1660 | 0.97 | 0.8340 | 1.97 | 0.9756 | 2.97 | 0.9985 |
| -2.98 | 0.0014 | -1.98 | 0.0239 | -0.98 | 0.1635 | 0.98 | 0.8365 | 1.98 | 0.9761 | 2.98 | 0.9986 |
| -2.99 | 0.0014 | -1.99 | 0.0233 | -0.99 | 0.1611 | 0.99 | 0.8389 | 1.99 | 0.9767 | 2.99 | 0.9986 |
| -3.00 | 0.0013 | -2.00 | 0.0228 | -1.00 | 0.1587 | 1.00 | 0.8413 | 2.00 | 0.9772 | 3.00 | 0.9987 |

This ($z = -1.67$) tells you that $15 million is 1.67 standard deviations below the mean (the forecast value).[9] Your problem has now been reduced to finding a standard normal probability:

Find the probability that a standard normal variable is less than $z = -1.67$.

From the table, you find the answer:

The probability of a really bad quarter is 0.0475, or about 5%.

Whew! It seems that a really bad quarter is not very likely. However, a 5% chance is an outside possibility that should not be disregarded altogether.

Figures 7.3.6 and 7.3.7 show this probability calculation, both in terms of sales dollars and in terms of standardized sales numbers (standard deviations above or below the mean).

This was an easy problem, since the answer was found directly from the standard normal probability table. Here is a question that requires a little more care:

Continuing with the sales-forecasting problem, find the probability of a "really good quarter," which is defined as sales in excess of $24 million.

The first step is to standardize the sales number: $24 million is $z = (24 - 20)/3 = 1.33$ standard deviations above the mean. Thus, you are asked to solve the following problem:

Find the probability that a standard normal variable *exceeds* $z = 1.33$.

Using the complement rule, we know that this probability is 1 minus the probability of being less than $z = 1.33$. Looking up 1.33 in the table, you find the answer:

Probability of a really good quarter $= 1 - 0.9082$
$\qquad\qquad\qquad\qquad\qquad\qquad = 0.0918$, or about 9%

This probability is illustrated, in standardized numbers, in Figure 7.3.8.

Here's another kind of problem:

Continuing with the sales-forecasting problem, find the probability of a "typical quarter," which is defined as sales between $16 million and $23 million.

Begin by standardizing both of these numbers, to see that your task is to solve the following problem:

Find the probability that a standard normal is *between* $z_1 = -1.33$ and $z_2 = 1.00$.

---

9. You know that this is below the mean because the standardized number $z = -1.67$ is negative. The standardized number $z$ will be positive for any number above the mean. The standardized number $z$ for the mean itself is 0.

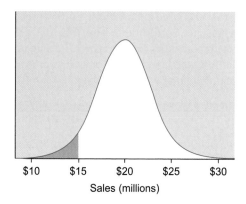

**FIGURE 7.3.6** The probability of a really bad quarter (sales less than $15 million) is represented by the shaded area under the curve. This is based on the forecast of $20 million and the standard deviation of $3 million. The answer is found by standardizing and then using the standard normal probability table.

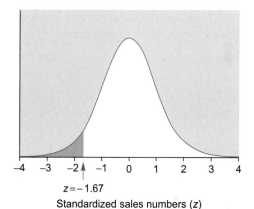

**FIGURE 7.3.7** The probability of a really bad quarter, in terms of standardized sales numbers. This is the probability that sales will be more than $z = -1.67$ standard deviations below the mean. The answer is 0.0475.

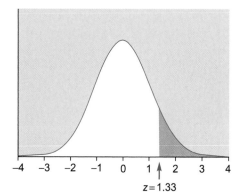

**FIGURE 7.3.8** The probability of a really good quarter, in terms of standardized sales numbers. The shaded area is 1 minus the unshaded area under the curve, which may be looked up in the table. The answer is 0.0918.

To solve this kind of problem, look up each standardized number in the table and find the difference between the probabilities for the answer. Be sure to subtract the smaller from the larger so that your answer is a positive number and therefore a "legal" probability!

$$\text{Probability of a typical quarter} = 0.8413 - 0.0918$$
$$= 0.7495, \text{ or about } 75\%$$

This probability is illustrated, in standardized numbers, in Figure 7.3.9.

Finally, here's yet another kind of problem:

Continuing with the sales-forecasting problem, find the probability of a "surprising quarter," which is defined as sales either less than $16 million or more than $23 million.

This asks for the probability of *not* being between two numbers. Using the complement rule, you may simply take 1 minus the probability found in the preceding example, which was the probability of being between these two values. The answer is therefore as follows:

$$\text{Probability of a surprising quarter} = 1 - 0.7495$$
$$= 0.2505, \text{ or about } 25\%$$

This probability is illustrated, in standardized numbers, in Figure 7.3.10.

To use Excel® to compute these first three probabilities, we use the function "NORMDIST(value,mean,standardDeviation,TRUE)" to find the probability that a normal distribution with specified mean and standard deviation is less than some value. There is no need to standardize because Excel will do this for you as part of the calculation. The first calculation is straightforward because it is a probability of being less. The second calculation is one minus the NORMDIST function because it is a probability of being greater. The third calculation is the difference of two

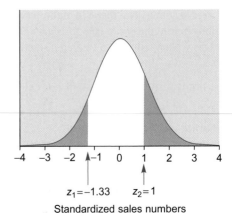

FIGURE 7.3.10   The probability of a surprising quarter, in terms of standardized sales numbers. The shaded area is found by looking up each standardized number in the table, finding the difference between the probabilities, and subtracting the result from 1. The answer is 0.2505.

NORMDIST calculations because it is the probability of being between two values. Here are the results:

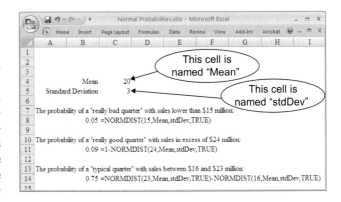

## The Four Different Probability Calculations

Here is a summary table of the four types of problems and how to solve them. The values $z$, $z_1$, and $z_2$ represent *standardized* numbers from the problem, found by subtracting the mean and dividing by the standard deviation. The table referred to is the standard normal probability table.

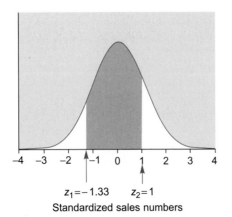

FIGURE 7.3.9   The probability of a typical quarter in terms of standardized sales numbers. The shaded area is found by looking up each standardized number in the table and then subtracting. Subtracting eliminates the *unshaded* area at the far left. The answer is 0.7495.

### Computing Probabilities for a Normal Distribution

| To Find the Probability of Being | Procedure |
|---|---|
| Less than $z$ | Look up $z$ in the table |
| More than $z$ | Subtract above answer from 1 |
| Between $z_1$ and $z_2$ | Look up $z_1$ and $z_2$ in the table, and subtract smaller probability from larger |
| Not between $z_1$ and $z_2$ | Subtract above answer (for "between $z_1$ and $z_2$") from 1 |

You may be wondering if there is a difference between the two events "sales exceeded $22 million" and "sales were at least $22 million." The term *exceeded* means *more than*, whereas the term *at least* means *more than or equal to*. In fact, for a normal distribution, there is *no difference* between the probabilities of these two events; the difference between the probabilities is just a geometric line, which represents no area under the normal curve.

## Be Careful: Things Need Not Be Normal!

If you have a normal distribution and you know the mean and standard deviation, you can find correct probabilities by standardizing and then using the standard normal probability table. Fortunately, if the distribution is only approximately normal, your probabilities will still be approximately correct.

However, if the distribution is very far from normal, then any probabilities you might compute based on the mean, the standard deviation, and the normal table could be very wrong indeed.

### Example
#### A Lottery (or Risky Project)

Consider a lottery (or a risky project, if you prefer) that pays back nothing 90% of the time, but pays $500 the remaining 10% of the time. The expected (mean) payoff is $50, and the standard deviation is $150 for this discrete random variable. Note that this does *not* represent a normal distribution; it's not even close because it's so discrete, with only two possible values.

What is the probability of winning at least $50? The correct answer is 10% because the *only* way to win

anything at all is to win the full amount, $500, which is at least $50.

What if you assumed a normal distribution with this same mean ($50) and standard deviation ($150)? How far from the correct answer (10%) would you be? Very far away, because the probability that a normally distributed random variable exceeds its mean is 0.5 or 50%.

This is a big difference: 10% (the correct answer) versus 50% (computed by wrongly assuming a normal distribution). Figure 7.3.11 shows the large difference between the actual discrete distribution and the normal distribution with the same mean and standard deviation. Always be careful about assuming a normal distribution!

## 7.4 THE NORMAL APPROXIMATION TO THE BINOMIAL

Remember the binomial distribution? It is the number of times that something happens out of $n$ independent tries. A binomial distribution can never be exactly normal, for two reasons. First, any normal distribution is free to produce observations with decimal parts (e.g., 7.11327), whereas the binomial number $X$ is restricted to whole numbers (e.g., 7). Second, a binomial distribution is skewed whenever $\pi$ is any number other than 0.5 (becoming more and more skewed when $\pi$ is close to 0 or to 1), whereas normal distributions are always perfectly symmetric.

However, a binomial distribution can be closely approximated by a normal distribution whenever the binomial $n$ is large and the probability $\pi$ is not too close to 0 or 1.[10,11] This fact will help you find probabilities (of being less, more, between, or not between given numbers) for a binomial distribution by replacing many complex and difficult calculations (using the earlier formula for individual binomial probabilities) with a single simpler calculation (using the normal distribution).

But how to choose a normal distribution that will be close to a given binomial distribution? A good choice is to use the normal distribution that has the same mean and standard deviation as the binomial distribution you wish to approximate. Since you already know how to find the mean and standard deviation for a binomial distribution (from Section 7.2), and you know how to find probabilities for a normal distribution with given mean

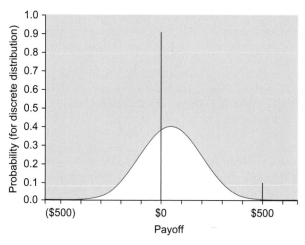

**FIGURE 7.3.11**  The discrete distribution of the payoff and the normal distribution having the same mean ($50) and standard deviation ($150). These distributions and their probabilities are very different. The discrete distribution gives the correct answers; the assumption of normality is wrong in this case.

10. When $\pi$ is close to 0 or 1, the approach to a normal distribution is slower as $n$ increases due to skewness of the binomial with rare or nearly definite events. The Poisson distribution, covered in a later section, is a good approximation to the binomial when $n$ is large and $\pi$ is close to 0.

11. The central limit theorem, to be covered in Chapter 8, tells how a normal distribution emerges when many independent random trials are combined by adding or averaging.

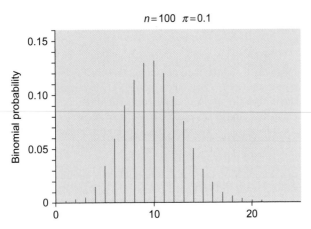

FIGURE 7.4.1  The probability distribution of a binomial with $n = 100$ and $\pi = 0.10$ is fairly close to normal.

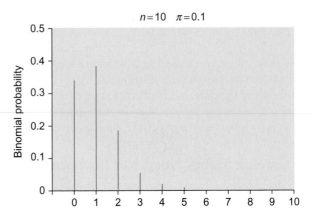

FIGURE 7.4.2  The probability distribution of a binomial with $n = 10$ and $\pi = 0.10$ is not very normal because $n$ is not large enough.

and standard deviation (from Section 7.3), it should not be difficult for you to compute these approximate binomial probabilities.

Here is convincing evidence of the binomial approximation to the normal. Suppose $n$ is 100 and $\pi$ is 0.10. The probability distribution, computed using the binomial formula, is shown in Figure 7.4.1. It certainly has the bell shape of a normal distribution. Although it is still discrete, with separate, individual bars, there are enough observations that the discreteness is not a dominant feature.

To approximate a binomial (which is discrete, only taking on whole-number values) by using a normal random variable (which is continuous), we will do better if we extend the limits by one-half in each direction in order to include all numbers that round to the whole number(s).[12] For example, to approximate the probability that a binomial $X$ is equal to 3, we would find the probability that a normal distribution (with the same mean and standard deviation) is between 2.5 and 3.5. We need to do this because the probability is zero that any normal random variable is equal to 3 and, actually, all values that the normal produced that were between 2.5 and 3.5 would round to 3. Similarly, to find the probability that a binomial is between 6 and 9, you would find the probability that a normal (with same mean and standard deviation) is between 5.5 and 9.5. The probability of *not* being between two numbers is, as usual, one minus the probability of being between them.

Using the Normal Approximation to the Binomial (Whole Numbers $a$ and $b$):

| The Probability That the Binomial Is: | Is Approximated by the Probability That the Corresponding Normal Is: |
|---|---|
| Exactly 8 | Between 7.5 and 8.5 |
| Exactly $a$ | Between $a - 0.5$ and $a + 0.5$ |
| Between 15 and 23 | Between 14.5 and 23.5 |
| Between $a$ and $b$ | Between $a - 0.5$ and $b + 0.5$ |

Compare Figure 7.4.1 ($n = 100$) to Figure 7.4.2 ($n = 10$) to see that, with smaller $n$, the distribution is not as normal. Furthermore, the discreteness is more important when $n$ is small.

**Example**
*High- and Low-Speed Microprocessors*

We often don't have as much control over a manufacturing process as we would like. Such is the case with sophisticated microprocessor chips, such as some of those used in microcomputers, which can have over a billion transistors placed on a chip of silicon smaller than a square inch. Despite careful controls, there is variation within the resulting chips: Some will run at higher speeds than others.

In the spirit of the old software saying "It's not a bug, it's a feature!" the chips are sorted according to the speed at which they will actually run and priced accordingly (with the faster chips commanding a higher price). The catalog lists two products: 2 gigahertz (slower) and 3 gigahertz (faster).

Your machinery is known to produce the slow chips 80% of the time, on average, and fast chips the remaining 20% of the time, with chips being slow or fast independently of one another. Today your goal is to ship 1,000 slow chips and 300 fast chips, perhaps with some chips left over. How many should you schedule for production?

If you schedule 1,300 total chips, you expect 80% (1,040 chips) to be slow and 20% (260 chips) to be fast. You would

---

12. We assume here that you are looking for probabilities about a binomial *number of occurrences X*. If, on the other hand, you need probabilities for a binomial *proportion* or *percentage p*, you should first convert to *X* and then proceed from there. For example, the probability of observing "at least 20% of 261" is the same as the probability of observing "at least 53 of 261" since you need at least $0.20 \times 261 = 52.2$ and can observe only whole numbers.

have enough slow ones, but not enough fast ones, on average.

Since you know that you are limited by the number of fast chips, you compute 300/0.20 = 1,500. This tells you that if you schedule 1,500 chips, you can expect 20% of these (or 300 chips) to be fast. So on average, you would just meet the goal. Unfortunately, this means that you have only about a 50% chance of meeting the goal for fast chips!

Suppose you schedule 1,650 chips for production. What is the probability that you will be able to meet the goal? To solve this, you first state it as a complete probability question:

Given a binomial random variable (the number of fast chips produced) with $n = 1,650$ total chips produced and $\pi = 0.20$ probability that a chip is fast, find the probability that this random variable is at least 300 but no more than 650.[13]

To solve this problem using the binomial distribution directly would require that you compute the probability for 300, for 301, for 302, and so on until you got tired. The normal approximation to the binomial allows you to solve it much faster using the standard normal probability table. You will need to know the mean and standard deviation of the number of fast chips produced:

$$\mu_{(\text{Number of fast chips})} = n\pi$$
$$= 1,650 \times 0.20 = 330$$

$$\sigma_{(\text{Number of fast chips})} = \sqrt{n\pi(1-\pi)}$$
$$= \sqrt{1,650 \times 0.20 \times 0.80} = 16.24807$$

You also need to standardize the bounds on the number of fast chips needed, 300 and 650 (after extending them by a half to 299.5 and 650.5), using the mean and standard deviation computed just above:

$$z_1 = \text{Standardized lower number of fast chips} = \frac{299.5 - 330}{16.24807}$$
$$= -1.88$$

$$z_2 = \text{Standardized upper number of fast chips} = \frac{650.5 - 330}{16.24807}$$
$$= 19.73$$

Looking up these standardized values in the standard normal probability table, you find 0.030 for $z_1 = -1.88$ and note that $z_2 = 19.73$ is way off the end of the table, so you use the value 1.[14] Subtracting these numbers to find the probability of being between the bounds, the answer is $1 - 0.030 = 0.970$. You conclude that if 1,650 chips are scheduled, you have a 97% chance of meeting the goal of shipping 300 fast chips and 1,000 slow ones.

---

13. The reason is that more than $1,650 - 1,000 = 650$ fast chips would imply fewer than 1,000 slow chips; thus, you would not be able to meet the goal for slow chips.
14. This makes sense; the probability that a standard normal variable will be less than $z_2 = 19.73$ standard deviations above its mean is essentially 1, since this nearly always happens.

One of the uses of probability is to help you understand what is happening "behind the scenes" in the real world. Let's see what might really be happening in an opinion poll by using a *What if* scenario analysis.

## Example
### Polling the Electorate

Your telephone polling and research firm was hired to conduct an opinion poll to see if a new municipal bond initiative is likely to be approved by the voters in the next election. You decided to interview 800 randomly selected representative people who are likely to vote, and you found that 437 intend to vote in favor. Here's the *What if*: If the entire electorate were, in fact, evenly divided on the issue, what is the probability that you would expect to see this many or more of your sample in favor?

**Your coworker:** "It looks pretty close: 437 out of 800 is pretty close to 50–50, which would be 400 out of 800."
**You:** "But 437 seems lots bigger than 400 to me. Let's find out if the extra 37 could reasonably be just randomness."
**Your coworker:** "OK. Let's assume that each person is as likely to be in favor as not. Then we can compute the chances of seeing 437 or more."
**You:** "OK. If the chances are more than 5% or 10%, then the extra 37 could reasonably be just randomness. But if the chances are really small, say under 5% or under 1%, then it would seem that more than just randomness is involved."

To do the calculation, let $X$ represent the following binomial random variable: the number of people (out of 800 interviewed) who say they intend to vote in favor. If we assume that people are evenly divided on the issue, then the probability for each person interviewed is $\pi = 0.50$ that they intend to vote in favor. Now let's find the mean and standard deviation of $X$ using formulas for the binomial distribution:

$$\mu_X = n\pi = (800)(0.50) = 400$$

$$\sigma_X = \sqrt{n\pi(1-\pi)}$$
$$= \sqrt{(800)(0.50)(1 - 0.50)} = 14.14214$$

Now, to find the probability that $X$ is at least 437, we extend the limit by one-half to find the probability that $X$ is at least 436.5 and then use the fact that $X$ is approximately normal. That is, we now find the probability that a normally distributed random variable with mean 400 and standard deviation 14.14214 is larger than 436.5 by standardizing as follows:

$$z = \text{Standardized value} = \frac{436.5 - \mu_X}{\sigma_X}$$
$$= \frac{436.5 - 400}{14.14214} = 2.58$$

Using the normal probability tables, you find that the probability of seeing this large a margin (or larger) in favor

*(Continued)*

**Example—cont'd**

in the sample, assuming that the population is evenly divided, is 1 – 0.995 = 0.005. This is very unlikely: a probability of about a half a percent, or 1 out of 200.

You asked, what if the population were evenly divided, and found the answer: A sample percentage of 54.6% (this is 437/800) or more is highly unlikely. The conclusion is that you have evidence against the *What if* scenario of evenly divided voters. It looks good for the initiative!

## 7.5 TWO OTHER DISTRIBUTIONS: THE POISSON AND THE EXPONENTIAL

Many other probability distributions are useful in statistics. This section provides brief descriptions of two such distributions with an indication of how they might fit in with some general classes of business applications. The *Poisson distribution* is often useful as a model of the *number of events* that occur during a fixed time, such as arrivals. The *exponential distribution* can work well as a model of the *amount of time*, such as that required to complete an operation. These distributions work well together (as the *Poisson process*) with the Poisson representing the number of events and the exponential describing the time between events.

### The Poisson Distribution

The Poisson distribution, like the binomial, is a counted number of times something happens. The difference is that there is no specified number *n* of possible tries. Here is one way that it can arise. If an event happens independently and randomly over time, and the mean rate of occurrence is constant over time, then the number of occurrences in a fixed amount of time will follow the **Poisson distribution**.[15] The Poisson is a *discrete* distribution and depends only on the mean number of occurrences expected.

Here are some random variables that might follow a Poisson distribution:

1. The number of orders your firm receives tomorrow.
2. The number of people who apply for a job tomorrow to your human resources division.
3. The number of defects in a finished product.
4. The number of calls your firm receives next week for help concerning an "easy-to-assemble" toy.
5. A binomial number *X* when *n* is large and $\pi$ is small.

The following figures show what the Poisson probabilities look like for a system expecting a mean of 0.5 occurrence (Figure 7.5.1), 2 occurrences (Figure 7.5.2), and 20

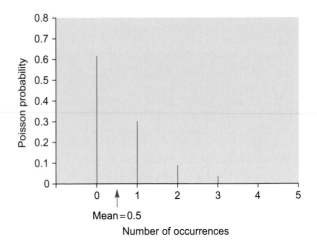

**FIGURE 7.5.1** The Poisson distribution with 0.5 occurrences expected is a skewed distribution. There is a high probability, 0.607, that no occurrences will happen at all.

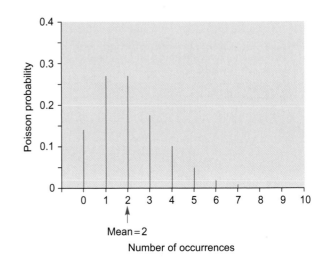

**FIGURE 7.5.2** The Poisson distribution with two occurrences expected. The distribution is still somewhat skewed.

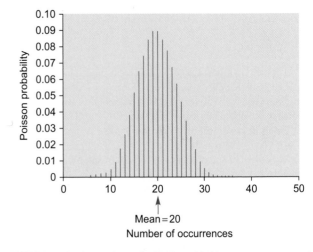

**FIGURE 7.5.3** The Poisson distribution with 20 occurrences expected. The distribution, although still discrete, is now fairly close to normal.

---

15. *Poisson* is a French name, pronounced (more or less) "pwah-*soh*."

occurrences (Figure 7.5.3). Note from the bell shape of Figure 7.5.3 that the Poisson distribution is approximately normal when many occurrences are expected.

There are three important facts about a Poisson distribution. These facts, taken together, tell you how to find probabilities for a Poisson distribution when you know only its mean.

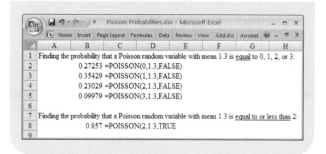

### For a Poisson Distribution

1. The standard deviation is always equal to the square root of the mean: $\sigma = \sqrt{\mu}$.
2. The exact probability that a Poisson random variable $X$ with mean $\mu$ is equal to $a$ is given by the formula

$$P(X = a) = \frac{\mu^a}{a!}e^{-\mu}$$

   where $e = 2.71828\ldots$ is a special number.[16]
3. If the mean is large, then the Poisson distribution is approximately normal.

16. This special mathematical number also shows up in continuously compounded interest formulas.

### Example
#### How Many Warranty Returns?

Because your firm's quality is so high, you expect only 1.3 of your products to be returned, on average, each day for warranty repairs. What are the chances that no products will be returned tomorrow? That one will be returned? How about two? How about three?

Since the mean (1.3) is so small, exact calculations are needed. Here are the details:

$$P(X = 0) = \frac{1.3^0}{0!}e^{-1.3} = \frac{1}{1} \times 0.27253 = 0.27253$$

$$P(X = 1) = \frac{1.3^1}{1!}e^{-1.3} = \frac{1.3}{1} \times 0.27253 = 0.35429$$

$$P(X = 2) = \frac{1.3^2}{2!}e^{-1.3} = \frac{1.69}{2} \times 0.27253 = 0.23029$$

$$P(X = 3) = \frac{1.3^3}{3!}e^{-1.3} = \frac{2.197}{6} \times 0.27253 = 0.09979$$

From these basic probabilities, you could add up the appropriate probabilities for 0, 1, and 2 to also find the probability that two items *or fewer* will be returned. The probability is, then, $0.27253 + 0.35429 + 0.23029 = 0.857$, or 85.7%.

To use Excel to compute these probabilities, you could use the function "POISSON(value,mean,FALSE)" to find the probability that a Poisson random variable is exactly equal to some value, and you could use "POISSON(value, mean,TRUE)" to find the probability that a Poisson random variable is *less than or equal* to the value. Here are the results:

### Example
#### How Many Phone Calls?

Your firm handles 460 calls per day, on average. Assuming a Poisson distribution, find the probability that you will be overloaded tomorrow, with 500 or more calls received.

The mean, $\mu = 460$, is given. The standard deviation is $\sigma = 21.44761$. You may use the normal approximation because the mean (460) is so large. Since the normal distribution is continuous, any value over 499.5 will round to 500 or more. The standardized number of calls is

$$z = \frac{499.5 - \mu}{\sigma} = \frac{499.5 - 460}{21.44761} = 1.84$$

When you use the standard normal probability table, the answer is a probability of $1 - 0.967 = 0.033$, so you may expect to be overloaded tomorrow with probability only about 3% (not very likely but within possibility).

## The Exponential Distribution

The **exponential distribution** is the very skewed continuous distribution shown in Figure 7.5.4. Its rise is vertical at 0, on the left, and it descends gradually, with a long tail on the right.

The following is a situation in which the exponential distribution is appropriate. If events happen independently

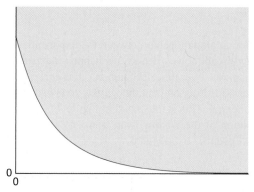

The exponential distribution

**FIGURE 7.5.4** The exponential distribution is a very skewed distribution that is often used to represent waiting times between events.

and randomly with a constant rate over time, the *waiting time* between successive events follows an exponential distribution.[17]

Here are some examples of random variables that might follow an exponential distribution:

1. Time between customer arrivals at an auto repair shop.
2. The amount of time your copy machine works between visits by the repair people.
3. The length of time of a typical telephone call.
4. The time until a TV system fails.
5. The time it takes to provide service for one customer.

The exponential distribution *has no memory* in the surprising sense that after you have waited awhile without success for the next event, your mean waiting time remaining until the next event is no shorter than it was when you started! This makes sense for waiting times, since occurrences are independent of one another and "don't know" that none have happened recently.

What does this property say about telephone calls? Suppose you are responsible for a switching unit for which the average call lasts five minutes. Consider all calls received at a given moment. On average, you expect them to last five minutes, with the individual durations following the exponential distribution. After one minute passes, some of these calls have ended. However, the calls that remain are all expected, on average, to last *five minutes more*. The reason is that the shorter calls have already been eliminated. While this may be difficult to believe, it has been confirmed (approximately) using real data.

Here are the basic facts for an exponential distribution. Note that there is no "normal approximation" because the exponential distribution is *always* very skewed.

---

**For an Exponential Distribution**

1. The standard deviation is always equal to the mean: $\sigma = \mu$.
2. The exact probability that an exponential random variable $X$ with mean $\mu$ is less than $a$ is given by the formula

$$P(X \leq a) = 1 - e^{-a/\mu}$$

---

There is a relationship between the exponential and the Poisson distributions when events happen independently at a constant rate over time. The *number of events* in any fixed time period is Poisson, and the *waiting time between events* is exponential. This is illustrated in Figure 7.5.5. In fact, the distribution of the waiting time from any fixed time until the next event is exponential.

---

17. Note that this implies that the total number of events follows a Poisson distribution.

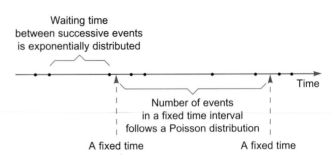

**FIGURE 7.5.5**   The relationship between the exponential and the Poisson distributions when events happen over time independently and at a constant rate.

---

**Example**
*Customer Arrivals*

Suppose customers arrive independently at a constant mean rate of 40 per hour. To find the probability that at least one customer arrives in the next five minutes, note that this is the probability that the exponential waiting time until the next customer arrives is less than five minutes. Since 40 customers arrive each hour, on average, the mean of this exponential random variable is $\mu = 1/40 = 0.025$ hours, or $0.025 \times 60 = 1.5$ minutes. The probability is then $P(X \leq 5) = 1 - e^{-5/1.5} = 0.964$, which may be computed in Excel® using the formula "=1−EXP(−5/1.5)". So the chances are high (96.4%) that at least one customer will arrive in the next 5 minutes.

---

## 7.6  END-OF-CHAPTER MATERIALS

### Summary

A **random variable** is a specification or description of a numerical result from a random experiment. A particular value taken on by a random variable is called an **observation**. The pattern of probabilities for a random variable is called its **probability distribution**. Random variables are either **discrete** (if you can list all possible outcomes) or **continuous** (if any number in a range is possible). Some random variables are actually discrete, but you can work with them as though they were continuous.

For a discrete random variable, the probability distribution is a list of the possible values together with their probabilities of occurrence. The mean or expected value and the standard deviation are computed as follows.

For a discrete random variable:

$$\mu = E(X) = \text{Sum of (value times probability)} = \sum XP(X)$$

$$\sigma = \sqrt{\text{Sum of (squared deviation times probability)}}$$

$$= \sqrt{\sum (X - \mu)^2 P(X)}$$

The interpretations are familiar. The **mean** or **expected value** indicates the typical or average value, and the

**standard deviation** indicates the risk in terms of approximately how far from the mean you can expect to be.

A random variable $X$ has a **binomial distribution** if it represents the *number of occurrences* of an event out of $n$ trials, provided (1) for each of the $n$ trials, the event always has the same probability $\pi$ of happening, and (2) the trials are independent of one another. The **binomial proportion** is $p = X/n$, which also represents a percentage. The mean and standard deviation of a binomial or binomial proportion may be found as follows:

**Mean and Standard Deviation for a Binomial Distribution**

| | Number of Occurrences, $X$ | Proportion or Percentage, $p = X/n$ |
|---|---|---|
| Mean | $E(X) = \mu_X = n\pi$ | $E(p) = \mu_p = \pi$ |
| Standard deviation | $\sigma_X = \sqrt{n\pi(1-\pi)}$ | $\sigma_p = \sqrt{\dfrac{\pi(1-\pi)}{n}}$ |

The probability that a binomial random variable $X$ is equal to some given number $a$ (from 0 to $n$) is given by the following formula. Binomial probability that $X$ equals $a$:

$$P(X = a) = \binom{n}{a} \pi^a (1-\pi)^{n-a}$$

$$= \frac{n!}{a!(n-a)!} \pi^a (1-\pi)^{n-a}$$

$$= \frac{1 \times 2 \times 3 \times \cdots \times n}{(1 \times 2 \times 3 \times \cdots \times a)[1 \times 2 \times 3 \times \cdots \times (n-a)]} \pi^a (1-\pi)^{n-a}$$

The notation $n!$ is $n$ factorial, the product of the numbers from 1 to $n$, with $0! = 1$ by definition. The notation

$$\binom{n}{a} = \frac{n!}{a!(n-a)!}$$

is the *binomial coefficient*, read aloud as "$n$ choose $a$."

The **normal distribution**, a continuous distribution, is represented by the familiar bell-shaped curve. The probability that a normal random variable will be between any two values is equal to the area under the normal curve between these two values. There is a normal distribution for each combination of a mean $\mu$ and a (positive) standard deviation $\sigma$. The **standard normal distribution** is a normal distribution with mean $\mu = 0$ and standard deviation $\sigma = 1$. You may think of the standard normal distribution as representing the number of standard deviations above or below the mean. The **standard normal probability table** gives the probability that a standard normal random variable $Z$ is *less than* any given number $z$.

To solve word problems involving normal probabilities, first identify the mean $\mu$, standard deviation $\sigma$, and the probability asked for. Convert to a **standardized number** $z$ (the number of standard deviations above the mean, or below the mean if the standardized number is negative) by subtracting the mean and dividing by the standard deviation:

$$z = \text{Standardized number} = \frac{\text{Number} - \text{Mean}}{\text{Standard deviation}}$$

$$= \frac{\text{Number} - \mu}{\sigma}$$

Finally, look up the standardized number or numbers in the standard normal probability table and use the following summary table (where $z$, $z_1$, and $z_2$ are standardized numbers) to find the final answer:

**Computing Probabilities for a Normal Distribution**

| To Find the Probability of Being | Procedure |
|---|---|
| Less than $z$ | Look up $z$ in the table |
| More than $z$ | Subtract above answer from 1 |
| Between $z_1$ and $z_2$ | Look up $z_1$ and $z_2$ in the table, and subtract smaller probability from larger |
| Not between $z_1$ and $z_2$ | Subtract above answer (for "between $z_1$ and $z_2$") from 1 |

Probabilities for a binomial distribution may be approximated using the normal distribution with the same mean and standard deviation, provided $n$ is large and $\pi$ is not too close to 0 or 1. Since the normal distribution is continuous, extend the limits by one-half in each direction (for example, the probability that a binomial is exactly a whole number $a$ is approximated by the probability that the corresponding normal is between $a - 0.5$ and $a + 0.5$).

If occurrences happen independently and randomly over time, and the average rate of occurrence is constant over time, then the number of occurrences that happen in a fixed amount of time will follow the **Poisson distribution**, a discrete random variable. The standard deviation is the square root of the mean. If the mean is large, the Poisson distribution is approximately normal and the standard normal probability table may be used. Exact Poisson probabilities may be found using the following formula:

$$P(X = a) = \frac{\mu^a}{a!} e^{-\mu}$$

The **exponential distribution** is a very skewed continuous distribution useful for understanding such variables as waiting times and durations of telephone calls. It has no "memory," in the sense that after you have waited awhile without success for the next event, your average waiting time until the next event is no shorter than it was when you started. Its standard deviation is always equal to its mean. The probability that an exponential random variable $X$ with mean $\mu$ is less than or equal to $a$ is $P(X \leq a) = 1 - e^{-a/\mu}$. There is no normal approximation for an exponential random variable.

# Key Words

binomial distribution, *159*
binomial proportion, *159*
continuous random variable, *156*
discrete random variable, *156*
exponential distribution, *177*
mean or expected value, *156*
normal distribution, *165*
observation, *155*
Poisson distribution, *176*
probability distribution, *155*
random variable, *155*
standard deviation, *157*
standard normal distribution, *166*
standard normal probability table, *166*
standardized number, *166*

## Questions

1. **a.** What is a random variable?
   **b.** What is the difference between a random variable and a number?
2. **a.** What is a discrete random variable?
   **b.** What is a continuous random variable?
   **c.** Give an example of a discrete random variable that is continuous for practical purposes.
3. **a.** What is the probability distribution of a discrete random variable?
   **b.** How do you find the mean of a discrete random variable? How do you interpret the result?
   **c.** How do you find the standard deviation of a discrete random variable? How do you interpret the result?
4. **a.** How do you tell if a random variable has a binomial distribution?
   **b.** What is a binomial proportion?
   **c.** What are $n$, $\pi$, $X$, and $p$?
5. For a binomial distribution:
   **a.** Why don't you just construct the probability tree to find the probabilities?
   **b.** How do you find the mean and the standard deviation?
   **c.** How do you find the probability that $X$ is equal to some number?
   **d.** How do you find the exact probability that $X$ is greater than or equal to some number?
   **e.** If $n$ is a large number, how do you find the approximate probability that $X$ is greater than or equal to some number?
6. **a.** What is a factorial?
   **b.** Find 3!, 0!, and 15!.
   **c.** What is a binomial coefficient? What does it represent in the formula for a binomial probability?
   **d.** Find the binomial coefficient "8 choose 5."

7. **a.** What is a normal distribution?
   **b.** Identify all of the different possible normal distributions.
   **c.** What does the area under the normal curve represent?
   **d.** What is the standard normal distribution? What is it used for?
   **e.** What numbers are found in the standard normal probability table?
   **f.** Find the probability that a standard normal random variable is less than –1.65.
   **g.** How do you standardize a number?
8. **a.** What kinds of situations give rise to a Poisson distribution?
   **b.** Is the Poisson a discrete or a continuous distribution?
   **c.** What is the standard deviation of a Poisson distribution?
   **d.** How do you find probabilities for a Poisson distribution if the mean is large?
   **e.** How do you find exact probabilities for a Poisson distribution?
9. **a.** What kinds of situations give rise to an exponential distribution?
   **b.** What is meant by the fact that an exponential random variable has no memory?
   **c.** Can the standard normal probability table be used to find probabilities for an exponential distribution? Why or why not?
   **d.** How do you find probabilities for an exponential distribution?

## Problems

*Problems marked with an asterisk (*) are solved in the Self Test in Appendix C*

1. A call option on common stock is being evaluated. If the stock goes down, the option will expire worthless. If the stock goes up, the payoff depends on just how high the stock goes. For simplicity, the payoffs are modeled as a discrete distribution with the probability distribution in Table 7.6.1. Even though options markets, in fact, behave more like a continuous random variable, this discrete approximation will give useful approximate results. Answer the following questions based on the discrete probability distribution given.
   **a.*** Find the mean, or expected value, of the option payoff.
   **b.*** Describe briefly what this expected value represents.
   **c.*** Find the standard deviation of the option payoff.
   **d.*** Describe briefly what this standard deviation represents.
   **e.*** Find the probability that the option will pay at least $20.
   **f.** Find the probability that the option will pay less than $30.

**TABLE 7.6.1  Probability Distribution of Payoff**

| Payoff | Probability |
|--------|-------------|
| $0 | 0.50 |
| 10 | 0.25 |
| 20 | 0.15 |
| 30 | 0.10 |

**TABLE 7.6.2  Probability Distribution of Downtime**

| Problem | Downtime (minutes) | Probability |
|---------|--------------------|-------------|
| Minor | 5 | 0.60 |
| Substantial | 30 | 0.30 |
| Catastrophic | 120 | 0.10 |

2. The length of time a system is "down" (that is, broken) is described (approximately) by the probability distribution in Table 7.6.2. Assume that these downtimes are exact. That is, there are three types of easily recognized problems that always take this long (5, 30, or 120 minutes) to fix.
   a. What kind of probability distribution does this table represent?
   b. Find the mean downtime.
   c. Find the standard deviation of the downtime.
   d. What is the probability that the downtime will be greater than 10 minutes, according to this table?
   e. What is the probability that the downtime is literally within one standard deviation of its mean? Is this about what you would expect for a normal distribution?
3. An investment will pay $105 with probability 0.7, and $125 with probability 0.3. Find the risk (as measured by standard deviation) for this investment.
4. On a given day, assume that there is a 30% chance you will receive no orders, a 50% chance you will receive one order, a 15% chance of two orders, and a 5% chance of three orders. Find the expected number of orders and the variability in the number of orders.
5. A new project has an uncertain cash flow. A group meeting has resulted in a consensus that a reasonable way to view the possible risks and rewards is to say that the project will pay $50,000 with probability 0.2, will pay $100,000 with probability 0.3, will pay $200,000 with probability 0.4, and will pay $400,000 with probability 0.1. How much risk is involved here? Please give both the name and the numerical value of your answer.
6. Your company is hoping to fill a key technical position and has advertised in hopes of obtaining qualified applicants. Because of the demanding qualifications, the pool of qualified people is limited and Table 7.6.3 shows your subjective probabilities for each outcome.
   a. Find the probability of obtaining at least one applicant.
   b. Find the probability of obtaining two or more applicants.
   c. Find the mean number of applicants.
   d. Find the standard deviation of the number of applicants and write a sentence interpreting its meaning.
7. You work for the loan department of a large bank. You know that one of your customers has been having trouble with the recession and may not be able to make the loan payment that is due next week. You believe there is a 60% chance that the payment of $50,000 will be made in full, a 30% chance that only half will be paid, and a 10% chance that no payment will be made at all.
   a. Find the expected loan payment.
   b. Find the degree of risk for this situation.
8. You are planning to invest in a new high-tech company, and figure your rate of return over the coming year as in Table 7.6.4 (where 100% says that you doubled your money, –50% says you lost half, etc.).
   a. Find the mean rate of return and explain what it represents.
   b. Find the standard deviation of the rate of return and explain what it represents.
   c. Find the probability that you will earn more than 40%, according to the table.
   d. How would you measure the risk of this investment?

**TABLE 7.6.3  Probabilities for Qualified Technical Applicants**

| Number of Applicants | Probability |
|----------------------|-------------|
| 0 | 0.30 |
| 1 | 0.55 |
| 2 | 0.10 |
| 3 | 0.05 |

**TABLE 7.6.4  Rates of Return and Probabilities for Four Scenarios**

| Rate of Return | Probability |
|----------------|-------------|
| 100% | 0.20 |
| 50 | 0.40 |
| 0 | 0.25 |
| –50 | 0.15 |

9. You can invest in just one of four projects on a lot of land you own. For simplicity, you have modeled the payoffs (as net present value in today's dollars) of the projects as discrete distributions. By selling the land, you can make $60,000 for sure. If you build an apartment, you estimate a payoff of $130,000 if things go well (with probability 0.60) and $70,000 otherwise. If you build a single-family house, the payoff is $100,000 (with probability 0.60) and $60,000 otherwise. Finally, you could build a gambling casino which would pay very well—$500,000—but with a probability of just 0.10 since the final government permits are not likely to be granted; all will be lost otherwise.

   a. Find the expected payoff for each of these four projects. In terms of just the expected payoff, rank these projects in order from best to worst.

   b. Find the standard deviation for each of these four projects. In terms of risk only, rank the projects from best to worst.

   c. Considering both the expected payoff and the risk involved, can any project or projects be eliminated from consideration entirely?

   d. How would you decide among the remaining projects? In particular, does any single project dominate the others completely?

10. Your quality control manager has identified the four major problems, the extent to which each one occurs (i.e., the probability that this problem occurs per item produced), and the cost of reworking to fix each one (see Table 7.6.5). Assume that only one problem can occur at a time.

   a. Compute the expected rework cost for each problem separately. For example, the expected rework cost for "broken case" is 0.04 × 6.88. Compare the results and indicate the most serious problem in terms of expected dollar costs.

   b. Find the overall expected rework cost due to all four problems together.

   c. Find the standard deviation of rework cost (don't forget the nonreworked items).

   d. Write a brief memo, as if to your supervisor, describing and analyzing the situation.

**TABLE 7.6.5 Quality Control Problems: Type, Extent, and Cost**

| Problem | Probability | Rework Cost |
| --- | --- | --- |
| Broken case | 0.04 | $6.88 |
| Faulty electronics | 0.02 | 12.30 |
| Missing connector | 0.06 | 0.75 |
| Blemish | 0.01 | 2.92 |

11. Suppose that 8% of the loans you authorize as vice president of the consumer loan division of a neighborhood bank will never be repaid. Assume further that you authorized 284 loans last year and that loans go sour independently of one another.

   a. How many of these loans, authorized by you, do you expect will never be repaid? What percentage do you expect?

   b. Find the usual measure of the level of uncertainty in the number of loans you authorized that will never be repaid. Briefly interpret this number.

   c. Find the usual measure of the level of uncertainty in the percentage of loans you authorized that will never be repaid. Briefly interpret this number.

12. Your company is planning to market a new reading lamp and has segmented the market into three groups—avid readers, regular readers, and occasional readers—and currently assumes that 25% of avid readers, 15% of regular readers, and 10% of occasional readers will want to buy the new product. As part of a marketing survey, 400 individuals will be randomly selected from the population of regular readers. Using the current assumptions, find the mean and standard deviation of the percentage among those surveyed who will want to buy the new product.

13. A company is conducting a survey of 235 people to measure the level of interest in a new product. Assume that the probability of a randomly selected person's being "very interested" is 0.88 and that people are selected independently of one another.

   a. Find the standard deviation of the percentage who will be found by the survey to be very interested.

   b. How much uncertainty is there in the number of people who will be found to be very interested?

   c. Find the expected number of people in the sample who will say that they are very interested.

   d. Find the expected percentage that the survey will identify as being very interested.

14. An election coming up next week promises to be very close. In fact, assume that 50% are in favor and 50% are against. Suppose you conduct a poll of 791 randomly selected likely voters. Approximately how different will the percent in favor (from the poll) be from the 50% in the population you are trying to estimate?

15. Repeat the previous problem, but now assume that 85% are in favor in the population. Is the uncertainty larger or smaller than when 50% was assumed? Why?

16. You have just performed a survey interviewing 358 randomly selected people. You found that 94 of them are interested in possibly purchasing a new cable TV service. How much uncertainty is there in this number "94" as compared to the average number you would expect to find in such a survey? (You may assume that exactly 25% of all people you might have interviewed would have been interested.)

17. You are planning to make sales calls at eight firms today. As a rough approximation, you figure that each call has a 20% chance of resulting in a sale and that firms make

their buying decisions without consulting each other. Find the probability of having a really terrible day with no sales at all.

18. It's been a bad day for the market, with 80% of securities losing value. You are evaluating a portfolio of 15 securities and will assume a binomial distribution for the number of securities that lost value.

   a.* What assumptions are being made when you use a binomial distribution in this way?

   b.* How many securities in your portfolio would you expect to lose value?

   c.* What is the standard deviation of the number of securities in your portfolio that lose value?

   d.* Find the probability that all 15 securities lose value.

   e.* Find the probability that exactly 10 securities lose value.

   f. Find the probability that 13 or more securities lose value.

19. Your firm has decided to interview a random sample of 10 customers in order to determine whether or not to change a consumer product. Your main competitor has already done a similar but much larger study and has concluded that exactly 86% of consumers approve of the change. Unfortunately, your firm does not have access to this information (but you may use this figure in your computations here).

   a. What is the name of the probability distribution of the number of consumers who will approve of the change in your study?

   b. What is the expected number of people, out of the 10 you will interview, who will approve of the change?

   c. What is the standard deviation of the number of people, out of the 10 you will interview, who will approve of the change?

   d. What is the expected percentage of people, out of the 10 you will interview, who will approve of the change?

   e. What is the standard deviation of the percentage of people, out of the 10 you will interview, who will approve of the change?

   f. What is the probability that exactly eight of your interviewed customers will approve of the change?

   g. What is the probability that eight or more of your interviewed customers will approve of the change?

20. Suppose that the number of hits on your company's website, from noon to 1 p.m. on a typical weekday, follows a normal distribution (approximately) with a mean of 190 and a standard deviation of 24.

   a. Find the probability that the number of hits is more than 160.

   b. Find the probability that the number of hits is less than 215.

   c. Find the probability that the number of hits is between 165 and 195.

   d. Find the probability that the number of hits is not between 150 and 225.

21. Find the probability that you will see moderate improvement in productivity, meaning an increase in productivity between 6 and 13. You may assume that the productivity increase follows a normal distribution with a mean of 10 and a standard deviation of 7.

22. Under usual conditions, a distillation unit in a refinery can process a mean of 135,000 barrels per day of crude petroleum, with a standard deviation of 6,000 barrels per day. You may assume a normal distribution.

   a. Find the probability that more than 135,000 barrels will be produced on a given day.

   b. Find the probability that more than 130,000 barrels will be produced on a given day.

   c. Find the probability that more than 150,000 barrels will be produced on a given day.

   d. Find the probability that less than 125,000 barrels will be produced on a given day.

   e. Find the probability that less than 100,000 barrels will be produced on a given day.

23. The quality control section of a purchasing contract for valves specifies that the diameter must be between 2.53 and 2.57 centimeters. Assume that the production equipment is set so that the mean diameter is 2.56 centimeters and the standard deviation is 0.01 centimeter. What percent of valves produced, over the long run, will be within these specifications, assuming a normal distribution?

24. Assume that the stock market closed at 13,246 points today. Tomorrow you expect the market to rise a mean of 4 points, with a standard deviation of 115 points. Assume a normal distribution.

   a. Find the probability that the stock market goes down tomorrow.

   b. Find the probability that the market goes up more than 50 points tomorrow.

   c. Find the probability that the market goes up more than 100 points tomorrow.

   d. Find the probability that the market goes down more than 150 points tomorrow.

   e. Find the probability that the market changes by more than 200 points in either direction.

25. Based on recent experience, you expect this Saturday's total receipts to have a mean of $2,353.25 and a standard deviation of $291.63 and to be normally distributed.

   a. Find the probability of a typical Saturday, defined as total receipts between $2,000 and $2,500.

   b. Find the probability of a terrific Saturday, defined as total receipts over $2,500.

   c. Find the probability of a mediocre Saturday, defined as total receipts less than $2,000.

26. The amount of ore (in tons) in a segment of a mine is assumed to follow a normal distribution with mean 185 and standard deviation 40. Find the probability that the amount of ore is less than 175 tons.

27. You are a farmer about to harvest your crop. To describe the uncertainty in the size of the harvest, you feel that it may be described as a normal distribution with a mean

value of 80,000 bushels and a standard deviation of 2,500 bushels. Find the probability that your harvest will exceed 84,000 bushels.

28. Assume that electronic microchip operating speeds are normally distributed with a mean of 2.5 gigahertz and a standard deviation of 0.4 gigahertz. What percentage of your production would you expect to be "superchips" with operating speeds of 3 gigahertz or more?

29. Although you don't know the exact total amount of payments you will receive next month, based on past experience you believe it will be approximately $2,500 more or less than $13,000, and will follow a normal distribution. Find the probability that you will receive between $10,000 and $15,000 next month.

30. A new project will be declared "successful" if you achieve a market share of 10% or more in the next two years. Your marketing department has considered all possibilities and decided that it expects the product to attain a market share of 12% in this time. However, this number is not certain. The standard deviation is forecast to be 3%, indicating the uncertainty in the 12% forecast as 3 percentage points. You may assume a normal distribution.
    a.* Find the probability that the new project is successful.
    b. Find the probability that the new project fails.
    c. Find the probability that the new project is wildly successful, defined as achieving at least a 15% market share.
    d. To assess the precision of the marketing projections, find the probability that the attained market share falls close to the projected value of 12%, that is, between 11% and 13%.

31. A manufacturing process produces semiconductor chips with a known failure rate of 6.3%. Assume that chip failures are independent of one another. You will be producing 2,000 chips tomorrow.
    a. What is the name of the probability distribution of the number of defective chips produced tomorrow?
    b. Find the expected number of defective chips produced.
    c. Find the standard deviation of the number of defective chips.
    d. Find the (approximate) probability that you will produce fewer than 130 defects.
    e. Find the (approximate) probability that you will produce more than 120 defects.
    f. You just learned that you will need to ship 1,860 working chips out of tomorrow's production of 2,000. What are the chances that you will succeed? Will you need to increase the scheduled number produced?
    g. If you schedule 2,100 chips for production, what is the probability that you will be able to ship 1,860 working ones?

32. A union strike vote is scheduled tomorrow, and it looks close. Assume that the number of votes to strike follows a binomial distribution. You expect 300 people to vote,

and you have projected a probability of 0.53 that a typical individual will vote to strike.
    a. Identify $n$ and $\pi$ for this binomial random variable.
    b. Find the mean and standard deviation of the number who will vote to strike.
    c. Find the (approximate) probability that a strike will result (i.e., that a majority will vote to strike).

33. Reconsider the previous problem and answer each part, but assume that 1,000 people will vote. (The probability for each one remains unchanged.)

34. Assume that if you were to interview the entire population of Detroit, exactly 18.6% would say that they are ready to buy your product. You plan to interview a representative random sample of 250 people. Find the (approximate) probability that your observed sample percentage is overoptimistic, where this is defined as the observed percentage exceeding 22.5%.

35. Suppose 15% of the items in a large warehouse are defective. You have chosen a random sample of 250 items to examine in detail. Find the (approximate) probability that more than 20% of the sample is defective.

36. You are planning to interview 350 consumers randomly selected from a large list of likely sales prospects, in order to assess the value of this list and whether you should assign salespeople the task of contacting them all. Assuming that 13% of the large list will respond favorably, find (approximate) probabilities for the following:
    a. More than 10% of randomly selected consumers will respond favorably.
    b. More than 13% of randomly selected consumers will respond favorably.
    c. More than 15% of randomly selected consumers will respond favorably.
    d. Between 10% and 15% of randomly selected consumers will respond favorably.

37. You have just sent out a test mailing of a catalog to 1,000 people randomly selected from a database of 12,320 addresses. You will go ahead with the mass mailing to the remaining 11,320 addresses provided you receive orders from 2.7% or more from the test mailing within two weeks. Find the (approximate) probability that you will do the mass mailing under each of the following scenarios:
    a. Assume that, in reality, exactly 2% of the population would send in an order within two weeks.
    b. Assume that, in reality, exactly 3% of the population would send in an order within two weeks.
    c. Assume that, in reality, exactly 4% of the population would send in an order within two weeks.

38. You expect a mean of 1,671 warranty repairs next month, with the actual outcome following a Poisson distribution.
    a. Find the standard deviation of the number of such repairs.
    b. Find the (approximate) probability of more than 1,700 such repairs.

39. If tomorrow is a typical day, your human resources division will expect to receive résumés from 175 job applicants. You may assume that applicants act independently of one another.
    a. What is the name of the probability distribution of the number of résumés received?
    b. What is the standard deviation of the number of résumés received?
    c. Find the approximate probability that you will receive more than 185 résumés.
    d. Find the approximate probability of a slow day, with 160 or fewer résumés received.

40. On a typical day, your clothing store takes care of 2.6 "special customers" on average. These customers are taken directly to a special room in the back, are assigned a full-time server, are given tea (or espresso) and scones, and have clothes brought to them. You may assume that the number who will arrive tomorrow follows a Poisson distribution.
    a. Find the standard deviation of the number of special customers.
    b. Find the probability that no special customers arrive tomorrow.
    c. Find the probability exactly 4 special customers will arrive tomorrow.

41. In order to earn enough to pay your firm's debt this year, you will need to be awarded at least 2 contracts. This is not usually a problem, since the yearly average is 5.1 contracts. You may assume a Poisson distribution.
    a. Find the probability that you will not earn enough to pay your firm's debt this year.
    b. Find the probability that you will be awarded exactly 3 contracts.

42. Customers arrive at random times, with an exponential distribution for the time between arrivals. Currently the mean time between customers is 6.34 minutes.
    a. Since the last customer arrived, three minutes have gone by. Find the mean time until the next customer arrives.
    b. Since the last customer arrived, 10 minutes have gone by. Find the mean time until the next customer arrives.

43. In the situation described in the previous problem, a customer has just arrived.
    a. Find the probability that the time until the arrival of the next customer is less than 3 minutes.
    b. Find the probability that the time until the arrival of the next customer is more than 10 minutes.
    c. Find the probability that the time until the arrival of the next customer is between 5 and 6 minutes.

44. A TV system is expected to last for 50,000 hours before failure. Assume an exponential distribution for the time until failure.
    a. Is the distribution skewed or symmetric?
    b. What is the standard deviation of the length of time until failure?
    c. The system has been working continuously for the past 8,500 hours and is still on. What is the expected time from now until failure? (Be careful!)

45. Assuming the appropriate probability distribution for the situation described in the preceding problem:
    a. Find the probability that the system will last 100,000 hours or more (twice the average lifetime).
    b. The system is guaranteed to last at least 5,000 hours. What percentage of production is expected to fail during the guarantee period?

46. Compare the "probability of being within one standard deviation of the mean" for the exponential and normal distributions.

## Database Exercises

*Problems marked with an asterisk (*) are solved in the Self Test in Appendix C.*

Refer to the employee database in Appendix A.

1. View each column as a collection of independent observations of a random variable.
   a. In each case, what kind of variable is represented, continuous or discrete? Why?
   b.* Consider the event "annual salary is above $40,000." Find the value of the binomial random variable $X$ that represents the number of times this event occurred. Also find the binomial proportion $p$ and say what it represents.
   c. What fraction of employees are male? Interpret this number as a binomial proportion. What is $n$?

2. You have a position open and are trying to hire a new person. Assume that the new person's experience will follow a normal distribution with the mean and (sample) standard deviation of your current employees.
   a. Find the probability that the new person will have more than six years of experience.
   b. Find the probability that the new person will have less than three years of experience.
   c. Find the probability that the new person will have between four and seven years of experience.

3. Suppose males and females are equally likely and that the number of each gender follows a binomial distribution. (Note that the database contains observations of random variables, not the random variables themselves.)
   a. Find $n$ and $\pi$ for the binomial distribution of the number of males.
   b. Find $n$ and $\pi$ for the binomial distribution of the number of females.
   c. Find the observed value of $X$ for the number of females.
   d. Use the normal approximation to the binomial distribution to find the probability of observing this many females (your answer to part c) or fewer in the database.

## Projects

1. Choose a continuous random quantity that you might deal with in your current or future work as an executive. Model it as a normally distributed random variable and

estimate (i.e., guess) the mean and standard deviation. Identify three events of interest to you relating to this random variable and compute their probabilities. Briefly discuss what you have learned.

2. Choose a discrete random quantity (taking on from 3 to 10 different values) that you might deal with in your current or future work as an executive. Estimate (i.e., guess) the probability distribution. Compute the mean and standard deviation. Identify two events of interest to you relating to this random variable and compute their probabilities. Briefly discuss what you have learned.

3. Choose a binomial random quantity that you might deal with in your current or future work as an executive. Estimate (i.e., guess) the value of $n$ and $\pi$. Compute the mean and standard deviation. Identify two events of interest to you relating to this random variable and compute their probabilities. Briefly discuss what you have learned.

4. On the Internet, find and record observations on at least five different random variables such as stock market indices, interest rates, corporate sales, or any business-related topic of interest to you.

## Case

### The Option Value of an Oil Lease

There's an oil leasing opportunity that looks too good to be true, and it probably *is* too good to be true: An estimated 1,500,000 barrels of oil sitting underground that can be leased for three years for just $1,300,000. It looks like a golden opportunity: Pay just over a million, bring the oil to the surface, sell it at the current spot price of $76.45 per barrel, and retire.

However, upon closer investigation, you come across the facts that explain why nobody else has snapped up this "opportunity." Evidently, it is difficult to remove the oil from the ground due to the geology and the remote location. A careful analysis shows that estimated costs of extracting the oil are a whopping $120,000,000. You conclude that by developing this oil field, you would actually *lose* money. Oh well.

During the next week, although you are busy investigating other capital investment opportunities, your thoughts keep returning to this particular project. In particular, the fact that the lease is so cheap and that it lasts for three

years inspires you to do a *What if* scenario analysis, recognizing that there is no obligation to extract the oil and that it could be extracted fairly quickly (taking a few months) at any time during the three-year lease. You are wondering: What if the price of oil rises enough during the three years for it to be profitable to develop the oil field? If so, then you would extract the oil. But if the price of oil didn't rise enough, you would let the term of the lease expire in three years, leaving the oil still in the ground. You would let the future price of oil determine whether or not to exercise the option to extract the oil.

But such a proposition is risky! How much risk? What are the potential rewards? You have identified the following basic probability structure for the source of uncertainty in this situation:

| Future Price of Oil | Probability |
|---|---|
| 60 | 0.10 |
| 70 | 0.15 |
| 80 | 0.20 |
| 90 | 0.30 |
| 100 | 0.15 |
| 110 | 0.10 |

### Discussion Questions

1. How much money would you make if there were no costs of extraction? Would this be enough to retire?

2. Would you indeed lose money if you leased and extracted immediately, considering the costs of extraction? How much money?

3. Continue the scenario analysis by computing the future net payoff implied by each of the future prices of oil. To do this, multiply the price of oil by the number of barrels; then subtract the cost of extraction. If this is negative, you simply won't develop the field, so change negative values to zero. (At this point, do not subtract the lease cost, because we are assuming that it has already been paid.)

4. Find the average future net payoff, less the cost of the lease. How much, on average, would you gain (or lose) by leasing this oil field? (You may ignore the time value of money.)

5. How risky is this proposition?

6. Should you lease or not?

# Statistical Inference

The real power of statistics comes from applying the concepts of *probability* to situations where you have *data*. The results, called *statistical inference*, give you exact probability statements about the world based on a set of data. Even a relatively small set of data will work just fine if you're careful. This is how, for example, those political and marketing polls can claim to know what "all Americans" think or do based on interviews of a carefully selected sample. One of the best ways to select a smaller representative sample from a larger group is to use a *random sample*, which will be covered in Chapter 8. The *confidence interval* provides an exact probability statement about an unknown quantity and will be presented in Chapter 9. When you need to decide between two possibilities, you can use *hypothesis testing* (Chapter 10) to tell you what the data have to say about the situation. In an environment of uncertainty, where perfect answers are unavailable, statistical inference at least gives you answers that have some *known* error rates.

## Chapter 8

# Random Sampling

## Planning Ahead for Data Gathering

It's 9:00 a.m. on Wednesday, and you've just read a memo from The Boss asking how your firm's customers would react to a new proposed discount pricing schedule. Your report will be needed in time for tomorrow's 10:00 meeting of the board of directors. What should you do? Some things are clear. For example, you will need to speak with some customers. Thinking it over, you decide that it will take 10 minutes just to interview one customer over the phone. With the time and people you can spare, and with 1,687 customers listed in your database, you could not possibly call every single one to get a reaction. What will you do?

The answer is, of course, that you will draw a *sample* from the *population* of all customers in the database. That is, you will call a manageable number of selected customers. Then you will cross your fingers and hope that the sample is *representative* enough of the larger population so that your report will convey facts needed at the board meeting. After all, the board is interested in how *all* customers would react, not just those in your smaller sample. But crossing your fingers is not enough, by itself, to ensure that the sample is representative. Depending on how you choose the sample, you have some control over whether or not the sample will be useful.

But how will you draw the sample? You might make a list of customers you need to talk with for other reasons and ask them for a reaction to the proposed discount schedule.

However, you rightly reject this idea because this list is not representative, since it consists largely of "squeaky wheels" who require more hand-holding and are less self-sufficient than customers in general. Even worse, this group tends to place smaller orders of lower-tech equipment compared with your typical customers. To get a representative group of customers, you will need a different way of drawing a sample.

*Random sampling* will help you in two ways. First, it will ensure a representative sample (at least in an average sense) through selection without regard to any characteristic that might *bias* the sample and confuse the issue.[1] Second, by using a carefully controlled statistical procedure, you will also be able to describe approximately *how different* your results (from the sample) are from the characteristics of the population. Thus, through the use of random sampling, your survey results will be approximately correct (compared with what you would find by interviewing all customers), and you will know if your results are close enough for comfort.

In particular, if you are interested in estimating the mean of a large population by using the average of a random sample,

---

1. This parenthetical qualification is needed because there may not be a smaller sample that is *exactly* representative of the population. This would happen, for example, if each member of the population were unique.

this sample average has (approximately) a *normal distribution* with smaller and smaller variability (and therefore greater precision) as the sample size *n* increases. The *central limit theorem*, a mathematical fact, guarantees this. An estimate of (approximately) how far the observed sample average is from the unknown population mean is provided by the *standard error of the average*, which plays a crucial role in statistical inference because it measures the quality of your information, telling how good or bad your estimate is.

## 8.1 POPULATIONS AND SAMPLES

A **population** is a collection of units (people, objects, or whatever) that you are interested in knowing about. A **sample** is a smaller collection of units selected from the population. Usually, you have detailed information about individuals in the sample, but not for those in the population. There are many different ways a sample can be selected; naturally, some methods are much better than others for a given purpose. Here are some examples of populations and samples:

1. *The population:* The approximately 386,000 residents of Tulsa, Oklahoma, where your firm is considering opening a fast-food Mexican restaurant.
   a. A sample might be obtained by hiring people with clipboards to interview every 35th shopper they see at a local mall. Although this sample would tell you about some shoppers, you would have no information about the rest of the population.
   b. Another way to obtain a sample would be to interview (by telephone) every 2,000th person in the phone book. This systematic sample would tell you something about people who are available to answer their telephones.
   c. Yet another way to draw a sample would be to interview customers as they leave a local McDonald's restaurant. This sample would tell you about a certain group of people who eat fast food.
2. *The population:* The 826 boxes of miscellaneous hardware that just arrived at your shipping dock. You will want to spot-check the invoice against the contents of selected boxes and also note any unacceptable items.
   a. A very convenient sample might be obtained by using the nearest 10 boxes and examining their contents. But this sample is hardly representative and, if your suppliers figured out your selection method, you could be taken advantage of.
   b. Another way to choose a sample would be to select three large, three medium, and three small boxes to examine. This seems to be an attempt to broaden the sampling method, but it also may not be representative of the boxes in general (which might be nearly all large boxes).

c. Yet another way would be to use the invoice itself and select a random sample of the boxes listed. You would then find and open these boxes. This is an appropriate sample. By starting from the invoice, you are ensuring the correctness of this document. By choosing a random sample, your suppliers cannot guess beforehand which boxes you will examine.

3. *The population:* Your suppliers (there are 598 of them). You are considering a new system that would involve paying a higher price for supplies in return for guarantees of higher quality and a faster response time.[2] The system would be worthwhile only if enough of your suppliers were interested.
   a. One sample would be just your five key suppliers. Although it is good to include these important firms, you might want to include some of the others also.
   b. A different sample could be obtained by delegating the selection to one of your workers with a memo that says, "Please get me a list of 10 suppliers for the just-in-time project." This method of delegating without control leaves you with an unknown quantity since you don't know the selection criterion used. You may suspect that the "quickest" or "most convenient" sample was used, and this may not be representative.
   c. Yet another way to choose a sample would be to take your 5 key suppliers and include 10 more chosen by controlled delegation (say, in compliance with a memo reading "Please get me a list of 10 random nonkey suppliers, using the random number table"). This would be a useful sample since it includes all of the most important players as well as a selection of the others.

## What Is a Representative Sample?

The process of sampling is illustrated in Figure 8.1.1. From the larger population, a sample is chosen to be measured and analyzed in detail. We hope that the sample will be **representative**, meaning that each characteristic (and combination of characteristics) arises the same percentage of the time in the sample as in the population. A sample that is not representative in an important way is said to show **bias**. For example, if the sample has a much greater proportion of males than does the population, you could say that the sample shows a gender bias or that the sample is biased toward males.

---

2. You may recognize this system as relating to the "just-in-time" method of supplying a factory. Instead of having raw materials sit around in inventory, eating up real estate space and incurring interest costs on the money paid for them, these materials arrive in the right place at the factory just as they are needed.

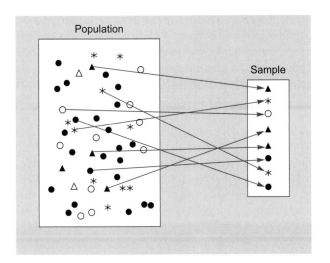

Population

Sample

**FIGURE 8.1.1**   A sample is a collection drawn from a larger population. This sample appears to be fairly representative but is not perfectly so because neither of the two open triangles was selected to be in the sample.

Since each individual may be unique, there may be no sample that is completely representative. But how do you tell when a sample is representative enough? By deliberately *not* sampling based on any measurable characteristic, a randomly selected statistical sample will be free (on average) from bias and therefore representative (on average). Furthermore, the randomness introduced in a controlled way into a statistical sample will let you make probability statements about the results (beginning with specification of *confidence intervals*, discussed in the next chapter). Thus, a careful statistical sample will be nearly representative, *and* you will be able to compute just how representative it is.

Once you have carefully identified the most appropriate population for your purpose, the next step is to figure out how to access it. You keep track of the population by creating or identifying the **frame**, which tells you how to gain access to the population units by number. For our purposes, the frame consists of a list of population units numbered from 1 to $N$, where $N$ is the number of units in the population. To access the 137th population unit, for example, you would locate it in the list to find information (such as name, serial number, or customer number) on how to measure it.

There are two basic kinds of samples. After a unit is chosen from the population to be in the sample, either it is put back (*replaced*) into the population so that it may be sampled again, or else it is not replaced. **Sampling without replacement** occurs if units cannot be chosen more than once in the sample, that is, if all units in the sample must be different. **Sampling with replacement** takes place if a population unit can appear more than once in the sample. Note that these properties are determined by the *process* used to choose the sample, and not by the *results* of that process. For a small sample chosen from a large population, there is very little difference between these two methods. From

now on in this book, we will work primarily with samples having distinct units chosen *without replacement*.

We will use the following standard notation for the number of units in the population (a property of the population) and for the number of units to be selected for the sample (which depends on how many you decide to select):

> **Notation for Number of Elementary Units**
>
> $N$ = Size of population
> $n$ = Size of sample

A sample that includes the entire population (i.e., $n = N$) is called a **census**. But even if you can examine the entire population, you may well decide not to. When weighing costs against benefits, you may decide that it isn't worth the time and trouble to examine all units.

## A Sample Statistic and a Population Parameter

A **sample statistic** (or just **statistic**) is defined as any number computed from your sample data. Examples include the sample average, median, sample standard deviation, and percentiles. A statistic is a *random variable* because it is based on data obtained by random sampling, which is a random experiment. Therefore, a statistic is *known* and *random*.

A **population parameter** (or just **parameter**) is defined as any number computed for the entire population. Examples include the population mean and population standard deviation. A parameter is a *fixed number* because no randomness is involved. However, you will not usually have data available for the entire population. Therefore, a parameter is *unknown* and *fixed*.

There is often a natural correspondence between statistics and parameters. For each population parameter (a number you would like to know but can't know exactly), there is a sample statistic computed from data that represents your best information about the unknown parameter. Such a sample statistic is called an **estimator** of the population parameter, and the actual number computed from the data is called an **estimate** of the population parameter. For example, the sample average is an estimator of the population mean, and in a particular case, the estimate might be 18.3. The **error of estimation** is defined as the estimator (or estimate) minus the population parameter; it is usually unknown.

An **unbiased estimator** is neither systematically too high nor too low compared with the corresponding population parameter. This is a desirable property for an estimator. Technically, an estimator is unbiased if its mean value (the mean of its sampling distribution) is equal to the population parameter.

Many commonly used statistical estimators are unbiased or approximately unbiased. For example, the sample average

$\overline{X}$ is an unbiased estimator of the population mean $\mu$. Of course, for any *given* set of data, $\overline{X}$ will (usually) be high or low relative to the population mean, $\mu$. If you were to repeat the sampling process many times, computing a new $\overline{X}$ for each sample, the results would average out close to $\mu$ and thus would not be *systematically* too high or too low.

The sample standard deviation $S$ is (perhaps surprisingly) a biased estimator of the population standard deviation $\sigma$, although it is approximately unbiased. Its square, the sample variance $S^2$, is an unbiased estimator of the population variance $\sigma^2$. For a binomial situation, the sample proportion $p$ is an unbiased estimator of the population proportion $\pi$.

## 8.2 THE RANDOM SAMPLE

A **random sample** or **simple random sample** is selected such that (1) each population unit has an *equal probability of being chosen*, and (2) units are *chosen independently*, without regard to one another. By making population units equally likely to be chosen, random sampling is as fair and unbiased as possible. By ensuring independent selection, random sampling aims at gathering as much independent information as possible.[3] Because personal tastes and human factors are removed from the selection process, the resulting sample is more likely to be fair and representative than if you assigned someone to choose an "arbitrary" sample.

An equivalent way to define *random sample* is to say that of all possible samples that might have been chosen, one was chosen at random. This definition, however, is more difficult to work with since the number of samples can be huge. For example, there are 17,310,309,456,440 different samples of $n = 10$ units that could be chosen from a population of just $N = 100$ units.[4] But this way of thinking about a random sample shows how fair it is, since random sampling does not favor any particular sample over the others.

How is a random sample better than an arbitrary sample? By choosing a random sample, you are assured that the theory of mathematical statistics is on your side. You are not just "hoping for the best" but are genuinely assured that the sample is representative, at least on average, of all population characteristics (even those characteristics that might not yet have occurred to you and those that are difficult or impossible to measure!). In addition, by choosing a random sample, you put in place the foundations for correctness of the conclusions (statistical inferences) you will draw about the population based on the data obtained from the sample. On the other hand, for example, if you select a nonrandom sample to be representative with respect to (a) the number of men and women, (b) family status, and (c) income, the resulting sample might be quite different from the population regarding some important characteristic, such as Internet usage or the willingness to order from catalogs. This could easily result in unfortunate business decisions, because random sampling was not used.

### Selecting a Random Sample

One way to choose a random sample is to use a table of random digits to represent the number of each selected population unit. The unit itself is then found with the help of the frame (this is the purpose of the frame: to go from a number to the actual population unit). A **table of random digits** is a list in which the digits 0 through 9 each occur with probability 1/10, independently of each other. Here are the details for choosing a random sample of size $n$ without replacement:

> **Selecting a Random Sample without Replacement**
>
> 1. Establish a frame so that the members of the population are numbered from 1 through $N$.
> 2. Select a place to begin reading from the table of random digits. This might be done randomly, for example, by tossing a coin.
> 3. Starting at the selected place, read the digits successively in the usual way (i.e., from left to right and continuing on the next line).
> 4. Organize these digits in groups whose size is the number of digits in the number $N$ itself. For example, with a population of $N = 5,387$, read the random digits four at a time (since 5,387 has four digits). Or, if the population had $N = 3,163,298$ units, you would read the random digits in groups of seven.
> 5. Proceed as follows until you have a sample of $n$ units:
>     a. If the random number is between 1 and $N$ and has not yet been chosen, include it in the sample.
>     b. If the random number is 0 or is larger than $N$, discard it because there is no corresponding unit to be chosen.
>     c. If the random number has already been chosen, discard it because you are sampling without replacement.

For example, let's choose a random sample of 9 customers from a list of 38 beginning with 69506 in row 11, column 3 of Table 8.2.1, the Table of Random Digits.

---

3. Although independence is a technical concept, it has important practical consequences as well. Here is an example to help you see the problems involved when units are *not* selected independently: In a hospital with 20 wards and 50 patients in each ward, you could select a sample (*not* a random sample) by choosing a ward at random and interviewing all patients in that ward. Note that each patient has an equal chance (1 in 20) of being interviewed. However, since patients are selected as a group instead of independently, your sample will not include important information about diversity within the hospital.

4. In case you're interested, the formula for the number of distinct samples that could be chosen without replacement is $\binom{N}{n} = N! / [n!(N-n)!]$, which you may recognize as a part of the binomial probability formula.

**TABLE 8.2.1 Table of Random Digits**

| | 1 | 2 | 3 | 4 | 5 | 6 | 7 | 8 | 9 | 10 |
|---|---|---|---|---|---|---|---|---|---|---|
| 1 | 51449 | 39284 | 85527 | 67168 | 91284 | 19954 | 91166 | 70918 | 85957 | 19492 |
| 2 | 16144 | 56830 | 67507 | 97275 | 25982 | 69294 | 32841 | 20861 | 83114 | 12531 |
| 3 | 48145 | 48280 | 99481 | 13050 | 81818 | 25282 | 66466 | 24461 | 97021 | 21072 |
| 4 | 83780 | 48351 | 85422 | 42978 | 26088 | 17869 | 94245 | 26622 | 48318 | 73850 |
| 5 | 95329 | 38482 | 93510 | 39170 | 63683 | 40587 | 80451 | 43058 | 81923 | 97072 |
| 6 | 11179 | 69004 | 34273 | 36062 | 26234 | 58601 | 47159 | 82248 | 95968 | 99722 |
| 7 | 94631 | 52413 | 31524 | 02316 | 27611 | 15888 | 13525 | 43809 | 40014 | 30667 |
| 8 | 64275 | 10294 | 35027 | 25604 | 65695 | 36014 | 17988 | 02734 | 31732 | 29911 |
| 9 | 72125 | 19232 | 10782 | 30615 | 42005 | 90419 | 32447 | 53688 | 36125 | 28456 |
| 10 | 16463 | 42028 | 27927 | 48403 | 88963 | 79615 | 41218 | 43290 | 53618 | 68082 |
| 11 | 10036 | 66273 | 69506 | 19610 | 01479 | 92338 | 55140 | 81097 | 73071 | 61544 |
| 12 | 85356 | 51400 | 88502 | 98267 | 73943 | 25828 | 38219 | 13268 | 09016 | 77465 |
| 13 | 84076 | 82087 | 55053 | 75370 | 71030 | 92275 | 55497 | 97123 | 40919 | 57479 |
| 14 | 76731 | 39755 | 78537 | 51937 | 11680 | 78820 | 50082 | 56068 | 36908 | 55399 |
| 15 | 19032 | 73472 | 79399 | 05549 | 14772 | 32746 | 38841 | 45524 | 13535 | 03113 |
| 16 | 72791 | 59040 | 61529 | 74437 | 74482 | 76619 | 05232 | 28616 | 98690 | 24011 |
| 17 | 11553 | 00135 | 28306 | 65571 | 34465 | 47423 | 39198 | 54456 | 95283 | 54637 |
| 18 | 71405 | 70352 | 46763 | 64002 | 62461 | 41982 | 15933 | 46942 | 36941 | 93412 |
| 19 | 17594 | 10116 | 55483 | 96219 | 85493 | 96955 | 89180 | 59690 | 82170 | 77643 |
| 20 | 09584 | 23476 | 09243 | 65568 | 89128 | 36747 | 63692 | 09986 | 47687 | 46448 |
| 21 | 81677 | 62634 | 52794 | 01466 | 85938 | 14565 | 79993 | 44956 | 82254 | 65223 |
| 22 | 45849 | 01177 | 13773 | 43523 | 69825 | 03222 | 58458 | 77463 | 58521 | 07273 |
| 23 | 97252 | 92257 | 90419 | 01241 | 52516 | 66293 | 14536 | 23870 | 78402 | 41759 |
| 24 | 26232 | 77422 | 76289 | 57587 | 42831 | 87047 | 20092 | 92676 | 12017 | 43554 |
| 25 | 87799 | 33602 | 01931 | 66913 | 63008 | 03745 | 93939 | 07178 | 70003 | 18158 |
| 26 | 46120 | 62298 | 69126 | 07862 | 76731 | 58527 | 39342 | 42749 | 57050 | 91725 |
| 27 | 53292 | 55652 | 11834 | 47581 | 25682 | 64085 | 26587 | 92289 | 41853 | 38354 |
| 28 | 81606 | 56009 | 06021 | 98392 | 40450 | 87721 | 50917 | 16978 | 39472 | 23505 |
| 29 | 67819 | 47314 | 96988 | 89931 | 49395 | 37071 | 72658 | 53947 | 11996 | 64631 |
| 30 | 50458 | 20350 | 87362 | 83996 | 86422 | 58694 | 71813 | 97695 | 28804 | 58523 |
| 31 | 59772 | 27000 | 97805 | 25042 | 09916 | 77569 | 71347 | 62667 | 09330 | 02152 |
| 32 | 94752 | 91056 | 08939 | 93410 | 59204 | 04644 | 44336 | 55570 | 21106 | 76588 |
| 33 | 01885 | 82054 | 45944 | 55398 | 55487 | 56455 | 56940 | 68787 | 36591 | 29914 |
| 34 | 85190 | 91941 | 86714 | 76593 | 77199 | 39724 | 99548 | 13827 | 84961 | 76740 |
| 35 | 97747 | 67607 | 14549 | 08215 | 95408 | 46381 | 12449 | 03672 | 40325 | 77312 |

*(Continued)*

### TABLE 8.2.1  Table of Random Digits—cont'd

|    | 1 | 2 | 3 | 4 | 5 | 6 | 7 | 8 | 9 | 10 |
|----|-----|-----|-----|-----|-----|-----|-----|-----|-----|-----|
| 36 | 43318 | 84469 | 26047 | 86003 | 34786 | 38931 | 34846 | 28711 | 42833 | 93019 |
| 37 | 47874 | 71365 | 76603 | 57440 | 49514 | 17335 | 71969 | 58055 | 99136 | 73589 |
| 38 | 24259 | 48079 | 71198 | 95859 | 94212 | 55402 | 93392 | 31965 | 94622 | 11673 |
| 39 | 31947 | 64805 | 34133 | 03245 | 24546 | 48934 | 41730 | 47831 | 26531 | 02203 |
| 40 | 37911 | 93224 | 87153 | 54541 | 57529 | 38299 | 65659 | 00202 | 07054 | 40168 |
| 41 | 82714 | 15799 | 93126 | 74180 | 94171 | 97117 | 31431 | 00323 | 62793 | 11995 |
| 42 | 82927 | 37884 | 74411 | 45887 | 36713 | 52339 | 68421 | 35968 | 67714 | 05883 |
| 43 | 65934 | 21782 | 35804 | 36676 | 35404 | 69987 | 52268 | 19894 | 81977 | 87764 |
| 44 | 56953 | 04356 | 68903 | 21369 | 35901 | 86797 | 83901 | 68681 | 02397 | 55359 |
| 45 | 16278 | 17165 | 67843 | 49349 | 90163 | 97337 | 35003 | 34915 | 91485 | 33814 |
| 46 | 96339 | 95028 | 48468 | 12279 | 81039 | 56531 | 10759 | 19579 | 00015 | 22829 |
| 47 | 84110 | 49661 | 13988 | 75909 | 35580 | 18426 | 29038 | 79111 | 56049 | 96451 |
| 48 | 49017 | 60748 | 03412 | 09880 | 94091 | 90052 | 43596 | 21424 | 16584 | 67970 |
| 49 | 43560 | 05552 | 54344 | 69418 | 01327 | 07771 | 25364 | 77373 | 34841 | 75927 |
| 50 | 25206 | 15177 | 63049 | 12464 | 16149 | 18759 | 96184 | 15968 | 89446 | 07168 |

**Start with random number table:**

69506      19610      01479      92338      55140      81097      73071      61544      85356      51400

**Arrange in groups of two (because 38 has two digits):**

69  50  61  96  10  01  47  99  23  38  55  14  08  10  97  73  07  16  15  44  85  35  65  14  00

**Eliminate numbers larger than 38 or smaller than 1:**

10  01         23  38      14  08  10            07  16  15            35         14

**Eliminate numbers previously listed:**

10  01         23  38      14  08            07  16  15            35

**Choose the first nine numbers:**

10   1         23  38      14   8            07  16  15

**FIGURE 8.2.1**  Selecting a random sample without replacement of *n* = 9 units from a population of *N* = 38 units. The random digits are used in groups of two (since 38 has two digits) starting in row 11, column 3 of Table 8.2.1. Numbers are discarded for being either greater than 38 or smaller than 1. The number 10 is discarded the second time. Stop when you have found *n* units.

Since $N = 38$ has two digits, arrange the sequence of random digits in groups of two as follows: 69 50 61 96 10 01 47 99 23 38…. Ignore the first ones because 69, 50, 61, and 96 are all larger than $N = 38$. Consequently, the first sample number selected is 10. The rest of the selection is illustrated in Figure 8.2.1. When the number 10 comes up a second time, do not include it in

the sample again (because you are choosing a sample without replacement); instead, continue until $n = 9$ units are chosen.[5]

---

5. If you wish to select a sample *with* replacement, most of the steps are the same except that in step 5a, you would include all random numbers between 1 and *N*, and step 5c should be omitted.

## Sampling by Shuffling the Population

Another way to choose a random sample from a population is easily implemented on a spreadsheet program. The idea here is to shuffle the population items into a completely random order and then select as many as you wish. This is just like shuffling a card deck and then dealing out as many cards as are needed.

In one column, list the numbers from 1 through $N$; there is usually a command for doing this automatically. In the next column, use the random number function to place uniform random numbers from 0 to 1 alongside your first column. Next, sort both columns in order according to the random number column. The result so far is that the population has been thoroughly shuffled into a random ordering. Finally, select the first $n$ items from the shuffled population to determine your sample.

To use Excel® to select a random sample of $n = 3$ from a population of size $N = 10$, you could type $= $ RAND() in the top cell of the random number column, press Enter, and then copy the result down the column to produce a column of random numbers. After selecting both columns (frame numbers and random numbers, including these headers), use Sort from the Sort & Filter area of Excel's Data Ribbon, being sure to sort by the random numbers. After the columns are sorted randomly, you may take the first three frame numbers to obtain your random sample, which results in selection of items 8, 6, and 1 in this example because these three had the smallest random numbers associated with them (note that Excel calculates new random numbers after you sort, so your random numbers will not be in order after sorting, but the frame numbers will be sorted according to the random numbers before they were recalculated).

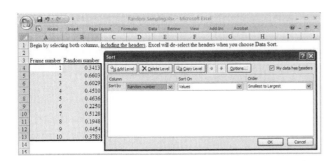

The resulting random sample has the same desirable properties attainable using the random number table.

### Example
*Auditing*

Microsoft Corporation reported revenues of $58.4 billion, with net income of $14.6 billion, for 2009. The number of individual transactions must have been huge, and the reporting system

must be carefully monitored in order for us to have faith in numbers like these. The auditors, the accounting firm Deloitte & Touche LLP, reported their opinion[6] as follows:

*In our opinion, such consolidated financial statements present fairly, in all material respects, the financial position of Microsoft Corporation and subsidiaries as of June 30, 2009 and 2008, and the results of their operations and their cash flows for each of the three years in the period ended June 30, 2009, in conformity with accounting principles generally accepted in the United States of America.*

To back up their opinion, they also reported (in part) as follows:

*We conducted our audits in accordance with the standards of the Public Company Accounting Oversight Board (United States). Those standards require that we plan and perform the audit to obtain reasonable assurance about whether the financial statements are free of material misstatement. An audit includes examining, on a test basis, evidence supporting the amounts and disclosures in the financial statements.... We believe that our audits provide a reasonable basis for our opinion.*

An auditing problem like this one involves statistics because it requires analysis of large amounts of data about transactions. Although all large transactions are checked in detail, many auditors rely on statistical sampling as a way of spot-checking long lists of smaller transactions.[7]

Say a particular list of transactions (perhaps out of many such lists) has been generated and is numbered from 1 through 7,329. You have been asked to draw a random sample of 20 accounts starting from row 23, column 8 of the table of random digits. When you arrange the random digits in groups of four and place brackets around large numbers to be discarded, your initial list looks like this:

2387 0784 0241 [7592] 6232 [7742] 2762 [8957] 5874 2831 [8704] 7200 [9292] 6761 2017 4355 4877 [9933] 6020 1931 6691 3630 0803 [7459] 3939 0717 [8700] 0318 1584 6120...

Selecting the first $n = 20$ available numbers, you end up with a sample including the following transaction numbers:

2387, 784, 241, 6232, 2762, 5874, 2831, 7200, 6761, 2017, 4355, 4877, 6020, 1931, 6691, 3630, 803, 3939, 717, 318

Placing these numbers in order will make it easier to look up the actual transactions for verification. Your final ordered list of sample transactions is

241, 318, 717, 784, 803, 1931, 2017, 2387, 2762, 2831, 3630, 3939, 4355, 4877, 5874, 6020, 6232, 6691, 6761, 7200

You would then look up these transactions in detail and verify their accuracy. The information learned by sampling from this list of transactions would be combined with other information learned by sampling from other lists and from complete examination of large, crucial transactions.

6. Microsoft 2009 Annual Report, accessed at http://www.microsoft.com/msft/reports/ar09/index.html on July 10, 2010.
7. A review of the wide variety of techniques that can be used is provided by A. J. Wilburn, *Practical Sampling for Auditors* (New York: Marcel Dekker, 1984).

**Example**
*A Pilot Study of Large Insurance Firms*

You have a new product that is potentially very useful to insurance companies. To formulate a marketing strategy in the early stages, you have decided to gather information about these firms. The problem is that the product is not yet entirely developed and you're not even sure how to gather the information! Therefore, you decide to run a **pilot study**, which is a small-scale version of a study designed to help you identify problems and fix them before the real study is run. For your pilot study, you have decided to use three of these firms ($n = 3$), selected at random.

First, you construct the frame, shown in Table 8.2.2. When you read two digits at a time (since $N = 32$ has two digits), starting from row 39, column 6 in the table of random digits, and place brackets around the ones to be discarded, the initial list is

[48], [93], [44], 17, 30, [47], [83], 12, [65], 31, 02, 20, [33], [79], 11, [93], 22, [48], [71], [53], . . .

When you select the first $n = 3$ available numbers, your sample will include firms with the following numbers:

17, 30, 12

Looking back at the frame to see which firms these are and arranging them in alphabetical order, you can see your sample of $n = 3$ firms for the pilot study will consist of Massachusetts Mutual Life Insurance, Travelers Cos., and Guardian Life Ins. Co. of America.

**TABLE 8.2.2 The Frame**

| | |
|---|---|
| 1 | AFLAC |
| 2 | Allstate |
| 3 | American Family Insurance Group |
| 4 | American International Group |
| 5 | Auto-Owners Insurance |
| 6 | Berkshire Hathaway |
| 7 | Chubb |
| 8 | Erie Insurance Group |
| 9 | Fidelity National Financial |
| 10 | First American Corp. |
| 11 | Genworth Financial |
| 12 | Guardian Life Ins. Co. of America |
| 13 | Hartford Financial Services |
| 14 | Liberty Mutual Insurance Group |
| 15 | Lincoln National |
| 16 | Loews |
| 17 | Massachusetts Mutual Life Insurance |
| 18 | MetLife |
| 19 | Mutual of Omaha Insurance |
| 20 | Nationwide |
| 21 | New York Life Insurance |
| 22 | Northwestern Mutual |
| 23 | Pacific Life |
| 24 | Principal Financial |
| 25 | Progressive |
| 26 | Prudential Financial |
| 27 | Reinsurance Group of America |
| 28 | State Farm Insurance Cos. |
| 29 | TIAA-CREF |
| 30 | Travelers Cos. |
| 31 | United Services Automobile Association |
| 32 | Unum Group |

## 8.3  THE SAMPLING DISTRIBUTION AND THE CENTRAL LIMIT THEOREM

Any statistic you measure based on a random sample of data will have a probability distribution called the **sampling distribution** of that statistic. Through understanding of this sampling distribution, you will be able to make the leap from information about a sample (what you already have) to information about the population (what you would like to know). Fortunately, in many cases, the sampling distribution of a statistic such as the sample average is approximately normally distributed even though individuals may not follow a normal distribution. This result, called the *central limit theorem*, will simplify statistical inference because you already know how to find probabilities for a normal distribution.

When you complete a survey of randomly selected consumers and find that, on average, they plan to spend $21.26 on groceries per trip, the number 21.26 may not look random to you. *But the result of your survey is random*. Let's be careful. The number 21.26 itself is not random. Instead, it's "the average spending on groceries per trip for these randomly selected consumers" that is the random variable. Looking at the situation this way, it is clear why it's random: Each time the random experiment is run, a new random sample of consumers would be interviewed, and the result would be different each time.

The way to *think* about a statistic can be very different from the way to *work* with one. To understand the concepts involved here, imagine repeating a study lots of times. This is necessary in order to see where randomness comes from; after all, if you just do the study once, the results will look like fixed numbers. But don't lose sight of the fact that, due to the constraints of real life, when you actually do a study, you (usually) just do it once. The idea of repeating it over and over is just a way of understanding the actual result by placing it in perspective along with all of the other possible results. With this in mind, examine the idea of a sampling distribution shown in Figure 8.3.1.

There are two reasons that the normal distribution is so special. First, many data sets follow a normal distribution (although in business, we often have to work with logarithms of the data to eliminate skewness). Second, even when a distribution is not normal, the distribution of an *average* or a *sum* of numbers from this distribution will be closer to a normal distribution.

It is important to distinguish an *individual* measurement from an *average* or *sum* of measurements, which combines many individuals. Although the individuals retain whatever distribution they happen to have, the process of combining many individuals into an average or sum results in a more normal distribution.

To understand this, recognize that the process of obtaining a random sample and computing the average is itself a random experiment, and the average is a random variable. Therefore, it makes sense to speak of the distribution of the average or the distribution of the sum as either having a normal distribution or not. Since we are dealing with probabilities and not statistics, we are free to imagine repeating the data-gathering process many times, producing multiple observations of the average. A histogram of these observations represents (approximately) its distribution.

The **central limit theorem** specifies that, for a random sample of $n$ observations from a population, the following statements are true:

1. Distributions become more and more normal as $n$ gets large, for both the *average* and the *sum*.
2. The means and standard deviations of the distributions of the average and the sum are as follows, where $\mu$ is the mean of the individuals and $\sigma$ is the standard deviation of these individuals in the population.

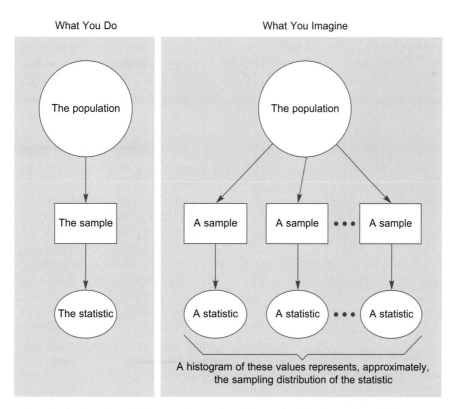

What You Do           What You Imagine

A histogram of these values represents, approximately, the sampling distribution of the statistic

**FIGURE 8.3.1** When you *imagine* that the entire study is repeated many times, the *sampling distribution* of a statistic corresponds to the histogram of the statistic's values. In reality, of course, you just have one sample and one value of the statistic. This one value is interpreted with respect to all of the other outcomes that *might* have happened, as represented by the sampling distribution.

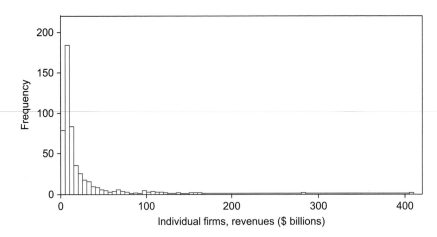

**FIGURE 8.3.2**   A histogram of sales of the largest 500 U.S. corporations. The standard deviation is $32.0 billion.

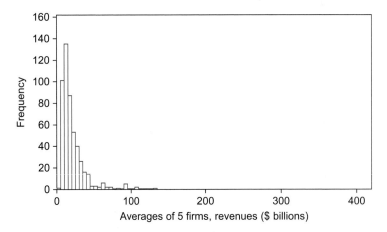

**FIGURE 8.3.3**   A histogram of averages of 5 randomly selected firms (repeated many times, with replacement), representing the sampling distribution of averages of 5 firms. Note that the skewness is reduced compared to the previous figure and that the standard deviation is reduced by a little more than half.

### Mean and Standard Deviation for Averages and Sums

|  | Random Variable | |
|---|---|---|
|  | **Average** | **Sum Total** |
| Mean | $\mu_{\bar{x}} = \mu$ | $\mu_{sum} = n\mu$ |
| Standard deviation | $\sigma_{\bar{x}} = \dfrac{\sigma}{\sqrt{n}}$ | $\sigma_{sum} = \sigma\sqrt{n}$ |

The central limit theorem gives you all the information needed to compute probabilities for a sum or an average based on a random sample. If $n$ is large enough, you may assume a normal distribution and use the standard normal probability tables.[8] To standardize the numbers, you may use the appropriate mean and standard deviation from the preceding table. Then all you have to do is remember how to compute probabilities for a normal distribution!

Figure 8.3.2 shows a histogram of the revenues of the top 500 U.S. corporations.[9] As you can see, the distribution is quite skewed. Figure 8.3.3 shows the distribution of the *averages of 5 firms* taken from this list of 500 (i.e., the average sales of 5 randomly selected firms was computed many times) and represents the sampling distribution of averages of 5 firms. The distribution is still skewed, but less so, and it is more normal than the distribution of the individual firms (the very long tail to the right in the hundreds of billions is gone, but we still have a long tail out to about $100 billion or so). Figure 8.3.4 uses a larger sample size, $n = 25$. Note how the means stay about the same throughout, whereas the

8. How large is large enough? If the distribution of individuals is not too skewed, $n = 30$ is generally sufficient. However, if the distribution is extremely skewed or has large outliers, $n$ may have to be much larger. If the distribution is fairly close to normal already, then $n$ can be much smaller than 30, say, 20, 10, or even 5. Of course, if the distribution was normal to begin with, then $n = 1$ is enough.

9. Data for these Fortune 500 firms were accessed at http://money.cnn .com/magazines/fortune/fortune500/2010/full_list on July 12, 2010.

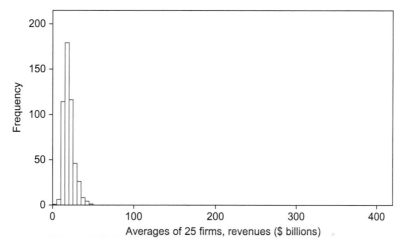

**FIGURE 8.3.4**   A histogram of averages of 25 randomly selected firms (repeated many times, with replacement), representing the sampling distribution of averages of 25 firms. The distribution is now fairly normal, although some skewness remains. The standard deviation has been reduced to approximately $\frac{1}{\sqrt{25}} = \frac{1}{5}$ of the initial standard deviation.

standard deviations get smaller according to the "divide by the square root of *n*" rule. Note also the progression in the figures from skewness to normality.

How does the central limit theorem work? The idea is that the extreme values in the data are averaged with each other. The long tail on the right of Figure 8.3.2, due to skewness, moves inward because the very large firms are averaged with some of the others. This explains the reduction in skewness.

Why does the distribution move toward a normal one? The complete answer relies on advanced mathematical statistics and will not be presented here. However, it is a general theoretical result, demanding only that $\sigma$ be finite and nonzero.

### Example
#### How Much Do Shoppers Spend?

At your supermarket, the typical shopper spends $18.93 with a standard deviation of $12.52. You are wondering what would happen in a typical morning hour with 400 typical shoppers, assuming that each one shops independently. Thus, $\mu = 18.93$, $\sigma = 12.52$, and $n = 400$. The central limit theorem can be used to tell you all about your total sales for the hour and, in particular, how likely it is that you will exceed $8,000 in total sales for all 400 shoppers.

First, for the *total* sales for this hour, which is the sum of the purchases of all 400 shoppers, the expected value is

$$
\begin{aligned}
\mu_{(total\,sales)} &= n\mu \\
&= 400 \times 18.93 \\
&= \$7,572.00
\end{aligned}
$$

Next, you wonder how much variability you can expect in total sales. This is an *hour-to-hour* variability for total sales, as compared to the *shopper-to-shopper* variability of $\sigma = \$12.52$. The answer is

$$
\begin{aligned}
\sigma_{(total\,sales)} &= \sigma\sqrt{n} \\
&= 12.52\sqrt{400} \\
&= 12.52 \times 20 \\
&= \$250.40
\end{aligned}
$$

In summary, for these 400 shoppers, you expect total sales of about $7,572.00, with a standard deviation of $250.40.

The central limit theorem also tells you that total sales will be approximately normally distributed. Since you have the mean and standard deviation, you can compute probabilities using the standard normal table. Note that the standard normal table might *not* apply to individual shoppers but (using the central limit theorem) would apply to total sales.

What is the probability that total sales will exceed $8,000 for these 400 shoppers? First, standardize this number using the appropriate mean and standard deviation (for total sales):

$$
\begin{aligned}
\text{Standardized total sales} &= \frac{8,000 - \mu_{(total\,sales)}}{\sigma_{(total\,sales)}} \\
&= \frac{8,000 - 7,572}{250.40} \\
&= 1.71
\end{aligned}
$$

When you look up 1.71 in the standard normal probability table, you find the final answer is $1 - 0.9564 = 0.0436$, or about a 4% chance of exceeding $8,000 in total sales. This probability is shown in Figure 8.3.5.

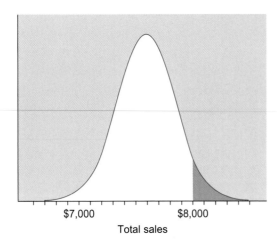

**FIGURE 8.3.5**   The probability that total sales will exceed $8,000 is 0.0436. The mean and standard deviation were computed using the central limit theorem.

**Example**
*Consistency in Bubble Gum Production*

It's OK if each individual piece of bubble gum isn't *exactly* 0.20 ounce, as promised on the package, so long as the average weight isn't too low (to avoid loss of good will, not to mention consumer and government lawsuits) or too high (to avoid unnecessary costs). In your production facility, you know from experience that individual pieces of gum have a standard deviation of 0.074 ounce, representing variability about their mean value of 0.201 ounce. Any bags of 30 pieces that have an average weight per piece lower than 0.18 ounce will be rejected. What fraction of bags will be rejected this way?

We will assume that pieces are independently produced (which may not be reasonable, since a problem could affect a number of pieces at a time).

First, you need to find the mean and standard deviation of the average of $n = 30$ pieces, where each piece has a mean weight of $\mu = 0.201$ ounce and a standard deviation of $\sigma = 0.074$:

$$\mu_{(\text{average weight})} = \mu$$
$$= 0.201 \text{ ounce}$$

$$\sigma_{(\text{average weight})} = \frac{\sigma}{\sqrt{30}}$$

$$= \frac{0.074}{5.477226}$$

$$= 0.01351 \text{ ounce}$$

Next, convert 0.18 ounce to a standardized number:

$$z = \text{Standardized average weight limit} = \frac{0.18 - \mu_{(\text{average weight})}}{\sigma_{(\text{average weight})}}$$

$$= \frac{0.18 - 0.201}{0.01351}$$

$$= -1.55$$

Finally, the probability of rejecting a bag is the probability that a standard normal variable is less than −1.55. From the standard normal probability table, the answer is 0.06, meaning about 6% of the bags will be rejected. This probability is shown in Figure 8.3.6.

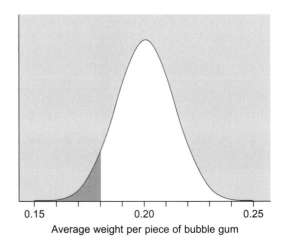

**FIGURE 8.3.6**   The probability that the average weight per piece will be lower than 0.18 ounce is 0.06. The mean and standard deviation were computed using the central limit theorem.

## 8.4 A STANDARD ERROR IS AN ESTIMATED STANDARD DEVIATION

Unfortunately, in real life you usually can't work directly with a sampling distribution because it is determined by properties of the entire population, and you have information only about a sample. Every (reasonable) distribution has a standard deviation, so the sampling distribution of any statistic has a standard deviation also. If you knew this standard deviation, you would know approximately how far the sample statistic is from its mean value (a population parameter). This would then help you know more about the population since, in addition to having a "best guess" (your statistic), you would have an indication of how good this guess is. Unfortunately, you don't know this standard deviation exactly because it depends on the population.

The solution is to use the sample information to guess, or estimate, the standard deviation of the sampling distribution of the statistic. The resulting approximation to the standard deviation of the statistic, based only on sample data, is called the **standard error of the statistic**. You interpret this standard error just as you would any standard deviation. The standard error indicates approximately how far the observed value of the statistic is from its mean. Literally, it indicates (approximately) the standard deviation you would find if you took a very large number of samples, found the sample average for each one, and worked with these sample averages as a data set.

Why do we use two terms (standard deviation and standard error) since a standard error is just a kind of standard deviation? This is done primarily to emphasize that the *standard error* indicates the amount of uncertainty in a *summary number* (a statistic) representing the entire sample. By contrast, the term *standard deviation* is usually used to indicate the amount of variability among *individuals* (elementary units), specifically indicating how far individuals are typically from the average.

## How Close Is the Sample Average to the Population Mean?

The sample average, $\overline{X}$, is a statistic since it is computed from the sample data. The **standard error of the average** (or just **standard error**, for short) estimates the sampling variability of the sample average, indicating approximately how far it is from the population mean. From the central limit theorem (Section 8.3), you know that the standard deviation of the sample average is $\sigma_{\overline{X}} = \sigma/\sqrt{n}$. In Section 8.3 we assumed that we knew values for population parameters such as $\sigma$ because we were doing probability. Now that we are doing statistics again, $\sigma$ is unknown, and so is the standard deviation of the sample average. But we have an estimate of the standard deviation: $S$, the standard deviation of the sample, from Chapter 5. If we replace $\sigma$ with $S$, the result is an

indication of the uncertainty in $\overline{X}$. Here are the standard deviation of $\overline{X}$ (which is exact) and the standard error of $\overline{X}$ (which is estimated and, therefore, only approximate):

> **Standard Deviation of the Average**
>
> $$\sigma_{\overline{X}} = \frac{\sigma}{\sqrt{n}}$$
>
> **Standard Error of the Average**
>
> $$S_{\overline{X}} = \frac{S}{\sqrt{n}}$$

This standard error indicates approximately how far the sample average, $\overline{X}$, is from the population mean, $\mu$. Since $\overline{X}$ is often your best information about $\mu$, this standard error tells you roughly how far off you are when you use the best available sample information (for example, average spending of 100 random consumers) in place of the unavailable population information (mean spending for the entire city). The standard error of the average is in the same measurement units (dollars, miles per gallon, people, or whatever) as the data values.

Note the difference between $S$ and $S_{\overline{X}}$. The standard deviation, $S$, indicates approximately how far *individuals* are from the average, whereas the standard error, $S_{\overline{X}}$, indicates approximately how far the *average*, $\overline{X}$, is from the population mean, $\mu$. This is illustrated in Figures 8.4.1 and 8.4.2. You would expect the sample average, $\overline{X}$, to be within two

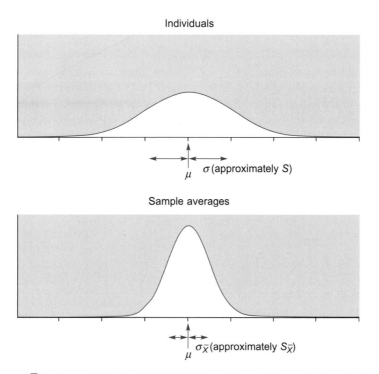

**FIGURE 8.4.1**  The sample average, $\overline{X}$, has less variability than individual $X$ values. The standard error, $S_{\overline{X}}$, is smaller than the standard deviation, $S$, and would become even smaller (indicating greater precision of $\overline{X}$) for larger $n$ than the sample of four used here.

What You Do

What You Imagine

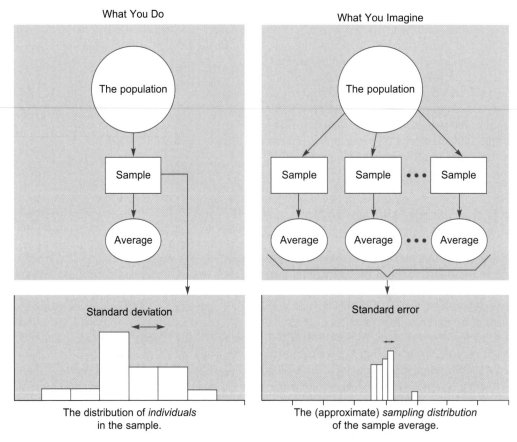

FIGURE 8.4.2 The random experiments and histograms of the resulting data for individuals (left) and sample averages (right). Note how the standard error of the average is much smaller than the standard deviation (about 1/3 the size with sample size $n = 10$ used here). The standard deviation indicates the variability of *individuals* about their average, whereas the standard error indicates the variability of the *sampling distribution* of the sample average.

standard errors of the population mean $\mu$ about 95% of the time. Following is a summary table to help you distinguish between variability of individuals and variability of averages, for both the population and a sample:

**Variability: Individuals and Averages, Population and Sample**

|  | For the Population | For a Sample |
| --- | --- | --- |
| Variability of individuals | $\sigma$ | $S$ |
| Variability of $\overline{X}$, the average of $n$ | $\sigma_{\overline{X}} = \sigma/\sqrt{n}$ | $S_{\overline{X}} = S/\sqrt{n}$ |

Why should you ever look at more than one individual? Because individuals are more variable, more random, and less precise than the sample average. The reason is that the standard error is $S$ divided by $\sqrt{n}$ and is thus smaller than $S$ whenever $n$ is 2 or more.

Why should you sample more rather than fewer individuals? Because the error $(\overline{X} - \mu)$ is typically smaller

when information from more individuals is combined in a sample. The standard error indicates the approximate size of this error. Since the standard error gets smaller as $n$ gets larger (all else equal), your information about the unknown $\mu$ improves as sample size grows because $\overline{X}$ will typically be closer to $\mu$.

**Example**
*Shopping Trips*

Suppose $n = 200$ randomly selected shoppers interviewed in a mall say that they plan to spend an average of $\overline{X} = \$19.42$ today with a standard deviation of $S = \$8.63$. This tells you that shoppers typically plan to spend about $19.42, and that a typical *individual* shopper plans to spend about $8.63 more or less than this amount. So far, this is no more and no less than a description of the individuals interviewed.

In fact, you can do more than just describe the sample data. You can say something about the unknown population mean, $\mu$, which is the mean amount that *all* shoppers in the mall today plan to spend, including those you did not interview. The standard error is

$$S_{\overline{X}} = \frac{S}{\sqrt{n}}$$

$$= \frac{\$8.63}{\sqrt{200}}$$

$$= \frac{\$8.63}{14.14213562}$$

$$= \$0.610$$

This tells you that when you use the sample average, $19.42, as an estimate of the unknown value $\mu$ for all shoppers, your error is only about $0.610. Note how much smaller the standard error ($0.610) is than the standard deviation ($8.63).

If you had interviewed only one person and (foolishly) tried to use the answer as an estimate of spending of all shoppers, your error would have been approximately $8.63. When you go to the extra trouble of sampling $n = 200$ shoppers and combine this information by using the sample average, your error is reduced—considerably—to approximately $0.610.

## Correcting for Small Populations

When the population is small so that the sample is a major fraction of the population, the standard error formula should be reduced by applying the **finite-population correction factor** $\sqrt{(N-n)/N}$ to obtain the **adjusted standard error**.

When you sample nearly all of the population, your information about the population is very good. In fact, when you sample all of the population (so that $n = N$), your information is perfect and the standard error should be 0. The following formula is used to adjust the standard error to make it more accurate:[10]

**Adjusted Standard Error**

(Finite-population correction factor) × (Standard error)

$$= \sqrt{\frac{N-n}{N}} \times S_{\overline{X}}$$

$$= \sqrt{\frac{N-n}{N}} \times \frac{S}{\sqrt{n}}$$

When the sample size is close to the population size, the term $N - n$ is small and the adjusted standard error is also small, reflecting the high quality of this nearly complete sample. When the population size, $N$, is large, the finite-

**FIGURE 8.4.3**  When a small amount is sampled from a large population, the size of the population does not affect the standard error. The two urns are the same except for size. If a few balls are drawn at random from each one, the distributions of the number of black balls in the samples should be similar. Each sample is providing information about the percentage of black ones.

population correction factor is nearly 1 and thus will not change the standard error very much.[11]

You may be wondering why the population size, $N$, doesn't seem to matter for a large population, since the standard error depends only on the sample information, $n$ and $S$. This is reasonable because the standard error reflects the randomness *in the sampling process* rather than any particular characteristic of the population. With a small sample from a large population, the sample values cannot "interfere" with each other very much due to non-replacement, and sample properties (such as the variability in the average) will look pretty much the same even if you double the population size (holding its characteristics the same). On the other hand, with a small population, nonreplacement affects the sample more strongly by limiting the selection, an effect that changes with the population size. Figure 8.4.3 shows a situation in which it is reasonable that the sample values do not depend on the (large) population size.

You may not always want to apply the finite-population correction factor, even in cases where you seem to be entitled to use it. Sometimes the population frame you sample from is not the population you are really interested in. If you are willing to assume that your frame represents a random sample from a much larger population, and you

---

10. The theoretical justification for this formula may be found, for example, in W. G. Cochran, *Sampling Techniques*, 3rd ed. (New York: Wiley, 1977), Equation 2.20, p. 26; or in L. Kish, *Survey Sampling* (New York: Wiley, 1967), Equation 2.2.2, p. 41.

11. When $N$ is large and $n$ is a small fraction of it, the finite-population correction factor reduces the standard error by approximately half of this fraction, that is, by $n/(2N)$. If, say, you are sampling 8% of a large population, then the correction will reduce the standard error by about 4%. (The exact correction in this case is 4.08%, fairly close to the 4% approximation.)

want to learn about this much larger population, then it is better not to use the correction factor. An **idealized population** might be defined as the much larger, sometimes imaginary, population that your sample represents. When you are interested in the idealized population, you do not use the finite-population correction factor. On the other hand, if you just want to learn about the population frame and not go beyond it, then the correction factor will work to your advantage by expressing the lower variability of this system.

For example, suppose from a list of 300 recent customers you have selected a random list of 50 to be interviewed about customer satisfaction. If you are interested only in the 300 recent customers on your list, you may feel free to reduce your initial standard error downward by 8.71%. However, if you wish to learn about *customers in general*, a potentially very large group that is represented by your convenient list of 300, you would not perform the adjustment. If you did (wrongly) adjust in this case, you would be deceiving yourself into thinking that the results are more precise than they really are.

When in doubt, the conservative, safe choice is *not to* use the finite-population correction factor.

---

**Example**

*Quality of the Day's Production*

Of the 48 truckloads that left your factory today loaded with newly manufactured goods, you had 10 selected at random for detailed quality inspection. During inspection, a number from 1 to 20 is assigned to the shipment, with 20 being "perfect" and 1 being "@X%#!!!." As it turned out, the measurements were 19, 20, 20, 17, 20, 20, 15, 18, 20, and 15. The average for the day's sample is 18.4, and the standard deviation is 2.065591 for individual shipments. The (uncorrected) standard error is 0.653.

If you are interested in how close the *day's sample average* of 18.4 is to the *day's average* for all 48 shipments (which is unknown, since you measured only 10 truckloads), you may use the finite-population correction factor. In fact, you probably should use this factor because you have sampled a large portion (10/48 = 20.8%) of the population. The adjusted standard error is then

$$\text{Adjusted standard error} = \sqrt{\frac{N-n}{N}} \times S_{\bar{X}}$$

$$= \sqrt{\frac{48-10}{48}} \times \frac{2.065591}{\sqrt{10}}$$

$$= 0.889757 \times 0.653197$$

$$= 0.581$$

The adjustment process has reduced the original standard error by about 11.0% (= 1 − 0.889757), from 0.653 to 0.581.

If, on the other hand, you are interested in how close the day's sample average of 18.4 is to the *quality in general* of your factory, you would not use the correction factor but

would use the larger (uncorrected) standard error of 0.653. In essence, you would try to generalize to a very large *idealized population* of all of the truckloads that *might* have been produced today under the current conditions in your factory.

---

## The Standard Error of the Binomial Proportion

For a binomial situation, there are two standard errors: one for the count $X$ and one for the proportion $p$. The standard error $S_X$ indicates the uncertainty or variability in the observed count and is easily computed from information in the sample. Similarly, the standard error $S_p$ indicates the uncertainty in the observed proportion. These are based on the (population) standard deviations $\sigma_X$ and $\sigma_p$ from Chapter 7, replacing the unknown population proportion $\pi$ by its sample estimate $p$. This is a common process: using the best information we have from the sample ($p$, in this case) in place of information about the population (e.g., $\pi$) that we need but don't have. Here are the formulas for population standard deviations and the standard errors (estimated from sample data) for a binomial situation:[12]

| **Binomial Standard Deviation and Standard Error** | | |
|---|---|---|
| | Number of Occurrences, $X$ | Proportion or Percentage, $p = X/n$ |
| Standard deviation (for the population) | $\sigma_X = \sqrt{n\pi(1-\pi)}$ | $\sigma_p = \sqrt{\dfrac{\pi(1-\pi)}{n}}$ |
| Standard error (estimated from a sample) | $S_X = \sqrt{np(1-p)}$ | $S_p = \sqrt{\dfrac{p(1-p)}{n}}$ |

For example, if we found 8 machines out of 50 to be defective, then the observed binomial proportion $p$ would be 0.16 or 16%, with uncertainty $S_p = 0.0518$ or 5.18 percentage points. The observed count $X$ would be 8, with uncertainty $S_X = 2.59$.

---

**Example**

*A Consumer Survey*

You have surveyed 937 people and found that 302, or 32.2%, of these would consider purchasing your product. You are wondering just how reliable these numbers are. In

---

12. Note that these formulas take you directly to the standard errors for a binomial situation. There is no need to first compute a standard deviation, $S$, and then divide by the square root of $n$, as you would do for a list of numbers (a nonbinomial situation).

particular, how far are they from their values for the entire, much larger population? The standard error would provide a good answer.

You may assume that this is a binomial situation because you have selected people independently and randomly from the population. At this point, you know that $n = 937$, $X = 302$, and $p = 0.322$, or 32.2%. However, you don't know $\pi$, which is the percentage for the entire population (which you would like to know but can't). The standard error (the estimated standard deviation) may be used here because you do have an *estimate* of $\pi$, namely, the observed value of 0.322.

|  | Number of People $X$ | Proportion or Percentage $p = X/n$ |
|---|---|---|
| Standard error (for binomial distribution) | $S_X = \sqrt{np(1-p)}$ $= \sqrt{937 \times 0.322(1-0.322)}$ $= 14.3$ people | $S_p = \sqrt{\dfrac{p(1-p)}{n}}$ $= \sqrt{\dfrac{0.322(1-0.322)}{937}}$ $= 0.0153$ or $1.53\%$ |

The observed number, 302 people, is about 14.3 people away from (above or below) the unknown value you would expect, on average, for this kind of study for this population. The observed proportion of people, 32.2%, is approximately 1.53 percentage points different from the unknown, true percentage in the entire population.

## 8.5 OTHER SAMPLING METHODS

The random sample is not the only way to select a sample from a population. There are many other methods, each with advantages and disadvantages. Some, such as the *stratified random sample*, use the principles of random sampling carefully. Others, such as *systematic sampling*, use very different methods and provide a fragile foundation, if any, for your statistical analysis.

One important consideration is to weigh the amount of hostile scrutiny the results will be subject to against the cost of obtaining the data. For an internal study in a friendly working environment, without much in the way of office politics (if there is such a place!), you may not need the rigor and care of a random sample. However, for an external study to be used by neutral or possibly even hostile parties, such as in a lawsuit, where people on the other side may choose to question your wisdom, it will be worth your while to pay attention to detail and use careful, randomized sampling methods.

### The Stratified Random Sample

Sometimes a population contains clear, known, easily identified groups. If you choose a random sample from such a population as a whole, each segment or *stratum* may be under- or overrepresented in the sample as compared to the population.[13] This may contribute some extra randomness to the results since you would not be using the known information about these groups.

A **stratified random sample** is obtained by choosing a random sample separately from each of the strata (segments or groups) of the population. If the population is similar (homogeneous) within each stratum but differs markedly from one segment to another, stratification can increase the precision of your statistical analysis. Stratification can also make administration easier since you may be able to delegate the selection process to your field offices.

You are free to choose any sample size for each individual stratum. There is no requirement that you sample the same number from each stratum or that you allocate your sample size according to population percentages. This allows you to determine sample sizes according to costs and benefits. Some strata may be more costly to sample than others, and you will therefore tend to use smaller sample sizes for them. Some strata will be known to have more variability than others, and for these you will therefore tend to use larger samples.

Table 8.5.1 shows the details and notation for the population sizes, sample sizes, sample averages, and sample standard deviations for each stratum.

It remains to show how to combine the estimates from each stratum into an estimate for the entire population and how to obtain the standard error of the resulting estimate. Each stratum provides a random sample and a sample average. To find the stratified sampling estimate of the population mean, combine the sample averages using a weighted average based on the population size of each stratum. In this way, the larger segments of the population exert their rightful influence on the results. First, multiply each sample average by the corresponding population size; then sum these and divide by the total population size.

**TABLE 8.5.1** Notation for Stratified Sampling

| Stratum | Population Size | Sample Size | Sample Average | Sample Standard Deviation |
|---|---|---|---|---|
| 1 | $N_1$ | $n_1$ | $\overline{X}_1$ | $S_1$ |
| 2 | $N_2$ | $n_2$ | $\overline{X}_2$ | $S_2$ |
| . | . | . | . | . |
| . | . | . | . | . |
| . | . | . | . | . |
| $L$ | $N_L$ | $n_L$ | $\overline{X}_L$ | $S_L$ |

---

13. *Stratum* is the singular and *strata* the plural for referring to segments or layers within a population. Perhaps you have seen stratified rock, with its pronounced layers.

## Computing the Average for a Stratified Sample

$$\overline{X} = \frac{N_1\overline{X}_1 + N_2\overline{X}_2 + \cdots + N_L\overline{X}_L}{N_1 + N_2 + \cdots + N_L}$$

$$= \frac{1}{N}\sum_{i=1}^{L} N_i\overline{X}_i$$

where $N$ is the total population size $N_1 + N_2 + \cdots + N_L$.

How much variability is there in the resulting combined estimate? The answer, as always, is given by its standard error. This is found by combining the standard deviations from the individual strata in the following way. For each stratum, multiply the square of the population size by the square of the standard deviation and divide by the sample size. Add these up, take the square root, and divide by the total population size.

## Standard Error for a Stratified Sample

$$S_{\overline{X}} = \frac{1}{N}\sqrt{\frac{N_1^2 S_1^2}{n_1} + \frac{N_2^2 S_2^2}{n_2} + \cdots + \frac{N_L^2 S_L^2}{n_L}}$$

$$= \frac{1}{N}\sqrt{\sum_{i=1}^{L} \frac{N_i^2 S_i^2}{n_i}}$$

If the population sizes for some strata are small, so that more than just a small fraction is sampled, then the finite-population correction factor may be applied, resulting in a more accurate standard error. For each stratum, multiply the population size by the square of the standard deviation and by the difference between the population and sample sizes, and then divide by the sample size. Add these up, take the square root, and divide by the total population size.

## Adjusted Standard Error for a Stratified Sample

$$\left(\begin{array}{c}\text{Adjusted}\\\text{standard error}\end{array}\right)$$

$$= \frac{1}{N}\sqrt{\frac{N_1(N_1-n_1)S_1^2}{n_1} + \frac{N_2(N_2-n_2)S_2^2}{n_2} + \cdots + \frac{N_L(N_L-n_L)S_L^2}{n_L}}$$

$$= \frac{1}{N}\sqrt{\sum_{i=1}^{L} \frac{N_i(N_i-n_i)S_i^2}{n_i}}$$

## Example

### Adjusting for Sophistication of the Consumer

In order to develop a marketing strategy for your high-tech video and audio products, you need good information about potential customers. These people can be divided in a natural way into two groups according to whether they are sophisticated or naive about how the technology

works. The sophisticated group wants to know detailed facts or "specs" on the products; the naive group needs information at a much more basic level.

To find out the dollar amount a typical potential consumer plans to spend on products like yours this year, you decide to use a stratified random sampling plan. This is reasonable because you expect sophisticated users to plan larger expenditures. By stratifying, you may be able to reduce the overall variability (of low and high expenditures) in the situation.

Your sampling frame, a list of names and addresses from a marketing firm, has 14,000 potential consumers. These consumers have already been classified; there are 2,532 sophisticated and 11,468 naive consumers. You decide to select 200 sophisticated and 100 naive users for detailed interviews, concentrating on the sophisticated segment because of their larger expected purchases. The results come out as follows:

| Stratum | Population Size | Sample Size | Sample Average | Sample Standard Deviation |
|---|---|---|---|---|
| Naive | $N_1 = 11,468$ | $n_1 = 100$ | $\overline{X}_1 = \$287$ | $S_1 = \$83$ |
| Sophisticated | $N_2 = 2,532$ | $n_2 = 200$ | $\overline{X}_2 = \$1,253$ | $S_2 = \$454$ |

These estimates for the two strata are quite interesting in their own right. The results certainly confirm your suspicion that sophisticated users plan to spend more!

To come up with a single average value per potential consumer for the entire population, find the weighted average:

$$\overline{X} = \frac{N_1\overline{X}_1 + N_2\overline{X}_2}{N_1 + N_2}$$

$$= \frac{11,468 \times 287 + 2,532 \times 1,253}{11,468 + 2,532}$$

$$= \frac{6,463,912}{14,000} = \$462$$

The resulting average, $462, is much closer to the naive expenditure of $287 than to the sophisticated expenditure of $1,253. The reason is that the naive segment is a much larger part of the population. Even though you sampled twice as much from the sophisticated segment, this simply increases your knowledge about these consumers and (properly) does not increase the influence of this segment of the population.

How much uncertainty is involved in this estimate of $462 per person? The standard error is found to be

$$S_{\overline{X}} = \frac{1}{N}\sqrt{\frac{N_1^2 S_1^2}{n_1} + \frac{N_2^2 S_2^2}{n_2}}$$

$$= \frac{1}{14,000}\sqrt{\frac{11,468^2 \times 83^2}{100} + \frac{2,532^2 \times 454^2}{200}}$$

$$= \frac{1}{14,000}\sqrt{9,060,070,003 + 6,607,073,115}$$

$$= \$8.94$$

How can you be so precise as to know the population mean expenditure to within approximately $8.94 when the individual variation from one person to another is around $83 or $454 depending on the group? The answer is that you are estimating a *mean* and not the behavior of the individuals. In fact, looking just at the naive segment, which is most of your market here, the standard error is $83/10, or just above $8.

What have you gained by stratifying instead of sampling 300 at random from the entire population? You have controlled for much of the variation. Instead of having the large numbers (for sophisticated users) and the small numbers (for naive ones) lumped together in a single, highly variable sample, you have separated out this variation according to its known sources. As a result, your answers are much more precise. A careful calculation (details not presented here) suggests that without stratification, your standard error could have been three times larger than the $8.94 you found in this case. And don't forget that a threefold reduction in standard error is ordinarily achievable only with a ninefold increase in the sample size (since $9 = 3^2$). By stratifying and sampling 300, you have achieved results comparable to a simple random sample of about 900. Stratification can be an important cost saver!

In other applications, stratification may help you more or less than in this example. Stratification will help you more when individuals are similar within each stratum but the strata are different from one another. That is to say, the strata do divide up the population into helpful, meaningful pieces.

If you recompute the standard error more carefully using the finite-population correction factor, you find that the standard error is reduced from $8.94 to $8.77. However,

we will assume you decide to stay with the uncorrected, larger $8.94 to be conservative (so that you are not exaggerating the precision of your results) and because you are interested in possibly generalizing beyond the frame of 14,000 consumers to a much larger idealized population that your frame represents.

### Example
#### The Price of a Typical Suit in a Department Store

Consider a store with two departments: general sales (which sells budget suits) and top fashion (selling high-ticket, expensive clothing). The general sales department has a higher volume but a lower price per suit. The top fashion department has fewer customers but a higher price per suit. To summarize sales patterns in the store as a whole, management would like a single number that represents the price paid for a typical suit of clothes.

Figure 8.5.1 shows the basic situation: 90% of suits are bought in general sales (where suits cost $60), and 10% are from top fashion (where suits cost $450). At the top of the figure is shown a representative sample of 10 customers and the average price of $99.00 (which represents nine suits at $60 and one suit at $450).

At the bottom of Figure 8.5.1 are the results of an unrepresentative sample of one customer from each department. Taking a simple average of the two suits, one at $60 and the other at $450, results in $255, which is clearly wrong. The problem is that the one customer in general sales actually represents much more of the population than does the one customer in top fashion.

Stratified sampling corrects this problem with the weighted average formula. Giving 90% of the weight to the
(Continued)

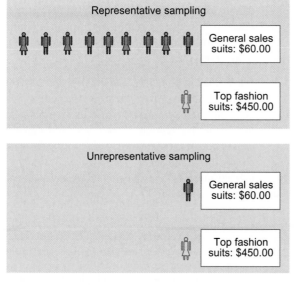
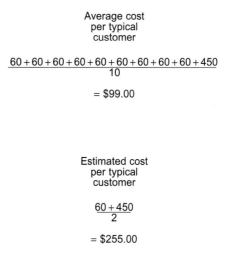

**FIGURE 8.5.1**  Careful stratified sampling methods can correct for the problems of unrepresentative sampling. The simple average for the unrepresentative sample, $255, is wrong. However, a weighted average for this same sample will give the correct answer, $99.

**FIGURE 8.5.2**   A systematic sample is made through regular selection from the population. In this case, every fifth population unit is selected, beginning with number 3 of the frame.

**Example—cont'd**

general sales customer and the remaining 10% to the top fashion customer, the weighted average does indeed give the correct answer:[14]

Weighted Average = $0.90 \times \$60 + 0.10 \times \$450 = \$99$

This shows why the average computed through the use of stratified sampling methods is correct. The weights in the weighted average reflect the importance in the population of each stratum in the sample.

_____

14. The weighted average was covered in Chapter 4.

## The Systematic Sample

A **systematic sample** is obtained by selecting a single, random starting place in the frame and then taking units separated by a fixed interval. It is easy and convenient to select a sample by taking, say, every fifth unit from the frame, as illustrated in Figure 8.5.2. It is even possible to introduce some randomness to this sampling method by selecting the starting place at random. But this systematic sampling method has some serious problems because it is impossible to assess its precision. If you wish to select a systematic sample of $n$ from a population of $N$, your interval between

selected items will be $N/n$.[15] If you select the starting place as a random digit between 1 and $N/n$, the sample average will be a reasonable estimate of the population mean in the sense that it will be *unbiased;* that is, it will not be regularly too high or too low. This is the good news.

The bad news is that you cannot know *how good* your estimate is. When you ask, "What's the standard error?" the answer is, "Who knows? The sample is not sufficiently random." In the words of W. Edwards Deming (who is famous for, among other things, bringing quality to Japanese products):

*One method of sampling, used much in previous years, by me as well as by others, was to take a random start and every kth sampling unit thereafter (a patterned or systematic sample).... As there is no replication, there is no valid way to compute an unbiased estimate of the variance of an estimate made by this procedure.... The replicated method [random sampling] is so simple to apply that there is no point in taking a chance with an estimate that raises questions.*[16]

_____

15. If $N/n$ is not a whole number, there are some small technical difficulties that require attention. See Chapter 4 of Kish, *Survey Sampling,* for a detailed discussion of this and other aspects of systematic sampling.

16. W. Edwards Deming, *Sample Design in Business Research* (New York: Wiley, 1960), p. 98.

One way in which systematic sampling can fail is when the list is ordered in an important, meaningful way. In this case, your random start determines how large your estimate will be so that a low starting number, for example, guarantees a low estimate.

A more serious failure of systematic sampling occurs if there is a repetitive pattern in the frame that matches your sampling interval. For example, if every 50th car that is produced gets special care and attention along the assembly line, and if you just happen to select every such 50th car to be in your systematic sample, your results will be completely useless in terms of representing the quality of *typical* cars.

So the reviews of systematic sampling are mixed. You might feel justified in using a systematic sample if (1) you are reasonably sure that there is no important ordering in the frame, (2) there are no important repetitive patterns in the frame, (3) you don't need to assess the quality of your estimate, and (4) you are sure that nobody will challenge your wisdom in selecting a systematic instead of a random sample.

Since a proper random sample will usually not cost very much more than a systematic sample, you may wonder why systematic samples are still used in some areas of business. So do I.

## 8.6 END-OF-CHAPTER MATERIALS

### Summary

Sampling is used to learn about a system that is too large and costly to study in its entirety. A **population** is the collection of units (people, objects, or whatever) that you are interested in knowing about. A **sample** is a smaller collection of units selected from the population. A sample is **representative** if each characteristic (and combination of characteristics) arises the same percent of the time in the sample as in the population. A sample that is not representative in an important way is said to show **bias**. The **frame** tells you how to gain access to the population units by number from 1 to the population size, $N$. A sample is said to be chosen **without replacement** if units cannot be selected more than once to be in the sample. A sample is said to be chosen **with replacement** if a population unit can appear more than once in the sample. A sample that includes the entire population ($n = N$) is called a **census**.

A **statistic**, or **sample statistic**, is defined as any number computed from your sample data. A **parameter**, or **population parameter**, is defined as any number computed for the entire population. An **estimator** is a sample statistic used as a guess for the value of a population parameter. The actual number computed from the data is called an **estimate**. The **error of estimation** is defined as the estimator (or estimate) minus the population parameter and is usually

unknown. An estimator is **unbiased** if it is correct on the average, that is, neither systematically too high nor too low, as compared to the corresponding population parameter. A **random sample** or **simple random sample** is selected such that (1) each population unit has an *equal probability of being chosen*, and (2) units are *chosen independently*, without regard to one another. A **table of random digits** is a list in which the digits 0 through 9 each occur with probability 1/10 independently of each other. Using such a table to select successive distinct population units is one way to select a random sample without replacement.

A **pilot study** is a small-scale version of a study, designed to help you identify problems and fix them before the real study is run.

The **central limit theorem** says that, for a random sample of $n$ observations from a population, the following statements are true:

1. Distributions become more and more normal as $n$ gets large, for both the *average* and the *sum*.
2. The means and standard deviations of the distributions of the average and the sum are as follows, where $\mu$ is the mean of the individuals and $\sigma$ is the standard deviation of these individuals in the population:

|  | Random Variable | |
| --- | --- | --- |
|  | Average | Sum Total |
| Mean | $\mu_{\overline{X}} = \mu$ | $\mu_{\text{sum}} = n\mu$ |
| Standard deviation | $\sigma_{\overline{X}} = \dfrac{\sigma}{\sqrt{n}}$ | $\sigma_{\text{sum}} = \sigma\sqrt{n}$ |

Applying the central limit theorem, you can find probabilities for a sum or an average from a random sample by using the standard normal probability table, finding the appropriate mean and standard deviation in the preceding summary table.

Anything you measure, based on a random sample of data, will have a probability distribution called the **sampling distribution** of that statistic. The **standard error of the statistic**, an estimate of the standard deviation of its sampling distribution, indicates approximately how far from its mean value (a population parameter) the statistic is. The **standard error of the average** (or just **standard error**, for short) indicates approximately how far the (random, observed) sample average $\overline{X}$ is from the (fixed, unknown) population mean $\mu$:

$$\text{Standard error} = S_{\overline{X}} = S/\sqrt{n}$$

The standard error gets smaller as the sample size $n$ grows (all else equal), reflecting the greater information and precision achieved with a larger sample.

When the population is small, so that the sample is an important fraction of the population, the standard error formula can be reduced by applying the **finite-population**

**correction factor** to obtain the **adjusted standard error** as follows:

(Finite-population correction factor) × (Standard error)

$$= \sqrt{\frac{N-n}{N}} \times S_{\overline{X}}$$

$$= \sqrt{\frac{N-n}{N}} \times \frac{S}{\sqrt{n}}$$

An **idealized population** can be defined as the much larger, sometimes imaginary, population that your sample represents. When you are interested in the idealized population, you do *not* use the finite-population correction factor. On the other hand, if you just want to learn about the population frame and not go beyond it, then the correction factor will work to your advantage by expressing the lower variability of this system. When in doubt, the safe choice is *not* to use the finite-population correction factor.

For a binomial distribution, the (population) standard deviations and the (sample) standard errors for both $X$ (the number) and $p = X/n$ (the proportion) are as follows:

|  | Binomial Number of Occurrences, $X$ | Binomial Proportion or Percentage, $p = X/n$ |
|---|---|---|
| Standard deviation (for the population) | $\sigma_X = \sqrt{n\pi(1-\pi)}$ | $\sigma_p = \sqrt{\dfrac{\pi(1-\pi)}{n}}$ |
| Standard error (estimated from a sample) | $S_X = \sqrt{np(1-p)}$ | $S_p = \sqrt{\dfrac{p(1-p)}{n}}$ |

The standard error $S_p$ indicates the uncertainty or variability in the observed proportion, $p$, and the standard error $S_X$ indicates the uncertainty in the observed count, $X$.

A **stratified random sample** is obtained by choosing a random sample separately from each of the strata (segments or groups) of the population. If the population is similar within each stratum but differs markedly from one to another, stratification can increase the precision of your statistical analysis. For a population with $L$ strata and $N_i$ units in the $i$th stratum, denote the sample size by $n_i$, the sample average by $\overline{X}_i$, and the standard deviation by $S_i$. To combine these averages into a single number that reflects the entire population, take a weighted average. Here are formulas for this weighted average and its standard error:

$$\overline{X} = \frac{N_1\overline{X}_1 + N_2\overline{X}_2 + \cdots + N_L\overline{X}_L}{N_1 + N_2 + \cdots + N_L} = \frac{1}{N}\sum_{i=1}^{L} N_i\overline{X}_i$$

$$S_{\overline{X}} = \frac{1}{N}\sqrt{\frac{N_1^2 S_1^2}{n_1} + \frac{N_2^2 S_2^2}{n_2} + \cdots + \frac{N_L^2 S_L^2}{n_L}} = \frac{1}{N}\sqrt{\sum_{i=1}^{L} \frac{N_i^2 S_i^2}{n_i}}$$

$$\left(\begin{array}{c}\text{Adjusted} \\ \text{standard error}\end{array}\right)$$

$$= \frac{1}{N}\sqrt{\frac{N_1(N_1-n_1)S_1^2}{n_1} + \frac{N_2(N_2-n_2)S_2^2}{n_2} + \cdots + \frac{N_L(N_L-n_L)S_L^2}{n_L}}$$

$$= \frac{1}{N}\sqrt{\sum_{i=1}^{L} \frac{N_i(N_i-n_i)S_i^2}{n_i}}$$

Use the adjusted standard error as a finite-population correction when you have sampled more than just a small fraction of any of the strata.

A **systematic sample** is obtained by selecting a single random starting place in the frame and then taking units separated by a regular interval. Although the sample average from a systematic sample is an unbiased estimator of the population mean (i.e., not regularly too high or too low), there are some serious problems with this technique. You cannot know how good the estimate is, since there is no reliable standard error for it. Problems can be particularly serious when population items are ordered in a meaningful way in the frame or when there is a repetitive pattern among population units in the frame. Since a proper random sample does not usually cost very much more than a systematic sample, you may wish to avoid using systematic samples.

## Key Words

adjusted standard error, *203*
bias, *190*
census, *191*
central limit theorem, *197*
error of estimation, *191*
estimate, *191*
estimator, *191*
finite-population correction factor, *203*
frame, *191*
idealized population, *204*
pilot study, *196*
population, *190*
population parameter, *191*
random sample or simple random sample, *192*
representative, *190*
sample, *190*
sample statistic, *191*
sampling distribution, *196*
sampling with replacement, *191*
sampling without replacement, *191*
standard error of the average, *201*
standard error of the statistic, *200*
stratified random sample, *205*
systematic sample, *208*
table of random digits, *192*
unbiased estimator, *191*

## Questions

1. a. What is a population?
   b. What is a sample? Why is sampling useful?
   c. What is a census? Would you always want to do a census if you had the resources?
2. a. What is a representative sample?
   b. What is a biased sample?
   c. How can a representative sample be chosen?
3. What is a frame? What is its role in sampling?
4. a. What is a random sample?
   b. Why is a random sample approximately representative?
   c. What is the difference between a random sample selected with and one selected without replacement?
   d. What is a table of random digits? How is it used in sample selection?
   e. What other method can be used to select a random sample using a spreadsheet program?
5. a. What is a pilot study?
   b. What can go wrong if you don't do a pilot study?
6. a. What is a statistic?
   b. What is a parameter?
7. a. What is an estimator?
   b. What is an estimate?
   c. A sample standard deviation is found to be 13.8. Is this number an estimator or an estimate of the population standard deviation?
   d. What is the error of estimation? When you estimate an unknown number, do you know the size of this error or not?
8. a. What is the sampling distribution of a statistic?
   b. What is the standard deviation of a statistic?
9. a. What is the central limit theorem?
   b. Does the central limit theorem specify that individual cases follow a normal distribution?
   c. How do you interpret the idea that the average has a normal distribution?
   d. What is the mean of a sum of independent observations of a random variable? What is its standard deviation?
   e. What is the mean of an average of independent observations of a random variable? What is its standard deviation?
10. a. What is the standard error of a statistic?
    b. In what way does the standard error indicate the quality of the information provided by an estimate?
    c. What typically happens to the standard error as the sample size, $n$, increases?
11. a. What is the finite-population correction factor?
    b. What is the adjusted standard error?
    c. What is an idealized population?
    d. In what way are your results more limited if you use the finite-population correction factor than if you don't?
12. What do the standard errors $S_x$ and $S_p$ indicate for a binomial situation?

13. a. What is a stratified random sample?
    b. What are the benefits of stratification?
    c. When is stratification most likely to be helpful?
14. a. What is a systematic sample?
    b. What are the main problems with systematic samples?
    c. Why is there no reliable standard error available for use with a sample average computed from a systematic sample?

## Problems

*Problems marked with an asterisk (*) are solved in the Self Test in Appendix C.*

1. Your automatic transmission factory has had some problems with quality. You have decided to gather information from tomorrow's production for careful evaluation. For each of the following sampling methods, say if the procedure is good, acceptable, or unreasonable. Give a reason for your choice.
   a.* The first 5 transmissions produced.
   b. The 18 transmissions that are sitting outside the plant because they never worked.
   c. Every 20th transmission produced.
   d. A random sampling taken at the end of the day using the day's production as a frame.
   e. All obviously defective transmissions together with a random sampling of the apparently normal ones.
2. Which of the following samples is likely to be the most representative of the population of all registered voters in the United States?
   a. A sample of 200 people at a Denver shopping mall.
   b. A sample of 200 of your friends and their friends.
   c. A sample of 200 people, chosen by dialing telephone numbers at random.
   d. A sample of 200 people selected at random from all students at the University of Nebraska.
3. Which of the following samples is likely to be the most representative of the population of all employees at IBM?
   a. The 10 oldest and most experienced researchers at the Thomas J. Watson Research Center.
   b. A random sample of 10 computer repair specialists.
   c. The 10 employees selected as "most typical" by middle management.
   d. A random sample of 10 chosen from a list of all employees at IBM.
4. Consider an election poll designed so that each household has an equal chance of being selected and one registered voter is interviewed from each selected household. Analyze what would happen if single-voter households were more likely to vote Democratic than households containing more than one registered voter. In particular: Would the "percent voting Democrat in the sample" be an unbiased estimator for the percent of all registered voters? If not, would it overestimate or underestimate this true percentage?

5. You have chosen a sample of 25 supermarkets out of the 684 you have responsibility for. These 25 have been inspected, and the number of violations of company policy has been recorded. For each of the following quantities, state whether it is a statistic or a parameter.

a.* The average number of violations among the 25 supermarkets you inspected.

b.* The mean number of violations you would have recorded had you inspected all 684 supermarkets you are responsible for.

c. The variability from one supermarket to another in the population.

d. The variability from one supermarket to another, as measured by the standard deviation you computed.

e. The standard deviation of your sample average.

f. The standard error of your sample average.

6. Select a random sample of three without replacement from the following (very small) population of firms: IBM, GM, Ford, Shell, HP, Boeing, and ITT. Use the following sequence of random digits: 5887053671352339.

7. Select a random sample of four without replacement from the following metal products corporations: Gillette, Crown Cork & Seal, MASCO, Tyco Laboratories, Illinois Tool Works, McDermott, Ball, Stanley Works, Harsco, Hillenbrand Industries, Newell, Snap-on Tools, Danaher, Silgan, Robertson-Ceco, and Barnes Group. Begin in row 28, column 7 of the table of random digits.

8.* Draw a random sample of 3 account numbers from a population of 681 accounts receivable documents, starting with row 6, column 2 of the table of random digits.

9. Draw a random sample of 4 firms from a population of 86 suppliers, starting with row 30, column 4 of the table of random digits.

10. Draw a random sample of 5 contracts from a population of 362 recent contracts with cost overruns, starting with row 13, column 5 of the table of random digits.

11. Draw a random sample of 8 invoices from a population of 500 overdue billings, starting with row 17, column 5 of the table of random digits.

12. The mean account balance is $500 and the standard deviation is $120 for a large population of bank accounts. Find the standard deviation of the average balance of groups of eight accounts (chosen independently of one another).

13. The population mean productivity is 35, the population standard deviation is 10, and the sample size is 15. Find the standard deviation of the total amount represented by a random sample.

14. You have eight machines operating independently. The mean production rate for each machine is 20.3 tons per day, and the standard deviation is 1.4 tons per day. Approximately how much uncertainty is there in the average daily production for the eight machines? Please report the usual summary measure.

15. You have estimated the inventory value of your competition as $384,000 but later learn that the true inventory value was $416,000. Find the estimation error.

16. At a medical clinic, patient office visits are found to last a mean time of 17 minutes, with a standard deviation of 10 minutes. Assume that the probability distribution of time is independent from one patient to another.

a. What is the approximate probability that the 25 patients scheduled for tomorrow will require more than 7 hours altogether?

b. What is the approximate probability that the average time for 25 patients will be more than 20 minutes?

17. Deposits have a mean of $125 and a standard deviation of $36. Find the standard deviation of the average amount of 12 randomly selected deposits.

18. You have interviewed 369 people out of a population of 30,916 and found that 51.8% expect to vote for the challenger in the upcoming election. Find the standard error of this estimate.

19. You have a factory with 40 production machines that are essentially identical, each producing at a mean daily rate of 90 products with a standard deviation of 35. You may assume that they produce independently of one another. Consider the random variable "the average daily production per machine tomorrow."

a. Find the mean of this random variable. Compare it to the mean for a single machine.

b. Find the standard deviation of this random variable. Compare it to the standard deviation for a single machine.

c. What is the approximate probability distribution of this random variable? How do you know?

d. Find the approximate probability that your average daily production per machine will be between 95 and 100 products tomorrow.

20. Breakfast cereal is packed into packages labeled "net weight 20 ounces, packed by weight not by volume; some settling may occur during shipment." However, weights of individual packages are not really all exactly equal to 20 ounces—although they are close, they do have some randomness. Based on past observation, assume that the mean weight is 20.04 ounces, the standard deviation is 0.15 ounce, and the distribution is approximately normal. Consider the average weight of 30 packages selected independently at random.

a. What is the mean weight of this random variable?

b. How variable is this random variable?

c. What is the approximate probability that the average weight is less than 20 ounces?

21. A farmer has 5 identical cornfields, each of which independently produces a normally distributed harvest with a mean of 80,000 bushels and a standard deviation of 15,000 bushels. Find the probability that the average Tharvest for the 5 fields will exceed 88,000 bushels.

22.* The population mean is $65 and the population standard deviation is $30. Find the probability that the average of 35 randomly selected transactions is between $55 and $60. You may assume that the population is approximately normally distributed.

23. You have analyzed a project using four scenarios, with the results shown in Table 8.6.1. Suppose you actually have 40 projects just like this one and they pay off

**TABLE 8.6.1 Probability and Profit for Four Scenarios**

| Scenario | Probability | Profit or Loss ($ millions) |
|---|---|---|
| Really bad | 0.10 | −10 |
| So-so | 0.15 | 2 |
| Pretty good | 0.50 | 5 |
| Great | 0.25 | 15 |

independently of one another. Find the probability that your average profit per project will be between $5 million and $6 million.

24. A typical incoming telephone call to your catalog sales force results in a mean order of $28.63 with a standard deviation of $13.91. You may assume that orders are received independently of one another.
   a. Based only on this information, can you find the probability that a single incoming call will result in an order of more than $40? Why or why not?
   b. An operator is expected to handle 110 incoming calls tomorrow. Find the mean and standard deviation of the resulting total order.
   c. What is the approximate probability distribution of the total order to be received by the operator in part b tomorrow? How do you know?
   d. Find the (approximate) probability that the operator in part b will generate a total order of more than $3,300 tomorrow.
   e. Find the (approximate) probability that the operator in part b will generate an average order between $27 and $29 tomorrow.

25. Your restaurant will serve 50 dinner groups tonight. Assume that the mean check size of dinner groups in general is $60, the standard deviation is $40, and the distribution is slightly skewed with a longer tail toward high values.
   a. Find the mean and standard deviation for the total of all 50 checks.
   b. Find the mean and standard deviation for the average of all 50 checks.
   c. What further assumption is needed in order for you to conclude that the total of all 50 checks is approximately normally distributed?
   d. Find the probability that the total of all 50 checks is more than $3,100, assuming a normal distribution.
   e. Find the probability that the average of all 50 checks is between $58 and $65, assuming a normal distribution.

26. Your customers' average order size is $2,601, with a standard deviation of $1,275. You are wondering what would happen if exactly 45 typical customers independently placed orders tomorrow.
   a.* Find the mean of tomorrow's total orders.
   b.* Find the standard deviation of tomorrow's total orders.

c.* Next (for the rest of this problem) assume that tomorrow's total orders follow a normal distribution. Why is this assumption reasonable, even if individual customer orders are somewhat skewed?
   d.* Find the probability that total orders will be at or above your break-even point of $105,000.
   e. Find the probability of a truly amazing day, with total orders exceeding $135,000.
   f. Find the probability of a typical day, with total orders between $110,000 and $125,000.
   g. Find the probability of a surprising day, with total orders either below $100,000 or above $135,000.
   h. What are the chances that tomorrow's average order per customer will be between $2,450 and $2,750?

27. You have a factory with 40 production machines that are essentially identical, each producing at a mean daily rate of 100 products with a standard deviation of 15. You may assume that they produce independently of one another. Consider the average daily production per machine tomorrow, which is a random variable.
   a. Find the mean of this random variable. Compare it to the mean for a single machine.
   b. Find the standard deviation of this random variable. Compare it to the standard deviation for a single machine.
   c. What is the approximate probability distribution of this random variable? How do you know?
   d. Find the (approximate) probability that your average daily production per machine will be more than 102 products tomorrow.
   e. Find the (approximate) probability that your average daily production per machine will be between 97 and 103 products tomorrow.

28. Consider the profits as a percent of revenue for a group of companies involved in petroleum and/or mining, as shown in Table 8.6.2.

**TABLE 8.6.2 Profit for Petroleum and/or Mining Firms**

| Firm | Profit as a Percent of Revenue |
|---|---|
| Anadarko Petroleum | −1.50% |
| Apache | −3.30 |
| Baker Hughes | 4.36 |
| Cameron International | 9.10 |
| Chesapeake Energy | −75.70 |
| Chevron | 6.41 |
| ConocoPhillips | 3.48 |
| Consol Energy | 11.68 |

*(Continued)*

**TABLE 8.6.2  Profit for Petroleum and/or Mining Firms—cont'd**

| Firm | Profit as a Percent of Revenue |
|---|---|
| Devon Energy | −27.67 |
| El Paso | −11.64 |
| Enbridge Energy Partners | 5.36 |
| Energy Transfer Equity | 8.17 |
| Enterprise GP Holdings | 0.80 |
| EOG Resources | 11.42 |
| Exxon Mobil | 6.77 |
| Freeport-McMoRan Copper & Gold | 18.28 |
| Halliburton | 7.80 |
| Hess | 2.50 |
| Holly | 0.41 |
| Kinder Morgan | 6.90 |
| Marathon Oil | 2.96 |
| Murphy Oil | 4.38 |
| National Oilwell Varco | 11.56 |
| Newmont Mining | 16.76 |
| Occidental Petroleum | 18.77 |
| Oneok | 2.75 |
| Peabody Energy | 7.10 |
| Plains All American Pipeline | 3.13 |
| Smith International | 1.81 |
| Spectra Energy | 17.95 |
| Sunoco | −1.11 |
| Tesoro | −0.84 |
| Valero Energy | −2.83 |
| Western Refining | −5.15 |
| XTO Energy | 22.27 |

**Source:** Data from the Fortune 500, accessed from http://money.cnn.com/magazines/fortune/fortune500/2010/index.html on July 12, 2010.

a. Construct a sampling frame, viewing this list as a population of large petroleum and/or mining-related firms.

b. Draw a random sample of 10 firms, starting from row 13, column 2 of the table of random digits.

c. Compute the sample average.

d. Compute the standard error of the average, both with and without use of the finite-population correction factor.

e. Write a paragraph explaining and interpreting the standard error.

f. Compute the population mean. (*Note:* In real life, you usually can't do this!)

g. Write a brief paragraph explaining the relationship among the sample average, population mean, and standard error.

29. Consider the percent change in revenues for the five largest soap and cosmetics firms as shown in Table 8.6.3. View this list as a (very small!) population with just $N = 5$ units. Consider drawing samples of size $n = 2$.

   a. Make a list of all possible samples of size 2 that might be chosen. (*Hint:* There are 10 such samples.) For each sample, find the average.

   b. Construct a histogram of the 10 sample averages from part a. This is the sampling distribution of the sample average.

   c. Select a random sample of size 2 from the population by starting with the number in row 26, column 4 of the table of random digits. Find the sample average.

   d. Indicate where the average (from part c) falls with respect to the sampling distribution (from part b).

   e. Write a paragraph explaining how "drawing a random sample from the population and finding the average" gives essentially the same result as "drawing a number from the sampling distribution of the average."

30. Economists often make forecasts of future conditions. Consider the U.S. unemployment rate for June 2010 as predicted in December 2009 as part of a survey of economists, as shown in Table 8.6.4.

   a. Find the average and the standard deviation. Briefly interpret these values.

   b. Find the standard error of the average. Interpret this number carefully, by viewing this list as a random sample from a much larger list of economists that might have been selected.

   c. Six months after these predictions were made, the actual unemployment rate[17] was recorded as 9.5%. Compare the average forecast to this actual outcome.

**TABLE 8.6.3  Percent Change in Revenues for Fortune 500 Soap and Cosmetics Companies**

| Company | Revenues % Change |
|---|---|
| Procter & Gamble | 5% |
| Colgate-Palmolive | 3 |
| Avon Products | 7 |
| Estee Lauder | 10 |
| Clorox | 2 |

**Source:** Data are from www.fortune.com, accessed on December 3, 2001.

**TABLE 8.6.4  Economists' Forecasts of the U.S. Unemployment Rate**

| Economist | Forecast of Rate in June 2010, as Made in December 2009 |
|---|---|
| Paul Ashworth | 9.7% |
| Nariman Behravesh | 10.2 |
| Richard Berner/ David Greenlaw | 10.1 |
| Ram Bhagavatula | 9.5 |
| Jay Brinkmann | 10.2 |
| Joseph Carson | 9.5 |
| Mike Cosgrove | 9.8 |
| Lou Crandall | 9.8 |
| J. Dewey Daane | 9.9 |
| Richard DeKaser | 10.1 |
| Douglas Duncan | 10.1 |
| Brian Fabbri | 10.8 |
| Maria Fiorini Ramirez/ Joshua Shapiro | 10.5 |
| Stephen Gallagher | 10.2 |
| Ethan Harris | 9.5 |
| Maury Harris | 10.4 |
| Jan Hatzius | 10.2 |
| Tracy Herrick | 9.6 |
| Gene Huang | 9.5 |
| William B. Hummer | 9.7 |
| Dana Johnson | 10.1 |
| Kurt Karl | 10.1 |
| Bruce Kasman | 10.0 |
| Paul Kasriel | 10.5 |
| Joseph A. LaVorgna | 9.6 |
| Edward Leamer/David Shulman | 10.3 |
| John Lonski | 9.9 |
| Dean Maki | 9.6 |
| David Malpass | 10.3 |
| Jim Meil | 9.7 |
| Mark Nielson, Ph.D. | 9.0 |
| Michael P. Niemira | 9.5 |
| Nicholas S. Perna | 9.6 |
| Joel Prakken/ Chris Varvares | 10.0 |
| Arun Raha | 9.8 |
| David Resler | 10.0 |
| John Ryding/ Conrad DeQuadros | 9.8 |
| Ian Shepherdson | 10.2 |
| John Silvia | 10.5 |
| Allen Sinai | 10.0 |
| James F. Smith | 9.0 |
| Sung Won Sohn | 9.8 |
| Neal Soss | 10.3 |
| Stephen Stanley | 9.9 |
| Susan M. Sterne | 9.8 |
| Diane Swonk | 10.0 |
| Bart van Ark | 10.5 |
| Brian S. Wesbury/ Robert Stein | 9.4 |
| William T. Wilson | 10.1 |
| David Wyss | 10.5 |
| Lawrence Yun | 10.0 |

**Source:** Accessed from http://online.wsj.com/article/SB1260393982 47384245.html on July 12, 2010.

d. How many standard errors away from the sample average is the actual outcome (9.5%)? Would you ordinarily be surprised by such an extreme difference?

e. Explain why the forecast error (average forecast minus actual outcome) need not be approximately equal in size to the standard error. Do this by identifying the population mean and showing that the actual outcome is not the same object.

31. Here is a list of the dollar amounts of recent billings:

$994, $307, $533, $443, $646, $148, $307, $524, $71, $973, $710, $342, $494

a. Find the average sale. What does this number represent?

b. Find the standard deviation. What does this number represent?

c. Find the standard error. What does this number represent?

d. You are anticipating sending another 500 billings similar to these next month. What total amount should you forecast for these additional billings?

32. The sample average age is 69.8 and the sample standard deviation is 9.2, based on a sample of 200 individuals in a retirement community. Your friend claims that "the sample average is approximately 9.2 away from the population mean." Is your friend correct? Why or why not?

33. Find the standard error of the average for the following data set representing quality of agricultural produce:

16.7, 17.9, 23.5, 13.8, 15.9, 15.2, 12.9, 15.7

34. A random sample of 50 recent patient records at a clinic shows that the average billing per visit was $53.01 and the standard deviation was $16.48.
    a.* Find the standard error of the average and interpret it.
    b. You feel that this standard error is too large for reasonable budgeting purposes. If the standard deviation were the same (which it would be, approximately), find the standard error you would expect to see for a sample of size 200.

35. Find the average and the standard error for the amounts that your regular customers spent on your products last month, viewing the data from problem 2 of Chapter 4 as a sample of customer orders.

36. Find the average and the standard error for the strength of cotton yarn used in a weaving factory, based on the data in problem 23 of Chapter 4.

37. Find the average and the standard error for the weight of candy bars before intervention, based on the data in problem 13 of Chapter 5.

38. A survey of 823 randomly selected adults in the United States finds that 63% support current government policies. Find the usual measure that indicates approximately how far this sample percentage is from the value that would have been found if all adults in the United States had been interviewed.

39. In a study of brand recognition, out of 763 people chosen at random, 152 were unable to identify your product.
    a. Estimate the percentage of the population (from which this sample was taken) who would be unable to identify your product.
    b. Find the standard error of the estimate found in part a and briefly interpret its meaning.

40. Based on careful examination of a sample of size 868 taken from 11,013 inventory items in a warehouse, you learn that 3.6% are not ready to be shipped.
    a. Find the standard error associated with this estimated percentage and indicate its meaning.
    b. Would you be surprised to learn that in fact 4% of the 11,013 inventory items are not ready to be shipped? Why or why not?
    c. Would you be surprised to learn that in fact 10% of the 11,013 inventory items are not ready to be shipped? Why or why not?

41. From a list of the 729 people who went on a cruise, 25 were randomly selected for interview. Of these, 21 said that they were "very happy" with the accommodations.
    a. What percent of the sample said that they were "very happy"?

    b. If you had been able to interview all 729 people, approximately how different a percentage would you expect to find as compared to your answer to part a? To answer this, please indicate which statistical quantity you are using, and compute its value.

42. A poll involved interviews with 1,487 people and found that 42.3% of those interviewed were in favor of the candidate in question. The election will be held in three weeks.
    a. Approximately what percentage of the entire population would say they were in favor of the candidate, if they were interviewed under the same conditions?
    b. Give at least two reasons why the actual outcome of the election could differ from this 42.3% by more than the standard error.

43. The accounts of a firm have been classified into 56 large accounts, 956 medium-sized accounts, and 16,246 small accounts. Each account has a book value (which is provided to you) representing the amount of money that is supposed to be in the account. Each account also has an audit value (which requires time and effort to track down) representing the amount of money that is really in the account. You are working with the auditors in preparing financial statements. The group has decided to examine all 56 large accounts, 15% of the medium-sized accounts, and 2% of the small accounts. The total error (book value minus audit value) was found to be $15,018.00 for the large accounts, $1,165.00 for the sampled medium-sized accounts, and $792.00 for the sampled small accounts. The standard deviations of the errors were $968.62, $7.12, and $5.14, respectively. (*Hint:* Do not confuse the error, which is measured for each account, with the standard error of an average.)
    a. Find the sample average error per account in each of the three strata (groups) of accounts.
    b. Combine these three averages to find the stratified sampling average estimate of the population mean error per account.
    c. Find the standard error of your estimate in part b both with and without use of the finite-population correction factor. Why are the answers so different in this case?
    d. Interpret the (corrected) standard error value in terms of the (unknown) population mean error per account.

44. Randomly selected consumers in four cities have been interviewed as part of a study by a shoe retailer. Each consumer reported the number of pairs of shoes in his or her closet (each line represents one city) (Table 8.6.5):
    a. Estimate the mean number of pairs of shoes for the population representing all four cities combined.
    b. Find the standard error for your answer to part a.
    c. Find the standard error with and without use of the finite-population correction factor, and compare. Why are they so similar?

17. Accessed from the Bureau of Labor Statistics at http://www.bls.gov/cps/ on July 12, 2010.

**TABLE 8.6.5 Sample Results for Shoes in Four Cities**

| Population Size | Sample Size | Sample Average (number of pairs of shoes) | Sample Standard Deviation |
|---|---|---|---|
| 3,638,815 | 200 | 13.77 | 13.57 |
| 6,899,665 | 200 | 12.72 | 12.11 |
| 9,608,853 | 250 | 8.79 | 12.34 |
| 709,212 | 200 | 10.43 | 14.99 |

## Database Exercises

*Problems marked with an asterisk (\*) are solved in the Self Test in Appendix C.*

Please refer to the employee database in Appendix A. For now, view this database as the population of interest.

1. Show that this database is arranged in the form of a frame. In particular, how would you use it to gain access to population information for a particular employee?

2. Draw a random sample without replacement of 10 employees, using the table of random digits, starting in row 23, column 7.
   a.\* List the employee numbers for your sample.
   b. Find the average salary for your sample and interpret this number.
   c. Find the standard deviation of salary for your sample and interpret this number.
   d. Find the standard error of salary for your sample and interpret this number. In particular, in what way is it different from the standard deviation found in the previous part of this exercise?

3. Continuing with the sample from the preceding exercise:
   a. Find the population mean for salary. (*Note:* In real life, you usually cannot find the population mean. We are peeking "behind the scenes" here.)
   b. Compare this population mean to the sample average for salary. In particular, how many standard errors apart are they?
   c. Find the population standard deviation for salary and interpret this number.
   d. Compare this population standard deviation to the sample standard deviation for salary.
   e. Find the population standard deviation for the average salary for a sample and interpret this number. Compare it to the standard error from the sample salary data.
   f. Arrange the numbers you have computed in a table where the columns are "population" and "sample" and the rows are "sample average and population mean," "standard deviation of individuals," and "standard deviation and standard error of sample averages of 10 employees."

4. Do exercise 2 using the ages instead of the salaries.
5. Do exercise 2 using the experiences instead of the salaries.
6. Do exercise 3 using the ages instead of the salaries.

7. Do exercise 3 using the experiences instead of the salaries.

8. Continuing with the sample from exercise 2:
   a.\* Find the binomial $X$ for the gender variable (counting the number of females) and interpret it.
   b.\* Find the standard error of $X$ and interpret it.
   c. Find the population mean for the binomial $X$.
   d. How far is the observed $X$ for your sample from its population mean?
   e. How does this difference compare to the standard error of $X$?

9. Do exercise 8 using the binomial proportion $p$ in place of $X$.

## Projects

1. Financial information about individual firms is now often available on the Internet, either as summaries (e.g., at http://www.finance.yahoo.com, where you can enter a company name to find its stock market symbol and then choose "Key Statistics" when its information comes up) or linked to the firm's home page (often under a heading such as "investor relations"). Select an important number such as "profit margin" that can be meaningfully compared across large and small firms.
   a. Identify a population of firms of interest to you and create a sampling frame.
   b. Select a random sample of 10 firms. Find the data for these firms.
   c. Compute the average and the standard error.
   d. Indicate (approximately) how far your average is from the mean value for all firms listed in your frame.
   e. Write a paragraph summarizing what you have learned about statistics and about the firms in your population.

2. Your firm is planning the marketing strategy for a "new and improved" consumer product. Your advertising company has shown you five TV ads, and you must choose two of these. Having seen them, you feel that some ads appeal to women more than to men. Before your company commits $1.8 million to this campaign, your supervisor would like to know more about consumer reaction to these ads. Write a one-page memo to your supervisor suggesting how to go about

gathering this needed information. Be sure to cover the following topics: random sample, stratified random sample, pilot study.

3. Identify a situation relating to your work or business interests in which statistical sampling might be (or has been) useful.
   a. Describe the population and indicate how a sample could be chosen.
   b. Identify a population parameter of interest and indicate how a sample statistic could shed light on this unknown.
   c. Explain the concept of the sampling distribution of this statistic for your particular example.

## Case

### Can This Survey Be Saved?

"What's troubling me is that you can't just pick a new random sample just because somebody didn't like the results of the first survey. Please tell me more about what's been done." Your voice is clear and steady, trying to discover what actually happened and, you hope, to identify some useful information without the additional expense of a new survey.

"It's not that we didn't like the *results* of the first survey," responded R. L. Steegmans, "it's that only 54% of the membership responded. We hadn't even looked at their planned spending when the decision [to sample again] was made. Since we had (naively) planned on receiving answers from nearly all of the 400 people initially selected, we chose 200 more at random and surveyed them also. That's the second sample." At this point, sensing that there's more to the story, you simply respond "Uh huh. …" Sure enough, more follows:

"Then E. S. Eldredge had this great idea of following up on those who didn't respond. We sent them another whole questionnaire, together with a crisp dollar and a letter telling them how important their responses are to the planning of the industry. Worked pretty well. Then, of course, we had to follow up the second sample as well."

"Let me see if I understand," you reply. "You have two samples: one of 400 people and one of 200. For each, you have the initial responses and follow-up responses. Is that it?"

"Well, yes, but there was also the pilot study—12 people in offices downstairs and across the street. We'd kinda like to include them, average them, with the rest because we worked so hard on that at the start, and it seems a shame to throw them away. But all we really want is to know average spending to within about a hundred dollars."

At this point, you feel that you have enough of the background information to evaluate the situation and to either recommend an estimate or an additional survey. Here are additional details for the survey of the 8,391 overall membership in order to determine planned spending over the next quarter.

|  | Pilot Study | First Sample | Second Sample | Both Samples | All Combined |
|---|---|---|---|---|---|
| **Initial mailing** | | | | | |
| Mailed | 12 | 400 | 200 | 600 | 612 |
| Responses | 12 | 216 | 120 | 336 | 348 |
| Average | $39,274.89 | $3,949.40 | $3,795.55 | $3,894.45 | $5,114.47 |
| Std. dev. | $9,061.91 | $849.26 | $868.39 | $858.02 | $6,716.42 |
| **Follow-up mailing** | | | | | |
| Mailed | 0 | 184 | 80 | 264 | 264 |
| Responses | 0 | 64 | 18 | 82 | 82 |
| Average | | $1,238.34 | $1,262.34 | $1,243.60 | $1,243.60 |
| Std. dev. | | $153.19 | $156.59 | $153.29 | $153.29 |
| **Initial and follow-up** | | | | | |
| Mailed | 12 | 400 | 200 | 600 | 612 |
| Responses | 12 | 280 | 138 | 418 | 430 |
| Average | $39,274.89 | $3,329.73 | $3,465.13 | $3,374.43 | $4,376.30 |
| Std. dev. | $9,061.91 | $1,364.45 | $1,179.50 | $1,306.42 | $6,229.77 |

### Discussion Questions

1. Do you agree that drawing a second sample was a good idea?
2. Do you agree that the follow-up mailings were a good idea?
3. How might you explain differences among averages in the results?
4. Are there useful results here? Which ones are useful? Are they sufficient or is further study needed?

## Chapter 9

# Confidence Intervals

## Admitting That Estimates Are Not Exact

There are two paths to larger profits: increasing the revenue or decreasing the costs. In the medical care business, revenue can be difficult to control because insurance companies and the government determine the maximum amount they are willing to reimburse for a given diagnosis. Consider a hospital whose managers are attempting to find a reasonable answer to the question, How much can we expect to earn or lose per patient, over the long run, for cardiac surgery? Careful analysis of a sample of medical and financial records for 35 cardiac surgery patients revealed that the average profit was $390.26, with a standard deviation of $450.56. So far, you know a lot about these particular 35 patients. But what do you really know about cardiac patients in general? After all, you are concerned about future performance, not just these particular individuals. You remember that the standard error, $450.56/\sqrt{35} = 76.16$, tells you approximately how far the sample average, $390.26, is from the mean for the entire population you sampled from. But now you need to go further, since "approximate" is not good enough. You would like to create a definite, authoritative, exact statement. The confidence interval is designed to do just that. In this case, the confidence interval (as computed by the methods to be described soon) will specify that:

We are 95% sure that the mean profit per patient, for the population from which these 35 were sampled, is somewhere between $235.51 and $545.01.

This is indeed an exact statement about a population, specifying a region around the sample average of $390.26 in order to reflect the randomness of sampling. We have not surveyed the entire population (of many more cardiac patients) and we may not need to. Nonetheless, we can make a statement about what we would find if we somehow could spend the money necessary to comb through old records in order to find out. The confidence interval provides an important link between the affordable survey of these 35 patients and the larger population, going well beyond the approximate interpretation of the standard error.[1]

The practical interpretation of this confidence interval tells us that when we try to generalize beyond these

---

1. What if all patient records are already computerized? Then you will be able to analyze larger sample sizes more easily. The issue now becomes the discrepancy between the particular patients you have seen (the sample) and the experience you are likely to have in the future. A confidence interval can still be useful because it indicates the random component of the difference. However, systematic changes in your marketplace or technology should be considered as well.

particular patients, the sample average of $390.26 is not as exact as it appeared initially. It could reasonably be in error in either direction by over $100 per patient. Why is the error so large? Because of variability (profit was larger for some patients than for others) and the small sample size (a study of more than 35 would be expected to estimate the population mean more closely).

A confidence interval can also be used to indicate how closely a computed percentage (for a sample) reflects the percentage you are really interested in (for a population). For example, the results of a market survey of 150 people selected randomly from your targeted group indicate that 46, or 30.7%, are aware of your brand name. You don't believe, even for a minute, that exactly 30.7% of the entire targeted group is aware of your brand name because you know that the randomness of sampling introduces an error of approximately one standard error. In this case, the standard error is $S_p = 3.76$ percentage points, indicating the approximate difference between the sample and the population percentages. The confidence interval formalizes this notion of *approximate difference*, as computed by the methods to be described soon, leading to the following exact statement:

> We are 95% sure that the percentage of our targeted group (the population) who are aware of our brand is somewhere between 23.3% and 38.0%.

The objective is to get rid of as much uncertainty as possible and to make the statement as exact as possible. Probability is necessary in order to make an exact statement in the face of uncertainty. Statistics is needed in order to take advantage of the information in your sample data. This process of generalizing from sample data to probability-based statements about the population is called **statistical inference**. In particular, a **confidence interval** is an interval computed from the data in such a way that there is a *known probability* of including the (unknown) population parameter of interest, where this probability is interpreted with respect to a random experiment that begins with the selection of a random sample. Thus, specifying the confidence interval is the best you can do under uncertainty: It is an exact probability statement in place of vague observations such as, We're not sure, but … or, It's probably pretty close.

Confidence intervals come in great variety: Here is a brief preview of the coming attractions. You can choose the probability of the statement, called the **confidence level**. By tradition, this is set at 95%, but it is common to find 90%, 99%, and even 99.9% levels used. The trade-off for a higher confidence level is a larger, less useful interval. A confidence interval for a population percentage can be computed easily using the standard error for a binomial distribution. Depending on the question of interest, you may also decide whether the interval is two-sided (it's between this and that) or one-sided (e.g., it's at least as big as this). As always, you must watch out for the technical assumptions—in this case, normality and random sampling—lurking in the background, which, if not satisfied, will invalidate your confidence interval statements. And be careful to distinguish the 95% probability for the *process* of generating the confidence interval from the 95% confidence you have for a particular interval after it is computed.

There is an approximate, all-purpose confidence interval statement that applies in many situations. Once you have estimated a population parameter using an appropriate unbiased estimator and found the appropriate standard error of this estimator, the confidence interval statement (in generic form) is as follows:

> **Approximate Confidence Interval Statement**
>
> We are 95% sure that the population parameter is somewhere between the estimator *minus* two standard errors and the estimator *plus* two standard errors.

You may remember that a normal variable will be within two standard deviations from the mean approximately 95% of the time; this is where these values come from (although somewhat indirectly) in this generic confidence interval statement.

How widely applicable is the notion of confidence interval? Essentially every number you see reported in the newspapers, in your confidential strategic internal memos, on television, and by media on the Internet is an estimate of an important number. Essentially all estimators have their own "personal" standard errors, indicating their precision. Once you have these two numbers (estimate and standard error), you can use the approximate confidence interval statement. But there are some details, improvements, and warnings to be observed for some cases, which follow.

## 9.1 THE CONFIDENCE INTERVAL FOR A POPULATION MEAN OR A POPULATION PERCENTAGE

We have just drawn a sample of data and computed the sample average, $\overline{X}$, in order to estimate the population mean. Let's pretend for a moment that we know the value of the (usually unknown) population mean, $\mu$, so that the situation is as in Figure 9.1.1. The distance between the sample average and the population mean (the estimation error) is *the same* whether you measure it starting from one or from the other. This says that measuring in terms of standard errors from the population mean will give the same result as measuring in terms of standard errors from the sample average. This is no trivial result. Since the sample average is known, you can measure in terms of a *known*

Sampling distribution of $\overline{X}$

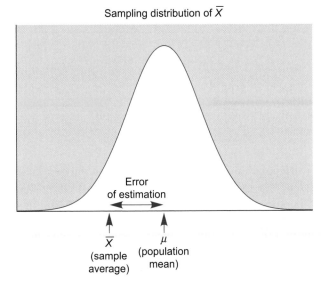

Error of estimation

$\overline{X}$
(sample average)

$\mu$
(population mean)

**FIGURE 9.1.1** The sampling distribution of $\overline{X}$ is centered at $\mu$. The distance between the sample average and the population mean (the error of estimation) is the same whether you measure starting from one or from the other. The confidence interval will be constructed by measuring from $\overline{X}$, which is known, instead of from $\mu$, which is unknown.

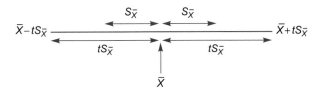

**FIGURE 9.1.2** The 95% confidence interval extends $t$ (approximately 2) standard errors, on either side of the sample average.

quantity (the standard error) from another *known* quantity (the sample average), and the same basic relationships hold as if you were measuring from the (unknown) population mean.

The intuitive reasoning behind the confidence interval is as follows. Recall the fact that, for a normal distribution, the probability is approximately 0.95 of falling within two standard deviations from the mean.[2] This leads to the following probability statement:

The probability that the sample average is within 1.960 "standard deviations of the sample average" from the population mean is 0.95.

However, this statement involves measuring from the *unknown* population mean, $\mu$. To avoid this problem, you can measure from the sample average, which leads to the following equivalent probability statement:

The probability that the population mean is within 1.960 "standard deviations of the sample average" from sample average is 0.95.

This statement still involves an unknown population parameter, since the standard deviation of the sample average is $\sigma_{\overline{X}} = \sigma/\sqrt{n}$. Statistics often proceeds by substituting what you know (an estimate) for what you don't know. The statement will still be approximately correct if you substitute the standard error $S_{\overline{X}} = S/\sqrt{n}$, which represents your best information about the standard deviation of the sample

average. This leads to the following approximate probability statement:

The probability that the population mean is within 1.960 *standard errors* from the sample average is *approximately* 0.95.

Unfortunately, this is only an approximate probability statement. To make it exact, we will use the $t$ table, discovered by Student and published in 1908.[3] By using the critical value from the $t$ table in place of 1.960, we obtain an *exact* probability statement instead of an approximate one:

The probability that the population mean is within [critical value from the $t$ table] standard errors from the sample average is 0.95.

The price you pay for substituting a sample estimate (the standard error $S_{\overline{X}}$ estimated from the data) in place of an unknown population parameter ($\sigma_{\overline{X}}$) is that the critical value from the $t$ table will be larger than 1.960, giving a wider, less precise interval. When the sample size, $n$, is small, this value will be larger than 1.960. When $n$ is larger than 40, we will feel free to use 1.960 in hand calculation as an approximation, although computer programs often use the slightly larger, exact critical $t$ value.

To obtain a practical confidence interval from this probability statement, we need to change "probability 0.95" to "95% confidence." This final step is necessary because the confidence interval, in practice, is stated in terms of numbers instead of random variables (more on this in Section 9.3). The final confidence interval (see Figure 9.1.2) is as follows:

**Exact Confidence Interval Statement for a Population Mean**

We are 95% sure that the population mean is somewhere between the estimator minus $t$ standard errors and the estimator plus $t$ standard errors. That is, we are 95% sure that the population mean $\mu$ is somewhere between

$$\overline{X} - tS_{\overline{X}} \quad \text{and} \quad \overline{X} + tS_{\overline{X}}$$

*(Continued)*

---

2. You can verify from the normal probability tables that the probability is *exactly* 0.95 of being within 1.960... standard deviations.

3. "Student" is the name used by W. S. Gossett, who was an executive (Head Brewer) at Guinness. He invented this important technique to help in controlling and improving the brewing process.

**Exact Confidence Interval Statement for a Population Mean—cont'd**

where $t$ is taken from the $t$ table for two-sided 95% confidence. There is a 5% chance that the population mean is actually outside the confidence interval. This distance $tS_{\overline{X}}$ that the confidence interval extends in each direction is called the **margin of error** because it indicates how far away from the estimated mean $\overline{X}$ we could reasonably find the population mean $\mu$.

This reasoning also applies to the problem of estimating an unknown population percentage based on the percentage in a sample from a survey, for example. The reason is that the sample percentage $p$ is itself an average $\overline{X}$ where each data value is 0 or 1 depending on whether the feature being studied is present or not (and the population percentage $\pi$ is also equal to the population mean $\mu$). For example, the results of a small survey of five people asked "Do you like this color for the product?" would be 0, 1, 0, 0, 1 if only the second and last respondents like the color. Then $p = \overline{X} = 0.4$, or 40%. By the central limit theorem (which justifies the normal approximation to the binomial), if $n$ is large and $\pi$ is not too close to 0 or 1, $p$ will be approximately normally distributed and the confidence interval will be correct.

This is a binomial situation, in which you are estimating the unknown probability of occurrence, $\pi$, after having observed $X$ occurrences out of $n$ trials (with $n$ large). Remember from Chapter 8 that we use the sample proportion $p = X/n$ to estimate $\pi$ and that the standard error of $p$ is $S_p = \sqrt{p(1-p)/n}$. According to the confidence interval statement, when you use $\pi$ as the population parameter (in place of $\mu$), $p$ as the estimator (in place of $\overline{X}$), and $S_p$ as the standard error (in place of $S_{\overline{X}}$), your confidence interval (see Figure 9.1.3) is as follows:[4]

**Confidence Interval Statement for a Binomial Situation ($n$ Large)**

In particular, we are 95% sure that the population percentage $\pi$ is somewhere between

$$p - tS_p \quad \text{and} \quad p + tS_p$$

where $t$ is taken from the $t$ table.

In general, the width of a confidence interval is determined primarily by the sample size $n$ and the uncertainty

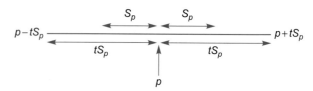

**FIGURE 9.1.3**   For a binomial proportion $\pi$, the 95% confidence interval extends $t$ (approximately 2) standard errors, $S_p$, on either side of the sample proportion $p$.

in the population. All else equal, if you have a larger sample size $n$, then the confidence interval will be smaller, indicating that there is less uncertainty when you have more information because the standard error involves dividing by $\sqrt{n}$. In addition, if there is less uncertainty in the sample, then the confidence interval will be smaller (this can happen if the standard deviation $S$ is smaller or, in the case of a binomial distribution, if the percentage $p$ is close to 0 or to 1).

## The $t$ Table and the $t$ Distribution

The $t$ **table**, presented here as Table 9.1.1, can be used in many situations to find the multiplier for the confidence interval. Initially, the headings "confidence level" and "two-sided" at the top concern us the most. For constructing an ordinary two-sided 95% confidence interval, which is the most common case, use the featured column in the table. The one-sided case will be covered later in this chapter, and hypothesis testing will be covered in the next chapter.

In statistics, the general concept of **degrees of freedom** represents the number of independent pieces of information in your standard error. For a single sample, the number of degrees of freedom is $n - 1$ (1 less than the number of observations) because the average is subtracted when the standard deviation is computed.[5] For example, with $n = 10$ observations, you have 9 degrees of freedom and would use 2.262 from the $t$ table to form an ordinary two-sided 95% confidence interval. If $\sigma_{\overline{X}}$ is known exactly, use the $t$ value 1.960 corresponding to an infinite number of degrees of freedom because you have perfect knowledge about the variability. When the sample size $n$ is more than 40, you may feel free to use the $t$ value (for example, 1.960 for 95% confidence) for an infinite sample size as an acceptable approximation (when you do this, your answer may be slightly different from direct computer calculations, which often use the exact $t$ value). Any $t$ value from the bottom

---

4. Note that because $n$ is large here, the $t$ table values will match those for normal probabilities. Although you may use the normal tables directly, we use $t$ for two reasons: (1) the estimated standard deviation $S_p$ is used in place of its true standard deviation, and (2) so that the procedures are similar for a sample average $\overline{X}$ and a sample proportion $p$, as they should be since $p$ and $\overline{X}$ are identical if we code each elementary unit's response as 0 or 1. The standard error computations lead to similar results because with this coding, $S_{\overline{X}} = S_p \sqrt{n/(n-1)}$.

5. After the average has been subtracted from the data values, one degree of freedom has indeed been lost because the resulting deviations add up to 0 and, therefore, only $n - 1$ deviations are free to vary because the last one must be equal to negative the sum of the others. Another way to see that information is lost is to note that, given only the residuals, we would have no idea of how large the average was.

## TABLE 9.1.1 The *t* Table

| Confidence Level | | | | | | | |
|---|---|---|---|---|---|---|---|
| Two-sided | 80% | **90%** | **95%** | 98% | **99%** | 99.8% | **99.9%** |
| One-sided | **90%** | **95%** | 97.5% | **99%** | 99.5% | **99.9%** | 99.95% |
| **Hypothesis Test Level** | | | | | | | |
| Two-sided | 0.20 | **0.10** | **0.05** | 0.02 | **0.01** | 0.002 | **0.001** |
| One-sided | **0.10** | **0.05** | 0.025 | **0.01** | 0.005 | **0.001** | 0.0005 |

| For One Sample | In General | | | | | | | |
|---|---|---|---|---|---|---|---|---|
| *n* | Degrees of Freedom | | | | | | | |
| | | | | | Critical Values | | | |
| 2 | 1 | 3.078 | 6.314 | 12.706 | 31.821 | 63.657 | 318.309 | 636.619 |
| 3 | 2 | 1.886 | 2.920 | 4.303 | 6.965 | 9.925 | 22.327 | 31.599 |
| 4 | 3 | 1.638 | 2.353 | 3.182 | 4.541 | 5.841 | 10.215 | 12.924 |
| 5 | 4 | 1.533 | 2.132 | 2.776 | 3.747 | 4.604 | 7.173 | 8.610 |
| 6 | 5 | 1.476 | 2.015 | 2.571 | 3.365 | 4.032 | 5.893 | 6.869 |
| 7 | 6 | 1.440 | 1.943 | 2.447 | 3.143 | 3.707 | 5.208 | 5.959 |
| 8 | 7 | 1.415 | 1.895 | 2.365 | 2.998 | 3.499 | 4.785 | 5.408 |
| 9 | 8 | 1.397 | 1.860 | 2.306 | 2.896 | 3.355 | 4.501 | 5.041 |
| 10 | 9 | 1.383 | 1.833 | 2.262 | 2.821 | 3.250 | 4.297 | 4.781 |
| 11 | 10 | 1.372 | 1.812 | 2.228 | 2.764 | 3.169 | 4.144 | 4.587 |
| 12 | 11 | 1.363 | 1.796 | 2.201 | 2.718 | 3.106 | 4.025 | 4.437 |
| 13 | 12 | 1.356 | 1.782 | 2.179 | 2.681 | 3.055 | 3.930 | 4.318 |
| 14 | 13 | 1.350 | 1.771 | 2.160 | 2.650 | 3.012 | 3.852 | 4.221 |
| 15 | 14 | 1.345 | 1.761 | 2.145 | 2.624 | 2.977 | 3.787 | 4.140 |
| 16 | 15 | 1.341 | 1.753 | 2.131 | 2.602 | 2.947 | 3.733 | 4.073 |
| 17 | 16 | 1.337 | 1.746 | 2.120 | 2.583 | 2.921 | 3.686 | 4.015 |
| 18 | 17 | 1.333 | 1.740 | 2.110 | 2.567 | 2.898 | 3.646 | 3.965 |
| 19 | 18 | 1.330 | 1.734 | 2.101 | 2.552 | 2.878 | 3.610 | 3.922 |
| 20 | 19 | 1.328 | 1.729 | 2.093 | 2.539 | 2.861 | 3.579 | 3.883 |
| 21 | 20 | 1.325 | 1.725 | 2.086 | 2.528 | 2.845 | 3.552 | 3.850 |
| 22 | 21 | 1.323 | 1.721 | 2.080 | 2.518 | 2.831 | 3.527 | 3.819 |
| 23 | 22 | 1.321 | 1.717 | 2.074 | 2.508 | 2.819 | 3.505 | 3.792 |
| 24 | 23 | 1.319 | 1.714 | 2.069 | 2.500 | 2.807 | 3.485 | 3.768 |
| 25 | 24 | 1.318 | 1.711 | 2.064 | 2.492 | 2.797 | 3.467 | 3.745 |
| 26 | 25 | 1.316 | 1.708 | 2.060 | 2.485 | 2.787 | 3.450 | 3.725 |
| 27 | 26 | 1.315 | 1.706 | 2.056 | 2.479 | 2.779 | 3.435 | 3.707 |
| 28 | 27 | 1.314 | 1.703 | 2.052 | 2.473 | 2.771 | 3.421 | 3.690 |
| 29 | 28 | 1.313 | 1.701 | 2.048 | 2.467 | 2.763 | 3.408 | 3.674 |

*(Continued)*

**TABLE 9.1.1  The *t* Table—cont'd**

| 30 | 29 | 1.311 | 1.699 | 2.045 | 2.462 | 2.756 | 3.396 | 3.659 |
|----|----|-------|-------|-------|-------|-------|-------|-------|
| 31 | 30 | 1.310 | 1.697 | 2.042 | 2.457 | 2.750 | 3.385 | 3.646 |
| 32 | 31 | 1.309 | 1.696 | 2.040 | 2.453 | 2.744 | 3.375 | 3.633 |
| 33 | 32 | 1.309 | 1.694 | 2.037 | 2.449 | 2.738 | 3.365 | 3.622 |
| 34 | 33 | 1.308 | 1.692 | 2.035 | 2.445 | 2.733 | 3.356 | 3.611 |
| 35 | 34 | 1.307 | 1.691 | 2.032 | 2.441 | 2.728 | 3.348 | 3.601 |
| 36 | 35 | 1.306 | 1.690 | 2.030 | 2.438 | 2.724 | 3.340 | 3.591 |
| 37 | 36 | 1.306 | 1.688 | 2.028 | 2.434 | 2.719 | 3.333 | 3.582 |
| 38 | 37 | 1.305 | 1.687 | 2.026 | 2.431 | 2.715 | 3.326 | 3.574 |
| 39 | 38 | 1.304 | 1.686 | 2.024 | 2.429 | 2.712 | 3.319 | 3.566 |
| 40 | 39 | 1.304 | 1.685 | 2.023 | 2.426 | 2.708 | 3.313 | 3.558 |
|    | Infinity | 1.282 | 1.645 | 1.960 | 2.326 | 2.576 | 3.090 | 3.291 |

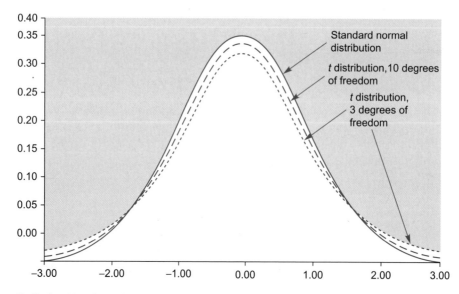

**FIGURE 9.1.4**   The *t* distribution. Note that as the sample size (and hence the degrees of freedom) gets larger, the shape is closer to a standard normal distribution. Because the *t* distribution has longer tails than the standard normal, you have to move farther out in the tails to capture 95% (or exclude 5%) of the probability. This is why the *t* table values are larger when the degrees of freedom are smaller.

of the table (for an infinite sample size) is often called a *z* value because it corresponds to probabilities for a standard normal distribution.

Where does the *t* table come from? Statisticians have defined the *t distribution* so that it matches the sampling distribution of $(\overline{X} - \mu)/S_{\overline{X}}$ when sampling from a normal distribution with mean $\mu$. (This ratio tells you how many standard errors, $S_{\overline{X}}$, the sample average $\overline{X}$ is above the population mean.) For a large sample size (with many degrees of freedom), the denominator is nearly the same as $\sigma_{\overline{X}}$ and the *t* distribution is nearly standard normal. This is why you find familiar numbers for the normal distribution (such as 1.960) at the bottom of the table. However, with smaller sample sizes, the distribution is not normal (see Figure 9.1.4). The effect of the denominator $S_{\overline{X}}$ is to spread out the *t* distribution and give it longer tails than the normal. This is why the numbers are larger at the top of the table than at the bottom.

## The Widely Used 95% Confidence Interval

Why are most confidence intervals computed at the 95% confidence level? One answer is that tradition has settled on this as a reasonable choice. The 95% level represents a compromise between trying to have as much confidence as possible and using a reasonably small interval.

The 100% confidence interval, unfortunately, is not very useful because it is too large. Consider the following exchange:

**The Boss:** Jones, how much do you think a typical consumer will be willing to spend for our new brand of toothpaste?
**Jones:** We estimate that the typical consumer will be willing to spend $2.45 per tube.
**The Boss:** How exact is that estimate? What do we really believe?
**Jones:** The analysis division is 100% sure that the typical consumer will be willing to spend between $0 and $35 million per tube.

Wait a minute! This is ridiculous. But the point is, to be 100% confident, you have to consider *every* remote, unrealistic possibility. When you back off from 100% to a confidence level that is still large but leaves some room for error, the result is a realistic and useful interval. Let's try it again:

**The Boss:** Jones, how much do you think a typical consumer will be willing to spend for our new brand of toothpaste?
**Jones:** We estimate that the typical consumer will be willing to spend $2.45 per tube.
**The Boss:** How exact is that estimate? What do we really believe?
**Jones:** The analysis division is 95% sure that the typical consumer will be willing to spend between $2.36 and $2.54 per tube.

Over the years, the 95% confidence level has emerged as a convenient, round number that is close but not too close to 100%. Other confidence levels are also used in practice, such as 90%, 99%, and even 99.9%; these will be considered after some examples.

### Example
#### Controlling the Average Thickness of Paper

The controls on the machinery in your paper factory have to be carefully adjusted so that the paper has the right thickness. Measurements of the thickness of selected sheets from an initial run of 0.004-inch paper are shown in Table 9.1.2.

Note that the average thickness was 0.004015 inch, which is about one-third of a percent larger than the 0.004 inch that these sheets are supposed to be. Although some variation from the ideal has to be tolerated in nearly any process, your immediate concern is to determine the state of the machinery. You do not believe for a minute that the average, 0.004015 inch, represents the machinery output perfectly.

**TABLE 9.1.2 Thicknesses of Selected Sheets of Paper (inches)**

|  |  |
| --- | --- |
|  | 0.00385 |
|  | 0.00358 |
|  | 0.00372 |
|  | 0.00418 |
|  | 0.00380 |
|  | 0.00399 |
|  | 0.00424 |
|  | 0.00375 |
|  | 0.00449 |
|  | 0.00422 |
|  | 0.00407 |
|  | 0.00434 |
|  | 0.00381 |
|  | 0.00421 |
|  | 0.00397 |
| Average | 0.0040146667 |
| Standard deviation | 0.0002614210 |
| Standard error | 0.0000674986 |
| $n$ | 15 |

The confidence interval will allow you to generalize from the 15 selected sheets you measured to the population, which, in this case, may be thought of as either the real population (all of the paper produced in the current run) or the idealized population (all of the paper the machine might produce under the current circumstances). This idealized population represents the current state of the machinery.

With a sample size $n = 15$, you have $n - 1 = 14$ degrees of freedom, for which the $t$ table value for a two-sided 95% confidence interval is 2.145. The confidence interval (assuming a normal distribution) extends from

$$\bar{X} - tS_{\bar{X}} = 0.0040146667 - (2.145)(0.0000674986)$$
$$= 0.00387$$

to

$$\bar{X} + tS_{\bar{X}} = 0.0040146667 + (2.145)(0.0000674986)$$
$$= 0.00416$$

The final confidence interval statement is therefore:

We are 95% sure that the machinery is currently producing paper with a mean thickness between 0.00387 inch and 0.00416 inch.

*(Continued)*

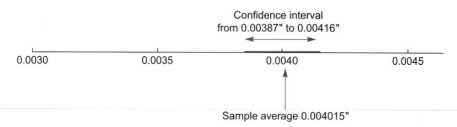

**FIGURE 9.1.5** The confidence interval for mean paper thickness, based on a sample of $n = 15$ sheets with $\overline{X} = 0.004015$ inch and $S_{\overline{X}} = 0.0000675$ inch.

**Example—cont'd**

This confidence interval is illustrated in Figure 9.1.5. Your end result is now an exact statement with a known level of confidence about the general state of your machinery (or about the larger supply of paper from which you sampled) based on a small amount of sample data.

What can you do if the statement is not precise enough and you want to pin it down to a smaller interval than 0.00387 to 0.00416? You would have to do something to decrease $S_{\overline{X}}$, your standard error.[6] There are two ways to accomplish this. First, by increasing the sample size $n$, you will decrease the standard error if all other factors are held the same (because of the denominator in $S_{\overline{X}} = S/\sqrt{n}$). Second, if you can decrease the variability in the production process by finding the causes of important sources of variation and correcting them, then your standard error will decrease, even with the same sample size (because of the numerator in $S_{\overline{X}} = S/\sqrt{n}$).

---

6. Although you can decrease $t$ somewhat, this is not likely to help very much unless your initial sample size was incredibly small.

Here is how you might use Excel® to find this confidence interval. First (if it is not yet named), give the data column the name "Thickness" by selecting the column of numbers and then choosing Define Name from Excel's Formula Ribbon. Next, use Excel's AVERAGE, STDEV, and COUNT functions to compute $\overline{X}$, $S$, and $n$, respectively, and name the cells so they can be easily used. The 95% confidence interval formula is then computed as $\overline{X} \pm tS/\sqrt{n}$ where we use Excel's TINV function to find the $t$ value.[7]

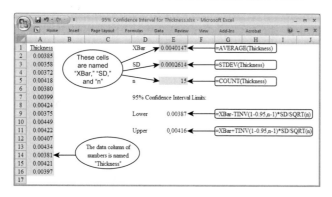

---

7. Excel's TINV function is shown using "1 − 0.95" because it needs "one minus the confidence level" instead of the confidence level itself. The term $n − 1$ is used because TINV needs the number of degrees of freedom.

**Example**
*Opinion Polling and Health Care Reform (A Binomial Situation)*

A major reorganization of health care in the United States, the Patient Protection and Affordable Care Act, was signed into law on March 23, 2010, by President Obama. Because health care is an important component of our economy—health expenditures represented 17.6% of total economic activity[8]—many people have strong opinions from various points of view, including as health care providers, as health care consumers, and as taxpayers. Some insights into these opinions are provided by Rasmussen Reports as summarized in Table 9.1.3 and Figure 9.1.6, indicating

**TABLE 9.1.3** Percentage of Voters Who Favor Repeal of the Health Care Law

| Poll Ending | Favor Repeal |
|---|---|
| July 11, 2010 | 53% |
| July 1, 2010 | 60 |
| June 26, 2010 | 52 |
| June 20, 2010 | 55 |
| June 12, 2010 | 58 |
| June 6, 2010 | 58 |
| May 29, 2010 | 60 |
| May 23, 2010 | 63 |
| May 15, 2010 | 56 |
| May 10, 2010 | 56 |
| May 1, 2010 | 54 |
| April 25, 2010 | 58 |
| April 17, 2010 | 56 |
| April 11, 2010 | 58 |
| April 3, 2010 | 54 |
| March 28, 2010 | 54 |
| March 24, 2010 | 55 |

Source: Rasmussen Reports, accessed at http://www.rasmussenreports.com/public_content/politics/current_events/healthcare/health_care_law on July 12, 2010.

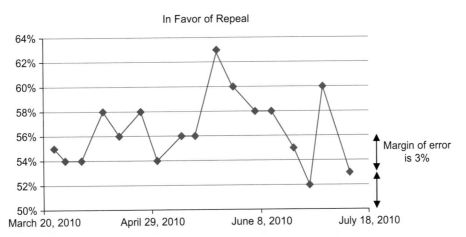

**FIGURE 9.1.6**   The percentage of voters who favor repeal of the health care law has risen and fallen over time since the bill was signed into law. However, some of these movements could be due to sampling error, as indicated by the margin of error (plus or minus 3 percentage points) as indicated at the right.

the percentage of voters who favor repeal of this law. While it looks as though this percentage has moved up and down over time, some of these movements reflect actual changes in opinion, whereas others may reflect the statistical noise due to sampling (note the margin of error, plus or minus 3 percentage points). In particular, how much should we believe in our ability to measure opinions by polling only a sample of the population? Statistics, probability, and confidence intervals help us understand the errors introduced when sampling is used because it is not possible to measure the entire population.

Polls like this one provide crucial information for those involved in politics and its effects on the economy, and help the rest of us feel informed about what is going on in the world. However, for various reasons, poll results are not as exact as they may seem. The way the question is asked can influence the results (e.g., "Should the new health law be repealed?" as compared to "Do you favor keeping the new health law?"). If there are several possible answers, the order in which questions are presented can make a difference. Sometimes, a question may be stated either positively or negatively (compare "Action would improve the quality of life: Should the government do it?" to "Action would require higher taxes: Should the government leave things as they are?"). The nature of any previous questions asked during the interview can also have an effect by "setting the stage" and giving the person positive or negative ideas. Finally, there are statistical sampling errors (perhaps the easiest to control and understand, using confidence intervals) that reflect the fact that a small sample cannot perfectly mirror the population being studied.

Along with the results and analysis, the report and methodology presenting these Rasmussen Reports polls included a look behind the scenes at some details of the design and analysis of such a nationwide poll (italics have been removed on the part that relates to confidence intervals):[9]

*Data for Rasmussen Reports survey research is collected using an automated polling methodology....For tracking surveys such as the Rasmussen Reports daily Presidential Tracking Poll or the Rasmussen Consumer Index, the automated technology insures*

*that every respondent hears exactly the same question, from the exact same voice, asked with the exact same inflection every single time....Calls are placed to randomly-selected phone numbers through a process that insures appropriate geographic representation....After the calls are completed, the raw data is processed through a weighting program to insure that the sample reflects the overall population in terms of age, race, gender, political party, and other factors. The processing step is required because different segments of the population answer the phone in different ways. For example, women answer the phone more than men, older people are home more and answer more than younger people, and rural residents typically answer the phone more frequently than urban residents.*

*The survey of 1,000 Likely Voters was conducted on July 10–11, 2010 by Rasmussen Reports. The margin of sampling error is +/– 3 percentage points with a 95% level of confidence.*

The first paragraph above gives general information about their methods: how they ensure consistency, that they use random sampling of phone numbers, and that they apply a weighting adjustment to control for some known sources of bias. The second paragraph gives the sample size $n = 1,000$ and the 3 percentage point margin of error for the 95% confidence interval. For example, the 95% confidence interval for the July 11 poll result of 53% would extend from 50% to 56%, which we find by simply adding and subtracting the margin of error: 53% ± 3%.

Now focus attention on the sample percentage, 53%, of likely voters who favored repeal in the July 11 poll. This is the exact sample percentage, but is only an estimate of the population percentage. The two-sided 95% confidence interval is computed by adding and subtracting the margin of error $tS_p$ for a binomial percentage. Using $t = 1.960$ from the $t$ table, we find

$$tS_p = t\sqrt{\frac{p(1-p)}{n}}$$

$$= 1.960\sqrt{\frac{0.53 \times (1-0.53)}{1,000}}$$

$$= 0.0309 \text{ or } 3.09\%$$

*(Continued)*

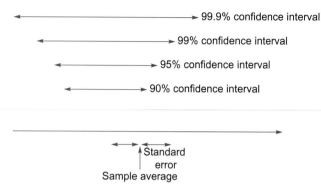

FIGURE 9.1.7   The relative sizes of 90%, 95%, 99%, and 99.9% confidence intervals from a large sample. The more confidence you wish, the larger the interval must be in order to cover your demands.

## Example—cont'd

As claimed, this margin of error, 3.09%, is indeed "3 percentage points" after rounding to the nearest percentage point. The 95% confidence interval is from

$$p - tS_p = 0.53 - 0.03 = 0.50 \quad \text{or} \quad 50\%$$

to

$$p + tS_p = 0.53 + 0.03 = 0.56 \quad \text{or} \quad 56\%$$

The final confidence interval statement is therefore as follows:

> We are 95% sure that, among all likely voters in the United States, some number between 50% and 56% would have said that they favor repeal at the time this poll was taken.

Your end result is now an exact confidence interval statement about all likely voters in the United States in July 2010, based on the exact results from a smaller sample. Or so it seems. If you want to be extra cautious in the inferences here, you may wish to redefine the population to be all likely voters in the United States who could have been reached by telephone during the time of the survey and who were willing to communicate their views. If you are very careful, you will note that there is no sampling frame that can be used to identify "likely voters," and you would look to see how it is decided which individuals, reached at random, will qualify. Rasmussen Reports indicates in its methodology section that

> *For political surveys, census bureau data provides a starting point and a series of screening questions are used to determine likely voters. The questions involve voting history, interest in the current campaign, and likely voting intentions.*

Will the health care law be repealed? Only time will tell. These opinion polls, with their confidence intervals to remind us of their lack of complete certainty, provide a guide to voter opinion and how these opinions change over time. A separate question (from whether or not voters favor repeal) is whether or not repeal is likely to happen and, of course, Rasmussen Reports has also done a poll on this topic. But don't forget that, while opinions play an important role in providing information for the political process, actual political events such as votes taken by the Congress are decided (according to our Constitution) based on voting by our representatives and not on opinion polls.

8. This was computed from health expenditures of $2.509 trillion and gross domestic product (GDP) of $14.256 trillion for 2009. Health expenditures are from U.S. Census Bureau, *Statistical Abstract of the United States: 2010* (129th edition), Washington, DC, 2009, Table 127, accessed from http://www.census.gov/compendia/statab/cats/health_nutrition.html. GDP is from U.S. Bureau of Economic Analysis, National Economic Accounts, accessed from http://www.bea.gov/national/#gdp on July 13, 2010.
9. Polling report accessed at http://www.rasmussenreports.com/public_content/politics/current_events/healthcare/health_care_law on July 12, 2010. Information about methodology accessed at http://www.rasmussenreports.com/public_content/about_us/methodology on July 13, 2010.

## Other Confidence Levels

Although the most commonly used confidence level is 95%, other confidence levels are appropriate for use in special situations. The basic principle here is a trade-off of the size of the interval (a smaller interval suggests more precision and is therefore better) against the probability of including the population parameter (a higher probability of being correct is also better). In some situations you may need so much precision that you are willing to widen the interval in order to be correct more often. In other situations you may have a stronger need for a smaller interval and be willing to be wrong more often in order to achieve it. The standard 95% confidence level represents a common trade-off of these two factors, but it is not the only reasonable choice.

There is a tendency to prefer round numbers for confidence levels (avoiding confusing statements such as "being 92.649% confident," for example). The *t* table presents values for constructing 90%, 95%, 99%, and 99.9% confidence intervals (in boldface at the top of Table 9.1.1) as well as a few other levels, which are listed primarily to help with one-sided intervals introduced in Section 9.4.

How much smaller is a confidence interval when you go for a lower confidence level? For a large sample, the relative sizes of confidence intervals are shown in Figure 9.1.7.

## Example
### *Average Selling Price as Determined through Rebates*

You probably know about rebates. They can look a lot like a discount when you buy a product, but you need to put together a bunch of paperwork (such as the sales receipt; the label, which is permanently glued to the product; and the promise of your first-born child—just kidding), spend some pocket change to mail the letter, and, finally, wait a while to receive a check for a dollar, which you have to cash at your bank.

One feature of rebates, from the manufacturer's point of view, is that they provide some useful information. Suppose your firm has a rebate program on a particular battery package with a list price of $2.99, and you would like to find out how much the public is *really* paying for these products after discounts and store sales. A carefully designed and randomized survey would provide good information about this, but it would not be justifiable on a cost basis just now. So you decide to analyze the sales receipts people have been sending in with their rebate requests.

First of all, what kind of sample is this? It's not random in any real sense of the word. Consumers who send in for rebates are not a representative cross-section of all consumers. For example, they might be better organized (so that they can keep track of everything and send it in) and poorer (so that the rebate money is worth the trouble to them) than the public at large. Nonetheless, you decide that you want to know about the population of consumers who are likely to send in rebate requests, and you decide to view the ones you receive as a random sample from this idealized population.

So far, $n = 15,603$ receipts have been received and are available for analysis. The summary values are

### Summary of Sales Receipts

| | |
|---|---|
| $n$ | 15,603 |
| Average sales price | $2.387 |
| Standard deviation | $0.318 |
| Standard error | $0.00255 |

Wow! Just look at the size of that standard error! It's so small because the sample is so large. Your estimate ($2.39) is *very* close to the population mean.

The 95% confidence interval, computed using $t = 1.960$, goes from $2.382 to $2.392. Evidently, the mean price in your idealized population is within about half a penny from the estimated $2.387.

With precision like this, very little will be lost by making a statement at a much higher confidence level. For example, let's use the highest confidence level in the table. To achieve 99.9% confidence, use $t = 3.291$ in place of 1.960 to compute the confidence interval, which extends from

$$\overline{X} - tS_{\overline{X}} = 2.387 - (3.291)(0.00255) = \$2.379$$

to

$$\overline{X} + tS_{\overline{X}} = 2.387 + (3.291)(0.00255) = \$2.395$$

The final confidence interval statement is therefore as follows:

We are 99.9% sure that the mean purchase price for your batteries by consumers motivated to send in rebate requests is somewhere between $2.379 and $2.395.

Compare this to the 95% confidence interval. Although the 99.9% confidence interval is slightly larger, it is still very close to the estimated mean price (about a penny away from $2.387). Because the level of variability is low in this case (as measured by the standard error), you can make a very exact statement that is correct with very high probability.

### Example
#### Yield of a Manufacturing Process

Now that you have brought a new chemical processing facility into production, top management wants to know the dependable long-term capabilities of the system. The processes are delicate, and no matter how carefully things are controlled, there is still some variation from day to day and even from hour to hour in the amount produced. Let's construct a confidence interval for the long-term yield (viewing this number as the population mean) based on measured yields for a sample of time periods.

Table 9.1.4 shows the raw data, consisting of 12 measurements of the yield of the facility, together with the usual summary measures. As you can see, there is much variability here.

Because the variability is so high, you are concerned that the confidence interval will be larger than you would like. You have talked things over with others at work, and it seems that a 90% confidence interval would be acceptable.

(Continued)

### TABLE 9.1.4 Yields of a Chemical Processing Facility (tons)

| | |
|---|---|
| | 71.7 |
| | 46.0 |
| | 103.9 |
| | 54.4 |
| | 43.3 |
| | 68.1 |
| | 73.4 |
| | 45.1 |
| | 45.6 |
| | 44.9 |
| | 77.8 |
| | 50.5 |
| Average | 60.3917 |
| Standard deviation | 18.7766 |
| Standard error | 5.4203 |
| $n$ | 12 |

**Example—cont'd**

The *t* value for a two-sided 90% confidence interval with *n* − 1 = 11 degrees of freedom is 1.796. The confidence interval therefore extends from

$$\overline{X} - tS_{\overline{X}} = 60.3917 - (1.796)(5.4203) = 50.7$$

to

$$\overline{X} + tS_{\overline{X}} = 60.3917 + (1.796)(5.4203) = 70.1$$

The final confidence interval statement is therefore as follows:

We are 90% sure that the mean long-term yield of this highly variable process is somewhere between 50.7 and 70.1 tons.

Compared with the 95% confidence interval, which extends from 48.5 to 72.3 tons, this 90% interval is only slightly shorter. The effect is not dramatic. By being wrong an additional 5% of the time, you have gained only slightly more precision as compared to the standard 95% interval.

## 9.2 ASSUMPTIONS NEEDED FOR VALIDITY

How can you be sure that your confidence levels are accurate? That is, when you claim 95% confidence, how can you be sure that the population mean will really be in the interval with 95% probability? Some technical assumptions are required in order for the statistical theory to apply to your particular case. If the assumptions apply, your confidence intervals will be correctly specified. If the assumptions do not apply to a situation, the confidence statement may be wrong.

When we say that the confidence interval statement is correct, what are we really saying? Suppose you have a procedure for constructing a 95% confidence interval. If the procedure is correct and you imagine repeating it many times (constructing many confidence intervals), you would find that approximately 95% of the confidence intervals include the population mean. This does not ensure that the population mean will definitely be in your interval, just that it is very likely to be there.

When we say that the confidence interval statement is *wrong*, in the case of incorrect assumptions, what are we really saying? Simply that the probability of including the population mean is *not necessarily* equal to the 95% level (or other confidence level) that you claimed. Your procedure might claim to have 95% confidence, but in reality it may have a much smaller confidence level, even as low as 50%, 10%, or smaller. Such a confidence interval is nearly worthless even though it might *appear* to be just fine. On the other hand, your confidence level could actually be *larger* than the 95% you claimed, if the assumptions

are not satisfied. Unfortunately, in some cases you don't know whether the true confidence level is higher or lower than your claimed 95%.

The two **assumptions required for the confidence interval** are (1) a random sample and (2) a normal distribution. These must both be satisfied for the confidence statement to be valid. We will consider each assumption in turn.

### Random Sampling

**Assumption 1 Required for the Confidence Interval**
The data set is a random sample from the population of interest.

The confidence interval is a statement about a population mean based on sample data. Naturally, there must be a strong relationship between your data and the population mean. A random sample ensures that your data represent the population and that each observation conveys new, independent information. Without a random sample, you would not be able to make exact probability statements about the results. If your sample consists only of your friends, for example, you cannot expect any confidence interval you compute to reflect a cross-section of all of society.

One interpretation of this assumption is that you must select a random sample from a carefully identified population frame, as discussed in Chapter 8. Certainly, the result of such efforts will satisfy the assumption. But this assumption is not as restrictive as it might seem; there is an alternative way to satisfy the random sampling assumption using an idealized population.

If you have some data and would like to construct a confidence interval, but the data do not really represent a deliberately chosen, random sample from a precisely specified population, you could try to construct an idealized population. Ask yourself what your data do represent. If you can identify a larger group and are willing to assume that your data are a lot like a random sample from this larger group, then you may legitimately construct a confidence interval to tell you about the unknown mean of this idealized population.[10]

For example, suppose you have some data on people who have recently come in to apply for employment. Strictly speaking, this group is *not* a random sample from any population because no randomization has been applied in their selection. It is not enough to observe that they look like a random sample or that they look like a diverse group.

---

10. However, if others disagree with you as to the identification of the idealized population, as the other side might in a lawsuit, then you have a problem. Since this is a conceptual problem, not a purely statistical problem, I can't help you.

The fact remains that, strictly speaking, they are not a random sample. However, if you are willing to view them as representatives of a larger population of people seeking employment and willing to take the time to try a firm like yours, then you may construct a confidence interval. This confidence interval goes beyond the particular people who applied for employment and tells you about others like them in your idealized population.

Here is an example to show how a confidence interval can fail if the data are not a random sample from the intended population.

### Example
#### Forecasting Interest Rates

Businesses depend on economic forecasts of future conditions as a way of dealing with uncertainty in the strategic planning process. These predictions are often viewed as the "best possible" information available. This may be so, yet how many of us know how dependable these forecasts really are? From time to time, the *Wall Street Journal* publishes past forecasts of selected economists together with the actual outcomes to see how well the predictions did the job.

A histogram of predictions of the long-term interest rate on 30-year U.S. Treasury bonds in the middle of 2001 (on June 29), as forecast six months in advance (about January 1, 2001) by 53 economists, is shown in Figure 9.2.1.[11] The two-sided 95% confidence interval based on these forecasts extends from 5.27% to 5.44%, which does not include the actual outcome of 5.70%. Were we just unlucky (in the sense that the 95% confidence interval will fail to cover the population mean 5% of the time), or is it unreasonable to expect the confidence interval to cover a situation like this? The answer is that we were not unlucky; in fact, the confidence interval is not being correctly interpreted in this situation because the data set was not sampled from the population it is being compared to.

Consider predictions of the short-term interest rate done at the same time as reported in the same source. The histogram of predictions of the short-term three-month U.S. treasury bill interest rate in mid-2001, as forecast six months in advance by 51 economists, is shown in Figure 9.2.2. The two-sided 95% confidence interval based on these forecasts extends from 5.25% to 5.46% and also does not include the actual outcome of 3.60%. In fact, the economists' predictions are disturbingly far from the actual outcome in this case.

Should you be troubled by the fact that a confidence interval does not include its intended number? Not necessarily; after all, the intervals are only guaranteed to be correct about 95% of the time. However, you should be surprised if the actual number is extremely far from the confidence interval. In this case (the short-term rates), the standard error is 0.054%, and the outcome of 3.60% (the intended number) is (5.36–3.60)/0.054 = 32.6 standard errors away from the sample average (5.36%). These 32.6 standard errors represent a very large distance and demand an explanation.

The explanation is simple. The random sampling assumption is not satisfied here; therefore, the confidence interval statement is not guaranteed to be correct. We should have been more careful and skeptical in using the economists' predictions as an indication of the future.

What *do* these economic predictions represent? About the best we can do with predictions made by a sample of economists is to view them as a random sample from the *idealized population* of similar forecasts by economists at the same time period. Our confidence interval, then, tells us about the mean consensus of this group of economists at this time. It does *not* tell us directly about the future interest rate because this future rate is not the mean of the population being sampled.

In between the forecasts and the actual outcome, there was a sudden, unforeseen downward shift in short-term interest rates. This is something that can and does happen

*(Continued)*

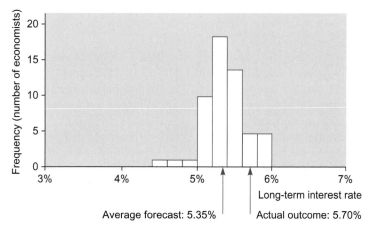

FIGURE 9.2.1  A histogram of six-month-ahead forecasts of the long-term interest rate on 30-year U.S. Treasury bonds, made around January 2001, compared to the actual outcome six months later on June 29, 2001. The two-sided 95% confidence interval about the average forecast does not include the actual outcome.

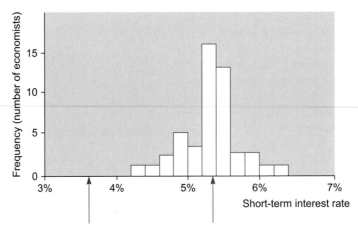

Actual outcome: 3.60%    Average forecast: 5.36%

**FIGURE 9.2.2**  A histogram of six-month-ahead forecasts of the short-term interest rate on six-month U.S. Treasury bills, made around January 2001, compared to the actual outcome six months later on June 29, 2001. The two-sided 95% confidence interval about the average forecast again does not include the actual outcome. The economists' consensus was not as accurate in this case.

**Example—cont'd**

in economics. However, in *statistics*, if your assumptions are satisfied, the correctness of your statements is guaranteed.

It often helps to clarify your thinking by separating the subject matter (in this case, economics) from the statistical principles. Then the best you can do is make a limited, exact statistical statement and interpret it according to the accepted ways of the subject matter.

___

11. Data are from the *Wall Street Journal* "Mid-Year Forecasting Survey," dated July 2, 2001, accessed on the *Wall Street Journal Interactive Edition* at http://interactive.wsj.com/documents/forecast-2001-07-02.htm on July 10, 2001.

## Normal Distribution

**Assumption 2 Required for the Confidence Interval**

The quantity being measured is normally distributed.

The detailed theory behind the confidence interval is based on the assumption that the quantity being measured is normally distributed *in the population*. Such a simplifying assumption makes it possible to work out all of the equations and compute the *t* table (which has already been done for you). Fortunately, in practice this requirement is much less rigid for two reasons.

First of all, you could never really tell whether or not the population is perfectly normal, since all you have is the sample with its randomness. In practice, therefore, you would look at a histogram of the data to see if the distribution is *approximately* normal, that is, not too skewed and with no extreme outliers.

Second, the central limit theorem often comes to the rescue. Since statistical inference is based primarily on the sample average, $\overline{X}$, what you need primarily is that the sampling distribution of $\overline{X}$ be approximately normal. The central limit theorem tells you that if *n* is large, $\overline{X}$ will be approximately normally distributed even if the individuals in the population (and the sample) are not.

Thus, the practical rule here may be summarized as follows:

**Assumption 2 (in Practice)**

Look at a histogram of the data. If it looks approximately normal, then you're OK (i.e., the confidence interval statement is approximately valid). If the histogram is slightly skewed, then you're OK provided the sample size is not too small. If the histogram is moderately skewed or has very few moderate outliers, then you're OK provided the sample size is large. If the histogram is extremely skewed or has extreme outliers, then you may be in trouble.

For a binomial situation, the central limit theorem implies that the sample percentage *p* is approximately normally distributed when *n* is large (provided the population percentage is not too close to 0% or 100%). This shows how the assumption of a normal distribution can be (approximately) satisfied for a binomial situation.

What can you do if the normal distribution assumption is not satisfied at all, due, say, to extreme skewness? One approach is to transform the data (perhaps with logarithms) to bring about a normal distribution; keep in mind that the resulting confidence interval would then be for the mean of the population *logarithm* values. Another possibility is to use *nonparametric methods*, to be described in Chapter 16.

### Example

#### Data Mining to Understand the Average Donation Amount

Consider the donations database with 20,000 entries on the companion site. The total amount given by these 20,000 people in response to the current mailing was $15,592.07, with 989 making a current donation and 19,011 not donating at this time. Thus, the average donation is $0.7796035, or about 78 cents per person. Certainly, the amount donated will vary according to the circumstances of a particular mailing. One source of variation is pure statistical variation, leading to the following question: If we were to send a mailing to a similar (but much larger) group of people that these 20,000 people represent (viewing these 20,000 as a random sample from the larger group), how much, on average, should we expect to receive from each person in the new mailing? An answer may be found using the confidence interval.

The standard deviation of the 20,000 donations is $4.2916438, and the standard error is $0.0303465, leading to a 95% confidence interval extending from $0.720122 to $0.839085. If we plan a new mailing to 500,000 people, then we would expect to receive donations totaling between

$360,061 and $419,543 (obtained by multiplying the ends of the confidence interval by 500,000 people).

What about the assumptions for validity of this confidence interval from about 72 to 84 cents for the population mean donation amount? The first assumption requires that the data be a random sample from the population of interest, and this would be true (for example) if the 20,000 were initially selected randomly from the 500,000 as part of a pilot study to see if it would be worthwhile mailing to all 500,000 at this time.[12] The second assumption requires that the quantity being measured be normally distributed; this assumption does not appear to be satisfied, as is seen from the very nonnormal histogram for the 20,000 donation amounts in Figure 9.2.3. However, the confidence interval is OK in this case, even though the distribution of individual donations is very skewed, because the sample size is large enough to make the distribution of "averages of 20,000 donations" approximately normal. To show that the distribution of "averages of 20,000 donations" is normally distributed, Figure 9.2.4 shows a histogram of 500 "bootstrap

*(Continued)*

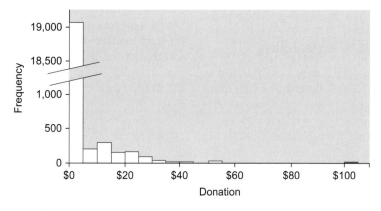

**FIGURE 9.2.3**  A histogram of the 20,000 individual donation amounts shows a highly skewed and very nonnormal distribution. However, assumption 2 for validity of the confidence interval may still be satisfied because the sample average might be approximately normal.

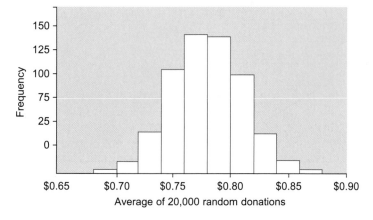

**FIGURE 9.2.4**  A histogram of *averages of 20,000 donations* shows that the average of 20,000 donations is very nearly normally distributed (due to the central limit theorem) even though individual donation amounts are highly skewed. In this case, 500 averages are shown, with each average chosen by random sampling (with replacement, according to the bootstrap technique) from the database of 20,000 donation amounts.

**Example—cont'd**

samples" with each bootstrap sample of size 20,000 chosen by sampling with replacement from the database of 20,000 donation amounts. Even though the individual donation amounts are highly skewed, the averages of 20,000 donations are actually very close to a normal distribution because of the central limit theorem.

12. Even if a random sample had not been chosen, it might still be instructive to consider the purely statistical variation in this average amount, as represented by the confidence interval.

## 9.3 INTERPRETING A CONFIDENCE INTERVAL

What are you really communicating when you say that, based on weights from a sample of the day's production, you are 95% sure that the mean weight of all soap boxes produced today is between 15.93 and 16.28 ounces? It looks like a probability statement, but it must be interpreted carefully. The mean weight of all soap boxes produced today is some fixed, unknown number. It is either in the interval, or it isn't. In this light, where does the probability come from?

### Which Event Has a 95% Probability?

In order for there to be a probability, there must be a random experiment. The probability refers to the entire *process* rather than just to the particular result. By saying that you are 95% sure that the population mean weight is between 15.93 and 16.28 ounces, you are making a statement about the exact numerical results based on the data. However, the 95%

probability comes from the process itself, which views the numbers as *random*. A careful probability statement might be: "There is a 95% probability for the event 'the population mean weight is within the confidence interval' for the random experiment 'randomly choose some boxes and compute the confidence interval.'" Each time you collect data and compute a 95% confidence interval, you are performing a random experiment that has a probability for every event. The probability that the unknown population mean falls within a computed interval is 0.95.

This subtlety is partly a question of timing of information. You might reasonably say there is a 55% chance that a stock market index will go up tomorrow. However, when tomorrow afternoon comes along and you see that the market did indeed go up, there is no remaining uncertainty or probability, the market did go up. Yet there *was* uncertainty before the fact. The one difference between this stock market example and the usual confidence interval statement is that when you compute a confidence interval you either include the population mean within the confidence interval or you don't, yet you may never know whether or not you did!

One useful way to interpret the 95% probability is to imagine repeating the sampling process over and over to obtain multiple confidence intervals, each one based on a different random sample. The notion of relative frequency and the law of large numbers (from Chapter 6) tell you that about 95% of these random, known intervals include the fixed, unknown population mean. This is illustrated in Figure 9.3.1. Note that each sample has its own average, $\overline{X}$, so some intervals are shifted to the right or left with respect to the others. Also, each one has its own standard error, $S_{\overline{X}}$, so some intervals are larger or smaller than others.

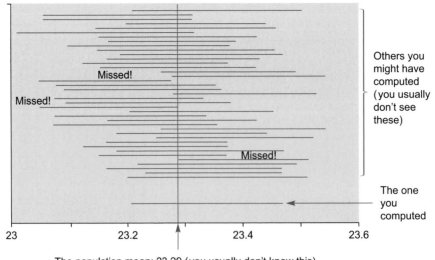

The population mean: 23.29 (you usually don't know this)

**FIGURE 9.3.1** What if you had used a different random sample? This figure shows how different the resulting confidence intervals can be from one random sample to another (independently chosen) random sample from the same population. Over the long run, 95% of these confidence intervals will include the unknown mean, provided the assumptions are satisfied.

Note that the confidence intervals that "missed" the population mean were still fairly close.

## Your Lifetime Track Record

Of course, you ordinarily compute just *one* confidence interval for the population mean in a given situation. However, since many such studies will be independent of each other (i.e., the sampling done for each study will ordinarily be chosen independently), you can interpret the meaning of "95% confidence" in terms of your lifetime track record. If you compute many 95% confidence intervals over your lifetime, and if the required assumptions are satisfied for each one, then approximately 95% of these confidence intervals will contain their respective population means.

Looking back over your life from the golf course at the retirement home, you get that satisfying feeling that 95% of the time your confidence intervals were correct. Unfortunately, you also get that sinking feeling that 5% of them were wrong. And, to top things off, you may *never know* which cases were right and which were wrong! Such are the ways of statistical inference.

## 9.4 ONE-SIDED CONFIDENCE INTERVALS

In some cases it may not be necessary to specify that the population mean is probably *between* two confidence interval numbers. It may suffice to say that the population mean is *at least as large as* some number or (in other situations) to say that the population mean is *no larger than* some number. A **one-sided confidence interval** states with known confidence that the population mean is either *at least* or *no larger than* some computed number, depending on which side is relevant to your needs. If you are careful, constructing a one-sided confidence interval can provide a more effective statement than use of a two-sided interval would.

For example, you may be interested only in something being *big enough:* We are 95% sure that sales will be at least $560,000. Or you might be interested only in something being *small enough:* We are 95% sure that our defect rate is no larger than 1 in 10,000 units produced. Situations like these can benefit from a one-sided confidence interval statement.

## Be Careful! You Can't Always Use a One-Sided Interval

There is one important criterion you must satisfy to use a one-sided confidence interval:

**Criterion for Using a One-Sided Confidence Interval**

In order to use a one-sided interval, you must be sure that *no matter how the data had come out,* you would still have used a one-sided interval on the same side ("at least" or "no larger than"). If, had the data come out differently, you might have used a one-sided interval *on the other side,* you should use a two-sided confidence interval instead. If in doubt, use a two-sided interval.

Suppose your break-even cost is $18 per item produced, you have the basic data from a sample, and you are ready to compute a confidence interval. You might be tempted to proceed as follows: If the estimated cost is *high* (more than $18), you will state that you are 95% sure that costs are *at least...,* but if the estimated cost is *low* (less than $18), you will state instead that you are 95% sure that costs are *no more than....* Don't be tempted! Because switching the side of the interval based on the data is not allowed (by the preceding criterion), you should compute a two-sided interval instead (you are 95% sure that costs are between ... and ...). There are two good reasons for this. First of all, you are interested in both sides—sometimes one, sometimes the other. Second, switching sides of a one-sided confidence interval can invalidate your probability statement so that your true confidence level might be much lower than the 95% claimed.[13]

## Computing the One-Sided Interval

To compute a one-sided interval, first find the *t* value from Table 9.1.1 using the "one-sided" confidence level heading at the top. (The row is the same as for a two-sided interval since the number of degrees of freedom is still $n - 1$.) For example, to compute a 95% one-sided confidence interval with a sample size of $n = 23$, you would use $t = 1.717$. For a 99.9% one-sided confidence interval with $n = 35$, you would use $t = 3.348$.

Next, choose *one* of the following one-sided confidence interval statement types:

We are 95% sure that the population mean is *at least as large* as $\overline{X} - t_{\text{one-sided}} S_{\overline{X}}$

or

We are 95% sure that the population mean is *not larger than* $\overline{X} + t_{\text{one-sided}} S_{\overline{X}}$

An easy way to remember whether to add or subtract is to be sure that the sample average, $\overline{X}$, is included in your one-sided confidence interval. (It should be, after all, since it is your best estimate of the population mean.)

---

13. In the worst case of switching, you might end up with a 90% one-sided confidence interval when you are claiming 95% confidence. This happens if you switch sides according to whether $\overline{X}$ is above or below $\mu$ because you suffer from the 5% errors of *both* intervals, adding up to a total error rate of 10% instead of the 5% you thought you had.

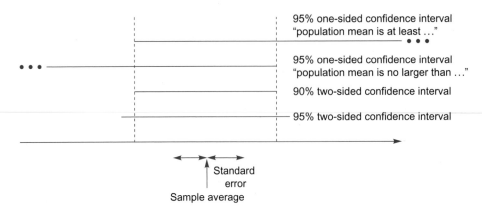

**FIGURE 9.4.1** Both kinds of one-sided confidence intervals are illustrated at the top. One-sided confidence intervals always include the sample average, starting from a point on one side and continuing indefinitely on the other. Note that the endpoint for the 95% *one-sided* confidence interval is the same as one of the endpoints for the 90% *two-sided* confidence interval.

Thus, when the one-sided interval extends upward to larger values ("at least"), it must start *below* the sample average; and when the one-sided interval extends downward to smaller values ("no larger than"), it must start *above* the sample average.

Figure 9.4.1 illustrates this and also compares one- and two-sided intervals. The beginning point of a *one-sided 95%* confidence interval is the same as one of the endpoints of a *two-sided 90%* confidence interval. The idea here is that there are two ways in which a two-sided interval might be wrong: Either the population mean is too big, or else it is too small. A one-sided interval sharing an endpoint with a two-sided interval can be wrong only half as often.

The one-sided confidence interval allows you to concentrate your attention on the most interesting cases. If you care only about errors on one side and don't care at all about errors on the other side, then the one-sided interval can begin *closer to the sample average* (and will therefore seem more precise) than a two-sided confidence interval. For example, for a large sample with an average of 19.0 and a standard error of 8.26, rather than saying it's between 2.81 and 35.2, you could say that it's at least 5.41. Knowing that it's at least 5.41 provides more information than knowing it's at least 2.81. You can claim a stronger lower bound because you are not claiming any upper bound at all.

**Example**

*The Savings of a New System*

You are evaluating a new automated production system and have decided to buy it if it can be demonstrated to save enough money per item produced. You have arranged for it to be installed on the premises so that you can try it out for a week. It will be programmed to produce a cross-section of typical products, and the cost savings will be determined for each item produced.

What is the population here? It is an idealized population of all of the items the system *might* produce under conditions similar to the ones you tested under. Statistical inference can help you here by extending your information from the particular items you did produce to the mean of the much larger group of items that you might produce in the indefinite future under similar conditions.

Should a one-sided confidence interval be used here? Yes, because regardless of how the data come out, you are interested only in whether you will save enough money. Your final statement will be of the form: We are 95% sure that the mean cost savings per item produced over the long run will be *at least*....

For a sample size of $n = 18$ items produced, with an average savings $\overline{X} = \$39.21$ and a standard error $S_{\overline{X}} = \$6.40$, the 95% one-sided confidence interval will extend indefinitely to larger values starting from

$$\overline{X} - t_{\text{one-sided}}S_{\overline{X}} = 39.21 - (1.740)(6.40) = 28.07$$

Therefore, your final one-sided confidence statement is

We are 95% sure that the mean cost savings are at least $28.07 per item produced.

Note that the one-sided confidence interval includes the sample average $\overline{X} = \$39.21$, as it must. That is, the sample average of $39.21 satisfies the confidence interval statement by being at least as large as $28.07. It would have been wrong to have used the other endpoint. By using this way of checking as a guide, you will always make the correct one-sided statement.

To compute a one-sided confidence interval at a different level, simply substitute the appropriate $t$ value from the table. For example, the 99% one-sided confidence interval statement uses $t = 2.567$. Compared to the 95% interval statement, this is a weaker statement about cost savings that you are more confident about:

We are 99% sure that the mean cost savings are at least $22.78 per item produced.

## Example
### Travel Costs

In an effort to prepare a realistic travel budget, you have examined the costs of typical trips made in the recent past. In an effort to ensure that the budget will cover the demands of the coming year, you would like to arrive at a maximum dollar figure for mean cost per trip. This allows you to say that the mean cost *is no more than* this figure. Since you are interested only in this one side, you may use a one-sided confidence interval. You choose the 95% level of confidence.

Working from a list of 83 recent trips, you find that the average cost was $1,286 with a standard error of $71.03. The one-sided confidence interval will include all values from $0 (since you know the cost can't be negative) to the upper bound,

$$\overline{X} + t_{\text{one-sided}}S_{\overline{X}} = 1,286 + (1.645)(71.03) = \$1,403$$

Here is your final one-sided confidence statement:

We are 95% sure that the mean travel expense per trip is no larger than $1,403.

Checking to make sure that adding (instead of subtracting) is correct here, you note that the sample average ($1,286) is indeed within the confidence interval ($1,286 is no larger than $1,403).

This confidence interval is of limited use because the data are not really a random sample from the population of interest. You would like to predict *future* travel costs, but your sample data are from the past. The confidence interval takes past variability in travel costs into account, which is useful information for you. However, it does not (and cannot) take into account future trends in travel costs.

## 9.5  PREDICTION INTERVALS

The confidence interval tells you where the *population mean* is, with known probability. This is fine if you are seeking a summary measure for a large population. If, on the other hand, you want to know about the observed value for an *individual case*, this confidence interval is not appropriate. Instead, you need a much wider interval that reflects not just the estimated uncertainty $S_{\overline{X}} = S/\sqrt{n}$ of $\overline{X}$ (which may be very small when $n$ is large) but also the estimated uncertainty $S$ of an individual observation.

The **prediction interval** allows you to use data from a sample to predict a new observation with known probability, provided you obtain this additional observation in the same way as you obtained your past data. The situation is as follows: You have a random sample of $n$ units from a population and have measured each one to obtain $X_1$, $X_2, \ldots, X_n$. You would now like to make a prediction about an *additional* unit randomly selected from the same population.

The uncertainty measure to use here is the **standard error for prediction**, a measure of variability of the distance between the sample average and the new observation. Two kinds of randomness are combined: for the sample average and for the new observation. This standard error for prediction is found by multiplying the standard deviation by the square root of $(1 + 1/n)$:

### Standard Error for Prediction

$$S\sqrt{1 + \frac{1}{n}}$$

The standard error for prediction is even larger than the estimator $S$ of the variability of individuals in the population. This is appropriate because the prediction interval must combine the uncertainty of individuals in the population (as measured by $S$) together with the uncertainty of the sample average (as measured by $S_{\overline{X}} = S/\sqrt{n}$).

Once you have an estimator ($\overline{X}$) and the standard error for prediction, you can form the prediction interval in much the same way as you form an ordinary confidence interval. The $t$ value is found in the table in just the same way for a given prediction confidence level and sample size $n$ (not including the additional observation, of course). Only the standard error is different; be sure to use the standard error for prediction in place of the standard error of the average.

### The Prediction Interval for a New Observation
**Two-sided**
We are 95% sure that the new observation will be between

$$\overline{X} - tS\sqrt{1 + 1/n} \quad \text{and} \quad \overline{X} + tS\sqrt{1 + 1/n}$$

**One-sided**
We are 95% sure that the new observation will be at least

$$\overline{X} - t_{\text{one-sided}}S\sqrt{1 + 1/n}$$

or

We are 95% sure that the new observation will be no larger than

$$\overline{X} + t_{\text{one-sided}}S\sqrt{1 + 1/n}$$

What does the figure 95% signify here? It is a probability according to the following random experiment: Get a random sample, find the prediction interval, get a new random observation, and see if the new observation falls in the interval. Note in particular that the 95% probability refers to drawing a new *sample* as well as a new observation. This is only natural; since one sample differs from another, the proportion of new observations that falls within the prediction interval will also vary from one sample to another.

Averaged over the randomness of the initial sample, the resulting probability is 95% (or some other specified confidence level).

The following table summarizes when to use a prediction interval instead of a confidence interval.

| When You Need to Learn About | Use |
| --- | --- |
| The population mean | Confidence interval |
| A new observation like the others | Prediction interval |

---

### Example
#### How Long until Your Order Is Filled?

How long should you wait before ordering new supplies for production inventory? If you order too soon, you pay interest on the capital used to buy them while they sit around costing you rent for the warehouse space they occupy. If you order too late, then you risk being without necessary parts and bringing part of the production line to a halt.

The past eight times that your supplier has said, "They'll be there in two weeks," you made a note of how many business days it actually took for them to arrive. These numbers were as follows:

$$10, 9, 7, 10, 3, 9, 12, 5$$

The average is $\overline{X} = 8.125$ days, and the standard deviation is $S = 2.94897$ days. The standard error of the average is $S_{\overline{X}} = 1.04262$ days, but we do not need it. The standard error for prediction is

$$
\begin{aligned}
\text{Standard error for prediction} &= S\sqrt{1 + 1/n} \\
&= 2.94897\sqrt{1 + 1/8} \\
&= 2.94897\sqrt{1.125} \\
&= 3.12786
\end{aligned}
$$

For a two-sided 95% prediction interval, the $t$ value from the table for $n = 8$ is $t = 2.365$. The prediction interval extends from

$$
\begin{aligned}
\overline{X} - t(\text{Standard error for prediction}) &= 8.125 - (2.365)(3.12786) \\
&= 0.728
\end{aligned}
$$

to

$$
\begin{aligned}
\overline{X} + t(\text{Standard error for prediction}) &= 8.125 + (2.365)(3.12786) \\
&= 15.52
\end{aligned}
$$

You will be assuming that the delivery times are approximately normally distributed, that the $n = 8$ delivery times observed represent a random sample from the idealized population of "typical delivery times," and that the next delivery time is randomly selected from this same population. The final prediction interval statement is as follows:

> We are 95% sure that the next delivery time will be somewhere between 0.7 and 15.5 days.

Why does this prediction interval extend over such a large range? This reflects the underlying uncertainty of the

situation. In the past, based on your eight observations, the delivery times have been quite variable. Naturally, this makes exact predictions difficult.

If you merely want to be assured that the next delivery time will not be *too late*, you may construct a one-sided prediction interval using $t = 1.895$ from the one-sided 95% confidence column in the $t$ table. The upper limit is then

$$
\begin{aligned}
\overline{X} + t(\text{Standard error for prediction}) &= 8.125 + (1.895)(3.12786) \\
&= 14.1
\end{aligned}
$$

You may then make the following one-sided prediction interval statement:

> We are 95% sure that the next delivery time will be no more than 14.1 days.

If you are willing to accept a 90% one-sided prediction interval, then the upper limit (using $t = 1.415$) would be

$$
\begin{aligned}
\overline{X} + t(\text{Standard error for prediction}) &= 8.125 + (1.415)(3.12786) \\
&= 12.6
\end{aligned}
$$

You would then make the following one-sided prediction interval statement:

> We are 90% sure that the next delivery time will be no more than 12.6 days.

---

## 9.6 END-OF-CHAPTER MATERIALS

### Summary

The process of generalizing from sample data to make probability-based statements about the population is called **statistical inference**. A **confidence interval** is an interval computed from the data in such a way that there is a *known probability* of including the (unknown) population parameter of interest, where this probability is interpreted with respect to a random experiment that begins with the selection of a random sample. The probability that the population parameter is included within the confidence interval is called the **confidence level**, which is set by tradition at 95%, although levels of 90%, 99%, and 99.9% are also commonly used. The higher the confidence level, the larger (and usually less useful) the confidence interval. The approximate all-purpose confidence interval statement goes as follows:

> We are 95% sure that the population parameter is somewhere between the estimator *minus* two of the estimator's standard errors and the estimator *plus* two of its standard errors.

This is a restatement of the fact that, for a normal distribution, you expect to be within 1.960 (approximately 2) standard deviations from the mean with probability 0.95.

The two-sided 95% confidence interval statement for the population mean goes as follows:

We are 95% sure that the population mean, $\mu$, is somewhere between $\overline{X} - tS_{\overline{X}}$ and $\overline{X} + tS_{\overline{X}}$ where $t$ is taken from the $t$ table.

For a binomial situation ($n$ large), this leads to the following interval:

We are 95% sure that $\pi$ is somewhere between $p - tS_p$ and $p + tS_p$, where $t$ is taken from the $t$ table.

To achieve a confidence level other than 95%, simply substitute the appropriate $t$ value in the interval statement. The *t* **table** is used to adjust for the added uncertainty due to the fact that an estimator (the standard error) is being used in place of the unknown exact variability for the population. When you work with a single sample of size $n$, your **degrees of freedom** number is $n - 1$, which represents the number of independent pieces of information in your standard error (because the average is subtracted when the standard deviation is computed). If the standard error is known exactly, use the $t$ value for an infinite number of degrees of freedom.

The two **assumptions required for the confidence interval** statement to be valid are (1) the data are a random sample from the population of interest, and (2) the quantity being measured is normally distributed. The first assumption ensures that the data properly represent the unknown parameter, and the second assumption forms the basis for the probability calculations underlying the $t$ table. In practice, because the confidence interval is based largely on the sample average, $\overline{X}$, the central limit theorem allows you to relax assumption 2 so that even for a moderately skewed distribution, the assumption will be satisfied provided the sample size is large enough.

The reason we say 95% sure or 95% confident is that once the numbers have been computed for the confidence interval, they are not random anymore; the event that has probability 0.95 must include the randomness of the sampling process. The relative frequency interpretation is that if you were to repeat the sampling process over and over, computing a confidence interval each time, about 95% of the random, known intervals would include the fixed, unknown population mean. Similarly, your lifetime track record for confidence intervals computed under correct assumptions should include about 95% successes (that is, intervals containing the unknown parameter) and about 5% mistakes. However, you will not generally know which ones were right and which were wrong!

A **one-sided confidence interval** specifies with known confidence that the population mean is either *at least* or *no larger than* a computed number. You compute the endpoint of the one-sided confidence interval in the same way as for the two-sided interval, except for substituting the one-sided

$t$ value for the two-sided value and choosing the endpoint so that your one-sided interval includes the sample average, $\overline{X}$. To use a one-sided interval, you must be sure that *no matter how the data had come out* you would still have used a one-sided interval on the same side (above or below). Otherwise, your confidence interval statement may not be valid. If in doubt, use a two-sided interval. The one-sided confidence interval statements take the following form:

We are 95% sure that the population mean is *at least as large as* $\overline{X} - t_{\text{one-sided}}S_{\overline{X}}$.

or

We are 95% sure that the population mean is *no larger than* $\overline{X} + t_{\text{one-sided}}S_{\overline{X}}$.

The **prediction interval** allows you to use data from a sample to predict a new observation with known probability, provided you obtain this additional observation in the same way as you obtained your data. The uncertainty measure to use is the **standard error for prediction**, $S\sqrt{1 + 1/n}$, a measure of variability of the distance between the sample average and the new observation. The prediction interval is then constructed in the same way as a confidence interval; simply substitute the standard error for prediction for the standard error of the average. The prediction interval formula for a new observation (two-sided) is

We are 95% sure that the new observation will be between $\overline{X} - tS\sqrt{1 + 1/n}$ and $\overline{X} + tS\sqrt{1 + 1/n}$.

The prediction intervals for a new observation (one-sided) are

We are 95% sure that the new observation will be at least $\overline{X} - tS_{\text{one-sided}}\sqrt{1 + 1/n}$.

or

We are 95% sure that the new observation will be no larger than $\overline{X} + tS_{\text{one-sided}}\sqrt{1 + 1/n}$.

Prediction intervals at levels other than 95% are available by selecting the appropriate $t$ value from the table. Remember that the confidence interval tells you about the population mean, whereas the prediction interval tells you about a new, single observation selected at random from the same population.

## Key Words

## Questions

1. In what important way does statistical inference go beyond summarizing the data?

2. What does a confidence interval tell you about the population that an estimated value alone does not?

3. Which fact about a normal distribution leads to the factor 2 (or 1.960) in the generic confidence interval statement?

4. Why is it correct to say, "We are 95% sure that the population mean is between $15.85 and $19.36" but not proper to say, "The probability is 0.95 that the population mean is between $15.85 and $19.36"?

5. Why does the *t* table give numbers larger than 1.960 for a two-sided 95% confidence interval?

6. **a.** How many degrees of freedom are there for a single sample of size *n*?

   **b.** What accounts for the degree of freedom lost?

   **c.** How many degrees of freedom should you use if the standard error is known exactly?

7. **a.** What confidence levels other than 95% are in common use?

   **b.** What would you do differently to compute a 99% confidence interval instead of a 95% interval?

   **c.** Which is larger, a two-sided 90% confidence interval or a two-sided 95% confidence interval?

8. **a.** Describe the two assumptions needed for the confidence interval statement to be valid.

   **b.** For each assumption, give an example of what could go wrong if it weren't satisfied.

   **c.** How does the central limit theorem help satisfy one of these assumptions?

   **d.** Under what circumstances would the central limit theorem not guarantee that the second assumption is satisfied?

9. **a.** What is the relative frequency interpretation of the correctness of a confidence interval?

   **b.** What is the "lifetime track record" interpretation of the correctness of many confidence intervals?

10. **a.** Why must a one-sided confidence interval always include the sample average?

    **b.** Must a one-sided confidence interval always include the population mean?

11. **a.** What additional criterion must be satisfied for a one-sided confidence interval to be valid (in addition to the two assumptions needed for a two-sided confidence interval)?

    **b.** If in doubt, should you use a one-sided or a two-sided confidence interval?

12. **a.** What is the difference between a prediction interval and a confidence interval?

    **b.** Which type of interval should you use to learn about the mean spending habits of your typical customer?

    **c.** Which type of interval should you use to learn about the spending habits of an individual customer?

13. **a.** What is the standard error for prediction?

    **b.** Why is the standard error for prediction even larger than the standard deviation *S*?

14. **a.** What would you change in the computation of a two-sided 95% prediction interval to find a two-sided 99% prediction interval instead?

    **b.** What would you change in the computation of a two-sided 95% prediction interval to find a one-sided 95% prediction interval instead?

    **c.** What would you change in the computation of a two-sided 95% prediction interval to find a one-sided 90% prediction interval instead?

## Problems

*Problems marked with an asterisk (\*) are solved in the Self Test in Appendix C.*

1.\* Your agricultural firm is considering the purchase of some farmland, and an indication of the quality of the land will be helpful. A random sample of 62 selected locations planted with corn indicates an average yield of 103.6 bushels per acre, with a standard deviation of 9.4 bushels per acre. Find the two-sided 95% confidence interval for the mean yield for the entire area under consideration.

2. Your company prepares and distributes frozen foods. The package claims a net weight of 24.5 ounces. A random sample of today's production was weighed, and the results were summarized as follows: average = 24.41 ounces, standard deviation = 0.11 ounce, sample size = 5 packages. Find the two-sided 95% confidence interval for the mean weight you would have found had you weighed all packages produced today.

3. Your hospital is negotiating with medical insurance providers, who would like to reduce the amount they pay as reimbursement for hospital stays. For a particular procedure, they would like to reduce payment by $300 and have patients go home one day earlier. To see what effect this would have on hospital costs, a random sample of 50 patients who were recently admitted for this procedure was analyzed. Had they left one day earlier, the average savings would have been $322.44, and the standard deviation was found to be $21.71. Find the two-sided 95% confidence interval for the mean savings, per patient, for the larger population of recent patients.

4. Your quality control department has just analyzed the contents of 20 randomly selected barrels of materials to be used in manufacturing plastic garden equipment. The results found an average of 41.93 gallons of usable material per barrel, with a standard error of 0.040 gallon per barrel. Find the two-sided 95% confidence interval for the population mean.

5.\* Intensities have been measured for eight flashlights. Find the appropriate *t* value from the *t* table to use for each of the following confidence interval calculations:

a. Two-sided 95% confidence.
b. Two-sided 99% confidence.
c. Two-sided 99.9% confidence.
d. Two-sided 90% confidence.

6. Cost observations have been provided for 21 production situations. Find the appropriate $t$ value from the $t$ table to use for each of the following confidence interval calculations:

a. Two-sided 95% confidence.
b. Two-sided 99% confidence.
c. Two-sided 99.9% confidence.
d. Two-sided 90% confidence.

7. Vaccine responses have been observed for 1,859 people. Find the appropriate $t$ value from the $t$ table to use for each of the following confidence interval calculations:

a. Two-sided 95% confidence.
b. Two-sided 99% confidence.
c. Two-sided 99.9% confidence.
d. Two-sided 90% confidence.

8. Main-course taste scores have been recorded for 48 restaurant diners. Find the appropriate $t$ value from the $t$ table to use for each of the following confidence interval calculations:

a. Two-sided 95% confidence.
b. Two-sided 99% confidence.
c. Two-sided 99.9% confidence.
d. Two-sided 90% confidence.

9. Production yield data with 17 degrees of freedom have been collected. Find the appropriate $t$ value from the $t$ table to use for each of the following confidence interval calculations:

a. One-sided 95% confidence.
b. One-sided 99% confidence.
c. One-sided 99.9% confidence.
d. One-sided 90% confidence.

10. Consumer preferences have been observed in a situation for which the standard error is known. Find the appropriate $t$ value from the $t$ table to use for each of the following confidence interval calculations:

a. Two-sided 95% confidence.
b. Two-sided 99% confidence.
c. Two-sided 99.9% confidence.
d. Two-sided 90% confidence.
e. One-sided 95% confidence.
f. One-sided 99% confidence.

11. A random sample of eight customers was interviewed in order to find the number of personal computers they planned to order next year. The results were 22, 18, 24, 47, 64, 32, 45, and 35. You are interested in knowing about the larger population that these customers represent.

a. Find the usual summary measure of the variability of individuals.
b. Approximately how far is the sample average from the population mean?
c. Find the 95% confidence interval for the population mean.

d. Find the 99% confidence interval for the population mean.

12. View the 989 donors in the donations database (out of 20,000 people represented on the companion site) as a random sample from a much larger population of people who would make a donation in response to the mailing. Note that the column of 989 donation amounts has been named "donation_D1."

a. Find the 95% confidence interval for the population mean donation amount.
b. Find the 99% confidence interval for the population mean donation amount.

13. View the 20,000 people represented in the donations database (on the companion site) as a sample from a much larger population. Of these 20,000 people, 989 made a donation in response to the current mailing.

a. Find the 95% confidence interval for the population percentage who would make a donation.
b. Find the 99% confidence interval for the population percentage who would make a donation.

14. Your bakery produces loaves of bread with "1 pound" written on the label. Here are weights of randomly sampled loaves from today's production:

1.02, 0.97, 0.98, 1.10, 1.00, 1.02,
0.98, 1.03, 1.03, 1.05, 1.02, 1.06

Find the 95% confidence interval for the mean weight of all loaves produced today.

15. A market survey has shown that people will spend an average of $15.48 each for your product next year, based on a sample survey of 483 people. The standard deviation of the sample was $2.52. Find the two-sided 95% confidence interval for next year's mean expenditure per person in the larger population.

16. The following quotes for cleaning cost have been obtained from a random sample of 12 providers chosen from a much larger population, prior to awarding a contract for these services:

$114, 154, 142, 132, 127, 145,
135, 138, 126, 142, 135, 124

a. Approximately how far is the average of these 12 quotes from the unknown mean for the entire population of providers?
b. Find the 95% confidence interval for the population mean quote.
c. Find the 99% confidence interval for the population mean quote.
d. Find the one-sided 95% confidence interval that claims that the population mean quote is no larger than some value.
e. Find the one-sided 99% confidence interval that claims that the population mean quote is no larger than some value.

17. Your company is planning to market a new reading lamp and has segmented the market into three groups: avid readers, regular readers, and occasional readers. As part of a marketing survey, 400 individuals have been randomly selected from the population of regular readers, and 58 said that they would like to purchase such a product. Find the 95% confidence interval

for the percentage of the population of regular readers who would express such interest in buying the new product.

18. A recent survey of 252 customers, selected at random from a database with 12,861 customers, found that 208 are satisfied with the service they are receiving. Find the 99% confidence interval for the percentage satisfied for all customers in the database.

19. In a sample of 258 individuals selected randomly from a city of 750,339 people, 165 were found to be supportive of a new public works project. Find the 99.9% confidence interval for the support level percentage in the entire city.

20. Out of 763 people chosen at random, 152 were unable to identify your product.
    a. Estimate the percentage of the population (from which this sample was taken) who would be unable to identify your product.
    b. Find the standard error of the estimate found in part a.
    c. Find the two-sided 95% confidence interval for the population percentage.
    d. Find the one-sided 99% confidence interval that claims that the population percent is no more than some amount.
    e. Why is this statistical inference approximately valid even though the population distribution is not normal?

21. A nationwide poll claims that the margin of error is no more than 3 percentage points in either direction (i.e., plus or minus) at the 95% confidence level.
    a. Verify this claim in a particular case by computing the $t$ table value times the standard error of the binomial fraction $p$ for the case of 309 out of 1,105 registered voters reporting that they are in favor of a particular candidate.
    b. Find the 95% confidence interval for the percentage of registered voters who favor the candidate as indicated in part a.

22. A nationwide poll claims that the margin of error is no more than 4.3 percentage points for questions asked of half the sample.
    a. Verify this claim in a particular case by computing the $t$ table value times the standard error of the binomial fraction $p$ for the case of a candidate having 46% of the 553 registered women voters in favor.
    b. Find the 95% confidence interval for the percentage of registered women voters who favor the candidate as indicated in part a.

23. A survey of 21 business intelligence analysts, who had been in their current positions from 10 to 20 years, revealed $90,734 as the average salary.[14] Assume a random sample with a standard deviation of $15,000.
    a. Find the 95% confidence interval for the population mean salary.
    b. Find the 99% confidence interval for the population mean salary.
    c. Complete the following sentence: We are 95% sure that the population mean salary is at least _____.

24. A survey of eight vice presidents of information technology and information systems, who had been in their current positions for 10 years or less, revealed $174,813 as the average salary.[15] Assume a random sample with a standard deviation of $25,000.
    a. Find the 95% confidence interval for the population mean salary.
    b. Find the 99% confidence interval for the population mean salary.
    c. Complete the following sentence: We are 95% sure that the population mean salary is at least _____.

25. Your firm is in the market to hire an experienced vice president of information technology or information systems. If one is chosen at random from the population represented in the preceding problem, complete the following sentence: We are 99% sure that the chosen manager's salary is at least _____.

26. Comparison shopping is available on the Internet, and this is useful because the exact same item is available at a variety of prices. Table 9.6.1 shows results from MySimon for prices of the Eureka 4750A Bagged Upright Vacuum at 15 stores. Find the 95% confidence interval for the average price in the population that these particular stores represent.

### TABLE 9.6.1 Prices of the Eureka 4750A Bagged Upright Vacuum Cleaner

| Store | Price |
| --- | --- |
| AJ Madison | $56.05 |
| Amazon Marketplace | 69.43 |
| Beach Camera | 54.95 |
| Compuplus.com | 57.99 |
| CPO Eureka | 59.99 |
| Discount Office Items | 54.90 |
| eBay | 59.99 |
| eVacuumStore | 54.99 |
| Gettington | 69.95 |
| GoVacuum | 59.99 |
| Home Depot | 79.99 |
| OneCall | 59.99 |
| PlumberSurplus.com | 56.11 |
| QVC | 69.84 |
| TheWiz.com | 65.57 |

**Source:** MySimon, accessed at http://www.mysimon.com/prices/eureka-4750a-bagged-upright-vacuum on July 13, 2010.

27. Based on the following daily percent changes of the S&P 500 stock market index for June 2010 (accessed at http://finance.yahoo.com on July 13, 2010), find the 95% confidence interval for the population mean daily change. (This is not, strictly speaking, a random sample from a population. However, the random walk theory of the stock market suggests that the changes of the market do actually behave like a random sample. The population would represent all daily market changes that might happen under conditions that prevailed during that time.)

    −1.01%, −3.10%, −0.20%, 0.29%, −1.68%, −0.30%, −1.61%, −0.39%, 0.13%, 0.13%, −0.06%, 2.35%, −0.18%, 0.44%, 2.95%, −0.59%, 1.10%, −1.35%, −3.44%, 0.41%, 2.58%, −1.72%

28. During a one-week experiment, motion was added to in-store sales displays at a random sample of your firm's stores nationwide. The resulting sales increases for these products (compared to the week before) averaged $441.84, with a standard deviation of $68.91. There were 18 stores participating in the experiment.[16]

    a. Find the 95% confidence interval for the population mean sales increase.

    b. Complete the following sentence: We are 95% confident that the population mean sales increase is at least _____.

    c. The manager of one of your firm's stores would like to assess the possible sales increases. This store was not part of the survey. Assuming conditions are similar to those of the experiment, complete the following sentence: We are 95% sure that the one-week savings for this store when motion is added will be between _____ and _____.

29.* Table 9.6.2 shows the performance of stocks recommended by securities firms.

    a. Compute the average and briefly describe its meaning.

    b. Compute the standard deviation and briefly describe its meaning.

    c. Compute the standard error of the average and briefly describe its meaning.

    d. Find the two-sided 95% confidence interval for the mean performance of stocks recommended by similar brokerage firms during this time period, viewing the data set as a random sample from this idealized population.

    e. Find the two-sided 90% confidence interval and compare it to the 95% confidence interval.

    f. Find the one-sided 99% confidence interval statement to the effect that the mean performance was at least as good as some number.

    g. Suppose you decide that, had the data come out with an average performance loss, you would have used a one-sided confidence interval statement that the mean performance was no larger than some number (in place of your answer to part f). In this case, and using the same data table, is your answer to part f a

**TABLE 9.6.2 Performance of Stocks Recommended by Securities Firms, One-Year Rate of Return**

| Firm | Performance |
| --- | --- |
| A.G. Edwards | −0.3% |
| Bear Stearns | −14.3 |
| Credit Suisse F.B. | −4.2 |
| Edward Jones | −5.3 |
| First Union Sec. | −38.3 |
| Goldman Sachs | 9.6 |
| J.P. Morgan Sec. | 16.2 |
| Lehman Bros. | −22.0 |
| Merrill Lynch | 12.3 |
| Morgan Stanley D.W. | 1.0 |
| PaineWebber | −29.1 |
| Prudential Securities | −22.5 |
| Raymond James | 43.9 |
| Salomon S.B. | −15.7 |
| U.S. Bancorp Piper | −5.4 |

**Source:** G. Jasen, "Raymond James Was Top Picker of Stocks in the Fourth Quarter," *Wall Street Journal*, February 12, 2001, p. C1. The source of the data is Zacks Investment Research.

valid confidence interval statement? Why or why not? If not, what should you do instead?

30. An election poll shows your favorite candidate ahead with 52.4% of the vote, based on interviews with 921 randomly selected people.

    a.* Find the two-sided 95% confidence interval for the percentage of the population in favor of your candidate.

    b. Since this candidate has been your favorite for a long time now, and you want her to win, you are interested only in knowing that she has at least some percentage of the votes. In this case, would it be valid to make a one-sided confidence interval statement?

    c. Find the one-sided 95% confidence interval that is appropriate, given the information in part b.

    d. Find the similarly appropriate one-sided 90% confidence interval.

    e. Write a brief paragraph describing how these confidence intervals shed important light on your candidate's chances. In particular, how much more do you know now as compared to knowing only the 52.4% figure?

31. A market survey has shown that people will spend an average of $2.34 each for your product next year,

based on a sample survey of 400 people. The standard deviation of the sample was $0.72. Find the two-sided 95% confidence interval for next year's mean expenditure per person in the larger population.

32. A survey of your customers shows, to your surprise, that 42 out of 200 randomly selected customers were not satisfied with after-sale support and service.

   a. Find the summary statistics: the sample size, $n$; the sample percentage, $p$; and the standard error, $S_p$.

   b. Find the two-sided 95% confidence interval for the percent dissatisfied among all of your customers (i.e., not just those surveyed).

   c. Your population consists of 28,209 customers. Convert the percentages representing the endpoints of the confidence interval in part b to numbers of people in the population. State and interpret your result as a confidence interval for the population number of dissatisfied customers.

33. A sample of 93 coils of sheet steel showed that the average length was 101.37 meters, with a standard deviation of 2.67 meters.

   a. Interpret the standard deviation in words; in particular, what is it measuring the variability of?

   b. Find the standard error. Interpret this number in words and distinguish it from the standard deviation you explained in part a.

   c. Find the two-sided 95% confidence interval for the mean length of coils in the larger population. Write a brief paragraph explaining its meaning.

   d. Find the two-sided 95% prediction interval for the length of the next coil to be produced. Write a brief paragraph explaining its meaning and distinguish this prediction interval from the confidence interval you found for part c. In particular, why is the prediction interval so much wider than the confidence interval?

   e. Your integrity requires that you guarantee coils to be at least a certain length. Does this information make it appropriate for you to compute one-sided intervals? Why or why not?

   f. Find the appropriate one-sided 99% confidence interval and explain its meaning.

   g. Find the appropriate one-sided 99% prediction interval and explain its meaning.

34. As a basis for a brochure describing the speed of a new computer system, you have measured how long it takes the machine to complete a particular benchmark database program. Since the state of the disks in the database is constantly changing as records are added, changed, and deleted, there is some variation in the test results. Here are times, in minutes, for 14 independent repetitions of this testing procedure:

   5, 6, 8, 11, 5, 8, 11, 10, 6, 10, 5, 9, 5, 5

   a. Find the two-sided 95% confidence interval and describe its meaning in terms of the long-run mean performance of the system.

   b. Find the appropriate one-sided 95% confidence interval, assuming you wish to show off how

fast the system is (so that lower numbers are better).

   c. Find the appropriate one-sided 90% confidence interval.

   d. Find the appropriate one-sided 99% confidence interval.

   e. Write a brief paragraph for an advertising brochure describing one (or more) of the preceding results. Be honest, but put your "best foot forward," and write in ordinary English. Include a technical footnote if necessary so that technically knowledgeable people can tell what you really did.

35. Samples of rock taken from various places in a proposed mine have been analyzed. For each sample, a "rate of return" number has been computed that represents the profit obtained (by selling the refined metal at the current market price) as a percentage of the cost of extraction. This measure reflects the difficulty in removing the ore, the difficulty in processing it, and the yield of the finished product, all in meaningful economic terms. You may assume that the samples were drawn at random and represent the conditions under which actual production would take place, if it is economically viable. Economic viability will require that the return be high enough to justify the costs of operation. For the 13 samples obtained, the rates of return were as follows:

   8.1%, 6.2%, 19.8%, –4.3%, 5.1%, 0.2%, –10.4%, 11.8%, 2.0%, 4.7%, –3.2%, 8.9%, –6.2%

   a. Find the summary statistics: $n$, the average, the standard deviation, and the standard error. Write a brief paragraph describing the situation as if you were explaining it to the board of directors.

   b. Identify the population and the population mean. Why is the population mean important to the management and owners of the proposed mine?

   c. Find the appropriate one-sided 99% confidence interval. Write a brief paragraph summarizing its meaning.

   d. Write a brief paragraph outlining the situation and making recommendations on possible action to top management.

36. You are concerned about waste in the newspaper publishing process. Previously, no measurements have been taken, although it is clear that frequent mistakes often require many pounds of newsprint to be thrown away. To judge the severity of the problem and to help you decide if action is warranted, you have begun collecting data. You will take action only if the amount of waste is large enough. So far, on 27 selected mornings, the weight of wastepaper has been recorded. The average is 273.1 pounds per day, with a standard deviation of 64.2 pounds.

   a. Is it appropriate for you to compute a one-sided confidence interval for this situation? Why or why not?

   b. Find the most useful one-sided 99% confidence interval. Why did you choose the side you did?

c. Express your confidence interval in terms of pounds per year, assuming operations continue 365 days per year.

d. Find a one-sided 99% prediction interval for tomorrow's waste. Compare and contrast this result to your confidence interval in part b.

37. So far at your new job, you have landed nine sales contracts with an average price of $3,782 and a standard deviation of $1,290.

a. Identify a reasonable idealized population that this sample represents.

b. If the distribution of sales prices is heavily skewed, would it be appropriate to construct the usual two-sided 95% confidence interval? Why or why not?

c. Assume now that the distribution of sales prices is only slightly skewed and not too different from a normal distribution. Compute the usual two-sided 95% confidence interval, and interpret it carefully in terms of your long-term prospects at this job. Be sure to address both the useful information and the limitations of the confidence interval in this situation.

d. Find the two-sided 90% prediction interval for the sales price of the next contract you land, assuming that conditions will remain essentially unchanged.

38. A random sample of 50 recent patient records at a clinic shows that the average billing per visit was $53.01 and the standard deviation was $16.48.

a. Find the 95% confidence interval for the mean and interpret it.

b. Find the 99% confidence interval.

c. Find the one-sided 95% confidence interval specifying at least some level of billing.

39. Find the 95% confidence interval for the amounts that your regular customers spent on your products last month, viewing the data from Table 4.3.1 of Chapter 4 as a random sample of customer orders.

40. Find the 99% confidence interval for the strength of cotton yarn used in a weaving factory based on the data in problem 23 of Chapter 4.

41. Find the 99.9% confidence interval for the weight of candy bars before intervention, based on the data in Table 5.5.4 of Chapter 5.

42. Find the one-sided 95% confidence interval for the weight of candy bars after intervention, based on the data in Table 5.5.4, indicating that the population mean weight is no more than some amount.

43. From a list of the 729 people who went on a cruise, 130 were randomly selected for interview. Of these, 112 said that they were very happy with the accommodations. Find the 95% confidence interval for the population percentage who would have said they were very happy with the accommodations.

44. Consider the quality scores measured for a random sample of agricultural produce:

16.7, 17.9, 23.5, 13.8, 15.9, 15.2, 12.9, 15.7

a. Find the 95% confidence interval for the population mean quality.

b. Find the 95% prediction interval for the quality of the next measurement.

c. Find the 99% confidence interval for the population mean quality.

d. Find the 99% prediction interval for the quality of the next measurement.

45. The amount of caffeine (milligrams) in randomly sampled cups of coffee was as follows:

112.8, 86.4, 45.9, 110.3, 100.3, 93.3,
101.9, 115.7, 92.5, 117.3, 105.6, 81.6

a. Find the one-sided 99% confidence interval for the population mean caffeine content of a cup of coffee that claims "at least...."

b. Find the one-sided 99% prediction interval for the caffeine content of the next cup of coffee, again claiming "at least...."

14. *Computerworld's Smart Salary Tool 2010*, accessed at http://www.computerworld.com/s/salary-survey/tool/2010/ on July 13, 2010.
15. Ibid.
16. Situations like this have been studied by Bennett-Chaikin, Inc., as reported in an ad for Menasha Corporation in *Advertising Age*, August 21, 1995, p. 17.

## Database Exercises

*Problems marked with an asterisk (*) are solved in the Self Test in Appendix C.*

Refer to the employee database in Appendix A.

1.* View this database as a population. Consider the following sample of five employee numbers from this database: 24, 54, 17, 34, and 53.

a. Find the average, standard deviation, and standard error for annual salary based on this sample.

b. Find the 95% confidence interval for the population mean salary.

c. Draw a graph in the style of Figure 9.1.5 indicating the sample average and confidence interval.

2. Now look at the entire population of salaries, which you can't usually do in real life.

a. Find the population mean and standard deviation, and compare them to the sample estimates from the previous problem.

b. Draw a graph for this situation in the style of Figure 9.1.1. Be sure to use $\sigma_{\overline{X}}$ as the standard deviation of the sampling distribution.

c. Is the population mean in the confidence interval (from exercise 1) in this case? Will it always be in the interval, for all random samples? Why or why not?

3. Repeat exercise 1, parts b and c, using a 99% confidence interval. Is the population mean annual salary in the interval?

4. Repeat exercise 1, parts b and c, using a 90% confidence interval. Is the population mean annual salary in the interval?

5. Repeat exercise 1 using a 95% confidence interval for a different random sample: employee numbers 4, 47, 45, 12, and 69. Also, answer the following:
   d. In real life, what (if anything) could you do about the fact that the population mean is not in the confidence interval?
   e. Also compute 99% and 99.9% intervals. At what confidence level (if any) is the confidence interval large enough to include the population mean?

6. Consider the following random sample of 15 employee numbers from this database: 66, 37, 56, 11, 32, 23, 53, 43, 55, 25, 7, 26, 36, 22, and 20.
   a. Find the percentage of women for this sample.
   b. Find the standard error for the percentage of women and interpret it.
   c. Why should you be hesitant to use this sample and the methods of this chapter to compute a confidence interval for the percentage of women?

7. Viewing the database in Appendix A as a random sample from a much larger population, consider the annual salary values.
   a. Find the 95% confidence interval.
   b. Find the 99% confidence interval.

8. Viewing the database in Appendix A as a random sample from a much larger population, consider the age values.
   a. Find the 95% confidence interval.
   b. Find the 90% confidence interval.

9. Viewing the database in Appendix A as a random sample from a much larger population, consider the experience values.
   a. Find the 95% confidence interval.
   b. Find the 99.9% confidence interval.

10. Viewing the database in Appendix A as a random sample from a much larger population, consider the percentage of women. Find the 95% confidence interval.

11. Viewing the database in Appendix A as a random sample from a much larger population, consider the percentage who are advanced (at training level B or C). Find the 99% confidence interval.

12. Viewing the database in Appendix A as a random sample from a much larger population of employees:
    a. Find the 95% one-sided confidence interval for the population mean annual salary specifying that salaries are at least some amount.
    b. Find the 99% one-sided confidence interval for part a.
    c. Find the 95% one-sided confidence interval for the population mean experience specifying that experience is at least some amount.
    d. Find the 99% one-sided confidence interval for part c.

13. Viewing the database in Appendix A as a random sample from the idealized population of potential employees you might hire next:
    a. Find the 95% prediction interval for the experience of your next hire. Why is this interval so much wider than the confidence interval for the population mean experience?
    b. Find the 95% prediction interval for the age of your next hire.

## Projects

1. Obtain an estimated value and its standard error (either from data or by educated guess) for each of two situations important to your business interests. For each case, find a confidence interval and write a sentence interpreting it. Explain your reasons if you use a confidence level other than 95% or if you use a one-sided confidence interval.

2. Obtain an estimated value and a standard deviation (either from data or by educated guess) for each of two situations important to your business interests. For each case, find a prediction interval and write a sentence interpreting it. Explain your reasons if you use a prediction level other than 95% or if you use a one-sided prediction interval.

3. Find a report of an opinion poll on the Internet or in a newspaper. Write a paragraph summarizing one of the poll's results. Be sure to mention sample size, the percentage, and the standard error. Compute your own two-sided 95% confidence interval. Compare your results to the margin of error, if this is reported in your source.

## Case

### Promising Results from a Specialty Catalog Survey

The preliminary survey results just came back on the specialty catalog project, and they look great! The average planned order size was $42.33, well above the $15 that was hoped for. The group leader will probably be delighted—after all, $42.33 for each of the 1,300,000 target addresses comes out to over $55 million in average sales!

As part of the preparation for the meeting, one of your responsibilities is to look through the fine print of how the survey was done. The initial memo included few details beyond the $42.33 figure. After some calls, you locate the employee who did most of the work. Here is what you learn. A random sample was drawn from a proprietary database of 600,000 addresses of well-off people who purchase luxury items by mail, and 600 catalogs were mailed together with the questionnaire. You also learn that 55 of the 600 surveys were returned. Of these, 13 indicated that "Yes, I will place an order for items totaling $_____ before the end of the year." These amounts were $9.97, $12.05, $29.27, $228.26, $6.10, $87.35, $27.48, $8.86, $19.95, $13.29, $44.06, $11.27, and $52.39.

Well, you now know that there is substantial variability in order size. The 95% confidence interval about the mean extends from $5.82 to $78.84. Multiplying each of these by the size of the target mailing (1,300,000), you compute bounds from $7.6 million to $102.5 million. So even after taking randomness into account, it seems to look as though there is real money to be made here. Or is there?

**Discussion Questions**

1. Is it proper to multiply the average order size, $42.33, by the number of addresses (1,300,000) in the target mailing?

2. Is it better, as suggested, to multiply the endpoints of the confidence interval by the target mailing size?

3. Would it be better to multiply by the size of the frame used to select the random sample?

4. Should anything else trouble you about this situation?

5. What is your best estimate, with confidence limits, for potential catalog sales?

# Chapter 10

# Hypothesis Testing
## Deciding between Reality and Coincidence

**Chapter Outline**

Oh no. Not again. Your high-pressure sales contact is on the line, trying to sell you that miracle yield-enhancing additive to increase the productivity of your refinery. It looks like a good deal, but you're just not sure. You've been trying it out for a week (free, of course, for now), and—sure enough—the yield is up. But it's not up a whole lot, and, naturally, the process is variable, so it's hard to tell whether or not there's anything important going on. What you need is an objective assessment, but you know that what you'll get from your contact on the phone is just another sales pitch: "The yield is up, isn't it? Well, what did I tell you? If you sign up today, we'll throw in a free engraved pen-and-pencil set! Blah blah blah." So you give the secret signal to your secretary, who says that you're in a meeting just now and will call back later.

Here's what's troubling you. Sure, the yield is up. But even if you do nothing special at all, you know that the yield fluctuates from day to day and from week to week about its long-run mean value. So the yield is up for one of two reasons: Either the additive is really working, or it's just a coincidence. After all, regardless of the additive, there's about a 50–50 chance that the week's yield would

be higher than the long-term mean and about a 50–50 chance for it to be lower.

Look at this situation from the salesperson's point of view. Suppose for a moment that the additive is actually worthless and has no effect whatsoever on the yield. Next, convince managers at 100 different companies to try it out for a week. About 50 of these managers will find that their yield went down—no need to follow up those cases. But the other 50 or so will find slightly higher yields. Maybe some of these will even pay big money to continue using this worthless product.

What you need is a way of using the information gathered so far about the yield to help you determine if (on the one hand) it could reasonably be *just coincidence* that the yield was higher last week or if (on the other hand) you have convincing evidence that the additive really works. This is what hypothesis testing can help you do.

**Hypothesis testing** uses data to decide between two possibilities (called *hypotheses*).[1] It can tell you whether

---

1. The singular is one *hypothesis*, and the plural is two *hypotheses* (pronounced *hypothesees*).

the results you're witnessing are just coincidence (and could reasonably be due to chance) or are likely to be real. Some people think of hypothesis testing as a way of using statistics to make decisions. Taking a broader view, an executive might look at hypothesis testing as *one component* of the decision-making process. Hypothesis testing by itself probably shouldn't be used to tell you whether to buy a product or not; nonetheless, it provides critically important information about how substantial and effective the product is.

## 10.1 HYPOTHESES ARE NOT CREATED EQUAL!

A **hypothesis** is a statement about how the world is. It is a statement about the *population*. A hypothesis is not necessarily true; it can be either right or wrong, and you use the sample data to help you decide. When you know everything, there is no need for statistical hypothesis testing. When there is uncertainty, statistical hypothesis testing will help you learn as much as possible from the information available to you.

You will ordinarily work with a *pair* of hypotheses at a time. The data will help you decide which of the two will prevail. But the two hypotheses are not interchangeable; each one plays a different, special role.

### The Null Hypothesis

The **null hypothesis**, denoted $H_0$, represents the *default* possibility that you will accept *unless you have convincing evidence to the contrary*. This is a very favored position. If your data are sketchy or too variable, you will end up accepting the null hypothesis because it has the "benefit of the doubt." In fact, you can end up accepting the null hypothesis without really proving anything at all, putting you in a fairly weak position. Thus, it can make an important difference which of your two hypotheses you refer to as the null hypothesis.

The null hypothesis is often the *more specific* hypothesis of the two. For example, the null hypothesis might claim that the population mean is exactly equal to some known reference value or that an observed difference is just due to random chance. To see that the hypothesis of random chance is indeed more specific, note that *non*random things can have very many different kinds of structure, but randomness implies a lack of structure.

### The Research Hypothesis

The **research hypothesis**, denoted $H_1$, is to be accepted only if there is convincing statistical evidence that would

rule out the null hypothesis as a reasonable possibility. The research hypothesis is also called the **alternative hypothesis**. Accepting the research hypothesis represents a much stronger position than accepting the null hypothesis because it requires convincing evidence.

People are often interested in establishing the research hypothesis as their hidden agenda, and they set up an appropriate null hypothesis solely for the purpose of refuting it. The end result would be to show that "it's not just random, and so here's my explanation...." This is an accepted way of doing research. Since people have fairly creative imaginations, the research community has found that by requiring that the null hypothesis of pure randomness be rejected before publication of a research finding, they can effectively screen many wild ideas that have no basis in fact. This approach does not *guarantee* that all research results are true, but it does screen out many incorrect ideas.

In deciding which hypothesis should be the research hypothesis, ask yourself, Which one has the *burden of proof*? That is, determine which hypothesis requires the more convincing evidence before you decide to believe in it. This one will be the research hypothesis. Don't neglect your own self-interest! Feel free to shift the burden of proof onto those trying to sell you things. Make them prove their claims!

### What Will the Result Tell You?

There are two possible outcomes of a hypothesis test. By convention, they are described as follows:

---

**Results of a Hypothesis Test**

Either:  Accept the null hypothesis, $H_0$, as a reasonable possibility.  A weak conclusion; not a significant result.

Or:  Reject the null hypothesis, $H_0$, and accept the research hypothesis, $H_1$.  A strong conclusion; a significant result.

---

Note that we *never* speak of rejecting the research hypothesis. The reason has to do with the favored status of the null hypothesis as default. Accepting the null hypothesis merely implies that you don't have enough evidence to decide against it. When we decide to "accept" a null hypothesis, $H_0$, we should not necessarily believe that it is true. While accepting it as a reasonably possible scenario that could have generated the data, we nonetheless recognize that there are many other such believable scenarios *close to* the null hypothesis that also might have generated the data. For example, when we accept the null hypothesis that claims the population mean is $2,000, we have not usually ruled out the possibility that this mean is $2,001 or $1,999. For this reason, some statisticians prefer to say that we "fail to reject" the null hypothesis rather than simply say that we "accept" it.

It may help you to think of the hypotheses in terms of a criminal legal case. The null hypothesis is "innocent," and the research hypothesis is "guilty." Since our legal system is based on the principle of "innocent until proven guilty," this assignment of hypotheses makes sense. Accepting the null hypothesis of innocence says that there was not enough evidence to convict; it does not prove that the person is truly innocent. On the other hand, rejecting the null hypothesis and accepting the research hypothesis of guilt says that there is enough evidence to rule out innocence as a possibility and to convincingly establish guilt. We do not have to rule out guilt in order to find someone innocent, but we do have to rule out innocence in order to find someone guilty.

## Examples of Hypotheses

Following are some examples of null and research hypotheses about the population. Note in each case that they cannot both be true, and that the data will be used to decide which one to accept.

1. *The situation*: A randomly selected group of 200 people view an advertisement, and the number of people who buy the product during the next week is recorded.
   *The null hypothesis*: The ad has no effect. That is, the percentage of buyers among those in the general population who viewed the ad is *exactly equal* to the baseline rate for those who did *not* view the ad in the general population. This baseline rate is known to be 19.3%, based on extensive past experience.
   *The research hypothesis*: The ad has an effect. That is, the percentage of buyers among those in the general population who viewed the ad is *different from* the baseline rate of 19.3% representing those buyers who did *not* view the ad in the general population.
   *Discussion*: Note that these hypotheses are statements about the general *population*, not about the 200 people in the sample. The sample evidence accumulated by observing the behavior of 200 randomly selected people will help decide which hypothesis to accept. Since the null hypothesis gives an exact value for the percentage, it is more specific than the research hypothesis, which specifies a large range (i.e., any percentage different from 19.3%). Also note that when you decide that an ad is effective, you will be making a strong statement since this is the research hypothesis. It is as if you are saying "OK. If this ad works as well as we all think it does, let's give it a chance to prove it to us. Or, on

the other hand, if it will be a disaster to sales, let's find that out also."

2. *The situation*: You are evaluating the yield-enhancing additive described at the start of this chapter.
   *The null hypothesis*: The additive has no effect on the long-run yield, an amount known from past experience.
   *The research hypothesis*: The additive has some effect on the long-run yield.
   *Discussion*: The null hypothesis is more specific. Both hypotheses refer to the population (long-run yield) and not just to the particular results of last week (the sample). Your default is that the additive has no effect, and to convince you otherwise will require a conclusive demonstration. The burden of proof is on them (the manufacturers of the additive) to show effectiveness. It's not up to you to prove to them that it's *not* effective.

3. *The situation*: Your firm is being sued for gender discrimination, and you are evaluating the documents filed by the other side. They include a statistical hypothesis test based on salaries of men and women that finds a "highly significant difference" on average between men's and women's salaries.
   *The null hypothesis*: Men's and women's salaries are equal except for random variation. That is, the population from which the men's salaries were sampled has the same mean as the population from which the women's salaries were sampled. Another way to view this idea is that the actual salary differences between men and women are not unreasonably different from what you might get if you were to put all salaries into a hat, mix them up well, and deal them out to people without regard to gender.
   *The research hypothesis*: The population means of men's and women's salaries are different (even before random variation is added).
   *Discussion*: Note the use of idealized populations here. Since these employees are not a random sample in any real sense, the hypotheses refer to an idealized population for each gender (one population of similar men's salaries, the other for the women). With the null hypothesis, the two populations have equal means, while with the research hypothesis, the means are different for the two genders. Your firm is in trouble since the null hypothesis has been rejected and the research hypothesis has been accepted. This is a strong conclusion that goes against you. But all is not necessarily lost. Don't forget that statistical methods generally tell you about the numbers only and not about *why* the numbers are this way. The salary differential might be due directly to gender discrimination, or it might be due to other factors, such as education, experience, and ability. A statistical hypothesis test that addresses only gender and salary cannot tell which factors

are responsible.[2] Also, the hypothesis test results could be wrong, since errors can happen whenever statistical methods are used.

## 10.2 TESTING THE POPULATION MEAN AGAINST A KNOWN REFERENCE VALUE

The simplest case of hypothesis testing involves testing the population mean against a known reference value. This **reference value** is a known, fixed number $\mu_0$ that does not come from the sample data. The hypotheses are as follows:

> ### The Null and the Research Hypothesis
>
> $$H_0 : \mu = \mu_0$$
>
> The null hypothesis $H_0$ claims that the unknown population mean, $\mu$, is *exactly equal* to the known reference value, $\mu_0$.
>
> $$H_1 : \mu \neq \mu_0$$
>
> The research hypothesis $H_1$ claims that the unknown population mean, $\mu$, is *not equal* to the known reference value, $\mu_0$.

This is a **two-sided test** because the research hypothesis includes values for the population mean $\mu$ on both sides (smaller and larger) of the reference value, $\mu_0$.[3] Note that there are actually *three* different numbers involved here that have something to do with an average or mean value:

$\mu$ is the unknown population mean, which you are interested in learning about.

$\mu_0$ is the known reference value you are testing against.

$\overline{X}$ is the known sample average you will use to decide which hypothesis to accept. Of these three numbers, this is the only one that is at all random because it is computed from the sample data. Note that $\overline{X}$ estimates, and hence represents, $\mu$.

The hypothesis test proceeds by comparing the two known numbers $\overline{X}$ and $\mu_0$ against each other. If they are more different than random chance could reasonably account for, then the null hypothesis $\mu = \mu_0$ will be rejected because $\overline{X}$ provides information about the unknown mean, $\mu$. If $\overline{X}$ and $\mu_0$ are fairly close to each other, then the null hypothesis $\mu = \mu_0$ will be accepted. But how close is

close? Where will we draw the line? Closeness must be based on $S_{\overline{X}}$, since this standard error tells you about the randomness in $\overline{X}$. Thus, if $\overline{X}$ and $\mu_0$ are a sufficient number of standard errors apart, then you have convincing evidence against $\mu$ being equal to $\mu_0$.

There are two different ways of carrying out the hypothesis test and getting the results. The first method uses confidence intervals, which we covered in the preceding chapter. This is the easier method because (a) you already know how to construct and interpret a confidence interval, and (b) the confidence interval is directly meaningful because it is in the same units as your data (e.g., dollars, people, defect rates). The second method (based on the *t* statistic) is more traditional but less intuitive since it requires that you calculate something new that is not in the same units as your data and that must be compared to the appropriate critical value from the *t* table before you know the result.

It really doesn't matter which of the two methods (confidence interval or *t* statistic) you use for hypothesis testing since they always give the same answer in the end. You may want to use the confidence interval method most of the time since it is quicker and easier, and it provides more information about the situation. However, you will also want to know how to use the *t* statistic method because it is so commonly used in practice. Since the two methods give the same result, either one may be called a *t test*.

### Using Confidence Intervals: The Easy Way

Here is how to test the null hypothesis $H_0 : \mu = \mu_0$ against the research hypothesis $H_1 : \mu \neq \mu_0$ based on a random sample from the population. First, construct the 95% confidence interval based on $\overline{X}$ and $S_{\overline{X}}$ in the usual way (see Chapter 9). Then look to see whether or not the reference value, $\mu_0$, is in the interval. If $\mu_0$ is outside the confidence interval, it is not a reasonable value for the population mean, $\mu$, and you will accept the research hypothesis; otherwise, you will accept the null hypothesis. This is illustrated in Figure 10.2.1. There are a number of equivalent ways of describing the result of such a hypothesis test. Your decision in each case may be stated as indicated in Table 10.2.1.

Why does this method work? Remember that the confidence interval statement says that the probability that $\mu$ is in the (random) confidence interval is 0.95. Assume for a moment that the null hypothesis is true, so that $\mu = \mu_0$ exactly. Then the probability that $\mu_0$ is in the confidence interval is also 0.95. This says that when the null hypothesis is true, you will make the correct decision in approximately 95% of all cases and be wrong only about 5% of the time. In this sense you now have a decision-making process with exact, controlled probabilities. For a more detailed discussion of the various types of errors in hypothesis testing, see Section 10.3.

---

2. In a later chapter, you will learn how *multiple regression* can adjust for other factors (such as education and experience) and can provide an *adjusted estimate* of the effect of gender on salary while holding these other factors constant.

3. You will learn about *one-sided* hypothesis testing in Section 10.4.

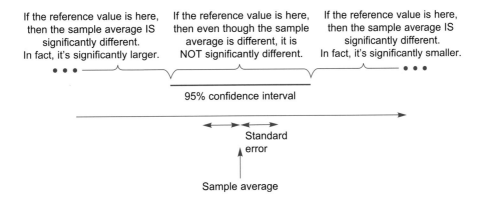

FIGURE 10.2.1   A hypothesis test for the population mean can be decided based on the confidence interval. The question is whether or not the population mean could reasonably be equal to a given reference value. If the reference value is in the interval, then it is reasonably possible. If the reference value is outside the interval, then you would decide that it is not the population mean.

## TABLE 10.2.1 Deciding a Hypothesis Test about the Population Mean

**If the reference value, $\mu_0$, is in the confidence interval from $\overline{X} - tS_{\overline{X}}$ to $\overline{X} + tS_{\overline{X}}$, then**

Accept the null hypothesis, $H_0$, as a reasonable possibility.

Do not accept the research hypothesis, $H_1$.

The sample average, $\overline{X}$, is *not significantly different* from the reference value, $\mu_0$.

The observed difference between the sample average, $\overline{X}$, and the reference value, $\mu_0$, could reasonably be due to random chance alone.

The result is *not statistically significant*. (All of the preceding statements are equivalent.)

**If the reference value, $\mu_0$, is not in the confidence interval from $\overline{X} - tS_{\overline{X}}$ to $\overline{X} + tS_{\overline{X}}$, then**

Accept the research hypothesis, $H_1$.

Reject the null hypothesis, $H_0$.

The sample average, $\overline{X}$, is *significantly different* from the reference value, $\mu_0$.

The observed difference between the sample average, $\overline{X}$, and the reference value, $\mu_0$, could not reasonably be due to random chance alone.

The result is *statistically significant*. (All of the preceding statements are equivalent.)

## Example
### Does the "Yield-Increasing" Additive Really Work?

Recall the (supposedly) yield-increasing additive you were considering purchasing at the start of this chapter.

## TABLE 10.2.2 Basic Facts for the "Yield-Increasing" Additive

| | | |
|---|---|---|
| Average daily yield over the past week | $\overline{X}$ | 39.6 tons |
| Standard error | $S_{\overline{X}}$ | 4.2 tons |
| Sample size | $n$ | 7 days |
| Your known mean daily long-term yield (without additive) | $\mu_0$ | 32.1 tons |

Suppose that the basic facts of the matter are as shown in Table 10.2.2. Your data set consists of $n = 7$ observations of the yield taken while the additive was in use. Your population therefore should be all possible daily yields using the additive; in particular, the population mean, $\mu$, should be the long-term mean yield achieved while using the additive (this is unknown and therefore not listed in the table). The sample average, $\overline{X}$, provides your best estimate of $\mu$.

Indeed, it looks as if the additive is working well. The average daily yield achieved with the additive ($\overline{X} = 39.6$ tons) is 7.5 tons higher than the mean daily long-term yield ($\mu_0 = 32.1$ tons) you expect without the additive. This is no surprise. In hypothesis testing, the reference value is almost never *exactly* equal to the observed value ($\overline{X}$ here). The question is if they are more different than random chance alone would reasonably allow. A histogram of the data, with the sample average and the reference value indicated, is shown in Figure 10.2.2.

In preparation for hypothesis testing, you identify the hypotheses, which may be stated directly in terms of the known value, $\mu_0 = 32.1$ tons (There is no reason to continue to use the symbolic notation $\mu_0$ instead of its known value in

*(Continued)*

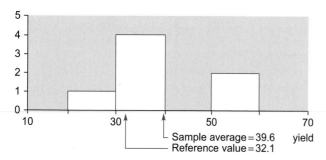

**FIGURE 10.2.2** A histogram of the seven yields obtained with the additive. The sample average summarizes the available data and is higher than the reference value. But is it significantly higher? The result of a hypothesis test will tell whether this sample histogram *could reasonably have come* from a population distribution whose mean is the reference value.

**TABLE 10.2.3  Hypothesis Test Result for the "Yield-Increasing" Additive**

Since the reference value, $\mu_0 = 32.1$ tons, is in the confidence interval from 29.3 to 49.9 tons

Accept the null hypothesis, $H_0: \mu = 32.1$ tons, as a reasonable possibility.

Do *not* accept the research hypothesis, $H_1: \mu \neq 32.1$ tons.

The sample average yield, $\overline{X} = 39.6$, is *not significantly different* from the reference value, $\mu_0 = 32.1$.

The observed difference between the sample average yield, $\overline{X} = 39.6$, and the reference value, $\mu_0 = 32.1$, could reasonably be due to random chance alone.

The result is *not statistically significant*. (All of the preceding statements are equivalent.)

**Example—cont'd**

the formal hypothesis statements.) The hypotheses are as follows:

$$H_0: \mu = 32.1 \text{ tons}$$

The null hypothesis claims that the unknown long-term mean daily yield with the additive, $\mu$, is exactly *equal* to the known reference value, $\mu_0 = 32.1$ tons (without the additive).

$$H_1: \mu \neq 32.1 \text{ tons}$$

The research hypothesis claims that the unknown long-term mean daily yield with the additive, $\mu$, is *not equal to* the known reference value, $\mu_0 = 32.1$ tons (without the additive).

Next, to facilitate the hypothesis test, compute the confidence interval in the usual way using $t = 2.447$ from the $t$ table for $n - 1 = 6$ degrees of freedom:

We are 95% sure that the long-term mean daily yield with the additive is somewhere between 29.3 and 49.9 tons.

Finally, to perform the actual hypothesis test, simply check whether or not the reference value, $\mu_0 = 32.1$ tons, is in the confidence interval.[4] It is in the interval because 32.1 is indeed between 29.3 and 49.9. That is, $29.3 \leq 32.1 \leq 49.9$ is a true statement. Your hypothesis test result is therefore as shown in Table 10.2.3.

The sample average daily yield with the additive, $\overline{X} = 39.6$ tons, is not significantly different from the long-term mean daily yield without the additive, $\mu_0 = 32.1$ tons. This result is inconclusive and ambiguous. You do not have convincing evidence in favor of the additive. When you next talk to the high-pressure sales contact who is trying hard to sell you the stuff, you'll have the confidence to say that *even though the yield is up, it is not up significantly,* and you are not yet convinced that the additive is worthwhile.

Does this test prove that the additive is ineffective? No. It *might* be effective; you just don't have convincing evidence one way or the other.

What else might be done to resolve the issue? Your sales contact might suggest that you use it for another month—free of charge, of course—to see if the additional information will be convincing enough. Or you might suggest this solution to your contact, if you have the nerve.

---

4. It would be silly to check whether or not $\overline{X}$ is in the interval, since, of course, $\overline{X}$ will always be in the confidence interval. The question here is if the known *reference value, $\mu_0$,* is in the confidence interval.

**Example**
*Should Your Company Sponsor the Olympics?*

Why do some companies choose to pay hundreds of millions of dollars each in order to be one of the official sponsors of the Olympic Games? Research by Miyazaki and Morgan points out some of the pluses (the opportunity for marketing visibility and enhancement of the corporate image) and minuses (many consumers cannot correctly identify official sponsors and the high cost).[5] In addition, Miyazaki and Morgan performed an "event study" to see whether the market value (as measured by the stock price) of companies tends to increase or decrease significantly around the time of the official Olympic sponsorship announcement in major print media (such as the *Wall Street Journal* or the *New York Times*).

In the financial markets, over the short term, the stock price of a company generally moves up and down more or less at random, in accordance with the random walk theory of efficient markets. Therefore, the null hypothesis says that the change in a company's market value near the time of the official Olympic sponsorship announcement will be zero on average. If Olympic sponsorship adds value, then we would expect to find market value significantly increased; if sponsorship hurts value, then we would find a decrease in market value on average for these companies.

A test statistic called "C.A.R." (which stands for "Cumulative Abnormal Return") is used to measure the amount of value added to the company, as a percentage over a set time period. Here are some of the results from Miyazaki and Morgan's research, for which some companies showed an increase in value and others showed a decrease:

**Average Change in Company Market Value from Four Days before an Official Olympic Sponsorship Announcement until the Day of the Announcement, as Measured by C.A.R, along with Its Standard Error and Sample Size**

| | | |
|---|---|---|
| Average change in market value | $\overline{X}$ | 1.24 |
| Standard error | $S_{\overline{X}}$ | 0.59 |
| Sample size (number of firms) | $n$ | 27 |

The question here is: Does sponsoring the Olympics enhance a company's value? The answer will be found by performing a hypothesis test. Why not just use the fact that the average company increased its value by 1.24 (in percentage points) to say that Olympic sponsorship enhances value? Because this is a result for a *sample* of 27 companies, and it may or may not represent the larger population of Olympic sponsoring companies in general. In order to infer the effect of sponsorship on value in general, based on the average from a sample, we will use a hypothesis test.

Since we will want to be convinced before concluding that Olympic sponsorship has any effect (positive or negative), this has the burden of proof and will be the research hypothesis. The null hypothesis will claim that Olympic sponsorship has no effect. If we let $\mu$ denote the mean percentage change for the larger population of sponsoring firms (where that population consists of companies that are similar to those included in the study sample, viewing this sample as a random sample from the population), the hypotheses are as follows:

$$H_0 : \mu = 0$$

The null hypothesis claims that the unknown mean effect $\mu$ of Olympic sponsorship on company value is exactly *equal* to the known reference value $\mu_0 = 0$.

$$H_1 : \mu \neq 0$$

The research hypothesis claims that the unknown mean effect $\mu$ of Olympic sponsorship on company value is *not equal* to the known reference value $\mu_0 = 0$.

Next, to facilitate the hypothesis test, compute the confidence interval in the usual way using $t = 2.056$ from the $t$ table with $n = 27$ companies:

We are 95% sure that the mean effect $\mu$ of Olympic sponsorship on company value is somewhere between 0.03 and 2.45.

Finally, to perform the hypothesis test, simply check whether or not the reference value $\mu_0 = 0$ is in the interval. It is *not* in the interval because 0 is not between 0.03 and 2.45. Your $t$ test result is therefore as shown in Table 10.2.4.

Based on the performance of company stock, the announcement of an Olympic sponsorship has a *statistically significant positive effect* on the value of the company.[6] The result is conclusive. You do have convincing evidence that sponsoring the Olympics, in general, enhances the value of a company. Even though the effect may be small (only 1.24 percentage points), the hypothesis test has declared that it cannot be dismissed as a mere random stock price fluctuation.

Does this test absolutely prove that, if the larger population of companies sponsoring the Olympics could be studied, the resulting mean change in company stock value would be positive? Not really. Absolute proof is generally impossible in the presence of even a small amount of randomness. You have convincing evidence but not absolute proof. This says that you might be making an error in rejecting the null hypothesis and accepting the research hypothesis here, although an error is not very likely. These ever-present errors will be discussed in Section 10.3.

---

5. A. D. Miyazaki and A. G. Morgan, "Assessing Market Value of Event Sponsoring: Corporate Olympic Sponsorships," *Journal of Advertising Research*, January–February 2001, pp. 9–15. The "event study" methodology they used also includes careful adjustments for overall stock-market movements and for the risk level of each company, where the event is the announcement of an Olympic sponsorship.
6. You may claim a *positive* effect because (1) the result is statistically significant and (2) the effect as measured by $\overline{X}$ is positive (i.e., $\overline{X}$ is a positive number, larger than $\mu_0 = 0$).

**TABLE 10.2.4 Hypothesis Test Result for the Value of Becoming an Official Olympic Sponsor**

**Since the reference value, $\mu_0 = 0$, is not in the confidence interval from 0.03 to 2.45**

Accept the research hypothesis, $H_1 : \mu \neq 0$.

Reject the null hypothesis, $H_0 : \mu = 0$.

The sample average score, $\overline{X} = 1.24$, is *significantly different* from the reference value, $\mu_0 = 0$.

The observed difference between the sample average score, $\overline{X} = 1.24$, and the reference value, $\mu_0 = 0$, could not reasonably be due to random chance alone.

The result is *statistically significant*. (All of the preceding statements are equivalent.)

If you have a binomial situation, it is straightforward to test whether the population percentage $\pi$ is equal to a given reference value $\pi_0$, provided $n$ is not too small. The situation and procedure are not very different from testing a population mean because once you have an estimator and its standard error, the confidence interval and $t$ statistic are formed in the same way. These similarities are shown in the following table:

|  | Normal | Binomial |
|---|---|---|
| Population mean | $\mu$ | $\pi$ |
| Reference value | $\mu_0$ | $\pi_0$ |
| Null hypothesis | $H_0 : \mu = \mu_0$ | $H_0 : \pi = \pi_0$ |
| Research hypothesis | $H_1 : \mu \neq \mu_0$ | $H_1 : \pi \neq \pi_0$ |
| Data | $X_1, X_2, \ldots, X_n$ | $X$ occurrences out of $n$ trials |
| Estimator | $\overline{X}$ | $p = X/n$ |
| Standard error | $S_{\overline{X}} = S/\sqrt{n}$ | $S_p = \sqrt{p(1-p)/n}$ |
| Confidence interval | From $\overline{X} - tS_{\overline{X}}$ to $\overline{X} + tS_{\overline{X}}$ | From $p - tS_p$ to $p + tS_p$ |
| $t$ statistic | $t = (\overline{X} - \mu_0)/S_{\overline{X}}$ | $t = (p - \pi_0)/S_p$ |

### Example
#### Pushing the Limits of Production (A Binomial Situation)

One of the mysteries of producing electronic computer chips is that you can't tell for sure how good the results are until you test them. At that point, you find that some are unacceptable, others are fine, and some are especially good. These especially good ones are separated and sold at a premium as "extra fast" because they process information more quickly than the others.

Your goal has been to improve the production process to the point where more than 10% of the long-run production can be sold as extra-fast chips. Based on a sample of 500 recently produced chips, you plan to perform a hypothesis test to see if the 10% goal has been exceeded, if you are far short, or if it's too close to call.

Because of recent improvements to the production process, you are hopeful. There were 58 extra-fast chips, giving an estimated rate of 11.6%, which exceeds 10%. But did you *significantly* exceed the goal, or were you just lucky? You'd like to know before celebrating.

Let's model this as a binomial situation in which each chip is either extra fast or not. The binomial probability $\pi$ represents the probability of being extra fast. The sample size is $n = 500$, the observed count is $X = 58$, and the sample proportion is $p = 11.6\%$. The reference value is $\pi_0 = 10\%$.

Hypothesis testing for a binomial (with sufficiently large $n$) is really no different from testing with quantitative data. After all, in each case you have an estimate ($\overline{X}$ or $p$), a standard error ($S_{\overline{X}}$ or $S_p$), and a reference value ($\mu_0$ or $\pi_0$). Here are the formal hypothesis statements for this binomial situation:

$$H_0 : \pi = 10\%$$

The null hypothesis claims that extra-fast chips represent 10% of production.

$$H_1 : \pi \neq 10\%$$

The research hypothesis claims that the rate is different from 10%: either higher (Hooray! Time to celebrate!) or lower (Uh-oh, time to make some adjustments).

The 95% confidence interval is computed in the usual way for a binomial, based on the standard error $S_p = \sqrt{p(1-p)/n} = 0.0143$ and 1.960 from the $t$ table. The interval is found to extend from 8.8% to 14.4%:

You are 95% sure that extra-fast chips are being produced at a rate somewhere between 8.8% and 14.4% of total production.

Finally, to perform the hypothesis test, simply see whether or not the reference value $\pi_0 = 10\%$ is in the interval. It *is* in the interval because 10% is between 8.8% and 14.4%. Your $t$ test result is therefore as shown in Table 10.2.5.

The observed rate of production of extra-fast chips is not statistically significantly different from 10%. You do not have enough information to tell whether the rate is conclusively either higher or lower. The result is inconclusive. Although 11.6% looked like a good rate (and actually exceeds the goal of 10%), it is not significantly different from the goal. Since 11.6% may be just randomly different from the 10% goal, you do not have strong evidence that the goal has been reached.

Remember that you are doing statistical inference. You're not just interested in these particular 500 chips. You'd like to know about the long-run production rate for many more chips, with the machinery running as it is now. Statistical inference has told you that the rate is so

### TABLE 10.2.5 Hypothesis Test Result for Chip Production

**Since the reference value, $\pi_0 = 10\%$, is in the confidence interval from 8.8% to 14.4%**

Accept the null hypothesis, $H_0 : \pi = 10\%$, as a reasonable possibility.

Do not accept the research hypothesis, $H_1 : \pi \neq 10\%$.

The sample proportion, $p = 11.6\%$, is *not significantly different* from the reference value, $\pi_0 = 10\%$.

The observed difference between the sample proportion, $p = 11.6\%$, and the reference value, $\pi_0 = 10\%$, could reasonably be due to random chance alone.

The result is *not statistically significant*. (All of the preceding statements are equivalent.)

close to 10% that you can't tell whether or not the goal has been reached yet.

You might decide to collect more data from tomorrow's production to see if the added information will allow you to show that the goal has been reached (by accepting the research hypothesis, you hope, with an observed rate *higher* than 10%). On the other hand, rather than just squeak by, you might want to hedge your bets by instituting some more improvements.

## The *t* Statistic: Another Way, Same Answer

Another way to carry out a two-sided test for a population mean is to first compute the *t* statistic, which is defined as $t = (\overline{X} - \mu_0)/S_{\overline{X}}$, and then use the *t* table to decide which hypothesis to accept. The answer will always be the same as from the confidence interval method, so it doesn't matter which method you use. The hypothesis testing procedure for comparing the population mean to a reference value based on $\overline{X}$ and $S_{\overline{X}}$ (using either method) is called **Student's *t* test** or simply the *t* test. The name *Student* was used by W. S. Gossett, Head Brewer for Guinness, when he published the first paper to use the *t* table (which he invented) in place of the normal probability table, correcting for the use of the sample standard deviation, *S*, in place of the unknown population standard deviation, $\sigma$, when the sample size, *n*, is small.[7]

In general, hypothesis tests proceed by first computing a number called a **test statistic** based on the data that provides the best information for discriminating between the two hypotheses. Next, this test statistic (e.g., the *t* statistic) is compared to the appropriate **critical value** taken from a standard table of critical values (e.g., the *t* table) to determine which hypothesis should be accepted. In situations that are more complex than just testing a population mean, it can require some creative effort (1) to come up with a test statistic that uses the sample information most efficiently and (2) to find the appropriate critical value. Either this critical value is found by theory (as is the case with the *t* table), or, increasingly in modern times, computers are used to create a new, special critical value for each particular situation.

There are two different values referred to as *t*. The **critical *t* value** is the number $t_{\text{table}}$ found in the *t* table and does not reflect the sample data in any way. The *t* **statistic**, on the other hand, is the test statistic and represents how many standard errors there are separating $\mu_0$ and $\overline{X}$:

7. Student, "The Probable Error of a Mean," *Biometrika* 6 (1908), pp. 1–25.

**The *t* Statistic**

For univariate data:

$$t_{\text{statistic}} = \frac{\overline{X} - \mu_0}{S_{\overline{X}}}$$

For a binomial situation:

$$t_{\text{statistic}} = \frac{p - \pi_0}{S_p}$$

The *t* test uses both of these *t* numbers, comparing the *t* statistic computed from the data to the *t* value found in the *t* table. The result of the test is as stated in Table 10.2.6.

The *absolute value* of a number, denoted by enclosing the number between two vertical bars, is defined by removing the minus sign, if any. For example, $|3| = 3$, $|-17| = 17$, and $|0| = 0$. A useful rule of thumb is that if the *t* statistic is larger in absolute value than 2, reject the null hypothesis; otherwise, accept it. This would apply when *n* is larger than about 40, using 2 as an approximation to the *t* value of 1.960. It is thus easy to scan a column of *t* statistics and tell which

**TABLE 10.2.6 Using the *t* Statistic to Decide a Hypothesis Test**

| |
|---|
| **If the *t* statistic is *smaller* in absolute value than the *t* value from the *t* table ($|t_{\text{statistic}}| < t_{\text{table}}$), then** |
| Accept the null hypothesis, $H_0$, as a reasonable possibility. |
| Do *not* accept the research hypothesis, $H_1$. |
| The sample average, $\overline{X}$, is *not significantly different* from the reference value, $\mu_0$. |
| The observed difference between the sample average, $\overline{X}$, and the reference value, $\mu_0$, could reasonably be due to random chance alone. |
| The result is *not statistically significant*. (All of the preceding statements are equivalent.) |
| **If the *t* statistic is *larger* in absolute value than the *t* value from the *t* table ($|t_{\text{statistic}}| > t_{\text{table}}$), then** |
| Accept the research hypothesis, $H_1$. |
| Reject the null hypothesis, $H_0$. |
| The sample average, $\overline{X}$, is *significantly different* from the reference value, $\mu_0$. |
| The observed difference between the sample average, $\overline{X}$, and the reference value, $\mu_0$, could not reasonably be due to random chance alone. |
| The result is *statistically significant*. (All of the preceding statements are equivalent.) |

are significant. For example, 6.81, –4.97, 13.83, 2.46, and –5.81 are significant $t$ statistics, whereas 1.23, –0.51, 0.02, –1.86, and 0.75 are not significant $t$ statistics. (A negative value for the $t$ statistic tells you that the sample average, $\overline{X}$, is smaller than the reference value, $\mu_0$.)

You might wonder what to do if the $t$ statistic is *exactly equal* to the $t$ value in the $t$ table. This would happen when $\mu_0$ falls exactly at an endpoint of the confidence interval. How would you decide? Fortunately, this almost never happens. You might compute more decimal digits to decide, or you might conclude that your result is "significant, but just borderline."

Although the $t$ statistic may be easily compared to the value 2 (or to the more exact number from the $t$ table) to decide significance, remember that it is not in the same measurement units as the data. Since the measurement units in the numerator and denominator of the $t$ statistic cancel each other, the result is a pure number without measurement units. It represents the distance between $\overline{X}$ and $\mu_0$ in *standard errors* rather than in dollars, miles per gallon, people, or whatever units your data set represents.

Other than this, there is nothing really different between the $t$ statistic and confidence interval approaches. To verify this, reconsider the preceding examples.

For the example of the "yield-increasing" additive, the sample average is $\overline{X} = 39.6$ tons, the standard error is $S_{\overline{X}} = 4.2$ tons, the sample size is $n = 7$, and the reference value is $\mu_0 = 32.1$ tons. The reference value *is* in the confidence interval, which extends from 29.3 to 49.9. Based on this, you accept the null hypothesis. If you had computed the $t$ statistic instead, you would have found

$$t_{\text{statistic}} = \frac{\overline{X} - \mu_0}{S_{\overline{X}}}$$

$$= \frac{39.6 - 32.1}{4.2}$$

$$= 1.785714$$

Since the absolute value of the $t$ statistic, 1.785714, is less than the $t$ table value of 2.447, you accept the null hypothesis. Thus, the $t$ statistic approach gives you the same end result as the confidence interval approach.

Consider, as an example, a survey in which managers were asked to rate the effect of employee stock ownership on product quality, for which the sample average score is $\overline{X} = 0.35$, the standard error is $S_{\overline{X}} = 0.14$, the sample size is $n = 343$, and the reference value is $\mu_0 = 0$ which expresses a neutral opinion (neither positive nor negative, on average).[8] The reference value is *not* in the confidence interval, which extends from 0.08 to 0.62. Based on this,

---

8. P. B. Voos, "Managerial Perceptions of the Economic Impact of Labor Relations Programs," *Industrial and Labor Relations Review* 40 (1987), pp. 195–208.

you accept the research hypothesis. Had you computed the $t$ statistic instead, you would have found

$$t_{\text{statistic}} = \frac{\overline{X} - \mu_0}{S_{\overline{X}}}$$

$$= \frac{0.35 - 0}{0.14}$$

$$= 2.50$$

Since the absolute value of the $t$ statistic, 2.50, is greater than the $t$ table value of 1.960, you accept the research hypothesis. The $t$ statistic approach gives you the same end result as the confidence interval approach, as it always must do.

For the binomial example involving the limits of production, there are $X = 58$ extra-fast chips out of the sample size $n = 500$, the binomial proportion is $p = 0.116$, the standard error is $S_p = 0.0143$, and the reference value is $\pi_0 = 0.10$. The reference value *is* in the confidence interval, which extends from 0.088 to 0.144. Based on this, you accept the null hypothesis. If you had computed the $t$ statistic instead, you would have found

$$t_{\text{statistic}} = \frac{p - \pi_0}{S_p}$$

$$= \frac{0.116 - 0.10}{0.0143}$$

$$= 1.12$$

Since the absolute value of the $t$ statistic, 1.12, is less than the $t$ table value of 1.960, you accept the null hypothesis, reaching the same conclusion as with the confidence interval approach.

## 10.3  INTERPRETING A HYPOTHESIS TEST

Now that you know the mechanics involved in performing a hypothesis test and conventional ways to describe the result, it is time to learn the probability statement behind it all. Just as in the case of confidence intervals, since it is not possible to be correct 100% of the time, you end up with a statement involving the unknown population mean that is correct 95% (or 90% or 99% or 99.9%) of the time.

By convention, the formal details of hypothesis testing are set up in terms of the various *errors* that can be made. The result of a hypothesis test is acceptance of one of the hypotheses based on information from the sample data. You might be right and you might be wrong since the hypotheses are statements about the *population*, for which you have incomplete information. Generally, you will not know for sure if you are right or wrong in your choice. Of course, you hope that you are correct; however, depending on the situation, there may or may not be a useful probability statement to reassure you.

Each type of error is based on a different assumption about which hypothesis is *really* true. Of course, in reality, you won't ordinarily know which hypothesis is true, even after you've made a decision to accept one. However, to understand the results of your hypothesis test, it is helpful to put it in perspective with respect to all of the different ways the test could have come out.

## Errors: Type I and Type II

If the null hypothesis is really true (even though, in reality, you won't know for sure if it is or not) but you wrongly decide to reject it and accept the research hypothesis instead, then you have committed a **type I error**, pronounced "type one error." The probability of a type I error occurring (when the null hypothesis is true) is controlled by convention at the 5% level:

$$P(\text{type I error when } H_0 \text{ is true}) = 0.05$$

It is possible to control the probability of a type I error because the null hypothesis is very specific, so there is an exact probability. For example, when you assume that the null hypothesis $H_0: \mu = \mu_0$ is true, you are assuming that you know the value of the population mean. Once you know the population mean, probabilities can be easily calculated.

Testing at other levels (10%, 1%, or 0.1%, say) can be done by using a different $t$ value from the $t$ table—for example, by working with a different confidence interval (90%, 99%, or 99.9%, respectively). If you are not willing to be wrong 5% of the time when the null hypothesis is true, you might test at the 1% level instead (using the appropriate value in the $t$ table column with 2.576 at the bottom) so that your probability of committing a type I error (when the null hypothesis is true) would only be 1%.

If the research hypothesis is really true (even though, again, you won't usually know for sure if it is or not), but you wrongly decide to accept the null hypothesis instead, you have committed a **type II error**. The probability of a type II error occurring cannot be easily controlled:

$$P(\text{type II error when } H_1 \text{ is true}) \text{ is not easily controlled}$$

It is difficult to control the probability of a type II error because, depending on the true value of $\mu$, this probability will vary.[9] Suppose $\mu$ is very close to $\mu_0$. Then, due to randomness in the data, it will be very difficult to tell them apart. For example, suppose the null hypothesis claims that $\mu$ is 15.00000, but $\mu$ is actually 15.00001. Then, although the research hypothesis is technically true (since 15.00000 ≠ 15.00001), in practical terms you will have

| | Your Decision | |
| --- | --- | --- |
| | Accept null hypothesis | Accept research hypothesis |
| **The Truth** — Null hypothesis | Correct decision | Type I error (controlled at level 0.05 or other level) |
| **The Truth** — Research hypothesis | Type II error (not easily controlled) | Correct decision |

**FIGURE 10.3.1** Your decision to accept one of the two hypotheses may or may not be correct. Depending on which hypothesis is really true, there are two types of errors. Only the type I error is easily controlled, conventionally at the 5% level.

much trouble telling them apart and the probability of a type II error will be approximately 95%. On the other hand, if $\mu$ is far from 15, the probability of a type II error will be nearly 0, a pleasing situation. Thus, since the probability of a type II error depends so heavily on the true value of $\mu$, it is difficult to control. These errors are illustrated in Figure 10.3.1.

## Assumptions Needed for Validity

You may have already suspected that some assumptions must be satisfied for the results of the hypothesis test to be valid. Since the test can be done based on the confidence interval, the assumptions for hypothesis testing are the same as the assumptions needed for confidence intervals. The **assumptions for hypothesis testing** are (1) the data set is a random sample from the population of interest, and (2) either the quantity being measured is approximately normal, or else the sample size is large enough that the central limit theorem ensures that the sample average is approximately normally distributed.

What happens if these assumptions are not satisfied? Consider the probability of a type I error (wrongly rejecting the null hypothesis when it is actually true). This error probability will no longer be controlled at the low, manageable level of 5% (or other claimed level of your choice). Instead, the true error probability could be much higher or lower than 5%. A finding of significance then loses much of its prestige since the event "wrongly finding significance" is now a more common occurrence.

If the data set is not a random sample from the population of interest, there is essentially nothing that statistics can do for you because the required information is just not in the data.

---

9. In principle, the probability can be computed for each value for $\mu$. The resulting table or graph provides the basis for what is called the *power* of the test. This is basically a *what if* analysis, giving the type II error properties of the test under each possible value of $\mu$.

Suppose your data set is a random sample, but the distribution is not normal. If your data distribution is so far from normal that you are concerned, you might try transforming the data (for example, using logarithms if all numbers are positive) to obtain a more normal distribution. If you decide to transform, note that you would no longer be testing the mean of the population but the mean of the *logarithm* of the population instead. Another solution would be to use a nonparametric test, to be explained in Chapter 16.

## Hypotheses Have No Probabilities of Being True or False

Perhaps you've noticed that we have never said that a hypothesis is "probably" either true or false. We've always been careful either to accept or to reject a hypothesis, making a definite, exact decision each time. We talk about the errors we might make and *their* probabilities, but never about the probability of a hypothesis being true or false. The reason is simple:

There is nothing random about a hypothesis!

The null hypothesis is either true or false, depending on the value of the population mean, $\mu$. There is no randomness involved in the population mean by itself. Similarly, the research hypothesis is also either true or false, and although you do not know which, there is no randomness involved in the hypothesis itself. The randomness comes only from the random sampling process, which gives you the data to help you decide.

Thus, your *decision* is known and random, just like a sample statistic, since it is based on the data. However, the true hypothesis is fixed but unknown, just like a population parameter.

## Statistical Significance and Test Levels

By convention, a result is defined to be **statistically significant** if you accept the research hypothesis using a test at the 5% level (for example, based on a standard 95% confidence interval). Note that this is probably not the same use of the word *significant* that you are used to. Ordinarily, something "significant" has special importance. This is not necessarily so in statistics.

To illustrate, a lawyer once came to me deeply concerned because the other side in a lawsuit had found a *statistically significant difference* between measurements made on a door that had been involved in an accident and other, similar doors in the same building. Oh no! But after the special statistical meaning of the word *significant* was pointed out, the attorney was relieved to find that the other side had *not* shown that the door was

extremely different from the others. They had only demonstrated that there was a *statistically detectable* difference. In fact, the difference was quite small. But with enough careful measurements, it was detectable! It was not just randomly different from the other doors (the null hypothesis); it was systematically different (the research hypothesis). Although the difference was statistically significant, it was not large enough to matter very much. The situation is analogous to snowflakes; each one is truly different from the others, yet for many purposes they are essentially identical.

The moral of this story is that you should not automatically be impressed when someone boasts of a "significant result." If the word is being used in its statistical sense, it says only that random chance has been ruled out. It still remains to examine the data to see if the effect is strong enough to be important to you. Statistical methods work only with the numbers; it is up to you to use knowledge from other fields to decide the importance and relevance of the statistical results.

There is another reason you should not be overly impressed by statistically significant results. Over your lifetime, approximately 5% of the test results you will see *for situations in which the null hypothesis is really true* will be found (wrongly, due to random error) to be significant. This implies that about 1 in every 20 uninteresting situations will be declared significant by mistake (i.e., by type I error). A pharmaceutical researcher once noticed that about 5% of the drugs being tested for a particularly difficult disease were found to have a significant effect. Since this is about the fraction of drugs that would be found *by mistake* to be effective, *even if none were in fact effective*, this observation suggests that the entire program might not be successful in finding a cure.

Depending on which column you use in the *t* table, you can perform a hypothesis test at the 10%, 5%, 1%, or 0.1% level. This **test level** or **significance level** is the probability of a type I error when the null hypothesis is in fact true.[10] When you reject the null hypothesis and accept the research hypothesis, you may claim that your result is *significant* at the 10%, 5%, 1%, or 0.1% level, depending on the *t* table column you used. The smaller the test level for which you can find significance, the more impressive the result. For example, finding a result that is significant at the 1% level is more impressive than finding significance at the 10% or 5% level; your type I error probability is smaller and your evidence against the null hypothesis is stronger. By convention,

---

10. In more general situations, including one-sided testing, the test level or significance level is defined more carefully as the *maximum* type I error probability, maximized over all possibilities included within the null hypothesis.

the following phrases may be used to describe your results:

| | |
|---|---|
| Not significant | Not significant at the conventional 5% level |
| Significant | Significant at the conventional 5% level |
| Highly significant | Significant at the 1% level |
| Very highly significant | Significant at the 0.1% level |

What should you do if you find significance at more than one level? Celebrate! Seriously, however, the smaller the test level at which you find significance, the stronger your evidence is against the null hypothesis. You would therefore report only the *smaller* of the significance levels for which you find significance. For example, if you find significance at both the 5% and 1% levels, it would be sufficient to report only that your result is highly significant.

Whenever you find significance at one level, you will necessarily find significance at all *larger* levels.[11] Thus, a highly significant result (i.e., significant at the 1% level) must *necessarily* (i.e., provably, using mathematics) be significant at the 5% and 10% levels. However, it might or might not be significant at the 0.1% level.

## *p*-Values

Every hypothesis test has a ***p*-value**, which tells you how surprised you would be to learn that the null hypothesis had produced the data, with smaller *p*-values indicating more surprise and leading to rejection of $H_0$. By convention, we reject $H_0$ whenever the *p*-value is less than 0.05. The *p*-value tells you the probability, assuming that the null hypothesis is true, that such data (or data showing even more differences from $H_0$) would be observed. Because small *p*-values are unlikely to arise when $H_0$ is true, they lead to rejection of $H_0$. For example, if $p < 0.001$, data with such large differences from $H_0$ occur less often than once in 1,000 random samples. Rather than suppose that rare 1-in-1,000 events can reasonably happen (because they don't, at least not very often), it is simpler to decide that $H_0$ is false and should be rejected. By convention, *p*-values are reported as follows:

| As Reported | Interpretation |
|---|---|
| Not significant ($p > 0.05$) | Not significant at the conventional 5% level |
| Significant ($p < 0.05$) | Significant at the conventional 5% level but not at the 1% level |
| Highly significant ($p < 0.01$) | Significant at the 1% level but not at the 0.1% level |
| Very highly significant ($p < 0.001$) | Significant at the 0.1% level |

In some fields of study, you are permitted to report a result that is significant at the 10% level. This represents an error probability of 1 in 10 of rejecting a null hypothesis that is really true; many other fields consider this an unacceptably high error rate. However, some fields recognize that the unpredictability and variability of their data make it difficult to obtain a (conventionally) significant result at the 5% level. If you are working in such a field, you may use the following *p*-value statement as a possible alternative:

Significant at the 10% level but not at the conventional 5% level ($p < 0.10$)

A *p*-value statement is often found inserted into text, as in "The style of music was found to have a significant effect ($p < 0.05$) on purchasing behavior." You may also see *p*-value statements included as footnotes either to text or to a table, as in "Productivity improved significantly[12] as a result of the new exercise program."

Most statistical software packages report an exact *p*-value as the result of a hypothesis test. For testing at the 5% level, if this *p*-value is any number less than 0.05, then the result is significant (e.g., $p = 0.0358$ corresponds to a significant test result, whereas $p = 0.2083$ is not significant because 0.0358 is less than 0.05, whereas 0.2083 is not). Note that the *p*-value is a statistic (not a population parameter) because it can be computed based on the data (and the reference value).

Consider the example of testing whether or not the observed average yield $\overline{X} = 39.6$ is significantly different from the reference value $\mu_0 = 32.1$ tons (based on $n = 7$ observations with standard error $S_{\overline{X}} = 4.2$). The result might be reported by the computer as follows:

| | n | Mean | STDEV | SE Mean | t | p-Value |
|---|---|---|---|---|---|---|
| Yield | 7 | 39.629 | 11.120 | 4.203 | 1.79 | 0.12 |

Since the computed *p*-value (0.12) is more than the conventional 5% test level (i.e., $0.12 > 0.05$) we have the test result "not significant ($p > 0.05$)." This result (not significant) may also be obtained by comparing the computed *t* statistic (1.79) to the *t* table value (from the *t* table). The exact *p*-value here tells you that there is a 12% chance of seeing such a large difference (between observed mean and reference value) under the assumption that the population mean is equal to the reference value $\mu_0 = 32.1$. By convention, a 12% chance is not considered out of the ordinary, but chances of 5% or less are considered

---

11. Note that a *larger* level of significance is actually a *less* impressive result. For example, being significant at the 1% level is highly significant, whereas being significant at the (larger) 5% level is (merely) significant.

12. ($p < 0.05$).

unlikely. Alternatively, you might first ask the computer for the 95% confidence interval:

| | N | Mean | STDEV | SE Mean | 95.0 Percent C.I. |
|---|---|---|---|---|---|
| Yield | 7 | 39.63 | 11.12 | 4.20 | (29.34, 49.92) |

From this output, you can see that the test is not significant because the reference value $\mu_0 = 32.1$ is within the confidence interval (29.34 to 49.92).

Next, consider testing whether or not managers in general view employee stock ownership as worthwhile for improving product quality, as measured on a scale from –2 (strongly not worthwhile) to +2 (strongly worthwhile). The computer output might look like this:

| | n | Mean | STDEV | SE Mean | t | p-Value |
|---|---|---|---|---|---|---|
| Score | 343 | 0.350 | 2.593 | 0.140 | 2.50 | 0.013 |

This output tells you that the $p$-value is $p = 0.013$. Thus, the result is significant at the 5% level (since $p < 0.05$) but is not significant at the 1% level (since $p > 0.01$). The conclusion is that managers perceive employee stock ownership as significantly worthwhile ($p < 0.05$).

In a binomial situation, note that there may be two different quantities referred to as $p$ by convention. One is the observed percentage in the sample, $p = X/n$. The other is the $p$-value computed for a hypothesis test involving a particular reference value. While this may be confusing, it is standard statistical notation.

## 10.4 ONE-SIDED TESTING

All of the tests we have done so far are two-sided tests because they test the null hypothesis $H_0 : \mu = \mu_0$ against the research hypothesis $H_1 : \mu \neq \mu_0$. This research hypothesis is two-sided because it allows for possible values for the population mean both above and below the reference value, $\mu_0$.

However, you may not really be interested in testing whether the population mean is *different* from the reference value. You may have a special interest in it being *larger* (in some cases) or *smaller* (in other cases) than the reference value. For example, you might purchase a system only if possible long-term savings are *significantly larger* than some special number (the reference value, $\mu_0$). Or you might be interested in claiming that your quality is high because the defect rate is *significantly smaller* than some impressively small number.

You don't need to use a one-sided test to be able to claim that the sample average is significantly larger or significantly smaller than the reference value; you may be able to use a two-sided test for this. If the two-sided test comes out significant (i.e., you accept the research hypothesis), then you may base your claim of significance on whether the sample average, $\overline{X}$, is larger than or smaller than the reference value:

---

**Using a Two-Sided Test but Reporting a One-Sided Conclusion[13]**

| | |
|---|---|
| If the two-sided test is significant and $\overline{X} > \mu_0$ | The sample average, $\overline{X}$, is significantly larger than the reference value, $\mu_0$ |
| If the two-sided test is significant and $\overline{X} < \mu_0$ | The sample average, $\overline{X}$, is significantly smaller than the reference value, $\mu_0$ |

13. Remember, the *two-sided* conclusion might be "$\overline{X}$ is significantly different from $\mu_0$."

---

However, it may be advantageous to use a one-sided test. If you meet the requirements, you might be able to report a significant result using a one-sided test that would not be significant had you used a two-sided test. How is this possible? By focusing on just one side and ignoring the other, the one-sided test can better detect a difference on that side. The trade-off is that the one-sided test is incapable of detecting a difference, no matter how large, on the other side.

A **one-sided $t$ test** is set up with the null hypothesis claiming that $\mu$ is on one side of $\mu_0$ and the research hypothesis claiming that it is on the other side. (We always include the case of $\mu = \mu_0$ in the null hypothesis, which is the default. This ensures that when you accept the research hypothesis and find significance, you have a stronger conclusion: either "significantly larger than" or "significantly smaller than.")[14] The hypotheses for the two different kinds of one-sided tests are as follows:

---

**One-Sided Testing to See If $\mu$ Is Smaller Than $\mu_0$**

$$H_0 : \mu \geq \mu_0$$

The null hypothesis claims that the unknown population mean, $\mu$, is *at least as large* as the known reference value, $\mu_0$.

$$H_1 : \mu < \mu_0$$

The research hypothesis claims that the unknown population mean, $\mu$, is *smaller* than the known reference value, $\mu_0$.

---

**One-Sided Testing to See If $\mu$ Is Larger Than $\mu_0$**

$$H_0 : \mu \leq \mu_0$$

The null hypothesis claims that the unknown population mean, $\mu$, is *not larger* than the known reference value, $\mu_0$.

$$H_1 : \mu > \mu_0$$

The research hypothesis claims that the unknown population mean, $\mu$, is *larger* than the known reference value, $\mu_0$.

---

14. Technically speaking, the case of equality could be included in the research hypothesis. However, the mechanics of the test would be identical and significance would be found in exactly the same cases, but your conclusion would seem weaker.

There is an important criterion you must satisfy before using a one-sided hypothesis test; it is essentially the same criterion that must be met for a one-sided confidence interval:

> In order to use a one-sided test, you must be sure that *no matter how the data had come out*, you would still have used a one-sided test on the same side ("larger than" or "smaller than") as you will use. If, had the data come out different, you might have used a one-sided test *on the other side* instead of the side you plan to use, you should use a two-sided test instead. If in doubt, use a two-sided test.

In particular, using a one-sided test can leave you open to criticism. Since the decision of what's interesting can be a subjective one, your decision to focus only on what's interesting to you may conflict with the opinions of others you want to convince. If you need to convince people who might have a very different viewpoint (for example, regulators or opposing lawyers), you should consider using a two-sided test and giving the one-sided conclusion. On the other hand, if you need only convince "friendly" people, with interests similar to yours (for example, within your department or firm), and you satisfy the preceding criterion, you will want to take advantage of one-sided testing.

The research hypothesis will be accepted only if there is convincing evidence against the null hypothesis. This says that you will accept the research hypothesis only when the sample average, $\overline{X}$, and the reference value, $\mu_0$, have the same relationship as described in the research hypothesis *and* are far enough apart (namely, $t_{\text{table}}$ or more standard errors apart, which represents the extent of reasonable variation from a mean value). There are two different ways to implement a one-sided test, namely, using a one-sided confidence interval or using the $t$ statistic.

### Example
#### Launching a New Product

Suppose a break-even analysis for a new consumer product suggests that it will be successful if more than 23% of consumers are willing to try it. This 23% is the reference value, $\mu_0$; it comes from a theoretical analysis, not from a random sample of data. To decide whether or not to launch the product, you have gathered some data from randomly selected consumers and have computed a one-sided confidence interval. Based just on the data, you expect 44.1% of consumers to try it ($\overline{X} = 44.1\%$), and your one-sided confidence interval statement is that you are 95% sure that *at least* 38.4% of consumers will be willing to try it. Since your break-even point, $\mu_0 = 23\%$, is well outside of this confidence interval (and, hence, it is *not* reasonable to suppose that the mean could be 23%), you do have convincing

**TABLE 10.4.1** Testing the Percentage of Consumers Who Are Willing to Try a New Product (Confidence Interval Approach)

| | | |
|---|---|---|
| Null hypothesis | $H_0: \mu \le \mu_0$ | $H_0: \mu \le 23\%$ |
| Research hypothesis | $H_1: \mu > \mu_0$ | $H_1: \mu > 23\%$ |
| Average | $\overline{X}$ | 44.1% |
| Standard error | $S_{\overline{X}}$ | 3.47% |
| Sample size | $n$ | 205 |
| Reference value | $\mu_0$ | 23% |
| Confidence interval | $\overline{X} - t_{\text{table}} S_{\overline{X}}$ | "We are 95% sure that the population mean is at least 38.4%" |
| Decision | Accept $H_1$ | "We expect significantly more than 23% of consumers to try our product"* |

*\* Significant ($p < 0.05$) using a one-sided test.*

evidence that the population mean is greater than 23%. A summary of the situation is shown in Table 10.4.1.[15]

The decision is made to accept $H_1$ because the reference value is not in the confidence interval (i.e., 23% is not "at least 38.4%").

Since the reference value, 23%, is so far below the confidence interval, you should try for a more impressive significance level. In fact, you can claim 99.9% confidence that the population mean is at least 33.4% using the $t$ table value 3.090 for this one-sided confidence interval. Since the reference value, 23%, is outside even this confidence interval, the result is *very highly significant* ($p < 0.001$).

---

15. Because this is a binomial situation, you may substitute $p$ in place of $\overline{X}$, $S_p$ in place of $S_{\overline{X}}$, $\pi$ in place of $\mu$, and $\pi_0$ in place of $\mu_0$ throughout. To see that this example is also correct as stated here, note that $\overline{X} = p$ for the data set $X_1, X_2, ..., X_n$ where each number is either 0 or 1 according to the response of each consumer.

### How to Perform the Test

Table 10.4.2 shows how to perform a one-sided test, giving complete instructions for both types of testing situations (i.e., to see if $\overline{X}$ is significantly larger than, or significantly smaller than, $\mu_0$), performed using either the confidence interval or the $t$ statistic method, and giving both types of possible conclusions (significant or not) and their interpretations. A useful guiding principle is that it's significant if the reference value, $\mu_0$, is not within the one-sided confidence interval constructed to match the direction of the research hypothesis.

## TABLE 10.4.2 One-Sided Testing

**One-Sided Testing to See If $\mu$ Is Larger than $\mu_0$**

The hypotheses being tested are $H_0 : \mu \leq \mu_0$ against $H_1 : \mu > \mu_0$.

The confidence interval statement is "We are 95% sure that the population mean is *at least* as large as $\overline{X} - t_{\text{table}} S_{\overline{X}}$."

The $t$ statistic is $t_{\text{statistic}} = (\overline{X} - \mu_0)/S_{\overline{X}}$ (*Note:* Do not use absolute values for one-sided testing).

**Is $\overline{X} - t_{\text{table}} S_{\overline{X}} \leq \mu_0$? This is the confidence interval approach, asking: Is the reference value, $\mu_0$, *inside* the confidence interval? Equivalently, with the $t$ statistic approach: Is $t_{\text{statistic}} \leq t_{\text{table}}$? If so, then**

Accept the null hypothesis, $H_0$, as a reasonable possibility.

Do *not* accept the research hypothesis, $H_1$.

The sample average, $\overline{X}$, is *not significantly larger* than the reference value, $\mu_0$.

If $\overline{X}$ is larger than $\mu_0$, the observed difference could reasonably be due to random chance alone.

The result is *not statistically significant.*

**Is $\overline{X} - t_{\text{table}} S_{\overline{X}} > \mu_0$? This is the confidence interval approach, asking: Is the reference value, $\mu_0$, *outside* the confidence interval? Equivalently, with the $t$ statistic approach: Is $t_{\text{statistic}} > t_{\text{table}}$? If so, then**

Accept the research hypothesis, $H_1$.

Reject the null hypothesis, $H_0$.

The sample average, $\overline{X}$, is *significantly larger* than the reference value, $\mu_0$.

The observed difference between the sample average, $\overline{X}$, and the reference value, $\mu_0$, could *not* reasonably be due to random chance alone.

The result is *statistically significant.*

**One-Sided Testing to See If $\mu$ Is Smaller than $\mu_0$**

The hypotheses being tested are $H_0 : H_0 : \mu \geq \mu_0$ against $H_1 : \mu < \mu_0$

The confidence interval statement is "We are 95% sure that the population mean is *not larger* than $\overline{X} + t_{\text{table}} S_{\overline{X}}$."

The $t$ statistic is $t_{\text{statistic}} = (\overline{X} - \mu_0)/S_{\overline{X}}$ (*Note:* Do not use absolute values for one-sided testing).

**Is $\overline{X} + t_{\text{table}} S_{\overline{X}} \geq \mu_0$? This is the confidence interval approach, asking: Is the reference value, $\mu_0$, *inside* the confidence interval? Equivalently, with the $t$ statistic approach: Is $t_{\text{statistic}} \geq -t_{\text{table}}$? If so, then**

Accept the null hypothesis, $H_0$, as a reasonable possibility.

Do *not* accept the research hypothesis, $H_1$.

The sample average, $\overline{X}$, is *not significantly smaller* than the reference value, $\mu_0$.

If $\overline{X}$ is smaller than $\mu_0$, the observed difference could reasonably be due to random chance alone.

The result is *not statistically significant.*

**Is $\overline{X} + t_{\text{table}} S_{\overline{X}} < \mu_0$? This is the confidence interval approach, asking: Is the reference value, $\mu_0$, *outside* the confidence interval? Equivalently, with the $t$ statistic approach: Is $t_{\text{statistic}} < -t_{\text{table}}$? If so, then**

Accept the research hypothesis, $H_1$.

Reject the null hypothesis, $H_0$.

The sample average, $\overline{X}$, is *significantly smaller* than the reference value, $\mu_0$.

The observed difference between the sample average, $\overline{X}$, and the reference value, $\mu_0$, could *not* reasonably be due to random chance alone.

The result is *statistically significant.*

If you use the confidence interval method, remember that there are two different one-sided confidence interval statements. You want to choose the one that matches the side of the claim of the research hypothesis. For example, if your research hypothesis is $H_1 : \mu > \mu_0$, your one-sided confidence interval will consist of all values for $\mu$ that are *at least as large* as the appropriate computed number, $\overline{X} - t_{\text{table}} S_{\overline{X}}$, using the one-sided $t$ value from the $t$ table.

Figure 10.4.1 shows that in order for you to decide that $\overline{X}$ is significantly larger than $\mu_0$, the distance between them must be sufficiently large to ensure that it is not just due to random chance. Figure 10.4.2 gives the corresponding picture for a one-sided test on the other side.

If you use the $t$ statistic, the test is decided by comparing $t_{\text{statistic}} = (\overline{X} - \mu_0)/S_{\overline{X}}$ either to the $t$ table value, $t_{\text{table}}$, or to its negative, $-t_{\text{table}}$, depending on the side being tested (specifically, depending on whether the research hypothesis is $H_1 : \mu > \mu_0$ or $H_1 : \mu < \mu_0$). The idea behind the calculation is that the test is significant if the data correspond to the side of the research hypothesis and the $t$ statistic is large in magnitude (which ensures that the difference between $\overline{X}$ and $\mu_0$ is larger than ordinary randomness). Note that the $t$ statistic is the same for one- and for two-sided testing, but that you use it differently to decide significance.

### Example
#### Launching a New Product, Revisited

For the consumer product launch example considered earlier, the launch will be profitable only if more than 23% of consumers try the product. The appropriate facts and results are as shown in Table 10.4.3. We now perform the same one-sided test we did earlier, except that we use the $t$ statistic method this time.

The decision is made to accept $H_1$ because $t_{\text{statistic}} > t_{\text{table}}$; that is, we have $6.08 > 1.645$, using the appropriate criterion from Table 10.4.2 for this research hypothesis ($H_1 : \mu > \mu_0$).

*(Continued)*

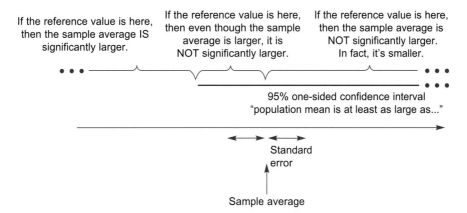

**FIGURE 10.4.1** Using a one-sided test to see if $\mu$ is larger than the reference value, $\mu_0$. The one-sided confidence interval uses the same side as the research hypothesis (namely, those reasonably possible values of $\mu$ that are at least as large as the endpoint value of the interval). Only if the reference value, $\mu_0$, is well below the sample average will you decide that the sample average is significantly larger.

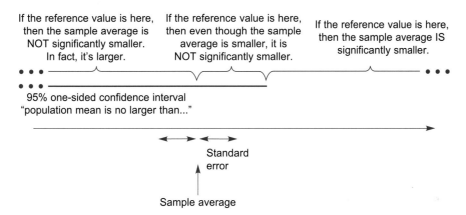

**FIGURE 10.4.2** Using a one-sided test to see if $\mu$ is smaller than the reference value, $\mu_0$. The one-sided confidence interval uses the same side as the research hypothesis (namely, those reasonably possible values of $\mu$ that are smaller than or equal to the endpoint value of the interval). Only if $\mu_0$ is well above the sample average will you decide that the sample average is significantly smaller.

**TABLE 10.4.3 Testing the Percentage of Consumers Who Are Willing to Try a New Product (*t* Statistic Approach)**

| Null hypothesis | $H_0 : \mu \leq \mu_0$ | $H_0 : \mu \leq 23\%$ |
|---|---|---|
| Research hypothesis | $H_1 : \mu > \mu_0$ | $H_1 : \mu > 23\%$ |
| Average | $\overline{X}$ | 44.1% |
| Standard error | $S_{\overline{X}}$ | 3.47% |
| Reference value | $\mu_0$ | 23% |
| *t* statistic | $t_{\text{statistic}} = \dfrac{\overline{X} - \mu_0}{S_{\overline{X}}}$ | $\dfrac{44.1 - 23}{3.47} = 6.08$ |
| Critical value | $t_{\text{table}}$ | 1.645 |
| Decision | Accept $H_1$ | "We expect significantly more than 23% of consumers to try our product"* |

*\* Significant (p < 0.05) using a one-sided test.*

## Example—cont'd

In the end, the result is the same (i.e., significant) whether you use the one-sided confidence interval or the one-sided *t* test. In fact, since the *t* statistic exceeds even the one-sided *t* table value (3.090) for testing at level 0.001, you may claim a *very highly significant* result (*p*, 0.001).

## Example
### *Will Costs Go Down?*

You have tested a new system that is supposed to reduce the *variable cost* or *unit cost* of production (i.e., the cost of producing each extra unit, ignoring fixed costs such as rent, which would be the same whether or not you produced the extra unit). Since the new system would involve some expenditures, you will be willing to use it only if you can be convinced that the variable cost will be less than $6.27 per unit produced.

Based on a careful examination of 30 randomly selected units that were produced using the new system, you found an average variable cost of $6.05. It looks as if your variable cost is, on average, less than the target of $6.27. But is it *significantly* less? That is, can you expect the long-run mean cost to be below $6.27, or is this just a random fluke of the particular 30 units you examined? Based on the information so far, you cannot say because you don't yet know how random the process is. Is $6.05 less than $6.27? Yes, of course. Is $6.05 *significantly* less than $6.27? You can tell only by comparing the difference to the randomness of the process, using the standard error and the *t* table.

So you go back and find the standard deviation, and then you compute the standard error, $0.12. Table 10.4.4 shows a summary of the situation and the results of a one-sided test (with both methods shown) of whether or not your costs are significantly lower than the required amount.

Using the confidence interval approach, you are 95% sure that the mean variable cost is less than $6.25. You are even more sure that it is less than the required amount, $6.27. Hence, the result is significant. Or you could simply note that the reference value ($6.27) is not in the confidence interval, which extends only up to $6.25.

When you use the *t* statistic approach, the result is significant because $t_{statistic} < -t_{table}$; that is, we have $-1.833 < -1.699$, using the appropriate criterion from Table 10.4.2 for this research hypothesis ($H_1: \mu < \mu_0$).

You are entitled to use a one-sided test in this case because you are really interested in just that one side. If you can be convinced that the mean variable cost is less than $6.27, then the system is worth considering. If not, then you are not interested. By using a one-sided test in this way, you are admitting that, had the system been really bad, you would not have been able to say that "variable costs are significantly *more than*…"; you would only be able to say that "they are *not significantly less*."

Had you decided to use a two-sided test, which is valid but less efficient, you would actually have found that the

### TABLE 10.4.4 Testing the Long-Run Variable Cost

| Reference value | $\mu_0$ | $6.27 |
|---|---|---|
| Null hypothesis | $H_0: \mu \geq \mu_0$ | $H_0: \mu \geq \$6.27$ |
| Research hypothesis | $H_1: \mu < \mu_0$ | $H_1: \mu < \$6.27$ |
| Average | $\overline{X}$ | $6.05 |
| Standard error | $S_{\overline{X}}$ | $0.12 |
| Sample size | $n$ | 30 |
| Confidence interval | $\overline{X} + t_{table} S_{\overline{X}}$ | "You are 95% sure that the long-run mean variable cost is less than $6.25" |
| *t* statistic | $t_{statistic} = \dfrac{\overline{X} - \mu_0}{S_{\overline{X}}}$ | $\dfrac{6.05 - 6.27}{0.12} = -1.833$ |
| Critical value | $-t_{table}$ | $-1.699$ |
| Decision | Accept $H_1$ | "Variable costs under the new system are significantly less than $6.27"* |

*\* Significant (p < 0.05) using a one-sided test.*

result is *not* significant! The two-sided confidence interval extends from $5.80 to $6.30, which *does* contain the reference value. The *t* statistic is still $-1.833$, but the two-sided *t* value from the table is 2.045, which is now larger than the absolute value (1.833) of the *t* statistic. Thus, this example shows that you can have a significant one-sided result but a nonsignificant two-sided result. This can happen only when the one-sided significance is borderline, passing the test with just a small margin, as in this example.

Should you buy the system? This is a business strategy question, not a statistical question. Use the results of the hypothesis test as one of your inputs, but consider all the other factors, such as availability of investment capital, personnel implications, and interactions with other projects. And don't forget this: Although hypothesis testing has led you to accept the research hypothesis that the variable costs are less than your threshold, this has *not* been absolutely proven; there is still room for error. You can't give a number for the probability that your hypothesis testing decision is wrong because you don't know which hypothesis is really true. The best you can say is that *if* the new system has variable costs of exactly $6.27, then you would wrongly decide significance (as you may have here) only 5% of the time.

## Example
### *Can You Create Value by Changing Your Firm's Name?*

When a large firm changes its name, it's a major event. The budgets for advertising the change of name and for setting up the new image can be enormous. Why do firms do it? According to the theory of financial budgeting, firms should undertake only projects that increase the value of the firm to the owners, the shareholders. If it is reasonable for a firm to spend those resources to change its name, then you should observe an increase in the firm's value as measured by the price of its stock.

A study of the change in firm value around the time of a name change announcement might use a one-sided statistical hypothesis test to see if the stock price really did go up. One of the difficulties of measuring this kind of market price reaction is that the stock market as a whole responds to many forces, and the name change should be evaluated in light of what the stock price did compared to what it should have done based on the entire stock market during that time. So if the stock market was up, you would have to find the firm's stock up an *even larger percentage* than you would ordinarily expect on such a day before deciding that the announcement has had an effect.

This kind of *event study*, using an adjustment for large-scale market forces, involves computing an *abnormal return*, which represents the rate of return an investor would have received by holding the firm's stock, minus the rate of return the investor would have expected from an investment of similar risk (but involving no name change) during the same time period.

Thus, a positive abnormal return would show that the name change caused a price rise even larger than we would have expected in the absence of the name change. This could happen because the stock market was generally up and the firm's stock was up much more. Or it could happen because the stock market was down but the firm's stock was down less than expected given the market as a whole.

One study looked at 58 corporations that changed their names and reported the methods as follows:[16]

*In order to test if [the abnormal return due to the name change] is different from zero, the test statistic employed is the ratio of the average abnormal returns … to their standard deviation…. This test statistic … is distributed standard normal if* n *is large enough.*

The study's authors are saying that they used a *t* statistic to test the sample average against the reference value, $\mu_0 = 0$, of no abnormal returns due to a name change. The "standard deviation" they refer to is the standard error of this estimated quantity. Since their sample size was large enough, the *t* table values (for $n = $ infinity) are the same as the standard normal values.

The results of this study were given as follows:

*The mean abnormal return was found to be 0.61%, with a corresponding t statistic … of 2.15. Thus, if the null hypothesis is that the residual returns are drawn from a population with a nonpositive mean, the one-sided null hypothesis can be rejected.*

Since the study's authors rejected the null hypothesis, they accepted the research hypothesis. They showed that the stock price rises significantly as a result of a name change. Does this say that you should rush out and change

your firm's name as often as possible? Well, not really. They discussed the implications as follows:

*Our findings are that, for most of the firms, name changes are associated with improved performance, and that the greatest improvement tends to occur in firms that produce industrial goods and whose performance prior to the change was relatively poor…. Our findings do not support, however, the contention that the new name per se [i.e., by itself] will enhance demand for the firm's products. Rather, it seems that the act of a name change serves as a signal that other measures to improve performance such as changes in product offerings and organizational changes will be seriously and successfully undertaken.*

Note that with a *t* statistic of 2.15, they found a significant result at the 5% level (since it exceeds 1.645). However, since the one-sided *t* table value at the 1% level is 2.326, their result is significant but not highly significant.

16. D. Horsky and P. Swyngedouw, "Does It Pay to Change Your Company's Name? A Stock Market Perspective," *Marketing Science* 6 (1987), pp. 320–35.

## 10.5 TESTING WHETHER OR NOT A NEW OBSERVATION COMES FROM THE SAME POPULATION

By now, you probably have the idea that if you can construct a confidence interval, you can do a hypothesis test. This is correct. Based on the prediction interval in Chapter 9 for a new observation (instead of for the population mean), you may now quickly test whether or not this new observation came from the same population as the sample data. The null hypothesis, $H_0$, claims that the new observation comes from the same normally distributed population as your sample, and the research hypothesis, $H_1$, claims that it does not. The data set is assumed to be a random sample.

The test is fairly simple, now that you know the basics of hypothesis testing and confidence intervals. Find the prediction interval (a special kind of confidence interval) based on the sample (but not using the new observation) using the *standard error for prediction*, $S\sqrt{1 + 1/n}$, as explained in Chapter 9. Then get the new observation. If the new observation is *not* in the interval, you will conclude that it is significantly different from the others.

If you want to use the *t* test method, simply compute your *t* statistic as the new observation minus the sample average, divided by the standard error for prediction. Then proceed just as before, comparing the *t* statistic to the critical value from the *t* table (with $n - 1$ degrees of freedom). The *t* statistic for testing a new observation is

$$t_{\text{statistic}} = \frac{X_{\text{new}} - \overline{X}}{S\sqrt{1 + 1/n}}$$

If you want a one-sided test to claim that the new observation is either significantly larger or significantly smaller than the average of the others, simply find the appropriate one-sided prediction interval or compare the $t$ statistic to the one-sided value in the $t$ table.

---

### Example
#### Is This System under Control?

You are scratching your head. Usually, these art objects of molded porcelain that come out of the machine weigh about 30 pounds each. Of course, there is some variation; they don't all weigh *exactly* 30 pounds each—in fact, these "one-of-a-kind" objects are not supposed to be identical. But this is ridiculous! A piece that just came out weighs 38.31 pounds, way over the expected weight. You are wondering if the process has gone *out of control*, or if this is just a random occurrence to be expected every now and again. You would rather not adjust the machinery, since this involves shutting down the assembly line and finding the trouble; but if the line is out of control, the sooner you fix the problem, the better.

The null hypothesis claims that the system is still under control, that is, that the most recent piece is the same as the others except for the usual random variation. The research hypothesis claims that the system is out of control, and the most recent piece is significantly different from the others. Here is the information for the most recent piece as well as for a sample of ordinary pieces:

| | |
|---|---|
| Sample size, $n$ | 19 |
| Sample average, $\overline{X}$ | 31.52 pounds |
| Standard deviation, $S$ | 4.84 pounds |
| New observation, $X_{new}$ | 38.31 pounds |

The standard error for prediction is

$$\text{Standard error for prediction} = S\sqrt{1 + \frac{1}{n}}$$

$$= 4.84\sqrt{1 + \frac{1}{19}}$$

$$= 4.965735$$

It would not be fair to use a one-sided test here because you would most certainly be interested in items that are greatly underweight as well as those that are overweight; either way, the system would be considered out of control. The two-sided prediction interval, based on the $t$ value of 2.101 from the $t$ table, extends from $31.52 - 2.101 \times 4.97 = 21.1$ to $31.52 + 2.101 \times 4.97 = 42.0$.

We are 95% sure that a new observation, taken from the same population as the sample, will be somewhere between 21.1 and 42.0 pounds.

Since the new observation, at 38.31 pounds, is within this prediction interval, it seems to be within the range of reasonable variation. Although it is near the high end, it is *not significantly different* from the others.

---

The $t$ statistic is less (in absolute value) than the critical value, 2.101, confirming your decision to accept the null hypothesis:

$$t_{\text{statistic}} = \frac{38.31 - 31.52}{4.965735}$$

$$= 1.367$$

In retrospect, you probably should not have been surprised at a piece weighing 38.31 pounds. Since the sample standard deviation is 4.84 pounds, you expect individuals to be about this far from the average. This piece is not even two standard deviations away and is therefore (even according to this approximate rule) within the reasonable 95% region. This quick look is just an approximation; when you use the standard error for prediction, your answer is exact because you took into account both the variation in your sample and the variation of the new observation in a mathematically correct way.

## 10.6  TESTING TWO SAMPLES

To test whether or not two samples are significantly different, on average, all you need to know are (1) the appropriate standard error to use for evaluating the average difference and (2) its degrees of freedom. The problem will then be essentially identical to the methods you already know: You will be testing an observed quantity (the observed average difference) against a known reference value (zero, indicating no difference) using the appropriate standard error and critical value from the $t$ table.

You will see this method repeated over and over in statistics. Whenever you have an estimated quantity and its own standard error, you can easily construct confidence intervals and do hypothesis testing. The applications get more complex (and more interesting), but the methods are just the same as the procedures you already know. Let's generalize this method now to the two-sample case.

### The Paired $t$ Test

The **paired $t$ test** is used to test whether two columns of numbers are different, on average, *when there is a natural pairing between the two columns*. This is appropriate, for example, for "before/after" studies, where you have a measurement (such as a score on a test or a rating) for each person or thing both before and after some intervention (seeing an advertisement, taking a medication, adjusting the gears, etc.).

In fact, you already know how to do a paired $t$ test because it can be changed into a familiar one-sample problem by working with the *differences*, for example, "after" minus "before," instead of with the two samples individually. It is crucial that the data be paired; otherwise, it would not be clear for which pairs to find differences.

It is not enough to have the averages and standard deviations for each of the two groups. This would lack any indication of the important information conveyed by the pairing of the observations. Instead, you will work with the average and the standard deviation of the *differences*.

A paired *t* test can be very effective even when individuals show lots of variation from one to another. Since it concentrates on *changes*, it can ignore the (potentially confusing) variation in *absolute* levels of individuals. For example, individuals could be very different from one another, and yet the changes could be very similar (e.g., everyone receives a $100 pay raise). The paired *t* test is not distracted by this individual variability in its methods to detect a systematic change.

Again, some assumptions are required for validity of the paired *t* test. The first assumption is that the elementary units being measured are a random sample selected from the population of interest. Each elementary unit produces two measurements. Next, look at the data set consisting of the differences between these two sets of measurements. The second assumption is that these differences are (at least approximately) normally distributed.

**Example**
*Reactions to Advertising*

An advertisement is being tested to determine if it is effective in creating the intended mood of relaxation. A sample of 15 people has been tested just before and just after viewing the ad. Their questionnaire included many items, but the one being considered now asked them to describe their current feelings on a scale from 1 (very tense) to 5 (completely relaxed). The results are shown in Table 10.6.1. (Note in particular that the average relaxation score increased by 0.67, going from 2.80 before to 3.47 after.)

This looks a lot like a two-sample problem, but, in a way, it isn't. It can be viewed as a one-sample problem based on the changes in the relaxation scores. For example, person 1 went from a 3 to a 2, for a change of –1 in relaxation score. (By convention, we compute the differences as "after" minus "before" so that increases end up as positive numbers and decreases as negatives.) Computing the difference for each person, you end up with a familiar one-sample problem, as shown in Table 10.6.2.

You know exactly how to attack this kind of one-sample problem. Using the two-sided *t* value of 2.145 from the *t* table, together with a sample average of $\overline{X} = 0.6667$ and a standard error of $S_{\overline{X}} = 0.2702$, you find

You are 95% sure that the mean change in relaxation score for the larger population is somewhere between 0.087 and 1.25.

**TABLE 10.6.1  Relaxation Scores**

|  | Before | After |
|---|---|---|
| Person 1 | 3 | 2 |
| Person 2 | 2 | 2 |
|  | 2 | 2 |
|  | 4 | 5 |
|  | 2 | 4 |
|  | 2 | 1 |
| . | 1 | 1 |
| . | 3 | 5 |
| . | 3 | 4 |
|  | 2 | 4 |
|  | 5 | 5 |
|  | 2 | 3 |
|  | 4 | 5 |
|  | 3 | 5 |
| Person 15 | 4 | 4 |
| Sample size | 15 | 15 |
| Average | 2.8000 | 3.4667 |
| Standard deviation | 1.0823 | 1.5055 |

What is the reference value, $\mu_0$, here? It is $\mu_0 = 0$ because a change of zero indicates *no effect* on relaxation in the population due to viewing the advertisement.

The hypothesis test merely involves seeing whether or not $\mu_0 = 0$ is in the confidence interval. It isn't, so the result is significant; that is, 0 is not a reasonable value for the change in the population based on your data:

Viewing of the advertisement resulted in a significant increase in relaxation ($p < 0.05$, two-sided test).

A two-sided test is needed here because it is also of interest to find out if an advertisement caused a significant *reduction* in relaxation. Having found significance with the two-sided test, you may state the result as a one-sided conclusion.

For completeness, here are the underlying hypotheses: The null hypothesis, $H_0: \mu = 0$, claims that the population mean change in relaxation from before to after is zero; that is, there is no change in mean relaxation. The research hypothesis, $H_1: \mu \neq 0$, claims that there is indeed a change in mean relaxation from before to after.

## TABLE 10.6.2 Change in Score

| | After – Before |
|---|---|
| Person 1 | –1 |
| Person 2 | 0 |
| | 0 |
| | 1 |
| | 2 |
| . | –1 |
| . | 0 |
| . | 2 |
| | 1 |
| | 2 |
| | 0 |
| | 1 |
| | 1 |
| | 2 |
| Person 15 | 0 |
| | |
| Sample size | 15 |
| Average | 0.6667 |
| Standard deviation | 1.0465 |
| Standard error | 0.2702 |

### Example

*Data Mining to Compare Current to Previous Donations*

After people skip a charitable contribution, does their next donation tend to be smaller or larger than the amount they used to give? To answer this, consider the donations database with 20,000 entries on the companion website, which we will view as a random sample from a much larger database. Recall that, at the time of the mailing, these people had given in the past but not recently. Focusing attention on the 989 donors (who did make a donation in response to the current mailing), we know the average size of their previous donations (this variable is named "AvgGift_D1" in the database for these 989 current donors) and the actual size of their current donation (named "Donation_D1").

The current donation (on average, $15.765 for the 989 current donors) is larger than the average of the previous donations (on average, $11.915, for the current donors). This suggests that it is indeed worthwhile to continue to ask for donations even when people do not respond initially, by showing that (at least among those who do respond, after a while) the donation amount increases, on

average. But does it increase significantly? A hypothesis test will give the answer. This is a paired situation because each person has a current donation amount and a past average donation amount, and we are interested in their difference. The standard error of the differences is $0.2767, and the 99.9% confidence interval statement (using 3.291 from the $t$ table and the average difference of $15.765 – $11.915 = $3.850) is

> We are 99.9% sure that the population mean increase in donation amount, for past donors who lapsed but then resumed donating, is somewhere between $2.94 and $4.76.

Because the reference value 0 (representing no difference, on average) is not in this confidence interval, we conclude that the difference is significant at the 0.1% level, and the difference is very highly significant ($p < 0.001$). The conclusion of this paired hypothesis test may be stated as

> Past donors who lapsed but then resumed donating showed a very highly significant increase ($p < 0.001$) in their donation amount, on average, as compared to their previous average donation.

Alternatively, the $t$ statistic is $(15.765 – 11.915)/0.2767 = 13.9$, which is considerably larger than the $t$ table value, 3.291, for testing at the 0.1% level, confirming the differences as very highly significant.

Often, when large amounts of data are available, differences are found to be very highly significant. This happens because a larger sample size (all else equal) leads to a smaller standard error after dividing the standard deviation by the square root of the large sample size $n$. In fact, in this example the $p$-value is much smaller than 0.001. When we use statistical software, the exact $p$-value is $p = 2.48E – 40$, where the "$E – 40$" in computer output tells us to move the decimal point 40 places (to the left since –40 is a negative number) so we actually have the very small $p$-value

$$p = 0.0000000000000000000000000000000000000000248$$

which would be reported by some statistical software as "$p = 0.0000$," obtained by rounding.

## The Unpaired *t* Test

The **unpaired *t* test** is used to test whether two *independent* columns of numbers are different, on the average. Such columns have no natural pairings. For example, you might have data on firms in each of two industry groups, or you might want to compare samples from two different production lines. These cases cannot be reduced to just a single column of numbers; you'll have to deal with both samples.

Once you find the appropriate standard error for this situation, the rest is easy. You will have an estimate (the difference between the two sample averages), its "personal"

**TABLE 10.6.3 Notation for Two Samples**

|  | Sample 1 | Sample 2 |
|---|---|---|
| Sample size | $n_1$ | $n_2$ |
| Average | $\overline{X}_1$ | $\overline{X}_2$ |
| Standard deviation | $S_1$ | $S_2$ |
| Standard error | $S_{\overline{X}_1}$ | $S_{\overline{X}_2}$ |
| Average difference | | $\overline{X}_2 - \overline{X}_1$ |

Small-sample situation (equal variabilities assumed):

$$S_{\overline{X}_2 - \overline{X}_1} = \sqrt{\frac{(n_1 - 1)S_1^2 + (n_2 - 1)S_2^2}{n_1 + n_2 - 2}\left(\frac{1}{n_1} + \frac{1}{n_2}\right)}$$

Degrees of freedom $= n_1 + n_2 - 2$

Be careful to use the correct variability measure for each formula, either the sample standard deviation or the standard error for the sample; the large-sample formula shows how to use either one. If, in the small-sample case, you are given the standard errors instead of the standard deviations, convert them to standard deviations by multiplying by the square root of the sample size for each sample. Note that in both formulas the standard deviations are squared before being combined.[18]

For the large-sample standard error formula, the estimated *variances* of the estimators $\overline{X}_1$ and $\overline{X}_2$ are added to derive the estimated variance of the difference. Taking the square root, you find the estimated standard deviation of the difference, which gives you the standard error of the difference.

For the small-sample standard error formula, the first fraction inside the square root sign combines the standard deviations using a weighted average (weighted according to the number of degrees of freedom for each one). The rest of the formula converts from the variation of *individuals* to the variation of the *average difference* by summing the reciprocal sample sizes, doing twice what you would do once to find an ordinary standard error.

The hypotheses being tested are $H_0 : \mu_1 = \mu_2$ against $H_1 : \mu_1 \neq \mu_2$. These may be written equivalently as $H_0 : \mu_1 - \mu_2 = 0$ against $H_1 : \mu_1 - \mu_2 \neq 0$. The assumptions needed in order for an unpaired two-sample $t$ test to be valid include the usual ones, plus one new one for the small-sample case only. First, each sample is assumed to be a random sample from its population. (There are two populations here, with each sample representing one of them independently of the other.) Second, each sample average is assumed to be approximately normally distributed, as we have required before. Finally, for the small-sample case only, it is also assumed that the *standard deviations are equal* in the two populations: $\sigma_1 = \sigma_2$. That is, the two populations differ (if at all) only in mean value and not in terms of the variability of individuals from the mean for their population.

standard error, and the appropriate number of degrees of freedom. The rest, constructing the confidence interval and performing the hypothesis test, should be routine for you by now.

We have two samples, sample 1 and sample 2. The summary statistics for each sample will be denoted in a natural way, as shown in Table 10.6.3.

Here's what's new. The **standard error of the difference** indicates the sampling variability of the *difference* between the two sample averages. There are two different formulas: a large-sample formula, to be used whenever both sample sizes are 30 or larger, and a small-sample formula that is based on the assumption that the two populations have the same variability.[17] The large-sample formula works even when the variabilities are unequal by directly combining the two standard errors, $S_{\overline{X}_1}$ and $S_{\overline{X}_2}$. The small-sample formula includes a weighted average of the sample standard deviations to estimate the population variability (assumed equal in the two populations). The small-sample standard error has $n_1 + n_2 - 2$ degrees of freedom: We start with the combined sample size, $n_1 + n_2$, and then subtract 1 for each sample average that was estimated. Here are formulas for the standard error of the difference for each case:

**Standard Error of the Difference**

Large-sample situation ($n_1 \geq 30$ and $n_2 \geq 30$):

$$S_{\overline{X}_2 - \overline{X}_1} = \sqrt{S_{\overline{X}_1}^2 + S_{\overline{X}_2}^2} = \sqrt{\frac{S_1^2}{n_1} + \frac{S_2^2}{n_2}}$$

$$S_{p_2 - p_1} = \sqrt{S_{p_1}^2 + S_{p_2}^2} \text{ (for two binomials)}$$

Degrees of freedom $=$ infinity

---

17. Solutions are available for the small-sample problem when variabilities are unequal, but they are more complex. One approach is presented in G. W. Snedecor and W. G. Cochran, *Statistical Methods*, 6th ed. (Ames: Iowa State University Press, 1976), p. 115.

18. Thus, the *variances* are averaged here, as happens in so many formulas like this one. This has led theoretical statisticians to concentrate their attention on the variance. However, anyone who wants to interpret such numbers in their meaningful measurement units will have to take the square root. This is why we work with the standard deviation rather than with the variance in this book. Note that their information is equivalent because either may be converted to the other.

## Example
### Gender Discrimination and Salaries

Your firm is being sued for gender discrimination, and you are evaluating the documents filed by the other side. They have included a statistical hypothesis test, based on salaries of men and women, that finds a "highly significant difference," on average, between men's and women's salaries. Table 10.6.4 shows a summary of their results.

### TABLE 10.6.4 Salaries Arranged by Gender

| | Women | Men |
|---|---|---|
| | $21,100 | $38,700 |
| | 29,700 | 30,300 |
| | 26,200 | 32,800 |
| | 23,000 | 34,100 |
| | 25,800 | 30,700 |
| | 23,100 | 33,300 |
| | 21,900 | 34,000 |
| | 20,700 | 38,600 |
| | 26,900 | 36,900 |
| | 20,900 | 35,700 |
| | 24,700 | 26,200 |
| | 22,800 | 27,300 |
| | 28,100 | 32,100 |
| | 25,000 | 35,800 |
| | 27,100 | 26,100 |
| | | 38,100 |
| | | 25,500 |
| | | 34,000 |
| | | 37,400 |
| | | 35,700 |
| | | 35,700 |
| | | 29,100 |
| Sample size | 15 | 22 |
| Average | $24,467 | $33,095 |
| Standard deviation | $2,806 | $4,189 |
| Standard error | $724 | $893 |
| Average difference | $8,628 | |

There are 15 women and 22 men in this department; the average yearly salaries are $24,476 for women and $33,095 for men. On average, men earn $8,628 more than women. This is a plain, clear fact. However, the issue is whether or not this difference is within the usual random variation. Essentially, no matter how you divide this group of 37 people into two groups of sizes 15 and 22, you will find different average salaries. The question is whether such a large difference as found here could reasonably be the result of a *random* allocation of salaries to men and to women, or if there is a need for some other explanation for the apparent inequity.

Each standard deviation ($2,806 for women, $4,189 for men) indicates that individuals within each group differ from their group average by roughly this amount. There is a bit more variation among the men than the women, but not enough to keep us from going ahead with a two-sample unpaired *t* test.

The standard errors ($724 for women, $893 for men) indicate about how far the group averages are from the means for their respective idealized populations. For example, if you view these particular 15 women as a random sample from the idealized population of women in similar circumstances, then the average women's salary of $24,467 (random, due to the fact that only 15 have been examined) is approximately $724 away from the idealized population mean.

This is clearly a two-sample *unpaired* situation. Although you might want to subtract Mary's salary from Jim's, there is no systematic way to complete the process because these are really two separate, unpaired groups.

To evaluate the average difference of $8,628 to see if it could be reasonably due to randomness, you need its standard error and number of degrees of freedom. Here are computations for the small-sample formula:

$$S_{\bar{X}_2 - \bar{X}_1} = \sqrt{\frac{(n_1 - 1)S_1^2 + (n_2 - 1)S_2^2}{n_1 + n_2 - 2} \left(\frac{1}{n_1} + \frac{1}{n_2}\right)}$$

$$= \sqrt{\frac{(15 - 1)2,806^2 + (22 - 1)4,189^2}{15 + 22 - 2} \left(\frac{1}{15} + \frac{1}{22}\right)}$$

$$= \sqrt{\frac{(14)2,806^2 + (22)4,189^2}{35}(0.066667 + 0.045455)}$$

$$= \sqrt{13.678.087 \times 0.112121}$$

$$= \$1,238$$

$$\left(\begin{array}{c}\text{Degrees of} \\ \text{freedom}\end{array}\right) = n_1 + n_2 - 2 = 15 + 22 - 2 = 35$$

The 99.9% confidence interval is based on the *t* value 3.591 from the *t* table. Be sure to use the *degrees of freedom* column in the *t* table since more than one sample is involved. The confidence interval extends from $8,628 - 3.591 \times 1,238$ to $8,628 + 3.591 \times 1,238$:

You are 99.9% sure that the population mean salary difference is somewhere between $4,182 and $13,074.

This confidence interval does *not* include the reference value of 0, where such a reference value corresponds to no mean difference in salary between men and women in the population. Your hypothesis testing decision therefore is as follows:

> The average difference between men's and women's salaries is very highly significant ($p < 0.001$).

This result is also supported by the fact that the $t$ statistic is $8,628/1,238 = 6.97$, well above the $t$ table value of 3.591 for testing at the 0.001 significance level.

What can you conclude from this? First of all, the salary allocation between men and women is not just random. Well, it *could* be random, but only if you are willing to admit that a rare, less than 1-in-1,000 event has happened (since this is the meaning of the significance level 0.001). Second, if the salary allocation is not random, there must be some other explanation. At this point, an individual may give his or her own favorite reason as though it were completely proven by the test result. However, it is one thing to say that there is a reason and another to be able to say *what* the reason is. Statistics has ruled out random chance as a reasonable possibility. That's all. If you want to propose a reason for the observed salary difference, you are entitled to do so, but don't expect the field of statistics to back you up. Having set the stage for an explanation, the field of statistics then exits, riding off into the sunset like the Lone Ranger.

So what might cause the salary difference? One explanation is that management, in its outdated, selfish, and illegal ways, has deliberately decided to pay people less if they are women than if they are men, looking only at the person's gender. However, it is not the only plausible explanation. The salary difference might be due to some other factor that (1) determines salary and (2) is linked to gender. In its defense, the firm might argue that it pays solely on the basis of *education* and *experience*, and it is not to be blamed for the fact that its pool of applicants consisted of better-educated and more-experienced men as compared to the women. This argument basically shifts the blame from the firm to society in general.

This is a complicated issue. Fortunately (for the author!) the resolution of the question one way or the other will not be attempted in this book. It can be dodged by pointing out that it is not a statistical question and should be decided using expertise from another field of human endeavor. But stay tuned. This question will reappear in Chapter 12 on multiple regression (with more data) in our continuing efforts to understand the interactions among gender, salary, education, and experience.

The field of statistics can be very helpful in providing exact answers in the presence of uncertainty, but the answers are limited in their scope and much further work and thought may be required before you reach a final explanation.

## Example
### *Your Productivity versus Theirs*

You have a friendly rivalry going with the manager of the other division over employee productivity. Actually, it's not entirely friendly because you both report to the same boss, who allocates resources based on performance. You would not only like to have the higher productivity, but would like it to be *significantly* higher so that there's no question about whose employees produce more.[19]

Here are summary measures of employee productivity in the two divisions:

|  | Your Division | Your Rival's Division |
|---|---|---|
| Sample size | 53 | 61 |
| Average | 88.23 | 83.70 |
| Standard deviation | 11.47 | 9.21 |
| Standard error | 1.58 | 1.18 |
| Average difference | | 4.53 |

To evaluate the average difference of 4.53 to see if it could be reasonably due to randomness, you need its standard error. Following are computations for the large-sample formula, which is appropriate because both sample sizes are at least 30:

$$S_{\bar{X}_2 - \bar{X}_1} = \sqrt{S_{\bar{X}_1}^2 + S_{\bar{X}_2}^2}$$
$$= \sqrt{1.58^2 + 1.18^2}$$
$$= \sqrt{2.48 + 1.39}$$
$$= 1.97$$

The 95% confidence interval is based on the $t$ value 1.960 from the $t$ table. The confidence interval extends from $4.53 - 1.960 \times 1.97$ to $4.53 + 1.960 \times 1.97$:

> You are 95% sure that the population mean productivity difference is somewhere between 0.67 and 8.39.

This confidence interval does *not* include the reference value of 0, which would indicate no mean difference in productivity between the two divisions in the (idealized) population. Thus, your hypothesis testing decision is as follows:

> The average difference between your employee productivity and that of your rival is statistically significant.

The $t$ statistic approach would, of course, have given the same answer. The $t$ statistic here is $4.53/1.97 = 2.30$, which exceeds the $t$ table critical value of 1.960.

The one-sided conclusion to a significant two-sided test may be used here. Because your division's productivity is higher, it follows that your division had *significantly higher productivity* than your rival's. Congratulations!

19. In reality, even after the hypothesis test is done, there will still be *some* questions because there is always some possibility of an error (type I or type II). What you hope to demonstrate is that the superior average productivity of your employees is not likely to be due to randomness alone.

# END-OF-CHAPTER MATERIALS

## Summary

**Hypothesis testing** uses data to decide between two possibilities (called *hypotheses*); it is often used to distinguish structure from mere randomness and should be viewed as a helpful input to executive decision making. A **hypothesis** is a statement about the population that may be either right or wrong; the data will help you decide which one (of two hypotheses) to accept as true. The **null hypothesis**, denoted $H_0$, represents the *default*, often a very specific case, such as pure randomness. The **research hypothesis** or **alternative hypothesis**, $H_1$, has the burden of proof, requiring convincing evidence against $H_0$ for its acceptance. Accepting the null hypothesis is a weak conclusion, whereas rejecting the null and accepting the research hypothesis is a strong conclusion and a significant result.

For testing whether or not the population mean, $\mu$, is equal to a reference value, $\mu_0$, the hypotheses are $H_0 : \mu = \mu_0$ versus $H_1 : \mu \neq \mu_0$. The **reference value**, $\mu_0$, is a known, fixed number that does not come from the sample data. This is a **two-sided test** because the research hypothesis allows possible population mean values on both sides of the reference value. This test of a population mean is known as the *t* **test** or **Student's *t* test**. The outcome of the test is determined by checking if the sample average, $\overline{X}$, is farther from the reference value, $\mu_0$, than random chance would allow if the population mean, $\mu$, were actually equal to $\mu_0$. Thus, the distance from $\overline{X}$ to $\mu_0$ is compared with the standard error, $S_{\overline{X}}$, using the *t* table. The test may be based either on the two-sided confidence interval (from Chapter 9) or on the *t* **statistic**, defined as follows:

$$t_{\text{statistic}} = \frac{\overline{X} - \mu_0}{S_{\overline{X}}}$$

Here is how the two-sided *t* test is decided, using your choice of the confidence interval approach or the *t* statistic approach (which always give identical results):

If the reference value, $\mu_0$, is in the two-sided confidence interval, or (equivalently) $|t_{\text{statistic}}| < t_{\text{table}}$, then accept the null hypothesis, $H_0$, as a reasonable possibility. The sample average, $\overline{X}$, is *not significantly different* from $\mu_0$. The observed difference between $\overline{X}$ and $\mu_0$ could reasonably be just random. The result is *not statistically significant*.

If the reference value, $\mu_0$, is *not* in the two-sided confidence interval, or (equivalently) $|t_{\text{statistic}}| > t_{\text{table}}$, then accept the research hypothesis, $H_1$, and reject the null hypothesis, $H_0$. The sample average, $\overline{X}$, is *significantly different* from $\mu_0$. The observed difference between $\overline{X}$ and $\mu_0$ could *not* reasonably be just random. The result is *statistically significant*.

By deciding the hypothesis test in this way, you are accepting the null hypothesis ($\mu = \mu_0$) whenever $\mu_0$ appears to be a reasonably possible value for $\mu$. When the null hypothesis is true, your probability of deciding correctly is equal to the confidence level (95% or other) for the column you used in the *t* table.

Table 10.7.1 shows a summary of the situation for testing either the mean of a normal distribution or the probability of occurrence for a binomial distribution.

The *t* statistic is an example of the general concept of a **test statistic**, which is the most helpful number that can be computed from your data for the purpose of deciding between two given hypotheses. The test statistic is compared to the appropriate **critical value** found in a standard statistical table; for example, a *t* value found in the *t* table is a **critical *t* value**. A useful rule of thumb is that if the *t* statistic is larger in absolute value than 2, you reject the null hypothesis; otherwise, you accept it.

Depending on which is (in reality) the true hypothesis, there are two types of errors that you might make. The **type I error** is committed when the null hypothesis is true, but you reject it and declare that your result is statistically significant. The probability of committing a type I error (when the null hypothesis is true) is controlled by your choice of column in the *t* table, conventionally the 5% level. The **type II error** is committed when the research hypothesis is true, but you accept the null hypothesis instead and declare the result *not* to be significant. The probability of committing a type II error (when the research hypothesis is true) is not easily controlled but can (depending on the true value of $\mu$) be anywhere between 0 and the confidence level of the test (e.g., 95%).

**TABLE 10.7.1 Testing Either the Mean of a Normal Distribution or the Probability of Occurrence for a Binomial Distribution**

|  | Normal | Binomial |
|---|---|---|
| Population mean | $\mu$ | $\pi$ |
| Reference value | $\mu_0$ | $\pi_0$ |
| Null hypothesis | $H_0 : \mu = \mu_0$ | $H_0 : \pi = \pi_0$ |
| Research hypothesis | $H_1 : \mu \neq \mu_0$ | $H_1 : \pi \neq \pi_0$ |
| Data | $X_1, X_2, \ldots, X_n$ | $X$ occurrences out of $n$ trials |
| Estimator | $\overline{X}$ | $p = X/n$ |
| Standard error | $S_{\overline{X}} = S/\sqrt{n}$ | $S_p = \sqrt{p(1-p)/n}$ |
| Confidence interval | From $\overline{X} - tS_{\overline{X}}$ to $\overline{X} + tS_{\overline{X}}$ | From $p - tS_p$ to $p + tS_p$ |
| *t* statistic | $t = (\overline{X} - \mu_0)/S_{\overline{X}}$ | $t = (p - \pi_0)/S_p$ |

Note that each type of error is based on an assumption about which hypothesis is true. Since each hypothesis is either true or false depending on the population (*not* on the data), there is no notion of the probability of a hypothesis being true.

The **assumptions for hypothesis testing** are (1) the data set is a random sample from the population of interest, and (2) either the quantity being measured is approximately normal, or else the sample size is large enough that the central limit theorem ensures that the sample average is approximately normally distributed.

The **test level** or **significance level** is the probability of accepting the research hypothesis when the null hypothesis is really true (i.e., committing a type I error). By convention, this level is set at 5% but may reasonably be set at 1% or 0.1% (or even 10% for some fields of study) by using the appropriate column in the *t* table. The ***p*-value** tells you how surprised you would be to learn that the null hypothesis had produced the data, with smaller *p*-values indicating more surprise and leading to rejection of $H_0$. By convention, we reject $H_0$ whenever the *p*-value is less than 0.05. A result is **statistically significant** ($p < 0.05$) if it is significant at the 5% level. Other terms used are *highly significant* ($p < 0.01$), *very highly significant* ($p < 0.001$), and *not significant* ($p > 0.05$).

A **one-sided test** is set up with the null hypothesis claiming that $\mu$ is on one side of $\mu_0$ and the research hypothesis claiming that it is on the other side. To use a one-sided test, you must be sure that *no matter how the data had come out* you would still have used a one-sided test on the same side ("larger than" or "smaller than"). If in doubt, use a two-sided test; if it is significant, you are then entitled to state the *one*-sided conclusion. The test may be performed either by constructing the appropriate one-sided confidence interval (matching the claim of the research hypothesis) or by using the *t* statistic. A significant result (accepting the research hypothesis) will be declared whenever the reference value $\mu_0$ does *not* fall in the confidence interval. This will happen whenever $\overline{X}$ is on the side of $\mu_0$ claimed in the research hypothesis and the absolute value of the *t* statistic is larger than the *t* value from the *t* table. A significant result will occur whenever $t_{statistic} > t_{table}$ (if testing $H_1: \mu > \mu_0$) or $t_{statistic} < -t_{table}$ (if testing $H_1: \mu < \mu_0$).

For the one-sided *t* test to see if $\mu$ is *larger* than $\mu_0$, the hypotheses are $H_0: \mu \leq \mu_0$ and $H_1: \mu > \mu_0$. The confidence interval includes all values *at least* as large as $\overline{X} - t_{table}S_{\overline{X}}$.

If $\mu_0$ is in the confidence interval or (equivalently) $t_{statistic} \leq t_{table}$, then accept the null hypothesis, $H_0$, as a reasonable possibility. The sample average, $\overline{X}$, is *not significantly larger* than $\mu_0$. If $\overline{X}$ is larger than $\mu_0$, the observed difference could reasonably be just random. The result is *not statistically significant*.

If $\mu_0$ is *not* in the confidence interval or (equivalently) $t_{statistic} > t_{table}$, then accept the research hypothesis, $H_1$, and reject the null hypothesis, $H_0$. The sample average, $\overline{X}$, is *significantly larger* than $\mu_0$. The observed difference could *not* reasonably be just random. The result is *statistically significant*.

For the one-sided *t* test to see if $\mu$ is *smaller* than $\mu_0$, the hypotheses are $H_0: \mu \geq \mu_0$ and $H_1: \mu < \mu_0$. The confidence interval includes all values *no larger* than $\overline{X} + t_{table}S_{\overline{X}}$.

If $\mu_0$ is in the confidence interval or (equivalently) $t_{statistic} \geq -t_{table}$, then accept the null hypothesis, $H_0$, as a reasonable possibility. The sample average, $\overline{X}$, is *not significantly smaller* than $\mu_0$. If $\overline{X}$ is smaller than $\mu_0$, the observed difference could reasonably be just random. The result is *not statistically significant*.

If $\mu_0$ is *not* in the confidence interval or (equivalently) $t_{statistic} < -t_{table}$, then accept the research hypothesis, $H_1$, and reject the null hypothesis, $H_0$. The sample average, $\overline{X}$, is *significantly smaller* than $\mu_0$. The observed difference could *not* reasonably be just random. The result is *statistically significant*.

Whenever you have an estimator (such as $\overline{X}$), the appropriate standard error for that estimator (such as $S_{\overline{X}}$), and a critical value from the appropriate table (such as the *t* table), you may construct one- or two-sided confidence intervals (at various confidence levels) and perform one- or two-sided hypothesis tests (at various significance levels).

For the test of whether a new observation came from the same population as a sample, the null hypothesis claims that it did, and the research hypothesis claims otherwise. Using the standard error for prediction, $S\sqrt{1 + 1/n}$, to construct the prediction interval, accept the null hypothesis if the new observation falls within the interval, and accept the research hypothesis and declare significance otherwise. Or compute the *t* statistic using the following equation, and compare it to the *t* value in the *t* table:

For Testing a New Observation

$$t_{statistic} = \frac{X_{new} - \overline{X}}{S\sqrt{1 + 1/n}}$$

Whichever method you choose (confidence interval or *t* statistic), you have available all of the significance levels, *p*-value statements, and one- or two-sided testing procedures as before.

The **paired *t* test** is used to test whether or not two samples have the same population mean value when there is a natural pairing between the two samples—for example, "before" and "after" measurements on the same people. By working with the differences ("after" minus "before"), we reduce such a problem to the familiar one-sample *t* test, using $\mu_0 = 0$ as the reference value expressing the null hypothesis of no difference in means.

The **unpaired $t$ test** is used to test whether or not two samples have the same population mean value when there is *no* natural pairing between the two samples; that is, each is an independent sample from a different population. For a two-sided test, the null hypothesis claims that the mean difference is 0. To construct confidence intervals for the mean difference and to perform the hypothesis test, you need the **standard error of the difference** (which gives the estimated standard deviation of the sample average difference) and its degrees of freedom.

For a large-sample situation ($n_1 \geq 30$ and $n_2 \geq 30$):

$$S_{\overline{X}_2 - \overline{X}_1} = \sqrt{S_{\overline{X}_1}^2 + S_{\overline{X}_2}^2} = \sqrt{\frac{S_1^2}{n_1} + \frac{S_2^2}{n_2}}$$

$$S_{p_2 - p_1} = \sqrt{S_{p_1}^2 + S_{p_2}^2} \text{ (for two binomials)}$$

Degrees of freedom = infinity

For a small-sample situation (assuming equal variabilities):

$$S_{\overline{X}_2 - \overline{X}_1} = \sqrt{\frac{(n_1 - 1)S_1^2 + (n_2 - 1)S_2^2}{n_1 + n_2 - 2}\left(\frac{1}{n_1} + \frac{1}{n_2}\right)}$$

Degrees of freedom = $n_1 + n - 2$

Based on the average difference, its standard error, its number of degrees of freedom, and the reference value (0), you can construct confidence intervals and perform hypothesis tests in the usual way. Note that, in addition to the usual assumptions of random samples and normal distributions, the small-sample situation also requires that the population variabilities be equal ($\sigma_1 = \sigma_2$).

## Key Words

**assumptions for hypothesis testing**, *259*
**critical $t$ value**, *257*
**critical value**, *257*
**hypothesis**, *250*
**hypothesis testing**, *249*
**null hypothesis**, *250*
**one-sided $t$ test**, *262*
**paired $t$ test**, *268*
**$p$-value**, *261*
**reference value**, *252*
**research hypothesis** or **alternative hypothesis**, *274*
**standard error of the difference**, *271*
**statistically significant**, *260*
**$t$ statistic**, *257*
**$t$ test** or **Student's $t$ test**, *274*
**test level** or **significance level**, *260*
**test statistic**, *257*
**two-sided test**, *252*
**type I error**, *259*
**type II error**, *259*
**unpaired $t$ test**, *270*

## Questions

1. a. What is the purpose of hypothesis testing?
   b. How is the result of a hypothesis test different from a confidence interval statement?
2. a. What is a hypothesis? In particular, is it a statement about the population or the sample?
   b. How is the role of the null hypothesis different from that of the research hypothesis? Which one usually includes the case of pure randomness? Which one has the burden of proof? Which one has the benefit of the doubt?
   c. Suppose you decide in favor of the null hypothesis. Is this a weak or a strong conclusion?
   d. Suppose you decide in favor of the research hypothesis. Is this a weak or a strong conclusion?
   e. "A null hypothesis can never be disproved." Comment.
3. a. Briefly describe the steps involved in performing a two-sided test concerning a population mean based on a confidence interval.
   b. Briefly describe the steps involved in performing a two-sided test concerning a population mean based on the $t$ statistic.
4. a. What is Student's $t$ test?
   b. Who was Student? What was his contribution?
5. a. What is the reference value? Does it come from the sample data? Is it known or unknown?
   b. What is the $t$ statistic? Does it depend on the reference value?
   c. Does the confidence interval change depending on the reference value?
6. a. What, in general, is a test statistic?
   b. Which test statistic would you use for a two-sided $t$ test?
   c. What, in general, is a critical value?
   d. Which critical value would you use for a two-sided $t$ test?
7. a. What assumptions must be satisfied for a two-sided $t$ test to be valid?
   b. Consider each assumption in turn. What happens if the assumption is not satisfied? What, if anything, can be done to fix the problem?
8. a. What is a type I error? Can it be controlled? Why or why not?
   b. What is a type II error? Can it be controlled? Why or why not?
   c. When, if ever, is it correct to say that "the null hypothesis is true with probability 0.95"?
   d. What can you say about your lifetime track record in terms of correct decisions to accept a true null hypotheses?
9. What $p$-value statement is associated with each of the following outcomes of a hypothesis test?
   a. Not significant.
   b. Significant.
   c. Highly significant.
   d. Very highly significant.

10. a. What is a one-sided test?
   b. What are the hypotheses for a one-sided test?
   c. When are you allowed to perform a one-sided test? What should you do if you're not sure if it's allowed?
   d. If you perform a one-sided test when it's really not permitted, what is the worst that can happen?
   e. Under what conditions are you permitted to make a one-sided statement based on a two-sided test?

11. a. How is a one-sided test performed based on a confidence interval?
   b. How is a one-sided test performed based on the t statistic?

12. Suppose you have an estimator and would like to test whether or not the population mean value equals 0. What do you need in addition to the estimated value?

13. What standard error would you use to test whether a new observation came from the same population as a sample? (Give both its name and the formula.)

14. a. What is a paired t test?
   b. Identify the two hypotheses involved in a paired t test.
   c. What is the "pairing" requirement? Give a concrete example.
   d. How is a paired t test similar to and different from an ordinary t test for just one sample?

15. a. What is an unpaired t test?
   b. Identify the two hypotheses involved in an unpaired t test.
   c. What is the "independence" requirement? Give a concrete example.
   d. How is an unpaired t test similar to and different from an ordinary t test for just one sample?
   e. When is each standard error appropriate? (Answer both in words and using a formula.)
   f. What new assumption is needed for the unpaired t test to be valid for small samples? What can you do if this assumption is grossly violated?

16. a. Describe the general process of constructing confidence intervals and performing hypothesis tests using the rule of thumb when you have an estimator and its standard error.
   b. If you also know the number of degrees of freedom and can use the t table, how would your answer change to be more exact?

## Problems

*Problems marked with an asterisk (*) are solved in the Self Test in Appendix C.*

1.* To help your restaurant marketing campaign target the right age levels, you want to find out if there is a statistically significant difference, on the average, between the age of your customers and the age of the general population in town, 43.1 years. A random sample of 50 customers shows an average age of 33.6 years with a standard deviation of 16.2 years.

   a. Identify the null and research hypotheses for a two-sided test using both words and mathematical symbols.
   b. Perform a two-sided test at the 5% significance level and describe the result.

2.* a. Perform a two-sided test at the 1% significance level for the previous problem and describe the result.
   b. State the p-value as either $p > 0.05$, $p < 0.05$, $p < 0.01$, or $p < 0.001$.

3. Part of the assembly line will need adjusting if the consistency of the injected plastic becomes either too viscous or not viscous enough as compared with a value (56.00) your engineers consider reasonable. You will decide to adjust only if you are convinced that the system is "not in control," that is, there is a real need for adjustment. The average viscosity for 13 recent measurements was 51.22 with a standard error of 3.18.

   a. Identify the null and research hypotheses for a two-sided test, using both words and mathematical symbols.
   b. Perform a two-sided test at the 5% significance level and describe the result.
   c. Perform a two-sided test at the 1% significance level and describe the result.
   d. State the p-value as either $p > 0.05$, $p < 0.05$, $p < 0.01$, or $p < 0.001$.

4. a. Why is a two-sided test appropriate for the previous problem?
   b. State the one-sided result of the two-sided test at the 5% level, if appropriate.

5. Some of your advertisements seem to get no reaction, as though they are being ignored by the public. You have arranged for a study to measure the public's awareness of your brand before and after viewing a TV show that includes the advertisement in question. You wish to see if the ad has a statistically significant effect as compared with zero, representing no effect. Your brand awareness, measured on a scale from 1 to 5, was found to have increased an average of 0.22 point when 200 people were shown an advertisement and questioned before and after. The standard deviation of the increase was 1.39 points.

   a. Identify the null and research hypotheses for a two-sided test, using both words and mathematical symbols.
   b. Perform a two-sided test at the 5% significance level and describe the result.
   c. Perform a two-sided test at the 1% significance level and describe the result.
   d. State the p-value as either $p > 0.05$, $p < 0.05$, $p < 0.01$, or $p < 0.001$.

6. a. Why is a two-sided test appropriate for the previous problem?
   b. State the one-sided result of the two-sided test at the 5% level, if appropriate.

7. In a random sample of 725 selected for interview from your database of 13,916 customers, 113 said they are dissatisfied with your company's service.

a. Find the best estimate of the percentage of all customers in your entire database who are dissatisfied.

b. Find the standard error of your estimate of the percentage of all customers who are dissatisfied.

c. Find the best estimate of the overall number of dissatisfied customers within your database.

d. Find the 95% confidence interval for the percentage of dissatisfied customers.

e. The company's goal has been to keep the percentage of dissatisfied customers at or below 10%. Could this reasonably still be the case, or do you have convincing evidence that the percentage is larger than 10%? Justify your answer.

8. Your factory's inventory level was determined at 12 randomly selected times last year, with the following results:

   313, 891, 153, 387, 584, 162, 742, 684, 277, 271, 285, 845

a. Find the typical inventory level throughout the whole year, using the standard statistical summary.

b. Identify the population.

c. Find the 95% confidence interval for the population mean inventory level.

d. Is the average of the measured inventory levels significantly different from 500, which is the number used for management budgeting purposes? Justify your answer.

9. Your bakery produces loaves of bread with "1 pound" written on the label. Here are weights of randomly sampled loaves from today's production:

   1.02, 0.97, 0.98, 1.10, 1.00, 1.02, 0.98, 1.03, 1.03, 1.05, 1.02, 1.06

a. Find the 95% confidence interval for the mean weight of all loaves produced today.

b. Find the reference value for testing the average of the actual weights against the claim on the label.

c. Find the hypotheses, $H_0$ and $H_1$.

d. Perform the hypothesis test (two-sided, level 0.05) and report the result.

e. What error, if any, might you have committed?

10. View the 20,000 people in the donations database on the companion site as a random sample from a much larger group of potential donors. Determine whether or not the amount donated in response to the current mailing (named "Donation" in the database), on average, is enough to cover the per-person cost (assumed to be 38 cents) of preparing materials and mailing them. In particular, can you conclude that it was significantly worthwhile to solicit a donation from this group?

11. Suppose that the target response rate was 4% when the current mailing was sent to the 20,000 people in the donations database on the companion site.

a. Find the actual response rate represented by the 989 donations received in response to this mailing to 20,000 people.

b. How does the actual response rate compare to the target? Give a statement that includes information about statistical significance (or lack of significance).

12. If the list price of the Eureka 4750A Bagged Upright Vacuum cleaner is $79.99, is the average price, based on the data from Table 9.6.1, significantly different from a 10% discount?

13. At a recent meeting, it was decided to go ahead with the introduction of a new product if "interested consumers would be willing, on average, to pay $20.00 for the product." A study was conducted, with 315 random interested consumers indicating that they would pay an average of $18.14 for the product. The standard deviation was $2.98.

a. Identify the reference value for testing the mean for all interested consumers.

b. Identify the null and research hypotheses for a two-sided test using both words and mathematical symbols.

c. Perform a two-sided test at the 5% significance level and describe the result.

d. Perform a two-sided test at the 1% significance level and describe the result.

e. State the $p$-value as either $p > 0.05$, $p < 0.05$, $p < 0.01$, or $p < 0.001$.

14. a. Why might a one-sided test be appropriate for the preceding problem?

b. Identify the null and research hypotheses for a one-sided test, using both words and mathematical symbols.

c. Perform a one-sided test at the 5% significance level and describe the result.

15. The $p$-value is 0.0371. What conclusion can you reach and what error might have been made?

16. Do initial public offerings (IPOs) of stock significantly increase in value, on average, in the short term? Test using the data from Table 4.3.7 that show performance of initial offerings as percent increases from the offer price, with most newly traded companies increasing in value while some lost money. Please give the $p$-value (as either $p > 0.05$, $p < 0.05$, $p < 0.01$, or $p < 0.001$) as part of your answer.

17. A recent poll of 809 randomly selected registered voters revealed that 426 plan to vote for your candidate in the coming election.

a. Is the observed percentage more than 50%?

b. Is the observed percentage significantly more than 50%? How do you know? Base your answer on a two-sided test.

18. Test whether or not the population percentage could reasonably be 20%, based on the observed 18.4% who like your products, from a random sample of 500 consumers.

19. As part of a decision regarding a new product launch, you want to test whether or not a large enough percentage (10% or more) of the community would be interested in purchasing it. You will launch the product only if you find convincing evidence of such demand. A survey of 400 randomly selected people in the community finds that 13.0% are willing to try your proposed new product.

a. Why is a one-sided test appropriate here?
b. Identify the null and research hypotheses for a one-sided test using both words and mathematical symbols.
c. Perform the test at the 5% significance level and describe the result.
d. Perform the test at the 1% significance level and describe the result.
e. State the *p*-value as either $p > 0.05$, $p < 0.05$, $p < 0.01$, or $p < 0.001$.

20. You are considering a new delivery system and wish to test whether delivery times are significantly different, on average, than your current system. It is well established that the mean delivery time of the current system is 2.38 days. A test of the new system shows that, with 48 observations, the average delivery time is 1.91 days with a standard deviation of 0.43 day.
a. Identify the null and research hypotheses for a two-sided test, using both words and mathematical symbols.
b. Perform a two-sided test at the 5% significance level and describe the result.
c. Perform a two-sided test at the 1% significance level and describe the result.
d. State the *p*-value as either $p > 0.05$, $p < 0.05$, $p < 0.01$, or $p < 0.001$.
e. Summarize the results in a brief memo to management.

21. You work for a company that prepares and distributes frozen foods. The package claims a net weight of 14.5 ounces. A random sample of today's production was weighed, producing the following data set:
    14.43, 14.37, 14.38, 14.29, 14.60, 14.45, 14.16, 14.52, 14.19, 14.04, 14.31
A sample was also selected from yesterday's production. The average was 14.46 and the standard deviation was 0.31.
a. Estimate the mean weight you would have found had you been able to weigh all packages produced today.
b. For a typical individual package produced yesterday, approximately how different was the actual weight from yesterday's average?
c. Find the 95% confidence interval for the mean weight for all packages produced today.
d. Identify the hypotheses you would work with to test whether or not your claimed weight is correct, on average, today.
e. Is there a significant difference between claimed and actual mean weight today? Justify your answer.

22. Although your product, a word game, has a list price of $12.95, each store is free to set the price as it wishes. You have just completed a quick survey, and the marked prices at a random sample of stores that sell the product were as follows:
    $12.95, 9.95, 8.95, 12.95, 12.95, 9.95, 9.95, 9.98, 13.00, 9.95

a. Estimate the mean selling price you would have found had you been able to survey all stores selling your product.
b. For a typical store, approximately how different is the actual selling price from the average?
c. Find the 95% confidence interval for the mean selling price for all stores selling your product.
d. Your marketing department believes that games generally sell at a mean discount of 12% from the list price. Identify the hypotheses you would work with to test the population mean selling price against this belief.
e. Test the hypotheses from part d.

23. Some frozen food dinners were randomly selected from this week's production and destroyed in order to measure their actual calorie content. The claimed calorie content is 200. Here are the calorie counts for each dinner:
    221, 198, 203, 223, 196, 202, 219, 189, 208, 215, 218, 207
a. Estimate the mean calorie content you would have found had you been able to measure all packages produced this week.
b. Approximately how different is the average calorie content (for the sample) from the mean value for all dinners produced this week?
c. Find the 99% confidence interval for the mean calorie content for all packages produced this week.
d. Is there a significant difference between claimed and measured calorie content? Justify your answer.

24. Consider the dollar value (in thousands) of gifts returned to each of your department stores after the holiday season (Table 10.7.2):
a. Compute the standard deviation.
b. Interpret the standard deviation as a measure of the variation from one store to another.
c. Compute the standard error of the average and briefly describe its meaning.
d. Find the two-sided 95% confidence interval for the mean value of returned merchandise for all downtown stores.

**TABLE 10.7.2 Dollar Value of Returned Gifts**

| Store | Returned |
|-------|----------|
| A | 13 |
| B | 8 |
| C | 36 |
| D | 18 |
| E | 6 |
| F | 21 |

e.   The Association of Downtown Merchants had been expecting an average value of $10,000 of returned merchandise per store, since this has been typical in the past. Test to see if this year's average differs significantly from their expectation.

25. Here are the satisfaction scores given by 12 randomly selected customers:

   89, 98, 96, 65, 99, 81, 76, 51, 82, 90, 96, 76

   Does the observed average score differ significantly from the target score of 80? Justify your answer.

26. Regulations require that your factory provide convincing evidence that it discharges less than 25 milligrams of a certain pollutant each week, on average, over the long run. A recent sample shows weekly amounts of 13, 12, 10, 8, 22, 14, 10, 15, 9, 10, 6, and 12 milligrams released.

   a.   Have you complied with the regulations? Explain your answer based on a one-sided hypothesis test at the 5% level.

   b.   Report the $p$-value as either $p > 0.05$, $p < 0.05$, $p < 0.01$, or $p < 0.001$. In particular, is the result highly significant?

   c.   Identify the underlying hypotheses and assumptions involved in these tests.

   d.   All else equal, would the use of a two-sided test, instead of a one-sided test, result in more or fewer instances of "out-of-compliance" findings? Explain.

27. A manufacturing process is considered to be "in control" if the long-run mean weight of components produced is 0.20 kilograms, even though individual components may vary from this mean. Here are weights of a random sample of recently produced components:

   0.253, 0.240, 0.247, 0.183, 0.247, 0.223, 0.252, 0.195, 0.235, 0.241, 0.251, 0.261, 0.194, 0.236, 0.256, and 0.241

   Does this process seem to be in control? Justify your answer.

28. Production yields vary and can be high or low on a given day. If they're high, you want to find out why so that yields could be similarly increased on other days. If they're low, you want to fix the problem. You have just learned that today's production yields seem to be lower than usual. Should you use a one-sided test or a two-sided test to investigate? Why?

29. A recent poll of 1,235 randomly sampled likely voters shows your favorite candidate ahead, with 52.1% in favor. There are two candidates. Use hypothesis testing to infer to the larger group of all likely voters.

   a.   Carefully identify the two-sided hypotheses.

   b.   Perform the hypothesis test at level 0.05 and give the result.

   c.   Make a careful, exact statement summarizing the result of the test and what it means.

   d.   Repeat parts b and c assuming that the percentage is 58.3% instead of 52.1%.

   e.   Explain why a one-sided test would be inappropriate here by showing that each of the three possible outcomes of a two-sided test would be of interest.

30. Managers perceived employee stock ownership as having a significant positive effect on product quality. As part of that same study, managers were also asked to rate the effect of employee stock ownership on unit labor cost.[20] This effect, on a scale from –2 (large negative effect) to 2 (large positive effect), was 0.12 with a standard error of 0.11, based on a sample of 343 managers.

   a.   Find the 95% confidence interval and state carefully what this represents. Keep in mind that these are opinions of randomly selected managers.

   b.   Is there a significant relationship between employee stock ownership and the unit cost of labor as perceived by managers? Why or why not?

   c.   Identify the null and research hypotheses.

   d.   Which hypothesis has been accepted? Is this a weak or a strong conclusion?

   e.   Has the accepted hypothesis been absolutely proven? If not, what type of error may have been made?

31. The goal of your marketing campaign is for more than 25% of supermarket shoppers to recognize your brand name. A recent survey of 150 random shoppers found that 21.3% recognized your brand name.

   a.   It might be argued that the burden of proof is to show that more than 25% of shoppers recognize your brand name. Identify the appropriate one-sided hypotheses in this case and perform the test at level 0.05.

   b.   On the other hand, it might be argued that you would be interested in knowing about all three possibilities: significantly more than 25% (indicating success), significantly less than 25% (indicating failure), and not significantly different from 25% (indicating that there is not enough information to say for sure). Identify the appropriate two-sided hypotheses in this case and perform the test at level 0.05.

   c.   For the two-sided test, write a brief paragraph describing the result, the error that might have been made, and its implications for your marketing strategy.

32. You are supervising an audit to decide whether or not any errors in the recording of account transactions are "material errors." Each account has a reported balance, whose accuracy can be verified only by careful and costly investigation; the account's error is defined as the difference between the reported balance and the actual balance. Note that the error is zero for any account that is correctly reported. In practical terms, for this situation involving 12,000 accounts, the total error is material only if it is at least $5,000. The average error amount for 250 randomly selected accounts was found to be $0.25, and the standard deviation of the error amount was $193.05. You may assume that your reputation as an auditor is on the line, so you want to be fairly certain before declaring that the total error is not material.

   a.   Find the estimated total error based on your sample and compare it to the material amount.

b. Identify the null and research hypotheses for a one-sided test of the population mean error per account and explain why a one-sided test is appropriate here.

c. Find the appropriate one-sided 95% confidence interval statement for the population mean error per account.

d. Find the $t$ statistic.

e. Which hypothesis is accepted as a result of a one-sided test at the 5% level?

f. Write a brief paragraph explaining the results of this audit.

33. Dishwasher detergent is packaged in containers that claim a weight of 24 ounces. Although there is some variation from one package to another, your policy is to ensure that the mean weight for each day's production is slightly over 24 ounces. A random sample of 100 packages from today's production indicates an average of 24.23 ounces with a standard deviation of 0.15 ounce.

a. Find the $p$-value (as either $p > 0.05$, $p < 0.05$, $p < 0.01$, or $p < 0.001$) for a one-sided hypothesis test to check if the population mean weight is above the claimed weight.

b. Write a brief paragraph summarizing your test and its results.

c. Is your conclusion a strong one or a weak one? Why?

34. Do employees take more sick leave in the year before retirement? They may well have an incentive to do so if their accumulated paid sick leave (the number of days they are entitled to be away with full pay) is about to expire. Indeed, this appears to happen with government workers. One evaluation of this issue looked at statistics gathered by the U.S. General Accounting Office (GAO).[21] The study concluded

*[What if] the bulge in sick days was just an aberration in the GAO sample rather than a real symptom of goofing off? In zeroing in on this question, we note that the 714 retirees in the GAO sample averaged 30 sick days in their last year instead of the "expected" 14 days. So in a work year of 251 days (average for federal employees), the retirees were finding themselves indisposed 12.0% of the time instead of 5.6%. Could that happen by chance? The science of statistics tells us that the probability of any such swing in so large a sample is low. To be precise, one in 200,000.*

a. Identify the population and the sample.

b. Identify the hypotheses being tested, in terms of percent of time indisposed.

c. Identify the $p$-value here.

d. Which hypothesis (if any) has been rejected? Which has been accepted?

e. How significant (statistically) is the result?

35. Selected mutual funds that practice socially aware investing, with year-to-date rates of return, are shown in Table 10.7.3. On average, these funds lost value in the first half of 2010, in the sense that their average rate of return was negative. However, the Standard & Poor's 500 stock market index lost 9.03% of its value during the same period, so this was a difficult time for the market in general.

**TABLE 10.7.3 Performance of Socially Aware Investment Funds**

| Fund | Rate of Return |
|---|---|
| Calvert Global Alternative Energy Fund A | −26.99% |
| Calvert Global Water Fund | −9.28% |
| Calvert New Vision Small Cap A | −5.42% |
| Calvert Social Investment Balanced A | −2.16% |
| Calvert World Values International A | −11.07% |
| Domini Social Equity A | 4.38% |
| Gabelli SRI Green Fund Inc A | −16.32% |
| Green Century Balanced | −4.00% |
| Legg Mason Prt Social Awareness Fund A | −4.27% |
| Neuberger Berman Socially Resp Inv | −0.53% |
| Pax World Global Green Fund—Individual Investor | −10.35% |
| Sentinel Sustainable Core Opportunities Fund | −7.38% |
| TIAA-CREF Social Choice Eq Retail | −5.98% |
| Walden Social Balanced Fund | −2.75% |
| Winslow Green Growth Fund | −15.59% |

**Source:** Social Investment Forum, accessed at http://www.socialinvest.org/resources/mfpc/ on July 14, 2010. Their source is Bloomberg.

a. On average, as a group, did socially aware mutual funds lose significantly more than the market index? Please use the market index as the reference value.

b. Find the $p$-value for this test (as either $p > 0.05$, $p < 0.05$, $p < 0.01$, or $p < 0.001$). In particular, is it highly significant?

c. Identify the underlying hypotheses and assumptions involved in part a.

d. Under these assumptions, the hypothesis test makes a clear and correct statement. However, are the assumptions realistic? Be sure to address independence (note that some of these funds are part of the same group).

e. Why is a two-sided test appropriate in this case? (*Hint:* You may wish to consider how the situation would have appeared if these funds had performed better than the market, on average.)

36. World investments markets were highly volatile in 1998. Table 10.7.4 shows one-year rates of return on closed-end mutual funds that specialize in income from international sources.

### TABLE 10.7.4  Performance of Closed-End World Income Funds: One-Year Market Return

| Fund | Return | Fund | Return |
|------|--------|------|--------|
| ACM Mgd $-x | −27.7% | Global Partners -x | −16.7% |
| Alliance Wld $ | −17.1 | Kleinwort Aust | −5.4 |
| Alliance Wld $ 2 | −27.0 | Morg St Em Debt -x | −24.9 |
| BlckRk North Am -x | 3.9 | Morgan St Glbl -x | −27.3 |
| Dreyfus Str Govt | 4.0 | Salomon SBG -x | −0.6 |
| Emer Mkts Float | −19.7 | Salomon SBW -x | −18.9 |
| Emer Mkts Inc -x | −18.4 | Scudder Glbl High Inc -x | −53.8 |
| Emer Mkts Inc II -x | −16.9 | Strategic Gl Inc | 5.8 |
| First Aust Prime -x | −5.3 | Templeton Em Inc | −12.8 |
| First Commonwlth -x | −3.5 | Templtn Gl Govt | −1.1 |
| Global HI Inc $ | −10.7 | Templtn Glbl Inc | 2.2 |
| Global Income Fund -x | −17.3 | Worldwide $Vest -x | −48.2 |

**Source:** From "Quarterly Closed-End Funds Review," *Wall Street Journal*, January 7, 1999, p. R14. Overall performance measures are from "Mutual-Fund Performance Yardsticks," p. R3.

a.  Do the rates of return of these closed-end world income funds, as a group, differ significantly on average from the 2.59% overall performance representing all world mutual funds over the same time period? If so, were these closed-end funds significantly better or significantly worse? In your calculations, you may assume that the overall performance is measured without randomness.

b.  Do the rates of return of these closed-end world income funds, as a group, differ significantly on average from the −26.83% overall performance representing all emerging markets' mutual funds over the same time period? If so, were these closed-end funds significantly better or significantly worse? In your calculations, you may assume that the overall performance is measured without randomness.

37.  Your broker achieved a rate of return of 18.3% on your portfolio last year. For a sample of 25 other brokers in the area, according to a recent news article, the average rate of return was 15.2% with a standard deviation of 3.2% (as percentage points).

a.  To test whether your broker significantly outperformed this group, identify the idealized population and the hypotheses being tested. In particular, are you testing against a mean or against a new observation?

b.  Find the standard error for prediction.

c.  Find the two-sided 95% prediction interval for a new observation.

d.  Did your broker outperform this group?

e.  Did your broker significantly outperform this group?

f.  Find the $t$ value and the $p$-value (as either $p > 0.05$, $p < 0.05$, $p < 0.01$, or $p < 0.001$) for this two-sided test.

38.  Last year you received an average of 129.2 complaints (i.e., individual items to be fixed under warranty) per new car sold, with a standard deviation of 42.1 complaints based on 3,834 new cars sold. This year you have set up a quality assurance program to fix some of these problems before the car is delivered. So far this year, you have had an average of just 93.4 complaints per new car sold with a standard deviation of 37.7, based on 74 cars sold so far.

a.  To see if your new quality assurance program is working, what hypothesis testing method would you use?

b.  Identify the populations, samples, and hypotheses.

c.  Perform a two-sided test at the 5% level and report the results.

39.  Why do firms change ownership? One possible reason for acquisitions is that the new owners expect to be able to manage the operations more efficiently than the current management. This theory leads to testable hypotheses. For example, it predicts that productivity should increase following a takeover and also that firms changing ownership should have lower productivity than firms in general. A study of this situation examined the productivity year by year for some firms that changed ownership and other firms that did not change owners.[22] In particular, they reported

*These numbers display a very clear pattern. Plants that changed owners … tended to be less efficient … than nonchangers…. But the differences … [after the change] were declining in*

*magnitude.... This signifies that the productivity of ... changers relative to that of ... nonchangers was both low and declining before the ownership change, and increasing (albeit still low) after the ownership change. With one exception, all of the productivity differences are highly statistically significant.*

a. In the last line of the preceding quote, explain what is implied by "highly statistically significant."

b. Consider the comparison of average productivity of firms that changed ownership (at the time of the change) to average productivity of firms that did not change ownership. Identify all elements of this hypothesis testing situation, in particular: the hypotheses, the sample data, the type of test used, and the assumptions being made.

c. One result they reported was "at the time of ownership change, productivity level was 3.9% lower as compared to plants that did not change ownership. The $t$ statistic is 9.10." Perform a hypothesis test based on this information and state your conclusion.

d. Why have they gone to the trouble of doing statistical hypothesis tests? What have they gained over and above simply observing and describing the productivity differences in their data?

40.* Stress levels were recorded during a true answer and a false answer given by each of six people in a study of lie-detecting equipment, based on the idea that the stress involved in telling a lie can be measured. The results are shown in Table 10.7.5.

**TABLE 10.7.5 Vocal Stress Level**

| Person | True Answer | False Answer |
|--------|-------------|--------------|
| 1 | 12.8 | 13.1 |
| 2 | 8.5 | 9.6 |
| 3 | 3.4 | 4.8 |
| 4 | 5.0 | 4.6 |
| 5 | 10.1 | 11.0 |
| 6 | 11.2 | 12.1 |

a. Was everyone's stress level higher during a false answer than during a true answer?

b. Find the average stress levels for true and for false answers. Find the average change in stress level (false minus true).

c. Find the appropriate standard error for the average difference. In particular, is this a paired or an unpaired situation?

d. Find the 95% two-sided confidence interval for the mean difference in stress level.

e. Test to see if the average stress levels are significantly different. If they are significantly different, are they significantly higher or lower when a false answer is given?

f. Write a paragraph interpreting the results of this test. In particular, is this a conclusion about these six people or about some other group? Also, how can you find a significant difference when some individuals had higher stress and some had lower stress for the false answer?

41. A group of experts has rated your winery's two best varietals. Ratings are on a scale from 1 to 20, with higher numbers being better. The results are shown in Table 10.7.6.

a. Is this a paired or unpaired situation? Why?

b. Find the average rating for each varietal and the average difference in ratings (Chardonnay minus Cabernet Sauvignon).

c. Find the appropriate standard error for the average difference.

d. Find the 95% two-sided confidence interval for the mean difference in rating.

e. Test to see if the average ratings are significantly different. If they are significantly different, which varietal is superior?

f. Write a paragraph interpreting the results of this test.

42.* To understand your competitive position, you have examined the reliability of your product as well as the reliability of your closest competitor's product. You have subjected each product to abuse that represents about a year's worth of wear-and-tear per day. Table 10.7.7 shows the data indicating how long each item lasted.

**TABLE 10.7.6 Wine-Tasting Scores**

| Expert | Chardonnay | Cabernet Sauvignon | Expert | Chardonnay | Cabernet Sauvignon |
|--------|------------|--------------------|--------|------------|--------------------|
| 1 | 17.8 | 16.6 | 6 | 19.9 | 18.8 |
| 2 | 18.6 | 19.9 | 7 | 17.1 | 18.9 |
| 3 | 19.5 | 17.2 | 8 | 17.3 | 19.5 |
| 4 | 18.3 | 19.0 | 9 | 18.0 | 16.2 |
| 5 | 19.8 | 19.7 | 10 | 19.8 | 18.6 |

TABLE 10.7.7 Days until Failure

| Your Products | Competitor's |
|---|---|
| 1.0 | 0.2 |
| 8.9 | 2.8 |
| 1.2 | 1.7 |
| 10.3 | 7.2 |
| 4.9 | 2.2 |
| 1.8 | 2.5 |
| 3.1 | 2.6 |
| 3.6 | 2.0 |
| 2.1 | 0.5 |
| 2.9 | 2.3 |
| 8.6 | 1.9 |
| 5.3 | 1.2 |
| | 6.6 |
| | 0.5 |
| | 1.2 |

TABLE 10.7.8 Monthly Day Care Rates in North Seattle*

| Laurelhurst Area | Non-Laurelhurst Area |
|---|---|
| $400 | $500 |
| 625 | 425 |
| 440 | 300 |
| 550 | 350 |
| 600 | 550 |
| 500 | 475 |
| | 325 |
| | 350 |
| | 350 |

*I am grateful to Ms. Colleen Walker for providing this data set.

TABLE 10.7.9 Preference Levels for Six Individuals in Each of Two Cities

| Milwaukee | Green Bay |
|---|---|
| 3 | 4 |
| 2 | 5 |
| 1 | 4 |
| 1 | 3 |
| 3 | 2 |
| 2 | 4 |

a. Find the average time to failure for your and your competitor's products. Find the average difference (yours minus your competitor's).
b. Find the appropriate standard error for this average difference. In particular, is this a paired or an unpaired situation? Why?
c. Find the two-sided 99% confidence interval for the mean difference in reliability.
d. Test at the 1% level if there is a significant difference in reliability between your products and your competitor's.
e. Find the p-value for the difference in reliability (as either $p > 0.05$, $p < 0.05$, $p < 0.01$, or $p < 0.001$).
f. Write a brief paragraph, complete with footnote(s), that might be used in an advertising brochure showing off your products.

43. Child care is one of life's necessities for working parents. Monthly rates per child at a sample of family day care centers in the North Seattle area are shown in Table 10.7.8. The Laurelhurst area is considered to be a highly desirable neighborhood, and real estate prices are higher in this area. Perform a one-sided hypothesis test at the 5% level to see if day care prices are also higher in the Laurelhurst area.

44. An advertising study interviewed six randomly selected people in each of two cities, recording each person's level of preference for a new product (Table 10.7.9).
a. Is this a paired or an unpaired two-sample problem?
b. Find the average preference level for each city.

c. Find the standard error of the difference between these average preference levels. (Note that these are small samples.)
d. Find the 95% two-sided confidence interval for the mean difference in preference between these two cities (Green Bay minus Milwaukee).
e. Test whether the apparent difference in preference is significant at the 5% test level.

45. There are two manufacturing processes, old and new, that produce the same product. The defect rate has been measured for a number of days for each process, resulting in the following summaries (Table 10.7.10).
a. By how much would we estimate that the defect rate would improve if we switched from the old to the new process?
b. What is the standard error of your answer to part a?
c. Your firm is interested in switching to the new process only if it can be demonstrated convincingly that the new process improves quality.

**TABLE 10.7.10 Defect Rate Summaries for Two Manufacturing Processes**

|  | Old | New |
|---|---|---|
| Average defect rate | 0.047 | 0.023 |
| Standard deviation | 0.068 | 0.050 |
| Sample size (days) | 50 | 44 |

State the null and research hypotheses for this situation.

d. Find the appropriate one-sided 95% confidence interval for the (population) long-term reduction in the defect rate.

e. Is the improvement (as estimated in part a) statistically significant?

46. To help you decide which of your two current suppliers deserves the larger contract next year, you have rated a random sample of plastic cases from each one. The data are a composite of several measurements, with higher numbers indicating higher quality (Table 10.7.11).

a. Find the average quality for each supplier.

b. Find the standard deviation of quality for each supplier.

c. Find the average difference in quality (International minus Custom) and its standard error.

d. Find the two-sided 95% confidence interval for the quality difference.

e. Is there a significant difference in quality? How do you know?

47. Consider the weights for two samples of candy bars, before and after intervention, from Table 5.5.4.

a. Is this a paired or an unpaired situation?

b. Find the 95% confidence interval for the population mean difference in weight per candy bar (after minus before).

c. Did intervention produce a significant change in weight? How do you know?

48. Your Detroit division produced 135 defective parts out of the total production of 983 last week. The Kansas City division produced 104 defectives out of 1,085 produced during the same time period.

a. Find the percent defective for each division and compare them.

b. Find the difference between these two percentages (Detroit minus Kansas City) and interpret it.

c. Find the standard error for this difference using the large-sample formula.

d. Find the 95% confidence interval for the difference.

e. Test to see if these two divisions differ significantly in terms of quality of production, based on the defect rate.

49. You are analyzing the results of a consumer survey of a product, rated on a scale from 1 to 10. For the 130 consumers who described themselves as "outgoing," the average rating was 8.36, and the standard deviation was 1.82. For the 218 "shy" consumers, the average was 8.78, and the standard deviation was 0.91.

a. Test to see if there is a significant difference between the ratings of outgoing and shy consumers.

b. Report the test results using $p$-value notation (as either $p > 0.05$, $p < 0.05$, $p < 0.01$, or $p < 0.001$).

50. Repeat the previous problem for a different product. For 142 outgoing consumers, the average rating was 7.28, and the standard deviation was 2.18. For 277 shy consumers, the average rating was 8.78, and the standard deviation was 1.32.

51. Repeat problem 49 for yet another product. For 158 outgoing consumers, the average rating was 7.93, and the standard deviation was 2.03. For 224 shy consumers, the average rating was 8.11, and the standard deviation was 1.55.

52. A cup of coffee is found to have only 72.8 milligrams of caffeine. Test (at the 5% level) whether the beans used could have come from the same population as those that generated the data in problem 45 of Chapter 9.

20. P. B. Voos, "Managerial Perceptions of the Economic Impact of Labor Relations Programs," *Industrial and Labor Relations Review* 40 (1987), pp. 195–208.

21. D. Seligman, "Sick in Washington," *Fortune*, March 28, 1988, p. 155.

22. F. Lichtenberg, "Productivity Improvements from Changes in Ownership," *Mergers & Acquisitions* 23 (1988), pp. 48–50.

**TABLE 10.7.11 Supplier Quality**

| Custom Cases Corp. | International Plastics, Inc. |
|---|---|
| 54.3 | 93.6 |
| 58.8 | 69.7 |
| 77.8 | 87.7 |
| 81.1 | 96.0 |
| 54.2 | 82.2 |
| 78.3 | |

### Database Exercises

Refer to the employee database in Appendix A. View this data set as a random sample from a much larger population of employees.

1.* Is the average annual salary significantly different from $40,000?

2. You would like to claim that the population has significantly more than five years of experience. Can you support this claim?

3. Test to see if the gender ratio differs significantly from 50%.
4. Test to see if the population mean annual salary for men differs from that for women.
5. Test to see if the population mean age for men differs from that for women.
6. Test to see if the average annual salary for training level A differs significantly from that for levels B and C combined.
7. Test to see if the population mean age for training level A differs from that for levels B and C combined.

## Projects

1. Identify a decision process within your work or business interests that could be resolved based on data.
   a. Describe the null and research hypotheses.
   b. Compute (or use an educated guess for) an appropriate estimate and its standard error.
   c. Find a confidence interval.
   d. Test the hypothesis.
   e. Interpret and explain your results.
2. Find a news item (from the Internet, a newspaper, a magazine, radio, or television) that reaches a conclusion based on data.
   a. Identify the null and research hypotheses.
   b. Identify the population and the sample, to the extent that you can from the information given. Was any important information omitted?
   c. What was the result of their hypothesis test?
   d. Is their conclusion a weak one or a strong one?
   e. Discuss and interpret their claims.

## Case

### So Many Ads, So Little Time

It's almost decision time, and the stakes are huge. With astronomical TV advertising costs per minute of airtime, it's been worthwhile to do some preliminary work so that nothing is wasted. In particular, you've been helping manage an effort to produce 22 ads for a personal hygiene product, even though only just a few will ever actually be shown to the general public. They have all been tested and ranked using the responses of representative consumers who were each randomly selected and assigned to view one ad, answering questions before and after. A composite score from 0 to 10 points, representing both recall and persuasion, has been produced for each consumer in the sample.

At your firm, the ads traditionally have been ranked using the average composite results, and the highest have run on nationwide TV. Recently, however, statistical hypothesis testing has been used to make sure that the ad or ads to be run are significantly better than a minimum score of 3.5 points.

Everything looks straightforward this time, with the two best ads scoring significantly above the minimum. The decision meeting should be straightforward, with Country Picnic the favorite for the most airtime and Coffee Break as an alternate. Following are the summaries, sorted in descending order by average composite score. The number of consumers viewing the ad is $n$. The $p$-values are from one-sided hypothesis tests against the reference value 3.5, computed separately for each ad.

| Ad | n | avg | stDev | stdErr | t | p |
|---|---|---|---|---|---|---|
| Country Picnic | 49 | 3.95 | 0.789 | 0.113 | 3.985 | 0.0001 |
| Coffee Break | 51 | 3.70 | 0.744 | 0.104 | 1.921 | 0.0302 |
| Anniversary | 51 | 3.66 | 0.934 | 0.131 | 1.214 | 0.1153 |
| Ocean Breeze | 49 | 3.63 | 0.729 | 0.104 | 1.255 | 0.1078 |
| Friends at Play | 56 | 3.62 | 0.896 | 0.120 | 0.969 | 0.1683 |
| Tennis Match | 56 | 3.60 | 0.734 | 0.098 | 1.037 | 0.1521 |
| Walking Together | 51 | 3.57 | 0.774 | 0.108 | 0.687 | 0.2476 |
| Swimming Pool | 52 | 3.56 | 0.833 | 0.116 | 0.532 | 0.2984 |
| Shopping | 49 | 3.54 | 0.884 | 0.126 | 0.355 | 0.3619 |
| Jogging | 47 | 3.54 | 0.690 | 0.101 | 0.423 | 0.3372 |
| Family Scene | 54 | 3.54 | 0.740 | 0.101 | 0.404 | 0.3438 |
| Mountain Retreat | 49 | 3.53 | 0.815 | 0.116 | 0.298 | 0.3836 |
| Cool & Comfortable | 52 | 3.52 | 0.780 | 0.108 | 0.195 | 0.4229 |
| Coffee Together | 53 | 3.52 | 0.836 | 0.115 | 0.148 | 0.4415 |
| City Landscape | 47 | 3.51 | 0.756 | 0.110 | 0.058 | 0.4770 |
| Friends at Work | 53 | 3.50 | 0.674 | 0.093 | 0.020 | 0.4919 |
| Sailing | 48 | 3.49 | 0.783 | 0.113 | 20.055 | 0.5219 |
| Desert Oasis | 55 | 3.48 | 0.716 | 0.097 | 20.226 | 0.5890 |
| Birthday Party | 50 | 3.48 | 0.886 | 0.125 | 20.175 | 0.5693 |
| Weekend Brunch | 53 | 3.45 | 0.817 | 0.112 | 20.437 | 0.6681 |
| Home from Work | 55 | 3.35 | 0.792 | 0.107 | 21.430 | 0.9207 |
| Windy | 47 | 3.34 | 0.678 | 0.099 | 21.593 | 0.9410 |

Thinking it over, you have some second thoughts. Because you want to really understand what the decision is based on, and because you remember material about errors in hypothesis testing from a course taken long ago, you wonder. The probability of a type I error is 0.05, so you expect to find about one ad in 20 to be significantly good even if it isn't. That says that sometimes none would be significant, yet other times more than one could reasonably be significant.

Your speculation continues: Could it be that decisions are being made on the basis of pure randomness? Could it be that consumers, on average, rate these ads equally good? Could it be that all you have here is the randomness of the particular consumers who were chosen for each ad?

You decide to run a computer simulation model, setting the population mean score for all ads to exactly 3.5. Hitting the recalculation button on the spreadsheet 10 times, you observe that 3 times no ads are significant, 5 times one ad is significant, once two ads are, and once three ads are significant. Usually, the significant ads are different each time. Even more troubling, the random simulated results look a lot like the real ones that are about to be used to make real decisions.

**Discussion Questions**

1. Choose two ads, one that is significant and one that is not. Verify significance based on the average, standard error, and $n$, to make sure that they are correct. Is it appropriate to use one-sided tests here?

2. If the type I error is supposed to be controlled at 5%, how is it that in the computer simulation model, type I errors occurred 70% of the time?

3. Could it reasonably be that no ads are worthwhile, in a study for which 2 of 22 are significant?

4. What is your interpretation of the effectiveness of the ads in this study? What would you recommend in this situation?

# Regression and Time Series

At this point, you know the basics: how to look at data, compute and interpret probabilities, draw a random sample, and do statistical inference. Now it's a question of applying these concepts to see the relationships hidden within the more complex situations of real life. Chapter 11 shows you how statistics can summarize the relationship between two factors based on a bivariate data set with two columns of numbers. The *correlation* will tell you how strong the relationship is, and *regression* will help you predict one factor from the other. Perhaps the most important statistical method is *multiple regression,* covered in Chapter 12, which lets you use all of the factors you have available in order to predict (that is, reduce the uncertainty of) some important but unknown number. Since *communication* is such an important business skill, Chapter 13 will show you how to effectively tell others all about the useful things you have learned from a multiple regression analysis. While the basic concepts stay the same, new ways of applying statistical methods are needed for *time-series analysis,* presented in Chapter 14, in order to extract the extra information contained in the time sequence of the observations.

## Chapter 11

# Correlation and Regression

## Measuring and Predicting Relationships

The world is filled with relationships: between attitude and productivity, between corporate strategy and market share, between government intervention and the state of the economy, between quantity and cost, between sales and earnings, and so on.

Up to now you have been concerned primarily with statistical summaries such as the average and variability, which are usually sufficient when you have *univariate* data (i.e., just *one* measurement, such as salary) for each elementary unit (e.g., employee). When you have *bivariate* data (e.g., salary and education), you can always study each measurement individually as though it were part of a univariate data set. But the real payoff comes from studying both measurements together to see the relationship between the two.

There are three basic goals to keep in mind when studying relationships in bivariate data:

**One**: *Describing and understanding the relationship.* This is the most general goal, providing background information to help you understand how the world works. When you are studying a complex system, knowing which factors interact most strongly with each other (and which other ones do not affect one another) will help you gain the perspective necessary for long-range planning and other strategies. Included in this concept is the idea of using one measurement to explain something about another measurement.

**Two:** *Forecasting and predicting a new observation.* Once you understand the relationship, you can use information about one of the measurements to help you do a better job of predicting the other. For example, if you know that orders are up this quarter, you can expect to see an increase in sales. If you have analyzed the relationship between orders and sales in the past, you may be able to come up with a good forecast of future sales based on current orders.

**Three:** *Adjusting and controlling a process.* When you *intervene* in a process (e.g., by adjusting the production level or providing an additive or service), you have to choose the extent of your intervention. If there is a direct relationship between intervention and result and you understand it, this knowledge can help you make the best possible adjustments.

Bivariate data can have many different kinds of structures; some are easy to work with, and others are more difficult. By exploring your data using a *scatterplot*, you can gain additional insights beyond the few conventional statistical summaries. There are two basic tools for summarizing bivariate data: *correlation* analysis summarizes the strength of the relationship (if any) between the two factors, while *regression* analysis shows you how to predict or control one of the variables using the other one. Hypothesis testing looks at the relationship that appears to exist in the data and lets you decide either that the relationship is significant or that it could reasonably be due to randomness alone.

## 11.1 EXPLORING RELATIONSHIPS USING SCATTERPLOTS AND CORRELATIONS

Whenever you have bivariate data, you should draw a *scatterplot* so that you can really *see* the structure. Just as the histogram shows structure in univariate data (normal, skewed, outliers, etc.), the scatterplot will show you everything that is going on in bivariate data. If there are some problems in your data, such as outliers or other unexpected features, often the only way to uncover them is by looking at a scatterplot.

The *correlation* is a summary measure of the strength of the relationship. Like all statistical summaries, the correlation is both helpful and limited. If the scatterplot shows either a well-behaved *linear* relationship (to be defined soon) or no relationship at all, then the correlation provides an excellent summarization of the relationship. But if there are problems (to be defined soon) such as a *nonlinear* relationship, *unequal variability*, *clustering*, or *outliers* in the data, the correlation can be misleading.

By itself, the correlation is limited because its interpretation depends on the type of relationship in the data. This is why the scatterplot is so important: It will either confirm the usual interpretation of the correlation or show that there are problems with the data that render the correlation number misleading.

## The Scatterplot Shows You the Relationship

A **scatterplot** displays each case (or elementary unit) using two axes to represent the two factors. If one variable is seen as causing, affecting, or influencing the other, then it is called *X* and defines the horizontal axis. The variable that might respond or be influenced is called *Y* and defines the vertical axis. If neither clearly influences nor causes the other, you may choose either factor to be *X* and the other to be *Y*.

For the small bivariate data set in Table 11.1.1, the scatterplot is shown in Figure 11.1.1. Since we ordinarily think of effort influencing results, it is natural to display contacts made (effort) on the horizontal axis and the resulting sales on the vertical axis. Sometimes it's helpful to have the points labeled as in Figure 11.1.1; at other times these labels may cause too much clutter. A more conventional scatterplot for this data set is shown in Figure 11.1.2. It is conventional to say that these are plots of sales "against"

**TABLE 11.1.1  First-Quarter Performance**

|         | Contacts | Sales     |
|---------|----------|-----------|
| Bill    | 147      | $126,300  |
| Martha  | 223      | 182,518   |
| Colleen | 163      | 141,775   |
| Gary    | 172      | 138,282   |

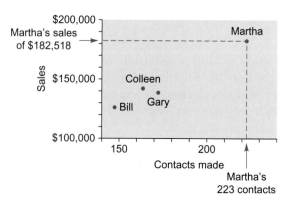

**FIGURE 11.1.1**   The scatterplot displays one point for each row in your bivariate data set. Each point is labeled here to show where it came from. Martha's exceptional performance is highlighted to show her 223 contacts, which resulted in sales of $182,518 for the quarter.

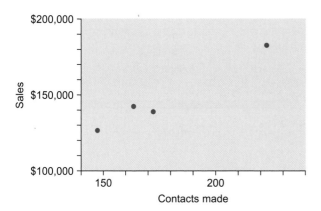

FIGURE 11.1.2   The scatterplot from Figure 11.1.1, without any extra information. You can see the distribution of contacts (along the horizontal axis), the distribution of sales (along the vertical axis), and the generally increasing relationship between contacts and sales (i.e., the points rise upward to the right).

(or "versus") contacts made to indicate which variable is on the vertical axis ($Y$) and which is horizontal ($X$); by convention we say that we plot $Y$ against $X$.

From either figure you can see information about each individual variable and about the relationship between variables. First, the distribution of the number of contacts (looking down at the horizontal scale) goes from about 150 to about 220, with a typical value of around 170. Second, the distribution of sales (looking at the vertical axis) goes from about $130,000 to about $180,000, with a typical value perhaps around $150,000. Finally, the relationship between contacts and sales appears to be a positive one: The pattern of points tilts upward and to the right. This tells you that those with more contacts (the data points to the right in the figure) also tended to have more sales (since these points are higher up in the figure). Although there appears to be such an increasing relationship overall, it does not apply to every case. This is typical of statistical analysis where you look for the trends of the "big picture," revealing patterns that are useful but usually not perfect.

## Example
### Measuring Internet Site Usage

Many Internet companies are in the business of "selling eye-balls" in the sense that they earn money by charging advertisers for access to their visitors, who cannot help looking at the ads. For example, you can see an ad for LendingTree at the right of MSN's search screen that changed to an ad for Education Degree Source the next time I visited the site.

The DoubleClick company, an early pioneer in Internet advertising that was bought by Google for $3.1 billion in 2008, helps advertisers plan their marketing strategy and also provides ratings for the top Internet sites. There are several different ways to measure a site's popularity, and Table 11.1.2 shows results for the top 25 web properties for the month of May 2010. The number of *Users* is the total number of distinct visitors, not counting repeat visits from the same person, which can be determined by so-called cookies, which are small files stored on the user's computer that are read each time the user visits a page but that do not necessarily reveal any personal information (such as name, address, or phone number). The *Reach* tells the percentage of Internet users who visited the site. The *Page Views* variable shows the total number of times a user visited a page (for example, if a person visited Yahoo News and Yahoo Business News, this would be counted as two page views but just one user for Yahoo).

To enhance its advertising revenues, an Internet site needs to attract many visitors to view multiple pages to engage users over an extended time period. Two measures of activity (Users and Page Views from Table 11.1.2) are plotted in the bivariate scatterplot in Figure 11.1.3. Note that the original data set is multivariate, but that we have focused attention on a bivariate data set formed by just two of its columns. The visual impression is that the sites spread out at high values of both Users and Page Views, and that Facebook is the leader in both measures.

To use Excel® to create such a scatterplot, you might begin by selecting both columns of numbers (with the horizontal $X$ axis data to the left, including the labels at the top if you wish). Then you click on Scatter from the Insert Ribbon's Charts area and choose "Scatter with Only Markers." Here is how it looks after you select the data and

*(Continued)*

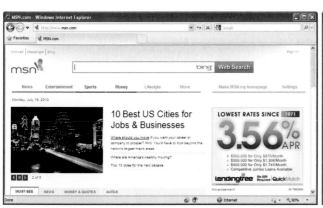

**TABLE 11.1.2 Top 25 Most-Visited Internet Sites**

| Site | Category | Users (millions) | Reach | Page Views (billions) | Has Advertising |
|---|---|---|---|---|---|
| facebook.com | Social Networks | 540 | 34.80% | 630.00 | Yes |
| yahoo.com | Web Portals | 490 | 31.50% | 70.00 | Yes |
| live.com | Search Engines | 370 | 24.00% | 40.00 | Yes |
| wikipedia.org | Dictionaries & Encyclopedias | 340 | 21.70% | 7.80 | No |
| msn.com | Web Portals | 280 | 18.00% | 11.00 | Yes |
| microsoft.com | Software | 230 | 14.70% | 3.40 | Yes |
| baidu.com | Web Portals | 170 | 11.20% | 27.00 | Yes |
| qq.com | Email & Messaging | 130 | 8.30% | 18.00 | Yes |
| wordpress.com | Blogging Resources & Services | 120 | 7.60% | 1.30 | Yes |
| mozilla.com | Internet Clients & Browsers | 110 | 6.80% | 1.90 | Yes |
| bing.com | Search Engines | 110 | 7.00% | 3.00 | Yes |
| sina.com.cn | Web Portals | 110 | 7.00% | 3.00 | Yes |
| ask.com | Search Engines | 98 | 6.30% | 1.90 | Yes |
| twitter.com | Email & Messaging | 97 | 6.20% | 5.80 | No |
| adobe.com | Programming | 96 | 6.20% | 1.10 | Yes |
| yahoo.co.jp | Web Portals | 81 | 5.20% | 30.00 | Yes |
| 163.com | Web Portals | 81 | 5.20% | 2.30 | Yes |
| amazon.com | Shopping | 81 | 5.20% | 3.60 | Yes |
| taobao.com | Auctions | 80 | 5.10% | 4.40 | Yes |
| youku.com | Video Clips & Movie Downloads | 74 | 4.70% | 1.30 | Yes |
| myspace.com | Social Networks | 73 | 4.70% | 17.00 | Yes |
| ebay.com | Auctions | 72 | 4.60% | 9.50 | Yes |
| apple.com | Computers & Electronics | 67 | 4.30% | 0.96 | Yes |
| sohu.com | Web Portals | 66 | 4.30% | 1.70 | Yes |
| hotmail.com | Email & Messaging | 60 | 3.90% | 1.00 | No |

**Source:** DoubleClick ad planner by Google for May 2010, accessed at http://www.google.com/adplanner/static/top1000/ on July 15, 2010. It is interesting that Google seems to have removed itself from the list, although information from Nielsen (accessed at http://en-us.nielsen.com/content/nielsen/en_us/insights/rankings/internet.html on July 19, 2010) shows that during this time period Google had 365 million Unique Audience (Users).

**Example—cont'd**

begin to insert a chart, together with the finished chart in the worksheet.

Because Facebook's Page Views were so much larger than any other site (it's clearly an outlier), it is difficult to compare the other sites to one another. As we found with outliers in Chapter 3, it can be useful to also analyze a data set with the outlier removed, as shown in Figure 11.1.4. Now we can see that Yahoo and Live dominate the others in both Users and Page Views, but that Wikipedia (which comes next in the Users ranking) tends to have fewer Page Views than you might expect for its number of Users. On the other hand, we can see clearly from the scatterplot that while Yahoo's Japan site yahoo.co.jp comes next (after Yahoo and Live) in the Page Views ranking, it seems to have fewer Users than we might expect for this large number of Page Views (it has 17% of the Users but 43% of the page views).

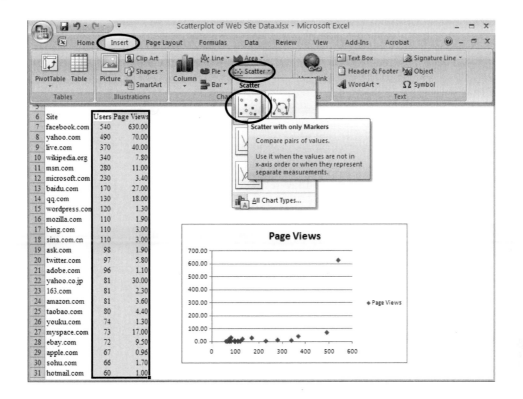

| Site | Users | Page Views |
|------|-------|-----------|
| facebook.com | 540 | 630.00 |
| yahoo.com | 490 | 70.00 |
| live.com | 370 | 40.00 |
| wikipedia.org | 340 | 7.80 |
| msn.com | 280 | 11.00 |
| microsoft.com | 230 | 3.40 |
| baidu.com | 170 | 27.00 |
| qq.com | 130 | 18.00 |
| wordpress.con | 120 | 1.30 |
| mozilla.com | 110 | 1.90 |
| bing.com | 110 | 3.00 |
| sina.com.cn | 110 | 3.00 |
| ask.com | 98 | 1.90 |
| twitter.com | 97 | 5.80 |
| adobe.com | 96 | 1.10 |
| yahoo.co.jp | 81 | 30.00 |
| 163.com | 81 | 2.30 |
| amazon.com | 81 | 3.60 |
| taobao.com | 80 | 4.40 |
| youku.com | 74 | 1.30 |
| myspace.com | 73 | 17.00 |
| ebay.com | 72 | 9.50 |
| apple.com | 67 | 0.96 |
| sohu.com | 66 | 1.70 |
| hotmail.com | 60 | 1.00 |

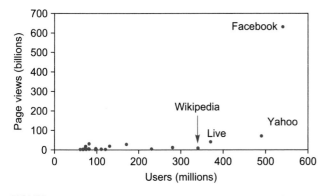

**FIGURE 11.1.3** A scatterplot of two measures of the extent of Internet site usage for $n = 25$ websites (with the top four identified). We clearly see that, while Facebook has the most users and most page views, its lead in terms of Page Views is much stronger than its lead in terms of Users.

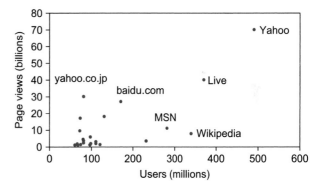

**FIGURE 11.1.4** After removing (and making careful note of) the outlier, Facebook, we can see more of the details in the rankings of the rest of the Sites because the Page Views scale can be expanded.

## Correlation Measures the Strength of the Relationship

The **correlation** or **correlation coefficient**, denoted $r$, is a pure number between –1 and 1 summarizing the strength of the relationship in the data. A correlation of 1 indicates a perfect straight-line relationship, with higher values of one variable associated with perfectly predictable higher values of the other. A correlation of –1 indicates a perfect negative straight-line relationship, with one variable *decreasing* as the other increases.

The usual interpretation of intermediate correlations between –1 and 1 is that the size (absolute value) of the correlation indicates the strength of the relationship, and the sign (positive or negative) indicates the direction (increasing or decreasing). The usual interpretation of a correlation of 0 is that there is no relationship, just randomness. However, these interpretations must be used with caution since curves (nonlinear structure) and outliers can distort the usual interpretation of the correlation. A quick look at the scatterplot will either confirm or rule out these possibly nasty possibilities. Table 11.1.3 indicates how to interpret the correlation in each case. Remember that the correlation shows you how close the points are to being exactly on a tilted straight line. It does *not* tell you how steep that line is.

To find the correlation in Excel®, you can use the CORREL function after naming your two columns of numbers (for example, by selecting each column of numbers in turn and choosing Define Name from Excel's Formula Ribbon), as shown here to find the correlation of 0.985 between contacts and sales.

## The Formula for the Correlation

The correlation is computed based on the data using a straightforward but time-consuming formula. Computers

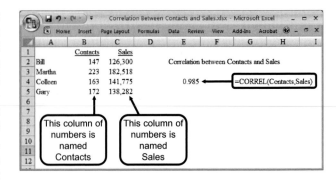

and many calculators can quickly compute the correlation for you. The formula is included in this section not so much for your use but rather to provide some insight into how it works.

The formula for the correlation coefficient is based on bivariate data consisting of the two measurements $(X_1, Y_1)$ made on the first elementary unit through the measurements $(X_n, Y_n)$ made on the last one. For example, $X_1$ might be sales of IBM and $Y_1$ might be net income of IBM; $X_n$ could be sales of Ford and $Y_n$ could be net income of Ford. Looking at each column of numbers separately, you

**TABLE 11.1.3 Interpreting the Correlation Coefficient**

| Correlation | Usual Interpretation | Some Other Possibilities |
|---|---|---|
| 1 | Perfect positive relationship. All data points must fall exactly on a line that tilts *upward* to the right. | None. |
| Close to 1 | Strong positive relationship. Data points bunch tightly, but with some random scatter, about a line that tilts *upward* to the right. | Data points fall exactly on an upward-sloping *curve* (nonlinear structure).<br><br>Data points mostly have no relationship, but one *outlier* has distorted the correlation. |
| Close to 0 but positive | Slight positive relationship. Data points form a random cloud with a slight *upward* tilt toward the right. | *Clustering* has distorted the correlation. |
| 0 | *No relationship,* just a random cloud tilting neither up nor down toward the right. | Data points fall exactly on a *curve* tilting up on one side and down on the other.<br><br>Data points fall exactly on a line, but one *outlier* has distorted the correlation.<br><br>*Clustering* has distorted the correlation. |
| Close to 0 but negative | Slight negative relationship. Data points form a random cloud with a slight *downward* tilt toward the right. | Data points fall exactly on a downward-sloping *curve* (nonlinear structure). |
| Close to −1 | Strong negative relationship. Data points bunch tightly, but with some random scatter, about a line that tilts *downward* to the right. | Data points mostly have no structure, but one *outlier* has distorted the correlation.<br><br>*Clustering* has distorted the correlation. |
| −1 | Perfect negative relationship. All data points must fall exactly on a line that tilts *downward* to the right. | None. |
| Undefined | Data points fall exactly on a horizontal or vertical line. | Not enough data (less than $n = 2$ distinct pairs of $X$ and $Y$ values). |

could compute the usual sample standard deviation for just the $X$ values to find $S_X$; similarly, $S_Y$ represents the standard deviation of just the $Y$ values.[1] The formula for the correlation also includes a sum of cross products involving $X$ and $Y$, which captures their interdependence, divided by $n - 1$ (as was used in the standard deviation calculation):

**Formula for the Correlation Coefficient**

$$r = \frac{\frac{1}{n-1}\sum_{i=1}^{n}(X_i - \overline{X})(Y_i - \overline{Y})}{S_X S_Y}$$

The terms in the numerator summation involve the interaction of the two variables and determine whether the correlation will be positive or negative. For example, if there is a strong positive (increasing) relationship, each term in the sum will be positive: If a point has a high $X$ and a high $Y$, the product will be positive; if a point has a low $X$ and a low $Y$, the product will again be positive because the two elements in the product will be negative (since $X$ and $Y$ are both below their respective averages), and a negative times a negative gives a positive number. Similarly, if there is a strong negative relationship, all terms in the sum in the numerator will be negative, resulting in a negative correlation.

The denominator merely scales the numerator so that the resulting correlation will be an easily interpreted pure number between $-1$ and $1$. Since the numerator involves the product of the two variables, it is reasonable to convert to a pure number through dividing by a product of terms involving these variables. If you did not divide, the numerator by itself would be difficult to interpret because its measurement units would be unfamiliar. For example, if $X$ and $Y$ are both measured as dollar amounts, the numerator would be in units of "squared dollars," whatever they are.

The numerator in the formula for the correlation coefficient, which is difficult to interpret due to its measurement units, is known as the **covariance** of $X$ and $Y$. Although it is used occasionally (for example, in finance theory to describe the covariation of one stock market price with another), it is probably easier to use the correlation instead. The correlation and the covariance both represent the same information (provided you also know the individual standard deviations), but the correlation presents that information in a more accessible form.

Note also that the roles of $X$ and $Y$ are interchangeable in the formula; it is *symmetric* in $X$ and $Y$. Thus, the correlation

of $X$ with $Y$ is the same as the correlation of $Y$ with $X$; it makes no difference which one comes first. This is true for correlation but not for regression (to be covered in Section 11.2).

## The Various Types of Relationships

The following sections list the various types of relationships you might find when you look at a scatterplot of a bivariate data set. At least one example will be provided for each kind of relationship, together with the scatterplot, the correlation coefficient, and some discussion. The types of relationships include linear (straight line) relationship, no relationship, nonlinear (curved) relationship, unequal variability, clustering (groupings), and bivariate outlier.

## Linear Relationship

Some kinds of bivariate data sets are easier to analyze than others. Those with a *linear relationship* are the easiest. This relationship plays the same special role for bivariate data that the normal distribution plays for univariate data. A bivariate data set shows a **linear relationship** if the scatterplot shows points bunched randomly around a straight line.[2] The points might be tightly bunched and fall almost exactly on a line, or they might be wildly scattered, forming a cloud of points. But the relationship will not be strongly curved, will not be funnel-shaped, and will not have any extreme outliers.

> **Example**
> *Economic Activity and Population of the States*
>
> One way to measure the amount of economic activity is to use GDP, the gross domestic product consisting of the value of all goods and services produced, and each state of the United States has its GDP as measured by the Survey of Current Business of the U.S. Bureau of Economic Analysis. One reason that a state might have more economic activity than another may be that it has more people contributing to its economy. On the other hand, a state with a busier economy (perhaps due to natural resources) than another might tend to attract more people than another state. Either way, we might expect to see a relationship between GDP and population of the states, as listed in Table 11.1.4, and we do find a strong linear relationship as shown in Figure 11.1.5.
>
> *(Continued)*

---

1. Note that $S_X$ and $S_Y$ are standard deviations representing the variability of *individuals* and should not be confused with standard errors $S_{\overline{X}}$ and $S_{\overline{Y}}$ representing the variability of the sample averages $\overline{X}$ and $\overline{Y}$, respectively.

2. A bivariate data set is said to have a *bivariate normal distribution* if it shows a linear relationship and, in addition, each of the individual variables has a normal distribution. A more careful technical definition would also require that for each $X$ value, the $Y$ values be normally distributed with constant variation.

### Example—cont'd

The scatterplot (Figure 11.1.5) shows linear structure because the points could be described as following a straight line but with some scatter. The relationship is positive because states with more people (to the right) generally also have more economic activity (toward the top). The high correlation, $r = 0.989$, summarizes the fact that there is strong positive association but not a perfect relationship. There is some randomness, which could make a difference in ranking these states.

**TABLE 11.1.4** Population and Economic Activity (GDP) of the States

| State | Population (millions) | GDP ($ billions) |
| --- | --- | --- |
| Alabama | 4.71 | 170 |
| Alaska | 0.70 | 48 |
| Arizona | 6.60 | 249 |
| Arkansas | 2.89 | 98 |
| California | 36.96 | 1,847 |
| Colorado | 5.02 | 249 |
| Connecticut | 3.52 | 216 |
| Delaware | 0.89 | 62 |
| Florida | 18.54 | 744 |
| Georgia | 9.83 | 398 |
| Hawaii | 1.30 | 64 |
| Idaho | 1.55 | 53 |
| Illinois | 12.91 | 634 |
| Indiana | 6.42 | 255 |
| Iowa | 3.01 | 136 |
| Kansas | 2.82 | 123 |
| Kentucky | 4.31 | 156 |
| Louisiana | 4.49 | 222 |
| Maine | 1.32 | 50 |
| Maryland | 5.70 | 273 |
| Massachusetts | 6.59 | 365 |
| Michigan | 9.97 | 383 |
| Minnesota | 5.27 | 263 |
| Mississippi | 2.95 | 92 |
| Missouri | 5.99 | 238 |
| Montana | 0.97 | 36 |

| | | |
| --- | --- | --- |
| Nebraska | 1.80 | 83 |
| Nevada | 2.64 | 131 |
| New Hampshire | 1.32 | 60 |
| New Jersey | 8.71 | 475 |
| New Mexico | 2.01 | 80 |
| New York | 19.54 | 1,144 |
| North Carolina | 9.38 | 400 |
| North Dakota | 0.65 | 31 |
| Ohio | 11.54 | 472 |
| Oklahoma | 3.69 | 146 |
| Oregon | 3.83 | 162 |
| Pennsylvania | 12.60 | 553 |
| Rhode Island | 1.05 | 47 |
| South Carolina | 4.56 | 156 |
| South Dakota | 0.81 | 37 |
| Tennessee | 6.30 | 252 |
| Texas | 24.78 | 1,224 |
| Utah | 2.78 | 110 |
| Vermont | 0.62 | 25 |
| Virginia | 7.88 | 397 |
| Washington | 6.66 | 323 |
| West Virginia | 1.82 | 62 |
| Wisconsin | 5.65 | 240 |
| Wyoming | 0.54 | 35 |

**Source:** U.S. Census Bureau, *Statistical Abstract of the United States: 2010* (129th edition), Washington, DC, 2009, accessed at http://www.census.gov/compendia/statab/rankings.html on July 16, 2010.

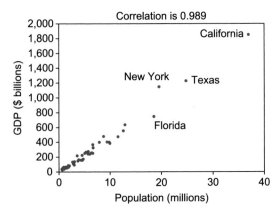

**FIGURE 11.1.5** A linear relationship in the scatterplot of GDP (economic activity) and population of the $n = 50$ states. Note the strong positive association, summarized by the high correlation of $r = 0.989$.

## Example
### Mergers

Investment bankers earn large fees for making arrangements and giving advice relating to mergers and acquisitions when one firm joins with or purchases another. Who are the big players? How many deals and how much money are involved to produce these huge fees? Some answers are provided by the bivariate data set in Table 11.1.5 for 2008 oil and gas company transactions.

The scatterplot, in Figure 11.1.6, shows a linear relationship, but one with substantially more scatter, or randomness, than in the previous example. There is an increasing trend, with the more successful firms being involved in more deals (toward the right) that involved more money (toward the top). The randomness involves substantial dollar amounts; for example, among firms involved with around 18 deals, the dollar amount of these deals differed by tens of billions of dollars. The correlation is $r = 0.581$, summarizing this increasing trend in the presence of noticeable randomness.

**TABLE 11.1.5 Top Merger and Acquisition Advisors for Oil and Gas Companies**

| | Number of Transactions | Deal Value (billions) |
|---|---|---|
| Goldman Sachs & Co | 18 | 24.6 |
| Scotia Waterous | 29 | 20.1 |
| JP Morgan | 19 | 18.6 |
| Deutsche Bank AG | 8 | 11.8 |
| TD Securities | 7 | 11.3 |
| Credit Suisse | 7 | 10.1 |
| Macquarie Group Ltd | 11 | 7.3 |
| Tristone Capital Inc | 17 | 6.8 |
| Merrill Lynch & Co | 14 | 6.6 |

**Source:** "Goldman Sachs, Scotia Waterous Were Top M&A Advisors in 2008" in *Oil and Gas Financial Journal*, accessed at http://www.ogfj.com/index/article-display/361732/articles/oil-gas-financial-journal/volume-6/issue-5/features/goldman-sachs-scotia-waterous-were-top-mampa-advisors-in-2008.html on July 15, 2010.

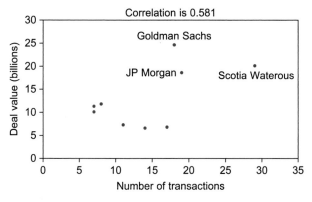

**FIGURE 11.1.6**   A linear relationship between the dollar amount and number of deals involved for the largest advisors for mergers and acquisitions. The correlation, $r = 0.581$, summarizes the increasing trend (successful advisors had lots of deals involving lots of money) partially obscured by randomness.

## Example
### Mortgage Rates and Fees

When you take out a mortgage, there are many different kinds of costs. Often the two largest are the *interest rate* (a yearly percentage that determines the size of your monthly payment) and the number of *points* (a one-time percentage charged to you at the time the loan is made). Some financial institutions let you "buy down" the interest rate by paying a higher initial loan fee (as points), suggesting that there should be a relationship between these two costs. The relationship should be *negative*, or *decreasing*, since a higher loan fee should go with a lower interest rate.

Table 11.1.6 shows a bivariate data set consisting of points and interest rates for lenders of 30-year fixed-rate mortgage loans.

The scatterplot, in Figure 11.1.7, shows a linear relationship with scatter and a *decreasing* association between points and interest rate. The negative correlation, $r = -0.401$, confirms the decreasing relationship we had expected. This correlation is consistent with the moderate association in the scatterplot, along with randomness.

Where did all the data go? There are 27 financial institutions in the bivariate data listing, but the number of points in the scatterplot looks much smaller than that. The reason is that some combinations are used by several institutions (for example, a point of 1% with an interest rate of 4.25%). These overlapping data points look deceptively like just one point in a simple scatterplot such as Figure 11.1.7. By adding a little bit of extra randomness or "jitter" (just for purposes of *looking* at the data, not analyzing it!), you can separate these overlapping points and see your data more clearly.[3] The resulting "jittered" scatterplot is shown in Figure 11.1.8.

3. For a general introduction to the many different techniques for looking at data (including jittering, on p. 135), see J. M. Chambers, W. S. Cleveland, B. Kleiner, and P. A. Tukey, *Graphical Methods for Data Analysis* (New York: Wadsworth, 1983).

## No Relationship

A bivariate data set shows **no relationship** if the scatterplot is just random, with no tilt (i.e., slanting neither upward nor downward as you move from left to right). The case of no

### TABLE 11.1.6 Mortgage Costs

| Institution | Points | Interest Rate |
|---|---|---|
| AimLoan.com | 0.000% | 4.500% |
| AimLoan.com | 0.532 | 4.375 |
| AimLoan.com | 1.856 | 4.125 |
| Bank of America | 1.125 | 4.750 |
| Bellevue Center Financial | 0.000 | 4.375 |
| Bellevue Center Financial | 1.000 | 4.250 |
| Cornerstone Mortgage Group | 0.000 | 4.375 |
| Cornerstone Mortgage Group | 1.000 | 4.250 |
| Cornerstone Mortgage Group | 2.000 | 4.250 |
| First Savings Bank Northwest | 0.000 | 5.375 |
| Interstate Mortgage Service, Inc | 0.000 | 4.375 |
| Interstate Mortgage Service, Inc | 0.750 | 4.250 |
| Interstate Mortgage Service, Inc | 1.010 | 4.250 |
| KeyBank | 0.500 | 4.750 |
| Mortgage Capital Associates | 0.000 | 4.500 |
| Mortgage Capital Associates | 1.000 | 4.375 |
| Mortgage Capital Associates | 2.000 | 4.375 |
| Oxford Lending Group, LLC | 0.000 | 4.500 |
| Oxford Lending Group, LLC | 1.000 | 4.375 |
| Oxford Lending Group, LLC | 2.000 | 4.125 |
| Quicken Loans | 0.000 | 4.750 |
| Quicken Loans | 1.000 | 4.625 |
| Quicken Loans | 1.875 | 4.375 |
| The Money Store | 0.000 | 4.375 |
| The Money Store | 1.000 | 4.250 |
| The Money Store | 2.000 | 4.125 |
| Washington Federal | 1.000 | 5.250 |

**Source:** Data are for a loan to borrow $300,000 in Seattle for 30 years at a fixed rate, accessed at http://rates.interest.com on July 15, 2010. Discount and origination points have been combined.

relationship is a special linear relationship that is neither increasing nor decreasing. Such a scatterplot may look like a cloud that is either circular or oval-shaped (the oval points either up and down or left to right; it is not tilted). In fact, by changing the scale for one or the other of your variables, you can make a data set with no relationship have either a circular or an oval-shaped scatterplot.

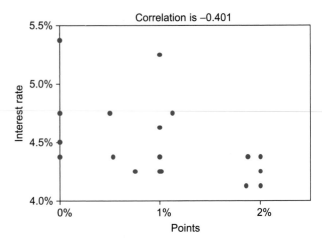

**FIGURE 11.1.7** A decreasing linear relationship between loan fee and interest rate for mortgages. The correlation, –0.401, summarizes this decreasing relationship: Higher fees tend to go with lower interest rates but not perfectly so, reflecting a moderate association, along with randomness.

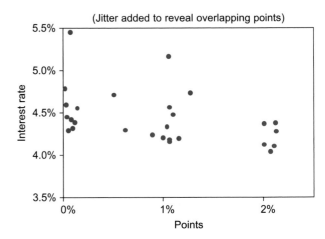

**FIGURE 11.1.8** The previous scatterplot with "jitter" added to separate the overlapping points and show the data set more clearly.

### Example
#### Short-Term "Momentum" and the Stock Market

Does the stock market have "momentum"? That is, is the market likely to keep going up this week because it went up last week? If there is a relationship between current market performance and the recent past, you would expect to find it in a scatterplot. After all, this is our best statistical tool for seeing the relationship, if any, between market behavior last week (one variable) and market behavior this week (the other variable).

The bivariate data set consists of weekly rates of return for the S&P 500 Stock Market Index, that is, the percent changes (increases or decreases) from one week to the next.[4] Although this seems to be a univariate time series, you can put essentially the same data in the two columns, offsetting the columns by one row so that each value for this week's close (on the left in the table) can be found in the next row (one week later) as last week's close (on the right). This is shown in Table 11.1.7.

The scatterplot, in Figure 11.1.9, shows no relationship! There is a lot of random scatter but no trend either upward (which would have suggested momentum) or downward (which would have suggested that the market "overreacted" one week and then corrected itself the next) as you move from left to right in the picture. The correlation, $r = 0.026$, is close to 0, confirming the lack of a strong relationship.[5]

A scatterplot such as this one is consistent with the ideas of market efficiency and random walks. Market efficiency says that all information that is available or can be anticipated is immediately reflected in market prices. Since traders anticipate future changes in market prices, there can be no systematic relationships, and only randomness (i.e., a *random walk*) can remain. A random walk generates a time series of data with no relationship between previous behavior and the next step or change.[6]

By changing the scale of the horizontal or vertical axis, you can make the cloud of points look more like a line. However, because the line is either *horizontal* or *vertical*, with no tilt, it still indicates no relationship. These cases are shown in Figures 11.1.10 and 11.1.11.

---

4. The formula for daily percent return is (This week's price – Last week's price)/(Last week's price).

5. A correlation coefficient such as this, computed for a time series and its own previous values, is called the *autocorrelation* of the series because it measures the correlation of the series with itself. You might say that this time series is not strongly autocorrelated because the autocorrelation is close to zero.

6. There is an entire book on this subject: B. G. Malkiel, *A Random Walk Down Wall Street* (New York: W. W. Norton, 2007).

---

**TABLE 11.1.7  Weekly Percent Change in the S&P 500 Stock Market Index**

| Date | This Week | Last Week |
|------|-----------|-----------|
| 1/4/2010 | 2.68% | –1.01% |
| 1/11/2010 | –0.78 | 2.68 |
| 1/19/2010 | –3.90 | –0.78 |
| 1/25/2010 | –1.64 | –3.90 |
| 2/1/2010 | –0.72 | –1.64 |
| 2/8/2010 | 0.87 | –0.72 |
| 2/16/2010 | 3.13 | 0.87 |
| 2/22/2010 | –0.42 | 3.13 |
| 3/1/2010 | 3.10 | –0.42 |
| 3/8/2010 | 0.99 | 3.10 |
| 3/15/2010 | 0.86 | 0.99 |
| 3/22/2010 | 0.58 | 0.86 |
| 3/29/2010 | 0.99 | 0.58 |
| 4/5/2010 | 1.38 | 0.99 |
| 4/12/2010 | –0.19 | 1.38 |
| 4/19/2010 | 2.11 | –0.19 |
| 4/26/2010 | –2.51 | 2.11 |
| 5/3/2010 | –6.39 | –2.51 |
| 5/10/2010 | 2.23 | –6.39 |
| 5/17/2010 | –4.23 | 2.23 |
| 5/24/2010 | 0.16 | –4.23 |
| 6/1/2010 | –2.25 | 0.16 |
| 6/7/2010 | 2.51 | –2.25 |
| 6/14/2010 | 2.37 | 2.51 |
| 6/21/2010 | –3.65 | 2.37 |
| 6/28/2010 | –4.59 | –3.65 |

**Source:** Calculated from adjusted closing prices accessed at http://finance.yahoo.com on July 16, 2010.

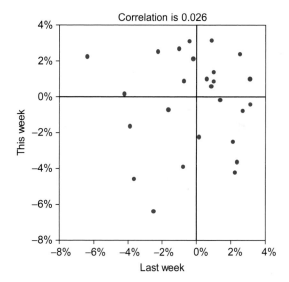

**FIGURE 11.1.9**   There is essentially no relationship discernable between this week's and last week's stock market performance. The correlation, $r = 0.026$, is close to 0, summarizing the lack of a relationship. If last week was a "good" week, then this week's performance will look about the same as if last week had been a "bad" week.

## Nonlinear Relationship

Other kinds of bivariate data sets are not so simple to analyze. A bivariate data set has a **nonlinear relationship** if the scatterplot shows points bunched around a *curved* rather than a straight line. Since there are so many different kinds of curves that can be drawn, the analysis is more complex.

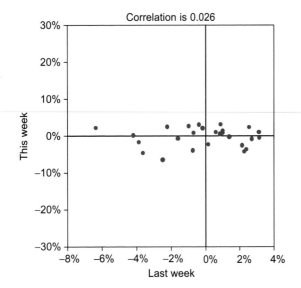

**FIGURE 11.1.10** There is no relationship here, even though the scatterplot looks like a distinct line, because the line is horizontal, with no tilt. This is the same data set as in Figure 11.1.9, but with the *Y* scale expanded to flatten the appearance of the plot.

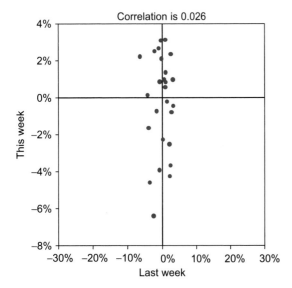

**FIGURE 11.1.11** No relationship is apparent here either, although the scatterplot looks like a distinct line because the line is vertical, with no tilt. In this case the *X* axis has been expanded from Figure 11.1.9.

Correlation and regression analysis must be used with care on nonlinear data sets. For some problems, you may want to transform one or both of the variables to obtain a linear relationship. This simplifies the analysis (since correlation and regression are more easily applied to a linear relationship) provided the results are transformed back to the original data, if appropriate.[7]

7. Transformations in regression will be considered further in Chapter 12.

### Example
### *Index Options*

If you buy a *call option*, you have the right (but not the obligation) to buy some asset (it might be a lot of land, 100 shares of Google stock, etc.) at a set price (the *strike price* or *exercise price*) whenever you want until the option expires. Businesses use options to hedge (i.e., reduce) risks at a much lower price compared to buying and perhaps later selling the asset itself. Options on stocks can be used either to reduce the risk of a portfolio or to create a portfolio of high risk with high expected return.

The higher the strike price, the less the option is worth. For example, an option to buy a candy bar for $2,000 is worthless, but an option to buy it at $0.50 would have some value. In fact, if candy bar prices are stable at $0.85, then the option would be worth $0.35 = $0.85 – $0.50. However, for most markets, uncertainty about the future adds to the value of the option. For example, on July 19, 2010, an option to buy Google stock at a strike price of $470 anytime during the next two months, when the stock was trading at $466, was worth about $20 per share. Why would you want to buy stock at $470 through the option when you could buy it for $466 right then? You wouldn't, but you could hold on to the option and still buy the stock for $470 when the market price for the stock rises to $480 (if it does). This market volatility (the *possibility* of a rise in prices) accounts for an important part of the value of an option.

So we expect to find a negative relationship between the *strike price* specified in the option contract and the *call price* at which the option contract itself is traded. Table 11.1.8 shows a bivariate data set for a popular set of index options, based on Standard & Poor's 500 stock market index.

The scatterplot, in Figure 11.1.12, shows a nonlinear relationship. The relationship is clearly negative, since higher strike prices are associated with lower call prices; the correlation of $r = -0.903$ confirms a strong negative relationship. Since the relationship is nearly perfect and there is almost no randomness, you might expect a correlation closer to –1. However, this could happen only if the points were exactly

**TABLE 11.1.8 S&P 500 Index Call Options**

| Strike Price | Call Price |
| --- | --- |
| 940 | 132.60 |
| 950 | 119.25 |
| 970 | 105.00 |
| 980 | 100.22 |
| 990 | 85.45 |
| 1000 | 77.30 |
| 1010 | 68.50 |

## TABLE 11.1.8 S&P 500 Index Call Options—cont'd

| Strike Price | Call Price |
|---|---|
| 1020 | 64.75 |
| 1030 | 57.90 |
| 1040 | 47.75 |
| 1050 | 40.00 |
| 1060 | 34.15 |
| 1070 | 29.00 |
| 1080 | 23.00 |
| 1090 | 18.00 |
| 1100 | 14.60 |
| 1110 | 10.90 |
| 1120 | 7.80 |
| 1130 | 5.70 |
| 1140 | 3.97 |
| 1150 | 2.75 |
| 1160 | 1.80 |
| 1170 | 1.55 |
| 1180 | 0.75 |
| 1190 | 0.65 |
| 1200 | 0.40 |
| 1210 | 0.50 |
| 1240 | 0.20 |
| 1300 | 0.05 |

**Source:** Data are for August expiration, accessed at the *Wall Street Journal's* Data Center from http://www.wsj.com on July 19, 2010. The index itself was trading at 1065.

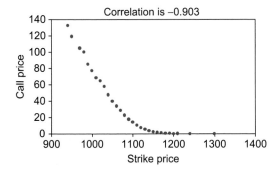

**FIGURE 11.1.12**   A nonlinear relationship between the price of an option and the strike price. You can see the expected negative relationship, but it is *nonlinear* because the line is curved. The correlation, $r = -0.903$, expresses a strong negative relationship. Due to the curvature, the correlation cannot be exactly $-1$ even though this is nearly a perfect relationship, with almost no random scatter.

on a *straight line*. Since the points are exactly on a curve, the correlation must be different from $-1$ because the correlation measures only *linear* association.

Advanced statistical methods, based on assumptions of an underlying normal distribution and a random walk for the stock price, have allowed analysts to compute an appropriate value for a call option price.[8] This complex and advanced theory is based on a careful computation of the expectation (mean value) of the random variable representing the option's ultimate payoff value and is computed using probabilities for a normal distribution.

8. An overview of the theory and practice of options is provided by J. C. Cox and M. Rubenstein, *Options Markets* (Englewood Cliffs, NJ: Prentice Hall, 1985).

### Example
### *Yield and Temperature*

You can have a strong nonlinear relationship even if the correlation is *nearly zero!* This can happen if the strong relationship is neither increasing nor decreasing, as might happen if there is an optimal, or best possible, value. Consider the data taken as part of an experiment to find the temperature that produces the largest output yield for an industrial process, shown in Table 11.1.9.

The scatterplot, in Figure 11.1.13, shows a strong nonlinear relationship with some random scatter. The correlation, $r = -0.0155$, is essentially useless for summarizing this kind of nonlinear relationship: It can't decide whether the relationship is increasing or decreasing because it is doing both!

This scatterplot will be very useful to your firm, since it tells you that to maximize your output yield, you should set the temperature of the process at around 700 degrees. The yield falls off if the temperature is either too cold or too hot. This useful information has come to you from looking at the strong relationship between yield and temperature on the scatterplot.

Remember: A correlation value near zero might mean there is no relationship in your data, but it also might mean the relationship is nonlinear with no overall trend up or down.

### TABLE 11.1.9 Temperature and Yield for an Industrial Process

| Temperature | Yield | Temperature | Yield |
|---|---|---|---|
| 600 | 127 | 750 | 153 |
| 625 | 139 | 775 | 148 |
| 650 | 147 | 800 | 146 |
| 675 | 147 | 825 | 136 |
| 700 | 155 | 850 | 129 |
| 725 | 154 | | |

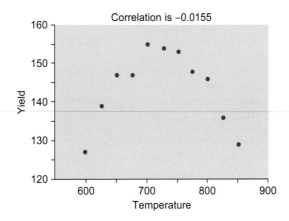

**FIGURE 11.1.13**   A nonlinear relationship between the output yield and the temperature of an industrial process. Although there is a strong relationship here, it is nonlinear. The correlation, $r = -0.0155$, merely tells you that, *overall*, the trend is neither up nor down.

## Unequal Variability

Another technical difficulty that, unfortunately, arises quite often in business and economic data is that the vertical variability in a plot of the data may depend on where you are on the horizontal scale. When you measure large businesses (or other kinds of elementary units), you find lots of variability, perhaps millions or billions of dollars' worth, but when you measure small businesses you might find variability only in the tens of thousands. A scatterplot is said to have **unequal variability** when the variability on the vertical axis changes dramatically as you move horizontally across the scatterplot.[9]

The problem with unequal variability is that the places with high variability, which represent the *least precise* information, tend to influence statistical summaries the most. So if you have a scatterplot with extremely unequal variability, the correlation coefficient (and other summaries of the relationship) will probably be unreliable.

This problem can often be solved by transforming the data, perhaps by using logarithms. It is fortunate that such a transformation, when applied to each variable, often solves several problems. Not only will the variability be equalized, but the variables will also be more normally distributed in many cases. Logarithms (either natural base *e* or common base 10; pick one and stick with it) tend to work well with dollar amounts. The square root transformation often works well with count data, which are measures of the number of things or the number of times something happened.

---

9. The technical words *heteroscedastic* (adjective) and *heteroscedasticity* (noun) also describe unequal variability. They are also spelled *heteroskedastic* and *heteroskedasticity*.

**Example**
*Employees and Sales*

What is the right staffing level for a company? While there are many considerations, generally a more successful firm with more business activity will require more staff to take care of these operations; however, some companies are by nature more labor-intensive than others. As a result, we might expect to see a positive relationship, with considerable randomness, between the number of employees and the total sales of companies. This information is shown in Table 11.1.10 for a group of major Northwest companies.

The scatterplot, in Figure 11.1.14, shows a generally increasing relationship: Firms with more employees tended to produce higher sales levels. However, the variability is substantially unequal, as may be seen from the "funnel shape" of the data, opening out to the right. The smaller firms are crowded together at the lower left, indicating much less variability in sales, whereas the larger firms at the right show much greater variability in the sales they achieve with their larger staffing. Figure 11.1.15 indicates exactly which variabilities are unequal: the variabilities measured vertically, for employees.

Could transformation take care of this unequal variability problem? Let's try natural logarithms. Alaska Air has 14,485 employees, so the logarithm is 9.581. Sales for Alaska Air are 3,663, so the logarithm is 8.206. Results of taking the log of each data value (using Excel's =LN function) are shown in Table 11.1.11.

The scatterplot, shown in Figure 11.1.16, shows a very nice *linear* relationship between the logarithms of employees and the logarithms of sales. The scatterplot would have looked the same had you used common logarithms (base 10) or if you had transformed the sales numbers as dollars instead of as $ millions. The unequal variability problem disappears when we use the logarithmic scale.

**TABLE 11.1.10 Employees and Sales for Northwest Companies**

|  | Employees | Sales ($ millions) |
|---|---|---|
| Alaska Air Group | 3,663 | 14,485 |
| Amazon.com | 19,166 | 13,900 |
| Avista | 1,677 | 1,995 |
| Cascade Corp | 534 | 2,100 |
| Coinstar | 912 | 1,900 |
| Coldwater Creek | 1,024 | 11,577 |
| Columbia Sportswear | 1,318 | 2,810 |
| Costco Wholesale | 72,483 | 127,000 |

### TABLE 11.1.10 Employees and Sales for Northwest Companies—cont'd

| | Employees | Sales ($ millions) |
|---|---|---|
| Esterline Technologies | 1,483 | 8,150 |
| Expedia | 2,937 | 6,600 |
| Expeditors International | 5,634 | 11,600 |
| F5 Networks | 650 | 1,068 |
| FEI | 599 | 1,683 |
| Flir Systems | 1,077 | 1,419 |
| Greenbrier | 1,290 | 3,661 |
| Idacorp | 960 | 1,976 |
| Intermec | 891 | 2,407 |
| Itron | 1,910 | 2,400 |
| Lithia Motors | 2,138 | 6,261 |
| Micron Technology | 5,841 | 23,500 |
| Microsoft | 60,420 | 71,000 |
| MWI Veterinary Supply | 831 | 719 |
| Nike | 18,627 | 28,000 |
| Nordstrom | 8,573 | 52,900 |
| Northwest Natural Gas | 1,038 | 1,211 |
| Northwest Pipe | 440 | 1,185 |
| Paccar | 14,973 | 21,000 |
| Plum Creek Timber | 1,614 | 2,000 |
| Portland General Electric | 1,745 | 2,635 |
| Precision Castparts | 6,852 | 16,063 |
| Puget Energy | 3,358 | 2,400 |
| RealNetworks | 605 | 1,649 |
| Schnitzer Steel Industries | 3,642 | 3,252 |
| StanCorp Financial Group | 2,667 | 3,280 |
| Starbucks | 10,383 | 145,800 |
| Sterling Financial | 787 | 2,405 |
| TriQuint Semiconductor | 573 | 1,780 |
| Umpqua Holdings | 541 | 1,530 |
| Washington Federal | 730 | 765 |
| Weyerhaeuser | 8,018 | 46,700 |

**Source:** Data are from the *Seattle Times*, accessed March 27, 2010 at http://seattletimes.nwsource.com/flatpages/businesstechnology/2009northwestcompaniesdatabase.html.

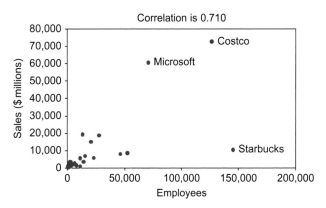

**FIGURE 11.1.14** Unequal variability in the relationship between sales and employees. The large players (to the right) show much more variability in sales levels than the smaller players (at the left).

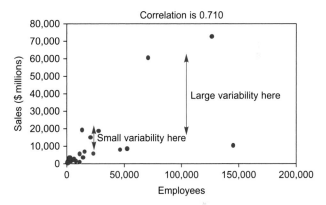

**FIGURE 11.1.15** The scatterplot of sales against employees, with the unequal variabilities clearly indicated. Note that we are referring to the *vertical* variability in *Y* (the sales levels) being different at different horizontal positions (different employee levels).

### TABLE 11.1.11 Employees and Sales for Northwest Companies (Natural Log Scale)

| | Log of Employees | Sales (log of $ millions) |
|---|---|---|
| Alaska Air Group | 8.206 | 9.581 |
| Amazon.com | 9.861 | 9.540 |
| Avista | 7.425 | 7.598 |
| Cascade Corp | 6.280 | 7.650 |
| Coinstar | 6.816 | 7.550 |
| Coldwater Creek | 6.931 | 9.357 |

*(Continued)*

**TABLE 11.1.11 Employees and Sales for Northwest Companies (Natural Log Scale)—cont'd**

| | Log of Employees | Sales (log of $ millions) |
|---|---|---|
| Columbia Sportswear | 7.184 | 7.941 |
| Costco Wholesale | 11.191 | 11.752 |
| Esterline Technologies | 7.302 | 9.006 |
| Expedia | 7.985 | 8.795 |
| Expeditors International | 8.637 | 9.359 |
| F5 Networks | 6.477 | 6.974 |
| FEI | 6.395 | 7.428 |
| Flir Systems | 6.982 | 7.258 |
| Greenbrier | 7.162 | 8.205 |
| Idacorp | 6.867 | 7.589 |
| Intermec | 6.792 | 7.786 |
| Itron | 7.555 | 7.783 |
| Lithia Motors | 7.668 | 8.742 |
| Micron Technology | 8.673 | 10.065 |
| Microsoft | 11.009 | 11.170 |
| MWI Veterinary Supply | 6.723 | 6.578 |
| Nike | 9.832 | 10.240 |
| Nordstrom | 9.056 | 10.876 |
| Northwest Natural Gas | 6.945 | 7.099 |
| Northwest Pipe | 6.087 | 7.077 |
| Paccar | 9.614 | 9.952 |
| Plum Creek Timber | 7.386 | 7.601 |
| Portland General Electric | 7.465 | 7.877 |
| Precision Castparts | 8.832 | 9.684 |
| Puget Energy | 8.119 | 7.783 |
| RealNetworks | 6.405 | 7.408 |
| Schnitzer Steel Industries | 8.200 | 8.087 |
| StanCorp Financial Group | 7.889 | 8.096 |
| Starbucks | 9.248 | 11.890 |
| Sterling Financial | 6.668 | 7.785 |
| TriQuint Semiconductor | 6.351 | 7.484 |
| Umpqua Holdings | 6.293 | 7.333 |
| Washington Federal | 6.593 | 6.640 |
| Weyerhaeuser | 8.989 | 10.751 |

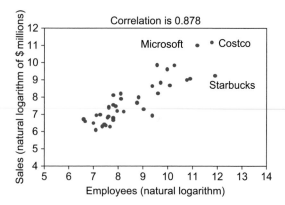

**FIGURE 11.1.16**  Transforming to a linear relationship between the natural logarithms of employees and of sales. By transforming in this way, we eliminated the problem of unequal variability. This data set, on the log scale, has a linear relationship.

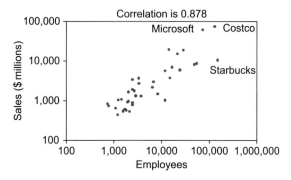

**FIGURE 11.1.17**  Again the transformed (logarithms) of employees and of sales, this time produced by reformatting the axes from Figure 11.1.15 (in Excel, right-click on each of the axes in turn, choose Format Axis at the bottom of the context-sensitive menu that appears, and check the box next to Logarithmic scale near the middle of the Axis Options). In this case, the scale shows actual numbers of employees and sales ($ millions) on a proportionate scale.

It is common for the correlation to increase (as it does here, from 0.710 untransformed to 0.878 for the logs) when a good transformation is used, but this may not always happen. However, it is generally true that because the correlation on the original scale is so sensitive to those few very large firms, the correlation number after transforming is a more *reliable* indication of the relationship.

Excel provides an easy way to show a chart on the log scale without having to transform the data. Right-click on each of the axes in turn, choose Format Axis at the bottom of the context-sensitive menu that appears, and check the box next to Logarithmic scale near the middle of the Axis Options. The result is shown in Figure 11.1.17. Note that the scale shows actual numbers of employees and actual $ millions for sales, with a scale that has equal increments for each multiple of 10 (that is, from 100 to 1,000 employees is the same distance on the log scale as 1,000 to 10,000 employees) because the log scale is a proportionate scale.

# Clustering

A bivariate data set is said to show **clustering** if there are separate, distinct groups in the scatterplot. This can be a problem if your data are clustered but you are not aware of it because the usual statistical summaries of the relationship are not sophisticated enough to respond to this kind of relationship. It is up to you to recognize clustering and to respond, for example, by separating the data set into two or more data sets, one for each cluster.

A typical problem with clustering is that within each cluster there is a clear relationship, but the correlation coefficient suggests that there is no relationship. Even worse, the correlation coefficient can suggest that the overall relationship is *opposite* to the relationship within each cluster! Always look at a scatterplot to see if you have this problem; the correlation alone can't tell you.

### Example
#### Inflation-Protected Bonds

U.S. Treasury securities are among the least risky investments, in terms of the likelihood of your receiving the promised payments.[10] In addition to the primary market auctions by the treasury, there is an active secondary market in which all outstanding issues can be traded. You would expect to see an increasing relationship between the *coupon* of the bond, which indicates the size of its periodic payment (cash twice a year), and the current selling price. Table 11.1.12 shows a bivariate data set of coupons and bid prices for long-maturity U.S. Treasury securities maturing during or after the year 2020. There are two types of securities: ordinary (notes and bonds) and TIPS (Treasury Inflation-Protected Securities).

The scatterplot in Figure 11.1.18 shows clustering. The ordinary bonds form one cluster with a very strong linear relationship. After careful investigation, you would find out that the special bonds in the cluster to the left are *inflation-protected* Treasury bonds (TIPS). These inflation-indexed bonds form a cluster with a very different relationship between coupon and price (although the slopes are similar for the two clusters, the TIPS prices are generally higher at a given coupon rate). The overall correlation, $r = 0.899$, indicates the strength of the relationship among *all* of the data points in all clusters. The relationship among the ordinary bonds is very much stronger, with a correlation of $r = 0.977$, found by leaving the inflation-indexed bonds out of the calculation.

What might have happened if you had not identified the clusters? You might have misjudged the strength of the relationship, concluding that the relationship between coupon and price is merely "quite strongly related" with a correlation of 0.899, instead of identifying the true relationship for ordinary bonds, which is "nearly perfectly related with a correlation of 0.977. If you were using this data set to compute prices or to decide which ordinary bonds to trade, your results would have been compromised by the presence of the inflation-protected bonds, which are, in a sense, a different type of

**TABLE 11.1.12** U.S. Treasury Securities

| Coupon Rate | Bid Price |
|---|---|
| 8.750 | 149.19 |
| 8.750 | 149.66 |
| 8.500 | 146.47 |
| 8.125 | 145.59 |
| 8.125 | 146.03 |
| 8.000 | 145.22 |
| 7.875 | 142.94 |
| 7.625 | 143.34 |
| 7.625 | 146.75 |
| 7.500 | 144.91 |
| 7.250 | 139.09 |
| 7.125 | 138.25 |
| 6.875 | 138.56 |
| 6.750 | 138.66 |
| 6.625 | 137.47 |
| 6.500 | 135.72 |
| 6.375 | 134.50 |
| 6.250 | 129.47 |
| 6.250 | 134.63 |
| 6.125 | 131.38 |
| 6.125 | 132.34 |
| 6.000 | 128.72 |
| 5.500 | 123.28 |
| 5.375 | 122.34 |
| 5.250 | 119.97 |
| 5.250 | 119.91 |
| 5.000 | 118.16 |
| 4.750 | 113.78 |
| 4.625 | 111.28 |
| 4.500 | 109.56 |
| 4.500 | 109.25 |
| 4.500 | 109.03 |
| 4.375 | 107.09 |
| 4.375 | 106.88 |
| 4.375 | 107.06 |

*(Continued)*

**TABLE 11.1.12** U.S. Treasury Securities—cont'd

| Coupon Rate | Bid Price |
|---|---|
| 4.250 | 104.72 |
| 3.625 | 105.41 |
| 3.500 | 104.41 |
| 3.500 | 91.91 |
| 3.875 | 132.20 |
| 3.625 | 127.25 |
| 3.375 | 128.02 |
| 2.500 | 111.21 |
| 2.375 | 109.29 |
| 2.375 | 109.20 |
| 2.125 | 106.26 |
| 2.000 | 104.20 |
| 1.750 | 100.08 |
| 1.375 | 101.27 |
| 1.250 | 100.10 |

**Source:** *Wall Street Journal* Market Data Center, accessed at http://online
.wsj.com/mdc/page/marketsdata.html on July 15, 2010. Their source is
Thomson Reuters. The bid prices are listed per "face value" of $100 to
be paid at maturity. Half of the coupon is paid every six months. For
example, the first one listed pays $4.375 (half of the 8.750 coupon, as a
percentage of $100 of face value) every six months until maturity, at
which time it pays an additional $100. The TIPS cash flows are linked to
the inflation level and are therefore more complex. The two types of

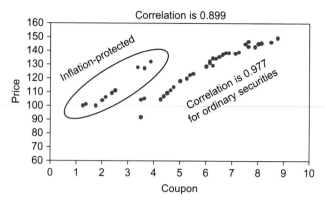

**FIGURE 11.1.18** Clustering in the relationship between bid price and
coupon payment for Treasury securities. Pricing of an ordinary bond differs
from that of an *inflation-protected* bond, so a separate relationship applies
for each cluster. The overall correlation, $r = 0.899$, does not take the relationships within each cluster into account. The correlation for the cluster of
ordinary bonds is much higher, at $r = 0.977$.

## Bivariate Outliers

A data point in a scatterplot is a **bivariate outlier** if it does
not fit the relationship of the rest of the data. An outlier can
distort statistical summaries and make them very misleading. You should always watch out for outliers in bivariate
data by looking at a scatterplot. If you can justify removing
an outlier (for example, by finding that it should not have
been there in the first place), then do so. If you have to
leave it in, at least be aware of the problems it can cause
and consider reporting statistical summaries (such as the
correlation coefficient) both with and without the outlier.

An outlier can distort the correlation to make it seem
that there is a strong relationship when, in fact, there is
nothing but randomness and one outlier. An outlier can
also distort the correlation to make it seem that there is
*no* relationship when, in fact, there is a strong relationship
and one outlier. How can you protect yourself from these
traps? By looking at a scatterplot, of course.

security altogether. It is somewhat confusing to have them
listed alongside the others in bond price listings.

What are inflation-protected securities, and why are they
priced so differently? As the U.S. Bureau of the Public Debt
explains[11]

*Treasury Inflation-Protected Securities, or TIPS, provide protection
against inflation. The principal of a TIPS increases with inflation
and decreases with deflation, as measured by the Consumer
Price Index. When a TIPS matures, you are paid the adjusted
principal or original principal, whichever is greater.*

Thus, the higher the inflation rate while you hold the bond,
the more you receive. This is why the inflation-indexed
bonds sell for more than ordinary bonds. According to the
scatterplot (looking at the relationship for ordinary bonds),
these bonds should be worth about $30 less due to their
low coupon payment. It is the upward adjustment for inflation, which makes it likely to pay more than an ordinary
bond, that pushes the price up so high.

---

10. However, there is still *interest rate risk* if you decide to sell the bond
before maturity because as interest rates change over time, so does the
price of the bond.
11. Accessed at http://www.treasurydirect.gov/indiv/products/prod_tips_
glance.htm on July 15, 2010.

### Example
#### Number Produced and Cost

Consider the number of items produced each week in a
factory together with the total cost for that week. There should
be a fairly strong relationship here. On high-volume weeks,
lots of items will be produced, requiring lots of costly input
materials. However, there can be surprises in the data. For
the data set shown in Table 11.1.13, the correlation is
negative, at $r = -0.623$. How could it be negative?

The scatterplot shown in Figure 11.1.19 has an extreme
outlier. This explains the negative correlation even though
the rest of the data show some positive association (which is
difficult to tell because the outlier causes the rest of the data
to be squashed together). Outliers should be investigated. In
fact, what happened here is that there was a fire in the factory.

Lots of the input materials were ruined, and these showed up as costs for that week. The output was low because production was halted at 11 a.m., and not even all of the production to that point could be used.

Is it permissible to omit the outlier? Probably so in this case. Certainly so if you are interested in the relationship for "ordinary" weeks and are willing to treat a disaster as a special case outside the usual circumstances. Indeed, if you omit the outlier, the correlation becomes strongly positive, at $r = 0.869$, indicating a fairly strong increasing relationship between inputs and outputs.

The data set without the outlier is shown in Figure 11.1.20. Note that, without the outlier present, the scale can be expanded, and more detail can be seen in the rest of the data.

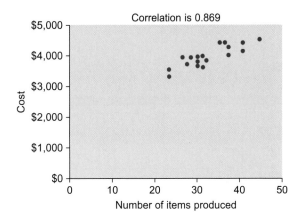

**FIGURE 11.1.20** The same data set with the outlier omitted to show the relationship for ordinary (i.e., nondisaster) weeks. The correlation is now a reasonably strong and positive $r = 0.869$, indicating an increasing relationship.

**TABLE 11.1.13 Weekly Production**

| Number Produced | Cost | Number Produced | Cost |
|---|---|---|---|
| 22 | $3,470 | 30 | $3,589 |
| 30 | 3,783 | 38 | 3,999 |
| 26 | 3,856 | 41 | 4,158 |
| 31 | 3,910 | 27 | 3,666 |
| 36 | 4,489 | 28 | 3,885 |
| 30 | 3,876 | 31 | 3,574 |
| 22 | 3,221 | 37 | 4,495 |
| 45 | 4,579 | 32 | 3,814 |
| 38 | 4,325 | 41 | 4,430 |
| 3 | 14,131 | | |

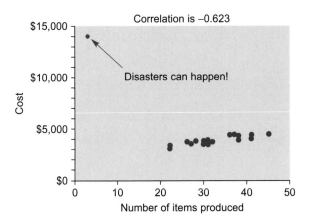

**FIGURE 11.1.19** An outlier has distorted the correlation. Instead of revealing a generally increasing relationship between number and cost, the correlation $r = -0.623$ suggests that there is a decreasing relationship, with higher production requiring a *lower* cost (which is not reasonable).

## Correlation Is Not Causation

We often think of correlation and causation as going together. This thinking is reasonable because when one thing *causes* another, the two tend to be associated and therefore correlated (e.g., effort and results, inspection and quality, investment and return, environment and productivity).

However, there can be correlation without causation. Think of it this way: The correlation is just a number that reveals whether large values of one variable tend to go with large (or with small) values of the other. The correlation cannot explain *why* the two are associated. Indeed, the correlation provides no sense of whether the investment is producing the return or vice versa! The correlation just indicates that the numbers seem to go together in some way.

One possible basis for correlation without causation is that there is some hidden, unobserved, *third factor* that makes one of the variables *seem* to cause the other when, in fact, each is being caused by the missing variable. The term **spurious correlation** refers to a high correlation that is actually due to some third factor. For example, you might find a high correlation between hiring new managers and building new facilities. Are the newly hired managers "causing" new plant investment? Or does the act of constructing new buildings "cause" new managers to be hired? Probably there is a third factor, namely, *high long-term demand* for the firm's products, that is causing both.

### Example
#### Food Store and Restaurant Spending

You will find a very high correlation ($r = 0.986$) between the amount of money spent in food stores ("food and beverage stores") and the amount spent in restaurants ("food services and drinking places") based on data by state in the United States shown in Figure 11.1.21, and this correlation is very

*(Continued)*

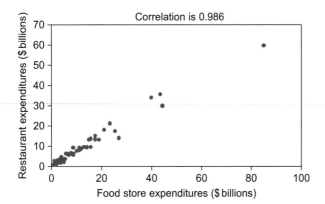

**FIGURE 11.1.21** Correlation without direct causation of restaurant spending by food store spending, by state. There is a very strong positive relationship, $r = 0.986$, suggesting that high restaurant spending goes with high food store spending, despite the fact that these are (to some extent) economic substitutes and the fact that people who dine out more often might be expected to *reduce* their food store spending as a consequence. High food store spending does not directly "cause" high restaurant spending; instead, the relationship is indirectly caused by variations in state populations: States with more people tend to have higher spending in both restaurants and food stores.

**Example—cont'd**

highly significant ($p < 0.001$).[12] To make sense out of this, first ask: "Does spending more money in food stores 'cause' a person to spend more in restaurants?" I really don't think so. As for me, when I spend more in food stores, I tend to eat at restaurants a little *less* often because I am well stocked at home. Next, could causation work the other way around; that is, does spending more money in restaurants "cause" people to spend more in food stores? For similar reasons, the answer is probably "no" here as well because a person spending more in restaurants likely will not need to spend as much in food stores. Economists would consider food stores and restaurants to be substitutes, to some extent.

If neither variable (food store spending or restaurant spending) directly causes the other to be high or low, then can we identify a third factor that influences both? How about state population?[13] Its correlations are also very high: $r = 0.980$ between population and food store spending and $r = 0.992$ between population and restaurant spending. A very reasonable explanation is that states with larger populations tend to have more spent in food stores *and* more in restaurants simply because they have more people! The connection between food store and restaurant spending is indirect but has a simple explanation in terms of this third factor that reasonably helps explains both.

12. Based on 2008 data for the 50 states and the District of Columbia from Table 1025 of U.S. Census Bureau, *Statistical Abstract of the United States: 2010* (129th edition), Washington, DC, 2009, accessed from http://www.census.gov/compendia/statab/cats/wholesale_retail_trade.html on July 19, 2010. Significance testing for bivariate data will be covered in Section 11.2.
13. Based on 2008 population data from Table 12 of U.S. Census Bureau, *Statistical Abstract of the United States: 2010* (129th edition), Washington, DC, 2009, accessed from http://www.census.gov/compendia/statab/cats/population.html on July 5, 2010.

## 11.2 REGRESSION: PREDICTION OF ONE VARIABLE FROM ANOTHER

Regression analysis is explaining or predicting one variable from the other, using an estimated straight line that summarizes the relationship between the variables. By convention, the variable being predicted is denoted $Y$, and the variable that helps with the prediction is $X$. It makes a big difference which one you denote as $Y$ and which as $X$, since *X predicts Y* and *Y is predicted by X*. Table 11.2.1 shows some of the standard ways used to refer to the role of each variable, together with some examples.

### A Straight Line Summarizes a Linear Relationship

**Linear regression analysis** is explaining or predicting one variable from the other when the two have a linear relationship. Just as you use the average to summarize a single variable, you can use a straight line to summarize a linear relationship between two variables. Just as there is variability about the average (for univariate data), there is also variability about the straight line (for bivariate data). Just like the average, the straight line is a useful but imperfect summary due to this randomness.

Figure 11.2.1 shows the straight-line summary for population and economic activity of the states, an example of a linear relationship given earlier in the chapter in Table 11.1.4 and Figure 11.1.5. Note how the line summarizes the increasing relationship. It captures the basic structure in the data, leaving the points to fluctuate randomly around it.

**TABLE 11.2.1 Variables in Regression Analysis**

| | X | Y |
|---|---|---|
| Roles | Predictor | Predicted |
| | Independent variable | Dependent variable |
| | Explanatory variable | Explained variable |
| | Stimulus | Response |
| | Exogenous (from outside) | Endogenous (from inside) |
| Examples | Sales | Earnings |
| | Number produced | Cost |
| | Effort | Results |
| | Investment | Outcome |
| | Experience | Salary |
| | Temperature of process | Output yield |

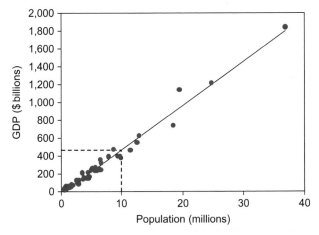

FIGURE 11.2.1   The regression line summarizes the relationship between population and economic activity of the states. This line shows how to explain the economic activity ($Y$) from the number of people ($X$) living in each state. For example, based on the line, for a state with 10 (million) people, we would expect the GDP to be about $473 (billion) per year.

After a brief discussion of straight lines, you will be shown how to compute and interpret the regression line, how to measure how well it works, how to do inference about a population relationship based on a sample, and how to be careful in problematic situations.

To use Excel® to add the least-squares line to a graph of the data, simply right-click *on a data point* in the chart, then select Add Trendline from the context-sensitive menu that appears, and finally, specify Linear as the Trend/Regression Type (and select Display equation on chart if you wish, near

the bottom) before clicking Close. The initial step of right-clicking on a data point is shown here, followed by the dialog box, and finally the end result after the line (with equation) has been added.

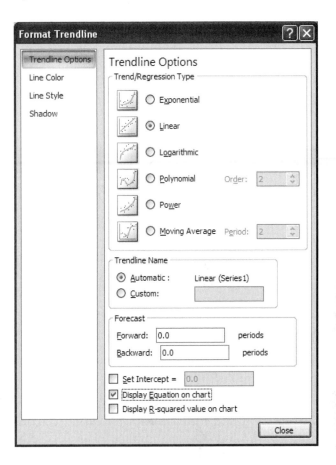

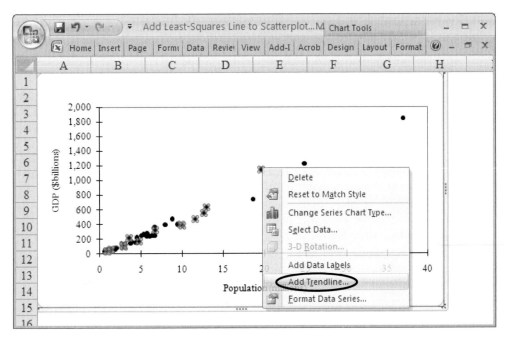

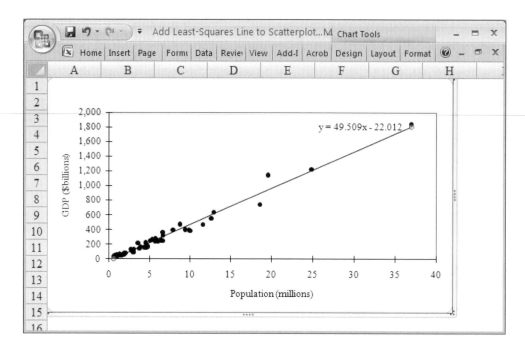

## Straight Lines

A straight line is described by two numbers: the *slope*, *b*, and the *intercept*, *a*. The **slope** indicates how steeply the line rises (or falls, if *b* is negative). As you move horizontally to the right exactly 1 unit (measured in *X* units), the line will rise (or fall, if $b < 0$) vertically a distance *b* units (measured in *Y* units). The **intercept** is simply the (vertical) value for *Y* when *X* is 0. In cases where it is absurd for *X* to be 0, the intercept should be viewed as a technical necessity for specifying the line and should not be interpreted directly.[14] The equation for a straight line is as follows:

> **Equation for a Straight Line**
>
> $$Y = (\text{Intercept}) + (\text{Slope})(X)$$
> $$= a + bX$$

The slope and intercept are illustrated in Figures 11.2.2, 11.2.3, and 11.2.4.

## Finding a Line Based on Data

How should you find the best summary line to predict *Y* from *X* based on a bivariate data set? One well-established approach is to find the line that has the smallest prediction error overall, in some sense. The conventional way to do this is to use the **least-squares line**, which has the smallest sum of squared vertical prediction errors compared to all

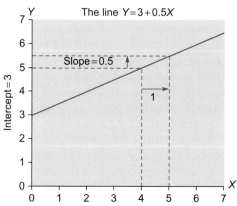

**FIGURE 11.2.2**   The straight line $Y = 3 + 0.5X$ starts at the intercept ($a = 3$), when *X* is 0, and rises 0.5 (one slope value, $b = 0.5$) for each distance of 1 moved to the right.

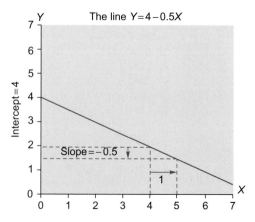

**FIGURE 11.2.3**   A line with negative slope. The straight line $Y = 4 - 0.5X$ starts at the intercept ($a = 4$), when *X* is 0, and falls 0.5 (since the slope value is negative, $b = -0.5$) for each distance of 1 moved to the right.

---

14. It is possible to define the line in terms of the slope together with the value of *Y* at $\overline{X}$ so that the two numbers that specify the line are both always meaningful. However, this is rarely done at present.

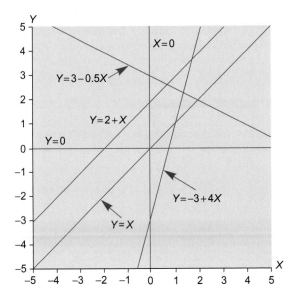

FIGURE 11.2.4 An assortment of straight lines and their equations, showing the slope and intercept. The vertical line is the only one that cannot be written in the form $Y = a + bX$.

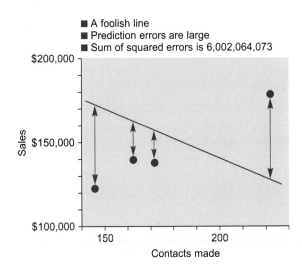

FIGURE 11.2.6 A foolish choice of line will have large prediction errors and will not be the least-squares line.

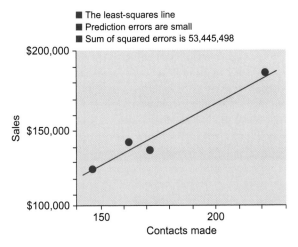

FIGURE 11.2.5 The least-squares line has the smallest sum of squared prediction errors of all possible lines. The prediction errors are measured vertically.

other lines that could possibly be drawn. These prediction errors, the sum of whose squares is to be minimized, are shown in Figure 11.2.5, for the least-squares line, and in Figure 11.2.6, for a foolish choice of line, for the sales data from Table 11.1.1.

The least-squares line can easily be found. Computers and many calculators can automatically find the least-squares slope, $b$, and intercept, $a$. The slope is also called the **regression coefficient** of $Y$ on $X$, and the intercept is also called the **constant term** in the regression. The slope, $b$, is found as the correlation, $r$, times the ratio of

standard deviations, $S_Y/S_X$ (which is in appropriate units of $Y$ per $X$). The intercept, $a$, is determined so that the line goes through the most reasonable landmark, namely $(\overline{X}, \overline{Y})$. The formulas are as follows:

**The Least-Squares Slope and Intercept**

$$\text{Slope} = b = r\frac{S_Y}{S_X}$$

$$\text{Intercept} = a = \overline{Y} - b\overline{X} = \overline{Y} - r\frac{S_Y}{S_X}\overline{X}$$

**The Least-Squares Line**

$$(\text{Predicted value of } Y) = a + bX$$

$$= \left(\overline{Y} - r\frac{S_Y}{S_X}\overline{X}\right) + \left(r\frac{S_Y}{S_X}\right)X$$

Don't expect to find all of the points exactly on the line. Think of the line as summarizing the overall relationship in the data. Think of the data as the line together with randomness. Your **predicted value** for $Y$ given a value of $X$ will be the height of the line at $X$, which you find by using the equation of the least-squares line. You can find the predicted value either for a data point or for a new value of $X$. Each of your data points has a **residual**, which tells you how far the point is above (or below, if negative) the line. These residuals allow you to make adjustments, comparing actual values of $Y$ to what you would expect them to be for corresponding values of $X$. The formula for the residual for the data point $(X, Y)$ is

$$\text{Residual} = (\text{Actual } Y) - (\text{Predicted } Y) = Y - (a + bX)$$

## Example
### Fixed and Variable Costs

Recall the production data from an earlier example, but with the outlier removed. Table 11.2.2 shows the data with $X$ and $Y$ indicated and summary statistics included. It is natural for $X$ to be the number produced and $Y$ to be the cost because a manager often needs to anticipate costs based on currently scheduled production.

The slope represents the *variable cost* (the marginal cost of producing one more item) and may be found from these summaries as follows:

$$\begin{aligned}\text{Variable cost} &= b\\ &= rS_Y/S_X\\ &= (0.869193)(389.6131)/6.5552\\ &= \$51.66\end{aligned}$$

The other term, the intercept, represents the *fixed cost*. These are baseline costs such as rent that are incurred even if no items are produced. This intercept term may be found as follows:[15]

$$\begin{aligned}\text{Fixed cost} &= a\\ &= \overline{Y} - b\overline{X}\\ &= 3{,}951.06 - (51.66)(32.5)\\ &= \$2{,}272\end{aligned}$$

The least-squares line may be written as follows:

Predicted
cost = Fixed cost + (Variable cost)(Number produced)
$$= \$2{,}272 + \$51.66(\text{Number produced})$$

The least-squares line is shown with the data in Figure 11.2.7.

You might use this estimated relationship to help with budgeting. If you anticipate the need to produce 36 of these items next week, you can predict your cost using the relationship in the past data as summarized by the least-squares line. Your forecast will be as follows:

$$\begin{aligned}\text{Predicted cost for producing 36 items} &= a + (b)(36)\\ &= \$2{,}272 + (\$51.66)(36)\\ &= \$4{,}132\end{aligned}$$

Your forecast of the cost is the height of the line at production equal to 36 items, as shown in Figure 11.2.8. Naturally you don't expect the cost to be exactly $4,132. However, you may reasonably expect the cost to just randomly differ from your best guess of $4,132.

---

15. To interpret the calculated intercept term as a fixed cost requires the assumption that the linear relationship continues to hold even outside the range of the data because we are extending the line (extrapolating beyond the data) to reach the $Y$ axis where $X = 0$.

### TABLE 11.2.2   Weekly Production

| Number | |
|---|---|
| Produced, $X$ | Cost, $Y$ |
| 22 | $3,470 |
| 30 | 3,783 |
| 26 | 3,856 |
| 31 | 3,910 |
| 36 | 4,489 |
| 30 | 3,876 |
| 22 | 3,221 |
| 45 | 4,579 |
| 38 | 4,325 |
| 30 | 3,589 |
| 38 | 3,999 |
| 41 | 4,158 |
| 27 | 3,666 |
| 28 | 3,885 |
| 31 | 3,574 |
| 37 | 4,495 |
| 32 | 3,814 |
| 41 | 4,430 |
| Average | $\overline{X} = 32.50$ | $\overline{Y} = \$3{,}951.06$ |
| Standard deviation | $S_X = 6.5552$ | $S_Y = \$389.6131$ |
| Correlation | $r = 0.869193$ | |

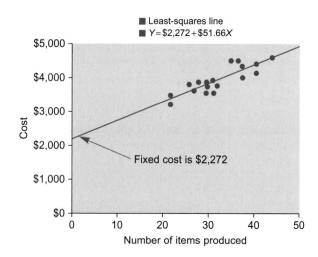

**FIGURE 11.2.7**   The least-squares line summarizes the production cost data by estimating a fixed cost (the intercept, $a = \$2{,}272$) and a variable per-unit cost (the slope, $b = \$51.66$ per item produced).

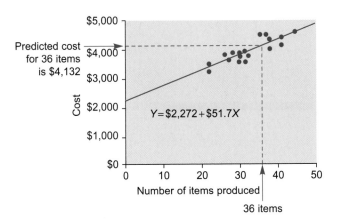

FIGURE 11.2.8   The least-squares line may be used to forecast, or predict, the expected value for Y given a new value for X. In this case, you are expecting to produce 36 items next week. The least-squares line suggests that your expected cost will be $4,132. Of course, the real cost will come in with some randomness, just like the other points.

### Example
#### Territory and Sales

Your sales managers are a varied lot. Sure, some work harder than others, and some bring in more sales than others, but it's not so simple as that. Somebody assigned each one to a territory, and some territories provide more opportunity for business than others. In addition to just looking at how much each person sold (which is important, of course), you have decided to try to *adjust for territory size* to find out who is doing well and who is doing poorly. It might turn out that some of the good performers are not really doing well because you would expect higher sales for a territory that large. You also might discover some hidden talent: people with smaller sales levels who are above average for a territory that small. A regression analysis will help you make this adjustment. The data set is shown in Table 11.2.3.

The least-squares line is

Expected sales = $1,371,744 + $0.23675045 (Tettitory)

By inserting each sales manager's territory size into this equation, you find the expected sales based on territory size. For example, Anson's expected sales are $1,371,744 + (0.23675045) × (4,956,512) = $2,545,000 (rounding the answer to the nearest thousand). Anson's actual sales (about $2,687,000) are $142,000 higher than his expected sales. Thus, Anson has a residual value of $142,000, possibly indicating added value. The expected sales levels and residuals may be found for each sales manager and are shown in Table 11.2.4.

The residuals are interesting. The largest one, $791,000, indicates that Bonnie pulled in about $0.79 million more in sales than you would have expected for a territory that size. Although her actual sales were not the highest, when you take account of the size of her territory (fairly small, actually), her results are very impressive. Another residual is

fairly large, at $538,000, telling you that Clara's impressive sales of $5,149,127 (the highest of all) were not just due to her large territory. Indeed, she pulled in about $0.5 million more than you would have expected for that territory size. However, the smallest residual, −$729,000, is negative, suggesting that Rod may not be pulling his weight. You would have expected about $0.73 million more from him based on the size of his territory. The data, least-squares line, and notes on these three special managers are shown in Figure 11.2.9.

Be careful not to interpret these results too literally. Although these three special cases might well indicate two stars and a trouble spot, there could be other explanations. Perhaps Rod has had trouble because his territory is in a depressed area of the country, in which case his low adjusted total should not be attributed to his personal performance. Perhaps a more careful regression analysis could be done, taking other important factors into account.

## How Useful Is the Line?

You have already seen that the least-squares line does not usually describe the data perfectly. It is a useful summary of the main trend, but it does not capture the random variation of the data points about the line. This raises the question: How useful is the regression line? The answer is based on two important measures: the *standard error of estimate* (an absolute measure of how big the prediction errors are) and $R^2$ (a relative measure of how much has been explained).

## The Standard Error of Estimate: How Large Are the Prediction Errors?

The **standard error of estimate**, denoted $S_e$ here (but often denoted $S$ in computer printouts), tells you approximately how large the prediction errors (residuals) are for your data set, in the same units as Y. How well can you predict Y? The answer is, to within about $S_e$ above or below.[16] Since you usually want your forecasts and predictions to be as accurate as possible, you would be glad to find a *small* value for $S_e$. You can interpret $S_e$ as a standard deviation in the sense that, if you have a normal distribution for the prediction errors, then you will expect about 2/3 of the data points to fall within a distance $S_e$ either above or below the regression line. Also, about 95% of the data values should fall within $2S_e$, and so forth. This is illustrated in Figure 11.2.10 for the production cost example.

---

16. A more careful, exact answer will be provided in a later section for predicting a new value of Y given a value for X.

### TABLE 11.2.3 Territory and Performance of Salespeople

|          | Territory (population size) | Sales (past year) |         | Territory (population size) | Sales (past year) |
|----------|----------------------------|-------------------|---------|----------------------------|-------------------|
| Anson    | 4,956,512                  | $2,687,224        | Clara   | 13,683,663                 | $5,149,127        |
| Ashley   | 8,256,603                  | 3,543,166         | Brittany| 3,580,058                  | 2,024,809         |
| Jonathan | 9,095,310                  | 3,320,214         | Ian     | 2,775,820                  | 1,711,720         |
| Rod      | 12,250,809                 | 3,542,722         | Bonnie  | 4,637,015                  | 3,260,464         |
| Nicholas | 4,735,498                  | 2,251,482         |         |                            |                   |

### TABLE 11.2.4 Territory, Actual Performance, Expected Performance, and Residuals

|          | Territory   | Actual Sales | Expected Sales (rounded) | Residuals (rounded) |
|----------|-------------|--------------|--------------------------|---------------------|
| Anson    | 4,956,512   | $2,687,224   | $2,545,000               | $142,000            |
| Ashley   | 8,256,603   | 3,543,166    | 3,326,000                | 217,000             |
| Jonathan | 9,095,310   | 3,320,214    | 3,525,000                | −205,000            |
| Rod      | 12,250,809  | 3,542,722    | 4,272,000                | −729,000            |
| Nicholas | 4,735,498   | 2,251,482    | 2,493,000                | −241,000            |
| Clara    | 13,683,663  | 5,149,127    | 4,611,000                | 538,000             |
| Brittany | 3,580,058   | 2,024,809    | 2,219,000                | −195,000            |
| Ian      | 2,775,820   | 1,711,720    | 2,029,000                | −317,000            |
| Bonnie   | 4,637,015   | 3,260,464    | 2,470,000                | 791,000             |

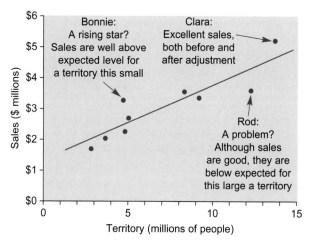

FIGURE 11.2.9   By comparing each data point to the regression line, you can evaluate performance after adjusting for some other factor. In this case, points above the line (with positive residuals) represent managers with higher sales than you would have expected for their size of territory. Points below the line indicate lower sales than expected.

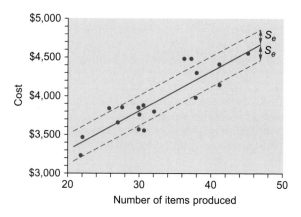

FIGURE 11.2.10   The standard error of estimate, $S_e$ indicates approximately how much error you make when you use the predicted value for $Y$ (on the least-squares line) instead of the actual value of $Y$. You may expect about 2/3 of the data points to be within $S_e$ above or below the least-squares line for a data set with a normal linear relationship, such as this one.

The standard error of estimate may be found using the following formulas:

### Standard Error of Estimate

$$S_e = S_Y \sqrt{(1 - r^2)\frac{n - 1}{n - 2}} \qquad \text{(for computation)}$$

$$= \sqrt{\frac{1}{n - 2}\sum_{i=1}^{n}[Y_i - (a + bX_i)]^2} \qquad \text{(for interpretation)}$$

The first formula shows how $S_e$ is computed by reducing $S_Y$ according to the correlation and sample size. Indeed, $S_e$ will usually be smaller than $S_Y$ because the line $a + bX$ summarizes the relationship and therefore comes closer to the $Y$ values than does the simpler summary, $\overline{Y}$. The second formula shows how $S_e$ can be interpreted as the estimated standard deviation of the residuals: The squared prediction errors are averaged by dividing by $n - 2$ (the appropriate number of degrees of freedom when two numbers, $a$ and $b$, have been estimated), and the square root undoes the earlier squaring, giving you an answer in the same measurement units as $Y$.

For the production cost data, the correlation was found to be $r = 0.869193$, the variability in the individual cost numbers is $S_Y = \$389.6131$, and the sample size is $n = 18$. The standard error of estimate is therefore

$$S_e = S_Y \sqrt{(1 - r^2)\frac{n - 1}{n - 2}}$$

$$= 389.6131 \sqrt{(1 - 0.869193^2)\frac{18 - 1}{18 - 2}}$$

$$= 389.6131 \sqrt{(0.0244503)\frac{17}{16}}$$

$$= 389.6131 \sqrt{0.259785}$$

$$= \$198.58$$

This tells you that, for a typical week, the actual cost was different from the predicted cost (on the least-squares line) by about $198.58. Although the least-squares prediction line takes full advantage of the relationship between cost and number produced, the predictions are far from perfect.

## $R^2$: How Much Is Explained?

$R^2$, pronounced "r squared" and also called the **coefficient of determination**, tells you how much of the variability of $Y$ is explained by $X$.[17] It is found by simply squaring the

correlation, $r$ (i.e., $R^2 = r^2$). This leaves $1 - R^2$ of the variation in $Y$ unexplained. Ordinarily, larger values of $R^2$ are considered better because they indicate a stronger relationship between $X$ and $Y$ that can be used for prediction or other purposes. However, in practice, a small $R^2$ does not necessarily say that $X$ is not helpful in explaining $Y$; instead, a small $R^2$ may merely signal that $Y$ is also partly determined by other important factors.

For example, the correlation of the production cost data set is $r = 0.869193$. Thus, the $R^2$ value is

$$R^2 = 0.869193^2 = 0.755 \text{ or } 75.5\%$$

This tells you that, based on the $R^2$ value, 75.5% of the variation in weekly cost is explained by the number produced each week. The remainder, 24.5% of the total cost variation, must be due to other causes.

Think of it this way: There is variation in cost from one week to another (summarized by $S_Y$). Some of this variation is due to the fact that production is higher in some weeks (resulting in a higher cost) and lower in others. Thus, the number produced "explains" part of the week-to-week variation in cost. But it doesn't explain all of this variation. There are other factors (such as occasional breakdowns, overtime, and mistakes) that also contribute to the variation in cost. The $R^2$ value tells you that 75.5% of the variation in cost is explained by the production level; the remaining 24.5% of the variation is still unexplained.

## Confidence Intervals and Hypothesis Tests for Regression

Up to now you have been summarizing the *data:* estimating the strength of the relationship using the correlation coefficient, estimating the relationship using the least-squares line, and estimating the accuracy of the line using the standard error of estimate and $R^2$. Now it's time to go beyond merely summarizing the sample data and start doing statistical inference about the larger population you really want to understand. But what is the appropriate population to consider for a regression problem? The conventional answer is provided by the *linear model.*

## The Linear Model Assumption Defines the Population

For statistical inference to be valid, the data set must be a random sample from the population of interest. As always, this ensures that the data set represents the population in an exact, controlled way. We will also need a technical assumption that will justify using the $t$ table, which is based on a normal distribution. For this purpose, we will assume that the bivariate data are independently chosen

17. Literally, $R^2$ is the proportion of the *variance* of $Y$ that has been explained by $X$. For technical reasons (the total *squared* error can be decomposed into two *squared* components: explained and unexplained), the variance (the squared standard deviation) has traditionally been used.

**lear model** which states that the observed value of *Y* is equal to the straight-line population relationship plus a random error that has a normal distribution:

---

**Linear Model for the Population**

$$Y = (\alpha + \beta X) + \varepsilon$$
$$= (\text{Population relationship}) + \text{Randomness}$$

where $\varepsilon$ has a normal distribution with mean 0 and constant standard deviation $\sigma$.

---

These assumptions help ensure that the data set consists of independent observations having a linear relationship with equal variability and approximately normal randomness.

The population relationship is given by two parameters: $\alpha$ is the population intercept (or constant term), and $\beta$ is the population slope. Another population parameter, $\sigma$, indicates the amount of uncertainty in the situation. If your data were a census of the entire population, then your least-squares line would be the population relationship. Ordinarily, however, you use the least-squares intercept, $a$, as an *estimator* of $\alpha$; the least-squares slope, $b$, as an *estimator* of $\beta$; and the standard error of estimate, $S_e$, as an *estimator* of $\sigma$. Of course, there are errors involved in this estimating since $a$, $b$, and $S_e$ are based on a smaller sample and not on the entire population. Table 11.2.5 shows a summary of these population parameters and sample statistics.

The linear model is the basic assumption required for statistical inference in regression and correlation analysis. Confidence intervals and hypothesis tests based on the slope coefficient will assume that the linear model holds in the population. In particular, these confidence intervals and hypothesis tests will not be valid if the relationship is nonlinear or has unequal variability. It is up to you to watch for problems; if the linear model is not appropriate for your data, then the inferences from regression analysis could be wrong.

---

**TABLE 11.2.5  Population Parameters and Sample Statistics**

| | Population (parameters: fixed and unknown) | Sample (estimators: random and known) |
|---|---|---|
| Intercept | $\alpha$ | $a$ |
| Slope | $\beta$ | $b$ |
| Regression line | $Y = \alpha + \beta X$ | $Y = a + bX$ |
| Uncertainty | $\sigma$ | $S_e$ |

---

## Standard Errors for the Slope and Intercept

You may suspect that there are standard errors lurking in the background, since there are population parameters and sample estimators. Once you know the standard errors and degrees of freedom, you will be able to construct confidence intervals and hypothesis tests using the familiar methods of Chapters 9 and 10.

The **standard error of the slope coefficient**, $S_b$, indicates approximately how far the estimated slope, $b$ (the regression coefficient computed from the sample), is from the population slope, $\beta$, due to the randomness of sampling. Note that $S_b$ is a sample statistic. The formula for $S_b$ is as follows:

---

**Standard Error of the Regression Coefficient**

$$S_b = \frac{S_e}{S_X \sqrt{n-1}} \quad \text{Degrees of freedom} = n - 2$$

---

This formula says that the uncertainty in $b$ is proportional to the basic uncertainty ($S_e$) in the situation, but (1) $S_b$ will be smaller when $S_X$ is large (since the line is better defined when the $X$ values are more spread out) and (2) $S_b$ will be smaller when the sample size $n$ is large (because there is more information). It is very common to see a term such as the square root of $n$ in the denominator of a standard error formula, expressing the effect of additional information.

The degrees of freedom number for this standard error is $n - 2$, since two numbers, $a$ and $b$, have been estimated to find the regression line.

For the production cost example (without the outlier!), the correlation is $r = 0.869193$, the sample size is $n = 18$, and the slope (variable cost) is $b = 51.66$ for the sample. The population is an idealized one: All of the weeks that might have happened under the same basic circumstances as the ones you observed. You might think of the population slope, $\beta$, as the slope you would compute if you had a lot more data. The standard error of $b$ is

$$S_b = \frac{S_e}{S_X \sqrt{n-1}}$$
$$= \frac{198.58}{6.5552\sqrt{18-1}}$$
$$= \frac{198.58}{27.0278}$$
$$= 7.35$$

The intercept term, $a$, was also estimated from the data. Therefore, it too has a standard error indicating its estimation uncertainty. The **standard error of the intercept term**, $S_a$, indicates approximately how far your estimate *a* is from $\alpha$, the true population intercept term. This standard error, whose computation follows, also has $n - 2$ degrees of freedom and is a sample statistic:

### Standard Error of the Intercept Term

$$S_a = S_e \sqrt{\frac{1}{n} + \frac{\overline{X}^2}{S_X^2 (n-1)}} \qquad \text{Degrees of freedom} = n - 2$$

This formula states that the uncertainty in *a* is proportional to the basic uncertainty $(S_e)$, that it is small when the sample size *n* is large, that it is large when $\overline{X}$ is large (either positive or negative) with respect to $S_X$ (because the *X* data would be far from 0 where the intercept is defined), and that there is a $1/n$ baseline term because *a* would be the average of *Y* if $\overline{X}$ were 0.

For the production cost example, the intercept, *a* = \$2,272, indicates your estimated fixed costs. The standard error of this estimate is

$$S_a = S_e \sqrt{\frac{1}{n} + \frac{\overline{X}^2}{S_X^2 (n-1)}}$$

$$= 198.58 \sqrt{\frac{1}{18} + \frac{32.50^2}{6.5552^2 (18-1)}}$$

$$= 198.58 \sqrt{0.0555556 + \frac{1,056.25}{730.50}}$$

$$= 198.58 \sqrt{1.5015}$$

$$= 243.33$$

## Confidence Intervals for Regression Coefficients

This material on confidence intervals should now be familiar. You take an estimator (such as *b*), its own personal standard error (such as $S_b$), and the *t* value from the *t* table (for $n - 2$ degrees of freedom). The two-sided confidence interval extends from $b - tS_b$ to $b + tS_b$. The one-sided confidence interval claims either that the population slope, $\beta$, is at least $b - tS_b$ or that the population slope, $\beta$, is no more than $b + tS_b$ (using the one-sided *t* values, of course). You may wish to reread the summary of Chapter 9 for a review of the basics of confidence intervals; the only difference here is that you are estimating a population *relationship* rather than just a population mean.

Similarly, inference for the population intercept term, $\alpha$, is based on the estimator *a* and its standard error, $S_a$.

### Confidence Intervals

For the population slope, $\beta$:

$$\text{From } b - tS_b \text{ to } b + tS_b$$

For the population intercept, $\alpha$:

$$\text{From } a - tS_a \text{ to } a + tS_a$$

### Example
*Variable Costs of Production*

For the production cost data, the estimated slope is *b* = 51.66, the standard error is $S_b$ = 7.35, and the two-sided *t* table value for $n - 2 = 16$ degrees of freedom is 2.120 for 95% confidence. The 95% confidence interval for $\beta$ extends from 51.66 − (7.35)(2.120) = 36.08 to 51.66 + (7.35)(2.120) = 67.24. Your confidence interval statement is as follows:

> We are 95% confident that the long-run (population) variable costs are somewhere between \$36.08 and \$67.24 per item produced.

As often happens, the confidence interval reminds you that the estimate (\$51.66) is not nearly as exact as it appears. Viewing your data as a random sample from the population of production and cost experiences that might have happened under similar circumstances, you find that with just 18 weeks' worth of data, there is substantial uncertainty in the variable cost.

A one-sided confidence interval provides a reasonable upper bound that you might use for budgeting purposes. This reflects the fact that you don't know what the variable costs really are; you just have an *estimate* of them. In this example, the one-sided *t* value from the table is 1.746, so your upper bound is 51.66 + (7.35)(1.746) = 64.49. The one-sided confidence interval statement is as follows:

> We are 95% confident that the long-run (population) variable costs are no greater than \$64.49 per item produced.

Note that this bound (\$64.49) is smaller than the upper bound of the two-sided interval (\$67.24) because you are interested only in the upper side. That is, because you are not interested at all in the lower side (and will not take any error on that side into account), you can obtain an upper bound that is closer to the estimated value of \$51.66.

## Testing Whether the Relationship Is Real or Coincidence

This chapter is about the relationship between *X* and *Y*. The correlation summarizes the strength of the relationship, and the regression equation exploits the relationship to explain *Y* from *X*. However, as often happens in statistics, you can summarize a relationship whether or not it is really there. It is the job of hypothesis testing to tell you if (on the one hand) the relationship that *appears* to be in your data could reasonably be pure coincidence or if (on the other hand) there is actually a significant association between X and Y.

The null hypothesis claims that there is *no relationship* between X and Y, that the apparent relationship in the data is just an artifact of the random pairing of X and Y values. The only way the linear model, $Y = \alpha + \beta X + \varepsilon$, can have Y *not* depend on X is if $\beta = 0$, so that X disappears and the linear model reduces to $Y = \alpha + \varepsilon$. Another way to say that there is no relationship is to say that X and Y are *independent* of each other.

The research hypothesis claims that there *is a relationship* between X and Y, not just randomness. This will happen whenever $\beta \neq 0$, so that the X term remains in the linear model for Y. Here are both hypotheses in mathematical form:

---

**Hypotheses for Testing Significance of a Relationship**

$$H_0 : \beta = 0$$

$$H_1 : \beta \neq 0$$

---

The test itself is performed in the usual way; again, there is nothing new here.[18] You might use the confidence interval approach and see whether or not the reference value, 0, is in the interval, deciding significance (accepting $H_1$) if it is not. Or you might construct the t statistic $b/S_b$ and compare it to the t value from the table, deciding significance ($H_1$) if the absolute value of the t statistic is larger.

For the variable production costs example, the confidence interval extends from \$36.08 to \$67.24. Since the reference value, 0, is *not* in this confidence interval, you may conclude that you *do* have significant variable costs. That is, based on your data, there is indeed a relationship between the number produced and the cost each week (since zero is not one of the reasonable values in the confidence interval). The apparent association (higher numbers produced tend to cost more) could not reasonably be due to randomness alone.

Of course, the t statistic approach gives the same answer. The t statistic is $t = b/S_b = 51.66/7.35 = 7.03$. Since the absolute value of the t statistic (7.03) is greater than the t table value (2.120) with $n - 2 = 16$ degrees of freedom for testing at the 5% level, you would conclude that the slope (51.66) is indeed significantly different from 0.

---

18. You may wish to review the summary of Chapter 10 to refresh your memory on the basics of hypothesis testing.

## Other Methods of Testing the Significance of a Relationship

There are other methods for testing the significance of a relationship. Although they may appear different at first glance, they always give the same answer as the method just described, based on the regression coefficient. These alternate tests are based on other statistics—for example, the correlation, r, instead of the slope coefficient, b. But since the basic question is the same (is there a relationship or not?), the answers will be the same also. This can be mathematically proven.

There are two ways to perform the significance test based on the correlation coefficient. You could look up the correlation coefficient in a special table, or else you could transform the correlation coefficient to find the t statistic $t = r\sqrt{(n-2)/(1-r^2)}$ to be compared to the t table value with $n - 2$ degrees of freedom. In the end, the methods yield the same answer as testing the slope coefficient. In fact, the t statistic defined from the correlation coefficient is the same number as the t statistic defined from the slope coefficient ($t = b/S_b$).

This implies that you may conclude that there is significant correlation or that the correlation is not significant based on a test of significance of the regression coefficient, b. In fact, you may conclude that there is significant positive correlation if the relationship is significant and $b > 0$. Or, if the relationship is significant and $b < 0$, you may conclude that there is a significant negative correlation.

There is a significance test called the F test for overall significance of a regression relationship. This test will be covered in the next chapter, on multiple regression. Although this test may also look different at first, in the end it is the same as testing the slope coefficient when you have just X and Y as the only variables in the analysis.

## Computer Results for the Production Cost Data

Many of these results are available from computer analysis for the production cost data. First is the prediction equation (or "regression equation"). Next are the coefficients ("coeff") $a = 2272.1$ and $b = 51.661$ with their standard errors ("stdev") $S_a = 243.3$ and $S_b = 7.347$, their t statistics $t_a = 9.34$ and $t_b = 7.03$, and their p-values (both of which are very highly significant because $p < 0.001$ in both cases). The next line indicates the standard error of estimate, $S_e = 198.6$, and the $R^2 = 0.755$.

The regression equation is:

Cost = 2272 + 51.7 Production

| Predictor | COEFF | STDEV | t Ratio | p |
|---|---|---|---|---|
| Constant | 2272.1 | 243.3 | 9.34 | 0.000 |
| Production | 51.661 | 7.347 | 7.03 | 0.000 |

S = 198.6, R-sq = 75.5%, R-sq(adj) = 74.0%

## Example
### Momentum in the Stock Market Revisited

Earlier in the chapter, the stock market's weekly percent changes were used as an example of the apparent lack of relationship between $X$ = last week's change and $Y$ = this week's change, as percentage changes in the S&P 500 stock market index. Let's now use regression to estimate the relationship between last week's and this week's changes and then use hypothesis testing to see whether or not the relationship is significant. The data set, with least-squares line, is shown in Figure 11.2.11.

The least-squares line is

This week $= -0.00277 + 0.02792$ (Last week)

For example, on May 17, 2010, we have $X$ = 2.23% = 0.0223 and $Y$ = −4.23% = −0.0423. The predicted value for $Y$ on this day is −0.00277 + 0.02792 (0.0223) = −0.00215 or −0.215%.

Should you believe this prediction equation? It pretends to help you forecast today's market behavior based on yesterday's (assuming that market behavior continues to act as though it is drawn from the same population). The key is the slope coefficient, $b$ = 0.02792, which says that only about 3% of last week's rise (or fall) will continue this week, on average. However, how accurately has this coefficient been estimated? The answer is provided by the confidence interval based on the estimate ($b$ = 0.02792), its standard error ($S_b$ = 0.21577), and the $t$ table value 2.064 with 26 − 2 = 24 degrees of freedom. The confidence interval is:

We are 95% sure that the population slope $\beta$ is between −0.417 and 0.473.

This is a wide interval; in fact, it contains 0, which would indicate no relationship. Thus, we conclude that since 0 is contained in the interval, the apparent slope is *not significant*. There is no significant association between yesterday's and today's market performance. You might also say that the slope coefficient is not significantly different from 0.

The $t$ statistic approach gives the same answer, of course. The standard error of the regression coefficient is $S_b$ = 0.21577, so the $t$ statistic is

$$t = b/S_b = 0.02792/0.21577 = 0.129$$

Such a small $t$ statistic is not significant (compare to the $t$ table value of 2.064).

To perform regression analysis with Excel®, first give a name to each column of numbers (if not yet named) using Excel's Define Name from the Formulas Ribbon. Then look in the Data Ribbon for Data Analysis in the Analysis area,[19] and then select Regression. In the resulting dialog box, you may specify the range name for the $Y$ variable ("This_Week" in this example) and for the $X$ variable ("Last_Week"). Click Output Range in the dialog box and specify where in the worksheet you want the results to be placed; then click OK. Here is the dialog box and its results, which include the $R^2$ value of 0.000697 or 0.0697%, the standard error of estimate, $S_e$ of 0.027018, as well as $b$ = 0.02792, $S_b$ = 0.21577, $t$ = 0.129, and the $p$-value of 0.8981.

---

19. If you cannot find Data Analysis in the Analysis area of Excel's Data Ribbon, click the Office button at the very top left, choose Excel Options at the bottom, select Add-Ins at the left, click Go at the bottom, and make sure the Analysis ToolPak is checked. If the Analysis ToolPak was not installed when Excel® was installed on your computer, you may need to run the Microsoft Office Setup Program.

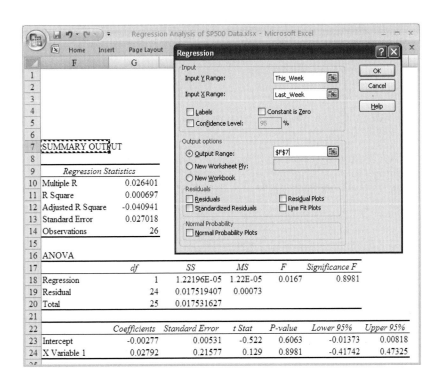

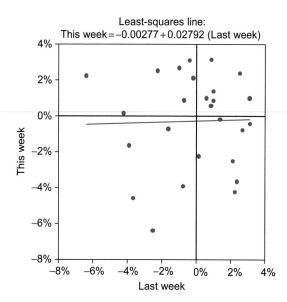

**FIGURE 11.2.11** Weekly percent changes ($X$ = Last week and $Y$ = This week) for the first half of 2010. The least-squares line is nearly horizontal but has a slight tilt to it. Since the small tilt could be due to randomness, the hypothesis test concludes that there is no significant relationship between last week's and this week's market performance.

**Example**

*Mining the Donations Database to Predict Dollar Amounts*

What determines the amount of a donation to a worthy cause? Why do some people donate larger amounts than others in response to a mailing? We know the donation amount for each of the 989 people who gave in response to the mailing, out of the 20,000 people in the donations database on the companion site. We also have information (known before the mailing) about each person's donation history and neighborhood characteristics that might help explain the amount of the donation. Regression analysis can help us search for connections between donation amount and the information known before the mailing. Patterns identified in such a regression analysis can be very useful in targeting marketing efforts because regression can predict the response, under similar conditions, *before* the next mailing is sent.

One reasonable place to begin explaining donation size is with the person's income (or wealth) because people who can afford it may give larger amounts. However, we don't have each person's income (such personal information is difficult to collect for a large database). In place of unavailable personal information, we can use the average (per capita) income for each donor's neighborhood.[20] Figure 11.2.12 shows a scatterplot with regression line to predict the donation amount from the neighborhood per capita income. There seems to be a positive relationship with considerable randomness, with people in higher-income neighborhoods donating larger amounts on average. The apparent positive relationship is very highly significant ($p < 0.001$; the more exact value is $p = 0.00000002$).

How useful is this very highly significant result? With the large sample sizes available in data mining applications, it is often possible to detect statistical significance for a small effect (i.e., a small effect that can be distinguished from randomness but may or may not be useful). In this case, neighborhood income explains only 3.2% of the variation in donations (using the coefficient of determination, $R^2 = 0.032$), which is small but to be expected given that donors' incomes vary even within a neighborhood (where they cannot be explained by average neighborhood income). But how important is the effect of income on donation, practically speaking, in terms of dollar amounts? The regression coefficient, $b = 0.000203$, says that for each $10,000 increase in neighborhood income, we expect to see an increase of $0.000203 \times 10,000 = \$2.03$ in the average donation size. The 95% confidence interval for this average donation increase extends from $1.33 to $2.73, per $10,000 of neighborhood income. From this, we conclude that income does have a useful effect because this average difference will be multiplied by thousands of potential donors, leading to thousands of dollars overall. We have found a useful connection between neighborhood income and donation amount but should not lose sight of the importance of donations even from the poorer neighborhoods: The intercept $a = \$12.37$ gives an indication of the average donation size for very low-income neighborhoods. Note also in Figure 11.2.12 that some of the highest donations ($100) did not come from wealthy neighborhoods.

What about age? Might it be true that neighborhoods with more people aged 60 to 64 will tend to give more (or give less) than others? To answer this, consider the scatterplot with regression line (Figure 11.2.13) to predict the donation amount from the percentage of people in the neighborhood who are between 60 and 64 years of age. Even with the large sample size (989 donors), there is no significant relationship. The neighborhood share in this age group explains only $R^2 = 0.0003\%$ of the variability of donations, and the $p$-value for testing the regression coefficient is

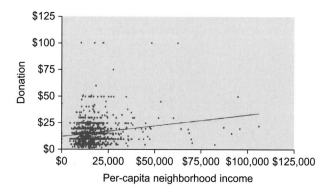

**FIGURE 11.2.12** Scatterplot with least-squares line to predict donation amount from neighborhood per capita income for donors. There seems to be a positive relationship with considerable randomness, with people from higher-income neighborhoods donating larger amounts on average.

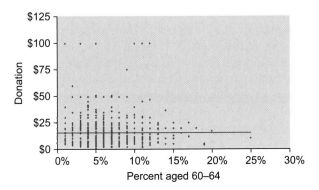

FIGURE 11.2.13  Scatterplot with least-squares line to predict donation amount from the neighborhood percentage of people 60–64 years old. There is no significant relationship.

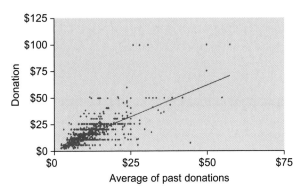

FIGURE 11.2.14  Scatterplot with least-squares line to predict donation amount from the average of previous donations from that donor. The relationship here is much stronger, with $R^2 = 51.3\%$ of the variation in the current donation being explained from the average of past donations.

0.960, leading to the conclusion that the relationship is not significant ($p > 0.05$).

Finally, consider a variable that is specific to the donor (not just to the neighborhood). We do have the average past donation amount for each donor, and (after omitting a single outlier: a donor with a previous average of $200 from a single past gift, who gave $100 this time around), you can see in Figure 11.2.14 that this well-targeted variable does a much better job of explaining donation amounts. In fact, the coefficient of determination shows that about half ($R^2 = 51.3\%$) of the variability of donations is explained by the past average. The regression coefficient, $b = \$1.21$, gives the amount of additional current donation (with standard error $S_b = \$0.0374$) per additional dollar of average past donations and is very highly significant ($p < 0.001$).[21]

Here are some of the lessons learned so far from mining this database for relationships to donation amount. First, donors from wealthier neighborhoods do tend to donate more on average than others, but many substantial donations come from less-wealthy neighborhoods that should not be ignored. Second, a quick look at (one aspect of) the age in donors' neighborhoods showed that this is not helpful in explaining donation amounts. Finally, the best explanation found so far uses information on the donor (not just for the neighborhood) and shows that donors who gave large amounts in the past tend to continue to do so. While it is not surprising that we find a relationship between past and current gifts, all the components of regression analysis (including the scatterplot, line, equation, and inference) have been helpful in understanding the nature and quantitative extent of the relationship.

20. Average income for each person's neighborhood is much easier to obtain than the actual income of each person. Neighborhood information can be automatically added to a database, for example, by using postal (ZIP) codes to link to a database of estimated average neighborhood incomes obtained through sampling (such information is available, e.g., from the U.S. Census Bureau at http://factfinder.census.gov).

21. With an effect this strong and a sample size this large, the $p$-value has 155 zeros after the decimal point! $p = 0.0000000000 \ldots 000000000025$. This is *very* highly unlikely to have occurred by chance, if there were no true relationship in the population!

## Other Tests of a Regression Coefficient

In some applications you will want to test whether the slope is significantly different from some *reference value* $\beta_0$ representing an external standard for comparison. The reference value does not come from the same data set used for regression. For example, you might test to see whether your recent variable costs (the slope from a regression of $Y$ = cost on $X$ = units produced) differ significantly from the budgeting assumptions (the reference value) you've used in the past.

The test of significance of the relationship between $X$ and $Y$ covered in the previous section is actually a test of whether the observed slope, $b$, is significantly different from the reference value $\beta_0 = 0$, which expresses the condition of no relationship. In this section we will allow $\beta_0$ to be nonzero. The test proceeds in the usual way. The hypotheses and results are as follows.

**Null and Research Hypotheses for Testing a Slope Coefficient**

Two-sided testing:

$$H_0 : \beta = \beta_0$$

$$H_1 : \beta \neq \beta_0$$

One-sided testing:

$$H_0 : \beta \leq \beta_0$$

$$H_1 : \beta > \beta_0$$

or

$$H_0 : \beta \geq \beta_0$$

$$H_1 : \beta < \beta_0$$

**Results of the Test**

If $\beta_0$ is *not* in the confidence interval for the slope, then the result is *significant*. For a two-sided test, use a two-sided interval and conclude that $b$ is significantly different from $\beta_0$. If $b$ is larger than $\beta_0$, you may conclude that it is significantly larger; otherwise, it is significantly smaller. For a one-sided test, use a one-sided confidence interval and conclude that $b$ is either significantly larger or significantly smaller than $\beta_0$, as appropriate.

If $\beta_0$ is in the confidence interval for the slope, then the result is *not significant*. For a two-sided test, use a two-sided interval and conclude that $b$ is not significantly different from $\beta_0$. For a one-sided test, use a one-sided confidence interval and conclude that $b$ is either not significantly larger or not significantly smaller than $\beta_0$, as appropriate.

Of course, the $t$ test may be used. The $t$ statistic is defined as follows:

$$t = \frac{b - \beta_0}{S_b}$$

Using the $t$ statistic, you would test these hypotheses about a population slope, $\beta$, just as you did in Chapter 10 for one- and two-sided testing of a population mean, $\mu$.

For the variable production costs example, suppose your budgeting process assumes a variable cost of $100.00 per item produced. The 95% confidence interval computed earlier extends from $36.08 to $67.24 per unit. Since the reference value, $\beta_0 = \$100.00$, is not in the confidence interval, you may conclude that the estimated variable costs, $b = \$51.66$, are significantly different from your budgeting assumption. In fact, since the estimated costs are smaller, you may conclude that actual variable costs are *significantly under budget*.

Continuing this example, suppose your intelligence sources indicate that one of your competitors bids on projects based on a variable cost of $60.00 per item produced. Since this reference value, $\beta_0 = \$60.00$, is in the confidence interval for your variable costs, there is *no significant difference*. You may conclude that your variable costs do not differ significantly from your competitor's. Even though your estimated variable costs ($51.66) are lower, this might reasonably be due to random chance rather than to any actual cost advantage.

## A New Observation: Uncertainty and the Confidence Interval

When you use regression to make a prediction about the value of a new observation, you want to know the uncertainty involved. You may even want to construct a confidence interval that you know has a 95% likelihood of containing the next observed value.

In this situation you know the value $X_0$, and you have predicted the value $a + bX_0$ for $Y$. There are now two sources of uncertainty that must be combined in order for you to find the standard error for this prediction. First of all, since $a$ and $b$ are estimated, the prediction $a + bX_0$ is uncertain. Second, there is always the randomness, $\varepsilon$, from the linear model (with standard deviation estimated by standard error $S_e$) to be considered when you work with a single observation. The result of combining these uncertainties is the standard error of $Y$ given $X_0$, denoted $S_{Y|X_0}$.[22] Following are the formula, together with its degrees of freedom, and the resulting confidence interval statement:

**Standard Error of a New Observation of $Y$ Given $X_0$**

$$S_{Y|X_0} = \sqrt{S_e^2\left(1 + \frac{1}{n}\right) + S_b^2\left(X_0 - \overline{X}\right)^2}$$

Degrees of freedom $= n - 2$

**Confidence Interval for a New Observation of $Y$ Given $X_0$**

From $(a + bX_0) - tS_{Y|X_0}$   to   $(a + bX_0) + tS_{Y|X_0}$

The standard error depends on $S_e$ (the basic uncertainty in the situation), on $S_b$ (the uncertainty in the slope used for prediction), and on the distance from $X_0$ to $\overline{X}$. The standard error of a new observation will be smaller when $X_0$ is close to $\overline{X}$ because this is where you know the most. The standard error of a new observation will be large when $X_0$ is far from $\overline{X}$ because the information you have (the observed $X$ values) is not near enough to the information you need ($X_0$). This behavior is shown in Figure 11.2.15 for the production cost data set.

For the production cost example, suppose you have scheduled $X_0 = 39$ units for production next week. Using the prediction equation, you have estimated the cost as $a + bX_0 = 2,272 + (51.66)(39) = \$4,287$. The uncertainty in this estimated cost for next week is

$$\begin{aligned}
S_{Y|X_0} &= \sqrt{S_e^2\left(1 + \frac{1}{n}\right) + S_b^2\left(X_0 - \overline{X}\right)^2} \\
&= \sqrt{198.58^2\left(1 + \frac{1}{18}\right) + 7.35^2\left(39 - 32.50\right)^2} \\
&= \sqrt{(39,434)(1.055556) + (54.0225)(42.25)} \\
&= \sqrt{43,907} \\
&= \$209.54
\end{aligned}$$

---

22. Note the use of the word *given*; this situation is similar to that of *conditional probability*, where you used additional information to update a probability. When you know the value $X_0$, you have additional information that can be used to decrease the uncertainty in $Y$ from the (unconditional) standard deviation $S_Y$ to the conditional standard error $S_{Y|X_0}$.

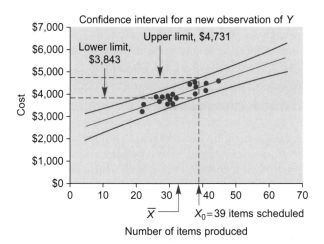

**FIGURE 11.2.15**  The confidence interval of a new observation for $Y$ when $X_0$ is known depends on how far $X_0$ is from $\overline{X}$. The interval is smallest, indicating slightly greater precision, near $\overline{X}$, where you have the most information from the data.

The 95% confidence interval, using the $t$ value (2.120) from the $t$ table for $n - 2 = 16$ degrees of freedom, extends from $4,287 - (2.120)(209.54)$ to $4,287 + (2.120)(209.54)$. Therefore,

> We have 95% confidence that next week's production cost (forecast as $4,287 based on production of 39 units) will be somewhere between $3,843 and $4,731.

This confidence interval takes into account all of the statistical sources of error: the smallish sample size, the estimation of the least-squares line for prediction, and the estimated additional uncertainty of a new observation. If the linear model is an appropriate description of your cost structure, then the confidence interval will be correct. This statistical method, however, cannot and does not take into account other sources of error, such as a fire at the plant (such a large error could not reasonably come from the same normal distribution as the randomness in your data), an unforeseen shift in the cost structure, or the additional unforeseen costs of doubling or tripling production.

## The Mean of $Y$: Uncertainty and the Confidence Interval

If you are interested in the *mean* value of $Y$ at a given value $X_0$, you need the appropriate standard error, denoted $S_{\text{predicted }Y|X_0}$ (to be defined soon) in order to construct confidence intervals and perform hypothesis tests. This procedure is much like that of the previous section, where $S_{Y|X_0}$ was used for statistical inference involving a new observation of $Y$ given $X_0$, except that the standard errors are different.

How do these two situations (mean of $Y$ versus single observation of $Y$, given $X_0$) compare with each other? You use the *same estimated value* in both cases, namely, the predicted value $a + bX_0$ from the least-squares line. However, because individual observations are more variable than statistical summaries, $S_{Y|X_0}$ is larger than $S_{\text{predicted }Y|X_0}$. The reason is that an individual observation of $Y$ (which is $\alpha + \beta X_0 + \varepsilon$ from the linear model) includes the random error term, $\varepsilon$, whereas the mean of $Y$ (which is $\alpha + \beta X_0$) does not.

Think of it this way. After looking at incomes ($X$) and sporting goods expenditures ($Y$), you may find that you have a very good idea of how much a typical person who earns $35,000 a year will spend on sporting goods, on average. This is the *average* amount spent on sporting goods by all people earning approximately $35,000; it is well estimated in a large sample because it is an average value and will be close to the mean spent by all people in the population who earn around $35,000. However, *individuals* differ substantially from one another—after all, not everyone plays racquetball with clients at lunchtime. The variability of individuals is *not* averaged out in a large sample; it's still there no matter how large $n$ is.

Given that $X$ is equal to a known value, $X_0$, the mean value for $Y$ is $\alpha + \beta X_0$. Note that this mean value is a population parameter because it is unknown and *fixed*, not random. The mean value for $Y$ given $X_0$ is estimated by the predicted value $a + bX_0$, which is *random* because the least-squares estimates $a$ and $b$ are computed from the random sample of data. The extent of this randomness is summarized by the following formula for the standard error of the predicted value (the mean value) of $Y$ given $X_0$.

---

**Standard Error of the Predicted (Mean) Value of $Y$ Given $X_0$**

$$S_{\text{predicted }Y|X_0} = \sqrt{S_e^2 \left(\frac{1}{n}\right) + S_b^2 (X_0 - \overline{X})^2}$$

Degrees of freedom $= n - 2$

**Confidence Interval for the Predicted (Mean) Value of $Y$ Given $X_0$**

From $(a + bX_0) - tS_{\text{predicted }Y|X_0}$ to $(a + bX_0) + tS_{\text{predicted }Y|X_0}$

---

This standard error depends on $S_e$ (the basic uncertainty in the situation), on $S_b$ (the uncertainty in the slope used for prediction), and on the distance from $X_0$ to $\overline{X}$. It will be smaller when $X_0$ is close to $\overline{X}$ because this is where you know the most. The standard error of the predicted (mean) value will be large when $X_0$ is far from $\overline{X}$ because the information you have (the observed $X$ values) is not near enough to the information you need ($X_0$). This behavior is shown in Figure 11.2.16 for the production cost data set.

Suppose you have set the production schedule at $X_0 = 39$ units for the indefinite future, and you want a good estimate of the long-term mean weekly production cost. Using the prediction equation, you have estimated this cost as $a + bX_0 = 2,272 + (51.66)(39) = \$4,287$. The uncertainty in this estimated long-term weekly cost is

$$S_{\text{predicted } Y|X_0} = \sqrt{S_e^2\left(\frac{1}{n}\right) + S_b^2\,(X_0 - \overline{X})^2}$$

$$= \sqrt{198.58^2\left(\frac{1}{18}\right) + 7.35^2(39 - 32.50)^2}$$

$$= \sqrt{(39,434)(0.055556) + (54.0225)(42.25)}$$

$$= \sqrt{4,473.25}$$

$$= \$66.88$$

Note how much smaller this standard error is (\$66.88, for the mean) compared to the standard error for an individual week (\$209.54) from the previous section.[23]

The 95% confidence interval, using the $t$ value (2.120) from the $t$ table for $n - 2 = 16$ degrees of freedom, extends from $\$4,287 - (2.120)(66.88)$ to $\$4,287 + (2.120)(66.88)$. In other words,

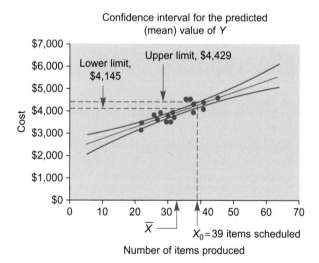

**Confidence interval for the predicted (mean) value of Y**

FIGURE 11.2.16   The confidence interval of the predicted (mean) value $Y$ when $X_0$ is known depends on how far $X_0$ is from $\overline{X}$. This interval is smaller than that for an individual observation of $Y$ because of the extra randomness of individuals (compare to Figure 11.2.15).

23. Due to round-off error, this result is off by a penny. If you use more precision in $S_b$ (using 7.3472 instead of 7.35), you will find the more accurate answer, \$66.87.

We have 95% confidence that the long-run mean weekly production cost (forecast as \$4,287, based on production of 39 units scheduled every week) will be somewhere between \$4,145 and \$4,429.

This confidence interval takes into account only the statistical error in estimating the predicted value, \$4,287, from the least-squares line based on this relatively small random sample of data. If the linear model is an appropriate description of your cost structure, the confidence interval will be correct. Again, however, this statistical method cannot take into account other, unforeseeable sources of error.

## Regression Can Be Misleading

Although regression is one of the most powerful and useful methods of statistics, there are problems to watch out for. Since inference from a regression analysis is based on the linear model, the results may be invalid if the linear model fails to hold in the population. Your error rate might be much higher than the 5% claimed, your confidence may be much lower than the 95% you think you have, or your predictions might simply be worse than they would be if the problems were addressed.

Since you have a limited amount of data, you have little information about cases of which your data are unrepresentative. Since your regression is based on the observed situation, it cannot necessarily anticipate the results of some intervention that produces a new situation with new dynamics. Furthermore, on a fairly technical note, it can make a big difference whether you are predicting $Y$ from $X$ or predicting $X$ from $Y$.

These are some of the problems you should be aware of in order to make the best use of your own statistical analyses as well as those of others. Following is some further information about the pitfalls of regression.

## The Linear Model May Be Wrong

Recall the linear model for the population

$$Y = (\alpha + \beta X) + \varepsilon$$
$$= (\text{Population relationship}) + \text{Randomness}$$

where $\varepsilon$ has a normal distribution with mean 0 and constant standard deviation. There are several ways in which this relationship might fail to hold in the population.

If the true relationship is *nonlinear*, the estimated straight line will not do a good job of predicting $Y$, as shown in Figures 11.2.17 and 11.2.18. Most computer programs will not object to using the least-squares estimation method in such situations, and few will even alert you to the problem. It's up to you to explore the data to discover trouble.

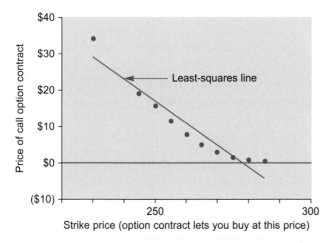

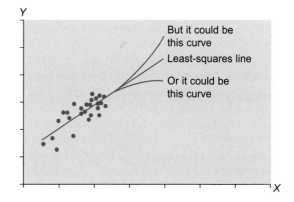

FIGURE 11.2.17  A nonlinear relationship cannot be well predicted by a line. Regression based on the linear model would predict negative stock index option prices at high strike prices, which is financially impossible.

FIGURE 11.2.19  Extrapolating beyond the range of the data is risky. Although the population might follow a straight line, you don't have enough information to rule out other possibilities. The two curved lines shown are also nearly straight in the region where you have information, and they cannot be ruled out.

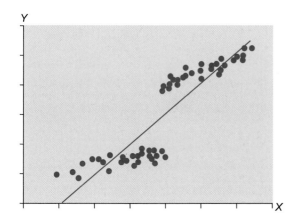

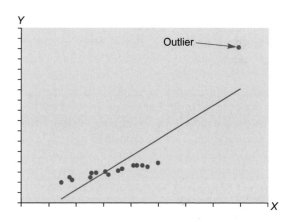

FIGURE 11.2.18  Nonlinearity may involve a "threshold effect," again resulting in poor predictions. In this particular case, it looks like a clustering problem. You might get much better results by fitting a separate regression line to each cluster.

FIGURE 11.2.20  An outlier can seriously distort the results of a least-squares regression analysis. Your prediction ability for the typical cases is seriously harmed by the outlier's presence.

The process of **extrapolation**, namely, predicting beyond the range of your data, is especially risky because you cannot protect yourself by exploring the data. Figure 11.2.19 illustrates the problem.

A single outlier can ruin everything, as in Figure 11.2.20. The linear model's assumption of a normal distribution for the randomness says that an outlier far from the population line is highly unlikely. The least-squares line will try hard to accommodate the outlier, and this interferes with its ability to predict the typical, nonoutlier cases. So-called robust regression methods provide some solutions to this problem.[24]

Finally, if there is *unequal variability* in your data, the inference will be unreliable. Too much importance will be given to the high-variability part of the data, and too little importance will be given to the more reliable low-variability part. There are two solutions to this problem: (1) Transform the data to equalize the variability and achieve a straight-line fit, or (2) use the advanced technique of *weighted* regression analysis to rebalance the importance of the observations.

Next, let's see what can go wrong in regression and its interpretation even when your data set does follow a linear model.

## Predicting Intervention from Observed Experience Is Difficult

When you use regression to make predictions based on data, you are assuming that the new observation being predicted will arise from the same basic system that generated

---

24. See, for example, D. C. Hoaglin, F. Mosteller, and J. W. Tukey, *Understanding Robust and Exploratory Data Analysis* (New York: Wiley, 1983).

the data. If the system changes, either through its own evolution or due to an external intervention, the predictions may no longer apply.

For example, you might estimate a regression line to predict the dollar volume of new orders from the number of telephone calls answered in a department. The slope would give you an indication of the average value of each call. Should you undertake a marketing program to encourage customers to call? If you do this, by intervening in the system, you may change the mixture of calls received. The marketing program may generate new calls that are primarily for information and are less likely to generate dollars immediately. You may very well experience an increase in orders through this campaign; the point is that the slope (based on past data) may not apply to the new system.

## The Intercept May Not Be Meaningful

When you are regressing cost data ($Y$) on number of units produced ($X$), the intercept term gives you the fixed costs, which are very meaningful. But in other situations, the intercept term may have no useful meaning. You may still need it for technical reasons, in order to get your best predictions, but there may be no practical interpretation.

For example, consider a regression of salary ($Y$) on age ($X$). The slope indicates the incremental (extra) salary you can expect from an additional year of age, on average. Some kind of intercept term is needed to establish a baseline so that actual salaries can be predicted, for example, from the equation $a + bX$. Although $a$ is useful here, its interpretation is difficult. Literally, it is the expected salary you would pay a person whose age is $X = 0$: a newborn baby!

This is not really a problem. Just don't feel that you need to interpret $a$ in these difficult cases.[25]

## Explaining Y from X versus Explaining X from Y

It really *does* matter which of your variables is being predicted from the other: Predicting $Y$ from $X$ is different from predicting $X$ from $Y$, and each approach requires a different regression line. This seems reasonable because the errors are different in each case. For example, predicting productivity from experience involves making prediction errors in productivity units, whereas predicting experience from productivity involves making prediction errors in experience units. Of course, if all of your data points fall exactly on a line (so that the correlation is 1 or −1), this line will work for predicting either variable from the other.

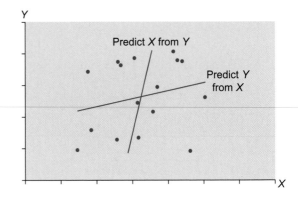

FIGURE 11.2.21 The two regression lines: one to predict $Y$ from $X$ (the usual procedure) and the other to predict $X$ from $Y$. Since there is lots of randomness here, the lines are very different. Each one is close to predicting the average value ($\overline{X}$ or $\overline{Y}$) of the respective variable (i.e., a horizontal or vertical line).

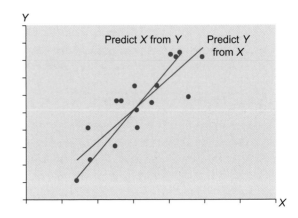

FIGURE 11.2.22 The two regression lines are similar when there is less randomness and the data points are fairly close to a line. When the data points fall exactly on a line, the two regression lines will coincide.

However, in the usual case, there is some randomness or uncertainty, which tends to push your predicted values toward the average of the particular variable being predicted (either $X$ or $Y$). In the extreme case, pure randomness, your best predictor of $Y$ from $X$ is $\overline{Y}$, and your best predictor of $X$ from $Y$ is $\overline{X}$. Remember the slope formula, $b = rS_Y/S_X$? This indicates that the line gets flatter (less steep) when there is more uncertainty (correlation, $r$, closer to 0).

Figures 11.2.21 and 11.2.22 show the two regression lines. Note that when the data points fall closer to a line, the two regression lines are closer together because the line is better defined by the data.

## A Hidden "Third Factor" May Be Helpful

This last consideration is more of a suggestion for improvement than a problem. Although the least-squares line is the best way to predict $Y$ from $X$, there is always the possibility that you could do a better job in predicting $Y$ if you had

---

25. One way around the problem is to use a so-called centercept instead of the intercept. The line would then be expressed as $Y = c + b(X - \overline{X})$. The centercept $c$ is the expected value of $Y$ for the most typical value of $X$, namely, $\overline{X}$, so interpretation is no problem. The slope is the same as before.

more information. That is, perhaps $X$ doesn't contain enough information about $Y$ to do the best job, and maybe you could find another variable (a third factor) that would improve the predictions.

If you can substitute a different variable for $X$, you can perform another regression analysis to predict the same $Y$. A comparison of the $R^2$ terms (or the $S_e$ terms) from each regression would provide some indication of which explanatory variable is most useful in predicting $Y$.

If you wish to combine the information in *two or more X variables*, you will need to use *multiple regression*, a very important method in business and research, which is the topic of the next chapter.

## 11.3 END-OF-CHAPTER MATERIALS

### Summary

The three basic goals of bivariate data analysis of $(X, Y)$ pairs are (1) describing and understanding the relationship, (2) forecasting and predicting a new observation, and (3) adjusting and controlling a process. *Correlation analysis* summarizes the strength of the relationship, and *regression analysis* is used to predict or explain one variable from the other (usually $Y$ from $X$).

Bivariate data are explored using the **scatterplot** of $Y$ against $X$, providing a visual picture of the relationship in the data. The **correlation** or **correlation coefficient** ($r$) is a pure number between $-1$ and $1$ summarizing the strength of the relationship. A correlation of 1 indicates a perfect straight-line relationship with upward tilt; a correlation of $-1$ indicates a perfect straight-line relationship with downward (negative) tilt. The correlation tells you how close the points are to being exactly on a tilted straight line, but it does not tell you how steep that line is. The formula for the correlation coefficient is

$$r = \frac{\frac{1}{n-1} \sum_{i=1}^{n} (X_i - \overline{X})(Y_i - \overline{Y})}{S_X S_Y}$$

The **covariance** of $X$ and $Y$ is the numerator in the formula for the correlation coefficient. Because its measurement units are difficult to interpret, it is probably easier to work with the correlation coefficient instead.

There are a number of relationships you might see when exploring a bivariate scatterplot. The easiest to analyze is a **linear relationship**, where the scatterplot shows points bunched randomly around a straight line with constant scatter. A scatterplot shows **no relationship** if it is just random, tilting neither upward nor downward as you move from left to right. There is a **nonlinear relationship** if the points bunch around a *curved* rather than a straight line. Since there are so many different kinds of curves that can be

drawn, the analysis is more complex, but a transformation may help straighten out the relationship. You have the problem of **unequal variability** when the vertical variability changes dramatically as you move horizontally across the scatterplot. This causes correlation and regression analysis to be unreliable; these problems may be fixed by using either transformations or a so-called weighted regression. You have **clustering** if there are separate, distinct groups in the scatterplot; you may wish to analyze each group separately. A data point is a **bivariate outlier** if it does not fit with the relationship of the rest of the data; outliers can distort statistical summaries.

Correlation is not causation. The correlation coefficient summarizes the association in the numbers but cannot explain it. Correlation might be due to the $X$ variable affecting $Y$, or the $Y$ variable might affect $X$, or there might be a hidden third factor affecting both $X$ and $Y$ so that they seem associated. The term **spurious correlation** refers to a high correlation that is actually due to some third factor.

Regression analysis is explaining or predicting one variable from the other. **Linear regression analysis** is predicting one variable from the other using a straight line. The **slope**, $b$, is in measurement units of $Y$ per unit $X$ and indicates how steeply the line rises (or falls, if $b$ is negative). The **intercept**, $a$, is the value for $Y$ when $X$ is 0. The equation for a straight line is

$$Y = (\text{Intercept}) + (\text{Slope})(X)$$
$$= a + bX$$

The **least-squares line** has the smallest sum of squared vertical prediction errors of all possible lines and is used as the best predictive line based on data. The slope $b$ is also called the **regression coefficient** of $Y$ on $X$, and the intercept $a$ is also called the **constant term** in the regression. Here are the equations for the least-squares slope and intercept:

$$\text{Slope} = b = r \frac{S_Y}{S_X}$$

$$\text{Intercept} = a = \overline{Y} - b\overline{X} = \overline{Y} - r \frac{S_Y}{S_X} \overline{X}$$

The formula for the least-squares line is

$$(\text{Predicted value of } Y) = a + bX$$

$$= \left( \overline{Y} - r \frac{S_Y}{S_X} \overline{X} \right) + \left( r \frac{S_Y}{S_X} \right) X$$

The **predicted value** for $Y$ given a value of $X$ is found by substituting the value of $X$ into the equation for the least-squares line. Each of the data points has a **residual**, a prediction error that tells you how far the point is above or below the line.

There are two measures of how useful the least-squares line is. The **standard error of estimate**, denoted $S_e$, tells you approximately how large the prediction errors (residuals) are for your data set *in the same units as Y*. The formulas are

$$S_e = S_Y \sqrt{(1 - r^2) \frac{n-1}{n-2}} \qquad \text{(for computation)}$$

$$= \sqrt{\frac{1}{n-2} \sum_{i=1}^{n} [Y_i - (a + bX_i)]^2} \qquad \text{(for interpretation)}$$

The $R^2$ value, also called the **coefficient of determination**, is the square of the correlation and tells you what percentage of the variability of $Y$ is explained by $X$.

Confidence intervals and hypothesis tests for the regression coefficient require assumptions about the data set to help ensure that it consists of independent observations having a linear relationship with equal variability and approximately normal randomness. First, the data must be a random sample from the population of interest. Second, the **linear model** specifies that the observed value for $Y$ is equal to the population relationship plus a random error that has a normal distribution. There are population parameters, corresponding to the least-squares slope and intercept term computed from the sample, in the linear model

$$Y = (\alpha + \beta X) + \varepsilon$$
$$= (\text{Population relationship}) + \text{Randomness}$$

where $\varepsilon$ has a normal distribution with mean 0 and constant standard deviation $\sigma$.

Inference (use of confidence intervals and hypothesis tests) for the coefficients of the least-squares line is based on their standard errors (as always) using the $t$ table with $n-2$ degrees of freedom. The **standard error of the slope coefficient**, $S_b$, indicates approximately how far the estimated slope, $b$ (the regression coefficient computed from the sample), is from the population slope, $\beta$, due to the randomness of sampling. The **standard error of the intercept term**, $S_a$, indicates approximately how far the estimated $a$ is from $\alpha$, the true population intercept term. Here are the formulas:

Standard Error of the Regression Coefficient

$$S_b = \frac{S_e}{S_X \sqrt{n-1}}$$

Standard Error of the Intercept Term

$$S_a = S_e \sqrt{\frac{1}{n} + \frac{\overline{X}^2}{S_X^2 (n-1)}}$$

Confidence Interval for the Population Slope, $\beta$

$$\text{From } b - tS_b \text{ to } b + tS_b$$

Confidence Interval for the Population Intercept, $\alpha$

$$\text{From } a - tS_a \text{ to } a + tS_a$$

One way to test whether the apparent relationship between $X$ and $Y$ is real or just coincidence is to test $\beta$ against the reference value $\beta_0 = 0$. There is a significant relationship if 0 is not in the confidence interval based on $b$ and $S_b$, or if the absolute value of $t = b/S_b$ exceeds the $t$ value in the $t$ table. This test is equivalent to testing the significance of the correlation coefficient and is also the same as the $F$ test in multiple regression (see next chapter) when you have only one $X$ variable. Of course, either coefficient ($a$ or $b$) may be tested against any appropriate reference value using a one- or a two-sided test, as appropriate, and using the same methods you learned in Chapter 10 for testing a population mean.

For predicting a new observation of $Y$ given that $X = X_0$, the uncertainty of prediction is estimated by the standard error $S_{Y|X_0}$, which also has $n-2$ degrees of freedom. This allows you to construct confidence intervals and hypothesis tests for a new observation. An alternative formula gives the standard error $S_{\text{predicted } Y|X_0}$ for predicting the *mean* value of $Y$ given $X_0$:

$$S_{Y|X_0} = \sqrt{S_e^2 \left(1 + \frac{1}{n}\right) + S_b^2 (X_0 - \overline{X})^2}$$

$$S_{\text{predicted } Y|X_0} = \sqrt{S_e^2 \left(\frac{1}{n}\right) + S_b^2 (X_0 - \overline{X})^2}$$

The confidence interval for a new observation of $Y$ given $X_0$ is

$$\text{From } (a + bX_0) - tS_{Y|X_0} \text{ to } (a + bX_0) + tS_{Y|X_0}$$

and the confidence interval for the predicted (mean) value of $Y$ given $X_0$ is

$$\text{From } (a + bX_0) - tS_{\text{predicted } Y|X_0} \text{ to } (a + bX_0) + tS_{\text{predicted } Y|X_0}$$

Regression analysis has its problems. If the linear model does not adequately describe the population, your predictions and inferences may be faulty. Exploring the data will help you determine if the linear model is appropriate by showing you problems such as nonlinearity, unequal variability, or outliers. The process of **extrapolation**, predicting beyond the range of the data, is especially risky because you cannot protect yourself by exploring the data.

Even if the linear model is appropriate, there are still problems. Since regression predicts based on past data, it cannot perfectly anticipate the effects of an intervention that changes the structure of a system. In some cases, the intercept term, $a$, is difficult to interpret, although it may be a necessary part of the least-squares prediction equation. Be sure to choose carefully the variable you wish to predict because predicting $Y$ from $X$ requires a different line than predicting $X$ from $Y$, especially when there is substantial randomness in the data. Finally, there may be a third factor that would help you do a better job of predicting $Y$ than using $X$ alone; the next chapter will discuss this.

## Key Words

bivariate outlier, *308*
clustering, *307*
coefficient of determination, $R^2$, *317*
constant term, *a*, *313*
correlation coefficient, *r*, *295*
covariance, *297*
extrapolation, *327*
intercept, *312*
least-squares line, $Y = a + bX$, *312*
linear model, *318*
linear regression analysis, *310*
linear relationship, *297*
no relationship, *299*
nonlinear relationship, *301*
predicted value, *313*
regression coefficient, *b*, *313*
residual, *313*
scatterplot, *292*
slope, *312*
spurious correlation, *309*
standard error of estimate, $S_e$, *315*
standard error of the intercept term, $S_a$, *318*
standard error of the slope coefficient, $S_b$, *318*
unequal variability, *304*

### Questions

1. What is new and different about analysis of bivariate data compared to univariate data?
2. Distinguish correlation and regression analysis.
3. Which activity (correlation or regression analysis) is involved in each of the following situations?
   a. Investigating to see whether there is any measurable connection between advertising expenditures and sales.
   b. Developing a system to predict portfolio performance based on changes in a major stock market index.
   c. Constructing a budgeting tool to express costs in terms of the number of items produced.
   d. Examining data to see how strong the connection is between employee morale and productivity.
4. For each of the following summaries, first indicate what the usual interpretation would be. Then indicate whether there are any other possibilities.
   a. $r = 1$.
   b. $r = 0.85$.
   c. $r = 0$.
   d. $r = -0.15$.
   e. $r = -1$.
5. a. What is the covariance between $X$ and $Y$?
   b. Which is easier to interpret, the covariance or the correlation? Why?

6. Draw a scatterplot to illustrate each of the following kinds of structure in bivariate data. There is no need to work from data for this question; you may draw the points directly.
   a. No relationship between $X$ and $Y$.
   b. Linear relationship with strong positive correlation.
   c. Linear relationship with weak negative correlation.
   d. Linear relationship with correlation –1.
   e. Positive association with unequal variability.
   f. Nonlinear relationship.
   g. Clustering.
   h. Positive association with an outlier.
7. a. If large values of $X$ cause the $Y$ values to be large, would you expect the correlation to be positive, negative, or zero? Why?
   b. If you find a strong positive correlation, does this prove that large values of $X$ cause the $Y$ values to be large? If not, what other possibilities are there?
8. a. What is so special about the least-squares line that distinguishes it from all other lines?
   b. How does the least-squares line "know" that it is predicting $Y$ from $X$ instead of the other way around?
   c. It is reasonable to summarize the "most typical" data value as having $\overline{X}$ as its $X$ value and $\overline{Y}$ as its $Y$ value. Show that the least-squares line passes through this most typical point.
   d. Suppose the standard deviations of $X$ and of $Y$ are held fixed while the correlation decreases from one positive number to a smaller one. What happens to the slope coefficient, $b$?
9. Define the predicted value and the residual for a given data point.
10. For each of the following situations tell whether the predicted value or the residual would be most useful.
    a. For budgeting purposes you need to know what number to place under "cost of goods sold" based on the expected sales figure for the next quarter.
    b. You would like to see how your divisions are performing after adjusting for how well you expect them to do given the resources they consume.
    c. To help you set the salary of a new employee, you want to know a reasonable pay figure for a person with the same experience.
    d. As part of a payroll policy analysis report, you wish to show how much more (or less) each employee is paid compared to the expected salary for a person with the same experience.
11. Distinguish the standard error of estimate and the coefficient of determination.
12. a. Which is better, a lower or a higher value for $R^2$?
    b. Which is better, a lower or a higher value for $S_e$?
13. a. What is the linear model?
    b. Which two parameters define the population straight-line relationship?
    c. What sample statistics are used to estimate the three population parameters $\alpha$, $\beta$, and $\sigma$?

d. Is the slope of the least-squares line, computed from a sample of data, a parameter or a statistic? How do you know?

14. Identify and write a formula for each of the following quantities, which are useful for accomplishing statistical inference.
   a. The standard error used for the regression coefficient.
   b. The standard error of the intercept term.
   c. The standard error of a new observation.
   d. The number of degrees of freedom for each of these standard errors.

15. Statistical inference in regression is based on the linear model. Name at least three problems that can arise when the linear model fails to hold.

16. What is extrapolation? Why is it especially troublesome?

17. Using a least-squares line, you have predicted that the cost of goods sold will rise to $8.33 million at the end of next quarter based on expected sales of $38.2 million. Your friend in the next office remarks, "Isn't it also true that a cost of goods sold of $8.33 million implies an expected sales level of $38.2 million?" Is this conclusion correct? Why or why not? (*Hint:* Which is *X* and which is *Y* in each case, and which is being predicted from the other?)

18. a. Give an example in which the intercept term, *a*, has a natural interpretation.
   b. Give an example in which the intercept term, *a*, does not have a natural interpretation.

## Problems

*Problems marked with an asterisk (*) are solved in the Self Test in Appendix C.*

1.* Consider the data set in Table 11.3.1, representing the ages (in years) and maintenance costs (in thousands of dollars per year) for five similar printing presses.
   a. Draw a scatterplot of this data set. What kind of relationship do you see?
   b. Find the correlation between age and maintenance cost. What do you learn from it?
   c. Find the least-squares regression equation that predicts maintenance cost from the age of the machine. Draw this line on a scatterplot of the data.

### TABLE 11.3.1 Age (in years) and Maintenance Cost for Printing Presses

| Age | Maintenance Cost |
| --- | --- |
| 2 | 6 |
| 5 | 13 |
| 9 | 23 |
| 3 | 5 |
| 8 | 22 |

d. What would you expect the annual maintenance to be for a press that is seven years old?
   e. What is a typical size for the prediction errors?
   f. How much of the variation in maintenance cost can be attributed to the fact that some presses are older than others?
   g. Does age explain a significant amount of the variation in maintenance cost? How do you know?
   h. Your conservative associate has suggested that you use $20,000 for planning purposes as the extra annual maintenance cost per additional year of life for each machine. Perform a hypothesis test at the 5% level to see if the extra annual cost is significantly different from your associate's suggestion.

2. A linear regression analysis has produced the following equation relating profits to hours of managerial time spent developing the past year's projects at a firm:
   $$\text{Profits} = -\$957 + \$85 \times \text{Number of hours}$$
   a. According to this estimated relationship, how large would the profits (or losses) be if no time were spent in planning?
   b. On the average, an extra 10 hours spent planning resulted in how large an increase in project profits?
   c. Find the break-even point, which is the number of hours for which the estimated profits would be zero.
   d. If the correlation is $r = 0.351$, what percentage of the variation in profits is explained by the time spent?
   e. How much of the variation in profits is left unexplained by the number of hours spent? Write a paragraph explaining how much faith you should have in this prediction equation and discussing other factors that might have an impact on profits.

3. Table 11.3.2 shows the on-time performance of nine airlines, both for one month (May 2010) and for the preceding four (January to April 2010). These numbers represent percentages of flights that arrived on time. We will investigate the consistency of performance by examining the relationship, if any, between the one month and the preceding four.
   a. Draw a scatterplot of this data set and comment on the relationship.
   b. Find the correlation between performance for one month and for the preceding four. Is there a strong relationship?
   c. Find the coefficient of determination and say what it represents.
   d. Find the linear regression equation to predict the performance for May from that of the previous four months.
   e. Find the predicted value and residual value for American Airlines. Say what each one represents.
   f. Find the standard error of estimate. What does this measure?
   g. Find the standard error of the regression coefficient.
   h. Find the 95% confidence interval for the regression coefficient.

## TABLE 11.3.2 Airline On-Time Performance

| Airline | January–April | May |
|---|---|---|
| Alaska | 87.58% | 91.50% |
| American | 78.18 | 76.58 |
| Continental | 80.32 | 82.48 |
| Delta | 80.34 | 75.64 |
| Frontier | 81.54 | 80.20 |
| JetBlue | 75.12 | 82.70 |
| Southwest | 81.21 | 80.35 |
| United | 83.98 | 84.77 |
| US Airways | 81.16 | 85.29 |

**Source:** Data are from U.S. Department of Transportation, Research and Innovation Technology Administration, Bureau of Transportation Statistics, accessed at http://www.transtats.bts.gov/ot_delay/OT_DelayCause1.asp on July 20, 2010.

    i.  Test at the 5% level to see whether or not there is a significant relationship between performance during these two time periods. What does this tell you about consistency in airline performance?

4. Closed-end funds sell shares in a fixed basket (portfolio) of securities (as distinguished from ordinary mutual funds, which continuously buy and sell shares of securities). Consider the net asset value and the market price for Sector Equity Funds, as shown in Table 11.3.3. While you might expect each fund to sell (the market price) at the same price as the sum of its components (the net asset value), there is usually some discrepancy.

    a.  How strong is the relationship between the net asset value and the market price for these closed-end funds?

    b.  Are the net asset value and the market price significantly related, or is it as though the market prices were randomly assigned to funds? How do you know?

    c.  Find the least-squares line to predict market price from net asset value.

    d.  Does the slope of the least-squares line differ significantly from 1? Interpret your answer in terms of this question: Could it be that a one-point increase in net asset value translates, on average, into a one-point increase in market price?

5. Consider the number of transactions and the total dollar value of merger and acquisition deals in the oil and gas industry, from Table 11.1.5.

    a.  Find the regression equation for predicting the dollar value from the number of transactions.

    b.  What is the estimated dollar value attributable to a single additional transaction for these investment bankers, on average?

    c.  Draw a scatterplot of the data set with the regression line.

## TABLE 11.3.3 Sector Equity Closed-End Funds

| Fund | Net Asset Value | Market Price |
|---|---|---|
| ASA Limited (ASA) | 28.36 | 26.20 |
| BlackRock EcoSolutions (BQR) | 9.55 | 9.54 |
| ClearBridge Energy MLP (CEM) | 19.48 | 20.40 |
| Cohen & Steers Infrastrc (UTF) | 16.41 | 13.67 |
| Cushing MLP Tot Ret (SRV) d | 7.21 | 8.39 |
| Diamond Hill Finl Trends (DHFT) | 10.26 | 8.52 |
| DWS Enh Commodity Strat (GCS) | 8.55 | 8.32 |
| Energy Income & Growth (FEN) | 23.48 | 24.60 |
| Evergreen Util & Hi Inc (ERH) | 11.07 | 10.90 |
| Fiduciary/Clay MLP Opp (FMO) | 17.79 | 19.56 |
| First Opportunity Fund (FOFI) | 7.92 | 6.00 |
| Gabelli Utility Trust (GUT) | 4.75 | 7.53 |
| H&Q Healthcare Investors (HQH) | 13.60 | 11.23 |
| ING Risk Mgd Nat Res (IRR) | 14.02 | 14.63 |
| J Hancock Bank & Thrift (BTO) | 17.28 | 14.63 |
| Kayne Anderson Enrgy TR (KYE) | 22.84 | 24.21 |
| Macquarie Gl Infrstrc TR (MGU) | 17.75 | 14.43 |
| MLP & Strat Eqty (MTP) | 17.12 | 16.91 |
| Petroleum & Resources (PEO) | 23.85 | 20.69 |
| Reaves Utility Income (UTG) a | 18.33 | 19.57 |

**Source:** Data are from the *Wall Street Journal*, accessed from http://online.wsj.com/mdc on July 20, 2010. Their source is Lipper, Inc.

    d.  Find the expected dollar amount for Goldman Sachs and the residual value. Interpret both of these values in business terms.

    e.  Find the standard error of the slope coefficient. What does this number indicate?

    f.  Find the 95% confidence interval for the expected marginal value of an additional transaction to these firms. (This is economics language for the slope.)

    g.  Test at the 5% level to see if there is a significant relationship between the number of transactions and the dollar value.

    h.  Your investment banking firm is aiming to be in the top group next year, with 25 transactions. Assuming that you will be "just like the big ones," compute a 95% confidence interval for the dollar amount you will handle.

6. Consider the slightly scary topic of business bankruptcies. Table 11.3.4 shows data for each state on the number of failed businesses and the population in millions.

**TABLE 11.3.4 Business Bankruptcies by State**

| State | Bankruptcies | Population |
|---|---|---|
| Alabama | 395 | 4.662 |
| Alaska | 78 | 0.686 |
| Arizona | 691 | 6.500 |
| Arkansas | 429 | 2.855 |
| California | 4,697 | 36.757 |
| Colorado | 756 | 4.939 |
| Connecticut | 366 | 3.501 |
| Delaware | 361 | 0.873 |
| District of Columbia | 42 | 0.592 |
| Florida | 2,759 | 18.328 |
| Georgia | 1,714 | 9.686 |
| Hawaii | 51 | 1.288 |
| Idaho | 167 | 1.524 |
| Illinois | 1,178 | 12.902 |
| Indiana | 692 | 6.377 |
| Iowa | 270 | 3.003 |
| Kansas | 244 | 2.802 |
| Kentucky | 390 | 4.269 |
| Louisiana | 571 | 4.411 |
| Maine | 154 | 1.316 |
| Maryland | 405 | 5.634 |
| Massachusetts | 334 | 6.498 |
| Michigan | 1,394 | 10.003 |
| Minnesota | 656 | 5.220 |
| Mississippi | 298 | 2.939 |
| Missouri | 534 | 5.912 |
| Montana | 71 | 0.967 |
| Nebraska | 221 | 1.783 |
| Nevada | 395 | 2.600 |
| New Hampshire | 318 | 1.316 |
| New Jersey | 925 | 8.683 |
| New Mexico | 185 | 1.984 |
| New York | 1,534 | 19.490 |
| North Carolina | 738 | 9.222 |
| North Dakota | 54 | 0.641 |
| Ohio | 1,436 | 11.486 |

| | | |
|---|---|---|
| Oklahoma | 392 | 3.642 |
| Oregon | 283 | 3.790 |
| Pennsylvania | 1,078 | 12.448 |
| Rhode Island | 122 | 1.051 |
| South Carolina | 196 | 4.480 |
| South Dakota | 95 | 0.804 |
| Tennessee | 653 | 6.215 |
| Texas | 2,728 | 24.327 |
| Utah | 302 | 2.736 |
| Vermont | 52 | 0.621 |
| Virginia | 798 | 7.769 |
| Washington | 565 | 6.549 |
| West Virginia | 169 | 1.814 |
| Wisconsin | 525 | 5.628 |
| Wyoming | 42 | 0.533 |

**Source:** Data are from U.S. Census Bureau, *Statistical Abstract of the United States: 2010* (129th edition), Washington, DC, 2009. Bankruptcies for 2008 were accessed at http://www.census.gov/compendia/statab/cats/business_enterprise/establishments_employees_payroll.html on July 5, 2010. Populations for 2008 are from Table 12, accessed from http://www.census.gov/compendia/statab/cats/population.html on July 5, 2010.

a. Construct a scatterplot of business bankruptcies ($Y$) against population ($X$). Describe the relationship that you see. Does there appear to be some association?

b. Does the linear model appear to hold? Why or why not?

c. Find the logarithm of each data value, both for population and for business bankruptcies. You may choose either base 10 or base $e$, but use only one type.

d. Construct a scatterplot of the logarithms and describe the relationship.

e. Find the equation of the regression line to predict the log of business bankruptcies from the log of population.

f. Find the two-sided 95% confidence interval for the slope coefficient of the log relationship.

g. Test at the 5% level to see whether there is a significant relationship between the logs of bankruptcies and of population. Explain why the result is reasonable.

h. If the slope for the logs were exactly 1, then business bankruptcies would be proportional to population. A value larger than 1 would say that large states have proportionately more bankruptcies, and a slope less than 1 would suggest that the smaller states have proportionately more bankruptcies. Test at the 5% level to see whether the population slope for the logs is significantly different from 1 or not, and briefly discuss your conclusion.

7. Consider the daily percent changes of McDonald's stock price and those of the Dow Jones Industrial Average for trading days in the months of January and February 2010, as shown in Table 11.3.5.

   a. Draw a scatterplot of McDonald's daily percent changes against the Dow Jones percent changes.

   b. Describe the relationship you see in this scatterplot.

   c. Find the correlation between these percent changes. Does this agree with your impression from the scatterplot?

**TABLE 11.3.5  Daily Changes in Stock Market Prices, January and February 2010**

| Day | Dow Jones | McDonald's |
|---|---|---|
| 26-Feb | 0.04% | −0.82% |
| 25-Feb | −0.51 | −0.51 |
| 24-Feb | 0.89 | 0.61 |
| 23-Feb | −0.97 | 0.16 |
| 22-Feb | −0.18 | 0.05 |
| 19-Feb | 0.09 | 0.39 |
| 18-Feb | 0.81 | 0.35 |
| 17-Feb | 0.39 | 0.39 |
| 16-Feb | 1.68 | 0.67 |
| 12-Feb | −0.44 | −0.32 |
| 11-Feb | 1.05 | 0.85 |
| 10-Feb | −0.20 | −0.49 |
| 9-Feb | 1.52 | 1.03 |
| 8-Feb | −1.04 | −0.72 |
| 5-Feb | 0.10 | −1.07 |
| 4-Feb | −2.61 | −1.76 |
| 3-Feb | −0.26 | 1.84 |
| 2-Feb | 1.09 | 0.22 |
| 1-Feb | 1.17 | 2.34 |
| 29-Jan | −0.52 | −0.64 |
| 28-Jan | −1.13 | −1.41 |
| 27-Jan | 0.41 | −0.13 |
| 26-Jan | −0.03 | 1.13 |
| 25-Jan | 0.23 | −0.48 |
| 22-Jan | −2.09 | 0.30 |
| 21-Jan | −2.01 | 0.30 |
| 20-Jan | −1.14 | −0.75 |
| 19-Jan | 1.09 | 1.93 |
| 15-Jan | −0.94 | −0.58 |
| 14-Jan | 0.28 | 0.10 |
| 13-Jan | 0.50 | −0.11 |
| 12-Jan | −0.34 | 0.55 |
| 11-Jan | 0.43 | 0.77 |
| 8-Jan | 0.11 | −0.10 |
| 7-Jan | 0.31 | 0.74 |
| 6-Jan | 0.02 | −1.36 |
| 5-Jan | −0.11 | −0.77 |

**Source:** http://finance.yahoo.com, accessed March 5, 2010.

   d. Find the coefficient of determination. (You may just square the correlation.) Interpret this number as "variation explained." In financial terms, it represents the proportion of nondiversifiable risk in McDonald's. For example, if it were 100%, McDonald's stock would track the market perfectly, and diversification would introduce nothing new.

   e. Find the proportion of diversifiable risk. This is just $1 − R^2$ (or 100% minus the percentage of nondiversifiable risk). This indicates the extent to which you can diversify away the risk of McDonald's stock by investing part of your portfolio in the Dow Jones Industrial stocks.

   f. Find the regression equation to explain the percent change in McDonald's stock from the percent change in the Dow Jones Index. Identify the stock's so-called beta, a measure used by market analysts, which is equal to the slope of this line. According to the capital asset pricing model, stocks with large beta values tend to give larger expected returns (on average, over time) than stocks with smaller betas.

   g. Find the 95% confidence interval for the slope coefficient.

   h. Test at the 5% level to see whether or not the daily percent changes of McDonald's and of the Dow Jones Index are significantly associated.

   i. Test at the 5% level to see whether the beta of McDonald's is significantly different from 1, which represents the beta of a highly diversified portfolio.

8. This problem continues the analysis of McDonald's and Dow Jones stock market data.

   a. Find the 95% confidence interval for the percent change in McDonald's stock on a day in which the Dow Jones Index is unchanged.

b. Find the 95% confidence interval for the mean percent change in McDonald's stock for the idealized population of all days in which the Dow Jones Index is unchanged.

c. Find the 95% confidence interval for the percent change in McDonald's stock on a day in which the Dow Jones Index is up 1.5%.

d. Find the 95% confidence interval for the mean percent change in McDonald's stock for the idealized population of all days in which the Dow Jones Index is up 1.5%.

9. In the territory versus sales example (based on the data from Table 11.2.3), the least-squares line to predict sales based on the population of the territory was found to be

Expected sales = $1,371,744 + $0.23675045 (Population)

a. Interpret the slope coefficient as a number with a simple and direct business meaning.

b. What proportion of the variation in sales from one agent to another is attributable to territory size? What proportion is due to other factors?

c. Does territory size have a significant impact on sales? How do you know?

d. Find the $p$-value (as either $p > 0.05$, $p < 0.05$, $p < 0.01$, or $p < 0.001$) for the significance of the slope coefficient.

10. Using the donations database on the companion site, and using only people who made a donation in response to the current mailing, consider predicting the amount of a donation (named "Donation_D1" in the worksheet) from the percentage of households in the neighborhood with three or more cars (named "Cars_D1").

a. Find the regression equation and the coefficient of determination.

b. Is there a significant relationship?

11. Using the donations database on the companion site, and using only people who made a donation in response to the current mailing, consider predicting the amount of a donation (named "Donation_D1" in the worksheet) from the percentage of households in the neighborhood that are self-employed (named "SelfEmployed_D1").

a. Find the regression equation and the coefficient of determination.

b. Is there a significant relationship?

12. The least-squares prediction equation is, Predicted costs = 35.2 + 5.3 (Items), with predicted costs measured in dollars. Find the predicted value and residual for a situation with costs of $600 and 100 items.

13. Find the $t$ table value that would be used to construct the confidence interval for the slope coefficient in a regression analysis for each of the following situations:

a. For 95% confidence, based on a sample size of $n = 298$.

b. For 99% confidence, based on a sample size of $n = 15$.

c. For 95% confidence, based on a sample size of $n = 25$.

d. For 99.9% confidence, based on a sample size of $n = 100$.

14. Consider the expense ratio and the total one-year rate of return on the W&R family of mutual funds in Table 11.3.6.

a. What percentage of the variation in rate of return is explained by expense ratio? Please provide both the name of the measure and its numeric value.

b. Based on this information, how different is the rate of return for Core EqC from what you would expect for a fund with its expense ratio? Please provide both the name of the measure and its numeric value.

c. Find the equation to predict rate of return from expense ratio. Please enter rates of return as percentage points (so 11.2% would be entered as 11.2).

d. Is the regression coefficient significant in the equation to predict rate of return from expense ratio? Please give the result and a brief justification.

15. In the presidential election of 2000, a number of events occurred between the initial vote count of November 7 and the count as certified by the Florida Secretary of State following counting of absentee ballots, a machine recount, and a Florida Supreme Court decision to require some hand recounts. Table 11.3.7 shows results in selected Florida counties indicating the initial count, the certified count, and the change in votes for Gore.

a. Based on this information, how strong is the relationship between the vote on November 7 and the change (from November 7 to the certified totals)? Please give both the name of the measure and its value.

b. Find the regression equation to predict the change from the number of votes cast on November 7.

c. Based on this information, how large a change would you expect to see in a county that recorded 250,000 votes for Gore on November 7?

d. How much larger (or smaller) was the change in Duval County as compared to what you would expect for the number of votes counted in Duval County on November 7, based on this information?

## TABLE 11.3.6 Expense Ratio and One-Year Rate of Return

| Fund | Expense Ratio | Return |
|---|---|---|
| AssetsS | 2.24 | 11.2% |
| Core EqC | 1.98 | 1.2 |
| HilncC | 2.17 | −1.0 |
| IntGthC | 2.37 | −41.6 |
| LtdTrmC | 1.81 | 9.2 |
| MuniC | 1.98 | 9.9 |
| ScTechC | 2.20 | −50.3 |
| SmCapGrC | 2.11 | −33.3 |

Source: Wall Street Journal, March 5, 2001, p. R17.

## TABLE 11.3.7 Votes for Albert Gore, Jr.

| County | November 7 | Certified | Change |
|---|---|---|---|
| Broward | 386,518 | 387,760 | 1,242 |
| Palm Beach | 268,945 | 269,754 | 809 |
| Dade | 328,702 | 328,867 | 165 |
| Volusia | 97,063 | 97,313 | 250 |
| Orange | 140,115 | 140,236 | 121 |
| Duval | 107,680 | 108,039 | 359 |
| Brevard | 97,318 | 97,341 | 23 |
| Hillsborough | 169,529 | 169,576 | 47 |

Source: www.cnn.com in Fall 2000.

   e.  How much of the variability in certified totals is explained by the initial count on November 7? Please give the name of the usual measure with its value.
   f.  Is there a significant relationship between the vote on November 7 and the certified total? Please give the result with justification.
16. How predictable are advertising budgets from year to year? Consider the 2008 and 2009 advertising spending of selected firms as reported in Table 11.3.8.

## TABLE 11.3.8 Total U.S. Advertising Spending (millions)

| Advertiser | Budget 2008 | Budget 2009 |
|---|---|---|
| AT&T | 3,073.0 | 2,797.0 |
| Citigroup | 970.7 | 560.4 |
| Coca-Cola Co | 752.1 | 721.5 |
| eBay | 429.1 | 365.2 |
| General Electric Co | 2,019.3 | 1,575.7 |
| Hewlett-Packard Co | 412.1 | 340.0 |
| MasterCard | 437.5 | 343.5 |
| Microsoft Corp | 802.3 | 1,058.6 |
| Procter & Gamble Co | 4,838.1 | 4,188.9 |
| Safeway | 420.2 | 430.6 |
| Sony Corp | 1,464.9 | 1,219.3 |
| U.S. Government | 1,195.6 | 1,034.1 |
| Walmart Stores | 1,659.8 | 1,729.5 |
| Walt Disney Co | 2,217.6 | 2,003.8 |

Source: *Advertising Age.* Data for 2009 accessed at http://adage.com/marketertrees2010/ on July 7, 2010. Data for 2008 accessed at http://adage.com/marketertrees09/ on July 20, 2010.

   a.  Summarize the strength of the year-to-year relationship in advertising budget by computing and interpreting the correlation coefficient and the coefficient of variation.
   b.  Draw a scatterplot with a least-squares line to predict spending in 2009 from spending in 2008.
   c.  Estimate the regression equation to predict spending in 2009 from spending in 2008.
   d.  Find the predicted value and residual value for Disney and interpret the residual value.
   e.  With this time period involving an economic recession, it is of interest to see whether advertising budgets were expanding (regression coefficient larger than 1) or shrinking (regression coefficient less than 1) from year to year. State your conclusion based on a hypothesis test of the regression coefficient against this reference value.
17. Gaining visibility for your products can be expensive, and television advertising during the Super Bowl is a good example, with a cost of nearly $2 million for a 30-second message. This high cost is due, in part, to the large number of Super Bowl viewers. Table 11.3.9 shows the market share and ad cost for selected broadcasts.
   a.  Estimate the dollar cost of an additional unit of market share from this data set.
   b.  Find the standard error for the additional unit cost from the previous part.
   c.  Find the 95% confidence interval for the dollar cost of an additional unit of market share.
   d.  Is there a significant relationship between market share and ad cost?
18. Consider the weight and price of gold coins from Table 11.3.10.
   a.  How strong is the association between weight and price for these coins? Please give both a number and its interpretation in words.

## TABLE 11.3.9 Market Share and 30-Second Advertising Cost (millions) for Selected Television Broadcasts

| Show | Share | Cost |
|---|---|---|
| Super Bowl | 40 | $1.90 |
| Academy Awards | 26 | 1.60 |
| NCAA Final Four | 16 | 0.90 |
| Prime-Time Winter Olympics | 15 | 0.60 |
| Grammy Awards | 17 | 0.57 |
| Barbara Walters Pre-Oscar | 12 | 0.55 |
| *Survivor* finale | 15 | 0.53 |
| Golden Globe Awards | 15 | 0.45 |
| *E.R.* | 17 | 0.37 |
| *Friends* | 17 | 0.35 |

Source: V. O'Connell, "Super Bowl Gets Competition," *Wall Street Journal,* January 28, 2002, p. B1.

## TABLE 11.3.10 Gold Coins

| Name | Weight (troy ounce) | Price |
|---|---|---|
| Maple Leaf | 1.00 | $1,197.00 |
| Mex. | 1.20 | 1,439.75 |
| Aus. | 0.98 | 1,142.50 |
| American Eagle | 1.00 | 1,197.00 |
| American Eagle | 0.50 | 629.50 |
| American Eagle | 0.25 | 340.50 |
| American Eagle | 0.10 | 158.25 |

b. Find the regression equation to predict price from weight.

c. Interpret the slope coefficient as a meaningful price.

d. Within approximately how many dollars are the predicted from the actual prices?

e. Find the 95% confidence interval for the slope coefficient.

f. Is the slope coefficient significantly different from 0? How do you know?

19. Are top executives of larger companies paid significantly more than those of smaller companies? Consider data on CEO pay (dollars) and market capitalization (the total market value of stock, in $ millions) for a sample of companies, as shown in Table 11.3.11.

a. How strong is the association between CEO pay and market capitalization? Please give both a number and its interpretation in words.

b. Find the regression equation to explain CEO pay using market capitalization.

c. Find and interpret the residual value for Red Lion Hotels, predicting CEO pay from market capitalization.

d. Find and interpret the 95% confidence interval for the slope coefficient.

e. Is there a significant relationship between CEO pay and market capitalization? How do you know?

## TABLE 11.3.11 CEO Pay and Market Capitalization

| Company Name | CEO | CEO Pay | Market Cap |
|---|---|---|---|
| Red Lion Hotels | Anupam Narayan | 688,085 | 43 |
| F5 Networks | John McAdam | 4,336,857 | 1,821 |
| InfoSpace | James F. Voelker | 6,224,477 | 261 |
| SeaBright Insurance Holdings | John G. Pasqualetto | 1,949,555 | 251 |
| Fisher Communications | Colleen B. Brown | 1,102,630 | 180 |
| Esterline Technologies | Robert W. Cremin | 5,063,367 | 1,125 |
| Washington Banking | John L. Wagner | 362,883 | 83 |
| Columbia Sportswear | Timothy P. Boyle | 827,799 | 1,197 |
| American Ecology | Stephen A. Romano | 501,210 | 370 |
| Cascade Financial | Carol K. Nelson | 302,380 | 66 |
| Merix | Michael D. Burger | 1,406,758 | 6 |
| Coinstar | David W. Cole | 1,994,972 | 551 |
| Intermec | Patrick J. Byrne | 2,888,301 | 820 |
| Jones Soda | Stephen C. Jones | 384,207 | 8 |
| Rentrak | Paul A. Rosenbaum | 501,828 | 124 |
| Coeur d'Alene Mines | Dennis E. Wheeler | 2,167,021 | 485 |
| Key Technology | David M. Camp, Ph.D. | 997,818 | 99 |
| Pacific Continental | Hal M. Brown | 528,341 | 180 |
| Cardiac Science | John R. Hinson | 838,023 | 172 |
| Washington Federal | Roy M. Whitehead | 720,415 | 1,316 |

**Source:** Data on market capitalization accessed March 27, 2010 at http://seattletimes.nwsource.com/flatpages/businesstechnology/2009northwestcompaniesdatabase .html. Data on CEO compensation accessed March 27, 2010 at http://seattletimes.nwsource.com/flatpages/businesstechnology/ceopay2008.html.

20. For each of the scatterplots in Figures 11.3.1 through 11.3.4, say whether the correlation is closest to 0.9, 0.5, 0.0, −0.5, or −0.9.

21. Consider the retail price of regular gasoline at selected locations and times shown in Table 11.3.12.
    a. How strong is the association between prices in 2010 and prices a year earlier? Please give both a number and its interpretation in words.
    b. Find the regression equation to predict the later from the earlier prices.
    c. Find the residual value for Florida (predicting later from earlier prices).
    d. Find the 95% confidence interval for the slope coefficient.
    e. Is the slope coefficient significantly different from 0? How do you know?

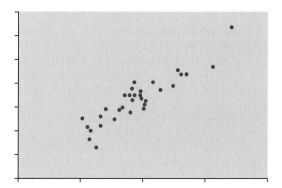

FIGURE 11.3.1

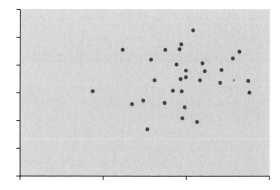

FIGURE 11.3.2

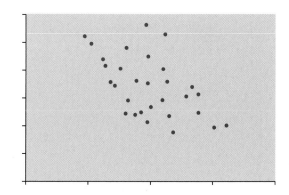

FIGURE 11.3.3

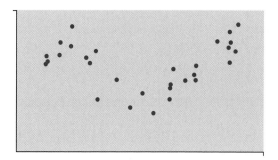

FIGURE 11.3.4

TABLE 11.3.12 Gasoline Prices

| Location | Price on 7/21/2010 | Price Year Before |
|---|---|---|
| Florida | 2.647 | 2.491 |
| Minnesota | 2.670 | 2.311 |
| Nebraska | 2.765 | 2.388 |
| Ohio | 2.705 | 2.311 |
| Texas | 2.557 | 2.305 |

Source: AAA's Daily Fuel Gauge Report, accessed at http://www.fuelgaugereport.com/sbsavg.html on July 21, 2010.

22. High salaries for presidents and high executives of charitable organizations have been in the news from time to time. Consider the information in Table 11.3.13 for the United Way in 10 major cities.
    a. What percent of the variation in presidents' salaries is explained by the fact that some raised more money per capita than others? Please give both the number and the usual statistical name for this concept.
    b. Find the regression equation to predict salary from money raised per capita.
    c. Find the residual value for Seattle, predicting salary from money raised per capita.
    d. Find the usual summary measure of the typical error made when using the regression equation to predict salaries from money raised per capita.
    e. Is there a significant relationship between president's salary and per capita money raised? How do you know?

23. Table 11.3.14 gives mailing-list size (thousands of names) and sales (thousands of dollars) for a group of catalogs.
    a. How strong is the association between these two variables? Find the appropriate summary measure and interpret it.
    b. Find the equation to predict sales from the size of the mailing list.
    c. What level of sales would you expect for a catalog mailed to 5,000 people?
    d. What percent of the variation in list size can be explained by the fact that some generated more sales than others?

### TABLE 11.3.13 Charitable Organizations

| City | Salary of President | Money Raised (per capita) | City | Salary of President | Money Raised (per capita) |
|---|---|---|---|---|---|
| Atlanta | $161,396 | $17.35 | Houston | $146,641 | $15.89 |
| Chicago | 189,808 | 15.81 | Kansas City | 126,002 | 23.87 |
| Cleveland | 171,798 | 31.49 | Los Angeles | 155,192 | 9.32 |
| Denver | 108,364 | 15.51 | Minneapolis | 169,999 | 29.84 |
| Detroit | 201,490 | 16.74 | Seattle | 143,025 | 24.19 |

### TABLE 11.3.14 Mailing Lists

| List Size | Sales | List Size | Sales |
|---|---|---|---|
| 168 | 5,178 | 249 | 7,325 |
| 21 | 2,370 | 43 | 2,449 |
| 94 | 3,591 | 589 | 15,708 |
| 39 | 2,056 | 41 | 2,469 |

    e. Is there a significant relationship between list size and sales? How do you know?

24. Table 11.3.15 compares short-term bond funds, showing the average maturity (in years until the fund's bonds mature) and the rate of return as a percentage.

### TABLE 11.3.15 Short-Term Bond Funds

| Fund | Maturity | Return |
|---|---|---|
| Strong Short-Term Bond Fund | 1.11 | 7.43% |
| DFA One-Year Fixed-Income Portfolio | 0.76 | 5.54 |
| Scudder Target Government Zero-Coupon 1990 | 2.3 | 5.01 |
| IAI Reserve Fund | 0.4 | 4.96 |
| Scudder Target Fund General 1990 | 1.9 | 4.86 |
| Vanguard Fixed-Income Short-Term Bond Portfolio | 2.3 | 4.86 |
| Criterion Limited-Term Institutional Trust | 1.3 | 4.8 |
| Franklin Series Trust Short-Int. U.S. Govt. | 2 | 4.64 |
| Benham Target Maturities Trust-Series 1990 | 2.3 | 4.62 |
| Delaware Treasury Reserves Investors Series | 2.84 | 4.35 |

    a. Find the correlation between maturity and return and interpret it.

    b. Find the least-squares regression equation to predict return from maturity.

    c. What rate of return would you expect for a fund with a current maturity of exactly one year?

    d. Find the standard error of prediction (for predicting "return" at a given maturity level) and explain its meaning.

    e. Is there a significant relationship between maturity and return? How do you know?

25. From Table 11.3.16, consider the daily production and the number of workers assigned for each of a series of days.

    a. Find the regression equation for predicting production from the number of workers.

    b. What is the estimated production amount attributable to a single additional worker?

    c. Draw a scatterplot of the data set with the regression line.

    d. Find the expected production and the residual value for the first data pair. Interpret both of these values in business terms.

    e. Find the standard error of the slope coefficient. What does this number indicate?

    f. Find the 95% confidence interval for the expected marginal value of an additional worker. (This is economics language for the slope.)

    g. Test at the 5% level to see if there is a significant relationship between production level and the number of workers.

### TABLE 11.3.16 Daily Production

| Workers | Production | Workers | Production |
|---|---|---|---|
| 7 | 483 | 9 | 594 |
| 6 | 489 | 9 | 575 |
| 7 | 486 | 6 | 464 |
| 8 | 562 | 9 | 647 |
| 8 | 568 | 8 | 595 |
| 9 | 559 | 6 | 499 |

26. Given the correlation $r = -0.603$ and the least-squares prediction equation $Y = 38.2 - 5.3X$, find the predicted value for $Y$ when $X$ is 15.

27. Given the correlation $r = 0.307$ and the least-squares prediction equation $Y = 55.6 + 18.2X$, find the predicted value for $Y$ when $X$ is $25.

28. One day your factory used $385 worth of electricity to produce 132 items. On a second day, $506 worth of electricity was consumed to produce 183 items. On a third day, the numbers were $261 and 105. How much electricity do you estimate it would take to produce 150 items?

29. Which of the following correlation coefficients corresponds to a moderately strong relationship with higher $X$ values associated with higher $Y$ values: $r = 1$, $r = 0.73$, $r = 0.04$, $r = -0.83$, or $r = -0.99$?

30. On Monday, your business produced 7 items which cost you $18. On Tuesday, you produced 8 costing $17. On Wednesday, you produced 18 costing $32. On Thursday, you produced 3 items costing $16. Using a linear regression model accounting for fixed and variable costs, give your estimate of Friday's costs for producing 10 items.

31. One weekend when you reduced prices 5%, your store had $58,000 worth of sales. The next weekend, with a 15% reduction, your sales were $92,000. The weekend after that, with a 17.5% reduction, your sales were $95,000. Based on all this information, how much would you expect to sell next weekend with a 10% price reduction?

32. Identify the structure of the scatterplot in Figure 11.3.5.

33. Identify the structure of the scatterplot in Figure 11.3.6.

34. Consider the international currency markets and, in particular, whether geographical proximity implies association with respect to market movements. Because

the United Kingdom kept the pound and did not convert to the euro, we can examine changes of these important currencies of Europe. Data on daily percentage changes in the price of a dollar, in each of these currencies, are shown in Table 11.3.17.

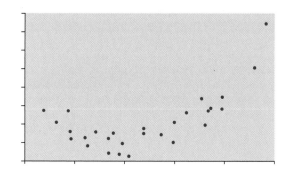

FIGURE 11.3.5

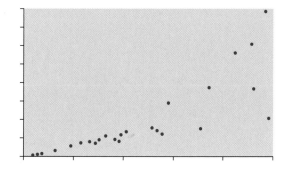

FIGURE 11.3.6

**TABLE 11.3.17  Change in the Price of a Dollar in Selected Currencies**

| Date | Euro | Yen | Pound |
|------|------|-----|-------|
| 6/21/2010 | -0.15% | -0.06% | -0.09% |
| 6/22/2010 | 0.02 | 0.41 | -0.09 |
| 6/23/2010 | 0.78 | -0.43 | 0.39 |
| 6/24/2010 | 0.15 | -0.50 | -0.78 |
| 6/25/2010 | -0.37 | -0.75 | -0.49 |
| 6/26/2010 | -0.06 | -0.06 | 0.04 |
| 6/27/2010 | -0.32 | -0.24 | -0.64 |
| 6/28/2010 | -0.01 | 0.01 | 0.00 |
| 6/29/2010 | 0.27 | 0.08 | -0.09 |
| 6/30/2010 | 1.04 | -0.78 | 0.02 |
| 7/1/2010 | -0.35 | -0.07 | 0.45 |
| 7/2/2010 | -0.91 | -0.79 | -0.16 |
| 7/3/2010 | -1.40 | -0.09 | -1.04 |
| 7/4/2010 | -0.20 | -0.01 | -0.05 |
| 7/5/2010 | 0.01 | 0.00 | 0.02 |
| 7/6/2010 | 0.21 | 0.02 | 0.30 |
| 7/7/2010 | -0.50 | -0.14 | -0.20 |
| 7/8/2010 | -0.04 | -0.41 | 0.12 |
| 7/9/2010 | -0.50 | 1.15 | -0.03 |
| 7/10/2010 | 0.05 | 0.25 | 0.21 |
| 7/11/2010 | 0.18 | 0.10 | 0.42 |
| 7/12/2010 | 0.01 | 0.01 | 0.00 |
| 7/13/2010 | 0.43 | 0.04 | 0.30 |
| 7/14/2010 | -0.38 | -0.28 | -0.57 |
| 7/15/2010 | -0.75 | 0.19 | -0.94 |
| 7/16/2010 | -0.84 | -0.91 | -0.66 |
| 7/17/2010 | -0.86 | -1.09 | -0.11 |
| 7/18/2010 | 0.13 | -0.25 | 0.43 |
| 7/19/2010 | 0.01 | -0.01 | 0.02 |
| 7/20/2010 | -0.17 | 0.26 | 0.20 |
| 7/21/2010 | 0.18 | 0.25 | 0.12 |

**Source:** Exchange rate information provided by www.OANDA.com – the currency site.

a. Create a scatterplot of the euro's against the pound's percentage changes.

b. Given that the pound and the euro are both used in Europe, we might expect that their values, with respect to the dollar, would tend to move together. To evaluate this, compute and interpret the correlation between the pound's percentage changes and the euro's.

c. Test at the 5% level to see if there is a significant link between movements in the pound and the euro.

d. On a day when the euro's percentage change is half of a percentage point, what would you expect the pound's percentage change to be?

35. Now consider also the daily percentage changes in the price of a dollar in Japanese yen from Table 11.3.17, along with the euro.

a. Create a scatterplot of the euro's against the yen's percentage changes.

b. Given that the yen and the euro are far apart geographically, we might expect that their values, with respect to the dollar, would tend to move

together only weakly if at all. To evaluate this, compute and interpret the correlation between the yen's percentage changes and the euro's.

c. Compare the correlation between the yen and the euro to the correlation between the pound and the euro, and interpret these results with respect to geographical closeness.

d. Test at the 5% level to see if there is a significant link between movements in the yen and the euro.

36. Many companies do not restrict themselves to operating inside any particular country, instead choosing to participate in the global economy, and stock market movements should reflect this reality. Consider data on the monthly percentage changes in stock market indexes (and one company), as shown in Table 11.3.18. In particular, the Hang Seng Index is for the Hong Kong Stock Exchange, the FTSE 100 Index is for companies on the London Stock Exchange, and the S&P 500 Index is primarily for the United States (companies are traded on either the NYSE Euronext or the NASDAQ Stock Exchanges).

**TABLE 11.3.18 Monthly Percentage Changes for Stock Market Indexes and for Microsoft**

| Date | Hang Seng Index | FTSE 100 Index | S&P 500 Index | Microsoft |
|---|---|---|---|---|
| 4/1/2010 | 4.32% | 2.56% | 3.61% | 5.39% |
| 3/1/2010 | 3.06 | 6.07 | 5.88 | 2.16 |
| 2/1/2010 | 2.42 | 3.20 | 2.85 | 2.21 |
| 1/4/2010 | −8.00 | −4.15 | −3.70 | −7.55 |
| 12/1/2009 | 0.23 | 4.28 | 1.78 | 3.66 |
| 11/2/2009 | 0.32 | 2.90 | 5.74 | 6.51 |
| 10/1/2009 | 3.81 | −1.74 | −1.98 | 7.81 |
| 9/1/2009 | 6.24 | 4.58 | 3.57 | 4.34 |
| 8/3/2009 | −4.13 | 6.52 | 3.36 | 5.39 |
| 7/2/2009 | 11.94 | 8.45 | 7.41 | −1.02 |
| 6/1/2009 | 1.14 | −3.82 | 0.02 | 13.74 |
| 5/1/2009 | 17.07 | 4.10 | 5.31 | 3.78 |
| 4/1/2009 | 14.33 | 8.09 | 9.39 | 10.28 |
| 3/2/2009 | 5.97 | 2.51 | 8.54 | 13.79 |
| 2/2/2009 | −3.51 | −7.70 | −10.99 | −4.93 |
| 1/2/2009 | −7.71 | −6.42 | −8.57 | −12.06 |
| 12/1/2008 | 3.59 | 3.41 | 0.78 | −3.81 |
| 11/3/2008 | −0.58 | −2.04 | −7.48 | −8.85 |
| 10/2/2008 | −22.47 | −10.71 | −16.94 | −16.33 |
| 9/1/2008 | −15.27 | −13.02 | −9.08 | −2.20 |
| 8/1/2008 | −6.46 | 4.15 | 1.22 | 6.51 |

### TABLE 11.3.18 Monthly Percentage Changes for Stock Market Indexes and for Microsoft—cont'd

| Date | Hang Seng Index | FTSE 100 Index | S&P 500 Index | Microsoft |
|---|---|---|---|---|
| 7/2/2008 | 2.85 | −3.80 | −0.99 | −6.50 |
| 6/2/2008 | −9.91 | −7.06 | −8.60 | −2.86 |
| 5/1/2008 | −4.75 | −0.56 | 1.07 | −0.33 |
| 4/1/2008 | 12.72 | 6.76 | 4.75 | 0.48 |

**Source:** Accessed at finance.yahoo.com on April 15, 2010.

a. Draw a scatterplot of the S&P 500 Index against the FTSE 100 Index and a scatterplot of the S&P 500 Index against the Hang Seng Index, and comment on any association you see in these country indexes.

b. Find and interpret the correlation between the S&P 500 Index and the FTSE 100 Index.

c. What percentage of the variation in the S&P 500 Index is explained by variations in the Hang Seng Index? How does this compare to the percentage of the variation in the S&P 500 Index explained by the FTSE 100? Base your answers on two separate regressions.

d. Test to see if there is a significant relationship between the S&P 500 and the FTSE 100 Indexes, and if there is a significant relationship between the S&P 500 and the Hang Seng.

e. Write a paragraph explaining and interpreting your results.

37. Microsoft is a company that sells its products in many countries all over the world. Use the data from Table 11.3.18 to explore how its market price movements relate to more general movements in the global economy.

a. Is Microsoft significantly related to the S&P 500 Index of the U.S. stock market? Please support your result with the *t* statistic and a *p*-value statement.

b. Is Microsoft significantly related to the FTSE 100 Index of the London Stock Exchange? Please support your result with the *t* statistic and a *p*-value statement.

c. Is Microsoft significantly related to the Hang Seng Index of the Hong Kong Stock Exchange? Please support your result with the *t* statistic and a *p*-value statement.

d. Compare the coefficients of determination from three separate regressions, where each regression predicts Microsoft stock movements from one of the indexes, to make a statement about Microsoft and the global economy.

38. Your firm is having a quality problem with the production of plastic automotive parts: There are too many defectives. One of your engineers thinks the reason is that the temperature of the process is not controlled carefully enough. Another engineer is sure that it's the assembly line being shut down too often for unrelated reasons. You have decided to analyze the problem and have come up with figures for the percent defective each day recently, the standard deviation of temperature measured hourly each day (as a measure of temperature control), and the number of assembly line stoppages each day. The raw data set is shown in Table 11.3.19.

a. Find the correlation of the defect rate with the temperature variability.

b. Find the correlation of the defect rate with stoppages.

### TABLE 11.3.19 Defects and Possible Causes

| Defect Rate | Temperature Variability | Stoppages | Defect Rate | Temperature Variability | Stoppages |
|---|---|---|---|---|---|
| 0.1% | 11.94 | 5 | 0.0% | 10.10 | 2 |
| 0.1 | 9.33 | 4 | 5.2 | 13.08 | 2 |
| 8.4 | 21.89 | 0 | 4.9 | 17.19 | 0 |
| 0.0 | 8.32 | 1 | 0.1 | 10.76 | 1 |
| 4.5 | 14.55 | 0 | 6.8 | 13.73 | 3 |
| 2.6 | 12.08 | 8 | 4.8 | 12.42 | 2 |
| 3.2 | 12.16 | 0 | 0.0 | 12.83 | 2 |
| 0.0 | 12.56 | 2 | 0.9 | 5.78 | 5 |

c. Which possible cause, temperature variability or stoppages, accounts for more of the variation in defect rate from day to day? How do you know?

d. Test each of these correlations for statistical significance.

e. Draw a scatterplot of defect rate against stoppages. Write a brief paragraph interpreting the scatterplot and correlation.

f. Draw a scatterplot of defect rate against temperature variability. Write a brief paragraph interpreting the scatterplot and correlation.

g. Write a paragraph summarizing what you have learned and proposing a plan for action.

## Database Exercises

Refer to the employee database in Appendix A.

1. Consider annual salary as the Y variable and experience as the X variable.
   a. Draw a scatterplot and describe the relationship.
   b. Find the correlation coefficient. What does it tell you? Is it appropriate, compared to the scatterplot?
   c. Find the least-squares regression line to predict Y from X and draw it on a scatterplot of the data.
   d. Find the standard error of estimate. What does it tell you?
   e. Find the standard error of the slope coefficient.
   f. Find the 95% confidence interval for the slope coefficient.
   g. Test at the 5% level to see if the slope is significantly different from 0. Interpret the result.
   h. Test at the 1% level to see if the slope is significantly different from 0.
   i. Test at the 5% level to see if the correlation coefficient is significantly different from 0.

2.* a. What fraction of the variation in salaries can be explained by the fact that some employees have more experience than others?
   b. What salary would you expect for an individual with eight years of experience?
   c. Find the 95% confidence interval for the salary of a new individual (from the same population from which the data were drawn) who has eight years of experience.
   d. Find the 95% confidence interval for the mean salary for those individuals in the population who have eight years of experience.

3. a. What salary would you expect for an individual with three years of experience?
   b. Find the 95% confidence interval for the salary of a new individual (from the same population from which the data were drawn) who has three years of experience.
   c. Find the 95% confidence interval for the mean salary of those individuals in the population who have three years of experience.

4. a. What salary would you expect for an individual with no (zero years of) experience?

b. Find the 95% confidence interval for the salary of a new individual (from the same population from which the data were drawn) who has no experience.

c. Find the 95% confidence interval for the mean salary of those individuals in the population who have no experience.

5. Consider annual salary as the Y variable and age as the X variable.
   a. Draw a scatterplot and describe the relationship.
   b. Find the correlation coefficient. What does it tell you? Is it appropriate, compared to the scatterplot?
   c. Find the least-squares regression line to predict Y from X and draw it on a scatterplot of the data.
   d. Find the standard error of estimate. What does it tell you?
   e. Find the standard error of the slope coefficient.
   f. Find the 95% confidence interval for the slope coefficient.
   g. Test at the 5% level to see if the slope is significantly different from 0. Interpret the result.
   h. Test at the 1% level to see if the slope is significantly different from 0.

6. a. What fraction of the variation in salaries can be explained by the fact that some employees are older than others?
   b. What salary would you expect for a 42-year-old individual?
   c. Find the 95% confidence interval for the salary of a new individual (from the same population from which the data were drawn) who is 42 years old.
   d. Find the 95% confidence interval for the mean salary of all 42-year-olds in the population.

7. a. What salary would you expect for a 50-year-old individual?
   b. Find the 95% confidence interval for a new individual (from the same population from which the data were drawn) who is 50 years old.
   c. Find the 95% confidence interval for the mean salary of all 50-year-olds in the population.

8. Consider experience as the Y variable and age as the X variable.
   a. Draw a scatterplot and describe the relationship.
   b. Find the correlation coefficient. What does it tell you? Is it appropriate, compared to the scatterplot?
   c. Find the least-squares regression line to predict Y from X and draw it on a scatterplot of the data.
   d. Find the standard error of estimate. What does it tell you?
   e. Find the standard error of the slope coefficient.
   f. Find the 95% confidence interval for the slope coefficient.
   g. Test at the 5% level to see if the slope is significantly different from 0. Interpret the result.
   h. Test at the 1% level to see if the slope is significantly different from 0.

9. a. What fraction of the variation in experience can be explained by the fact that some employees are older than others?

**b.** How much experience would you expect for a 42-year-old individual?

**c.** Find the 95% confidence interval for the experience of a new individual (from the same population from which the data were drawn) who is 42 years old.

**d.** Find the 95% confidence interval for the mean experience of all 42-year-olds in the population.

**10. a.** How much experience would you expect for a 50-year-old individual?

**b.** Find the 95% confidence interval for the experience of a new individual (from the same population from which the data were drawn) who is 50 years old.

**c.** Find the 95% confidence interval for the mean experience of all 50-year-olds in the population.

## Projects

Find a bivariate data set relating to your work or business interests on the Internet, in a newspaper, or in a magazine, with a sample size of $n = 15$ or more.

**a.** Give your choice of dependent variable ($Y$) and independent variable ($X$) and briefly explain your reasons.

**b.** Draw a scatterplot and comment on the relationship.

**c.** Compute the correlation coefficient and briefly interpret it.

**d.** Square the correlation coefficient and briefly interpret it.

**e.** Compute the least-squares regression equation and draw the line on a scatterplot of your data.

**f.** For two elementary units in your data set, compute predicted values for $Y$ and residuals.

**g.** Find a confidence interval for the slope coefficient.

**h.** Test whether or not anything is being explained by your regression equation.

**i.** Choose a value of $X$. Find the expected value of $Y$ for this $X$. Find the confidence interval for the $Y$ value of an individual with this $X$ value. Find the confidence interval for the population mean $Y$ value for individuals with this $X$ value. Summarize and interpret these results.

**j.** Comment on what you have learned by applying correlation and regression analysis to your data set.

## Case

### Just One More Production Step: Is It Worthwhile?

The "techies" (scientists) in the laboratory have been lobbying you, and management in general, to include just one more laboratory step. They think it's a good idea, although you have some doubt because one of them is known to be good friends with the founder of the startup biotechnology company that makes the reagent used in the reaction. But if adding this step works as expected, it could help immensely in reducing production costs. The trouble is, the test results just came back and they don't look so good. Discussion at the upcoming meeting between the technical staff and management will be spirited, so you've decided to take a look at the data.

Your firm is anticipating government approval from the Food and Drug Administration (FDA) to market a new medical diagnostic test made possible by monoclonal antibody technology, and you are part of the team in charge of production. Naturally, the team has been investigating ways to increase production yields or lower costs.

The proposed improvement is to insert yet another reaction as an intermediate purifying procedure. This is good because it focuses resources down the line on the particular product you want to produce. But it shares the problem of any additional step in the laboratory: one more manipulation, one more intervention, one more way for something to go wrong. In this particular case, it has been suggested that, while small amounts of the reagent may be helpful, trying to purify too well will actually decrease the yield and increase the cost.

The design of the test was to have a series of test production runs, each with a different amount of purifier, including one test run with the purification step omitted entirely (i.e., 0 purifier). The order of the tests was randomized so that any time trends would not be mistakenly interpreted as being due to purification. Here are the data and the regression results:

| Amount of Purifier | Observed Yield | Amount of Purifier | Observed Yield |
|---|---|---|---|
| 0 | 13.39 | 6 | 37.07 |
| 1 | 11.86 | 7 | 51.07 |
| 2 | 27.93 | 8 | 51.69 |
| 3 | 35.83 | 9 | 31.37 |
| 4 | 28.52 | 10 | 21.26 |
| 5 | 41.21 | | |

### Summary Output

Regression Statistics

| | |
|---|---|
| Multiple R | 0.516 |
| R Square | 0.266 |
| Adjusted R Square | 0.184 |
| Standard Error | 12.026 |
| Observations | 11 |

#### ANOVA

| | df | SS | MS | F | Significance F |
|---|---|---|---|---|---|
| Regression | 1 | 471.339 | 471.339 | 3.259 | 0.105 |
| Residual | 9 | 1301.553 | 144.615 | | |
| Total | 10 | 1772.872 | | | |

| | Coefficients | Std Err | t | p | Low 95% | Up 95% |
|---|---|---|---|---|---|---|
| Intercept | 21.577 | 6.783 | 3.181 | 0.011 | 6.232 | 36.922 |
| Purifier | 2.070 | 1.147 | 1.805 | 0.105 | –0.524 | 4.664 |

### Discussion Questions

1. Does the amount of purifier have a significant effect on yield according to this regression analysis? Based on this alone, would you be likely to recommend including a purifying step in the production process?

2. What would you recommend? Are there any other considerations that might change your mind?

## Chapter 12

# Multiple Regression

## Predicting One Variable from Several Others

### Chapter Outline

The world is multivariate. In realistic business problems, you have to consider data on more than just one or two factors. But don't despair; the next step, *multiple regression*, is a relatively easy procedure that will build on your ability to deal with the simpler cases of univariate and bivariate data. In fact, all of the basic ideas are in place already: average, variability, correlation, prediction, confidence intervals, and hypothesis tests.

Explaining or predicting a single $Y$ variable from *two or more $X$* variables is called **multiple regression**. Predicting a single $Y$ variable from a single $X$ variable is called *simple regression* and was covered in the preceding chapter. The goals when using multiple regression are the same as

with simple regression. Here is a review of those goals with some examples:

**One:** Describing and understanding the relationship.

a. Consider the relationship between salary ($Y$) and some basic characteristics of employees, such as gender ($X_1$, represented as 0 or 1 to distinguish male and female), years of experience ($X_2$), and education ($X_3$). Describing and understanding how these $X$ factors influence $Y$ could provide important evidence in a gender discrimination lawsuit. The regression coefficient for gender would give an estimate of how large the salary gap is between men and women after adjustment for age and experience. Even if your firm is not currently being sued, you might want to run such a multiple regression analysis so that any small problems can be fixed before they become large ones.

**b.** If your firm makes bids on projects, then for the projects you win, you will have data available for the actual cost ($Y$), the estimated direct labor cost ($X_1$), the estimated materials cost ($X_2$), and supervisory function costs ($X_3$). Suppose you suspect your bids are unrealistically low. By figuring out the relationship between actual cost and the estimates made earlier during the bidding process, you will be able to tell which, if any, of the estimates are systematically too low or too high in terms of their contribution to the actual cost.

**Two:** Forecasting (predicting) a new observation.

**a.** An in-depth understanding of your firm's cost structure would be useful for many purposes. For instance, you would have a better idea of how much extra to budget during the rush season (e.g., for overtime). If the business is changing, you may be able to anticipate the effects of changes on your costs. One way to understand your cost structure is through multiple regression of costs ($Y$) on each potentially useful factor you can think of, such as the number of items produced ($X_1$), the number of workers ($X_2$), and the amount of overtime ($X_3$). Results of an analysis such as this can tell you much more than just "add a bunch for overtime." It can respond to hidden costs that tend to increase along with overtime, giving you the best predictions of actual cost based on the available information.

**b.** Your firm's monthly sales (a time series) might be explained by an overall trend with seasonal effects. One way to analyze and forecast would be to use multiple regression to explain sales ($Y$) based on a trend (for example, $X_1 = 1, 2, 3, \ldots$, indicating months since the start) and a variable for each month (for example, $X_2$ would be 1 for January and 0 otherwise, $X_3$ would represent February, and so forth). You could use multiple regression to forecast sales several months ahead, as well as to understand your long-term trend and to see which months tend to be higher than others.

**Three:** Adjusting and controlling a process.

**a.** The wood pulp goes in one end, and paper comes flying out of the other, ready to be rolled and shipped. How do you control such a large piece of machinery? Just reading the instruction manual is often not good enough; it takes experience to get the thickness just right and to dry it sufficiently without wasting energy dollars. When this "experience"

consists of data, a multiple regression analysis can help you find out which combination of adjustments (the $X$ variables) produces the result (the $Y$ variable) you want.

**b.** "Hedging" in securities markets consists of setting up a portfolio of securities (often futures and options) that matches the risk of some asset as closely as possible. If you hold an inventory, you should consider hedging its risk. Banks use treasury futures and options contracts to hedge the interest rate risk of their deposit accounts and loans. Agricultural industries use hedging to decrease their risk due to commodity price fluctuations. The process of choosing a hedge portfolio may be accomplished using multiple regression analysis. Based on past data, you would attempt to explain the price movements of your asset ($Y$) by the price movements of securities ($X_1$, $X_2$, and so forth). The regression coefficients would tell you how much of each security to include in the hedge portfolio to get rid of as much risk as possible. You would be using multiple regression to adjust and control your risk exposure.

## 12.1 INTERPRETING THE RESULTS OF A MULTIPLE REGRESSION

What will the computer analysis look like, and how can you interpret it? First of all, we will provide an overview of the input and main results. A more detailed explanation will follow.

Let $k$ stand for the number of explanatory variables ($X$ variables); this can be any manageable number. Your elementary units are often called *cases*; they might be customers, firms, or items produced.[1] The input data for a typical multiple regression analysis is shown in Table 12.1.1.

There will be an **intercept** or **constant term**, $a$, that gives the predicted value for $Y$ when *all* $X$ variables are 0. Also, there will be a **regression coefficient** for each $X$ variable, indicating the effect of that $X$ variable on $Y$ while holding the other $X$ variables fixed; the regression coefficient $b_j$ for the $j$th $X$ variable indicates how much larger you expect $Y$ to be for a case that is identical to another except for being one unit larger in $X_j$. Taken together, these regression coefficients give you the **prediction equation** or **regression equation**, (Predicted $Y$) $= a + b_1 X_1 + b_2 X_2 + \cdots + b_k X_k$, which may be used for prediction or control. These coefficients ($a$, $b_1$, $b_2$, $\ldots$, $b_k$) are traditionally computed using the method of least squares, which minimizes

---

1. For technical reasons, you must have at least one more case than you have $X$ variables; that is, $n \geq k + 1$. For practical reasons, you should probably have many more.

**TABLE 12.1.1 Input Data for a Multiple Regression**

| | Y (dependent variable to be explained) | $X_1$ (first independent or explanatory variable) | $X_2$ (second independent or explanatory variable) | ... | $X_k$ (last independent or explanatory variable) |
|---|---|---|---|---|---|
| Case 1 | 10.9 | 2.0 | 4.7 | ... | 12.5 |
| Case 2 | 23.6 | 4.0 | 3.4 | ... | 12.3 |
| . | . | . | . | . | . |
| . | . | . | . | . | . |
| . | . | . | . | . | . |
| Case n | 6.0 | 0.5 | 3.1 | ... | 7.0 |

the sum of the squared prediction errors. The **prediction errors** or **residuals** are defined as $Y$ – (Predicted $Y$).

Just as for simple regression, with only one $X$, the **standard error of estimate**, $S_e$, indicates the approximate size of the prediction errors. Also as for simple regression, $R^2$ is the **coefficient of determination**, which indicates the percentage of the variation in $Y$ that is "explained by" or "attributed to" all of the $X$ variables.[2]

Inference will begin with an overall test, called the **$F$ test**, to see if the $X$ variables explain a significant amount of the variation in $Y$. If your regression is *not* significant, you are not permitted to go further (essentially there is nothing there, end of story). On the other hand, if the regression *is* significant, you may proceed with statistical inference using **$t$ tests for individual regression coefficients**, which show you whether an $X$ variable has a significant impact on $Y$ *holding all other X variables fixed*. Confidence intervals and hypothesis tests for an individual regression coefficient will be based on its standard error, of course. There is a standard error for each regression coefficient; these are denoted $S_{b_1}, S_{b_2}, \ldots, S_{b_k}$. Table 12.1.2 shows a list of the results of a multiple regression analysis.

### Example
*Magazine Ads*

The price of advertising is different from one consumer magazine to another. What causes these differences in price? Probably something relating to the value of the ad to the advertiser. Magazines that reach more readers (all else equal) should be able to charge more for an ad. Also,

magazines that reach a better-paid reading audience should probably be able to charge more. Although there may be other important factors, let's look at these two together with one more, gender difference, to see if magazines charge more based on the percentage of men or women among the readers. Multiple regression will provide some answers and can help explain the impact of audience size, income, and gender on advertising prices.

Table 12.1.3 shows the multivariate data set to be analyzed. The page costs for a "four-color, one-page ad run once" will be the $Y$ variable to be explained. The explanatory variables are $X_1$, audience (projected readers, in thousands); $X_2$, percent male among the projected readership; and $X_3$, median household income. The sample size is $n = 45$.

Table 12.1.4 shows the computer output from a multiple regression analysis from MINITAB®. Other statistical software packages will give much the same basic information. For example, Excel® can perform multiple regression (look in the Data Ribbon for Data Analysis in the Analysis area;[3] then select Regression). Figure 12.1.1a shows Excel's regression dialog box and Figure 12.1.lb shows Excel's regression results. These results will be interpreted in the following sections.

---

3. If you cannot find Data Analysis in the Analysis area of Excel's Data Ribbon, click the Office button at the very top left, choose Excel Options at the bottom, select Add-Ins at the left, click Go at the bottom, and make sure the Analysis ToolPak is checked. If the Analysis ToolPak was not installed when Excel® was installed on your computer, you may need to run the Microsoft Office Setup Program.

### Regression Coefficients and the Regression Equation

The intercept or constant term, $a$, and the regression coefficients $b_1$, $b_2$, and $b_3$ are found by the computer using the method of least squares. Among all possible regression equations with various values for these coefficients, these

---

2. However, it is not just the square of the correlation of $Y$ with *one* of the $X$ variables. Instead, it is the square of $r$, the correlation of $Y$ with the least-squares predictions (based on the regression equation), which uses *all* of the $X$ variables.

## 12.1.2 Results of a Multiple Regression Analysis

| Name | Result | Description |
|---|---|---|
| Intercept or constant term | $a$ | Predicted value for $Y$ when every $X$ is 0 |
| Regression coefficients | $b_1, b_2, \ldots, b_k$ | The effect of each $X$ on $Y$, holding all other $X$ variables constant |
| Prediction equation or regression equation | Predicted $Y = a + b_1X_1 + b_2X_2 + \cdots + b_kX_k$ | Predicted value for $Y$ given the values for the $X$ variables |
| Prediction errors or residuals | $Y -$ Predicted $Y$ | Error made by using the prediction equation instead of the actual value of $Y$ for each case |
| Standard error of estimate | $S_e$ or $S$ | Approximate size of prediction errors (typical difference between actual $Y$ and predicted $Y$ from regression equation) |
| Coefficient of determination | $R^2$ | Percentage of variability in $Y$ explained by the $X$ variables as a group |
| $F$ test | Significant or not significant | Tests whether the $X$ variables, as a group, can predict $Y$ better than just randomly; essentially a test to see if $R^2$ is larger than pure randomness would produce |
| $t$ tests for individual regression coefficients | Significant or not significant, for each $X$ variable | Tests whether a particular $X$ variable has an effect on $Y$, holding the other $X$ variables constant; should be performed only if the $F$ test is significant |
| Standard errors of the regression coefficients | $S_{b_1}, S_{b_2}, \ldots, S_{b_k}$ | Indicates the estimated sampling standard deviation of each regression coefficient; used in the usual way to find confidence intervals and hypothesis tests for individual regression coefficients |
| Degrees of freedom for standard errors of the regression coefficients | $n - k - 1$ | Used to find the $t$ table value for confidence intervals and hypothesis tests for individual regression coefficients |

## TABLE 12.1.3 Advertising Costs and Characteristics of Magazines

|  | Y<br>Page Costs<br>(color ad) | X₁<br>Audience<br>(thousands) | X₂<br>Percent<br>Male | X₃<br>Median Household<br>Income |
|---|---|---|---|---|
| AAA Westways | $53,310 | 8,740 | 47.0% | $92,600 |
| AARP The Magazine | 532,600 | 35,721 | 39.7 | 58,990 |
| Allure | 131,721 | 6,570 | 9.0 | 65,973 |
| Architectural Digest | 119,370 | 4,988 | 42.0 | 100,445 |
| Audubon | 25,040 | 1,924 | 39.0 | 73,446 |
| Better Homes & Gardens | 468,200 | 38,946 | 19.6 | 67,637 |
| Bicycling | 55,385 | 2,100 | 73.2 | 74,175 |
| Bon Appétit | 143,612 | 8,003 | 25.0 | 91,849 |
| Brides | 82,041 | 5,800 | 10.0 | 56,718 |
| Car and Driver | 187,269 | 2,330 | 72.8 | 141,873 |
| Conde Nast Traveler | 118,657 | 3,301 | 45.0 | 110,037 |
| Cosmopolitan | 222,400 | 18,331 | 15.8 | 57,298 |
| Details | 69,552 | 1,254 | 69.0 | 82,063 |
| Discover | 57,300 | 7,140 | 61.0 | 61,127 |

**TABLE 12.1.3 Advertising Costs and Characteristics of Magazines—cont'd**

| | Y<br>Page Costs<br>(color ad) | X₁<br>Audience<br>(thousands) | X₂<br>Percent<br>Male | X₃<br>Median Household<br>Income |
|---|---|---|---|---|
| *Every Day with Rachael Ray* | 139,000 | 6,860 | 12.0% | 70,162 |
| *Family Circle* | 254,600 | 21,062 | 10.0 | 52,502 |
| *Fitness* | 142,300 | 6,196 | 23.7 | 70,442 |
| *Food & Wine* | 86,000 | 8,034 | 37.0 | 84,750 |
| *Golf Magazine* | 141,174 | 5,608 | 83.0 | 96,659 |
| *Good Housekeeping* | 344,475 | 24,484 | 12.2 | 60,981 |
| *GQ (Gentlemen's Quarterly)* | 143,681 | 6,360 | 77.0 | 75,103 |
| *Kiplinger's Personal Finance* | 54,380 | 2,407 | 62.0 | 101,900 |
| *Ladies' Home Journal* | 254,000 | 13,865 | 5.9 | 55,249 |
| *Martha Stewart Living* | 157,700 | 11,200 | 11.0 | 74,436 |
| *Midwest Living* | 125,100 | 3,913 | 25.0 | 69,904 |
| *Money* | 201,800 | 7,697 | 63.0 | 98,057 |
| *More* | 148,400 | 1,389 | 0.0 | 93,550 |
| *O, The Oprah Magazine* | 150,730 | 15,575 | 12.0 | 72,953 |
| *Parents* | 167,800 | 15,300 | 17.6 | 59,616 |
| *Prevention* | 134,900 | 10,403 | 16.0 | 66,799 |
| *Reader's Digest* | 171,300 | 31,648 | 38.0 | 62,076 |
| *Readymade* | 32,500 | 1,400 | 16.0 | 52,894 |
| *Road & Track* | 109,373 | 1,492 | 75.0 | 143,179 |
| *Self* | 166,773 | 6,078 | 6.6 | 85,671 |
| *Ser Padres* | 74,840 | 3,444 | 28.0 | 37,742 |
| *Siempre Mujer* | 48,300 | 1,710 | 17.0 | 46,041 |
| *Sports Illustrated* | 352,800 | 21,000 | 80.0 | 72,726 |
| *Teen Vogue* | 115,897 | 5,829 | 9.0 | 56,608 |
| *The New Yorker* | 135,263 | 4,611 | 49.9 | 91,359 |
| *Time* | 287,440 | 20,642 | 52.0 | 73,946 |
| *TV Guide* | 134,700 | 14,800 | 45.0 | 49,850 |
| *Vanity Fair* | 165,600 | 6,890 | 23.0 | 74,765 |
| *Vogue* | 151,133 | 12,030 | 12.0 | 68,667 |
| *Wired* | 99,475 | 2,789 | 75.5 | 91,056 |
| *Woman's Day* | 259,960 | 20,325 | 0.0 | 58,053 |
| Average | 160,397 | 10,226 | 34.7 | 75,598 |
| Standard deviation | 105,639 | 9,298 | 25.4 | 22,012 |

Sample size: *n* = 45.

**Source:** Individual magazine websites, accessed January and July 2010.

**TABLE 12.1.4** MINITAB Computer Output for Multiple Regression Analysis of Magazine Ads

**The regression equation is**

Page = –22385 + 10.5 Audience – 20779 Male + 1.09 Income

| Predictor | Coeff | SE Coeff | t | p |
|---|---|---|---|---|
| Constant | –22385 | 36060 | –0.62 | 0.538 |
| Audience | 10.5066 | 0.9422 | 11.15 | 0.000 |
| Male | –20779 | 37961 | –0.55 | 0.587 |
| Income | 1.0920 | 0.4619 | 2.36 | 0.023 |

$S = 53812.4$    R-Sq = 75.8%    R-Sq(adj) = 74.1%

**Analysis of Variance**

| Source | DF | SS | MS | F | p |
|---|---|---|---|---|---|
| Regression | 3 | 3.72299E+11 | 1.24100E+11 | 42.86 | 0.000 |
| Residual Error | 41 | 1.18727E+11 | 2895770619 | | |
| Total | 44 | 4.91025E+11 | | | |

| Source | DF | Seq SS |
|---|---|---|
| Audience | 1 | 3.54451E+11 |
| Male | 1 | 1662287268 |
| Income | 1 | 16185501423 |

**Unusual Observations**

| Obs | Audience | Page | Fit | SE Fit | Residual | St Resid |
|---|---|---|---|---|---|---|
| 1 | 8740 | 53310 | 160794 | 10265 | –107484 | –2.03R |
| 2 | 35721 | 532600 | 409087 | 24551 | 123513 | 2.58R |
| 31 | 31648 | 171300 | 370017 | 20995 | –198717 | –4.01R |

R denotes an observation with a large standardized residual.

are the ones that make the sum of squared prediction errors the smallest possible for these particular magazines. The regression equation, or prediction equation, is

$$\text{Predicted page costs} = a + b_1 X_1 + b_2 X_2 + b_3 X_3$$
$$= -22{,}385 + 10.50658 \,(\text{Audience})$$
$$- 20{,}779 \,(\text{Percent male})$$
$$+ 1.09198 \,(\text{Median income})$$

The intercept, $a = -\$22{,}385$, suggests that the typical charge for a one-page color ad in a magazine with no paid subscribers, no men among its readership, and no income among its readers is –\$22,385, suggesting that such an ad has negative value. However, there are no such magazines in this data set, so it may be best to view the intercept, $a$, as a possibly helpful step in getting the best predictions and not interpret it too literally.

## Interpreting the Regression Coefficients

The regression coefficients are interpreted as the effect of each variable on page costs, if all of the other explanatory variables are held constant. This is often "adjusting for" or "controlling for" the other explanatory variables. Because of this, the regression coefficient for an $X$ variable may change (sometimes considerably) when other $X$ variables are included or dropped from the analysis. In particular, each regression coefficient gives you the average increase in page costs per increase of 1 in its $X$ variable, where 1 refers to one unit of whatever that $X$ variable measures.

The regression coefficient for audience, $b_1 = 10.50658$, indicates that, all else equal, a magazine with an extra 1,000 readers (since $X_1$ is given in thousands in the original data set) will charge an extra \$10.51 (on average) for a one-page ad. You can also think of it as meaning that each extra reader is worth \$0.0105, just over a penny per person. So if a different magazine had the same percent male readership and the same median income but 3,548 more people in its audience, you would expect the page costs to be $10.50658 \times 3.548 = \$37.28$ higher (on average) due to the larger audience.

The regression coefficient for percent male, $b_2 = -20,779$, indicates that, all else equal, a magazine with an extra 1% of male readers would charge \$208 less (on average) for a full-page color ad, where we have divided 20,779 by 100 because 1 percentage point is one hundredth of the unit for percentages (because 100% has the value 1). This suggests that women readers are more valuable than men readers. Statistical inference will confirm or deny this hypothesis by comparing the size of this effect (i.e., −\$208) to what you might expect to find here due to random coincidence alone.

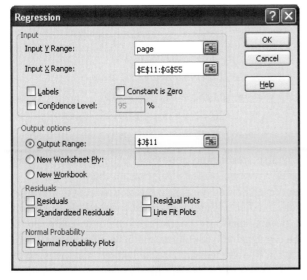

**FIGURE 12.1.1a** Excel's® Regression dialog box. You may give a range name for the $Y$ variable ("page" here), but the $X$ variables must be in adjacent columns: You might drag the mouse cursor across the columns (just the data, not including any titles above them) or type in the cell address.

| | J | K | L | M | N | O | P |
|---|---|---|---|---|---|---|---|
| 11 | SUMMARY OUTPUT | | | | | | |
| 12 | | | | | | | |
| 13 | *Regression Statistics* | | | | | | |
| 14 | Multiple R | 0.871 | | | | | |
| 15 | R Square | 0.758 | | | | | |
| 16 | Adjusted R Square | 0.741 | | | | | |
| 17 | Standard Error | 53812.365 | | | | | |
| 18 | Observations | 45 | | | | | |
| 19 | | | | | | | |
| 20 | ANOVA | | | | | | |
| 21 | | *df* | *SS* | *MS* | *F* | *Significance F* | |
| 22 | Regression | 3 | 3.72299E+11 | 1.24E+11 | 42.855 | 1.04673E-12 | |
| 23 | Residual | 41 | 1.18727E+11 | 2.9E+09 | | | |
| 24 | Total | 44 | 4.91025E+11 | | | | |
| 25 | | | | | | | |
| 26 | | *Coefficients* | *Standard Error* | *t Stat* | *P-value* | *Lower 95%* | *Upper 95%* |
| 27 | Intercept | -22385.094 | 36059.902 | -0.621 | 0.538 | -95209.542 | 50439.353 |
| 28 | X Variable 1 | 10.507 | 0.942 | 11.151 | 5.46E-14 | 8.604 | 12.409 |
| 29 | X Variable 2 | -20779.017 | 37961.310 | -0.547 | 0.587 | -97443.437 | 55885.404 |
| 30 | X Variable 3 | 1.092 | 0.462 | 2.364 | 0.023 | 0.159 | 2.025 |
| 31 | | | | | | | |

**FIGURE 12.1.1b** Excel's® regression results for magazine ads.

The regression coefficient for income, $b_3 = 1.09198$, indicates that, all else equal, a magazine with an extra dollar of median income among its readers would charge about $1.09 more (on average) for a full-page ad. The sign (positive) is reasonable because people with more income can spend more on advertised products. If a magazine had the same audience and percent male readership but had a median income $4,000 higher, you would expect the page costs to be $1.09198 \times 4,000 = \$4,368$ higher (on average) due to the higher income level.

Remember that regression coefficients indicate the effect of one $X$ variable on $Y$ while all other $X$ variables are held constant. This should be taken literally. For example, the regression coefficient $b_3$ indicates the effect of median income on page costs, computed while holding audience and percent male readership fixed. In this example, higher median income levels tend to result in higher page costs at fixed levels of audience and of male readership (due to the fact that $b_3$ is a positive number).

What would the relationship be if the other variables, audience and percent male, were *not* held constant? This may be answered by looking at the ordinary correlation coefficient (or regression coefficient predicting $Y$ from this $X$ alone), computed for just the two variables page costs and median income. In this case, higher median income is actually associated with *lower* page costs (the correlation of page costs with median income is negative: $-0.148$)! How can this be? One reasonable explanation is that magazines targeting a higher median income level cannot support a large audience due to a relative scarcity of rich people in the general population, and this is supported by the negative correlation ($-0.377$) of audience size with median income. If this audience decrease is large enough, it can offset the effect of higher income per reader.

## Predictions and Prediction Errors

The prediction equation or regression equation is defined as follows: Predicted $Y = a + b_1 X_1 + b_2 X_2 + \cdots + b_k X_k$. For the magazine ads example, to find a predicted value for page costs based on the audience, percent male readership, and median income for a magazine similar to those in the data set, substitute the $X$ values into the prediction equation as follows:

Predicted page costs
$$= a + b_1 X_1 + b_2 X_2 + b_3 X_3$$
$$= -22,385 + 10.50658 X_1 - 20,779 X_2 + 1.09198 X_3$$
$$= -22,385 + 10.50658 \,(\text{Audience})$$
$$\quad - 20,779 \,(\text{Percent male}) + 1.09198 \,(\text{Median income})$$

For example, suppose you planned to launch a new magazine, *Popular Statistics*, that would reach an audience of 900,000 with a readership that is 55% women and that has a median

income of $80,000. Be sure to put these numbers into the regression equation in the same form as the original data set: $X_1 = 900$ (audience in thousands), $X_2 = 0.45$ (percent male expressed as a decimal), and $X_3 = \$80,000$ (median income). The predicted value for this situation is

Predicted page costs for *Popular Statistics*
$$= -22,385 + 10.50658 \,(\text{Audience})$$
$$\quad - 20,779 \,(\text{Percent male})$$
$$\quad + 1.09198 \,(\text{Median income})$$
$$= -22,385 + 10.50658(900)$$
$$\quad - 20,779(0.45) + 1.09198(80,000)$$
$$= \$65,079$$

Of course, you would not expect page costs to be exactly $65,079 for two reasons. First of all, there is random variation even among the magazines for which you have data, so the predictions are not perfect even for these. Second, predictions can only be useful to the extent that the predicted magazine is similar to the magazines in the original data set. For a new magazine, the advertising rates may be determined differently than for the well-established magazines used to compute the regression equation.

You can also use the equation to find the predicted page costs for the magazines in the original data set. For example, *Martha Stewart Living* has $X_1 = 11,200$ (indicating an audience of 11.2 million readers), $X_2 = 11.0\%$ (indicating the percentage of men among its readers), and $X_3 = \$74,436$ (indicating the median annual income for its readers). The predicted value is

Predicted page costs for *Martha Stewart Living*
$$= -22,385 + 10.50658 \,(\text{Audience})$$
$$\quad - 20,779 \,(\text{Precent male})$$
$$\quad + 1.09198 \,(\text{Median income})$$
$$= -22,385 + 10.50658(11,200)$$
$$\quad - 20,779(0.110) + 1.09198(74,436)$$
$$= \$174,286$$

The residual, or prediction error, is defined as $Y -$ (Predicted $Y$). For a magazine in the original data set, this is the actual minus the predicted page costs. For *Martha Stewart Living*, the actual page costs are $157,700, compared to the predicted value of $174,286. Thus, the prediction error is $157,700 - 174,286 = -16,586$. A negative residual like this indicates that the actual page costs are lower than predicted—about $17,000 lower for this magazine. For many of us, this would be a lot of money; it is a good idea to look at some of the other prediction errors to see how well or poorly the predictions do. How can *Martha Stewart Living* charge so much less than we would expect? Basically, because the prediction used only $k = 3$ of the many factors that influence the cost of advertising (and many of these factors are not well understood and cannot easily be measured).

Table 12.1.5 shows the actual values and the predicted values (also called the *expected values* or *fitted values*) for page costs together with the prediction errors for each of the magazines in the original data set.

## How Good Are the Predictions?

This section will be primarily a review, since the standard error of estimate, $S_e$, and the coefficient of determination, $R^2$, have much the same interpretation for multiple regression as they had for simple regression in the preceding chapter. The only difference is that your predictions are now based on more than one $X$ variable. They are so similar because you are still predicting just one $Y$.

## Typical Prediction Error: Standard Error of Estimate

Just as for simple regression, with only one $X$, the standard error of estimate indicates the approximate size of the prediction errors. For the magazine ads example, $S_e = \$53,812$. This tells you that actual page costs for these magazines are typically within about $53,812 from the predicted page costs, in the sense of a standard deviation. That is, if the error distribution is normal, then you would expect about 2/3 of the actual page costs to be within $S_e$ of the predicted page costs, about 95% to be within $2S_e$, and so forth.

The standard error of estimate, $S_e = \$53,812$, indicates the remaining variation in page costs after you have used the $X$ variables (audience, percent male, and median income) in the regression equation to predict page costs

### TABLE 12.1.5 Predicted and Residual Values for Magazine Ads

|  | Page Costs (actual) | Page Costs (predicted) | Prediction Errors (residuals) |
|---|---|---|---|
| AAA Westways | $53,310 | $160,794 | –$107,484 |
| AARP The Magazine | 532,600 | 409,087 | 123,513 |
| Allure | 131,721 | 116,814 | 14,907 |
| Architectural Digest | 119,370 | 130,979 | –11,609 |
| Audubon | 25,040 | 69,927 | –44,887 |
| Better Homes & Gardens | 468,200 | 456,590 | 11,610 |
| Bicycling | 55,385 | 65,466 | –10,081 |
| Bon Appétit | 143,612 | 156,802 | –13,190 |
| Brides | 82,041 | 98,410 | –16,369 |
| Car and Driver | 187,269 | 141,891 | 45,378 |
| Conde Nast Traveler | 118,657 | 123,105 | –4,448 |
| Cosmopolitan | 222,400 | 229,496 | –7,096 |
| Details | 69,552 | 66,064 | 3,488 |
| Discover | 57,300 | 106,706 | –49,406 |
| Every Day with Rachael Ray | 139,000 | 123,812 | 15,188 |
| Family Circle | 254,600 | 254,158 | 442 |
| Fitness | 142,300 | 114,710 | 27,590 |
| Food & Wine | 86,000 | 146,882 | –60,882 |
| Golf Magazine | 141,174 | 124,839 | 16,335 |
| Good Housekeeping | 344,475 | 298,913 | 45,562 |
| GQ (Gentlemen's Quarterly) | 143,681 | 110,448 | 33,233 |
| Kiplinger's Personal Finance | 54,380 | 101,294 | –46,914 |
| Ladies' Home Journal | 254,000 | 182,394 | 71,606 |

*(Continued)*

**TABLE 12.1.5  Predicted and Residual Values for Magazine Ads—cont'd**

|  | Page Costs (actual) | Page Costs (predicted) | Prediction Errors (residuals) |
|---|---|---|---|
| Martha Stewart Living | 157,700 | 174,286 | −16,586 |
| Midwest Living | 125,100 | 89,866 | 35,234 |
| Money | 201,800 | 152,470 | 49,330 |
| More | 148,400 | 94,363 | 54,037 |
| O, The Oprah Magazine | 150,730 | 218,425 | −67,695 |
| Parents | 167,800 | 199,808 | −32,008 |
| Prevention | 134,900 | 156,533 | −21,633 |
| Reader's Digest | 171,300 | 370,017 | −198,717 |
| Readymade | 32,500 | 46,759 | −14,259 |
| Road & Track | 109,373 | 134,055 | −24,682 |
| Self | 166,773 | 133,654 | 33,119 |
| Ser Padres | 74,840 | 49,195 | 25,645 |
| Siempre Mujer | 48,300 | 42,325 | 5,975 |
| Sports Illustrated | 352,800 | 261,045 | 91,755 |
| Teen Vogue | 115,897 | 98,803 | 17,094 |
| The New Yorker | 135,263 | 115,454 | 19,809 |
| Time | 287,440 | 264,434 | 23,006 |
| TV Guide | 134,700 | 178,197 | −43,497 |
| Vanity Fair | 165,600 | 126,868 | 38,732 |
| Vogue | 151,133 | 176,499 | −25,366 |
| Wired | 99,475 | 90,661 | 8,814 |
| Woman's Day | 259,960 | 254,554 | 5,406 |

for each magazine. Compare this to the ordinary univariate standard deviation, $S_Y = \$105,639$ for the page costs, computed by ignoring all the other variables. This standard deviation, $S_Y$, indicates the remaining variation in page costs after you have used only $\overline{Y}$ to predict the page costs for each magazine. Note that $S_e = \$53,812$ is smaller than $S_Y = \$105,639$; your errors are typically smaller if you use the regression equation instead of just $\overline{Y}$ to predict page costs. This suggests that the $X$ variables are helpful in explaining page costs.

Think of the situation this way. If you knew nothing of the $X$ variables, you would use the average page costs ($\overline{Y} = 160,397$) as your best guess, and you would be wrong by about $S_Y = \$105,639$. But if you knew the audience, percent male readership, and median reader income, you could use the regression equation to find a prediction for page costs that would be wrong by only $S_e = \$53,812$. This

reduction in prediction error (from \$105,639 to \$53,812) is one of the helpful payoffs from running a regression analysis.

## Percent Variation Explained: $R^2$

The coefficient of determination, $R^2$, indicates the percentage of the variation in $Y$ that is explained by or attributed to all of the $X$ variables.

For the magazine ads example, the coefficient of determination, $R^2 = 0.758$ or 75.8%, tells you that the explanatory variables (the $X$ variables audience, percent male, and median income) have explained 75.8% of the variation in page costs.[4] This leaves 24.2% of the variation unaccounted

---

4. Technically, it is the fraction of the *variance* (the squared standard deviation) of $Y$ that is explained by the $X$ variables.

for and attributable to other factors. This is a fairly large $R^2$ number; many research studies find much smaller numbers yet still provide useful predictions. You usually want $R^2$ to be as large a value as possible, since higher numbers tell you that the relationship is a strong one. The highest possible value is $R^2 = 100\%$, which happens only when all prediction errors are 0 (and is usually a signal to look for a mistake somewhere!).

## Inference in Multiple Regression

The regression output so far is a fairly complete description of these particular ($n = 45$) magazines, but statistical inference will help you generalize to the idealized population of similar conceivable magazines. Rather than just observe that there is an average decrease in page costs of $208 per percentage point increase in male readership, you can infer about a large population of similar magazines that could reasonably have produced this data, to see whether there is *necessarily* any connection between gender and page costs, or if the –208 regression coefficient could reasonably be just randomness. Could it be that this effect of percent male readership on page costs is just a random number, rather than indicating a systematic relationship? Statistical inference will provide an answer.

For reference, Table 12.1.6 shows the portion of the computer output from Table 12.1.4 that deals with statistical inference by providing *p*-values for the overall $F$ test as well as for each independent ($X$) variable. We will discuss each item in the following sections, after indicating the population you will be inferring about.

## Assumptions

For simplicity, assume that you have a random sample from a much larger population. Assume also that the population has a linear relationship with randomness, as expressed by the **multiple regression linear model**, which specifies that the observed value for $Y$ is equal to the population relationship plus a random error that has a normal distribution. These random errors are also assumed to be independent from one case (elementary unit) to another.

---

**The Multiple Regression Linear Model for the Population**

$$Y = (\alpha + \beta_1 X_1 + \beta_2 X_2 + \dots + \beta_k X_k) + \varepsilon$$
$$= (\text{Population relationship}) + \text{Randomness}$$

where $\varepsilon$ has a normal distribution with mean 0 and constant standard deviation $\sigma$, and this randomness is independent from one case to another.

---

The population relationship is given by $k + 1$ parameters: $\alpha$ is the population intercept (or constant term), and $\beta_1, \beta_2, \dots, \beta_k$ are the population regression coefficients, indicating the mean effect of each $X$ on $Y$ (in the population), holding all other $X$ constant. A summary of population and sample quantities is shown in Table 12.1.7. If your data were a census of the entire population, then your least-squares regression coefficients would be the same as those in the population relationship. Ordinarily, however, you will use the least-squares intercept, $a$, as an

---

**TABLE 12.1.6 Statistical Inference for Magazine Ads**

| Predictor | Coeff | SE Coeff | t | p |
|---|---|---|---|---|
| Constant | –22385 | 36060 | –0.62 | 0.538 |
| Audience | 10.5066 | 0.9422 | 11.15 | 0.000 |
| Male | –20779 | 37961 | –0.55 | 0.587 |
| Income | 1.0920 | 0.4619 | 2.36 | 0.023 |

$S = 53812.4$   R-Sq = 75.8%   R-Sq(adj) = 74.1%

**Analysis of Variance**

| Source | DF | SS | MS | F | p |
|---|---|---|---|---|---|
| Regression | 3 | 3.72299E+11 | 1.24100E+11 | 42.86 | 0.000 |
| Residual Error | 41 | 1.18727E+11 | 2895770619 | | |
| Total | 44 | 4.91025E+11 | | | |

**TABLE 12.1.7 Population and Sample Quantities in Multiple Regression**

| | Population (parameters: fixed and unknown) | Sample (estimators: random and known) |
|---|---|---|
| Intercept or constant | $\alpha$ | $a$ |
| Regression coefficients | $\beta_1$ | $b_1$ |
| | $\beta_2$ | $b_2$ |
| | . | . |
| | . | . |
| | . | . |
| | $\beta_k$ | $b_k$ |
| Uncertainty in $Y$ | $\sigma$ | $S_e$ |

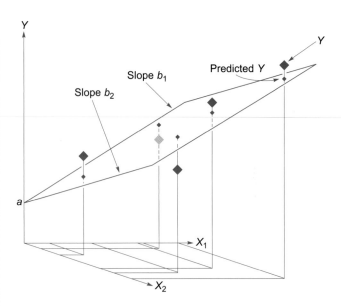

**FIGURE 12.1.2**  When two explanatory $X$ variables are used to predict $Y$, the prediction equation can be visualized as a flat plane chosen to be closest to the data points in a three-dimensional space. The intercept term $a$ is the place where this prediction plane hits the $Y$ axis. The regression coefficients $b_1$ and $b_2$ show the tilt of the prediction plane in two of its directions.

*estimator* of $\alpha$, and the least-squares regression coefficients, $b_1, b_2, \ldots, b_k$, as *estimators* of $\beta_1, \beta_2, \ldots, \beta_k$, respectively. Of course, there are errors involved in estimating since the sample is much smaller than the entire population.

How can you picture a multiple regression linear relationship using a scatterplot? Each time a new explanatory $X$ variable is added, we get one more dimension. For example, with just one $X$ variable in the previous chapter, we had the prediction line in a flat two-dimensional space. With two $X$ variables, we have a flat prediction plane in the three-dimensional space defined by $X_1$, $X_2$, and $Y$, as shown in Figure 12.1.2. One assumption of multiple regression analysis is that the relationship in the population is basically flat, not curved.

## Is the Model Significant? The *F* Test or $R^2$ Test

Inference begins with the $F$ test to see if the $X$ variables explain a significant amount of the variation in $Y$. The $F$ test is used as a gateway to statistical inference: If it is significant, then there is a relationship, and you may proceed to investigate and explain it. If it is not significant, you might as well just have a bunch of unrelated random numbers; essentially nothing can be explained. Don't forget that whenever you accept the null hypothesis, it is a *weak* conclusion. You have not proven that there is no relationship; you merely lack convincing evidence that there is a relationship. There could be a relationship but, due to randomness or small sample size, you are unable to detect it with the data with which you are working.

The null hypothesis for the $F$ test claims that there is *no* predictive relationship between the $X$ variables and $Y$ in the population. That is, $Y$ is pure randomness and has no regard for the values of the $X$ variables. Looking at the multiple regression linear model, you can see this claim is equivalent to $Y = \alpha + \varepsilon$, which happens whenever *all* of the population regression coefficients are 0.

The research hypothesis for the $F$ test claims that there is some predictive relationship between the $X$ variables and $Y$ in the population. Thus, $Y$ is more than just pure randomness and must depend on at least one of the $X$ variables. Thus, the research hypothesis claims that *at least one* of the regression coefficients is not 0. Note that it is not necessary for every $X$ variable to affect $Y$; it is enough for there to be just one.

**Hypotheses for the *F* Test**

$$H_0: \beta_1 = \beta_2 = \cdots = \beta_k = 0$$

$$H_1: \text{At least one of } \beta_1, \beta_2, \ldots, \beta_k \neq 0$$

The easiest way to perform the $F$ test is to look for the appropriate $p$-value in the computer analysis and to interpret the resulting significance level as we did in Chapter 10. If the $p$-value is more than 0.05, then the result is not significant. If the $p$-value is less than 0.05, then the result is significant. If $p < 0.01$, then it is highly significant, and so forth.

Another way to perform the $F$ test is to compare the $R^2$ value (the percent of the variation in $Y$ that is explained by the $X$ variables) to a table of $R^2$ values for the appropriate test level (5% or other). If the $R^2$ value is large enough, then the regression is significant; more than just a random amount of the variation in $Y$ is explained. The table is indexed by $n$ (the number of cases) and $k$ (the number of $X$ variables).

The traditional way of performing the $F$ test is harder to interpret, but it always gives the same result as the $R^2$ table. The $F$ test is traditionally done by computing the $F$ statistic and comparing it to the critical value in the $F$ table at the appropriate test level.[5] Two different degrees of freedom numbers are used: the numerator degrees of freedom $k$ (the number of $X$ variables used to explain $Y$) and the denominator degrees of freedom $n - k - 1$ (a measure of the randomness of the residuals after estimation of the $k + 1$ coefficients $a, b_1, \ldots, b_k$).

However, the $F$ statistic is an unnecessary complication because the $R^2$ value may be tested directly. Furthermore, $R^2$ is more directly meaningful than the $F$ statistic because $R^2$ tells you the percent of the variation in $Y$ that is accounted for (or explained by) the $X$ variables, while $F$ has no simple, direct interpretation in terms of the data. Whether you use the $F$ or the $R^2$ approach, your answer (significant or not) will be the same at each test level.

Why is the more complex $F$ statistic traditionally used when $R^2$, which is more directly meaningful, can be used instead? Perhaps the reason is just tradition, and perhaps it's that the $F$ tables have long been available. The use of a meaningful number (such as $R^2$) provides more insight into the situation and seems preferable, especially in the field of business. All three methods (the $p$-value, the $R^2$ value, and the $F$ statistic) must lead to the same answer.

### Result of the $F$ Test (Decided Using the $p$-Value)

If the $p$-value is *larger* than 0.05, then the model is *not significant* (you accept the null hypothesis that the $X$ variables *do not* help predict $Y$).

If the $p$-value is *smaller* than 0.05, then the model is *significant* (you reject the null hypothesis and accept the research hypothesis that the $X$ variables do help predict $Y$).

### Result of the $F$ Test (Decided Using $R^2$)

If the $R^2$ value is *smaller* than the critical value in the $R^2$ table, then the model is *not significant*.

If the $R^2$ value is *larger*, then the model is *significant*. This answer will always be the same as the result found using the $p$-value.

### Result of the $F$ Test (Decided Using $F$ Directly)

If the $F$ value is *smaller* than the critical value in the $F$ table, then the model is *not significant*.

If the $F$ value is *larger*, then the model is *significant*.

Remember that the statistical meaning of *significant* is slightly different from its everyday usage. When you find a significant regression model, you know that the relationship between the $X$ variables and $Y$ is stronger than you would ordinarily expect from randomness alone. That is, you can tell that a relationship is there. It may or may not be a strong or useful relationship in any practical sense—you must look separately at this issue—but it is strong enough that it does not look purely random.

For the magazine advertising cost example, the prediction equation *does* explain a significant proportion of the variation in page costs, as is indicated by the $p$-value of 0.000 to the right of the $F$ value of 42.86 in the computer printout.[6] This says that page costs do depend systematically on (at least one of) these factors and are not just random. You don't yet know which one or more of the $X$ variables are responsible for this prediction of $Y$, but you know that there is at least one.

To verify—using $R^2$—that the regression equation is really significant, note that the coefficient of determination is $R^2 = 0.758$ or 75.8%. The $R^2$ table for testing at the 5% level with $n = 45$ magazines and $k = 3$ variables (see Table 12.1.8) shows the critical value 0.172 or 17.2%. In order for the equation to be significant at the conventional 5% level, the $X$ variables would need to explain only 17.2% of the variation in page costs ($Y$). Since they explain more, the regression is significant.

Looking at the 1% and 0.1% $R^2$ tables (Tables 12.1.9 and 12.1.10) for $n = 45$ and $k = 3$, you find critical values of 23.9% and 32.4%, respectively. Since the observed $R^2 = 75.8\%$ exceeds both of these, you may conclude that these $X$ variables (audience, percent male, and median income) have a *very highly significant* impact on $Y$ (page costs).

---

5. In case you are curious, the $F$ statistic gets its name from Sir Ronald A. Fisher and is defined as the "explained mean square" divided by the "unexplained mean square." Large values of $F$ suggest that the regression model is significant because you have explained a lot relative to the amount of unexplained randomness. Large values of $R^2$ also suggest significance. The link between $F$ and $R^2$ is the fact that $F = (n - k - 1)[1/(1 - R^2) - 1]/k$ and $R^2 = 1 - 1/[1 + kF/(n - k - 1)]$, so that larger values of $F$ go with larger values of $R^2$ and vice versa. This is why tests for large $F$ are exactly equivalent to tests for large values of $R^2$.

6. When a $p$-value is listed as 0.000, it may be interpreted as $p < 0.0005$, because a $p$-value larger than or equal to 0.0005 would be rounded to show as at least 0.001.

**TABLE 12.1.8** $R^2$ Table: Level 5% Critical Values (Significant)

| Number of Cases (n) | Number of X Variables (k) | | | | | | | | | |
|---|---|---|---|---|---|---|---|---|---|---|
| | 1 | 2 | 3 | 4 | 5 | 6 | 7 | 8 | 9 | 10 |
| 3 | 0.994 | | | | | | | | | |
| 4 | 0.902 | 0.997 | | | | | | | | |
| 5 | 0.771 | 0.950 | 0.998 | | | | | | | |
| 6 | 0.658 | 0.864 | 0.966 | 0.999 | | | | | | |
| 7 | 0.569 | 0.776 | 0.903 | 0.975 | 0.999 | | | | | |
| 8 | 0.499 | 0.698 | 0.832 | 0.924 | 0.980 | 0.999 | | | | |
| 9 | 0.444 | 0.632 | 0.764 | 0.865 | 0.938 | 0.983 | 0.999 | | | |
| 10 | 0.399 | 0.575 | 0.704 | 0.806 | 0.887 | 0.947 | 0.985 | 0.999 | | |
| 11 | 0.362 | 0.527 | 0.651 | 0.751 | 0.835 | 0.902 | 0.954 | 0.987 | 1.000 | |
| 12 | 0.332 | 0.486 | 0.604 | 0.702 | 0.785 | 0.856 | 0.914 | 0.959 | 0.989 | 1.000 |
| 13 | 0.306 | 0.451 | 0.563 | 0.657 | 0.739 | 0.811 | 0.872 | 0.924 | 0.964 | 0.990 |
| 14 | 0.283 | 0.420 | 0.527 | 0.618 | 0.697 | 0.768 | 0.831 | 0.885 | 0.931 | 0.967 |
| 15 | 0.264 | 0.393 | 0.495 | 0.582 | 0.659 | 0.729 | 0.791 | 0.847 | 0.896 | 0.937 |
| 16 | 0.247 | 0.369 | 0.466 | 0.550 | 0.624 | 0.692 | 0.754 | 0.810 | 0.860 | 0.904 |
| 17 | 0.232 | 0.348 | 0.440 | 0.521 | 0.593 | 0.659 | 0.719 | 0.775 | 0.825 | 0.871 |
| 18 | 0.219 | 0.329 | 0.417 | 0.494 | 0.564 | 0.628 | 0.687 | 0.742 | 0.792 | 0.839 |
| 19 | 0.208 | 0.312 | 0.397 | 0.471 | 0.538 | 0.600 | 0.657 | 0.711 | 0.761 | 0.807 |
| 20 | 0.197 | 0.297 | 0.378 | 0.449 | 0.514 | 0.574 | 0.630 | 0.682 | 0.731 | 0.777 |
| 21 | 0.187 | 0.283 | 0.361 | 0.429 | 0.492 | 0.550 | 0.604 | 0.655 | 0.703 | 0.749 |
| 22 | 0.179 | 0.270 | 0.345 | 0.411 | 0.471 | 0.527 | 0.580 | 0.630 | 0.677 | 0.722 |
| 23 | 0.171 | 0.259 | 0.331 | 0.394 | 0.452 | 0.507 | 0.558 | 0.607 | 0.653 | 0.696 |
| 24 | 0.164 | 0.248 | 0.317 | 0.379 | 0.435 | 0.488 | 0.538 | 0.585 | 0.630 | 0.673 |
| 25 | 0.157 | 0.238 | 0.305 | 0.364 | 0.419 | 0.470 | 0.518 | 0.564 | 0.608 | 0.650 |
| 26 | 0.151 | 0.229 | 0.294 | 0.351 | 0.404 | 0.454 | 0.501 | 0.545 | 0.588 | 0.629 |
| 27 | 0.145 | 0.221 | 0.283 | 0.339 | 0.390 | 0.438 | 0.484 | 0.527 | 0.569 | 0.609 |
| 28 | 0.140 | 0.213 | 0.273 | 0.327 | 0.377 | 0.424 | 0.468 | 0.510 | 0.551 | 0.590 |
| 29 | 0.135 | 0.206 | 0.264 | 0.316 | 0.365 | 0.410 | 0.453 | 0.495 | 0.534 | 0.573 |
| 30 | 0.130 | 0.199 | 0.256 | 0.306 | 0.353 | 0.397 | 0.439 | 0.480 | 0.518 | 0.556 |
| 31 | 0.126 | 0.193 | 0.248 | 0.297 | 0.342 | 0.385 | 0.426 | 0.466 | 0.503 | 0.540 |
| 32 | 0.122 | 0.187 | 0.240 | 0.288 | 0.332 | 0.374 | 0.414 | 0.452 | 0.489 | 0.525 |
| 33 | 0.118 | 0.181 | 0.233 | 0.279 | 0.323 | 0.363 | 0.402 | 0.440 | 0.476 | 0.511 |
| 34 | 0.115 | 0.176 | 0.226 | 0.271 | 0.314 | 0.353 | 0.391 | 0.428 | 0.463 | 0.497 |
| 35 | 0.111 | 0.171 | 0.220 | 0.264 | 0.305 | 0.344 | 0.381 | 0.417 | 0.451 | 0.484 |
| 36 | 0.108 | 0.166 | 0.214 | 0.257 | 0.297 | 0.335 | 0.371 | 0.406 | 0.440 | 0.472 |
| 37 | 0.105 | 0.162 | 0.208 | 0.250 | 0.289 | 0.326 | 0.362 | 0.396 | 0.429 | 0.461 |

**TABLE 12.1.8** $R^2$ Table: Level 5% Critical Values (Significant)—cont'd

| Number of Cases | Number of X Variables (k) | | | | | | | | | |
|---|---|---|---|---|---|---|---|---|---|---|
| (n) | 1 | 2 | 3 | 4 | 5 | 6 | 7 | 8 | 9 | 10 |
| 38 | 0.103 | 0.157 | 0.203 | 0.244 | 0.282 | 0.318 | 0.353 | 0.386 | 0.418 | 0.449 |
| 39 | 0.100 | 0.153 | 0.198 | 0.238 | 0.275 | 0.310 | 0.344 | 0.377 | 0.408 | 0.439 |
| 40 | 0.097 | 0.150 | 0.193 | 0.232 | 0.268 | 0.303 | 0.336 | 0.368 | 0.399 | 0.429 |
| 41 | 0.095 | 0.146 | 0.188 | 0.226 | 0.262 | 0.296 | 0.328 | 0.359 | 0.390 | 0.419 |
| 42 | 0.093 | 0.142 | 0.184 | 0.221 | 0.256 | 0.289 | 0.321 | 0.351 | 0.381 | 0.410 |
| 43 | 0.090 | 0.139 | 0.180 | 0.216 | 0.250 | 0.283 | 0.314 | 0.344 | 0.373 | 0.401 |
| 44 | 0.088 | 0.136 | 0.176 | 0.211 | 0.245 | 0.276 | 0.307 | 0.336 | 0.365 | 0.393 |
| 45 | 0.086 | 0.133 | 0.172 | 0.207 | 0.239 | 0.271 | 0.300 | 0.329 | 0.357 | 0.384 |
| 46 | 0.085 | 0.130 | 0.168 | 0.202 | 0.234 | 0.265 | 0.294 | 0.322 | 0.350 | 0.377 |
| 47 | 0.083 | 0.127 | 0.164 | 0.198 | 0.230 | 0.259 | 0.288 | 0.316 | 0.343 | 0.369 |
| 48 | 0.081 | 0.125 | 0.161 | 0.194 | 0.225 | 0.254 | 0.282 | 0.310 | 0.336 | 0.362 |
| 49 | 0.079 | 0.122 | 0.158 | 0.190 | 0.220 | 0.249 | 0.277 | 0.304 | 0.330 | 0.355 |
| 50 | 0.078 | 0.120 | 0.155 | 0.186 | 0.216 | 0.244 | 0.272 | 0.298 | 0.323 | 0.348 |
| 51 | 0.076 | 0.117 | 0.152 | 0.183 | 0.212 | 0.240 | 0.267 | 0.293 | 0.318 | 0.342 |
| 52 | 0.075 | 0.115 | 0.149 | 0.180 | 0.208 | 0.235 | 0.262 | 0.287 | 0.312 | 0.336 |
| 53 | 0.073 | 0.113 | 0.146 | 0.176 | 0.204 | 0.231 | 0.257 | 0.282 | 0.306 | 0.330 |
| 54 | 0.072 | 0.111 | 0.143 | 0.173 | 0.201 | 0.227 | 0.252 | 0.277 | 0.301 | 0.324 |
| 55 | 0.071 | 0.109 | 0.141 | 0.170 | 0.197 | 0.223 | 0.248 | 0.272 | 0.295 | 0.318 |
| 56 | 0.069 | 0.107 | 0.138 | 0.167 | 0.194 | 0.219 | 0.244 | 0.267 | 0.290 | 0.313 |
| 57 | 0.068 | 0.105 | 0.136 | 0.164 | 0.190 | 0.215 | 0.240 | 0.263 | 0.285 | 0.308 |
| 58 | 0.067 | 0.103 | 0.134 | 0.161 | 0.187 | 0.212 | 0.236 | 0.258 | 0.281 | 0.303 |
| 59 | 0.066 | 0.101 | 0.131 | 0.159 | 0.184 | 0.208 | 0.232 | 0.254 | 0.276 | 0.298 |
| 60 | 0.065 | 0.100 | 0.129 | 0.156 | 0.181 | 0.205 | 0.228 | 0.250 | 0.272 | 0.293 |
| Multiplier 1 | 3.84 | 5.99 | 7.82 | 9.49 | 11.07 | 12.59 | 14.07 | 15.51 | 16.92 | 18.31 |
| Multiplier 2 | 2.15 | −0.27 | −3.84 | −7.94 | −12.84 | −18.24 | −23.78 | −30.10 | −36.87 | −43.87 |

**TABLE 12.1.9** $R^2$ Table: Level 1% Critical Values (Highly Significant)

| Number of Cases | Number of X Variables (k) | | | | | | | | | |
|---|---|---|---|---|---|---|---|---|---|---|
| (n) | 1 | 2 | 3 | 4 | 5 | 6 | 7 | 8 | 9 | 10 |
| 3 | 1.000 | | | | | | | | | |
| 4 | 0.980 | 1.000 | | | | | | | | |
| 5 | 0.919 | 0.990 | 1.000 | | | | | | | |
| 6 | 0.841 | 0.954 | 0.993 | 1.000 | | | | | | |

(Continued)

**TABLE 12.1.9** $R^2$ Table: Level 1% Critical Values (Highly Significant)—cont'd

| Number of Cases | Number of X Variables (k) | | | | | | | | | |
|---|---|---|---|---|---|---|---|---|---|---|
| (n) | 1 | 2 | 3 | 4 | 5 | 6 | 7 | 8 | 9 | 10 |
| 7 | 0.765 | 0.900 | 0.967 | 0.995 | 1.000 | | | | | |
| 8 | 0.696 | 0.842 | 0.926 | 0.975 | 0.996 | 1.000 | | | | |
| 9 | 0.636 | 0.785 | 0.879 | 0.941 | 0.979 | 0.997 | 1.000 | | | |
| 10 | 0.585 | 0.732 | 0.830 | 0.901 | 0.951 | 0.982 | 0.997 | 1.000 | | |
| 11 | 0.540 | 0.684 | 0.784 | 0.859 | 0.916 | 0.958 | 0.985 | 0.997 | 1.000 | |
| 12 | 0.501 | 0.641 | 0.740 | 0.818 | 0.879 | 0.928 | 0.963 | 0.987 | 0.998 | 1.000 |
| 13 | 0.467 | 0.602 | 0.700 | 0.778 | 0.842 | 0.894 | 0.936 | 0.967 | 0.988 | 0.998 |
| 14 | 0.437 | 0.567 | 0.663 | 0.741 | 0.806 | 0.860 | 0.906 | 0.943 | 0.971 | 0.989 |
| 15 | 0.411 | 0.536 | 0.629 | 0.706 | 0.771 | 0.827 | 0.875 | 0.915 | 0.948 | 0.973 |
| 16 | 0.388 | 0.508 | 0.598 | 0.673 | 0.738 | 0.795 | 0.844 | 0.887 | 0.923 | 0.953 |
| 17 | 0.367 | 0.482 | 0.570 | 0.643 | 0.707 | 0.764 | 0.814 | 0.858 | 0.896 | 0.929 |
| 18 | 0.348 | 0.459 | 0.544 | 0.616 | 0.678 | 0.734 | 0.784 | 0.829 | 0.869 | 0.904 |
| 19 | 0.331 | 0.438 | 0.520 | 0.590 | 0.652 | 0.707 | 0.757 | 0.802 | 0.843 | 0.879 |
| 20 | 0.315 | 0.418 | 0.498 | 0.566 | 0.626 | 0.681 | 0.730 | 0.775 | 0.816 | 0.854 |
| 21 | 0.301 | 0.401 | 0.478 | 0.544 | 0.603 | 0.656 | 0.705 | 0.750 | 0.791 | 0.829 |
| 22 | 0.288 | 0.384 | 0.459 | 0.523 | 0.581 | 0.633 | 0.681 | 0.726 | 0.767 | 0.805 |
| 23 | 0.276 | 0.369 | 0.442 | 0.504 | 0.560 | 0.612 | 0.659 | 0.703 | 0.744 | 0.782 |
| 24 | 0.265 | 0.355 | 0.426 | 0.487 | 0.541 | 0.591 | 0.638 | 0.681 | 0.721 | 0.759 |
| 25 | 0.255 | 0.342 | 0.410 | 0.470 | 0.523 | 0.572 | 0.618 | 0.660 | 0.700 | 0.738 |
| 26 | 0.246 | 0.330 | 0.396 | 0.454 | 0.506 | 0.554 | 0.599 | 0.641 | 0.680 | 0.717 |
| 27 | 0.237 | 0.319 | 0.383 | 0.440 | 0.490 | 0.537 | 0.581 | 0.622 | 0.661 | 0.698 |
| 28 | 0.229 | 0.308 | 0.371 | 0.426 | 0.475 | 0.521 | 0.564 | 0.605 | 0.643 | 0.679 |
| 29 | 0.221 | 0.298 | 0.359 | 0.413 | 0.461 | 0.506 | 0.548 | 0.588 | 0.625 | 0.661 |
| 30 | 0.214 | 0.289 | 0.349 | 0.401 | 0.448 | 0.492 | 0.533 | 0.572 | 0.609 | 0.644 |
| 31 | 0.208 | 0.280 | 0.338 | 0.389 | 0.435 | 0.478 | 0.519 | 0.557 | 0.593 | 0.627 |
| 32 | 0.201 | 0.272 | 0.329 | 0.378 | 0.423 | 0.465 | 0.505 | 0.542 | 0.578 | 0.612 |
| 33 | 0.195 | 0.264 | 0.319 | 0.368 | 0.412 | 0.453 | 0.492 | 0.529 | 0.563 | 0.597 |
| 34 | 0.190 | 0.257 | 0.311 | 0.358 | 0.401 | 0.442 | 0.479 | 0.515 | 0.550 | 0.583 |
| 35 | 0.185 | 0.250 | 0.303 | 0.349 | 0.391 | 0.430 | 0.468 | 0.503 | 0.537 | 0.569 |
| 36 | 0.180 | 0.244 | 0.295 | 0.340 | 0.381 | 0.420 | 0.456 | 0.491 | 0.524 | 0.556 |
| 37 | 0.175 | 0.237 | 0.287 | 0.332 | 0.372 | 0.410 | 0.446 | 0.480 | 0.512 | 0.543 |
| 38 | 0.170 | 0.231 | 0.280 | 0.324 | 0.363 | 0.400 | 0.435 | 0.469 | 0.501 | 0.531 |
| 39 | 0.166 | 0.226 | 0.274 | 0.316 | 0.355 | 0.391 | 0.426 | 0.458 | 0.490 | 0.520 |
| 40 | 0.162 | 0.220 | 0.267 | 0.309 | 0.347 | 0.382 | 0.416 | 0.448 | 0.479 | 0.509 |
| 41 | 0.158 | 0.215 | 0.261 | 0.302 | 0.339 | 0.374 | 0.407 | 0.439 | 0.469 | 0.498 |

**TABLE 12.1.9** $R^2$ Table: Level 1% Critical Values (Highly Significant)—cont'd

| Number of Cases | Number of X Variables (k) | | | | | | | | | |
|---|---|---|---|---|---|---|---|---|---|---|
| (n) | 1 | 2 | 3 | 4 | 5 | 6 | 7 | 8 | 9 | 10 |
| 42 | 0.155 | 0.210 | 0.255 | 0.295 | 0.332 | 0.366 | 0.399 | 0.430 | 0.459 | 0.488 |
| 43 | 0.151 | 0.206 | 0.250 | 0.289 | 0.325 | 0.358 | 0.390 | 0.421 | 0.450 | 0.478 |
| 44 | 0.148 | 0.201 | 0.244 | 0.283 | 0.318 | 0.351 | 0.382 | 0.412 | 0.441 | 0.469 |
| 45 | 0.145 | 0.197 | 0.239 | 0.277 | 0.311 | 0.344 | 0.375 | 0.404 | 0.432 | 0.460 |
| 46 | 0.141 | 0.193 | 0.234 | 0.271 | 0.305 | 0.337 | 0.367 | 0.396 | 0.424 | 0.451 |
| 47 | 0.138 | 0.189 | 0.230 | 0.266 | 0.299 | 0.330 | 0.360 | 0.389 | 0.416 | 0.443 |
| 48 | 0.136 | 0.185 | 0.225 | 0.261 | 0.293 | 0.324 | 0.353 | 0.381 | 0.408 | 0.435 |
| 49 | 0.133 | 0.181 | 0.221 | 0.256 | 0.288 | 0.318 | 0.347 | 0.374 | 0.401 | 0.427 |
| 50 | 0.130 | 0.178 | 0.217 | 0.251 | 0.283 | 0.312 | 0.341 | 0.368 | 0.394 | 0.419 |
| 51 | 0.128 | 0.175 | 0.213 | 0.246 | 0.278 | 0.307 | 0.335 | 0.361 | 0.387 | 0.412 |
| 52 | 0.125 | 0.171 | 0.209 | 0.242 | 0.273 | 0.301 | 0.329 | 0.355 | 0.381 | 0.405 |
| 53 | 0.123 | 0.168 | 0.205 | 0.238 | 0.268 | 0.296 | 0.323 | 0.349 | 0.374 | 0.398 |
| 54 | 0.121 | 0.165 | 0.201 | 0.233 | 0.263 | 0.291 | 0.318 | 0.343 | 0.368 | 0.391 |
| 55 | 0.119 | 0.162 | 0.198 | 0.229 | 0.259 | 0.286 | 0.312 | 0.337 | 0.362 | 0.385 |
| 56 | 0.117 | 0.160 | 0.194 | 0.226 | 0.254 | 0.281 | 0.307 | 0.332 | 0.356 | 0.379 |
| 57 | 0.115 | 0.157 | 0.191 | 0.222 | 0.250 | 0.277 | 0.302 | 0.326 | 0.350 | 0.373 |
| 58 | 0.113 | 0.154 | 0.188 | 0.218 | 0.246 | 0.272 | 0.297 | 0.321 | 0.345 | 0.367 |
| 59 | 0.111 | 0.152 | 0.185 | 0.215 | 0.242 | 0.268 | 0.293 | 0.316 | 0.339 | 0.361 |
| 60 | 0.109 | 0.149 | 0.182 | 0.211 | 0.238 | 0.264 | 0.288 | 0.311 | 0.334 | 0.356 |
| Multiplier 1 | 6.63 | 9.21 | 11.35 | 13.28 | 15.09 | 16.81 | 18.48 | 20.09 | 21.67 | 23.21 |
| Multiplier 2 | −5.81 | −15.49 | −25.66 | −36.39 | −47.63 | −59.53 | −71.65 | −84.60 | −97.88 | −111.76 |

**TABLE 12.1.10** $R^2$ Table: Level 0.1% Critical Values (Very Highly Significant)

| Number of Cases | Number of X Variables (k) | | | | | | | | | |
|---|---|---|---|---|---|---|---|---|---|---|
| (n) | 1 | 2 | 3 | 4 | 5 | 6 | 7 | 8 | 9 | 10 |
| 3 | 1.000 | | | | | | | | | |
| 4 | 0.998 | 1.000 | | | | | | | | |
| 5 | 0.982 | 0.999 | 1.000 | | | | | | | |
| 6 | 0.949 | 0.990 | 0.999 | 1.000 | | | | | | |
| 7 | 0.904 | 0.968 | 0.993 | 0.999 | 1.000 | | | | | |
| 8 | 0.855 | 0.937 | 0.977 | 0.995 | 1.000 | 1.000 | | | | |
| 9 | 0.807 | 0.900 | 0.952 | 0.982 | 0.996 | 1.000 | 1.000 | | | |

*(Continued)*

**TABLE 12.1.10** $R^2$ Table: Level 0.1% Critical Values (Very Highly Significant)—cont'd

| Number of Cases | Number of X Variables (k) | | | | | | | | | |
|---|---|---|---|---|---|---|---|---|---|---|
| (n) | 1 | 2 | 3 | 4 | 5 | 6 | 7 | 8 | 9 | 10 |
| 10 | 0.761 | 0.861 | 0.922 | 0.961 | 0.985 | 0.996 | 1.000 | 1.000 | | |
| 11 | 0.717 | 0.822 | 0.889 | 0.936 | 0.967 | 0.987 | 0.997 | 1.000 | 1.000 | |
| 12 | 0.678 | 0.785 | 0.856 | 0.908 | 0.945 | 0.972 | 0.989 | 0.997 | 1.000 | 1.000 |
| 13 | 0.642 | 0.749 | 0.822 | 0.878 | 0.920 | 0.952 | 0.975 | 0.990 | 0.997 | 1.000 |
| 14 | 0.608 | 0.715 | 0.790 | 0.848 | 0.894 | 0.930 | 0.958 | 0.978 | 0.991 | 0.998 |
| 15 | 0.578 | 0.684 | 0.759 | 0.819 | 0.867 | 0.906 | 0.938 | 0.962 | 0.980 | 0.992 |
| 16 | 0.550 | 0.654 | 0.730 | 0.790 | 0.840 | 0.881 | 0.916 | 0.944 | 0.966 | 0.982 |
| 17 | 0.525 | 0.627 | 0.702 | 0.763 | 0.813 | 0.856 | 0.893 | 0.923 | 0.949 | 0.968 |
| 18 | 0.502 | 0.602 | 0.676 | 0.736 | 0.787 | 0.831 | 0.869 | 0.902 | 0.930 | 0.953 |
| 19 | 0.480 | 0.578 | 0.651 | 0.711 | 0.763 | 0.807 | 0.846 | 0.880 | 0.910 | 0.935 |
| 20 | 0.461 | 0.556 | 0.628 | 0.688 | 0.739 | 0.784 | 0.824 | 0.859 | 0.890 | 0.917 |
| 21 | 0.442 | 0.536 | 0.606 | 0.665 | 0.716 | 0.761 | 0.801 | 0.837 | 0.869 | 0.897 |
| 22 | 0.426 | 0.517 | 0.586 | 0.644 | 0.694 | 0.739 | 0.780 | 0.816 | 0.849 | 0.878 |
| 23 | 0.410 | 0.499 | 0.567 | 0.624 | 0.674 | 0.718 | 0.759 | 0.795 | 0.829 | 0.859 |
| 24 | 0.395 | 0.482 | 0.548 | 0.605 | 0.654 | 0.698 | 0.739 | 0.775 | 0.809 | 0.839 |
| 25 | 0.382 | 0.466 | 0.531 | 0.587 | 0.635 | 0.679 | 0.719 | 0.756 | 0.790 | 0.821 |
| 26 | 0.369 | 0.452 | 0.515 | 0.570 | 0.618 | 0.661 | 0.701 | 0.737 | 0.771 | 0.802 |
| 27 | 0.357 | 0.438 | 0.500 | 0.553 | 0.601 | 0.644 | 0.683 | 0.719 | 0.753 | 0.784 |
| 28 | 0.346 | 0.425 | 0.486 | 0.538 | 0.585 | 0.627 | 0.666 | 0.702 | 0.735 | 0.767 |
| 29 | 0.335 | 0.412 | 0.472 | 0.523 | 0.569 | 0.611 | 0.649 | 0.685 | 0.718 | 0.750 |
| 30 | 0.325 | 0.401 | 0.459 | 0.510 | 0.555 | 0.596 | 0.634 | 0.669 | 0.702 | 0.733 |
| 31 | 0.316 | 0.389 | 0.447 | 0.496 | 0.541 | 0.581 | 0.619 | 0.654 | 0.686 | 0.717 |
| 32 | 0.307 | 0.379 | 0.435 | 0.484 | 0.527 | 0.567 | 0.604 | 0.639 | 0.671 | 0.702 |
| 33 | 0.299 | 0.369 | 0.424 | 0.472 | 0.515 | 0.554 | 0.590 | 0.625 | 0.657 | 0.687 |
| 34 | 0.291 | 0.360 | 0.414 | 0.460 | 0.503 | 0.541 | 0.577 | 0.611 | 0.643 | 0.673 |
| 35 | 0.283 | 0.351 | 0.404 | 0.450 | 0.491 | 0.529 | 0.564 | 0.598 | 0.629 | 0.659 |
| 36 | 0.276 | 0.342 | 0.394 | 0.439 | 0.480 | 0.517 | 0.552 | 0.585 | 0.616 | 0.646 |
| 37 | 0.269 | 0.334 | 0.385 | 0.429 | 0.469 | 0.506 | 0.540 | 0.573 | 0.604 | 0.633 |
| 38 | 0.263 | 0.326 | 0.376 | 0.420 | 0.459 | 0.495 | 0.529 | 0.561 | 0.591 | 0.620 |
| 39 | 0.257 | 0.319 | 0.368 | 0.411 | 0.449 | 0.485 | 0.518 | 0.550 | 0.580 | 0.608 |
| 40 | 0.251 | 0.312 | 0.360 | 0.402 | 0.440 | 0.475 | 0.508 | 0.539 | 0.569 | 0.597 |
| 41 | 0.245 | 0.305 | 0.352 | 0.393 | 0.431 | 0.465 | 0.498 | 0.529 | 0.558 | 0.586 |
| 42 | 0.240 | 0.298 | 0.345 | 0.385 | 0.422 | 0.456 | 0.488 | 0.518 | 0.547 | 0.575 |
| 43 | 0.235 | 0.292 | 0.338 | 0.378 | 0.414 | 0.447 | 0.479 | 0.509 | 0.537 | 0.564 |
| 44 | 0.230 | 0.286 | 0.331 | 0.370 | 0.406 | 0.439 | 0.470 | 0.499 | 0.527 | 0.554 |

**TABLE 12.1.10** $R^2$ Table: Level 0.1% Critical Values (Very Highly Significant)—cont'd

| Number of Cases | Number of X Variables (k) | | | | | | | | | |
|---|---|---|---|---|---|---|---|---|---|---|
| (n) | 1 | 2 | 3 | 4 | 5 | 6 | 7 | 8 | 9 | 10 |
| 45 | 0.225 | 0.280 | 0.324 | 0.363 | 0.398 | 0.431 | 0.461 | 0.490 | 0.518 | 0.544 |
| 46 | 0.220 | 0.275 | 0.318 | 0.356 | 0.391 | 0.423 | 0.453 | 0.482 | 0.509 | 0.535 |
| 47 | 0.216 | 0.269 | 0.312 | 0.349 | 0.383 | 0.415 | 0.445 | 0.473 | 0.500 | 0.526 |
| 48 | 0.212 | 0.264 | 0.306 | 0.343 | 0.377 | 0.408 | 0.437 | 0.465 | 0.491 | 0.517 |
| 49 | 0.208 | 0.259 | 0.301 | 0.337 | 0.370 | 0.401 | 0.429 | 0.457 | 0.483 | 0.508 |
| 50 | 0.204 | 0.255 | 0.295 | 0.331 | 0.363 | 0.394 | 0.422 | 0.449 | 0.475 | 0.500 |
| 51 | 0.200 | 0.250 | 0.290 | 0.325 | 0.357 | 0.387 | 0.415 | 0.442 | 0.467 | 0.492 |
| 52 | 0.197 | 0.246 | 0.285 | 0.320 | 0.351 | 0.381 | 0.408 | 0.435 | 0.460 | 0.484 |
| 53 | 0.193 | 0.242 | 0.280 | 0.314 | 0.345 | 0.374 | 0.402 | 0.428 | 0.453 | 0.477 |
| 54 | 0.190 | 0.237 | 0.276 | 0.309 | 0.340 | 0.368 | 0.395 | 0.421 | 0.446 | 0.469 |
| 55 | 0.186 | 0.233 | 0.271 | 0.304 | 0.334 | 0.362 | 0.389 | 0.414 | 0.439 | 0.462 |
| 56 | 0.183 | 0.230 | 0.267 | 0.299 | 0.329 | 0.357 | 0.383 | 0.408 | 0.432 | 0.455 |
| 57 | 0.180 | 0.226 | 0.262 | 0.294 | 0.324 | 0.351 | 0.377 | 0.402 | 0.426 | 0.448 |
| 58 | 0.177 | 0.222 | 0.258 | 0.290 | 0.319 | 0.346 | 0.371 | 0.396 | 0.419 | 0.442 |
| 59 | 0.174 | 0.219 | 0.254 | 0.285 | 0.314 | 0.341 | 0.366 | 0.390 | 0.413 | 0.436 |
| 60 | 0.172 | 0.215 | 0.250 | 0.281 | 0.309 | 0.336 | 0.361 | 0.384 | 0.407 | 0.429 |
| Multiplier 1 | 10.83 | 13.82 | 16.27 | 18.47 | 20.52 | 22.46 | 24.32 | 26.12 | 27.88 | 29.59 |
| Multiplier 2 | −31.57 | −54.02 | −75.12 | −96.26 | −117.47 | −138.94 | −160.86 | −183.33 | −206.28 | −229.55 |

Using $p$-value notation, you could say that the regression is very highly significant ($p < 0.001$).

To verify this very high level of significance using $F$ directly, compare the $F$ statistic 42.86 from the computer printout to the $F$ table value for testing at the 0.1% level (see Table D–11 in Appendix D) of between 7.054 and 6.171 for $k = 3$ numerator degrees of freedom and $n - k - 1 = 41$ denominator degrees of freedom. (Since 41 does not appear in the table, we know that the $F$ table value is between the values 7.054 for 30 denominator degrees of freedom and 6.171 for 60 denominator degrees of freedom.) Because the $F$ statistic (42.86) is larger than the $F$ table value (between 7.054 and 6.171), we again conclude that the result is very highly significant ($p < 0.001$).

## Tables of Critical Values for Testing $R^2$

Tables 12.1.8 through 12.1.11 can be used for testing whether or not the model is significant (the $F$ test). These tables are provided for test levels 0.05 (significant), 0.01 (highly significant), 0.001 (very highly significant), and 0.1. For each test level, your regression is significant if $R^2$ exceeds the value in the table for your number of X variables ($k$) and number of cases ($n$). For example, if you have run a regression with $k = 2$ explanatory X variables and $n = 35$ cases, then it is significant at level 0.05 if $R^2$ exceeds the critical value 0.171 (from the 5% table).

In practice, most computer programs will perform the $F$ test for you and report whether or not it is significant and, if so, at what level. These $R^2$ tables would not be needed in such cases. Their use is twofold: (1) to find significance when you have an $R^2$ value reported without significance test information and (2) to show you how strongly the significance level depends on $n$ and $k$. The critical $R^2$ value required for significance is smaller (less demanding) when $n$ is larger because you have more information. However, the critical $R^2$ value required for significance is larger (more demanding) when $k$ is larger because of the effort involved in estimating the extra regression coefficients.

**TABLE 12.1.11** $R^2$ Table: Level 10% Critical Values

| Number of Cases (n) | Number of X Variables (k) | | | | | | | | | |
|---|---|---|---|---|---|---|---|---|---|---|
| | 1 | 2 | 3 | 4 | 5 | 6 | 7 | 8 | 9 | 10 |
| 3 | 0.976 | | | | | | | | | |
| 4 | 0.810 | 0.990 | | | | | | | | |
| 5 | 0.649 | 0.900 | 0.994 | | | | | | | |
| 6 | 0.532 | 0.785 | 0.932 | 0.996 | | | | | | |
| 7 | 0.448 | 0.684 | 0.844 | 0.949 | 0.997 | | | | | |
| 8 | 0.386 | 0.602 | 0.759 | 0.877 | 0.959 | 0.997 | | | | |
| 9 | 0.339 | 0.536 | 0.685 | 0.804 | 0.898 | 0.965 | 0.998 | | | |
| 10 | 0.302 | 0.482 | 0.622 | 0.738 | 0.835 | 0.914 | 0.970 | 0.998 | | |
| 11 | 0.272 | 0.438 | 0.568 | 0.680 | 0.775 | 0.857 | 0.925 | 0.974 | 0.998 | |
| 12 | 0.247 | 0.401 | 0.523 | 0.628 | 0.721 | 0.803 | 0.874 | 0.933 | 0.977 | 0.998 |
| 13 | 0.227 | 0.369 | 0.484 | 0.584 | 0.673 | 0.753 | 0.825 | 0.888 | 0.940 | 0.979 |
| 14 | 0.209 | 0.342 | 0.450 | 0.545 | 0.630 | 0.708 | 0.779 | 0.842 | 0.899 | 0.946 |
| 15 | 0.194 | 0.319 | 0.420 | 0.510 | 0.592 | 0.667 | 0.736 | 0.799 | 0.857 | 0.907 |
| 16 | 0.181 | 0.298 | 0.394 | 0.480 | 0.558 | 0.630 | 0.697 | 0.759 | 0.816 | 0.868 |
| 17 | 0.170 | 0.280 | 0.371 | 0.453 | 0.527 | 0.596 | 0.661 | 0.721 | 0.778 | 0.830 |
| 18 | 0.160 | 0.264 | 0.351 | 0.428 | 0.499 | 0.566 | 0.628 | 0.687 | 0.742 | 0.794 |
| 19 | 0.151 | 0.250 | 0.332 | 0.406 | 0.474 | 0.538 | 0.598 | 0.655 | 0.709 | 0.760 |
| 20 | 0.143 | 0.237 | 0.316 | 0.386 | 0.452 | 0.513 | 0.571 | 0.626 | 0.679 | 0.729 |
| 21 | 0.136 | 0.226 | 0.301 | 0.368 | 0.431 | 0.490 | 0.546 | 0.599 | 0.650 | 0.699 |
| 22 | 0.129 | 0.215 | 0.287 | 0.352 | 0.412 | 0.469 | 0.523 | 0.575 | 0.624 | 0.671 |
| 23 | 0.124 | 0.206 | 0.275 | 0.337 | 0.395 | 0.450 | 0.502 | 0.552 | 0.600 | 0.646 |
| 24 | 0.118 | 0.197 | 0.263 | 0.323 | 0.379 | 0.432 | 0.482 | 0.530 | 0.577 | 0.622 |
| 25 | 0.113 | 0.189 | 0.253 | 0.310 | 0.364 | 0.415 | 0.464 | 0.511 | 0.556 | 0.599 |
| 26 | 0.109 | 0.181 | 0.243 | 0.298 | 0.350 | 0.400 | 0.447 | 0.492 | 0.536 | 0.579 |
| 27 | 0.105 | 0.175 | 0.234 | 0.287 | 0.338 | 0.386 | 0.431 | 0.475 | 0.518 | 0.559 |
| 28 | 0.101 | 0.168 | 0.225 | 0.277 | 0.326 | 0.372 | 0.417 | 0.459 | 0.501 | 0.541 |
| 29 | 0.097 | 0.162 | 0.218 | 0.268 | 0.315 | 0.360 | 0.403 | 0.444 | 0.484 | 0.523 |
| 30 | 0.094 | 0.157 | 0.210 | 0.259 | 0.305 | 0.348 | 0.390 | 0.430 | 0.469 | 0.507 |
| 31 | 0.091 | 0.152 | 0.203 | 0.251 | 0.295 | 0.337 | 0.378 | 0.417 | 0.455 | 0.492 |
| 32 | 0.088 | 0.147 | 0.197 | 0.243 | 0.286 | 0.327 | 0.366 | 0.405 | 0.442 | 0.478 |
| 33 | 0.085 | 0.142 | 0.191 | 0.236 | 0.277 | 0.317 | 0.356 | 0.393 | 0.429 | 0.464 |
| 34 | 0.082 | 0.138 | 0.185 | 0.229 | 0.269 | 0.308 | 0.346 | 0.382 | 0.417 | 0.451 |
| 35 | 0.080 | 0.134 | 0.180 | 0.222 | 0.262 | 0.300 | 0.336 | 0.371 | 0.406 | 0.439 |
| 36 | 0.078 | 0.130 | 0.175 | 0.216 | 0.255 | 0.291 | 0.327 | 0.361 | 0.395 | 0.427 |

**TABLE 12.1.11** $R^2$ Table: Level 10% Critical Values—cont'd

| Number of Cases | | | | Number of X Variables (k) | | | | | | |
|---|---|---|---|---|---|---|---|---|---|---|
| (n) | 1 | 2 | 3 | 4 | 5 | 6 | 7 | 8 | 9 | 10 |
| 37 | 0.075 | 0.127 | 0.170 | 0.210 | 0.248 | 0.284 | 0.318 | 0.352 | 0.385 | 0.416 |
| 38 | 0.073 | 0.123 | 0.166 | 0.205 | 0.241 | 0.276 | 0.310 | 0.343 | 0.375 | 0.406 |
| 39 | 0.071 | 0.120 | 0.162 | 0.199 | 0.235 | 0.269 | 0.302 | 0.334 | 0.366 | 0.396 |
| 40 | 0.070 | 0.117 | 0.157 | 0.194 | 0.229 | 0.263 | 0.295 | 0.326 | 0.357 | 0.387 |
| 41 | 0.068 | 0.114 | 0.154 | 0.190 | 0.224 | 0.257 | 0.288 | 0.319 | 0.348 | 0.378 |
| 42 | 0.066 | 0.111 | 0.150 | 0.185 | 0.219 | 0.250 | 0.281 | 0.311 | 0.340 | 0.369 |
| 43 | 0.065 | 0.109 | 0.146 | 0.181 | 0.214 | 0.245 | 0.275 | 0.304 | 0.333 | 0.361 |
| 44 | 0.063 | 0.106 | 0.143 | 0.177 | 0.209 | 0.239 | 0.269 | 0.297 | 0.325 | 0.353 |
| 45 | 0.062 | 0.104 | 0.140 | 0.173 | 0.204 | 0.234 | 0.263 | 0.291 | 0.318 | 0.345 |
| 46 | 0.060 | 0.102 | 0.137 | 0.169 | 0.200 | 0.229 | 0.257 | 0.285 | 0.312 | 0.338 |
| 47 | 0.059 | 0.099 | 0.134 | 0.166 | 0.196 | 0.224 | 0.252 | 0.279 | 0.305 | 0.331 |
| 48 | 0.058 | 0.097 | 0.131 | 0.162 | 0.191 | 0.220 | 0.247 | 0.273 | 0.299 | 0.324 |
| 49 | 0.057 | 0.095 | 0.128 | 0.159 | 0.188 | 0.215 | 0.242 | 0.268 | 0.293 | 0.318 |
| 50 | 0.055 | 0.093 | 0.126 | 0.156 | 0.184 | 0.211 | 0.237 | 0.263 | 0.287 | 0.312 |
| 51 | 0.054 | 0.092 | 0.123 | 0.153 | 0.180 | 0.207 | 0.233 | 0.258 | 0.282 | 0.306 |
| 52 | 0.053 | 0.090 | 0.121 | 0.150 | 0.177 | 0.203 | 0.228 | 0.253 | 0.277 | 0.300 |
| 53 | 0.052 | 0.088 | 0.119 | 0.147 | 0.174 | 0.199 | 0.224 | 0.248 | 0.272 | 0.295 |
| 54 | 0.051 | 0.086 | 0.116 | 0.144 | 0.170 | 0.196 | 0.220 | 0.244 | 0.267 | 0.290 |
| 55 | 0.050 | 0.085 | 0.114 | 0.142 | 0.167 | 0.192 | 0.216 | 0.239 | 0.262 | 0.284 |
| 56 | 0.049 | 0.083 | 0.112 | 0.139 | 0.164 | 0.189 | 0.212 | 0.235 | 0.257 | 0.279 |
| 57 | 0.049 | 0.082 | 0.110 | 0.137 | 0.162 | 0.185 | 0.209 | 0.231 | 0.253 | 0.275 |
| 58 | 0.048 | 0.080 | 0.108 | 0.134 | 0.159 | 0.182 | 0.205 | 0.227 | 0.249 | 0.270 |
| 59 | 0.047 | 0.079 | 0.107 | 0.132 | 0.156 | 0.179 | 0.202 | 0.223 | 0.245 | 0.266 |
| 60 | 0.046 | 0.078 | 0.105 | 0.130 | 0.153 | 0.176 | 0.198 | 0.220 | 0.241 | 0.261 |
| Multiplier 1 | 2.71 | 4.61 | 6.25 | 7.78 | 9.24 | 10.65 | 12.02 | 13.36 | 14.68 | 15.99 |
| Multiplier 2 | 3.12 | 3.08 | 2.00 | 0.32 | −1.92 | −4.75 | −7.59 | −11.12 | −14.94 | −19.05 |

If you have more than 60 cases, you may find critical values using the two multipliers at the bottom of the $R^2$ table according to the following formula:

Critical Values for $R^2$ when $n > 60$

$$\text{Critical value} = \frac{\text{Multiplier 1}}{n} + \frac{\text{Multiplier 2}}{n^2}$$

For example, with $n = 135$ cases and $k = 6$ explanatory $X$ variables, to test at level 0.05 you would use the two multipliers 12.59 and −18.24 at the bottom of the $k = 6$ column in the 5% table. Using the formula, you would find the critical value for $R^2$ to be

$$\begin{aligned}\text{Critical value} &= \frac{\text{Multiplier 1}}{n} + \frac{\text{Multiplier 2}}{n^2} \\ &= \frac{12.59}{135} + \frac{-18.24}{135^2} \\ &= 0.09326 - 0.00100 \\ &= 0.0923\end{aligned}$$

If the $R^2$ for your data set (from the computer printout) exceeds this value (0.0923, or 9.23%), the $F$ test is significant; otherwise, it is not.

## Which Variables Are Significant? A *t* Test for Each Coefficient

If the $F$ test is significant, you know that one or more of the $X$ variables is helpful in predicting $Y$, and you may proceed with statistical inference using $t$ tests for individual regression coefficients to find out which one (or more) of the $X$ variables is useful. These $t$ tests show you whether an $X$ variable has a significant impact on $Y$, holding all other $X$ variables fixed. Don't forget that whenever you accept the null hypothesis, it is a *weak* conclusion and you have not proven that an $X$ is not useful—you merely lack convincing evidence that there is a relationship. There could be a relationship, but due to randomness or small sample size, you might be unable to detect it with the data with which you are working.

If the $F$ test is not significant, then you are *not* permitted to use $t$ tests on the regression coefficients. In rare cases, these $t$ tests can be significant even though the $F$ test is not. The $F$ test dominates in these cases, and you must conclude that nothing is significant. To do otherwise would raise your type I error rate above the claimed level (5%, for example).

The $t$ test for each coefficient is based on the estimated regression coefficient and its standard error and uses the $t$ table critical value for $n - k - 1$ degrees of freedom. The confidence interval for a particular population regression coefficient, say, the $j$th one, $b_j$, extends in the usual way.

> ### Confidence Interval for the *j*th Regression Coefficient, $\beta_j$
>
> From $b_j - tS_{b_j}$   to   $b_j + tS_{b_j}$
>
> where $t$ is from the $t$ table for $n - k - 1$ degrees of freedom.

The $t$ test is significant if the reference value 0 (indicating no effect) is *not* in the confidence interval. There is nothing really new here; this is just the usual generic procedure for a two-sided test.

Alternatively, you could compare the $t$ statistic $b_j/S_{b_j}$ to the $t$ table value, deciding significance if the absolute value of the $t$ statistic is larger. As you look down a column of $t$ values, one for each coefficient, there is an easy, approximate way to tell which coefficients are significant: These are the ones that are 2 or larger in absolute value, since the $t$ table value is approximately 2 for a test at the 5% level (provided $n$ is large enough). As always, the $t$ statistic and the confidence interval approaches must always give the same result (significant or not) for each test.

What exactly is being tested here? The $t$ test for $\beta_j$ decides whether or not $X_j$ has a significant effect on $Y$ in the population *with all other X variables held constant*. This is not a test for the correlation of $X_j$ with $Y$, which would ignore all the other $X$ variables. Rather, it is a test for the effect of $X_j$ on $Y$ after adjustment has been made for all other factors. In studies of salary designed to identify possible gender discrimination, it is common to adjust for education and experience. Although the men in a department might be paid more than women, on average, it is also important to see whether these differences can be explained by factors other than gender. When you include all these factors in a multiple regression (by regressing $Y$ = salary on $X_1$ = gender, $X_2$ = education, and $X_3$ = experience), the regression coefficient for gender will represent the effect of gender on salary after adjustment for education and experience.[7]

Here are the formulas for the hypotheses for the significance test of the $j$th regression coefficient:

> ### Hypotheses for the *t* Test of the *j*th Regression Coefficient, $\beta_j$
>
> $$H_0: \beta_j = 0$$
>
> $$H_1: \beta_j \neq 0$$

For the magazine advertising example, the $t$ test will have $n - k - 1 = 45 - 3 - 1 = 41$ degrees of freedom. The two-sided critical value from the $t$ table is 1.960 (or, more accurately, 2.020).[8] Table 12.1.12 provides the appropriate information from the computer results of Table 12.1.6.

Two of the $X$ variables are significant because their $p$-values are less than 0.05. Another (equivalent) way to verify significance is to see which $t$ statistics are larger than the $t$ table value 2.020. Yet another (equivalent) way

**TABLE 12.1.12  Multiple Regression Computer *t*-Test Results**

| Predictor | Coeff | SE Coeff | t | p |
|-----------|-------|----------|------|-------|
| Constant | –22385 | 36060 | –0.62 | 0.538 |
| Audience | 10.5066 | 0.9422 | 11.15 | 0.000 |
| Male | –20779 | 37961 | –0.55 | 0.587 |
| Income | 1.0920 | 0.4619 | 2.36 | 0.023 |

---

7. The gender variable, $X_1$, might be represented as 0 for a woman and 1 for a man. In this case, the regression coefficient would represent the extra pay, on average, for a man compared to a woman with the same education and experience. If the gender variable were represented as 1 for a woman and 0 for a man, the regression coefficient would be the extra pay for a woman compared to a man with the same characteristics otherwise. Fortunately, the conclusions will be identical regardless of which representation is used.

8. Don't forget that using the $t$ value for an infinite number of degrees of freedom whenever there are 40 or more degrees of freedom is just an approximation. In this case, the true $t$ table value is 2.020; 1.960 is just a convenient approximation.

to verify significance is to see which of the 95% confidence intervals for the regression coefficients do not include 0. As we suspected originally, audience size has a lot to do with advertising costs. With such a high $t$ value, 11.15, the effect of audience on page costs is very highly significant (holding percent male and median income constant). The effect of median income on page costs is also significant (holding audience and percent male constant).

Evidently, the percentage of male readership does not significantly affect page costs (holding audience and median income constant), since this $t$ test is not significant. It is possible that this percentage has an impact on page costs only through median income, which may be higher for males. Thus, after you adjust for median income, it is reasonable that the variable for percent male would provide no further information for determining page costs. Although the estimated effect of percent male is –20,779, it is only randomly different from 0. Strictly speaking, you are not allowed to interpret this –20,779 coefficient; since it is not significant, you have no "license" to explain it. It may *look* like –20,779, but it is not distinguishable from $0.00$; you can't even really tell if it is positive or negative!

The constant, $a = -22,385$, is not significant. It is not significantly different from zero. No claim can be made as to whether the population parameter, $\alpha$, is positive or negative, since it might reasonably be zero. In cost accounting applications, $a$ often estimates the fixed cost of production. The confidence intervals and hypothesis tests would show you whether or not there is a significant fixed component in your cost structure.

## Other Tests for a Regression Coefficient

You may also perform other tests with regression coefficients, just as you have with mean values. If there is a reference value for one of the regression coefficients (which does not come from the data), you may test whether or not the estimated regression coefficient differs significantly from the reference value. Simply see if the reference value is in the confidence interval and decide "significantly different" if it is not, as usual. Or use the $t$ statistic $(b_j - \text{Reference value})/S_{b_j}$ deciding "significantly different" if its absolute value is larger than the $t$ table value for $n - k - 1$ degrees of freedom.

Suppose you had believed (before coming across this data set) that the additional cost per reader was $5.00 per thousand people. To test this assumption, use $5.00 as the reference value. Since the confidence interval (e.g., from Figure 12.1.1b) for audience, $8.604 to $12.409, excludes the reference value, you may conclude that the effect of audience on ad costs, adjusting for percent male readership and median income, is significantly more than $5.00 per thousand. Note that this is a one-sided conclusion based on a two-sided test. The two-sided test is appropriate here because the estimate might have come out on the other side of $5.00.

One-sided confidence intervals may be computed in the usual way for one (or more) regression coefficients, giving you a one-sided statement about the population regression coefficient (or coefficients) of interest. Be sure to use the one-sided $t$ values from the $t$ table for $n - k - 1$ degrees of freedom, and be sure that your confidence interval for $\beta_j$ includes the regression coefficient $b_j$.

For example, the regression coefficient for income is $b_3 = 1.09198$, indicating that (all else equal) an extra dollar of median income would boost the price of a full-page ad by $1.09198, on average. The standard error is $S_{b_3}$ and the one-sided $t$ table value for $n - k - 1 = 41$ is $t = 1.683$ (this is the more correct value than the approximation 1.645 from our $t$ table), so the one-sided interval will extend upward from $b_3 - tS_{b_3} = 1.09198 - 1.683 \times 0.46189 = 0.315$. Your conclusion is

> You are 95% sure that an extra dollar of median income will boost the average page cost by at least $0.315.

The 31.5 cents defining the one-sided confidence bound is much lower than the estimated value of $1.09198 because you have allowed for the random errors of estimation. By using a one-sided interval instead of a two-sided interval, you can claim 31.5 cents instead of the smaller value of 15.9 cents, which defines the lower endpoint of the two-sided interval.

One-sided tests may be done for regression coefficients in the usual way, provided you are interested in only *one side* of the reference value and would not change the side of interest had the estimates come out differently.

## Which Variables Explain the Most?

Which $X$ variable or variables explain the most about $Y$? This is a good question. Unfortunately, there is no completely satisfying answer because relationships among the $X$ variables can make it fundamentally impossible to decide precisely which $X$ variable is really responsible for the behavior of the $Y$ variable. The answer depends on how you view the situation (in particular, whether or not you can change the $X$ variables individually). The answer also depends on how the $X$ variables interrelate (or correlate) with each other. Following are two useful but incomplete answers to this tough question.

### Comparing the Standardized Regression Coefficients

Since the regression coefficients $b_1, \ldots, b_k$ may all be in different measurement units, direct comparison is difficult; a small coefficient may actually be more important than a larger one. This is the classic problem of "trying to compare apples and oranges." The *standardized regression coefficients* eliminate this problem by expressing the coefficients in terms of a single, common set of statistically reasonable units so that comparison may at least be attempted.

The regression coefficient $b_i$ indicates the effect of a change in $X_i$ on $Y$ with all of the other $X$ variables unchanged. The measurement units of regression coefficient $b_i$ are units of $Y$ per unit of $X_i$. For example, if $Y$ is the dollar amount of sales and $X_1$ is the number of people in the sales force, $b_1$ is in units of dollars of sales per person. Suppose that the next regression coefficient, $b_2$, is in units of dollars of sales per number of total miles traveled by the sales force. The question of which is more important to sales, staffing level or travel budget, cannot be answered by comparing $b_1$ to $b_2$ because dollars per person and dollars per mile are not directly comparable.

The **standardized regression coefficient**, found by multiplying the regression coefficient $b_i$ by $S_{X_i}$ and dividing it by $S_Y$, represents the expected change in $Y$ (in standardized units of $S_Y$ where each "unit" is a statistical unit equal to one standard deviation) due to an increase in $X_i$ of one of its standardized units (i.e., $S_{X_i}$), with all other $X$ variables unchanged.[9] The absolute values of the standardized regression coefficients may be compared, giving a rough indication of the relative importance of the variables.[10] Each standardized regression coefficient is in units of standard deviations of $Y$ per standard deviation of $X_i$. These are just the ordinary sample standard deviations for each variable that you learned in Chapter 5. Use of these units is natural because they set the measurement scale according to the actual variation in each variable in your data set.

> ### Standardized Regression Coefficients
>
> $$b_i \frac{S_{X_i}}{S_Y}$$
>
> Each regression coefficient is adjusted according to a ratio of ordinary sample standard deviations. The absolute values give a rough indication of the relative importance of the $X$ variables.

To standardize the regression coefficients for the magazine ads example, you first need the standard deviation for each variable:

### Standard Deviations

| Page Costs | Audience | Percent Male | Median Income |
|---|---|---|---|
| $S_Y = 105,639$ | $S_{X_1} = 9,298$ | $S_{X_2} = 25.4\%$ | $S_{X_3} = 22,012$ |

---

9. Standardized regression coefficients are sometimes referred to as *beta coefficients*. We will avoid this term because it could easily be confused with the population regression coefficients (also $\beta$, or beta) and with the nondiversifiable component of risk in finance (which is called the *beta* of a stock and is an ordinary, unstandardized sample regression coefficient, where $X$ is the percent change in a market index and $Y$ is the percent change in the value of a stock certificate).
10. Recall that the absolute value simply ignores any minus sign.

You also need the regression coefficients:

### Regression Coefficients

| Audience | Percent Male | Median Income |
|---|---|---|
| $b_1 = 10.50658$ | $b_2 = -20,779$ | $b_3 = 1.09198$ |

Finally, the standardized regression coefficients may be computed:

### Standardized Regression Coefficients

| Audience | Percent Male | Median Income |
|---|---|---|
| $b_1 S_{X_1}/S_Y$ | $b_2 S_{X_2}/S_Y$ | $b_3 S_{X_3}/S_Y$ |
| $= 10.50658$ | $= -20,779$ | $= 1.09198$ |
| $\times 9,298/105,639$ | $\times 0.254/105,639$ | $\times 22,012/105,639$ |
| $= 0.925$ | $= -0.050$ | $= 0.228$ |

Here's the direct interpretation for one of these standardized coefficients: The value 0.925 for audience says that an increase in audience of one of its standard deviations (9,298, in thousands of readers) will result in an expected increase in page costs of 0.925 of its standard deviations ($105,639). That is, an audience increase of 9,298 (one standard deviation) will result in an expected page-cost increase of about $97,700 computed as $0.925 \times 105,639$ (slightly less, 0.925 or 92.5%, than one standard deviation of page costs).

More importantly, these standardized regression coefficients may now be compared. The largest in absolute value is 0.925 for audience, suggesting that this is the most important of the three $X$ variables. Next is median income, with 0.228. The smallest absolute value is $|-0.050| = 0.050$ for percent male.

It would be wrong to compare the regression coefficients directly, without first standardizing. Note that percent male has the largest regression coefficient (in absolute value), $|-20,779| = 20,779$. However, because it is in different measurement units from the other regression coefficients, a direct comparison does not make sense.

The absolute values of the standardized regression coefficients may be compared, providing a *rough* indication of importance of the variables. Again, the results are not perfect because relationships among the $X$ variables can make it fundamentally impossible to decide which $X$ variable is really responsible for the behavior of the $Y$ variable.

### Comparing the Correlation Coefficients

You might not really be interested in the regression coefficients from a multiple regression, which represent the effects of each variable with all others fixed. If you simply want to see how strongly each $X$ variable affects $Y$, allowing the other $X$ variables to "do what comes naturally" (i.e., deliberately *not* holding them fixed), you may compare the *absolute values of the correlation coefficients* for $Y$ with each $X$ in turn.

The correlation clearly measures the strength of the relationship (as was covered in Chapter 11), but why use the absolute value? Remember, a correlation near 1 or −1 indicates a strong relationship, and a correlation near 0 suggests no relationship. The absolute value of the correlation gives the *strength* of the relationship without indicating its *direction.*

Multiple regression *adjusts* or *controls* for the other variables, whereas the correlation coefficient does not.[11] If it is important that you adjust for the effects of other variables, then multiple regression is your answer. If you don't need to adjust, the correlation approach may meet your needs.

Here are the correlation coefficients of $Y$ with each of the $X$ variables for the magazine ads example. For example, the correlation of page costs with median income is −0.148.

### Correlation with Page Costs

| Audience | Percent Male | Median Income |
|----------|--------------|---------------|
| 0.850    | −0.126       | −0.148        |

In terms of the relationship to page costs, without adjustments for the other $X$ variables, audience has by far the highest absolute value of correlation, 0.850. Next in absolute value of correlation is median income, with $|-0.148| = 0.148$. Percent male has the smallest absolute value, $|-0.126| = 0.126$. It looks as if only audience is important in determining page costs. In fact, neither of the other two variables (by itself, without holding the others constant) explains a significant amount of page costs.

The multiple regression gives a different picture because it controls for other variables. After you adjust for audience, the multiple regression coefficient for median income indicates a significant effect on page costs. Here's how to interpret this: The adjustment for audience controls for the fact that higher incomes go with smaller audiences. The audience effect is removed, leaving only the pure income effect, which can be detected because it is no longer masked by the competing audience effect.

Although the correlation coefficients indicate the *individual* relationships with $Y$, the standardized regression coefficients from a multiple regression can provide you with important additional information.

## 12.2 PITFALLS AND PROBLEMS IN MULTIPLE REGRESSION

Unfortunately, multiple regression does not always work out as well in real life as it does in the textbooks. This section includes a checklist of potential problems and some suggestions about how to fix them (when possible).

There are three main kinds of problems. Following is a quick overview of each; more details will follow.

1. The problem of **multicollinearity** arises when some of your explanatory ($X$) variables are too similar. Although they do a good job of explaining and predicting $Y$ (as indicated by a high $R^2$ and a significant $F$ test), the individual regression coefficients are poorly estimated. The reason is that there is not enough information to decide which one (or more) of the variables is doing the explaining. One solution is to omit some of the variables in an effort to end the confusion. Another solution is to redefine some of the variables (perhaps using ratios) to distinguish them from one another.

2. The problem of **variable selection** arises when you have a long list of potentially useful explanatory $X$ variables and you are trying to decide which ones to include in the regression equation. On one hand, if you have too many $X$ variables, the unnecessary ones will degrade the quality of your results (perhaps due to multicollinearity). Some of the information in the data is wasted on the estimation of unnecessary parameters. On the other hand, if you omit a necessary $X$ variable, your predictions will lose quality because helpful information is being ignored. One solution is to think hard about *why* each $X$ variable is important and to make sure that each one you include is performing a potentially useful function. Another solution is to use an automated procedure that automatically tries to select the most useful variables for you.

3. The problem of **model misspecification** refers to the many different potential incompatibilities between your application and the multiple regression linear model, which is the underlying basis and framework for a multiple regression analysis. Your particular application might or might not conform to the assumptions of the multiple regression linear model. By exploring the data, you can be alerted to some of the potential problems with nonlinearity, unequal variability, or outliers. However, you may or may not have a problem: Even though the histograms of some variables may be skewed, and even though some scatterplots may be nonlinear, the multiple regression linear model might still hold. There is a *diagnostic plot* that helps you decide when the problem is serious enough to need fixing. A possible solution is creation of new $X$ variables, constructed from the current ones, and/or transformation of some or all of the variables. Another serious problem arises if you have a *time series*, so that the independence assumption of the multiple regression linear model is not satisfied. The time-series problem is complex; however, you may be able to do multiple regression using *percent changes* from one time period to the next in place of the original data.

---

11. There is an advanced statistical concept, the *partial correlation coefficient*, which is not covered in this book. It gives the correlation between two variables while controlling or adjusting for one or more additional variables.

## Multicollinearity: Are the Explanatory Variables Too Similar?

When some of your explanatory ($X$) variables are similar to one another, you may have a *multicollinearity* problem because it is difficult for multiple regression to distinguish between the effect of one variable and the effect of another. The consequences of multicollinearity can be *statistical* or *numerical*:

1. *Statistical* consequences of multicollinearity include difficulties in testing individual regression coefficients due to inflated standard errors. Thus, you may be unable to declare an $X$ variable significant even though (by itself) it has a strong relationship with $Y$.
2. *Numerical* consequences of multicollinearity include difficulties in the computer's calculations due to numerical instability. In extreme cases, the computer may try to divide by zero and thus fail to complete the analysis. Or, even worse, the computer may complete the analysis but then report meaningless, wildly incorrect numbers.[12]

Multicollinearity may or may not be a problem, depending on the purpose of your analysis and the extent of the multicollinearity. Small to moderate amounts of multicollinearity are usually not a problem. Extremely strong multicollinearity (for example, including the same variable twice) will always be a problem and may cause serious errors (numerical consequences). Fortunately, if your purpose is primarily to predict or forecast $Y$, strong multicollinearity may not be a problem because a careful multiple regression program can still produce the best (least-squares) forecasts of $Y$ based on all of the $X$ variables. However, if you want to use the individual regression coefficients to explain how $Y$ is affected by each $X$ variable, then the statistical consequences of multicollinearity will probably cause trouble because these effects cannot be separated. Table 12.2.1 summarizes the impact of multicollinearity on the regression analysis.

How can you tell if you have a multicollinearity problem? One simple way is to look at the ordinary bivariate correlations of each pair of variables.[13] The **correlation matrix** is a table giving the correlation between every pair of variables in your multivariate data set. The higher the correlation coefficient between one $X$ variable and another, the more

**TABLE 12.2.1 Effects of Multicollinearity in Regression**

| Extent of Multicollinearity | Effect on the Regression Analysis |
| --- | --- |
| Little | Not a problem |
| Moderate | Not usually a problem |
| Strong | Statistical consequences: Often a problem if you want to estimate effects of individual $X$ variables (i.e., regression coefficients); may not be a problem if your goal is just to predict or forecast $Y$ |
| Extremely strong | Numerical consequences: Always a problem; computer calculations may even be wrong due to numerical instability |

multicollinearity you have. The reason is that a high correlation (close to 1 or to $-1$) indicates strong association and tells you that these two $X$ variables are measuring something similar, bringing overlapping information to the analysis.

The primary statistical effect of multicollinearity is to *inflate the standard errors of some or all of the regression coefficients* ($S_{b_j}$). This is only natural: If two $X$ variables contain overlapping information, it is difficult to compute the effect of each individually. A high standard error is the computer's way of saying, "I found the regression coefficient for you, but it's not very precise because I can't tell whether this variable or some other one is really responsible." The result is that the confidence intervals for the regression coefficients become larger, and the $t$ tests are less likely to be significant.

With strong multicollinearity, you may find a regression that is very highly significant (based on the $F$ test) but for which not even one of the $t$ tests of the individual $X$ variables is significant. The computer is telling you that the $X$ variables taken as a group explain a lot about $Y$, but it is impossible to single out any particular $X$ variables as being responsible. Remember that the $t$ test for a particular variable $X_i$ measures its effect with the others held fixed. Thus, the $t$ test for $X_i$ measures only the *additional* information conveyed by $X_i$ over and above that of the other variables. If some other variable is very similar, then $X_i$ really isn't bringing significant new information into the regression.

One solution is to omit $X$ variables that duplicate the information already available in other $X$ variables. For example, if your $X$ variables include three different size measures, consider either eliminating two of them or else combining them into a single size measure (for example, using their average).

Another solution is to redefine some of the variables so that each $X$ variable has a clear, unique role in explaining $Y$. One common way to apply this idea to a group of similar $X$ variables is to choose a single representative $X$ variable

---

12. Dividing by zero is mathematically impossible; for example, 5/0 is undefined. However, due to small round-off errors made in computation, the computer may divide 5.0000000000968 by 0.0000000000327 instead. Rather than stopping and reporting trouble, the computer would use the inappropriate huge result of this division, 152,905,198,779.72.

13. Unfortunately, a complete diagnosis of multicollinearity is much more difficult than this because the $X$ variables must be considered all at once, not just two at a time. Full technical details may be found, for example, in D. A. Belsley, E. Kuh, and R. E. Welsch, *Regression Diagnostics: Identifying Influential Data and Sources of Collinearity* (New York: Wiley, 1980).

(choose one or form an index from all of them) and express other variables as ratios (per capita values can do this) with respect to your representative $X$ variable. For example, you could explain sales ($Y$) using population ($X_1$) and total income ($X_2$) for each region; however, these variables are multicollinear (that is, population and total income are highly correlated). You could correct the problem by explaining sales ($Y$) using population ($X_1$) and per capita income (the new $X_2$ variable). In effect, you would let population be the representative variable, indicating the overall size of the territory, and you would redefine income to convey new information (about how well off people are) instead of repeating old information (about how large the territory is).

### Example
#### Predicting Market Value from Assets and Employees

What is the equity market value of a firm, and how is it determined? It is the total value of all outstanding common stock, easily found by multiplying the total shares outstanding by the current price per share. It is determined by supply and demand in the stock market. Financial theorists tell us that it represents the present value of the (uncertain, risky) future cash flows of the firm. But how does it relate to other features of a firm? Let's use multiple regression to find out.

Consider the information shown in Table 12.2.2 on market value (the $Y$ variable, to be explained) and some explanatory $X$ variables (the value of assets owned and the number of employees) for the aerospace and defense companies in the Fortune 500.

You should anticipate a multicollinearity problem with this data set because every $X$ is basically a size variable. The $X$ variables bring in similar, overlapping information; that is, large firms tend to be large in every aspect: market value, assets, and employees. Small firms tend to be small in every aspect. Table 12.2.3 summarizes the multiple regression results.

Note that the regression is significant according to the $F$ test. Over three-quarters ($R^2 = 87.9\%$) of the variation in market value is explained by the $X$ variables taken together as a group, and this is very highly statistically significant. However, due to multicollinearity, no single $X$ variable is significant. Thus, market value is explained, but you can't say which $X$ is doing the explaining.

Some useful information about multicollinearity is provided by the correlation matrix, shown in Table 12.2.4,

which lists the correlation between every pair of variables in the multivariate data set. Note the extremely high correlations between the two $X$ variables: 0.944 between assets and employees. Such a high correlation suggests that, at least with respect to the numbers, these two $X$ variables are providing nearly identical information. No wonder the regression analysis can't tell them apart.

If you were to keep only one of the two $X$ variables, you would find a regression with a very highly significant $t$ test for that variable regardless of which $X$ variable you chose to keep. That is, each one does a good job of explaining market value by itself.

If you want to retain all of the information contained in both $X$ variables, a good way to proceed is to keep one of them as your representative size variable and define the other as a ratio. Let's keep assets as the representative size variable because it indicates the fixed investment required by the firm. The other variables will now be replaced by the ratio of employees to assets (indicating the number of employees per million dollars of assets). Now assets is the clear size variable, and the other brings in new information about efficiency in the use of employees. Table 12.2.5 shows the new data set.

Let's first check the correlation matrix, shown in Table 12.2.6, to look for multicollinearity problems. These correlations look much better. The correlation between the $X$ variables (–0.271) is no longer extremely high and is not statistically significant.

What should you expect to see in the multiple regression results? The regression should still be significant, and the $t$ test for assets should now be significant because it has no competing size variables. The uncertainty to be resolved is, knowing the assets, does it help you to know the employee ratio in order to explain market value? Table 12.2.7 shows the results.

These results confirm our expectations: The regression ($F$ test) is significant, and the $t$ test for assets is also significant now that the strong multicollinearity has been eliminated. We have also found that the other variable (employees per asset) is not significant.

Apparently, for this small ($n = 14$) group of large aerospace and defense firms, much of the variation in market value can be explained by the level of assets. Furthermore, information about human resources (employees) contributes little, if any, additional insight into the market value of these successful firms. Perhaps analysis with a larger sample of firms would be able to detect the impact of this variable.

### TABLE 12.2.2 Aerospace and Defense Companies in the Fortune 500

|  | Market Value (millions), $Y$ | Assets (millions), $X_1$ | Employees, $X_2$ |
|---|---|---|---|
| Alliant Techsystems | 2,724.7 | 3,593.2 | 19,000 |
| Boeing | 54,948.9 | 62,053.0 | 157,100 |
| General Dynamics | 29,670.0 | 31,077.0 | 91,700 |

*(Continued)*

**TABLE 12.2.2** Aerospace and Defense Companies in the Fortune 500—cont'd

|  | Market Value (millions), $Y$ | Assets (millions), $X_1$ | Employees, $X_2$ |
|---|---|---|---|
| Goodrich | 8,888.0 | 8,741.4 | 24,000 |
| Honeywell International | 34,099.0 | 36,004.0 | 122,000 |
| ITT | 9,646.1 | 11,129.1 | 40,200 |
| Lockheed Martin | 31,626.1 | 35,111.0 | 140,000 |
| L-3 Communications | 10,779.6 | 14,813.0 | 67,000 |
| Northrop Grumman | 19,840.6 | 30,252.0 | 120,700 |
| Precision Castparts | 17,542.6 | 6,721.4 | 20,600 |
| Raytheon | 21,717.7 | 23,607.0 | 75,000 |
| Rockwell Collins | 9,901.7 | 4,645.0 | 19,300 |
| Textron | 5,910.4 | 18,940.0 | 32,000 |
| United Technologies | 69,011.4 | 55,762.0 | 206,700 |

**Source:** Accessed at http://money.cnn.com/magazines/fortune/fortune500/2010/industries/157/index.html on July 23, 2010.

**TABLE 12.2.3** Regression Analysis of Aerospace and Defense Companies

**Multiple regression to predict Market Value from Assets and Employees.**

**The prediction equation is**

Market Value = −1125.58 + 0.5789 Assets + 0.1267 Employees

| | |
|---|---|
| 0.879 | $R$ squared |
| 7292.617 | Standard error of estimate |
| 14 | Number of observations |
| 39.830 | $F$ statistic |
| 0.0000092 | $p$-value |

|  | Coeff | 95% LowerCI | 95% UpperCI | StdErr | $t$ | $p$ |
|---|---|---|---|---|---|---|
| Constant | −1125.575 | −8532.015 | 6280.865 | 3365.057 | −0.334 | 0.744 |
| Assets | 0.579 | −0.156 | 1.314 | 0.334 | 1.733 | 0.111 |
| Employees | 0.127 | −0.098 | 0.351 | 0.102 | 1.242 | 0.240 |

**TABLE 12.2.4** Correlation Matrix for Aerospace and Defense Companies

|  | Market Value, $Y$ | Assets, $X_1$ | Employees, $X_2$ |
|---|---|---|---|
| Market Value, $Y$ | 1.000 | 0.928 | 0.920 |
| Assets, $X_1$ | 0.928 | 1.000 | 0.944 |
| Employees, $X_2$ | 0.920 | 0.944 | 1.000 |

**TABLE 12.2.5** Defining New $X$ Variables for Aerospace and Defense Companies, Using Ratio of Employees per Assets

| | Market Value (millions), $Y$ | Assets (millions), $X_1$ | Ratio of Employees to Assets, $X_2$ |
|---|---|---|---|
| Alliant Techsystems | 2,724.7 | 3,593.2 | 5.288 |
| Boeing | 54,948.9 | 62,053.0 | 2.532 |
| General Dynamics | 29,670.0 | 31,077.0 | 2.951 |
| Goodrich | 8,888.0 | 8,741.4 | 2.746 |
| Honeywell International | 34,099.0 | 36,004.0 | 3.389 |
| ITT | 9,646.1 | 11,129.1 | 3.612 |
| Lockheed Martin | 31,626.1 | 35,111.0 | 3.987 |
| L-3 Communications | 10,779.6 | 14,813.0 | 4.523 |
| Northrop Grumman | 19,840.6 | 30,252.0 | 3.990 |
| Precision Castparts | 17,542.6 | 6,721.4 | 3.065 |
| Raytheon | 21,717.7 | 23,607.0 | 3.177 |
| Rockwell Collins | 9,901.7 | 4,645.0 | 4.155 |
| Textron | 5,910.4 | 18,940.0 | 1.690 |
| United Technologies | 69,011.4 | 55,762.0 | 3.707 |

**TABLE 12.2.6** Correlation Matrix for New $X$ Variables for Aerospace and Defense Companies

| | Market Value, $Y$ | Assets, $X_1$ | Ratio of Employees to Assets, $X_2$ |
|---|---|---|---|
| Market value, $Y$ | 1.000 | 0.928 | −0.168 |
| Assets, $X_1$ | 0.928 | 1.000 | −0.271 |
| Ratio of employees to assets, $X_2$ | −0.168 | −0.271 | 1.000 |

## Variable Selection: Are You Using the Wrong Variables?

Statistical results depend heavily on the information you provide as data to be analyzed. In particular, you must use care in choosing the explanatory ($X$) variables for a multiple regression analysis. It is *not* a good idea to include too many $X$ variables "just to be safe" or because "each one seems like it might have some effect." If you do so, you may have trouble finding significance for the regression (the $F$ test), or, due to multicollinearity caused by unnecessary variables, you may have trouble finding significance for some of the regression coefficients.

What happens when you include one extra, irrelevant $X$ variable? Your $R^2$ will be slightly larger because a little more of $Y$ can be explained by exploiting the randomness of this new variable.[14] However, the $F$ test of significance of the regression takes this increase into account, so this larger $R^2$ is not an advantage.

In fact, it is a slight to moderate *disadvantage* to include such an extra $X$ variable. The estimation of an unnecessary parameter (the unneeded regression coefficient) leaves less information remaining in the standard error of estimate, $S_e$. For technical reasons, this leads to a less powerful $F$ test that is less likely to find significance when population $X$ variables do in fact help explain $Y$.

What happens when you omit a necessary $X$ variable? Important helpful information will be missing from the data set, and your predictions of $Y$ will not be as good as if this $X$ variable had been included. The standard error of estimate, $S_e$, will tend to be larger (indicating larger prediction errors), and $R^2$ will tend to be smaller (indicating that less of $Y$ has been explained). Naturally, if a crucial $X$

---

14. Although $R^2$ will always be either the same or larger, there is a similar quantity called *adjusted $R^2$* that may be either larger or smaller when an irrelevant $X$ variable is included. The adjusted $R^2$ will increase only if the $X$ variable explains more than a useless $X$ variable would be expected to randomly explain. The adjusted $R^2$ may be computed from the (ordinary, unadjusted) $R^2$ value using the formula $1 - (n - 1)(1 - R^2)/(n - k - 1)$.

**TABLE 12.2.7 Regression Analysis Using New *X* Variables for Aerospace and Defense Companies**

Multiple regression analysis to predict Market Value from Assets and Employees per Asset.

**The prediction equation is**

Market Value = –7756.40 + 0.9962 Assets + 1920.36 Employees per Asset

| 0.869 | *R* squared |
| 7573.221 | Standard error of estimate |
| 14 | Number of observations |
| 36.533 | *F* statistic |
| 0.000014 | *p*-value |

| | Coeff | 95% LowerCI | 95% UpperCI | StdErr | *t* | *p* |
|---|---|---|---|---|---|---|
| Constant | –7756.404 | –29408.961 | 13896.154 | 9837.666 | –0.788 | 0.4470987 |
| Assets | 0.996 | 0.735 | 1.257 | 0.118 | 8.407 | 0.0000041 |
| Employees per Asset | 1920.364 | –3400.177 | 7240.905 | 2417.345 | 0.794 | 0.4437581 |

variable is omitted, you may not even find a significant *F* test for the regression.

Your incentives here are to include just enough *X* variables (not too many, not too few) and to include the correct ones. If in doubt, you might include just a few of the *X* variables for which you are not sure. There is a subjective method for achieving this (based on a prioritized list of *X* variables), and there are many different automatic methods.

## Prioritizing the List of *X* Variables

One good way to proceed is to think hard about the problem, the data, and just what it is that you are trying to accomplish. Then produce a prioritized list of the variables as follows:

1. Select the *Y* variable you wish to explain, understand, or predict.
2. Select the single *X* variable that you feel is most important in determining or explaining *Y*. If this is difficult, because you feel that they are *all* so important, imagine that you are forced to choose.
3. Select the most important remaining *X* variable by asking yourself, "With the first variable taken into account, which *X* variable will contribute the most *new* information toward explaining *Y*?"
4. Continue selecting the most important remaining *X* variable in this way until you have a prioritized list of the *X* variables. At each stage, ask yourself, "With the

*X* variables previously selected taken into account, which remaining *X* variable will contribute the most *new* information toward explaining *Y*?"

Next, compute a regression with just those *X* variables from your list that you consider crucial. Also run a few more regressions including some (or all) of the remaining variables to see if they do indeed contribute toward predicting *Y*. Finally, choose the regression result you feel is most helpful.

Even though this procedure is fairly subjective (since it depends so heavily on your opinion), it has two advantages. First, when a choice is to be made between two *X* variables that are nearly equally good at predicting *Y*, you will have control over the selection (an automated procedure might make a less intuitive choice). Second, by carefully prioritizing your explanatory *X* variables, you gain further insight into the situation. Because it clarifies your thoughts, this exercise may be worth nearly as much as getting the multiple regression results!

## Automating the Variable Selection Process

If, instead of thinking hard about the situation, you want the selection of variables to be done for you based on the data, there are many different ways to proceed. Unfortunately, there is no single overall "best" answer to the automatic variable selection problem. Statistical research is still continuing on the problem. However, you can expect a good

**TABLE 12.2.8 List of All Possible Subsets of the X Variables When $k = 3$**

| 1 | None (just use $\overline{Y}$ to predict Y) |
|---|---|
| 2 | $X_1$ |
| 3 | $X_2$ |
| 4 | $X_3$ |
| 5 | $X_1, X_2$ |
| 6 | $X_1, X_3$ |
| 7 | $X_2, X_3$ |
| 8 | $X_1, X_2, X_3$ |

automatic method to provide you with a fairly short list of X variables that will do a fairly good job of predicting Y.

The best methods of automatic variable selection look at *all subsets* of the X variables. For example, if you have three explanatory X variables to choose from, then there are eight subsets to look at, as shown in Table 12.2.8. If you have 10 X variables, there will be 1,024 different subsets.[15] Even if you can compute that many regressions, how will you know which subset is best? A number of technical approaches have been suggested by statistical researchers based on formulas that trade off the additional information in a larger subset against the added difficulty in estimation.[16]

One widely used approach is called *stepwise selection.* At each step, a variable is either added to the list or removed from the list, depending on its usefulness. The process continues until the list stabilizes. This process is faster than looking at all subsets but may not work as well in some cases. Here are some further details of the stepwise selection procedure:

1. *Getting started.* Is there an X variable that is helpful in explaining Y? If not, stop now and report that no helpful X variable can be found. If there is a helpful one, put the most useful X variable on the list (this is the one with the highest absolute correlation with Y).
2. *Forward selection step.* Look at all of the X variables that are *not* on the list. In particular, look at the one that contributes the most *additional* explanation of Y. If it explains enough about Y, then include it on the list.

3. *Backward elimination steps.* Is there an X variable on the list that is no longer helpful (now that there are additional variables on the list)? If so, remove it, but be sure still to consider it for possible inclusion later. Continue eliminating X variables until none can be eliminated from the list.
4. *Repeat until done.* Repeat steps 2 and 3 until no variable can be added to or dropped from the list.

The end result of the stepwise selection process is usually a fairly useful, fairly short list of explanatory X variables to use in multiple regression analysis to explain Y.

## Model Misspecification: Does the Regression Equation Have the Wrong Form?

Even if you have a good list of X variables that contain the right information for explaining Y, there may still be problems. *Model misspecification* refers to all of the ways that the multiple regression linear model might fail to represent your particular situation. Here are some of the ways in which a regression model might be misspecified:

1. The expected response of Y to the X variables might be *nonlinear*. That is, the regression equation $a + b_1X_1 + b_2X_2 + \cdots + b_kX_k$ might not adequately describe the true relationship between Y and the X variables.
2. There might be *unequal variability* in Y. This would violate the assumption that the standard deviation, $\sigma$, in the multiple regression linear model is constant regardless of the values of the X variables.
3. There might be one or more *outliers* or *clusters*, which could seriously distort the regression estimates.
4. You might have a *time series*, in which case the randomness in the multiple regression linear model would not be independent from one time period to the next. In general, time-series analysis is complex (see Chapter 14). However, you may be able to work with a multiple regression using the *percent change* (from one time period to the next) in place of the original variables.

Some of these problems can be identified by exploring all of the possible scatterplots you can draw taking variables two at a time (for example, for $k = 3$, you would have six scatterplots: $[X_1, Y]$, $[X_2, Y]$, $[X_3, Y]$, $[X_1, X_2]$, $[X_1, X_3]$, $[X_2, X_3]$). For a complete analysis, all of these scatterplots should be at least briefly explored so that you are alerted to potential troubles. But keep in mind that these scatterplots may overstate the need for corrective action. For example, Y may show a curved relationship against $X_1$, yet this may not be a problem by itself.

Fortunately, there is a more direct method that often identifies serious problems. The *diagnostic plot* is a single scatterplot of residuals against predicted values that can

---

15. The general formula is that $2^k$ subsets can be formed from $k$ explanatory X variables.
16. One good measure for choosing the best subset of X variables in regression is *Mallows' $C_p$ statistic*. This and other approaches are discussed in N. R. Draper and H. Smith, *Applied Regression Analysis* (New York: Wiley, 1981), chapter 6; and in G. A. F. Seber, *Linear Regression Analysis* (New York: Wiley, 1977), chapter 12.

point out most serious problems involving nonlinearity, unequal variability, and outliers. Thus, you may use all of the scatterplots of the basic variables as background information and then use the diagnostic plot as a basis for deciding whether or not to change the analysis.

## Exploring the Data to See Nonlinearity or Unequal Variability

By looking at all of the scatterplots that can be made, one for each pair of variables, you can see much of the structure of the relationships among these variables. Often, you can make useful insights into your situation by looking at the data in this way. However, you still can't see *all* of the structure. For example, you would miss any combined effect of two variables on a third because you can see only two at a time.[17] Nonetheless, much useful background information can be obtained from the basic scatterplots.

Consider the earlier example of magazine ads, with page costs ($Y$) to be explained by audience ($X_1$), percent male readership ($X_2$), and median income ($X_3$). Let's look at the scatterplots defined by each of the four variables against each of the others (Figures 12.2.1 through 12.2.6).

The correlation matrix is also helpful, since it gives a summary of the strength and direction of the association in each of these scatterplots, as shown in Table 12.2.9.

How would you summarize the results of this exploration of scatterplots and examination of correlations? The strongest association is between audience and page costs (Figure 12.2.1); there is also strong association between median income and percent male (Figure 12.2.6). We also learn that the magazines with the largest audience and the largest page costs tend to appeal to the middle-income group, leading to some unequal variability patterns (Figures 12.2.3 and 12.2.5).

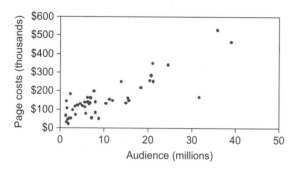

**FIGURE 12.2.1**　The scatterplot of $Y$ (page costs) against $X_1$ (audience) shows a fairly strong increasing relationship.

---

17. Some computer systems can rotate a scatterplot in real time so that you can visually inspect the three-dimensional plot of three variables at once! A collection of techniques for exploring multivariate data is given in J. M. Chambers, W. S. Cleveland, B. Kleiner, and P. A. Tukey, *Graphical Methods for Data Analysis* (Boston: Duxbury Press, 1983).

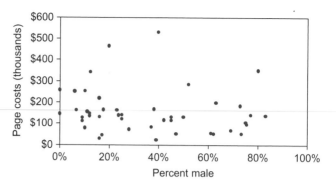

**FIGURE 12.2.2**　The scatterplot of $Y$ (page costs) against $X_2$ (percent male) shows very little if any structure.

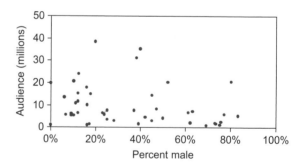

**FIGURE 12.2.3**　The scatterplot of $Y$ (page costs) against $X_3$ (median income) seems at first to show little if any structure. There may be a tendency for page costs to have higher variability within the lower-to-middle-income group. It may be difficult to charge high advertising (page) costs at the high end of the income scale because there are fewer such people.

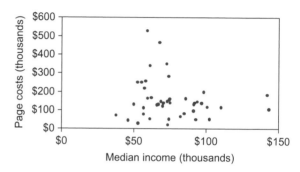

**FIGURE 12.2.4**　The scatterplot of $X_1$ (audience) against $X_2$ (percent male) shows relatively little, if any, structure. The slight downward tilt that you might see is not statistically significant ($p = 0.156$).

Is there a problem? The diagnostic plot will help you decide which problems (if any) require action and will show you whether or not the action works.

## Using the Diagnostic Plot to Decide If You Have a Problem

The **diagnostic plot** for multiple regression is a scatterplot of the prediction errors (residuals) against the predicted values

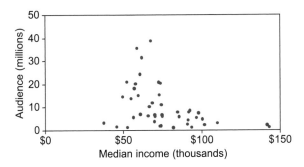

**FIGURE 12.2.5**  The scatterplot of $X_1$ (audience) against $X_3$ (median income) shows that the high-audience magazines tend to appeal to the low-to-middle-income group, but there is considerable variability within this group. The extremes (high-income and low-income) tend to have low audiences.

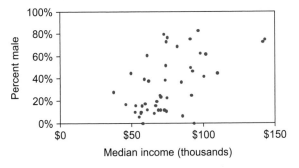

**FIGURE 12.2.6**  The scatterplot of $X_2$ (percent male) against $X_3$ (median income) suggests the existence of gender differences in income level. Magazines appealing to a high-income readership tend to have more males among their readers; magazines appealing to a low-income readership tend to have more females. Middle-income magazines show large variability in terms of gender.

**TABLE 12.2.9  Correlation Matrix for Magazine Ads Data Set**

|  | Page Costs, Y | Audience, $X_1$ | Percent Male, $X_2$ | Median Income $X_3$ |
|---|---|---|---|---|
| Page costs, Y | 1.000 | 0.850 | –0.126 | –0.148 |
| Audience, $X_1$ | 0.850 | 1.000 | –0.215 | –0.377 |
| Percent male, $X_2$ | –0.126 | –0.215 | 1.000 | 0.540 |
| Median income, $X_3$ | –0.148 | –0.377 | 0.540 | 1.000 |

and is used to see if the predictions can be improved by fixing problems in your data.[18] The residuals, $Y - [a + b_1 X_1 + b_2 X_2 + \cdots + b_k X_k]$, are plotted on the vertical axis, and the predicted values, $a + b_1 X_1 + b_2 X_2 + \cdots + b_k X_k$, go

18. The *predicted values* are also referred to as the *fitted values*.

on the horizontal axis. Because the methods for fixing problems are fairly complex (outlier removal, transformation, etc.), you fix a problem only if it is clear and extreme.

Do not intervene unless the diagnostic plot shows you a clear and definite problem.

You read a diagnostic plot in much the same way you would read any bivariate scatterplot (see Chapter 11). Table 12.2.10 shows how to interpret what you find.

Why does it work this way? The residuals represent the *unexplained* prediction errors for Y that could not be accounted for by the multiple regression linear model involving the X variables. The predicted values represent the *current explanation*, based on the X variables. If there is any strong relationship visible in the diagnostic plot, the current explanation can and should be improved by changing it to account for this visible relationship.

**TABLE 12.2.10  How to Interpret a Diagnostic Plot of Residuals against Predicted Values in Multiple Regression**

| Structure in the Diagnostic Plot | Interpretation |
|---|---|
| No relationship; just random, untilted scatter | Congratulations! No problems are indicated. Some improvements may still be possible, but the diagnostic plot cannot detect them. |
| Tilted linear relationship | Impossible by itself, since the least-squares regression equation should already have accounted for any purely linear relationship. |
| Tilted linear relationship with outlier(s) | The outlier(s) have distorted the regression coefficients and the predictions. The predictions for the well-behaved portion of the data can be improved if you feel that the outliers can be controlled (perhaps by transformation) or omitted.* |
| Curved relationship, typically U-shaped or inverted U-shaped | There is a nonlinear relationship in the data. Your predictions can be improved by either transforming, including an extra variable, or using nonlinear regression. |
| Unequal variability | Your prediction equation has been inefficiently estimated. Too much importance has been given to the less reliable portion of the data and too little to the most reliable part. This may be controlled by transforming Y (perhaps along with some of the X variables as well). |

*\* A transformation should never be used solely to control outliers. However, when you transform a distribution to reduce extreme skewness, you may find that the former outliers are no longer outliers in the transformed data set.*

Shown in Figure 12.2.7 is the diagnostic plot for the example of magazine ads, with page costs ($Y$) explained by audience ($X_1$), percent male readership ($X_2$), and median income ($X_3$). This plot shows unequal variability, with lower (vertical) variability on the left and greater variability on the right. There is potential here for improved regression results because the linear model assumption is not satisfied.

The histogram of audience size, shown in Figure 12.2.8, indicates strong skewness, while histograms of the other variables (not shown) do not. Although you don't necessarily have to transform $X$ variables just because of skewness, we will see what happens if we transform the audience variable, $X_1$.

Figure 12.2.9 shows the histogram of the natural logarithm of audience, log $X_1$ (use Excel's LN function).[19] The skewness is now mostly gone from the distribution. We will now see if transforming audience in this way will improve the regression results.

Table 12.2.11 shows the multiple regression results after the transformation to the log of audience. The variables are now page costs ($Y$) explained by the natural log of audience (the new $X_1$), percent male ($X_2$), and median income ($X_3$).

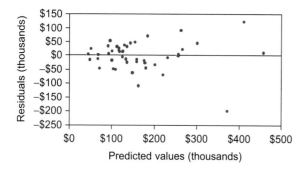

FIGURE 12.2.7 This diagnostic plot shows some possible unexplained structure remaining in the residuals: unequal variability, which you see in the pattern opening up as you move to the right. This is the diagnostic plot for the multiple regression of the basic variables page costs ($Y$) as explained by audience ($X_1$), percent male ($X_2$), and median income ($X_3$).

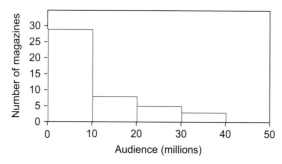

FIGURE 12.2.8 The histogram of audience ($X_1$) shows strong skewness.

19. For example, *Martha Stewart Living* has an audience of 11,200 (in thousands). The natural logarithm (sometimes denoted ln) of 11,200 is 9.324.

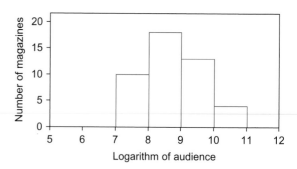

FIGURE 12.2.9 The histogram of the logarithm of audience does not show skewness.

In some ways, this is not better than before: The $R^2$ value has gone down (i.e., decreased, indicating a poorer explanation) to 57.4% from 75.8%, and the standard error of estimate has increased from $53,812 to $71,416. It is not clear that the transformation of audience has been useful in improving our understanding and prediction of page costs.

The diagnostic plot for this regression, shown in Figure 12.2.10, is certainly different from the diagnostic plot for the original data (Figure 12.2.7); in particular, the variability no longer seems to increase as you move to the right. However, a new possible problem has emerged: There appears to be nonlinearity in the data, curving up at both sides. There is potential here for an improved fit to the data.

Next, let's try transforming all of the variables that measure amounts (i.e., page costs and median income, as well as audience size) in the same way by using natural logarithms.[20] Table 12.2.12 shows the multiple regression results after transformation to the log of page costs, audience, and median income. The variables are now the log of page costs (the new $Y$) explained by the log of audience (the new $X_1$), percent male ($X_2$), and the log of median income (the new $X_3$). The $R^2$ value is up only slightly, indicating little improvement overall. The standard error of estimate is now on the log scale for page costs and so is not directly comparable to the previous values.[21] The diagnostic plot will tell you whether or not these transformations have helped.

The diagnostic plot for this regression, shown in Figure 12.2.11, indicates that the nonlinearity problem

20. If a variable, in another situation, contains both positive and negative values, transformation is difficult, and the logarithm cannot be used because it is undefined for zero or for negative values. In some situations, you may be able to redefine the variable so that it is always positive. For example, if it represents profit (= Income − Expenses), you might consider using the ratio Income/Expenses instead. The logarithm would then be log (Income/Expenses) = log (Income) − log (Expenses) and may be thought of as representing profit on a percentage scale rather than an absolute dollar scale.

21. Interpreting the results of a multiple regression when logarithms are used will be covered in a later section of this chapter.

**TABLE 12.2.11 Multiple Regression Output after Transforming to the Log of Audience**

**The regression equation is**

Page = –720580 + 91894 **log** Audience + 2042 Male + 0.906 Income

| Predictor | Coeff | SE Coeff | t | p |
|---|---|---|---|---|
| Constant | –720580 | 133894 | –5.38 | 0.000 |
| **log** Audience | 91894 | 12637 | 7.27 | 0.000 |
| Male | 2042 | 50531 | 0.040 | 0.968 |
| Income | 0.9059 | 0.6145 | 1.47 | 0.148 |

S = 71416.0   R-Sq = 57.4%   R-Sq(adj) = 54.3%

**Analysis of Variance**

| Source | DF | SS | MS | F | p |
|---|---|---|---|---|---|
| Regression | 3 | 2.81915E+11 | 93971814172 | 18.42 | 0.000 |
| Residual Error | 41 | 2.09110E+11 | 5100243105 | | |
| Total | 44 | 4.91025E+11 | | | |

| Source | DF | Seq SS |
|---|---|---|
| **log** Audience | 1 | 2.67048E+11 |
| Male | 1 | 3783749185 |
| Income | 1 | 11083990002 |

**Unusual Observations**

| Obs | log Audience | Page | Fit | SE Fit | Residual | St Resid |
|---|---|---|---|---|---|---|
| 1 | 9.1 | 53310 | 198262 | 15208 | –144952 | –2.08R |
| 2 | 10.5 | 532600 | 297038 | 23524 | 235562 | 3.49R |
| 6 | 10.6 | 468200 | 312404 | 23341 | 155796 | 2.31R |

R denotes an observation with a large standardized residual.

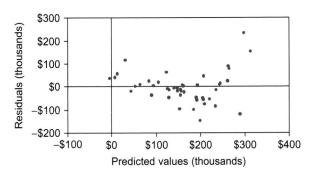

**FIGURE 12.2.10**   The diagnostic plot after transforming to the logarithm of audience. Nonlinearity may be a problem here, with a tendency to curve up at both sides.

has been fixed by transforming to the logarithms of page costs, audience, and median income.

## Using Percent Changes to Model an Economic Time Series

One assumption of the multiple regression linear model is that the random component ($\varepsilon$) is independent from one data value to the next. When you have time-series data, this assumption is often violated because changes are usually small from one period to the next, yet there can be large changes over longer periods of time.

**TABLE 12.2.12  Multiple Regression Output after Transforming to the Log of Page Costs, Audience, and Median Income**

**The regression equation is**

**log** Page = –2.05 + 0.581 **log** Audience – 0.258 Male + 0.786 **log** Income

| Predictor | Coeff | StDev | t | p |
|---|---|---|---|---|
| Constant | –2.047 | 3.185 | –0.64 | 0.524 |
| **log** Audience | 0.58090 | 0.06949 | 8.36 | 0.000 |
| Male | –0.2576 | 0.2816 | –0.91 | 0.366 |
| **log** Income | 0.7858 | 0.2686 | 2.93 | 0.006 |

$S = 0.400408$   R-Sq = 64.2%   R-Sq(adj) = 61.6%

**Analysis of Variance**

| Source | DF | SS | MS | F | p |
|---|---|---|---|---|---|
| Regression | 3 | 11.7996 | 3.9332 | 24.53 | 0.000 |
| Residual Error | 41 | 6.5734 | 0.1603 | | |
| Total | 44 | 18.3730 | | | |

| Source | DF | Seq SS |
|---|---|---|
| **log** Audience | 1 | 10.3812 |
| Male | 1 | 0.0461 |
| **log** Income | 1 | 1.3723 |

**Unusual Observations**

| Obs | log Audience | log Page | Fit | SE Fit | Residual | St Resid |
|---|---|---|---|---|---|---|
| 1 | 9.1 | 10.8839 | 12.0902 | 0.0884 | –1.2063 | –3.09R |
| 5 | 7.6 | 10.1282 | 11.0495 | 0.1055 | –0.9213 | –2.39R |
| 27 | 7.2 | 11.9077 | 11.1508 | 0.1866 | 0.7569 | 2.14R |

R denotes an observation with a large standardized residual.

Another way to understand the problem is to recognize that many economic time series increase over time, for example, gross national product, disposable income, and your firm's sales (we hope!). A multiple regression of one such variable ($Y$) on the others ($X$ variables) will have a high $R^2$ value, suggesting strong association. But if each series is *individually* increasing over time, in its own particular way and without regard to the others, this is deceiving. You should really conclude that there is meaningful association only if the *pattern* of increases over time for $Y$ can be predicted from those of the $X$ variables.

One way to solve this problem is to work with the *percent changes* of each variable, defined by the proportion (Current – Previous)/Previous, which represents the one-period growth rate of that variable. You lose nothing by doing this because the forecasting problem may be equivalently viewed either as predicting the *change* from the current level of $Y$ or as predicting the *future level* of $Y$.

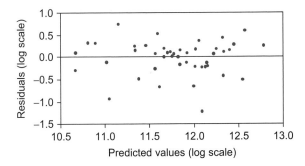

**FIGURE 12.2.11**  This is a very nice diagnostic plot: No further major problems are detectable in the data—neither unequal variability nor major curvature is present. There is no apparent relationship after transformation to the log of page costs ($Y$), the log of audience ($X_1$), and the log of median income ($X_3$). Only percent male ($X_2$) has been left untransformed.

Imagine a system that is more or less at equilibrium at each time period but that changes somewhat from one period to the next. What you really want to know is how to use information about the $X$ variables to predict the next value of your $Y$ variable. One problem is that your data set represents past history with $X$ values that are not currently reasonable as possibilities. By working with the percent changes, you make the past history much more relevant to your current experience. In other words, although your firm's sales are probably much different from what they were 5 years ago, the percent change in sales from one year to the next may well be similar from year to year. Or, if you are using gross national product (GNP) to predict something, although you may never see the same GNP as we had 10 years ago, you might easily see the same (at least approximately) growth rate (percent change) in the GNP.

Think of it this way: A system in equilibrium may tend to change in similar ways through time, even though it may find itself in entirely different circumstances as time goes by.

You may find that your $R^2$ suffers when you use percent changes instead of the original data values. In some cases the regression will no longer be significant. This might "look bad" at first (after all, everybody likes large $R^2$ values), but a closer look often shows that the original $R^2$ was overoptimistic, and the smaller value is closer to the truth.

### Example
#### Predicting Dividends from Sales of Nondurable and Durable Goods

How are dividends set by firms in the U.S. economy? If you are not careful, you might conclude that dividends respond in a very precise way to the level of sales of nondurable goods each year. If you are careful to use percent changes, then you will realize that dividends are not quite so simply explained.

Note that each of the columns in Table 12.2.13 shows a general increase over time. You should therefore expect to see strong correlations among these variables since high values of one are associated with high values of the others. This is indeed what you see in the correlation matrix, as shown in Table 12.2.14.

It would then be no surprise to find a very high $R^2$ value, suggesting that a whopping 98.6% of the variation in dividends is explained by sales of nondurables and durables. *But this would be wrong!* More precisely, in the historical context, it would be correct; however, it would not be nearly as useful in predicting future levels of dividends.

Table 12.2.15 shows the percent changes of these variables. For example, the 2007 value for dividends is $(788.7 - 702.1)/702.1 = 12.33\%$. (Note that data are missing for 2001 because there is no previous year given in the original data set.) The correlation matrix in Table 12.2.16 shows a much more modest association among the changes in these variables from one year to the next. In fact, with such a small sample size ($n = 6$ for the percent changes), not one of these pairwise correlations is even significant. The $R^2$ for the multiple regression of the percent changes has been reduced to 38.0%, and the $F$ test is not significant. This suggests that changes in sales levels of nondurables and durables cannot be used to help understand changes in dividends from one year to the next.

Economically speaking, the regression analysis using percent changes makes more sense. The level of dividends in the economy is a complex process involving the interaction of many factors. Due to our tax system and the investors' apparent dislike of sudden changes in dividend levels, we should not expect dividends to be almost completely explained just by sales levels.

**TABLE 12.2.13 Dividends, Sales of Nondurable Goods, and Sales of Durable Goods, 2001–2007**

| Year | Dividends (billions), $Y$ | Sales of Nondurable Goods (billions), $X_1$ | Sales of Durable Goods (billions), $X_2$ |
|------|------|------|------|
| 2001 | 370.9 | 2,017.1 | 883.7 |
| 2002 | 399.2 | 2,079.6 | 923.9 |
| 2003 | 424.7 | 2,190.2 | 942.7 |
| 2004 | 539.5 | 2,343.7 | 983.9 |
| 2005 | 577.4 | 2,514.1 | 1,020.8 |
| 2006 | 702.1 | 2,685.2 | 1,052.1 |
| 2007 | 788.7 | 2,833.0 | 1,082.8 |

**Source:** Data are from Tables 651 and 767 of U.S. Census Bureau, *Statistical Abstract of the United States: 2010* (129th edition), Washington, DC, 2009, accessed from http://www.census.gov/compendia/statab/ on July 23, 2010.

**TABLE 12.2.14 Correlation Matrix for Dividends, Sales of Nondurable Goods, and Sales of Durable Goods**

|  | Dividends, Y | Non-durables, $X_1$ | Durables, $X_2$ |
|---|---|---|---|
| Dividends, Y | 1.000 | 0.992 | 0.979 |
| Nondurables, $X_1$ | 0.992 | 1.000 | 0.992 |
| Durables, $X_2$ | 0.979 | 0.992 | 1.000 |

**TABLE 12.2.15 Yearly Percent Changes of Dividends, Sales of Nondurable Goods, and Sales of Durable Goods**

| Year | Dividends (yearly change), Y | Sales of Nondurable Goods (yearly change), $X_1$ | Sales of Durable Goods (yearly change), $X_2$ |
|---|---|---|---|
| 2001 | — | — | — |
| 2002 | 7.63% | 3.10% | 4.55% |
| 2003 | 6.39 | 5.32 | 2.03 |
| 2004 | 27.03 | 7.01 | 4.37 |
| 2005 | 7.03 | 7.27 | 3.75 |
| 2006 | 21.60 | 6.81 | 3.07 |
| 2007 | 12.33 | 5.50 | 2.92 |

**TABLE 12.2.16 Correlation Matrix for Percent Changes in Dividends, Sales of Nondurable Goods, and Sales of Durable Goods**

|  | Dividends, Y | Non-durables, $X_1$ | Durables, $X_2$ |
|---|---|---|---|
| Dividends, Y | 1.000 | 0.509 | 0.280 |
| Nondurables, $X_1$ | 0.509 | 1.000 | −0.128 |
| Durables, $X_2$ | 0.280 | −0.128 | 1.000 |

## 12.3 DEALING WITH NONLINEAR RELATIONSHIPS AND UNEQUAL VARIABILITY

The multiple regression techniques discussed so far are based on the multiple regression linear model, which has *constant variability*. If your data set does not have

such a linear relationship, as indicated by the diagnostic plot covered earlier, you have three choices. The first two use multiple regression and will be covered in this section.

1. *Transform some or all variables.* By transforming one or more of the variables (for example, using logarithms), you may be able to obtain a new data set that has a linear relationship. Remember that logarithms can be used only to transform positive numbers. If your data set shows unequal variability, you may be able to correct the problem by transforming Y and (perhaps) also transforming some of the X variables.

2. *Introduce a new variable.* By introducing an additional, necessary X variable (for example $X_1^2$, the square of $X_1$), you may be able to obtain a linear relationship between Y and the new set of X variables. This method can work well when you are seeking an *optimal* value for Y, for example, to maximize profits or production yield. In other situations, you might use products of the variables (for example, defining $X_5 = X_1 \times X_2$) so that the regression equation can reflect the *interaction* between these two variables.

3. *Use nonlinear regression.* There may be an important nonlinear relationship, perhaps based on some theory, that must be estimated directly. The advanced methods of *nonlinear regression* can be used in these cases if both the form of the relationship and the form of the randomness are known.[22]

## Transforming to a Linear Relationship: Interpreting the Results

There is a useful guideline for transforming your data. To keep things from getting too complicated, try to use the same transformation on all variables that are measured in the same units. For example, if you take the logarithm of sales (measured in dollars or thousands of dollars), you should probably also transform all other variables that measure dollar amounts in the same way. In this way, dollar amounts for all appropriate variables will be measured on a percentage scale (this is what logarithms do) rather than on an absolute dollar scale.

**Consistency Guideline for Transforming Multivariate Data**

Variables that are measured in the same basic units should probably all be transformed in the same way.

---

22. An introduction to nonlinear regression is provided in N. R. Draper and H. Smith, *Applied Regression Analysis,* 2nd ed. (New York: Wiley, 1981), chapter 10.

**TABLE 12.3.1 Interpreting a Multiple Regression When Transformation Has Been Used**

| | If $Y$ Is Not Transformed | If Natural Log of $Y$ Is Used |
|---|---|---|
| $R^2$ | *Usual interpretation:* The percent of variability of $Y$ explained by the (perhaps transformed) $X$ variables. | *Usual interpretation:* The percent of variability of (transformed) $Y$ that is explained by the (perhaps transformed) $X$ variables. |
| $S_e$ | *Usual interpretation:* Approximate size of prediction errors of $Y$. | *New interpretation:* The *coefficient of variation* of the prediction errors of $Y$ is given by* $\sqrt{2.71828^{S_e^2} - 1}$. |
| $b_i$ | *Usual interpretation:* The expected effect of a unit change in (perhaps transformed) $X_i$ on $Y$, all else equal. | *Similar interpretation:* The expected effect of a unit change in (perhaps transformed) $X_i$ on log $Y$. If $X_i$ has also been transformed using logarithms, then $b_i$ is also called the elasticity of $Y$ with respect to $X_i$: the expected effect (in percentage points of $Y$) of a 1% change in $X_i$, all else equal. |
| Significance test for $b_i$ | *Usual interpretation:* Does $X_i$ have an impact on $Y$, holding other $X$ variables fixed? | *Usual interpretation:* Does $X_i$ have an impact on $Y$, holding other $X$ variables fixed? |
| Prediction of $Y$ | *Usual procedure:* Use the regression equation to predict $Y$ from the $X$ variables, being sure to transform the $X$ variables first. | *New procedure:* Begin by using the regression equation to predict log $Y$ from the $X$ variables, being sure to transform the $X$ variables first. Then find the predicted value for $Y$ as follows:† $2.71828^{[(1/2)S_e^2 + \text{Predicted value for log }Y]}$ |

\* *Warning: This coefficient of variation may not be reliable for values larger than around 1 (or 100%) since the extreme skewness in these cases makes estimation of means and standard deviations very difficult.*

† *This predicts the expected (i.e., average or mean) value of Y for the given values of the X variables. To predict the median value of Y instead, use the following, simpler formula:* $2.71828^{(\text{Predicted value for log }Y)}$.

When you transform some or all of your variables and then perform a multiple regression analysis, some of the results will require a new interpretation. This section will show you how to interpret the results when either (1) $Y$ is left untransformed (that is, only some or all of the $X$ variables are transformed) or (2) $Y$ is transformed using the natural logarithm (regardless of whether none, some, or all of the $X$ variables are transformed). The $Y$ variable is special because it is the one being predicted, so transforming $Y$ redefines the meaning of a prediction error.

Table 12.3.1 is a summary of the interpretation of the basic numbers on the computer output: the coefficient of determination, $R^2$; the standard error of estimate, $S_e$; the regression coefficients, $b_i$; and the significance test for $b_i$ when you have used some transformations.[23] The procedure for producing predicted values for $Y$ using the regression equation is also included.

The $R^2$ value has the same basic interpretation, regardless of how you transform the variables.[24] It tells you how much of the variability of your current $Y$ (in whatever form, transformed or not) is explained by the current form of the $X$ variables.

The standard error of estimate, $S_e$, has a different interpretation depending on whether or not $Y$ is transformed. If $Y$ is not transformed, the usual interpretation (the typical size of the prediction errors) still applies because $Y$ itself is being predicted. However, if log $Y$ is used in the regression analysis, then $Y$ appears in the regression in percentage terms rather than as an absolute measurement. The appropriate measure of relative variability, from Chapter 5, is the *coefficient of variation* because the same percentage variability will be found at high predicted values of $Y$ as at smaller values. The formula for this coefficient of variation in Table 12.3.1 is based on theory for the lognormal distribution.[25]

The regression coefficients, $b_i$, have their usual interpretation if $Y$ is not transformed: They give the expected effect of an increase in $X_i$ on $Y$, where the increase in $X_i$ is one unit in whatever transformation was used on $X_i$. If $Y$ is transformed,

23. The relationships are easier to interpret if you use the *natural logarithm* for $Y$ (to the base $e = 2.71828 \ldots$, sometimes written ln) instead of the logarithm to the base 10.

24. We are assuming here that each transformation is "reasonable," in the sense that it does not change the relative ordering of the observations, and that it is a relatively "smooth" function.

25. A random variable is said to have a *lognormal distribution* if the distribution of its logarithm is normal. There are several excellent technical references for this distribution, including N. L. Johnson and S. Kotz, *Continuous Univariate Distributions* (New York: Wiley, 1970), chapter 14; and J. Aitchison and J. A. C. Brown, *The Lognormal Distribution* (London: Cambridge University Press, 1957). The lognormal distribution is also very important in the theory of pricing of financial options.

$b_i$ indicates the change in *transformed Y*. If you have used both the logarithm of $Y$ and the logarithm of $X_i$, then $b_i$ has the special economic interpretation of *elasticity*. The **elasticity** of $Y$ with respect to $X_i$ is the expected percentage change in $Y$ associated with a 1% increase in $X_i$, holding the other $X$ variables fixed; the elasticity is estimated using the regression coefficient from a regression using the natural logarithms of both $Y$ and $X_i$. Thus, an elasticity is just like a regression coefficient except that the changes are measured in percentages instead of the original units.

The significance test for a regression coefficient, $b_i$, retains its usual interpretation for any reasonable transformation. The basic question is, Does $X_i$ have a detectable impact on $Y$ (holding the other $X$ values fixed), or does $Y$ appear to behave just randomly with respect to $X_i$? Because the question has a yes or no answer, rather than a detailed description of the response, the basic question being tested is the same whether or not you use the logarithm transformation. Of course, the test proceeds in a different way in each case, and performance is best when you use the transformations that achieve a multiple regression linear model form for your data.

The predictions of $Y$ change considerably depending on whether or not you transform $Y$. If $Y$ is not transformed, the regression equation predicts $Y$ directly. Simply take the appropriately transformed values for each $X_i$, multiply each by its regression coefficient $b_i$, add them up, add $a$, and you have the predicted value for $Y$.

If $Y$ is transformed using natural logarithms, there is a correction for the skewness of the untransformed $Y$. Using the appropriately transformed values of the $X$ variables in the regression equation will get you a prediction of log $Y$. The new procedure for predicting (untransformed) $Y$ given in the preceding table does two things. First, by exponentiating, the prediction of log $Y$ is brought back to the original units of $Y$ and provides predicted (fitted) values for the median of $Y$. Second, if it is important to predict the average instead of the median of $Y$, the skewness correction (based on $S_e$) inflates this value to reflect the fact that an average value is larger than a median or mode for this kind of a skewed distribution.

### Example
#### Magazine Ads Transformed and Interpreted

Table 12.3.2 shows the multiple regression results for the magazine ads example after transformation to the log of page costs, audience, and median income. The variables are now the log of page costs (the new $Y$), explained by the log of audience (the new $X_1$), percent male ($X_2$), and the log of median income (the new $X_3$). Let's interpret these results.

The $R^2$ value, 64.2%, has its usual conceptual interpretation, even in terms of the untransformed variables. It tells you

that 64.2% of the variation in page costs from one magazine to another can be accounted for by knowing the values of audience, percent male, and median income of each magazine.[26] The concept of $R^2$ is the same whether or not you transform using logs, but the details are slightly different.

The standard error of estimate, $S_e = 0.4004$, has a new interpretation. To make sense of this number (which literally indicates the typical size of prediction errors on the log scale), you use this equation:

$$\sqrt{2.71828^{S_e^2} - 1} = \sqrt{2.71828^{(0.4004^2)} - 1}$$
$$= \sqrt{2.71828^{0.160320} - 1}$$
$$= \sqrt{1.1739 - 1}$$
$$= 0.417 \text{ or } 41.7\%$$

This tells you that your prediction error is typically about 41.7% of the predicted value. For example, if your predicted page costs are \$100,000, your variation is about 41.7% of this, or \$41,700, giving you a standard error of estimate for page costs that is applicable to such magazines. If your predicted page costs are \$250,000, taking 41.7%, you find \$104,250 as the appropriate standard error of estimate for these more expensive magazines. It makes sense that the standard error of estimate should depend on the size of the magazine because the pricier magazines have much more room for variability than the less expensive ones.

The regression coefficient $b_1 = 0.581$, for log audience, is an elasticity because the natural logarithm transformation was also used on $Y$. Thus, for every 1% increase in audience, you expect a 0.581% increase in page costs. This suggests that there are some declining returns to scale in that you achieve less than a full 1% increase in page costs for a 1% increase in audience. You might wonder whether these returns are significantly declining or if this $b_1 = 0.581$ is essentially equal to 1 except for randomness. The answer is found by noting that the reference value, 1, is outside the confidence interval for $b_1$ (which extends from 0.441 to 0.721), implying that the returns to scale are indeed significantly declining. Alternatively, you might discover this by computing the $t$ statistic, $t = (0.581 - 1)/0.0695 = -6.03$.

Does audience have a significant impact on page costs, holding percent male and median income fixed? The answer is yes, as given by the usual $t$ test for significance of $b_1$ in this multiple regression. This result is found by observing the $p$-value (listed as 0.000) in the computer output for the predictor "log audience."

Finally, let's obtain a predicted value for $Y$ for *Martha Stewart Living*. This will differ slightly from the predicted value computed much earlier in this chapter and will be slightly better because the data did not follow a multiple regression linear model before transformation. There are two steps to predicting $Y$: First, predict log $Y$ directly from the regression equation, and then combine with $S_e$ to obtain the predicted value.

The data values for *Martha Stewart Living* are $X_1 = 11,200$ (indicating an audience of 11.200 million readers), $X_2 = 11.0\%$ (indicating 11.0% men among its readers),

### TABLE 12.3.2 Multiple Regression Results Using the Logs of Page Costs, Audience, and Median Income

**The regression equation is**

log Page = –2.05 + 0.581 **log** Audience – 0.258 Male + 0.786 **log** Income

| Predictor | Coeff | StDev | t | p |
|-----------|-------|-------|---|---|
| Constant | –2.047 | 3.185 | –0.64 | 0.524 |
| **log** Audience | 0.58090 | 0.06949 | 8.36 | 0.000 |
| Male | –0.2576 | 0.2816 | –0.91 | 0.366 |
| **log** Income | 0.7858 | 0.2686 | 2.93 | 0.006 |

$S = 0.400408$   R-Sq = 64.2%   R-Sq(adj) = 61.6%

**Analysis of Variance**

| Source | DF | SS | MS | F | p |
|--------|----|----|----|---|---|
| Regression | 3 | 11.7996 | 3.9332 | 24.53 | 0.000 |
| Residual Error | 41 | 6.5734 | 0.1603 | | |
| Total | 44 | 18.3730 | | | |

| Source | DF | Seq SS |
|--------|----|--------|
| **log** Audience | 1 | 10.3812 |
| Male | 1 | 0.0461 |
| **log** Income | 1 | 1.3723 |

**Unusual Observations**

| Obs | log Audience | log Page | Fit | SE Fit | Residual | St Resid |
|-----|--------------|----------|-----|--------|----------|----------|
| 1 | 9.1 | 10.8839 | 12.0902 | 0.0884 | –1.2063 | –3.09R |
| 5 | 7.6 | 10.1282 | 11.0495 | 0.1055 | –0.9213 | –2.39R |
| 27 | 7.2 | 11.9077 | 11.1508 | 0.1866 | 0.7569 | 2.14R |

R denotes an observation with a large standardized residual.

and $X_3 = \$74,436$ (indicating the median household income for its readers). Transforming the audience and the median income values to logs in the regression equation, you find the predicted value for log (page costs) for *Martha Stewart Living*:

Predicted (Page costs)
= –2.047 + 0.58090 × log (Audience)
  – 0.2576 (Percent male) + 0.7858 × log (Median income)
= –2.047 + 0.58090 × log (11,200)
  – 0.2576(0.110) + 0.7858 × log (74,436)
= –2.047 + 0.58090 × 9.324 – 0.2576 × 0.110 + 0.7858 × 11.218
= 12.156

In order to find the predicted value for page costs, the next step is

$$\text{Predicted page costs} = e^{(1/2)S_e^2 + \text{Predicted value for log } Y}$$
$$= 2.71828^{[(1/2)0.4004^2 + 12.156]}$$
$$= 2.71828^{[(1/2)0.4004^2 + 12.156]}$$
$$= 2.71828^{12.236}$$
$$= \$206,074$$

This predicted value is about 31% larger ($48,374 larger) than the actual page costs for this magazine, $157,700.
*(Continued)*

**Example—cont'd**

However, we were a bit lucky to be this close. The appropriate standard error for comparing the actual to the predicted value is 41.7% of $206,074, which is $85,933. If you compute the predicted page costs for other magazines, you will find that they are typically not nearly this close to the actual values. For an idea of how this compares to some of the other magazines, the relative prediction errors for the first 10 magazines in the list are 262%, –41%, 5%, 26%, 172%, –18%, 19%, 34%, 39%, and –37%. So it seems that 41.7% is a reasonable summary of the typical size of the errors.

---

26. The particular measure of variability used here is variance of page costs on the natural log scale, as explained by a multiple regression linear model using log audience, percent male, and log median income.

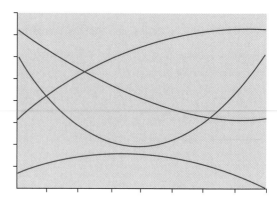

**FIGURE 12.3.1** Quadratic polynomials can be used to model a variety of curved relationships. Here is a selection of possibilities. Flipping any of these curves horizontally or vertically still gives you a quadratic polynomial.

## Fitting a Curve with Polynomial Regression

Consider a nonlinear *bivariate* relationship. If the scatterplot of $Y$ against $X$ shows a curved relationship, you may be able to use multiple regression by first introducing a new $X$ variable that is also curved with respect to $X$. The simplest choice is to introduce $X^2$, the square of the original $X$ variable. You now have a *multivariate* data set with three variables: $Y$, $X$, and $X^2$. You are using **polynomial regression** when you predict $Y$ using a single $X$ variable together with some of its powers ($X^2$, $X^3$, etc.). Let's consider just the case of $X$ with $X^2$.

With these variables, the usual multiple regression equation, $Y = a + b_1 X_1 + b_2 X_2$, becomes the *quadratic polynomial* $Y = a + b_1X + b_2X^2$.[27] This is still considered a linear relationship because the individual terms are added together. More precisely, you have a *linear* relationship between $Y$ and the pair of variables $(X, X^2)$ you are using to explain the *nonlinear* relationship between $Y$ and $X$.

At this point, you may simply compute the multiple regression of $Y$ on the two variables $X$ and $X^2$ (so that the number of variables rises to $k = 2$ while the number of cases, $n$, is unchanged). All of the techniques you learned earlier in this chapter can be used: predictions, residuals, $R^2$ and $S_e$ as measures of quality of the regression, testing of the coefficients, and so forth.

Figure 12.3.1 shows some of the variety of curves that quadratic polynomials can produce. If your scatterplot of $Y$ against $X$ resembles one of these curves, then the

---

27. The word *polynomial* refers to any sum of constants times nonnegative integer powers of a variable—for example, $3 + 5x - 4x^2 - 15x^3 + 8x^6$. The word *quadratic* indicates that no powers higher than 2 are used—for example, $7 - 4x + 9x^2$ or $9 - 3x^2$. Although higher powers can be used to model more complex nonlinear relationships, the results are often unstable when powers higher than 3 are used.

introduction of $X^2$ as a new variable will do a good job of explaining and predicting the relationship.

**Example**

*Optimizing the Yield of a Production Process*

Consider the data in Table 12.3.3, taken as part of an experiment to find the temperature that produces the largest yield for an industrial process. This data set could be very useful to your firm since it tells you that to maximize your output yield, you should set the temperature of the process at around 700 degrees. The yield apparently falls off if the temperature is either too cold or too hot.

The scatterplot, shown in Figure 12.3.2 with the least-squares line, shows how disastrous linear regression can be when it is inappropriately used to predict a nonlinear relationship. There is an abundance of structure here that could be used to predict yield from temperature and to determine the highest-yielding temperature, but a straight line just isn't going to do it for you!

Polynomial regression will correct this problem and also give you a good estimate of the optimal temperature that maximizes your yield. Table 12.3.4 shows the multivariate data set to use; note that only the last variable (the square of temperature) is new. Here is the prediction equation from multiple regression. It is graphed along with the data in Figure 12.3.3.

$$\text{Yield} = -712.10490 + (2.39119\,\text{Temperature})$$
$$- (0.00165\,\text{Temperature}^2)$$

The coefficient of determination for this multiple regression is $R^2 = 0.969$, indicating that a whopping 96.9% of the variation of yield has been explained by temperature and its square. (In fact, less than 1% had been explained by the straight line alone.) The standard error of estimate is $S_e = 1.91$, indicating that yield may be predicted to within a few units (compared to the much larger value of 10.23 for the straight line).

**TABLE 12.3.3 Temperature and Yield for an Industrial Process**

| Temperature, X | Yield, Y | Temperature, X | Yield, Y |
|---|---|---|---|
| 600 | 127 | 750 | 153 |
| 625 | 139 | 775 | 148 |
| 650 | 147 | 800 | 146 |
| 675 | 147 | 825 | 136 |
| 700 | 155 | 850 | 129 |
| 725 | 154 | | |

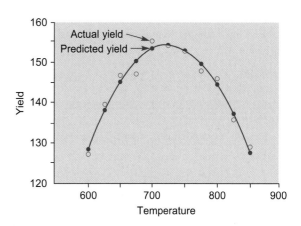

**FIGURE 12.3.3** The results of a quadratic polynomial regression to explain yield based on temperature and its square. The predictions are now excellent.

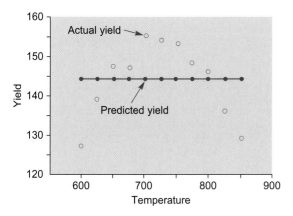

**FIGURE 12.3.2** The nonlinear relationship between output yield and the temperature of an industrial process is very badly described by the least-squares line. The predictions are unnecessarily far from the actual values.

How can you test whether the extra term (Temperature$^2$) was really necessary? The $t$ test for its regression coefficient ($b_2 = -0.00165$), based on a standard error of $S_{b_2} = 0.000104$ with 8 degrees of freedom, indicates that this term is *very highly significant*. Of course, this was obvious from the strong curvature in the scatterplot. Table 12.3.5 shows the results.

What is the best temperature to use in order to optimize yield? If the regression coefficient $b_2$ for your squared $X$ variable is *negative* (as it is here), then the quadratic polynomial has a *maximum* value at $-b_1/(2\ b_2)$.[28] For this example, the temperature that achieves the highest yield is

$$\text{Optimal temperature} = -b_1/(2b_2)$$
$$= -2.39119/[2(-0.00165)]$$
$$= 724.6$$

---

28. If $b_2$ is positive, then there is a *minimum* value at the same place: $-b_1/(2\ b_2)$.

A temperature setting of 725 degrees will be a good choice.

## Modeling Interaction between Two X Variables

In the multiple regression linear model, each of the $X$ variables is multiplied by its regression coefficient, and these are then added together with $a$ to find the prediction $a + b_1 X_1 + \cdots + b_k X_k$. There is no allowance for interaction among the variables. Two variables are said to show **interaction** if a change in both of them causes an expected shift in $Y$ that is different from the sum of the shifts in $Y$ obtained by changing each $X$ individually.

Many systems show interaction, especially if just the right combination of ingredients is required for success. For an extreme example, let $X_1 =$ gunpowder, $X_2 =$ heat,

**TABLE 12.3.4 Creating a New Variable (Squared Temperature) in Order to Do Polynomial Regression**

| Yield, Y | Temperature, $X_1 = X$ | Temperature Squared, $X_2 = X^2$ |
|---|---|---|
| 127 | 600 | 360,000 |
| 139 | 625 | 390,625 |
| 147 | 650 | 422,500 |
| 147 | 675 | 455,625 |
| 155 | 700 | 490,000 |
| 154 | 725 | 525,625 |
| 153 | 750 | 562,500 |
| 148 | 775 | 600,625 |
| 146 | 800 | 640,000 |
| 136 | 825 | 680,625 |
| 129 | 850 | 722,500 |

**TABLE 12.3.5** Multiple Regression Results, Using Squared Temperature as a Variable in Order to Do Polynomial Regression

$S = 1.907383$

$R^2 = 0.969109$

**Inference for Yield at the 5% level**

The prediction equation does explain a significant proportion of the variation in Yield.

$F = 125.4877$ with 2 and 8 degrees of freedom

|  | Effect on Yield | 95% Confidence Interval | | Hypothesis Test | StdErr of Coeff | t Statistic |
| --- | --- | --- | --- | --- | --- | --- |
| Variable | Coeff | From | To | Significant? | StdErr | t |
| Constant | −712.104 | −837.485 | −586.723 | Yes | 54.37167 | −13.0969 |
| Temperature | 2.391188 | 2.042414 | 2.739963 | Yes | 0.151246 | 15.80988 |
| Temperature2 | −0.00165 | −0.00189 | −0.00141 | Yes | 0.000104 | −15.8402 |

and $Y$ = reaction. A pound of gunpowder doesn't do much by itself; neither does a lighted match all by itself. But put these together and they interact, causing a very strong explosion as the reaction. In business, you have interaction whenever "the whole is more (or less) than the sum of its parts."

One common way to model interaction in regression analysis is to use a *cross-product*, formed by multiplying one $X$ variable by another to define a new $X$ variable that is to be included along with the others in your multiple regression. This cross-product will represent the interaction of those two variables. Furthermore, you will be able to test for the existence of interaction by using the $t$ test for significance of the regression coefficient for the interaction term.

If your situation includes an important interaction, but you don't provide for it in the regression equation, then your predictions will suffer. For example, consider predicting sales ($Y$) from business travel ($X_1$, miles) and contacts ($X_2$, number of people seen) for a group of salespeople. The usual regression equation that would be used to predict sales, $a + b_1$ (Miles) $+ b_2$ (Contacts), does not recognize the possibility of interaction between miles and contacts. The value of an additional mile of travel (by itself) is estimated as $b_1$, *regardless of the number of contacts seen*. Similarly, the value of an additional contact seen (by itself) is estimated as $b_2$, *regardless of the number of miles traveled*.

If you believed that there was some interaction between miles and contacts, so that salespeople with more contacts were able to make more productive use of their traveling miles, this model would be misspecified. One way to correct it would be to include a new $X$ variable, the cross-product $X_3 = X_1 \times X_2$ = Contacts × Miles. The resulting model is still a linear model and may be written in two equivalent ways:

$$\text{Predicted sales} = a + b_1(\text{Miles}) + b_2(\text{Contacts})$$
$$+ b_3(\text{Contacts} \times \text{Miles})$$
$$= a + [b_1 + b_3(\text{Contacts})](\text{Miles}) + b_2(\text{Contacts})$$

This says that an extra mile of travel counts more toward sales when the number of contacts is larger (provided $b_3 > 0$). Conveniently, you can use the $t$ test of $b_3$ to see if this effect is significant; if it is not, then you can omit the extra variable, $X_3$, and use the regression analysis of $Y$ on $X_1$ and $X_2$.

Another way in which interaction is modeled in regression analysis is to use transformation of some or all of the variables. Because logarithms convert multiplication to addition, the multiplicative equation with interaction

$$Y = AX_1^{b_1} X_2^{b_2}$$

is converted to a linear additive equation with no interaction by taking logs of all variables:

$$\log Y = \log A + b_1 \log X_1 + b_2 \log X_2$$

**Example**

*Mining the Donations Database to Predict Dollar Amounts from Combinations of the Other Variables*

Multiple regression is well suited to the task of predicting dollar amounts of gifts for the 989 people who donated in response to the mailing, out of the 20,000 people in the donations database on the companion site. For the X variables, we can use information (known before the mailing) about each person's donation history and neighborhood characteristics in order to explain the amount of the donation.

We find unequal variability in the diagnostic plot in Figure 12.3.4 for an initial regression analysis using all 24 $X$ variables (after excluding age because it has so many missing values).[29] To fix this problem, we will try using the logarithm transform on dollar amounts. After we take the logarithm of all variables measured in dollar amounts,[30] the diagnostic plot in Figure 12.3.5 is much better behaved. The regression results shown in Table 12.3.6 indicate that the $X$ variables explain 60.7% of the variability of (log) donations, and that four variables have significant $t$ tests: the log of the average past donation (AvgGift_Ln), the log of the lifetime total past donation, the number of recent gifts, and (curiously) the percentage of the neighborhood that is employed in sales.

With so many $X$ variables, there may be multicollinearity, and it is possible that more than these four significant $X$ variables are useful in predicting donation amounts. Stepwise regression performed using MINITAB does indeed include a fifth variable: The number of promotions mailed to this donor in the past is significant when unnecessary variables are omitted. Multiple regression results to predict the logarithm of donations, shown in Table 12.3.7, indicate that these five $X$ variables explain 60.3% of the variability of (log) donations (down only slightly from 60.7% when all 24 $X$ variables were used).

Standardized regression coefficients for these five $X$ variables, included in Table 12.3.7, indicate that the most influential explanatory variable by a large margin is the average of past donations (AvgGift, with standardized coefficient 0.618), which very sensibly says that people tend to continue to donate in their own personal style (smaller or larger amounts). The other variables have considerably less impact.

Curiously, the number of recent gifts has a significant *negative* impact on donation amount, holding the other four variables fixed. Indeed, larger donors do tend to give less frequently, as evidenced by the negative correlation −0.390 between donation amount and the number of recent gifts. Also of interest is the negative significant impact of the number of promotions received (holding the other four variables fixed) on donation amount.

To predict donation amounts using these results with the median method from the bottom of Table 12.3.1 when transformations are used, we use the prediction equation, which is

$$\text{Predicted Donation} = 2.71828^{\text{Predicted Log Donation}}$$

where

$$\text{Predicted Donation} = 0.526 + 0.746\,(\text{AvgGift\_Ln}) + 0.134(\text{Lifetime\_Ln})$$
$$- 0.00307\,(\text{Promotions}) - 0.112\,(\text{Recent Gifts}) + 0.828(\text{Sales})$$

The predicted donation may then be written as follows:

$$\text{Predicted Donation} = 1.692\,(\text{AvgGift})^{0.746}(\text{Lifetime})^{0.134}$$
$$\times 0.9969^{\text{Promotions}} 0.894^{\text{RecentGifts}} 2.228^{\text{Sales}}$$

where $2.71828^{0.526} = 1.692$ for the constant, the transformed $X$ variables (AvgGift_Ln and Lifetime_Ln) are exponentiated by their coefficients, and the nontransformed $X$ variables such as promotions appear in the exponent where, for example, $2.71828^{-0.00307} = 0.9969$. For example, considering a donor whose average of past gifts was $25, whose lifetime total of past gifts was $150, whose number of previous promotions was 35, whose number of recent gifts was 2, and for whom 20% = 0.20 of the neighborhood was employed in sales, the predicted donation would be $31.

Multiple regression has helped us identify the main variables that affect donation amounts and has provided a prediction equation that could be used in deciding who should receive the next mailing and how long to wait before asking again for money.

29. There is also an outlier, who happens to be an individual who made just one previous gift in the large amount of $200, with a current donation of $100. As you can see, the prediction equation is expecting about $150 this time around, and the donation was about $50 less.
30. The variables measured in dollar amounts are the current donation, the average past donation, the lifetime total past donation, the neighborhood median household income, and the neighborhood per capita income. Natural logarithms were used. For the income variables, the natural log of $10,000 plus the income was used in order to avoid problems with neighborhood incomes recorded as zero in the database.

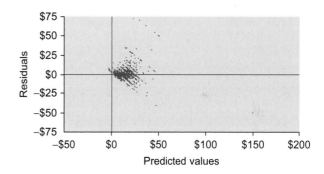

**FIGURE 12.3.4** Unequal variability is evident in the diagnostic plot for the multiple regression analysis of 989 donors to predict donation amount from 24 predictor variables relating to the past donation history and neighborhood of the donor.

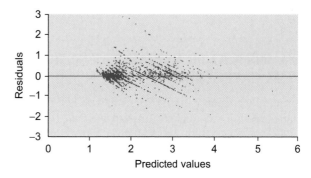

**FIGURE 12.3.5** After using the logarithm transform for all variables measured in dollar amounts, the diagnostic plot for 989 donors and 24 predictor variables no longer shows unequal variability.

**TABLE 12.3.6 Multiple Regression Results to Predict the Logarithm of the Donation Amount for 989 Donors with 24 Predictor Variables, Using the Logarithm Transformation for Dollar Amounts (the $R^2$ Is 60.7%)**

| | Coeff | LowerCI | UpperCI | StdErr | t | p |
|---|---|---|---|---|---|---|
| Constant | −0.173 | −1.662 | 1.316 | 0.759 | −0.228 | 0.820 |
| Age55_59 | 0.086 | −0.907 | 1.079 | 0.506 | 0.170 | 0.865 |
| Age60_64 | 0.535 | −0.411 | 1.481 | 0.482 | 1.109 | 0.268 |
| Avg Gift_Ln | 0.738 | 0.647 | 0.829 | 0.046 | 15.876 | 0.000*** |
| Cars | 0.075 | −0.265 | 0.416 | 0.174 | 0.434 | 0.664 |
| CatalogShopper | 0.012 | −0.081 | 0.105 | 0.047 | 0.251 | 0.802 |
| Clerical | −0.050 | −0.565 | 0.465 | 0.262 | −0.190 | 0.850 |
| Farmers | 0.129 | −0.517 | 0.775 | 0.329 | 0.391 | 0.696 |
| Gifts | −0.003 | −0.008 | 0.002 | 0.003 | −1.094 | 0.274 |
| HomePhone | 0.027 | −0.026 | 0.080 | 0.027 | 1.000 | 0.317 |
| Lifetime_Ln | 0.142 | 0.054 | 0.230 | 0.045 | 3.170 | 0.002** |
| MajorDonor | 0.283 | −0.306 | 0.872 | 0.300 | 0.944 | 0.346 |
| MedHouseInc_Ln | 0.155 | −0.048 | 0.358 | 0.103 | 1.503 | 0.133 |
| OwnerOccupied | −0.110 | −0.283 | 0.063 | 0.088 | −1.247 | 0.213 |
| PCOwner | −0.007 | −0.092 | 0.078 | 0.043 | −0.168 | 0.867 |
| PerCapIncome_Ln | −0.096 | −0.320 | 0.128 | 0.114 | −0.840 | 0.401 |
| Professional | 0.005 | −0.466 | 0.476 | 0.240 | 0.022 | 0.983 |
| Promotions | −0.003 | −0.006 | 0.000 | 0.002 | −1.943 | 0.052 |
| RecentGifts | −0.103 | −0.136 | −0.069 | 0.017 | −6.012 | 0.000*** |
| Sales | 0.738 | 0.160 | 1.317 | 0.295 | 2.505 | 0.012* |
| School | −0.001 | −0.030 | 0.027 | 0.015 | −0.083 | 0.934 |
| SelfEmployed | −0.030 | −0.653 | 0.592 | 0.317 | −0.095 | 0.924 |
| Technical | −0.340 | −1.439 | 0.759 | 0.560 | −0.607 | 0.544 |
| YearsSinceFirst | 0.006 | −0.013 | 0.025 | 0.010 | 0.608 | 0.543 |
| YearsSinceLast | 0.024 | −0.059 | 0.107 | 0.042 | 0.564 | 0.573 |

*$p < 0.05$, **$p < 0.01$, ***$p < 0.001$.

Note: Variables with "Ln" at the end have been transformed using logarithms.

## 12.4 INDICATOR VARIABLES: PREDICTING FROM CATEGORIES

Multiple regression is based on arithmetic and therefore requires meaningful numbers (quantitative data). What can you do if your variables are not all quantitative? An **indicator variable** (also called a *dummy variable*) is a quantitative variable using only the values 0 and 1 that is used to represent qualitative categorical data. For example, you might have a gender variable, which would be 1 for women and 0 for men (or the other way around). You can use one or more indicator variables as predictor ($X$) variables in your multiple regression analysis.[31]

---

31. If your response ($Y$) variable is qualitative, the situation is much more complex because the error term, $\varepsilon$, in the multiple regression linear model cannot have a normal distribution. If $Y$ has only two categories, you might use the *logit model* (*multiple logistic regression*) or the *probit model*. If your $Y$ variable has more than two categories, then the *multinomial logit model* or the *multinomial probit model* might be appropriate. Some helpful discussion is provided in J. Kmenta, *Elements of Econometrics* (New York: Macmillan, 1986), section 11–5.

TABLE 12.3.7 Multiple Regression Results, Including Standardized Regression Coefficients, to Predict the Logarithm of the Donation Amount for 989 Donors Using Five Predictor Variables Selected Using MINITAB's Stepwise Regression (the $R^2$ Is 60.3%)

| | Coeff | StdCoeff | LowerCI | UpperCI | StdErr | t | p |
|---|---|---|---|---|---|---|---|
| Constant | 0.526 | | 0.308 | 0.744 | 0.111 | 4.739 | 0.000*** |
| AvgGift_Ln | 0.746 | 0.618 | 0.676 | 0.816 | 0.036 | 21.006 | 0.000*** |
| Lifetime_Ln | 0.134 | 0.169 | 0.059 | 0.210 | 0.038 | 3.506 | 0.000*** |
| Promotions | −0.003 | −0.116 | −0.006 | −0.001 | 0.001 | −2.411 | 0.016* |
| RecentGifts | −0.112 | −0.198 | −0.141 | −0.082 | 0.015 | −7.429 | 0.000*** |
| Sales | 0.828 | 0.070 | 0.358 | 1.298 | 0.240 | 3.456 | 0.001*** |

*$p < 0.05$, **$p < 0.01$, ***$p < 0.001$.

Note: Variables with "Ln" at the end have been transformed using logarithms.

If a qualitative $X$ variable encompasses just two categories (such as men/women, buy/browse, or defective/conforming), you may represent it directly as an indicator variable. You may decide arbitrarily which of the two categories will be represented as 1 and which will be 0 (the baseline). Although the choice is arbitrary at this point, you must remember which alternative you chose in order to interpret the results later! Table 12.4.1 shows an example of a categorical variable that represents each person's gender, with the arbitrary choice of "woman" to be 1 and "man" to be 0.

If a qualitative $X$ variable has more than two categories, you will need to use more than one indicator variable to replace it. First, select one of the categories to use as the baseline value against which the effects of

the other categories will be measured. Do *not* use an indicator variable for the baseline category in the regression analysis because it will be represented by the constant term in the regression output. You will create a separate indicator variable for each of the nonbaseline categories. For each elementary unit (person, firm, or whatever) in the sample, you will have at most one value of 1 in the group of indicator variables; they will all be 0 if the elementary unit belongs to the baseline category. Remember the following rule.

**Rule for Using Indicator Variables**

The number of indicator variables used in multiple regression to replace a qualitative variable is *one less than* the number of categories. The remaining category defines the *baseline*. The baseline category is represented by the constant term in the regression output.

TABLE 12.4.1 An Indicator Variable Representing Gender

| Categorical Variable | Indicator Variable |
|---|---|
| Man | 0 |
| Man | 0 |
| Woman | 1 |
| Man | 0 |
| Woman | 1 |
| Woman | 1 |
| . | . |
| . | . |
| . | . |

Which category should be the baseline? You may choose the one you are most interested in comparing the other categories against.[32] You should probably choose a category that occurs fairly frequently.

Here is an example of a categorical variable that represents the nature of each item in a sample processed by a firm's mailroom. Four categories were used: business envelope, oversize envelope, small box, and large box. Since the vast majority of cases were business envelopes, this was chosen for convenience to be the baseline category.

---

32. What if you need to compare against more than one category? One simple solution is to run *several* multiple regression analyses, each one with a different category as the baseline.

TABLE 12.4.2 Using Three Indicator Variables to Represent Four Categories, Omitting "Business Envelope" as the Baseline Category

| Categorical Variable: Type of Item | Indicator Variables | | |
|---|---|---|---|
| | Oversize Envelope, $X_1$ | Small Box, $X_2$ | Large Box, $X_3$ |
| Business envelope | 0 | 0 | 0 |
| Small box | 0 | 1 | 0 |
| Business envelope | 0 | 0 | 0 |
| Business envelope | 0 | 0 | 0 |
| Large box | 0 | 0 | 1 |
| Oversize envelope | 1 | 0 | 0 |
| Business envelope | 0 | 0 | 0 |
| Oversize envelope | 1 | 0 | 0 |
| Business envelope | 0 | 0 | 0 |
| . | . | . | . |
| . | . | . | . |
| . | . | . | . |

This single qualitative variable (type of item) is to be used in a multiple regression analysis to help explain $Y$ = Processing time. Table 12.4.2 shows the three indicator variables that would be created and used along with the other $X$ variables.

## Interpreting and Testing Regression Coefficients for Indicator Variables

Once the categorical $X$ variables are replaced with indicator variables, the multiple regression can be performed in the usual way. Although the regression still has its usual interpretation, there are some special ways to think about the regression coefficients and their $t$ tests when you use indicator variables, as shown in Table 12.4.3. Remember that if $X_i$ is an indicator variable, it represents just one category of the original qualitative variable (namely, the category where it is 1).

TABLE 12.4.3 Interpreting the Regression Coefficient for an Indicator Variable $X_i$

| | |
|---|---|
| $b_i$ | The regression coefficient $b_i$ represents the *average difference* in $Y$ between the category represented by $X_i$ and the baseline category, holding all other $X$ variables fixed. If $b_i$ is a positive number, its category has a *higher* estimated average $Y$ than the baseline category; if $b_i$ is a negative number, then average $Y$ for its category is *lower* than for the baseline (all else equal). |
| Significance test for $b_i$ | In terms of the expected $Y$ value, holding all other $X$ variables fixed, is there any difference (other than randomness) between the category represented by $X_i$ and the baseline category? |

### Example
*Estimating the Impact of Gender on Salary after Adjusting for Experience*

Uh-oh. Your firm is worried about being sued for gender discrimination. There is a growing perception that males are being paid more than females in your department. A quick analysis of the 24 men and 26 women in the department shows that the average man was paid $4,214 more annually than the average woman. Furthermore, based on the standard error of $1,032, this is very highly statistically significant ($p < 0.001$).[33]

Does this imply discrimination against women? Well… not necessarily. The statistical results do summarize the salaries of the two groups and compare the difference against what would be expected due to randomness. Statistically, you may conclude that there are gender differences in salary that go well beyond randomness. However, statistics will not tell you the reason for these differences. Although there might be discrimination in hiring at your firm (either overt or subtle), there are other possibilities that could explain the gender differences. There may even be an economic basis for why men would be paid more in this particular situation.

At a meeting, someone suggests that the experience of your workers should also be considered as a possible explanation of the salary differences. The work of analyzing this possibility is delegated to you, and you decide to try multiple regression analysis as a way of understanding *the effect of gender on salary after adjusting for experience*. Multiple regression is the appropriate procedure because a regression coefficient is always adjusted for the other $X$ variables. The regression coefficient for the indicator variable representing gender will give you the expected salary difference between a man and a woman with the same experience.

Your multiple regression variables are salary ($Y$), experience ($X_1$), and gender ($X_2$). Gender will be represented as an indicator variable with Female = 1 and Male = 0. Table 12.4.4 shows the multivariate data set.

TABLE 12.4.4 Salary, Experience, and Gender for Employees

| Salary, Y | Experience (years), $X_1$ | Gender (1 = Female, 0 = Male), $X_2$ |
|---|---|---|
| $39,700 | 16 | 0 |
| 28,500 | 2 | 1 |
| 30,650 | 2 | 1 |
| 31,000 | 3 | 1 |
| 33,700 | 25 | 0 |
| 33,250 | 15 | 0 |
| 35,050 | 16 | 1 |
| 22,800 | 0 | 1 |
| 36,300 | 33 | 0 |
| 35,600 | 29 | 1 |
| 32,350 | 3 | 1 |
| 31,800 | 16 | 0 |
| 26,900 | 0 | 1 |
| 37,250 | 19 | 0 |
| 30,450 | 1 | 1 |
| 31,350 | 2 | 1 |
| 38,200 | 32 | 0 |
| 38,200 | 21 | 1 |
| 28,950 | 0 | 1 |
| 33,950 | 34 | 0 |
| 34,100 | 8 | 1 |
| 32,900 | 11 | 1 |
| 30,150 | 5 | 1 |
| 30,800 | 1 | 0 |
| 31,300 | 11 | 1 |
| 33,550 | 18 | 1 |
| 37,750 | 44 | 0 |
| 31,350 | 2 | 1 |
| 27,350 | 0 | 1 |
| 35,700 | 19 | 1 |
| 32,250 | 7 | 0 |
| 25,200 | 0 | 1 |
| 35,900 | 15 | 1 |
| 36,700 | 14 | 0 |
| 32,050 | 4 | 1 |
| 38,050 | 33 | 0 |
| 36,100 | 19 | 0 |
| 35,200 | 20 | 1 |
| 34,800 | 24 | 0 |
| 26,550 | 3 | 0 |
| 26,550 | 0 | 1 |
| 32,750 | 17 | 0 |
| 39,200 | 19 | 0 |
| 30,450 | 0 | 1 |
| 38,800 | 21 | 0 |
| 41,000 | 31 | 0 |
| 29,900 | 6 | 0 |
| 40,400 | 35 | 0 |
| 37,400 | 20 | 0 |
| 35,500 | 23 | 0 |
| **Average** | **33,313** | **13.98** | **52.0%** |
| **Standard deviation** | **4,188** | **11.87** | |

Sample size: $n = 50$

First, here are the results of data exploration. The scatterplot of salary against experience in Figure 12.4.1 shows a strong relationship (correlation $r = 0.803$). Employees with more experience are generally compensated more. There is a hint of nonlinearity, perhaps a "saturation effect" or an indication of "diminishing returns," in which an extra year of experience counts less and less as experience is accumulated. In any case, you can expect experience to account for much of the variation in salary.

The scatterplot of salary against gender, in Figure 12.4.2, confirms the fact that men are generally paid higher. However, this plot is much clearer when redone as two *box plots*, one for each gender, as in Figure 12.4.3. There is a clear relationship between gender and salary, with men paid higher on average. Although there is some overlap between the two box plots, the average salary difference is very highly significant (using the two-sample, unpaired $t$ test from Chapter 10).

The relationship between gender and experience, shown in Figure 12.4.4, shows that men have more experience on average than women. The lower part of the plot for women is not missing; it indicates that 25% of the women have little or no experience.

*(Continued)*

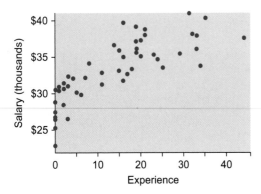

**FIGURE 12.4.1** The scatterplot of salary against experience shows a strong increasing relationship. The more experienced employees are compensated accordingly.

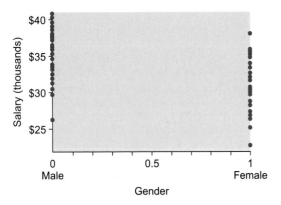

**FIGURE 12.4.2** The scatterplot of salary against gender is difficult to interpret because gender is an indicator variable. It's better to use box plots, as in Figure 12.4.3.

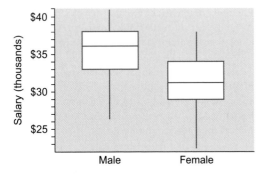

**FIGURE 12.4.3** Box plots of salary, one for each gender, provide a better way of exploring the relationship between gender and salary. Men are paid more on average, although there is considerable overlap in salary levels.

**Example—cont'd**

So far, what have you learned? There is a strong relationship between all pairs of variables. Extra experience is compensated, and being female is associated with a lower salary and less experience.

One important question remains: When you *adjust* for experience (in order to compare a man's salary to a woman's

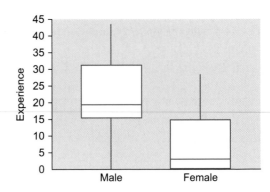

**FIGURE 12.4.4** On average, men have more experience than women. These are box plots representing the relationship between gender and experience.

with the same experience), is there a gender difference in salary? This information is not in the scatterplots because it involves all three variables simultaneously. The answer will be provided by the multiple regression. Table 12.4.5 shows the results.

The regression coefficient for gender, –488.08, indicates that the expected salary difference between a man and a woman with the same experience is $488.08, with the woman paid *less* than the man. The reason is that an increase of 1 in the indicator variable $X_2$ brings you from 0 (man) to 1 (woman) and results in a negative expected change (–$488.08) in salary.

Note that the regression coefficient for gender is not significant. It's not even close! The *t* test for significance of this coefficient is testing whether there is a difference between men and women at the same level of experience. This result tells you that, once experience is taken into account, there is no detectable difference between the average salary levels of men and women. The clear salary differences between men and women can be explained by differences in their experience. You have evidence that your firm may discriminate based on *experience* but not based solely on *gender*.

Does this analysis prove that there is no gender discrimination at your firm? Well…no. You may conclude only that there is no evidence of discrimination. Since accepting the null hypothesis (a finding of "not significant") leads to a weak conclusion (as discussed in Chapter 10), it is difficult to prove the absence of discrimination.

Does this analysis show that there is no gender discrimination in society at large? No, because the data reflect only one department at one firm and are hardly representative of this larger group.

But isn't the lower experience level of women due to past societal discrimination? This may well be true, but it cannot be determined based on the statistical analysis just performed. This data set has no information about the possible causes of the apparent salary discrimination other than that it may be explained by differences in experience.

33. This is the standard error of the difference for a two-sample, unpaired situation. So far, this is Chapter 10 material.

**TABLE 12.4.5 Multiple Regression Results for Employee Salary, Experience, and Gender**

**The regression equation is**

Salary

   = 29776

   + 271.15 * Experience

   −488.08 * Gender

The standard error of estimate

   $S = 2,538.76$

indicates the typical size of prediction errors in this data set.

The R-squared value

   $R^2 = 64.7\%$

indicates the proportion of the variance of Salary that is explained by the regression model.

**Inference for Salary at the 5% level**

The prediction equation DOES explain a significant proportion of the variation in Salary.

$F = 43.1572$ with 2 and 47 degrees of freedom

| Variable | Effect on Salary<br>Coeff | 95% Confidence Interval<br>From | To | Hypothesis Test<br>Significant? | StdErr of Coeff<br>StdErr | t Statistic<br>t |
|---|---|---|---|---|---|---|
| Constant | 29776 | 27867 | 31685 | Yes | 948.86 | 31.38 |
| Experience | 271.15 | 195.46 | 346.84 | Yes | 37.63 | 7.21 |
| Gender | −488.08 | −2269.06 | 1292.90 | No | 885.29 | −0.55 |

## Separate Regressions

A different approach to multiple regression analysis of multivariate data that includes a qualitative variable is to divide up the data set according to category and then perform a separate multiple regression for each category. For example, you might have two analyses: one for the men and another for the women. Or you might separately analyze the oil from the gas and from the nuclear power plants.

The use of indicator variables is just one step in the direction of separate regressions. With indicator variables, you essentially have a different constant term for each category but the same values for each regression coefficient. With separate regressions, you have a different constant term *and* a different regression coefficient for each category.

## 12.5 END-OF-CHAPTER MATERIALS

### Summary

Explaining or predicting a single Y variable from *two or more X* variables is called **multiple regression**. The goals of multiple regression are (1) to describe and understand the relationship, (2) to forecast (predict) a new observation, and (3) to adjust and control a process.

The **intercept** or **constant term**, *a*, gives the predicted (or "fitted") value for Y when *all X* variables are 0. The **regression coefficient** $b_j$, for the jth X variable, specifies the effect of $X_j$ on Y after adjusting for the other X variables; $b_j$ indicates how much larger you expect Y to be for a case that is identical to another except for being one unit larger in $X_j$. Taken together, these regression coefficients give you the **prediction equation** or **regression equation**, Predicted $Y = a + b_1 X_1 + b_2 X_2 + \ldots + b_k X_k$, which may be used for prediction or control. These coefficients ($a, b_1, b_2, \ldots, b_k$) are traditionally computed using the method of *least squares*, which minimizes the sum of the squared prediction errors. The **prediction errors** or **residuals** are given by $Y - (\text{Predicted } Y)$.

There are two ways of summarizing how good the regression analysis is. The **standard error of estimate**, $S_e$, indicates the approximate size of the prediction errors. The **coefficient of determination**, $R^2$, indicates the percentage of the variation in $Y$ that is "explained by" or "attributed to" the $X$ variables.

Inference begins with the **$F$ test**, an overall test to see if the $X$ variables explain a significant amount of the variation in $Y$. If your regression is *not* significant, you are not permitted to go further. If the regression is significant, you may proceed with statistical inference using *t* **tests for individual regression coefficients**. Confidence intervals and hypothesis tests for an individual regression coefficient will be based on its standard error, $S_{b_1}$, $S_{b_2}$, ..., or $S_{b_k}$. The critical value from the *t* table will have $n - k - 1$ degrees of freedom.

Inference is based on the **multiple regression linear model**, which specifies that the observed value for $Y$ is equal to the population relationship plus independent random errors that have a normal distribution

$$Y = (\alpha + \beta_1 X_1 + \beta_2 X_2 + \cdots + \beta_k X_k) + \varepsilon$$

$$= (\text{Population relationship}) + \text{Randomness}$$

where $\varepsilon$ has a normal distribution with mean 0 and constant standard deviation $\sigma$, and this randomness is independent from one case to another. For each population parameter ($\alpha$, $\beta_1$, $\beta_2$, ..., $\beta_k$, $\sigma$), there is a sample estimator ($a$, $b_1$, $b_2$, ..., $b_k$, $S_e$).

The hypotheses of the $F$ test are as follows:

$$H_0: \beta_1 = \beta_2 = \cdots = \beta_k = 0$$

$$H_1: \text{At least one of } \beta_1, \beta_2, ..., \beta_k \neq 0$$

The result of the $F$ test is given as follows:

If the $R^2$ value is *smaller* than the critical value in the table, the model is *not significant* (accept the null hypothesis that the $X$ variables *do not* help to predict $Y$).

If the $R^2$ value is *larger* than the value in the table, the model is *significant* (reject the null hypothesis and accept the research hypothesis that the $X$ variables *do* help to predict $Y$).

The confidence interval for an individual regression coefficient $b_j$ is

$$\text{From } b_j - tS_{b_j} \quad \text{to} \quad b_j + tS_{b_j}$$

where $t$ is from the $t$ table for $n - k - 1$ degrees of freedom. The hypotheses for the $t$ test of the $j$th regression coefficient are

$$H_0: \beta_j = 0$$

$$H_1: \beta_j \neq 0$$

There are two approaches to the difficult problem of deciding which $X$ variables are contributing the most to a regression equation. The **standardized regression coefficient**, $b_i S_{X_i}/S_y$, represents the expected change in $Y$ due to a change in $X_i$, measured in units of standard deviations of $Y$ per standard deviation of $X_i$, holding all other $X$ variables constant. If you don't want to adjust for all other $X$ variables (by holding them constant), you may compare the absolute values of the correlation coefficients for $Y$ with each $X$ instead.

There are some potential problems with a multiple regression analysis:

1. The problem of **multicollinearity** arises when some of your explanatory ($X$) variables are too similar to each other. The individual regression coefficients are poorly estimated because there is not enough information to decide *which one* (or more) of the variables is doing the explaining. You might omit some of the variables or redefine some of the variables (perhaps using ratios) to distinguish them from one another.

2. The problem of **variable selection** arises when you have a long list of potentially useful explanatory $X$ variables and would like to decide which ones to include in the regression equation. With too many $X$ variables, the quality of your results will decline because information is being wasted in estimating unnecessary parameters. If one or more important $X$ variables are omitted, your predictions will lose quality due to missing information. One solution is to include only those variables that are clearly necessary, using a prioritized list. Another solution is to use an automated procedure such as *all subsets* or *stepwise regression*.

3. The problem of **model misspecification** refers to the many different potential incompatibilities between your application and the multiple regression linear model. By exploring the data, you can be alerted to some of the potential problems with nonlinearity, unequal variability, or outliers. However, you may or may not have a problem: Even though the histograms of some variables may be skewed, and even though some scatterplots may be nonlinear, the multiple regression linear model might still hold. The *diagnostic plot* can help you decide when the problem is serious enough to need fixing. Another serious problem arises if you have a time series; it may help to do multiple regression using the percent changes from one time period to the next in place of the original data values for each variable.

The **diagnostic plot** for multiple regression is a scatterplot of the prediction errors (residuals) against the predicted values, and is used to decide whether you have any problems in your data that need to be fixed. Do not intervene unless the diagnostic plot shows you a clear and definite problem.

There are three ways of dealing with nonlinearity and/or unequal variability: (1) transform some or all variables, (2) introduce a new variable, or (3) use nonlinear regression. If you transform, each group of variables that is measured in the same basic units should probably be transformed in the same way. If you transform some of the $X$ variables but do not transform $Y$, then most of the interpretation of the results of a multiple regression analysis remains unchanged. If you use the natural logarithm of $Y$, then $R^2$ and significance tests for individual regression coefficients retain their usual interpretation, individual regression coefficients have similar interpretations, and a new interpretation is needed for $S_e$.

The **elasticity** of $Y$ with respect to $X_i$ is the expected *percentage* change in $Y$ associated with a 1% increase in $X_i$, holding the other $X$ variables fixed; the elasticity is estimated using the regression coefficient from a regression analysis using the natural logarithm of both $Y$ and $X_i$.

Another way to deal with nonlinearity is to use **polynomial regression** to predict $Y$ using a single $X$ variable together with some of its powers ($X^2$, $X^3$, etc.).

Two variables are said to show **interaction** if a change in both of them causes an expected shift in $Y$ that is different from the sum of the shifts in $Y$ obtained by changing each $X$ individually. Interaction is often modeled in regression analysis by using a *cross-product*, formed by multiplying one $X$ variable by another, which defines a new $X$ variable to be included along with the others in your multiple regression. Interaction can also be modeled by using transformations of some or all of the variables.

An **indicator variable** (also called a *dummy variable*) is a quantitative variable consisting only of the values 0 and 1 that is used to represent qualitative categorical data as an explanatory $X$ variable. The number of indicator variables used in multiple regression to replace a qualitative variable is *one less* than the number of categories. The remaining category defines the *baseline*. The baseline category is represented by the constant term in the regression output.

Instead of using indicator variables, you might compute separate regressions for each category. This approach provides a more flexible model with different regression coefficients for each $X$ variable for each category.

## Key Words

coefficient of determination, *349*
correlation matrix, *372*
diagnostic plot, *378*
elasticity, *386*
F test, *349*
indicator variable, *392*
interaction, *389*
intercept or constant term, *348*
model misspecification, *371*
multicollinearity, *371*
multiple regression, *347*
multiple regression linear model, *357*
polynomial regression, *388*
prediction equation or regression equation, *348*
prediction errors or residuals, *349*

regression coefficient, *348*
standard error of estimate, *349*
standardized regression coefficient, *370*
*t* tests for individual regression coefficients, *349*
variable selection, *371*

## Questions

1. For multiple regression, answer the following:
   a. What are the three goals?
   b. What kinds of data are necessary?
2. For the regression equation, answer the following:
   a. What is it used for?
   b. Where does it come from?
   c. What does the constant term tell you?
   d. What does a regression coefficient tell you?
3. Describe the two measures that tell you how helpful a multiple regression analysis is.
4. a. What does the result of the *F* test tell you?
   b. What are the two hypotheses of the *F* test?
   c. In order for the *F* test to be significant, do you need a high or a low value of $R^2$? Why?
5. a. What is the *t* test for an individual regression coefficient?
   b. In what way is such a test adjusted for the other $X$ variables?
   c. If the *F* test is not significant, are you permitted to go ahead and test individual regression coefficients?
6. a. How are the standardized regression coefficients computed?
   b. How are they useful?
   c. What are their measurement units?
7. a. What is multicollinearity?
   b. What are the harmful effects of extreme multicollinearity?
   c. How might moderate multicollinearity cause your *F* test to be significant, even though none of your *t* tests are significant?
   d. How can multicollinearity problems be solved?
8. a. If you want to be sure to get the best predictions, why not include among your $X$ variables every conceivably helpful variable you can think of?
   b. How can a prioritized list help you solve the variable selection problem?
   c. Briefly describe two automatic methods for variable selection?
9. a. What is the multiple regression linear model?
   b. List three ways in which the multiple regression linear model might fail to hold.
   c. What scatterplot can help you spot problems with the multiple regression linear model?
10. a. What are the axes in the diagnostic plot?
    b. Why is it good to find no structure in the diagnostic plot?
11. Why should variables measured in the same basic units be transformed in the same way?
12. a. What is an elasticity?
    b. Under what circumstances will a regression coefficient indicate the elasticity of $Y$ with respect to $X_i$?

13. How does polynomial regression help you deal with nonlinearity?
14. a. What is interaction?
    b. What can be done to include interaction terms in the regression equation?
15. a. What kind of variable should you create in order to include information about a categorical variable among your X variables? Please give the name of the variables and indicate how they are created.
    b. For a categorical variable with four categories, how many indicator variables would you create?
    c. What does the regression coefficient of an indicator variable indicate?

## Problems

*Problems marked with an asterisk (*) are solved in the Self Test in Appendix C.*

1.* Your firm is wondering about the results of magazine advertising as part of an assessment of marketing strategy. For each ad, you have information on its cost, its size, and the number of inquiries it generated. In particular, you are wondering if the number of leads generated by an ad has any connection with its cost and size. Identify the Y variable, the X variables, and the appropriate statistic or test.

2. It's budgeting time again, and you would like to know the expected payoff (in terms of dollars collected) of spending an extra dollar on collection of delinquent accounts, after adjusting for the size of the pool of delinquent accounts. Identify the Y variable, the X variables, and the appropriate statistic or test.

3. In order to substantiate a claim of damages, you need to estimate the revenues your firm lost when the opening of the new lumber mill was delayed for three months. You have access to data for similar firms on their total assets, lumber mill capacity, and revenues. For your firm, you know total assets and lumber mill capacity (if it had been working), but you wish to estimate the revenues. Identify the Y variable, the X variables, and the appropriate statistic or test.

4. Productivity is a concern. For each employee, you have data on productivity as well as other factors. You want to know how much these factors explain about the variation in productivity from one employee to another. Identify the Y variable, the X variables, and the appropriate statistic or test.

5.* Table 12.5.1 shows data on Picasso paintings giving the price, area of canvas, and year for each one.
    a. Find the regression equation to predict price from area and year.
    b. Interpret the regression coefficient for area.
    c. Interpret the regression coefficient for year.
    d. What would you expect the sales price to be for a 1954 painting with an area of 4,000 square centimeters?
    e. About how large are the prediction errors for these paintings?
    f. What percentage of the variation in prices of Picasso paintings can be attributed to the size of the painting and the year in which it was painted?
    g. Is the regression significant? Report the results of the appropriate test and interpret its meaning.

**TABLE 12.5.1 Price, Area, and Year for Picasso Paintings**

| Price (thousands) | Area (square cm) | Year | Price (thousands) | Area (square cm) | Year |
|---|---|---|---|---|---|
| $100 | 768 | 1911 | $360 | 1,141 | 1943 |
| 50 | 667 | 1914 | 150 | 5,520 | 1944 |
| 120 | 264 | 1920 | 65 | 5,334 | 1944 |
| 400 | 1,762 | 1921 | 58 | 1,656 | 1953 |
| 375 | 10,109 | 1921 | 65 | 2,948 | 1956 |
| 28 | 945 | 1922 | 95 | 3,510 | 1960 |
| 35 | 598 | 1923 | 210 | 6,500 | 1963 |
| 750 | 5,256 | 1923 | 32 | 1,748 | 1965 |
| 145 | 869 | 1932 | 55 | 3,441 | 1968 |
| 260 | 7,876 | 1934 | 80 | 7,176 | 1969 |
| 78 | 1,999 | 1940 | 18 | 6,500 | 1969 |
| 90 | 5,980 | 1941 | | | |

**Source:** Data are from E. Mayer, *International Auction Records*, vol. XVII (Caine, England: Hilmarton Manor Press, 1983), p. 1056-58.

   h.  Does area have significant impact on price, following adjustment for year? In particular, are larger paintings worth significantly more or significantly less on average than smaller paintings from the same year?
   i.  Does year have a significant impact on price, following adjustment for area? What does this tell you about older versus newer paintings?

6.  How are prices set for computer processor chips? At one time, the frequency was a good indicator of the processing speed; however, more recently manufacturers have developed alternative ways to deliver computing power because a high frequency tends to lead to overconsumption of power. Consider the multiple regression analysis in Table 12.5.2 to explain the price of eight chips from the two major manufacturers Intel (i3, i5, i7, i8) and AMD (Athlon and Phenom) based on their performance (as measured by the WorldBench score, where higher numbers are better), their frequency (gigahertz, or billions of cycles per second), and their power consumption (when idling, in watts).[34]

   a.  Approximately how much of the variation in price from one processor to another can be explained by the frequency, the power consumption, and the benchmark score?
   b.  Have frequency, power, and benchmark score taken together explained a significant proportion of the variability in price? How do you know?
   c.  Which, if any, of frequency, power, and benchmark score has a significant effect on price while controlling for the two others?
   d.  Find the predicted value of price and the residual value for the Phenom II X4 945 processor, given that its price is $140, its frequency is 3 gigahertz, its power is 98.1 watts, and its WorldBench score is 110.
   e.  What exactly do the regression coefficient 22.80 for the WorldBench benchmark test and its confidence interval tell you?
   f.  Approximately how accurately does the regression equation match the actual prices of these eight processor chips?
   g.  Write a paragraph summarizing what you can learn about the pricing structure of computer processors from this multiple regression analysis.

7.  One might expect the price of a tent to reflect various characteristics; for example, we might expect larger tents to cost more, all else equal (because they will hold more people) and heavier tents to cost less, all else equal (because they are harder to carry and therefore less desirable). Listings from a mail-order camping supply company provided the price, weight, and area of 30 tents. Results of a multiple regression analysis to predict price are shown in Table 12.5.3.

   a.  For tents of a given size (i.e., area), do heavier tents cost more or less on average than lighter tents?
   b.  What number from the computer output provides the answer to part a? Interpret this number and give its measurement units. Is it significant?
   c.  Is the result from part a consistent with expectations regarding tent pricing given at the start of the problem? Explain your answer.
   d.  For tents of a given weight, do larger tents cost more or less on average than smaller ones?

---

## TABLE 12.5.2 Multiple Regression Results for Computer Processor Chip Prices

**The regression equation is**

Price = −1,873 − 223.41 * (Frequency) + 2.37 * (Power) + 22.80 * WorldBench

$S = 99.827$

$R^2 = 94.3\%$

$F = 22.04$

$p = 0.00598$

| Variable | Effect on Price | 95% Confidence interval | | Hypothesis Test | StdErr of Coeff | t Statistic | p Value |
|---|---|---|---|---|---|---|---|
| | Coeff | From | To | Significant? | StdErr | t | p |
| Constant | −1,873 | −3441 | −305 | Yes | 565 | −3.317 | 0.029 |
| Frequency | −223.41 | −969.59 | 522.76 | No | 268.75 | −0.831 | 0.453 |
| Power | 2.37 | −8.29 | 13.04 | No | 3.84 | 0.618 | 0.570 |
| WorldBench | 22.80 | 13.37 | 32.23 | Yes | 3.40 | 6.713 | 0.003 |

## TABLE 12.5.3 Multiple Regression Results for Tent Pricing

**The regression equation is**

Price = 120 + 73.2 Weight − 7.52 Area

| Predictor | Coeff | StDev | *t*-ratio | *p* |
|---|---|---|---|---|
| Constant | 120.33 | 54.82 | 2.19 | 0.037 |
| Weight | 73.17 | 15.37 | 4.76 | 0.000 |
| Area | −7.517 | 2.546 | −2.95 | 0.006 |

*S* = 99.47  *R*-sq = 56.7%  *R*-sq(adj) = 53.5%

**Analysis of Variance**

| Source | DF | SS | MS | F | *p* |
|---|---|---|---|---|---|
| Regression | 2 | 349912 | 174956 | 17.68 | 0.000 |
| Error | 27 | 267146 | 9894 | | |
| Total | 29 | 617058 | | | |

e. What number from the computer output provides the answer to part d? Interpret this number and give its measurement units. Is it significant?

f. Is the result from part d consistent with expectations regarding tent pricing given at the start of the problem? Explain your answer.

8. Networked computers tend to slow down when they are overloaded. The response time is how long it takes from when you press the Enter key until the computer comes back with your answer. Naturally, when the computer is busier (either with users or with other work), you would expect it to take longer. This response time (in seconds) was measured at various times together with the number of users on the system and the load (the percent of the time that the machine is busy with high-priority tasks). The data are shown in Table 12.5.4.

a. Explore the data by commenting on the relationships in the three scatterplots you can produce by considering variables two at a time. In particular, do these relationships seem reasonable?

b. Compute the correlation matrix and compare it to the relationships you saw in the scatterplots.

c. Find the regression equation to predict response time from users and load. (You will probably need to use a computer for this and subsequent parts of this problem.)

d. To within approximately how many seconds can response time be predicted by users and load for this data set?

e. Is the *F* test significant? What does this tell you?

## TABLE 12.5.4 Computer Response Time, Number of Users, and Load Level

| Response Time | Users | Load |
|---|---|---|
| 0.31 | 1 | 20.2% |
| 0.69 | 8 | 22.7 |
| 2.27 | 18 | 41.7 |
| 0.57 | 4 | 24.6 |
| 1.28 | 15 | 20.0 |
| 0.88 | 8 | 39.0 |
| 2.11 | 20 | 33.4 |
| 4.84 | 22 | 63.9 |
| 1.60 | 13 | 35.8 |
| 5.06 | 26 | 62.3 |

f. Are the regression coefficients significant? Write a sentence for each variable, interpreting its adjusted effect on response time.

g. Note that the two regression coefficients are very different from each other. Compute the standardized regression coefficients to compare them and write a sentence about the relative importance of users and load in terms of effect on response time.

9. The unemployment rate can vary from one state to another, and in 2008 the standard deviation was 1.2% for the percent unemployed, which averaged 5.3% at the time. Table 12.5.5 shows these unemployment rates together with two possible explanatory variables: the educational level (percentage of college graduates for 2007) and the amount of federal spending (federal funds in dollars per capita, for 2007). To explore how well unemployment can be explained by these additional variables, please look at the multiple regression results from MINITAB in Table 12.5.6. Write a paragraph summarizing the strength of the connection. In particular, to what extent do education and

federal funding explain the cross-section of unemployment across the states?

10. Using the donations database on the companion site, and using only people who made a donation in response to the current mailing, consider predicting the amount of a donation (named "Donation_D1" in the worksheet) from the median years of school completed by adults in the neighborhood ("School_D1") and the number of promotions received before this mailing ("Promotions_D1").

    a. Find the regression equation and the coefficient of determination.

    b. Is there a significant relationship overall?

**TABLE 12.5.5** Unemployment Rate by State, with College Graduation Rate and Federal Spending

| State | Unemployment | College Grads | Federal Funds |
|---|---|---|---|
| Alabama | 5.6% | 21.4% | $9,571 |
| Alaska | 6.8 | 26.0 | 13,654 |
| Arizona | 5.9 | 25.3 | 7,519 |
| Arkansas | 5.2 | 19.3 | 7,655 |
| California | 7.1 | 29.5 | 7,006 |
| Colorado | 4.8 | 35.0 | 7,222 |
| Connecticut | 5.7 | 34.7 | 8,756 |
| Delaware | 5.0 | 26.1 | 6,862 |
| Florida | 6.1 | 25.8 | 7,905 |
| Georgia | 6.4 | 27.1 | 6,910 |
| Hawaii | 4.2 | 29.2 | 10,555 |
| Idaho | 5.4 | 24.5 | 6,797 |
| Illinois | 6.6 | 29.5 | 6,433 |
| Indiana | 6.0 | 22.1 | 6,939 |
| Iowa | 4.0 | 24.3 | 7,344 |
| Kansas | 4.5 | 28.8 | 7,809 |
| Kentucky | 6.3 | 20.0 | 8,945 |
| Louisiana | 5.0 | 20.4 | 16,357 |
| Maine | 5.4 | 26.7 | 8,350 |
| Maryland | 4.2 | 35.2 | 12,256 |
| Massachusetts | 5.3 | 37.9 | 8,944 |
| Michigan | 8.3 | 24.7 | 6,665 |
| Minnesota | 5.5 | 31.0 | 6,189 |
| Mississippi | 6.5 | 18.9 | 14,574 |
| Missouri | 6.1 | 24.5 | 8,952 |
| Montana | 5.2 | 27.0 | 8,464 |

*(Continued)*

**TABLE 12.5.5 Unemployment Rate by State, with College Graduation Rate and Federal Spending—cont'd**

| State | Unemployment | College Grads | Federal Funds |
|---|---|---|---|
| Nebraska | 3.3% | 27.5% | $7,895 |
| Nevada | 6.1 | 21.8 | 5,859 |
| New Hampshire | 3.8 | 32.5 | 6,763 |
| New Jersey | 5.4 | 33.9 | 7,070 |
| New Mexico | 4.4 | 24.8 | 10,784 |
| New York | 5.5 | 31.7 | 7,932 |
| North Carolina | 6.4 | 25.6 | 6,992 |
| North Dakota | 3.2 | 25.7 | 9,903 |
| Ohio | 6.5 | 24.1 | 7,044 |
| Oklahoma | 3.7 | 22.8 | 8,130 |
| Oregon | 6.4 | 28.3 | 6,391 |
| Pennsylvania | 5.3 | 25.8 | 8,324 |
| Rhode Island | 7.9 | 29.8 | 8,255 |
| South Carolina | 6.7 | 23.5 | 7,813 |
| South Dakota | 3.0 | 25.0 | 10,135 |
| Tennessee | 6.6 | 21.8 | 8,329 |
| Texas | 4.8 | 25.2 | 7,119 |
| Utah | 3.5 | 28.7 | 6,090 |
| Vermont | 4.9 | 33.6 | 8,496 |
| Virginia | 4.0 | 33.6 | 13,489 |
| Washington | 5.3 | 30.3 | 7,602 |
| West Virginia | 4.4 | 17.3 | 8,966 |
| Wisconsin | 4.7 | 25.4 | 6,197 |
| Wyoming | 3.0 | 23.4 | 10,082 |

**Source:** Data are from U.S. Census Bureau, *Statistical Abstract of the United States: 2010* (129th edition), Washington, DC, 2009, tables 228, 467, and 580, accessed at http://www.census.gov/compendia/statab/cats/federal_govt_finances_employment.html, http://www.census.gov/compendia/statab/cats/education.html, and http://www.census.gov/compendia/statab/cats/labor_force_employment_earnings.html on July 24, 2010.

c.  Which, if any, of the explanatory variables has a significant *t* test? What does this tell you?

11. Using the donations database on the companion site, and using only people who made a donation in response to the current mailing, consider predicting the amount of a donation (named "Donation_D1" in the worksheet) from the indicator variable that tells if the person is a catalog shopper ("CatalogShopper_D1"), the indicator variable that tells whether or not the person has a published home phone number ("HomePhone_D1"), and the number of recent gifts made ("RecentGifts_D1").

a.  Find the regression equation and the coefficient of determination.

b.  Is there a significant relationship overall?

c.  Which, if any, of the explanatory variables has a significant *t* test? What does this tell you?

12. There is considerable variation in the amount CEOs of different companies are paid, and some of it might be explained by differences in company characteristics. Consider the information in Table 12.5.7 on CEO salaries, sales, and return on equity (ROE) for selected northwest companies.

a.  What percentage of the variability in salary is explained by company sales and ROE?

b.  What is the estimated impact on salary, in additional dollars, of an increase in sales of 100 million dollars, holding ROE constant? Is this statistically significant?

**TABLE 12.5.6 Multiple Regression Results to Explain Unemployment Rate by State**

The regression equation is

Unemployment = 0.0713 – 0.000001 Federal Funds – 0.0357 College Grads

| Predictor | Coeff | SE Coeff | t | p |
|---|---|---|---|---|
| Constant | 0.07132 | 0.01278 | 5.58 | 0.000 |
| Federal Funds | –0.00000101 | 0.00000078 | –1.30 | 0.199 |
| College Grads | –0.03571 | 0.03749 | –0.95 | 0.346 |

$S = 0.0122090$   R-Sq = 4.7%   R-Sq(adj) = 0.7%

Analysis of Variance

| Source | DF | SS | MS | F | P |
|---|---|---|---|---|---|
| Regression | 2 | 0.0003476 | 0.0001738 | 1.17 | 0.321 |
| Residual Error | 47 | 0.0070058 | 0.0001491 | | |
| Total | 49 | 0.0073534 | | | |

| Source | DF | Seq SS |
|---|---|---|
| Federal Funds | 1 | 0.0002124 |
| College Grads | 1 | 0.0001352 |

**TABLE 12.5.7 CEO Salaries, Sales, and Return on Equity for Selected Northwest Companies**

| Company | Name | Salary | Sales (millions) | ROE |
|---|---|---|---|---|
| Alaska Air Group | William S. Ayer | $360,000 | $3,663 | –6.1% |
| Ambassadors Group | Jeffrey D. Thomas | 400,000 | 98 | 35.5 |
| American Ecology | Stephen A. Romano | 275,000 | 176 | 23.1 |
| Avista | Scott L. Morris | 626,308 | 1,677 | 8.7 |
| Blue Nile | Diane Irvine | 437,396 | 295 | 20.3 |
| Cardiac Science | John R. Hinson | 376,923 | 206 | 0.0 |
| Cascade Corp. | Robert C. Warren, Jr. | 540,000 | 534 | 17.1 |
| Cascade Microtech | Geoffrey Wild | 369,102 | 77 | 4.2 |
| Coeur d'Alene Mines | Dennis E. Wheeler | 587,633 | 189 | 19.2 |
| Coinstar | David W. Cole | 475,000 | 912 | 6.1 |
| Coldwater Creek | Dan Griesemer | 725,000 | 1,024 | 19.6 |
| Columbia Sportswear | Timothy P. Boyle | 804,231 | 1,318 | 15.6 |

(Continued)

**TABLE 12.5.7 CEO Salaries, Sales, and Return on Equity for Selected Northwest Companies—cont'd**

| Company | Name | Salary | Sales (millions) | ROE |
|---|---|---|---|---|
| Data I/O | Frederick R. Hume | 312,500 | 28 | 0.4 |
| Esterline Technologies | Robert W. Cremin | 849,231 | 1,483 | 8.4 |
| Expedia | Dara Khosrowshahi | 1,000,000 | 2,937 | 4.2 |
| Fisher Communications | Colleen B. Brown | 546,000 | 174 | 7.5 |
| Flir Systems | Earl R. Lewis | 823,206 | 1,077 | 26.7 |
| Flow International | Charles M. Brown | 384,624 | 244 | 12.6 |
| Hecla Mining | Phillips S. Baker, Jr. | 426,250 | 193 | 35.5 |
| InfoSpace | James F. Voelker | 403,077 | 157 | −2.2 |
| Jones Soda | Stephen C. Jones | 142,917 | 36 | 18.8 |
| Key Technology | David M. Camp, Ph.D. | 275,002 | 134 | −1.9 |
| Key Tronic | Jack W. Oehlke | 417,308 | 204 | 29.9 |
| LaCrosse Footwear | Joseph P. Schneider | 440,000 | 128 | 11.8 |
| Lattice Semiconductor | Bruno Guilmart | 307,506 | 222 | 0.6 |
| McCormick & Schmick's | Douglas L. Schmick | 415,385 | 391 | 8.8 |
| Micron Technology | Steven R. Appleton | 950,000 | 5,841 | 5.8 |
| MWI Veterinary Supply | James F. Cleary, Jr. | 300,000 | 831 | 12.8 |
| Nike | Mark G. Parker | 1,376,923 | 18,627 | 23.3 |
| Nordstrom | Blake W. Nordstrom | 696,111 | 8,573 | 31.8 |
| Northwest Pipe | Brian W. Dunham | 570,000 | 440 | 10.3 |
| Paccar | Mark C. Pigott | 1,348,846 | 14,973 | 35.8 |
| Plum Creek Timber | Rick R. Holley | 830,000 | 1,614 | 14.3 |
| Precision Castparts | Mark Donegan | 1,175,000 | 6,852 | 17.9 |
| RealNetworks | Robert Glaser | 236,672 | 605 | 16.0 |
| Red Lion Hotels | Anupam Narayan | 345,715 | 188 | −0.4 |
| SonoSite | Kevin M. Goodwin | 450,000 | 244 | 4.3 |
| Starbucks | Howard Schultz | 1,190,000 | 10,383 | 26.1 |
| Todd Shipyards | Stephen G. Welch | 340,653 | 139 | 9.9 |
| TriQuint Semiconductor | Ralph G. Quinsey | 414,953 | 573 | 4.7 |
| Umpqua Holdings | Raymond P. Davis | 714,000 | 541 | 8.9 |
| Weyerhaeuser | Daniel S. Fulton | 792,427 | 8,018 | 4.8 |
| Zumiez | Richard M. Brooks | 262,500 | 409 | 23.4 |

**Source:** *Seattle Times*, accessed March 27, 2010, at http://seattletimes.nwsource.com/flatpages/businesstechnology/2009northwestcompaniesdatabase.html and at http://seattletimes.nwsource.com/flatpages/businesstechnology/ceopay2008.html.

c. What is the estimated impact on salary, in additional dollars, of an increase in ROE of one percentage point, holding sales constant? Is this statistically significant?

d. Identify CEO and company corresponding to the smallest salary, the smallest predicted salary, and the smallest residual of salary (using sales and ROE as explanatory variables). Write a few sentences interpreting these results.

13. What explains the financial performance of brokerage houses? Table 12.5.8 shows the one-year performance of asset-allocation blends of selected brokerage houses, together with the percentages recommended in stocks and bonds at the end of the period.

a. What proportion of the variation in performance is explained by the recommended percentages in stocks and bonds?

b. Do the recommended percentages explain a significant amount of the variation in performance?

c. Does the recommended percentage for stocks have a significant impact on performance, adjusted for the recommended percentage for bonds?

d. Does the recommended percentage for bonds have a significant impact on performance, adjusted for the recommended percentage for stocks?

e. Which appears to have a greater impact on performance: stocks or bonds?

14. Table 12.5.9 shows some of the results of a multiple regression analysis to explain money spent on home food-processing equipment ($Y$) based on income ($X_1$),

education ($X_2$), and money spent on sporting equipment ($X_3$). All money variables represent total dollars for the past year; education is in years. There are 20 cases.

a. How much would you expect a person to spend on food-processing equipment if he or she earns $25,000 per year, has 14 years of education, and spent $292 on sporting equipment last year?

b. How successful is the regression equation in explaining food-processing expenditures? In particular, which statistic in the results should you look at, and is it statistically significant?

c. To within approximately what accuracy (in dollars per year) can predictions of food-processing expenditures be made for the people in this study?

d. For each of the three $X$ variables, state whether it has a significant effect on food-processing expenditures or not (after adjusting for the other $X$ variables).

15. Consider the multiple regression results shown in Table 12.5.10, which attempt to explain compensation of the top executives of 11 major energy corporations based on the revenues and the return on equity of the firms.[35] For example, the data for Consol Energy, Inc. consist of a compensation number of 13.75 (in millions of dollars) for the CEO J. Brett Harvey, an ROE number of 16.42% (which is the same number as 0.1642), and a revenue number of 4,570 (in millions of dollars).

a. To within approximately how many dollars can you predict the compensation of the CEO of these firms based on revenue and ROE?

**TABLE 12.5.8 Brokerage House Asset-Allocation One-Year Performance and Recommended Percentages in Stocks and Bonds**

| Brokerage Firm | Performance | Stocks | Bonds |
|---|---|---|---|
| Lehman Brothers | 14.62% | 80% | 10% |
| Morgan Stanley D.W. | 14.35 | 70 | 20 |
| Edward D. Jones | 13.86 | 71 | 24 |
| Prudential Securities | 13.36 | 75 | 5 |
| Goldman Sachs | 12.98 | 70 | 27 |
| Raymond James | 10.18 | 55 | 15 |
| A.G. Edwards | 10.04 | 60 | 35 |
| PaineWebber | 9.44 | 48 | 37 |
| Credit Suisse F.B. | 9.33 | 55 | 30 |
| J.P. Morgan | 9.13 | 50 | 25 |
| Bear Stearns | 8.75 | 55 | 35 |
| Salomon Smith Barney | 8.57 | 55 | 35 |
| Merrill Lynch | 5.15 | 40 | 55 |

**Source:** T. Ewing, "Bullish Stock Mix Paid Off in 4th Quarter," *Wall Street Journal*, March 17, 2000, p. C1.

### TABLE 12.5.9 Multiple Regression Results for Food-Processing Equipment

**The regression equation is**

$Y = -9.26 + 0.00137\ X_1 + 10.8\ X_2 + 0.00548\ X_3$

| Column | Coefficient | StDev of Coeff | t-ratio = Coeff/s.d. |
|---|---|---|---|
|  | −9.26247 | 13.37258 | −0.69264 |
| $X_1$ | 0.001373 | 0.000191 | 7.165398 |
| $X_2$ | 10.76225 | 0.798748 | 13.47389 |
| $X_3$ | 0.005484 | 0.025543 | 0.214728 |

$S = 16.11$
$R$-Squared = 94.2 percent

### TABLE 12.5.10 Multiple Regression Results for Executive Compensation

**The regression equation is**

Compensation = 0.908811 + 0.002309 * Revenue − 0.116328 * ROE.

$S = 4.6101\ R^2 = 0.5201$

$F = 4.3351$ with 2 and 8 degrees of freedom

$p = 0.0530$

| Variable | Effect on Compensation Coeff | 95% Confidence Interval From | To | StdErr of Coeff StdErr | t statistic t | p-Value p |
|---|---|---|---|---|---|---|
| Constant | 0.908811 | −4.7675 | 6.5852 | 2.4616 | 0.3692 | 0.722 |
| Revenue | 0.002309 | 0.0004 | 0.0042 | 0.0008 | 2.7434 | 0.025 |
| ROE | −0.116328 | −3.8677 | 3.6351 | 1.6268 | −0.0715 | 0.945 |

b.* Find the predicted compensation and the residual prediction error for the CEO of Consol Energy, Inc., expressing both quantities in dollars.

c. If ROE is interpreted as an indicator of the firm's performance, is there a significant link between performance and compensation (adjusting for firm sales)? How do you know?

d. Is there a significant link between revenue and compensation (adjusting for firm ROE)? How do you know?

e. What exactly does the regression coefficient 0.002309 for revenue tell you?

16. In many ways, nonprofit corporations are run much like other businesses. Charity organizations with larger operations would be expected to have a larger staff, although some have more overhead than others. Table 12.5.11 shows the number of paid staff members of charity organizations as well as the amounts of money (in millions of dollars) raised from public donations, government payments, and other sources of income.

a. Find the regression equation to predict staff levels from the contributions of each type for these charities. (You will probably need to use a computer for this.)

b. How many additional paid staff members would you expect to see, on average, working for a charity that receives $5 million more from public donations than another charity (all else equal)?

c. To within approximately how many people can the regression equation predict the staffing levels of these charities from their revenue figures?

d. Find the predicted staffing level and its residual for the American Red Cross.

e. What is the result of the F test? What does it tell you?

f. Does revenue from public donations have a significant impact on staffing level, holding other revenues fixed? How do you know?

**TABLE 12.5.11  Staff and Contribution Levels ($ millions) for Charities**

| Charity Organization | Staff | Public | Government | Other |
|---|---|---|---|---|
| Salvation Army | 29,350 | $473 | $92 | $300 |
| American Red Cross | 22,100 | 341 | 30 | 602 |
| Planned Parenthood | 8,200 | 67 | 106 | 101 |
| CARE | 7,087 | 45 | 340 | 12 |
| Easter Seals | 5,600 | 83 | 51 | 78 |
| Association of Retarded Citizens | 5,600 | 28 | 80 | 32 |
| Volunteers of America | 5,000 | 14 | 69 | 83 |
| American Cancer Society | 4,453 | 271 | 0 | 37 |
| Boys Clubs | 3,650 | 103 | 9 | 75 |
| American Heart Association | 2,700 | 151 | 1 | 27 |
| UNICEF | 1,652 | 67 | 348 | 48 |
| March of Dimes | 1,600 | 106 | 0 | 6 |
| American Lung Association | 1,500 | 80 | 1 | 17 |

**Source:** Data are from G. Kinkead, "America's Best-Run Charities," *Fortune*, November 9, 1987, p.146.

17. Consider the computer output in Table 12.5.12, part of an analysis to explain the final cost of a project based on management's best guess of labor and materials costs at the time the bid was placed, computed from 25 recent contracts. All variables are measured in dollars.
    a. What percentage of the variation in cost is explained by the information available at the time the bid is placed?
    b. Approximately how closely can we predict cost if we know the other variables?
    c. Find the predicted cost of a project involving $9,000 in labor and $20,000 in materials.
    d. Is the $F$ test significant? What does this tell you?
    e. Do materials have a significant impact on cost?

18. For the previous problem, interpret the regression coefficient for labor by estimating the average final cost associated with each dollar that management identified ahead of time as being labor related.

19. A coworker of yours is very pleased, having just found an $R^2$ value of 100%, indicating that the regression equation has explained all of the variability in $Y$ ("profits") based on the $X$ variables "revenues" and "costs." You then amaze this person by correctly guessing the values of the regression coefficients.

    a. Explain why the result ($R^2 = 100\%$) is reasonable—trivial, even—in this case.
    b. What are the values of the regression coefficients?

20. Quality control has been a problem with a new product assembly line, and a multiple regression analysis is being used to help identify the source of the trouble. The daily "percent defective" has been identified as the $Y$ variable, to be predicted from the following variables that were considered by some workers to be likely causes of trouble: the "percent overscheduled" (a measure of the extent to which the system is being worked over and above its capacity), the "buffer inventory level" (the extent to which stock builds up between workstations), and the "input variability" (the standard deviation of the weights for a key input component). Based on the multiple regression output in Table 12.5.13, where should management action be targeted? Explain your answer in the form of a memo to your supervisor.

21. By switching suppliers, you believe that the standard deviation of the key input component can be reduced from 0.62 to 0.38, on average. Based on the multiple regression output from the preceding problem, what size reduction in defect rate should you expect if you go ahead and switch suppliers? (The defect rate was

**TABLE 12.5.12 Regression Analysis for Final Cost of a Project**

**Correlations:**

|  | Cost | Labor |
|---|---|---|
| Labor | 0.684 | |
| Material | 0.713 | 0.225 |

**The regression equation is**

Cost = 13975 + 1.18 Labor + 1.64 Material

| Predictor | Coeff | StDev | t-ratio | p |
|---|---|---|---|---|
| Constant | 13975 | 4286 | 3.26 | 0.004 |
| Labor | 1.1806 | 0.2110 | 5.59 | 0.000 |
| Material | 1.6398 | 0.2748 | 5.97 | 0.000 |

$S = 3860$　$R\text{-sq} = 79.7\%$　$R\text{-sq(adj)} = 77.8\%$

**Analysis of Variance**

| Source | DF | SS | MS | F | p |
|---|---|---|---|---|---|
| Regression | 2 | 1286267776 | 643133888 | 43.17 | 0.000 |
| Error | 22 | 327775808 | 14898900 | | |
| Total | 24 | 1614043648 | | | |

| Source | DF | SEQ SS |
|---|---|---|
| Labor | 1 | 755914944 |
| Material | 1 | 530352896 |

measured in percentage points, so that "defect" = 5.3 represents a 5.3% defect rate.)

22. How do individual companies respond to economic forces throughout the globe? One way to explore this is to see how well rates of return for stock of individual companies can be explained by stock market indexes that reflect particular parts of the world. Table 12.5.14 shows monthly rates of return for two companies, Microsoft (headquartered in the United States) and China Telecom (in China), along with three indexes: the Hang Seng (Hong Kong), the FTSE100 (London), and the S&P 500 (United States).

a. Run a multiple regression to explain percentage changes in Microsoft stock from those of the three indexes. Which indexes, if any, show a significant t test? Report the p-value of each of these t tests. Is this consistent with where Microsoft is headquartered?

b. Run a multiple regression to explain percentage changes in China Telecom stock from those of the three indexes. Which indexes, if any, show a significant t test? Is this consistent with where China Telecom is headquartered?

c. Run an ordinary regression to explain percentage changes in Microsoft stock from the Hang Seng Index only. Is the regression significant? Report the overall p-value of the regression.

d. Reconcile the results of parts a and c, focusing in particular on whether the Hang Seng Index is significant in each regression. You may use the interpretation that when an explanatory variable is significant in a multiple regression, it says that this variable brings additional information about the Y variable over and above that brought by the other explanatory X variables.

**TABLE 12.5.13  Multiple Regression Results for New Product Assembly Line**

**The regression equation is**

Defect = −1.62 + 11.7 Sched + 0.48 Buffer + 7.29 Input

| Predictor | Coeff | StDev | t-ratio | p |
|---|---|---|---|---|
| Constant | −1.622 | 1.806 | −0.90 | 0.381 |
| Sched | 11.71 | 22.25 | 0.53 | 0.605 |
| Buffer | 0.479 | 2.305 | 0.21 | 0.838 |
| Input | 7.290 | 2.287 | 3.19 | 0.005 |

$S = 2.954$   $R\text{-sq} = 43.8\%$   $R\text{-sq(adj)} = 34.4\%$

**Analysis of Variance**

| Source | DF | SS | MS | F | P |
|---|---|---|---|---|---|
| Regression | 3 | 122.354 | 40.785 | 4.67 | 0.014 |
| Error | 18 | 157.079 | 8.727 | | |
| Total | 21 | 279.433 | | | |

**TABLE 12.5.14  Monthly Rates of Return for Two Companies along with Stock Market Indexes from Different Parts of the World**

| Date | Microsoft | China Telecom | Hang Seng Index | FTSE 100 Index | S&P 500 Index |
|---|---|---|---|---|---|
| 4/1/2010 | 5.39% | 5.92% | 4.32% | 2.56% | 3.61% |
| 3/1/2010 | 2.16 | 11.32 | 3.06 | 6.07 | 5.88 |
| 2/1/2010 | 2.21 | 7.26 | 2.42 | 3.20 | 2.85 |
| 1/4/2010 | −7.55 | −0.94 | −8.00 | −4.15 | −3.70 |
| 12/1/2009 | 3.66 | −6.82 | 0.23 | 4.28 | 1.78 |
| 11/2/2009 | 6.51 | 1.07 | 0.32 | 2.90 | 5.74 |
| 10/1/2009 | 7.81 | −7.02 | 3.81 | −1.74 | −1.98 |
| 9/1/2009 | 4.34 | −8.39 | 6.24 | 4.58 | 3.57 |
| 8/3/2009 | 5.39 | −0.69 | −4.13 | 6.52 | 3.36 |
| 7/2/2009 | −1.02 | 4.48 | 11.94 | 8.45 | 7.41 |
| 6/1/2009 | 13.74 | 4.56 | 1.14 | −3.82 | 0.02 |
| 5/1/2009 | 3.78 | −3.61 | 17.07 | 4.10 | 5.31 |
| 4/1/2009 | 10.28 | 22.14 | 14.33 | 8.09 | 9.39 |
| 3/2/2009 | 13.79 | 23.76 | 5.97 | 2.51 | 8.54 |
| 2/2/2009 | −4.93 | −7.35 | −3.51 | −7.70 | −10.99 |

*(Continued)*

**TABLE 12.5.14 Monthly Rates of Return for Two Companies along with Stock Market Indexes from Different Parts of the World—cont'd**

| Date | Microsoft | China Telecom | Hang Seng Index | FTSE 100 Index | S&P 500 Index |
|------|-----------|---------------|-----------------|----------------|---------------|
| 1/2/2009 | −12.06 | −5.04 | −7.71 | −6.42 | −8.57 |
| 12/1/2008 | −3.81 | 0.00 | 3.59 | 3.41 | 0.78 |
| 11/3/2008 | −8.85 | 7.94 | −0.58 | −2.04 | −7.48 |
| 10/2/2008 | −16.33 | −13.81 | −22.47 | −10.71 | −16.94 |
| 9/1/2008 | −2.20 | −19.85 | −15.27 | −13.02 | −9.08 |
| 8/1/2008 | 6.51 | −6.45 | −6.46 | 4.15 | 1.22 |
| 7/2/2008 | −6.50 | 0.28 | 2.85 | −3.80 | −0.99 |
| 6/2/2008 | −2.86 | −23.57 | −9.91 | −7.06 | −8.60 |
| 5/1/2008 | −0.33 | 5.28 | −4.75 | −0.56 | 1.07 |
| 4/1/2008 | 0.48 | 9.26 | 12.72 | 6.76 | 4.75 |
| 3/3/2008 | 4.37 | −14.41 | −6.09 | −3.10 | −0.60 |
| 2/1/2008 | −16.25 | 2.00 | 3.73 | 0.08 | −3.48 |
| 1/2/2008 | −8.44 | −7.80 | −15.67 | −8.94 | −6.12 |
| 12/3/2007 | 5.95 | −3.34 | −2.90 | 0.38 | −0.86 |
| 11/1/2007 | −8.39 | −8.10 | −8.64 | −4.30 | −4.40 |
| 10/1/2007 | 24.93 | 14.89 | 15.51 | 3.94 | 1.48 |
| 9/3/2007 | 2.56 | 31.48 | 13.17 | 2.59 | 3.58 |
| 8/1/2007 | −0.58 | 0.83 | 3.45 | −0.89 | 1.29 |
| 7/3/2007 | −1.61 | −2.46 | 6.49 | −3.75 | −3.20 |
| 6/1/2007 | −3.98 | 9.97 | 5.52 | −0.20 | −1.78 |
| 5/1/2007 | 2.86 | 13.89 | 1.55 | 2.67 | 3.25 |
| 4/2/2007 | 7.40 | −1.35 | 2.62 | 2.24 | 4.33 |
| 3/1/2007 | −1.05 | 6.83 | 0.76 | 2.21 | 1.00 |
| 2/1/2007 | −8.39 | −5.38 | −2.26 | −0.51 | −2.18 |
| 1/2/2007 | 3.34 | −10.94 | 0.71 | −0.28 | 1.41 |

Source: Data are from http://finance.yahoo.com/, accessed on April 15, 2010.

23. In a multiple regression, what would you suspect is the problem if the $R^2$ is large and significant, but none of the X variables has a t test that is significant?

24. Consider Table 12.5.15, showing the partial results from a multiple regression analysis that explains the annual sales of 25 grocery stores by some of their characteristics. The variable "mall" is 1 if the store is in a shopping mall and 0 otherwise. The variable "customers" is the number of customers per year.

   a. To within approximately how many dollars can you predict sales with this regression model?

   b. Find the predicted sales for a store that is in a shopping mall and has 100,000 customers per year.

   c. Does each of the explanatory variables have a significant impact on sales? How do you know?

   d. What, exactly, does the regression coefficient for customers tell you?

   e. Does the location (mall or not) have a significant impact on sales, comparing two stores with the same number of customers? Give a brief explanation of why this might be the case.

## TABLE 12.5.15 Multiple Regression Results for Grocery Stores' Annual Sales

**The regression equation is**

Sales = –36589 + 209475 Mall + 10.3 Customers

| Predictor | Coeff | StDev | t-ratio | p |
|---|---|---|---|---|
| Constant | –36589 | 82957 | –0.44 | 0.663 |
| Mall | 209475 | 77040 | 2.72 | 0.013 |
| Customers | 10.327 | 4.488 | 2.30 | 0.031 |

$S = 183591$   R-sq = 39.5%   R-sq(adj) = 34.0%

f.  Approximately how much extra in annual sales comes to a store in a mall, as compared to a similar store not located in a mall?

25. Setting prices is rarely an easy task. A low price usually results in higher sales, but there will be less profit per sale. A higher price produces higher profit per sale, but sales will be lower overall. Usually, a firm wants to choose the price that will maximize the total profit, but there is considerable uncertainty about the demand. Table 12.5.16 shows hypothetical results of a study of profits in comparable test markets of equal sizes, where only the price was changed.

    a.  Find the regression equation of the form Predicted Profit = $a + b$(Price).

    b.  Test to see whether or not the regression is significant. Is this result reasonable?

    c.  To within approximately how many dollars can profit be predicted from price in this way?

    d.  Examine a diagnostic plot to see if there is any further structure remaining that would help you explain profit based on price. Describe the structure that you see.

## TABLE 12.5.16 Price and Profit in Test Markets

| Price | Profit |
|---|---|
| $8 | $6,486 |
| 9 | 10,928 |
| 10 | 15,805 |
| 11 | 13,679 |
| 12 | 12,758 |
| 13 | 9,050 |
| 14 | 5,702 |
| 15 | –109 |

e.  Create another X variable using the squared price values and find the multiple regression equation to predict profit from price and squared price.

f.  To within approximately how many dollars can profit be predicted from price using these two X variables?

g.  Test to see whether a significant proportion of the variation in profit can be explained by price and squared price taken together.

h.  Find the price at which the predicted profit is maximized. Compare this to the price at which the observed profit was the highest.

26. Table 12.5.17 shows the results of a multiple regression analysis designed to explain the salaries of chief executive officers based on the sales of their firm and the industry group.[36] The Y variable represents CEO salary (in thousands of dollars). The $X_1$ variable is the firm's sales (in millions of dollars). $X_2$, $X_3$, and $X_4$ are indicator variables representing the industry groups aerospace, banking, and natural resources, respectively (the natural resources group includes the large oil companies). The indicator variable for the baseline group, automotive, has been omitted. There are $n = 49$ observations in this data set.

    a.  Do sales and industry groups have a significant impact on CEO salary?

    b.  What is the estimated effect of an additional million dollars of sales on CEO salary, adjusted for industry group?

    c.  Is the salary difference you estimated due to sales in part b statistically significant? What does this tell you in practical terms about salary differences?

    d.  According to the regression coefficient, how much more or less is the CEO of a bank paid compared to the CEO of an automotive firm of similar size?

    e.  Is the salary difference comparing banking to automotive that you estimated in part d statistically significant? What does this tell you in practical terms about salary differences?

27. Consider the magazine advertising page cost data from Table 12.1.3.

## TABLE 12.5.17 Multiple Regression Results for CEO Salaries

**The regression equation is**

Salary = 931.8383

+ 0.01493 * Sales

– 215.747 * Aerospace

– 135.550 * Bank

– 303.774 * Natural Resources

$S = 401.8215$

$R^2 = 0.423469$

| Variable | Coeff | StdErr |
|---|---|---|
| Constant | 931.8383 | 163.8354 |
| Sales | 0.014930 | 0.003047 |
| Aerospace | –215.747 | 222.3225 |
| Bank | –135.550 | 177.0797 |
| Natural Resources | –303.774 | 187.4697 |

| 1999 | 4.97 | 4.64 | 5.65 |
|---|---|---|---|
| 2000 | 6.24 | 5.82 | 6.03 |
| 2001 | 3.88 | 3.40 | 5.02 |
| 2002 | 1.67 | 1.61 | 4.61 |
| 2003 | 1.13 | 1.01 | 4.02 |
| 2004 | 1.35 | 1.37 | 4.27 |
| 2005 | 3.22 | 3.15 | 4.29 |
| 2006 | 4.97 | 4.73 | 4.80 |
| 2007 | 5.02 | 4.36 | 4.63 |
| 2008 | 1.92 | 1.37 | 3.66 |

**Source:** Data are from U.S. Census Bureau, *Statistical Abstract of the United States: 2010* (129th edition), Washington, DC, 2009. Fed Funds and Treasury bill data from Table 1160, accessed at http://www.census.gov/compendia/statab/cats/banking_finance_insurance.html on July 24, 2010. The 10-year Treasury bond data are from Table 1161, accessed at http://www.census.gov/compendia/statab/cats/banking_finance_insurance.html on July 24, 2010.

a. Which *X* variable is the least helpful in explaining page costs? How do you know?

b. Rerun the regression analysis omitting this *X* variable.

c. Compare the following results without the *X* variable to the results with the *X* variable: *F* test, $R^2$, regression coefficients, *t* statistics.

28. Consider the interest rates on securities with various terms to maturity, shown in Table 12.5.18.

a. Find the regression equation to predict the long-term interest rate (Treasury bonds) from the two shorter-term rates.

b. Create a new variable, "interaction," by multiplying the two shorter-term rates together. Find the regression equation to predict the long-term interest rate (Treasury bonds) from both of the shorter-term rates and the interaction.

c. Test whether there is any interaction between the two shorter-term interest rates that would enter into the relationship between short-term and long-term interest rates.

34. Data analyzed are from D. Murphy, "Chip Showdown: Confused about Which Processor to Pick for Your New System?" *PCWorld*, August 2010, p. 91.

35. CEO salary data are from http://www.aflcio.org/corporatewatch/paywatch/ceou/industry.cfm, accessed on July 4, 2010. Revenue and ROE are from http://www.finance.yahoo.com, accessed on July 24, 2010.

36. The data used are from "Executive Compensation Scoreboard," *Business Week*, May 2, 1988, p. 57.

## TABLE 12.5.18 Interest Rates

| Year | Federal Funds (overnight) | Treasury Bills (3-month) | Treasury Bonds (10-year) |
|---|---|---|---|
| 1992 | 3.52% | 3.43% | 7.01% |
| 1993 | 3.02 | 3.00 | 5.87 |
| 1994 | 4.21 | 4.25 | 7.09 |
| 1995 | 5.83 | 5.49 | 6.57 |
| 1996 | 5.30 | 5.01 | 6.44 |
| 1997 | 5.46 | 5.06 | 6.35 |
| 1998 | 5.35 | 4.78 | 5.26 |

## Database Exercises

Refer to the employee database in Appendix A.

1.* Consider the prediction of annual salary from age and experience.

a. Find and interpret the regression equation and regression coefficients.

b. Find and interpret the standard error of estimate.

c. Find and interpret the coefficient of determination.

d. Is the model significant? What does this tell you?

e. Test each regression coefficient for significance and interpret the results.

f. Find and interpret the standardized regression coefficients.

g. Examine the diagnostic plot and report serious problems, if there are any.

2. Continue using predictions of annual salary based on age and experience.

**a.*** Find the predicted annual salary and prediction error for employee 33 and compare the result to the actual annual salary.

**b.** Find the predicted annual salary and prediction error for employee 52 and compare the result to the actual annual salary.

**c.** Find the predicted annual salary and prediction error for the highest-paid employee and compare the result to the actual annual salary. What does this comparison tell you?

**d.** Find the predicted annual salary and prediction error for the lowest-paid employee and compare the result to the actual annual salary. What does this comparison tell you?

**3.** Consider the prediction of annual salary from age alone (as compared to exercise 1, where experience was also used as an *X* variable).

**a.** Find the regression equation to predict annual salary from age.

**b.** Using results from part a of exercise 1 and this exercise, compare the effect of age on annual salary with and without an adjustment for experience.

**c.** Test whether age has a significant impact on annual salary with and without an adjustment for experience. Briefly discuss your results.

**4.** Now examine the effect of gender on annual salary, with and without adjusting for age and experience.

**a.** Find the average annual salary for men and for women and compare them.

**b.** Using a two-sided test at the 5% level, test whether men are paid significantly more than women. (You may wish to refer back to Chapter 10 for the appropriate test to use.)

**c.** Find the multiple regression equation to predict annual salary from age, experience, and gender, using an indicator variable for gender that is 1 for a woman.

**d.** Examine and interpret the regression coefficient for gender.

**e.** Does gender have a significant impact on annual salary after adjustment for age and experience?

**f.** Compare and discuss your results from parts b and e of this exercise.

**5.** Now examine the effect of training level on annual salary, with and without adjusting for age and experience.

**a.** Find the average annual salary for each of the three training levels and compare them.

**b.** Find the multiple regression equation to predict annual salary from age, experience, and training level, using indicator variables for training level. Omit the indicator variable for level A as the baseline.

**c.** Examine and interpret the regression coefficient for each indicator variable for training level.

**d.** Does training level appear to have a significant impact on annual salary after adjustment for age and experience?

**e.** Compare and discuss the average salary differential between training levels A and C, both with and without adjusting for age and experience.

**6.** Consider predicting annual salary from age, experience, and an interaction term.

**a.** Create a new variable, "interaction," by multiplying age by experience for each employee.

**b.** Find the regression equation to predict annual salary from age, experience, and interaction.

**c.** Test whether you have a significant interaction by using a *t* test for the regression coefficient of the interaction variable.

**d.** What is the average effect on annual salary of an extra year's experience for a 40-year-old employee?

**e.** What is the average effect on annual salary of an extra year's experience for a 50-year-old employee?

**f.** Interpret the interaction between age and experience by comparing your answers to parts d and e of this exercise.

## Project

Find a multivariate data set relating to your work or business interests on the Internet, in your library, in a newspaper, or in a magazine, with a sample size of *n* = 25 or more, for which the *F* test is significant and at least one of the *t* tests is significant.

**a.** Give your choice of dependent variable (*Y*) and briefly explain your reasons.

**b.** Examine and comment on the scatterplots defined by plotting *Y* against each *X* variable.

**c.** Compute and briefly interpret the correlation matrix.

**d.** Report the regression equation.

**e.** For two elementary units in your data set, compute predicted values for *Y* and residuals.

**f.** Interpret each regression coefficient and its confidence interval.

**g.** Which regression coefficients are significant? Which (if any) are not? Are these results reasonable?

**h.** Comment on what you have learned from multiple regression analysis about the effects of the *X* variables on *Y*.

## Case

### Controlling Quality of Production

Everybody seems to disagree about just why so many parts have to be fixed or thrown away after they are produced. Some say that it's the temperature of the production process, which needs to be held constant (within a reasonable range). Others claim that it's clearly the density of the product, and that if we could only produce a heavier material, the problems would disappear. Then there is Ole, who has been warning everyone forever to take care not to push the equipment beyond its limits. This problem would be the easiest to fix, simply by slowing down the production rate; however, this would increase costs. Interestingly, many of the workers on the morning shift think that the problem is "those inexperienced workers in the afternoon," who, curiously, feel the same way about the morning workers.

Ever since the factory was automated, with computer network communication and bar code readers at each station, data have been piling up. You've finally decided to have a look. After your assistant aggregated the data by four-hour blocks and then typed in the AM/PM variable, you found the following note on your desk with a printout of the data already loaded into the computer network:

*Whew! Here are the variables:*

- *Temperature actually measures temperature variability as a standard deviation during the time of measurement.*
- *Density indicates the density of the final product.*
- *Rate indicates the rate of production.*
- *AM/PM is an indicator variable that is 1 during morning production and is 0 during the afternoon.*
- *Defect is the average number of defects per 1,000 produced.*

| Temperature | Density | Rate | AM/PM | Defect |
|---|---|---|---|---|
| 0.97 | 32.08 | 177.7 | 0 | 0.2 |
| 2.85 | 21.14 | 254.1 | 0 | 47.9 |
| 2.95 | 20.65 | 272.6 | 0 | 50.9 |
| 2.84 | 22.53 | 273.4 | 1 | 49.7 |
| 1.84 | 27.43 | 210.8 | 1 | 11.0 |
| 2.05 | 25.42 | 236.1 | 1 | 15.6 |
| 1.50 | 27.89 | 219.1 | 0 | 5.5 |
| 2.48 | 23.34 | 238.9 | 0 | 37.4 |
| 2.23 | 23.97 | 251.9 | 0 | 27.8 |
| 3.02 | 19.45 | 281.9 | 1 | 58.7 |
| 2.69 | 23.17 | 254.5 | 1 | 34.5 |
| 2.63 | 22.70 | 265.7 | 1 | 45.0 |
| 1.58 | 27.49 | 213.3 | 0 | 6.6 |
| 2.48 | 24.07 | 252.2 | 0 | 31.5 |
| 2.25 | 24.38 | 238.1 | 0 | 23.4 |
| 2.76 | 21.58 | 244.7 | 1 | 42.2 |
| 2.36 | 26.30 | 222.1 | 10 | 13.4 |
| 1.09 | 32.19 | 181.4 | 1 | 0.0 |
| 2.15 | 25.73 | 241.0 | 0 | 20.6 |
| 2.12 | 25.18 | 226.0 | 0 | 15.9 |
| 2.27 | 23.74 | 256.0 | 0 | 44.4 |
| 2.73 | 24.85 | 251.9 | 1 | 37.6 |
| 1.46 | 30.01 | 192.8 | 1 | 2.2 |
| 1.55 | 29.42 | 223.9 | 1 | 1.5 |
| 2.92 | 22.50 | 260.0 | 0 | 55.4 |
| 2.44 | 23.47 | 236.0 | 0 | 36.7 |
| 1.87 | 26.51 | 237.3 | 0 | 24.5 |
| 1.45 | 30.70 | 221.0 | 1 | 2.8 |
| 2.82 | 22.30 | 253.2 | 1 | 60.8 |
| 1.74 | 28.47 | 207.9 | 1 | 10.5 |

Naturally you decide to run a multiple regression to predict the defect rate from all of the explanatory variables, the idea being to see which (if any) are associated with the occurrence of defects. There is also the hope that if a variable helps predict defects, then you might be able to control (reduce) defects by changing its value. Here are the regression results as computed in your spreadsheet.[37]

### Summary Output
**Regression Statistics**

| | |
|---|---|
| Multiple $R$ | 0.948 |
| $R$ Square | 0.899 |
| Adjusted $R$ Square | 0.883 |
| Standard Error | 6.644 |
| Observations | 30 |

### ANOVA

| | df | SS | MS | F | P-value |
|---|---|---|---|---|---|
| Regression | 4 | 9825.76 | 2456.44 | 55.65 | 4.37E–12 |
| Residual | 25 | 1103.54 | 44.14 | | |
| Total | 29 | 10929.29 | | | |

| | Coeff | StdErr | t | p | Low95 | Up95 |
|---|---|---|---|---|---|---|
| Intercept | −28.756 | 64.170 | −0.448 | 0.658 | −160.915 | 103.404 |
| Temperature | 26.242 | 9.051 | 2.899 | 0.008 | 7.600 | 44.884 |
| Density | −0.508 | 1.525 | −0.333 | 0.742 | −3.649 | 2.633 |
| Rate | 0.052 | 0.126 | 0.415 | 0.682 | −0.207 | 0.311 |
| AM/PM | −1.746 | 0.803 | −2.176 | 0.039 | −3.399 | −0.093 |

At first, the conclusions appear obvious. But are they?

### Discussion Questions
1. What are the "obvious conclusions" from the hypothesis tests in the regression output?
2. Look through the data. Do you find anything that calls into question the regression results? Perform further analysis as needed.
3. What action would you recommend? Why?

37. Note that a number can be reported in scientific notation, so that 2.36E–5 stands for $(2.36)(10^{-5}) = 0.0000236$. Think of it as though the E–5 tells you to move the decimal point five places to the left.

## Chapter 13

# Report Writing

## Communicating the Results of a Multiple Regression

Communication is an essential management skill with many applications. You use communication strategies to motivate those who report to you, to convince your boss you've done a good job, to obtain resources needed for a new project, to persuade your potential customers, to bring your suppliers into line, and so on.

Statistical summaries can help you communicate the basic facts of a situation in the most objective and useful way.[1] They can help you make your point to an audience. You can gain credibility because it is clear that you have gone to some trouble to carefully survey the entire situation in order to present the "big picture." Here are some examples of reports that include statistical information:

**One:** *A market survey.* Your firm is in the process of deciding whether or not to go ahead and launch a new product. The market survey provides important background information about potential customers: their likes and dislikes, how much they might be willing to pay, what kind of support they expect, and so on. To help you understand these consumers, the report might include a multiple regression analysis to determine how eager they are to buy the product based on characteristics such

as income and industry group. The purpose of the report is to provide background information for decision making. The audience is middle- and upper-level management, including those who will ultimately make the decision.

**Two:** *Recommendations for improving a production process.* In the presence of anticipated domestic and international competition when your patent expires next year, you would like to remain in the market as a low-cost producer. This goal is reasonable because, after all, you have much more experience than anyone else. Based on data collected from actual operations, as well as experiments with alternative processes, the report summarizes the various approaches and their expected savings under various scenarios. There might be multiple regression results to identify the important factors and to suggest how to adjust them. The purpose of the report is to help reduce costs. The audience might be middle- and upper-level management, who would decide which suggestions to adopt.

**Three:** *A review of hiring and compensation practices.* Either as part of a periodic review or in reaction to accusations of unfairness, your firm is examining its human resources practices. A multiple regression analysis might be used to explain salary by age, experience, gender, qualifications, and so on. By comparing official

---

1. There are, of course, other uses for statistics. Consider, for example, D. Huff, *How to Lie with Statistics* (New York: Norton, 1993).

policies to the regression results, you can see whether or not the firm is effectively implementing its goals. By testing the regression coefficient for gender, you can see whether or not there is evidence of possible gender discrimination. The purposes of the study include helping the firm achieve its human resource management goals and, perhaps, defending the firm against accusations of discrimination. The audience might be middle- and upper-level management, who might make some adjustments in human resource management policies, or it might be the plaintiff and judge in a lawsuit.

**Four:** *Determination of cost structure.* To control costs, it helps to know what they are. In particular, as demand and production go up and down, what component of your total cost can be considered fixed cost, and what, then, are the variable costs per unit of each item you produce? A multiple regression analysis can provide estimates of your cost structure based on actual experience. The purpose of the study is to understand and control costs. The audience consists of those managers responsible for budgeting and for controlling costs.

**Five:** *Product testing.* Your firm, in your opinion, produces the best product within its category. One way to convince others of this is to report the statistical results from objective testing. Toothpaste firms have used this technique for years now ("…has been shown to be an effective decay-preventing dentifrice that can be of significant value …"). Various statistical hypothesis testing techniques might be used here, from an unpaired $t$ test to multiple regression. The purpose of the study is to prove superiority of your product. The audience consists of your potential customers. The form of the "report" might range from a quote on the package, to a paragraph in the information brochure, to a 200-page report filed with a government agency and made available upon request.

Let's assume that your primary consideration when using statistical results in a report is to *communicate*. Be kind to your readers and explain what you have learned using language that is easy for them to understand. They will probably be more impressed with your work if they understand it than if they get lost in technical terminology and details.

Reports are written for various reasons and for various audiences. Once you have identified these, you will find that the writing is easier because you can picture yourself addressing your audience for a real purpose. Defining your purpose helps you narrow down the topic so that you can concentrate on the relevant issues. Identifying your audience helps you select the appropriate writing style and level of detail.

## 13.1 HOW TO ORGANIZE YOUR REPORT

How you organize your report will depend on your purpose and the appropriate audience. In this section is an outline of the main parts of a statistical report, which you may modify to fit your particular purpose. There are six parts to this form of a typical report:

1. The *executive summary* is a paragraph at the beginning that describes the most important facts and conclusions from your work.
2. The *introduction* consists of several paragraphs in which you describe the background, the questions of interest, and the data with which you have worked.
3. The *analysis and methods section* lets you interpret the data by presenting graphic displays, statistical summary numbers, and results, which you explain as you go along.
4. The *conclusion and summary* move back to the big picture to give closure, pulling together all of the important thoughts you would like your readers to remember.
5. A *reference* is a note indicating the material you have taken from an outside source and giving enough information so that your audience can locate it. You might have references appear as notes on the appropriate page of a section, or you might gather them together in their own section.
6. The *appendix* should contain all supporting material that is important enough to include but not important enough to appear in the text of your report.

In addition, you may also wish to include a *title page* and a *table of contents*. The **title page** goes first and includes the title of the report, the name and title of the person you prepared it for, your name and title (as the preparer), and the date. The **table of contents** goes after the executive summary, giving an outline of the report together with page numbers.

The best organization is straightforward and direct. Remember, your purpose is to make it easy for the reader to understand what you've done. Don't play games with your audience by keeping them in suspense until the last page. Instead, tell them all of the most important results at the start and fill in details later. Your readers will appreciate this effort because they are as pressed for time as you are. This strategy will also help your message reach those people who read only a part of your paper.

Use an outline in planning your report, perhaps with one line representing each paragraph. This is an excellent way to keep the big picture in mind as you work out the exact wording.

### The Executive Summary Paragraph

The **executive summary** is a paragraph at the beginning that describes the most important facts and conclusions from your work, omitting unnecessary details. Writing in straightforward nontechnical language, you should orient the reader to the importance of the problem and explain your contribution to its understanding and solution. You are, in essence, reducing the entire report to a single paragraph.

Some people, especially those who are technically oriented, may complain that their hundreds of pages of analysis have already been reduced to 15 pages of report and that it would be impossible (and unfair) to reduce all of that precious work to a single paragraph. However, there are important people who will read no more than the executive summary, and if you want your message to reach them, you are doing yourself a favor by providing them the convenience of the executive summary.

Although the executive summary goes first, it is often easiest to write it last, after you have finished the rest of the report. Only at this time are you completely sure of just what it is you are summarizing!

## The Introduction Section

The **introduction** consists of several paragraphs in which you describe the background, the questions of interest, and the data set you have worked with. Write in nontechnical language, as if to an intelligent person who knows very little about the details of the situation. After reading the executive summary and introduction, your reader should be completely oriented to the situation. All that remains are the details.

It's OK to repeat material from the executive summary. You may even want to take some of the sentences directly from the executive summary and use them as initial topic sentences for paragraphs in the introduction.

## The Analysis and Methods Section

In the **analysis and methods** section, you interpret the data by presenting graphic displays, statistical summary numbers, and results, explaining them as you go along. This is your chance to give some of the details that have been only hinted at in the executive summary and introduction.

Be selective. You should probably leave out much of the analysis you have actually done on the problem. A careful analyst will explore many avenues "just in case," in order to check assumptions and validate the basic approach. But many of these results belong in a separate folder, in which you keep an archive of everything you looked at. From this folder, select only those items that are important and helpful to the story you are telling in the report. For example, if a group of scatterplots had ordinary linear structure, you might include just one of these with a comment that the others were much the same. Choose the materials that most strongly relate to your purpose.

To help your reader understand your points, you will want to include each graph on the page of text that discusses it. Many computers can help you do this. An alternative is to use a reducing copy machine to enable you to paste a small graph directly on the page of text. Here is a list of items you should consider including in the analysis and methods section, organized according to the five basic activities of statistics:

1. *Design.* If there are important aspects of how you obtained the data that could not be covered in the introduction, you can include them either here or in an appendix.
2. *Data exploration.* Tell your reader what you found in the data. You might want to include some graphs (histograms, box plots, or scatterplots) if they help the reader see what you're talking about. An extremely skewed histogram or a diagnostic plot with structure might be shown to justify using a transformation to help satisfy the underlying assumptions. If you have outlier trouble, now is your chance to mention it and justify how you dealt with it.
3. *Modeling.* Here is your opportunity to tell the reader in general terms how a particular modeling method, such as multiple regression, is useful to the analysis of your situation.
4. *Estimation.* Report the appropriate statistical summaries and explain what they say about the business situation you are examining. These might be averages (indicating typical value), standard deviations (perhaps indicating risk), correlations (indicating strength of relationship), or regression coefficients (indicating the adjusted effect of one factor on another). You will also want to include measures of the uncertainty in these estimates so that your readers can assess the quality of the information you are reporting. These would include standard errors and confidence intervals for the estimates, whenever possible and appropriate, as well as $R^2$ and the standard error of estimate for a regression analysis.
5. *Hypothesis tests.* If appropriate, tell your reader whether or not the estimated effects are "really there" by testing them, for example, against the reference value 0. Once you find a statistically significant result, you have license to explain it. If an estimate is *not* statistically significant, you are *not* permitted to explain it.[2] By testing, you reassure your reader by showing that your claims have a solid foundation.

---

2. For example, if a regression coefficient is computed as $-167.35$ but is not significantly different from 0, you really are not even sure that it is a negative number. Since the true effect (in the population) might be positive instead of negative, don't make the mistake of "explaining" why it is negative.

## The Conclusion and Summary Section

By this point, your reader is somewhat familiar with the details of your project. The **conclusion and summary** section now moves back to the big picture to give closure, pulling together all of the important thoughts you would like your readers to remember. Whereas the executive summary primarily provides the initial orientation, the conclusion and summary can draw on the details you have provided in the intervening sections. Keep in mind that while some readers get here by reading your pages in order, others may flip directly here to see how it ends.

In particular, be sure to tell exactly what the analysis has revealed about the situation. Why did you go to all of the trouble? What made it worth all that time? How have your questions been answered?

## Including References

Whenever you take text, data, or ideas from an outside source, you need to give proper credit. A **reference** is a note indicating the kind of material you have taken from an outside source and giving enough information so that your reader can obtain a copy. You might have each note appear as a footnote on the same page as its text reference, or you might gather them together in their own section.

Be sure to provide enough information in your reference so that an interested person can actually find the information you used. It is not enough to just give the author's name or to say "U.S. Department of Commerce." A full reference will also include such details as the date, volume, page, and publisher. Even a statement such as "the 2010 sales figure is from the firm's annual reports" does not provide enough information because, for example, the sales figures for 2010 may be revised in 2011 and appear as a different number in the annual report a year or two later.

Here are examples of common types of references:

1. For *quoted text taken from a book*, your reference should include the author(s), the year published, the title, the place of publication, and the publisher, as well as the page number for this material. Here is a sample paragraph and footnote:

> In order to give credit where credit is due, as well as to avoid being accused of plagiarism, you should provide adequate references in your report. In the words of Siegel:
>
> *Whenever you take text, data, or ideas from an outside source, it is necessary to give proper credit. A **reference** is a note indicating the kind of material you have taken from an outside source and giving enough information so that your reader can obtain a copy.*[3]
>
> ---
> 3. This quote is from A. F. Siegel, *Practical Business Statistics*, 6th ed. (Burlington, MA: Elsevier, 201), p. 420.

2. For an *idea taken from a book*, explained in your own words and not as a direct quote, proceed as follows:

> As explained by Hens and Schenk-Hoppé, liquidity in financial markets can be measured by the difference between the bid and the ask prices, and agents in these markets can choose to create additional liquidity or to make use of liquidity already available.[4] Moreover, each agent can make this decision individually without central control.
>
> ---
> 4. This material is covered in T. Hens and K. R. Schenk-Hoppé, *Handbook of Financial Markets: Dynamics and Evolution* (New York: Elsevier, 2009), p. 111.

3. For *data taken from a magazine article*, your note should include the title of the article, the author (if given), the magazine, the date, and the page. If the data are attributed to another source, you should mention that source as well. For example:

> The benchmark performance scores among the top Intel processors were higher than those of the best AMD processors, with Intel's chips achieving an impressive 147 (for the Core i7–980X) and 127 (for the Core i7–870) while AMD's best score was 118 (for the Phenom II X6 1090T) on PCWorld's WorldBench scale.[5] Of course, both companies can be expected to continue to improve their product lines.
>
> ---
> 5. Data are from D. Murphy, "Chip Showdown: Confused about which processor to pick for your new system?" *PCWorld*, August 2010, p. 91.

4. For material you obtained from electronic networks such as the Internet, your reference should include the author(s), the title, the date of posting and update (if available in the document), the date you accessed the material, and the electronic address known as the uniform resource locator or URL. Be sure to provide enough information so that, in case the URL address is changed or discontinued, your future readers will be able to perform a search to try to locate the material. Here is a sample paragraph and footnote in which the date is listed only once because the material was accessed on the date of posting:

> To some extent, the budgetary difficulties of the Greek government can be traced to the fact that unit labor costs in Greece rose 33% from 2001 to 2009 and this interfered with Greece's ability to rely on its export market.[6]
>
> ---
> 6. From J. Jubak, "Euro Crisis Is Tip of the Iceberg," MSN Money, accessed at http://articles.moneycentral.msn.com/Investing/JubaksJournal/euro-crisis-is-tip-of-the-iceberg.aspx on July 25, 2010.

5. For material you learned from an interview, letter, or telephone call, your reference is to a *personal communication*. Be sure to mention the person's name and title as well as the place and date of the communication. For example:

> One important aspect of the language in the audio of marketing materials is the intonation, which can "convey attitude and emotion, highlight new information, and indicate how the words are grouped into phrases."[7]
>
> ―――――――――――――――――――――
> 7. Ann Wennerstrom, Linguist, personal communication, July 25, 2010.

If you would like more complete information about references, *The Chicago Manual of Style* is an excellent source for further details.[8]

## The Appendix Section

The **appendix** contains all supporting material that is important enough to include but not important enough (perhaps due to space limitations) to be included in the text of your report. This would include a listing of your data (if it fits on a few pages) and their source. You might also include, if appropriate, some details of the design of your study, some extra graphs and tables, and further technical explanation for statements in the report itself. For clarity, you might put material into different appendix sections: Appendix A, Appendix B, and so on.

Using an appendix is your way of keeping everyone happy. The casual reader's thoughts will not be interrupted, and the more technical reader will have easy access to important material. For example:

> Since the gender effect was not statistically significant, the multiple regression analysis was repeated after removing the gender variable. The results did not change in any meaningful way (details may be found in Appendix C).

## 13.2  HINTS AND TIPS

This section gives some hints and tips to help you save time in producing an effective report.

## Think about Your Audience

Keep it brief. Remember that your readers probably don't have enough time to get things done either. You can do

―――――――――――――――――――――
8. See "Documentation I: Basic Patterns" and "Documentation II: Specific Content" in *The Chicago Manual of Style*, 15th ed. (Chicago: University of Chicago Press, 2003), Chapters 16 and 17.

them a favor by selecting and including only the most important results, graphs, and conclusions. If you must include lots of technical materials, try placing them in an appendix.

Make it clear. Be sure to use straightforward language and to provide enough introduction and orientation so that your readers don't get left behind.

Look it over. Read your rough draft, putting yourself in the place of your reader. Try to forget that you have been living and breathing this project for weeks, and pretend you have only heard vague things about it. See if you have provided enough connections with everyday life to bring your reader into the subject.

## What to Write First? Next? Last?

The order in which you do things can make a difference. You can't write the paper until you know what the results are. Many people write the introduction and executive summary *last* because only then do they know what they are introducing and summarizing.

Do the analysis first. Explore the data, look at the graphs, compute the estimates, and do the hypothesis tests. Perhaps even run the multiple regression with different $X$ variables. This stage will produce a file folder filled with more material than you could ever use. Save it all, just in case you'll need to check something out later.

Next, select the most important results from your analysis file. Now that you know how things have turned out, you are ready to make an outline of the analysis and methods section. Perhaps you could also outline some conclusions at this point.

All that remains is to create paragraphs from the lines on your outline, decide what to place in the appendix, choose the references, and write the introduction and executive summary. After you've done this rough draft, read it over with your audience in mind, and make the final changes. If possible, ask a friend to read it over, too. Print it out. You're done!

## Other Sources

There are many sources of information about good writing, covering language, style, and usage. For example:

1. Use a *dictionary* to check usage of a word you're not sure about. Computer word processors are often quicker for checking spelling (in Microsoft Word, you can check spelling in the Proofing group of the Review Ribbon).
2. Use a *thesaurus* to look for different words with a similar meaning. This can help you find just the right word or can help you avoid repeating the same term too many times. Many computer word processing

programs have a built-in thesaurus (in Microsoft Word, the thesaurus can be found in the Proofing group of the Review Ribbon).

3. A number of available books are filled with advice for managers who need to write reports that might involve some technical material, for example, Kuiper's *Contemporary Business Report Writing*, Miller's *Chicago Guide to Writing about Multivariate Analysis*, and Mamishev and Williams' *Technical Writing for Teams*.[9]

4. For more details about writing conventions, please consult *The Chicago Manual of Style*.

## 13.3 EXAMPLE: A QUICK PRICING FORMULA FOR CUSTOMER INQUIRIES

An example of a report based on a multiple regression analysis, following the organization plan suggested earlier in this chapter, is presented here. Note how the practical meaning rather than the details is emphasized throughout.

---

### A Quick Pricing Formula for Customer Inquiries

*Prepared for*
*B.S. Wennerstrom, Vice President of Sales*

*Prepared by*
*C.H. Siegel, Director of Research*
*Mount Olympus Design and Development Corporation*
*April 10, 2010*

#### Executive Summary

We are losing potential customers because we cannot respond immediately with a price quotation. Our salespeople report that by the time we call back with a price quote the next day, many of these contacts have already made arrangements to work with one of our competitors. Our proposed solution is to create a quick pricing formula. Potential customers for routine jobs will be able to obtain an approximate price over the phone. This should help keep them interested in us while they wait for the exact quote the next day. Not only will they know if we're "in the ballpark," but this will also help us appear more responsive to customer needs.

#### Introduction

Preparing a price quotation currently requires three to six hours of engineering work. When our customers call us about a new job, they want to know the price range so that they can "comparison shop." In the past, when we had fewer

---

competitors, this was not a problem. Even though our quality is superior, we are losing more and more jobs to competitors who can provide information more quickly.

We checked with Engineering and agree that the time delay is necessary if an exact quote is required. A certain amount of rough preliminary work is essential in order to determine the precise size of the layout and the power requirements, which, in turn, determine our cost.

We also checked with a few key customers. Although they do not require an exact price quote immediately, it would be a big help if we could give them a rough idea of the price during that initial contact. This would meet two of their needs: (1) They can be sure that we are competitive, and (2) this pricing information helps them with their design decisions because they can quickly evaluate several approaches.

Based on our own experience, we have created a formula that produces an approximate number for our cost remarkably quickly:

$$\text{Quick cost number} = \$1,356 + \$35.58(\text{Components}) + \$5.68(\text{Size})$$

The resulting "quick cost number" will, in most cases, differ from our own detailed cost computation by no more than $200. A quick telephone quotation could be given by simply adding in the appropriate (confidential) markup, depending on the customer's discount class.

We assembled data from recent detailed quotations, taken from our own internal computerized records that support the detailed price quote that we routinely commit ourselves to for a period of seven days. The key variables we analyzed include:

1. The **cost** computed by Engineering. This is internal confidential material. This is the variable to be predicted; it is the only variable not available during the initial phone conversation.

2. The **number of components** involved. This is a rough indication of the complexity of the design and is nearly always provided by the customer during the initial contact.

3. The **layout size**. This is a very rough indication of the size of the actual finished layout. It is provided by the customer as an initial starting point.

Of the 72 quotations given this quarter, we selected 56 as representative of the routine jobs we see most often. The cases that were rejected either required a special chemical process or coating or else used exotic components for which we have no stable source. The data set is included in the appendix.

#### Analysis and Methods

This section begins with a description of our typical price quotations and continues with the cost prediction formula (using multiple regression methodology) and its interpretation.

Here is a profile of our most typical jobs. From the histograms in Figure 13.3.1, you can see that our typical price quote involves a cost somewhere between $3,000 and

---

9. S. Kuiper, *Contemporary Business Report Writing*, 4th ed., (Mason, OH: Thomson South-Western, 2007); J. E. Miller, *The Chicago Guide to Writing about Multivariate Analysis* (Chicago: University of Chicago Press, 2005); A.V. Mamishev and S.D. Williams, *Technical Writing for Teams: The STREAM Tools Handbook* (New York: Wiley-IEEE Press, 2010).

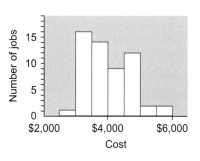
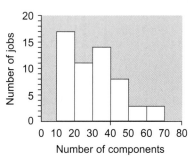
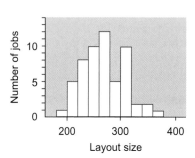

**FIGURE 13.3.1**  Histograms of the variables.

$5,000, with just a few above or below. The standard deviation of cost is $707, indicating the approximate error we would make if we were to (foolishly!) make quick quotes simply based on the average cost of $3,987. The number of components is typically 10 to 50, with just a few larger jobs. The layout size is typically anywhere from 200 to the low 300s. There are no outlier problems because all large or atypical jobs are treated separately as special cases, and the quick cost formula would not be appropriate.

Next, we considered the *relationship* between cost and each of the other variables. As you can see from the two scatterplots in Figure 13.3.2, there is a very strong relationship between the number of components and our cost (the correlation is 0.949) and a strong relationship between the layout size and our cost (correlation 0.857). These strong relationships suggest that we will indeed be able to obtain a useful prediction of cost based on these variables. The number of components and the layout size are moderately related (correlation 0.760; the scatterplot is in the appendix). Because this relationship is not perfect, the layout size may be bringing useful additional information into the picture. Furthermore, the relationships appear to be linear, suggesting that regression analysis is appropriate.

A multiple regression analysis to predict cost based on the other variables (number of components and layout size) produced the following prediction equation:

$$\text{Predicted cost} = \$1,356 + \$35.58(\text{Components}) + \$5.68(\text{Size})$$

This predicted cost can be easily computed from the customer's information provided over the telephone and represents our best prediction of the detailed cost figure obtainable (in the least-squares sense) from a linear model of this type. This is the "quick cost number" we are proposing.

This prediction equation appears very reasonable. The estimated fixed cost of $1,356 should more than cover our usual overhead expenses. The estimated cost per component of $35.58 is somewhat larger than what you might expect because the best prediction includes other factors (such as labor) that also increase with the complexity of the design. The $5.68 per unit of layout size is again higher than our actual cost because the layout size number also represents information about other costly aspects of the design.

Here is an example of the use of the prediction equation. Consider a customer who inquires about a job involving about 42 components and a layout size of around 315. Our cost may be estimated as follows:

$$\text{Predicted cost} = \$1,356 + \$35.58 \times 42 + \$5.68 \times 315$$
$$= \$4,640$$

If this customer ordinarily receives a 20% markup (according to confidential records easily available on the computer), then the price quote might be found as follows:

$$\text{Quick price quote} = \$4,640 \times 1.2$$
$$= \$5,568$$

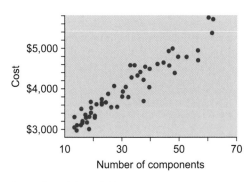
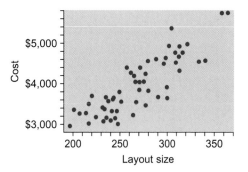

**FIGURE 13.3.2**  Scatterplots of the variables.

How accurate are these quick cost quotes? They should usually be within $200 of the detailed cost information (which is not immediately available) based on the standard error of estimate of $169.[10]

There are three approaches we might use to handle this remaining error. First, we could tell the customer that this number is only approximate and that the final figure will depend on the standard detailed cost calculation. Second, at the opposite extreme, we could give firm quotes instantly, perhaps after adding a few hundred dollars as a "safety margin." Finally, we could reserve the right to revise the quote but guarantee the customer that the price would not rise by more than some figure (perhaps $100).

How effective are number of components and layout size in predicting cost? Consider all of the variation in cost from one job to another; a very large fraction of this variation is explained by the number of components and the layout size ($R$-squared is 94.5%). This is extremely unlikely to have occurred by accident; the regression equation is very highly statistically significant.[11]

The average cost is $3,987. By using the prediction equation instead of this average cost, we reduce our error from $707 (the ordinary standard deviation of cost) to $169 (the standard error of estimate from the regression analysis).

Do we need both the number of components and the layout size in order to predict cost? Yes, because the additional contribution of each one (over and above the information provided by the other) is very highly statistically significant, according to the $t$-test of each regression coefficient.

We also checked for possible technical problems and didn't find any. For example, the diagnostic plot in the appendix shows no additional structure that could be used to improve the results further.

### Conclusion and Summary
We can offer our customers better service by providing instant price quotations for our most typical design jobs using the following predicted cost equation:

$$\text{Predicted cost} = \$1,356 + \$35.58(\text{Components}) + \$5.68(\text{Size})$$

After adding markup and, perhaps, a few hundred dollars to cover the prediction error, we could provide instant quotes in several different formats:

1. Provide the quote as an *approximate* price only. Tell the customer that the actual price will depend on the standard detailed cost calculation available the next day. In effect, the customer bears all the risk of the price uncertainty.
2. Provide a *firm* quote. We might add a little extra for this purpose in order to shift the price uncertainty risk from the customer to us.
3. Compromise. Provide an *approximate* quote, but limit the customer's risk by "capping" the price change. For

example, we might reserve the right to revise the price but promise not to raise it by more than $100. Consider also whether we should *lower* the price or not when the prediction is too high.

This predicted cost equation is based on our actual design experience and on conventional statistical methods. The multiple regression analysis is appropriate for our most typical design jobs, and the results are very highly statistically significant.

If we do implement a "quick price quote" policy, we should be aware of the following selection problem: As customers begin to "figure out" how we provide these quotes, they may bring unfairly complex layout problems to us. Since the prediction equation is based on our *typical* jobs, a serious change in the level of complexity could lead to incorrect pricing. We could respond by identifying the source of this additional complexity and updating our quick pricing model accordingly from time to time.

If this works well with customers, we might consider expanding the program to produce quick quotes on additional categories of design work.

### References
The data used were taken from confidential internal corporate records in the system as of 4/6/10.

The statistical software used was StatPad, a trademark of Skyline Technologies, Inc.

An explanation of general statistical principles in business is provided in A. F. Siegel, *Practical Business Statistics*, 6th ed. (New York: Elsevier, 2011).

### Appendix
On the following page is the data set that was analyzed. We included only routine designs, rejecting 16 designs that either required a special chemical process or coating or else used exotic components for which we have no stable source.

Below this data set is the computer printout of the multiple regression analysis:

Figure 13.3.3 shows the scatterplot of the two explanatory variables: number of components and layout size. The correlation is 0.760.

The diagnostic plot of cost prediction errors plotted against predicted cost (Figure 13.3.4) showed no structure, just a random scatter of data points. This suggests that there is no simple way to improve these predictions of cost based on the number of components and the layout size.

---

10. For the normal linear model, we would expect about 2/3 of the quick cost numbers for this data set to be within $169 of their respective detailed cost numbers. For similar jobs in the future, this error would be slightly higher due to uncertainties in the regression coefficients in the prediction equation. For *different* jobs in the future, it is difficult to tell the size of the error.
11. The *p*-value is less than 0.001, indicating the probability of finding such a strong predictive relationship if, in fact, there were just randomness, with no relationship.

| Number of Components | Layout Size | Cost | Number of Components | Layout Size | Cost |
|---|---|---|---|---|---|
| 27 | 268 | $4,064 | 42 | 288 | $4,630 |
| 23 | 243 | 3,638 | 24 | 244 | 3,659 |
| 30 | 301 | 3,933 | 37 | 220 | 3,712 |
| 16 | 245 | 3,168 | 23 | 235 | 3,677 |
| 37 | 265 | 4,227 | 32 | 250 | 3,826 |
| 14 | 233 | 3,105 | 28 | 267 | 3,576 |
| 20 | 247 | 3,352 | 38 | 309 | 4,547 |
| 33 | 334 | 4,581 | 49 | 317 | 4,806 |
| 23 | 290 | 3,708 | 56 | 313 | 4,717 |
| 35 | 314 | 4,360 | 14 | 196 | 2,981 |
| 31 | 270 | 4,058 | 46 | 314 | 4,949 |
| 17 | 236 | 3,162 | 13 | 248 | 3,032 |
| 47 | 322 | 5,008 | 39 | 274 | 4,051 |
| 52 | 309 | 4,790 | 33 | 262 | 4,283 |
| 34 | 341 | 4,593 | 17 | 264 | 3,250 |
| 25 | 271 | 3,869 | 44 | 277 | 4,280 |
| 26 | 252 | 3,572 | 44 | 299 | 4,665 |
| 19 | 300 | 3,677 | 17 | 233 | 3,389 |
| 16 | 224 | 3,211 | 17 | 206 | 3,296 |
| 30 | 280 | 3,840 | 61 | 358 | 5,732 |
| 36 | 257 | 4,428 | 56 | 302 | 4,963 |
| 61 | 306 | 5,382 | 15 | 241 | 3,109 |
| 19 | 216 | 3,518 | 48 | 271 | 4,419 |
| 16 | 277 | 3,496 | 39 | 297 | 4,524 |
| 46 | 280 | 4,588 | 19 | 232 | 3,425 |
| 60 | 366 | 5,752 | 20 | 201 | 3,368 |
| 18 | 217 | 3,042 | 13 | 213 | 3,295 |
| 18 | 242 | 3,358 | 21 | 237 | 3,605 |

Correlations:

|  | Cost | Components | Size |
|---|---|---|---|
| Cost | 1. | 0.949313 | 0.856855 |
| Components | 0.949313 | 1. | 0.759615 |
| Size | 0.856855 | 0.759615 | 1. |

The Regression Equation:

Cost = 1356.148
        +35.57949 * Components
        +5.678224 * Size
$S$ = 169.1943
$R2$ = 0.944757

Inference for Cost at the 5% Level:

The prediction equation DOES
explain a significant proportion
of the variation in Cost.
$F$ = 453.2050 with
2   and   53
degrees of freedom

| Variable | Effect on Cost Coeff | 95% Confidence Interval From | To | Hypothesis Test Significant? | StdErr of Coeff StdErr | t Statistic t |
|---|---|---|---|---|---|---|
| Constant | 1356.148 | 983.1083 | 1729.189 | Yes | 185.9845 | 7.291728 |
| Components | 35.57949 | 30.55847 | 40.60051 | Yes | 2.503300 | 14.21303 |
| Size | 5.678224 | 3.916505 | 7.439943 | Yes | 0.878329 | 6.464801 |

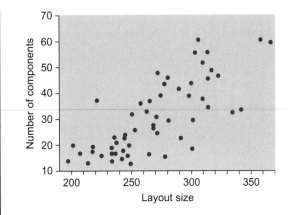

**FIGURE 13.3.3**  Scatterplots of number of components against layout size.

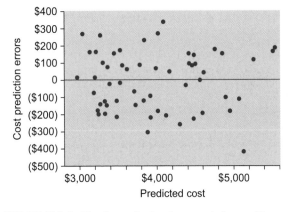

**FIGURE 13.3.4**  The diagnostic plot shows no obvious problems.

## 13.4  END-OF-CHAPTER MATERIALS

### Summary

Communication is an essential management skill, and statistical summaries can help you communicate the basic facts of a situation in a direct and objective way. Be kind to your readers and explain what you have learned using language that is easy for them to understand. Identifying your *purpose* helps you narrow down the topic so that you can concentrate on the relevant issues. Identifying your *audience* helps you select the appropriate writing style and level of detail.

The best organization is straightforward and direct. You want to make it quick and easy for your readers to hear your story. Develop an outline early on, perhaps with one line representing each paragraph of your paper. Here is one reasonable plan for organization:

1. The **executive summary** is a paragraph at the beginning that describes the most important facts and conclusions from your work.

2. The **introduction** consists of several paragraphs in which you describe the background, the questions of interest, and the data set you have worked with.
3. The **analysis and methods** section lets you interpret the data by presenting graphic displays, summaries, and results, which you explain as you go along.
4. The **conclusion and summary** move back to the big picture to give closure, pulling together all of the important thoughts you would like your readers to remember.
5. A **reference** is a note indicating the kind of material you have taken from an outside source and giving enough information so that your reader can obtain a copy. You might have these appear on the appropriate page of a section, or you might gather them together in their own section.
6. The **appendix** contains all supporting material that is important enough to include but not important enough to appear in the text of your report.

In addition, you may also wish to include a *title page* and a *table of contents* page. The **title page** goes first, including the title of the report, the name and title of the person you prepared it for, your name and title (as the preparer), and the date. The **table of contents** goes after the executive summary, showing an outline of the report together with page numbers.

Do the analysis first. Next, select the most important results from your analysis file. Then make an outline of the analysis and methods section and conclusions. Create paragraphs from the entries in your outline, decide what to place in the appendix, choose the references, and write the introduction and executive summary.

Be brief and clear. Read over your material with your audience in mind.

### Key Words

analysis and methods, *419*
appendix, *421*
conclusion and summary, *420*
executive summary, *418*
introduction, *419*
reference, *420*
table of contents, *418*
title page, *418*

### Questions

1. What is the primary purpose of writing a report?
2. Why is it necessary to identify the purpose and audience of a report?
3. How can an outline help you?

4. Give some reasons why you might want to include statistical results in a report.

5. Should you leave key results out of the executive summary, ending it with a sentence such as "We have examined these issues and have come up with some recommendations." Why or why not?

6. How can you use the executive summary and introduction to reach a diverse audience with limited time?

7. Is it OK to repeat material in the introduction that already appeared in the executive summary?

8. a. What kind of material appears in the analysis and methods section?

   b. Should you describe everything you have examined in the analysis and methods section? Why or why not?

9. Should you assume that everyone who reads your conclusion is already familiar with all of the details of the analysis and methods section?

10. a. Give two reasons for providing a reference when you make use of material from the Internet, a book, a magazine, or an other source.

    b. How can you tell if you have provided enough information in a reference?

    c. How would you reference material from a telephone call or an interview with an expert?

11. a. What material belongs in the appendix?

    b. How can an appendix help you satisfy both the casual and the dedicated reader?

12. What can you do to help those in your audience who are short of time?

13. When is the best time to write the introduction and executive summary, first or last? Why?

14. What is the relationship between the outline and the finished report?

15. How would you check the meaning of a word to be sure that you are using it correctly?

16. How can you find synonyms for a given word? Why might you want to?

## Problems

*Problems marked with an asterisk (\*) are solved in the Self Test in Appendix C.*

1. Your boss has just asked you to write a report. Identify the purpose and audience in each of the following situations:

   a.\* The firm is considering expansion of the shipping area. Background material is needed on the size of facilities at other firms.

   b. The new stereo system is almost ready to be shipped. It is clearly superior to everything else on the market. Your boss was contacted by a hi-fi magazine for some material to include in its "new products" column.

   c. Manufacturing equipment has been breaking down fairly often lately, and nobody seems to know why. Fortunately, information concerning the details of each breakdown is available.

   d. Your firm's bank is reluctant to lend any more money because it claims that your industry group's sales are too closely tied to the economy and are therefore too vulnerable to financial trouble in a recession. Your boss feels that data available for the firm's sales and for the U.S. gross national product might prove otherwise.

2. Fill in the blanks, using either *affect* or *effect*.

   a.\* The amount of overtime and the time of day had a significant _____ on accidents in the workplace.

   b.\* The amount of overtime and the time of day significantly _____ accidents in the workplace.

   c. Curiously, the amount of experience does not seem to _____ the productivity of these workers.

   d. Curiously, the amount of experience does not seem to have an _____ on the productivity of these workers.

3. Using your imagination, create an executive summary paragraph for a hypothetical report based on each of the situations in problem 1. (*Note:* Only one paragraph is required for each situation.)

4. For each of the following sentences, say which section (or sections) it might be located in.

   a.\* The scatterplot of defects against the rate of production shows a moderately strong relationship, with a correlation of 0.782, suggesting that our worst problems happen at those times of high demand, when we can least afford them.

   b. The third option, to sell the division to an independent outsider, should be held in reserve as a last resort, to be used only if the other two possibilities don't work out.

   c. These problems began five years ago, when the new power plant was installed, and have cost the firm at least $2 million over the past two years.

   d. Here is the data set, giving the prices for each product in each of the different markets.

5. Arrange selected information to form a proper reference for each of the following cases.

   a.\* The title of the article from the *Wall Street Journal* is "Tallying Up Viewers: Industry Group to Study How a Mobile Nation Uses Media." It appeared on July 26, 2010, which was a Monday. It was on page B4. It was written by Suzanne Vranica, the Advertising and Marketing Columnist for the newspaper. The article is about the search for "new ways of measuring audiences" using Apple Inc.'s iPhone.

   b. In order to be sure about a technical detail in a report about leasing and taxes, you have called an expert at the University of Washington. Professor Lawrence D. Schall confirms your suspicion that the laws are highly complex in this area and suggests that no simple approach to the problem will work in general.

   c. You have been using a book called *Quality Management: Tools and Methods for Improvement* as background reading for a report suggesting manufacturing improvements. The appropriate chapter is 8, and the authors are Howard Gitlow, Alan Oppenheim, and Rosa Oppenheim. The book was published in 1995 by Richard D. Irwin, Inc., located in Burr Ridge, Illinois. The book is copyrighted and has ISBN number 0–256–10665–7. It is dedicated to "the never-ending improvement of the species."

    **d.** The information is that consumer confidence had fallen, reaching 52.9. It was released on June 29, 2010. You found it on the Internet on July 26, 2010, at a page provided by The Conference Board and titled "The Conference Board Consumer Confidence Index® Drops Sharply." The URL address is http://www.conference-board.org/data/consumerconfidence.cfm.

**6.** What important information is missing from each of the following references?

    **a.*** Personal communication, 2010.

    **b.** *Business Week*, p. 80.

    **c.** *Basic Business Communication* (Burr Ridge, Ill.: Richard D. Irwin).

    **d.** James A. White, "Will the Real S&P 500 Please Stand Up? Investment Firms Disagree on Index," *Wall Street Journal*.

    **e.** Data were obtained from the White House Economic Statistics Briefing Room on the Internet.

## Database Exercises

Refer to the employee database in Appendix A.

    Write a three- to five-page report summarizing the relationship between gender and salary for these employees. Be sure to discuss the results of the following statistical analyses: (a) a two-sample *t* test of male salaries against female salaries and (b) a multiple regression to explain salary using age, experience, and an indicator variable for gender.

## Project

Perform a multiple regression analysis using business data of your choice from the Internet, the library, or your company and write up the results as a report to upper-level management, either as a background summary report or as a proposal for action. You should have a significant *F* test and at least one significant *t* test (so that you will be able to make some strong conclusions in your project). Your report should include five to seven pages plus an appendix and should be based on the following format:

**a.** *Introduction*: Describe the background and questions of interest and the data set clearly as if to an intelligent person who knows nothing about the details of the situation.

**b.** *Analysis and Methods*: Analyze the data, presenting displays and results, explaining as you go along. Consider including some of each of the following:

    **(1)** Explore the data using histograms or box plots for each variable and using scatterplots for each pair of variables.

    **(2)** Use a transformation (such as the logarithm) only if this would clearly help the analysis by dealing with a big problem in the diagnostic plot.

    **(3)** Compute the correlation of each pair of variables and interpret these values.

    **(4)** Report the multiple linear regression to predict one variable (chosen appropriately) from the others by explaining the regression equation and interpreting each regression coefficient. Comment on the quality of the regression analysis in terms of both prediction accuracy (standard error of estimate) and how well the relationship is explained (coefficient of determination). Report statistical significance using *p*-values both overall (*F* test) and for each regression coefficient (*t* tests). In particular, are your results reasonable?

**c.** *Conclusion and summary*: What has this analysis told you about the situation? How have your questions been answered? What have you learned?

**d.** *Appendix*: List the data, with their source indicated. (This does not count toward the page limit.)

# Time Series

## Understanding Changes over Time

A time series is different from cross-sectional data because *ordering of the observations conveys important information*. In particular, you are interested in more than just a typical value to summarize the entire series (the average, for example) or even the variability of the series (as described by, say, the standard deviation). You would like to know *what is likely to happen next*. Such a forecast must carefully extend the most recent behavior with respect to the patterns over time, which are evident in past behavior. Here are some examples of time-series situations:

**One:** In order to prepare a budget for next quarter, you need a good estimate of the expected sales. This forecast will be the basis for predicting the other numbers in the budget, perhaps using regression analysis. By looking at a time series of actual quarterly sales for the past few years, you should be able to come up with a forecast that represents your best guess based on the overall trend in sales (up, you hope) and taking into account any seasonal variation. For example, if there has always been a downturn from fourth quarter (which includes the

holiday shopping season) to first quarter, you will want your forecast to reflect the usual seasonal pattern.

**Two:** In order to decide whether or not to build that new factory, you need to know how quickly your market will grow. Analyzing the available time-series data on industry sales and prices will help you evaluate your chances for success. But don't expect to get exact answers. Predicting the future is a tricky and uncertain business, even with all of the computerized help you can get. Although time-series analysis will help you by providing a "reality check" to your decision making, substantial risk may still remain.

**Three:** By constantly monitoring time-series data related to your firm, both internal (sales, cost, etc.) and external (industrywide sales, imports, etc.), you will be in the best position to manage effectively. By anticipating future trends corresponding to those you spotted in the early stages, you will be ready to participate in growth areas or to move away from dead-end markets. By anticipating seasonal needs for cash, you can avoid the panic of having too little and the costs of having too much. By anticipating the need for inventory, you can minimize the losses due to unfilled orders (which help your competition) and the costs (interest and storage) of carrying too much. There is a tremendous amount of valuable information contained in these time-series data sets.

## 14.1 AN OVERVIEW OF TIME-SERIES ANALYSIS

Methods from previous chapters (confidence intervals and hypothesis tests, for example) must be modified before they will work with time-series data. Why? Because the necessary assumptions are not satisfied. In particular, a time series is *not a random sample* from a population.[1] Tomorrow's price, for example, is likely to be closer to today's than to last year's price; successive observations are *not* independent of one another. If you go ahead and compute confidence intervals and hypothesis tests in the usual way, the danger is that your error rate might be much higher than the 5% you might claim. Time-series analysis requires specialized methods that take into account the dependence among observations. The basic ideas and concepts of statistical inference are the same, but the methods are adapted to a new situation.

The primary goal of time-series analysis is to create forecasts of the future. This requires a model to describe your time series. A **model** (also called a **mathematical model** or a **process**) is a system of equations that can produce an assortment of artificial time-series data sets. Here are the basic steps involved in forecasting:

1. Select a family of time-series models.
2. Estimate the particular model (within this chosen family) that produces artificial data matching the essential features (but not the quirks and exceptions) of the actual time-series data set.
3. Your **forecast** will be the expected (i.e., mean) value of the future behavior of the estimated model. Note that you can predict the future for a mathematical model by using a computer, even though the future of the actual series is unavailable.
4. The **forecast limits** are the confidence limits for your forecast (if the model can produce them); if the model is correct, the future observation has a 95% probability, for example, of being within these limits. The limits are computed in the usual way from the standard error, which represents the variability of the future behavior of the estimated model.

Following these steps does more than just produce forecasts. By selecting an appropriate model that produces data sets that "look like" your actual series, you gain insight into the patterns of behavior of the series. This kind of deeper statistical understanding of how the world works will be useful to you as background information in decision making.

Although we all want dependable forecasts of the future, don't expect forecasts to be exactly right. The forecast accuracy we would really like to have is probably impossible because the truly unexpected, by definition, cannot be foreseen.[2] However, the need for forecasts is so strong that people are willing to try anything that *might* lead to a slight improvement, and sophisticated statistical methods have been developed to help fill this need. Although the results may be the best we can come up with based on the available information, they still might not suit your real needs very well.

There are many different approaches to time-series analysis. The methods of time-series analysis are varied and are still evolving. Following some examples of time-series data sets, we will discuss two of the most important methods for analyzing time series in business:

1. *Trend-seasonal analysis* is a direct, intuitive approach to estimating the basic components of a monthly or quarterly time series. These components include (1) the long-term trend; (2) the exactly repeating seasonal patterns; (3) the medium-term, wandering, cyclic ups and downs; and (4) the random, irregular "noise." Forecasts are obtained by imposing the usual seasonal patterns on the long-term trend.
2. *Box–Jenkins ARIMA processes* are flexible linear models that can precisely describe a wide variety of different time-series behaviors, including even the medium-term ups and downs of the so-called business cycle. Although these basic models are fairly simple to describe, their estimation requires extensive computer calculations. Forecasts and confidence limits are obtained by statistical theory based on the future behavior of the estimated model.

---

1. The exception is the *pure random noise process*, described in Section 14.3.

2. For example, on July 2, 2001, the *Wall Street Journal* reported the forecasts of 51 prominent economists together with the actual outcome. The average six-month-ahead forecast for the short-term U.S. Treasury bill interest rate was 5.36%. Six months later, the actual interest rate turned out to be 3.60%. When you consider that a difference of one quarter of a percentage point in interest rates can be worth $3,000 in present value to your typical first-time home purchaser, such a difference can be worth large amounts to industry and the economy. None of the economists, with all of their sophisticated forecasting methods, had expected such a steep plunge in interest rates, with the forecasts ranging from 4.30% to 6.40%, with a standard deviation of 0.38 percentage points.

## Example

### The Stock Market Is a Random Walk

Each day's closing value of a stock market index—for example, the Dow Jones Industrial Average—forms a time series of vital importance to many of us. Figure 14.1.1 shows a typical time-series graph, with the series itself plotted vertically as $Y$ against time (in number of trading days), which is plotted horizontally as $X$.

What information would be lost if you were to draw a histogram of the Dow Jones index values, compute the average, or find an ordinary confidence interval? You would lose information about the *ordering* of the observations; you would be treating the index values as if they had been arranged in an arbitrary sequence. Figure 14.1.2 demonstrates that essential information is lost when a random ordering is used, showing why special time-series methods are needed that will take advantage of this important information. A good time-series method for stock market data should recognize that the stock market usually changes by a small amount each day (relative to the day before) as it wanders through its ups and downs.

The financial theory of efficient markets argues that the stock market should follow a *random walk*, in which the daily changes amount to unpredictable, random noise.[3] Figure 14.1.3 shows that the daily changes (today's value minus yesterday's value) in stock market price over this time period are indeed random.

The *random walk* model is included within the Box–Jenkins ARIMA framework as a special case of a series that "knows" only where it is but not how it got there.

---

3. Because large investors act immediately on their available information, any *foreseeable* trends are already reflected in the stock price. The only changes that are possible are due to the *unforeseeable*, or randomness. A more careful analysis would work with the percent change rather than the change itself; this makes a difference only when a series has changed by an appreciable percentage of its value over the entire time period.

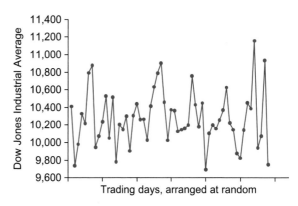

**FIGURE 14.1.2**  The results of randomly shuffling the order of the data in the time series from the preceding figure. Essential information is lost because all time trends disappear. Time-series analysis requires special methods that will *not* lose this important information.

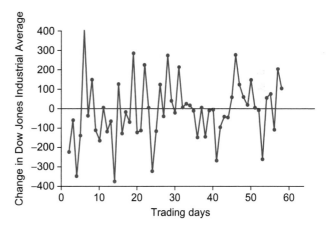

**FIGURE 14.1.3**  The daily changes in the Dow Jones Industrial Average, in proper order (no shuffling). This is basically a random series, supporting the *random walk* model for stock behavior.

## Example

### Warehouse Clubs and Superstores Sales Have Enjoyed Steady Growth

Warehouse clubs and superstores that sell a large variety of items (such as Walmart) have enjoyed substantial and steady growth. Table 14.1.1 shows U.S. sales at these stores from 1994 through 2008. The time-series plot in Figure 14.1.4 shows overall steady growth, with increased sales every year. The upward curvature suggests a constant growth rate, which would correspond to exponential growth. One way to estimate the rate of growth during this time period is to use the regression coefficient for predicting the natural logarithm of the time-series data ($Y$) from the time period ($X$). Table 14.1.2 shows the logarithms of the time-series values (these are natural logs of the dollar amounts in billions). A time-series plot of the logarithms, Figure 14.1.5, shows an approximately linear (i.e., straight-line) relationship, confirming the pattern of exponential growth at a constant rate for warehouse club and superstore sales.

*(Continued)*

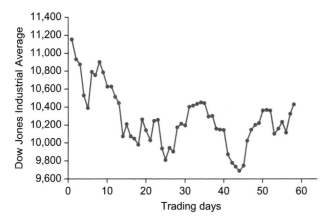

**FIGURE 14.1.1**   A time-series plot of the Dow Jones Industrial Average, daily from May 3 through July 23, 2010.

**Example—cont'd**

The estimated regression line is shown in Figure 14.1.6. The regression equation is

Predicted log of sales $= -240.54 + 0.13772 \times$ Year

Each additional year adds the regression coefficient 0.13772 to the previous predicted logarithm, so (using the exponential function to undo the logarithm) it multiplies the previous number of new orders by

$$e^{0.13772} = 2.71828^{0.13772} = 1.148$$

By subtracting 1, we find that the estimated growth rate in warehouse club and superstore sales is 14.8% per year from 1994 through 2008.

Growth rate $= 14.8\%$

Note that the data points appear to be evenly distributed above and below the regression line in Figure 14.1.6 but seem to wander above for a while and then stay below for a while. Such a tendency is called *serial correlation*. If you find that serial correlation is present, the least-squares line can still provide a good estimator of growth, but statistical inference (confidence intervals and hypothesis tests) would give incorrect results because serial correlation is not permitted in the linear regression model of Chapter 11.

**TABLE 14.1.1 Warehouse Club and Superstore Sales**

| Year | Sales (billions) |
|---|---|
| 1994 | $56.3 |
| 1995 | 63.3 |
| 1996 | 71.4 |
| 1997 | 81.9 |
| 1998 | 101.2 |
| 1999 | 123.6 |
| 2000 | 139.6 |
| 2001 | 164.7 |
| 2002 | 191.3 |
| 2003 | 216.3 |
| 2004 | 242.4 |
| 2005 | 270.2 |
| 2006 | 296.7 |
| 2007 | 323.8 |
| 2008 | 351.2 |

**Source:** Data for 2000–2008 are from Table 1017 of *U.S. Census Bureau, Statistical Abstract of the United States*: 2010 (129th edition), Washington, DC, 2009, accessed at http://www.census.gov/compendia/statab/cats/wholesale_retail_trade.html on July 26, 2010. Data for 1994–1999 are from Table 1020 of *U.S. Census Bureau, Statistical Abstract of the United States*: 2001, accessed at http://www.census.gov/prod/www/abs/statab2001_2005.html on July 26, 2010.

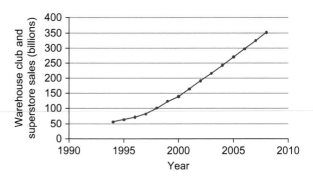

**FIGURE 14.1.4** Steady growth in warehouse club and superstore sales from 1994 to 2008.

**TABLE 14.1.2 Warehouse Club and Superstore Sales with Logarithms**

| Year, $X$ | Sales (billions) | Natural Logarithm of Sales, $Y$ |
|---|---|---|
| 1994 | $56.3 | 4.031 |
| 1995 | 63.3 | 4.148 |
| 1996 | 71.4 | 4.268 |
| 1997 | 81.9 | 4.405 |
| 1998 | 101.2 | 4.617 |
| 1999 | 123.6 | 4.817 |
| 2000 | 139.6 | 4.939 |
| 2001 | 164.7 | 5.104 |
| 2002 | 191.3 | 5.254 |
| 2003 | 216.3 | 5.377 |
| 2004 | 242.4 | 5.491 |
| 2005 | 270.2 | 5.599 |
| 2006 | 296.7 | 5.693 |
| 2007 | 323.8 | 5.780 |
| 2008 | 351.2 | 5.861 |

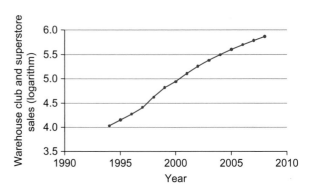

**FIGURE 14.1.5** The logarithms of warehouse club and superstore sales show linear (straight-line) growth over time. This indicates that sales have grown at a constant rate (i.e., exponential growth).

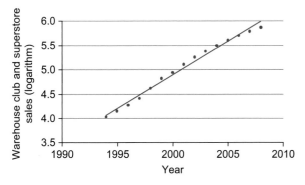

**FIGURE 14.1.6** The logarithms of warehouse club and superstore sales (*Y*) plotted against time (*X*), together with the least-squares regression line. The regression coefficient, a slope of 0.13772, is used to find the yearly growth rate of 14.8%.

### Example

#### Total Retail Sales Show Seasonal Variation

Table 14.1.3 shows the raw, unadjusted total U.S. retail sales, in millions, monthly from January 2006 through April 2010. The time-series plot of these sales figures, in Figure 14.1.7, does not show overall growth although there is substantial variation (bumpiness) from one month to the next. A close look at this variation reveals that it is not just random but shows a tendency to repeat itself from one year to the next. The highest points (that stand out from their neighbors) tend to be in December (just before the start of the next year); sales then drop to fairly low values in January and February. This kind of seasonal pattern matches our perception of holiday season shopping in the United States.

*(Continued)*

**TABLE 14.1.3 U.S. Retail Sales, Unadjusted**

| Year | Month | Sales (billions) |
|---|---|---|
| 2006 | January | $287 |
| 2006 | February | 283 |
| 2006 | March | 327 |
| 2006 | April | 317 |
| 2006 | May | 338 |
| 2006 | June | 332 |
| 2006 | July | 327 |
| 2006 | August | 340 |
| 2006 | September | 312 |
| 2006 | October | 314 |
| 2006 | November | 324 |
| 2006 | December | 381 |
| 2007 | January | 296 |
| 2007 | February | 291 |
| 2007 | March | 337 |
| 2007 | April | 323 |
| 2007 | May | 354 |
| 2007 | June | 339 |
| 2007 | July | 335 |
| 2007 | August | 350 |
| 2007 | September | 318 |
| 2007 | October | 332 |
| 2007 | November | 343 |
| 2007 | December | 389 |
| 2008 | January | 308 |
| 2008 | February | 309 |
| 2008 | March | 336 |
| 2008 | April | 332 |
| 2008 | May | 359 |
| 2008 | June | 342 |
| 2008 | July | 346 |
| 2008 | August | 345 |
| 2008 | September | 316 |
| 2008 | October | 314 |
| 2008 | November | 302 |
| 2008 | December | 350 |
| 2009 | January | 277 |
| 2009 | February | 268 |
| 2009 | March | 294 |
| 2009 | April | 296 |
| 2009 | May | 312 |
| 2009 | June | 311 |
| 2009 | July | 314 |
| 2009 | August | 320 |
| 2009 | September | 293 |
| 2009 | October | 306 |
| 2009 | November | 310 |
| 2009 | December | 370 |
| 2010 | January | 285 |
| 2010 | February | 282 |
| 2010 | March | 329 |
| 2010 | April | 326 |

**Source:** U.S. Census Bureau, *Monthly & Annual Retail Trade*, accessed at http://www.census.gov/retail/ on July 26, 2010.

**Example—cont'd**

The government also provides *seasonally adjusted* sales figures, removing the predictable changes from one month to the next, as shown in Table 14.1.4. When the predictable seasonal patterns are removed from the series, the result is a much smoother indication of the patterns of growth, decline, and more growth, as shown in Figure 14.1.8 (compare to previous figure). The remaining variation indicates fluctuations that were not consistent from one year to the next and therefore were not expected at that time of year. In particular, the seasonally adjusted series makes it very clear exactly how sales declined during the financial crisis in the fall of 2008, with the seasonal fluctuations removed.

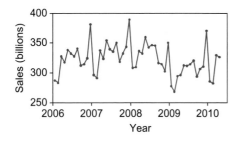

**FIGURE 14.1.7** U.S. retail sales, monthly from January 2006 through April 2010. Note the strong seasonal pattern that repeats each year.

**TABLE 14.1.4 U.S. Retail Sales, Unadjusted and Seasonally Adjusted**

| Year | Month | Sales (billions) | |
|------|-------|------------------|---|
| | | Unadjusted | Adjusted for Seasonal Variation |
| 2006 | January | $287 | $324 |
| 2006 | February | 283 | 321 |
| 2006 | March | 327 | 322 |
| 2006 | April | 317 | 325 |
| 2006 | May | 338 | 322 |
| 2006 | June | 332 | 323 |
| 2006 | July | 327 | 325 |
| 2006 | August | 340 | 325 |
| 2006 | September | 312 | 323 |
| 2006 | October | 314 | 323 |
| 2006 | November | 324 | 325 |
| 2006 | December | 381 | 329 |
| 2007 | January | 296 | 328 |
| 2007 | February | 291 | 329 |
| 2007 | March | 337 | 333 |
| 2007 | April | 323 | 331 |
| 2007 | May | 354 | 335 |
| 2007 | June | 339 | 331 |
| 2007 | July | 335 | 333 |
| 2007 | August | 350 | 333 |
| 2007 | September | 318 | 336 |
| 2007 | October | 332 | 337 |
| 2007 | November | 343 | 342 |
| 2007 | December | 389 | 339 |
| 2008 | January | 308 | 340 |
| 2008 | February | 309 | 336 |
| 2008 | March | 336 | 337 |
| 2008 | April | 332 | 338 |
| 2008 | May | 359 | 339 |
| 2008 | June | 342 | 339 |
| 2008 | July | 346 | 338 |
| 2008 | August | 345 | 335 |
| 2008 | September | 316 | 328 |
| 2008 | October | 314 | 317 |
| 2008 | November | 302 | 307 |
| 2008 | December | 350 | 297 |
| 2009 | January | 277 | 303 |
| 2009 | February | 268 | 303 |
| 2009 | March | 294 | 298 |
| 2009 | April | 296 | 298 |
| 2009 | May | 312 | 300 |
| 2009 | June | 311 | 305 |
| 2009 | July | 314 | 305 |
| 2009 | August | 320 | 313 |
| 2009 | September | 293 | 305 |
| 2009 | October | 306 | 310 |
| 2009 | November | 310 | 315 |
| 2009 | December | 370 | 316 |
| 2010 | January | 285 | 317 |
| 2010 | February | 282 | 318 |
| 2010 | March | 329 | 325 |
| 2010 | April | 326 | 327 |

**Source:** U.S. Census Bureau, *Monthly & Annual Retail Trade*, accessed at http://www.census.gov/retail/ on July 26, 2010.

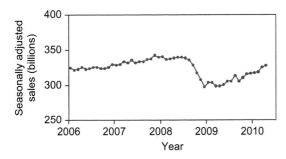

**FIGURE 14.1.8** Seasonally adjusted U.S. retail sales, monthly from January 2006 through April 2010. The seasonal pattern has been eliminated, and the impression emerges of steady growth followed by sharp decline and then growth again. The remaining variation indicates changes that were not expected at that time of year.

## Example
### Interest Rates

One way the U.S. government raises cash is by selling securities. Treasury bills are short-term securities, with one year or less until the time they mature, at which time they pay back the initial investment plus interest. Table 14.1.5 shows yields (interest rates) on three-month U.S. Treasury bills each year from 1970 through 2009.

The time-series plot, in Figure 14.1.9, indicates a general rise followed by a deep and prolonged fall in interest rates over this time period with substantial variation. There appears to be a cyclic pattern of rising and falling rates of increasing magnitude; however, it is difficult to use these patterns to predict future rates.

Unlike the example of warehouse club and superstore sales, interest rates are not expected to continue growing in the future. Unlike the example of total retail sales, the cycles of interest rates in Figure 14.1.9 do not show an exactly repeating pattern.

A time series that wanders about will often form trends and cycles that are not really expected to continue in the future. The Box–Jenkins ARIMA process approach (to be presented in Section 14.3) is especially well suited to this kind of time-series behavior because it takes into account the fact that a series will usually appear to produce cycles whenever it wanders about.

**TABLE 14.1.5  U.S. Treasury Bills (Three-Month Maturity)**

| Year | Yield |
|------|-------|
| 1970 | 6.39% |
| 1971 | 4.33 |
| 1972 | 4.06 |
| 1973 | 7.04 |
| 1974 | 7.85 |
| 1975 | 5.79 |
| 1976 | 4.98 |
| 1977 | 5.26 |
| 1978 | 7.18 |
| 1979 | 10.05 |
| 1980 | 11.39 |
| 1981 | 14.04 |
| 1982 | 10.60 |
| 1983 | 8.62 |
| 1984 | 9.54 |
| 1985 | 7.47 |
| 1986 | 5.97 |
| 1987 | 5.78 |
| 1988 | 6.67 |
| 1989 | 8.11 |
| 1990 | 7.50 |
| 1991 | 5.38 |
| 1992 | 3.43 |
| 1993 | 3.00 |
| 1994 | 4.25 |
| 1995 | 5.49 |
| 1996 | 5.01 |
| 1997 | 5.06 |
| 1998 | 4.78 |
| 1999 | 4.64 |
| 2000 | 5.82 |
| 2001 | 3.40 |
| 2002 | 1.61 |
| 2003 | 1.01 |
| 2004 | 1.37 |
| 2005 | 3.15 |
| 2006 | 4.73 |
| 2007 | 4.36 |
| 2008 | 1.37 |
| 2009 | 0.15 |

**Source:** Federal Reserve, accessed at http://www.federalreserve.gov/releases/h15/data/Annual/H15_TB_M3.txt on July 26, 2010.

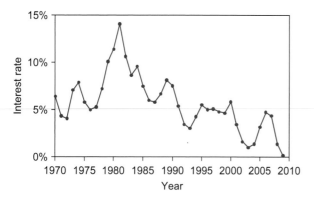

FIGURE 14.1.9   Interest rates on three-month U.S. Treasury bills from 1970 through 2009 have shown overall increasing, and then decreasing, trends with cyclic fluctuations. These cycles, however, have various lengths and are not expected to repeat exactly the way seasonal patterns do.

## 14.2  TREND-SEASONAL ANALYSIS

**Trend-seasonal analysis** is a direct, intuitive approach to estimating the four basic components of a monthly or quarterly time series: the long-term trend, the seasonal patterns, the cyclic variation, and the irregular component. The basic time-series model expresses the numbers in the series as the product obtained by multiplying these basic components together.

> **Trend-Seasonal Time-Series Model**
>
> Data = Trend × Seasonal × Cyclic × Irregular

Here are the definitions of these four components:

1. The long-term **trend** indicates the *very* long-term behavior of the time series, typically as a straight line or an exponential curve. This is useful in seeing the overall picture.
2. The exactly repeating **seasonal component** indicates the effects of the time of year. For example, heating demands are high in the winter months, sales are high in December, and agricultural sales are high at harvest time. Each time period during the year has its *seasonal index*, which indicates how much higher or lower this particular time usually is as compared to the others. For example, with quarterly data there would be a seasonal index for each quarter; an index of 1.235 for the fourth quarter says that sales are about 23.5% higher at this time compared to all quarters during the year. An index of 0.921 for the second quarter says that sales are 7.9% lower (since 1 − 0.921 = 0.079) at this time.
3. The medium-term **cyclic component** consists of the gradual ups and downs that do *not* repeat each year and so are excluded from the seasonal component. Since they are gradual, they are not random enough to

be considered part of the independent random error (the irregular component). The cyclic variation is especially difficult to forecast beyond the immediate future, yet it can be very important since basic business cycle phenomena (such as recessions) are considered to be part of the cyclic variation in economic performance.

4. The short-term, random **irregular component** represents the leftover, residual variation that can't be explained. It is the effect of those one-time occurrences that happen randomly, rather than systematically, over time. The best that can be done with the irregular component is to summarize how large it is (using a standard deviation, for example), to determine whether it changes over time, and to recognize that even in the best situation, a forecast can be no closer (on average) than the typical size of the irregular variation.

The four basic components of a time series (trend, seasonal, cyclic, and irregular components) can be estimated in different ways. Here is an overview of the **ratio-to-moving-average** method (to be presented in detail), which divides the series by a smooth moving average as follows:

1. A *moving average* is used to eliminate the seasonal effects by averaging over the entire year, reducing the irregular component and producing a combination of trend and cyclic components.
2. Dividing the series by the smoothed moving-average series gives you the *ratio-to-moving-average*, which includes both seasonal and irregular values. Grouping by time of year and then averaging within groups, you find the *seasonal index* for each time of year. Dividing each series value by the appropriate seasonal index for its time of year, you find *seasonally adjusted* values.
3. A regression of the seasonally adjusted series (*Y*) on time (*X*) is used to estimate the *long-term trend* as a straight line over time.[4] This trend has no seasonal variation and leads to a seasonally adjusted forecast.
4. Forecasting may be done by *seasonalizing the trend*. Taking predicted values from the regression equation (the trend) for future time periods and then multiplying by the appropriate seasonal index, you get forecasts that reflect both the long-term trend and the seasonal behavior.

The advantages of the ratio-to-moving-average method are its easy computation and interpretation. The main disadvantage is that the model is not completely specified; therefore, measures of uncertainty (such as forecast limits) are not easily found.[5]

---

4. For example, this time variable, *X*, might consist of the numbers 1, 2, 3, ....
5. In particular, the partially random structure of the cyclic component is not spelled out in detail. This problem is not solved by the multiple regression approach, which uses indicator variables to estimate seasonal indices.

The following example of a time series exhibits all of these components. We will refer back to this example through the remainder of this section.

### Example
#### Microsoft Revenues

Table 14.2.1 shows the quarterly revenues as reported by Microsoft Corporation. This time series shows some distinct seasonal patterns. For example (Figure 14.2.1), revenues always rise from third to fourth quarter and generally decline from fourth to first of the next year (this happens in all cases except for Q4 2006 to Q1 2007). Since the seasonal pattern is not repeated perfectly each year, there will be some cyclic and irregular behavior as well. Note also the long-term trend, represented by the general rise over time (except, perhaps, toward the end of the series).

The results of a trend-seasonal analysis are shown in Figure 14.2.2. The trend is a straight line, the seasonal index repeats exactly each year, the cyclic component wanders erratically, and the irregular component is basically random. Because the cyclic and irregular components are small compared to the seasonal variations, they have been enlarged in Figure 14.2.3 to show that they really are cyclic and irregular. The computations for this analysis will be explained soon; at this point you should understand how the basic components relate to the original time series.

## Trend and Cyclic: The Moving Average

Our objective is to identify the four basic components of a time series. We begin by averaging a year's worth of data at a time in order to eliminate the seasonal component and reduce the irregular component. A **moving average** is a new series created by averaging nearby observations of a time series and then moving along to the next time period; this produces a less bumpy series. A full year at a time is averaged so that the seasonal components always contribute in the same way regardless of where you are in the year.

$$\text{Moving average} = \text{Trend} \times \text{Cyclic}$$

Here is how to find the moving average for quarterly data at a given time period. Start with the value at this time, add it to the values of its neighbors, then add *half* the values of the next neighbors, and divide by 4. Such a weighted average is needed so that the span is symmetric around the base time and still captures exactly a year's worth of data.[6] If you have monthly data, average the series at the base time period together with the five nearest months on each side and half of the next one out on each side. The moving average is

### TABLE 14.2.1  Microsoft Revenues

| Year | Quarter | Revenues (billions) |
|------|---------|---------------------|
| 2003 | 1 | $7.835 |
| 2003 | 2 | 8.065 |
| 2003 | 3 | 8.215 |
| 2003 | 4 | 10.153 |
| 2004 | 1 | 9.175 |
| 2004 | 2 | 9.292 |
| 2004 | 3 | 9.189 |
| 2004 | 4 | 10.818 |
| 2005 | 1 | 9.620 |
| 2005 | 2 | 10.161 |
| 2005 | 3 | 9.741 |
| 2005 | 4 | 11.837 |
| 2006 | 1 | 10.900 |
| 2006 | 2 | 11.804 |
| 2006 | 3 | 10.811 |
| 2006 | 4 | 12.542 |
| 2007 | 1 | 14.398 |
| 2007 | 2 | 13.371 |
| 2007 | 3 | 13.762 |
| 2007 | 4 | 16.367 |
| 2008 | 1 | 14.454 |
| 2008 | 2 | 15.837 |
| 2008 | 3 | 15.061 |
| 2008 | 4 | 16.629 |
| 2009 | 1 | 13.648 |
| 2009 | 2 | 13.099 |
| 2009 | 3 | 12.920 |
| 2009 | 4 | 19.022 |
| 2010 | 1 | 14.503 |

**Source:** U.S. Securities and Exchange Commission, 10-K and 10-Q filings, accessed at http://www.sec.gov/cgi-bin/browse-edgar on May 7, 2009, April 26, 2010, and July 26, 2010.

---

6. By weighting the extremes by 1/2, you ensure that this quarter counts just the same in the moving average as the other quarters.

unavailable for the first two and last two quarters or, for a monthly series, for the first six and last six months.

For Microsoft, the moving average of revenue for the third quarter of 2003 is given by [(1/2)7.835 + 8.065 + 8.215 + 10.153 + (1/2)9.175]/4 = 8.7345. For the fourth quarter of

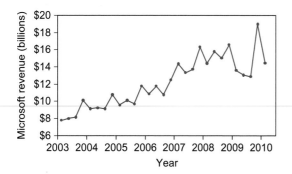

FIGURE 14.2.1   A time-series plot of quarterly revenues of Microsoft. Note the seasonal effects that repeat each year. You can also see an upward long-term trend throughout most of this time period, as well as some irregular behavior.

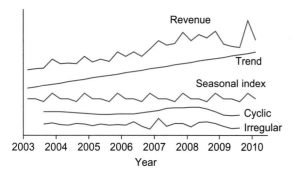

FIGURE 14.2.2   Quarterly revenues broken down into the four basic components: a straight-line trend, a seasonal index that repeats each year, a wandering cyclic component, and a random irregular component. These are shown on approximately the same scale as the original series.

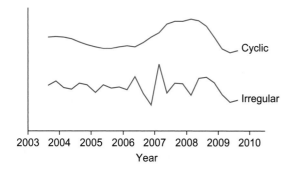

FIGURE 14.2.3   The cyclic and irregular components enlarged to show detail.

2003, the moving-average value is $[(1/2)8.065 + 8.215 + 10.153 + 9.175 + (1/2)9.292]/4 = 9.0554$. The moving-average values are shown in Table 14.2.2 and are displayed in the time-series plot of Figure 14.2.4.

## Seasonal Index: The Average Ratio-to-Moving-Average Indicates Seasonal Behavior

To isolate the seasonal behavior, start by taking the ratio of the original data to the moving average. (This is where the ratio-

### TABLE 14.2.2   Microsoft Revenues with Moving Average

| Year | Quarter | Revenues (billions) | Moving Average of Revenues (billions) |
|------|---------|---------------------|----------------------------------------|
| 2003 | 1 | $7.835 | (unavailable) |
| 2003 | 2 | 8.065 | (unavailable) |
| 2003 | 3 | 8.215 | $8.735 |
| 2003 | 4 | 10.153 | 9.055 |
| 2004 | 1 | 9.175 | 9.331 |
| 2004 | 2 | 9.292 | 9.535 |
| 2004 | 3 | 9.189 | 9.674 |
| 2004 | 4 | 10.818 | 9.838 |
| 2005 | 1 | 9.620 | 10.016 |
| 2005 | 2 | 10.161 | 10.212 |
| 2005 | 3 | 9.741 | 10.500 |
| 2005 | 4 | 11.837 | 10.865 |
| 2006 | 1 | 10.900 | 11.204 |
| 2006 | 2 | 11.804 | 11.426 |
| 2006 | 3 | 10.811 | 11.952 |
| 2006 | 4 | 12.542 | 12.585 |
| 2007 | 1 | 14.398 | 13.149 |
| 2007 | 2 | 13.371 | 13.996 |
| 2007 | 3 | 13.762 | 14.482 |
| 2007 | 4 | 16.367 | 14.797 |
| 2008 | 1 | 14.454 | 15.267 |
| 2008 | 2 | 15.837 | 15.463 |
| 2008 | 3 | 15.061 | 15.395 |
| 2008 | 4 | 16.629 | 14.952 |
| 2009 | 1 | 13.648 | 14.342 |
| 2009 | 2 | 13.099 | 14.373 |
| 2009 | 3 | 12.920 | 14.779 |
| 2009 | 4 | 19.022 | (unavailable) |
| 2010 | 1 | 14.503 | (unavailable) |

to-moving-average gets its name.) The result will include the seasonal and irregular components because the moving average cancels out the trend and cyclic components in the data:

$$(\text{Seasonal})(\text{Irregular}) = \frac{\text{Data}}{\text{Moving average}}$$

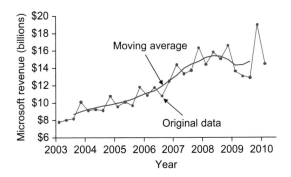

**FIGURE 14.2.4**   The moving average of revenues for Microsoft. The seasonal and irregular patterns have been eliminated, leaving only the trend and cyclic patterns.

| | Third Quarter |
|---|---|
| Year | Ratio-to-Moving-Average |
| 2003 | 0.941 |
| 2004 | 0.950 |
| 2005 | 0.928 |
| 2006 | 0.905 |
| 2007 | 0.950 |
| 2008 | 0.978 |
| 2009 | 0.874 |
| 2010 | (unavailable) |
| Average | 0.932 |

**TABLE 14.2.3** Computing the Third-Quarter Seasonal Index for Microsoft

Next, to eliminate the irregular component, you average these values for each season. The seasonal component will emerge because it is present each year, whereas the irregular will tend to be averaged away. The end results include a **seasonal index** for each time of the year, a factor that indicates how much larger or smaller this particular time period is compared to a typical period during the year. For example, a seasonal index of 1.088 for the fourth quarter indicates that the fourth quarter is generally 8.8% larger than a typical quarter. On the other hand, a third-quarter seasonal index of 0.932 would indicate that the third quarter is generally 6.8% lower.

$$\text{Seasonal index} = \text{Average of} \left( \frac{\text{Data}}{\text{Moving average}} \right)$$
$$\text{for that season}$$

For Microsoft, the first ratio-to-moving-average value is $8.215/8.7345 = 0.941$ for the third quarter of 2003. The third-quarter seasonal index is found by averaging these third-quarter ratios for all of the available years, as shown in Table 14.2.3.

Once each seasonal index has been found, it can be used throughout, even when the moving average is unavailable, because by definition, the seasonal pattern is exactly repeating. Table 14.2.4 shows the ratio-to-moving-average values and seasonal indexes for Microsoft. The typical yearly pattern is shown in Figure 14.2.5, and the repeating seasonal pattern is shown in Figure 14.2.6.

## Seasonal Adjustment: The Series Divided by the Seasonal Index

On July 21, 2010, the *Wall Street Journal* reported a seasonally adjusted statistic on its front page:

*On Tuesday, the U.S. Census Bureau said single-family housing starts in June fell by 0.7%, to a seasonally adjusted annual rate of 454,000.*

What is "seasonally adjusted," and how can there be a fall on a seasonally adjusted basis even though the actual value might have risen? **Seasonal adjustment** eliminates the expected seasonal component from a measurement (by dividing the series by the seasonal index for that period) so that one quarter or month may be directly compared to another (after seasonal adjustment) to reveal the underlying trends.

For retail sales, December is an especially good month. If sales are up in December as compared to November, it is no surprise; it is just the expected outcome. But if December sales are up even more than expected for this time of year, it may be time to bring out the champagne and visit that tropical island. To say, "December sales were higher than November's on a seasonally adjusted basis" is the same as saying, "December was up more than we expected." On the other hand, December sales could be way up but not as much as expected, so that December sales would actually be *down* on a seasonally adjusted basis.

To find a seasonally adjusted value, simply divide the original data value by the appropriate seasonal index for its month or quarter to remove the effect of this particular season:

$$\text{Seasonally adjusted value} = \left( \frac{\text{Data}}{\text{Seasonal index}} \right)$$
$$= \text{Trend} \times \text{Cyclic} \times \text{Irregular}$$

For Microsoft, the seasonally adjusted revenues for the second quarter of 2009 are the actual revenues (13.099, in

**TABLE 14.2.4 Microsoft Revenues and Seasonal Indexes**

| Year | Quarter | Revenues (billions) | Moving Average (billions) | Ratio-to-Moving-Average | Seasonal Index |
|------|---------|---------------------|---------------------------|-------------------------|----------------|
| 2003 | 1 | $7.835 | (unavailable) | (unavailable) | 0.985 |
| 2003 | 2 | 8.065 | (unavailable) | (unavailable) | 0.982 |
| 2003 | 3 | 8.215 | $8.735 | 0.941 | 0.932 |
| 2003 | 4 | 10.153 | 9.055 | 1.121 | 1.088 |
| 2004 | 1 | 9.175 | 9.331 | 0.983 | 0.985 |
| 2004 | 2 | 9.292 | 9.535 | 0.974 | 0.982 |
| 2004 | 3 | 9.189 | 9.674 | 0.950 | 0.932 |
| 2004 | 4 | 10.818 | 9.838 | 1.100 | 1.088 |
| 2005 | 1 | 9.620 | 10.016 | 0.960 | 0.985 |
| 2005 | 2 | 10.161 | 10.212 | 0.995 | 0.982 |
| 2005 | 3 | 9.741 | 10.500 | 0.928 | 0.932 |
| 2005 | 4 | 11.837 | 10.865 | 1.089 | 1.088 |
| 2006 | 1 | 10.900 | 11.204 | 0.973 | 0.985 |
| 2006 | 2 | 11.804 | 11.426 | 1.033 | 0.982 |
| 2006 | 3 | 10.811 | 11.952 | 0.905 | 0.932 |
| 2006 | 4 | 12.542 | 12.585 | 0.997 | 1.088 |
| 2007 | 1 | 14.398 | 13.149 | 1.095 | 0.985 |
| 2007 | 2 | 13.371 | 13.996 | 0.955 | 0.982 |
| 2007 | 3 | 13.762 | 14.482 | 0.950 | 0.932 |
| 2007 | 4 | 16.367 | 14.797 | 1.106 | 1.088 |
| 2008 | 1 | 14.454 | 15.267 | 0.947 | 0.985 |
| 2008 | 2 | 15.837 | 15.463 | 1.024 | 0.982 |
| 2008 | 3 | 15.061 | 15.395 | 0.978 | 0.932 |
| 2008 | 4 | 16.629 | 14.952 | 1.112 | 1.088 |
| 2009 | 1 | 13.648 | 14.342 | 0.952 | 0.985 |
| 2009 | 2 | 13.099 | 14.373 | 0.911 | 0.982 |
| 2009 | 3 | 12.920 | 14.779 | 0.874 | 0.932 |
| 2009 | 4 | 19.022 | (unavailable) | (unavailable) | 1.088 |
| 2010 | 1 | 14.503 | (unavailable) | (unavailable) | 0.985 |

billions of dollars) divided by the second-quarter seasonal index (0.982).

Seasonally adjusted revenues for second quarter 2009

$$= 13.099/0.982 = 13.34 \text{ (in \$ billions)}$$

Why is the seasonally adjusted result larger than the actual revenues? Because revenues are generally lower in the second quarter compared to a typical quarter in the year. In fact, you expect second-quarter revenues to be approximately 1.8% lower (based on the seasonal index of 0.982, subtracting it from 1). Dividing by the seasonal index removes this expected seasonal fluctuation, raising the second-quarter revenues to the status of a typical quarter.

In the next quarter (third quarter 2009), the seasonally adjusted revenues figure is 12.920/0.932 = 13.86. Note that

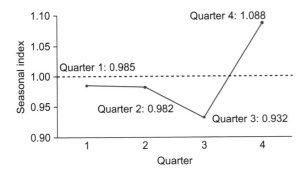

**FIGURE 14.2.5**  The seasonal indexes show that Microsoft revenues are typically highest in quarter 4 and lowest in quarter 3.

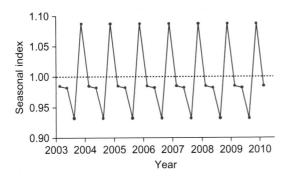

**FIGURE 14.2.6**  The seasonal component of Microsoft revenues, extracted from the original series, is exactly repeating each year.

revenues fell (from 13.099 to 12.920) from the second to the third quarter of 2009. However, on a seasonally adjusted basis, revenues actually increased from 13.34 to 13.86. This tells you that the drop, large as it seems, was actually *smaller than you would expect* for that time of year.

Note the strong increase in revenues from third to fourth quarter 2009 (from 12.920 to 19.022, in billions of dollars). On a seasonally adjusted basis, this is also an increase (from 13.86 to 17.49). Seasonal adjustment confirms your impression that this is a "real" and not just a seasonal increase in revenues, even after reducing the increase according to the anticipated rise at this time of year.

Table 14.2.5 shows the seasonally adjusted revenues for the entire time series. They are plotted in Figure 14.2.7 along with the original data. The seasonally adjusted series is somewhat smoother than the original data because the seasonal variation has been eliminated. However, some roughness remains because the irregular and cyclic components are present in the seasonally adjusted series, in addition to the trend.

## Long-Term Trend and Seasonally Adjusted Forecast: The Regression Line

When a time series shows an upward or downward long-term linear trend over time, regression analysis can be used to estimate this trend and to forecast the future. Although this leads to a useful forecast, an even more careful and complex method (an *ARIMA process*, for example) would pay more attention to the cyclic component than the method presented here.

Here's how the regression analysis works. Use the time period as the $X$ variable to predict the seasonally adjusted series as the $Y$ variable.[7] The resulting regression equation will represent the long-term trend. By substituting future time periods as new $X$ values, you will be able to forecast this long-term trend into the future.

Be careful how you represent the time periods. It is important that the numbers you choose be evenly spaced.[8] One easy way to do this is to use the numbers 1, 2, 3, ... to represent $X$ directly in terms of number of time periods (quarters or months). In this case, with 7 years of quarterly data (plus one extra), $X$ will use the numbers from 1 to 29.

Table 14.2.6 shows the data for the regression analysis (last two columns) to detect the long-term trend for Microsoft.

The regression equation, estimated using least squares, is

$$\text{Long-term trend} = 7.8027 + 0.2939(\text{Time period})$$

This suggests that Microsoft revenues have grown at an average rate of \$0.294 (in billions) per quarter.

It is easy to forecast this long-term trend by substituting the appropriate time period into the regression equation. For example, to find the trend value for the first quarter of 2013, use $X = 41$ to represent the time period that is three years (hence, 12 time periods) beyond the end of the series (which is $X = 29$). The forecast is then

$$\begin{aligned}
&(\text{Forecast trend value for first quarter 2013}) \\
&\quad = 7.8027 + 0.2939\,(\text{Time period}) \\
&\quad = 7.8027 + 0.2939 \times 41 \\
&\quad = \$19.85\ (\text{in billions})
\end{aligned}$$

Table 14.2.7 shows the predicted values, giving the long-term trend values and their (seasonally adjusted) forecasts for four years beyond the end of the data. Figure 14.2.8 shows how this trend line summarizes the seasonally adjusted series

---

7. If your series shows substantial exponential growth rather than a linear relationship, as a new startup firm might, you could use the *logarithm* of the seasonally adjusted series as your $Y$ variable and then transform back your predicted values (see Chapter 12) to make the forecast.

8. You would definitely *not* want to use 2003.1, 2003.2, 2003.3, 2003.4, 2004.1, ... because these numbers are not evenly spaced. You might use 2003.125, 2003.375, 2003.625, 2003.875, 2004.125, ... instead, which represents each time period as the halfway point of a quarter (adding 1/8, 3/8, 5/8, and 7/8 to each year). The first quarter of 2003 is represented by its midpoint, 2003.125, which is halfway between the beginning (2003.000) and the end (2003.250), as found by averaging them: (2003.000 + 2003.250)/2 = 2003.125.

**TABLE 14.2.5** Microsoft Revenues and Seasonally Adjusted Revenues

| Year | Quarter | Revenues (billions) | Seasonal Index | Seasonally Adjusted Revenues (billions) |
|------|---------|---------------------|----------------|-----------------------------------------|
| 2003 | 1 | $7.835 | 0.985 | $7.95 |
| 2003 | 2 | 8.065 | 0.982 | 8.21 |
| 2003 | 3 | 8.215 | 0.932 | 8.81 |
| 2003 | 4 | 10.153 | 1.088 | 9.34 |
| 2004 | 1 | 9.175 | 0.985 | 9.31 |
| 2004 | 2 | 9.292 | 0.982 | 9.46 |
| 2004 | 3 | 9.189 | 0.932 | 9.86 |
| 2004 | 4 | 10.818 | 1.088 | 9.95 |
| 2005 | 1 | 9.620 | 0.985 | 9.77 |
| 2005 | 2 | 10.161 | 0.982 | 10.34 |
| 2005 | 3 | 9.741 | 0.932 | 10.45 |
| 2005 | 4 | 11.837 | 1.088 | 10.88 |
| 2006 | 1 | 10.900 | 0.985 | 11.07 |
| 2006 | 2 | 11.804 | 0.982 | 12.02 |
| 2006 | 3 | 10.811 | 0.932 | 11.60 |
| 2006 | 4 | 12.542 | 1.088 | 11.53 |
| 2007 | 1 | 14.398 | 0.985 | 14.62 |
| 2007 | 2 | 13.371 | 0.982 | 13.61 |
| 2007 | 3 | 13.762 | 0.932 | 14.76 |
| 2007 | 4 | 16.367 | 1.088 | 15.05 |
| 2008 | 1 | 14.454 | 0.985 | 14.67 |
| 2008 | 2 | 15.837 | 0.982 | 16.12 |
| 2008 | 3 | 15.061 | 0.932 | 16.16 |
| 2008 | 4 | 16.629 | 1.088 | 15.29 |
| 2009 | 1 | 13.648 | 0.985 | 13.86 |
| 2009 | 2 | 13.099 | 0.982 | 13.34 |
| 2009 | 3 | 12.920 | 0.932 | 13.86 |
| 2009 | 4 | 19.022 | 1.088 | 17.49 |
| 2010 | 1 | 14.503 | 0.985 | 14.72 |

and extends to the right by extrapolation to indicate the seasonally adjusted forecasts.

## Forecast: The Seasonalized Trend

All you need to do now to forecast the future is to "seasonalize" the long-term trend by putting the expected seasonal variation back in. To do this, simply multiply the trend value by the appropriate seasonal index for the time period you are forecasting. This process is the reverse of seasonal adjustment. The resulting forecast includes the long-term trend and the seasonal variation:

Forecast = Trend × Seasonal index

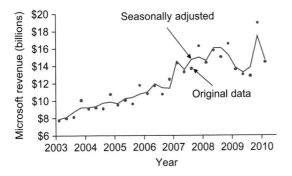

FIGURE 14.2.7 The seasonally adjusted series allows you to compare one quarter to another. By eliminating the *expected* seasonal changes, you have a clearer picture of where your business is heading.

To forecast the revenues of Microsoft for the first quarter of 2013, you would multiply the trend value of 19.85 (in billions of dollars, found by regression for the 41st time period) by the first-quarter seasonal index of 0.985:

(Revenue forecast for first quarter 2013)

$$= 19.85 \times 0.985 = \$19.55 \text{ (in billions)}$$

Table 14.2.8 shows the forecasts for four years beyond the end of the data. Figure 14.2.9 shows how this seasonalized trend summarizes the series and extends to the right by extrapolation to provide reasonable forecasts that include the expected seasonal behavior of revenues.

**TABLE 14.2.6  Microsoft Revenues with Regression Variables to Find the Long-Term Trend**

| Year | Quarter | Revenues (billions) | Seasonally Adjusted Revenues (billions), Y | Time Periods, X |
|------|---------|---------------------|--------------------------------------------|-----------------|
| 2003 | 1 | $7.835 | $7.95 | 1 |
| 2003 | 2 | 8.065 | 8.21 | 2 |
| 2003 | 3 | 8.215 | 8.81 | 3 |
| 2003 | 4 | 10.153 | 9.34 | 4 |
| 2004 | 1 | 9.175 | 9.31 | 5 |
| 2004 | 2 | 9.292 | 9.46 | 6 |
| 2004 | 3 | 9.189 | 9.86 | 7 |
| 2004 | 4 | 10.818 | 9.95 | 8 |
| 2005 | 1 | 9.620 | 9.77 | 9 |
| 2005 | 2 | 10.161 | 10.34 | 10 |
| 2005 | 3 | 9.741 | 10.45 | 11 |
| 2005 | 4 | 11.837 | 10.88 | 12 |
| 2006 | 1 | 10.900 | 11.07 | 13 |
| 2006 | 2 | 11.804 | 12.02 | 14 |
| 2006 | 3 | 10.811 | 11.60 | 15 |
| 2006 | 4 | 12.542 | 11.53 | 16 |
| 2007 | 1 | 14.398 | 14.62 | 17 |
| 2007 | 2 | 13.371 | 13.61 | 18 |
| 2007 | 3 | 13.762 | 14.76 | 19 |
| 2007 | 4 | 16.367 | 15.05 | 20 |
| 2008 | 1 | 14.454 | 14.67 | 21 |
| 2008 | 2 | 15.837 | 16.12 | 22 |
| 2008 | 3 | 15.061 | 16.16 | 23 |

*(Continued)*

**TABLE 14.2.6** Microsoft Revenues with Regression Variables to Find the Long-Term Trend—cont'd

| Year | Quarter | Revenues (billions) | Seasonally Adjusted Revenues (billions), Y | Time Periods, X |
|------|---------|---------------------|--------------------------------------------|-----------------|
| 2008 | 4 | 16.629 | 15.29 | 24 |
| 2009 | 1 | 13.648 | 13.86 | 25 |
| 2009 | 2 | 13.099 | 13.34 | 26 |
| 2009 | 3 | 12.920 | 13.86 | 27 |
| 2009 | 4 | 19.022 | 17.49 | 28 |
| 2010 | 1 | 14.503 | 14.72 | 29 |

**TABLE 14.2.7** Microsoft Revenues and Long-Term Trend Values

| Year | Quarter | Revenues (billions) | Seasonally Adjusted Revenues (billions), Y | Time Periods, X | Trend and Seasonally Adjusted Forecast (billions), Predicted Y |
|------|---------|---------------------|--------------------------------------------|-----------------|----------------------------------------------------------------|
| 2003 | 1 | $7.835 | $7.95 | 1 | $8.10 |
| 2003 | 2 | 8.065 | 8.21 | 2 | 8.39 |
| 2003 | 3 | 8.215 | 8.81 | 3 | 8.68 |
| 2003 | 4 | 10.153 | 9.34 | 4 | 8.98 |
| 2004 | 1 | 9.175 | 9.31 | 5 | 9.27 |
| 2004 | 2 | 9.292 | 9.46 | 6 | 9.57 |
| 2004 | 3 | 9.189 | 9.86 | 7 | 9.86 |
| 2004 | 4 | 10.818 | 9.95 | 8 | 10.15 |
| 2005 | 1 | 9.620 | 9.77 | 9 | 10.45 |
| 2005 | 2 | 10.161 | 10.34 | 10 | 10.74 |
| 2005 | 3 | 9.741 | 10.45 | 11 | 11.04 |
| 2005 | 4 | 11.837 | 10.88 | 12 | 11.33 |
| 2006 | 1 | 10.900 | 11.07 | 13 | 11.62 |
| 2006 | 2 | 11.804 | 12.02 | 14 | 11.92 |
| 2006 | 3 | 10.811 | 11.60 | 15 | 12.21 |
| 2006 | 4 | 12.542 | 11.53 | 16 | 12.50 |
| 2007 | 1 | 14.398 | 14.62 | 17 | 12.80 |
| 2007 | 2 | 13.371 | 13.61 | 18 | 13.09 |
| 2007 | 3 | 13.762 | 14.76 | 19 | 13.39 |
| 2007 | 4 | 16.367 | 15.05 | 20 | 13.68 |
| 2008 | 1 | 14.454 | 14.67 | 21 | 13.97 |
| 2008 | 2 | 15.837 | 16.12 | 22 | 14.27 |
| 2008 | 3 | 15.061 | 16.16 | 23 | 14.56 |

**TABLE 14.2.7 Microsoft Revenues and Long-Term Trend Values—cont'd**

| Year | Quarter | Revenues (billions) | Seasonally Adjusted Revenues (billions), Y | Time Periods, X | Trend and Seasonally Adjusted Forecast (billions), Predicted Y |
|------|---------|---------------------|---------------------------------------------|-----------------|-------------------------------------------------------------------|
| 2008 | 4 | 16.629 | 15.29 | 24 | 14.86 |
| 2009 | 1 | 13.648 | 13.86 | 25 | 15.15 |
| 2009 | 2 | 13.099 | 13.34 | 26 | 15.44 |
| 2009 | 3 | 12.920 | 13.86 | 27 | 15.74 |
| 2009 | 4 | 19.022 | 17.49 | 28 | 16.03 |
| 2010 | 1 | 14.503 | 14.72 | 29 | 16.32 |
| 2010 | 2 | | | 30 | 16.62 |
| 2010 | 3 | | | 31 | 16.91 |
| 2010 | 4 | | | 32 | 17.21 |
| 2011 | 1 | | | 33 | 17.50 |
| 2011 | 2 | | | 34 | 17.79 |
| 2011 | 3 | | | 35 | 18.09 |
| 2011 | 4 | | | 36 | 18.38 |
| 2012 | 1 | | | 37 | 18.68 |
| 2012 | 2 | | | 38 | 18.97 |
| 2012 | 3 | | | 39 | 19.26 |
| 2012 | 4 | | | 40 | 19.56 |
| 2013 | 1 | | | 41 | 19.85 |
| 2013 | 2 | | | 42 | 20.14 |
| 2013 | 3 | | | 43 | 20.44 |
| 2013 | 4 | | | 44 | 20.73 |
| 2014 | 1 | | | 45 | 21.03 |

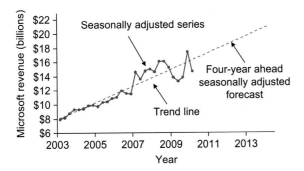

**FIGURE 14.2.8** The least-squares regression line used to predict the seasonally adjusted series from the time period can be extended to the right to provide seasonally adjusted forecasts.

Should you believe these forecasts? Keep in mind that nearly all forecasts are wrong. After all, by definition, the irregular component cannot be predicted. In addition, these trend-seasonal forecasts do not reflect the cyclic component. But they do seem to do a good job of capturing the long-term upward trend and the repeating seasonal patterns.

## 14.3 MODELING CYCLIC BEHAVIOR USING BOX–JENKINS ARIMA PROCESSES

The Box–Jenkins approach is one of the best methods we have for the important goals of *understanding* and *forecasting* a business time series. The resulting *ARIMA processes*

**TABLE 14.2.8 Microsoft Revenues and Forecasts**

| Year | Quarter | Revenues (billions) | Trend and Seasonally Adjusted Forecast (billions) | Seasonal Index | Seasonalized Trend and Forecast (billions) |
|------|---------|---------------------|---------------------------------------------------|----------------|--------------------------------------------|
| 2003 | 1 | $7.835 | $8.10 | 0.985 | $7.98 |
| 2003 | 2 | 8.065 | 8.39 | 0.982 | 8.24 |
| 2003 | 3 | 8.215 | 8.68 | 0.932 | 8.10 |
| 2003 | 4 | 10.153 | 8.98 | 1.088 | 9.76 |
| 2004 | 1 | 9.175 | 9.27 | 0.985 | 9.13 |
| 2004 | 2 | 9.292 | 9.57 | 0.982 | 9.40 |
| 2004 | 3 | 9.189 | 9.86 | 0.932 | 9.19 |
| 2004 | 4 | 10.818 | 10.15 | 1.088 | 11.04 |
| 2005 | 1 | 9.620 | 10.45 | 0.985 | 10.29 |
| 2005 | 2 | 10.161 | 10.74 | 0.982 | 10.55 |
| 2005 | 3 | 9.741 | 11.04 | 0.932 | 10.29 |
| 2005 | 4 | 11.837 | 11.33 | 1.088 | 12.32 |
| 2006 | 1 | 10.900 | 11.62 | 0.985 | 11.45 |
| 2006 | 2 | 11.804 | 11.92 | 0.982 | 11.70 |
| 2006 | 3 | 10.811 | 12.21 | 0.932 | 11.38 |
| 2006 | 4 | 12.542 | 12.50 | 1.088 | 13.60 |
| 2007 | 1 | 14.398 | 12.80 | 0.985 | 12.61 |
| 2007 | 2 | 13.371 | 13.09 | 0.982 | 12.86 |
| 2007 | 3 | 13.762 | 13.39 | 0.932 | 12.48 |
| 2007 | 4 | 16.367 | 13.68 | 1.088 | 14.88 |
| 2008 | 1 | 14.454 | 13.97 | 0.985 | 13.76 |
| 2008 | 2 | 15.837 | 14.27 | 0.982 | 14.01 |
| 2008 | 3 | 15.061 | 14.56 | 0.932 | 13.57 |
| 2008 | 4 | 16.629 | 14.86 | 1.088 | 16.16 |
| 2009 | 1 | 13.648 | 15.15 | 0.985 | 14.92 |
| 2009 | 2 | 13.099 | 15.44 | 0.982 | 15.17 |
| 2009 | 3 | 12.920 | 15.74 | 0.932 | 14.67 |
| 2009 | 4 | 19.022 | 16.03 | 1.088 | 17.43 |
| 2010 | 1 | 14.503 | 16.32 | 0.985 | 16.08 |
| 2010 | 2 | | 16.62 | 0.982 | 16.32 |
| 2010 | 3 | | 16.91 | 0.932 | 15.77 |
| 2010 | 4 | | 17.21 | 1.088 | 18.71 |
| 2011 | 1 | | 17.50 | 0.985 | 17.24 |

**TABLE 14.2.8** Microsoft Revenues and Forecasts—cont'd

| Year | Quarter | Revenues (billions) | Trend and Seasonally Adjusted Forecast (billions) | Seasonal Index | Seasonalized Trend and Forecast (billions) |
|------|---------|---------------------|--------------------------------------------------|----------------|--------------------------------------------|
| 2011 | 2 | | 17.79 | 0.982 | 17.48 |
| 2011 | 3 | | 18.09 | 0.932 | 16.86 |
| 2011 | 4 | | 18.38 | 1.088 | 19.99 |
| 2012 | 1 | | 18.68 | 0.985 | 18.39 |
| 2012 | 2 | | 18.97 | 0.982 | 18.63 |
| 2012 | 3 | | 19.26 | 0.932 | 17.96 |
| 2012 | 4 | | 19.56 | 1.088 | 21.27 |
| 2013 | 1 | | 19.85 | 0.985 | 19.55 |
| 2013 | 2 | | 20.14 | 0.982 | 19.79 |
| 2013 | 3 | | 20.44 | 0.932 | 19.05 |
| 2013 | 4 | | 20.73 | 1.088 | 22.55 |
| 2014 | 1 | | 21.03 | 0.985 | 20.71 |

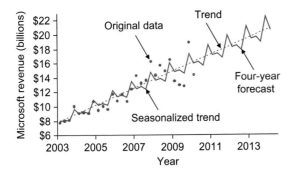

**FIGURE 14.2.9** Forecasts are made by multiplying the trend line by the seasonal index. The result includes the trend and seasonal components but not the cyclic and irregular behavior of the series.

are linear statistical models that can precisely describe many different time-series behaviors, including even the medium-term wandering of the so-called business cycle. Compared to the trend-seasonal approach of the previous section, the Box–Jenkins approach has a more solid statistical foundation but is somewhat less intuitive. As a result, you can obtain reasonable statistical measures of uncertainty (a standard error for the forecast, for example) once you have found an appropriate model within the Box–Jenkins family.

Here's an outline of the steps involved "behind the scenes" when you use Box–Jenkins methods to help you understand what the forecasts and their confidence intervals represent:

1. A fairly simple process is chosen from the Box–Jenkins family of ARIMA processes that generates data with the same overall look as your series, except for randomness. This involves selecting a particular type of model and estimating the parameters from your data. The resulting model will tell you useful facts such as (a) the extent to which each observation influences the future and (b) the extent to which each observation brings useful new information to help you forecast.

2. The forecast for any time is the expected (i.e., average or mean) future value of the estimated process at that time. Imagine the universe of all reasonably possible future behaviors of your series, starting with your data and extending it into the future according to the model selected in step 1. The formula for the forecast quickly computes the average of all of these future scenarios.

3. The standard error of a forecast for any time is the standard deviation of all reasonably possible future values for that time.

4. The forecast limits extend above and below the forecast value such that (if the model is correct) there is a 95% chance, for example, that the future value for any time will fall within the forecast limits. They are constructed so that for each future time period, 95% of the reasonably possible future behaviors of your series fall within the limits. This assumes that your series will continue to behave similarly to the estimated process.

The **Box–Jenkins ARIMA processes** are a family of linear statistical models based on the normal distribution that have the flexibility to imitate the behavior of many different real time series by combining *autoregressive (AR) processes, integrated (I) processes*, and *moving-average (MA) processes*.[9] The result is a **parsimonious model**, that is, one that uses just a few estimated parameters to describe the complex behavior of a time series. Although the theory and computations involved are complex, the models themselves are fairly simple and are quickly calculated using a computer.

We will begin by reviewing the random noise process and then describe how each component of an ARIMA process adds structure and smoothness to the model. We will cover only some of the basics of these complex models.[10]

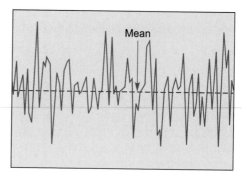

**FIGURE 14.3.1**   A random noise process consists of independent observations from a normal distribution. It is basically flat and bumpy, with constant variability.

## A Random Noise Process Has No Memory: The Starting Point

A **random noise process** consists of a random sample (independent observations) from a normal distribution with constant mean and standard deviation. There are no trends because, due to independence, the observations "have no memory" of the past behavior of the series.

The model for random noise says that at time $t$ the observed data, $Y_t$, will consist of a constant, $\mu$ (the long-term mean of the process), plus random noise, $\varepsilon_t$, with mean zero.

> **The Random Noise Process**
>
> Data = Mean value + Random noise
>
> $$Y_t = \mu + \varepsilon_t$$
>
> The long-term mean of $Y$ is $\mu$.

A random noise process tends to be basically flat (tilted neither up nor down), to be very irregular, and to have constant variability, as shown in Figure 14.3.1.

If you have a random noise process, the analysis is easy because the data form a random sample from a normal distribution—a situation you learned about in Chapters 9 and 10. The average is the best forecast for any future time period, and the ordinary prediction interval for a

new observation gives you the forecast limits for any future value of the series.

Most business and economic time-series data sets have some structure in addition to their random noise component. You may think of this structure in terms of the way each observation "remembers" the past behavior of the series. When this memory is strong, the series can be much smoother than a random noise process.

## An Autoregressive (AR) Process Remembers Where It Was

An observation of an **autoregressive process** (the *AR* in ARIMA) consists of a linear function of the previous observation plus random noise.[11] Thus, an autoregressive process remembers where it was and uses this information in deciding where to go next.

The model for an autoregressive process says that at time $t$ the data value, $Y_t$, consists of a constant, $\delta$ (delta), plus an autoregressive coefficient, $\varphi$ (phi), times the previous data value, $Y_{t-1}$, plus random noise, $\varepsilon_t$. Note that this is a linear regression model that predicts the current level ($Y = Y_t$) from the previous level ($X = Y_{t-1}$). In effect, the series moves a proportion $(1 - \varphi)$ back toward its long-run mean and then moves a random distance from there. By increasing $\varphi$ from 0 toward 1, you can make the process look smoother and less like random noise.[12] It is important that $\varphi$ be less than 1 (in absolute value) in order that the process be stable.

9. The word *process* here refers to any statistical procedure that produces time-series data.

10. For further details, see C. R. Nelson, *Applied Time Series Analysis for Managerial Forecasting* (San Francisco: Holden-Day, 1973); or G. E. P. Box and G. M. Jenkins, *Time Series Analysis: Forecasting and Control* (San Francisco: Holden-Day, 1976).

11. This is a *first-order* autoregressive process. In general, the observation might depend on *several* of the most recent observations, much like a multiple regression.

12. If the coefficient, $\varphi$, is negative, the autoregressive process can actually be *more* bumpy than random noise because it tends to be alternatively high and low. We will assume that $\varphi$ is positive so that an autoregressive process is smoother than random noise.

### The Autoregressive Process

> Data $= \delta + \varphi(\text{Previous value}) + \text{Random noise}$
>
> $Y_t = \delta + \varphi Y_{t-1} + \varepsilon_t$
>
> The long-term mean value of $Y$ is $\delta/(1-\varphi)$.

Because it has memory, an autoregressive process can stay high for a while, then stay low for a while, and so on, thereby generating a cyclic pattern of ups and downs about a long-term mean value, as shown in Figure 14.3.2. The particular process shown here has $\varphi = 0.8$, so that $Y_t = 0.8 Y_{t-1} + \varepsilon_t$, where $\varepsilon$ has standard deviation 1 and is the same noise as in Figure 14.3.1.

Autoregressive models often make sense for business data. They express the fact that where you go depends partly on where you are (as expressed by the autoregressive coefficient, $\varphi$) and partly on what happens to you along the way (as expressed by the random noise component).

Forecasting with an autoregressive process is done with predicted values from the estimated regression equation after going forward one unit in time, so that the predicted $Y_{t+1}$ is $\hat{\delta} + \hat{\varphi} Y_t$. (The "hats" over the coefficients indicate that they are estimated from the data rather than the population values.) The forecast is a compromise between the most recent data value and the long-term mean value of the series. The further into the future you look, the closer to the estimated long-term mean value your forecast will be because the process gradually "forgets" the distant past.

### Example
#### Forecasting the Unemployment Rate Using an Autoregressive Process

Table 14.3.1 shows the U.S. unemployment rate, recorded by year from 1960 through 2009. This data set is graphed in Figure 14.3.3.

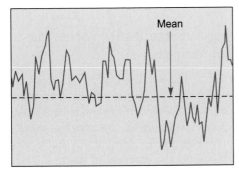

**FIGURE 14.3.2** An autoregressive process evolves as a linear regression equation in which the current value helps predict the next value. Note that the series is less bumpy than pure noise (compare to Figure 14.3.1) and that it can stray from its long-term mean value for extended periods.

An autoregressive (AR) model was estimated for this data set, using the method of least squares, with the results as shown in Table 14.3.2.[13] Note that the autoregressive coefficient and the mean are both statistically significant, based on $p$ value from the $t$ ratio.

These results give us an autoregressive (AR) model that produces time-series data that somewhat resemble the unemployment rate data, with the same kind of irregularity, smoothness, and cyclic behavior. This estimated AR model is as follows:

Data $= 0.01317 + 0.79012(\text{Previous value}) + \text{Random noise}$

$Y_t = 0.01317 + 0.79012\, Y_{t-1} + \varepsilon_t$

*(Continued)*

**TABLE 14.3.1 Unemployment Rate**

| Year | Unemployment Rate |
|------|-------------------|
| 1960 | 6.6% |
| 1961 | 6.0 |
| 1962 | 5.5 |
| 1963 | 5.5 |
| 1964 | 5.0 |
| 1965 | 4.0 |
| 1966 | 3.8 |
| 1967 | 3.8 |
| 1968 | 3.4 |
| 1969 | 3.5 |
| 1970 | 6.1 |
| 1971 | 6.0 |
| 1972 | 5.2 |
| 1973 | 4.9 |
| 1974 | 7.2 |
| 1975 | 8.2 |
| 1976 | 7.8 |
| 1977 | 6.4 |
| 1978 | 6.0 |
| 1979 | 6.0 |
| 1980 | 7.2 |
| 1981 | 8.5 |
| 1982 | 10.8 |
| 1983 | 8.3 |
| 1984 | 7.3 |

*(Continued)*

**TABLE 14.3.1** Unemployment Rate—cont'd

| Year | Unemployment Rate |
|------|-------------------|
| 1985 | 7.0 |
| 1986 | 6.6 |
| 1987 | 5.7 |
| 1988 | 5.3 |
| 1989 | 5.4 |
| 1990 | 6.3 |
| 1991 | 7.3 |
| 1992 | 7.4 |
| 1993 | 6.5 |
| 1994 | 5.5 |
| 1995 | 5.6 |
| 1996 | 5.4 |
| 1997 | 4.7 |
| 1998 | 4.4 |
| 1999 | 4.0 |
| 2000 | 3.9 |
| 2001 | 5.7 |
| 2002 | 6.0 |
| 2003 | 5.7 |
| 2004 | 5.4 |
| 2005 | 4.9 |
| 2006 | 4.4 |
| 2007 | 5.0 |
| 2008 | 7.4 |
| 2009 | 10.0 |

**Source:** Bureau of Labor Statistics, U.S. Department of Labor, accessed from http://www.bls.gov/cps/ on April 16, 2010.

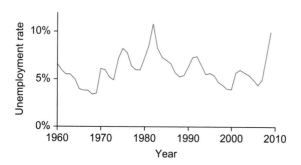

**FIGURE 14.3.3**   The U.S. unemployment rate from 1960 through 2009. Note the degree of smoothness (this is obviously *not* just random noise) and the tendency toward cyclic behavior.

**Example—cont'd**

where we have used the estimates from Table 14.3.2 to compute $\hat{\delta} = 0.06275 \times (1-0.79012) = 0.01317$. This estimated model suggests that the unemployment rate does not change by large amounts from one year to the next, since each year's data value is generated taking the previous year's level into account. Literally, each year's data value is found by first moving $(1 - 0.790) = 21.0\%$ of the way from the current unemployment rate toward the long-run mean value of 6.275% and then adding new random noise.

How closely do data from the estimated AR process mimic the unemployment rate? Figure 14.3.4 shows the actual unemployment rate together with two simulations created from the estimated AR process, starting at the same (6.6%) unemployment rate for 1960 but using different random noise. Think of these simulations as alternative scenarios of what might have happened instead of what actually did happen.

The real purpose of time-series analysis in business is to forecast. Table 14.3.3 shows forecasts of the unemployment rate, together with forecast limits, out to 2020 as computed based on the estimated AR model. Figure 14.3.5 shows that the forecast heads down from the last series value (10.0% for 2009) toward the long-term mean value of 6.275%. This forecast, the best that can be done based only on the data from Table 14.3.1 and this AR model, says that *on average* we expect the series to gradually forget that it was above its long-term mean and to revert back down. Of course, we really expect it to continue its cyclic and irregular behavior; this is the reason that the 95% forecast limits are so wide.

Figure 14.3.6 shows two simulations of the future, created from the estimated AR model using new, independent noise. The forecast represents the average of all such simulations of the future. The forecast limits enclose the middle 95% of all such simulations at each time period in the future.

13. The SPSS statistical software package was used. The goal of least-squares estimation is to make the noise component as small as possible, so that as much as possible of the structure of the series is captured by the autoregressive component of the model. When the noise is a random sample from a normal distribution, the powerful general method of maximum likelihood gives the same estimates as the method of least squares because of the exponential square term in the normal density function.

## A Moving-Average (MA) Process Has a Limited Memory

An observation of a **moving-average process** (the *MA* in ARIMA) consists of a constant, $\mu$ (the long-term mean of the process), plus independent random noise minus a fraction of the previous random noise.[14] A moving-average process does not remember exactly where it was, but it does remember the random noise component of where it was. Thus, its

14. This is a *first-order* moving-average process. In general, the observation might depend on several of the most recent random noise components, and the limited memory could be several steps long.

**TABLE 14.3.2 Estimates of an Autoregressive Model Fitted to the Unemployment Rate Data**

| Coefficient | Estimate | Standard Error | t Ratio | p |
|---|---|---|---|---|
| Autoregression ($\hat{\varphi}$) | 0.79012 | 0.10039 | 7.87 | 0.00000 |
| Mean ($\hat{\delta}/(1-\hat{\varphi})$) | 0.06275 | 0.00656 | 9.56 | 0.00000 |
| Standard deviation of random noise | 0.01045 | | | |

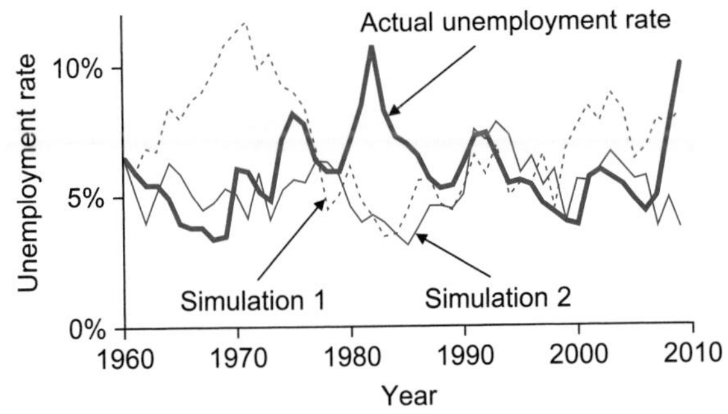

**FIGURE 14.3.4** Two simulations from the estimated AR process together with the actual unemployment rate. Note how the artificial simulations have the same basic character as the real data in terms of smoothness, irregularities, and cycles. This ability to behave like the real series is an important feature of Box–Jenkins analysis.

memory is limited to one step into the future; beyond that, it starts anew.

The model for a moving-average process says that at time $t$ the data value, $Y_t$, consists of a constant, $\mu$, plus random noise, $\varepsilon_t$, minus a fraction, $\theta$ (theta, the moving-average coefficient), of the previous random noise. By decreasing the coefficient, $\theta$, from 0 to $-1$, you can make the process look less like random noise, but it will be only slightly smoother.[15]

---

**The Moving-Average Process**

Data $= \mu + (\text{Random noise}) - \theta(\text{Previous random noise})$

$Y_t = \mu + \varepsilon_t - \theta\varepsilon_{t-1}$

The long-term mean value of $Y$ is $\mu$.

---

Because it has memory, a moving-average process can produce adjacent pairs of observations that are more likely to *both* be either high or low. However, because its memory is limited, the series is random again after only two steps. The result is a series that is not quite as random as a pure random noise series. Compare the moving-average process in Figure 14.3.7 to the pure random noise series of Figure 14.3.1 to see the decreased randomness. The particular process shown in Figure 14.3.7 has $\theta = -0.8$, so that $Y_t = \varepsilon_t + 0.8\varepsilon_{t-1}$, where $\varepsilon$ has standard deviation 1 and is the same noise as in Figure 14.3.1. In fact, this series *is* a moving average of random noise.

Pure moving-average models have only limited applicability for business data because of their limited memory (as expressed by the moving-average coefficient, $\theta$). They are best used in combination with autoregressive processes to permit a sharper focus on recent events than pure autoregressive processes allow.

Forecasting the next observation with a moving-average process is based on an estimate of the current random noise, $\hat{\varepsilon}$. Beyond the next observation, the best forecast is the estimated long-term mean, $\hat{\mu}$, because all but the immediate past has been forgotten.

## The Autoregressive Moving-Average (ARMA) Process Combines AR and MA

An observation of an **autoregressive moving-average (ARMA) process** consists of a linear function of the previous observation plus independent random noise minus a fraction of the previous random noise. This combines an autoregressive process with a moving-average process.[16] An autoregressive moving-average process remembers both where it was and the random noise component of where it was. Thus, its memory combines that of the autoregressive process with that of the moving-average process. The result is an autoregressive process with an improved short-term memory.

The model for an autoregressive moving-average (ARMA) process says that at time $t$ the data value, $Y_t$, consists of a constant, $\delta$, plus an autoregressive coefficient, $\varphi$, times the previous data value, $Y_{t-1}$, plus random noise, $\varepsilon_t$, minus a fraction, $\theta$, of the previous random noise. This is like a linear regression model except that the errors are

---

15. If the coefficient, $\theta$, is positive, the moving-average process can actually be somewhat *more* bumpy than random noise because it tends to be alternately high and low. We will assume that $\theta$ is negative so that a moving-average process is somewhat smoother than random noise.

16. This is a *first-order* autoregressive moving-average process. In general, the observation might depend on several of the most recent observations and random noise components.

**TABLE 14.3.3  Forecasts and Forecast Limits Given by the AR Model Fitted to the Unemployment Rate Data**

| Year | Forecast | 95% Forecast Limits | |
| --- | --- | --- | --- |
| | | Lower | Upper |
| 2010 | 9.218% | 7.100% | 11.337% |
| 2011 | 8.600 | 5.878 | 11.323 |
| 2012 | 8.112 | 5.058 | 11.167 |
| 2013 | 7.727 | 4.470 | 10.983 |
| 2014 | 7.422 | 4.037 | 10.807 |
| 2015 | 7.181 | 3.711 | 10.651 |
| 2016 | 6.991 | 3.464 | 10.518 |
| 2017 | 6.841 | 3.274 | 10.407 |
| 2018 | 6.722 | 3.128 | 10.316 |
| 2019 | 6.628 | 3.014 | 10.242 |
| 2020 | 6.554 | 2.926 | 10.182 |

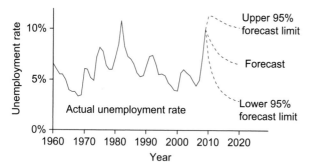

**FIGURE 14.3.5**  The unemployment rate, its forecast through 2020, and the 95% forecast limits, as computed based on the estimated AR model. The forecast says that the series, on average, will gradually forget that it is above its long-run mean. The forecast limits are wide enough to anticipate future cyclic and irregular behavior.

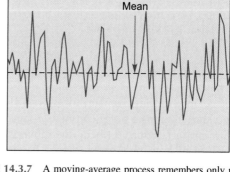

**FIGURE 14.3.7**  A moving-average process remembers only part of the previous noise. The result is slightly less irregular than pure random noise (compare to Figure 14.3.1). Two periods ahead, it becomes random again because it does not remember where it was.

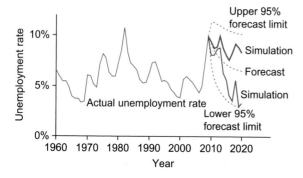

**FIGURE 14.3.6**  The unemployment rate, its forecast through 2020, the 95% forecast limits, and two simulations of the future. The forecast represents the average of all such simulations at each future time. The forecast limits enclose 95% of all such simulations at each future time.

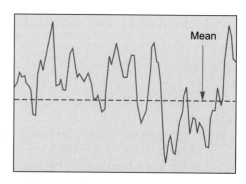

**FIGURE 14.3.8**  In an autoregressive moving-average (ARMA) process, both the current value and the current noise help determine the next value. The result is smoother due to the memory of the autoregressive process combined with the additional short-term (one-step-ahead) memory of the moving-average process.

not independent. As $\varphi$ goes from 0 toward 1, and as $\theta$ goes from 0 to –1, the resulting process looks smoother and less like random noise. It is important that $\varphi$ be less than 1 (in absolute value) in order that the process be stable.

> **The Autoregressive Moving-Average (ARMA) Process**
>
> Data $= \delta + \varphi$(Previous value) $+$ (Random noise)
> $\qquad - \theta$(Previous random noise)
>
> $Y_t = \delta + \varphi Y_{t-1} + \varepsilon_t - \theta \varepsilon_{t-1}$
>
> The long-term mean value of $Y$ is $\delta/(1 - \varphi)$.

Because of its memory, an ARMA process can stay high for a while, then stay low for a while, and so forth, generating a cyclic pattern of ups and downs about a long-term mean value, as shown in Figure 14.3.8. The particular process shown here has $\varphi = 0.8$ and $\theta = -0.8$, so that $Y_t = 0.8Y_{t-1} + \varepsilon_t + 0.8\varepsilon_{t-1}$ where $\varepsilon$ has standard deviation 1 and is the same noise as in Figure 14.3.1. Since the purely autoregressive process (Figure 14.3.2) shares this same random noise, a comparison of Figures 14.3.2 and 14.3.8 shows the contribution of the moving-average term to the smoothness of this ARMA process.

The combination of autoregressive and moving-average processes is a powerful and useful one for business data. By adjusting the coefficients ($\varphi$ and $\theta$), you can choose a model to match any of a wide variety of cyclic and irregular time-series data sets.

Forecasting the next observation with an ARMA process is done by combining the predicted value from the estimated autoregression equation ($Y_{t+1} = \hat{\delta} + \hat{\varphi} Y_t$ where "hats" again indicate estimates) with an estimate of the current random noise, $\varepsilon_t$. Beyond the next observation, the best forecast is based only on the previous forecast value. The further into the future you look, the closer to the estimated long-term mean value your forecast will be because the process gradually forgets the distant past.

## A Pure Integrated (I) Process Remembers Where It Was and Then Moves at Random

Each observation of a **pure integrated (I) process**, also called a **random walk**, consists of a random step away from the current observation. This process knows where it is but has forgotten how it got there. A random walk is said to be a **nonstationary process** because over time it tends to move farther and farther away from where it was. In contrast, the autoregressive, moving-average, and ARMA models each represent a **stationary process** because they tend to behave similarly over long time periods, staying relatively close to their long-run mean.

The model for a random walk says that at time $t$ the data value, $Y_t$, consists of a constant, $\delta$ (the "drift" term), plus the previous data value, $Y_{t-1}$, plus random noise, $\varepsilon_t$. Although this looks just like an autoregressive model with $\varphi = 1$, its behavior is very different.[17] The drift term, $\delta$, allows us to force the process to walk randomly upward on average over time (if $\delta > 0$) or downward (if $\delta < 0$). However, even if $\delta = 0$, the series will *appear* to have upward and downward trends over time.

> **The Pure Integrated (Random Walk) Process**
>
> Data $= \delta +$ (Previous value) $+$ (Random noise)
>
> $Y_t = \delta + Y_{t-1} + \varepsilon_t$
>
> Over time, $Y$ is not expected to stay close to any long-term mean value.

The easiest way to analyze pure integrated processes is to work with the series of *differences*, $Y_t - Y_{t-1}$, which follow a random noise process.[18]

> **The Pure Integrated (Random Walk) Process in Differenced Form**
>
> Data $-$ (Previous value) $= \delta +$ (Random noise)
>
> $Y_t - Y_{t-1} = \delta + \varepsilon_t$

Since there is no tendency to return to a long-run mean value, random walks can be deceptive, creating the appearance of trends where there really are none. The random walk in Figure 14.3.9 was created using $\delta = 0$, so there are *no real trends*, just random changes. The series did not "know" when it reached its highest point; it just continued at random in the same way from wherever it happened to be. The same random noise was used as in Figure 14.3.1, which represents the differences of the series in Figure 14.3.9.

Forecasting the next observation with a random walk is done by adding the estimated drift term, $\hat{\delta}$, to the current observation. For each additional time period you forecast into the future, an additional $\hat{\delta}$ is added. If there is no drift term (i.e., if you believe that $\delta = 0$), then the current value *is* the forecast of all future values. The forecast limits in either case will continue to widen over time (more than for ARMA processes) due to nonstationarity.

---

17. This is why we restricted $\varphi$ to be smaller than 1 in absolute value in the definition of autoregressive and ARMA processes. Remember that autoregressive and ARMA models are stationary, but the random walk is not. For an ARMA process, the long-run mean $\delta/(1 - \varphi)$ is undefined if $\varphi = 1$ due to division by zero.

18. For stock market and some other business data sets, it may be better to work with the *percent changes*, $(Y_t - Y_{t-1})/Y_{t-1}$. This is a variation on the idea of working with differences. Literally, percent changes are appropriate when the *logarithms* of the data follow a random walk with relatively small steps.

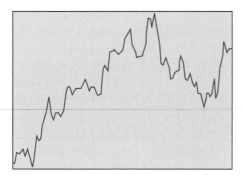

**FIGURE 14.3.9**  A pure integrated (I) process or random walk with no drift can appear to have trends when, in reality, there are none. The series remembers only where it is and takes totally random steps from there.

The random walk model is important on its own (as a stock market model, for example). It is also a key building block when used with ARMA models to create ARIMA models, with added flexibility to analyze more complex time-series data sets.

## The Autoregressive Integrated Moving-Average (ARIMA) Process Remembers Its Changes

If the changes or differences in a series are generated by an autoregressive moving-average (ARMA) process, then the series itself follows an **autoregressive integrated moving-average (ARIMA) process**. Thus, the *change* in the process consists of a linear function of the previous change, plus independent random noise, minus a fraction of the previous random noise.[19] This process knows where it is, remembers how it got there, and even remembers part of the previous noise component. Therefore, ARIMA processes can be used as a model for time-series data sets that are very smooth, changing direction slowly. These ARIMA processes are *nonstationary* due to the inclusion of an integrated component. Thus, over time, the series will tend to move farther and farther away from where it was.

The model for an autoregressive integrated moving-average process states that at time $t$ the data value's change, $Y_t - Y_{t-1}$, consists of a constant, $\delta$, plus an autoregressive coefficient, $\varphi$, times the previous change, $Y_{t-1} - Y_{t-2}$, plus random noise, $\varepsilon_t$, minus a fraction, $\theta$, of the previous random noise. This is like a linear regression model in terms of the *differences*, except that the errors are not independent. As $\varphi$ goes from 0 toward 1, and as $\theta$ goes from

0 to –1, the resulting process will be smoother. It is important that $\varphi$ be less than 1 (in absolute value) in order that the (differenced) process be stable.

---

**The Autoregressive Integrated Moving-Average (ARIMA) Process in Differenced Form**

Data change = $\delta + \varphi$(Previous change) + (Random noise) $- \theta$(Previous random noise)

$$Y_t - Y_{t-1} = \delta + \varphi(Y_{t-1} - Y_{t-2}) + \varepsilon_t - \theta\varepsilon_{t-1}$$

The long-term mean value of the *change* in Y is $\delta/(1 - \varphi)$. Over time, Y is not expected to stay close to any long-term mean value.

---

Figure 14.3.10 shows the ARIMA process created by summing (sometimes called *integrating*) the ARMA process of Figure 14.3.8. Since it shares the same random noise as the random walk of Figure 14.3.9, you can see how the autoregressive and moving-average components smooth out the changes while preserving the overall behavior of the series.

Forecasting with an ARIMA model is done by forecasting the changes of the ARMA model for the differences. Due to nonstationarity, the forecasts can tend indefinitely upward (or downward), and the forecast limits will widen as you extend your forecast further into the future.

When is differencing helpful? An ARIMA model (with differencing) will be useful for situations in which there is no tendency to return to a long-run mean value (for example, a stock's price, the U.S. GNP, the consumer price index, or your firm's sales). An ARMA model (which does not include differencing) will be useful for situations in which the series tends to stay near a long-term mean value (examples might include the unemployment rate, interest rates, changes in the price index, and your firm's debt ratio).

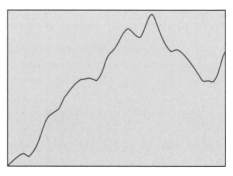

**FIGURE 14.3.10**  An autoregressive integrated moving-average (ARIMA) process remembers where it is, how it got there, and some of the previous noise. This results in a very smooth time-series model. Compare it to the random walk (with the same noise) in Figure 14.3.9.

---

19. This is a *first-order* autoregressive integrated moving-average process. In general, the change might depend on several of the most recent changes and random noise components.

More advanced ARIMA models can be created to include the seasonal behavior of quarterly and monthly series. The idea is to include last year's value in addition to last month's value in the model equations.

## 14.4 END-OF-CHAPTER MATERIALS

### Summary

A time series is different from cross-sectional data because the ordering of the observations conveys important information. Methods from previous chapters (confidence intervals and hypothesis tests, for example) must be modified before they will work with time-series data because a time series is usually not a random sample from a population.

The primary goal of time-series analysis is to create forecasts, that is, to *predict the future*. These are based on a **model** (also called a **mathematical model** or a **process**), which is a system of equations that can produce an assortment of different artificial time-series data sets. A **forecast** is the expected (i.e., mean) value of the future behavior of the estimated model. Like all estimates, forecasts are usually wrong. The **forecast limits** are the confidence limits for your forecast (if the model can produce them); if the model is correct for your data, then the future observation has a 95% probability, for example, of being within these limits.

**Trend-seasonal analysis** is a direct, intuitive approach to estimating the four basic components of a monthly or quarterly time series: the long-term trend, the seasonal patterns, the cyclic variation, and the irregular component. The long-term **trend** indicates the *very* long-term behavior of the time series, typically as a straight line or an exponential curve. The exactly repeating **seasonal component** indicates the effects of the time of year. The medium-term **cyclic component** consists of the gradual ups and downs that do not repeat each year. The short-term, random **irregular component** represents the leftover, residual variation that can't be explained. The formula for the trend-seasonal time-series model is

$$\text{Data} = \text{Trend} \times \text{Seasonal} \times \text{Cyclic} \times \text{Irregular}$$

The **ratio-to-moving-average** method divides the series by a smooth moving average as follows:

1. A **moving average** is a new series created by averaging nearby observations. We use a year of data in each average so that the seasonal component is eliminated:

$$\text{Moving average} = \text{Trend} \times \text{Cyclic}$$

2. Dividing the series by the smoothed moving-average series produces the ratio-to-moving-average, a combination of seasonal and irregular values. Grouping by time of year and then averaging produces the **seasonal index** for each

time of the year, indicating how much larger or smaller a particular time period is compared to a typical period during the year. **Seasonal adjustment** eliminates the expected seasonal component from an observation (by dividing the series by the seasonal index for that period) so that one quarter or month may be directly compared to another (after seasonal adjustment) to reveal the underlying trends:

$$(\text{Seasonal})(\text{Irregular}) = \frac{\text{Data}}{\text{Moving average}}$$

$$\text{Seasonal index} = \text{Average of} \left( \frac{\text{Data}}{\text{Moving average}} \right) \text{for that season}$$

$$\text{Seasonally adjusted value} = \left( \frac{\text{Data}}{\text{Seasonal index}} \right)$$

$$= \text{Trend} \times \text{Cyclic} \times \text{Irregular}$$

3. A regression of the seasonally adjusted series ($Y$) on time ($X$) is used to estimate the long-term trend as a straight line over time and to provide a seasonally adjusted forecast. This is appropriate only if the long-term trend in your series is linear.

4. Forecasting may be done by *seasonalizing the trend*, i.e., multiplying it by the appropriate seasonal index.

The **Box–Jenkins ARIMA processes** form a family of linear statistical models based on the normal distribution that have the flexibility to imitate the behavior of many different real time series by combining *autoregressive (AR) processes, integrated (I) processes*, and *moving-average (MA) processes*. The result is a **parsimonious model**, that is, one that uses just a few estimated parameters to describe the complex behavior of a time series. Here is an outline of the steps involved:

1. A process is selected from the Box–Jenkins family of ARIMA processes that generates data with the same overall look as your series, except for randomness.

2. The forecast at any time is the expected (i.e., average or mean) future value of the estimated process at that time.

3. The standard error of a forecast at any time is the standard deviation of the future value of the estimated process at that time.

4. The forecast limits extend above and below the forecast value, so that there is a 95% chance, for example, that the future value of the estimated process will fall within the forecast limits at any time. This assumes that the future behavior of the series will be similar to that of the estimated process.

A **random noise process** consists of a random sample (independent observations) from a normal distribution with constant mean and standard deviation. The average is the best forecast for any future time period, and the ordinary prediction interval for a new observation gives you the forecast limits for any future value of the series. The formula for the random noise process is

$$\text{Data} = \text{Mean value} + \text{Random noise}$$

$$Y_t = \mu + \varepsilon_t$$

The long-term mean of $Y$ is $\mu$.

An observation of an **autoregressive process** consists of a linear function of the previous observation plus independent random noise. Forecasting is done using predicted values from the estimated regression equation $\hat{\delta} + \hat{\varphi}Y_t$. The forecast is a compromise between the most recent data value and the long-term mean value of the series. The further into the future you look, the closer to the long-term mean value your forecast will be. The formula is

$$\text{Data} = \delta + \varphi(\text{Previous value}) + \text{Random noise}$$

$$Y_t = \delta + \varphi Y_{t-1} + \varepsilon_t$$

The long-term mean value of $Y$ is $\delta/(1 - \varphi)$.

An observation of a **moving-average process** consists of a constant, $\mu$ (the long-term mean of the process), plus independent random noise minus a fraction of the previous random noise:

$$\text{Data} = \mu + \varphi(\text{Random noise}) - \theta(\text{Previous random noise})$$

$$Y_t = \mu + \varepsilon_t - \theta\varepsilon_{t-1}$$

where the long-term mean value of $Y$ is $\mu$. This produces a moving average of two observations at a time from a random noise process. Forecasting the next observation is based on an estimate of the current random noise, $\varepsilon_t$; beyond this, the best forecast is the estimated long-term mean.

An observation of an **autoregressive moving-average (ARMA) process** consists of a linear function of the previous observation, plus independent random noise, minus a fraction of the previous random noise. This combines an autoregressive with a moving-average process:

$$\text{Data} = \delta + \varphi(\text{Previous value}) + (\text{Random noise})$$

$$- \theta(\text{Previous random noise})$$

$$Y_t = \delta + \varphi Y_{t-1} + \varepsilon_t - \theta\varepsilon_{t-1}$$

where the long-term mean value of $Y$ is $\delta/(1 - \varphi)$. Forecasting the next observation is done by combining the predicted value from the estimated autoregression equation $\hat{\delta} + \hat{\varphi}Y_t$ with an estimate of the current random noise, $\hat{\varepsilon}_t$. Beyond this, the best forecast is based only on the previous forecast value. The further in the future you look, the closer to the long-term mean value your forecast will be.

An observation of a **pure integrated (I) process**, also called a **random walk**, consists of a random step away from the current observation. A random walk is said to be a **nonstationary process** because it tends to move farther and farther away from where it was. In contrast, the autoregressive, moving-average, and ARMA models are **stationary processes** because they tend to behave similarly over long time periods, staying relatively close to their long-run means. Forecasting the next observation with a random walk is done by adding an estimate of $\delta$, the drift term, to the current observation for each additional period in the future. For the pure integrated (random walk) process,

$$\text{Data} = \delta + (\text{Previous value}) + (\text{Random noise})$$

$$Y_t = \delta + Y_{t-1} + \varepsilon_t$$

Over time, $Y$ is not expected to stay close to any long-term mean value. For the pure integrated (random walk) process in differenced form,

$$\text{Data} - (\text{Previous value}) = \delta + (\text{Random noise})$$

$$Y_t - Y_{t-1} = \delta + \varepsilon_t$$

If the changes or differences of a series are generated by an autoregressive moving-average (ARMA) process, then the series follows an **autoregressive integrated moving-average (ARIMA) process**. These are nonstationary processes: Over time, the series will tend to move farther and farther away from where it was. Forecasting with an ARIMA model is done by forecasting the changes of the ARMA model for the differences. Due to nonstationarity, the forecasts can tend indefinitely upward (or downward), and the forecast limits widen as you extend further into the future. Here is the formula for the ARIMA process in differenced form:

$$\text{Data change} = \delta + \varphi(\text{Previous change}) + (\text{Random noise})$$

$$- \theta(\text{Previous random noise})$$

$$Y_t - Y_{t-1} = \delta + \varphi(Y_{t-1} - Y_{t-2}) + \varepsilon_t - \theta\varepsilon_{t-1}$$

The long-term mean value of the *change* in $Y$ is $\delta/(1 - \varphi)$. Over time, $Y$ is not expected to stay close to any long-term mean value.

More advanced ARIMA models can be created to include the seasonal behavior of quarterly and monthly series.

# Key Words

autoregressive (AR) process, *448*
autoregressive integrated moving-average (ARIMA) process, *454*
autoregressive moving-average (ARMA) process, *451*
Box–Jenkins ARIMA processes, *448*
cyclic component, *436*
forecast, *430*

## Questions

1. a. How is a time series different from cross-sectional data?
   b. What information is lost when you look at a histogram for time-series data?
2. a. What is a forecast?
   b. What are the forecast limits?
   c. What role does a mathematical model play in forecasting?
   d. Why doesn't trend-seasonal analysis produce forecast limits?
3. a. Name the four basic components of a monthly or quarterly time series, from the trend-seasonal approach.
   b. Carefully distinguish the cyclic and the irregular components.
4. a. How is the moving average different from the original series?
   b. For trend-seasonal analysis, why do we use exactly one year of data at a time in the moving average?
   c. Which components remain in the moving average? Which are reduced or eliminated?
5. a. How do you compute the ratio-to-moving-average? Which components does it represent?
   b. What do you do to the ratio-to-moving-average to produce a seasonal index? Why does this work?
   c. What does a seasonal index represent?
   d. How do you seasonally adjust a time-series value? How do you interpret the result?
6. a. How is a linear trend estimated in trend-seasonal analysis?
   b. What kind of forecast does the linear trend represent?
   c. What do you do to produce a forecast from the linear trend?
   d. Which components are represented in this forecast? Which are missing?
7. a. How is the flexibility of the Box–Jenkins ARIMA process approach helpful in time-series analysis?
   b. What is parsimony?

c. How does the forecast relate to the actual future behavior of the estimated process?
   d. How do the forecast limits relate to the actual future behavior of the estimated process?
8. a. Define the random noise process in terms of the relationship between successive observations.
   b. Comment on the following: If it's a random noise process, then special time-series methods aren't needed to analyze it.
   c. What are the forecast and forecast limits for a random noise process?
9. a. Define a first-order autoregressive process in terms of the relationship between successive observations.
   b. What are the X and Y variables in the regression model to predict the next observation in a first-order autoregressive process?
   c. Describe the forecasts of an autoregressive process in terms of the most recent data observation and the long-run mean value for the estimated model.
10. a. Define a first-order moving-average process in terms of the relationship between successive observations.
    b. What is a moving-average process a moving average of?
    c. Describe the forecasts for two or more periods into the future of a first-order moving-average process in terms of the long-run mean value for the estimated model.
11. a. Define a first-order ARMA process in terms of the relationship between successive observations.
    b. What parameter value would you set equal to zero in an ARMA process in order to have an autoregressive process?
    c. What parameter value would you set equal to zero in an ARMA process in order to have a moving-average process?
    d. Describe the forecasts for the distant future based on an ARMA process.
12. a. Define a random walk in terms of the relationship between successive observations.
    b. Carefully distinguish a random noise process from a random walk.
    c. Comment on the following: If it's a random walk, then special time-series methods aren't needed to analyze it.
    d. What is the effect of the drift term in a random walk?
    e. Describe the forecasts for a random walk process.
13. Distinguish stationary and nonstationary time-series behavior.
14. For each of the following, say whether it is stationary or nonstationary:
    a. Autoregressive process.
    b. Random walk.
    c. Moving-average process.
    d. ARMA process.
15. a. Define a first-order ARIMA process in terms of the relationship between successive observations.
    b. What parameter values would you set equal to zero in an ARIMA process in order to have a random walk?

c. How can you construct an ARMA process from an ARIMA process?

d. Describe the forecasts for the distant future based on an ARIMA process.

16. What kinds of additional terms are needed to include seasonal behavior in advanced ARIMA models?

## Problems

*Problems marked with an asterisk (*) are solved in the Self Test in Appendix C.*

1. For each of the following, tell whether or not you would expect it to have a strong seasonal component and why.
   a. Sales of colorful wrapping paper, recorded monthly.
   b. The number of air travelers to Hawaii from Chicago, recorded monthly.
   c. The S&P 500 stock market index, recorded daily. Assume that the stock market is efficient, so that any foreseeable trends have already been eliminated through the action of large investors attempting to profit from them.

2. You have suspected for some time that production problems tend to flare up in the wintertime, during the first quarter of each year. A trend-seasonal analysis of the defect rates indicates seasonal indices of 1.00, 1.01, 1.03, and 0.97 for quarters 1, 2, 3, and 4, respectively. Does this analysis support the view that defects are highest in the first quarter? If yes, justify your answer. If no, is there a quarter you should look at instead?

3. A bank had 38,091 ATM network transactions at its cash machines in January and had 43,182 in February. The seasonal indices are 0.925 for January and 0.986 for February.
   a. By what percent did ATM transactions increase from January to February?
   b. By what percent would you have expected ATM transactions to increase from January to February? (*Hint:* Use the seasonal indices.)
   c. Find seasonally adjusted transaction levels for each month.
   d. By what percent did ATM transactions increase (or decrease) from January to February on a seasonally adjusted basis?

4. At a meeting, everyone seems to be pleased by the fact that sales increased from $21,791,000 to $22,675,000 from the third to the fourth quarter. Given that the seasonal indices are 1.061 for quarter 3 and 1.180 for quarter 4, write a paragraph analyzing the situation on a seasonally adjusted basis. In particular, is this good news or bad news?

5. Which time-series method of analysis would be most appropriate to a situation in which forecasts and confidence limits are needed for a data set that shows medium-term cyclic behavior?

6. Which time-series method of analysis would be most appropriate to a situation in which prices are lower at harvest time in the fall but are typically higher the rest of the year and in which there is a need for a methodology that is relatively easy to understand?

7. Consider the Walt Disney Company's quarterly revenues as shown in Table 14.4.1.

### TABLE 14.4.1 Quarterly Revenues for Walt Disney Company and Subsidiaries

| Year | Revenues (millions) |
|------|---------------------|
| 1997 | $5,520 |
| 1997 | 5,194 |
| 1997 | 5,481 |
| 1997 | 6,278 |
| 1998 | 6,147 |
| 1998 | 5,248 |
| 1998 | 5,242 |
| 1998 | 6,339 |
| 1999 | 5,791 |
| 1999 | 5,531 |
| 1999 | 5,516 |
| 1999 | 6,597 |
| 2000 | 6,118 |
| 2000 | 6,053 |
| 2000 | 6,307 |
| 2000 | 6,940 |
| 2001 | 5,812 |
| 2001 | 5,975 |
| 2001 | 6,049 |
| 2001 | 7,433 |

Source: Annual reports on the Internet, accessed at http://disney .go.com/corporate/investors/financials/annual.html on February 10, 2002.

a. Draw a time-series plot for this data set. Describe any trend and seasonal behavior that you see.
b. Find the moving average values and plot them on the same graph as the original data. Comment on what you see.
c. Find the seasonal index for each quarter. In particular, how much higher is the fourth quarter than a typical quarter during the year?
d. Find the seasonally adjusted values and plot them with the original data. Comment on what you see.
e. From fourth quarter 1997 to first quarter 1998, revenues fell from 6,278 to 6,147. What happened on a seasonally adjusted basis?

f. Find the regression equation to predict the long-term trend in seasonally adjusted sales for each time period, using 1, 2, ... for the X variable.

g. Compute the seasonally adjusted forecast for the fourth quarter of 2003.

h. Compute the forecast for the first quarter of 2004.

8. Consider Intel's Revenues in Table 14.4.2.

**TABLE 14.4.2 Quarterly Revenues for Intel**

| Year | Net Sales (billions) | Year | Net Sales (billions) |
|------|------|------|------|
| 2006 | $8.940 | 2007 | $10.090 |
| 2006 | 8.009 | 2007 | 10.712 |
| 2006 | 8.739 | 2008 | 9.673 |
| 2006 | 9.694 | 2008 | 9.470 |
| 2007 | 8.852 | 2008 | 10.217 |
| 2007 | 8.680 | 2008 | 8.226 |

**Source:** U.S. Securities and Exchange Commission, 10-K filings, accessed at http://www.sec.gov/cgi-bin/browse-edgar on October 13, 2010.

a. Construct a time-series plot for this data set. Describe the seasonal and cyclic behavior that you see, as well as any evidence of irregular behavior.

b. Which quarter (1, 2, 3, or 4) appears to be Intel's best in terms of sales level, based on your plot in part a?

c. Is the seasonal pattern (in your graph for part a) consistent across the entire time period?

d. Calculate the moving average (using one year of data at a time) for this time series. Construct a time-series plot with both the data and the moving average.

e. Describe the cyclic behavior revealed by the moving average.

f. Find the seasonal index for each quarter. Do these values appear reasonable compared to the time-series plot of the data?

g. Find the seasonally adjusted sales corresponding to each of the original sales values. Construct a time-series plot of this seasonally adjusted series.

h. Do you see an overall linear long-term trend up or down in these sales data? Would it be appropriate to use a regression line for forecasting this series?

i. Intel's revenue fell from 8.852 to 8.680 from the first to the second quarter of 2007. What happened to revenue on a seasonally adjusted basis?

9.* Table 14.4.3 shows the quarterly net sales of Mattel, a major designer, manufacturer, and marketer of toys. Because of seasonal gift giving, you might expect fourth-quarter sales to be much higher, generally, than those of the other three quarters of the year.

**TABLE 14.4.3 Quarterly Net Sales for Mattel**

| Year | Net Sales (millions) | Year | Net Sales (millions) |
|------|------|------|------|
| 2005 | $783.120 | 2006 | $1,790.312 |
| 2005 | 886.823 | 2006 | 2,108.842 |
| 2005 | 1,666.145 | 2007 | 940.265 |
| 2005 | 1,842.928 | 2007 | 1,002.625 |
| 2006 | 793.347 | 2007 | 1,838.574 |
| 2006 | 957.655 | 2007 | 2,188.626 |

**Source:** U.S. Securities and Exchange Commission, 10-K filings, accessed at http://www.sec.gov/cgi-bin/browse-edgar on October 14, 2010.

a. Construct a time-series plot for this data set. Describe any trend and seasonal behavior that you see in the plot.

b. Calculate the moving average (using one year of data at a time) for this time series. Construct a time-series plot with both the data and the moving average.

c. Find the seasonal index for each quarter. Do these values appear reasonable when you look at the time-series plot of the data?

d. Which is Mattel's best quarter (1, 2, 3, or 4)? On average, how much higher are sales as compared to a typical quarter during the year?

e. Find the seasonally adjusted sales corresponding to each of the original sales values.

f. From the third to the fourth quarter of 2007, sales went up from 1,838.574 to 2,188.626. What happened on a seasonally adjusted basis?

g. From the second to the third quarter of 2007, Mattel's sales rose considerably, from 1,002.625 to 1,838.574. What happened on a seasonally adjusted basis?

h. Find the regression equation to predict the long-term trend in seasonally adjusted sales for each time period, using 1, 2, ... for the X variable.

i. Compute the seasonally adjusted forecast for the fourth quarter of 2009.

j. Compute the forecast for the fourth quarter of 2009.

k. Compare the forecast from part i to Mattel's actual net sales of 1,955.128 for the fourth quarter of 2009. Is your result consistent with the possibility that the recession from December 2007 through June 2009 might have disrupted the sales pattern?

10. Amazon.com is an e-commerce firm that has shown considerable growth since its founding in 1995, and its quarterly sales are shown in Table 14.4.4. Their 2009 annual report includes a section titled "Seasonality" that states: "Our business is affected by seasonality, which historically has resulted in higher sales volume during our fourth quarter, which ends December 31."

### TABLE 14.4.4 Quarterly Sales for Amazon.com

| Year | Quarter | Revenue (billions) |
|------|---------|--------------------|
| 2003 | 1 | $1.084 |
| 2003 | 2 | 1.100 |
| 2003 | 3 | 1.134 |
| 2003 | 4 | 1.946 |
| 2004 | 1 | 1.530 |
| 2004 | 2 | 1.387 |
| 2004 | 3 | 1.462 |
| 2004 | 4 | 2.541 |
| 2005 | 1 | 1.902 |
| 2005 | 2 | 1.753 |
| 2005 | 3 | 1.858 |
| 2005 | 4 | 2.977 |
| 2006 | 1 | 2.279 |
| 2006 | 2 | 2.139 |
| 2006 | 3 | 2.307 |
| 2006 | 4 | 3.986 |
| 2007 | 1 | 3.015 |
| 2007 | 2 | 2.886 |
| 2007 | 3 | 3.262 |
| 2007 | 4 | 5.673 |
| 2008 | 1 | 4.135 |
| 2008 | 2 | 4.063 |
| 2008 | 3 | 4.264 |
| 2008 | 4 | 6.704 |
| 2009 | 1 | 4.889 |
| 2009 | 2 | 4.651 |
| 2009 | 3 | 5.449 |
| 2009 | 4 | 9.519 |
| 2010 | 1 | 7.131 |
| 2010 | 2 | 6.566 |

Source: U.S. Securities and Exchange Commission, 10-Q filings, accessed at http://www.sec.gov/cgi-bin/browse-edgar on July 26, 2010.

a. Construct a time-series plot for this data set. Do you agree that there are seasonal factors present here?

b. Calculate the moving average (using one year of data at a time) for this time series. Construct a time-series plot with both the data and the moving average.

c. Describe any cyclic behavior that you see in the moving average.

d. Find the seasonal index for each quarter. Do these values appear reasonable when you look at the time-series plot of the data?

e. Which is Amazon.com's best quarter (1, 2, 3, or 4)? On average, how much higher are sales then as compared to a typical quarter during the year?

f. Which is Amazon.com's worst quarter (1, 2, 3, or 4)? On average, how much lower are sales then as compared to a typical quarter during the year?

g. Find the seasonally adjusted sales corresponding to each of the original sales values. Construct a time-series plot of this seasonally adjusted series.

h. Describe the behavior of the seasonally adjusted series. In particular, identify any variations in growth rate that are visible over this time period.

11. Consider PepsiCo's quarterly net sales as shown in Table 14.4.5.

### TABLE 14.4.5 Quarterly Net Sales for PepsiCo

| Year | Net Sales (millions) |
|------|----------------------|
| 1996 | $4,053 |
| 1996 | 5,075 |
| 1996 | 5,159 |
| 1996 | 6,050 |
| 1997 | 4,213 |
| 1997 | 5,086 |
| 1997 | 5,362 |
| 1997 | 6,256 |
| 1998 | 4,353 |
| 1998 | 5,258 |
| 1998 | 5,544 |
| 1998 | 7,193 |
| 1999 | 5,114 |
| 1999 | 4,982 |
| 1999 | 4,591 |
| 1999 | 5,680 |
| 2000 | 4,191 |
| 2000 | 4,928 |
| 2000 | 4,909 |
| 2000 | 6,410 |

Source: Annual reports on the Internet accessed at www.pepsico.com on February 12, 2002.

a. Draw a time-series plot for this data set. Describe any trend and seasonal behavior that you see.

b. Plot the moving average values on the same graph as the original data. Comment on what you see.

c. Find the seasonal index for each quarter. Which is generally the best quarter for PepsiCo? About how much larger are net sales in this quarter, as compared to a typical quarter?

d. Plot the seasonally adjusted series with the original data.

e. Find the regression equation to predict the long-term trend in seasonally adjusted sales for each time period, using 1, 2, … for the X variable.

f. Does PepsiCo show a significant trend (either up or down) over this time period as indicated by the regression analysis in the previous part of this problem?

12. Based on past data, your firm's sales show a seasonal pattern. The seasonal index for November is 1.08, for December it is 1.38, and for January it is 0.84. Sales for November were $285,167.

a. Would you ordinarily expect an increase in sales from November to December in a typical year? How do you know?

b.* Find November's sales, on a seasonally adjusted basis.

c.* Take the seasonally adjusted November figure and seasonalize it using the December index to find the expected sales level for December.

d. Sales for December have just been reported as $430,106. Is this higher or lower than expected, based on November's sales?

e. Find December's sales, on a seasonally adjusted basis.

f. Were sales up or down from November to December, on a seasonally adjusted basis? What does this tell you?

g. Using the same method as in part c, find the expected level for January sales based on December's sales.

13. The number of diners per quarter eating at your après-ski restaurant has been examined using trend-seasonal analysis. The quarterly seasonal indexes are 1.45, 0.55, 0.72, and 1.26 for quarters 1, 2, 3, and 4, respectively. A linear trend has been estimated as 5,423 + 408 (Quarter number), where the quarter number starts at 1 in the first quarter of 2007 and increases by 1 each successive quarter.

a.* Find the seasonally adjusted forecast value for the first quarter of 2014.

b. Find the seasonally adjusted forecast value for the second quarter of 2014.

c. Why is the seasonally adjusted forecast larger in the second quarter, in which you would expect fewer skiers coming to dinner?

d.* Find the forecast value for the first quarter of 2014.

e. Find the forecast value for the second quarter of 2014.

f. On a seasonally adjusted basis, according to this estimated linear trend, how many more diners do you expect to serve each quarter compared to the previous quarter?

g. Your strategic business plan includes a major expansion project when the number of diners reaches 80,000 per year. In which calendar year will this first happen, according to your forecasts? (Hint: Compute and add the four forecasts for each year to find yearly totals for 2015 and 2016.)

14. Consider the time series of quarterly sales in thousands shown in Table 14.4.6. The seasonal indices are 0.89 for quarter 1, 0.88 for 2, 1.27 for 3, and 0.93 for 4.

a. Find the seasonally adjusted sales corresponding to each sales value.

b. In which quarter is the most business generally done?

c. As indicated in the data, sales increased from 817 to 1,073 in 2010 from quarters 2 to 3. What happened during this period on a seasonally adjusted basis?

**TABLE 14.4.6 Quarterly Sales**

| Quarter | Year | Sales (thousands) | Quarter | Year | Sales (thousands) |
|---------|------|-------------------|---------|------|-------------------|
| 1 | 2007 | $438 | 1 | 2009 | $676 |
| 2 | 2007 | 432 | 2 | 2009 | 645 |
| 3 | 2007 | 591 | 3 | 2009 | 1,084 |
| 4 | 2007 | 475 | 4 | 2009 | 819 |
| 1 | 2008 | 459 | 1 | 2010 | 710 |
| 2 | 2008 | 506 | 2 | 2010 | 817 |
| 3 | 2008 | 736 | 3 | 2010 | 1,073 |
| 4 | 2008 | 542 | | | |

d. As indicated in the data, sales decreased from 1,084 to 819 in 2009 from quarters 3 to 4. What happened during this period on a seasonally adjusted basis?

e. The exponential trend values for the four quarters of 2014 are 1,964, 2,070, 2,183, and 2,301. Seasonalize these trend forecasts to obtain actual sales forecasts for 2014.

15. Which type of time-series analysis would provide the simplest results for studying demand for heating oil, which tends to be highest in the winter?

16. Your seasonally adjusted monthly sales forecast is $382,190 + $4,011(Month number), where the month number is 1 for January 2006 and increases by 1 each month. The seasonal index for February sales is 0.923, and it is 1.137 for April. What you need now is a forecast for cost of goods sold in order to plan ahead for filling future orders. You have found that monthly sales have been a good predictor of monthly cost of goods sold and have estimated the following regression equation:

Predicted cost of good sold $= \$106{,}582 + 0.413$ (Sales)

a. Find the seasonally adjusted forecast of monthly sales for February 2013.

b. Find the forecast of monthly sales for February 2013.

c. Find the forecast of cost of goods sold for February 2013.

d. Find the forecast of cost of goods sold for April 2014.

17. For each of the following, tell whether it is likely to be stationary or nonstationary and why.

a. The price per share of IBM stock, recorded daily.

b. The prime rate, recorded weekly. This is the interest rate that banks charge their best customers for their loans.

c. The thickness of paper, measured five times per minute as it is being produced and rolled, assuming that the process is in control.

d. The price of a full-page advertisement in TV Guide, recorded each year.

18. Table 14.4.7 shows basic computer results from a Box–Jenkins analysis of the daily percentage changes in the Dow Jones Industrial stock market index from July 31 to October 9, 1987, prior to the crash of 1987.

a. What kind of process has been estimated?

b. Write the model in a way that shows how the next observation is determined from the previous one. Use the actual estimated coefficients.

c. Which estimated coefficients are statistically significant?

d. Using 0 in place of all estimated coefficients that are not statistically significant, write down the model that shows how the next observation is determined from the previous one. What kind of process is this?

e. Write a brief paragraph summarizing your results as support for the random walk theory of market behavior.

19. Help-wanted advertising is used by businesses when they are caught short and need workers quickly, as might be expected to happen when business activity picks up after a downturn. Table 14.4.8 shows a Box–Jenkins analysis of the help-wanted advertising index, while Figure 14.4.1 shows the data series with the Box–Jenkins forecasts.[20]

a. What kind of process has been estimated?

b. Which estimated coefficients (if any) are significant?

c. Based on the figure, would you be surprised if the help-wanted advertising index dropped to 15 in the year 1995?

d. Based on the figure, would you be surprised if the help-wanted advertising index rose to 120 in the year 1996?

e. The forecasts in the figure appear to level off after about 1995. Does this tell you that the help-wanted advertising index will stop changing from year to year in the future? Explain.

**TABLE 14.4.8 Results of a Box–Jenkins Analysis of the Help-Wanted Advertising Index**

| Final Estimates of Parameters | | | |
|---|---|---|---|
| Type | Estimate | StDev | t Ratio |
| AR 1 | 0.5331 | 0.1780 | 2.99 |
| MA 1 | −0.8394 | 0.1170 | −7.18 |
| Constant | 50.653 | 5.302 | 9.55 |
| Mean | 108.48 | 11.36 | |

No. of Obs.: 29

Residuals: SS = 6161.77 (Backforecasts excluded)

MS = 236.99 DF = 26

**TABLE 14.4.7 Results of a Box–Jenkins Analysis of Daily Percentage Changes in the Dow Jones Index**

| Coefficient | Estimate | Standard Error | t Ratio |
|---|---|---|---|
| Autoregression | −0.3724 | 1.7599 | −0.21 |
| Moving average | −0.4419 | 1.6991 | −0.26 |
| Constant | −0.000925 | 0.002470 | −0.37 |
| Mean | −0.000674 | 0.001799 | −0.37 |
| Standard deviation of random noise | 0.011950 | | |

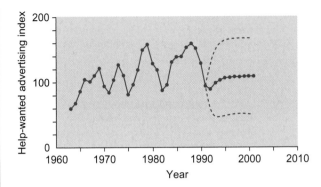

FIGURE 14.4.1 The help-wanted advertising index from 1963 to 1991, with forecasts and 95% intervals for 10 years further based on a Box–Jenkins time-series model.

**TABLE 14.4.9 Basic Results of a Box–Jenkins Analysis of U.S. Treasury Bill Interest Rates**

| Coefficient | Estimate | Standard Error | t Ratio | p-Value |
|---|---|---|---|---|
| Autoregression | 0.6901 | 0.1347 | 5.12 | 0.000 |
| Moving average | −0.7438 | 0.1164 | −6.39 | 0.000 |
| Constant | 1.6864 | 0.3953 | 4.27 | 0.000 |
| Mean | 5.441 | 1.275 | | |

**TABLE 14.4.10 Resulting Forecasts from a Box–Jenkins Analysis of U.S. Treasury Bill Interest Rates**

| Year | Forecast | 95% Forecast Limits | |
|---|---|---|---|
| | | Lower | Upper |
| 2010 | 1.366 | −1.427 | 4.158 |
| 2011 | 2.629 | −2.253 | 7.511 |
| 2012 | 3.501 | −2.110 | 9.110 |
| 2013 | 4.102 | −1.823 | 10.027 |
| 2014 | 4.517 | −1.553 | 10.587 |
| 2015 | 4.804 | −1.334 | 10.941 |
| 2016 | 5.001 | −1.168 | 11.170 |
| 2017 | 5.138 | −1.047 | 11.322 |
| 2018 | 5.232 | −0.960 | 11.423 |
| 2019 | 5.297 | −0.898 | 11.491 |
| 2020 | 5.341 | −0.855 | 11.538 |

**TABLE 14.4.11 Results of Box–Jenkins Analysis of Dividends as a Percentage of Profits**

| Final Estimates of Parameters | | | | |
|---|---|---|---|---|
| Type | Estimate | StDev | t Ratio | p-Value |
| AR1 | −0.0197 | 0.2869 | −0.07 | 0.946 |
| Constant | 0.009964 | 0.009142 | 1.09 | 0.292 |

Differencing: 1 regular difference

Number of observations: Original series 19, after differencing 18

Residuals: SS = 0.0236910 (backforecasts excluded)

MS = 0.0014807 DF = 16

20. Tables 14.4.9 and 14.4.10 show basic computer results from a Box–Jenkins analysis of yields on three-month U.S. Treasury bills each year from 1970 through 2009.
    a. What kind of process has been fitted?
    b. Write the model in a way that shows how the next observation is determined from the previous one.
    c. Which estimated coefficients are statistically significant?
    d. Draw a time-series plot of the original data (from Table 14.1.5), the forecasts, and the forecast limits.
    e. Comment on these forecasts and forecast limits.
21. Dividends, paid by corporations from their profits to their shareholders, have fluctuated over time as a percentage of profits and this percentage has risen to some extent over the years. Table 14.4.11 shows the computer results of a Box–Jenkins analysis of dividends as a percentage of corporate profits, annually from 1990 to 2008, while Figure 14.4.2 shows the data series with the Box–Jenkins forecasts.[21]
    a. What kind of component (autoregressive or moving-average) does the estimated model include?
    b. How many differences are used in the model?
    c. Is the model component that you identified in part a significant?

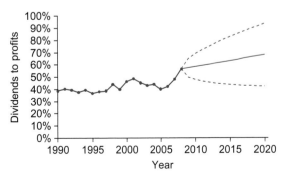

FIGURE 14.4.2 Dividends as a percentage of profits, from 1990 to 2008, with forecasts and 95% forecast intervals through 2020 based on a Box-Jenkins time-series model.

**d.** Is the constant term significant?

**e.** Based on the figure, would you be surprised if dividends dropped to 30% of profits in 2015?

**f.** Based on the figure, would you be surprised if dividends rose to 65% of profits in 2015?

20. Data are from U.S. Department of Commerce, Bureau of Economic Analysis, *Business Statistics 1963–91* (Washington, DC: U.S. Government Printing Office, June 1992), p. 59.
21. Data are from Table 767 of *U.S. Census Bureau, Statistical Abstract of the United States*: 2010 (129th Edition) Washington, DC, 2009, accessed at http://www.census.gov/compendia/statab/cats/business_enterprise.html on July 28, 2010. Box–Jenkins analysis is from MINITAB statistical software.

## Projects

**1.** Select a firm of interest to you and obtain at least three continuous years of quarterly sales figures from the firm's annual reports at your library or on the Internet.

**a.** Draw a time-series graph and comment on the structure you see.

**b.** Compute the one-year moving average, draw it on your graph, and comment.

**c.** Compute the seasonal indexes, graph them, and comment.

**d.** Compute and graph the seasonally adjusted series; then comment on what you see. In particular, what new information have you gained through seasonal adjustment?

**e.** Compute the trend line and seasonalize it to find forecasts for two years into the future. Graph these forecasts along with the original data. Comment on how reasonable these forecasts seem to you.

**2.** Find yearly data on a business or economic time series of interest to you for at least 20 continuous years. (This project requires access to computer software that can estimate ARIMA models.)

**a.** Graph this time series and comment on the structure you see.

**b.** Does the series appear to be stationary or nonstationary? If it is extremely nonstationary (ending up far away from where it started, for example), graph the differences to see if they appear to be stationary.

**c.** Fit a first-order autoregressive process to your series (or to the differences, if the series was nonstationary). Based on the $t$ statistic, is the autoregressive coefficient significant?

**d.** Fit a first-order moving-average process to your series (or to the differences, if the series was nonstationary). Based on the $t$ statistic, is the moving-average coefficient significant?

**e.** Fit a first-order ARMA process to your series (or to the differences, if the series was nonstationary). Based on the $t$ statistic, which coefficients are significant?

**f.** Based on the results for these three models, which one do you select? You may want to exclude components that are not significant.

**g.** Now work with the original series (even if you have been using differences). Estimate your chosen model, including an integrated (I) component if you have been differencing, and find forecasts and forecast limits.

**h.** Graph the forecasts and forecast limits along with the original data and comment.

**i.** Comment on the model selection procedure. (Keep in mind that the selection procedure is much more complicated when higher-order processes are used.)

# Methods and Applications

The selected topics of these last four chapters will show you how the ideas and methods of statistics can be applied to some special situations. The *analysis of variance* (ANOVA for short) is a collection of methods that compare the extent of variability due to various sources in order to perform hypothesis tests for complex situations. Chapter 15 introduces these methods and shows how ANOVA can be used to test whether or not several samples come from similar populations. In Chapter 16, *nonparametric* statistical methods (based on ranks) will show you how to test hypotheses in some difficult situations, for example, when the distributions are not normal or the data are merely *ordinal* (ordered categories) instead of quantitative (meaningful numbers). When you have only *nominal* data (unordered categories), the special testing methods of *chi-squared analysis* (see Chapter 17) are needed because you can't do arithmetic or ranking on these categories. Finally, Chapter 18 covers the basic statistical methods of *quality control:* how to decide which problems to solve, how to manage when there is variability in production, and how to decide when to fix things and when to leave well enough alone.

## Chapter 15

# ANOVA

Testing for Differences among Many Samples, and Much More

The **analysis of variance** (or **ANOVA**, for short) provides a general framework for statistical hypothesis testing based on careful examination of the different sources of variability in a complex situation. Here are some examples of situations in which ANOVA would be helpful:

**One:** In order to cut costs, you have tested five additives that claim to improve the yield of your chemical production process. You have completed 10 production runs for each additive and another 10 using no additive. This is an example of a *one-way design*, since a single factor ("additive") appears at several different levels. The result is a data set consisting of six lists of yields. The usual variability from one run to another makes it difficult to tell whether any improvements were just due to luck or whether the additives were truly better than the others. What you need is a test of the following null hypothesis: These six lists are really all the same, and any differences are just random. You can't just use a *t* test, from Chapter 10, because you have more than two samples.[1] Instead, the one-way analysis of variance will tell you whether there are any significant (i.e., systematic or nonrandom) differences among these additives. If significant differences exist, you may examine them in detail. Otherwise, you may

conclude that there are no detectable systematic differences among the additives.

**Two:** It occurs to you that you could run *combinations* of additives for the situation just described. With five additives, there are $2^5 = 32$ possible combinations (including no additives), and you have run each combination twice.[2] This is an example of a *factorial design* with five factors (the additives) each at two levels (either used or not used). The analysis of variance for this data set would indicate (a) whether or not each additive has a significant effect on yield and (b) whether there are any significant interactions resulting from combinations of additives.

---

1. You might consider using an unpaired *t* test to compare the additives two at a time. However, there are 15 such tests, and this *group* of tests is no longer valid because the probability of error is no longer controlled. In particular, assuming validity of the null hypothesis of no differences, the probability of wrongly declaring *some pair* of yields to be significantly different could be much higher than the 5% error rate that applies to an *individual* test. The *F* test will keep the error rate at 5%, and you may use modified *t* tests if the *F* test is significant.
2. It is a good idea to run each case more than once, if you can afford it, because you will then have more information about the variability in each situation.

**Three:** As part of a marketing study, you have tested three types of media (newspaper, radio, and television) in combination with two types of ads (direct and indirect approach). Each person in the study was exposed to one combination, and a score was computed reflecting the effectiveness of the advertising. This is an example of a *two-way design* (the factors are "media" and "type of ad"). The analysis of variance will tell you (a) whether the media types have significantly different effectiveness, (b) whether the two types of ads have significantly different effectiveness, and (c) whether there are any significant interactions between media and type of ad.

The analysis of variance uses an **F test** based on the **F statistic**, a ratio of two variance measures, to perform each hypothesis test.[3] The numerator represents the variability due to the special, interesting effect being tested, and the denominator represents a baseline measure of randomness. If the ratio is larger than the value in the *F* table, the effect is significant.

The **one-way analysis of variance**, in particular, is used to test whether or not the averages from several different situations are significantly different from one another. This is the simplest kind of analysis of variance. Although more complex situations require more complicated calculations, the general ANOVA idea remains the same: to test significance by comparing one source of variability (the one being tested) against another source of variability (the underlying randomness of the situation).

## 15.1  USING BOX PLOTS TO LOOK AT MANY SAMPLES AT ONCE

Since the purpose of the analysis of variance is only to test hypotheses, it is up to you to remember to explore your data. You should examine statistical summaries (average and standard deviation, for example) and histograms or box plots for each list of numbers in your data set. The analysis of variance might tell you that there are significant differences, but you would also have to examine ordinary statistical summaries to actually see those estimated differences.

Box plots are particularly well suited to the task of comparing several distributions because unnecessary details are omitted, allowing you to concentrate on the essentials. Here is a checklist of things to look for when using box

plots or histograms to compare similar measurements across a variety of situations:

1. Do the box plots look reasonable? You might as well spot trouble *before* you spend a lot more time working with the data set. For example, you might discover that you've called up the wrong data set. (Do the numbers seem to be much too big or too small? Are these last year's data?) You might also spot some outliers that could be examined and, if they are errors, corrected.

2. Do the centers (medians) appear different from one box plot to another? This provides an initial, informal assessment for which the analysis of variance will provide an exact, formal answer. Also, do the centers show any patterns of special interest?

3. Is the variability reasonably constant from one box plot to another? This is important because the analysis of variance will assume that these variabilities are equal in the population. If, for example, the higher boxes (with larger medians) are systematically wider (indicating more variability), then the analysis of variance may give incorrect answers.[4]

### Example
#### Comparing the Quality of Your Suppliers' Products
Your firm currently purchases the same electronic components from three different suppliers, and you are concerned. Although some of your associates contend that this arrangement allows your firm to get good prices and fast delivery times, you are troubled by the fact that products must be designed to work with the *worst* combination of components that might be installed. The overriding concern is with costs and benefits. In particular, would the firm be better off negotiating an exclusive contract with just one supplier to obtain faster delivery of high-quality components at a higher price? As part of the background information on this question, you have been looking at the quality of components delivered by each supplier.

The data set has just arrived. You had asked your firm's QA (quality assurance) department to check out 20 components from each supplier, randomly selected from recent deliveries. The QA staff actually tested 21 components for each, but not all produced a reliable measurement. The quality score is a composite of several different measurements, on a scale of 0 to 100, indicating the extent of agreement with the

---

3. Recall that the variance is the square of the standard deviation. This is the accepted way to proceed with ANOVA. Had the history of statistics developed differently, we might be comparing ratios of standard deviations to tables containing the square roots of our current *F* tables. However, the variance method is firmly established by tradition, and we will use it here.

4. This problem of unequal variability can often be fixed by transforming the original data values, for example, using logarithms if all of your data values are positive numbers. Examine the box plots of the transformed data to see if the problem has been fixed. If the analysis of variance finds significant differences on the log scale, you would conclude that the original groups also show significant differences. Thus, the interpretation of the results of an analysis of variance remains much the same even when you transform, provided that you use the same transformation on all of your data.

specifications of the component and your firm's needs. Higher scores are better, and a score of 75 or higher is sufficient for many applications. Table 15.1.1 shows the data set, together with some basic statistical summaries.

On average, Consolidated has the highest quality score (87.7), followed by Amalgamated (82.1), and finally Bipolar (80.7). The box plots in Figure 15.1.1 also suggest that Consolidated's products are generally higher in quality than the others, although there is considerable overlap and the highest quality component actually came from Amalgamated (with a score of 97). Much additional information is also provided by these box plots: No component achieved perfect quality (a score of 100), and the variability is similar from one supplier to another (as indicated by the size of each box).

Although Consolidated appears to have the highest quality, you wonder if that could just be due to the random selection of these particular components. If you could examine many more components from each supplier, would Consolidated still be the best, on average? The analysis of variance provides an answer without requiring the prohibitive cost of obtaining more data.

**TABLE 15.1.1 Quality Scores for Suppliers' Products**

| | Amalgamated | Bipolar | Consolidated |
|---|---|---|---|
| | 75 | 94 | 90 |
| | 72 | 87 | 86 |
| | 87 | 80 | 92 |
| | 77 | 86 | 75 |
| | 84 | 80 | 79 |
| | 82 | 67 | 94 |
| | 84 | 86 | 95 |
| | 81 | 82 | 85 |
| | 78 | 86 | 86 |
| | 97 | 82 | 92 |
| | 85 | 72 | 92 |
| | 81 | 77 | 85 |
| | 95 | 87 | 87 |
| | 81 | 68 | 86 |
| | 72 | 80 | 92 |
| | 89 | 76 | 85 |
| | 84 | 68 | 93 |
| | 73 | 86 | 89 |
| | | 74 | 83 |
| | | 86 | |
| | | 90 | |
| Average | $\overline{X}_1 = 82.055556$ | $\overline{X}_2 = 80.666667$ | $\overline{X}_3 = 87.684211$ |
| Standard deviation | $S_1 = 7.124706$ | $S_2 = 7.598245$ | $S_3 = 5.228688$ |
| Sample size | $n_1 = 8$ | $n_2 = 21$ | $n_3 = 19$ |

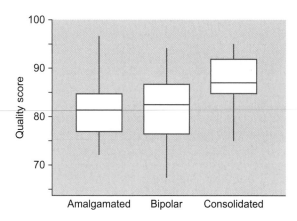

**FIGURE 15.1.1** Box plots of the quality of components purchased from each of your three suppliers. Amalgamated and Bipolar are quite similar (although the downside appears worse for Bipolar). Consolidated appears to have consistently higher quality, although there is considerable overlap in the distribution of these quality scores.

## 15.2 THE *F* TEST TELLS YOU IF THE AVERAGES ARE SIGNIFICANTLY DIFFERENT

The *F* test for the one-way analysis of variance will tell you whether the averages of several independent samples are significantly different from one another. This replaces the unpaired *t* test (from Chapter 10) when you have more than two samples and gives the identical result when you have exactly two samples.

### The Data Set and Sources of Variation

The data set for the one-way analysis of variance consists of *k* independent univariate samples, each one with the same measurement units (e.g., dollars or miles per gallon). The sample sizes may be different from one sample to another; this is permitted. The data set will then be of the form in Table 15.2.1.

One way to identify sources of variation is to ask the question, Why are these data values different from one another? There are two sources of variation here, so there are two answers:

1. One source of variation is the fact that the populations may be different from one another. For example, if sample 2 involves a particularly effective treatment, then the numbers in sample 2 will generally be different (higher) than the numbers in the other samples. This source is called the *between-sample variability*. The larger the between-sample variability, the more evidence you have of differences from one population to another.
2. The other source of variation is the fact that there is (usually) diversity within every sample. For example, you wouldn't expect all of the numbers in sample 2 to be the same. This source is called the *within-sample variability*. The larger the within-sample variability, the more random your situation is and the harder it is to tell whether or not the populations are actually different.

**TABLE 15.2.1 Data Set for One-Way ANOVA**

|  | Sample 1 | Sample 2 | ... | Sample *k* |
|---|---|---|---|---|
|  | $X_{1,1}$ | $X_{2,1}$ | ... | $X_{k,1}$ |
|  | $X_{1,2}$ | $X_{2,2}$ | ... | $X_{k,2}$ |
|  | . | . | . |  |
|  | . | . | . |  |
|  | . | . | . |  |
|  | $X_{1,n_1}$ | $X_{2,n_2}$ | ... | $X_{k,n_k}$ |
| Average | $\overline{X}_1$ | $\overline{X}_2$ | ... | $\overline{X}_k$ |
| Standard deviation | $S_1$ | $S_2$ | ... | $S_k$ |
| Sample size | $n_1$ | $n_2$ | ... | $n_k$ |

**Sources of Variation for a One-Way Analysis of Variance**
Between-sample variability (from one sample to another).
Within-sample variability (inside each sample).

For the supplier quality example, the two sources of variation are (1) the possibly different quality levels of the three suppliers and (2) the possibly different quality scores on different components from the same supplier.

The *F* test will be based on a ratio of measures of these sources of variation. But first we will consider the foundations of this hypothesis test.

### The Assumptions

The assumptions underlying the *F* test in the one-way analysis of variance provide a solid framework for an exact probability statement based on observed data.

**Assumptions for a One-Way Analysis of Variance**
1. The data set consists of *k* random samples from *k* populations.
2. Each population has a normal distribution, and the standard deviations of the populations are identical, so that $\sigma_1 = \sigma_2 = \cdots = \sigma_k$. This allows you to use standard statistical tables for hypothesis testing.

Note that there are no assumptions about the mean values of the normal distributions in the populations. The means are permitted to take on any values; the hypothesis-testing procedures will deal with them.

For the supplier quality example, the assumptions are that the QA department's data represent three random samples, one from the population of quality scores of each supplier. Furthermore, the distribution of quality scores is assumed to be normal for each supplier, and all three suppliers are assumed to have the same variability (population standard deviation) of quality score. The box plots examined earlier suggest that the assumptions about normal distributions and equal variability are reasonable and roughly satisfied by this data set.

## The Hypotheses

The null hypothesis for the $F$ test in the one-way analysis of variance claims that the $k$ populations (represented by the $k$ samples) all have the same mean value. The research hypothesis claims that they are *not* all the same, that is, at least two population means are different.

> **Hypotheses for a One-Way Analysis of Variance**
>
> **Null Hypothesis:**
>
> $$H_0 : \mu_1 = \mu_2 = \cdots = \mu_k \text{ (the means are all equal)}$$
>
> **Research Hypothesis:**
>
> $$H_1 : \mu_i \neq \mu_j \text{ for at least one pair of populations}$$
>
> (the means are *not* all equal)

Because the standard deviations are assumed to be equal in all of the populations, the null hypothesis actually claims that *the populations are identical* (in distribution) to each other. The research hypothesis claims that some differences exist, regardless of the number of populations that actually differ. That is, the research hypothesis includes the cases in which just one population is different from the others, in which several are different, and in which all are different.

For the supplier quality example, the null hypothesis claims that the three suppliers have identical quality characteristics: Their components have the same distribution of quality scores (the same normal distribution with the same mean and standard deviation). The research hypothesis claims that there are some supplier differences in terms of mean quality level (they might all three be different, or two might be the same with the third one either higher or lower).

## The $F$ Statistic

The $F$ statistic for a one-way analysis of variance is the ratio of variability measures for the two sources of variation: the between-sample variability divided by the within-sample variability. Think of the $F$ statistic as measuring how many times more variable the sample averages are compared to what you would expect if they were just randomly different. The $F$ test is performed by computing the $F$ statistic and comparing it to the value in the $F$ table. There are several computations involved.

Because the null hypothesis claims that all population means are equal, we will need an estimate of this mean value that combines all of the information from the samples.

The **grand average** is the average of all of the data values from all of the samples combined. It may also be viewed as a *weighted average* of the sample averages, where the larger samples have more weight.

> **The Total Sample Size, $n$, and The Grand Average, $\overline{X}$**
>
> $$n = n_1 + n_2 + \cdots + n_k$$
> $$= \sum_{i=1}^{k} n_i$$
> $$\overline{X} = \frac{n_1 \overline{X}_1 + n_2 \overline{X}_2 + \cdots + n_k \overline{X}_k}{n}$$
> $$= \frac{1}{n} \sum_{i=1}^{k} n_i \overline{X}_i$$

For the supplier quality example, the total sample size and grand average are as follows:

$$n = n_1 + n_2 + \cdots + n_k$$
$$= 18 + 21 + 19$$
$$= 58$$

$$\overline{X} = \frac{n_1 \overline{X}_1 + n_2 \overline{X}_2 + \cdots + n_k \overline{X}_k}{n}$$
$$= \frac{(18 \times 82.055556) + (21 \times 80.666667) + (19 \times 87.684211)}{58}$$
$$= \frac{4{,}837}{58}$$
$$= 83.396552$$

The **between-sample variability** measures how different the sample averages are from one another. This would be zero if the sample averages were all identical, and it would be large if they were very different. It is basically a measure of the variability of the sample averages.[5] Here is the formula:

> **The Between-Sample Variability for One-Way Analysis of Variance**
>
> Between-sample variability
>
> $$= \frac{n_1 (\overline{X}_1 - \overline{X})^2 + n_2 (\overline{X}_2 - \overline{X})^2 + \cdots + n_k (\overline{X}_2 - \overline{X})^2}{k - 1}$$
>
> $$= \frac{1}{k - 1} \sum_{i=1}^{k} n_i (\overline{X}_i - \overline{X})^2$$
>
> Degrees of freedom $= k - 1$

The degrees of freedom number expresses the fact that you are measuring the variability of $k$ averages. One degree of freedom is lost (as for an ordinary standard deviation) because the grand average was estimated.

---

5. The formula for the between-sample variability may be viewed as the result of replacing all data values by their sample averages, combining all of these to form one large data set, computing the ordinary sample standard deviation, squaring to get the variance, and multiplying by the scaling factor $(n - 1)/(k - 1)$.

For the supplier quality example, the between-sample variability (with $k - 1 = 3 - 1 = 2$ degrees of freedom) is computed as follows:

Between-sample variability

$$= \frac{n_1(\overline{X}_1 - \overline{X})^2 + n_2(\overline{X}_2 - \overline{X})^2 + \cdots + n_k(\overline{X}_k - \overline{X})^2}{k - 1}$$

$$= \frac{18(82.055556 - 83.396552)^2 + 21(80.666667 - 83.396552)^2}{3 - 1}$$

$$= \frac{32.369 + 156.498 + 349.296}{2}$$

$$= 269.08$$

The **within-sample variability** measures how variable each sample is. Because the samples are assumed to have equal variability, there is only one measure of within-sample variability. This would be zero if each sample consisted of its sample average repeated many times, and it would be large if each sample contained a wide diversity of numbers. The square root of the within-sample variability provides an estimator of the population standard deviations. Here is the formula:

**The Within-Sample Variability for One-Way Analysis of Variance**

Within-sample variability

$$= \frac{(n_1 - 1)(S_1)^2 + (n_2 - 1)(S_2)^2 + \cdots + (n_k - 1)(S_k)^2}{n - k}$$

$$= \frac{1}{n - k} \sum_{i=1}^{k} (n_i - 1)(S_i)^2$$

Degrees of freedom = $n - k$

The degrees of freedom number expresses the fact that you are measuring the variability of all $n$ data values about their sample averages but have lost $k$ degrees of freedom because $k$ different sample averages were estimated.

For the supplier quality example, the within-sample variability (with $n - k = 58 - 3 = 55$ degrees of freedom) is computed as follows:

Within-Sample Variability

Within-sample variability

$$= \frac{(n_1 - 1)(S_1)^2 + (n_2 - 1)(S_2)^2 + \cdots + (n_k - 1)(S_k)^2}{n - k}$$

$$= \frac{(18 - 1)(7.124706)^2 + (21 - 1)(7.598245)^2 + (19 - 1)(5.228688)^2}{58 - 3}$$

$$= \frac{(17 \times 50.7614) + (20 \times 57.7333) + (18 \times 27.3392)}{55}$$

$$= \frac{862.944 + 1154.667 + 492.105}{55}$$

$$= 45.63$$

The $F$ statistic is the ratio of these two variability measures, indicating the extent to which the sample averages differ from one another (the numerator) with respect to the overall level of variability in the samples (the denominator).

**The $F$ Statistic for One-Way Analysis of Variance**

$$F = \frac{\text{Between-sample variability}}{\text{Within-sample varibaility}}$$

Degrees of freedom = $k - 1$ (numerator) and $n - k$ (denominator)

Note that the $F$ statistic has *two* numbers for degrees of freedom. It inherits the degrees of freedom of both of the variability measures it is based on.

For the supplier quality example, the $F$ statistic (with 2 and 55 degrees of freedom) is computed as follows:

$$F = \frac{\text{Between-sample variability}}{\text{Within-sample varibaility}}$$

$$= \frac{269.08}{45.63}$$

$$= 5.897$$

This tells you that the between-sample variability (due to differences among suppliers) is 5.897 times the within-sample variability. That is, there is 5.897 times as much variability among suppliers as you would expect, based only on the variability of individual suppliers. Is this large enough to indicate significant supplier differences? A statistical table is needed.

## The $F$ Table

The $F$ **table** is a list of critical values for the distribution of the $F$ statistic when the null hypothesis is true, so that the $F$ statistic exceeds the $F$ table value a controlled percentage of the time (5%, for example) when the null hypothesis is true. To find the critical value, use your numbers of degrees of freedom to find the row and column in the $F$ table corresponding to the level at which you are testing (e.g., 5%). Tables 15.2.2 through 15.2.5 give $F$ table critical values for testing at the 5%, 1%, 0.1%, and 10% levels, respectively.

For the supplier quality example, the degrees of freedom are $k - 1 = 2$ (for between-sample variability) and $n - k = 55$ (for within-sample variability). The critical value for testing at the usual 5% level, found in the $F$ table, is somewhere between 3.316 and 3.150 (these are the respective values for 30 and for 60 within-sample degrees of freedom, which bracket the unlisted value for

**TABLE 15.2.2** *F* Table: Level 5% Critical Value (Significant)

Numerator Degrees of Freedom (*k* – 1 for between-sample variability in one-way ANOVA)

| Denominator Degrees of Freedom (*n* – *k* for within-sample variability in one-way ANOVA) | 1 | 2 | 3 | 4 | 5 | 6 | 7 | 8 | 9 | 10 | 12 | 15 | 20 | 30 | 60 | 120 | Infinity |
|---|---|---|---|---|---|---|---|---|---|---|---|---|---|---|---|---|---|
| 1 | 161.45 | 199.50 | 215.71 | 224.58 | 230.16 | 233.99 | 236.77 | 238.88 | 240.54 | 241.88 | 243.91 | 245.95 | 248.01 | 250.10 | 252.20 | 253.25 | 254.32 |
| 2 | 18.513 | 19.000 | 19.164 | 19.247 | 19.296 | 19.330 | 19.353 | 19.371 | 19.385 | 19.396 | 19.413 | 19.429 | 19.446 | 19.462 | 19.479 | 19.487 | 19.496 |
| 3 | 10.128 | 9.552 | 9.277 | 9.117 | 9.013 | 8.941 | 8.887 | 8.845 | 8.812 | 8.786 | 8.745 | 8.703 | 8.660 | 8.617 | 8.572 | 8.549 | 8.526 |
| 4 | 7.709 | 6.944 | 6.591 | 6.388 | 6.256 | 6.163 | 6.094 | 6.041 | 5.999 | 5.964 | 5.912 | 5.858 | 5.803 | 5.746 | 5.688 | 5.658 | 5.628 |
| 5 | 6.608 | 5.786 | 5.409 | 5.192 | 5.050 | 4.950 | 4.876 | 4.818 | 4.772 | 4.735 | 4.678 | 4.619 | 4.558 | 4.496 | 4.431 | 4.398 | 4.365 |
| 6 | 5.987 | 5.143 | 4.757 | 4.534 | 4.387 | 4.284 | 4.207 | 4.147 | 4.099 | 4.060 | 4.000 | 3.938 | 3.874 | 3.808 | 3.740 | 3.705 | 3.669 |
| 7 | 5.591 | 4.737 | 4.347 | 4.120 | 3.972 | 3.866 | 3.787 | 3.726 | 3.677 | 3.637 | 3.575 | 3.511 | 3.445 | 3.376 | 3.304 | 3.267 | 3.230 |
| 8 | 5.318 | 4.459 | 4.066 | 3.838 | 3.687 | 3.581 | 3.500 | 3.438 | 3.388 | 3.347 | 3.284 | 3.218 | 3.150 | 3.079 | 3.005 | 2.967 | 2.928 |
| 9 | 5.117 | 4.256 | 3.863 | 3.633 | 3.482 | 3.374 | 3.293 | 3.230 | 3.179 | 3.137 | 3.073 | 3.006 | 2.936 | 2.864 | 2.787 | 2.748 | 2.707 |
| 10 | 4.965 | 4.103 | 3.708 | 3.478 | 3.326 | 3.217 | 3.135 | 3.072 | 3.020 | 2.978 | 2.913 | 2.845 | 2.774 | 2.700 | 2.621 | 2.580 | 2.538 |
| 12 | 4.747 | 3.885 | 3.490 | 3.259 | 3.106 | 2.996 | 2.913 | 2.849 | 2.796 | 2.753 | 2.687 | 2.617 | 2.544 | 2.466 | 2.384 | 2.341 | 2.296 |
| 15 | 4.543 | 3.682 | 3.287 | 3.056 | 2.901 | 2.790 | 2.707 | 2.641 | 2.588 | 2.544 | 2.475 | 2.403 | 2.328 | 2.247 | 2.160 | 2.114 | 2.066 |
| 20 | 4.351 | 3.493 | 3.098 | 2.866 | 2.711 | 2.599 | 2.514 | 2.447 | 2.393 | 2.348 | 2.278 | 2.203 | 2.124 | 2.039 | 1.946 | 1.896 | 1.843 |
| 30 | 4.171 | 3.316 | 2.922 | 2.690 | 2.534 | 2.421 | 2.334 | 2.266 | 2.211 | 2.165 | 2.092 | 2.015 | 1.932 | 1.841 | 1.740 | 1.683 | 1.622 |
| 60 | 4.001 | 3.150 | 2.758 | 2.525 | 2.368 | 2.254 | 2.167 | 2.097 | 2.040 | 1.993 | 1.917 | 1.836 | 1.748 | 1.649 | 1.534 | 1.467 | 1.389 |
| 120 | 3.920 | 3.072 | 2.680 | 2.447 | 2.290 | 2.175 | 2.087 | 2.016 | 1.959 | 1.910 | 1.834 | 1.750 | 1.659 | 1.554 | 1.429 | 1.352 | 1.254 |
| Infinity | 3.841 | 2.996 | 2.605 | 2.372 | 2.214 | 2.099 | 2.010 | 1.938 | 1.880 | 1.831 | 1.752 | 1.666 | 1.571 | 1.459 | 1.318 | 1.221 | 1.000 |

**TABLE 15.2.3** *F* Table: Level 1% Critical Values (Highly Significant)

Numerator Degrees of Freedom ($k - 1$ for between-sample variability in one-way ANOVA)

| Denominator Degrees of Freedom ($n - k$ for within-sample variability in one-way ANOVA) | 1 | 2 | 3 | 4 | 5 | 6 | 7 | 8 | 9 | 10 | 12 | 15 | 20 | 30 | 60 | 120 | Infinity |
|---|---|---|---|---|---|---|---|---|---|---|---|---|---|---|---|---|---|
| 1 | 4052.2 | 4999.5 | 5403.4 | 5624.6 | 5763.7 | 5859.0 | 5928.4 | 5891.1 | 6022.5 | 6055.8 | 6106.3 | 6157.3 | 6208.7 | 6260.6 | 6313.0 | 6339.4 | 6365.9 |
| 2 | 98.501 | 98.995 | 99.159 | 99.240 | 99.299 | 99.333 | 99.356 | 99.374 | 99.388 | 99.399 | 99.416 | 99.432 | 99.449 | 99.466 | 99.482 | 99.491 | 99.499 |
| 3 | 34.116 | 30.816 | 29.456 | 28.709 | 28.236 | 27.910 | 27.671 | 27.488 | 27.344 | 27.228 | 27.051 | 26.871 | 26.689 | 26.503 | 26.315 | 26.220 | 26.125 |
| 4 | 21.197 | 18.000 | 16.694 | 15.977 | 15.522 | 15.207 | 14.976 | 14.799 | 14.659 | 14.546 | 14.374 | 14.198 | 14.020 | 13.838 | 13.652 | 13.558 | 13.463 |
| 5 | 16.258 | 13.274 | 12.060 | 11.392 | 10.967 | 10.672 | 10.455 | 10.289 | 10.158 | 10.051 | 9.888 | 9.722 | 9.553 | 9.379 | 9.202 | 9.112 | 9.021 |
| 6 | 13.745 | 10.925 | 9.780 | 9.148 | 8.746 | 8.466 | 8.260 | 8.102 | 7.976 | 7.874 | 7.718 | 7.559 | 7.396 | 7.229 | 7.057 | 6.969 | 6.880 |
| 7 | 12.246 | 9.547 | 8.451 | 7.847 | 7.460 | 7.191 | 6.993 | 6.840 | 6.719 | 6.620 | 6.469 | 6.314 | 6.155 | 5.992 | 5.823 | 5.737 | 5.650 |
| 8 | 11.258 | 8.649 | 7.591 | 7.006 | 6.632 | 6.371 | 6.178 | 6.029 | 5.911 | 5.814 | 5.667 | 5.515 | 5.359 | 5.198 | 5.032 | 4.946 | 4.859 |
| 9 | 10.561 | 8.021 | 6.992 | 6.422 | 6.057 | 5.802 | 5.613 | 5.467 | 5.351 | 5.257 | 5.111 | 4.962 | 4.808 | 4.649 | 4.483 | 4.398 | 4.311 |
| 10 | 10.044 | 7.559 | 6.552 | 5.994 | 5.636 | 5.386 | 5.200 | 5.057 | 4.942 | 4.849 | 4.706 | 4.558 | 4.405 | 4.247 | 4.082 | 3.996 | 3.909 |
| 12 | 9.330 | 6.927 | 5.953 | 5.412 | 5.064 | 4.821 | 4.640 | 4.499 | 4.388 | 4.296 | 4.155 | 4.010 | 3.858 | 3.701 | 3.535 | 3.449 | 3.361 |
| 15 | 8.683 | 6.359 | 5.417 | 4.893 | 4.556 | 4.318 | 4.142 | 4.004 | 3.895 | 3.805 | 3.666 | 3.522 | 3.372 | 3.214 | 3.047 | 2.959 | 2.868 |
| 20 | 8.096 | 5.849 | 4.938 | 4.431 | 4.103 | 3.871 | 3.699 | 3.564 | 3.457 | 3.368 | 3.231 | 3.088 | 2.938 | 2.778 | 2.608 | 2.517 | 2.421 |
| 30 | 7.562 | 5.390 | 4.510 | 4.018 | 3.699 | 3.473 | 3.304 | 3.173 | 3.067 | 2.979 | 2.843 | 2.700 | 2.549 | 2.386 | 2.208 | 2.111 | 2.006 |
| 60 | 7.077 | 4.977 | 4.126 | 3.649 | 3.339 | 3.119 | 2.953 | 2.823 | 2.718 | 2.632 | 2.496 | 2.352 | 2.198 | 2.028 | 1.836 | 1.726 | 1.601 |
| 120 | 6.851 | 4.786 | 3.949 | 3.480 | 3.174 | 2.956 | 2.792 | 2.663 | 2.559 | 2.472 | 2.336 | 2.191 | 2.035 | 1.860 | 1.656 | 1.533 | 1.381 |
| Infinity | 6.635 | 4.605 | 3.782 | 3.319 | 3.017 | 2.802 | 2.639 | 2.511 | 2.407 | 2.321 | 2.185 | 2.039 | 1.878 | 1.696 | 1.473 | 1.325 | 1.000 |

**TABLE 15.2.4**  *F* Table: Level 0.1% Critical Values (Very Highly Significant)

| Denominator Degrees of Freedom ($n - k$ for within-sample variability in one-way ANOVA) | Numerator Degrees of Freedom ($k - 1$ for between-sample variability in one-way ANOVA) | | | | | | | | | | | | | | | | |
|---|---|---|---|---|---|---|---|---|---|---|---|---|---|---|---|---|---|
| | 1 | 2 | 3 | 4 | 5 | 6 | 7 | 8 | 9 | 10 | 12 | 15 | 20 | 30 | 60 | 120 | Infinity |
| 1 | 405284 | 500000 | 540379 | 562500 | 576405 | 585937 | 592873 | 598144 | 602284 | 605621 | 610668 | 615764 | 620908 | 626099 | 631337 | 633972 | 636629 |
| 2 | 998.50 | 999.00 | 999.17 | 999.25 | 999.30 | 999.33 | 999.36 | 999.38 | 999.39 | 999.40 | 999.42 | 999.43 | 999.45 | 999.47 | 999.48 | 999.49 | 999.50 |
| 3 | 167.03 | 148.50 | 141.11 | 137.10 | 134.58 | 132.85 | 131.58 | 130.62 | 129.86 | 129.25 | 128.32 | 127.37 | 126.42 | 125.45 | 124.47 | 123.97 | 123.47 |
| 4 | 74.137 | 61.246 | 56.177 | 53.436 | 51.712 | 50.525 | 49.658 | 48.996 | 48.475 | 48.053 | 47.412 | 46.761 | 46.100 | 45.429 | 44.746 | 44.400 | 44.051 |
| 5 | 47.181 | 37.122 | 33.202 | 31.085 | 29.752 | 28.834 | 28.163 | 27.649 | 27.244 | 26.917 | 26.418 | 25.911 | 25.395 | 24.869 | 24.333 | 24.060 | 23.785 |
| 6 | 35.507 | 27.000 | 23.703 | 21.924 | 20.803 | 20.030 | 19.463 | 19.030 | 18.688 | 18.411 | 17.989 | 17.559 | 17.120 | 16.672 | 16.214 | 15.981 | 15.745 |
| 7 | 29.245 | 21.689 | 18.772 | 17.198 | 16.206 | 15.521 | 15.019 | 14.634 | 14.330 | 14.083 | 13.707 | 13.324 | 12.932 | 12.530 | 12.119 | 11.909 | 11.697 |
| 8 | 25.415 | 18.494 | 15.829 | 14.392 | 13.485 | 12.858 | 12.398 | 12.046 | 11.767 | 11.540 | 11.194 | 10.841 | 10.480 | 10.109 | 9.727 | 9.532 | 9.334 |
| 9 | 22.857 | 16.387 | 13.902 | 12.560 | 11.714 | 11.128 | 10.698 | 10.368 | 10.107 | 9.894 | 9.570 | 9.238 | 8.898 | 8.548 | 8.187 | 8.001 | 7.813 |
| 10 | 21.040 | 14.905 | 12.553 | 11.283 | 10.481 | 9.926 | 9.517 | 9.204 | 8.956 | 8.754 | 8.445 | 8.129 | 7.804 | 7.469 | 7.122 | 6.944 | 6.762 |
| 12 | 18.643 | 12.974 | 10.804 | 9.633 | 8.892 | 8.379 | 8.001 | 7.710 | 7.480 | 7.292 | 7.005 | 6.709 | 6.405 | 6.090 | 5.762 | 5.593 | 5.420 |
| 15 | 16.587 | 11.339 | 9.335 | 8.253 | 7.567 | 7.092 | 6.741 | 6.471 | 6.256 | 6.081 | 5.812 | 5.535 | 5.248 | 4.950 | 4.638 | 4.475 | 4.307 |
| 20 | 14.818 | 9.953 | 8.098 | 7.095 | 6.460 | 6.018 | 5.692 | 5.440 | 5.239 | 5.075 | 4.823 | 4.562 | 4.290 | 4.005 | 3.703 | 3.544 | 3.378 |
| 30 | 13.293 | 8.773 | 7.054 | 6.124 | 5.534 | 5.122 | 4.817 | 4.581 | 4.393 | 4.239 | 4.000 | 3.753 | 3.493 | 3.217 | 2.920 | 2.759 | 2.589 |
| 60 | 11.973 | 7.767 | 6.171 | 5.307 | 4.757 | 4.372 | 4.086 | 3.865 | 3.687 | 3.541 | 3.315 | 3.078 | 2.827 | 2.555 | 2.252 | 2.082 | 1.890 |
| 120 | 11.378 | 7.321 | 5.781 | 4.947 | 4.416 | 4.044 | 3.767 | 3.552 | 3.379 | 3.237 | 3.016 | 2.783 | 2.534 | 2.262 | 1.950 | 1.767 | 1.543 |
| Infinity | 10.827 | 6.908 | 5.422 | 4.617 | 4.103 | 3.743 | 3.475 | 3.266 | 3.097 | 2.959 | 2.742 | 2.513 | 2.266 | 1.990 | 1.660 | 1.447 | 1.000 |

**TABLE 15.2.5** *F* Table: Level 10% Critical Values

| Denominator Degrees of Freedom ($n-k$ for within-sample variability in one-way ANOVA) | \multicolumn{17}{c}{Numerator Degrees of Freedom ($k-1$ for between-sample variability in one-way ANOVA)} | | | | | | | | | | | | | | | | |
|---|---|---|---|---|---|---|---|---|---|---|---|---|---|---|---|---|---|
| | 1 | 2 | 3 | 4 | 5 | 6 | 7 | 8 | 9 | 10 | 12 | 15 | 20 | 30 | 60 | 120 | Infinity |
| 1 | 39.863 | 49.500 | 53.593 | 55.833 | 57.240 | 58.204 | 58.906 | 59.439 | 59.858 | 60.195 | 60.705 | 61.220 | 61.740 | 62.265 | 62.794 | 63.061 | 63.328 |
| 2 | 8.526 | 9.000 | 9.162 | 9.243 | 9.293 | 9.326 | 9.349 | 9.367 | 9.381 | 9.392 | 9.408 | 9.425 | 9.441 | 9.458 | 9.475 | 9.483 | 9.491 |
| 3 | 5.538 | 5.462 | 5.391 | 5.343 | 5.309 | 5.285 | 5.266 | 5.252 | 5.240 | 5.230 | 5.216 | 5.200 | 5.184 | 5.168 | 5.151 | 5.143 | 5.134 |
| 4 | 4.545 | 4.325 | 4.191 | 4.107 | 4.051 | 4.010 | 3.979 | 3.955 | 3.936 | 3.920 | 3.896 | 3.870 | 3.844 | 3.817 | 3.790 | 3.775 | 3.761 |
| 5 | 4.060 | 3.780 | 3.619 | 3.520 | 3.453 | 3.405 | 3.368 | 3.339 | 3.316 | 3.297 | 3.268 | 3.238 | 3.207 | 3.174 | 3.140 | 3.123 | 3.105 |
| 6 | 3.776 | 3.463 | 3.289 | 3.181 | 3.108 | 3.055 | 3.014 | 2.983 | 2.958 | 2.937 | 2.905 | 2.871 | 2.836 | 2.800 | 2.762 | 2.742 | 2.722 |
| 7 | 3.589 | 3.257 | 3.074 | 2.961 | 2.883 | 2.827 | 2.785 | 2.752 | 2.725 | 2.703 | 2.668 | 2.632 | 2.595 | 2.555 | 2.514 | 2.493 | 2.471 |
| 8 | 3.458 | 3.113 | 2.924 | 2.806 | 2.726 | 2.668 | 2.624 | 2.589 | 2.561 | 2.538 | 2.502 | 2.464 | 2.425 | 2.383 | 2.339 | 2.316 | 2.293 |
| 9 | 3.360 | 3.006 | 2.813 | 2.693 | 2.611 | 2.551 | 2.505 | 2.469 | 2.440 | 2.416 | 2.379 | 2.340 | 2.298 | 2.255 | 2.208 | 2.184 | 2.159 |
| 10 | 3.285 | 2.924 | 2.728 | 2.605 | 2.522 | 2.461 | 2.414 | 2.377 | 2.347 | 2.323 | 2.284 | 2.244 | 2.201 | 2.155 | 2.107 | 2.082 | 2.055 |
| 12 | 3.177 | 2.807 | 2.606 | 2.480 | 2.394 | 2.331 | 2.283 | 2.245 | 2.214 | 2.188 | 2.147 | 2.105 | 2.060 | 2.011 | 1.960 | 1.932 | 1.904 |
| 15 | 3.073 | 2.695 | 2.490 | 2.361 | 2.273 | 2.208 | 2.158 | 2.119 | 2.086 | 2.059 | 2.017 | 1.972 | 1.924 | 1.873 | 1.817 | 1.787 | 1.755 |
| 20 | 2.975 | 2.589 | 2.380 | 2.249 | 2.158 | 2.091 | 2.040 | 1.999 | 1.965 | 1.937 | 1.892 | 1.845 | 1.794 | 1.738 | 1.677 | 1.643 | 1.607 |
| 30 | 2.881 | 2.489 | 2.276 | 2.142 | 2.049 | 1.980 | 1.927 | 1.884 | 1.849 | 1.819 | 1.773 | 1.722 | 1.667 | 1.606 | 1.538 | 1.499 | 1.456 |
| 60 | 2.791 | 2.393 | 2.177 | 2.041 | 1.946 | 1.875 | 1.819 | 1.775 | 1.738 | 1.707 | 1.657 | 1.603 | 1.543 | 1.476 | 1.395 | 1.348 | 1.291 |
| 120 | 2.748 | 2.347 | 2.130 | 1.992 | 1.896 | 1.824 | 1.767 | 1.722 | 1.684 | 1.652 | 1.601 | 1.545 | 1.482 | 1.409 | 1.320 | 1.265 | 1.193 |
| Infinity | 2.706 | 2.303 | 2.084 | 1.945 | 1.847 | 1.774 | 1.717 | 1.670 | 1.632 | 1.599 | 1.546 | 1.487 | 1.421 | 1.342 | 1.240 | 1.169 | 1.000 |

55 degrees of freedom). For testing at the 1% level, the critical value is between 5.390 and 4.977. While interpolation using reciprocal degrees of freedom can give an approximate value, computer software can give you the exact value. Of course, in practice, computer output will give you the exact $p$-value.

## The Result of the F Test

The $F$ test is performed by comparing the $F$ statistic (computed from your data) to the critical value from the $F$ table as shown in Table 15.2.6. The result is *significant* if the $F$ statistic is *larger* because this indicates greater differences among the sample averages. Remember that, as is usually the case with hypothesis testing, when you accept the null hypothesis, you have a weak conclusion in the sense that you should *not* believe that the null hypothesis has been shown to be true. Your conclusion when accepting the null hypothesis is really that there is not enough evidence to reject it.

To test the supplier quality example at the 5% level, the $F$ statistic (5.897) is compared to the $F$ table critical value (somewhere between 3.316 and 3.150). Since the $F$ statistic is larger, the result is significant:

> There are significant differences among your suppliers in terms of average quality level ($p < 0.05$).

---

**TABLE 15.2.6  Finding the Result of the F Test**

**If the F statistic is *smaller* than the critical value from the F table:**

Accept the null hypothesis, $H_0$, as a reasonable possibility.

Do *not* accept the research hypothesis, $H_1$.

The sample averages are *not significantly different* from each other.

The observed differences among the sample averages could reasonably be due to random chance alone.

The result is *not statistically significant*. (All of the above statements are equivalent to one another.)

**If the F statistic is *larger* than the critical value from the F table:**

Accept the research hypothesis, $H_1$.

Reject the null hypothesis, $H_0$.

The sample averages are *significantly different* from each other.

The observed differences among the sample averages could *not* reasonably be due to random chance alone.

The result is *statistically significant*. (All of the above statements are equivalent to one another.)

---

To test supplier quality at the 1% level, the $F$ statistic (5.897) is compared to the $F$ table critical value (somewhere between 5.390 and 4.977). Since the $F$ statistic is larger, the result is *highly* significant, a stronger result than before:

> The supplier differences are highly significant ($p < 0.01$).

## Computer Output: The One-Way ANOVA Table

The following computer output shows an ANOVA table for this example, using a standard format for reporting ANOVA results. The sources are the *factor* (this is the supplier effect, indicating the extent to which Amalgamated, Bipolar, and Consolidated vary systematically from one another), the *error* (the random variation within a supplier), and the *total* variation. The degrees of freedom (DF) are in the next column, followed by the sums of squares (SS). Dividing SS by DF, we find the mean squares (MS), which are the between-sample and the within-sample variabilities. Dividing the *factor* MS by the *error* MS produces the $F$ statistic in the next column, followed by its significance level ($p$-value) in the last column, indicating that the supplier differences are highly significant.

**Analysis of Variance**

| Source | DF | SS | MS | F | p |
|--------|-----|--------|-------|------|-------|
| Factor | 2 | 538.2 | 269.1 | 5.90 | 0.005 |
| Error | 55 | 2509.7 | 45.6 | | |
| Total | 57 | 3047.9 | | | |

To perform a one-way ANOVA analysis with Excel®, first highlight your data by dragging across the columns.[6] Then look in the Data Ribbon for Data Analysis in the Analysis area,[7] select Anova: Single Factor, and click OK. In the resulting dialog box, be sure that your input range is correctly indicated, click Output Range, and specify where in the worksheet you want the results to be placed. Then click OK to see the results. Following are the initial worksheet, the dialog boxes, and the results for the quality score example. Note that the results include the sample sizes (18, 21, and 19), the averages (82.06, 80.67, and 87.68), the between-sample variability (MS

---

6. Your data should be arranged as adjacent columns. Although they do not have to have the same length (i.e., some columns can be shorter than others), you should (1) be sure to highlight down to the end of the longest column and (2) be sure that any highlighted cells after the last data value in any column are truly empty.

7. If you cannot find Data Analysis in the Analysis area of Excel's Data Ribbon, click the Office button at the top left, choose Excel Options at the bottom, select Add-Ins at the left, click Go at the bottom, and make sure the Analysis ToolPak is checked. If the Analysis ToolPak was not installed when Excel® was installed on your computer, you may need to run the Microsoft Office Setup Program.

between groups of 269.08), the within-sample variability (MS within groups of 45.63), the $F$ statistic of 5.897, its $p$-value 0.00478, and the critical $F$ value from the $F$ table of 3.165.

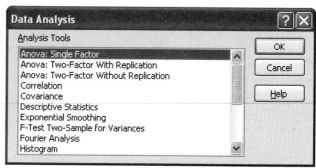

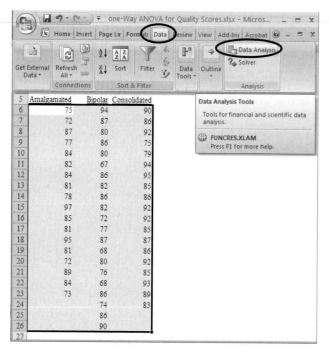

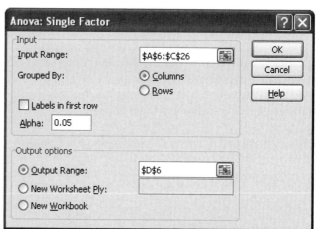

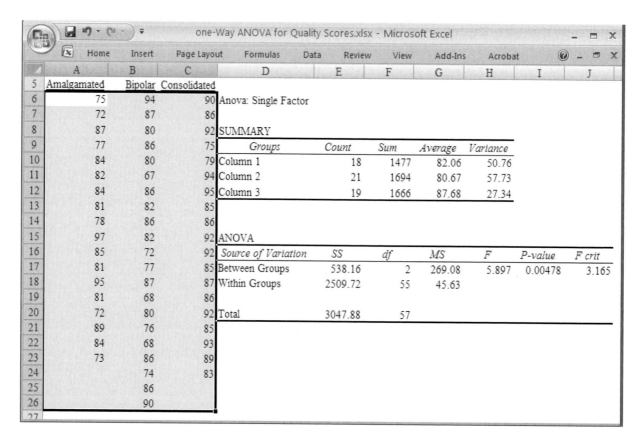

**Anova: Single Factor**

**SUMMARY**

| Groups | Count | Sum | Average | Variance |
|---|---|---|---|---|
| Column 1 | 18 | 1477 | 82.06 | 50.76 |
| Column 2 | 21 | 1694 | 80.67 | 57.73 |
| Column 3 | 19 | 1666 | 87.68 | 27.34 |

**ANOVA**

| Source of Variation | SS | df | MS | F | P-value | F crit |
|---|---|---|---|---|---|---|
| Between Groups | 538.16 | 2 | 269.08 | 5.897 | 0.00478 | 3.165 |
| Within Groups | 2509.72 | 55 | 45.63 | | | |
| Total | 3047.88 | 57 | | | | |

## 15.3 THE LEAST-SIGNIFICANT-DIFFERENCE TEST: WHICH PAIRS ARE DIFFERENT?

What if you want to know *which* sample averages are significantly different from others? The *F* test doesn't give you this information; it merely tells you whether or not there are differences. There are a number of different solutions to this problem. The method presented here, the **least-significant-difference test**, is based on *t* tests for the average difference between pairs of samples.

There is a strict rule that must be obeyed in order that the probability of a type I error remain at a low rate of 5% (or other chosen level). The problem is that there are many *t* tests (one for each pair of samples), and even though the *individual* error rate for each one remains at 5%, the *group* error rate, for all pairs, can be much higher.[8]

### Strict Requirement for Testing Individual Pairs

If your *F* test is not significant, you may *not* test particular samples to see if they are different from each other. The *F* test has already told you that there are *no* significant differences whatsoever. If the *F* test is not significant, but some *t* test appears to be significant, the *F* test overrides; the *t* test is not really significant.

If your *F* test is significant, you may go ahead and test all of the samples against one another to find out which particular pairs of samples are different.

The *t* test for deciding whether or not two particular samples are different is based on three numbers:

1. The *average difference* between those samples, found by subtracting one average from the other. (It doesn't matter which you subtract from which, as long as you remember how you did it.)
2. The *standard error* for this average difference.
3. The number of *degrees of freedom*, which is $n - k$ regardless of which groups are being compared because the standard error uses the information from all of the samples.

The standard error is computed as follows:

### Standard Error for the Average Difference between Two Samples

$$\text{Standard error} = \sqrt{(\text{Within-sample variability}) \left( \frac{1}{n_i} + \frac{1}{n_j} \right)}$$

when $n_i$ and $n_j$ are the sample sizes of the two samples being compared.

Note that this standard error may change depending on which pairs of samples you are comparing. The reason is that the variability of the sample averages being compared depends on the sample sizes.

The test then proceeds using the *t* table in the usual way, either by constructing a confidence interval for the population mean difference and seeing if it includes the reference value 0 (for no difference) or else by computing the *t* statistic (dividing the average difference by the standard error) and comparing the result to the *t* table value.

For the supplier quality example, there are three pairs of suppliers to be compared: Amalgamated to Bipolar, Amalgamated to Consolidated, and Bipolar to Consolidated. Are you permitted to test these pairs against each other? Yes, because the *F* test shows that there are significant differences in mean quality score from one supplier to another.

Here are the calculations for comparing Amalgamated to Bipolar, using the approximate *t* table value of 1.960 for $n - k = 58 - 3 = 55$ degrees of freedom:[9]

$$\text{Average difference} = 80.667 - 82.056 = -1.389$$

$$\begin{aligned}
\text{Standard error} &= \sqrt{(\text{Within-sample variability}) \left( \frac{1}{n_2} + \frac{1}{n_1} \right)} \\
&= \sqrt{(45.63) \left( \frac{1}{21} + \frac{1}{18} \right)} \\
&= \sqrt{45.63 \times 0.103175} \\
&= 2.170
\end{aligned}$$

The 95% confidence interval for the population mean difference extends from

$$-1.389 - (1.960 \times 2.170) = -5.64$$

to

$$-1.389 + (1.960 \times 2.170) = 2.86$$

The *t* statistic is

$$t = -\frac{1.389}{2.170} = -0.640$$

Bipolar has a lower quality score than Amalgamated, −1.389 points difference on average, but the difference is *not statistically significant*. You are 95% sure that the difference is somewhere between −5.64 and 2.86. Because this confidence interval includes the possibility of zero difference, you accept the null hypothesis that there is no difference in the population mean quality scores of Amalgamated and Bipolar. You could also perform the *t* test by

---

8. The group error rate is the probability that *any one or more* of the *t* tests wrongly declares significance when, in fact, there are no population mean differences.

9. This *t* table value, 1.960, is for an infinite number of degrees of freedom and is often used as an approximation when the number of degrees of freedom is 40 or more. Slightly different answers will be obtained if the more exact value, 2.004, is used.

observing that the *t* statistic (−0.640) is smaller in magnitude than the *t* table value of 1.960.

However, Consolidated *does* have significantly higher quality than either Amalgamated or Bipolar. Here are the calculations for comparing Amalgamated to Consolidated:

$$\text{Average difference} = 87.684 - 82.056 = 5.628$$

$$\text{Standard error} = \sqrt{(45.63)\left(\frac{1}{19} + \frac{1}{18}\right)} = 2.222$$

The 95% confidence interval for the population mean difference extends from

$$5.628 - (1.960 \times 2.222) = 1.27$$

to

$$5.628 + (1.960 \times 2.222) = 9.98$$

and the *t* statistic is

$$t = \frac{5.628}{2.222} = 2.533$$

The computations for comparing Bipolar to Consolidated are

$$\text{Average difference} = 87.684 - 80.677 = 7.017$$

$$\text{Standard error} = \sqrt{(45.63)\left(\frac{1}{19} + \frac{1}{21}\right)} = 2.139$$

The 95% confidence interval for the population mean difference extends from

$$7.017 - (1.960 \times 2.139) = 2.82$$

to

$$7.017 + (1.960 \times 2.139) = 11.21$$

and the *t* statistic is

$$t = \frac{7.017}{2.139} = 3.281$$

Here is a summary of what the analysis of variance has told you about the quality of these three suppliers:

1. There are significant differences among the suppliers. The *F* test decided that their population mean quality scores are not all identical.
2. Consolidated has significantly superior quality compared to each of the other suppliers (based on the least-significant-difference test).
3. The other two suppliers, Amalgamated and Bipolar, are not significantly different from each other in terms of average quality level.

## 15.4 MORE ADVANCED ANOVA DESIGNS

When your data set has more structure than just a single collection of samples, the analysis of variance can often be adapted to help answer the more complex questions that can be asked.

In order for ANOVA to be the appropriate analysis, your data set should still consist of a collection of samples with one basic measurement for each elementary unit, just as it was for the one-way analysis of variance. What's new is that there is now some structure or pattern to the arrangement of these samples. For example, while salary data for the four groups "white male," "white female," "minority male," and "minority female" could be analyzed using one-way ANOVA to see whether salaries differ significantly from one group to another, a two-way ANOVA would also allow you to ask questions about a gender difference and a minority difference.

You will still need to satisfy the basic assumptions. First, each sample is assumed to be a random sample from the population to which you wish to generalize. Second, each population is assumed to follow a normal distribution, and the standard deviations of these populations are assumed to be identical.

There is another way to look at the kind of data structure needed in order for ANOVA to be appropriate: You need a multivariate data set in which exactly one variable is quantitative (the basic measurement), and all others are qualitative. The qualitative variables, taken together, define the grouping of the quantitative observations into samples.

### Variety Is the Spice of Life

There is tremendous diversity in the world of ANOVA because there are many different ways the samples might relate to one another. What distinguishes one kind of analysis from another is the *design*, that is, the way in which the data were collected. When you're using advanced ANOVA, it's up to you to be sure that the computer uses an ANOVA model that is appropriate for your data; in many cases there's no way the computer could choose the correct model based only on the data set. Here are some highlights of these more advanced ANOVA methods.

### Two-Way ANOVA

When your samples form a table, with rows for one factor and columns for another, you may ask three basic kinds of questions: (1) Does the first factor make any difference? (2) Does the second factor make any difference? and (3) Does the effect of the first factor depend on the second factor, or do the two factors act independently of one another? The first two questions refer to the *main effects* of each factor by itself, while the third question refers to the *interaction* of the factors with one another.

Here is an example for which two-way ANOVA would be appropriate. The first factor is *shift*, indicating whether the day shift, the night shift, or the swing shift was on duty at the time the part was manufactured. The second factor is *supplier*, indicating which of your four suppliers provided the raw materials. The measurement is the overall *quality score* of the manufactured product. The first question concerns the main effect of shift: Do the quality scores differ significantly from one shift to another? The second question concerns the main effect of supplier: Do the quality scores differ significantly from one supplier to another? The third question concerns the interaction of shift with supplier, for example: Do the quality scores for the three shifts show different patterns depending on which supplier's raw materials are being used?

For a concrete example of interaction, suppose that the night shift's quality scores are usually higher than for the other shifts, except that the night crew has trouble working with raw materials from one particular supplier (and the other shifts have no such trouble). The interaction here is due to the fact that a supplier's raw materials affect the shifts differently. For there to be no interaction, *all* shifts would have to have similar trouble with that supplier's raw materials.

Figure 15.4.1 shows how interaction might look when you plot these quality scores as line graphs against the shift (day, night, or swing), with one line for each supplier (A, B, or C). Because the lines do not move up and down together, there appears to be interaction. Figure 15.4.2 shows how the quality scores might look if there were absolutely no interaction whatsoever. Of course, in real life there is usually randomness in data, so there will nearly always appear to be some interaction. The purpose of ANOVA's significance test for interaction is to test whether or not an apparent interaction is significant (in the statistical sense of being more than just randomly different from the "no interaction" case).

## Three-Way and More

When there are three or more factors defining your samples, the analysis of variance still examines the *main effects* of each factor (to see if it makes a difference) and the

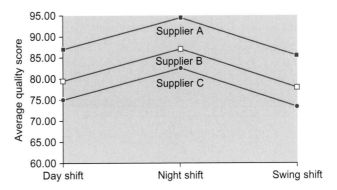

FIGURE 15.4.2  *If there were no interaction whatsoever,* this is how a plot of the averages might look. Note how the lines move up and down together. In this case, the night shift gets the highest quality score regardless of which supplier's materials are being used (being careful to compare quality scores for one supplier at a time). Note also that every shift has similar trouble with supplier C's materials.

*interactions* among factors (to see how the factors relate to one another). What's new is that there are more kinds of interactions, each to be examined separately, than for just a two-way ANOVA with only two factors. There are *two-way interactions* that consider the factors two at a time, *three-way interactions* that consider combinations of three factors at once, and so forth up until you look at the highest-level interaction of all factors at once.

## Analysis of Covariance (ANCOVA)

The analysis of covariance combines regression analysis with ANOVA. For example, in addition to the basic data for ANOVA, you might have an important additional quantitative variable. Instead of either doing ANOVA while ignoring this additional variable, or doing regression while ignoring the groups, ANCOVA will do both at once. You may think of the analysis either in terms of the relationship between separate regression analyses performed for each sample or in terms of an ANOVA that has been adjusted for differences in the additional variable.

## Multivariate Analysis of Variance (MANOVA)

When you have more than one quantitative response variable, you may use the multivariate analysis of variance to study the differences in all responses from one sample to another. If, for example, you had three quantitative ratings measured for each finished product (How nice does it look? How well does it work? How noisy is it?), then you could use MANOVA to see whether these measures differ significantly according to the main effects of shift (day, night, or swing) and supplier.

## How to Read an ANOVA Table

The general form of the traditional ANOVA table is shown in Table 15.4.1. While such a table is useful for testing

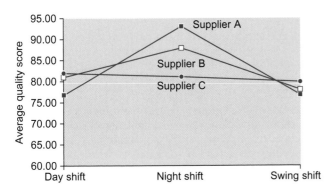

FIGURE 15.4.1  A plot of the averages that shows interaction because the lines do not all move up and down together. Note, in particular, how the night shift (in the middle) generally has better quality scores than the other shifts, except when it works with materials from supplier C.

**TABLE 15.4.1 General Form of the Traditional ANOVA Table**

| Source of Variation | Sum of Squares | Degrees of Freedom | Mean Square | F Value | p-Value |
|---|---|---|---|---|---|
| Source 1 | $SS_1$ | $df_1$ | $MS_1 = SS_1/df_1$ | $F_1 = MS_1/MS_e$ | $p_1$ |
| Source 2 | $SS_2$ | $df_2$ | $MS_2 = SS_2/df_2$ | $F_2 = MS_2/MS_e$ | $p_2$ |
| . | . | . | . | . | . |
| . | . | . | . | . | . |
| . | . | . | . | . | . |
| Source $k$ | $SS_k$ | $df_k$ | $MS_k = SS_k/df_k$ | $F_k = MS_k/MS_e$ | $p_k$ |
| Error or residual | $SS_e$ | $df_e$ | $MS_e = SS_e/df_e$ | | |

hypotheses about your population means (once you know how to read it!), it has two serious deficiencies. First, it tells you nothing in terms of the original measurements; you will need to examine a separate table of average values to find out, for example, whether quality was higher for the day or the night shift. Second, most of the ANOVA table has no direct practical interpretation: For many applications only the first (source of variation) and last (*p*-value) columns are useful; the others are merely computational steps along the way to the *p*-values that give you the results of the significance tests. Nonetheless, it is traditional to report this table to substantiate your claims of statistical significance in ANOVA.

Each hypothesis is tested using an *F* test, which compares the mean square for that source of variation (which is large when that source of variation makes a difference in your quantitative measurement) to the error mean square, asking the question, How much stronger than purely random is this particular source? To find out whether or not that source of variation (the *i*th one, say) has a significant effect, compare the computed *F* value ($F_i$) in the table to the *F* table value with numerator degrees of freedom ($df_i$) for that source and denominator degrees of freedom ($df_e$) for the error. Or else look directly at the *p*-value ($p_i$) and decide "significant" if it is small enough, for example, if $p < 0.05$.

**Example**
*The Effect of Price Changes and Product Type on Grocery Sales*

We expect sales to go up when an item is temporarily "on sale" at a price lower than usual. If the product is one that consumers can easily stock up on at home, you would expect to see even higher sales than for a more perishable or less frequently consumed product. These questions are addressed in a study by Litvack, Calantone, and Warshaw.[10] They used a two-way ANOVA with the following basic structure:

*The first factor is defined by the two product types: stock-up and nonstock-up items. Stock-up items are those that consumers can easily buy in quantity and store at home, such as dog food, tissues, and canned fish. Nonstock-up items included mustard, cheese, and breakfast cereals.*

*The second factor is defined by the three price manipulations: lowered 20%, unchanged, and raised 20% as compared to the usual price at each store.*

*The measurement is defined as the change in sales, in number of units sold per $1 million of grocery sales for each store. Note that by dividing in this way, they have adjusted for the different sizes of one store as compared to another, making it appropriate to analyze smaller and larger grocery stores together.*

The change in sales was measured for a variety of products of each type and a variety of price manipulations, resulting in an ANOVA table like Table 15.4.2. The *p*-value 0.5694 for product type indicates *no significant differences* on average between stock-up items and nonstock-up items. This result is somewhat surprising because we expected to find a difference; however, please read on for further results. The *p*-value 0.0001 for price manipulation shows that there are *very highly significant differences* on average among lowered, unchanged, and raised prices; that is, the price change had a significant impact on sales.

The interaction term is highly significant ($p = 0.0095$ is less than 0.01). This says that a product's sales reaction to price depended on whether it was a stock-up item or a nonstock-up item. That is, stock-up items reacted differently to price changes than the others did.

How can it be that the main effect for product type was not significant, but the interaction of product type with price manipulation was significant? Remember that the main effect looks only at the *average* for each item type, while the interaction looks at all combinations of product type and pricing manipulation.

Although the ANOVA table provides useful results for significance tests, much important information is missing. In particular: What was the effect of a 20% price reduction on sales of a typical stock-up item? The answer to such

**TABLE 15.4.2 ANOVA Table for Product and Price Effects on Sales**

| Source of Variation | Sum of Squares | Degrees of Freedom | Mean Square | F Value | p-Value |
|---|---|---|---|---|---|
| Product type | 0.469 | 1 | 0.469 | 0.32 | 0.5694 |
| Price manipulation | 33.284 | 2 | 16.642 | 11.50 | 0.0001 |
| Interaction | | | | | |
|    Product type × Price manipulation | 13.711 | 2 | 6.856 | 4.74 | 0.0095 |
| Error or residual | 377.579 | 261 | 1.447 | | |

**TABLE 15.4.3 Percent Change in Standardized Sales**

| | Price Manipulation | | |
|---|---|---|---|
| Product Type | Lowered 20% | Unchanged | Raised 20% |
| Stock-up | 54.95 | 1.75 | –24.10 |
| Nonstock-up | 10.55 | 6.95 | –7.60 |

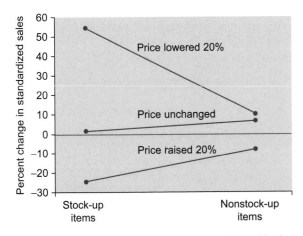

**FIGURE 15.4.3** The average change in sales for the six combinations of product type with price manipulation. The ANOVA table tests hypotheses about the six population means that these averages represent.

practical questions cannot be found in the ANOVA table! To answer questions like this, you will need to examine the average values, as shown in Table 15.4.3 and displayed in Figure 15.4.3. Each average represents 12 products and 4 stores. Note that sales of stock-up items (on the left in the figure) are *not* uniformly higher or lower than for nonstock-up items (on the right); this helps explain why the main effect for product type was not significant. Note also that sales went way up only when prices were lowered for stock-up items; this happened only when the right combination of both factors was present and is therefore part of the interaction term that is indeed significant. It is also clear that sales

dropped off when prices were raised and were highest when prices were lowered, leading to the significant main effect for price manipulation.

Don't be fooled or intimidated into thinking that an ANOVA table (such as Table 15.4.2) is supposed to tell the whole story. If it's not already provided to you, be sure to ask to see the average values (such as Table 15.4.3 and Figure 15.4.3) so that you can understand what's really going on!

10. D. S. Litvack, R. J. Calantone, and P. R. Warshaw, "An Examination of Short-Term Retail Grocery Price Effects," *Journal of Retailing* 61 (1985), pp. 9–25.

### Example
*Jokes in the Workplace*

What kinds of jokes are unacceptable in the workplace? Why do some people take offense at some kinds of jokes while others do not? These matters were studied using ANOVA by Smeltzer and Leap.[11] As an executive, you may be involved in these issues beyond your personal appreciation for humor because "joking may impact on civil and human rights litigation and on the quality of work life."

The study involved a three-way ANOVA. The three factors were gender (male or female), race (black or white), and experience (inexperienced with less than one full year or experienced). Considering all combinations of these three factors, each with two categories, there are eight different types of employees (male black inexperienced, male black experienced, male white inexperienced, and so forth). Let's look at how the 165 people in their study each rated five sexist jokes on a seven-point scale indicating how inappropriate these jokes were in the workplace.

The ANOVA results are shown in Table 15.4.4. The sums of squares, the mean squares, and the residual results are not needed. All of the usual hypothesis tests can be performed using the p-values provided.

Of the main effects, only race is significant. This says that whites and blacks had different opinions, overall, on the appropriateness of sexist jokes in the workplace. As always, the ANOVA table doesn't tell you which group felt they

*(Continued)*

**TABLE 15.4.4 ANOVA Table for Appropriateness of Sexist Jokes in the Workplace**

| Source of Variation | Degrees of Freedom | F Value | p-Value |
|---|---|---|---|
| Main effects | | | |
| Gender | 1 | 2.83 | 0.09 |
| Race | 1 | 15.59 | 0.0001 |
| Experience | 1 | 0.54 | 0.46 |
| Two-way interactions | | | |
| Gender × Race | 1 | 6.87 | 0.009 |
| Gender × Experience | 1 | 0.00 | 1.0 |
| Race × Experience | 1 | 2.54 | 0.11 |
| Three-way interaction | | | |
| Gender × Race × Experience | 1 | 1.44 | 0.23 |

**Example—cont'd**

were more appropriate; you would have to examine the average responses for each group to find out this information. The differences were not large (5.4 for whites as compared to 4.36 for blacks, according to the study), but they are highly unlikely to be this different due to random chance alone.

Of the two-way interactions, only gender × race is significant. This indicates that the difference in attitudes between men and women depended on whether they were black or white. The three-way interaction was not significant, indicating that there are no further detailed distinctions among attitudes that are discernible in this data set.

11. L. R. Smeltzer and T. L. Leap, "An Analysis of Individual Reactions to Potentially Offensive Jokes in Work Settings," *Human Relations* 41 (1988), pp. 295–304.

## 15.5   END-OF-CHAPTER MATERIALS

### Summary

The **analysis of variance** (or **ANOVA** for short) provides a general framework for statistical hypothesis testing based on careful examination of the different sources of variability in a complex situation. The analysis of variance uses an **F test** based on the **F statistic**, a ratio of two variance measures, to perform each hypothesis test. The numerator represents the variability due to that special, interesting effect being tested, and the denominator represents a baseline measure of randomness. If the ratio is larger than the value in the *F* table, the effect is significant. The **one-way analysis of variance**, in particular, is used to test whether or not the averages from several different situations are significantly different from one another.

Don't forget to explore your data. Box plots help you compare several distributions at once, so that you can see the structure in your data, identify problems (if any), and check assumptions required for the analysis of variance, such as normal distributions and equal variability.

The data set for the one-way analysis of variance consists of $k$ independent univariate samples, each using the same measurement units. The one-way analysis of variance compares two sources of variation:

Between-sample variability (from one sample to another).

Within-sample variability (inside each sample).

There are two assumptions that must be satisfied for the result of a one-way analysis of variance to be valid:

1. The data set consists of $k$ random samples from $k$ populations.
2. Each population has a normal distribution, and the standard deviations of the populations are identical, so that $\sigma_1 = \sigma_2 = \cdots = \sigma_k$.

The null hypothesis claims that there are no differences from one population to another, and the research hypothesis claims that some differences exist:

$$H_0 : \mu_1 = \mu_2 = \cdots = \mu_k \text{ (the means are all equal)}$$

$$H_1 : \mu_i \neq \mu_j \text{ for at least one pair of populations}$$

$$\text{(the means are \textit{not} all equal)}$$

The **grand average** is the average of all of the data values from all of the samples combined:

$$\overline{X} = \frac{n_1\overline{X}_1 + n_2\overline{X}_2 + \cdots + n_k\overline{X}_k}{n}$$

$$= \frac{1}{n} \sum_{i=1}^{k} n_i \overline{X}_i$$

where the total sample size is $n = n_1 + n_2 + \cdots + n_k$.

The **between-sample variability** measures how different the sample averages are from one another, and the **within-sample variability** measures how variable each sample is:

Between-sample variability

$$= \frac{n_1(\overline{X}_1 - \overline{X})^2 + n_2(\overline{X}_2 - \overline{X})^2 + \cdots + n_k(\overline{X}_k - \overline{X})^2}{k-1}$$

$$= \frac{1}{k-1} \sum_{i=1}^{k} n_i(\overline{X}_i - \overline{X})^2$$

Degrees of freedom $= k - 1$

Within-sample variability

$$= \frac{(n_1 - 1)(S_1)^2 + (n_2 - 1)(S_2)^2 + \cdots + (n_k - 1)(S_k)^2}{n - k}$$

$$= \frac{1}{n - k} \sum_{i=1}^{k} (n_i - 1)(S_i)^2$$

Degrees of freedom $= n - k$

The $F$ statistic is the ratio of these two variability measures, indicating the extent to which the sample averages differ from one another (the numerator) with respect to the overall level of variability in the samples (the denominator):

$$F = \frac{\text{Between-sample variability}}{\text{Within-sample variability}}$$

Degrees of freedom $= k - 1$ (numerator) and $n - k$ (denominator)

The **$F$ table** is a list of critical values for the distribution of the $F$ statistic such that the $F$ statistic exceeds the $F$ table value a controlled percentage of the time (5%, for example) when the null hypothesis is true. The $F$ test is performed by comparing the $F$ statistic (computed from your data) to the critical value from the $F$ table.

The $F$ test tells you only whether or not there are differences. If the $F$ test finds significance, the **least-significant-difference test** may be used to compare each pair of samples to see which ones are significantly different from each other. This test is based on the average difference for the two groups being compared, the standard error of this average difference, and the number of degrees of freedom $(n - k)$:

$$\text{Standard error} = \sqrt{(\text{Within-sample variability}) \left( \frac{1}{n_i} + \frac{1}{n_j} \right)}$$

where $n_i$ and $n_j$ are the sizes of the two samples being compared.

There are many advanced ANOVA techniques, including two-way and higher designs. In order for ANOVA to be the appropriate analysis, your data set should consist of a collection of samples, with one basic measurement for each elementary unit, just as it was for the one-way analysis of variance. Remember that the ANOVA table does not tell the whole story; always ask to see the average values so that you can understand what's really going on.

## Key Words

analysis of variance (ANOVA), *467*
between-sample variability, *471*
*F* statistic, *468*
*F* table, *472*

*F* test, *468*
grand average, *471*
least-significant-difference test, *479*
one-way analysis of variance, *468*
within-sample variability, *472*

## Questions

1. Explain in what sense the analysis of variance involves actually analyzing variance—in particular, what variances are analyzed and why?
2. a.  What kind of data set should be analyzed using the one-way analysis of variance?
   b.  Why shouldn't you use the unpaired *t* test instead of the one-way analysis of variance?
3. Name and interpret the two sources of variation in the one-way analysis of variance.
4. Which assumption helps the data be representative of the population?
5. What assumptions are required concerning the distribution of each population?
6. Do the sample sizes have to be equal in the one-way analysis of variance?
7. a.  State the hypotheses for the one-way analysis of variance.
   b.  Is the research hypothesis very specific about the nature of any differences?
8. Describe and give a formula for each of the following quantities, which are used in performing a one-way analysis of variance:
   a.  Total sample size, *n*.
   b.  Grand average, $\overline{X}$.
   c.  Between-sample variability and its degrees of freedom.
   d.  Within-sample variability and its degrees of freedom.
   e.  *F* statistic and its degrees of freedom.
   f.  *F* table (describe only).
9. When may you use the least-significant-difference test to compare individual pairs of samples? When is this not permitted?
10. Why can the standard error of the average difference be a different number depending on which samples you are comparing?

## Problems

*Problems marked with an asterisk (*) are solved in the Self Test in Appendix C.*

1.* Three advertisements have been tested, each one using a different random sample of consumers from the same city. Scores indicating the effectiveness of the advertisement were analyzed; the results are shown in Table 15.5.1.
   a.  Which advertisement appears to have the highest effectiveness? Which appears to have the lowest?
   b.  Find the total sample size, *n*, the grand average, $\overline{X}$, and the number of samples, *k*.

**TABLE 15.5.1 Analysis of Ad Effectiveness**

|                            | Ad 1 | Ad 2 | Ad 3 |
| -------------------------- | ---- | ---- | ---- |
| Average                    | 63.2 | 68.1 | 53.5 |
| Standard deviation         | 7.9  | 11.3 | 9.2  |
| Sample size (consumers)    | 101  | 97   | 105  |

**TABLE 15.5.2 Analysis of Waste Measurements**

|                            | Sludge Away | Cleen Up | No Yuk |
| -------------------------- | ----------- | -------- | ------ |
| Average                    | 245.97      | 210.92   | 240.45 |
| Standard deviation         | 41.05       | 43.52    | 35.91  |
| Sample size (batches)      | 10          | 10       | 10     |

    **c.** Find the between-sample variability and its degrees of freedom.

    **d.** Find the within-sample variability and its degrees of freedom.

**2.** Refer to the data for problem 1.

    **a.** Find the $F$ statistic and its numbers of degrees of freedom.

    **b.** Interpret the $F$ statistic in terms of how many times more volatile one source of variability is than another.

    **c.** Find the critical value from the $F$ table at the 5% level.

    **d.** Report the result of the $F$ test at the 5% level.

    **e.** Summarize what this test has told you about any differences among these ads for consumers in general in this city.

**3.** Refer to the data for problem 1.

    **a.** Find the critical value from the $F$ table at the 0.1% level and report the result of the $F$ test at this level.

    **b.** Summarize what this test has told you about any differences among these ads for consumers in general in this city.

**4.** Refer to the data for problem 1.

    **a.** Find the average difference between the effectiveness of ad 1 and that of ad 2 (computed as ad 2 minus ad 1).

    **b.*** Find the standard error for this average difference.

    **c.** How many degrees of freedom does this standard error have?

    **d.** Find the 99.9% confidence interval for the population mean difference in effectiveness between ad 1 and ad 2.

    **e.** Are the effectiveness scores of ad 1 and ad 2 very highly significantly different? How do you know?

**5.** Refer to the data for problem 1.

    **a.** Find the average difference and its standard error for every pair of advertisements (computed as ad 2 minus ad 1, ad 1 minus ad 3, and ad 2 minus ad 3).

    **b.** Test every pair of advertisements at the 1% level and report the results.

**6.** Three companies are trying to sell you their additives to reduce waste in a chemical manufacturing process. You're not sure their products are appropriate because your process is different from the industry standard (it's a proprietary trade secret). You have arranged to get a small supply of each additive, provided free, for testing purposes. Table 15.5.2 shows the summaries of the waste measurements when each additive was used, all else kept equal.

    **a.** Which additive appears to leave the highest amount of waste? Which appears to leave the lowest?

    **b.** Find the total sample size, $n$, the grand average, $\overline{X}$, and the number of samples, $k$.

    **c.** Find the between-sample variability and its degrees of freedom.

    **d.** Find the within-sample variability and its degrees of freedom.

**7.** Refer to the data for problem 6.

    **a.** Find the $F$ statistic and its numbers of degrees of freedom.

    **b.** Interpret the $F$ statistic in terms of how many times more volatile one source of variability is than another.

    **c.** Find the critical value from the $F$ table at the 5% level.

    **d.** Report the result of the $F$ test at the 5% level.

    **e.** Summarize what this $F$ test has told you about the comparative abilities of these additives to reduce waste.

**8.** Refer to the data for problem 6. Would it be appropriate to use the least-significant-difference test to find out whether Cleen Up has significantly lower waste than Sludge Away (at the 5% level)? Why or why not?

**9.** Refer to the data for problem 6.

    **a.** Find the critical value from the $F$ table at the 10% level and report the result of the $F$ test at this level.

    **b.** Summarize what this $F$ test has told you about the comparative abilities of these additives to reduce waste.

**10.** Refer to the data for problem 6. Select the two additives with the largest average difference in waste and answer the following. (Use the least-significant-difference test method for this problem, subtracting smaller from larger even if you feel that it is not appropriate to do so.)

    **a.** Find the size of the average difference for this pair.

    **b.** Find the standard error of this average difference.

    **c.** How many degrees of freedom does this standard error have?

    **d.** Find the two-sided 90% confidence interval for the mean difference.

    **e.** Based on the average difference, the standard error, the degrees of freedom, and the $t$ table, do these two additives appear to be significantly different at the 10% test level?

**f.** Can you conclude that the two additives are really significantly different at the 10% level? Why or why not? (Be careful. You may wish to consider the result of the *F* test from the preceding problem.)

11. In an attempt to regain control of your time, you have been recording the time required, in minutes, to respond to each telephone call for the day. Before you make changes (such as referring certain types of calls to subordinates), you would like to have a better understanding of the situation. With calls grouped by type, the results for call lengths were as shown in Table 15.5.3.

**a.** Draw box plots on the same scale for these four types of calls and describe the structure you see.

**b.** Compute the average and standard deviation for each type of call.

**c.** Which type of call appears to have the highest average length? Which has the lowest?

**d.** Are the assumptions of normal distribution and equal variability for the one-way analysis of variance satisfied for this data set? Why or why not?

**e.** Find the natural logarithm of each data value and draw box plots for these logarithms.

**f.** Is the assumption of equal variability better satisfied using logarithms than using the original data?

12. Refer to the data for problem 11. Continue using the logarithms of the lengths of calls.

**a.** Find the total sample size, $n$, the grand average, $\overline{X}$, and the number of samples, $k$.

**b.** Find the between-sample variability and its degrees of freedom.

**c.** Find the within-sample variability and its degrees of freedom.

13. Refer to the data for problem 11. Continue using the logarithms of the lengths of calls.

**a.** Find the *F* statistic and its numbers of degrees of freedom.

**b.** Find the critical value from the *F* table at the 5% level.

**c.** Report the result of the *F* test at the 5% level.

**d.** Summarize what this test has told you about any differences among these types of calls.

14. Refer to the data for problem 11. Continue using the logarithms of the lengths of calls.

**a.** Find the average difference and its standard error for every pair of types of calls (subtracting smaller from larger in each case).

**b.** Which pairs of types of calls are significantly different from each other, in terms of average logarithm of length?

15. Use multiple regression with indicator variables, instead of one-way ANOVA, to test whether the quality data in Table 15.1.1 show significant differences from one supplier to another. (You may wish to review the material on indicator variables from Chapter 12.)

**a.** Create the *Y* variable by listing all quality scores in a single, long column. Do this by stacking Amalgamated's scores on top of Bipolar's on top of Consolidated's.

**b.** Create two indicator variables, one for Amalgamated and one for Bipolar.

**c.** Run a multiple regression analysis.

**d.** Compare the *F* statistic from the multiple regression to the *F* statistic from the one-way ANOVA. Comment.

**e.** Compare the regression coefficients for the indicator variables to the average differences in quality scores from one supplier to another. Comment.

**f.** Do these two methods—multiple regression and one-way ANOVA—give different answers or are they in complete agreement? Why do you think it works this way?

16. Table 15.5.4 shows the average quality scores for production, averaged according to which supplier (A, B, or C) provided the materials and which shift (day, night, or swing) was active at the time the part was produced; this is followed by the computer version of the ANOVA table. For this experiment, there were five observations for each combination (a combination specifies both supplier and shift). Note that the last row (and column) represents the averages of the numbers in their column (and row), respectively (e.g., 82.42 is the average quality for Supplier A across all shifts).

**TABLE 15.5.3 Lengths of Telephone Calls**

| Information | Sales | Service | Other |
|---|---|---|---|
| 0.6 | 5.1 | 5.2 | 6.3 |
| 1.1 | 1.7 | 2.9 | 1.2 |
| 1.0 | 4.4 | 2.6 | 3.1 |
| 1.9 | 26.6 | 1.2 | 2.5 |
| 3.8 | 7.4 | 7.0 | 3.0 |
| 1.6 | 1.4 | 14.2 | 2.6 |
| 0.4 | 7.0 | 8.4 | 0.8 |
| 0.6 | 3.9 | 0.6 | |
| 2.2 | 3.1 | 26.7 | |
| 12.3 | 1.2 | 7.7 | |
| 4.2 | 1.9 | 4.8 | |
| 2.8 | 17.3 | 7.2 | |
| 1.4 | 7.8 | 2.7 | |
| | 4.3 | 3.4 | |
| | 3.4 | 13.3 | |
| | 1.3 | | |
| | 2.0 | | |

**TABLE 15.5.4** Average Quality Scores and ANOVA Table

|  | Day Shift | Night Shift | Swing Shift | Average |
|---|---|---|---|---|
| Supplier A | 77.06 | 93.12 | 77.06 | 82.42 |
| Supplier B | 81.14 | 88.13 | 78.11 | 82.46 |
| Supplier C | 82.02 | 81.18 | 79.91 | 81.04 |
| Average | 80.08 | 87.48 | 78.36 | 81.97 |

**Analysis of Variance for Quality**

| Source | DF | SS | MS | F | p |
|---|---|---|---|---|---|
| shift | 2 | 704.07 | 352.04 | 11.93 | 0.000 |
| supplier | 2 | 19.60 | 9.80 | 0.33 | 0.720 |
| shift*supplier | 4 | 430.75 | 107.69 | 3.65 | 0.014 |
| Error | 36 | 1062.05 | 29.50 | | |
| Total | 44 | 2216.47 | | | |

**TABLE 15.5.5** Effects of Competition-Cooperation and Value Dissensus on Performance

| Source of Variation | Sum of Squares | Degrees of Freedom | Mean Square | F Value | p-Value |
|---|---|---|---|---|---|
| Competition-cooperation (A) | 3,185.77 | 1 | 3,185.77 | 4.00 | 0.049682 |
| Value dissensus (B) | 58.04 | 1 | 54.04 | 0.07 | 0.792174 |
| A × B | 424.95 | 1 | 424.95 | 0.53 | 0.469221 |
| Error | 51,729.98 | 65 | 795.85 | | |
| Total | 55,370.49 | 68 | | | |

a. Compare the overall average for supplier A to that for suppliers B and C. Does it appear that there are large differences (more than two or three quality points) among suppliers?

b. Are the average supplier scores significantly different? How do you know?

17. Compare the overall average for the day shift to that for the night and swing shifts (refer to Table 15.5.4). Does it appear that there are large differences (more than two or three quality points) among shifts? Are these differences significant? How do you know?

18. Is there a significant interaction between supplier and shift in Table 15.5.4? Justify and interpret your answer.

19. Which is better: competition or cooperation? And does the answer depend on whether the participants share the same values? A study by Cosier and Dalton sheds light on these issues.[12] One of their ANOVA tables provides the basis for Table 15.5.5.

a.* The average performance was higher for the cooperation group than for the competition group. Was it significantly higher? How do you know?

b. The average performance was higher when value dissensus was low. Does value dissensus have a significant impact on performance? How do you know?

c. Is the interaction significant? What does this tell you?

20. Are prices really higher in department stores as compared to off-price stores? Kirby and Dardis examined prices of 20 items (shirts, pants, etc.) for 13 weeks and found that prices are indeed 40% higher in department stores.[13] The ANOVA table, adapted from their report, is shown in Table 15.5.6.

a. Are the higher prices (40% higher at department stores on average) significantly higher? How do you know?

b. What kind of ANOVA is this?

c. Identify the three factors in this analysis. How many categories are there for each one?

d. What does the p-value for main effect B tell you?

e. Are there significant differences in pricing from one week to another? How do you know?

TABLE 15.5.6 Analysis of Variance for Prices by Store Type, Item, and Week

| Source of Variation | Sum of Squares | Degrees of Freedom | Mean Square | F Value | p-Value |
|---|---|---|---|---|---|
| Store type (A) | 1,794,577,789 | 1 | 1,794,577,789 | 1,121.52 | 0.000000 |
| Item (B) | 25,726,794,801 | 19 | 1,354,041 | 864.21 | 0.000000 |
| Week (C) | 246,397,563 | 12 | 2,053,313 | 12.83 | 0.000000 |
| Two-way interaction | | | | | |
| A × B | 970,172,936 | 19 | 510,617 | 31.91 | 0.000000 |
| A × C | 69,197,628 | 12 | 5,766,469 | 3.60 | 0.000027 |
| B × C | 320,970,292 | 228 | 140,776 | 0.88 | 0.884253 |
| Three-way interaction | | | | | |
| A × B × C | 264,279,428 | 228 | 115,912 | 0.72 | 0.998823 |
| Residual | 1,664,128,185 | 1040 | 1,600,123 | | |
| Total | 31,056,518,626 | 1559 | | | |

f. Consider the interaction between type of store and item. Is it significant? What does this tell you?

g. Consider the interaction between type of store and week. Is it significant? What does this tell you?

h. Consider the interaction between item and week. Is it significant? What does this tell you?

i. Usually, we examine p-values only to see if they are small enough to declare significance. However, the three-way interaction p-value appears suspiciously large, suggesting that there may be significantly less randomness than was expected for this model. Which technical assumption of hypothesis testing may not be satisfied here?

21. Camera angle can make a difference in advertising; it can even affect the viewer's evaluation of a product. A research article reported a main effect for camera angle ($F_{2,29} = 14.48$, $p < 0.001$) based on an analysis of variance.[14] The average score was 4.51 for eye-level camera angle, 5.49 for a low angle looking up, and 3.61 for a high angle looking down. Higher scores represent more positive evaluations of the product (a personal computer). Are there significant differences among these three camera angles? If so, which angle appears to be best?

22. Another experiment in the report by Meyers-Levy and Peracchio involved the evaluation of bicycle pictures taken with various camera angles, as evaluated by two groups of individuals with different levels of motivation. (The high-motivation group believed they had a reasonable chance to win a bicycle.) Evaluation scores, on average, were higher when the camera angle was upward or at eye level, and lower when the bicycle was viewed looking down. These differences were larger for the low-motivation group. The ANOVA results of the evaluation scores included an examination of the main effect for camera angle

($F_{2,106} = 7.00$, $p < 0.001$), the main effect for motivation ($F_{1,106} = 3.78$, $p < 0.05$), and their interaction ($F_{2,106} = 3.83$, $p < 0.03$).

a. Are there significant differences in the average evaluation scores of the low-motivation and the high-motivation groups? Justify your answer.

b. Does the information provided here from the analysis of variance tell you whether it was the low-motivation group or the high-motivation group that gave higher evaluations, on average?

c. Is there a significant interaction between camera angle and motivation? Justify your answer.

d. Can you conclude that the camera angle makes more of a difference when marketing to the low-motivation group than to the high-motivation group, or are the effects of camera angle basically similar for the two groups, except for randomness? Explain your answer.

12. R. A. Cosier and D. R. Dalton, "Competition and Cooperation: Effects of Value Dissensus and Predisposition to Help," *Human Relations* 41 (1988), pp. 823–39.

13. G. H. Kirby and R. Dardis, "Research Note: A Pricing Study of Women's Apparel in Off-Price and Department Stores," *Journal of Retailing* 62 (1986), pp. 321–30.

14. J. Meyers-Levy and L. A. Perrachio, "Getting an Angle in Advertising: The Effect of Camera Angle on Product Evaluations," *Journal of Marketing Research* 29 (1992), pp. 454–61.

### Database Exercises

Refer to the employee database in Appendix A.

1. Break down the annual salaries into three groups according to training level (A, B, or C).

   a.* Draw box plots to compare these three groups. Comment on what you see.

   **b.\*** Find the average for each training level, and comment.

   **c.\*** Find the between-sample and the within-sample variabilities and their respective degrees of freedom.

   **d.** Find the $F$ statistic and its numbers of degrees of freedom.

   **e.** Perform the $F$ test at level 0.05 and report the results.

   **f.** Report the results of the least-significant-difference test, if appropriate.

   **g.** Summarize what you have learned about the database from this problem.

**2.** Answer the parts of exercise 1 using age in place of annual salary.

**3.** Answer the parts of exercise 1 using experience in place of annual salary.

## Projects

**1.** Find a quantity of interest to you and look up its value on the Internet or in your library for at least 10 firms in each of at least three industry groups. You should thus have at least 30 numbers.

   **a.** Draw box plots, one for each industry group, and summarize your data set. Be sure to use the same scale, to facilitate comparison.

   **b.** Find the average and standard deviation for each industry group.

   **c.** Comment on whether the assumptions for the one-way analysis of variance appear to be (1) satisfied, (2) somewhat satisfied, or (3) not at all satisfied by your data. Correct any serious problem, if possible, by transforming.

   **d.** Find the between-sample variability and its degrees of freedom.

   **e.** Find the within-sample variability and its degrees of freedom.

   **f.** Find the $F$ statistic and its numbers of degrees of freedom.

   **g.** Find the appropriate critical value in the $F$ table at your choice of significance level.

   **h.** Perform the $F$ test and report the results.

   **i.** If the $F$ test is significant, perform the least-significant-difference test for each pair of industry groups and summarize any differences you find.

**2.** From your library, choose a few scholarly journals in a business field of interest to you (the reference librarian may be able to help you). Skim the articles in several issues to locate one that uses the analysis of variance. Write a page summarizing the following:

   **a.** What is the main question being addressed?

   **b.** What kind of data have been analyzed? How were they obtained?

   **c.** Find a hypothesis test that has been performed. Identify the null and research hypotheses. State the results of the test.

## Chapter 16

# Nonparametrics

Testing with Ordinal Data or Nonnormal Distributions

## Chapter Outline

Have you been at all troubled by the assumptions required for statistical inference? Perhaps you should be. In particular, it should be disturbing that the population distribution is required to be *normal* when this can be so difficult to verify based on sample data. Sure, the central limit theorem helps you sometimes, but when your sample size is not large enough or when you have strong skewness or outliers, it would be nice if you had an alternative. You do.

**Nonparametric methods** are statistical procedures for hypothesis testing that do not require a normal distribution (or any other particular shape of distribution) because they are based on counts or ranks (the smallest observation has rank 1, the next is 2, then 3, etc.) instead of the actual data values. These methods still require that you have a random sample from the population, to ensure that your data provide useful information. Because they are based on ranking and do not require computing sums of data values, many nonparametric methods work with *ordinal data* as well as with quantitative data. Here is a summary of two nonparametric approaches:

### The Nonparametric Approach Based on Counts

1. Count the number of times some event occurs in the data set.
2. Use the binomial distribution to decide whether this count is reasonable or not under the null hypothesis.

### The Nonparametric Approach Based on Ranks

1. Create a new data set using the rank of each data value. The **rank** of a data value indicates its position after you order the data set. For example, the data set (35, 95, 48, 38, 57) would become (1, 5, 3, 2, 4) because 35 is the smallest (it has rank 1), 95 is the largest (with rank 5), 48 is the third smallest (rank 3), and so on.
2. Ignore the original data and concentrate on the rank ordering.
3. Use statistical formulas and tables created especially for testing ranks.

**Parametric methods** are statistical procedures that require a completely specified model. Most of our statistical inference so far has required parametric models (including *t* tests, regression tests, and the *F* test). For example, the linear model for regression specifies the prediction equation as well as the exact form of the random noise. By contrast, *non*parametric methods are more flexible and do not require an exact specification of the situation.

The biggest surprise about nonparametric methods is a pleasant one: You lose very little when you fail to take advantage of a normal distribution (when you have one), and you can win very big when your distribution is not normal. Thus, using a nonparametric method is like taking out an insurance policy: You pay a small premium, but you will receive a lot if problems do arise.

One way to measure the effectiveness of different statistical tests is in terms of their efficiency. One test is said to be more **efficient** than another if it makes better use of the information in the data.[1] Thus, nonparametric methods are nearly as efficient as parametric ones when you have a normal distribution and can be much more efficient when you don't. Here is a summary of the advantages of the nonparametric approach:

**Advantages of Nonparametric Testing**

1. No need to assume normality; can be used even if the distribution is not normal.
2. Avoids many problems of transformation; can be used even if data cannot easily be transformed to be normal and, in fact, gives the same result whether you transform or not.
3. Can even be used to test ordinal data because ranks can be found based on the natural ordering.
4. Can be much more efficient than parametric methods when distributions are not normal.

There is only one disadvantage of nonparametric methods, and it is relatively small:

**Disadvantage of Nonparametric Testing**

Less statistically efficient than parametric methods when distributions are normal; however, this efficiency loss is often slight.

In this chapter, you will learn about the one-sample problem (testing the median), as well as the paired and unpaired two-sample problems (testing for a difference).

## 16.1 TESTING THE MEDIAN AGAINST A KNOWN REFERENCE VALUE

On the one hand, when you have an ordinary univariate sample of data from a population, you might use the average and standard error to test a hypothesis about the population mean (the *t* test). And this is fine if the distribution is normal.

On the other hand, the nonparametric approach, because it is based on the rank ordering of the data, tests the population *median*. The median is the appropriate summary because it is defined in terms of ranks. (Remember that the median has rank $(1 + n)/2$ for a sample of size $n$.)

How can we get rid of the normal distribution assumption? It's easy once you realize that half of the population is below the median and half is above, if the population distribution is continuous.[2] There is a binomial probability distribution inherent here, since the data set is a random sample of independent observations. Using probability language from Chapters 6 and 7, we know the number of data values below the population median is the number of "below-median" events that occur in $n$ independent trials, where each event has probability 1/2. Therefore:

The number of sample data values below a continuous population's median follows a binomial distribution where $\pi = 0.5$ and $n$ is the sample size.

## The Sign Test

The *sign test* makes use of this binomial distribution. To test whether or not the population median could reasonably be $65,536, for example, you could see how many sample values fall below $65,536 and determine if this is a reasonable observation from a binomial distribution. The **sign test** decides whether the population median is equal to a given reference value based on the number of sample values that fall below that reference value. No arithmetic is performed on the data values, only comparing and counting. Here is the procedure:

**The Sign Test**

1. Count the number of data values that are *different* from the reference value, $\theta_0$. This number is $m$, the **modified sample size**.
2. Find the limits in the table for this modified sample size.
3. Count how many data values fall below the reference value, $\theta_0$, and compare this number to the limits in the table.[3]
4. If the count from step 3 falls *outside* the limits of the table, the difference is statistically significant. If it falls *at or within* the limits, the difference is not statistically significant.

3. If you prefer, you may count how many fall *above* the reference value. The result of the test will be the same.

## The Hypotheses

First, assume that the population distribution is continuous. The null hypothesis for the sign test claims that the population median, $\theta$, is exactly equal to some specified reference value, $\theta_0$. (As usual, this reference value is assumed to be known precisely and was not computed

---

from the current data set.) The research hypothesis claims the contrary: The population median is not equal to this reference value.

**Hypothesis for the Sign Test for the Median of a Continuous Population Distribution**

$$H_0: \theta = \theta_0$$

$$H_1: \theta \neq \theta_0$$

where $\theta$ is the (unknown) population median and $\theta_0$ is the (known) reference value being tested.

In general, even if the distribution is not continuous, the sign test will decide whether or not your reference value, $\theta_0$, divides the population exactly in half:[4]

**Hypotheses for the Sign Test in General**

$H_0$: The probability of being above $\theta_0$ is equal to the probability of being below $\theta_0$ in the population

$H_1$: These probabilities are not equal

where $\theta_0$ is the (known) reference value being tested.

## The Assumption

There is an assumption required for validity of the sign test. One of the strengths of this nonparametric method is that so little is required for it to be valid.

**Assumption Required for the Sign Test**

The data set is a random sample from the population of interest.

Table 16.1.1 lists the ranks for the sign test. If $m$ is larger than 100, you would find the table values for level 0.05 by rounding $(m - 1.960\sqrt{m})/2$ and $(m + 1.960\sqrt{m})/2$ to the nearest whole numbers. For example, for $m = 120$, these formulas give 49.3 and 70.7, which round to the table values 49 and 71. For level 0.01, you would round $(m - 2.576\sqrt{m})/2$ and $(m + 2.576\sqrt{m})/2$.

---

4. This is slightly different from $\theta_0$ being the median. For example, the (very small) population consisting of the numbers (11, 12, 13, 13, 14) has a median of 13. However, there are two values below but just one above 13. Thus, the null hypothesis of the sign test specifies *more* than just that the population median be $\theta_0$.

**TABLE 16.1.1 Ranks for the Sign Test**

| Modified Sample Size, $m$ | 5% Test Level Sign Test Is Significant If Number Is Either | | | 1% Test Level Sign Test Is Significant If Number Is Either | | |
|---|---|---|---|---|---|---|
| | Less than | or | More than | Less than | or | More than |
| 6 | 1 | | 5 | — | | — |
| 7 | 1 | | 6 | — | | — |
| 8 | 1 | | 7 | 1 | | 7 |
| 9 | 2 | | 7 | 1 | | 8 |
| 10 | 2 | | 8 | 1 | | 9 |
| 11 | 2 | | 9 | 1 | | 10 |
| 12 | 3 | | 9 | 2 | | 10 |
| 13 | 3 | | 10 | 2 | | 11 |
| 14 | 3 | | 11 | 2 | | 12 |
| 15 | 4 | | 11 | 3 | | 12 |
| 16 | 4 | | 12 | 3 | | 13 |
| 17 | 5 | | 12 | 3 | | 14 |
| 18 | 5 | | 13 | 4 | | 14 |
| 19 | 5 | | 14 | 4 | | 15 |
| 20 | 6 | | 14 | 4 | | 16 |
| 21 | 6 | | 15 | 5 | | 16 |
| 22 | 6 | | 16 | 5 | | 17 |
| 23 | 7 | | 16 | 5 | | 18 |
| 24 | 7 | | 17 | 6 | | 18 |
| 25 | 8 | | 17 | 6 | | 19 |
| 26 | 8 | | 18 | 7 | | 19 |
| 27 | 8 | | 19 | 7 | | 20 |
| 28 | 9 | | 19 | 7 | | 21 |
| 29 | 9 | | 20 | 8 | | 21 |
| 30 | 10 | | 20 | 8 | | 22 |
| 31 | 10 | | 21 | 8 | | 23 |
| 32 | 10 | | 22 | 9 | | 23 |
| 33 | 11 | | 22 | 9 | | 24 |
| 34 | 11 | | 23 | 10 | | 24 |
| 35 | 12 | | 23 | 10 | | 25 |
| 36 | 12 | | 24 | 10 | | 26 |
| 37 | 13 | | 24 | 11 | | 26 |

*(Continued)*

## TABLE 16.1.1 Ranks for the Sign Test—cont'd

| Modified Sample Size, m | 5% Test Level Sign Test Is Significant If Number Is Either | | | 1% Test Level Sign Test Is Significant If Number Is Either | | |
|---|---|---|---|---|---|---|
| | Less than | or | More than | Less than | or | More than |
| 38 | 13 | | 25 | 11 | | 27 |
| 39 | 13 | | 26 | 12 | | 27 |
| 40 | 14 | | 26 | 12 | | 28 |
| 41 | 14 | | 27 | 12 | | 29 |
| 42 | 15 | | 27 | 13 | | 29 |
| 43 | 15 | | 28 | 13 | | 30 |
| 44 | 16 | | 28 | 14 | | 30 |
| 45 | 16 | | 29 | 14 | | 31 |
| 46 | 16 | | 30 | 14 | | 32 |
| 47 | 17 | | 30 | 15 | | 32 |
| 48 | 17 | | 31 | 15 | | 33 |
| 49 | 18 | | 31 | 16 | | 33 |
| 50 | 18 | | 32 | 16 | | 34 |
| 51 | 19 | | 32 | 16 | | 35 |
| 52 | 19 | | 33 | 17 | | 35 |
| 53 | 19 | | 34 | 17 | | 36 |
| 54 | 20 | | 34 | 18 | | 36 |
| 55 | 20 | | 35 | 18 | | 37 |
| 56 | 21 | | 35 | 18 | | 38 |
| 57 | 21 | | 36 | 19 | | 38 |
| 58 | 22 | | 36 | 19 | | 39 |
| 59 | 22 | | 37 | 20 | | 39 |
| 60 | 22 | | 38 | 20 | | 40 |
| 61 | 23 | | 38 | 21 | | 40 |
| 62 | 23 | | 39 | 21 | | 41 |
| 63 | 24 | | 39 | 21 | | 42 |
| 64 | 24 | | 40 | 22 | | 42 |
| 65 | 25 | | 40 | 22 | | 43 |
| 66 | 25 | | 41 | 23 | | 43 |
| 67 | 26 | | 41 | 23 | | 44 |
| 68 | 26 | | 42 | 23 | | 45 |
| 69 | 26 | | 43 | 24 | | 45 |
| 70 | 27 | | 43 | 24 | | 46 |
| 71 | 27 | | 44 | 25 | | 46 |
| 72 | 28 | | 44 | 25 | | 47 |
| 73 | 28 | | 45 | 26 | | 47 |
| 74 | 29 | | 45 | 26 | | 48 |
| 75 | 29 | | 46 | 26 | | 49 |
| 76 | 29 | | 47 | 27 | | 49 |
| 77 | 30 | | 47 | 27 | | 50 |
| 78 | 30 | | 48 | 28 | | 50 |
| 79 | 31 | | 48 | 28 | | 51 |
| 80 | 31 | | 49 | 29 | | 51 |
| 81 | 32 | | 49 | 29 | | 52 |
| 82 | 32 | | 50 | 29 | | 53 |
| 83 | 33 | | 50 | 30 | | 53 |
| 84 | 33 | | 51 | 30 | | 54 |
| 85 | 33 | | 52 | 31 | | 54 |
| 86 | 34 | | 52 | 31 | | 55 |
| 87 | 34 | | 53 | 32 | | 55 |
| 88 | 35 | | 53 | 32 | | 56 |
| 89 | 35 | | 54 | 32 | | 57 |
| 90 | 36 | | 54 | 33 | | 57 |
| 91 | 36 | | 55 | 33 | | 58 |
| 92 | 37 | | 55 | 34 | | 58 |
| 93 | 37 | | 56 | 34 | | 59 |
| 94 | 38 | | 56 | 35 | | 59 |
| 95 | 38 | | 57 | 35 | | 60 |
| 96 | 38 | | 58 | 35 | | 61 |
| 97 | 39 | | 58 | 36 | | 61 |
| 98 | 39 | | 59 | 36 | | 62 |
| 99 | 40 | | 59 | 37 | | 62 |
| 100 | 40 | | 60 | 37 | | 63 |

## Example

### Comparing Local to National Family Income

Your upscale restaurant is considering franchises in new communities. One of the ways you screen is by looking at *median* family income, because the *mean* family income might be high due to just a few families. A survey of one

### TABLE 16.1.2  Incomes of Sampled Families

| | | | |
|---|---|---|---|
| $39,465 | $96,270 | $16,477* | $138,933 |
| 80,806 | 85,421 | 5,921* | 70,547 |
| 267,525 | 56,240 | 187,445 | 81,802 |
| 163,819 | 14,706* | 83,414 | 78,464 |
| 58,525 | 54,348 | 36,346 | |
| 25,479* | 7,081* | 19,605* | |
| 29,341 | 137,414 | 156,681 | |

*\* Income below $27,735.*

community estimated the median family income as $70,547, and you are wondering whether this is significantly higher than the national median family income of $27,735.[5] It certainly *appears* that this community has a higher median income, but with a sample of only 25 families, you would like to be careful before coming to a conclusion. Table 16.1.2 shows the data set, indicating those families with incomes below $27,735.

Your reference value is $\theta_0 = \$27,735$, a number that is not from the data set itself. Here are the steps involved in performing the sign test:

1. All 25 families have incomes different from this reference value, so the modified sample size is $m = 25$, the same as the actual sample size.
2. The limits from the table for testing at the 5% level are 8 and 17 for $m = 25$.
3. There are six families with incomes below the reference value.
4. Since the number 6 falls outside the limits (i.e., it is less than 8), you reject the null hypothesis and conclude that the result is statistically significant:

   The observed median family income of $70,547 for this community is significantly different from the national median family income of $27,735.

Your suspicions have been confirmed: This is indeed an upscale community. The median family income in the community is significantly higher than the national median.[6]

---

5. As reported for 1985 in U.S. Bureau of the Census, *Statistical Abstract of the United States, 1987* (Washington, DC, 1986), p. 437.
6. This is the one-sided conclusion to a two-sided test, as described in Chapter 10.

## 16.2  TESTING FOR DIFFERENCES IN PAIRED DATA

When your data set consists of *paired* observations, arranged as two columns, you can create a single sample that represents the changes or the differences between

them. This is appropriate, for example, for before/after studies, where you have a measurement for each person or thing both before and after some intervention (seeing an advertisement, taking a medication, adjusting the gears, etc.). That is how the paired *t* test worked in Chapter 10. Here we will illustrate the nonparametric solution.

## Using the Sign Test on the Differences

The nonparametric procedure for testing whether two columns of values are significantly different, the **sign test for the differences**, applies the sign test (from the previous section) to a single column representing the differences between the two columns. The reference value, $\theta_0$, will be 0, representing "no net difference" in the population. The sign test will then determine whether the changes are balanced (so that there are as many increases as decreases, except for randomness) or systematically different (for example, significantly more increases than decreases).

Table 16.2.1 shows how the data set for a typical application would look. In some applications, column 1 (*X*) would represent "before" and column 2 (*Y*) "after." It is important that there be a natural pairing so that each row represents two observations (in the same measurement units) for the *same* person or thing. Here is how to perform the test:

### The Sign Test for the Differences

1. Count the number of data values that change between columns 1 and 2. This number is *m*, the modified sample size.
2. Find the limits in the table for this modified sample size.
3. Count how many data values went down (that is, have a smaller value in column 2 compared to column 1) and compare this count to the limits in the table.[7]
4. If this count falls *outside* the limits from the table, then the two samples are significantly different. If it falls *at* or *within* the limits, then the two samples are not statistically different.

---

7. If you prefer, you may count how many went *up* instead. The result of the test will be the same.

Note that only the direction (up or down) of the change matters (from column 1 to column 2), not the actual size of the change. This implies that you can use this test on ordinal as well as on quantitative data. But some sense of ordering is required so that you can know the direction (up or down) of the change.

| TABLE 16.2.1 Paired Observations | | |
|---|---|---|
| Elementary Units | Column 1 | Column 2 |
| 1 | $X_1$ | $Y_1$ |
| 2 | $X_2$ | $Y_2$ |
| . | . | . |
| . | . | . |
| . | . | . |
| $n$ | $X_n$ | $Y_n$ |

| TABLE 16.2.2 Level of Creativity | | | |
|---|---|---|---|
| Ad 1 | Ad 2 | Ad 1 | Ad 2 |
| 4 | 2 | 5 | 4 |
| 2 | 4 | 3 | 4 |
| 4 | 5 | 3 | 5 |
| 4 | 4 | 4 | 5 |
| 4 | 4 | 5 | 5 |
| 2 | 5 | 4 | 5 |
| 3 | 3 | 5 | 4 |
| 4 | 5 | 2 | 5 |
| 3 | 5 | | |

## The Hypotheses

The null hypothesis claims that just as many units go up (comparing the paired data values $X$ and $Y$) as down in the population. Any net movement up or down in the sample would just be random under this hypothesis. The research hypothesis claims that the probabilities of going up and down are different.

### Hypotheses for the Sign Test for the Differences

$H_0$: The probability of $X < Y$ equals the probability of $Y < X$. That is, the probability of going up equals the probability of going down.

$H_1$: The probability of $X < Y$ is *not* equal to the probability of $Y < X$. The probabilities of going up and down are unequal.

## The Assumption

As with other, similar tests, there is an assumption required for validity of the sign test for the differences.

### Assumption Required for the Sign Test for the Differences

The data set is a random sample from the population of interest. Each elementary unit in this population has both values $X$ and $Y$ measured for it.

### Example
#### Rating Two Advertisements

Two advertisements were shown to each member of a group of 17 people. Each ad was scored by each person on a scale from 1 to 5 indicating the creativity of the advertisement. The results are shown in Table 16.2.2.

1. The number of data values that went either up or down (from ad 1 to ad 2) is 13. That is, 13 people gave different scores to the two ads, and the remaining 4 gave the same score to both. Thus, the modified sample size is $m = 13$.
2. The limits in the table for testing at the 5% level at $m = 13$ are 3 and 10.
3. Three people gave ad 2 a lower score than ad 1. This is within the limits of the table (you would have to find either *less than 3* or *more than 10* such people for the result to be significant).
4. You therefore accept the null hypothesis that the creativity ratings for these two ads are similar.

The result is not significant. Even though 3 out of 17 people rated ad 1 higher in creativity, this could reasonably be due to random chance and not to any particular quality of the advertisements.

## 16.3 TESTING TO SEE IF TWO UNPAIRED SAMPLES ARE SIGNIFICANTLY DIFFERENT

Now suppose that you have two independent (unpaired) samples and want to test whether or not they could have come from populations with the same distribution. On the one hand, the unpaired $t$ test from Chapter 10 assumes that the distributions are normal (with the same standard deviation for small samples) and then tests whether the means are identical. On the other hand, the nonparametric approach assumes only that you have two random samples from two populations and then tests whether the populations' distributions are identical. Table 16.3.1 shows how a typical data set with two unpaired samples would look.

**TABLE 16.3.1 Two Unpaired Samples**

| Sample 1 ($n_1$ observations from population 1) | Sample 2 ($n_2$ observations from population 2) |
|---|---|
| $X_{1,1}$ | $X_{2,1}$ |
| $X_{1,2}$ | $X_{2,2}$ |
| . | . |
| . | . |
| . | . |
| $X_{1,n_1}$ | $X_{2,n_2}$ |

Note: The two sample sizes, $n_1$ and $n_2$, may be different.

## The Procedure Is Based on the Ranks of All of the Data

The procedure is to first *put both samples together* and define an overall set of ranks. If one sample has systematically smaller values than the other, its ranks will be smaller, too. By comparing the overall ranks of one sample to those of the other sample, you can test whether they are systematically or just randomly different.

There are several different ways to get the same basic answer to this problem. The **Wilcoxon rank-sum test** and the **Mann–Whitney $U$ test** are two different ways to compute the same result of a nonparametric test for two unpaired samples; they both lead to the same conclusion. The Wilcoxon rank-sum test is based on the sum of the overall ranks in one of the samples, and the Mann–Whitney $U$ test is based on the number of ways you can find a value in one sample that is bigger than a value in the other sample.

An easier approach is to work with the *average* rank of the two samples. This test is algebraically equivalent to the others (in the sense that the Wilcoxon test, the Mann–Whitney test, and the average rank difference test as presented here always lead to the same result) and makes clear the main idea of what is happening with many nonparametric methods: Although you work with the ranks instead of the data values, the basic ideas of statistics remain the same.[8] Here's how to perform this test:

---

8. For example, here is the formula to express the average difference in ranks in terms of the $U$ statistic: $(n_1 + n_2)(U - n_1 n_2/2)/(n_1 n_2)$. The Mann–Whitney $U$ statistic is defined as $n_1 n_2 + n_1(n_1 + 1)/2$ minus the sum of the overall ranks in the first sample.

### The Nonparametric Test for Two Unpaired Samples

1. Put both samples together and sort them to obtain the *overall ranks*. If you have repeated numbers (ties), use the average of their ranks so that equal numbers are assigned equal ranks.
2. Find the average overall rank for each sample, $\overline{R}_1$ and $\overline{R}_2$.
3. Find the difference between these average overall ranks, $\overline{R}_2 - \overline{R}_1$.
4. Find the standard error for the average difference in the ranks:[9]

$$\text{Standard error} = (n_1 + n_2)\sqrt{\frac{n_1 + n_2 + 1}{12 n_1 n_2}}$$

5. Divide the average difference (from step 3) by its standard error (from step 4) to find the test statistic:

$$\text{Test statistic} = \frac{\overline{R}_2 - \overline{R}_1}{(n_1 + n_2)\sqrt{\dfrac{n_1 + n_2 + 1}{12 n_1 n_2}}}$$

6. If the test statistic is larger than 1.960 in magnitude, the two samples are *significantly different*. If the test statistic is smaller than 1.960 in magnitude, the two samples are *not significantly different*.[10]

---

9. This standard error is exact in the absence of ties. There is no need for estimation because it can be computed directly from the properties of randomly shuffled ranks under the null hypothesis.

10. To use a different test level, you would substitute the appropriate $t$ table value with an infinite number of degrees of freedom in place of 1.960 here. For example, to test at the 1% level, use 2.576 in place of 1.960.

## The Hypotheses

The null hypothesis claims that the two samples were drawn from populations with the *same distribution;* the research hypothesis claims that these population distributions are different.

### Hypotheses for Testing Two Unpaired Samples

$H_0$: The two samples come from populations with the same distribution.

$H_1$: The two samples come from populations with different distributions.

## The Assumptions

There are assumptions required for validity of the test for two unpaired samples. In addition to the usual requirement of random sampling, if you want to use the $t$ table value with an infinite number of degrees of freedom, the sample sizes must be large enough.

### Assumptions Required for Testing Two Unpaired Samples

1. Each sample is a random sample from its population.
2. More than 10 elementary units have been chosen from each population, so that $n_1 > 10$ and $n_2 > 10$.

### Example
#### Fixed-Rate and Adjustable-Rate Mortgage Applicants

The bank is planning a marketing campaign for home equity loans. Some people in the meeting feel that variable-rate mortgages appeal more to the lower-income applicants because they can qualify for a larger loan amount and therefore can afford to buy a more expensive house. Others feel that the higher risk of a variable-rate mortgage appeals more to the higher-income applicants because they have a larger "cushion" to use in case their payments go up in the future. Which group is correct? You have just compiled some data on the incomes of recent mortgage applicants, which are shown in Table 16.3.2.

Note the outlier ($240,000). This alone would call into question the use of a two-sample *t* test. You called to check and found that the number is correct. Should you delete the outlier? Probably not, because it really does represent a high-income family applying for a fixed-rate loan, and this is useful information to you.

Here comes nonparametrics to the rescue! Because it works with ranks instead of the actual data values, it wouldn't matter if the highest income were $1 trillion; it would still be treated as simply the largest value.

A look at this two-sample data set using box plots is provided in Figure 16.3.1. This figure suggests that the variable-rate group has higher incomes, but this is unclear due to the considerable overlap between the two groups.

You will have to rank-order *all* of the income data. To do this, create a new database structure with one column for income (listing *both* columns of the original data set) and another as mortgage type, as shown in Table 16.3.3.

Now you are ready to sort the database in order by income; the results are shown in Table 16.3.4. You will want to indicate the mortgage type along with each income level in the sorting process; this is an easy task on a computer spreadsheet. After sorting, list the ranks as the numbers 1, 2, 3, and so forth.

If you have ties (two or more income levels that are the same), use the average rank for these levels. For example, in Table 16.3.4 the income level $36,500 occurs three times (at ranks 12, 13, and 14), so the average rank for $36,500, (12 + 13 + 14)/3 = 13, is listed for all three occurrences. There are two incomes at $57,000, so the average rank, (18 + 19)/2 = 18.5, is listed for both.

At this point you are ready to compute the average ranks. From the column of ranks, first select only the numbers representing fixed-rate mortgages and compute their average rank:

$$\frac{\left(\begin{array}{c}1+3+4+5+6+7+9+13+16+17\\+18.5+18.5+20+23+25+30\end{array}\right)}{16} = \frac{216}{16} = 13.50$$

**TABLE 16.3.2 Incomes of Mortgage Applicants**

| Fixed-Rate | Variable-Rate |
|---|---|
| $34,000 | $37,500 |
| 25,000 | 86,500 |
| 41,000 | 36,500 |
| 57,000 | 65,500 |
| 79,000 | 21,500 |
| 22,500 | 36,500 |
| 30,000 | 99,500 |
| 17,000 | 36,000 |
| 36,500 | 91,000 |
| 28,000 | 59,500 |
| 240,000 | 31,000 |
| 22,000 | 88,000 |
| 57,000 | 35,500 |
| 68,000 | 72,000 |
| 58,000 | |
| 49,500 | |

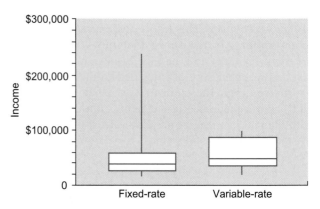

**FIGURE 16.3.1**   Although the highest income is in the fixed-rate group, it appears that incomes are actually higher in general for the variable-rate group. However, there is considerable overlap in income scores between these two groups. The presence of the large outlier would be a problem for a two-sample *t* test, but it is no problem for a nonparametric test.

## TABLE 16.3.3 Initial Database before Sorting

| Income | Mortgage Type | Income | Mortgage Type |
|--------|---------------|--------|---------------|
| $34,000 | Fixed | $49,500 | Fixed |
| 25,000 | Fixed | 37,500 | Variable |
| 41,000 | Fixed | 86,500 | Variable |
| 57,000 | Fixed | 36,500 | Variable |
| 79,000 | Fixed | 65,500 | Variable |
| 22,500 | Fixed | 21,500 | Variable |
| 30,000 | Fixed | 36,500 | Variable |
| 17,000 | Fixed | 99,500 | Variable |
| 36,500 | Fixed | 36,000 | Variable |
| 28,000 | Fixed | 91,000 | Variable |
| 240,000 | Fixed | 59,500 | Variable |
| 22,000 | Fixed | 31,000 | Variable |
| 57,000 | Fixed | 88,000 | Variable |
| 68,000 | Fixed | 35,500 | Variable |
| 58,000 | Fixed | 72,000 | Variable |

Next, calculate the average rank for variable-rate mortgages:

$$\frac{\left(\begin{array}{c}2+8+10+11+13+13+15+21\\+22+24+26+27+28+29\end{array}\right)}{14} = \frac{249}{14} = 17.7857$$

Table 16.3.5 shows what has been accomplished. The original income figures are listed in the original order, and the ranks have been assigned. Note that our outlier ($240,000) has the highest rank (30) but that its *rank* is not an outlier. Note also that (as was suggested by the box plots) the apparently lower income level corresponding to fixed-rate mortgages has a lower average rank.

But are incomes for fixed-rate and variable-rate applicants different? That is, are these average ranks (13.50 and 17.79) significantly different? The standard error of the average difference in the ranks is needed next:

$$\text{Standard error} = (n_1 + n_2)\sqrt{\frac{n_1 + n_2 + 1}{12 n_1 n_2}}$$

$$= (16 + 14)\sqrt{\frac{16 + 14 + 1}{12 \times 16 \times 14}}$$

$$= 30\sqrt{\frac{31}{2,688}}$$

$$= 3.2217$$

*(Continued)*

## TABLE 16.3.4 Database after Sorting Income, Then Including Ranks

| Income | Mortgage Type | Ranks by Income (ties, in bold, have been averaged) | Income | Mortgage Type | Ranks by Income (ties, in bold, have been averaged) |
|--------|---------------|------|--------|---------------|------|
| $17,000 | Fixed | 1 | $41,000 | Fixed | 16 |
| 21,500 | Variable | 2 | 49,500 | Fixed | 17 |
| 22,000 | Fixed | 3 | 57,000 | Fixed | **18.5** |
| 22,500 | Fixed | 4 | 57,000 | Fixed | **18.5** |
| 25,000 | Fixed | 5 | 58,000 | Fixed | 20 |
| 28,000 | Fixed | 6 | 59,500 | Variable | 21 |
| 30,000 | Fixed | 7 | 65,500 | Variable | 22 |
| 31,000 | Variable | 8 | 68,000 | Fixed | 23 |
| 34,000 | Fixed | 9 | 72,000 | Variable | 24 |
| 35,500 | Variable | 10 | 79,000 | Fixed | 25 |
| 36,000 | Variable | 11 | 86,500 | Variable | 26 |
| 36,500 | Fixed | **13** | 88,000 | Variable | 27 |
| 36,500 | Variable | **13** | 91,000 | Variable | 28 |
| 36,500 | Variable | **13** | 99,500 | Variable | 29 |
| 37,500 | Variable | 15 | 240,000 | Fixed | 30 |

**Example—cont'd**

The test statistic is the difference in ranks divided by the standard error:

$$\text{Test statistic} = \frac{\text{Average difference in ranks}, \overline{R}_2 - \overline{R}_1}{\text{Standard error}}$$

$$= \frac{17.7857 - 13.5000}{3.2217}$$

$$= 1.3303$$

Because the magnitude of this test statistic (you would now ignore a minus sign, if any), 1.3303, is less than 1.960, the two samples are not significantly different. The observed differences between incomes of fixed-rate and variable-rate mortgage applicants are *not* statistically significant.

**TABLE 16.3.5 Income and Ranks for Mortgage Applicants**

| | Fixed-Rate | | Variable-Rate | |
|---|---|---|---|---|
| Income | Rank | | Income | Rank |
| $34,000 | 9 | | $37,500 | 15 |
| 25,000 | 5 | | 86,500 | 26 |
| 41,000 | 16 | | 36,500 | 13 |
| 57,000 | 18.5 | | 65,500 | 22 |
| 79,000 | 25 | | 21,500 | 2 |
| 22,500 | 4 | | 36,500 | 13 |
| 30,000 | 7 | | 99,500 | 29 |
| 17,000 | 1 | | 36,000 | 11 |
| 36,500 | 13 | | 91,000 | 28 |
| 28,000 | 6 | | 59,500 | 21 |
| 240,000 | 30 | | 31,000 | 8 |
| 22,000 | 3 | | 88,000 | 27 |
| 57,000 | 18.5 | | 35,500 | 10 |
| 68,000 | 23 | | 72,000 | 24 |
| 58,000 | 20 | | | |
| 49,500 | 17 | | | |
| Average rank | $\overline{R}_1 = 13.50$ | | | $\overline{R}_2 = 17.7857$ |

*Sample sizes: $n_1 = 16$, $n_2 = 14$.*

## 16.4 END-OF-CHAPTER MATERIALS

### Summary

**Nonparametric methods** are statistical procedures for hypothesis testing that do not require a normal distribution (or any other particular shape of distribution) because they are based on counts or ranks instead of the actual data values. Many nonparametric methods work with ordinal data as well as with quantitative data. To use the nonparametric approach based on counts:

1. Count the number of times some event occurs in the data set.
2. Use the binomial distribution to decide whether or not this count is reasonable under the null hypothesis.

To use the nonparametric approach based on ranks:

1. Create a new data set using the rank of each data value. The **rank** of a data value indicates its position after you order the data set. For example, the data set (35, 95, 48, 38, 57) would become (1, 5, 3, 2, 4) because 35 is the smallest (it has rank 1), 95 is the largest (with rank 5), 48 is ranked as the third-smallest (rank 3), and so on.
2. Ignore the original data values and concentrate only on the rank ordering.
3. Use statistical formulas and tables created especially for testing ranks.

**Parametric methods** are statistical procedures that require a completely specified model, as do most of the methods considered before this chapter. One issue is the efficiency of nonparametric tests as compared to parametric ones. One test is said to be more **efficient** than another if it makes better use of the information in the data. Nonparametric tests have many advantages:

1. There is no need to assume a normal distribution.
2. Many problems of transformation are avoided. In fact, the test gives the same result whether you transform or not.
3. Even ordinal data can be tested because ranks can be found based on the natural ordering.
4. Such tests can be much more efficient than parametric methods when distributions are not normal.

The only disadvantage of nonparametric testing is that it is less statistically efficient than parametric methods when distributions are normal; however, the efficiency lost is often slight.

The **sign test** decides whether the population median is equal to a given reference value based on the number of sample values that fall below that reference value. Instead of assuming a normal distribution, the theory is based on the fact that the number of sample data values below a continuous population's median follows a binomial distribution, where $\pi = 0.5$ and $n$ is the sample size.

To perform the sign test:

1. Count the number of data values that differ from the reference value, $\theta_0$. This number is $m$, the **modified sample size**.
2. Find the limits in the table for $m$.
3. Count how many data values fall below the reference value and compare this number to the limits in the table.
4. If the count from step 3 falls outside the limits from the table, the difference is statistically significant. If it falls at or within the limits, the difference is not statistically significant.

The hypotheses for the sign test for the median of a continuous population distribution are

$$H_0: \theta = \theta_0$$

$$H_1: \theta \neq \theta_0$$

where $\theta$ is the (unknown) population median and $\theta_0$ is the (known) reference value being tested.

The hypotheses for the sign test in general are

$H_0$: The probability of being above $\theta_0$ is equal to the probability of being below $\theta_0$ in the population

$H_1$: These probabilities are not equal

where $\theta_0$ is the (known) reference value being tested. The data set is assumed to be a random sample from the population of interest.

When you have paired observations ("before and after," for example), you may apply the sign test to the differences or changes. This nonparametric procedure for testing whether the two columns are significantly different is called the **sign test for the differences**. The procedure is as follows:

1. Count the number of data values that differ between columns 1 and 2. This number is $m$, the modified sample size.
2. Find the limits in the table for $m$.
3. Count how many data values went down (that is, have a smaller value in column 2 than in column 1) and compare this count to the limits in the table.
4. If this count falls outside the limits from the table, then the two samples are significantly different. If it falls at or within the limits, then the two samples are not statistically different.

The hypotheses for the sign test for the differences are

$H_0$: The probability of $X < Y$ equals the probability of $Y < X$. That is, the probability of going up equals the probability of going down.

$H_1$: The probability of $X < Y$ is *not* equal to the probability of $Y < X$. The probabilities of going up and down are unequal.

The data set is assumed to be a random sample from the population of interest, where each elementary unit in this population has both $X$ and $Y$ values measured for it.

If you have two independent (unpaired) samples and wish to test for differences, there is a nonparametric procedure that substitutes for the unpaired $t$ test.

The **Wilcoxon rank-sum test** and the **Mann–Whitney $U$ test** are two different ways to compute the same result of a nonparametric test for two unpaired samples. The Wilcoxon rank-sum test is based on the sum of the overall ranks in one of the samples, and the Mann–Whitney $U$ test is based on the number of ways you can find a value in one sample that is bigger than a value in the other sample. An easier approach is to work with the *average rank* of the two samples:

1. Put both samples together and sort them to obtain the overall ranks. In case of ties, use the average of their ranks.
2. Find the average overall rank for each sample, $\overline{R}_1$ and $\overline{R}_2$.
3. Find the difference between these overall ranks, $\overline{R}_2 - \overline{R}_1$.
4. Find the standard error for the average difference in the ranks:

$$\text{Standard error} = (n_1 + n_2)\sqrt{\frac{n_1 + n_2 + 1}{12 n_1 n_2}}$$

5. Divide the difference from step 3 by its standard error from step 4 to find the test statistic:

$$\text{Test statistic} = \frac{\overline{R}_2 - \overline{R}_1}{(n_1 + n_2)\sqrt{\dfrac{n_1 + n_2 + 1}{12 n_1 n_2}}}$$

6. If the test statistic is larger than 1.960 in magnitude, the two samples are significantly different. If the test statistic is smaller than 1.960 in magnitude, the two samples are not significantly different.

The hypotheses for testing two unpaired samples are

$H_0$: The two samples come from populations with the same distribution.

$H_1$: The two samples come from populations with different distributions.

The assumptions that must be satisfied for testing two unpaired samples are

1. Each sample is a random sample from its population.
2. More than 10 elementary units have been chosen from each population, so that $n_1 > 10$ and $n_2 > 10$.

# Key Words

efficient, *492*
Mann–Whitney *U* test, *497*
modified sample size, *492*
nonparametric methods, *491*
parametric methods, *491*
rank, *491*
sign test, *492*
sign test for the differences, *495*
Wilcoxon rank-sum test, *497*

## Questions

1. **a.** What is a nonparametric statistical method?
   **b.** For the nonparametric approach based on counts, what is being counted? What probability distribution is used to make the decision?
   **c.** For the nonparametric approach based on ranks, what information is disregarded from the data set? What is substituted in its place?
2. **a.** What is a parametric statistical method?
   **b.** Name some parametric methods you have used.
3. **a.** List the advantages of nonparametric testing over parametric methods, if any.
   **b.** List the disadvantages, if any, of nonparametric testing compared with parametric methods. How serious are these shortcomings?
4. **a.** How should you interpret this statement: One test is more efficient than another?
   **b.** If the distribution is normal, which would be more efficient, a parametric test or a nonparametric test?
   **c.** If the distribution is far from normal, which would be likely to be more efficient, a parametric test or a nonparametric test?
5. **a.** For a continuous population, which measure of typical value is the sign test concerned with?
   **b.** What probability distribution does this test rely on?
   **c.** Suppose the population is discrete and an appreciable fraction of the population is equal to its median. How do the hypotheses for the sign test change as compared to the case of a continuous population?
6. **a.** Can the sign test be used with quantitative data? Why or why not?
   **b.** Can the sign test be used with ordinal data? Why or why not?
   **c.** Can the sign test be used with nominal data? Why or why not?
7. **a.** What assumption must be met for the sign test to be valid?
   **b.** What assumption is not required for the sign test but would be required for a *t* test to be valid?
8. Describe the similarities and differences between the sign test and the *t* test.
9. **a.** What kind of data set is appropriate for the sign test for the differences?
   **b.** What hypotheses are being tested?
   **c.** What assumption is required?

10. **a.** Can the sign test for the differences be used with quantitative data? Why or why not?
    **b.** Can the sign test for the differences be used with ordinal data? Why or why not?
    **c.** Can the sign test for the differences be used with nominal data? Why or why not?
11. Describe the similarities and differences between the sign test for the differences and the paired *t* test.
12. **a.** Describe the data set consisting of two unpaired samples.
    **b.** What hypotheses would ordinarily be tested for such data?
13. **a.** What is the difference (if any) between the Wilcoxon rank-sum test and the Mann–Whitney *U* test?
    **b.** What is the relationship between the Wilcoxon rank-sum test, the Mann–Whitney *U* test, and the test based on the difference between the average overall ranks in each sample?
14. **a.** What happens if there is an outlier in one of the samples in a test of two unpaired samples? For each case (very large or very small outlier), say what the rank of the outlier would be.
    **b.** Which statistical method (nonparametric or parametric) is more sensitive to an outlier? Why?
15. **a.** Can the nonparametric test for two unpaired samples be used with quantitative data? Why or why not?
    **b.** Can the nonparametric test for two unpaired samples be used with ordinal data? Why or why not?
    **c.** Can the nonparametric test for two unpaired samples be used with nominal data? Why or why not?
16. Describe the similarities and differences between the nonparametric test for two unpaired samples and the unpaired *t* test.

## Problems

*Problems marked with an asterisk (*) are solved in the Self Test in Appendix C.*

1. For each of the following situations, say whether parametric or nonparametric methods would be preferred. Give a reason for your choice and indicate how serious a problem it would be to use the other method.
   **a.** Your data set consists of bond ratings, of which AAA is a higher grade than AA, which is higher than A, and so on.
   **b.** Your data set consists of profits as a percentage of sales, and there is an outlier due to one firm that is involved in a serious lawsuit. You feel that the outlier must be left in because this lawsuit represents one of the risks of this industry group.
   **c.** Your data set consists of the weights of faucet washers being produced by a manufacturing system that is currently under control. The histogram looks very much like a normal distribution.

2. Consider the profits of the building materials firms in the Fortune 500, given in Table 16.4.1.
   a. Draw a histogram of these profit percentages. Describe the distribution.
   b. Find the average and the median. Explain why they are either similar or different.
   c. Use the *t* test to see if the mean profit (for the idealized population of similar firms operating under similar circumstances) is significantly different from the reference value –5% (i.e., a 5% loss). (Use the *t* test for now, even if you feel it is inappropriate.)
   d.* Use the sign test to see whether the median profit of this idealized population differs significantly from a 5% loss.
   e. Compare these two testing approaches to this data set. In particular, which of these two tests (*t* test or sign test) is appropriate here? Are both appropriate? Why?

3. Consider the profits of the aerospace firms in the Fortune 500, shown in Table 16.4.2.
   a. Draw a histogram of these profit percentages. Describe the distribution.
   b. Find the average and the median. Explain why they are either similar or different.
   c. Use the *t* test to see if the mean profit (for the idealized population of similar firms operating under similar circumstances) is significantly different from zero. (Use the *t* test for now, even if you feel it is inappropriate.)
   d. Use the sign test to see whether the median profit of this idealized population is significantly different from zero.

**TABLE 16.4.1  Profits of Building Materials Firms**

| Firm | Profits (as a percentage of sales) | Firm | Profits (as a percentage of sales) |
|------|------|------|------|
| American Standard | –1 | Norton | 7 |
| Owens-Illinois | –2 | Lafarge | 7 |
| Owens-Corning Fiberglas | 7 | Certainteed | 4 |
| USG | 4 | National Gypsum | –7 |
| Manville | –59 | Anchor Glass | –1 |
| Corning Glass Works | 10 | Calmat | 9 |
| Nortek | 1 | Southdown | 9 |

**Source:** Data are from *Fortune*, April 24, 1989, pp. 380–81.

**TABLE 16.4.2  Profits of Aerospace Firms**

| Firm | Profits (as a percentage of sales) | Firm | Profits (as a percentage of sales) |
|------|------|------|------|
| United Technologies | 4 | Martin Marietta | 6 |
| Boeing | 4 | Grumman | 2 |
| McDonnell Douglas | 2 | Gencorp | 3 |
| Rockwell International | 7 | Sequa | 4 |
| Allied-Signal | 4 | Colt Industries | 5 |
| Lockheed | 6 | Sundstrand | –5 |
| General Dynamics | 4 | Rohr Industries | 4 |
| Textron | 3 | Kaman | 3 |
| Northrop | 2 | | |

**Source:** Data are from *Fortune*, April 24, 1989, p. 380.

e. Compare these two testing approaches to this data set. In particular, which of these two tests (*t* test or sign test) is appropriate here? Are both appropriate? Why?

4. Of the 35 people in your sales force, more than half have productivity above the national median. The exact numbers are 23 above and 12 below. Are you just lucky, or is your sales force significantly more productive than the national median? How do you know?

5. Last year your department handled a median of 63,821 calls per day. (This is the median of the total calls handled each day during the year.) So far this year, more than half of the days have had total calls above this level (there were 15 days above and only 9 days below). Do you have the right to claim that you are overloaded compared to last year? Explain why or why not.

6. An advertisement is being tested to see if it is effective in creating the intended mood of relaxation. A sample of 15 people was tested just before and just after viewing the ad. Their questionnaire included many items, but the one being considered now is: Please describe your current feeling on a scale from 1 (very tense) to 5 (completely relaxed). The results are shown in Table 16.4.3.

a. How many people reported higher relaxation after viewing the ad than before? How many reported lower relaxation? How many were unchanged?

b.* Find the modified sample size.

c.* Perform the nonparametric sign test for the differences.

d. Briefly summarize this result in terms of the effect (if any) of the advertisement.

7. Stress levels were recorded during a true answer and a false answer given by each of six people in a study of lie-detecting equipment, based on the idea that the stress involved in telling a lie can be measured. The results are shown in Table 16.4.4.

a. Was everyone's stress level higher during a false answer than during a true answer?

b. How many had more stress during a true answer? During a false answer?

c. Find the modified sample size.

d. Use the nonparametric sign test for the differences to tell whether there is a significant difference in stress level between true and false answers.

8. Your human resources department has referred 26 employees for alcohol counseling. While the work habits of 15 improved, 4 actually got worse, and the remaining 7 were unchanged. Use the sign test for the differences to tell whether significantly more people improved than got worse.

9. Use the data sets from problems 2 and 3, on profit as a percent of sales for building materials firms and for aerospace firms.

a. Find the median profit for each industry group and compare them.

b. Combine the two data sets into a single column of profit percentages next to a column indicating industry group.

c. Sort the profit percentages, carrying along the industry group information. List the ranks in a third column, averaging appropriately when there are ties.

d. List the overall ranks for each industry group.

TABLE 16.4.3 Effects of Advertisement on Mood

| Person | Relaxation Scores | |
| | Before | After |
| --- | --- | --- |
| 1 | 3 | 2 |
| 2 | 2 | 2 |
| 3 | 2 | 2 |
| 4 | 4 | 5 |
| 5 | 2 | 4 |
| 6 | 2 | 1 |
| 7 | 1 | 1 |
| 8 | 3 | 5 |
| 9 | 3 | 4 |
| 10 | 2 | 4 |
| 11 | 5 | 5 |
| 12 | 2 | 3 |
| 13 | 4 | 5 |
| 14 | 3 | 5 |
| 15 | 4 | 4 |

TABLE 16.4.4 Stress Levels

| Person | Vocal Stress Level | |
| | True Answer | False Answer |
| --- | --- | --- |
| 1 | 12.8 | 13.1 |
| 2 | 8.5 | 9.6 |
| 3 | 3.4 | 4.8 |
| 4 | 5.0 | 4.6 |
| 5 | 10.1 | 11.0 |
| 6 | 11.2 | 12.1 |

e. Find the average rank for each industry group; also find the difference between these average ranks (subtracting the smaller from the larger).

f. Find the appropriate standard error for this difference in average rank.

g. Find the test statistic for the nonparametric test for two unpaired samples.

h. What is your conclusion from this test regarding profits in these two industry groups?

10. Your firm is being sued for gender discrimination, and you are evaluating the documents filed by the other side. Their data set is shown in Table 16.4.5.

a. Draw box plots for this data set on the same scale and comment on their appearance.

b.* Use a nonparametric method to test whether these salary distributions are significantly different.

c. Briefly summarize your conclusions based on the result of this test.

11. To understand your competitive position, you have examined the reliability of your product as well as the reliability of your closest competitor's product. You have subjected each product to abuse that represents about a year's worth of wear and tear per day. Table 16.4.6 gives the data indicating how long each item lasted.

a. Find the median time to failure for your and your competitor's products. Find the difference in medians (subtracting the smaller from the larger).

b. Find the nonparametric test statistic to determine whether your reliability differs significantly from your competitor's.

c. State the result of this nonparametric test.

d. Write a brief paragraph, complete with footnote(s), that might be used in an advertising brochure showing off your products.

12. Would there be any problem with a nonparametric analysis (two unpaired samples) of data in Table 10.7.8 listing day care rates comparing those of the well-to-do Laurelhurst area to other parts of Seattle? Why or why not?

13. Are tasting scores significantly different for the Chardonnay and Cabernet Sauvignon wines listed in Table 10.7.6? Is this a paired or unpaired situation?

14. The number of items returned for each of the past 9 days was 13, 8, 36, 18, 6, 4, 39, 47, and 21. Test to see if the median number returned is significantly different from 40 and find the $p$-value (as either $p > 0.05$, $p < 0.05$, or $p < 0.01$).

## TABLE 16.4.5 Gender Discrimination Data

| Salaries | |
|---|---|
| Women | Men |
| $21,100 | $38,700 |
| 29,700 | 30,300 |
| 26,200 | 32,800 |
| 23,000 | 34,100 |
| 25,800 | 30,700 |
| 23,100 | 33,300 |
| 21,900 | 34,000 |
| 20,700 | 38,600 |
| 26,900 | 36,900 |
| 20,900 | 35,700 |
| 24,700 | 26,200 |
| 22,800 | 27,300 |
| 28,100 | 32,100 |
| 25,000 | 35,800 |
| 27,100 | 26,100 |
| | 38,100 |
| | 25,500 |
| | 34,000 |
| | 37,400 |
| | 35,700 |
| | 35,700 |
| | 29,100 |

## TABLE 16.4.6 Reliability of Products under Abuse

| Days until Failure | |
|---|---|
| Your Products | Competitor's Products |
| 1.0 | 0.2 |
| 8.9 | 2.8 |
| 1.2 | 1.7 |
| 10.3 | 7.2 |
| 4.9 | 2.2 |
| 1.8 | 2.5 |
| 3.1 | 2.6 |
| 3.6 | 2.0 |
| 2.1 | 0.5 |
| 2.9 | 2.3 |
| 8.6 | 1.9 |
| 5.3 | 1.2 |
| | 6.6 |
| | 0.5 |
| | 1.2 |

**TABLE 16.4.7** Prescription Drug Prices per 100 Tablets

| Drug | United States | Canada |
|------|--------------|--------|
| Ativan | 49.43 | 6.16 |
| Ceclor | 134.18 | 84.14 |
| Coumadin | 36.70 | 19.59 |
| Dilantin | 15.03 | 4.67 |
| Feldene | 167.54 | 123.61 |
| Halcion | 47.69 | 16.09 |
| Lopressor | 35.71 | 15.80 |
| Naprosyn | 72.36 | 42.64 |
| Pepcid | 103.74 | 76.22 |
| Premarin | 26.47 | 10.10 |

**Source:** Data are from the *Wall Street Journal*, February 16, 1993, p. A9. Their source is Prime Institute, University of Minnesota.

15. Perform a nonparametric analysis of prescription drug prices in the United States and Canada, as reported in Table 16.4.7.
    a. Is this a paired or unpaired situation?
    b. Are prices significantly higher in the United States? How do you know?

### Database Exercises

Refer to the employee database in Appendix A.

1. Use a nonparametric test to see whether the median age of employees differs significantly from 40 years.

2. Use a nonparametric test to see whether the median experience of employees differs significantly from three years.

3. Use a nonparametric test to see whether the distribution of annual salaries for men differs significantly from the distribution for women.

### Projects

1. Find a univariate data set related to your work or business interests on the Internet, from your company, or at your library, and select a reasonable reference value to test against.
   a. Draw a histogram of your data and comment on its appearance.
   b. Perform the *t* test and report the conclusion.
   c. Perform the sign test and report the conclusion.
   d. Compare your two test results. If they are different, tell which one should be believed (if an appropriate choice can be made).
2. Continue with the same data set from the preceding project but insert a single, additional data value that is an extreme outlier. Repeat parts a–d of the preceding project for this new data set. Also, write a paragraph describing your experience of how the *t* test and the sign test each respond to an outlier.
3. Find two unpaired univariate data sets related to your work or business interests on the Internet, from your company, or at your library. It should make sense to test whether the two samples are significantly different.
   a. Draw box plots for your data on the same scale and comment on their appearance.
   b. Write down the null and research hypotheses for the appropriate nonparametric test.
   c. Perform the appropriate nonparametric test and report the conclusion.
   d. Write a paragraph summarizing what the test result has told you about this business situation.

## Chapter 17

# Chi-Squared Analysis

Testing for Patterns in Qualitative Data

How can you do statistical inference on qualitative data, where each observation gives you a category (such as a color or an energy source) instead of a number? You already know the answer for two cases. First, if you have *attribute data* (i.e., qualitative data with just two categories), the binomial distribution and its normal approximation can provide you with confidence intervals and hypothesis tests for the population percentage. Second, if you have *ordinal data* (with a natural ordering), the nonparametric methods of Chapter 16 can be used. However, if you have *nominal data* (with no natural ordering) and more than two categories (or more than one variable), you will need other methods. Here are some examples:

**One:** No manufacturing process is perfect, and yours is no exception. When defects occur, they are grouped into categories of assignable causes. The overall percent defective comes from an *attribute variable* and may be analyzed with the binomial distribution (if you can assume independence). The percent of defective items can be computed for each assignable cause. You would find the percent due to a bad chip, the percent due to a bad soldering connection, the percent due to a bad circuit board, and so on. As these percentages fluctuate from week to week, you would like to know whenever the system goes out of control, changing more than just randomly from its state before.

**Two:** Opinion polls are a useful source of information for many of us. In addition to the political details provided by the media's polls, many firms use polls to learn how their customers (actual and potential) feel about how things are and how they might be. This information is useful in planning strategy for marketing and new product introduction. Many opinion polls produce qualitative data, such as the categories "yes," "no," and "no opinion." Another type of qualitative data would result from the selection of a preferred product from a list of leading brands. You might use statistical inference to compare the opinions of two groups of people and test whether they differ significantly. Or you might test a single group against a known standard.

The **chi-squared tests** provide hypothesis tests for qualitative data, where you have categories instead of numbers. With nominal qualitative data, you can only count (since ordering and arithmetic cannot be done). Chi-squared tests are therefore based on counts that represent the number of items in the sample falling into each category. The **chi-squared statistic** measures the difference between the *actual* counts and the *expected* counts (assuming validity of the null hypothesis) as follows:

---

**The Chi-Squared Statistic**

Chi-squared statistic =

$$\text{Sum of } \frac{(\text{Observed count} - \text{Expected count})^2}{\text{Expected count}}$$

$$= \sum \frac{(O_i - E_i)^2}{E_i}$$

where the sum extends over all categories or combinations of categories. The definition of *expected count* will depend on the particular form of the null hypothesis being tested.

---

Based on the chi-squared statistic as a measure of how close the data set conforms to the null hypothesis, the chi-squared test can decide whether or not the null hypothesis is a reasonable possibility.

## 17.1 SUMMARIZING QUALITATIVE DATA BY USING COUNTS AND PERCENTAGES

Here is a typical qualitative data set, given in the usual way as a list of measurements for each elementary unit in the sample. The elementary units here are people who came to an auto showroom, and the measurement is the type of vehicle they are looking for:

Pickup, Economy car, Economy car, Family sedan, Pickup, Economy car, Sports car, Economy car, Family sedan, Pickup, Economy car, Family sedan, Van, Van, Economy car, Family sedan, Pickup, Sports car, Family sedan, Family sedan, Economy car, Van, Economy car, Family sedan, Sports car, Economy car, Economy car, Van, Van, …

Because a list like this goes on and on, you will probably want to work with a summary table of counts (frequencies) or percentages. This type of table preserves the information from the data while presenting it in a more useful way in a smaller space. An example is shown in Table 17.1.1.

Summary tables of counts or percentages are also helpful for analysis of *bivariate* qualitative data, where you have more than one measurement. When Gallup researchers investigated trends in Americans' feelings about saving and spending money, they asked several

**TABLE 17.1.1 Vehicle Desired**

| Type | Count (frequency) | Percent of Total |
|---|---|---|
| Family sedan | 187 | (187/536 =) 34.9% |
| Economy car | 206 | 38.4 |
| Sports car | 29 | 5.4 |
| Van | 72 | 13.4 |
| Pickup | 42 | 7.8 |
| Total | 536 | 100.0% |

**TABLE 17.1.2 Spending versus Saving**

| | Age 18 to 29 | Age 30 to 49 | Age 50+ | Overall |
|---|---|---|---|---|
| Prefer spending | 43% | 38% | 29% | 35% |
| Prefer saving | 56 | 59 | 66 | 62 |
| Not sure* | 1 | 3 | 5 | 3 |
| Total | 100% | 100% | 100% | 100% |

* The "Not sure" category was computed based on the other numbers available.

questions of each person responding.[1] Two qualitative variables were

1. The answer to the question: "Thinking about money for a moment, are you the type of person who more enjoys spending money or who more enjoys saving money," where the order of "spending" and "saving" was switched to guard against bias.
2. A classification by age into one of three groups: 18 to 29 years, 30 to 49 years, or 50+ years.

Since each person provides a category as a response to each of these variables, the actual results of the poll looked something like this, person by person:

Prefers spending age 18 to 29, Prefers saving age 50+, Prefers saving age 30 to 49, Not sure age 50+, …

The results for the 1,025 adults polled are summarized, in part, in Table 17.1.2.

---

1. D. Jacobe, "Spending Less Becoming New Norm for Many Americans, An Increasing Percentage of Americans Say They More Enjoy Saving Than Spending," February 25, 2010, accessed at http://www.gallup.com/poll/126197/spending-less-becoming-new-norm-americans.aspx on July 29, 2010.

Whenever you see a table of counts or percentages, it may help to imagine the underlying qualitative data set it came from. The next step is to test various hypotheses about these counts and percentages.

## 17.2 TESTING IF POPULATION PERCENTAGES ARE EQUAL TO KNOWN REFERENCE VALUES

You already know how to test a single percentage against a reference value using the binomial distribution (see Chapter 10). However, another method is needed for testing an entire table of percentages against another table of reference values. A common application of such a test is to find out whether your recent experience (summarized by counts and percentages) is typical relative to your past experience (the reference values).

### The Chi-Squared Test for Equality of Percentages

The **chi-squared test for equality of percentages** is used to determine whether a table of *observed* counts or percentages could reasonably have come from a population with known percentages (the reference values). Here is a summary of the situation and its solution:

> **The Chi-Squared Test for Equality of Percentages**
>
> *The data:* A table indicating the count for each category for a single qualitative variable.
>
> *The hypotheses:*
>
> $H_0$: The population percentages are equal to a set of known, fixed reference values.
> $H_1$: The population percentages are not equal to this set of reference values; at least one category is different.
>
> *The expected counts:* For each category, multiply the population reference proportion by the sample size, $n$.
>
> *The assumptions:*
>
> 1. The data set is a random sample from the population of interest.
> 2. At least five counts are expected in each category.
>
> *The chi-squared statistic:*
>
> Chi-squared statistic =
>
> $$\text{Sum of } \frac{(\text{Observed count} - \text{Expected count})^2}{\text{Expected count}}$$
>
> $$= \sum \frac{(O_i - E_i)^2}{E_i}$$
>
> *Degrees of freedom:* Number of categories minus 1.
> *The chi-squared test result:* Significant if the chi-squared statistic is larger than the value from Table 17.2.1; not significant otherwise.

**TABLE 17.2.1  Critical Values for Chi-Squared Tests**

| Degrees of Freedom | 10% Level | 5% Level | 1% Level | 0.1% Level |
|---|---|---|---|---|
| 1 | 2.706 | 3.841 | 6.635 | 10.828 |
| 2 | 4.605 | 5.991 | 9.210 | 13.816 |
| 3 | 6.251 | 7.815 | 11.345 | 16.266 |
| 4 | 7.779 | 9.488 | 13.277 | 18.467 |
| 5 | 9.236 | 11.071 | 15.086 | 20.515 |
| 6 | 10.645 | 12.592 | 16.812 | 22.458 |
| 7 | 12.017 | 14.067 | 18.475 | 24.322 |
| 8 | 13.362 | 15.507 | 20.090 | 26.124 |
| 9 | 14.684 | 16.919 | 21.666 | 27.877 |
| 10 | 15.987 | 18.307 | 23.209 | 29.588 |
| 11 | 17.275 | 19.675 | 24.725 | 31.264 |
| 12 | 18.549 | 21.026 | 26.217 | 32.909 |
| 13 | 19.812 | 22.362 | 27.688 | 34.528 |
| 14 | 21.064 | 23.685 | 29.141 | 36.123 |
| 15 | 22.307 | 24.996 | 30.578 | 37.697 |
| 16 | 23.542 | 26.296 | 32.000 | 39.252 |
| 17 | 24.769 | 27.587 | 33.409 | 40.790 |
| 18 | 25.989 | 28.869 | 34.805 | 42.312 |
| 19 | 27.204 | 30.144 | 36.191 | 43.820 |
| 20 | 28.412 | 31.410 | 37.566 | 45.315 |
| 21 | 29.615 | 32.671 | 38.932 | 46.797 |
| 22 | 30.813 | 33.924 | 40.289 | 48.268 |
| 23 | 32.007 | 35.172 | 41.638 | 49.728 |
| 24 | 33.196 | 36.415 | 42.980 | 51.179 |
| 25 | 34.382 | 37.652 | 44.314 | 52.620 |
| 26 | 35.563 | 38.885 | 45.642 | 54.052 |
| 27 | 36.741 | 40.113 | 46.963 | 55.476 |
| 28 | 37.916 | 41.337 | 48.278 | 56.892 |
| 29 | 39.087 | 42.557 | 49.588 | 58.301 |
| 30 | 40.256 | 43.773 | 50.892 | 59.703 |
| 31 | 41.422 | 44.985 | 52.191 | 61.098 |
| 32 | 42.585 | 46.194 | 53.486 | 62.487 |
| 33 | 43.745 | 47.400 | 54.776 | 63.870 |
| 34 | 44.903 | 48.602 | 56.061 | 65.247 |

*(Continued)*

**TABLE 17.2.1 Critical Values for Chi-Squared Tests—cont'd**

| Degrees of Freedom | 10% Level | 5% Level | 1% Level | 0.1% Level |
|---|---|---|---|---|
| 35 | 46.059 | 49.802 | 57.342 | 66.619 |
| 36 | 47.212 | 50.998 | 58.619 | 67.985 |
| 37 | 48.363 | 52.192 | 59.893 | 69.346 |
| 38 | 49.513 | 53.384 | 61.162 | 70.703 |
| 39 | 50.660 | 54.572 | 62.428 | 72.055 |
| 40 | 51.805 | 55.758 | 63.691 | 73.402 |
| 41 | 52.949 | 56.942 | 64.950 | 74.745 |
| 42 | 54.090 | 58.124 | 66.206 | 76.084 |
| 43 | 55.230 | 59.304 | 67.459 | 77.419 |
| 44 | 56.369 | 60.481 | 68.710 | 78.749 |
| 45 | 57.505 | 61.656 | 69.957 | 80.077 |
| 46 | 58.641 | 62.830 | 71.201 | 81.400 |
| 47 | 59.774 | 64.001 | 72.443 | 82.720 |
| 48 | 60.907 | 65.171 | 73.683 | 84.037 |
| 49 | 62.038 | 66.339 | 74.919 | 85.351 |
| 50 | 63.167 | 67.505 | 76.154 | 86.661 |
| 51 | 64.295 | 68.669 | 77.386 | 87.968 |
| 52 | 65.422 | 69.832 | 78.616 | 89.272 |
| 53 | 66.548 | 70.993 | 79.843 | 90.573 |
| 54 | 67.673 | 72.153 | 81.069 | 91.872 |
| 55 | 68.796 | 73.311 | 82.292 | 93.167 |
| 56 | 69.919 | 74.468 | 83.513 | 94.461 |
| 57 | 71.040 | 75.624 | 84.733 | 95.751 |
| 58 | 72.160 | 76.778 | 85.950 | 97.039 |
| 59 | 73.279 | 77.931 | 87.166 | 98.324 |
| 60 | 74.397 | 79.082 | 88.379 | 99.607 |
| 61 | 75.514 | 80.232 | 89.591 | 100.888 |
| 62 | 76.630 | 81.381 | 90.802 | 102.166 |
| 63 | 77.745 | 82.529 | 92.010 | 103.442 |
| 64 | 78.860 | 83.675 | 93.217 | 104.716 |
| 65 | 79.973 | 84.821 | 94.422 | 105.988 |
| 66 | 81.085 | 85.965 | 95.626 | 107.258 |
| 67 | 82.197 | 87.108 | 96.828 | 108.526 |
| 68 | 83.308 | 88.250 | 98.028 | 109.791 |
| 69 | 84.418 | 89.391 | 99.228 | 111.055 |
| 70 | 85.527 | 90.531 | 100.425 | 112.317 |
| 71 | 86.635 | 91.670 | 101.621 | 113.577 |
| 72 | 87.743 | 92.808 | 102.816 | 114.835 |
| 73 | 88.850 | 93.945 | 104.010 | 116.091 |
| 74 | 89.956 | 95.081 | 105.202 | 117.346 |
| 75 | 91.061 | 96.217 | 106.393 | 118.599 |
| 76 | 92.166 | 97.351 | 107.583 | 119.850 |
| 77 | 93.270 | 98.484 | 108.771 | 121.100 |
| 78 | 94.374 | 99.617 | 109.958 | 122.348 |
| 79 | 95.476 | 100.749 | 111.144 | 123.594 |
| 80 | 96.578 | 101.879 | 112.329 | 124.839 |
| 81 | 97.680 | 103.010 | 113.512 | 126.083 |
| 82 | 98.780 | 104.139 | 114.695 | 127.324 |
| 83 | 99.880 | 105.267 | 115.876 | 127.565 |
| 84 | 100.980 | 106.395 | 117.057 | 129.804 |
| 85 | 102.079 | 107.522 | 118.236 | 131.041 |
| 86 | 103.177 | 108.648 | 119.414 | 132.277 |
| 87 | 104.275 | 109.773 | 120.591 | 133.512 |
| 88 | 105.372 | 110.898 | 121.767 | 134.745 |
| 89 | 106.469 | 112.022 | 122.942 | 135.978 |
| 90 | 107.565 | 113.145 | 124.116 | 137.208 |
| 91 | 108.661 | 114.268 | 125.289 | 138.438 |
| 92 | 109.756 | 115.390 | 126.462 | 139.666 |
| 93 | 110.850 | 116.511 | 127.633 | 140.893 |
| 94 | 111.944 | 117.632 | 128.803 | 142.119 |
| 95 | 113.038 | 118.752 | 129.973 | 143.344 |
| 96 | 114.131 | 119.871 | 131.141 | 144.567 |
| 97 | 115.223 | 120.990 | 132.309 | 145.789 |
| 98 | 116.315 | 122.108 | 133.476 | 147.010 |
| 99 | 117.407 | 123.225 | 134.642 | 148.230 |
| 100 | 118.498 | 124.342 | 135.807 | 149.449 |

If the chi-squared statistic is *larger* than the critical value from the chi-squared table for the appropriate number of degrees of freedom, you have evidence that the observed counts are very different from those expected for your reference percentages. You would then reject the null hypothesis and accept the research hypothesis, concluding that the observed sample percentages are *significantly different* from the reference values.

If the chi-squared statistic is *smaller* than the critical value from the chi-squared table, then the observed data are not very different from what you would expect based on the reference percentages. You would accept the null hypothesis (as a reasonable possibility) and conclude that the observed sample percentages are *not significantly different* from the reference values.

As a rule of thumb, there should be at least five counts expected in each category because the chi-squared test is an approximate, not an exact, test. The approximation is close enough for practical purposes when this rough guideline is followed, but it may be in error when you have too few expected counts in some category. The risk is that your type I error probability will not be controlled at the 5% level (or other chosen level).

## Example
### Quality Problems Categorized by Their Causes

As part of your firm's commitment to total quality control, defects are carefully monitored because this provides useful information for quality improvement. Each defective component is checked to see whether the problem is a bad chip, a bad soldering joint, or a bad circuit board. Based on past data from this assembly line, you know what percentages to expect (the reference percentages) when the process is under control. By comparing the current results to these reference percentages, you can test whether or not the process is currently under control.[2]

Table 17.2.2 shows the data set, representing the problems from the previous week. Table 17.2.3 shows the reference values, based on the past year's experience when the assembly line seemed to be working properly. Although the chip problems are very close to the reference value (16.0% observed chip defects in the past week compared to the reference value of 15.2%), the others are fairly different (70.0% compared to 60.5% for a bad soldering joint, for example). The question is whether or not these differences are significant. That is, could the defect rates have reasonably been produced by randomly sampling from a population whose percentages correspond to the reference values? Or are the differences so great that they could not reasonably be due to random chance alone? The chi-squared test for equality of percentages will provide an answer. Here are the hypotheses:

$H_0$: The process is still in control. (The observed defect rates are equal to the reference values.)
$H_1$: The process is not in control. (The observed defect rates are not equal to the reference values.)

Table 17.2.4 shows the expected counts, found by multiplying the reference percentages (15.2%, 60.5%, and 24.3%) by the sample size, $n = 50$. Note that it's perfectly all right for the expected counts to include a decimal part; this is necessary so that the computed percentages give you the reference percentages exactly. Note also that the total expected count matches the total actual count, namely, the sample size, $n = 50$.

As for the assumptions, this data set is sampled from the idealized population of all components that would be built under similar circumstances. Thus, your actual experience is viewed as a random sample from this idealized population of what *might* have happened. The second assumption is also satisfied because all expected counts are at least 5 (i.e., 7.6, 30.3, and 12.2 are all 5 or larger). Note that this assumption is checked by looking at the *expected,* not the *observed,* counts.

The chi-squared statistic is computed by combining the observed and expected counts for all categories as follows (be sure to omit the total because it is not a category):

Chi-squared statistic =

$$\text{Sum of } \frac{(\text{Observed count} - \text{Expected count})^2}{\text{Expected count}}$$

$$= \frac{(8 - 7.60)^2}{7.60} + \frac{(35 - 30.25)^2}{30.25} + \frac{(7 - 12.15)^2}{12.15}$$

$$= \frac{0.1600}{7.60} + \frac{22.5625}{30.25} + \frac{26.5225}{12.15}$$

$$= 2.950$$

The number of degrees of freedom is 1 less than the number of categories. There are three categories here (chip, solder, and board); therefore,

$$\text{Degrees of freedom} = 3 - 1 = 2$$

Looking in the chi-squared table for 2 degrees of freedom, you see the critical value for testing at the 5% level is 5.991. Since the chi-squared statistic (2.950) is smaller, you accept the null hypothesis. It is reasonably possible that the assembly line is producing defects that match the "in control" reference proportions and that the discrepancy is just due to the usual random chance for a sample of size 50. You do not have convincing evidence that the process is out of control, so you accept the possibility that it is still in control.

The observed percentages are not significantly different from the reference percentages. Based on this, there is no evidence that the process has gone out of control.

---

2. Of course, you would also, separately, keep a close eye on the overall percent defective. The analysis here helps you identify trouble due to a specific cause. The topic of quality control will be covered in Chapter 18.

**TABLE 17.2.2 The Observed Data: Defective Components from the Previous Week**

| Problem | Observed Count (frequency) | Percent of Total |
|---------|---------------------------|------------------|
| Chip | 8 | (8/50 =) 16.0% |
| Solder | 35 | 70.0 |
| Board | 7 | 14.0 |
| | | |
| Total | 50 | 100.0% |

**TABLE 17.2.3 The Reference Values: Defective Components from the Previous Year When "In Control"**

| Problem | Percent of Total |
|---------|------------------|
| Chip    | 15.2%            |
| Solder  | 60.5             |
| Board   | 24.3             |
| Total   | 100.0%           |

**TABLE 17.2.4 The Expected Counts: Hypothetical Numbers of Defective Components to Match "In Control" Reference Percentages**

| Problem | Expected Count       |
|---------|----------------------|
| Chip    | $(0.152 \times 50 =)$ 7.60 |
| Solder  | 30.25                |
| Board   | 12.15                |
| Total   | 50.00                |

## 17.3  TESTING FOR ASSOCIATION BETWEEN TWO QUALITATIVE VARIABLES

Now suppose you have *two* qualitative variables; that is, your data set consists of *bivariate qualitative data*. After you have examined each variable separately by examining counts and percentages, you may be interested in the *relationship* (if any) between these two variables. In particular, you may be interested in whether there is any relationship at all between the two. Here are some applications for such a test:

1.  One variable represents the favorite recreational activity of each person (from a list including sports, reading, TV, etc.). The other variable is each person's favorite breakfast cereal. By understanding the relationship between these two variables, you are better equipped to plan marketing strategy. If you are in cereals, this would help you decide what kind of material to put on cereal boxes. If you are in recreation, this would help you decide which cereal companies to approach about joint marketing plans.
2.  One variable represents the cause of the defective component. The other variable represents the manager in charge when the component was produced. The relationship, if any, will help you focus your efforts where they will do the most good by identifying the specific managers who should devote more attention to solving the problems. If a cause shows up for all

managers, you have a systematic problem and should look at the larger system (not the individual managers) for a solution. If a cause shows up for just one manager, you could begin by delegating its solution to that manager.

## The Meaning of Independence

Two qualitative variables are said to be **independent** if knowledge about the value of one variable does not help you predict the other; that is, the *probabilities* for one variable are the same as the *conditional probabilities* given the other variable. Each variable has its own population percentages, which represent the probabilities of occurrence for each category. The **conditional population percentages** are the probabilities of occurrence for one variable when you restrict attention to just one category of the other variable. These restricted population percentages represent the conditional probabilities for one variable given this category of the other.

For example, suppose the population percentage of "paint-flake" defects is 3.1% overall. When Jones is the manager on duty, however, the conditional population percentage of paint-flake defects is 11.2%. In this case, knowledge about one variable (the particular manager) helps you predict the outcome of the other (the defect type) because 3.1% and 11.2% are different. Paint-flake defects are more likely when Jones is on duty and less likely when someone else is in charge. Therefore, these two variables are *not independent*.

You should note that the real-life situation is somewhat more involved than this example because you have to work with *sample percentages* as estimates of population probabilities; you won't be able to just look at the percentages and see if they are different because they will (nearly) always be different due to random chance. The chi-squared test for independence will tell you when the differences go *beyond* what is reasonable due to random chance alone.

## The Chi-Squared Test for Independence

The **chi-squared test for independence** is used to decide whether or not two qualitative variables are independent, based on a table of observed counts from a bivariate qualitative data set. It is computed from a table that gives the counts you would expect if the two variables were independent. Here is a summary of the situation and its solution:

**The Chi-Squared Test for Independence**

*The data:* A table indicating the counts for each combination of categories for two qualitative variables, summarizing a bivariate data set.

*The hypotheses:*

$H_0$: The two variables are independent of one another. That is, the probabilities for either variable are equal to the conditional probabilities given the other variable.

$H_1$: The two variables are associated; they are not independent of one another. There is at least one category of one variable whose probability is not equal to the conditional probability given some category of the other variable.

*The expected table:* For each combination of categories, one for each variable, multiply the count for one category by the count for the other category and then divide by the total sample size, $n$:

$$\text{Expected count} = \frac{\left(\begin{array}{c}\text{Count for category}\\\text{for one variable}\end{array}\right)\left(\begin{array}{c}\text{Count for category}\\\text{for other variable}\end{array}\right)}{n}$$

*The assumptions:*

1. The data set is a random sample from the population of interest.
2. At least five counts are expected in each combination of categories.

*The chi-squared statistic:*

$$\text{Chi-squared statistic} =$$

$$\text{Sum of } \frac{(\text{Observed count} - \text{Expected count})^2}{\text{Expected count}}$$

where the sum extends over all combinations of categories.

*Degrees of freedom:*

$$\left(\begin{array}{c}\text{Number of categories}\\\text{for first variable}\end{array} -1\right)\left(\begin{array}{c}\text{Number of categories}\\\text{for second variable}\end{array} -1\right)$$

*The chi-squared test result:* Significant association if the chi-squared statistic is larger than the table value; no significant association otherwise.

If the chi-squared statistic is larger than the critical value from the chi-squared table (Table 17.2.1), you have evidence that the observed counts are very different from those that would be expected if the variables were independent. You would reject the null hypothesis of independence and accept the research hypothesis. You would conclude that the variables show *significant association*; that is, they *are not independent* of each other.

If the chi-squared statistic is smaller than the critical value from the chi-squared table, then the observed data are not very different from what you would expect if the variables were independent of each other in the population. You would accept the null hypothesis of independence as a reasonable possibility. You would conclude that the variables do not show significant association. This is a weak conclusion because the null hypothesis of independence has been accepted; you *accept* independence, but you have *not proven* independence.

Why is the expected table computed this way? Remember from probability theory (Chapter 6) that when two events are independent, the probability that they will *both* occur is equal to the product of their probabilities.

The equation that defines the expected count expresses independence, in effect, by multiplying these probabilities.[3]

## Example

### Is Your Market Segmented?

You are trying to set strategy for a marketing campaign for a new product line consisting of three rowing machines. The basic model is made of industrial-strength chrome with black plastic fittings and a square seat. The designer model comes in a variety of colors and with a sculpted seat. The complete model adds a number of accessories to the designer model (computerized display, running water, sound effects, etc.).

To help your team write the information brochure and press releases, you need to know which model each type of customer will prefer so that, for example, you don't go overboard[4] about how practical a model is when your market actually consists of impulsive types.

A marketing firm has gathered data in a small test market. For each purchase, two qualitative variables were recorded. One is the model (basic, designer, or complete), and the other is the type of consumer (summarized for this purpose as either practical or impulsive). Table 17.3.1 shows the data set, presented as a table of counts for these $n = 221$ customers. For example, of the 221 purchases, 22 were basic machines purchased by practical customers.

The *overall percentages*, obtained by dividing each count by the total sample size, $n$, indicate what percent fall into each category of each variable and each combination of categories (one category for each variable). From Table 17.3.2, you see that your largest group of sales is designer models to impulsive shoppers (39.8% of all sales). The next largest group is complete models to practical shoppers (24.4% of total sales). Looking at the totals by model type (the rightmost column), you see that the basic model is the slowest seller (only 21.3% of all rowing machines sold were the basic model). The reason may be that your customers are upscale and able to pay more to get the higher-end models. (That's good news—congratulations!)

The *percentages by model*, obtained by dividing each count by the total count for that model, indicate the percentages of each type of customer for each model. These are the *conditional sample percentages given the model* and estimate the corresponding conditional population percentages

*(Continued)*

---

3. If you divide both sides of the equation for the expected count by $n$, it is easier to see this. The result is $\frac{\text{Expected count}}{n} = \frac{\left(\begin{array}{c}\text{Count for category}\\\text{for one variable}\end{array}\right)}{n} \times \frac{\left(\begin{array}{c}\text{Count for category}\\\text{for other variable}\end{array}\right)}{n}$. Since dividing by $n$ gives you a proportion that estimates the probability, the equation states that the probability of a combination of a particular category for one variable with a particular category for the other variable is equal to the product of these probabilities. This is the same as the definition of independence for probabilities (see Chapter 6).

**Example—cont'd**

(the conditional probabilities). This shows you the customer profile for each type of machine. From Table 17.3.3, you can see that the basic model is purchased in roughly equal proportions by each type of customer (46.8% practical versus 53.2% impulsive). The designer model's customers are almost exclusively impulsive, whereas the complete model's customers are much more likely to be practical than impulsive.

The *percentages by customer type,* obtained by dividing each count by the total count for that customer type, indicate the percentages of each model type bought by each type of customer. These are the *conditional sample percentages given the customer type* and estimate the corresponding conditional population percentages (the conditional probabilities). This gives you a profile of the model preferences for each customer type. From Table 17.3.4, you can see that practical customers strongly prefer the complete model (60.7% of them purchased it), and impulsive shoppers strongly prefer the designer model (66.7% of them purchased it). However, you should not ignore the other choices (such as basic model purchases by practical consumers) since they still represent a sizable minority fraction of your market.

Do these two variables look like they are independent? No. We have already noted several facts that indicate some kind of relationship between customer type and model preference. For example, looking at the table of percentages by customer type, you see that although 33.0% of all customers purchased complete units, a much larger percentage (60.7%) of practical customers purchased these units. If they were independent, you would expect the practical consumers to show the same buying patterns as anyone else. Knowledge of customer *does* seem to help you predict the model purchased, suggesting that the two factors are not independent.

If they were independent, the percentages by customer type would be the same in all three columns: Practical shoppers and impulsive shoppers would have the same model purchase profile as all shoppers together. Similarly, if they were independent, the percentages by model would be the same for all four rows: The basic model, the designer model, and the complete model would have the same customer profile as that for all models together.

The *expected table* (obtained by multiplying the total count for each customer type by the total count for each model type and then dividing by total sample size, $n = 221$) indicates the counts you would expect if customer type were independent of model purchased. Table 17.3.5 shows that the expected table keeps the same total count for each customer type as was actually observed (89 practical and 132 impulsive). The model purchases are also the same (47 basic, 101 designer, and 73 complete). But the counts inside the table have been rearranged to show what you would expect (on average) assuming independence.

**TABLE 17.3.1 Counts: Rowing Machine Purchases**

|  | Practical | Impulsive | Total |
|---|---|---|---|
| Basic | 22 | 25 | 47 |
| Designer | 13 | 88 | 101 |
| Complete | 54 | 19 | 73 |
|  |  |  |  |
| Total | 89 | 132 | 221 |

**TABLE 17.3.2 Overall Percentages: Rowing Machine Purchases**

|  | Practical | Impulsive | Total |
|---|---|---|---|
| Basic | (22/221 =) 10.0% | 11.3% | 21.3% |
| Designer | 5.9 | 39.8 | 45.7 |
| Complete | 24.4 | 8.6 | 33.0 |
|  |  |  |  |
| Total | 40.3% | 59.7% | 100.0% |

**TABLE 17.3.3 Percentages by Model: Rowing Machine Purchases**

|  | Practical | Impulsive | Total |
|---|---|---|---|
| Basic | (22/47 =) 46.8% | 53.2% | 100.0% |
| Designer | 12.9 | 87.1 | 100.0 |
| Complete | 74.0 | 26.0 | 100.0 |
|  |  |  |  |
| Total | 40.3% | 59.7% | 100.0% |

**TABLE 17.3.4 Percentages by Customer Type: Rowing Machine Purchases**

|  | Practical | Impulsive | Total |
|---|---|---|---|
| Basic | (22/89 =) 24.7% | 18.9% | 21.3% |
| Designer | 14.6 | 66.7 | 45.7 |
| Complete | 60.7 | 14.4 | 33.0 |
|  |  |  |  |
| Total | 100.0% | 100.0% | 100.0% |

Note that you would expect 40.67 purchases of designer machines by practical consumers (based on 89 practical shoppers and 101 designer machines sold out of the 221 total). However, you actually sold only 13 (from the original table of counts), far fewer than the 40.67 expected under the assumption of independence.

Are model purchases independent of customer type? Are the differences between the original observed counts and these expected counts greater than would be reasonable due to random chance if they were, in fact, independent? The chi-squared test will decide.

The hypotheses are as follows:

$H_0$: Customer type is independent of the model purchased.
$H_1$: Customer type is not independent of the model purchased.

The null hypothesis claims that customers have the same preferences (percent of each model purchased) regardless of their type (practical or impulsive). The research hypothesis claims that these preferences are different.

Let's check the assumptions. Is the data set a random sample from the population of interest? Not really, but it may be close enough. This depends in part on how carefully the marketing study was designed—that is, how representative the test market is of your overall customers. Keep in mind that statistical inference can generalize only to the larger population (real or idealized) that is *represented* by your sample. For now, let's assume it is a random sample of purchases in cities similar to your test area, in stores like those that were included in the test. The second assumption is satisfied because every entry in the expected table is 5 or greater.

The chi-squared statistic is the sum of (Observed − Expected)$^2$/Expected. Table 17.3.6 shows these values (note that the row and column of totals are omitted). The chi-squared statistic is the sum of these values:

Chi-squared statistic =

$$\text{Sum of } \frac{(\text{Observed count} - \text{Expected count})^2}{\text{Expected count}}$$

$$= 0.50 + 18.83 + 20.58 + 0.34 + 12.69 + 13.88$$

$$= 66.8$$

The number of degrees of freedom is 2 because there are three categories of rowing machines and two categories of customers:

$$\text{Degrees of freedom} = (3-1)(2-1)$$

$$= 2 \times 1$$

$$= 2$$

Looking in the chi-squared table for 2 degrees of freedom at the 5% test level, you find the critical value 5.991. Because the chi-squared statistic (66.8) is larger than this critical value, there is significant association between these

### TABLE 17.3.5 Expected Counts: Rowing Machine Purchases

|  | Practical | Impulsive | Total |
|---|---|---|---|
| Basic | (89 × 47/221 =) 18.93 | 28.07 | 47.00 |
| Designer | 40.67 | 60.33 | 101.00 |
| Complete | 29.40 | 43.60 | 73.00 |
| Total | 89.00 | 132.00 | 221.00 |

### TABLE 17.3.6 (Observed–Expected)$^2$/Expected: Rowing Machine Purchases

|  | Practical | Impulsive |
|---|---|---|
| Basic | ([22 − 18.93]$^2$/18.93 =) 0.50 | 0.34 |
| Designer | 18.83 | 12.69 |
| Complete | 20.58 | 13.88 |

qualitative variables. In fact, because it is so much larger, let's test at the 0.1% level. This critical value (still with 2 degrees of freedom) is 13.816. The chi-squared statistic (66.8) is still much higher. Your conclusion is therefore:

The association between customer type and model purchased is very highly significant ($p < 0.001$).

Because, assuming independence, the probability that you would see data with this much association or more is so small ($p < 0.001$), you have very strong evidence against the null hypothesis of independence. You may now plan the marketing campaign with some assurance that the different models really do appeal differently to different segments of the market.

Excel® can help you compute the $p$-value of the chi-squared test for independence using the CHITEST function, but you have to compute the table of expected counts first. The results are shown below: first the original table of counts, next the table of expected counts,[5] and finally the CHITEST function, which uses both the original table and the table of expected counts. The resulting CHITEST $p$-value is 3.07823E-15, which represents the very small number 0.00000000000000307823 because the scientific notation "E-15" tells you to move the decimal point 15 places to the left. Clearly, the result is very highly significant because this $p$-value is less than 0.001.

Begin by selecting the cell where you want the $p$-value to go. Then choose Insert Function from the Function Library of

(Continued)

**Example—cont'd**

the Formulas Ribbon, select Statistical as the function category, and choose CHITEST as the function name. A dialog box will then pop up after you click OK, allowing you first to drag across your table of counts. Then click in the Expected_range box and drag across your table of expected counts, and finally press Enter to complete the process. Here's how it looks:

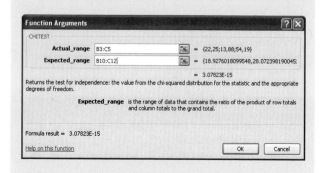

4. Sorry about the pun.
5. To create a formula for expected counts that will copy correctly to fill the entire table, note the use of "absolute addressing" using dollar signs in the formula "=B$6*$D3/$D$6" to find the expected 18.93 purchases of basic machines by practical consumers. This formula can be copied and pasted to fill the table while always taking row totals from row 6 (hence, the reference B$6), always taking the column totals from column D (hence, the reference $D3), and always taking the overall total from cell D6 (hence, the reference $D$6).

## 17.4 END-OF-CHAPTER MATERIALS

### Summary

Qualitative data are summarized using counts and percentages. The **chi-squared tests** provide hypothesis tests for qualitative data, where you have categories instead of numbers. The **chi-squared statistic** measures the difference

between the *actual* counts and the *expected* counts (based on the null hypothesis):

$$\text{Chi-squared statistic} = \text{Sum of } \frac{(\text{Observed count} - \text{Expected count})^2}{\text{Expected count}}$$

where the sum extends over all categories or combinations of categories. The definition of *expected count* will depend on the particular form of the null hypothesis being tested.

The **chi-squared test for equality of percentages** is used to decide whether a table of observed counts or percentages (summarizing a single qualitative variable) could reasonably have come from a population with known percentages (the reference values). The hypotheses are

$H_0$: The population percentages are equal to a set of known, fixed reference values.

$H_1$: The population percentages are not equal to this set of reference values. At least one category is different.

The assumptions are

1. The data set is a random sample from the population of interest.
2. At least five counts are expected in each category.

For the chi-squared statistic, the expected count for each category is the population reference percentage multiplied by the sample size, $n$. The degrees of freedom equal the number of categories minus 1.

If the chi-squared statistic is larger than the critical value from the chi-squared table at the appropriate degrees of freedom, you have evidence that the observed counts are very different from those expected for your reference percentages. You would reject the null hypothesis and accept the research hypothesis. The observed sample percentages are significantly different from the reference values.

If the chi-squared statistic is smaller than the critical value from the chi-squared table, then the observed data are not very different from what you would expect based on the reference percentages. You would accept the null hypothesis as a reasonable possibility. The observed sample percentages are *not* significantly different from the reference values.

When you have *bivariate qualitative data*, you may wish to test whether or not the two variables are associated. Two qualitative variables are said to be **independent** if knowledge about the value of one variable does not help you predict the other; that is, the probabilities for one variable are the same as the conditional probabilities given the other variable. The **conditional population percentages** are the probabilities of occurrence for one variable when you restrict attention to just one category of the other variable. Your sample data set provides estimates of these population percentages and conditional population percentages.

One way to summarize bivariate qualitative data is to use *overall percentages*, which give you the relative frequency of each *combination* of categories, one for each variable. Another approach is to use the *percentages by one of the variables* to obtain a profile of estimated conditional probabilities for the other variable *given* each category of the first variable.

The **chi-squared test for independence** is used to decide whether or not two qualitative variables are independent, based on a table of observed counts from a bivariate qualitative data set. It is computed from a table that gives the counts you would expect if the two variables were independent. The hypotheses are

$H_0$: The two variables are independent of one another. That is, the probabilities for either variable are equal to the conditional probabilities given the other variable.
$H_1$: The two variables are associated; they are not independent of one another. There is at least one category of one variable whose probability is not equal to the conditional probability given some category for the other variable.

The expected table is constructed as follows: For each combination of categories, one for each variable, multiply the count for one category by the count for the other category and then divide by the total sample size, $n$:

$$\text{Expected count} = \frac{\left(\begin{array}{c}\text{Count for category}\\ \text{for one variable}\end{array}\right)\left(\begin{array}{c}\text{Count for category}\\ \text{for other variable}\end{array}\right)}{n}$$

The assumptions are

1. The data set is a random sample from the population of interest.
2. At least five counts are expected in each combination of categories.

When calculating the chi-squared statistic in the test for independence, the degrees of freedom number is

$$\left(\begin{array}{c}\text{Number of categories}\\ \text{for first variable}\end{array} - 1\right)\left(\begin{array}{c}\text{Number of categories}\\ \text{for second variable}\end{array} - 1\right).$$

If the chi-squared statistic is larger than the critical value from the chi-squared table, you have evidence that the observed counts are very different from those that would be expected if the variables were independent. You would reject the null hypothesis of independence and accept the research hypothesis, concluding that the variables show significant association.

If the chi-squared statistic is smaller than the critical value from the chi-squared table, the observed data are not very different from what you would expect if they were independent in the population. You would accept the null hypothesis of independence (as a reasonable possibility) and conclude that the variables do not show significant association. This is a weak conclusion because the null hypothesis of independence has been accepted; you *accept* independence but have *not proven* independence.

## Key Words

chi-squared statistic, *507*
chi-squared test for equality of percentages, *509*
chi-squared test for independence, *512*
chi-squared tests, *507*
conditional population percentages, *512*
independent, *512*

### Questions

1. For what kind of variables are chi-squared tests useful?
2. a. What does the chi-squared statistic measure, in terms of the relationship between the observed data and the null hypothesis?
   b. Do you reject the null hypothesis for large or for small values of the chi-squared statistic? Why?
3. What is the purpose of the chi-squared test for equality of percentages?
4. a. For what kind of data set is the chi-squared test for equality of percentages appropriate?
   b. What are the reference values for this test?
   c. What are the hypotheses?
   d. How are the expected counts obtained? What do they represent?
   e. What assumptions are required in order that this test be valid?
5. a. For the chi-squared test for equality of percentages, what do you conclude if the chi-squared statistic is larger than the value in the chi-squared table?
   b. What do you conclude if the chi-squared statistic is smaller than the value in the chi-squared table?
6. a. What is meant by independence of two qualitative variables?
   b. What is the relationship between conditional probabilities and independence of qualitative variables?
7. What is the purpose of the chi-squared test for independence?
8. a. For what kind of data set is the chi-squared test for independence appropriate?
   b. What are the reference values for this test, if any?
   c. What are the hypotheses?
   d. How are the expected counts obtained? What do they represent?
   e. What assumptions are required in order that this test be valid?
9. a. For the chi-squared test for independence, what do you conclude if the chi-squared statistic is larger than the value in the chi-squared table?
   b. What do you conclude if the chi-squared statistic is smaller than the value in the chi-squared table?

10. Why is it much more difficult to establish independence than it is to establish dependence (lack of independence)?

## Problems

*Problems marked with an asterisk (\*) are solved in the Self Test in Appendix C.*

1. **a.** If an observed count is 3 and the expected count is 8.61, is there any problem with going ahead with a chi-squared test?
   **b.** If an observed count is 8 and the expected count is 3.29, is there any problem with going ahead with a chi-squared test?
2. For each potential customer entering an auto showroom, the type of vehicle desired is recorded. Table 17.4.1 shows data for the past week, together with percentages for the past year at this time.
   **a.** Find the percentages for last week's vehicles.
   **b.** Compare last week's percentages to last year's percentages. Describe any differences you see in terms that would be useful to an automobile salesperson.
   **c.\*** Assuming last year's percentages continued to apply, how many of these 536 people would you expect to be looking for an economy car? Compare this to the observed number of such people.
   **d.\*** Find the expected count for each type of vehicle, assuming last year's percentages still apply.
   **e.\*** Find the chi-squared statistic, viewing last year's percentages as exact.
   **f.** Discuss the assumptions required for the chi-squared test to be valid. In particular, what population are you inferring about?
   **g.** How many degrees of freedom are there for this chi-squared test?
   **h.** Find the appropriate chi-squared table value for the 5%, 1%, and 0.1% levels.
   **i.** Perform the chi-squared test at each of these levels and report the results.

**j.** State your conclusions (with a *p*-value reported as $p > 0.05$, $p < 0.05$, $p < 0.01$, or $p < 0.001$), discussing any shifts in consumer preferences.

3. Last year at this time, your firm's incoming telephone calls followed the percentages shown in Table 17.4.2.
   **a.** Find the percentages for the first day of this month and compare them to last year's percentages for this month.
   **b.** Find the expected number of calls of each type for the first day of this month, assuming the population percentages are given by last year's total for the month.
   **c.** Find the chi-squared statistic and the number of degrees of freedom.
   **d.** Report the results of the chi-squared test at the 5% level.
   **e.** Summarize what the chi-squared test has told you about any change in the pattern of calls compared to last year at this time.

4. Out of 267 roller skates randomly selected for close examination, 5 were found to have a loose rivet, and 12 were not cleaned according to specifications.
   **a.** Use the formula for computing an expected count. How many roller skates would you expect to have both problems if the problems are independent?
   **b.** Find the estimated probability of a loose rivet, using the relative frequency.
   **c.** Similarly, find the estimated probability of a cleaning problem.
   **d.** Use the probability formula from Chapter 6 to find the estimated probability of a loose rivet and a cleaning problem, assuming they are independent, using your estimated probabilities from parts b and c.
   **e.** Convert your probability from part d to an expected count by multiplying it by the sample size.
   **f.** Compare your answers to parts a (from the expected count formula) and e (from the independence formula). Comment on why they both came out this way.

**TABLE 17.4.1 Vehicle Desired**

| Type | Last Week's Count | Last Year's Percentages |
|---|---|---|
| Family sedan | 187 | 25.8% |
| Economy car | 206 | 46.2 |
| Sports car | 29 | 8.1 |
| Van | 72 | 12.4 |
| Pickup | 42 | 7.5 |
| | | |
| Total | 536 | 100.0% |

**TABLE 17.4.2 Incoming Calls**

| Type | Count (first day of the month) | Percent of Total (this month last year) |
|---|---|---|
| Reservation | 53 | 33.2% |
| Information | 54 | 38.1 |
| Service request | 28 | 12.5 |
| Cancellation | 18 | 9.7 |
| Other | 7 | 6.5 |
| | | |
| Total | 160 | 100.0% |

**TABLE 17.4.3 Survey Responses Regarding Future Business Conditions**

|  | Managers | Other Employees | Total |
|---|---|---|---|
| Better | 23 | 185 | |
| Same | 37 | 336 | |
| Worse | 11 | 161 | |
| Not sure | 15 | 87 | |
| | | | |
| Total | | | |

5. Your firm is considering expansion to a nearby city. A survey of employees in that city, asked to respond to the question "Will business conditions in this area get better, stay the same, or get worse?" produced the data set shown in Table 17.4.3.
   a. Fill in the "Total" row and column.
   b. Find the table of overall percentages. Interpret these as estimates of probabilities in the population. In particular, what probabilities do they represent?
   c. Find the table of percentages by type of employee. Interpret these as estimates of probabilities in the population. In particular, what probabilities do they represent?
   d. Find the table of percentages by response. Interpret these as estimates of probabilities in the population. In particular, what probabilities do they represent?
   e. Does the response appear to be independent of the employee's classification? Why or why not?

6. Refer to the data for problem 5.
   a. What does the null hypothesis of independence claim, in practical terms, for this situation?
   b. How many managers responding "Worse" would you expect to find in this sample if response were independent of employee classification?

c. Find the table of expected counts, assuming independence.
d.* Find the chi-squared statistic.
e. How many degrees of freedom does the chi-squared test have?

7. Refer to the data for problem 5.
   a. Find the critical value from the chi-squared table for the 5% level and report the result of the chi-squared test.
   b. Find the critical value from the chi-squared table for the 1% level and report the result of the chi-squared test.
   c. Find the critical value from the chi-squared table for the 0.1% level and report the result of the chi-squared test.
   d. State your conclusions (with $p$-value reported as $p > 0.05$, $p < 0.05$, $p < 0.01$, or $p < 0.001$) and discuss the results in practical terms.

8. Consider the results of a small opinion poll concerning the chances of another stock market crash in the next 12 months comparable to the crash of 1987, shown in Table 17.4.4.
   a. Fill in the "Total" row and column.
   b. Find the table of overall percentages. Interpret these as estimates of probabilities in the population. In particular, what probabilities do they represent?
   c. Find the table of percentages by type of person (stockholder/nonstockholder). Interpret these as estimates of probabilities in the population. In particular, what probabilities do they represent?
   d. Find the table of percentages by response. Interpret these as estimates of probabilities in the population. In particular, what probabilities do they represent?
   e. Does the response appear to be independent of the stockholder/nonstockholder classification? Why or why not?

9. Refer to the data for problem 8.
   a. What does the null hypothesis of independence claim, in practical terms, for this situation?
   b. How many stockholders responding "Very likely" would you expect to find in this sample if response

**TABLE 17.4.4 Responses to the Opinion Poll on the Chances of Another Big Crash in the Stock Market**

|  | Stockholders | Nonstockholders | Total |
|---|---|---|---|
| Very likely | 18 | 26 | |
| Somewhat likely | 41 | 65 | |
| Not very likely | 52 | 68 | |
| Not likely at all | 19 | 31 | |
| Not sure | 8 | 13 | |
| | | | |
| Total | | | |

**Source:** Results of a much larger poll for this and related questions appeared in a *Business Week*/Harris Poll, *Business Week*, November 9, 1987, p. 36.

were independent of stockholder/nonstockholder classification?

   c. Find the table of expected counts, assuming independence.

   d. Find the chi-squared statistic.

   e. How many degrees of freedom does the chi-squared test have?

**10.** Refer to the data for problem 8.

   a. Find the critical value from the chi-squared table for the 5% level and report the result of the chi-squared test.

   b. Find the critical value from the chi-squared table for the 1% level and report the result of the chi-squared test.

   c. Find the critical value from the chi-squared table for the 0.1% level and report the result of the chi-squared test.

   d. State your conclusions (with $p$-value reported as $p > 0.05$, $p < 0.05$, $p < 0.01$, or $p < 0.001$) and discuss the results in practical terms.

**11.** A mail-order company is interested in whether or not "order rates" (the percent of catalogs mailed that result in an order) vary from one region of the country to another. Table 17.4.5 gives recent data on the number of catalogs mailed that produced an order, and the number that did not, according to region.

   a. Find the order rate (as a percentage) for each region. Which region appears to order at a higher rate on a per-catalog basis?

   b. Are the order rates significantly different between these two regions? How do you know?

**12.** A commercial bank is reviewing the status of recent real estate mortgage applications. Some applications have been accepted, some rejected, and some are pending while waiting for further information. The data are shown in Table 17.4.6 and graphed in Figure 17.4.1.

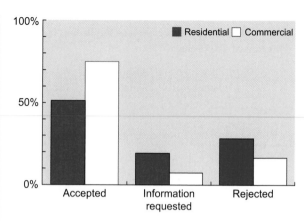

**FIGURE 17.4.1** The distribution of real estate loan application status for residential and for commercial mortgage applications.

   a. Write a paragraph, as if to your supervisor, describing Figure 17.4.1 and comparing the status of residential to commercial loan applications.

   b. Are the differences between residential and commercial customers significant? How do you know?

**13.** Does it really matter how you ask a question? A study was conducted that asked whether or not people would pay $30 to eat at a particular restaurant.[6] One group was told "there is a 50 percent chance that you will be satisfied," while the other was told "there is a 50 percent chance that you will be dissatisfied." The only difference in the wording is the use of *dissatisfied* in place of *satisfied*. The results were that 26% of the 240 people who were asked the "satisfied" question said they would eat there, as compared with only 11% of the 215 people who were asked the "dissatisfied" question. Is the difference (between 26% and 11%) significant, or is it possible to have differences this large due to randomness alone? How do you know?

**14.** The eastern factory had 28 accidents last year, out of a workforce of 673. The western factory had 31 accidents during this time period, out of 1,306 workers.

   a. Which factory had more accidents? Which factory had a greater accident rate?

   b. Is there a significant difference between the accident rates of these two factories? Justify your answer by reporting the chi-squared statistic and its degrees of freedom.

**15.** One group of households was asked how satisfied they were with their car, while the other group was asked how dissatisfied they were. Results are shown in Table 17.4.7.

   a. Which group was more likely to report that they were satisfied?

   b. Which group was more likely to report that they were dissatisfied?

   c. Are the differences significant? Justify your answer by reporting the chi-squared statistic and its degrees of freedom.

**16.** Here are the numbers of new customers who signed up for full service during each quarter of last year: 106, 108, 72, and 89.

---

**TABLE 17.4.5 Order Rates by Region**

|  | East | West |
| --- | --- | --- |
| Order produced | 926 | 352 |
| No order produced | 22,113 | 10,617 |

---

**TABLE 17.4.6 Status of Mortgage Applications**

|  | Residential | Commercial |
| --- | --- | --- |
| Accepted | 78 | 57 |
| Information requested | 30 | 6 |
| Rejected | 44 | 13 |

**TABLE 17.4.7 Number of Household Responses According to Question Asked**

| | Question Asked | |
| --- | --- | --- |
| | "Satisfied" | "Dissatisfied" |
| Very satisfied | 139 | 128 |
| Somewhat satisfied | 82 | 69 |
| Somewhat dissatisfied | 12 | 20 |
| Very dissatisfied | 10 | 23 |

**Source:** Data are adapted from R. A. Peterson and W. R. Wilson, "Measuring Customer Satisfaction: Fact and Artifact," *Journal of the Academy of Marketing Science* 20 (1992), pp. 61–71.

**TABLE 17.4.8 Newsletter Interest Level for Customers and Potential Customers**

| | Customer | Potential Customer |
| --- | --- | --- |
| Very interested | 49 | 187 |
| Somewhat interested | 97 | 244 |
| Not interested | 161 | 452 |

    a.  How many would you have expected to see in each quarter if these customers had signed up at exactly the same rate throughout the year?

    b.  Do the observed numbers differ significantly from those expected in part a? Justify your answer by reporting the chi-squared statistic and its degrees of freedom.

17.  Are your customers special? In particular, is their interest level in your promotional newsletter higher than for potential customers (who are not currently customers)? Justify your answer by reporting the chi-squared statistic and its degrees of freedom for the data set reported in Table 17.4.8 based on a random sample for each group.

6. R. A. Peterson and W. R. Wilson, "Measuring Customer Satisfaction: Fact and Artifact," *Journal of the Academy of Marketing Science* 20 (1992), pp. 61–71.

### Database Exercises

Refer to the employee database in Appendix A.

1.  Do training levels A, B, and C have approximately the same number of employees, except for randomness? (To answer this, test whether the percentage of employees at each training level differs significantly from the proportions 1/3, 1/3, and 1/3 for the three levels.)

2.  Is there evidence consistent with gender discrimination in training level? To answer this, proceed as follows:

    a.  Create a table of counts for the two qualitative variables "gender" and "training level."

    b.  Compute the overall percentage table and comment briefly.

    c.  Compute a table of percentages by gender; then comment.

    d.  Compute a table of percentages by training level; then comment.

    e.  Is it appropriate to use the chi-squared test for independence on this data set? Why or why not?

    f.  Omit training level C, restricting attention only to those employees at training levels A and B for the rest of this exercise. Compute the expected table.

    g.  Still omitting training level C, compute the chi-squared statistic.

    h.  Still omitting training level C, report the results of the chi-squared test for independence at the 5% level. Comment on these results.

### Projects

1.  Obtain data for a single (univariate) qualitative variable relating to your work or business interests on the Internet, in a newspaper, or in a magazine, together with a list of reference percentages for comparison.

    a.  Summarize these observations in a table of counts.

    b.  Summarize these observations in a table of percentages.

    c.  Compare the percentages for the data to your reference percentages and comment on the similarities and differences.

    d.  Find the table of expected counts, assuming that your reference percentages are correct for the population.

    e.  List the [(Observed − Expected)$^2$/Expected] values for each category. Find the largest and smallest values in this list, and explain why these particular categories are smallest and largest by comparing the observed and expected percentages for these categories.

    f.  Find the chi-squared statistic.

    g.  Perform the chi-squared test and find the *p*-value (reported as $p > 0.05$, $p < 0.05$, $p < 0.01$, or $p < 0.001$).

    h.  Comment on what the chi-squared test has told you about this business situation.

2.  Obtain data for two qualitative variables (a bivariate data set) on the Internet, in a newspaper, or in a magazine relating to your work or business interests.

    a.  Summarize these observations in a table of counts.

    b.  Summarize these observations in a table of overall percentages; then comment.

    c.  Summarize these observations in a table of percentages by one of your variables. Comment on the profiles of the other variable.

d. Repeat the preceding part using percentages by the other variable.

e. What does the null hypothesis of independence claim for this situation? Is it a reasonable possibility, in your opinion?

f. Find the table of expected counts, assuming that your two variables are independent.

g. List the [(Observed − Expected)$^2$/Expected] values for each combination of categories. Find the largest and smallest values in this table, and explain why these particular combinations of categories are smallest and largest by comparing the observed and expected counts for these categories.

h. Find the chi-squared statistic.

i. Perform the chi-squared test and find the $p$-value (reported as $p > 0.05$, $p < 0.05$, $p < 0.01$, or $p < 0.001$).

j. Comment on what the chi-squared test has told you about this business situation.

## Chapter 18

# Quality Control

## Recognizing and Managing Variation

**Statistical quality control** is the use of statistical methods for evaluating and improving the results of any activity. Whether the activity involves production or service, it can be monitored so that problems can be discovered and corrected before they become more serious. Statistics is well suited for this task because good management decisions must be based, in part, on *data*. Here are some examples to indicate the variety of ways that statistical quality control can be used to enhance the firm's bottom line:

**One:** Firms don't like it very much when a customer returns merchandise. In addition to losing the sale you thought you had, this customer might not be saying good things about your firm's products. But why not view this problem as an opportunity? By collecting data on the various reasons for returning the product, you obtain a wealth of information that is useful in a variety of areas. Analyzing this list will show you how to improve product quality by directing your attention to the most important problems. In addition, you may learn more about your customers. Such a proprietary database could be useful in marketing and sales efforts, as well as in the design of new products.

**Two:** Your package indicates that 16 ounces of dishwasher detergent are contained in every box. If you could do it inexpensively, you'd like to be sure that every box had *exactly* 16 ounces. However, the costs would be prohibitive. Some level of variation from one box to another will be tolerated, and you want to

control this level of variation. One goal is to avoid public relations trouble by making sure that no boxes are seriously underweight and that, on average, you have at least 16 ounces per box. Another goal is to hold costs down by not allowing too much extra in the boxes. By collecting and analyzing the net weights of these boxes (either every one or a random sample from each batch), you can monitor the process and its level of variability. When everything seems to be under control, you can leave the process alone. When you detect a problem, you can fix it by adjusting or replacing the machinery. Proper monitoring can even allow you to fix a problem before it occurs by enabling you to spot a trend that, if continued, will soon cause real trouble.

**Three:** The accounts receivable department has a very important function: to convert your firm's sales into cash. Any problems you might have with accounts receivable translate directly into lost value, for example, cash you would have received if the file hadn't been held up three weeks waiting for an internal review. And don't forget that cash received later is worth less due to the time value of money. By collecting data on the progress of a sample of normal and problem accounts, you can tell whether or not you have the process of bill collection "under control." By analyzing problems that show up over and over, you can identify those parts of this system that need to be redesigned. Perhaps you need a few extra workers over here.

Or maybe different steps could take place simultaneously—"in parallel"—so that one group doesn't have to wait for another group to finish before doing its work.

Quality control looks good on the cost-benefit scale. The costs of setting up a program of statistical quality control are typically small compared to the money saved as a result. When production of defective items is eliminated, you save because inspection of every item is unnecessary. You also save the cost of either reworking or junking defective items. Finally, the reputation for quality and dependability will help you land the contracts you need at favorable prices.

But don't expect statistical methods to do everything for you. They can only provide information; it's up to you to make good use of it. They might tell you that something funny probably happened around 10:30 AM because the packages are getting significantly heavier, but it's still up to you and your workers to adjust the machines involved.

And don't expect statistical methods to do the impossible. The capabilities of the system itself must be considered. If a drill bit is so old and worn out that it is incapable of drilling a smooth hole, no amount of statistical analysis by itself will correct the problem. Although a good quality control program will help you get the most out of what is available, you may discover that some modernization is necessary before acceptable results can be achieved.

The five basic activities of statistics all play important roles in quality control. The *design* phase involves identification of particular processes to look at and measurements to take. The *modeling* phase often remains in the background, allowing the calculations of control limits in a standard way based on assumptions such as independence and normal (or binomial) distributions. The other three activities are often assisted by a *control chart* to display the data. The *exploration* phase involves checking the data for particular kinds of problems in the process. The *estimation* phase involves characterizing the current state of the process and how well it's performing. The *hypothesis testing* phase involves deciding whether the process should be adjusted or left alone.

W. Edwards Deming brought statistical quality control methods to the Japanese in the 1950s and continued to help firms around the world with their quality control programs. Here are Deming's 14 points for managing continued improvement in quality, which summarize how a company should go about improving quality:[1]

1. *Create constancy of purpose toward improvement of product and service,* with the aim to become competitive, to stay in business, and to provide jobs.
2. *Adopt a new philosophy.* We are in a new economic age, created by Japan. We can no longer live with commonly accepted styles of American management, nor with commonly accepted levels of delays, mistakes, or defective products.
3. *Cease dependence on inspection to achieve quality.* Eliminate the need for inspection on a mass basis by building quality into the product in the first place.
4. *End the practice of awarding business on the basis of price tag.* Instead, minimize total cost.
5. *Improve constantly and forever the system of production and service to improve quality and productivity,* and thus constantly decrease costs.
6. *Institute training on the job.*
7. *Institute supervision:* the aim of supervision should be to help people and machines and gadgets do a better job. Supervision of management is in need of overhaul, as well as supervision of production workers.
8. *Drive out fear,* so that everyone may work effectively for the company.
9. *Break down barriers between departments.* People in research, design, sales, and production must work as a team to foresee problems of production and use that may be encountered with the product or service.
10. *Eliminate slogans, exhortations, and targets for the work force that ask for zero defects and new levels of productivity.* Such exhortations only create adversarial relationships. The bulk of the causes of low productivity belong to the system and thus lie beyond the power of the work force.
11. *Eliminate work standards that prescribe numerical quotas for the day.* Substitute aids and helpful supervision.
12. *Remove the barriers that rob the hourly worker of the right to pride of workmanship.* The responsibility of supervisors must be changed from sheer numbers to quality. This means abolishment of the annual rating, or merit rating, and management by objective.
13. *Institute a vigorous program of education and training.*
14. *Put everybody in the company to work to accomplish the transformation.*

## 18.1 PROCESSES AND CAUSES OF VARIATION

A **process** is any business activity that takes inputs and transforms them into outputs. A manufacturing process takes raw materials and turns them into products. Restaurants have processes that take food and energy and transform them into ready-to-eat meals. Offices have a variety of processes that transform some kind of information (perhaps papers

---

1. These 14 points are reprinted from *The ESB Journal*, Spring 1989, published by the Educational Service Bureau of Dow Jones & Co., Inc. Further discussion of these points may be found, for example, in W. Edwards Deming, *Out of the Crisis* (Cambridge, MA: MIT Center for Advanced Engineering Studies, 1986); and in H. Gitlow, A. Oppenheim, and R. Oppenheim, *Quality Management: Tools and Methods for Improvement*, 2nd ed. (Burr Ridge, IL: Richard D. Irwin, 1995).

with basic information) into other information (perhaps computerized records or paychecks).

A process can be made up of other processes, called *subprocesses*, each of which is a process in its own right. For example, airplane production might be viewed as a single process that takes various inputs (metal, plastic, wire, computerized information) and transforms them into airplanes. This enormous process consists of many subprocesses (for example, assembling the fuselage, connecting the wires to the cabin lights, testing the wing flaps). Each of these subprocesses consists of further subprocesses (placing one rivet, soldering one wire, checking the maximum extension). You are free to focus on whatever level of detail you want in order to achieve your purpose. The methods of statistical process control are adaptable to essentially any process or subprocesses.

**Statistical process control** is the use of statistical methods to monitor the functioning of a process so that you can adjust or fix it when necessary and can leave it alone when it's working properly. The goal is to detect problems and fix them *before* defective items are produced. By using sampled data to keep a process in a state of statistical control, you can ensure high-quality production without inspecting every single item!

Nearly every process shows some variation among its results. Some variations are small and unimportant, such as the exact number of chocolate chips varying by a few from one cookie to another. Other variations are more crucial, such as a metal box getting squashed and resembling a modern-art sculpture instead of what the customer wanted.

The difference between "what you wanted" and "what you got" could be due to any number of causes. Some of these causes are easier to find than others. Sometimes you find out the cause even before you get the data (for example, because you could smell the smoke). Some causes require some detective work to discover what (say, a machine that needs adjustment) and why (perhaps an employee needs new glasses). Yet other causes are not even worth the effort of investigation (such as, Why did the tape machine use an extra tiny fraction of an inch more for this box than for that one?).

Anytime you could reasonably find out why a problem occurred, you have an **assignable cause of variation**. Note that you may not actually know the cause; it is enough that you could reasonably find it out without too much expense. Here are some examples of assignable causes and possible solutions:

1. Dust gets into the "clean room" and interferes with the production of microchips and disk drives. Possible solutions include checking the cleaning apparatus, replacing filters and seals as needed, and reviewing employee procedures for entering and leaving the area.

2. Clerks fill out the forms incorrectly, placing the totals in the wrong boxes. Possible solutions include training the clerks, redesigning the forms, and doing both.

All causes of variation that would not be worth the effort to identify are grouped together as **random causes of variation**.[2] These causes should be viewed as defining the basic randomness of the situation, so that you know what to expect when things are in control and so that you don't expect too much. Here are some examples of variation due to random causes:

1. The bottle-filling machinery is accurate to a fraction of a drop of soda. Even when it is adjusted properly, there is still some small amount of variation in liquid from one bottle to the next.
2. The number of corn flakes in a cereal box varies from one box to the next. There is no reason to control this variability, provided it falls within acceptable limits and the total weight is close enough to the net weight promised on the package.

When you have identified and eliminated all of the assignable causes of variation, so that only random causes remain, your process is said to be **in a state of statistical control**, or, simply, **in control**. When a process is in control, all you have to do is monitor it to see if it stays in control. When a process goes *out of control*, there is an assignable cause and, therefore, a problem to be fixed.

Quality control programs are no longer completely internal to a company. More and more firms are demanding that their suppliers prove (by providing control charts) that the products they are buying were produced when the system was in control. If you are having problems with a supplier, you may want to consider this possibility.

## The Pareto Diagram Shows Where to Focus Attention

Suppose you have examined a group of defective components and have classified each one according to the cause of the defect. The **Pareto diagram** displays the causes of the various defects in order from most to least frequent so that you can focus attention on the most important problems. In addition to showing the number and percentage of

---

2. This is not exactly the same as the strict statistical definition of the word *random*. See, for example, the definition of *random sample* in Chapter 8. However, the Statistical Division of the American Society for Quality Control, in *Glossary and Tables for Statistical Quality Control* (Milwaukee, WI: American Society for Quality Control, 1983), p. 29, defines *random or chance causes* as follows: "Factors, generally numerous and individually of relatively small importance, which contribute to variation, but which are not feasible to detect or identify."

defectives due to each cause, the diagram indicates the *cumulative* percentage so that you can easily tell, for example, the percentage for the two (or three or four) biggest problems combined.

Here's how to create a Pareto diagram:

1. Begin with the number of defects (the frequency) for each problem cause. Find the total number of defects and the percentage of defects for each cause.
2. Sort the causes in descending order by frequency so that the biggest problems come first in the list.
3. Draw a bar graph of these frequencies.
4. Find the cumulative percentage for each cause by adding its percentage to all percentages above it in the list.
5. Draw a line (on top of the bar graph) indicating these cumulative percentages.
6. Add labels to the diagram. Indicate the name of each problem cause underneath its bar. On the left, indicate the vertical scale in terms of the number of defects. On the right, indicate the *cumulative percentage* of defects.

For example, consider problems encountered in the production of small consumer electronics components, as shown in Table 18.1.1. This same data set, sorted and with percentages computed, is shown in Table 18.1.2. The Pareto diagram, shown in Figure 18.1.1, indicates that the biggest problem is with power supplies, representing about twice the number of defects due to the next most important problem, which involves the plastic case. Looking at the cumulative percentage line, you see that these two major problems account for a large majority (85.9%) of the defects. The three biggest problems together account for nearly all (97.2%) of the defects.

In a Pareto diagram, the tallest bars will always be at the left (*indicating the most frequent problems*), and the shortest bars will be at the right. The line indicating the cumulative percentage will always go upward and will tend to level off toward the right.

**TABLE 18.1.1 Defect Causes, with Frequency of Occurrence**

| Cause of Problem | Number of Cases |
|---|---|
| Solder joint | 37 |
| Plastic case | 86 |
| Power supply | 194 |
| Dirt | 8 |
| Shock (was dropped) | 1 |
| | |
| Total | 326 |

**TABLE 18.1.2 Defect Causes Sorted by Frequency with Percentage of Occurrence and Cumulative Percentages**

| Cause of Problem | Number of Cases | Percent | Cumulative Percent |
|---|---|---|---|
| Power supply | 194 | 59.5% | 59.5% |
| Plastic case | 86 | 26.4 | 85.9 |
| Solder joint | 37 | 11.3 | 97.2 |
| Dirt | 8 | 2.5 | 99.7 |
| Shock (was dropped) | 1 | 0.3 | 100.0 |
| | | | |
| Total | 326 | 100.0% | |

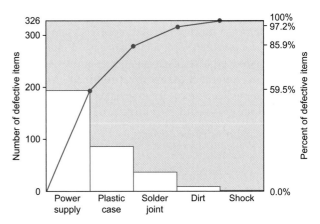

**FIGURE 18.1.1** The Pareto diagram for causes of defects in the production of small consumer electronics components. The bars show the importance (frequency) of each cause; for example, 86 or 26.4% of the problems involved the plastic case. The line shows the cumulative percent of the most important causes; for example, the top two causes (power supply and plastic case) together accounted for 85.9% of defects.

One useful function of the Pareto diagram is to introduce some objectivity into the discussion of what to do about quality. Rather than having employees choose problems to solve based just on what they know and on what they enjoy doing, you can use the Pareto diagram to help you concentrate their attention on those problems that are most crucial to the company.

## 18.2 CONTROL CHARTS AND HOW TO READ THEM

Once you have selected one of the many processes you are responsible for as a manager and have chosen one of the many possible measurements that can be made on that process, you will want to understand this information in order to know when to act and when *not* to act. A **control chart** displays successive measurements of a process together with a *center* *line* and *control limits*, which are computed to help you decide whether or not the process is in control. If you decide that the process is not in control, the control chart will help you identify the problem so that you can fix it.

All five basic activities of statistics are involved in the use of control charts for quality control. The *design* phase involves selecting the process and measurements to examine for producing the control chart. The *modeling* phase, often taken for granted, allows you to use standard tables or software to find the control limits, often by assuming that the data are approximately independent (even though, strictly speaking, you have a time-series data set) and that the distribution is somewhat normal (or binomial, when observing the number or percent of defects). Once you have a control chart, the other three activities come into play. The *exploration* phase involves looking at the chart in order to spot patterns, trends, and exceptions that tell you how the process is working and what it is likely to do in the future. The *estimation* phase involves computing summaries of the process, some of which are displayed in the control chart. Finally, the *hypothesis testing* phase involves using data (the measurements) to decide whether or not the process is in control. Here are the hypotheses being tested:

$H_0$: The process is in control.
$H_1$: The process is not in control.

Note that the default (the null hypothesis) is that the process is assumed to be in control. By setting it up this way, you ensure that the process will not be adjusted unless there is convincing evidence that there is a problem. Making adjustments can be very expensive due to lost production time, the cost of making the adjustments, and the possibility that adjusting will *add to* the variability of the system. You don't want to make adjustments unless you really need to. As they say, "If it ain't broke, don't fix it!"

The **false alarm rate** is how often you decide to fix the process when there really isn't any problem; this is the same concept as the type I error in hypothesis testing. Conventional control charts use a factor of three standard errors instead of two because the conventional 5% type I error rate of statistical hypothesis testing is unacceptably high for most quality control applications.[3]

The rest of this section will show you how to read a generic control chart. The details of how to construct the various types of control charts will be presented beginning in Section 18.3.

## The Control Limits Show If a Single Observation Is Out of Control

For a process that is in control, you should expect to see a plot that stays within the control limits, as shown in Figure 18.2.1. All observations are within the control limits; no observation falls outside. Although there is variation from one item to the next, the chart shows no clear patterns. The process is varying randomly within the control limits.

If any measurement falls outside the control limits, either above the upper control limit or below the lower limit, you have evidence that the process is not in control. An example is shown in Figure 18.2.2. In statistical terms, you reject the null hypothesis that the process is in control and accept the research hypothesis that the process is not in control.

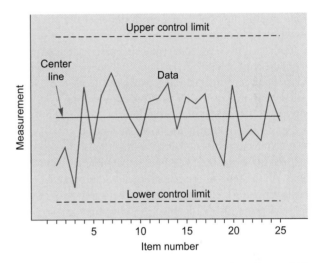

**FIGURE 18.2.1**   This control chart shows a process that is in control. The measurements fluctuate randomly within the control limits.

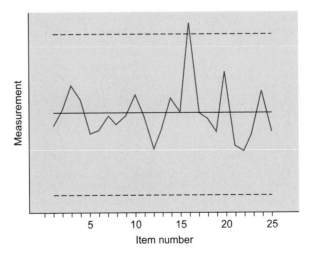

**FIGURE 18.2.2**   This process is not in control. The 16th measurement is outside the control limits (it is above the upper limit). An investigation of the circumstances of this particular item's production may help you avoid this kind of problem in the future.

---

3. From Chapter 10, on hypothesis testing, recall that the *t* table value for two-sided testing at the 5% level is 1.960, or approximately 2. When you use 3 instead of 1.960, the theoretical false alarm rate is reduced from 5% way down to 0.27% for large samples. With six standard deviations ("six sigmas") separating the short-term mean from the acceptable limits, and also allowing for a shift over time in the mean of 1.5 sigmas as the properties of the system change, the defect rate for a normal distribution becomes 0.0000034 or 3.4 per million.

In practical terms, you have a problem, and the control chart will help guide you toward a solution. The chart shows you which item is involved so that you can investigate by looking at the records for this item and talking with those responsible. You might quickly find an obvious assignable cause (such as a broken drill bit or an overheated solder bath). On the other hand, you might find that further investigation and testing are required before the problem can be identified. Or it could be the rare case of a false alarm, with no fixing needed.

## How to Spot Trouble Even within the Control Limits

One way in which you can spot a process that is out of control is to look for points that fall outside the control limits. However, even if all points are within the control limits, it still may be clear from looking at the chart that the process is not in control.

The idea is to consider the nature of combinations of points within the control limits. Even though all points are within the limits, if they tend to fall suspiciously close to one of the limits (as shown in Figure 18.2.3) or if they are moving systematically toward a limit (as shown in Figure 18.2.4), you would decide that the process is no longer in control.[4]

A sudden jump to another level in a control chart (such as occurred in Figure 18.2.3) suggests that there was an abrupt change in the process. Perhaps something fell into the gears and is slowing them down, or maybe a new worker just took charge and hasn't learned the system yet.

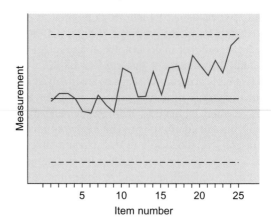

**FIGURE 18.2.4**   Even though all points are within the control limits, there is a troublesome trend here. Note that the sequence has a steady drift upward. There is no need to wait for the trend to continue through the upper control limit. You should decide now that the process is not in control.

A gradual drift upward or downward (as in Figure 18.2.4) suggests that something is progressively wearing out. Perhaps some component of the machinery has worn through its hard outer shell and has begun to slowly disintegrate, or maybe the chemical bath has begun to reach the end of its useful life and should be replaced.

Part of your job in interpreting control charts is to be a detective. By exploring the data and looking at the trends and patterns, you learn what kinds of things to look for in order to fix the problem and get the process back into control.

## 18.3   CHARTING A QUANTITATIVE MEASUREMENT WITH $\overline{X}$ AND $R$ CHARTS

For charting a quantitative measurement, it is conventional to choose a fairly small sample size ($n = 4$ or $n = 5$ are common choices) and then chart the results of many successive samples. Each point on the control chart will then represent a summary (perhaps center or variability) for the $n$ individual observations. It is advisable to choose a small $n$ because the process may change quickly and you would like to detect the change before very many defects are produced.

Perhaps the most common ways of charting a quantitative measurement are the $\overline{X}$ and $R$ charts. The **$\overline{X}$ chart** displays the *average* of each sample together with the appropriate center line and control limits so that you can monitor the level of the process. The **$R$ chart** displays the *range* (the largest value minus the smallest) for each sample together with center line and control limits so that you can monitor the variability of the process. These charts and others like them (including the percentage chart) are due to W. A. Shewhart.

Why is the range sometimes used in quality control charts instead of the (statistically) more conventional

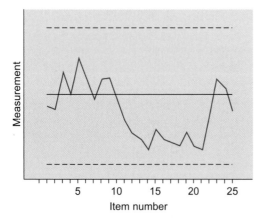

**FIGURE 18.2.3**   Even though all points are within the control limits, there is a troublesome pattern here. Note that a sequence of 11 points (items 12 through 22) fall close to, but inside, the lower control limit. Based on the evidence provided by the entire chart, you would decide that the process is not in control.

---

4. A collection of rules for deciding whether or not such a process is in control is given in Gitlow et al., *op cit.* In particular, you could decide that the process is not in control if eight or more consecutive points fall on one side of the center line or if eight or more consecutive points form a sequence that is always moving either upward or downward.

standard deviation as a measure of variability? Because the range is easier to compute. Back when computers were not widely available, this was an important advantage because it meant that workers could construct their own charts by hand. Although (as was shown in Chapter 5) the standard deviation is better than the range as a measure of variability, with small sample sizes (such as $n = 4$ or 5) the range is nearly as good as the standard deviation.

There are two ways to find the control limits and center line for each of these charts, depending on whether or not you have an external standard (for example, from past experience, engineering specifications, or customer requirement). If you don't have an external standard, the limits are computed based only on the data. If you do have an external standard, the limits are based only on this standard. Table 18.3.1 shows how to compute the center line and the control limits for the $\overline{X}$ and $R$ charts. The multipliers $A$, $A_2$, $d_2$, $D_1$, $D_2$, $D_3$, and $D_4$ are given in Table 18.3.2. The symbol $\overline{\overline{X}}$ is the average of all of the sample averages, and $\overline{R}$ is the average of the ranges for all of the samples.

For example, suppose there is no standard given, and the sample summaries are as shown in Table 18.3.3. With sample size $n = 4$, you will need the table values $A_2 = 0.729$, $D_3 = 0$, and $D_4 = 2.282$. For the $\overline{X}$ chart, the center line and control limits will be as follows:

Center line: $\qquad \overline{\overline{X}} = 21.84$
Lower control limit: $\quad \overline{\overline{X}} - A_2\overline{R} = 21.84 - (0.729)(1.20) = 20.97$
Upper control limit: $\quad \overline{\overline{X}} + A_2\overline{R} = 21.84 + (0.729)(1.20) = 22.71$

For the $R$ chart, these values are as follows:

Center line: $\qquad \overline{R} = 1.20$
Lower control limit: $\quad D_3\overline{R} = (0)(1.20) = 0$
Upper control limit: $\quad D_4\overline{R} = (2.282)(1.20) = 2.74$

Now suppose that there is a standard given in addition to the data in Table 18.3.3, so that you know (say, from past experience) that $\mu_0 = 22.00$ and $\sigma_0 = 0.50$. With sample size $n = 4$, you will need the table values $A = 1.500$,

---

**TABLE 18.3.1 Finding the Center Line and Control Limits for $\overline{X}$ and $R$ Charts**

|  |  | Center Line | Control Limits |
|---|---|---|---|
| $\overline{X}$ chart | Standard given ($\mu_0$ and $\sigma_0$) | $\mu_0$ | From $\mu_0 - A\sigma_0$ to $\mu_0 + A\sigma_0$ |
|  | No standard given | $\overline{\overline{X}}$ | From $\overline{\overline{X}} - A_2\overline{R}$ to $\overline{\overline{X}} + A_2\overline{R}$ |
| $R$ chart | Standard given ($\sigma_0$) | $d_2\sigma_0$ | From $D_1\sigma_0$ to $D_2\sigma_0$ |
|  | No standard given | $\overline{R}$ | From $D_3\overline{R}$ to $D_4\overline{R}$ |

---

**TABLE 18.3.2 Multipliers to Use for Constructing $\overline{X}$ and $R$ Charts**

| | Charts for Averages ($\overline{X}$ Chart): | | Charts for Ranges ($R$ Chart) | | | | |
|---|---|---|---|---|---|---|---|
| Sample Size | Factors for Control Limits | | Factor for Central Line | | Factors for Control Limits | | |
| $n$ | $A$ | $A_2$ | $d_2$ | $D_1$ | $D_2$ | $D_3$ | $D_4$ |
| 2 | 2.121 | 1.880 | 1.128 | 0.000 | 3.686 | 0.000 | 3.267 |
| 3 | 1.732 | 1.023 | 1.693 | 0.000 | 4.358 | 0.000 | 2.574 |
| 4 | 1.500 | 0.729 | 2.059 | 0.000 | 4.698 | 0.000 | 2.282 |
| 5 | 1.342 | 0.577 | 2.326 | 0.000 | 4.918 | 0.000 | 2.114 |
| 6 | 1.225 | 0.483 | 2.534 | 0.000 | 5.078 | 0.000 | 2.004 |
| 7 | 1.134 | 0.419 | 2.704 | 0.204 | 5.204 | 0.076 | 1.924 |
| 8 | 1.061 | 0.373 | 2.847 | 0.388 | 5.306 | 0.136 | 1.864 |
| 9 | 1.000 | 0.337 | 2.970 | 0.547 | 5.393 | 0.184 | 1.816 |

*(Continued)*

**TABLE 18.3.2 Multipliers to Use for Constructing $\overline{X}$ and $R$ Charts—cont'd**

| | Charts for Averages ($\overline{X}$ Chart): | | Charts for Ranges ($R$ Chart) | | | | |
| | | | Factor for Central Line | | Factors for Control Limits | | |
| Sample Size | Factors for Control Limits | | | | | | |
| $n$ | $A$ | $A_2$ | $d_2$ | $D_1$ | $D_2$ | $D_3$ | $D_4$ |
|---|---|---|---|---|---|---|---|
| 10 | 0.949 | 0.308 | 3.078 | 0.687 | 5.469 | 0.223 | 1.777 |
| 11 | 0.905 | 0.285 | 3.173 | 0.811 | 5.535 | 0.256 | 1.744 |
| 12 | 0.866 | 0.266 | 3.258 | 0.922 | 5.594 | 0.283 | 1.717 |
| 13 | 0.832 | 0.249 | 3.336 | 1.025 | 5.647 | 0.307 | 1.693 |
| 14 | 0.802 | 0.235 | 3.407 | 1.118 | 5.696 | 0.328 | 1.672 |
| 15 | 0.775 | 0.223 | 3.472 | 1.203 | 5.741 | 0.347 | 1.653 |
| 16 | 0.750 | 0.212 | 3.532 | 1.282 | 5.782 | 0.363 | 1.637 |
| 17 | 0.728 | 0.203 | 3.588 | 1.356 | 5.820 | 0.378 | 1.622 |
| 18 | 0.707 | 0.194 | 3.640 | 1.424 | 5.856 | 0.391 | 1.608 |
| 19 | 0.688 | 0.187 | 3.689 | 1.487 | 5.891 | 0.403 | 1.597 |
| 20 | 0.671 | 0.180 | 3.735 | 1.549 | 5.921 | 0.415 | 1.585 |
| 21 | 0.655 | 0.173 | 3.778 | 1.605 | 5.951 | 0.425 | 1.575 |
| 22 | 0.640 | 0.167 | 3.819 | 1.659 | 5.979 | 0.434 | 1.566 |
| 23 | 0.626 | 0.162 | 3.858 | 1.710 | 6.006 | 0.443 | 1.557 |
| 24 | 0.612 | 0.157 | 3.895 | 1.759 | 6.031 | 0.451 | 1.548 |
| 25 | 0.600 | 0.153 | 3.931 | 1.806 | 6.056 | 0.459 | 1.541 |

**Source:** These values are from ASTM-STP 15D, American Society for Testing and Materials.

**TABLE 18.3.3 Summaries of Measurements for Eight Samples of $n = 4$ Components Each**

| Sample Identification Number | Sample Average, $\overline{X}$ | Sample Range, $R$ |
|---|---|---|
| 1 | 22.3 | 1.8 |
| 2 | 22.4 | 1.2 |
| 3 | 21.5 | 1.1 |
| 4 | 22.0 | 0.9 |
| 5 | 21.1 | 1.1 |
| 6 | 21.7 | 0.9 |
| 7 | 22.1 | 1.5 |
| 8 | 21.6 | 1.1 |
| Average | $\overline{\overline{X}} = 21.84$ | $\overline{R} = 1.20$ |

$d_2 = 2.059$, $D_1 = 0$, and $D_2 = 4.698$. For the $\overline{X}$ chart, the center line and control limits will be as follows:

Center line:          $\mu_0 = 22.00$
Lower control limit:  $\mu_0 - A\sigma_0 = 22.00 - (1.500)(0.50) = 21.25$
Upper control limit:  $\mu_0 + A\sigma_0 = 22.00 + (1.500)(0.50) = 22.75$

For the $R$ chart, these values are as follows:

Center line:          $d_2\sigma_0 = (2.059)(0.50) = 1.03$
Lower control limit:  $D_1\sigma_0 = (0)(0.50) = 0$
Upper control limit:  $D_2\sigma_0 = (4.698)(0.50) = 2.35$

### Example
### Net Weight of Dishwasher Detergent

From each batch of 150 boxes of dishwasher detergent, 5 boxes are selected at random and the contents weighed. This information is then summarized in a control chart, which is examined to see whether or not the process needs to be adjusted. Measurements and summaries (average,

**TABLE 18.3.4** Net Weights of Sampled Boxes of Dishwasher Detergent, with Sample Summaries

| Sample Identification Number | Individual Measurements within Each Sample (net weight, in ounces) | | | | | Sample Summaries | | | |
| | 1 | 2 | 3 | 4 | 5 | Average $\overline{X}$ | Largest | Smallest | Range $R$ |
|---|---|---|---|---|---|---|---|---|---|
| 1 | 16.12 | 16.03 | 16.25 | 16.19 | 16.24 | 16.166 | 16.25 | 16.03 | 0.22 |
| 2 | 16.11 | 16.10 | 16.28 | 16.18 | 16.16 | 16.166 | 16.28 | 16.10 | 0.18 |
| 3 | 16.16 | 16.21 | 16.10 | 16.09 | 16.04 | 16.120 | 16.21 | 16.04 | 0.17 |
| 4 | 15.97 | 15.99 | 16.34 | 16.18 | 16.02 | 16.100 | 16.34 | 15.97 | 0.37 |
| 5 | 16.21 | 16.00 | 16.14 | 16.12 | 16.10 | 16.114 | 16.21 | 16.00 | 0.21 |
| 6 | 15.77 | 16.11 | 16.01 | 16.02 | 16.17 | 16.016 | 16.17 | 15.77 | 0.40 |
| 7 | 16.02 | 16.29 | 16.08 | 15.96 | 16.11 | 16.092 | 16.29 | 15.96 | 0.33 |
| 8 | 15.83 | 16.08 | 16.25 | 16.14 | 16.15 | 16.090 | 16.25 | 15.83 | 0.42 |
| 9 | 16.16 | 15.90 | 16.08 | 15.98 | 16.09 | 16.042 | 16.16 | 15.90 | 0.26 |
| 10 | 16.08 | 16.10 | 16.13 | 16.03 | 16.03 | 16.074 | 16.13 | 16.03 | 0.10 |
| 11 | 15.90 | 16.16 | 16.15 | 15.99 | 16.07 | 16.054 | 16.16 | 15.90 | 0.26 |
| 12 | 16.09 | 16.05 | 16.07 | 15.98 | 15.95 | 16.028 | 16.09 | 15.95 | 0.14 |
| 13 | 15.98 | 16.18 | 16.08 | 16.08 | 16.07 | 16.078 | 16.18 | 15.98 | 0.20 |
| 14 | 16.23 | 16.05 | 16.10 | 16.07 | 16.16 | 16.122 | 16.23 | 16.05 | 0.18 |
| 15 | 15.96 | 16.20 | 16.35 | 16.11 | 16.08 | 16.140 | 16.35 | 15.96 | 0.39 |
| 16 | 16.00 | 16.04 | 16.02 | 16.03 | 16.09 | 16.036 | 16.09 | 16.00 | 0.09 |
| 17 | 16.12 | 16.12 | 15.95 | 15.98 | 16.10 | 16.054 | 16.12 | 15.95 | 0.17 |
| 18 | 16.30 | 16.05 | 16.10 | 16.09 | 16.07 | 16.122 | 16.30 | 16.05 | 0.25 |
| 19 | 16.11 | 16.15 | 16.25 | 16.03 | 16.05 | 16.118 | 16.25 | 16.03 | 0.22 |
| 20 | 15.85 | 16.06 | 15.96 | 16.20 | 16.25 | 16.064 | 16.25 | 15.85 | 0.40 |
| 21 | 15.94 | 15.88 | 16.02 | 16.06 | 16.10 | 16.000 | 16.10 | 15.88 | 0.22 |
| 22 | 16.15 | 16.15 | 16.21 | 15.95 | 16.13 | 16.118 | 16.21 | 15.95 | 0.26 |
| 23 | 16.10 | 16.17 | 16.24 | 16.00 | 15.87 | 16.076 | 16.24 | 15.87 | 0.37 |
| 24 | 16.22 | 16.34 | 16.40 | 16.07 | 16.12 | 16.230 | 16.40 | 16.07 | 0.33 |
| 25 | 16.32 | 15.97 | 15.88 | 16.03 | 16.27 | 16.094 | 16.32 | 15.88 | 0.44 |
| | | | | | | $\overline{\overline{X}} = 16.093$ | | | $\overline{R} = 0.263$ |

largest, smallest, and range) are shown in Table 18.3.4 for 25 samples of 5 boxes each.

Since some of the packing equipment is fairly new, no outside standard is available. Here are the center line and control limits for the $\overline{X}$ and $R$ charts, using the table entries for sample size $n = 5$. For the $\overline{X}$ chart:

Center line: $\overline{\overline{X}} = 16.093$

Lower control limit: $\overline{\overline{X}} - A_2\overline{R} = 16.093 - (0.577)(0.263) = 15.941$

Upper control limit: $\overline{\overline{X}} + A_2\overline{R} = 16.093 + (0.577)(0.263) = 16.245$

*(Continued)*

## Example—cont'd

For the R chart:

Center line:            $\bar{R} = 0.263$
Lower control limit:    $D_3\bar{R} = (0)(0.263) = 0$
Upper control limit:    $D_4\bar{R} = (2.114)(0.263) = 0.556$

The $\bar{X}$ and $R$ charts (both shown in Figure 18.3.1) show a process that is in control. All observations in each chart fall within the control limits. There are no clear nonrandom patterns (such as an upward or downward trend or too many successive points falling too close to a limit). Does observation 24, at the right in the $\bar{X}$ chart, look suspicious to you because it's so close to the upper control limit? Perhaps it does, but because it's still within the limits, you would do better to resist the temptation to mess with the packaging process until the evidence is more clear. With a single isolated point near (but within) the control limits, you would still decide that the process is in control; but feel free to remain suspicious and to look for further evidence, clues, and patterns that would show that the process needs adjustment.

Figure 18.3.2 shows how this kind of control chart technology might be implemented on the shop floor for this example. This form, available from the American Society

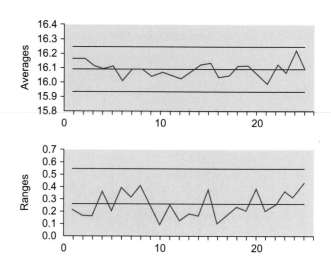

**FIGURE 18.3.1**   The $\bar{X}$ and $R$ charts for 25 samples of five boxes of dishwasher detergent. The process is in control because all points fall within the control limits, and the plots indicate randomness, with no clear trends or patterns.

for Quality Control, allows you to record background information, measurements, summaries, and control charts all in one place.

Here is how to use Excel® to draw an $\bar{X}$ chart for the detergent data. Begin with a column containing a list of the averages (of five observations each). Immediately to

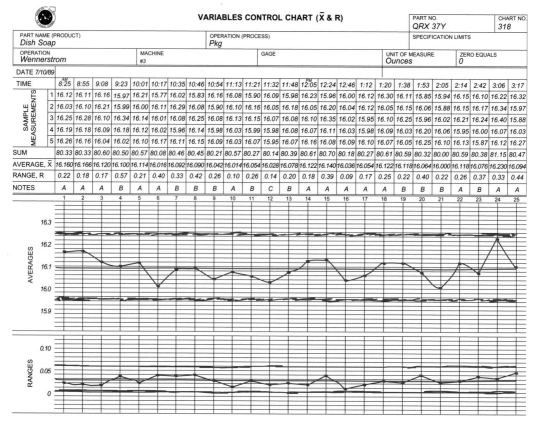

**FIGURE 18.3.2**   The background information, measurements, summaries, and control charts for the dishwasher detergent weights, as they might be recorded on the shop floor.

its right, create a column containing the average $\overline{\overline{X}} = 16.093$ of these averages repeated down the column. Next to it, create a column for the lower control limit $\overline{\overline{X}} - A_2\overline{R} = 15.941$ and one for the upper control limit $\overline{\overline{X}} + A_2\overline{R} = 16.245$. Now select all four of these columns (just the numbers) and, from the Chart area of Excel's® Insert Ribbon, choose Line; then select Line with Markers to create the $\overline{X}$ chart. You may then select and delete the

legend that might appear at the right in the chart, as well as the gridlines if you wish.

To use Excel® to draw an $R$ chart for the detergent data, proceed as for the $\overline{X}$ chart, but use the range values $R$ for the first column, their average $\overline{R} = 0.263$ for the second column, and the appropriate lower and upper control limits $D_3\overline{R} = 0$ and $D_4\overline{R} = 0.556$ for the third and fourth columns. Here are the charts in Excel®:

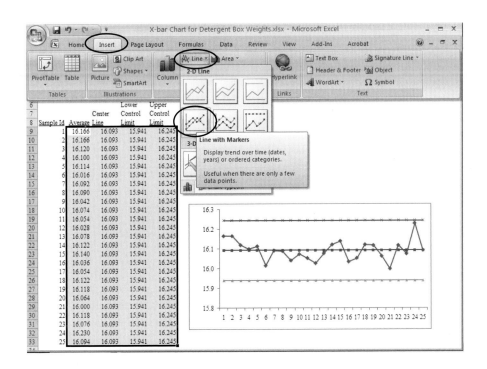

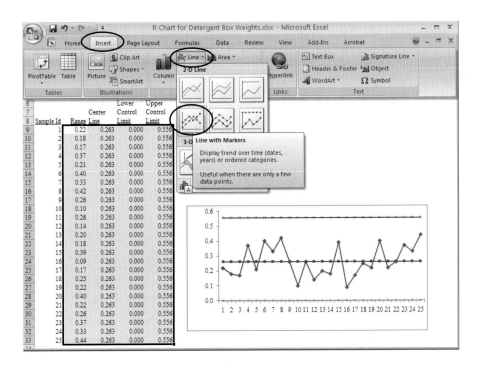

## 18.4 CHARTING THE PERCENT DEFECTIVE

Suppose that items are inspected and then classified as either defective or not. This is not a quantitative measurement, so a new control chart is needed. For each group of items, the *percent defective* can be computed and charted to give an idea of how widespread the problem is.

The **percentage chart** displays the percent defective together with the appropriate center line and control limits so that you can monitor the rate at which the process produces defective items. The control limits are set at three standard deviations, assuming a binomial distribution for the number of defective items. There is a rule of thumb to use for deciding on a sample size:

> **Selecting the Sample Size, *n*, for a Percentage Chart**
> You should expect at least five defective items in a sample.

This says that the sample size, *n*, will be much larger for the percentage chart than for the $\overline{X}$ and $R$ charts. For example, if you expect 10% defective items, your sample size should be at least $n = 5/0.10 = 50$. If you expect only 0.4% defectives, your sample size should be at least $n = 5/0.004 = 1,250$.

If you are fortunate enough to essentially *never* produce a defective item, don't despair. Although you can't find a sample size that satisfies this rule, you are nonetheless obviously in great shape. Congratulations on your dedication to quality.

Table 18.4.1 shows how to compute the center line and the control limits for the percentage chart. No special table of multipliers is needed because, as you may already have noticed, these formulas use the standard deviation of a binomial distribution. Recall that $p$ is the observed proportion or percentage in one sample. The symbol $\overline{p}$ represents the average of all of the sample proportions.

For example, suppose there is no standard given, and the sample summaries are as shown in Table 18.4.2. With

**TABLE 18.4.1 Finding the Center Line and Control Limits for the Percentage Chart**

| | Center Line | Control Limits |
|---|---|---|
| Standard given ($\pi_0$) | $\pi_0$ | From $\pi_0 - 3\sqrt{\dfrac{\pi_0(1-\pi_0)}{n}}$ to $\pi_0 + 3\sqrt{\dfrac{\pi_0(1-\pi_0)}{n}}$ |
| No standard given | $\overline{p}$ | From $\overline{p} - 3\sqrt{\dfrac{\overline{p}(1-\overline{p})}{n}}$ to $\overline{p} + 3\sqrt{\dfrac{\overline{p}(1-\overline{p})}{n}}$ |

**TABLE 18.4.2 Summaries of Measurements for 12 Samples of $n = 500$ Items Each**

| Sample Identification Number | Number of Defective Items, X | Sample Percentage, p |
|---|---|---|
| 1 | 10 | 2.0% |
| 2 | 11 | 2.2 |
| 3 | 10 | 2.0 |
| 4 | 12 | 2.4 |
| 5 | 7 | 1.4 |
| 6 | 14 | 2.8 |
| 7 | 13 | 2.6 |
| 8 | 11 | 2.2 |
| 9 | 6 | 1.2 |
| 10 | 12 | 2.4 |
| 11 | 11 | 2.2 |
| 12 | 13 | 2.6 |
| Average | 10.8333 | $\overline{p} = 2.1667\%$ |

sample size $n = 500$, the center line and control limits for the percentage chart will be as follows:

Center line:

$$\overline{p} = 0.121667 \text{ or } 2.1667\%$$

Lower control limit:

$$\overline{p} - 3\sqrt{\frac{\overline{p}(1-\overline{p})}{n}} = 0.021667 - 3\sqrt{\frac{0.021667(1-0.021667)}{500}}$$
$$= 0.021667 - (3)(0.006511)$$
$$= 0.0021 \text{ or } 0.21\%$$

Upper control limit:

$$\overline{p} + 3\sqrt{\frac{\overline{p}(1-\overline{p})}{n}} = 0.021667 + (3)(0.006511)$$
$$= 0.0412 \text{ or } 4.12\%$$

Now suppose that there is a standard given in addition to the data in Table 18.4.2, so that you know (say, from past experience) that the process produces defective items at a rate of $\pi_0 = 2.30\%$. With sample size $n = 500$, the center line and control limits for the percentage chart will be as follows:

Center line:

$$\pi_0 = 0.230 \text{ or } 2.30\%$$

Lower control limit:

$$\pi_0 - 3\sqrt{\frac{\pi_0(1-\pi_0)}{n}} = 0.0230 - 3\sqrt{\frac{0.0230(1-0.0230)}{500}}$$
$$= 0.0230 - (3)(0.0067039)$$
$$= 0.0029 \text{ or } 0.29\%$$

Upper control limit:

$$\pi_0 + 3\sqrt{\frac{\pi_0(1-\pi_0)}{n}} = 0.0230 + (3)(0.0067039)$$
$$= 0.0431 \text{ or } 4.31\%$$

### Example
#### Filling Out Purchase Orders

When purchase orders are entered into the computer and processed, errors are sometimes found that require special attention to correct them. Of course, "special attention" is expensive, and you would like the percentage of problems to be small. In order to keep an eye on this problem, you regularly look at a percentage chart.

For each batch of 300 purchase orders, the percent of errors is recorded, as shown in Table 18.4.3. You are not using any standard for the percentage chart, so the center line and control limits are as follows:

Center line:

$$\bar{p} = 0.515 \quad \text{or} \quad 5.15\%$$

Lower control limit:

$$\bar{p} - 3\sqrt{\frac{\bar{p}(1-\bar{p})}{n}} = 0.0515 - 3\sqrt{\frac{0.0515(1-0.0515)}{300}}$$
$$= 0.0515 - (3)(0.012760)$$
$$= 0.0132 \quad \text{or} \quad 1.32\%$$

Upper control limit:

$$\bar{p} + 3\sqrt{\frac{\bar{p}(1-\bar{p})}{n}} = 0.0515 + (3)(0.012760)$$
$$= 0.0898 \quad \text{or} \quad 8.98\%$$

The percentage chart (Figure 18.4.1) shows that the processing of purchase orders is *not* in control. Two batches (18 and 21) are outside the upper control limit due to excessive errors. Furthermore, the entire right-hand side of the chart appears to have a higher percentage than the left side. It appears that the process was in control, at a low error rate, but then some problem began around batch number 16 or 17 that increased the error rate.

The control chart has done its job: to alert you to problems and provide clues to help you solve them. Yes, there is a problem. The clue is a high error rate starting around batch 16 or 17.

An investigation of what was happening around the time of production of batch 16 shows that work began at that time on a new investment project. There were so many purchase orders to process that extra help was hired. Apparently, the higher error rates were due initially to the strain on the system (due to the large volume of purchase orders) and the fact that the new employees had not yet learned the system. Although the error rate does drop at the end of the chart (on the right), it still may be somewhat high. It may well be time to institute a quick course of further training for the new people in order to bring the error rate back down to a lower level.

**TABLE 18.4.3 Summaries of Errors in 25 Batches of $n = 300$ Purchase Orders**

| Batch Identification Number | Numbers of Errors, X | Sample Percentage, p |
|---|---|---|
| 1 | 5 | 1.7% |
| 2 | 11 | 3.7 |
| 3 | 7 | 2.3 |
| 4 | 14 | 4.7 |
| 5 | 5 | 1.7 |
| 6 | 11 | 3.7 |
| 7 | 11 | 3.7 |
| 8 | 10 | 3.3 |
| 9 | 14 | 4.7 |
| 10 | 8 | 2.7 |
| 11 | 5 | 1.7 |
| 12 | 16 | 5.3 |
| 13 | 12 | 4.0 |
| 14 | 9 | 3.0 |
| 15 | 13 | 4.3 |
| 16 | 17 | 5.7 |
| 17 | 20 | 6.7 |
| 18 | 30 | 10.0 |
| 19 | 23 | 7.7 |
| 20 | 25 | 8.3 |
| 21 | 35 | 11.7 |
| 22 | 24 | 8.0 |
| 23 | 22 | 7.3 |
| 24 | 23 | 7.7 |
| 25 | 16 | 5.3 |
| Average | 15.44 | $\bar{p} = 5.15\%$ |

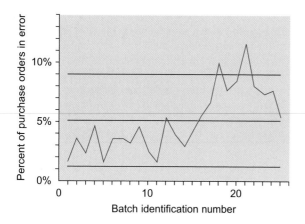

**FIGURE 18.4.1** The processing of purchase orders is not in control and needs some attention. Batches 18 and 21 are above the upper control limit, and there appears to have been a shift in level around batch number 16 or 17.

## 18.5 END-OF-CHAPTER MATERIALS

### Summary

**Statistical quality control** is the use of statistical methods for evaluating and improving the results of any activity. A **process** is any business activity that takes inputs and transforms them into outputs. A process can be made up of other processes, called *subprocesses*, each of which is a process in its own right. **Statistical process control** is the use of statistical methods to monitor the functioning of a process so that you can adjust or fix it when necessary and can leave it alone when it's working properly. Anytime you could reasonably find out why a problem occurred, you have an **assignable cause of variation**. All causes of variation that would not be worth the effort to identify are grouped together as **random causes of variation**. When you have identified and eliminated all of the assignable causes of variation, so that only random causes remain, your process is said to be **in a state of statistical control** or, simply, **in control**.

The **Pareto diagram** displays the causes of the various defects in order from most to least frequent so that you can focus attention on the most important problems. The tallest bars will always be at the left (indicating the most frequent problems) and the shortest bars will be at the right. The line indicating the cumulative percentage will always go upward and will tend to level off toward the right.

A **control chart** displays successive measurements of a process together with a *center line* and *control limits*, which are computed to help you decide whether or not the process is in control. If you decide that the process is not in control, the control chart helps you identify the problem so that you can fix it. The hypotheses being tested are

$H_0$: The process is in control.
$H_1$: The process is not in control.

The **false alarm rate** is how often you decide to fix the process when there really isn't any problem. This is the same concept as the type I error in hypothesis testing. Conventional control charts use a factor of three standard errors instead of two because the conventional 5% type I error rate of statistical hypothesis testing is unacceptably high for most quality control applications. A process that is in control will usually have a control chart plot that stays within the control limits. If any measurement falls outside of the control limits, either above the upper control limit or below the lower limit, you would decide that the process is not in control. A sudden jump to another level in a control chart or a gradual drift upward or downward can also indicate a process that is not in control, even if all points fall within the control limits.

It is conventional, for charting a quantitative measurement, to choose a fairly small sample size ($n = 4$ or $n = 5$ are common choices) and then chart the results of many successive samples. The $\overline{X}$ **chart** displays the *average* of each sample together with the appropriate center line and control limits so that you can monitor the level of the process. The $R$ **chart** displays the *range* (the largest value minus the smallest) for each sample together with center line and control limits so that you can monitor the variability of the process.

The **percentage chart** displays the percent defective together with the appropriate center line and control limits so that you can monitor the rate at which the process produces defective items. The sample size required is much larger than for charting a quantitative measurement. A common rule of thumb is that you should expect at least five defective items in a sample.

### Key Words

assignable cause of variation, *525*
control chart, *526*
false alarm rate, *527*
in a state of statistical control (in control), *525*
Pareto diagram, *525*
percentage chart, *534*
process, *524*
$R$ chart, *528*
random causes of variation, *525*
statistical process control, *525*
statistical quality control, *523*
$\overline{X}$ chart, *528*

### Questions

1. **a.** What is statistical quality control?
   **b.** Why are statistical methods so helpful for quality control?

2. **a.** What is a process?
   **b.** What is the relationship between a process and its subprocesses?
   **c.** What is statistical process control?
3. Can statistical process control be applied to business activities in general, or is it restricted to manufacturing?
4. Why should you monitor a process? Why not just inspect the results and throw away the defective ones?
5. **a.** What is an assignable cause of variation?
   **b.** What is a random cause of variation?
6. **a.** What do we mean when we say that a process is in a state of statistical control?
   **b.** What should you do when a process is not in control?
   **c.** What should you do when a process appears to be in control?
7. **a.** What information is displayed in a Pareto diagram?
   **b.** What makes the Pareto diagram useful as a management tool?
8. **a.** What is a control chart?
   **b.** Explain how control charts help you perform the five basic activities of statistics.
   **c.** What hypotheses are being tested when you use control charts?
   **d.** What is the false alarm rate? Is it conventional to set it at 5%?
9. **a.** Describe a typical control chart for a process that is in control.
   **b.** Describe three different ways in which a control chart could tell you that the process is not in control.
10. **a.** What is the purpose of the $\overline{X}$ chart?
    **b.** What is a typical sample size?
    **c.** How would you find the center line if you had no standard?
    **d.** How would you find the center line if you did have a standard?
    **e.** How would you find the control limits if you had no standard?
    **f.** How would you find the control limits if you did have a standard?
11. **a.** What is the purpose of the $R$ chart?
    **b.** What is a typical sample size?
    **c.** How would you find the center line if you had no standard?
    **d.** How would you find the center line if you did have a standard?
    **e.** How would you find the control limits if you had no standard?
    **f.** How would you find the control limits if you did have a standard?
12. **a.** What is the purpose of the percentage chart?
    **b.** How large should the sample size be?
    **c.** How would you find the center line if you had no standard?
    **d.** How would you find the center line if you did have a standard?
    **e.** How would you find the control limits if you had no standard?
    **f.** How would you find the control limits if you did have a standard?

## Problems

*Problems marked with an asterisk (\*) are solved in the Self Test in Appendix C.*

1. For each of the following situations, tell whether a Pareto diagram, an $\overline{X}$ chart, an $R$ chart, or a percentage chart would be the most helpful. Give a reason for your choice.
   **a.\*** Your workers all want to be helpful, but they can't seem to agree on which problems to solve first.
   **b.\*** Some of the engines come off the line with too much oil, and others come off with too little. Something has to be done to control the differences from one to the next.
   **c.** The gears being produced are all pretty much the same size in each batch, but they tend to be consistently too large compared to the desired specification.
   **d.** Management would like to track the rate at which defective candy coatings are produced.
   **e.** You would like to understand the limits of variation of the machinery so that you can set it to fill each bottle just a tiny bit more than the amount claimed on the label.
   **f.** Usually, you pay the bills on time, but a small fraction of them slip through the system and are paid late with a penalty. You'd like to keep an eye on this to see if things are getting worse.

2. A tractor manufacturing plant has been experiencing problems with the division that makes the transmissions. A Pareto diagram, shown in Figure 18.5.1, has been constructed based on recent experience.
   **a.** What is the most important problem, in terms of the number of transmissions affected? What percent of all difficulties does this problem represent?
   **b.** What is the next most important problem? What percent does it represent?
   **c.** What percent of defective transmissions do the top two problems, taken together, represent?
   **d.** What percent of problems do the top three problems together represent?
   **e.** Write a paragraph, as if to your supervisor, summarizing the situation and recommending action.

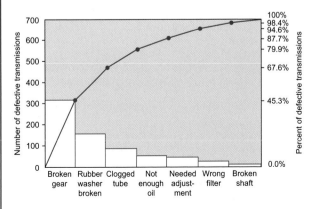

**FIGURE 18.5.1**   Pareto diagram for defective transmissions.

**TABLE 18.5.1 Frequency of Occurrence of Various Problems in Candy Manufacturing**

| Cause of Problem | Number of Cases |
|---|---|
| Miscellaneous | 22 |
| Not enough coating | 526 |
| Squashed | 292 |
| Too much coating | 89 |
| Two stuck together | 57 |

3. A candy manufacturer has observed the frequency of various types of problems that occur in the production of small chocolates with a hard candy coating. The basic data set is shown in Table 18.5.1.
   a. Arrange the problems in order from most to least frequent, and create a table showing number of cases, percent of total problem cases, and cumulative percent.
   b. Draw a Pareto diagram for this situation.
   c. What is the most important problem, in terms of the number of candies affected? What percent of all difficulties does this problem represent?
   d. What is the next most important problem? What percent does it represent?
   e. What percent of defective candies do the top two problems, taken together, represent?
   f. Write a paragraph, as if to your supervisor, summarizing the situation and recommending action.
4. A firm specializing in the processing of rebate certificates has tabulated the frequency of occurrence of various types of problems, as shown in Table 18.5.2.
   a. Arrange the problems in order from most to least frequent, and create a table indicating the number of cases, percent of total problem cases, and cumulative percent.
   b. Draw a Pareto diagram for this situation.
   c. What is the most important problem, in terms of the number of certificates affected? What percent of all difficulties does this problem represent?
   d. What is the next most important problem? What percent does it represent?

**TABLE 18.5.2 Frequency of Occurrence of Various Problems in Processing Rebate Certificates**

| Cause of Problem | Number of Cases |
|---|---|
| Blank | 53 |
| Illegible | 528 |
| Two numbers transposed | 184 |
| Wrong place on form | 330 |

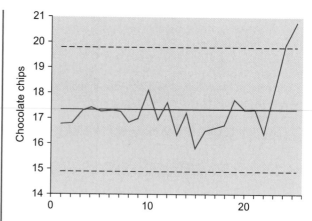

FIGURE 18.5.2   An $\bar{X}$ chart for the number of chocolate chips per cookie.

   e. What percent of defective certificates do the top two problems, taken together, represent?
   f. Write a paragraph, as if to your supervisor, summarizing the situation and recommending action.
5. Consider the $\bar{X}$ chart shown in Figure 18.5.2 showing the average number of chocolate chips per cookie.
   a. Describe in general terms what you see in the chart.
   b. Decide whether or not this process is in control. Give a reason for your answer.
   c. What action, if any, is warranted?
6. Find the center line and control limits for each of the following situations.
   a.* $\bar{X}$ chart, sample size $n = 6$, $\bar{\bar{X}} = 56.31$, $\bar{R} = 4.16$, no standard given.
   b. $R$ chart, sample size $n = 6$, $\bar{\bar{X}} = 56.31$, $\bar{R} = 4.16$, no standard given.
   c. $\bar{X}$ chart, sample size $n = 3$, $\bar{\bar{X}} = 182.3$, $\bar{R} = 29.4$, no standard given.
   d. $R$ chart, sample size $n = 3$, $\bar{\bar{X}} = 182.3$, $\bar{R} = 29.4$, no standard given.
   e. $\bar{X}$ chart, sample size $n = 5$, $\bar{\bar{X}} = 182.3$, $\bar{R} = 13.8$, standards are $\mu_0 = 100.0$ and $\sigma_0 = 5.0$.
   f. $R$ chart, sample size $n = 5$, $\bar{\bar{X}} = 182.3$, $\bar{R} = 13.8$, standards are $\mu_0 = 100.0$ and $\sigma_0 = 5.0$.
   g. $\bar{X}$ chart, sample size $n = 8$, standards are $\mu_0 = 2.500$ and $\sigma_0 = 0.010$.
   h. $R$ chart, sample size $n = 8$, standards are $\mu_0 = 2.500$ and $\sigma_0 = 0.010$.
7. Consider the data set shown in Table 18.5.3, representing the thicknesses of a protective coating.
   a. Find the average, $\bar{X}$, and the range, $R$, for each sample.
   b.* Find the overall average, $\bar{\bar{X}}$, and the average range, $\bar{R}$.
   c.* Find the center line for the $\bar{X}$ chart.
   d.* Find the control limits for the $\bar{X}$ chart.
   e. Draw the $\bar{X}$ chart.
   f. Comment on what you see in the $\bar{X}$ chart. In particular, is this process in control? How do you know?
   g. Write a paragraph, as if to your supervisor, summarizing the situation and defending the action you feel is appropriate.

**TABLE 18.5.3 Thicknesses of a Protective Coating: 25 Samples of Three Items Each**

| Sample Identification Number | Individual Measurements within Each Sample | | |
|---|---|---|---|
| | 1 | 2 | 3 |
| 1 | 12.51 | 12.70 | 12.57 |
| 2 | 12.60 | 12.53 | 12.39 |
| 3 | 12.40 | 12.81 | 12.56 |
| 4 | 12.44 | 12.57 | 12.60 |
| 5 | 12.78 | 12.61 | 12.58 |
| 6 | 12.75 | 12.43 | 12.61 |
| 7 | 12.53 | 12.51 | 12.68 |
| 8 | 12.64 | 12.49 | 12.51 |
| 9 | 12.57 | 12.74 | 12.81 |
| 10 | 12.70 | 12.87 | 12.95 |
| 11 | 12.74 | 12.80 | 12.86 |
| 12 | 12.90 | 12.83 | 12.91 |
| 13 | 13.05 | 13.00 | 13.02 |
| 14 | 12.88 | 12.88 | 13.11 |
| 15 | 13.03 | 12.85 | 13.05 |
| 16 | 12.96 | 12.88 | 12.95 |
| 17 | 12.91 | 12.75 | 13.01 |
| 18 | 12.95 | 13.03 | 12.89 |
| 19 | 13.17 | 12.81 | 13.17 |
| 20 | 13.17 | 13.05 | 12.97 |
| 21 | 12.95 | 13.04 | 12.80 |
| 22 | 13.04 | 13.25 | 12.95 |
| 23 | 13.12 | 13.07 | 13.11 |
| 24 | 12.83 | 13.13 | 13.31 |
| 25 | 13.24 | 13.18 | 13.13 |

8. Continue with the data set in Table 18.5.3, representing the thicknesses of a protective coating.
   a.* Find the center line for the $R$ chart.
   b.* Find the control limits for the $R$ chart.
   c.  Draw the $R$ chart.
   d.  Comment on what you see in the $R$ chart. In particular, is the variability of this process in control? How do you know?

9. Mr. K. R. Wood, president of Broccoli Enterprises, is interested in the data shown in Table 18.5.4, representing the lengths of broccoli trees after cutting.
   a.  Find the average, $\overline{X}$, and the range, $R$, for each sample.
   b.  Find the overall average, $\overline{\overline{X}}$, and the average range, $\overline{R}$.
   c.  Find the center line for the $\overline{X}$ chart.
   d.  Find the control limits for the $\overline{X}$ chart.
   e.  Draw the $\overline{X}$ chart.
   f.  Comment on what you see in the $\overline{X}$ chart. In particular, is this process in control? How do you know?
   g.  Write a paragraph summarizing the situation and defending the action you feel is appropriate.
10. Continue with the data set in Table 18.5.4, representing the lengths of broccoli trees after cutting.

**TABLE 18.5.4 Lengths of Trees of Broccoli: 20 Samples of Four Stems Each**

| Sample Identification Number | Individual Measurements within Each Sample | | | |
|---|---|---|---|---|
| | 1 | 2 | 3 | 4 |
| 1 | 8.60 | 8.47 | 8.44 | 8.51 |
| 2 | 8.43 | 8.42 | 8.62 | 8.46 |
| 3 | 8.65 | 8.32 | 8.65 | 8.51 |
| 4 | 8.39 | 8.54 | 8.50 | 8.41 |
| 5 | 8.49 | 8.53 | 8.61 | 8.46 |
| 6 | 8.63 | 8.46 | 8.64 | 8.54 |
| 7 | 8.47 | 8.63 | 8.54 | 8.55 |
| 8 | 8.52 | 8.50 | 8.31 | 8.63 |
| 9 | 8.35 | 8.43 | 8.51 | 8.61 |
| 10 | 8.31 | 8.65 | 8.46 | 8.40 |
| 11 | 8.58 | 8.43 | 8.55 | 8.45 |
| 12 | 8.28 | 8.57 | 8.58 | 8.48 |
| 13 | 8.45 | 8.52 | 8.52 | 8.54 |
| 14 | 8.38 | 8.48 | 8.41 | 8.57 |
| 15 | 8.56 | 8.60 | 8.58 | 8.51 |
| 16 | 8.39 | 8.47 | 8.59 | 8.41 |
| 17 | 8.53 | 8.58 | 8.54 | 8.42 |
| 18 | 8.78 | 8.52 | 8.46 | 8.50 |
| 19 | 8.48 | 8.49 | 8.74 | 8.59 |
| 20 | 8.46 | 8.47 | 8.70 | 8.32 |

a.  Find the center line for the R chart.
b.  Find the control limits for the R chart.
c.  Draw the R chart.
d.  Comment on what you see in the R chart. In particular, is the variability of this process in control? How do you know?

11. Find the center line and control limits for each of the following situations.
   a.* Percentage chart, sample size $n = 300$, $\overline{p} = 0.0731$.
   b.  Percentage chart, sample size $n = 450$, $\overline{p} = 0.1683$.
   c.* Percentage chart, sample size $n = 800$, $\overline{p} = 0.0316$, standard is $\pi_0 = 0.0350$.
   d.  Percentage chart, sample size $n = 1,500$, standard is $\pi_0 = 0.01$.

12. Consider the data set shown in Table 18.5.5, summarizing recent numbers of errors in batches of 500 invoices.
   a.  Find the percentage, $p$, for each batch.
   b.  Find the average percentage, $\overline{p}$.
   c.  Find the center line for the percentage chart.
   d.  Find the control limits for the percentage chart.
   e.  Draw the percentage chart.
   f.  Comment on what you see in the percentage chart. In particular, is this process in control? How do you know?
   g.  Write a paragraph, as if to your supervisor, summarizing the situation and defending the action you feel is appropriate.

13. No matter how closely the production process is monitored, some chips will work faster than others and be worth more in the marketplace. The goal is to make this number as high as possible, and improvements are being implemented continually. Consider the data set shown in Table 18.5.6, summarizing the number of memory chips that worked properly at the highest speed for each of 25 batches of 1,000 chips.

**TABLE 18.5.5 Summaries of Defective Invoices in 25 Batches of $n = 500$**

| Batch Identification Number | Number of Errors, X | Batch Identification Number | Number of Errors, X |
|---|---|---|---|
| 1 | 58 | 14 | 51 |
| 2 | 57 | 15 | 54 |
| 3 | 60 | 16 | 47 |
| 4 | 64 | 17 | 52 |
| 5 | 57 | 18 | 50 |
| 6 | 53 | 19 | 62 |
| 7 | 53 | 20 | 56 |
| 8 | 74 | 21 | 60 |
| 9 | 40 | 22 | 67 |
| 10 | 54 | 23 | 50 |
| 11 | 56 | 24 | 60 |
| 12 | 54 | 25 | 67 |
| 13 | 60 | | |

**TABLE 18.5.6 Number of Highest-Speed Memory Chips for 25 Batches of 1,000 Chips Each**

| Batch Identification Number | Number of Highest-Speed Chips, X | Proportion of Highest-Speed Chips, p |
|---|---|---|
| 1 | 75 | 0.075 |
| 2 | 61 | 0.061 |
| 3 | 62 | 0.062 |
| 4 | 70 | 0.070 |
| 5 | 60 | 0.060 |
| 6 | 56 | 0.056 |
| 7 | 61 | 0.061 |
| 8 | 65 | 0.065 |
| 9 | 54 | 0.054 |
| 10 | 71 | 0.071 |
| 11 | 84 | 0.084 |
| 12 | 84 | 0.084 |
| 13 | 110 | 0.110 |
| 14 | 71 | 0.071 |
| 15 | 103 | 0.103 |
| 16 | 103 | 0.103 |
| 17 | 80 | 0.080 |
| 18 | 90 | 0.090 |
| 19 | 84 | 0.084 |
| 20 | 88 | 0.088 |
| 21 | 111 | 0.111 |
| 22 | 118 | 0.118 |
| 23 | 147 | 0.147 |
| 24 | 136 | 0.136 |
| 25 | 123 | 0.123 |
| Average | 86.68 | 0.08668 |

a. Find the center line for the percentage chart.
b. Find the control limits for the percentage chart.
c. Draw the percentage chart.
d. Comment on what you see in the percentage chart. In particular, is this process in control? How do you know?
e. For the particular case of high-speed memory chips here, does the control chart show good or bad news?
f. Write a paragraph, as if to your supervisor, summarizing the situation.

14. Consider the data set shown in Table 18.5.7, indicating hourly summaries of the temperature for a baking oven measured four times per hour.

a. Draw an $\overline{X}$ and an $R$ chart for each day.
b. For each day, summarize the charts. In particular, was the process in control? How do you know?
c. For each day, tell what action is appropriate.
d. A new product requires that the temperature be constant to within plus or minus 10 degrees. Based on the "in control" control charts from part a, do you think that these ovens can be used for this purpose? Why or why not?

15. Find the probability that a particular set of eight consecutive points will fall on one side of the center line for a process that is in control. (*Hints:* For a process that is in control, assume that the probability is 0.5 that a point is on the same side as the first point, and

**TABLE 18.5.7 Average and Range of Temperatures Taken Four Times per Hour**

| Time | Monday $\overline{X}$ | Monday $R$ | Tuesday $\overline{X}$ | Tuesday $R$ | Wednesday $\overline{X}$ | Wednesday $R$ |
|---|---|---|---|---|---|---|
| 12:00 | 408.65 | 30.74 | 401.07 | 25.23 | 402.92 | 31.96 |
| 1:00 | 401.57 | 24.81 | 405.97 | 32.72 | 407.28 | 9.11 |
| 2:00 | 395.52 | 21.93 | 401.70 | 34.56 | 399.61 | 22.85 |
| 3:00 | 402.25 | 35.91 | 402.06 | 38.15 | 398.43 | 38.52 |
| 4:00 | 405.04 | 28.68 | 403.35 | 31.03 | 389.97 | 12.16 |
| 5:00 | 404.12 | 38.18 | 407.82 | 34.93 | 402.37 | 18.39 |
| 6:00 | 404.44 | 18.16 | 400.30 | 30.56 | 406.29 | 48.44 |
| 7:00 | 407.19 | 14.14 | 403.69 | 17.97 | 407.77 | 32.63 |
| 8:00 | 407.43 | 21.56 | 399.72 | 14.11 | 398.22 | 19.30 |
| 9:00 | 412.60 | 25.29 | 394.77 | 28.89 | 408.42 | 19.11 |
| 10:00 | 413.40 | 14.32 | 400.82 | 37.26 | 402.91 | 28.52 |
| 11:00 | 407.26 | 39.70 | 401.96 | 33.30 | 391.20 | 20.08 |
| 12:00 | 402.97 | 32.92 | 399.94 | 16.43 | 398.59 | 13.29 |
| 1:00 | 387.44 | 21.89 | 401.01 | 16.95 | 401.72 | 35.90 |
| 2:00 | 414.39 | 14.34 | 399.67 | 30.34 | 394.37 | 12.32 |
| 3:00 | 401.25 | 18.62 | 401.67 | 29.53 | 409.59 | 32.91 |
| 4:00 | 400.43 | 27.96 | 413.30 | 12.62 | 421.97 | 40.38 |
| 5:00 | 399.31 | 25.93 | 412.47 | 45.47 | 394.58 | 48.70 |
| 6:00 | 403.14 | 37.57 | 406.62 | 43.65 | 407.01 | 25.75 |
| 7:00 | 403.07 | 33.52 | 421.90 | 21.75 | 403.40 | 63.81 |
| 8:00 | 403.66 | 45.69 | 429.67 | 24.30 | 404.93 | 82.12 |
| 9:00 | 404.05 | 40.52 | 422.75 | 25.79 | 391.82 | 67.03 |
| 10:00 | 399.00 | 39.77 | 422.56 | 15.28 | 393.96 | 84.53 |
| 11:00 | 410.18 | 37.71 | 424.39 | 16.64 | 421.68 | 92.92 |

these are independent. You may therefore compute the probability for a binomial distribution with $\pi = 0.5$ and $n = 7$. You would use $n = 7$ instead of $n = 8$ because the first point of the sequence is free to fall on either side, so the situation is really determined by the seven other points.)

16. What problems, if any, are visible in the control charts in Figure 18.5.3? What action (if any) would you suggest?

17. What problems, if any, are visible in the control charts in Figure 18.5.4? What action (if any) would you suggest?

18. What problems, if any, are visible in the control charts in Figure 18.5.5? What action (if any) would you suggest?

19. What problems, if any, are visible in the control charts in Figure 18.5.6? What action (if any) would you suggest?

20. What problems, if any, are visible in the control charts in Figure 18.5.7? What action (if any) would you suggest?

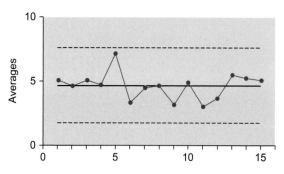

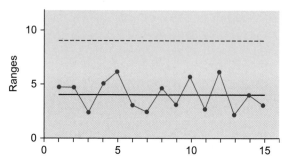

**FIGURE 18.5.3**

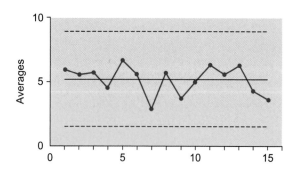

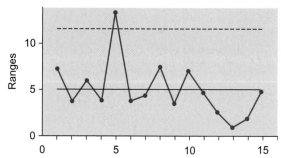

**FIGURE 18.5.4**

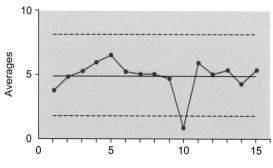

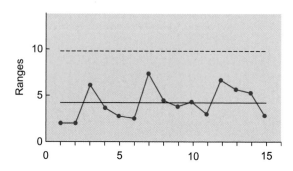

**FIGURE 18.5.5**

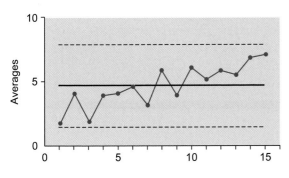

**FIGURE 18.5.6**

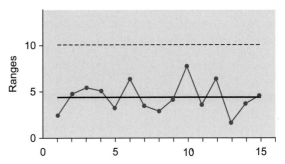

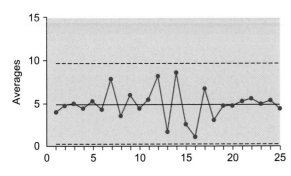

**FIGURE 18.5.7**

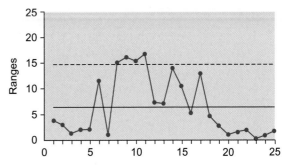

## Projects

1.  Obtain some qualitative data relating to quality showing how frequently different situations have occurred. Possible sources include the Internet, your firm, a local business, your own experiences, or the library. Draw the Pareto diagram and describe what you see. Write a one-page summary of the situation for management.

2.  Obtain some quantitative data relating to quality. Possible sources include the Internet, your firm, a local business, your own experiences, or the library. The data set should consist of at least 10 samples of from 3 to 20 observations each. Draw the $\overline{X}$ and $R$ charts and then describe what you see. Write a one-page summary of the situation for management.

# Employee Database

Following are the employee records of an administrative division:

| Employee Number[1] | Annual Salary | Gender | Age (years) | Experience (years) | Training Level[2] |
|---|---|---|---|---|---|
| 1 | $32,368 | F | 42 | 3 | B |
| 2 | 53,174 | M | 54 | 10 | B |
| 3 | 52,722 | M | 47 | 10 | A |
| 4 | 53,423 | M | 47 | 1 | B |
| 5 | 50,602 | M | 44 | 5 | B |
| 6 | 49,033 | M | 42 | 10 | A |
| 7 | 24,395 | M | 30 | 5 | A |
| 8 | 24,395 | F | 52 | 6 | A |
| 9 | 43,124 | M | 48 | 8 | A |
| 10 | 23,975 | F | 58 | 4 | A |
| 11 | 53,174 | M | 46 | 4 | C |
| 12 | 58,515 | M | 36 | 8 | C |
| 13 | 56,294 | M | 49 | 10 | B |
| 14 | 49,033 | F | 55 | 10 | B |
| 15 | 44,884 | M | 41 | 1 | A |
| 16 | 53,429 | F | 52 | 5 | B |
| 17 | 46,574 | M | 57 | 8 | A |
| 18 | 58,968 | F | 61 | 10 | B |
| 19 | 53,174 | M | 50 | 5 | A |
| 20 | 53,627 | M | 47 | 10 | B |
| 21 | 49,033 | M | 54 | 5 | B |
| 22 | 54,981 | M | 47 | 7 | A |
| 23 | 62,530 | M | 50 | 10 | B |
| 24 | 27,525 | F | 38 | 3 | A |
| 25 | 24,395 | M | 31 | 5 | A |
| 26 | 56,884 | M | 47 | 10 | A |
| 27 | 52,111 | M | 56 | 5 | A |
| 28 | 44,183 | F | 38 | 5 | B |
| 29 | 24,967 | F | 55 | 6 | A |
| 30 | 35,423 | F | 47 | 4 | A |
| 31 | 41,188 | F | 35 | 2 | B |
| 32 | 27,525 | F | 35 | 3 | A |
| 33 | 35,018 | M | 39 | 1 | A |
| 34 | 44,183 | M | 41 | 2 | A |
| 35 | 35,423 | M | 44 | 1 | A |
| 36 | 49,033 | M | 53 | 8 | A |
| 37 | 40,741 | M | 47 | 2 | A |
| 38 | 49,033 | M | 42 | 10 | A |
| 39 | 56,294 | F | 44 | 6 | C |
| 40 | 47,180 | F | 45 | 5 | C |
| 41 | 46,574 | M | 56 | 8 | A |
| 42 | 52,722 | M | 38 | 8 | C |
| 43 | 51,237 | M | 58 | 2 | B |
| 44 | 53,627 | M | 52 | 8 | A |
| 45 | 53,174 | M | 54 | 10 | A |
| 46 | 56,294 | M | 49 | 10 | B |
| 47 | 49,033 | F | 53 | 10 | B |
| 48 | 49,033 | M | 43 | 9 | A |
| 49 | 55,549 | M | 35 | 8 | C |
| 50 | 51,237 | M | 56 | 1 | C |
| 51 | 35,200 | F | 38 | 1 | B |
| 52 | 50,175 | F | 42 | 5 | A |
| 53 | 24,352 | F | 35 | 1 | A |
| 54 | 27,525 | F | 40 | 3 | A |
| 55 | 29,606 | F | 34 | 4 | B |
| 56 | 24,352 | F | 35 | 1 | A |
| 57 | 47,180 | F | 45 | 5 | B |
| 58 | 49,033 | M | 54 | 10 | A |
| 59 | 53,174 | M | 47 | 10 | A |
| 60 | 53,429 | F | 45 | 7 | B |
| 61 | 53,627 | M | 47 | 10 | A |
| 62 | 26,491 | F | 46 | 7 | A |
| 63 | 42,961 | M | 36 | 3 | B |
| 64 | 53,174 | M | 45 | 5 | A |
| 65 | 36,292 | M | 46 | 0 | A |
| 66 | 37,292 | M | 47 | 1 | A |
| 67 | 41,188 | F | 34 | 3 | B |
| 68 | 57,242 | F | 45 | 7 | C |
| 69 | 53,429 | F | 44 | 6 | C |
| 70 | 53,174 | M | 50 | 10 | B |
| 71 | 44,138 | F | 38 | 2 | B |

1. These numbers were assigned for the sole purpose of giving each employee a unique number.
2. The training is offered from time to time and is voluntary (it is not a job requirement). Employees who have not taken either training course are coded as "A." They become "B" after the first training course, and they change to "C" after the second and final course.

# Donations Database

Table B.1 shows part of the Donations Database on the companion website at http://www.elsevierdirect.com that gives information on 20,000 individuals at the time of a mailing, together with the amount (if any, in the first column) that each one donated as a result of that mailing. These individuals are "lapsed donors" who have donated before, but not in the past year. This database was adapted from a large data set originally used in The Second International Knowledge Discovery and Data Mining Tools Competition and is available as part of the UCI Knowledge Discovery in Databases Archive (Hettich, S. and Bay, S. D., 1999, The UCI KDD Archive, http://kdd.ics.uci.edu, Irvine, CA, University of California, Department of Information and Computer Science, now maintained as part of the UCI Machine Learning Archive at http://archive.ics.uci.edu/ml/). In the Excel file on the companion site, there are three worksheets:

- *Everyone:* The first workbook tab includes all 20,000 records. The Excel names for these columns are given in Table B.2. For example, "Donation" refers to the 20,000 donation amounts.

- *Donors only:* The second workbook tab includes records for only the 989 individuals (out of the original 20,000) who gave money in response to the current mailing. Excel names for these columns consist of the names in Table B.2 with "_D1" meaning "donors, yes" at the end. For example, "Donation _D1" refers to the 989 donation amounts for this group.

- *Nondonors only:* The third workbook tab includes records for only the 19,011 individuals (out of the original 20,000) who did not give money in response to the current mailing. Excel names for these columns consist of the names in Table B.2 with "_D0" meaning "donors, no" at the end. For example, "Donation_D0" refers to the 19,011 donation amounts for this group, which are all zero.

**TABLE B.1** The First and Last 10 Rows, Where Each Row Is One Person, in the Donations Database of 20,000 People*

| Donation | Life-time | Gifts | Years Since First | Years Since Last | Average Gift | Major Donor | Promos | Recent Gifts | Age | Home Phone | PC Owner | Catalog Shopper |
|---|---|---|---|---|---|---|---|---|---|---|---|---|
| $0.00 | $81.00 | 15 | 6.4 | 1.2 | $5.40 | 0 | 58 | 3 | | 0 | 0 | 0 |
| 15.00 | 15.00 | 1 | 1.2 | 1.2 | 15.00 | 0 | 13 | 1 | 33 | 1 | 0 | 1 |
| 0.00 | 15.00 | 1 | 1.8 | 1.8 | 15.00 | 0 | 16 | 1 | | 1 | 0 | 0 |
| 0.00 | 25.00 | 2 | 3.5 | 1.3 | 12.50 | 0 | 26 | 1 | 55 | 0 | 0 | 0 |
| 0.00 | 20.00 | 1 | 1.3 | 1.3 | 20.00 | 0 | 12 | 1 | 71 | 1 | 0 | 0 |
| 0.00 | 68.00 | 6 | 7.0 | 1.6 | 11.33 | 0 | 38 | 2 | 42 | 0 | 0 | 0 |
| 0.00 | 110.00 | 11 | 10.2 | 1.4 | 10.00 | 0 | 38 | 2 | 75 | 1 | 0 | 0 |
| 0.00 | 174.00 | 26 | 10.4 | 1.5 | 6.69 | 0 | 72 | 3 | | 0 | 0 | 0 |
| 0.00 | 20.00 | 1 | 1.8 | 1.8 | 20.00 | 0 | 15 | 1 | 67 | 1 | 0 | 0 |
| 14.00 | 95.00 | 7 | 6.1 | 1.3 | 13.57 | 0 | 56 | 2 | 61 | 0 | 0 | 0 |
| . | . | . | . | . | . | . | . | . | . | . | . | . |
| . | . | . | . | . | . | . | . | . | . | . | . | . |
| . | . | . | . | . | . | . | . | . | . | . | . | . |
| 0.00 | 25.00 | 2 | 1.5 | 1.1 | 12.50 | 0 | 18 | 2 | | 0 | 0 | 1 |
| 0.00 | 30.00 | 2 | 2.2 | 1.4 | 15.00 | 0 | 19 | 1 | 74 | 1 | 0 | 0 |
| 0.00 | 471.00 | 22 | 10.6 | 1.5 | 21.41 | 0 | 83 | 1 | 87 | 0 | 0 | 0 |
| 0.00 | 33.00 | 3 | 6.1 | 1.2 | 11.00 | 0 | 31 | 1 | 42 | 1 | 0 | 0 |
| 0.00 | 94.00 | 10 | 1.1 | 0.3 | 9.40 | 0 | 42 | 1 | 51 | 0 | 0 | 0 |
| 0.00 | 47.00 | 8 | 3.4 | 1.0 | 5.88 | 0 | 24 | 4 | 38 | 0 | 1 | 0 |
| 0.00 | 125.00 | 7 | 5.2 | 1.2 | 17.86 | 0 | 49 | 3 | 58 | 0 | 1 | 0 |
| 0.00 | 109.50 | 16 | 10.6 | 1.3 | 6.84 | 0 | 68 | 4 | 67 | 0 | 0 | 0 |
| 0.00 | 112.00 | 11 | 10.2 | 1.6 | 10.18 | 0 | 66 | 2 | 82 | 0 | 0 | 0 |
| 0.00 | 243.00 | 15 | 10.1 | 1.2 | 16.20 | 0 | 67 | 2 | 67 | 0 | 0 | 0 |

* The first column shows how much each person gave as a result of this mailing, while the other columns show information that was available before the mailing was sent. Data mining can use this information to statistically predict the mailing result, giving useful information about characteristics that are linked to the likelihood and amount of donations.

## TABLE B.1 cont'd

| Per Capita Income | Median Household Income | Professional | Technical | Sales | Clerical | Farmers | Self-Employed | Cars | Owner Occupied | Age 55–59 | Age 60–64 | School |
|---|---|---|---|---|---|---|---|---|---|---|---|---|
| $16,838 | $30,500 | 12% | 7% | 17% | 22% | 1% | 2% | 16% | 41% | 4% | 5% | 14.0 |
| 17,728 | 33,000 | 11 | 1 | 14 | 16 | 1 | 6 | 8 | 90 | 7 | 11 | 12.0 |
| 6,094 | 9,300 | 3 | 0 | 5 | 32 | 0 | 0 | 3 | 12 | 6 | 3 | 12.0 |
| 16,119 | 50,200 | 4 | 7 | 16 | 19 | 6 | 21 | 52 | 79 | 3 | 2 | 12.3 |
| 11,236 | 24,700 | 7 | 3 | 7 | 15 | 2 | 5 | 22 | 78 | 6 | 6 | 12.0 |
| 13,454 | 40,400 | 15 | 2 | 7 | 4 | 14 | 17 | 26 | 67 | 6 | 5 | 12.0 |
| 8,655 | 17,000 | 8 | 3 | 5 | 12 | 15 | 15 | 21 | 82 | 8 | 5 | 12.0 |
| 6,461 | 13,800 | 7 | 4 | 9 | 12 | 1 | 4 | 12 | 57 | 6 | 6 | 12.0 |
| 12,338 | 37,400 | 11 | 2 | 16 | 18 | 3 | 3 | 22 | 90 | 10 | 9 | 12.0 |
| 10,766 | 20,300 | 13 | 4 | 11 | 8 | 2 | 7 | 20 | 67 | 7 | 7 | 12.0 |
| . | . | . | . | . | . | . | . | . | . | . | . | . |
| . | . | . | . | . | . | . | . | . | . | . | . | . |
| . | . | . | . | . | . | . | . | . | . | . | . | . |
| 9,989 | 23,400 | 14 | 2 | 9 | 10 | 0 | 7 | 20 | 73 | 7 | 6 | 12.0 |
| 11,691 | 27,800 | 4 | 1 | 8 | 14 | 0 | 2 | 10 | 65 | 6 | 8 | 12.0 |
| 20,648 | 34,000 | 13 | 4 | 20 | 20 | 0 | 2 | 5 | 46 | 8 | 9 | 12.4 |
| 12,410 | 21,900 | 9 | 3 | 12 | 20 | 0 | 9 | 13 | 49 | 5 | 8 | 12.0 |
| 14,436 | 41,300 | 15 | 7 | 9 | 15 | 1 | 9 | 29 | 85 | 6 | 5 | 13.2 |
| 17,689 | 31,800 | 11 | 3 | 17 | 21 | 0 | 6 | 12 | 16 | 2 | 3 | 14.0 |
| 26,435 | 43,300 | 15 | 1 | 5 | 9 | 0 | 3 | 16 | 89 | 5 | 24 | 14.0 |
| 17,904 | 44,800 | 8 | 3 | 1 | 20 | 4 | 15 | 26 | 88 | 6 | 5 | 12.0 |
| 11,840 | 28,200 | 13 | 4 | 12 | 14 | 2 | 6 | 13 | 77 | 5 | 5 | 12.0 |
| 17,755 | 40,100 | 10 | 3 | 13 | 24 | 2 | 7 | 24 | 41 | 2 | 4 | 14.0 |

## TABLE B.2 Definitions for Variables in the Donations Database*

| Excel Range Name | Description |
| --- | --- |
| Donation | Donation amount in dollars in response to this mailing |
| Lifetime | Donation lifetime total before this mailing |
| Gifts | Number of lifetime gifts before this mailing |
| YearsSinceFirst | Years since first gift |
| YearsSinceLast | Years since most recent gift before this mailing |
| AvgGift | Average of gifts before this mailing |
| MajorDonor | Major donor indicator |
| Promotions | Number of promotions received before this mailing |
| RecentGifts | Frequency: (1, 2, 3, or 4, meaning 4 or more) is number of gifts in past two years (remember that these are lapsed donors who did not give during past year) |
| Age | Age in years (note that many are blank) |
| HomePhone | Published home phone number indicator |
| PCOwner | Home PC owner indicator |
| CatalogShopper | Shop by catalog indicator |
| PerCapIncome | Per capita neighborhood income |
| MedHouseInc | Median household neighborhood income |
| Professional | Percent professional in neighborhood |
| Technical | Percent technical in neighborhood |
| Sales | Percent sales in neighborhood |
| Clerical | Percent clerical in neighborhood |
| Farmers | Percent farmers in neighborhood |
| SelfEmployed | Percent self-employed in neighborhood |
| Cars | Percent households with 3+ vehicles |
| OwnerOccupied | Percent owner-occupied housing units in neighborhood |
| Age55–59 | Percent adults age 55–59 in neighborhood |
| Age60–64 | Percent adults age 60–64 in neighborhood |
| School | Median years of school completed by adults in neighborhood |

*The first group of variables represents information about the person who received the mailing; for example, the second variable, "Lifetime," shows the total dollar amount of all previous gifts by this person and "PC Owner" is 1 if he or she owns a PC and is 0 otherwise. The remaining variables, beginning with Per Capita Income and continuing through the last column, represent information about the person's neighborhood.

# Self Test: Solutions to Selected Problems and Database Exercises

## CHAPTER 1

### Problem

6. a. Exploring the data. Data are already available (being examined) so it is not designing the study. No further specifics are provided that would support estimation or hypothesis testing.
   b. Designing the study. Data are not yet available for any of the other activities of statistics.
   c. Modeling the data.
   d. Hypothesis testing. The two possibilities are that the salary pattern is merely random, or it is not.
   e. Estimation. The unknown quantity being guessed, based on data, is the size of next quarter's gross national product.

## CHAPTER 2

### Problem

11. a. The individual employee is the elementary unit for this data set.
    b. This is a multivariate data set, with three or more columns.
    c. Salary and years of experience are quantitative; gender and education are qualitative.
    d. Education is ordinal qualitative because the natural ordering HS, BA, MBA corresponds to more and more education.
    e. These are cross-sectional data, without a natural sequence.

### Database Exercises

1. a. This is a multivariate data set, with three or more columns.

   d. Training level is ordinal because it can be ranked in a meaningful way.
2. For gender:
   a. No, these categories cannot be added or subtracted as they are in the database.
   b. Yes, you can count how many men or women there are.
   c. No, there is no natural ordering.
   d. Yes, you can find the percentage of women or men.

## CHAPTER 3

### Problems

6. a.

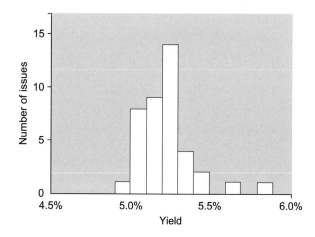

   b. Typical values are between 5.0% and 5.5%, approximately.
   c. Approximately normal (perhaps with one or two outliers at the right).

**19. a.**

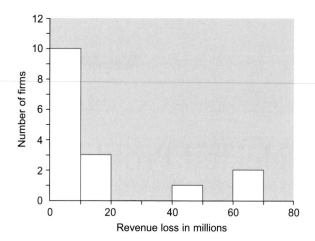

**b.** The distribution is skewed toward high values and shows two gaps with three outliers. In particular, 13 of the 16 data values are crammed into the first two columns of the display.

## Database Exercise

**2. a.**

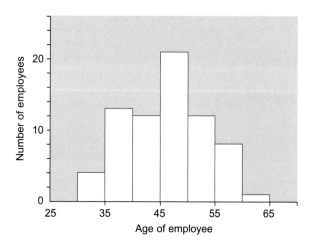

**b.** Approximately normal.

**c.** From the histogram we see that the youngest employee is between 30 and 35, and the oldest is between 60 and 65. The typical age is around 45 or 50 years old. The distribution shape is approximately normal, with concentration of employees near the middle of the distribution, with relatively few older or younger ones.

## CHAPTER 4

### Problems

**1. a.** Average is 15.6 defects per day.
   **b.** Median is 14 defects per day.

**c.**

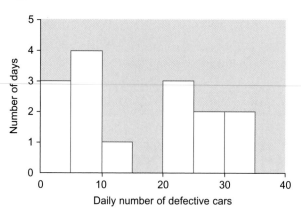

**d.** Mode is 7.5 defects per day. (With quantitative data, the mode is defined as the value at the highest point of the histogram, perhaps as the midpoint of the highest bar). With a different histogram (different bar widths, for example), a different value of the mode could be found.

**e.** Lower quartile is 6; upper quartile is 24.5 defects per day.

**f.** Smallest is 0; largest is 34 defects per day.

**g.**

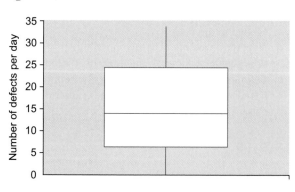

**h.**

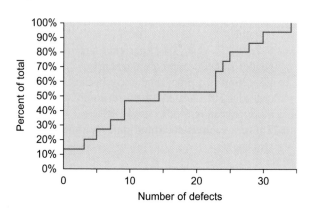

**i.** The 90th percentile is 30 defects per day.

**j.** The percentile ranking is approximately 87%, as can be seen from the cumulative distribution function, as follows:

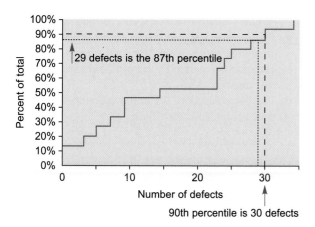

29 defects is the 87th percentile

90th percentile is 30 defects

8. The cost of capital is 14.6%, the weighted average of the rates of return (17%, 13%, and 11%) with weights equal to the respective market values:

$$\frac{4,500,000}{8,400,000} \times 0.17 + \frac{1,700,000}{8,400,000} \times 0.13 + \frac{2,200,000}{8,400,000}$$
$$\times 0.11 = 0.146$$

## Database Exercise

1. **a.** Average: $45,141.50.
   **b.** Median: $49,033.
   **c.**

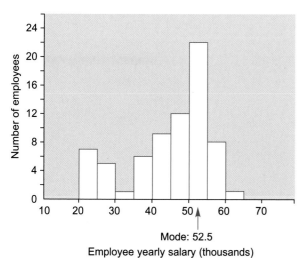

Mode: 52.5

Employee yearly salary (thousands)

Mode: Approximately $52,500; midpoint of highest bar of this histogram.

**d.** Average is lowest, median is next, and mode is highest. This is expected for a skewed distribution with a long tail toward low values.

From the average we know that the total of all the salaries paid divided by the number of employees is $45,141. If each employee received the average, the total payroll would be unchanged. The median shows that the same number of employees make more than $49,030 as get a salary of less than $49,030. From the mode, we see that a larger number of employees make from $50,000 to $55,000 per year than receive a salary in any other $5,000 segment of the salaries paid by this firm.

## CHAPTER 5

### Problems

1. **a.** The average budget size is $748.41 million.
   **b.** The standard deviation is $253.55 million.
   **c.** The standard deviation indicates, approximately, how far the individual budget amounts are from their average.
   **d.** The range is $810.4 million, computed as 1,167.0 − 356.6.
   **e.** The range is the largest minus the smallest. The firm with the largest budget has $810.4 million more to spend than the firm with the lowest budget.
   **f.** The coefficient of variation is 0.339, computed as 253.55/748.41. There are no units of measurement; i.e., this is a pure number and will be the same no matter which units are used in the calculation.
   **g.** A coefficient of variation of 0.339 indicates that the size of the advertising budget for these firms typically varies from the average price by 33.9% (that is, by 33.9% of the average).
   **h.** The variance is 64,288.36, measured in squared millions of dollars.
   **i.** There is no simple interpretation because the variance is measured in squared millions of dollars, which are beyond our ordinary business experience.
   **j.**

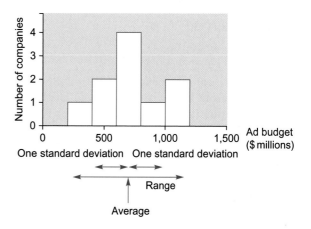

One standard deviation   One standard deviation

Range

Average

6. **a.** Average number of executives per firm: 10.4.

   **b.** The standard deviation, 7.19, indicates that these firms differ from the average by approximately 7.19 executives.

   **c.** There are 38 corporations (84.4%) within one standard deviation of the average (that is, from 3.21 to 17.59). This is more than the approximately two-thirds you would expect for a normal distribution.

   **d.** There are 43 corporations (95.6%) within two standard deviations from the average (from −3.97 to 24.77). This is quite close to the 95% you would expect for a normal distribution.

   **e.** There are 44 corporations (97.8%) within three standard deviations from the average (from −11.16 to 31.96). This is close to the 99.7% you would expect for a normal distribution.

   **f.**

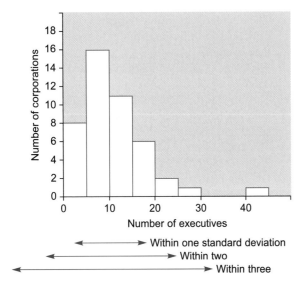

The histogram shows a possible outlier at 41. This has pulled the average to high values and has inflated the standard deviation. This may account for the larger than expected 84.4% within one standard deviation of the average.

## Database Exercise

1. **a.** The range is $38,555.

   **b.** The standard deviation is $10,806.

   **c.** The coefficient of variation is 0.239, or 23.9%.

   **d.** The gap from lowest to highest paid employee is $38,555 (the range). Employee salaries typically differ from the average by approximately $10,806 (the standard deviation), which is 23.9% of the average. The range is larger than the standard deviation because it measures the largest possible difference between two data values, instead of a typical difference from average.

# CHAPTER 6

## Problems

1. **a.** The random experiment is: You wait until the net earnings figure is announced and then observe it.

   **b.** The sample space consists of all dollar amounts, including positive, negative, and zero.

   **c.** The outcome will tell you Ford's net earnings for the past quarter.

   **d.** The list consists of all dollar amounts that exceed your computed dollar figure:

   Computed figure + .01, computed figure + .02, ···

   **e.** Subjective probability because it is based on opinion.

5. **a.** The probability is 35/118 = 0.297.

   **b.** The probability is (1 − 0.297) = 0.703.

8. **a.** The probability is 0.22. The event "big trouble" is the complement of the event "A and B." Using the relationship between *and* and *or*, we find

   Probability of (A and B) = 0.83 + 0.91 − 0.96 = 0.78.

   Using the complement rule, we find the answer:

   Probability of "big trouble" = 1 − 0.78 = 0.22.

   **b.** These events are not mutually exclusive because the probability of "A and B" is 0.78 (from part a).

   **c.** No. These are not independent events. The probability of meeting both deadlines, 0.78 (from part a), is not equal to the product of the probabilities, 0.83 × 0.91 = 0.755. This can also be seen using conditional probabilities.

18. **a.**

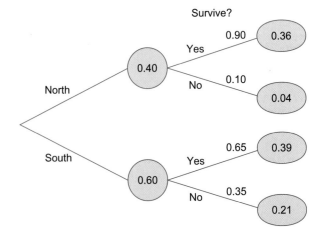

   **b.** The probability is 0.36 + 0.39 = 0.75 of surviving the first year.

   **c.** The probability is 0.39 of being built in the South and being successful.

**d.** The probability is 0.39/0.75 = 0.52. This is the conditional probability of South given survival and is equal to (Probability of "South and survival")/(Probability of survival).

**e.** The probability is 0.04/0.4 = 0.10, for failure given that it is built in the North. This is probability of "not surviving" and North)/(Probability of North).

## Database Exercise

**1. a.** Probability of selecting a woman is 28/71 = 0.394.
   **b.** Probability that the salary is over $35,000 is 58/71 = 0.817.
   **e.** Probability of over $35,000 given B is 0.310 /0.338 = 0.917.

## CHAPTER 7

### Problems

**1. a.** The mean payoff is $8.50.
   **b.** The expected option payoff, $8.50, indicates the typical or average value for the random payoff.
   **c.** The standard deviation is $10.14.
   **d.** This standard deviation, $10.14, gives us a measure of the risk of this investment. It summarizes the approximate difference between the actual (random) payoff and the expected payoff of $8.50.
   **e.** The probability is 0.15 + 0.10 = 0.25.

**18. a.** We are assuming that all of the $n = 15$ securities have the same probability of losing value and are independent of one another.
   **b.** You would expect 12 = (0.8)(15) securities to lose value.
   **c.** The standard deviation is $\sigma_X = 1.55$.
   **d.** The probability is 0.035

$$\frac{15!}{15! \times 0!} 0.8^{15}(1 - 0.8)^0 \,(\text{or use table})$$

   **e.** The probability is

$$0.103 = \frac{15!}{10! \times 5!} 0.8^{10}(1 - 0.8)^5$$
$$= 3003 \times 0.107374 \times 0.00032 \,(\text{or use table})$$

**30. a.** The probability is 0.75. The standardized number is $z = (0.10 - 0.12)/0.03 = -0.67$, which leads to 0.2514 in the standard normal probability table, for the event "being less than 10%." The answer, using the complement rule, is then $1 - 0.2514 = 0.75$.

## Database Exercise

**1. b.** $X = 52$; $p = 52/71 = 0.732$. Thus, 73.2% of the employees have salaries above $40,000.

## CHAPTER 8

### Problems

**1. a.** Unreasonable. This is an unrepresentative sample. The first transmissions of the day might get extra care.

**5. a.** Statistic. This is the average for the sample you have observed.
   **b.** Parameter. This is the mean for the entire population.

**8.** The sample consists of documents numbered 43, 427, and 336. Taking the random digits three at a time, we find 690, 043, 427, 336, 062, .... The first number is too big (690 > 681, the population size), but the next three can be used and do not repeat.

**22.** The probability is 0.14. The standard deviation of the average is $30/\sqrt{35} = 5.070926$ and the standardized numbers are $z_1 = (55 - 65)/5.070926 = -1.97$ and $z_2 = (60 - 65)/5.070926 = -0.99$. Looking up these standardized numbers in the standard normal table and subtracting, we find the answer $0.1611 - 0.0244 = 0.14$.

**26. a.** The mean is $2,601 \times 45 = \$117,045$.
   **b.** The standard deviation is $\$1,275\sqrt{45} = \$8,552.96$.
   **c.** Because of the central limit theorem.
   **d.** The probability is 0.92, using the standardized number $z = (105,000 - 117,045)/8,552.96 = -1.41$. Looking up this standardized number in the standard normal probability table, we find 0.0793, which represents the probability of being less than $105,000. The answer (being at or above) will then be $1 - 0.0793 = 0.92$.

**34. a.** The standard error of the average is $16.48/\sqrt{50} = \$2.33$. This indicates approximately how far the (unknown) population mean is from the average ($53.01) of the sample.

## Database Exercises

**2.** Arranging the random digits in groups of 2, we have the following:

14  53  62  38  70  78  40  24  17  59  26
23  27  74  22  76  28  95  75

Eliminating numbers that are more than 71 or less than 1:

14  53  62  38  70  40  24  17  59  26
23  27  22  28

The first 10 numbers have no duplicates and give us the following sample:

14, 53, 62, 38, 70, 40, 24, 17, 59, 26

If you want them in order by employee number:

14, 17, 24, 26, 38, 40, 53, 59, 62, 70

**a.** The employee numbers are 14, 53, 62, 38, 70, 40, 24, 17, 59, and 26.

8. a. The binomial $X$ is 5 females.
   b. The standard error is $\sqrt{10 \times 0.5 \times 0.5} = 1.58$, indicating that the observed binomial $X$ is approximately 1.58 above or below the mean number you would expect to find in a random sample of 10 from the same population.

## CHAPTER 9

## Problems

1. The 95% confidence interval extends from 101.26 to 105.94 bushels per acre (using 1.960 from the $t$ table). This is computed as $103.6 - (1.960)(9.4)/\sqrt{62} = 103.6 - 2.34 = 101.26$ and as $103.6 + (1.960)(9.4)/\sqrt{62} = 103.6 + 2.34 = 105.94$. (A more exact computer result is from 101.21 to 105.99.)

5. a. 2.365.
   b. 3.499.
   c. 5.408.
   d. 1.895.

29. a. The average, −4.94%, summarizes the performance of these stocks.
   b. The standard deviation, 0.2053 or 20.53%, summarizes difference from average. The performance of a typical stock in this list differed from the average value by about 20.53 percentage points.
   c. The standard error, $0.2053/\sqrt{15} = 0.0530$ or 5.30%, indicates the approximate difference (in percentage points) between the average (−4.94%) and the unknown mean for the (idealized) population of similar brokerage firms.
   d. The 95% confidence interval extends from −16.31% to 6.43% (using 2.145 from the $t$ table).
   e. The 90% confidence interval extends from −14.27% to 4.39% (using 1.761 from the $t$ table). The 90% two-sided confidence interval is smaller than the 95% two-sided confidence interval.
   f. We are 99% sure that the mean performance of the population of stocks is at least −18.85% (using 2.624 from the $t$ table).
   g. No, you must either use the same one side regardless of the data or else use the two-sided interval. Otherwise, you may not have the 99% confidence that you claim.

30. a. The 95% confidence interval extends from 49.2% to 55.6% (using 1.960 from the $t$ table). This is computed as

$$0.524 - 1.960\sqrt{0.524(1-0.524)/921}$$
$$= 0.524 - 0.032 = 0.492 \text{ and as}$$
$$0.524 + 1.960\sqrt{0.524(1-0.524)/921}$$
$$= 0.524 + 0.032 = 0.556.$$

## Database Exercise

1. a. The average is $34,031.80. The standard deviation is $10,472.93. The standard error is $4,683.64.
   b. The 95% confidence interval extends from $21,030 to $47,034 (using 2.776 from the $t$ table).
   c.

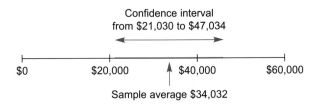

Confidence interval
from $21,030 to $47,034

$0        $20,000        $40,000        $60,000

Sample average $34,032

## CHAPTER 10

## Problems

1. a. The null hypothesis, $H_0: \mu = 43.1$, claims that the population mean age of customers is the same as that for the general population in town. The research hypothesis, $H_1: \mu \neq 43.1$, claims that they are different.
   b. Reject $H_0$ and accept $H_1$. The average customer age is significantly different from the general population. The 95% confidence interval extends from 29.11 to 38.09 (using 1.960 from the $t$ table) and does not include the reference value (43.1). The $t$ statistic is −4.15. (A more exact computer confidence interval is from 29.00 to 38.20.)

2. a. Reject $H_0$ and accept $H_1$. The average customer age is highly significantly different from the general population. The 99% confidence interval extends from 27.70 to 39.50 (using 2.576 from the $t$ table) and does not include the reference value (43.1). The $t$ statistic is −4.15, which is larger in absolute value than 2.576 from the $t$ table. (A more exact computer confidence interval is from 27.46 to 39.74.)
   b. $p < 0.001$. The $t$ table value for testing at the 0.001 level is 3.291.

40. a. No. Person #4 had a higher stress level with the true answer than with the false answer.
   b. Average stress levels are as follows: true (8.5), false (9.2); average change: 0.7.
   c. The standard error of the difference is $0.648074/\sqrt{6} = 0.264575$. This is a paired situation. There is a natural relationship between the two data sets since both measurements were made on the same subject.
   d. The 95% two-sided confidence interval extends from 0.02 to 1.38 (using 2.571 from the $t$ table). We are 95% certain that the population mean change in the vocal stress level (false minus true) is somewhere between 0.02 and 1.38.

e. These average stress levels are significantly different ($p < 0.05$) because the reference value (0, indicating no difference) is not in the confidence interval. The $t$ statistic is $0.7/0.264575 = 2.65$. The one-sided conclusion to this two-sided test says that stress is significantly higher when a false answer is given, as compared to a true answer.

f. The mean stress level is significantly higher when a false answer is given, as compared to a true answer. This is a conclusion about the mean stress levels in a large population, based on the six people measured as representatives of the larger group. The conclusion goes beyond these six people to the population (real or idealized) from which they may be viewed as a random sample. Although there was one person with lower stress for the false answer, the conclusion is about the mean difference in the population; it is not guaranteed to apply to each and every individual.

42. a. Average time to failure: yours, 4.475 days; your competitor's, 2.360. The average difference is 2.115 days.

b. The standard error is 1.0066. This is an unpaired situation. There is no natural relationship between the measurements made on the two samples since they are different objects. In addition, since there are different numbers of measurements (sample sizes) for the two samples, this could not be a paired situation.

c. The two-sided 99% confidence interval extends from –0.69 to 4.92 (using 2.787 from the $t$ table).

d. The difference in reliability is not significant at the 1% level. The reference value, 0, is in the 99% confidence interval from part c. The $t$ statistic is 2.10.

e. There is a significant difference in reliability ($p < 0.05$). This may be seen from the 95% confidence interval (from 0.04 to 4.19) or from the $t$ statistic (2.10), which exceeds the $t$ table value (2.060) for testing at the 5% level. From part d, we know that the test is not significant at the 1% level.

f. A study has shown that our products are significantly more reliable than our competitors ....[1]

## Database Exercise

1. Yes, the average annual salary ($45,142) is significantly different from $40,000 since this reference value is not in the 95% confidence interval (from $42,628 to $47,655, using 1.960 for the $t$ table value). Alternatively, the $t$ statistic is

$$(45,141.50 - 40,000)/1,282.42 = 4.01$$

Reject the null hypothesis and accept the research hypothesis that the population mean is different from

---

1. $p < 0.05$, using a two-sample unpaired $t$ test.

$40,000. (A more exact computer confidence interval is from $42,584 to $47,699.)

## CHAPTER 11

1. a.

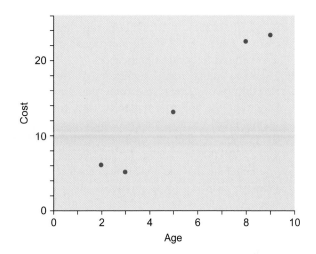

The scatterplot shows a linear structure (increasing relationship) with data values distributed about a straight line, with some randomness.

b. The correlation, $r$, between age and maintenance cost is 0.985. This correlation is very close to 1, indicating a strong positive relationship. It agrees with the scatterplot, which showed the maintenance cost increasing along a straight line, with increasing age.

c. Predicted cost = –1.0645 + 2.7527 Age

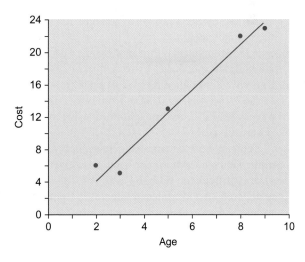

d. Predicted cost = $-1.06451 + (2.752688)(7) = 18.204$, in thousands of dollars, hence $18,204.

e. $S_e = 1.7248$, in thousands of dollars, or $1,725.

f. $R_2 = 96.9\%$ of the variation in maintenance cost can be attributed to the fact that some presses are older than others.

**g.** Yes, age does explain a significant amount of the variation in maintenance cost. This may be verified by testing whether the slope is significantly different from 0. The confidence interval for the slope, from 1.8527 to 3.6526, does not include 0; therefore, the slope is significantly different from 0. Alternatively, note that the $t$ statistic, $t = b/S_b = 2.7527/0.2828 = 9.73$, exceeds the $t$ table value (3.182) for $5 - 2 = 3$ degrees of freedom and may also be used to decide significance.

**h.** The extra annual cost is significantly different from $20,000. From part g, we know that we are 95% sure that the long-term yearly cost for annual maintenance per machine is somewhere between $1,853 and $3,653 per year. Since the reference value, $20,000, is not in the confidence interval, you conclude that the annual maintenance cost per year per printing press is significantly different from $20,000. In fact, it is significantly less than your conservative associate's estimate of $20,000. The $t$ statistic is $(2.752688 - 20)/0.282790 = -61.0$. Since the value for the $t$ statistic is larger than the critical value (12.294 with 3 degrees of freedom) at the 0.001 level, you reject the null hypothesis and accept the research hypothesis that the population maintenance cost is different from the reference value, and claim that the finding is very highly significant ($p < 0.001$).

## Database Exercise

**2. a.** $R^2 = 30.4\%$ of the variation in salaries can be explained by the years of experience found among these employees.

**b.** $49,285, computed as Predicted Salary $= 34,575.94 + 1,838.615$ Experience.

**c.** The 95% confidence interval extends from $31,304 to $67,265 (using 1.960 from the $t$ table and $S_{Y|X_0} = 9,173.75$ as the standard error of a new observation).

You are 95% certain that a new employee having eight years of experience would receive a yearly salary of between $31,304 and $67,265. (A more exact computer confidence interval is from $30,984 to $67,586.)

**d.** The 95% confidence interval extends from $46,707 to $51,863 (using 1.960 from the $t$ table and $S_{\text{predicted } Y|X_0} = 1,315.34$ as the standard error of the mean value of $Y$ given $X_0$).

You are 95% certain that the population mean salary level for employees with eight years of experience is between $46,707 and $51,863. (A more exact computer confidence interval is from $46,661 to $51,909.)

# CHAPTER 12

## Problems

**1.** Multiple regression would be used to predict $Y =$ Number of leads from $X_1 =$ Cost and $X_2 =$ Size. The appropriate test would be the $F$ test, which is the overall test for significance of the relationship.

**5. a.** Price $= 8,344.005 + 0.026260$ Area $- 4.26699$ Year.

**b.** The value of each additional square centimeter is $26.26. All else being equal (i.e., for a given year) for an increase in area of 1 square centimeter, the price of the painting would rise by ($1,000) (0.026260) = 26.26 on average.

**c.** Holding area constant, the regression coefficient for Year reveals that as the years increased the price a painting could command decreased by $4,266.99 a year on average. The earlier paintings are more valuable than the later ones.

**d.** Price $= 8,344.005 + (0.026260)(4,000) - (4.26699)$ (1954)
$= \$111.348$ (thousands) $= \$111,348$.

**e.** The prediction errors are about $153,111. The standard error of estimate, $S_e = 153.111$, indicates the typical size of prediction errors in this data set, in thousands of dollars (because $Y$ is in thousands).

**f.** $R^2 = 28.2\%$ of the variation in price of Picasso paintings can be attributed to the size of the painting and the year in which it was painted.

**g.** Yes, the regression is significant. The $F$ test done using $R^2$ is significant ($R^2 = 0.282$, with $n = 23$ cases and $k = 2$ $X$-variables, exceeds the table value of 0.259). This indicates that the variables, Area and Year taken together, explain a significant fraction of the variation in price from one painting to another.

The traditional $F$ test reports significance, $F = 3.93203$ with 2 and 20 degrees of freedom. (The $p$-value is 0.036.)

**h.** Yes, area does have a significant impact on price following adjustment for year. Larger paintings are worth significantly more than smaller ones from the same year.

The confidence interval extends from 0.000896 to 0.051623; it does not include the reference value of zero, so we accept the research hypothesis. Also, the $t$ statistic (2.16) is larger than the critical $t$ value (2.086) for 20 degrees of freedom. (The $p$-value is 0.043.)

**i.** Yes, year has a significant impact on price, adjusting for area. The impact of year on price is of a decrease of price, so that newer paintings are worth significantly less than older ones of the same size. The confidence interval extends from $-8.0048$ to $-0.5291$ and does not include the reference value of zero. Also, the

$t$ statistic of –2.38 is larger in absolute value than the critical $t$ table value of 2.086. (The $p$-value is 0.027.)

**15. b.** Predicted Compensation

$$= 0.9088 + 0.0023 * \text{Revenue} - 0.1163 * \text{ROE}$$
$$= 0.908811 + 0.002309 * 4{,}570 - 0.116328 * 0.1642$$
$$= 11.44 \,(\text{millions}) = \$11{,}440{,}000 \,(\text{rounded})$$
$$\text{Residual} = 13.75 - 11.44 = 2.31 \,(\text{millions})$$
$$= \$2{,}310{,}000 \,(\text{rounded}).$$

This executive is paid about \$2,310,000 more than we would expect for a firm with this level of revenues and ROE.

## Database Exercises

**1. a.** The regression equation is

$$\text{Salary} = 22{,}380.64 + 300.5515\,\text{Age}$$
$$+ 1{,}579.259\,\text{Experience}$$

This prediction equation gives you the expected (average) salary for a typical employee of a given age and experience. Each additional year of age adds \$301 to annual salary, on average, while each additional year of experience is valued at \$1,579.

**b.** $S_e = 8{,}910.19$. The standard error of estimate reveals that the predicted salary numbers differ from the actual salaries by approximately \$8,910.

**c.** $R^2 = 0.3395$. This says that 34.0% of the variation in salary can be attributed to age and experience. About 66% of the variation is due to other causes.

**d.** The model is significant. This tells you that age and experience, taken together, explain a significant proportion of the variation in salaries.

To carry out the $F$ test based on $R^2$, we need the critical value from the table. Since $n = 71 > 50$, we use the multipliers (for $k = 2$ variables) to determine the 5% critical value:

$$\text{Critical value} = 5.99/71 + (-0.27)/71^2$$
$$= 0.0843$$

Since the $R^2$ value (0.3395) exceeds this critical value, the $F$ test is significant.

**e.** Age does not have a significant effect on salary, holding experience constant. The confidence interval (from –14 to 615) contains the reference value 0. Also, the $t$ statistic, 1.91, is smaller than the critical value of 1.960.

Experience does have a significant influence on employee salaries, holding age constant. The confidence interval for experience (870 to 2,289) does not contain zero. The $t$ statistic for experience, 4.44, is larger than the critical value, 3.291, at the 0.1% level. This indicates that experience is very highly significant in predicting salary levels.

**f.** The standardized regression coefficient for age is 0.203. An increase in one standard deviation for age would result in an increase of 20.35% of one standard deviation of salary.

The standardized regression coefficient for experience is 0.474. An increase in one standard deviation for experience results in an increase of 47.4% of one standard deviation of salary.

This suggests that experience is more important than age in its effect on salary because the standardized regression coefficient for experience is larger. (The standard deviations are 7.315 for age, 3.241 for experience, and 10,806 for salary.)

**g.**

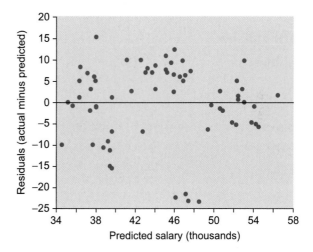

This is basically random: There is little, if any, structure in the diagnostic plot. However, there is one bunch of four data values at about \$46,000 to \$49,000 (predicted salary) that have exceptionally small residuals between –20 and –25. Perhaps this group is worthy of further study. They are all women in Training Group A who may be underpaid (since the residuals are negative) relative to their age and experience.

**2. a.** Predicted salary = 22,380.64 + (300.5515)(39) + (1,579.259)(1) = \$35,681.

Prediction error = Actual – Predicted = 35,018 – 35,681 = –663.

The predicted salary (\$35,681) is close to the actual salary (\$35,018). The prediction error, –663, suggests that this employee's salary is \$663 lower than you would expect for this age and experience.

## CHAPTER 13

## Problems

**1. a.** Purpose: to provide background information on the size of shipping facilities at other firms to help with expansion strategy. Audience: those executives who will be suggesting plans and making decisions about this expansion.

2. **a.** Effect. The usual statistical usage is that *effect* is the noun ("it has an effect …") and that *affect* is the verb ("it affects …").

   **b.** Affects.

4. **a.** Analysis and methods because it includes technical detail and its interpretation.

5. **a.** S. Vranica, "Tallying Up Viewers: Industry Group to Study How a Mobile Nation Uses Media," *Wall Street Journal*, July 26, 2010, p. B4.

6. **a.** The name and title of the person; also the place, month, and day.

# CHAPTER 14

## Problems

**9. a.**

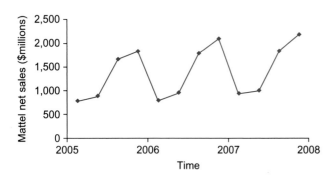

There is an overall increasing trend over time with strong seasonal variation.

**b.** The moving average is not available for the first two quarters of 2005. The first value is for third quarter 2005: $(783.120/2 + 886.823 + 1,666.145 + 1,842.928 + 793.347/2)/4 = \$1,296$. This and the other moving average values are as follows:

| Year | Net Sales (millions) | Moving Average |
|------|------|------|
| 2005 | 783.120 | (unavailable) |
| 2005 | 886.823 | (unavailable) |
| 2005 | 1,666.145 | 1,296 |
| 2005 | 1,842.928 | 1,306 |
| 2006 | 793.347 | 1,331 |
| 2006 | 957.655 | 1,379 |
| 2006 | 1,790.312 | 1,431 |
| 2006 | 2,108.842 | 1,455 |
| 2007 | 940.265 | 1,467 |
| 2007 | 1,002.625 | 1,483 |
| 2007 | 1,838.574 | (unavailable) |
| 2007 | 2,188.626 | (unavailable) |

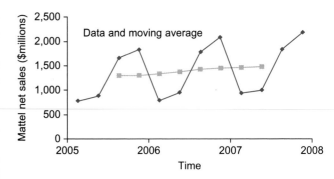

**c.** The seasonal indices for quarters 1 through 4 are: 0.6187, 0.6853, 1.2684, and 1.4302. Yes, these numbers seem reasonable: quarters 3 and 4 are considerably higher than the other two quarters.

**d.** Quarter 4 is the best. Sales are $1.4302 - 1 = 43.0\%$ higher as compared to a typical quarter.

**e.** Dividing each sales figure by the appropriate seasonal index:

| Year | Sales (millions) | Seasonally Adjusted |
|------|------|------|
| 2005 | 783.120 | 1,266 |
| 2005 | 886.823 | 1,294 |
| 2005 | 1,666.145 | 1,314 |
| 2005 | 1,842.928 | 1,289 |
| 2006 | 793.347 | 1,282 |
| 2006 | 957.655 | 1,397 |
| 2006 | 1,790.312 | 1,412 |
| 2006 | 2,108.842 | 1,474 |
| 2007 | 940.265 | 1,520 |
| 2007 | 1,002.625 | 1,463 |
| 2007 | 1,838.574 | 1,450 |
| 2007 | 2,188.626 | 1,530 |

**f.** On a seasonally adjusted basis, sales also went up from third to fourth quarter of 2007 (from 1,450 to 1,530).

**g.** On a seasonally adjusted basis, sales fell from the second to the third quarter of 2007 (from 1,463 to 1,450).

**h.** The regression equation using the time period 1, 2, 3, … for $X$ and the seasonally adjusted series for $Y$ is

Predicted Adjusted Net Sales
$$= 1,229.455 + 24.832 \,(\text{Time Period}).$$

**i.** Predicted Sales $= 1,726.087$, found by substituting 20 for Time Period.

**j.** The forecast is 2,469. The seasonally adjusted forecast from the previous part is seasonalized (by multiplying by the fourth quarter seasonal index) to find the forecast $(1,726.087)(1.4302) = 2,469$.

**k.** This forecast of 2,469 is higher than Mattel's actual net sales, 1,955, for fourth quarter 2009, and is consistent with the possibility that the recession disrupted sales patterns and accounts for this lower actual level.

12. **b.** 285,167/1.08 = 264,044.
   **c.** (264,043.5)(1.38) = 364,380.
13. **a.** 5,423 + (29)(408) = 17,255.
   **d.** (17,255)(1.45) = 25,020.

# CHAPTER 15

## Problems

1. **a.** Ad #2 appears to have the highest effectiveness (68.1). Ad #3 appears to have the lowest effectiveness (53.5).
   **b.** The total sample size is $n = 303$. The grand average is $\overline{X} = 61.4073$. The number of samples is $k = 3$.
   **c.** The between-sample variability is 5,617.30 with $k - 1 = 2$ degrees of freedom.
   **d.** The within-sample variability is 91.006 with $n - k = 300$ degrees of freedom.
4. **b.** The standard error for this average difference is 1.356192.
19. **a.** Yes. The difference in performance between the cooperation group and the competition group is statistically significant. The performance of the cooperation group is significantly higher than the performance of the competition group.

   You know the difference is significant because the $p$-value of 0.049682 (for the test "Competition-cooperation (A)" in the table) is smaller than 0.05, which indicates significance at the 5% level.

## Database Exercise

1. **a.**

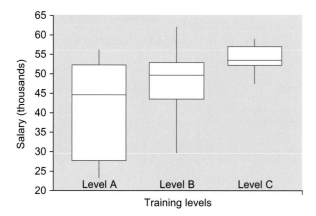

It appears that, typically, higher salaries go to those with more training. However, the highest salaries are in the B group of employees who have taken one training course, the lowest salaries in the A group having taken no training courses. In general,

salaries are larger in the B and C groups, who took more training courses.

The variabilities are unequal, with A showing the most and C the least variability.

**b.** The averages are A, $41,010.87; B, $48,387.17; and C, $53,926.89.

The average salary is seen to increase with increasing training. This is similar to the effect seen for the medians in the box plots.

**c.** Between-sample variability: 797,916,214, with $k - 1 = 2$ degrees of freedom.

Within-sample variability: 96,732,651, with $n - k = 68$ degrees of freedom.

# CHAPTER 16

## Problems

2. **d.** Reject the null hypothesis using the nonparametric test. The median profit of building material firms is significantly different from a loss of 5 percentage points. Here are the steps:
   1. The modified sample size is 14. Not one of the data values is equal to the reference value, −5.
   2. From the table, the sign test is significant at the 5% level if the number of ranked values counted is less than 3 or more than 11.
   3. The count is that 2 data values fall below the reference value, −5. They are Manville, −59, and National Gypsum, −7.
   4. The count falls outside the limits (at the 5% test level) from the sign test table. Therefore, you can claim that the median profit of building material firms is significantly different from a loss of 5 percentage points.
6. **b.** The modified sample size, the number who changed, is 8 + 2 = 10.
   **c.** Accept the null hypothesis; the difference is not significant. Here are the details:
   1. The modified sample size is 10.
   2. From the table, the sign test is significant at the 5% level if the number of ranked values counted is less than 2 or more than 8.
   3. The count gives 2 less than the reference value and 8 more than the reference value.
   4. Since the count falls at the limits given in the table for the sign test, the difference is not statistically significant.
10. **b.** Accept the research hypothesis because the test statistic, 4.53, is larger than 1.960. The difference between the salaries of men and women is significant, using the nonparametric test for two unpaired samples. Here are the details, starting with the two salary scales

combined with the ranks listed and averaged where appropriate:

| Rank | Salary | Gender |
|------|--------|--------|
| 1 | $20,700 | Woman |
| 2 | 20,900 | Woman |
| 3 | 21,100 | Woman |
| 4 | 21,900 | Woman |
| 5 | 22,800 | Woman |
| 6 | 23,000 | Woman |
| 7 | 23,100 | Woman |
| 8 | 24,700 | Woman |
| 9 | 25,000 | Woman |
| 10 | 25,500 | Man |
| 11 | 25,800 | Woman |
| 12 | 26,100 | Man |
| 13.5 | 26,200 | Woman |
| 13.5 | 26,200 | Man |
| 15 | 26,900 | Woman |
| 16 | 27,100 | Woman |
| 17 | 27,300 | Man |
| 18 | 28,100 | Woman |
| 19 | 29,100 | Man |
| 20 | 29,700 | Woman |
| 21 | 30,300 | Man |
| 22 | 30,700 | Man |
| 23 | 32,100 | Man |
| 24 | 32,800 | Man |
| 25 | 33,300 | Man |
| 26.5 | 34,000 | Man |
| 26.5 | 34,000 | Man |
| 28 | 34,100 | Man |
| 30 | 35,700 | Man |
| 30 | 35,700 | Man |
| 30 | 35,700 | Man |
| 32 | 35,800 | Man |
| 33 | 36,900 | Man |
| 34 | 37,400 | Man |
| 35 | 38,100 | Man |
| 36 | 38,600 | Man |
| 37 | 38,700 | Man |

Separating the two groups, we have the following:

| Rank | Salary | Gender |
|------|--------|--------|
| 1 | $20,700 | Woman |
| 2 | 20,900 | Woman |
| 3 | 21,100 | Woman |
| 4 | 21,900 | Woman |
| 5 | 22,800 | Woman |
| 6 | 23,000 | Woman |
| 7 | 23,100 | Woman |
| 8 | 24,700 | Woman |
| 9 | 25,000 | Woman |
| 11 | 25,800 | Woman |
| 13.5 | 26,200 | Woman |
| 15 | 26,900 | Woman |
| 16 | 27,100 | Woman |
| 18 | 28,100 | Woman |
| 20 | 29,700 | Woman |

| Rank | Salary | Gender |
|------|--------|--------|
| 10 | 25,500 | Man |
| 12 | 26,100 | Man |
| 13.5 | 26,200 | Man |
| 17 | 27,300 | Man |
| 19 | 29,100 | Man |
| 21 | 30,300 | Man |
| 22 | 30,700 | Man |
| 23 | 32,100 | Man |
| 24 | 32,800 | Man |
| 25 | 33,300 | Man |
| 26.5 | 34,000 | Man |
| 26.5 | 34,000 | Man |
| 28 | 34,100 | Man |
| 30 | 35,700 | Man |
| 30 | 35,700 | Man |
| 30 | 35,700 | Man |
| 32 | 35,800 | Man |
| 33 | 36,900 | Man |
| 34 | 37,400 | Man |
| 35 | 38,100 | Man |
| 36 | 38,600 | Man |
| 37 | 38,700 | Man |

The average rank for the women employees is 9.23333, the average rank for the men is 25.65909, the difference in the average ranks is 16.42576, the standard error is 3.624481, and the test statistic is 4.53.

# CHAPTER 17

## Problems

2. **c.** You would expect 247.63, or 248, people. This is 46.2% of the 536 people to be looking for an economy car: $0.462 \times 536 = 247.63$.

   **d.** For the expected count, multiply the population reference proportion by the sample size. For family sedan, the reference proportion is 0.258, and the sample size is 536. So the expected count is $0.258 \times 536 = 138.288$.

| Type | Last Year's Percentages | Expected Count |
|------|------------------------|----------------|
| Family sedan | 25.8% | 138.288 |
| Economy car | 46.2 | 247.632 |
| Sports car | 8.1 | 43.416 |
| Van | 12.4 | 66.464 |
| Pickup | 7.5 | 40.200 |
| Total | 100 | 536 |

   **e.** The chi-squared statistic is 29.49. Here are the observed and expected counts:

| Type | This Week's Count | Expected Count |
|------|-------------------|----------------|
| Family sedan | 187 | 138.288 |
| Economy car | 206 | 247.632 |
| Sports car | 29 | 43.416 |

| Type | This Week's Count | Expected Count |
|---|---|---|
| Van | 72 | 66.464 |
| Pickup | 42 | 40.200 |
| Total | 536 | 536 |

Chi-squared

$$= (187 - 138.288)^2 / 138.288$$
$$+ (206 - 247.632)^2 / 247.632$$
$$+ (29 - 43.416)^2 / 43.416$$
$$+ (72 - 66.464)^2 / 66.464$$
$$+ (42 - 40.200)^2 / 40.200$$
$$= 17.159 + 6.999 + 4.787 + 0.461 + 0.081$$
$$= 29.49.$$

6. d. Chi-squared = 5.224.

# CHAPTER 18

## Problems

1. a. Pareto diagram. The Pareto diagram displays the problems in the order from most to least frequent so that you can focus attention on the most important problems.

b. The $R$ chart. The $R$ chart enables you to monitor the variability of the process, so you can modify it, if that is necessary. This is a problem because the engines come out different from one another.

6. a. Center line $= \overline{\overline{X}} = 56.31$. The control limits extend from 54.30 to 58.32, computed as $\overline{\overline{X}} - A_2\overline{R}$ to $\overline{\overline{X}} + A_2\overline{R}$ where $A_2 = 0.483$.

7. b. $\overline{\overline{X}} = 12.8423$ and $\overline{R} = 0.208$.

c. The center line is 12.8423.

d. The control limits extend from 12.630 to 13.055, computed as $\overline{\overline{X}} - A_2\overline{R}$ to $\overline{\overline{X}} + A_2\overline{R}$ where $A_2 = 1.023$.

8. a. Center line $= \overline{R} = 0.208$.

b. The control limits extend from 0 to 0.535, computed as $D_3\overline{R}$ to $D_4\overline{R}$ where $D_3 = 0$ and $D_4 = 2.574$.

11. a. Center line $= \overline{p} = 0.0731$. The control limits extend from 2.80% to 11.82%, computed as

$$\overline{p} - 3\sqrt{\frac{\overline{p}(1-\overline{p})}{n}} \quad \text{to} \quad \overline{p} + 3\sqrt{\frac{\overline{p}(1-\overline{p})}{n}}$$

c. Center line $= \pi_0 = 0.0350$. The control limits extend from 1.55% to 5.45%, computed as

$$\pi_0 - 3\sqrt{\frac{\pi_0(1-\pi_0)}{n}} \quad \text{to} \quad \pi_0 + 3\sqrt{\frac{\pi_0(1-\pi_0)}{n}}$$

# Statistical Tables

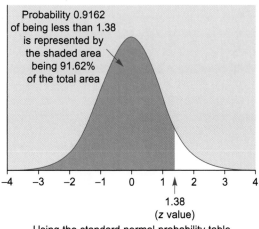

Probability 0.9162 of being less than 1.38 is represented by the shaded area being 91.62% of the total area

1.38
(z value)

Using the standard normal probability table

**TABLE D.1 Standard Normal Probability Table (See Figure on previous page)**

| z Value | Probability | z Value | Probability | z Value | Probability | Probability | z Value | Probability | z Value | Probability | z Value |
|---|---|---|---|---|---|---|---|---|---|---|---|
| -2.00 | 0.0228 | -1.00 | 0.1587 | 0.00 | 0.5000 | 0.5000 | 0.00 | 0.8413 | 1.00 | 0.9772 | 2.00 |
| -2.01 | 0.0222 | -1.01 | 0.1562 | -0.01 | 0.4960 | 0.5040 | 0.01 | 0.8438 | 1.01 | 0.9778 | 2.01 |
| -2.02 | 0.0217 | -1.02 | 0.1539 | -0.02 | 0.4920 | 0.5080 | 0.02 | 0.8461 | 1.02 | 0.9783 | 2.02 |
| -2.03 | 0.0212 | -1.03 | 0.1515 | -0.03 | 0.4880 | 0.5120 | 0.03 | 0.8485 | 1.03 | 0.9788 | 2.03 |
| -2.04 | 0.0207 | -1.04 | 0.1492 | -0.04 | 0.4840 | 0.5160 | 0.04 | 0.8508 | 1.04 | 0.9793 | 2.04 |
| -2.05 | 0.0202 | -1.05 | 0.1469 | -0.05 | 0.4801 | 0.5199 | 0.05 | 0.8531 | 1.05 | 0.9798 | 2.05 |
| -2.06 | 0.0197 | -1.06 | 0.1446 | -0.06 | 0.4761 | 0.5239 | 0.06 | 0.8554 | 1.06 | 0.9803 | 2.06 |
| -2.07 | 0.0192 | -1.07 | 0.1423 | -0.07 | 0.4721 | 0.5279 | 0.07 | 0.8577 | 1.07 | 0.9808 | 2.07 |
| -2.08 | 0.0188 | -1.08 | 0.1401 | -0.08 | 0.4681 | 0.5319 | 0.08 | 0.8599 | 1.08 | 0.9812 | 2.08 |
| -2.09 | 0.0183 | -1.09 | 0.1379 | -0.09 | 0.4641 | 0.5359 | 0.09 | 0.8621 | 1.09 | 0.9817 | 2.09 |
| -2.10 | 0.0179 | -1.10 | 0.1357 | -0.10 | 0.4602 | 0.5398 | 0.10 | 0.8643 | 1.10 | 0.9821 | 2.10 |
| -2.11 | 0.0174 | -1.11 | 0.1335 | -0.11 | 0.4562 | 0.5438 | 0.11 | 0.8665 | 1.11 | 0.9826 | 2.11 |
| -2.12 | 0.0170 | -1.12 | 0.1314 | -0.12 | 0.4522 | 0.5478 | 0.12 | 0.8686 | 1.12 | 0.9830 | 2.12 |
| -2.13 | 0.0166 | -1.13 | 0.1292 | -0.13 | 0.4483 | 0.5517 | 0.13 | 0.8708 | 1.13 | 0.9834 | 2.13 |
| -2.14 | 0.0162 | -1.14 | 0.1271 | -0.14 | 0.4443 | 0.5557 | 0.14 | 0.8729 | 1.14 | 0.9838 | 2.14 |
| -2.15 | 0.0158 | -1.15 | 0.1251 | -0.15 | 0.4404 | 0.5596 | 0.15 | 0.8749 | 1.15 | 0.9842 | 2.15 |
| -2.16 | 0.0154 | -1.16 | 0.1230 | -0.16 | 0.4364 | 0.5636 | 0.16 | 0.8770 | 1.16 | 0.9846 | 2.16 |
| -2.17 | 0.0150 | -1.17 | 0.1210 | -0.17 | 0.4325 | 0.5675 | 0.17 | 0.8790 | 1.17 | 0.9850 | 2.17 |
| -2.18 | 0.0146 | -1.18 | 0.1190 | -0.18 | 0.4286 | 0.5714 | 0.18 | 0.8810 | 1.18 | 0.9854 | 2.18 |
| -2.19 | 0.0143 | -1.19 | 0.1170 | -0.19 | 0.4247 | 0.5753 | 0.19 | 0.8830 | 1.19 | 0.9857 | 2.19 |
| -2.20 | 0.0139 | -1.20 | 0.1151 | -0.20 | 0.4207 | 0.5793 | 0.20 | 0.8849 | 1.20 | 0.9861 | 2.20 |
| -2.21 | 0.0136 | -1.21 | 0.1131 | -0.21 | 0.4168 | 0.5832 | 0.21 | 0.8869 | 1.21 | 0.9864 | 2.21 |
| -2.22 | 0.0132 | -1.22 | 0.1112 | -0.22 | 0.4129 | 0.5871 | 0.22 | 0.8888 | 1.22 | 0.9868 | 2.22 |
| -2.23 | 0.0129 | -1.23 | 0.1093 | -0.23 | 0.4090 | 0.5910 | 0.23 | 0.8907 | 1.23 | 0.9871 | 2.23 |
| -2.24 | 0.0125 | -1.24 | 0.1075 | -0.24 | 0.4052 | 0.5948 | 0.24 | 0.8925 | 1.24 | 0.9875 | 2.24 |
| -2.25 | 0.0122 | -1.25 | 0.1056 | -0.25 | 0.4013 | 0.5987 | 0.25 | 0.8944 | 1.25 | 0.9878 | 2.25 |

| z | Area | z | Area | z | Area | z | Area | z | Area | z | Area |
|---|---|---|---|---|---|---|---|---|---|---|---|
| 2.26 | 0.9881 | 1.26 | 0.8962 | 0.26 | 0.6026 | −0.26 | 0.3974 | −1.26 | 0.1038 | −2.26 | 0.0119 |
| 2.27 | 0.9884 | 1.27 | 0.8980 | 0.27 | 0.6064 | −0.27 | 0.3936 | −1.27 | 0.1020 | −2.27 | 0.0116 |
| 2.28 | 0.9887 | 1.28 | 0.8997 | 0.28 | 0.6103 | −0.28 | 0.3897 | −1.28 | 0.1003 | −2.28 | 0.0113 |
| 2.29 | 0.9890 | 1.29 | 0.9015 | 0.29 | 0.6141 | −0.29 | 0.3859 | −1.29 | 0.0985 | −2.29 | 0.0110 |
| 2.30 | 0.9893 | 1.30 | 0.9032 | 0.30 | 0.6179 | −0.30 | 0.3821 | −1.30 | 0.0968 | −2.30 | 0.0107 |
| 2.31 | 0.9896 | 1.31 | 0.9049 | 0.31 | 0.6217 | −0.31 | 0.3783 | −1.31 | 0.0951 | −2.31 | 0.0104 |
| 2.32 | 0.9898 | 1.32 | 0.9066 | 0.32 | 0.6255 | −0.32 | 0.3745 | −1.32 | 0.0934 | −2.32 | 0.0102 |
| 2.33 | 0.9901 | 1.33 | 0.9082 | 0.33 | 0.6293 | −0.33 | 0.3707 | −1.33 | 0.0918 | −2.33 | 0.0099 |
| 2.34 | 0.9904 | 1.34 | 0.9099 | 0.34 | 0.6331 | −0.34 | 0.3669 | −1.34 | 0.0901 | −2.34 | 0.0096 |
| 2.35 | 0.9906 | 1.35 | 0.9115 | 0.35 | 0.6368 | −0.35 | 0.3632 | −1.35 | 0.0885 | −2.35 | 0.0094 |
| 2.36 | 0.9909 | 1.36 | 0.9131 | 0.36 | 0.6406 | −0.36 | 0.3594 | −1.36 | 0.0869 | −2.36 | 0.0091 |
| 2.37 | 0.9911 | 1.37 | 0.9147 | 0.37 | 0.6443 | −0.37 | 0.3557 | −1.37 | 0.0853 | −2.37 | 0.0089 |
| 2.38 | 0.9913 | 1.38 | 0.9162 | 0.38 | 0.6480 | −0.38 | 0.3520 | −1.38 | 0.0838 | −2.38 | 0.0087 |
| 2.39 | 0.9916 | 1.39 | 0.9177 | 0.39 | 0.6517 | −0.39 | 0.3483 | −1.39 | 0.0823 | −2.39 | 0.0084 |
| 2.40 | 0.9918 | 1.40 | 0.9192 | 0.40 | 0.6554 | −0.40 | 0.3446 | −1.40 | 0.0808 | −2.40 | 0.0082 |
| 2.41 | 0.9920 | 1.41 | 0.9207 | 0.41 | 0.6591 | −0.41 | 0.3409 | −1.41 | 0.0793 | −2.41 | 0.0080 |
| 2.42 | 0.9922 | 1.42 | 0.9222 | 0.42 | 0.6628 | −0.42 | 0.3372 | −1.42 | 0.0778 | −2.42 | 0.0078 |
| 2.43 | 0.9925 | 1.43 | 0.9236 | 0.43 | 0.6664 | −0.43 | 0.3336 | −1.43 | 0.0764 | −2.43 | 0.0075 |
| 2.44 | 0.9927 | 1.44 | 0.9251 | 0.44 | 0.6700 | −0.44 | 0.3300 | −1.44 | 0.0749 | −2.44 | 0.0073 |
| 2.45 | 0.9929 | 1.45 | 0.9265 | 0.45 | 0.6736 | −0.45 | 0.3264 | −1.45 | 0.0735 | −2.45 | 0.0071 |
| 2.46 | 0.9931 | 1.46 | 0.9279 | 0.46 | 0.6772 | −0.46 | 0.3228 | −1.46 | 0.0721 | −2.46 | 0.0069 |
| 2.47 | 0.9932 | 1.47 | 0.9292 | 0.47 | 0.6808 | −0.47 | 0.3192 | −1.47 | 0.0708 | −2.47 | 0.0068 |
| 2.48 | 0.9934 | 1.48 | 0.9306 | 0.48 | 0.6844 | −0.48 | 0.3156 | −1.48 | 0.0694 | −2.48 | 0.0066 |
| 2.49 | 0.9936 | 1.49 | 0.9319 | 0.49 | 0.6879 | −0.49 | 0.3121 | −1.49 | 0.0681 | −2.49 | 0.0064 |
| 2.50 | 0.9938 | 1.50 | 0.9332 | 0.50 | 0.6915 | −0.50 | 0.3085 | −1.50 | 0.0668 | −2.50 | 0.0062 |
| 2.51 | 0.9940 | 1.51 | 0.9345 | 0.51 | 0.6950 | −0.51 | 0.3050 | −1.51 | 0.0655 | −2.51 | 0.0060 |
| 2.52 | 0.9941 | 1.52 | 0.9357 | 0.52 | 0.6985 | −0.52 | 0.3015 | −1.52 | 0.0643 | −2.52 | 0.0059 |
| 2.53 | 0.9943 | 1.53 | 0.9370 | 0.53 | 0.7019 | −0.53 | 0.2981 | −1.53 | 0.0630 | −2.53 | 0.0057 |

*(Continued)*

**TABLE D.1** Standard Normal Probability Table—cont'd

| z Value | Probability | z Value | Probability | z Value | Probability | z Value | Probability | z Value | Probability | z Value | Probability |
|---------|-------------|---------|-------------|---------|-------------|---------|-------------|---------|-------------|---------|-------------|
| -2.54 | 0.0055 | -1.54 | 0.0618 | -0.54 | 0.2946 | 0.54 | 0.7054 | 1.54 | 0.9382 | 2.54 | 0.9945 |
| -2.55 | 0.0054 | -1.55 | 0.0606 | -0.55 | 0.2912 | 0.55 | 0.7088 | 1.55 | 0.9394 | 2.55 | 0.9946 |
| -2.56 | 0.0052 | -1.56 | 0.0594 | -0.56 | 0.2877 | 0.56 | 0.7123 | 1.56 | 0.9406 | 2.56 | 0.9948 |
| -2.57 | 0.0051 | -1.57 | 0.0582 | -0.57 | 0.2843 | 0.57 | 0.7157 | 1.57 | 0.9418 | 2.57 | 0.9949 |
| -2.58 | 0.0049 | -1.58 | 0.0571 | -0.58 | 0.2810 | 0.58 | 0.7190 | 1.58 | 0.9429 | 2.58 | 0.9951 |
| -2.59 | 0.0048 | -1.59 | 0.0559 | -0.59 | 0.2776 | 0.59 | 0.7224 | 1.59 | 0.9441 | 2.59 | 0.9952 |
| -2.60 | 0.0047 | -1.60 | 0.0548 | -0.60 | 0.2743 | 0.60 | 0.7257 | 1.60 | 0.9452 | 2.60 | 0.9953 |
| -2.61 | 0.0045 | -1.61 | 0.0537 | -0.61 | 0.2709 | 0.61 | 0.7291 | 1.61 | 0.9463 | 2.61 | 0.9955 |
| -2.62 | 0.0044 | -1.62 | 0.0526 | -0.62 | 0.2676 | 0.62 | 0.7324 | 1.62 | 0.9474 | 2.62 | 0.9956 |
| -2.63 | 0.0043 | -1.63 | 0.0516 | -0.63 | 0.2643 | 0.63 | 0.7357 | 1.63 | 0.9484 | 2.63 | 0.9957 |
| -2.64 | 0.0041 | -1.64 | 0.0505 | -0.64 | 0.2611 | 0.64 | 0.7389 | 1.64 | 0.9495 | 2.64 | 0.9959 |
| -2.65 | 0.0040 | -1.65 | 0.0495 | -0.65 | 0.2578 | 0.65 | 0.7422 | 1.65 | 0.9505 | 2.65 | 0.9960 |
| -2.66 | 0.0039 | -1.66 | 0.0485 | -0.66 | 0.2546 | 0.66 | 0.7454 | 1.66 | 0.9515 | 2.66 | 0.9961 |
| -2.67 | 0.0038 | -1.67 | 0.0475 | -0.67 | 0.2514 | 0.67 | 0.7486 | 1.67 | 0.9525 | 2.67 | 0.9962 |
| -2.68 | 0.0037 | -1.68 | 0.0465 | -0.68 | 0.2483 | 0.68 | 0.7517 | 1.68 | 0.9535 | 2.68 | 0.9963 |
| -2.69 | 0.0036 | -1.69 | 0.0455 | -0.69 | 0.2451 | 0.69 | 0.7549 | 1.69 | 0.9545 | 2.69 | 0.9964 |
| -2.70 | 0.0035 | -1.70 | 0.0446 | -0.70 | 0.2420 | 0.70 | 0.7580 | 1.70 | 0.9554 | 2.70 | 0.9965 |
| -2.71 | 0.0034 | -1.71 | 0.0436 | -0.71 | 0.2389 | 0.71 | 0.7611 | 1.71 | 0.9564 | 2.71 | 0.9966 |
| -2.72 | 0.0033 | -1.72 | 0.0427 | -0.72 | 0.2358 | 0.72 | 0.7642 | 1.72 | 0.9573 | 2.72 | 0.9967 |
| -2.73 | 0.0032 | -1.73 | 0.0418 | -0.73 | 0.2327 | 0.73 | 0.7673 | 1.73 | 0.9582 | 2.73 | 0.9968 |
| -2.74 | 0.0031 | -1.74 | 0.0409 | -0.74 | 0.2296 | 0.74 | 0.7704 | 1.74 | 0.9591 | 2.74 | 0.9969 |
| -2.75 | 0.0030 | -1.75 | 0.0401 | -0.75 | 0.2266 | 0.75 | 0.7734 | 1.75 | 0.9599 | 2.75 | 0.9970 |
| -2.76 | 0.0029 | -1.76 | 0.0392 | -0.76 | 0.2236 | 0.76 | 0.7764 | 1.76 | 0.9608 | 2.76 | 0.9971 |
| -2.77 | 0.0028 | -1.77 | 0.0384 | -0.77 | 0.2206 | 0.77 | 0.7794 | 1.77 | 0.9616 | 2.77 | 0.9972 |
| -2.78 | 0.0027 | -1.78 | 0.0375 | -0.78 | 0.2177 | 0.78 | 0.7823 | 1.78 | 0.9625 | 2.78 | 0.9973 |
| -2.79 | 0.0026 | -1.79 | 0.0367 | -0.79 | 0.2148 | 0.79 | 0.7852 | 1.79 | 0.9633 | 2.79 | 0.9974 |

| z | Area | z | Area | z | Area | z | Area | z | Area | z | Area |
|---|---|---|---|---|---|---|---|---|---|---|---|
| -2.80 | 0.0026 | -1.80 | 0.0359 | -0.80 | 0.2119 | 0.80 | 0.7881 | 1.80 | 0.9641 | 2.80 | 0.9974 |
| -2.81 | 0.0025 | -1.81 | 0.0351 | -0.81 | 0.2090 | 0.81 | 0.7910 | 1.81 | 0.9649 | 2.81 | 0.9975 |
| -2.82 | 0.0024 | -1.82 | 0.0344 | -0.82 | 0.2061 | 0.82 | 0.7939 | 1.82 | 0.9656 | 2.82 | 0.9976 |
| -2.83 | 0.0023 | -1.83 | 0.0336 | -0.83 | 0.2033 | 0.83 | 0.7967 | 1.83 | 0.9664 | 2.83 | 0.9977 |
| -2.84 | 0.0023 | -1.84 | 0.0329 | -0.84 | 0.2005 | 0.84 | 0.7995 | 1.84 | 0.9671 | 2.84 | 0.9977 |
| -2.85 | 0.0022 | -1.85 | 0.0322 | -0.85 | 0.1977 | 0.85 | 0.8023 | 1.85 | 0.9678 | 2.85 | 0.9978 |
| -2.86 | 0.0021 | -1.86 | 0.0314 | -0.86 | 0.1949 | 0.86 | 0.8051 | 1.86 | 0.9686 | 2.86 | 0.9979 |
| -2.87 | 0.0021 | -1.87 | 0.0307 | -0.87 | 0.1922 | 0.87 | 0.8078 | 1.87 | 0.9693 | 2.87 | 0.9979 |
| -2.88 | 0.0020 | -1.88 | 0.0301 | -0.88 | 0.1894 | 0.88 | 0.8106 | 1.88 | 0.9699 | 2.88 | 0.9980 |
| -2.89 | 0.0019 | -1.89 | 0.0294 | -0.89 | 0.1867 | 0.89 | 0.8133 | 1.89 | 0.9706 | 2.89 | 0.9981 |
| -2.90 | 0.0019 | -1.90 | 0.0287 | -0.90 | 0.1841 | 0.90 | 0.8159 | 1.90 | 0.9713 | 2.90 | 0.9981 |
| -2.91 | 0.0018 | -1.91 | 0.0281 | -0.91 | 0.1814 | 0.91 | 0.8186 | 1.91 | 0.9719 | 2.91 | 0.9982 |
| -2.92 | 0.0018 | -1.92 | 0.0274 | -0.92 | 0.1788 | 0.92 | 0.8212 | 1.92 | 0.9726 | 2.92 | 0.9982 |
| -2.93 | 0.0017 | -1.93 | 0.0268 | -0.93 | 0.1762 | 0.93 | 0.8238 | 1.93 | 0.9732 | 2.93 | 0.9983 |
| -2.94 | 0.0016 | -1.94 | 0.0262 | -0.94 | 0.1736 | 0.94 | 0.8264 | 1.94 | 0.9738 | 2.94 | 0.9984 |
| -2.95 | 0.0016 | -1.95 | 0.0256 | -0.95 | 0.1711 | 0.95 | 0.8289 | 1.95 | 0.9744 | 2.95 | 0.9984 |
| -2.96 | 0.0015 | -1.96 | 0.0250 | -0.96 | 0.1685 | 0.96 | 0.8315 | 1.96 | 0.9750 | 2.96 | 0.9985 |
| -2.97 | 0.0015 | -1.97 | 0.0244 | -0.97 | 0.1660 | 0.97 | 0.8340 | 1.97 | 0.9756 | 2.97 | 0.9985 |
| -2.98 | 0.0014 | -1.98 | 0.0239 | -0.98 | 0.1635 | 0.98 | 0.8365 | 1.98 | 0.9761 | 2.98 | 0.9986 |
| -2.99 | 0.0014 | -1.99 | 0.0233 | -0.99 | 0.1611 | 0.99 | 0.8389 | 1.99 | 0.9767 | 2.99 | 0.9986 |
| -3.00 | 0.0013 | -2.00 | 0.0228 | -1.00 | 0.1587 | 1.00 | 0.8413 | 2.00 | 0.9772 | 3.00 | 0.9987 |

## TABLE D.2 Table of Random Digits

|    | 1     | 2     | 3     | 4     | 5     | 6     | 7     | 8     | 9     | 10    |
|----|-------|-------|-------|-------|-------|-------|-------|-------|-------|-------|
| 1  | 51449 | 39284 | 85527 | 67168 | 91284 | 19954 | 91166 | 70918 | 85957 | 19492 |
| 2  | 16144 | 56830 | 67507 | 97275 | 25982 | 69294 | 32841 | 20861 | 83114 | 12531 |
| 3  | 48145 | 48280 | 99481 | 13050 | 81818 | 25282 | 66466 | 24461 | 97021 | 21072 |
| 4  | 83780 | 48351 | 85422 | 42978 | 26088 | 17869 | 94245 | 26622 | 48318 | 73850 |
| 5  | 95329 | 38482 | 93510 | 39170 | 63683 | 40587 | 80451 | 43058 | 81923 | 97072 |
| 6  | 11179 | 69004 | 34273 | 36062 | 26234 | 58601 | 47159 | 82248 | 95968 | 99722 |
| 7  | 94631 | 52413 | 31524 | 02316 | 27611 | 15888 | 13525 | 43809 | 40014 | 30667 |
| 8  | 64275 | 10294 | 35027 | 25604 | 65695 | 36014 | 17988 | 02734 | 31732 | 29911 |
| 9  | 72125 | 19232 | 10782 | 30615 | 42005 | 90419 | 32447 | 53688 | 36125 | 28456 |
| 10 | 16463 | 42028 | 27927 | 48403 | 88963 | 79615 | 41218 | 43290 | 53618 | 68082 |
| 11 | 10036 | 66273 | 69506 | 19610 | 01479 | 92338 | 55140 | 81097 | 73071 | 61544 |
| 12 | 85356 | 51400 | 88502 | 98267 | 73943 | 25828 | 38219 | 13268 | 09016 | 77465 |
| 13 | 84076 | 82087 | 55053 | 75370 | 71030 | 92275 | 55497 | 97123 | 40919 | 57479 |
| 14 | 76731 | 39755 | 78537 | 51937 | 11680 | 78820 | 50082 | 56068 | 36908 | 55399 |
| 15 | 19032 | 73472 | 79399 | 05549 | 14772 | 32746 | 38841 | 45524 | 13535 | 03113 |
| 16 | 72791 | 59040 | 61529 | 74437 | 74482 | 76619 | 05232 | 28616 | 98690 | 24011 |
| 17 | 11553 | 00135 | 28306 | 65571 | 34465 | 47423 | 39198 | 54456 | 95283 | 54637 |
| 18 | 71405 | 70352 | 46763 | 64002 | 62461 | 41982 | 15933 | 46942 | 36941 | 93412 |
| 19 | 17594 | 10116 | 55483 | 96219 | 85493 | 96955 | 89180 | 59690 | 82170 | 77643 |
| 20 | 09584 | 23476 | 09243 | 65568 | 89128 | 36747 | 63692 | 09986 | 47687 | 46448 |
| 21 | 81677 | 62634 | 52794 | 01466 | 85938 | 14565 | 79993 | 44956 | 82254 | 65223 |
| 22 | 45849 | 01177 | 13773 | 43523 | 69825 | 03222 | 58458 | 77463 | 58521 | 07273 |
| 23 | 97252 | 92257 | 90419 | 01241 | 52516 | 66293 | 14536 | 23870 | 78402 | 41759 |
| 24 | 26232 | 77422 | 76289 | 57587 | 42831 | 87047 | 20092 | 92676 | 12017 | 43554 |
| 25 | 87799 | 33602 | 01931 | 66913 | 63008 | 03745 | 93939 | 07178 | 70003 | 18158 |
| 26 | 46120 | 62298 | 69126 | 07862 | 76731 | 58527 | 39342 | 42749 | 57050 | 91725 |
| 27 | 53292 | 55652 | 11834 | 47581 | 25682 | 64085 | 26587 | 92289 | 41853 | 38354 |
| 28 | 81606 | 56009 | 06021 | 98392 | 40450 | 87721 | 50917 | 16978 | 39472 | 23505 |
| 29 | 67819 | 47314 | 96988 | 89931 | 49395 | 37071 | 72658 | 53947 | 11996 | 64631 |
| 30 | 50458 | 20350 | 87362 | 83996 | 86422 | 58694 | 71813 | 97695 | 28804 | 58523 |
| 31 | 59772 | 27000 | 97805 | 25042 | 09916 | 77569 | 71347 | 62667 | 09330 | 02152 |
| 32 | 94752 | 91056 | 08939 | 93410 | 59204 | 04644 | 44336 | 55570 | 21106 | 76588 |
| 33 | 01885 | 82054 | 45944 | 55398 | 55487 | 56455 | 56940 | 68787 | 36591 | 29914 |
| 34 | 85190 | 91941 | 86714 | 76593 | 77199 | 39724 | 99548 | 13827 | 84961 | 76740 |
| 35 | 97747 | 67607 | 14549 | 08215 | 95408 | 46381 | 12449 | 03672 | 40325 | 77312 |

| 36 | 43318 | 84469 | 26047 | 86003 | 34786 | 38931 | 34846 | 28711 | 42833 | 93019 |
| 37 | 47874 | 71365 | 76603 | 57440 | 49514 | 17335 | 71969 | 58055 | 99136 | 73589 |
| 38 | 24259 | 48079 | 71198 | 95859 | 94212 | 55402 | 93392 | 31965 | 94622 | 11673 |
| 39 | 31947 | 64805 | 34133 | 03245 | 24546 | 48934 | 41730 | 47831 | 26531 | 02203 |
| 40 | 37911 | 93224 | 87153 | 54541 | 57529 | 38299 | 65659 | 00202 | 07054 | 40168 |
| 41 | 82714 | 15799 | 93126 | 74180 | 94171 | 97117 | 31431 | 00323 | 62793 | 11995 |
| 42 | 82927 | 37884 | 74411 | 45887 | 36713 | 52339 | 68421 | 35968 | 67714 | 05883 |
| 43 | 65934 | 21782 | 35804 | 36676 | 35404 | 69987 | 52268 | 19894 | 81977 | 87764 |
| 44 | 56953 | 04356 | 68903 | 21369 | 35901 | 86797 | 83901 | 68681 | 02397 | 55359 |
| 45 | 16278 | 17165 | 67843 | 49349 | 90163 | 97337 | 35003 | 34915 | 91485 | 33814 |
| 46 | 96339 | 95028 | 48468 | 12279 | 81039 | 56531 | 10759 | 19579 | 00015 | 22829 |
| 47 | 84110 | 49661 | 13988 | 75909 | 35580 | 18426 | 29038 | 79111 | 56049 | 96451 |
| 48 | 49017 | 60748 | 03412 | 09880 | 94091 | 90052 | 43596 | 21424 | 16584 | 67970 |
| 49 | 43560 | 05552 | 54344 | 69418 | 01327 | 07771 | 25364 | 77373 | 34841 | 75927 |
| 50 | 25206 | 15177 | 63049 | 12464 | 16149 | 18759 | 96184 | 15968 | 89446 | 07168 |

## TABLE D.3 Table of Binomial Probabilities*

| N | a | π = 0.05 Exact | Sum | π = 0.10 Exact | Sum | π = 0.20 Exact | Sum | π = 0.30 Exact | Sum | π = 0.40 Exact | Sum | π = 0.50 Exact | Sum | π = 0.60 Exact | Sum | π = 0.70 Exact | Sum | π = 0.80 Exact | Sum | π = 0.90 Exact | Sum | π = 0.95 Exact | Sum |
|---|---|---|---|---|---|---|---|---|---|---|---|---|---|---|---|---|---|---|---|---|---|---|---|
| 1 | 0 | 0.950 | 0.950 | 0.900 | 0.900 | 0.800 | 0.800 | 0.700 | 0.700 | 0.600 | 0.600 | 0.500 | 0.500 | 0.400 | 0.400 | 0.300 | 0.300 | 0.200 | 0.200 | 0.100 | 0.100 | 0.050 | 0.050 |
| 1 | 1 | 0.050 | 1.000 | 0.100 | 1.000 | 0.200 | 1.000 | 0.300 | 1.000 | 0.400 | 1.000 | 0.500 | 1.000 | 0.600 | 1.000 | 0.700 | 1.000 | 0.800 | 1.000 | 0.900 | 1.000 | 0.950 | 1.000 |
| 2 | 0 | 0.903 | 0.903 | 0.810 | 0.810 | 0.640 | 0.640 | 0.490 | 0.490 | 0.360 | 0.360 | 0.250 | 0.250 | 0.160 | 0.160 | 0.090 | 0.090 | 0.040 | 0.040 | 0.010 | 0.010 | 0.003 | 0.003 |
| 2 | 1 | 0.095 | 0.998 | 0.180 | 0.990 | 0.320 | 0.960 | 0.420 | 0.910 | 0.480 | 0.840 | 0.500 | 0.750 | 0.480 | 0.640 | 0.420 | 0.510 | 0.320 | 0.360 | 0.180 | 0.190 | 0.095 | 0.098 |
| 2 | 2 | 0.003 | 1.000 | 0.010 | 1.000 | 0.040 | 1.000 | 0.090 | 1.000 | 0.160 | 1.000 | 0.250 | 1.000 | 0.360 | 1.000 | 0.490 | 1.000 | 0.640 | 1.000 | 0.810 | 1.000 | 0.903 | 1.000 |
| 3 | 0 | 0.857 | 0.857 | 0.729 | 0.729 | 0.512 | 0.512 | 0.343 | 0.343 | 0.216 | 0.216 | 0.125 | 0.125 | 0.064 | 0.064 | 0.027 | 0.027 | 0.008 | 0.008 | 0.001 | 0.001 | 0.000 | 0.000 |
| 3 | 1 | 0.135 | 0.993 | 0.243 | 0.972 | 0.384 | 0.896 | 0.441 | 0.784 | 0.432 | 0.648 | 0.375 | 0.500 | 0.288 | 0.352 | 0.189 | 0.216 | 0.096 | 0.104 | 0.027 | 0.028 | 0.007 | 0.007 |
| 3 | 2 | 0.007 | 1.000 | 0.027 | 0.999 | 0.096 | 0.992 | 0.189 | 0.973 | 0.288 | 0.936 | 0.375 | 0.875 | 0.432 | 0.784 | 0.441 | 0.657 | 0.384 | 0.488 | 0.243 | 0.271 | 0.135 | 0.143 |
| 3 | 3 | 0.000 | 1.000 | 0.001 | 1.000 | 0.008 | 1.000 | 0.027 | 1.000 | 0.064 | 1.000 | 0.125 | 1.000 | 0.216 | 1.000 | 0.343 | 1.000 | 0.512 | 1.000 | 0.729 | 1.000 | 0.857 | 1.000 |
| 4 | 0 | 0.815 | 0.815 | 0.656 | 0.656 | 0.410 | 0.410 | 0.240 | 0.240 | 0.130 | 0.130 | 0.063 | 0.063 | 0.026 | 0.026 | 0.008 | 0.008 | 0.002 | 0.002 | 0.000 | 0.000 | 0.000 | 0.000 |
| 4 | 1 | 0.171 | 0.986 | 0.292 | 0.948 | 0.410 | 0.819 | 0.412 | 0.652 | 0.346 | 0.475 | 0.250 | 0.313 | 0.154 | 0.179 | 0.076 | 0.084 | 0.026 | 0.027 | 0.004 | 0.004 | 0.000 | 0.000 |
| 4 | 2 | 0.014 | 1.000 | 0.049 | 0.996 | 0.154 | 0.973 | 0.265 | 0.916 | 0.346 | 0.821 | 0.375 | 0.688 | 0.346 | 0.525 | 0.265 | 0.348 | 0.154 | 0.181 | 0.049 | 0.052 | 0.014 | 0.014 |
| 4 | 3 | 0.000 | 1.000 | 0.004 | 1.000 | 0.026 | 0.998 | 0.076 | 0.992 | 0.154 | 0.974 | 0.250 | 0.938 | 0.346 | 0.870 | 0.412 | 0.760 | 0.410 | 0.590 | 0.292 | 0.344 | 0.171 | 0.185 |
| 4 | 4 | 0.000 | 1.000 | 0.000 | 1.000 | 0.002 | 1.000 | 0.008 | 1.000 | 0.026 | 1.000 | 0.063 | 1.000 | 0.130 | 1.000 | 0.240 | 1.000 | 0.410 | 1.000 | 0.656 | 1.000 | 0.815 | 1.000 |
| 5 | 0 | 0.774 | 0.774 | 0.590 | 0.590 | 0.328 | 0.328 | 0.168 | 0.168 | 0.078 | 0.078 | 0.031 | 0.031 | 0.010 | 0.010 | 0.002 | 0.002 | 0.000 | 0.000 | 0.000 | 0.000 | 0.000 | 0.000 |
| 5 | 1 | 0.204 | 0.977 | 0.328 | 0.919 | 0.410 | 0.737 | 0.360 | 0.528 | 0.259 | 0.337 | 0.156 | 0.188 | 0.077 | 0.087 | 0.028 | 0.031 | 0.006 | 0.007 | 0.000 | 0.000 | 0.000 | 0.000 |
| 5 | 2 | 0.021 | 0.999 | 0.073 | 0.991 | 0.205 | 0.942 | 0.309 | 0.837 | 0.346 | 0.683 | 0.313 | 0.500 | 0.230 | 0.317 | 0.132 | 0.163 | 0.051 | 0.058 | 0.008 | 0.009 | 0.001 | 0.001 |
| 5 | 3 | 0.001 | 1.000 | 0.008 | 1.000 | 0.051 | 0.993 | 0.132 | 0.969 | 0.230 | 0.913 | 0.313 | 0.813 | 0.346 | 0.663 | 0.309 | 0.472 | 0.205 | 0.263 | 0.073 | 0.081 | 0.021 | 0.023 |
| 5 | 4 | 0.000 | 1.000 | 0.000 | 1.000 | 0.006 | 1.000 | 0.028 | 0.998 | 0.077 | 0.990 | 0.156 | 0.969 | 0.259 | 0.922 | 0.360 | 0.832 | 0.410 | 0.672 | 0.328 | 0.410 | 0.204 | 0.226 |
| 5 | 5 | 0.000 | 1.000 | 0.000 | 1.000 | 0.000 | 1.000 | 0.002 | 1.000 | 0.010 | 1.000 | 0.031 | 1.000 | 0.078 | 1.000 | 0.168 | 1.000 | 0.328 | 1.000 | 0.590 | 1.000 | 0.774 | 1.000 |
| 6 | 0 | 0.735 | 0.735 | 0.531 | 0.531 | 0.262 | 0.262 | 0.118 | 0.118 | 0.047 | 0.047 | 0.016 | 0.016 | 0.004 | 0.004 | 0.001 | 0.001 | 0.000 | 0.000 | 0.000 | 0.000 | 0.000 | 0.000 |
| 6 | 1 | 0.232 | 0.967 | 0.354 | 0.886 | 0.393 | 0.655 | 0.303 | 0.420 | 0.187 | 0.233 | 0.094 | 0.109 | 0.037 | 0.041 | 0.010 | 0.011 | 0.002 | 0.002 | 0.000 | 0.000 | 0.000 | 0.000 |
| 6 | 2 | 0.031 | 0.998 | 0.098 | 0.984 | 0.246 | 0.901 | 0.324 | 0.744 | 0.311 | 0.544 | 0.234 | 0.344 | 0.138 | 0.179 | 0.060 | 0.070 | 0.015 | 0.017 | 0.001 | 0.001 | 0.000 | 0.000 |
| 6 | 3 | 0.002 | 1.000 | 0.015 | 0.999 | 0.082 | 0.983 | 0.185 | 0.930 | 0.276 | 0.821 | 0.313 | 0.656 | 0.276 | 0.456 | 0.185 | 0.256 | 0.082 | 0.099 | 0.015 | 0.016 | 0.002 | 0.002 |
| 6 | 4 | 0.000 | 1.000 | 0.001 | 1.000 | 0.015 | 0.998 | 0.060 | 0.989 | 0.138 | 0.959 | 0.234 | 0.891 | 0.311 | 0.767 | 0.324 | 0.580 | 0.246 | 0.345 | 0.098 | 0.114 | 0.031 | 0.033 |
| 6 | 5 | 0.000 | 1.000 | 0.000 | 1.000 | 0.002 | 1.000 | 0.010 | 0.999 | 0.037 | 0.996 | 0.094 | 0.984 | 0.187 | 0.953 | 0.303 | 0.882 | 0.393 | 0.738 | 0.354 | 0.469 | 0.232 | 0.265 |
| 6 | 6 | 0.000 | 1.000 | 0.000 | 1.000 | 0.000 | 1.000 | 0.001 | 1.000 | 0.004 | 1.000 | 0.016 | 1.000 | 0.047 | 1.000 | 0.118 | 1.000 | 0.262 | 1.000 | 0.531 | 1.000 | 0.735 | 1.000 |
| 7 | 0 | 0.698 | 0.698 | 0.478 | 0.478 | 0.210 | 0.210 | 0.082 | 0.082 | 0.028 | 0.028 | 0.008 | 0.008 | 0.002 | 0.002 | 0.000 | 0.000 | 0.000 | 0.000 | 0.000 | 0.000 | 0.000 | 0.000 |
| 7 | 1 | 0.257 | 0.956 | 0.372 | 0.850 | 0.367 | 0.577 | 0.247 | 0.329 | 0.131 | 0.159 | 0.055 | 0.063 | 0.017 | 0.019 | 0.004 | 0.004 | 0.000 | 0.000 | 0.000 | 0.000 | 0.000 | 0.000 |

| n | x | | | | | | | | | | | | | | | | | | | | | | |
|---|---|---|---|---|---|---|---|---|---|---|---|---|---|---|---|---|---|---|---|---|---|---|---|
| 7 | 2 | 0.041 | 0.996 | 0.124 | 0.974 | 0.275 | 0.852 | 0.318 | 0.647 | 0.261 | 0.420 | 0.164 | 0.227 | 0.077 | 0.096 | 0.025 | 0.029 | 0.004 | 0.005 | 0.000 | 0.000 | 0.000 | 0.000 |
| 7 | 3 | 0.004 | 1.000 | 0.023 | 0.997 | 0.115 | 0.967 | 0.227 | 0.874 | 0.290 | 0.710 | 0.273 | 0.500 | 0.194 | 0.290 | 0.097 | 0.126 | 0.029 | 0.033 | 0.003 | 0.003 | 0.000 | 0.000 |
| 7 | 4 | 0.000 | 1.000 | 0.003 | 1.000 | 0.029 | 0.995 | 0.097 | 0.971 | 0.194 | 0.904 | 0.273 | 0.773 | 0.290 | 0.580 | 0.227 | 0.353 | 0.115 | 0.148 | 0.023 | 0.026 | 0.004 | 0.004 |
| 7 | 5 | 0.000 | 1.000 | 0.000 | 1.000 | 0.004 | 1.000 | 0.025 | 0.996 | 0.077 | 0.981 | 0.164 | 0.938 | 0.261 | 0.841 | 0.318 | 0.671 | 0.275 | 0.423 | 0.124 | 0.150 | 0.041 | 0.044 |
| 7 | 6 | 0.000 | 1.000 | 0.000 | 1.000 | 0.000 | 1.000 | 0.004 | 1.000 | 0.017 | 0.998 | 0.055 | 0.992 | 0.131 | 0.972 | 0.247 | 0.918 | 0.367 | 0.790 | 0.372 | 0.522 | 0.257 | 0.302 |
| 7 | 7 | 0.000 | 1.000 | 0.000 | 1.000 | 0.000 | 1.000 | 0.000 | 1.000 | 0.002 | 1.000 | 0.008 | 1.000 | 0.028 | 1.000 | 0.082 | 1.000 | 0.210 | 1.000 | 0.478 | 1.000 | 0.698 | 1.000 |
| 8 | 0 | 0.663 | 0.663 | 0.430 | 0.430 | 0.168 | 0.168 | 0.058 | 0.058 | 0.017 | 0.017 | 0.004 | 0.004 | 0.001 | 0.001 | 0.000 | 0.000 | 0.000 | 0.000 | 0.000 | 0.000 | 0.000 | 0.000 |
| 8 | 1 | 0.279 | 0.943 | 0.383 | 0.813 | 0.336 | 0.503 | 0.198 | 0.255 | 0.090 | 0.106 | 0.031 | 0.035 | 0.008 | 0.009 | 0.001 | 0.001 | 0.000 | 0.000 | 0.000 | 0.000 | 0.000 | 0.000 |
| 8 | 2 | 0.051 | 0.994 | 0.149 | 0.962 | 0.294 | 0.797 | 0.296 | 0.552 | 0.209 | 0.315 | 0.109 | 0.145 | 0.041 | 0.050 | 0.010 | 0.011 | 0.001 | 0.001 | 0.000 | 0.000 | 0.000 | 0.000 |
| 8 | 3 | 0.005 | 1.000 | 0.033 | 0.995 | 0.147 | 0.944 | 0.254 | 0.806 | 0.279 | 0.594 | 0.219 | 0.363 | 0.124 | 0.174 | 0.047 | 0.058 | 0.009 | 0.010 | 0.001 | 0.001 | 0.000 | 0.000 |
| 8 | 4 | 0.000 | 1.000 | 0.005 | 1.000 | 0.046 | 0.990 | 0.136 | 0.942 | 0.232 | 0.826 | 0.273 | 0.637 | 0.232 | 0.406 | 0.136 | 0.194 | 0.046 | 0.056 | 0.005 | 0.005 | 0.000 | 0.000 |
| 8 | 5 | 0.000 | 1.000 | 0.000 | 1.000 | 0.009 | 0.999 | 0.047 | 0.989 | 0.124 | 0.950 | 0.219 | 0.855 | 0.279 | 0.685 | 0.254 | 0.448 | 0.147 | 0.203 | 0.033 | 0.038 | 0.005 | 0.006 |
| 8 | 6 | 0.000 | 1.000 | 0.000 | 1.000 | 0.001 | 1.000 | 0.010 | 0.999 | 0.041 | 0.991 | 0.109 | 0.965 | 0.209 | 0.894 | 0.296 | 0.745 | 0.294 | 0.497 | 0.149 | 0.187 | 0.051 | 0.057 |
| 8 | 7 | 0.000 | 1.000 | 0.000 | 1.000 | 0.000 | 1.000 | 0.001 | 1.000 | 0.008 | 0.999 | 0.031 | 0.996 | 0.090 | 0.983 | 0.198 | 0.942 | 0.336 | 0.832 | 0.383 | 0.570 | 0.279 | 0.337 |
| 8 | 8 | 0.000 | 1.000 | 0.000 | 1.000 | 0.000 | 1.000 | 0.000 | 1.000 | 0.001 | 1.000 | 0.004 | 1.000 | 0.017 | 1.000 | 0.058 | 1.000 | 0.168 | 1.000 | 0.430 | 1.000 | 0.663 | 1.000 |
| 9 | 0 | 0.630 | 0.630 | 0.387 | 0.387 | 0.134 | 0.134 | 0.040 | 0.040 | 0.010 | 0.010 | 0.002 | 0.002 | 0.000 | 0.000 | 0.000 | 0.000 | 0.000 | 0.000 | 0.000 | 0.000 | 0.000 | 0.000 |
| 9 | 1 | 0.299 | 0.929 | 0.387 | 0.775 | 0.302 | 0.436 | 0.156 | 0.196 | 0.060 | 0.071 | 0.018 | 0.020 | 0.004 | 0.004 | 0.000 | 0.000 | 0.000 | 0.000 | 0.000 | 0.000 | 0.000 | 0.000 |
| 9 | 2 | 0.063 | 0.992 | 0.172 | 0.947 | 0.302 | 0.738 | 0.267 | 0.463 | 0.161 | 0.232 | 0.070 | 0.090 | 0.021 | 0.025 | 0.004 | 0.004 | 0.000 | 0.000 | 0.000 | 0.000 | 0.000 | 0.000 |
| 9 | 3 | 0.008 | 0.999 | 0.045 | 0.992 | 0.176 | 0.914 | 0.267 | 0.730 | 0.251 | 0.483 | 0.164 | 0.254 | 0.074 | 0.099 | 0.021 | 0.025 | 0.003 | 0.003 | 0.001 | 0.001 | 0.000 | 0.000 |
| 9 | 4 | 0.001 | 1.000 | 0.007 | 0.999 | 0.066 | 0.980 | 0.172 | 0.901 | 0.251 | 0.733 | 0.246 | 0.500 | 0.167 | 0.267 | 0.074 | 0.099 | 0.017 | 0.020 | 0.001 | 0.001 | 0.000 | 0.000 |
| 9 | 5 | 0.000 | 1.000 | 0.001 | 1.000 | 0.017 | 0.997 | 0.074 | 0.975 | 0.167 | 0.901 | 0.246 | 0.746 | 0.251 | 0.517 | 0.172 | 0.270 | 0.066 | 0.086 | 0.007 | 0.008 | 0.001 | 0.001 |
| 9 | 6 | 0.000 | 1.000 | 0.000 | 1.000 | 0.003 | 1.000 | 0.021 | 0.996 | 0.074 | 0.975 | 0.164 | 0.910 | 0.251 | 0.768 | 0.267 | 0.537 | 0.176 | 0.262 | 0.045 | 0.053 | 0.008 | 0.008 |
| 9 | 7 | 0.000 | 1.000 | 0.000 | 1.000 | 0.000 | 1.000 | 0.004 | 1.000 | 0.021 | 0.996 | 0.070 | 0.980 | 0.161 | 0.929 | 0.267 | 0.804 | 0.302 | 0.564 | 0.172 | 0.387 | 0.063 | 0.071 |
| 9 | 8 | 0.000 | 1.000 | 0.000 | 1.000 | 0.000 | 1.000 | 0.000 | 1.000 | 0.004 | 1.000 | 0.018 | 0.998 | 0.060 | 0.990 | 0.156 | 0.960 | 0.302 | 0.866 | 0.387 | 0.613 | 0.299 | 0.370 |
| 9 | 9 | 0.000 | 1.000 | 0.000 | 1.000 | 0.000 | 1.000 | 0.000 | 1.000 | 0.000 | 1.000 | 0.002 | 1.000 | 0.010 | 1.000 | 0.040 | 1.000 | 0.134 | 1.000 | 0.387 | 1.000 | 0.630 | 1.000 |
| 10 | 0 | 0.599 | 0.599 | 0.349 | 0.349 | 0.107 | 0.107 | 0.028 | 0.028 | 0.006 | 0.006 | 0.001 | 0.001 | 0.000 | 0.000 | 0.000 | 0.000 | 0.000 | 0.000 | 0.000 | 0.000 | 0.000 | 0.000 |
| 10 | 1 | 0.315 | 0.914 | 0.387 | 0.736 | 0.268 | 0.376 | 0.121 | 0.149 | 0.040 | 0.046 | 0.010 | 0.011 | 0.002 | 0.002 | 0.000 | 0.000 | 0.000 | 0.000 | 0.000 | 0.000 | 0.000 | 0.000 |
| 10 | 2 | 0.075 | 0.988 | 0.194 | 0.930 | 0.302 | 0.678 | 0.233 | 0.383 | 0.121 | 0.167 | 0.044 | 0.055 | 0.011 | 0.012 | 0.001 | 0.002 | 0.000 | 0.000 | 0.000 | 0.000 | 0.000 | 0.000 |
| 10 | 3 | 0.010 | 0.999 | 0.057 | 0.987 | 0.201 | 0.879 | 0.267 | 0.650 | 0.215 | 0.382 | 0.117 | 0.172 | 0.042 | 0.055 | 0.009 | 0.011 | 0.001 | 0.001 | 0.000 | 0.000 | 0.000 | 0.000 |
| 10 | 4 | 0.001 | 1.000 | 0.011 | 0.998 | 0.088 | 0.967 | 0.200 | 0.850 | 0.251 | 0.633 | 0.205 | 0.377 | 0.111 | 0.166 | 0.037 | 0.047 | 0.006 | 0.006 | 0.000 | 0.001 | 0.000 | 0.000 |
| 10 | 5 | 0.000 | 1.000 | 0.001 | 1.000 | 0.026 | 0.994 | 0.103 | 0.953 | 0.201 | 0.834 | 0.246 | 0.623 | 0.201 | 0.367 | 0.103 | 0.150 | 0.026 | 0.033 | 0.001 | 0.002 | 0.000 | 0.000 |
| 10 | 6 | 0.000 | 1.000 | 0.000 | 1.000 | 0.006 | 0.999 | 0.037 | 0.989 | 0.111 | 0.945 | 0.205 | 0.828 | 0.251 | 0.618 | 0.200 | 0.350 | 0.088 | 0.121 | 0.011 | 0.013 | 0.001 | 0.001 |

(Continued)

## TABLE D.3 Table of Binomial Probabilities*—cont'd

| N | a | π = 0.05 Exact | Sum | π = 0.10 Exact | Sum | π = 0.20 Exact | Sum | π = 0.30 Exact | Sum | π = 0.40 Exact | Sum | π = 0.50 Exact | Sum | π = 0.60 Exact | Sum | π = 0.70 Exact | Sum | π = 0.80 Exact | Sum | π = 0.90 Exact | Sum | π = 0.95 Exact | Sum |
|---|---|---|---|---|---|---|---|---|---|---|---|---|---|---|---|---|---|---|---|---|---|---|---|
| 10 | 7 | 0.000 | 1.000 | 0.000 | 1.000 | 0.001 | 1.000 | 0.009 | 0.998 | 0.042 | 0.988 | 0.117 | 0.945 | 0.215 | 0.833 | 0.267 | 0.617 | 0.201 | 0.322 | 0.057 | 0.070 | 0.010 | 0.012 |
| 10 | 8 | 0.000 | 1.000 | 0.000 | 1.000 | 0.000 | 1.000 | 0.001 | 1.000 | 0.011 | 0.998 | 0.044 | 0.989 | 0.121 | 0.954 | 0.233 | 0.851 | 0.302 | 0.624 | 0.194 | 0.264 | 0.075 | 0.086 |
| 10 | 9 | 0.000 | 1.000 | 0.000 | 1.000 | 0.000 | 1.000 | 0.000 | 1.000 | 0.002 | 1.000 | 0.010 | 0.999 | 0.040 | 0.994 | 0.121 | 0.972 | 0.268 | 0.893 | 0.387 | 0.651 | 0.315 | 0.401 |
| 10 | 10 | 0.000 | 1.000 | 0.000 | 1.000 | 0.000 | 1.000 | 0.000 | 1.000 | 0.000 | 1.000 | 0.001 | 1.000 | 0.006 | 1.000 | 0.028 | 1.000 | 0.107 | 1.000 | 0.349 | 1.000 | 0.599 | 1.000 |
| 11 | 0 | 0.569 | 0.569 | 0.314 | 0.314 | 0.086 | 0.086 | 0.020 | 0.020 | 0.004 | 0.004 | 0.000 | 0.000 | 0.000 | 0.000 | 0.000 | 0.000 | 0.000 | 0.000 | 0.000 | 0.000 | 0.000 | 0.000 |
| 11 | 1 | 0.329 | 0.898 | 0.384 | 0.697 | 0.236 | 0.322 | 0.093 | 0.113 | 0.027 | 0.030 | 0.005 | 0.006 | 0.001 | 0.001 | 0.000 | 0.000 | 0.000 | 0.000 | 0.000 | 0.000 | 0.000 | 0.000 |
| 11 | 2 | 0.087 | 0.985 | 0.213 | 0.910 | 0.295 | 0.617 | 0.200 | 0.313 | 0.089 | 0.119 | 0.027 | 0.033 | 0.005 | 0.006 | 0.001 | 0.001 | 0.000 | 0.000 | 0.000 | 0.000 | 0.000 | 0.000 |
| 11 | 3 | 0.014 | 0.998 | 0.071 | 0.981 | 0.221 | 0.839 | 0.257 | 0.570 | 0.177 | 0.296 | 0.081 | 0.113 | 0.023 | 0.029 | 0.004 | 0.004 | 0.000 | 0.002 | 0.000 | 0.000 | 0.000 | 0.000 |
| 11 | 4 | 0.001 | 1.000 | 0.016 | 0.997 | 0.111 | 0.950 | 0.220 | 0.790 | 0.236 | 0.533 | 0.161 | 0.274 | 0.070 | 0.099 | 0.017 | 0.022 | 0.002 | 0.012 | 0.000 | 0.000 | 0.000 | 0.000 |
| 11 | 5 | 0.000 | 1.000 | 0.002 | 1.000 | 0.039 | 0.988 | 0.132 | 0.922 | 0.221 | 0.753 | 0.226 | 0.500 | 0.147 | 0.247 | 0.057 | 0.078 | 0.010 | 0.050 | 0.000 | 0.001 | 0.000 | 0.000 |
| 11 | 6 | 0.000 | 1.000 | 0.000 | 1.000 | 0.010 | 0.998 | 0.057 | 0.978 | 0.147 | 0.901 | 0.226 | 0.726 | 0.221 | 0.467 | 0.132 | 0.210 | 0.039 | 0.161 | 0.002 | 0.003 | 0.001 | 0.002 |
| 11 | 7 | 0.000 | 1.000 | 0.000 | 1.000 | 0.002 | 1.000 | 0.017 | 0.996 | 0.070 | 0.971 | 0.161 | 0.887 | 0.236 | 0.704 | 0.220 | 0.430 | 0.111 | 0.383 | 0.016 | 0.019 | 0.014 | 0.015 |
| 11 | 8 | 0.000 | 1.000 | 0.000 | 1.000 | 0.000 | 1.000 | 0.004 | 0.999 | 0.023 | 0.994 | 0.081 | 0.967 | 0.177 | 0.881 | 0.257 | 0.687 | 0.221 | 0.678 | 0.071 | 0.090 | 0.087 | 0.102 |
| 11 | 9 | 0.000 | 1.000 | 0.000 | 1.000 | 0.000 | 1.000 | 0.001 | 1.000 | 0.005 | 0.999 | 0.027 | 0.994 | 0.089 | 0.970 | 0.200 | 0.887 | 0.295 | 0.914 | 0.213 | 0.303 | 0.329 | 0.431 |
| 11 | 10 | 0.000 | 1.000 | 0.000 | 1.000 | 0.000 | 1.000 | 0.000 | 1.000 | 0.001 | 1.000 | 0.005 | 1.000 | 0.027 | 0.996 | 0.093 | 0.980 | 0.236 | 1.000 | 0.384 | 0.686 | 0.329 | 0.569 |
| 11 | 11 | 0.000 | 1.000 | 0.000 | 1.000 | 0.000 | 1.000 | 0.000 | 1.000 | 0.000 | 1.000 | 0.000 | 1.000 | 0.004 | 1.000 | 0.020 | 1.000 | 0.086 | 1.000 | 0.314 | 1.000 | 0.569 | 1.000 |
| 12 | 0 | 0.540 | 0.540 | 0.282 | 0.282 | 0.069 | 0.069 | 0.014 | 0.014 | 0.002 | 0.002 | 0.000 | 0.000 | 0.000 | 0.000 | 0.000 | 0.000 | 0.000 | 0.000 | 0.000 | 0.000 | 0.000 | 0.000 |
| 12 | 1 | 0.341 | 0.882 | 0.377 | 0.659 | 0.206 | 0.275 | 0.071 | 0.085 | 0.017 | 0.020 | 0.003 | 0.003 | 0.000 | 0.000 | 0.000 | 0.000 | 0.000 | 0.000 | 0.000 | 0.000 | 0.000 | 0.000 |
| 12 | 2 | 0.099 | 0.980 | 0.230 | 0.889 | 0.283 | 0.558 | 0.168 | 0.253 | 0.064 | 0.083 | 0.016 | 0.019 | 0.002 | 0.003 | 0.000 | 0.000 | 0.000 | 0.000 | 0.000 | 0.000 | 0.000 | 0.000 |
| 12 | 3 | 0.017 | 0.998 | 0.085 | 0.974 | 0.236 | 0.795 | 0.240 | 0.493 | 0.142 | 0.225 | 0.054 | 0.073 | 0.012 | 0.015 | 0.001 | 0.002 | 0.001 | 0.001 | 0.000 | 0.000 | 0.000 | 0.000 |
| 12 | 4 | 0.002 | 1.000 | 0.021 | 0.996 | 0.133 | 0.927 | 0.231 | 0.724 | 0.213 | 0.438 | 0.121 | 0.194 | 0.042 | 0.057 | 0.008 | 0.009 | 0.003 | 0.004 | 0.000 | 0.000 | 0.000 | 0.000 |
| 12 | 5 | 0.000 | 1.000 | 0.004 | 0.999 | 0.053 | 0.981 | 0.158 | 0.882 | 0.227 | 0.665 | 0.193 | 0.387 | 0.101 | 0.158 | 0.029 | 0.039 | 0.016 | 0.019 | 0.000 | 0.000 | 0.000 | 0.000 |
| 12 | 6 | 0.000 | 1.000 | 0.000 | 1.000 | 0.016 | 0.996 | 0.079 | 0.961 | 0.177 | 0.842 | 0.226 | 0.613 | 0.177 | 0.335 | 0.079 | 0.118 | 0.053 | 0.073 | 0.000 | 0.001 | 0.000 | 0.000 |
| 12 | 7 | 0.000 | 1.000 | 0.000 | 1.000 | 0.003 | 0.999 | 0.029 | 0.991 | 0.101 | 0.943 | 0.193 | 0.806 | 0.227 | 0.562 | 0.158 | 0.276 | 0.133 | 0.205 | 0.004 | 0.004 | 0.000 | 0.000 |
| 12 | 8 | 0.000 | 1.000 | 0.000 | 1.000 | 0.001 | 1.000 | 0.008 | 0.998 | 0.042 | 0.985 | 0.121 | 0.927 | 0.213 | 0.775 | 0.231 | 0.507 | 0.236 | 0.442 | 0.021 | 0.026 | 0.002 | 0.002 |
| 12 | 9 | 0.000 | 1.000 | 0.000 | 1.000 | 0.000 | 1.000 | 0.001 | 1.000 | 0.012 | 0.997 | 0.054 | 0.981 | 0.142 | 0.917 | 0.240 | 0.747 | 0.283 | 0.725 | 0.085 | 0.111 | 0.017 | 0.020 |
| 12 | 10 | 0.000 | 1.000 | 0.000 | 1.000 | 0.000 | 1.000 | 0.000 | 1.000 | 0.002 | 1.000 | 0.016 | 0.997 | 0.064 | 0.980 | 0.168 | 0.915 | 0.206 | 0.931 | 0.230 | 0.341 | 0.099 | 0.118 |
| 12 | 11 | 0.000 | 1.000 | 0.000 | 1.000 | 0.000 | 1.000 | 0.000 | 1.000 | 0.000 | 1.000 | 0.003 | 1.000 | 0.017 | 0.998 | 0.071 | 0.986 | 0.069 | 1.000 | 0.377 | 0.718 | 0.341 | 0.460 |
| 12 | 12 | 0.000 | 1.000 | 0.000 | 1.000 | 0.000 | 1.000 | 0.000 | 1.000 | 0.000 | 1.000 | 0.000 | 1.000 | 0.002 | 1.000 | 0.014 | 1.000 | 0.069 | 1.000 | 0.282 | 1.000 | 0.540 | 1.000 |

| n | x |  |  |  |  |  |  |  |  |  |  |  |  |  |  |  |  |  |  |  |  |  |  |
|---|---|---|---|---|---|---|---|---|---|---|---|---|---|---|---|---|---|---|---|---|---|---|---|
| 13 | 0 | 0.513 | 0.513 | 0.254 | 0.254 | 0.055 | 0.055 | 0.010 | 0.010 | 0.001 | 0.001 | 0.000 | 0.000 | 0.000 | 0.000 | 0.000 | 0.000 | 0.000 | 0.000 | 0.000 | 0.000 | 0.000 | 0.000 |
| 13 | 1 | 0.351 | 0.865 | 0.367 | 0.621 | 0.179 | 0.234 | 0.054 | 0.064 | 0.011 | 0.013 | 0.002 | 0.002 | 0.000 | 0.000 | 0.000 | 0.000 | 0.000 | 0.000 | 0.000 | 0.000 | 0.000 | 0.000 |
| 13 | 2 | 0.111 | 0.975 | 0.245 | 0.866 | 0.268 | 0.502 | 0.139 | 0.202 | 0.045 | 0.058 | 0.010 | 0.011 | 0.001 | 0.001 | 0.000 | 0.000 | 0.000 | 0.000 | 0.000 | 0.000 | 0.000 | 0.000 |
| 13 | 3 | 0.021 | 0.997 | 0.100 | 0.966 | 0.246 | 0.747 | 0.218 | 0.421 | 0.111 | 0.169 | 0.035 | 0.046 | 0.006 | 0.008 | 0.001 | 0.001 | 0.000 | 0.000 | 0.000 | 0.000 | 0.000 | 0.000 |
| 13 | 4 | 0.003 | 1.000 | 0.028 | 0.994 | 0.154 | 0.901 | 0.234 | 0.654 | 0.184 | 0.353 | 0.087 | 0.133 | 0.024 | 0.032 | 0.003 | 0.004 | 0.000 | 0.000 | 0.000 | 0.000 | 0.000 | 0.000 |
| 13 | 5 | 0.000 | 1.000 | 0.006 | 0.999 | 0.069 | 0.970 | 0.180 | 0.835 | 0.221 | 0.574 | 0.157 | 0.291 | 0.066 | 0.098 | 0.014 | 0.018 | 0.001 | 0.001 | 0.000 | 0.000 | 0.000 | 0.000 |
| 13 | 6 | 0.000 | 1.000 | 0.001 | 1.000 | 0.023 | 0.993 | 0.103 | 0.938 | 0.197 | 0.771 | 0.209 | 0.500 | 0.131 | 0.229 | 0.044 | 0.062 | 0.006 | 0.007 | 0.000 | 0.000 | 0.000 | 0.000 |
| 13 | 7 | 0.000 | 1.000 | 0.000 | 1.000 | 0.006 | 0.999 | 0.044 | 0.982 | 0.131 | 0.902 | 0.209 | 0.709 | 0.197 | 0.426 | 0.103 | 0.165 | 0.023 | 0.030 | 0.001 | 0.001 | 0.000 | 0.000 |
| 13 | 8 | 0.000 | 1.000 | 0.000 | 1.000 | 0.001 | 1.000 | 0.014 | 0.996 | 0.066 | 0.968 | 0.157 | 0.867 | 0.221 | 0.647 | 0.180 | 0.346 | 0.069 | 0.099 | 0.006 | 0.006 | 0.000 | 0.000 |
| 13 | 9 | 0.000 | 1.000 | 0.000 | 1.000 | 0.000 | 1.000 | 0.003 | 0.999 | 0.024 | 0.992 | 0.087 | 0.954 | 0.184 | 0.831 | 0.234 | 0.579 | 0.154 | 0.253 | 0.028 | 0.034 | 0.003 | 0.003 |
| 13 | 10 | 0.000 | 1.000 | 0.000 | 1.000 | 0.000 | 1.000 | 0.001 | 1.000 | 0.006 | 0.999 | 0.035 | 0.989 | 0.111 | 0.942 | 0.218 | 0.798 | 0.246 | 0.498 | 0.100 | 0.134 | 0.021 | 0.025 |
| 13 | 11 | 0.000 | 1.000 | 0.000 | 1.000 | 0.000 | 1.000 | 0.000 | 1.000 | 0.001 | 1.000 | 0.010 | 0.998 | 0.045 | 0.987 | 0.139 | 0.936 | 0.268 | 0.766 | 0.245 | 0.379 | 0.111 | 0.135 |
| 13 | 12 | 0.000 | 1.000 | 0.000 | 1.000 | 0.000 | 1.000 | 0.000 | 1.000 | 0.000 | 1.000 | 0.002 | 1.000 | 0.011 | 0.999 | 0.054 | 0.990 | 0.179 | 0.945 | 0.367 | 0.746 | 0.351 | 0.487 |
| 13 | 13 | 0.000 | 1.000 | 0.000 | 1.000 | 0.000 | 1.000 | 0.000 | 1.000 | 0.000 | 1.000 | 0.000 | 1.000 | 0.001 | 1.000 | 0.010 | 1.000 | 0.055 | 1.000 | 0.254 | 1.000 | 0.513 | 1.000 |
| 14 | 0 | 0.488 | 0.488 | 0.229 | 0.229 | 0.044 | 0.044 | 0.007 | 0.007 | 0.001 | 0.001 | 0.000 | 0.000 | 0.000 | 0.000 | 0.000 | 0.000 | 0.000 | 0.000 | 0.000 | 0.000 | 0.000 | 0.000 |
| 14 | 1 | 0.359 | 0.847 | 0.356 | 0.585 | 0.154 | 0.198 | 0.041 | 0.047 | 0.007 | 0.008 | 0.001 | 0.001 | 0.000 | 0.000 | 0.000 | 0.000 | 0.000 | 0.000 | 0.000 | 0.000 | 0.000 | 0.000 |
| 14 | 2 | 0.123 | 0.970 | 0.257 | 0.842 | 0.250 | 0.448 | 0.113 | 0.161 | 0.032 | 0.040 | 0.006 | 0.006 | 0.001 | 0.001 | 0.000 | 0.000 | 0.000 | 0.000 | 0.000 | 0.000 | 0.000 | 0.000 |
| 14 | 3 | 0.026 | 0.996 | 0.114 | 0.956 | 0.250 | 0.698 | 0.194 | 0.355 | 0.085 | 0.124 | 0.022 | 0.029 | 0.003 | 0.004 | 0.000 | 0.000 | 0.000 | 0.000 | 0.000 | 0.000 | 0.000 | 0.000 |
| 14 | 4 | 0.004 | 1.000 | 0.035 | 0.991 | 0.172 | 0.870 | 0.229 | 0.584 | 0.155 | 0.279 | 0.061 | 0.090 | 0.014 | 0.018 | 0.001 | 0.002 | 0.000 | 0.000 | 0.000 | 0.000 | 0.000 | 0.000 |
| 14 | 5 | 0.000 | 1.000 | 0.008 | 0.999 | 0.086 | 0.956 | 0.196 | 0.781 | 0.207 | 0.486 | 0.122 | 0.212 | 0.041 | 0.058 | 0.007 | 0.008 | 0.000 | 0.000 | 0.000 | 0.000 | 0.000 | 0.000 |
| 14 | 6 | 0.000 | 1.000 | 0.001 | 1.000 | 0.032 | 0.988 | 0.126 | 0.907 | 0.207 | 0.692 | 0.183 | 0.395 | 0.092 | 0.150 | 0.023 | 0.031 | 0.002 | 0.002 | 0.000 | 0.000 | 0.000 | 0.000 |
| 14 | 7 | 0.000 | 1.000 | 0.000 | 1.000 | 0.009 | 0.998 | 0.062 | 0.969 | 0.157 | 0.850 | 0.209 | 0.605 | 0.157 | 0.308 | 0.062 | 0.093 | 0.009 | 0.012 | 0.000 | 0.000 | 0.000 | 0.000 |
| 14 | 8 | 0.000 | 1.000 | 0.000 | 1.000 | 0.002 | 1.000 | 0.023 | 0.992 | 0.092 | 0.942 | 0.183 | 0.788 | 0.207 | 0.514 | 0.126 | 0.219 | 0.032 | 0.044 | 0.001 | 0.001 | 0.000 | 0.000 |
| 14 | 9 | 0.000 | 1.000 | 0.000 | 1.000 | 0.000 | 1.000 | 0.007 | 0.998 | 0.041 | 0.982 | 0.122 | 0.910 | 0.207 | 0.721 | 0.196 | 0.416 | 0.086 | 0.130 | 0.008 | 0.009 | 0.000 | 0.000 |
| 14 | 10 | 0.000 | 1.000 | 0.000 | 1.000 | 0.000 | 1.000 | 0.001 | 1.000 | 0.014 | 0.996 | 0.061 | 0.971 | 0.155 | 0.876 | 0.229 | 0.645 | 0.172 | 0.302 | 0.035 | 0.044 | 0.004 | 0.004 |
| 14 | 11 | 0.000 | 1.000 | 0.000 | 1.000 | 0.000 | 1.000 | 0.000 | 1.000 | 0.003 | 0.999 | 0.022 | 0.994 | 0.085 | 0.960 | 0.194 | 0.839 | 0.250 | 0.552 | 0.114 | 0.158 | 0.026 | 0.030 |
| 14 | 12 | 0.000 | 1.000 | 0.000 | 1.000 | 0.000 | 1.000 | 0.000 | 1.000 | 0.001 | 1.000 | 0.006 | 0.999 | 0.032 | 0.992 | 0.113 | 0.953 | 0.250 | 0.802 | 0.257 | 0.415 | 0.123 | 0.153 |
| 14 | 13 | 0.000 | 1.000 | 0.000 | 1.000 | 0.000 | 1.000 | 0.000 | 1.000 | 0.000 | 1.000 | 0.001 | 1.000 | 0.007 | 0.999 | 0.041 | 0.993 | 0.154 | 0.956 | 0.356 | 0.771 | 0.359 | 0.512 |
| 14 | 14 | 0.000 | 1.000 | 0.000 | 1.000 | 0.000 | 1.000 | 0.000 | 1.000 | 0.000 | 1.000 | 0.000 | 1.000 | 0.001 | 1.000 | 0.007 | 1.000 | 0.044 | 1.000 | 0.229 | 1.000 | 0.488 | 1.000 |
| 15 | 0 | 0.463 | 0.463 | 0.206 | 0.206 | 0.035 | 0.035 | 0.005 | 0.005 | 0.000 | 0.000 | 0.000 | 0.000 | 0.000 | 0.000 | 0.000 | 0.000 | 0.000 | 0.000 | 0.000 | 0.000 | 0.000 | 0.000 |
| 15 | 1 | 0.366 | 0.829 | 0.343 | 0.549 | 0.132 | 0.167 | 0.031 | 0.035 | 0.005 | 0.005 | 0.000 | 0.000 | 0.000 | 0.000 | 0.000 | 0.000 | 0.000 | 0.000 | 0.000 | 0.000 | 0.000 | 0.000 |
| 15 | 2 | 0.135 | 0.964 | 0.267 | 0.816 | 0.231 | 0.398 | 0.092 | 0.127 | 0.022 | 0.027 | 0.003 | 0.004 | 0.000 | 0.000 | 0.000 | 0.000 | 0.000 | 0.000 | 0.000 | 0.000 | 0.000 | 0.000 |

*(Continued)*

## TABLE D.3 Table of Binomial Probabilities*—cont'd

| N | a | π = 0.05 Exact | Sum | π = 0.10 Exact | Sum | π = 0.20 Exact | Sum | π = 0.30 Exact | Sum | π = 0.40 Exact | Sum | π = 0.50 Exact | Sum | π = 0.60 Exact | Sum | π = 0.70 Exact | Sum | π = 0.80 Exact | Sum | π = 0.90 Exact | Sum | π = 0.95 Exact | Sum |
|---|---|---|---|---|---|---|---|---|---|---|---|---|---|---|---|---|---|---|---|---|---|---|---|
| 15 | 3 | 0.031 | 0.995 | 0.129 | 0.944 | 0.250 | 0.648 | 0.170 | 0.297 | 0.063 | 0.091 | 0.014 | 0.018 | 0.002 | 0.002 | 0.000 | 0.000 | 0.000 | 0.000 | 0.000 | 0.000 | 0.000 | 0.000 |
| 15 | 4 | 0.005 | 0.999 | 0.043 | 0.987 | 0.188 | 0.836 | 0.219 | 0.515 | 0.127 | 0.217 | 0.042 | 0.059 | 0.007 | 0.009 | 0.001 | 0.001 | 0.000 | 0.000 | 0.000 | 0.000 | 0.000 | 0.000 |
| 15 | 5 | 0.001 | 1.000 | 0.010 | 0.998 | 0.103 | 0.939 | 0.206 | 0.722 | 0.186 | 0.403 | 0.092 | 0.151 | 0.024 | 0.034 | 0.003 | 0.004 | 0.000 | 0.000 | 0.000 | 0.000 | 0.000 | 0.000 |
| 15 | 6 | 0.000 | 1.000 | 0.002 | 1.000 | 0.043 | 0.982 | 0.147 | 0.869 | 0.207 | 0.610 | 0.153 | 0.304 | 0.061 | 0.095 | 0.012 | 0.015 | 0.001 | 0.001 | 0.000 | 0.000 | 0.000 | 0.000 |
| 15 | 7 | 0.000 | 1.000 | 0.000 | 1.000 | 0.014 | 0.996 | 0.081 | 0.950 | 0.177 | 0.787 | 0.196 | 0.500 | 0.118 | 0.213 | 0.035 | 0.050 | 0.003 | 0.004 | 0.000 | 0.000 | 0.000 | 0.000 |
| 15 | 8 | 0.000 | 1.000 | 0.000 | 1.000 | 0.003 | 0.999 | 0.035 | 0.985 | 0.118 | 0.905 | 0.196 | 0.696 | 0.177 | 0.390 | 0.081 | 0.131 | 0.014 | 0.018 | 0.000 | 0.000 | 0.000 | 0.000 |
| 15 | 9 | 0.000 | 1.000 | 0.000 | 1.000 | 0.001 | 1.000 | 0.012 | 0.996 | 0.061 | 0.966 | 0.153 | 0.849 | 0.207 | 0.597 | 0.147 | 0.278 | 0.043 | 0.061 | 0.002 | 0.002 | 0.000 | 0.000 |
| 15 | 10 | 0.000 | 1.000 | 0.000 | 1.000 | 0.000 | 1.000 | 0.003 | 0.999 | 0.024 | 0.991 | 0.092 | 0.941 | 0.186 | 0.783 | 0.206 | 0.485 | 0.103 | 0.164 | 0.010 | 0.013 | 0.001 | 0.001 |
| 15 | 11 | 0.000 | 1.000 | 0.000 | 1.000 | 0.000 | 1.000 | 0.001 | 1.000 | 0.007 | 0.998 | 0.042 | 0.982 | 0.127 | 0.909 | 0.219 | 0.703 | 0.188 | 0.352 | 0.043 | 0.056 | 0.005 | 0.005 |
| 15 | 12 | 0.000 | 1.000 | 0.000 | 1.000 | 0.000 | 1.000 | 0.000 | 1.000 | 0.002 | 1.000 | 0.014 | 0.996 | 0.063 | 0.973 | 0.170 | 0.873 | 0.250 | 0.602 | 0.129 | 0.184 | 0.031 | 0.036 |
| 15 | 13 | 0.000 | 1.000 | 0.000 | 1.000 | 0.000 | 1.000 | 0.000 | 1.000 | 0.000 | 1.000 | 0.003 | 1.000 | 0.022 | 0.995 | 0.092 | 0.965 | 0.231 | 0.833 | 0.267 | 0.451 | 0.135 | 0.171 |
| 15 | 14 | 0.000 | 1.000 | 0.000 | 1.000 | 0.000 | 1.000 | 0.000 | 1.000 | 0.000 | 1.000 | 0.000 | 1.000 | 0.005 | 1.000 | 0.031 | 0.995 | 0.132 | 0.965 | 0.343 | 0.794 | 0.366 | 0.537 |
| 15 | 15 | 0.000 | 1.000 | 0.000 | 1.000 | 0.000 | 1.000 | 0.000 | 1.000 | 0.000 | 1.000 | 0.000 | 1.000 | 0.000 | 1.000 | 0.005 | 1.000 | 0.035 | 1.000 | 0.206 | 1.000 | 0.463 | 1.000 |
| 16 | 0 | 0.440 | 0.440 | 0.185 | 0.185 | 0.028 | 0.028 | 0.003 | 0.003 | 0.000 | 0.000 | 0.000 | 0.000 | 0.000 | 0.000 | 0.000 | 0.000 | 0.000 | 0.000 | 0.000 | 0.000 | 0.000 | 0.000 |
| 16 | 1 | 0.371 | 0.811 | 0.329 | 0.515 | 0.113 | 0.141 | 0.023 | 0.026 | 0.003 | 0.003 | 0.000 | 0.000 | 0.000 | 0.000 | 0.000 | 0.000 | 0.000 | 0.000 | 0.000 | 0.000 | 0.000 | 0.000 |
| 16 | 2 | 0.146 | 0.957 | 0.275 | 0.789 | 0.211 | 0.352 | 0.073 | 0.099 | 0.015 | 0.018 | 0.002 | 0.002 | 0.000 | 0.000 | 0.000 | 0.000 | 0.000 | 0.000 | 0.000 | 0.000 | 0.000 | 0.000 |
| 16 | 3 | 0.036 | 0.993 | 0.142 | 0.932 | 0.246 | 0.598 | 0.146 | 0.246 | 0.047 | 0.065 | 0.009 | 0.011 | 0.001 | 0.001 | 0.000 | 0.000 | 0.000 | 0.000 | 0.000 | 0.000 | 0.000 | 0.000 |
| 16 | 4 | 0.006 | 0.999 | 0.051 | 0.983 | 0.200 | 0.798 | 0.204 | 0.450 | 0.101 | 0.167 | 0.028 | 0.038 | 0.004 | 0.005 | 0.000 | 0.000 | 0.000 | 0.000 | 0.000 | 0.000 | 0.000 | 0.000 |
| 16 | 5 | 0.001 | 1.000 | 0.014 | 0.997 | 0.120 | 0.918 | 0.210 | 0.660 | 0.162 | 0.329 | 0.067 | 0.105 | 0.014 | 0.019 | 0.001 | 0.002 | 0.000 | 0.000 | 0.000 | 0.000 | 0.000 | 0.000 |
| 16 | 6 | 0.000 | 1.000 | 0.003 | 0.999 | 0.055 | 0.973 | 0.165 | 0.825 | 0.198 | 0.527 | 0.122 | 0.227 | 0.039 | 0.058 | 0.006 | 0.007 | 0.000 | 0.000 | 0.000 | 0.000 | 0.000 | 0.000 |
| 16 | 7 | 0.000 | 1.000 | 0.000 | 1.000 | 0.020 | 0.993 | 0.101 | 0.926 | 0.189 | 0.716 | 0.175 | 0.402 | 0.084 | 0.142 | 0.019 | 0.026 | 0.001 | 0.001 | 0.000 | 0.000 | 0.000 | 0.000 |
| 16 | 8 | 0.000 | 1.000 | 0.000 | 1.000 | 0.006 | 0.999 | 0.049 | 0.974 | 0.142 | 0.858 | 0.196 | 0.598 | 0.142 | 0.284 | 0.049 | 0.074 | 0.006 | 0.007 | 0.000 | 0.000 | 0.000 | 0.000 |
| 16 | 9 | 0.000 | 1.000 | 0.000 | 1.000 | 0.001 | 1.000 | 0.019 | 0.993 | 0.084 | 0.942 | 0.175 | 0.773 | 0.189 | 0.473 | 0.101 | 0.175 | 0.020 | 0.027 | 0.000 | 0.001 | 0.000 | 0.000 |
| 16 | 10 | 0.000 | 1.000 | 0.000 | 1.000 | 0.000 | 1.000 | 0.006 | 0.998 | 0.039 | 0.981 | 0.122 | 0.895 | 0.198 | 0.671 | 0.165 | 0.340 | 0.055 | 0.082 | 0.003 | 0.003 | 0.000 | 0.000 |
| 16 | 11 | 0.000 | 1.000 | 0.000 | 1.000 | 0.000 | 1.000 | 0.001 | 1.000 | 0.014 | 0.995 | 0.067 | 0.962 | 0.162 | 0.833 | 0.210 | 0.550 | 0.120 | 0.202 | 0.014 | 0.017 | 0.001 | 0.001 |
| 16 | 12 | 0.000 | 1.000 | 0.000 | 1.000 | 0.000 | 1.000 | 0.000 | 1.000 | 0.004 | 0.999 | 0.028 | 0.989 | 0.101 | 0.935 | 0.204 | 0.754 | 0.200 | 0.402 | 0.051 | 0.068 | 0.006 | 0.007 |
| 16 | 13 | 0.000 | 1.000 | 0.000 | 1.000 | 0.000 | 1.000 | 0.000 | 1.000 | 0.001 | 1.000 | 0.009 | 0.998 | 0.047 | 0.982 | 0.146 | 0.901 | 0.246 | 0.648 | 0.142 | 0.211 | 0.036 | 0.043 |
| 16 | 14 | 0.000 | 1.000 | 0.000 | 1.000 | 0.000 | 1.000 | 0.000 | 1.000 | 0.000 | 1.000 | 0.002 | 1.000 | 0.015 | 0.997 | 0.073 | 0.974 | 0.211 | 0.859 | 0.275 | 0.485 | 0.146 | 0.189 |
| 16 | 15 | 0.000 | 1.000 | 0.000 | 1.000 | 0.000 | 1.000 | 0.000 | 1.000 | 0.000 | 1.000 | 0.000 | 1.000 | 0.003 | 1.000 | 0.023 | 0.997 | 0.113 | 0.972 | 0.329 | 0.815 | 0.371 | 0.560 |

| n | 16 | 17 | 17 | 17 | 17 | 17 | 17 | 17 | 17 | 17 | 17 | 17 | 17 | 17 | 17 | 17 | 17 | 17 | 17 | 18 | 18 | 18 | 18 | 18 | 18 | 18 | 18 | 18 | 18 | 18 | 18 | 18 |
|---|---|---|---|---|---|---|---|---|---|---|---|---|---|---|---|---|---|---|---|---|---|---|---|---|---|---|---|---|---|---|---|---|
| x | 16 | 0 | 1 | 2 | 3 | 4 | 5 | 6 | 7 | 8 | 9 | 10 | 11 | 12 | 13 | 14 | 15 | 16 | 17 | 0 | 1 | 2 | 3 | 4 | 5 | 6 | 7 | 8 | 9 | 10 | 11 | 12 |
| | 1.000 | 0.000 | 0.000 | 0.000 | 0.000 | 0.000 | 0.000 | 0.000 | 0.000 | 0.000 | 0.000 | 0.000 | 0.000 | 0.001 | 0.009 | 0.050 | 0.208 | 0.582 | 1.000 | 0.000 | 0.000 | 0.000 | 0.000 | 0.000 | 0.000 | 0.000 | 0.000 | 0.000 | 0.000 | 0.000 | 0.001 | 0.006 |
| | 0.440 | 0.000 | 0.000 | 0.000 | 0.000 | 0.000 | 0.000 | 0.000 | 0.000 | 0.000 | 0.000 | 0.000 | 0.001 | 0.008 | 0.041 | 0.158 | 0.374 | 0.418 | 0.000 | 0.000 | 0.000 | 0.000 | 0.000 | 0.000 | 0.000 | 0.000 | 0.001 | 0.008 | 0.035 | 0.082 | 0.133 |
| | 1.000 | 0.000 | 0.000 | 0.000 | 0.000 | 0.000 | 0.000 | 0.001 | 0.003 | 0.013 | 0.040 | 0.105 | 0.225 | 0.403 | 0.611 | 0.798 | 0.923 | 0.981 | 0.998 | 1.000 | 0.000 | 0.000 | 0.000 | 0.000 | 0.001 | 0.006 | 0.021 | 0.060 | 0.141 | 0.278 | 0.466 | — |
| | 0.003 | 0.000 | 0.000 | 0.000 | 0.000 | 0.001 | 0.003 | 0.009 | 0.028 | 0.064 | 0.120 | 0.178 | 0.208 | 0.187 | 0.125 | 0.058 | 0.017 | 0.002 | — | 0.000 | 0.000 | 0.001 | 0.005 | 0.015 | 0.039 | 0.081 | 0.138 | 0.187 | — | — | — | — |
| | 1.000 | 0.000 | 0.000 | 0.000 | 0.003 | 0.011 | 0.035 | 0.092 | 0.199 | 0.359 | 0.552 | 0.736 | 0.874 | 0.954 | 0.988 | 0.998 | 1.000 | 1.000 | — | 0.000 | 0.001 | 0.006 | 0.020 | 0.058 | 0.135 | 0.263 | 0.437 | 0.626 | 0.791 | — | — | — |
| | 0.000 | 0.000 | 0.000 | 0.002 | 0.008 | 0.024 | 0.057 | 0.107 | 0.161 | 0.193 | 0.184 | 0.138 | 0.080 | 0.034 | 0.010 | 0.002 | 0.000 | 0.000 | — | 0.001 | 0.004 | 0.015 | 0.037 | 0.077 | 0.128 | 0.173 | 0.189 | 0.166 | — | — | — | — |
| | 1.000 | 0.000 | 0.001 | 0.006 | 0.025 | 0.072 | 0.166 | 0.315 | 0.500 | 0.685 | 0.834 | 0.928 | 0.975 | 0.994 | 0.999 | 1.000 | 1.000 | 1.000 | — | 0.000 | 0.001 | 0.004 | 0.015 | 0.048 | 0.119 | 0.240 | 0.407 | 0.593 | 0.760 | 0.881 | 0.952 | — |
| | 0.000 | 0.000 | 0.005 | 0.018 | 0.047 | 0.094 | 0.148 | 0.185 | 0.185 | 0.148 | 0.094 | 0.047 | 0.018 | 0.005 | 0.001 | 0.000 | 0.000 | — | — | 0.001 | 0.003 | 0.012 | 0.033 | 0.071 | 0.121 | 0.167 | 0.185 | 0.167 | 0.121 | 0.071 | — | — |
| | 1.000 | 0.002 | 0.012 | 0.046 | 0.126 | 0.264 | 0.448 | 0.641 | 0.801 | 0.908 | 0.965 | 0.989 | 0.997 | 1.000 | 1.000 | 1.000 | 1.000 | — | — | 0.001 | 0.008 | 0.033 | 0.094 | 0.209 | 0.374 | 0.563 | 0.737 | 0.865 | 0.942 | 0.980 | 0.994 | — |
| | 0.000 | 0.002 | 0.010 | 0.034 | 0.080 | 0.138 | 0.184 | 0.193 | 0.161 | 0.107 | 0.057 | 0.024 | 0.008 | 0.002 | 0.000 | 0.000 | — | — | — | 0.001 | 0.007 | 0.025 | 0.061 | 0.115 | 0.166 | 0.189 | 0.173 | 0.128 | 0.077 | 0.037 | 0.015 | — |
| | 1.000 | 0.019 | 0.077 | 0.202 | 0.389 | 0.597 | 0.775 | 0.895 | 0.960 | 0.987 | 0.997 | 0.999 | 1.000 | 1.000 | 1.000 | 1.000 | 1.000 | — | — | 0.014 | 0.060 | 0.165 | 0.333 | 0.534 | 0.722 | 0.859 | 0.940 | 0.979 | 0.994 | 0.999 | 1.000 | — |
| | 0.000 | 0.017 | 0.058 | 0.125 | 0.187 | 0.208 | 0.178 | 0.120 | 0.064 | 0.028 | 0.009 | 0.003 | 0.001 | 0.000 | 0.000 | 0.000 | — | — | — | 0.013 | 0.046 | 0.105 | 0.168 | 0.202 | 0.187 | 0.138 | 0.081 | 0.039 | 0.015 | 0.005 | 0.001 | — |
| | 1.000 | 0.023 | 0.118 | 0.310 | 0.549 | 0.758 | 0.894 | 0.962 | 0.989 | 0.997 | 1.000 | 1.000 | 1.000 | 1.000 | 1.000 | 1.000 | 1.000 | — | — | 0.099 | 0.271 | 0.501 | 0.716 | 0.867 | 0.949 | 0.984 | 0.996 | 0.999 | 1.000 | 1.000 | 1.000 | — |
| | 0.000 | 0.023 | 0.096 | 0.191 | 0.239 | 0.209 | 0.136 | 0.068 | 0.027 | 0.008 | 0.002 | 0.000 | 0.000 | 0.000 | 0.000 | 0.000 | — | — | — | 0.081 | 0.172 | 0.230 | 0.215 | 0.151 | 0.082 | 0.035 | 0.012 | 0.003 | 0.001 | 0.000 | 0.000 | — |
| | 1.000 | 0.167 | 0.482 | 0.762 | 0.917 | 0.978 | 0.995 | 0.999 | 1.000 | 1.000 | 1.000 | 1.000 | 1.000 | 1.000 | 1.000 | 1.000 | 1.000 | — | — | 0.450 | 0.734 | 0.902 | 0.972 | 0.994 | 0.999 | 1.000 | 1.000 | 1.000 | 1.000 | 1.000 | 1.000 | — |
| | 0.000 | 0.167 | 0.315 | 0.280 | 0.156 | 0.060 | 0.017 | 0.004 | 0.001 | 0.000 | 0.000 | 0.000 | 0.000 | 0.000 | 0.000 | 0.000 | — | — | — | 0.300 | 0.284 | 0.168 | 0.070 | 0.022 | 0.005 | 0.001 | 0.000 | 0.000 | 0.000 | 0.000 | 0.000 | — |
| | 1.000 | 0.418 | 0.792 | 0.950 | 0.991 | 0.999 | 1.000 | 1.000 | 1.000 | 1.000 | 1.000 | 1.000 | 1.000 | 1.000 | 1.000 | 1.000 | 1.000 | — | — | 0.774 | 0.942 | 0.989 | 0.998 | 1.000 | 1.000 | 1.000 | 1.000 | 1.000 | 1.000 | 1.000 | 1.000 | — |
| | 0.000 | 0.418 | 0.374 | 0.158 | 0.041 | 0.008 | 0.001 | 0.000 | 0.000 | 0.000 | 0.000 | 0.000 | 0.000 | 0.000 | 0.000 | 0.000 | — | — | — | 0.376 | 0.168 | 0.047 | 0.009 | 0.001 | 0.000 | 0.000 | 0.000 | 0.000 | 0.000 | 0.000 | 0.000 | — |
| | — | 0.397 | — | — | — | — | — | — | — | — | — | — | — | — | — | — | — | — | — | 0.397 | — | — | — | — | — | — | — | — | — | — | — | — |

*(Continued)*

## TABLE D.3 Table of Binomial Probabilities*—cont'd

| N | a | π = 0.05 | | π = 0.10 | | π = 0.20 | | π = 0.30 | | π = 0.40 | | π = 0.50 | | π = 0.60 | | π = 0.70 | | π = 0.80 | | π = 0.90 | | π = 0.95 | |
|---|---|---|---|---|---|---|---|---|---|---|---|---|---|---|---|---|---|---|---|---|---|---|---|
| | | Exact | Sum | Exact | Sum | Exact | Sum | Exact | Sum | Exact | Sum | Exact | Sum | Exact | Sum | Exact | Sum | Exact | Sum | Exact | Sum | Exact | Sum |
| 18 | 13 | 0.000 | 1.000 | 0.000 | 1.000 | 0.000 | 1.000 | 0.000 | 1.000 | 0.004 | 0.999 | 0.033 | 0.985 | 0.115 | 0.906 | 0.202 | 0.667 | 0.151 | 0.284 | 0.022 | 0.028 | 0.001 | 0.002 |
| 18 | 14 | 0.000 | 1.000 | 0.000 | 1.000 | 0.000 | 1.000 | 0.000 | 1.000 | 0.001 | 1.000 | 0.012 | 0.996 | 0.061 | 0.967 | 0.168 | 0.835 | 0.215 | 0.499 | 0.070 | 0.098 | 0.009 | 0.011 |
| 18 | 15 | 0.000 | 1.000 | 0.000 | 1.000 | 0.000 | 1.000 | 0.000 | 1.000 | 0.000 | 1.000 | 0.003 | 0.999 | 0.025 | 0.992 | 0.105 | 0.940 | 0.230 | 0.729 | 0.168 | 0.266 | 0.047 | 0.058 |
| 18 | 16 | 0.000 | 1.000 | 0.000 | 1.000 | 0.000 | 1.000 | 0.000 | 1.000 | 0.000 | 1.000 | 0.001 | 1.000 | 0.007 | 0.999 | 0.046 | 0.986 | 0.172 | 0.901 | 0.284 | 0.550 | 0.168 | 0.226 |
| 18 | 17 | 0.000 | 1.000 | 0.000 | 1.000 | 0.000 | 1.000 | 0.000 | 1.000 | 0.000 | 1.000 | 0.000 | 1.000 | 0.001 | 1.000 | 0.013 | 0.998 | 0.081 | 0.982 | 0.300 | 0.850 | 0.376 | 0.603 |
| 18 | 18 | 0.000 | 1.000 | 0.000 | 1.000 | 0.000 | 1.000 | 0.000 | 1.000 | 0.000 | 1.000 | 0.000 | 1.000 | 0.000 | 1.000 | 0.002 | 1.000 | 0.018 | 1.000 | 0.150 | 1.000 | 0.397 | 1.000 |
| 19 | 0 | 0.377 | 0.377 | 0.135 | 0.135 | 0.014 | 0.014 | 0.001 | 0.001 | 0.000 | 0.000 | 0.000 | 0.000 | 0.000 | 0.000 | 0.000 | 0.000 | 0.000 | 0.000 | 0.000 | 0.000 | 0.000 | 0.000 |
| 19 | 1 | 0.377 | 0.755 | 0.285 | 0.420 | 0.068 | 0.083 | 0.009 | 0.010 | 0.001 | 0.001 | 0.000 | 0.000 | 0.000 | 0.000 | 0.000 | 0.000 | 0.000 | 0.000 | 0.000 | 0.000 | 0.000 | 0.000 |
| 19 | 2 | 0.179 | 0.933 | 0.285 | 0.705 | 0.154 | 0.237 | 0.036 | 0.046 | 0.005 | 0.005 | 0.000 | 0.002 | 0.000 | 0.000 | 0.000 | 0.000 | 0.000 | 0.000 | 0.000 | 0.000 | 0.000 | 0.000 |
| 19 | 3 | 0.053 | 0.987 | 0.180 | 0.885 | 0.218 | 0.455 | 0.087 | 0.133 | 0.017 | 0.023 | 0.002 | 0.002 | 0.000 | 0.000 | 0.000 | 0.000 | 0.000 | 0.000 | 0.000 | 0.000 | 0.000 | 0.000 |
| 19 | 4 | 0.011 | 0.998 | 0.080 | 0.965 | 0.218 | 0.673 | 0.149 | 0.282 | 0.047 | 0.070 | 0.007 | 0.010 | 0.001 | 0.001 | 0.000 | 0.000 | 0.000 | 0.000 | 0.000 | 0.000 | 0.000 | 0.000 |
| 19 | 5 | 0.002 | 1.000 | 0.027 | 0.991 | 0.164 | 0.837 | 0.192 | 0.474 | 0.093 | 0.163 | 0.022 | 0.032 | 0.002 | 0.003 | 0.000 | 0.000 | 0.000 | 0.000 | 0.000 | 0.000 | 0.000 | 0.000 |
| 19 | 6 | 0.000 | 1.000 | 0.007 | 0.998 | 0.095 | 0.932 | 0.192 | 0.666 | 0.145 | 0.308 | 0.052 | 0.084 | 0.008 | 0.012 | 0.001 | 0.001 | 0.000 | 0.000 | 0.000 | 0.000 | 0.000 | 0.000 |
| 19 | 7 | 0.000 | 1.000 | 0.001 | 1.000 | 0.044 | 0.977 | 0.153 | 0.818 | 0.180 | 0.488 | 0.096 | 0.180 | 0.024 | 0.035 | 0.002 | 0.003 | 0.000 | 0.000 | 0.000 | 0.000 | 0.000 | 0.000 |
| 19 | 8 | 0.000 | 1.000 | 0.000 | 1.000 | 0.017 | 0.993 | 0.098 | 0.916 | 0.180 | 0.667 | 0.144 | 0.324 | 0.053 | 0.088 | 0.008 | 0.011 | 0.000 | 0.000 | 0.000 | 0.000 | 0.000 | 0.000 |
| 19 | 9 | 0.000 | 1.000 | 0.000 | 1.000 | 0.005 | 0.998 | 0.051 | 0.967 | 0.146 | 0.814 | 0.176 | 0.500 | 0.098 | 0.186 | 0.022 | 0.033 | 0.001 | 0.002 | 0.000 | 0.000 | 0.000 | 0.000 |
| 19 | 10 | 0.000 | 1.000 | 0.000 | 1.000 | 0.001 | 1.000 | 0.022 | 0.989 | 0.098 | 0.912 | 0.176 | 0.676 | 0.146 | 0.333 | 0.051 | 0.084 | 0.005 | 0.007 | 0.000 | 0.000 | 0.000 | 0.000 |
| 19 | 11 | 0.000 | 1.000 | 0.000 | 1.000 | 0.000 | 1.000 | 0.008 | 0.997 | 0.053 | 0.965 | 0.144 | 0.820 | 0.180 | 0.512 | 0.098 | 0.182 | 0.017 | 0.023 | 0.000 | 0.000 | 0.000 | 0.000 |
| 19 | 12 | 0.000 | 1.000 | 0.000 | 1.000 | 0.000 | 1.000 | 0.002 | 0.999 | 0.024 | 0.988 | 0.096 | 0.916 | 0.180 | 0.692 | 0.153 | 0.334 | 0.044 | 0.068 | 0.001 | 0.002 | 0.000 | 0.000 |
| 19 | 13 | 0.000 | 1.000 | 0.000 | 1.000 | 0.000 | 1.000 | 0.001 | 1.000 | 0.008 | 0.997 | 0.052 | 0.968 | 0.145 | 0.837 | 0.192 | 0.526 | 0.095 | 0.163 | 0.007 | 0.009 | 0.000 | 0.000 |
| 19 | 14 | 0.000 | 1.000 | 0.000 | 1.000 | 0.000 | 1.000 | 0.000 | 1.000 | 0.002 | 0.999 | 0.022 | 0.990 | 0.093 | 0.930 | 0.192 | 0.718 | 0.164 | 0.327 | 0.027 | 0.035 | 0.002 | 0.002 |
| 19 | 15 | 0.000 | 1.000 | 0.000 | 1.000 | 0.000 | 1.000 | 0.000 | 1.000 | 0.001 | 1.000 | 0.007 | 0.998 | 0.047 | 0.977 | 0.149 | 0.867 | 0.218 | 0.545 | 0.080 | 0.115 | 0.011 | 0.013 |
| 19 | 16 | 0.000 | 1.000 | 0.000 | 1.000 | 0.000 | 1.000 | 0.000 | 1.000 | 0.000 | 1.000 | 0.002 | 1.000 | 0.017 | 0.995 | 0.087 | 0.954 | 0.218 | 0.763 | 0.180 | 0.295 | 0.053 | 0.067 |
| 19 | 17 | 0.000 | 1.000 | 0.000 | 1.000 | 0.000 | 1.000 | 0.000 | 1.000 | 0.000 | 1.000 | 0.000 | 1.000 | 0.005 | 0.999 | 0.036 | 0.990 | 0.154 | 0.917 | 0.285 | 0.580 | 0.179 | 0.245 |
| 19 | 18 | 0.000 | 1.000 | 0.000 | 1.000 | 0.000 | 1.000 | 0.000 | 1.000 | 0.000 | 1.000 | 0.000 | 1.000 | 0.001 | 1.000 | 0.009 | 0.999 | 0.068 | 0.986 | 0.285 | 0.865 | 0.377 | 0.623 |
| 19 | 19 | 0.000 | 1.000 | 0.000 | 1.000 | 0.000 | 1.000 | 0.000 | 1.000 | 0.000 | 1.000 | 0.000 | 1.000 | 0.000 | 1.000 | 0.001 | 1.000 | 0.014 | 1.000 | 0.135 | 1.000 | 0.377 | 1.000 |
| 20 | 0 | 0.358 | 0.358 | 0.122 | 0.122 | 0.012 | 0.012 | 0.001 | 0.001 | 0.000 | 0.000 | 0.000 | 0.000 | 0.000 | 0.000 | 0.000 | 0.000 | 0.000 | 0.000 | 0.000 | 0.000 | 0.000 | 0.000 |
| 20 | 1 | 0.377 | 0.736 | 0.270 | 0.392 | 0.058 | 0.069 | 0.007 | 0.008 | 0.000 | 0.001 | 0.000 | 0.000 | 0.000 | 0.000 | 0.000 | 0.000 | 0.000 | 0.000 | 0.000 | 0.000 | 0.000 | 0.000 |
| 20 | 2 | 0.189 | 0.925 | 0.285 | 0.677 | 0.137 | 0.206 | 0.028 | 0.035 | 0.003 | 0.004 | 0.000 | 0.000 | 0.000 | 0.000 | 0.000 | 0.000 | 0.000 | 0.000 | 0.000 | 0.000 | 0.000 | 0.000 |

| n | x |  |  |  |  |  |  |  |  |  |  |  |  |  |  |  |  |  |  |  |  |  |  |
|---|---|---|---|---|---|---|---|---|---|---|---|---|---|---|---|---|---|---|---|---|---|---|---|
| 20 | 3 | 0.060 | 0.984 | 0.190 | 0.867 | 0.205 | 0.411 | 0.072 | 0.107 | 0.012 | 0.016 | 0.001 | 0.001 | 0.000 | 0.000 | 0.000 | 0.000 | 0.000 | 0.000 | 0.000 | 0.000 | 0.000 | 0.000 |
| 20 | 4 | 0.013 | 0.997 | 0.090 | 0.957 | 0.218 | 0.630 | 0.130 | 0.238 | 0.035 | 0.051 | 0.005 | 0.006 | 0.000 | 0.000 | 0.000 | 0.000 | 0.000 | 0.000 | 0.000 | 0.000 | 0.000 | 0.000 |
| 20 | 5 | 0.002 | 1.000 | 0.032 | 0.989 | 0.175 | 0.804 | 0.179 | 0.416 | 0.075 | 0.126 | 0.015 | 0.021 | 0.001 | 0.002 | 0.000 | 0.000 | 0.000 | 0.000 | 0.000 | 0.000 | 0.000 | 0.000 |
| 20 | 6 | 0.000 | 1.000 | 0.009 | 0.998 | 0.109 | 0.913 | 0.192 | 0.608 | 0.124 | 0.250 | 0.037 | 0.058 | 0.005 | 0.006 | 0.000 | 0.001 | 0.000 | 0.000 | 0.000 | 0.000 | 0.000 | 0.000 |
| 20 | 7 | 0.000 | 1.000 | 0.002 | 1.000 | 0.055 | 0.968 | 0.164 | 0.772 | 0.166 | 0.416 | 0.074 | 0.132 | 0.015 | 0.021 | 0.001 | 0.005 | 0.000 | 0.000 | 0.000 | 0.000 | 0.000 | 0.000 |
| 20 | 8 | 0.000 | 1.000 | 0.000 | 1.000 | 0.022 | 0.990 | 0.114 | 0.887 | 0.180 | 0.596 | 0.120 | 0.252 | 0.035 | 0.057 | 0.004 | 0.017 | 0.000 | 0.000 | 0.000 | 0.000 | 0.000 | 0.000 |
| 20 | 9 | 0.000 | 1.000 | 0.000 | 1.000 | 0.007 | 0.997 | 0.065 | 0.952 | 0.160 | 0.755 | 0.160 | 0.412 | 0.071 | 0.128 | 0.012 | 0.048 | 0.000 | 0.001 | 0.000 | 0.000 | 0.000 | 0.000 |
| 20 | 10 | 0.000 | 1.000 | 0.000 | 1.000 | 0.002 | 0.999 | 0.031 | 0.983 | 0.117 | 0.872 | 0.176 | 0.588 | 0.117 | 0.245 | 0.031 | 0.113 | 0.002 | 0.003 | 0.000 | 0.000 | 0.000 | 0.000 |
| 20 | 11 | 0.000 | 1.000 | 0.000 | 1.000 | 0.000 | 1.000 | 0.012 | 0.995 | 0.071 | 0.943 | 0.160 | 0.748 | 0.160 | 0.404 | 0.065 | 0.228 | 0.007 | 0.010 | 0.000 | 0.000 | 0.000 | 0.000 |
| 20 | 12 | 0.000 | 1.000 | 0.000 | 1.000 | 0.000 | 1.000 | 0.004 | 0.999 | 0.035 | 0.979 | 0.120 | 0.868 | 0.180 | 0.584 | 0.114 | 0.392 | 0.022 | 0.032 | 0.000 | 0.000 | 0.000 | 0.000 |
| 20 | 13 | 0.000 | 1.000 | 0.000 | 1.000 | 0.000 | 1.000 | 0.001 | 1.000 | 0.015 | 0.994 | 0.074 | 0.942 | 0.166 | 0.750 | 0.164 | 0.584 | 0.055 | 0.087 | 0.002 | 0.002 | 0.000 | 0.000 |
| 20 | 14 | 0.000 | 1.000 | 0.000 | 1.000 | 0.000 | 1.000 | 0.000 | 1.000 | 0.005 | 0.998 | 0.037 | 0.979 | 0.124 | 0.874 | 0.192 | 0.762 | 0.109 | 0.196 | 0.009 | 0.011 | 0.000 | 0.000 |
| 20 | 15 | 0.000 | 1.000 | 0.000 | 1.000 | 0.000 | 1.000 | 0.000 | 1.000 | 0.001 | 1.000 | 0.015 | 0.994 | 0.075 | 0.949 | 0.179 | 0.893 | 0.175 | 0.370 | 0.032 | 0.043 | 0.002 | 0.003 |
| 20 | 16 | 0.000 | 1.000 | 0.000 | 1.000 | 0.000 | 1.000 | 0.000 | 1.000 | 0.000 | 1.000 | 0.005 | 0.999 | 0.035 | 0.984 | 0.130 | 0.965 | 0.218 | 0.589 | 0.090 | 0.133 | 0.013 | 0.016 |
| 20 | 17 | 0.000 | 1.000 | 0.000 | 1.000 | 0.000 | 1.000 | 0.000 | 1.000 | 0.000 | 1.000 | 0.001 | 1.000 | 0.012 | 0.996 | 0.072 | 0.992 | 0.205 | 0.794 | 0.190 | 0.323 | 0.060 | 0.075 |
| 20 | 18 | 0.000 | 1.000 | 0.000 | 1.000 | 0.000 | 1.000 | 0.000 | 1.000 | 0.000 | 1.000 | 0.000 | 1.000 | 0.003 | 0.999 | 0.028 | 0.999 | 0.137 | 0.931 | 0.285 | 0.608 | 0.189 | 0.264 |
| 20 | 19 | 0.000 | 1.000 | 0.000 | 1.000 | 0.000 | 1.000 | 0.000 | 1.000 | 0.000 | 1.000 | 0.000 | 1.000 | 0.000 | 1.000 | 0.007 | 1.000 | 0.058 | 0.988 | 0.270 | 0.878 | 0.377 | 0.642 |
| 20 | 20 | 0.000 | 1.000 | 0.000 | 1.000 | 0.000 | 1.000 | 0.000 | 1.000 | 0.000 | 1.000 | 0.000 | 1.000 | 0.000 | 1.000 | 0.001 | 1.000 | 0.012 | 1.000 | 0.122 | 1.000 | 0.358 | 1.000 |

*Exact probabilities are found under the heading "Exact," while cumulative probabilities are found under the heading "Sum." For example, the probability that a binomial random variable with $\pi = 0.30$ and $n = 3$ is exactly equal to $a = 2$ is found to be 0.189, while the probability that this binomial random variable is less than or equal to $a = 2$ is found to be 0.973 (the sum of probabilities for $a = 0$, 1, and 2).

## TABLE D.4 The t Table

| Confidence Level | | | | | | | |
|---|---|---|---|---|---|---|---|
| Two-sided | 80% | 90% | 95% | 98% | 99% | 99.8% | 99.9% |
| One-sided | 90% | 95% | 97.5% | 99% | 99.5% | 99.9% | 99.95% |
| **Hypothesis Test Level** | | | | | | | |
| Two-sided | 0.20 | 0.10 | 0.05 | 0.02 | 0.01 | 0.002 | 0.001 |
| One-sided | 0.10 | 0.05 | 0.025 | 0.01 | 0.005 | 0.001 | 0.0005 |

| For One Sample | In General | | | | | | | |
|---|---|---|---|---|---|---|---|---|
| $n$ | Degrees of Freedom | | | | | | | |
| | | | | | Critical Values | | | |
| 2 | 1 | 3.078 | 6.314 | 12.706 | 31.821 | 63.657 | 318.309 | 636.619 |
| 3 | 2 | 1.886 | 2.920 | 4.303 | 6.965 | 9.925 | 22.327 | 31.599 |
| 4 | 3 | 1.638 | 2.353 | 3.182 | 4.541 | 5.841 | 10.215 | 12.924 |
| 5 | 4 | 1.533 | 2.132 | 2.776 | 3.747 | 4.604 | 7.173 | 8.610 |
| 6 | 5 | 1.476 | 2.015 | 2.571 | 3.365 | 4.032 | 5.893 | 6.869 |
| 7 | 6 | 1.440 | 1.943 | 2.447 | 3.143 | 3.707 | 5.208 | 5.959 |
| 8 | 7 | 1.415 | 1.895 | 2.365 | 2.998 | 3.499 | 4.785 | 5.408 |
| 9 | 8 | 1.397 | 1.860 | 2.306 | 2.896 | 3.355 | 4.501 | 5.041 |
| 10 | 9 | 1.383 | 1.833 | 2.262 | 2.821 | 3.250 | 4.297 | 4.781 |
| 11 | 10 | 1.372 | 1.812 | 2.228 | 2.764 | 3.169 | 4.144 | 4.587 |
| 12 | 11 | 1.363 | 1.796 | 2.201 | 2.718 | 3.106 | 4.025 | 4.437 |
| 13 | 12 | 1.356 | 1.782 | 2.179 | 2.681 | 3.055 | 3.930 | 4.318 |
| 14 | 13 | 1.350 | 1.771 | 2.160 | 2.650 | 3.012 | 3.852 | 4.221 |
| 15 | 14 | 1.345 | 1.761 | 2.145 | 2.624 | 2.977 | 3.787 | 4.140 |
| 16 | 15 | 1.341 | 1.753 | 2.131 | 2.602 | 2.947 | 3.733 | 4.073 |
| 17 | 16 | 1.337 | 1.746 | 2.120 | 2.583 | 2.921 | 3.686 | 4.015 |
| 18 | 17 | 1.333 | 1.740 | 2.110 | 2.567 | 2.898 | 3.646 | 3.965 |
| 19 | 18 | 1.330 | 1.734 | 2.101 | 2.552 | 2.878 | 3.610 | 3.922 |
| 20 | 19 | 1.328 | 1.729 | 2.093 | 2.539 | 2.861 | 3.579 | 3.883 |
| 21 | 20 | 1.325 | 1.725 | 2.086 | 2.528 | 2.845 | 3.552 | 3.850 |
| 22 | 21 | 1.323 | 1.721 | 2.080 | 2.518 | 2.831 | 3.527 | 3.819 |
| 23 | 22 | 1.321 | 1.717 | 2.074 | 2.508 | 2.819 | 3.505 | 3.792 |
| 24 | 23 | 1.319 | 1.714 | 2.069 | 2.500 | 2.807 | 3.485 | 3.768 |
| 25 | 24 | 1.318 | 1.711 | 2.064 | 2.492 | 2.797 | 3.467 | 3.745 |
| 26 | 25 | 1.316 | 1.708 | 2.060 | 2.485 | 2.787 | 3.450 | 3.725 |
| 27 | 26 | 1.315 | 1.706 | 2.056 | 2.479 | 2.779 | 3.435 | 3.707 |
| 28 | 27 | 1.314 | 1.703 | 2.052 | 2.473 | 2.771 | 3.421 | 3.690 |
| 29 | 28 | 1.313 | 1.701 | 2.048 | 2.467 | 2.763 | 3.408 | 3.674 |

## TABLE D.4  The *t* Table—cont'd

| Confidence Level | | | | | | | | |
|---|---|---|---|---|---|---|---|---|
| Two-sided | | 80% | 90% | 95% | 98% | 99% | 99.8% | 99.9% |
| One-sided | | 90% | 95% | 97.5% | 99% | 99.5% | 99.9% | 99.95% |
| **Hypothesis Test Level** | | | | | | | | |
| Two-sided | | 0.20 | 0.10 | 0.05 | 0.02 | 0.01 | 0.002 | 0.001 |
| One-sided | | 0.10 | 0.05 | 0.025 | 0.01 | 0.005 | 0.001 | 0.0005 |
| **For One Sample** | **In General** | | | | | | | |
| *n* | **Degrees of Freedom** | | | | | | | |
| | | | | Critical Values | | | | |
| 30 | 29 | 1.311 | 1.699 | 2.045 | 2.462 | 2.756 | 3.396 | 3.659 |
| 31 | 30 | 1.310 | 1.697 | 2.042 | 2.457 | 2.750 | 3.385 | 3.646 |
| 32 | 31 | 1.309 | 1.696 | 2.040 | 2.453 | 2.744 | 3.375 | 3.633 |
| 33 | 32 | 1.309 | 1.694 | 2.037 | 2.449 | 2.738 | 3.365 | 3.622 |
| 34 | 33 | 1.308 | 1.692 | 2.035 | 2.445 | 2.733 | 3.356 | 3.611 |
| 35 | 34 | 1.307 | 1.691 | 2.032 | 2.441 | 2.728 | 3.348 | 3.601 |
| 36 | 35 | 1.306 | 1.690 | 2.030 | 2.438 | 2.724 | 3.340 | 3.591 |
| 37 | 36 | 1.306 | 1.688 | 2.028 | 2.434 | 2.719 | 3.333 | 3.582 |
| 38 | 37 | 1.305 | 1.687 | 2.026 | 2.431 | 2.715 | 3.326 | 3.574 |
| 39 | 38 | 1.304 | 1.686 | 2.024 | 2.429 | 2.712 | 3.319 | 3.566 |
| 40 | 39 | 1.304 | 1.685 | 2.023 | 2.426 | 2.708 | 3.313 | 3.558 |
| | Infinity | 1.282 | 1.645 | 1.960 | 2.326 | 2.576 | 3.090 | 3.291 |

## TABLE D.5  $R^2$ Table: Level 5% Critical Values (Significant)

| Number of Cases | Number of *X* Variables (*k*) | | | | | | | | | |
|---|---|---|---|---|---|---|---|---|---|---|
| (*n*) | 1 | 2 | 3 | 4 | 5 | 6 | 7 | 8 | 9 | 10 |
| 3 | 0.994 | | | | | | | | | |
| 4 | 0.902 | 0.997 | | | | | | | | |
| 5 | 0.771 | 0.950 | 0.998 | | | | | | | |
| 6 | 0.658 | 0.864 | 0.966 | 0.999 | | | | | | |
| 7 | 0.569 | 0.776 | 0.903 | 0.975 | 0.999 | | | | | |
| 8 | 0.499 | 0.698 | 0.832 | 0.924 | 0.980 | 0.999 | | | | |
| 9 | 0.444 | 0.632 | 0.764 | 0.865 | 0.938 | 0.983 | 0.999 | | | |
| 10 | 0.399 | 0.575 | 0.704 | 0.806 | 0.887 | 0.947 | 0.985 | 0.999 | | |

*(Continued)*

**TABLE D.5** $R^2$ Table: Level 5% Critical Values (Significant)—cont'd

| Number of Cases (n) | Number of X Variables (k) | | | | | | | | | |
|---|---|---|---|---|---|---|---|---|---|---|
| | 1 | 2 | 3 | 4 | 5 | 6 | 7 | 8 | 9 | 10 |
| 11 | 0.362 | 0.527 | 0.651 | 0.751 | 0.835 | 0.902 | 0.954 | 0.987 | 1.000 | |
| 12 | 0.332 | 0.486 | 0.604 | 0.702 | 0.785 | 0.856 | 0.914 | 0.959 | 0.989 | 1.000 |
| 13 | 0.306 | 0.451 | 0.563 | 0.657 | 0.739 | 0.811 | 0.872 | 0.924 | 0.964 | 0.990 |
| 14 | 0.283 | 0.420 | 0.527 | 0.618 | 0.697 | 0.768 | 0.831 | 0.885 | 0.931 | 0.967 |
| 15 | 0.264 | 0.393 | 0.495 | 0.582 | 0.659 | 0.729 | 0.791 | 0.847 | 0.896 | 0.937 |
| 16 | 0.247 | 0.369 | 0.466 | 0.550 | 0.624 | 0.692 | 0.754 | 0.810 | 0.860 | 0.904 |
| 17 | 0.232 | 0.348 | 0.440 | 0.521 | 0.593 | 0.659 | 0.719 | 0.775 | 0.825 | 0.871 |
| 18 | 0.219 | 0.329 | 0.417 | 0.494 | 0.564 | 0.628 | 0.687 | 0.742 | 0.792 | 0.839 |
| 19 | 0.208 | 0.312 | 0.397 | 0.471 | 0.538 | 0.600 | 0.657 | 0.711 | 0.761 | 0.807 |
| 20 | 0.197 | 0.297 | 0.378 | 0.449 | 0.514 | 0.574 | 0.630 | 0.682 | 0.731 | 0.777 |
| 21 | 0.187 | 0.283 | 0.361 | 0.429 | 0.492 | 0.550 | 0.604 | 0.655 | 0.703 | 0.749 |
| 22 | 0.179 | 0.270 | 0.345 | 0.411 | 0.471 | 0.527 | 0.580 | 0.630 | 0.677 | 0.722 |
| 23 | 0.171 | 0.259 | 0.331 | 0.394 | 0.452 | 0.507 | 0.558 | 0.607 | 0.653 | 0.696 |
| 24 | 0.164 | 0.248 | 0.317 | 0.379 | 0.435 | 0.488 | 0.538 | 0.585 | 0.630 | 0.673 |
| 25 | 0.157 | 0.238 | 0.305 | 0.364 | 0.419 | 0.470 | 0.518 | 0.564 | 0.608 | 0.650 |
| 26 | 0.151 | 0.229 | 0.294 | 0.351 | 0.404 | 0.454 | 0.501 | 0.545 | 0.588 | 0.629 |
| 27 | 0.145 | 0.221 | 0.283 | 0.339 | 0.390 | 0.438 | 0.484 | 0.527 | 0.569 | 0.609 |
| 28 | 0.140 | 0.213 | 0.273 | 0.327 | 0.377 | 0.424 | 0.468 | 0.510 | 0.551 | 0.590 |
| 29 | 0.135 | 0.206 | 0.264 | 0.316 | 0.365 | 0.410 | 0.453 | 0.495 | 0.534 | 0.573 |
| 30 | 0.130 | 0.199 | 0.256 | 0.306 | 0.353 | 0.397 | 0.439 | 0.480 | 0.518 | 0.556 |
| 31 | 0.126 | 0.193 | 0.248 | 0.297 | 0.342 | 0.385 | 0.426 | 0.466 | 0.503 | 0.540 |
| 32 | 0.122 | 0.187 | 0.240 | 0.288 | 0.332 | 0.374 | 0.414 | 0.452 | 0.489 | 0.525 |
| 33 | 0.118 | 0.181 | 0.233 | 0.279 | 0.323 | 0.363 | 0.402 | 0.440 | 0.476 | 0.511 |
| 34 | 0.115 | 0.176 | 0.226 | 0.271 | 0.314 | 0.353 | 0.391 | 0.428 | 0.463 | 0.497 |
| 35 | 0.111 | 0.171 | 0.220 | 0.264 | 0.305 | 0.344 | 0.381 | 0.417 | 0.451 | 0.484 |
| 36 | 0.108 | 0.166 | 0.214 | 0.257 | 0.297 | 0.335 | 0.371 | 0.406 | 0.440 | 0.472 |
| 37 | 0.105 | 0.162 | 0.208 | 0.250 | 0.289 | 0.326 | 0.362 | 0.396 | 0.429 | 0.461 |
| 38 | 0.103 | 0.157 | 0.203 | 0.244 | 0.282 | 0.318 | 0.353 | 0.386 | 0.418 | 0.449 |
| 39 | 0.100 | 0.153 | 0.198 | 0.238 | 0.275 | 0.310 | 0.344 | 0.377 | 0.408 | 0.439 |
| 40 | 0.097 | 0.150 | 0.193 | 0.232 | 0.268 | 0.303 | 0.336 | 0.368 | 0.399 | 0.429 |
| 41 | 0.095 | 0.146 | 0.188 | 0.226 | 0.262 | 0.296 | 0.328 | 0.359 | 0.390 | 0.419 |
| 42 | 0.093 | 0.142 | 0.184 | 0.221 | 0.256 | 0.289 | 0.321 | 0.351 | 0.381 | 0.410 |
| 43 | 0.090 | 0.139 | 0.180 | 0.216 | 0.250 | 0.283 | 0.314 | 0.344 | 0.373 | 0.401 |
| 44 | 0.088 | 0.136 | 0.176 | 0.211 | 0.245 | 0.276 | 0.307 | 0.336 | 0.365 | 0.393 |

### TABLE D.5  $R^2$ Table: Level 5% Critical Values (Significant)—cont'd

| Number of Cases (n) | Number of X Variables (k) | | | | | | | | | |
|---|---|---|---|---|---|---|---|---|---|---|
| | 1 | 2 | 3 | 4 | 5 | 6 | 7 | 8 | 9 | 10 |
| 45 | 0.086 | 0.133 | 0.172 | 0.207 | 0.239 | 0.271 | 0.300 | 0.329 | 0.357 | 0.384 |
| 46 | 0.085 | 0.130 | 0.168 | 0.202 | 0.234 | 0.265 | 0.294 | 0.322 | 0.350 | 0.377 |
| 47 | 0.083 | 0.127 | 0.164 | 0.198 | 0.230 | 0.259 | 0.288 | 0.316 | 0.343 | 0.369 |
| 48 | 0.081 | 0.125 | 0.161 | 0.194 | 0.225 | 0.254 | 0.282 | 0.310 | 0.336 | 0.362 |
| 49 | 0.079 | 0.122 | 0.158 | 0.190 | 0.220 | 0.249 | 0.277 | 0.304 | 0.330 | 0.355 |
| 50 | 0.078 | 0.120 | 0.155 | 0.186 | 0.216 | 0.244 | 0.272 | 0.298 | 0.323 | 0.348 |
| 51 | 0.076 | 0.117 | 0.152 | 0.183 | 0.212 | 0.240 | 0.267 | 0.293 | 0.318 | 0.342 |
| 52 | 0.075 | 0.115 | 0.149 | 0.180 | 0.208 | 0.235 | 0.262 | 0.287 | 0.312 | 0.336 |
| 53 | 0.073 | 0.113 | 0.146 | 0.176 | 0.204 | 0.231 | 0.257 | 0.282 | 0.306 | 0.330 |
| 54 | 0.072 | 0.111 | 0.143 | 0.173 | 0.201 | 0.227 | 0.252 | 0.277 | 0.301 | 0.324 |
| 55 | 0.071 | 0.109 | 0.141 | 0.170 | 0.197 | 0.223 | 0.248 | 0.272 | 0.295 | 0.318 |
| 56 | 0.069 | 0.107 | 0.138 | 0.167 | 0.194 | 0.219 | 0.244 | 0.267 | 0.290 | 0.313 |
| 57 | 0.068 | 0.105 | 0.136 | 0.164 | 0.190 | 0.215 | 0.240 | 0.263 | 0.285 | 0.308 |
| 58 | 0.067 | 0.103 | 0.134 | 0.161 | 0.187 | 0.212 | 0.236 | 0.258 | 0.281 | 0.303 |
| 59 | 0.066 | 0.101 | 0.131 | 0.159 | 0.184 | 0.208 | 0.232 | 0.254 | 0.276 | 0.298 |
| 60 | 0.065 | 0.100 | 0.129 | 0.156 | 0.181 | 0.205 | 0.228 | 0.250 | 0.272 | 0.293 |
| Multiplier 1 | 3.84 | 5.99 | 7.82 | 9.49 | 11.07 | 12.59 | 14.07 | 15.51 | 16.92 | 18.31 |
| Multiplier 2 | 2.15 | −0.27 | −3.84 | −7.94 | −12.84 | −18.24 | −23.78 | −30.10 | −36.87 | −43.87 |

### TABLE D.6  $R^2$ Table: Level 1% Critical Values (Highly Significant)

| Number of Cases (n) | Number of X Variables (k) | | | | | | | | | |
|---|---|---|---|---|---|---|---|---|---|---|
| | 1 | 2 | 3 | 4 | 5 | 6 | 7 | 8 | 9 | 10 |
| 3 | 1.000 | | | | | | | | | |
| 4 | 0.980 | 1.000 | | | | | | | | |
| 5 | 0.919 | 0.990 | 1.000 | | | | | | | |
| 6 | 0.841 | 0.954 | 0.993 | 1.000 | | | | | | |
| 7 | 0.765 | 0.900 | 0.967 | 0.995 | 1.000 | | | | | |
| 8 | 0.696 | 0.842 | 0.926 | 0.975 | 0.996 | 1.000 | | | | |
| 9 | 0.636 | 0.785 | 0.879 | 0.941 | 0.979 | 0.997 | 1.000 | | | |
| 10 | 0.585 | 0.732 | 0.830 | 0.901 | 0.951 | 0.982 | 0.997 | 1.000 | | |
| 11 | 0.540 | 0.684 | 0.784 | 0.859 | 0.916 | 0.958 | 0.985 | 0.997 | 1.000 | |
| 12 | 0.501 | 0.641 | 0.740 | 0.818 | 0.879 | 0.928 | 0.963 | 0.987 | 0.998 | 1.000 |

*(Continued)*

**TABLE D.6** $R^2$ Table: Level 1% Critical Values (Highly Significant)—cont'd

| Number of Cases | Number of X Variables (k) | | | | | | | | | |
|---|---|---|---|---|---|---|---|---|---|---|
| (n) | 1 | 2 | 3 | 4 | 5 | 6 | 7 | 8 | 9 | 10 |
| 13 | 0.467 | 0.602 | 0.700 | 0.778 | 0.842 | 0.894 | 0.936 | 0.967 | 0.988 | 0.998 |
| 14 | 0.437 | 0.567 | 0.663 | 0.741 | 0.806 | 0.860 | 0.906 | 0.943 | 0.971 | 0.989 |
| 15 | 0.411 | 0.536 | 0.629 | 0.706 | 0.771 | 0.827 | 0.875 | 0.915 | 0.948 | 0.973 |
| 16 | 0.388 | 0.508 | 0.598 | 0.673 | 0.738 | 0.795 | 0.844 | 0.887 | 0.923 | 0.953 |
| 17 | 0.367 | 0.482 | 0.570 | 0.643 | 0.707 | 0.764 | 0.814 | 0.858 | 0.896 | 0.929 |
| 18 | 0.348 | 0.459 | 0.544 | 0.616 | 0.678 | 0.734 | 0.784 | 0.829 | 0.869 | 0.904 |
| 19 | 0.331 | 0.438 | 0.520 | 0.590 | 0.652 | 0.707 | 0.757 | 0.802 | 0.843 | 0.879 |
| 20 | 0.315 | 0.418 | 0.498 | 0.566 | 0.626 | 0.681 | 0.730 | 0.775 | 0.816 | 0.854 |
| 21 | 0.301 | 0.401 | 0.478 | 0.544 | 0.603 | 0.656 | 0.705 | 0.750 | 0.791 | 0.829 |
| 22 | 0.288 | 0.384 | 0.459 | 0.523 | 0.581 | 0.633 | 0.681 | 0.726 | 0.767 | 0.805 |
| 23 | 0.276 | 0.369 | 0.442 | 0.504 | 0.560 | 0.612 | 0.659 | 0.703 | 0.744 | 0.782 |
| 24 | 0.265 | 0.355 | 0.426 | 0.487 | 0.541 | 0.591 | 0.638 | 0.681 | 0.721 | 0.759 |
| 25 | 0.255 | 0.342 | 0.410 | 0.470 | 0.523 | 0.572 | 0.618 | 0.660 | 0.700 | 0.738 |
| 26 | 0.246 | 0.330 | 0.396 | 0.454 | 0.506 | 0.554 | 0.599 | 0.641 | 0.680 | 0.717 |
| 27 | 0.237 | 0.319 | 0.383 | 0.440 | 0.490 | 0.537 | 0.581 | 0.622 | 0.661 | 0.698 |
| 28 | 0.229 | 0.308 | 0.371 | 0.426 | 0.475 | 0.521 | 0.564 | 0.605 | 0.643 | 0.679 |
| 29 | 0.221 | 0.298 | 0.359 | 0.413 | 0.461 | 0.506 | 0.548 | 0.588 | 0.625 | 0.661 |
| 30 | 0.214 | 0.289 | 0.349 | 0.401 | 0.448 | 0.492 | 0.533 | 0.572 | 0.609 | 0.644 |
| 31 | 0.208 | 0.280 | 0.338 | 0.389 | 0.435 | 0.478 | 0.519 | 0.557 | 0.593 | 0.627 |
| 32 | 0.201 | 0.272 | 0.329 | 0.378 | 0.423 | 0.465 | 0.505 | 0.542 | 0.578 | 0.612 |
| 33 | 0.195 | 0.264 | 0.319 | 0.368 | 0.412 | 0.453 | 0.492 | 0.529 | 0.563 | 0.597 |
| 34 | 0.190 | 0.257 | 0.311 | 0.358 | 0.401 | 0.442 | 0.479 | 0.515 | 0.550 | 0.583 |
| 35 | 0.185 | 0.250 | 0.303 | 0.349 | 0.391 | 0.430 | 0.468 | 0.503 | 0.537 | 0.569 |
| 36 | 0.180 | 0.244 | 0.295 | 0.340 | 0.381 | 0.420 | 0.456 | 0.491 | 0.524 | 0.556 |
| 37 | 0.175 | 0.237 | 0.287 | 0.332 | 0.372 | 0.410 | 0.446 | 0.480 | 0.512 | 0.543 |
| 38 | 0.170 | 0.231 | 0.280 | 0.324 | 0.363 | 0.400 | 0.435 | 0.469 | 0.501 | 0.531 |
| 39 | 0.166 | 0.226 | 0.274 | 0.316 | 0.355 | 0.391 | 0.426 | 0.458 | 0.490 | 0.520 |
| 40 | 0.162 | 0.220 | 0.267 | 0.309 | 0.347 | 0.382 | 0.416 | 0.448 | 0.479 | 0.509 |
| 41 | 0.158 | 0.215 | 0.261 | 0.302 | 0.339 | 0.374 | 0.407 | 0.439 | 0.469 | 0.498 |
| 42 | 0.155 | 0.210 | 0.255 | 0.295 | 0.332 | 0.366 | 0.399 | 0.430 | 0.459 | 0.488 |
| 43 | 0.151 | 0.206 | 0.250 | 0.289 | 0.325 | 0.358 | 0.390 | 0.421 | 0.450 | 0.478 |
| 44 | 0.148 | 0.201 | 0.244 | 0.283 | 0.318 | 0.351 | 0.382 | 0.412 | 0.441 | 0.469 |
| 45 | 0.145 | 0.197 | 0.239 | 0.277 | 0.311 | 0.344 | 0.375 | 0.404 | 0.432 | 0.460 |
| 46 | 0.141 | 0.193 | 0.234 | 0.271 | 0.305 | 0.337 | 0.367 | 0.396 | 0.424 | 0.451 |
| 47 | 0.138 | 0.189 | 0.230 | 0.266 | 0.299 | 0.330 | 0.360 | 0.389 | 0.416 | 0.443 |

## TABLE D.6 $R^2$ Table: Level 1% Critical Values (Highly Significant)—cont'd

| Number of Cases | | | | | Number of $X$ Variables ($k$) | | | | | |
|---|---|---|---|---|---|---|---|---|---|---|
| ($n$) | 1 | 2 | 3 | 4 | 5 | 6 | 7 | 8 | 9 | 10 |
| 48 | 0.136 | 0.185 | 0.225 | 0.261 | 0.293 | 0.324 | 0.353 | 0.381 | 0.408 | 0.435 |
| 49 | 0.133 | 0.181 | 0.221 | 0.256 | 0.288 | 0.318 | 0.347 | 0.374 | 0.401 | 0.427 |
| 50 | 0.130 | 0.178 | 0.217 | 0.251 | 0.283 | 0.312 | 0.341 | 0.368 | 0.394 | 0.419 |
| 51 | 0.128 | 0.175 | 0.213 | 0.246 | 0.278 | 0.307 | 0.335 | 0.361 | 0.387 | 0.412 |
| 52 | 0.125 | 0.171 | 0.209 | 0.242 | 0.273 | 0.301 | 0.329 | 0.355 | 0.381 | 0.405 |
| 53 | 0.123 | 0.168 | 0.205 | 0.238 | 0.268 | 0.296 | 0.323 | 0.349 | 0.374 | 0.398 |
| 54 | 0.121 | 0.165 | 0.201 | 0.233 | 0.263 | 0.291 | 0.318 | 0.343 | 0.368 | 0.391 |
| 55 | 0.119 | 0.162 | 0.198 | 0.229 | 0.259 | 0.286 | 0.312 | 0.337 | 0.362 | 0.385 |
| 56 | 0.117 | 0.160 | 0.194 | 0.226 | 0.254 | 0.281 | 0.307 | 0.332 | 0.356 | 0.379 |
| 57 | 0.115 | 0.157 | 0.191 | 0.222 | 0.250 | 0.277 | 0.302 | 0.326 | 0.350 | 0.373 |
| 58 | 0.113 | 0.154 | 0.188 | 0.218 | 0.246 | 0.272 | 0.297 | 0.321 | 0.345 | 0.367 |
| 59 | 0.111 | 0.152 | 0.185 | 0.215 | 0.242 | 0.268 | 0.293 | 0.316 | 0.339 | 0.361 |
| 60 | 0.109 | 0.149 | 0.182 | 0.211 | 0.238 | 0.264 | 0.288 | 0.311 | 0.334 | 0.356 |
| Multiplier 1 | 6.63 | 9.21 | 11.35 | 13.28 | 15.09 | 16.81 | 18.48 | 20.09 | 21.67 | 23.21 |
| Multiplier 2 | −5.81 | −15.49 | −25.66 | −36.39 | −47.63 | −59.53 | −71.65 | −84.60 | −97.88 | −111.76 |

## TABLE D.7 $R^2$ Table: Level 0.1% Critical Values (Very Highly Significant)

| Number of Cases | | | | | Number of $X$ Variables ($k$) | | | | | |
|---|---|---|---|---|---|---|---|---|---|---|
| ($n$) | 1 | 2 | 3 | 4 | 5 | 6 | 7 | 8 | 9 | 10 |
| 3 | 1.000 | | | | | | | | | |
| 4 | 0.998 | 1.000 | | | | | | | | |
| 5 | 0.982 | 0.999 | 1.000 | | | | | | | |
| 6 | 0.949 | 0.990 | 0.999 | 1.000 | | | | | | |
| 7 | 0.904 | 0.968 | 0.993 | 0.999 | 1.000 | | | | | |
| 8 | 0.855 | 0.937 | 0.977 | 0.995 | 1.000 | 1.000 | | | | |
| 9 | 0.807 | 0.900 | 0.952 | 0.982 | 0.996 | 1.000 | 1.000 | | | |
| 10 | 0.761 | 0.861 | 0.922 | 0.961 | 0.985 | 0.996 | 1.000 | 1.000 | | |
| 11 | 0.717 | 0.822 | 0.889 | 0.936 | 0.967 | 0.987 | 0.997 | 1.000 | 1.000 | |
| 12 | 0.678 | 0.785 | 0.856 | 0.908 | 0.945 | 0.972 | 0.989 | 0.997 | 1.000 | 1.000 |
| 13 | 0.642 | 0.749 | 0.822 | 0.878 | 0.920 | 0.952 | 0.975 | 0.990 | 0.997 | 1.000 |
| 14 | 0.608 | 0.715 | 0.790 | 0.848 | 0.894 | 0.930 | 0.958 | 0.978 | 0.991 | 0.998 |
| 15 | 0.578 | 0.684 | 0.759 | 0.819 | 0.867 | 0.906 | 0.938 | 0.962 | 0.980 | 0.992 |
| 16 | 0.550 | 0.654 | 0.730 | 0.790 | 0.840 | 0.881 | 0.916 | 0.944 | 0.966 | 0.982 |
| 17 | 0.525 | 0.627 | 0.702 | 0.763 | 0.813 | 0.856 | 0.893 | 0.923 | 0.949 | 0.968 |

*(Continued)*

**TABLE D.7** $R^2$ Table: Level 0.1% Critical Values (Very Highly Significant)—cont'd

| Number of Cases | | | | | Number of $X$ Variables ($k$) | | | | | |
|---|---|---|---|---|---|---|---|---|---|---|
| ($n$) | 1 | 2 | 3 | 4 | 5 | 6 | 7 | 8 | 9 | 10 |
| 18 | 0.502 | 0.602 | 0.676 | 0.736 | 0.787 | 0.831 | 0.869 | 0.902 | 0.930 | 0.953 |
| 19 | 0.480 | 0.578 | 0.651 | 0.711 | 0.763 | 0.807 | 0.846 | 0.880 | 0.910 | 0.935 |
| 20 | 0.461 | 0.556 | 0.628 | 0.688 | 0.739 | 0.784 | 0.824 | 0.859 | 0.890 | 0.917 |
| 21 | 0.442 | 0.536 | 0.606 | 0.665 | 0.716 | 0.761 | 0.801 | 0.837 | 0.869 | 0.897 |
| 22 | 0.426 | 0.517 | 0.586 | 0.644 | 0.694 | 0.739 | 0.780 | 0.816 | 0.849 | 0.878 |
| 23 | 0.410 | 0.499 | 0.567 | 0.624 | 0.674 | 0.718 | 0.759 | 0.795 | 0.829 | 0.859 |
| 24 | 0.395 | 0.482 | 0.548 | 0.605 | 0.654 | 0.698 | 0.739 | 0.775 | 0.809 | 0.839 |
| 25 | 0.382 | 0.466 | 0.531 | 0.587 | 0.635 | 0.679 | 0.719 | 0.756 | 0.790 | 0.821 |
| 26 | 0.369 | 0.452 | 0.515 | 0.570 | 0.618 | 0.661 | 0.701 | 0.737 | 0.771 | 0.802 |
| 27 | 0.357 | 0.438 | 0.500 | 0.553 | 0.601 | 0.644 | 0.683 | 0.719 | 0.753 | 0.784 |
| 28 | 0.346 | 0.425 | 0.486 | 0.538 | 0.585 | 0.627 | 0.666 | 0.702 | 0.735 | 0.767 |
| 29 | 0.335 | 0.412 | 0.472 | 0.523 | 0.569 | 0.611 | 0.649 | 0.685 | 0.718 | 0.750 |
| 30 | 0.325 | 0.401 | 0.459 | 0.510 | 0.555 | 0.596 | 0.634 | 0.669 | 0.702 | 0.733 |
| 31 | 0.316 | 0.389 | 0.447 | 0.496 | 0.541 | 0.581 | 0.619 | 0.654 | 0.686 | 0.717 |
| 32 | 0.307 | 0.379 | 0.435 | 0.484 | 0.527 | 0.567 | 0.604 | 0.639 | 0.671 | 0.702 |
| 33 | 0.299 | 0.369 | 0.424 | 0.472 | 0.515 | 0.554 | 0.590 | 0.625 | 0.657 | 0.687 |
| 34 | 0.291 | 0.360 | 0.414 | 0.460 | 0.503 | 0.541 | 0.577 | 0.611 | 0.643 | 0.673 |
| 35 | 0.283 | 0.351 | 0.404 | 0.450 | 0.491 | 0.529 | 0.564 | 0.598 | 0.629 | 0.659 |
| 36 | 0.276 | 0.342 | 0.394 | 0.439 | 0.480 | 0.517 | 0.552 | 0.585 | 0.616 | 0.646 |
| 37 | 0.269 | 0.334 | 0.385 | 0.429 | 0.469 | 0.506 | 0.540 | 0.573 | 0.604 | 0.633 |
| 38 | 0.263 | 0.326 | 0.376 | 0.420 | 0.459 | 0.495 | 0.529 | 0.561 | 0.591 | 0.620 |
| 39 | 0.257 | 0.319 | 0.368 | 0.411 | 0.449 | 0.485 | 0.518 | 0.550 | 0.580 | 0.608 |
| 40 | 0.251 | 0.312 | 0.360 | 0.402 | 0.440 | 0.475 | 0.508 | 0.539 | 0.569 | 0.597 |
| 41 | 0.245 | 0.305 | 0.352 | 0.393 | 0.431 | 0.465 | 0.498 | 0.529 | 0.558 | 0.586 |
| 42 | 0.240 | 0.298 | 0.345 | 0.385 | 0.422 | 0.456 | 0.488 | 0.518 | 0.547 | 0.575 |
| 43 | 0.235 | 0.292 | 0.338 | 0.378 | 0.414 | 0.447 | 0.479 | 0.509 | 0.537 | 0.564 |
| 44 | 0.230 | 0.286 | 0.331 | 0.370 | 0.406 | 0.439 | 0.470 | 0.499 | 0.527 | 0.554 |
| 45 | 0.225 | 0.280 | 0.324 | 0.363 | 0.398 | 0.431 | 0.461 | 0.490 | 0.518 | 0.544 |
| 46 | 0.220 | 0.275 | 0.318 | 0.356 | 0.391 | 0.423 | 0.453 | 0.482 | 0.509 | 0.535 |
| 47 | 0.216 | 0.269 | 0.312 | 0.349 | 0.383 | 0.415 | 0.445 | 0.473 | 0.500 | 0.526 |
| 48 | 0.212 | 0.264 | 0.306 | 0.343 | 0.377 | 0.408 | 0.437 | 0.465 | 0.491 | 0.517 |
| 49 | 0.208 | 0.259 | 0.301 | 0.337 | 0.370 | 0.401 | 0.429 | 0.457 | 0.483 | 0.508 |
| 50 | 0.204 | 0.255 | 0.295 | 0.331 | 0.363 | 0.394 | 0.422 | 0.449 | 0.475 | 0.500 |
| 51 | 0.200 | 0.250 | 0.290 | 0.325 | 0.357 | 0.387 | 0.415 | 0.442 | 0.467 | 0.492 |
| 52 | 0.197 | 0.246 | 0.285 | 0.320 | 0.351 | 0.381 | 0.408 | 0.435 | 0.460 | 0.484 |
| 53 | 0.193 | 0.242 | 0.280 | 0.314 | 0.345 | 0.374 | 0.402 | 0.428 | 0.453 | 0.477 |
| 54 | 0.190 | 0.237 | 0.276 | 0.309 | 0.340 | 0.368 | 0.395 | 0.421 | 0.446 | 0.469 |

## TABLE D.7 $R^2$ Table: Level 0.1% Critical Values (Very Highly Significant)—cont'd

| Number of Cases (n) | Number of X Variables (k) | | | | | | | | | |
|---|---|---|---|---|---|---|---|---|---|---|
| | 1 | 2 | 3 | 4 | 5 | 6 | 7 | 8 | 9 | 10 |
| 55 | 0.186 | 0.233 | 0.271 | 0.304 | 0.334 | 0.362 | 0.389 | 0.414 | 0.439 | 0.462 |
| 56 | 0.183 | 0.230 | 0.267 | 0.299 | 0.329 | 0.357 | 0.383 | 0.408 | 0.432 | 0.455 |
| 57 | 0.180 | 0.226 | 0.262 | 0.294 | 0.324 | 0.351 | 0.377 | 0.402 | 0.426 | 0.448 |
| 58 | 0.177 | 0.222 | 0.258 | 0.290 | 0.319 | 0.346 | 0.371 | 0.396 | 0.419 | 0.442 |
| 59 | 0.174 | 0.219 | 0.254 | 0.285 | 0.314 | 0.341 | 0.366 | 0.390 | 0.413 | 0.436 |
| 60 | 0.172 | 0.215 | 0.250 | 0.281 | 0.309 | 0.336 | 0.361 | 0.384 | 0.407 | 0.429 |
| Multiplier 1 | 10.83 | 13.82 | 16.27 | 18.47 | 20.52 | 22.46 | 24.32 | 26.12 | 27.88 | 29.59 |
| Multiplier 2 | −31.57 | −54.02 | −75.12 | −96.26 | −117.47 | −138.94 | −160.86 | −183.33 | −206.28 | −229.55 |

## TABLE D.8 $R^2$ Table: Level 10% Critical Values

| Number of Cases (n) | Number of X Variables (k) | | | | | | | | | |
|---|---|---|---|---|---|---|---|---|---|---|
| | 1 | 2 | 3 | 4 | 5 | 6 | 7 | 8 | 9 | 10 |
| 3 | 0.976 | | | | | | | | | |
| 4 | 0.810 | 0.990 | | | | | | | | |
| 5 | 0.649 | 0.900 | 0.994 | | | | | | | |
| 6 | 0.532 | 0.785 | 0.932 | 0.996 | | | | | | |
| 7 | 0.448 | 0.684 | 0.844 | 0.949 | 0.997 | | | | | |
| 8 | 0.386 | 0.602 | 0.759 | 0.877 | 0.959 | 0.997 | | | | |
| 9 | 0.339 | 0.536 | 0.685 | 0.804 | 0.898 | 0.965 | 0.998 | | | |
| 10 | 0.302 | 0.482 | 0.622 | 0.738 | 0.835 | 0.914 | 0.970 | 0.998 | | |
| 11 | 0.272 | 0.438 | 0.568 | 0.680 | 0.775 | 0.857 | 0.925 | 0.974 | 0.998 | |
| 12 | 0.247 | 0.401 | 0.523 | 0.628 | 0.721 | 0.803 | 0.874 | 0.933 | 0.977 | 0.998 |
| 13 | 0.227 | 0.369 | 0.484 | 0.584 | 0.673 | 0.753 | 0.825 | 0.888 | 0.940 | 0.979 |
| 14 | 0.209 | 0.342 | 0.450 | 0.545 | 0.630 | 0.708 | 0.779 | 0.842 | 0.899 | 0.946 |
| 15 | 0.194 | 0.319 | 0.420 | 0.510 | 0.592 | 0.667 | 0.736 | 0.799 | 0.857 | 0.907 |
| 16 | 0.181 | 0.298 | 0.394 | 0.480 | 0.558 | 0.630 | 0.697 | 0.759 | 0.816 | 0.868 |
| 17 | 0.170 | 0.280 | 0.371 | 0.453 | 0.527 | 0.596 | 0.661 | 0.721 | 0.778 | 0.830 |
| 18 | 0.160 | 0.264 | 0.351 | 0.428 | 0.499 | 0.566 | 0.628 | 0.687 | 0.742 | 0.794 |
| 19 | 0.151 | 0.250 | 0.332 | 0.406 | 0.474 | 0.538 | 0.598 | 0.655 | 0.709 | 0.760 |
| 20 | 0.143 | 0.237 | 0.316 | 0.386 | 0.452 | 0.513 | 0.571 | 0.626 | 0.679 | 0.729 |
| 21 | 0.136 | 0.226 | 0.301 | 0.368 | 0.431 | 0.490 | 0.546 | 0.599 | 0.650 | 0.699 |
| 22 | 0.129 | 0.215 | 0.287 | 0.352 | 0.412 | 0.469 | 0.523 | 0.575 | 0.624 | 0.671 |
| 23 | 0.124 | 0.206 | 0.275 | 0.337 | 0.395 | 0.450 | 0.502 | 0.552 | 0.600 | 0.646 |
| 24 | 0.118 | 0.197 | 0.263 | 0.323 | 0.379 | 0.432 | 0.482 | 0.530 | 0.577 | 0.622 |

*(Continued)*

## TABLE D.8 $R^2$ Table: Level 10% Critical Values—cont'd

| Number of Cases (n) | Number of X Variables (k) | | | | | | | | | |
|---|---|---|---|---|---|---|---|---|---|---|
| | 1 | 2 | 3 | 4 | 5 | 6 | 7 | 8 | 9 | 10 |
| 25 | 0.113 | 0.189 | 0.253 | 0.310 | 0.364 | 0.415 | 0.464 | 0.511 | 0.556 | 0.599 |
| 26 | 0.109 | 0.181 | 0.243 | 0.298 | 0.350 | 0.400 | 0.447 | 0.492 | 0.536 | 0.579 |
| 27 | 0.105 | 0.175 | 0.234 | 0.287 | 0.338 | 0.386 | 0.431 | 0.475 | 0.518 | 0.559 |
| 28 | 0.101 | 0.168 | 0.225 | 0.277 | 0.326 | 0.372 | 0.417 | 0.459 | 0.501 | 0.541 |
| 29 | 0.097 | 0.162 | 0.218 | 0.268 | 0.315 | 0.360 | 0.403 | 0.444 | 0.484 | 0.523 |
| 30 | 0.094 | 0.157 | 0.210 | 0.259 | 0.305 | 0.348 | 0.390 | 0.430 | 0.469 | 0.507 |
| 31 | 0.091 | 0.152 | 0.203 | 0.251 | 0.295 | 0.337 | 0.378 | 0.417 | 0.455 | 0.492 |
| 32 | 0.088 | 0.147 | 0.197 | 0.243 | 0.286 | 0.327 | 0.366 | 0.405 | 0.442 | 0.478 |
| 33 | 0.085 | 0.142 | 0.191 | 0.236 | 0.277 | 0.317 | 0.356 | 0.393 | 0.429 | 0.464 |
| 34 | 0.082 | 0.138 | 0.185 | 0.229 | 0.269 | 0.308 | 0.346 | 0.382 | 0.417 | 0.451 |
| 35 | 0.080 | 0.134 | 0.180 | 0.222 | 0.262 | 0.300 | 0.336 | 0.371 | 0.406 | 0.439 |
| 36 | 0.078 | 0.130 | 0.175 | 0.216 | 0.255 | 0.291 | 0.327 | 0.361 | 0.395 | 0.427 |
| 37 | 0.075 | 0.127 | 0.170 | 0.210 | 0.248 | 0.284 | 0.318 | 0.352 | 0.385 | 0.416 |
| 38 | 0.073 | 0.123 | 0.166 | 0.205 | 0.241 | 0.276 | 0.310 | 0.343 | 0.375 | 0.406 |
| 39 | 0.071 | 0.120 | 0.162 | 0.199 | 0.235 | 0.269 | 0.302 | 0.334 | 0.366 | 0.396 |
| 40 | 0.070 | 0.117 | 0.157 | 0.194 | 0.229 | 0.263 | 0.295 | 0.326 | 0.357 | 0.387 |
| 41 | 0.068 | 0.114 | 0.154 | 0.190 | 0.224 | 0.257 | 0.288 | 0.319 | 0.348 | 0.378 |
| 42 | 0.066 | 0.111 | 0.150 | 0.185 | 0.219 | 0.250 | 0.281 | 0.311 | 0.340 | 0.369 |
| 43 | 0.065 | 0.109 | 0.146 | 0.181 | 0.214 | 0.245 | 0.275 | 0.304 | 0.333 | 0.361 |
| 44 | 0.063 | 0.106 | 0.143 | 0.177 | 0.209 | 0.239 | 0.269 | 0.297 | 0.325 | 0.353 |
| 45 | 0.062 | 0.104 | 0.140 | 0.173 | 0.204 | 0.234 | 0.263 | 0.291 | 0.318 | 0.345 |
| 46 | 0.060 | 0.102 | 0.137 | 0.169 | 0.200 | 0.229 | 0.257 | 0.285 | 0.312 | 0.338 |
| 47 | 0.059 | 0.099 | 0.134 | 0.166 | 0.196 | 0.224 | 0.252 | 0.279 | 0.305 | 0.331 |
| 48 | 0.058 | 0.097 | 0.131 | 0.162 | 0.191 | 0.220 | 0.247 | 0.273 | 0.299 | 0.324 |
| 49 | 0.057 | 0.095 | 0.128 | 0.159 | 0.188 | 0.215 | 0.242 | 0.268 | 0.293 | 0.318 |
| 50 | 0.055 | 0.093 | 0.126 | 0.156 | 0.184 | 0.211 | 0.237 | 0.263 | 0.287 | 0.312 |
| 51 | 0.054 | 0.092 | 0.123 | 0.153 | 0.180 | 0.207 | 0.233 | 0.258 | 0.282 | 0.306 |
| 52 | 0.053 | 0.090 | 0.121 | 0.150 | 0.177 | 0.203 | 0.228 | 0.253 | 0.277 | 0.300 |
| 53 | 0.052 | 0.088 | 0.119 | 0.147 | 0.174 | 0.199 | 0.224 | 0.248 | 0.272 | 0.295 |
| 54 | 0.051 | 0.086 | 0.116 | 0.144 | 0.170 | 0.196 | 0.220 | 0.244 | 0.267 | 0.290 |
| 55 | 0.050 | 0.085 | 0.114 | 0.142 | 0.167 | 0.192 | 0.216 | 0.239 | 0.262 | 0.284 |
| 56 | 0.049 | 0.083 | 0.112 | 0.139 | 0.164 | 0.189 | 0.212 | 0.235 | 0.257 | 0.279 |
| 57 | 0.049 | 0.082 | 0.110 | 0.137 | 0.162 | 0.185 | 0.209 | 0.231 | 0.253 | 0.275 |
| 58 | 0.048 | 0.080 | 0.108 | 0.134 | 0.159 | 0.182 | 0.205 | 0.227 | 0.249 | 0.270 |
| 59 | 0.047 | 0.079 | 0.107 | 0.132 | 0.156 | 0.179 | 0.202 | 0.223 | 0.245 | 0.266 |
| 60 | 0.046 | 0.078 | 0.105 | 0.130 | 0.153 | 0.176 | 0.198 | 0.220 | 0.241 | 0.261 |
| Multiplier 1 | 2.71 | 4.61 | 6.25 | 7.78 | 9.24 | 10.65 | 12.02 | 13.36 | 14.68 | 15.99 |
| Multiplier 2 | 3.12 | 3.08 | 2.00 | 0.32 | −1.92 | −4.75 | −7.59 | −11.12 | −14.94 | −19.05 |

**TABLE D.9** *F* Table: Level 5% Critical Value (Significant)

Numerator Degrees of Freedom ($k - 1$ for between-sample variability in one-way ANOVA)

| Denominator Degrees of Freedom ($n - k$ for within-sample variability in one-way ANOVA) | 1 | 2 | 3 | 4 | 5 | 6 | 7 | 8 | 9 | 10 | 12 | 15 | 20 | 30 | 60 | 120 | Infinity |
|---|---|---|---|---|---|---|---|---|---|---|---|---|---|---|---|---|---|
| 1 | 161.45 | 199.50 | 215.71 | 224.58 | 230.16 | 233.99 | 236.77 | 238.88 | 240.54 | 241.88 | 243.91 | 245.95 | 248.01 | 250.10 | 252.20 | 253.25 | 254.32 |
| 2 | 18.513 | 19.000 | 19.164 | 19.247 | 19.296 | 19.330 | 19.353 | 19.371 | 19.385 | 19.396 | 19.413 | 19.429 | 19.446 | 19.462 | 19.479 | 19.487 | 19.496 |
| 3 | 10.128 | 9.552 | 9.277 | 9.117 | 9.013 | 8.941 | 8.887 | 8.845 | 8.812 | 8.786 | 8.745 | 8.703 | 8.660 | 8.617 | 8.572 | 8.549 | 8.526 |
| 4 | 7.709 | 6.944 | 6.591 | 6.388 | 6.256 | 6.163 | 6.094 | 6.041 | 5.999 | 5.964 | 5.912 | 5.858 | 5.803 | 5.746 | 5.688 | 5.658 | 5.628 |
| 5 | 6.608 | 5.786 | 5.409 | 5.192 | 5.050 | 4.950 | 4.876 | 4.818 | 4.772 | 4.735 | 4.678 | 4.619 | 4.558 | 4.496 | 4.431 | 4.398 | 4.365 |
| 6 | 5.987 | 5.143 | 4.757 | 4.534 | 4.387 | 4.284 | 4.207 | 4.147 | 4.099 | 4.060 | 4.000 | 3.938 | 3.874 | 3.808 | 3.740 | 3.705 | 3.669 |
| 7 | 5.591 | 4.737 | 4.347 | 4.120 | 3.972 | 3.866 | 3.787 | 3.726 | 3.677 | 3.637 | 3.575 | 3.511 | 3.445 | 3.376 | 3.304 | 3.267 | 3.230 |
| 8 | 5.318 | 4.459 | 4.066 | 3.838 | 3.687 | 3.581 | 3.500 | 3.438 | 3.388 | 3.347 | 3.284 | 3.218 | 3.150 | 3.079 | 3.005 | 2.967 | 2.928 |
| 9 | 5.117 | 4.256 | 3.863 | 3.633 | 3.482 | 3.374 | 3.293 | 3.230 | 3.179 | 3.137 | 3.073 | 3.006 | 2.936 | 2.864 | 2.787 | 2.748 | 2.707 |
| 10 | 4.965 | 4.103 | 3.708 | 3.478 | 3.326 | 3.217 | 3.135 | 3.072 | 3.020 | 2.978 | 2.913 | 2.845 | 2.774 | 2.700 | 2.621 | 2.580 | 2.538 |
| 12 | 4.747 | 3.885 | 3.490 | 3.259 | 3.106 | 2.996 | 2.913 | 2.849 | 2.796 | 2.753 | 2.687 | 2.617 | 2.544 | 2.466 | 2.384 | 2.341 | 2.296 |
| 15 | 4.543 | 3.682 | 3.287 | 3.056 | 2.901 | 2.790 | 2.707 | 2.641 | 2.588 | 2.544 | 2.475 | 2.403 | 2.328 | 2.247 | 2.160 | 2.114 | 2.066 |
| 20 | 4.351 | 3.493 | 3.098 | 2.866 | 2.711 | 2.599 | 2.514 | 2.447 | 2.393 | 2.348 | 2.278 | 2.203 | 2.124 | 2.039 | 1.946 | 1.896 | 1.843 |
| 30 | 4.171 | 3.316 | 2.922 | 2.690 | 2.534 | 2.421 | 2.334 | 2.266 | 2.211 | 2.165 | 2.092 | 2.015 | 1.932 | 1.841 | 1.740 | 1.683 | 1.622 |
| 60 | 4.001 | 3.150 | 2.758 | 2.525 | 2.368 | 2.254 | 2.167 | 2.097 | 2.040 | 1.993 | 1.917 | 1.836 | 1.748 | 1.649 | 1.534 | 1.467 | 1.389 |
| 120 | 3.920 | 3.072 | 2.680 | 2.447 | 2.290 | 2.175 | 2.087 | 2.016 | 1.959 | 1.910 | 1.834 | 1.750 | 1.659 | 1.554 | 1.429 | 1.352 | 1.254 |
| Infinity | 3.841 | 2.996 | 2.605 | 2.372 | 2.214 | 2.099 | 2.010 | 1.938 | 1.880 | 1.831 | 1.752 | 1.666 | 1.571 | 1.459 | 1.318 | 1.221 | 1.000 |

**TABLE D.10** *F* Table: Level 1% Critical Values (Highly Significant)

| Denominator Degrees of Freedom ($n - k$ for within-sample variability in one-way ANOVA) | Numerator Degrees of Freedom ($k - 1$ for between-sample variability in one-way ANOVA) | | | | | | | | | | | | | | | | |
|---|---|---|---|---|---|---|---|---|---|---|---|---|---|---|---|---|---|
| | 1 | 2 | 3 | 4 | 5 | 6 | 7 | 8 | 9 | 10 | 12 | 15 | 20 | 30 | 60 | 120 | Infinity |
| 1 | 4052.2 | 4999.5 | 5403.4 | 5624.6 | 5763.7 | 5859.0 | 5928.4 | 5891.1 | 6022.5 | 6055.8 | 6106.3 | 6157.3 | 6208.7 | 6260.6 | 6313.0 | 6339.4 | 6365.9 |
| 2 | 98.501 | 98.995 | 99.159 | 99.240 | 99.299 | 99.333 | 99.356 | 99.374 | 99.388 | 99.399 | 99.416 | 99.432 | 99.449 | 99.466 | 99.482 | 99.491 | 99.499 |
| 3 | 34.116 | 30.816 | 29.456 | 28.709 | 28.236 | 27.910 | 27.671 | 27.488 | 27.344 | 27.228 | 27.051 | 26.871 | 26.689 | 26.503 | 26.315 | 26.220 | 26.125 |
| 4 | 21.197 | 18.000 | 16.694 | 15.977 | 15.522 | 15.207 | 14.976 | 14.799 | 14.659 | 14.546 | 14.374 | 14.198 | 14.020 | 13.838 | 13.652 | 13.558 | 13.463 |
| 5 | 16.258 | 13.274 | 12.060 | 11.392 | 10.967 | 10.672 | 10.455 | 10.289 | 10.158 | 10.051 | 9.888 | 9.722 | 9.553 | 9.379 | 9.202 | 9.112 | 9.021 |
| 6 | 13.745 | 10.925 | 9.780 | 9.148 | 8.746 | 8.466 | 8.260 | 8.102 | 7.976 | 7.874 | 7.718 | 7.559 | 7.396 | 7.229 | 7.057 | 6.969 | 6.880 |
| 7 | 12.246 | 9.547 | 8.451 | 7.847 | 7.460 | 7.191 | 6.993 | 6.840 | 6.719 | 6.620 | 6.469 | 6.314 | 6.155 | 5.992 | 5.823 | 5.737 | 5.650 |
| 8 | 11.258 | 8.649 | 7.591 | 7.006 | 6.632 | 6.371 | 6.178 | 6.029 | 5.911 | 5.814 | 5.667 | 5.515 | 5.359 | 5.198 | 5.032 | 4.946 | 4.859 |
| 9 | 10.561 | 8.021 | 6.992 | 6.422 | 6.057 | 5.802 | 5.613 | 5.467 | 5.351 | 5.257 | 5.111 | 4.962 | 4.808 | 4.649 | 4.483 | 4.398 | 4.311 |
| 10 | 10.044 | 7.559 | 6.552 | 5.994 | 5.636 | 5.386 | 5.200 | 5.057 | 4.942 | 4.849 | 4.706 | 4.558 | 4.405 | 4.247 | 4.082 | 3.996 | 3.909 |
| 12 | 9.330 | 6.927 | 5.953 | 5.412 | 5.064 | 4.821 | 4.640 | 4.499 | 4.388 | 4.296 | 4.155 | 4.010 | 3.858 | 3.701 | 3.535 | 3.449 | 3.361 |
| 15 | 8.683 | 6.359 | 5.417 | 4.893 | 4.556 | 4.318 | 4.142 | 4.004 | 3.895 | 3.805 | 3.666 | 3.522 | 3.372 | 3.214 | 3.047 | 2.959 | 2.868 |
| 20 | 8.096 | 5.849 | 4.938 | 4.431 | 4.103 | 3.871 | 3.699 | 3.564 | 3.457 | 3.368 | 3.231 | 3.088 | 2.938 | 2.778 | 2.608 | 2.517 | 2.421 |
| 30 | 7.562 | 5.390 | 4.510 | 4.018 | 3.699 | 3.473 | 3.304 | 3.173 | 3.067 | 2.979 | 2.843 | 2.700 | 2.549 | 2.386 | 2.208 | 2.111 | 2.006 |
| 60 | 7.077 | 4.977 | 4.126 | 3.649 | 3.339 | 3.119 | 2.953 | 2.823 | 2.718 | 2.632 | 2.496 | 2.352 | 2.198 | 2.028 | 1.836 | 1.726 | 1.601 |
| 120 | 6.851 | 4.786 | 3.949 | 3.480 | 3.174 | 2.956 | 2.792 | 2.663 | 2.559 | 2.472 | 2.336 | 2.191 | 2.035 | 1.860 | 1.656 | 1.533 | 1.381 |
| Infinity | 6.635 | 4.605 | 3.782 | 3.319 | 3.017 | 2.802 | 2.639 | 2.511 | 2.407 | 2.321 | 2.185 | 2.039 | 1.878 | 1.696 | 1.473 | 1.325 | 1.000 |

**TABLE D.11** $F$ Table: Level 0.1% Critical Values (Very Highly Significant)

| Denominator Degrees of Freedom ($n - k$ for within-sample variability in one-way ANOVA) | Numerator Degrees of Freedom ($k - 1$ for between-sample variability in one-way ANOVA) | | | | | | | | | | | | | | | | |
|---|---|---|---|---|---|---|---|---|---|---|---|---|---|---|---|---|---|
| | 1 | 2 | 3 | 4 | 5 | 6 | 7 | 8 | 9 | 10 | 12 | 15 | 20 | 30 | 60 | 120 | Infinity |
| 1 | 405284 | 500000 | 540379 | 562500 | 576405 | 585937 | 592873 | 598144 | 602284 | 605621 | 610668 | 615764 | 620908 | 626099 | 631337 | 633972 | 636629 |
| 2 | 998.50 | 999.00 | 999.17 | 999.25 | 999.30 | 999.33 | 999.36 | 999.38 | 999.39 | 999.40 | 999.42 | 999.43 | 999.45 | 999.47 | 999.48 | 999.49 | 999.50 |
| 3 | 167.03 | 148.50 | 141.11 | 137.10 | 134.58 | 132.85 | 131.58 | 130.62 | 129.86 | 129.25 | 128.32 | 127.37 | 126.42 | 125.45 | 124.47 | 123.97 | 123.47 |
| 4 | 74.137 | 61.246 | 56.177 | 53.436 | 51.712 | 50.525 | 49.658 | 48.996 | 48.475 | 48.053 | 47.412 | 46.761 | 46.100 | 45.429 | 44.746 | 44.400 | 44.051 |
| 5 | 47.181 | 37.122 | 33.202 | 31.085 | 29.752 | 28.834 | 28.163 | 27.649 | 27.244 | 26.917 | 26.418 | 25.911 | 25.395 | 24.869 | 24.333 | 24.060 | 23.785 |
| 6 | 35.507 | 27.000 | 23.703 | 21.924 | 20.803 | 20.030 | 19.463 | 19.030 | 18.688 | 18.411 | 17.989 | 17.559 | 17.120 | 16.672 | 16.214 | 15.981 | 15.745 |
| 7 | 29.245 | 21.689 | 18.772 | 17.198 | 16.206 | 15.521 | 15.019 | 14.634 | 14.330 | 14.083 | 13.707 | 13.324 | 12.932 | 12.530 | 12.119 | 11.909 | 11.697 |
| 8 | 25.415 | 18.494 | 15.829 | 14.392 | 13.485 | 12.858 | 12.398 | 12.046 | 11.767 | 11.540 | 11.194 | 10.841 | 10.480 | 10.109 | 9.727 | 9.532 | 9.334 |
| 9 | 22.857 | 16.387 | 13.902 | 12.560 | 11.714 | 11.128 | 10.698 | 10.368 | 10.107 | 9.894 | 9.570 | 9.238 | 8.898 | 8.548 | 8.187 | 8.001 | 7.813 |
| 10 | 21.040 | 14.905 | 12.553 | 11.283 | 10.481 | 9.926 | 9.517 | 9.204 | 8.956 | 8.754 | 8.445 | 8.129 | 7.804 | 7.469 | 7.122 | 6.944 | 6.762 |
| 12 | 18.643 | 12.974 | 10.804 | 9.633 | 8.892 | 8.379 | 8.001 | 7.710 | 7.480 | 7.292 | 7.005 | 6.709 | 6.405 | 6.090 | 5.762 | 5.593 | 5.420 |
| 15 | 16.587 | 11.339 | 9.335 | 8.253 | 7.567 | 7.092 | 6.741 | 6.471 | 6.256 | 6.081 | 5.812 | 5.535 | 5.248 | 4.950 | 4.638 | 4.475 | 4.307 |
| 20 | 14.818 | 9.953 | 8.098 | 7.095 | 6.460 | 6.018 | 5.692 | 5.440 | 5.239 | 5.075 | 4.823 | 4.562 | 4.290 | 4.005 | 3.703 | 3.544 | 3.378 |
| 30 | 13.293 | 8.773 | 7.054 | 6.124 | 5.534 | 5.122 | 4.817 | 4.581 | 4.393 | 4.239 | 4.000 | 3.753 | 3.493 | 3.217 | 2.920 | 2.759 | 2.589 |
| 60 | 11.973 | 7.767 | 6.171 | 5.307 | 4.757 | 4.372 | 4.086 | 3.865 | 3.687 | 3.541 | 3.315 | 3.078 | 2.827 | 2.555 | 2.252 | 2.082 | 1.890 |
| 120 | 11.378 | 7.321 | 5.781 | 4.947 | 4.416 | 4.044 | 3.767 | 3.552 | 3.379 | 3.237 | 3.016 | 2.783 | 2.534 | 2.262 | 1.950 | 1.767 | 1.543 |
| Infinity | 10.827 | 6.908 | 5.422 | 4.617 | 4.103 | 3.743 | 3.475 | 3.266 | 3.097 | 2.959 | 2.742 | 2.513 | 2.266 | 1.990 | 1.660 | 1.447 | 1.000 |

**TABLE D.12** *F* Table: Level 10% Critical Values

| Denominator Degrees of Freedom ($n - k$ for within-sample variability in one-way ANOVA) | Numerator Degrees of Freedom ($k - 1$ for between-sample variability in one-way ANOVA) | | | | | | | | | | | | | | | | |
|---|---|---|---|---|---|---|---|---|---|---|---|---|---|---|---|---|---|
| | 1 | 2 | 3 | 4 | 5 | 6 | 7 | 8 | 9 | 10 | 12 | 15 | 20 | 30 | 60 | 120 | Infinity |
| 1 | 39.863 | 49.500 | 53.593 | 55.833 | 57.240 | 58.204 | 58.906 | 59.439 | 59.858 | 60.195 | 60.705 | 61.220 | 61.740 | 62.265 | 62.794 | 63.061 | 63.328 |
| 2 | 8.526 | 9.000 | 9.162 | 9.243 | 9.293 | 9.326 | 9.349 | 9.367 | 9.381 | 9.392 | 9.408 | 9.425 | 9.441 | 9.458 | 9.475 | 9.483 | 9.491 |
| 3 | 5.538 | 5.462 | 5.391 | 5.343 | 5.309 | 5.285 | 5.266 | 5.252 | 5.240 | 5.230 | 5.216 | 5.200 | 5.184 | 5.168 | 5.151 | 5.143 | 5.134 |
| 4 | 4.545 | 4.325 | 4.191 | 4.107 | 4.051 | 4.010 | 3.979 | 3.955 | 3.936 | 3.920 | 3.896 | 3.870 | 3.844 | 3.817 | 3.790 | 3.775 | 3.761 |
| 5 | 4.060 | 3.780 | 3.619 | 3.520 | 3.453 | 3.405 | 3.368 | 3.339 | 3.316 | 3.297 | 3.268 | 3.238 | 3.207 | 3.174 | 3.140 | 3.123 | 3.105 |
| 6 | 3.776 | 3.463 | 3.289 | 3.181 | 3.108 | 3.055 | 3.014 | 2.983 | 2.958 | 2.937 | 2.905 | 2.871 | 2.836 | 2.800 | 2.762 | 2.742 | 2.722 |
| 7 | 3.589 | 3.257 | 3.074 | 2.961 | 2.883 | 2.827 | 2.785 | 2.752 | 2.725 | 2.703 | 2.668 | 2.632 | 2.595 | 2.555 | 2.514 | 2.493 | 2.471 |
| 8 | 3.458 | 3.113 | 2.924 | 2.806 | 2.726 | 2.668 | 2.624 | 2.589 | 2.561 | 2.538 | 2.502 | 2.464 | 2.425 | 2.383 | 2.339 | 2.316 | 2.293 |
| 9 | 3.360 | 3.006 | 2.813 | 2.693 | 2.611 | 2.551 | 2.505 | 2.469 | 2.440 | 2.416 | 2.379 | 2.340 | 2.298 | 2.255 | 2.208 | 2.184 | 2.159 |
| 10 | 3.285 | 2.924 | 2.728 | 2.605 | 2.522 | 2.461 | 2.414 | 2.377 | 2.347 | 2.323 | 2.284 | 2.244 | 2.201 | 2.155 | 2.107 | 2.082 | 2.055 |
| 12 | 3.177 | 2.807 | 2.606 | 2.480 | 2.394 | 2.331 | 2.283 | 2.245 | 2.214 | 2.188 | 2.147 | 2.105 | 2.060 | 2.011 | 1.960 | 1.932 | 1.904 |
| 15 | 3.073 | 2.695 | 2.490 | 2.361 | 2.273 | 2.208 | 2.158 | 2.119 | 2.086 | 2.059 | 2.017 | 1.972 | 1.924 | 1.873 | 1.817 | 1.787 | 1.755 |
| 20 | 2.975 | 2.589 | 2.380 | 2.249 | 2.158 | 2.091 | 2.040 | 1.999 | 1.965 | 1.937 | 1.892 | 1.845 | 1.794 | 1.738 | 1.677 | 1.643 | 1.607 |
| 30 | 2.881 | 2.489 | 2.276 | 2.142 | 2.049 | 1.980 | 1.927 | 1.884 | 1.849 | 1.819 | 1.773 | 1.722 | 1.667 | 1.606 | 1.538 | 1.499 | 1.456 |
| 60 | 2.791 | 2.393 | 2.177 | 2.041 | 1.946 | 1.875 | 1.819 | 1.775 | 1.738 | 1.707 | 1.657 | 1.603 | 1.543 | 1.476 | 1.395 | 1.348 | 1.291 |
| 120 | 2.748 | 2.347 | 2.130 | 1.992 | 1.896 | 1.824 | 1.767 | 1.722 | 1.684 | 1.652 | 1.601 | 1.545 | 1.482 | 1.409 | 1.320 | 1.265 | 1.193 |
| Infinity | 2.706 | 2.303 | 2.084 | 1.945 | 1.847 | 1.774 | 1.717 | 1.670 | 1.632 | 1.599 | 1.546 | 1.487 | 1.421 | 1.342 | 1.240 | 1.169 | 1.000 |

**TABLE D.13  Ranks for the Sign Test**

| Modified Sample Size, $m$ | 5% Test Level Sign Test Is Significant If Number Is Either Less than | or | More than | 1% Test Level Sign Test Is Significant If Number Is Either Less than | or | More than |
|---|---|---|---|---|---|---|
| 6 | 1 | | 5 | — | | — |
| 7 | 1 | | 6 | — | | — |
| 8 | 1 | | 7 | 1 | | 7 |
| 9 | 2 | | 7 | 1 | | 8 |
| 10 | 2 | | 8 | 1 | | 9 |
| 11 | 2 | | 9 | 1 | | 10 |
| 12 | 3 | | 9 | 2 | | 10 |
| 13 | 3 | | 10 | 2 | | 11 |
| 14 | 3 | | 11 | 2 | | 12 |
| 15 | 4 | | 11 | 3 | | 12 |
| 16 | 4 | | 12 | 3 | | 13 |
| 17 | 5 | | 12 | 3 | | 14 |
| 18 | 5 | | 13 | 4 | | 14 |
| 19 | 5 | | 14 | 4 | | 15 |
| 20 | 6 | | 14 | 4 | | 16 |
| 21 | 6 | | 15 | 5 | | 16 |
| 22 | 6 | | 16 | 5 | | 17 |
| 23 | 7 | | 16 | 5 | | 18 |
| 24 | 7 | | 17 | 6 | | 18 |
| 25 | 8 | | 17 | 6 | | 19 |
| 26 | 8 | | 18 | 7 | | 19 |
| 27 | 8 | | 19 | 7 | | 20 |
| 28 | 9 | | 19 | 7 | | 21 |
| 29 | 9 | | 20 | 8 | | 21 |
| 30 | 10 | | 20 | 8 | | 22 |
| 31 | 10 | | 21 | 8 | | 23 |
| 32 | 10 | | 22 | 9 | | 23 |
| 33 | 11 | | 22 | 9 | | 24 |
| 34 | 11 | | 23 | 10 | | 24 |
| 35 | 12 | | 23 | 10 | | 25 |
| 36 | 12 | | 24 | 10 | | 26 |
| 37 | 13 | | 24 | 11 | | 26 |
| 38 | 13 | | 25 | 11 | | 27 |
| 39 | 13 | | 26 | 12 | | 27 |
| 40 | 14 | | 26 | 12 | | 28 |
| 41 | 14 | | 27 | 12 | | 29 |
| 42 | 15 | | 27 | 13 | | 29 |
| 43 | 15 | | 28 | 13 | | 30 |
| 44 | 16 | | 28 | 14 | | 30 |
| 45 | 16 | | 29 | 14 | | 31 |
| 46 | 16 | | 30 | 14 | | 32 |
| 47 | 17 | | 30 | 15 | | 32 |
| 48 | 17 | | 31 | 15 | | 33 |
| 49 | 18 | | 31 | 16 | | 33 |
| 50 | 18 | | 32 | 16 | | 34 |
| 51 | 19 | | 32 | 16 | | 35 |
| 52 | 19 | | 33 | 17 | | 35 |
| 53 | 19 | | 34 | 17 | | 36 |
| 54 | 20 | | 34 | 18 | | 36 |
| 55 | 20 | | 35 | 18 | | 37 |
| 56 | 21 | | 35 | 18 | | 38 |
| 57 | 21 | | 36 | 19 | | 38 |
| 58 | 22 | | 36 | 19 | | 39 |
| 59 | 22 | | 37 | 20 | | 39 |
| 60 | 22 | | 38 | 20 | | 40 |
| 61 | 23 | | 38 | 21 | | 40 |
| 62 | 23 | | 39 | 21 | | 41 |
| 63 | 24 | | 39 | 21 | | 42 |
| 64 | 24 | | 40 | 22 | | 42 |
| 65 | 25 | | 40 | 22 | | 43 |
| 66 | 25 | | 41 | 23 | | 43 |
| 67 | 26 | | 41 | 23 | | 44 |
| 68 | 26 | | 42 | 23 | | 45 |
| 69 | 26 | | 43 | 24 | | 45 |
| 70 | 27 | | 43 | 24 | | 46 |
| 71 | 27 | | 44 | 25 | | 46 |
| 72 | 28 | | 44 | 25 | | 47 |
| 73 | 28 | | 45 | 26 | | 47 |
| 74 | 29 | | 45 | 26 | | 48 |
| 75 | 29 | | 46 | 26 | | 49 |
| 76 | 29 | | 47 | 27 | | 49 |

(Continued)

## TABLE D.13 Ranks for the Sign Test—cont'd

| Modified Sample Size, m | 5% Test Level | | 1% Test Level | |
|---|---|---|---|---|
| | Sign Test Is Significant If Number Is Either | | Sign Test Is Significant If Number Is Either | |
| | Less than or | More than | Less than or | More than |
| 77 | 30 | 47 | 27 | 50 |
| 78 | 30 | 48 | 28 | 50 |
| 79 | 31 | 48 | 28 | 51 |
| 80 | 31 | 49 | 29 | 51 |
| 81 | 32 | 49 | 29 | 52 |
| 82 | 32 | 50 | 29 | 53 |
| 83 | 33 | 50 | 30 | 53 |
| 84 | 33 | 51 | 30 | 54 |
| 85 | 33 | 52 | 31 | 54 |
| 86 | 34 | 52 | 31 | 55 |
| 87 | 34 | 53 | 32 | 55 |
| 88 | 35 | 53 | 32 | 56 |
| 89 | 35 | 54 | 32 | 57 |
| 90 | 36 | 54 | 33 | 57 |
| 91 | 36 | 55 | 33 | 58 |
| 92 | 37 | 55 | 34 | 58 |
| 93 | 37 | 56 | 34 | 59 |
| 94 | 38 | 56 | 35 | 59 |
| 95 | 38 | 57 | 35 | 60 |
| 96 | 38 | 58 | 35 | 61 |
| 97 | 39 | 58 | 36 | 61 |
| 98 | 39 | 59 | 36 | 62 |
| 99 | 40 | 59 | 37 | 62 |
| 100 | 40 | 60 | 37 | 63 |

## TABLE D.14 Critical Values for Chi-Squared Tests

| Degrees of Freedom | 10% Level | 5% Level | 1% Level | 0.1% Level |
|---|---|---|---|---|
| 1 | 2.706 | 3.841 | 6.635 | 10.828 |
| 2 | 4.605 | 5.991 | 9.210 | 13.816 |
| 3 | 6.251 | 7.815 | 11.345 | 16.266 |
| 4 | 7.779 | 9.488 | 13.277 | 18.467 |
| 5 | 9.236 | 11.071 | 15.086 | 20.515 |
| 6 | 10.645 | 12.592 | 16.812 | 22.458 |
| 7 | 12.017 | 14.067 | 18.475 | 24.322 |
| 8 | 13.362 | 15.507 | 20.090 | 26.124 |
| 9 | 14.684 | 16.919 | 21.666 | 27.877 |
| 10 | 15.987 | 18.307 | 23.209 | 29.588 |
| 11 | 17.275 | 19.675 | 24.725 | 31.264 |
| 12 | 18.549 | 21.026 | 26.217 | 32.909 |
| 13 | 19.812 | 22.362 | 27.688 | 34.528 |
| 14 | 21.064 | 23.685 | 29.141 | 36.123 |
| 15 | 22.307 | 24.996 | 30.578 | 37.697 |
| 16 | 23.542 | 26.296 | 32.000 | 39.252 |
| 17 | 24.769 | 27.587 | 33.409 | 40.790 |
| 18 | 25.989 | 28.869 | 34.805 | 42.312 |
| 19 | 27.204 | 30.144 | 36.191 | 43.820 |
| 20 | 28.412 | 31.410 | 37.566 | 45.315 |
| 21 | 29.615 | 32.671 | 38.932 | 46.797 |
| 22 | 30.813 | 33.924 | 40.289 | 48.268 |
| 23 | 32.007 | 35.172 | 41.638 | 49.728 |
| 24 | 33.196 | 36.415 | 42.980 | 51.179 |
| 25 | 34.382 | 37.652 | 44.314 | 52.620 |
| 26 | 35.563 | 38.885 | 45.642 | 54.052 |
| 27 | 36.741 | 40.113 | 46.963 | 55.476 |
| 28 | 37.916 | 41.337 | 48.278 | 56.892 |
| 29 | 39.087 | 42.557 | 49.588 | 58.301 |
| 30 | 40.256 | 43.773 | 50.892 | 59.703 |
| 31 | 41.422 | 44.985 | 52.191 | 61.098 |
| 32 | 42.585 | 46.194 | 53.486 | 62.487 |
| 33 | 43.745 | 47.400 | 54.776 | 63.870 |
| 34 | 44.903 | 48.602 | 56.061 | 65.247 |
| 35 | 46.059 | 49.802 | 57.342 | 66.619 |
| 36 | 47.212 | 50.998 | 58.619 | 67.985 |
| 37 | 48.363 | 52.192 | 59.893 | 69.346 |
| 38 | 49.513 | 53.384 | 61.162 | 70.703 |
| 39 | 50.660 | 54.572 | 62.428 | 72.055 |
| 40 | 51.805 | 55.758 | 63.691 | 73.402 |

**TABLE D.14** Critical Values for Chi-Squared Tests—cont'd

| Degrees of Freedom | 10% Level | 5% Level | 1% Level | 0.1% Level |
|---|---|---|---|---|
| 41 | 52.949 | 56.942 | 64.950 | 74.745 |
| 42 | 54.090 | 58.124 | 66.206 | 76.084 |
| 43 | 55.230 | 59.304 | 67.459 | 77.419 |
| 44 | 56.369 | 60.481 | 68.710 | 78.749 |
| 45 | 57.505 | 61.656 | 69.957 | 80.077 |
| 46 | 58.641 | 62.830 | 71.201 | 81.400 |
| 47 | 59.774 | 64.001 | 72.443 | 82.720 |
| 48 | 60.907 | 65.171 | 73.683 | 84.037 |
| 49 | 62.038 | 66.339 | 74.919 | 85.351 |
| 50 | 63.167 | 67.505 | 76.154 | 86.661 |
| 51 | 64.295 | 68.669 | 77.386 | 87.968 |
| 52 | 65.422 | 69.832 | 78.616 | 89.272 |
| 53 | 66.548 | 70.993 | 79.843 | 90.573 |
| 54 | 67.673 | 72.153 | 81.069 | 91.872 |
| 55 | 68.796 | 73.311 | 82.292 | 93.167 |
| 56 | 69.919 | 74.468 | 83.513 | 94.461 |
| 57 | 71.040 | 75.624 | 84.733 | 95.751 |
| 58 | 72.160 | 76.778 | 85.950 | 97.039 |
| 59 | 73.279 | 77.931 | 87.166 | 98.324 |
| 60 | 74.397 | 79.082 | 88.379 | 99.607 |
| 61 | 75.514 | 80.232 | 89.591 | 100.888 |
| 62 | 76.630 | 81.381 | 90.802 | 102.166 |
| 63 | 77.745 | 82.529 | 92.010 | 103.442 |
| 64 | 78.860 | 83.675 | 93.217 | 104.716 |
| 65 | 79.973 | 84.821 | 94.422 | 105.988 |
| 66 | 81.085 | 85.965 | 95.626 | 107.258 |
| 67 | 82.197 | 87.108 | 96.828 | 108.526 |
| 68 | 83.308 | 88.250 | 98.028 | 109.791 |
| 69 | 84.418 | 89.391 | 99.228 | 111.055 |
| 70 | 85.527 | 90.531 | 100.425 | 112.317 |
| 71 | 86.635 | 91.670 | 101.621 | 113.577 |
| 72 | 87.743 | 92.808 | 102.816 | 114.835 |
| 73 | 88.850 | 93.945 | 104.010 | 116.091 |
| 74 | 89.956 | 95.081 | 105.202 | 117.346 |
| 75 | 91.061 | 96.217 | 106.393 | 118.599 |
| 76 | 92.166 | 97.351 | 107.583 | 119.850 |
| 77 | 93.270 | 98.484 | 108.771 | 121.100 |
| 78 | 94.374 | 99.617 | 109.958 | 122.348 |
| 79 | 95.476 | 100.749 | 111.144 | 123.594 |
| 80 | 96.578 | 101.879 | 112.329 | 124.839 |
| 81 | 97.680 | 103.010 | 113.512 | 126.083 |
| 82 | 98.780 | 104.139 | 114.695 | 127.324 |
| 83 | 99.880 | 105.267 | 115.876 | 127.565 |
| 84 | 100.980 | 106.395 | 117.057 | 129.804 |
| 85 | 102.079 | 107.522 | 118.236 | 131.041 |
| 86 | 103.177 | 108.648 | 119.414 | 132.277 |
| 87 | 104.275 | 109.773 | 120.591 | 133.512 |
| 88 | 105.372 | 110.898 | 121.767 | 134.745 |
| 89 | 106.469 | 112.022 | 122.942 | 135.978 |
| 90 | 107.565 | 113.145 | 124.116 | 137.208 |
| 91 | 108.661 | 114.268 | 125.289 | 138.438 |
| 92 | 109.756 | 115.390 | 126.462 | 139.666 |
| 93 | 110.850 | 116.511 | 127.633 | 140.893 |
| 94 | 111.944 | 117.632 | 128.803 | 142.119 |
| 95 | 113.038 | 118.752 | 129.973 | 143.344 |
| 96 | 114.131 | 119.871 | 131.141 | 144.567 |
| 97 | 115.223 | 120.990 | 132.309 | 145.789 |
| 98 | 116.315 | 122.108 | 133.476 | 147.010 |
| 99 | 117.407 | 123.225 | 134.642 | 148.230 |
| 100 | 118.498 | 124.342 | 135.807 | 149.449 |

**TABLE D.15** Multipliers to Use for Constructing $\overline{X}$ and $R$ Charts

| Sample Size | Charts for Averages ($\overline{X}$ Chart): Factors for Control Limits | | Charts for Ranges ($R$ Chart) Factor for Central Line | Factors for Control Limits | | | |
|---|---|---|---|---|---|---|---|
| $n$ | $A$ | $A_2$ | $d_2$ | $D_1$ | $D_2$ | $D_3$ | $D_4$ |
| 2 | 2.121 | 1.880 | 1.128 | 0.000 | 3.686 | 0.000 | 3.267 |
| 3 | 1.732 | 1.023 | 1.693 | 0.000 | 4.358 | 0.000 | 2.574 |
| 4 | 1.500 | 0.729 | 2.059 | 0.000 | 4.698 | 0.000 | 2.282 |
| 5 | 1.342 | 0.577 | 2.326 | 0.000 | 4.918 | 0.000 | 2.114 |
| 6 | 1.225 | 0.483 | 2.534 | 0.000 | 5.078 | 0.000 | 2.004 |
| 7 | 1.134 | 0.419 | 2.704 | 0.204 | 5.204 | 0.076 | 1.924 |
| 8 | 1.061 | 0.373 | 2.847 | 0.388 | 5.306 | 0.136 | 1.864 |
| 9 | 1.000 | 0.337 | 2.970 | 0.547 | 5.393 | 0.184 | 1.816 |
| 10 | 0.949 | 0.308 | 3.078 | 0.687 | 5.469 | 0.223 | 1.777 |
| 11 | 0.905 | 0.285 | 3.173 | 0.811 | 5.535 | 0.256 | 1.744 |
| 12 | 0.866 | 0.266 | 3.258 | 0.922 | 5.594 | 0.283 | 1.717 |
| 13 | 0.832 | 0.249 | 3.336 | 1.025 | 5.647 | 0.307 | 1.693 |
| 14 | 0.802 | 0.235 | 3.407 | 1.118 | 5.696 | 0.328 | 1.672 |
| 15 | 0.775 | 0.223 | 3.472 | 1.203 | 5.741 | 0.347 | 1.653 |
| 16 | 0.750 | 0.212 | 3.532 | 1.282 | 5.782 | 0.363 | 1.637 |
| 17 | 0.728 | 0.203 | 3.588 | 1.356 | 5.820 | 0.378 | 1.622 |
| 18 | 0.707 | 0.194 | 3.640 | 1.424 | 5.856 | 0.391 | 1.608 |
| 19 | 0.688 | 0.187 | 3.689 | 1.487 | 5.891 | 0.403 | 1.597 |
| 20 | 0.671 | 0.180 | 3.735 | 1.549 | 5.921 | 0.415 | 1.585 |
| 21 | 0.655 | 0.173 | 3.778 | 1.605 | 5.951 | 0.425 | 1.575 |
| 22 | 0.640 | 0.167 | 3.819 | 1.659 | 5.979 | 0.434 | 1.566 |
| 23 | 0.626 | 0.162 | 3.858 | 1.710 | 6.006 | 0.443 | 1.557 |
| 24 | 0.612 | 0.157 | 3.895 | 1.759 | 6.031 | 0.451 | 1.548 |
| 25 | 0.600 | 0.153 | 3.931 | 1.806 | 6.056 | 0.459 | 1.541 |

**Source:** These values are from ASTM-STP 15D, American Society for Testing and Materials.

# A

**adjusted standard error**, *203* Used when the population is small, so that the sample is an important fraction of the population, and found by applying the finite-population correction factor to the standard error formula:

$$(\text{Finite-population correction factor}) \times (\text{Standard error})$$

$$= \sqrt{\frac{N-n}{N}} \times S_{\bar{X}}$$

$$= \sqrt{\frac{N-n}{N}} \times \frac{S}{\sqrt{n}}$$

**analysis and methods**, *419* The section of a report that lets you interpret the data by presenting graphic displays, summaries, and results, explaining as you go along.

**analysis of variance (ANOVA)**, *467* A general framework for statistical hypothesis testing based on careful examination of the different sources of variability in a complex situation.

**appendix**, *421* The section of a report that contains all supporting material important enough to include but not important enough to appear in the text of the report.

**assignable cause of variation**, *525* The basis for *why* a problem occurred, anytime this can be reasonably determined.

**assumptions for hypothesis testing**, *259* (1) The data set is a random sample from the population of interest, and (2) the quantity being measured is approximately normal.

**assumptions for the confidence interval**, *230* (1) The data are a random sample from the population of interest, and (2) the quantity being measured is approximately normal.

**autoregressive (AR) process**, *448* A time-series process in which each observation consists of a linear function of the previous observation plus independent random noise.

**autoregressive integrated moving-average (ARIMA) process**, *454* A time-series process in which the changes or differences are generated by an autoregressive moving-average (ARMA) process.

**autoregressive moving-average (ARMA) process**, *451* A time-series process in which each observation consists of a linear function of the previous observation, plus independent random noise, minus a fraction of the previous random noise.

**average**, *66* The most common method for finding a typical value for a list of numbers, found by adding up all the values and then dividing by the number of items; also called the *mean*.

# B

**Bayesian analysis**, *132* Statistical methods involving the use of subjective probabilities in a formal, mathematical way.

**between-sample variability**, *471* Used for ANOVA, a measure of how different the sample averages are from one another.

**biased sample**, *190* A sample that is not representative in an important way.

**bimodal distribution**, *47* A distribution with two clear and separate groups in a histogram.

**binomial distribution**, *159* The distribution of a random variable $X$ that represents the number of occurrences of an event out of $n$ trials, provided (1) for each of the $n$ trials, the event always has the same probability $\pi$ of happening, and (2) the trials are independent of one another.

**binomial proportion**, *159* The proportion $p = X/n$, which also represents a percentage.

**bivariate data**, *20* Data sets that have exactly two pieces of information recorded for each item.

**bivariate outlier**, *308* A data point in a scatterplot that does not fit the relationship of the rest of the data.

**Box-Jenkins ARIMA process**, *448* One of a family of linear statistical models based on the normal distribution that have the flexibility to imitate the behavior of many different real time series by combining autoregressive (AR) processes, integrated (I) processes, and moving-average (MA) processes.

**box plot**, *77* A plot displaying the five-number summary as a graph.

# C

**census**, *191* A sample that includes the entire population, so that $n = N$.

**central limit theorem**, *197* A rule stating that for a random sample of $n$ observations from a population, (1) the sampling distributions become more and more normal as $n$ gets large, for both the average and the sum, and (2) the means and standard deviations of the distributions of the average and the sum are as follows, where $\mu$ is the mean of the individuals and $\sigma$ is the standard deviation of these individuals:

$$\mu_{\bar{X}} = \mu \qquad \mu_{\text{sum}} = n\mu$$
$$\sigma_{\bar{X}} = \frac{\sigma}{\sqrt{n}} \qquad \sigma_{\text{sum}} = \sigma\sqrt{n}$$

**chi-squared statistic**, *507* A measure of the difference between the actual counts and the expected counts (assuming validity of the null hypothesis).

**chi-squared test for equality of percentages**, *509* A test used to determine whether a table of observed counts or percentages (summarizing a single qualitative variable) could reasonably have come from a population with known percentages (the reference values).

**chi-squared test for independence**, *512* A test used to determine whether or not two qualitative variables are independent, based on a table of observed counts from a bivariate qualitative data set.

**chi-squared tests**, *507* Tests that provide hypothesis tests for qualitative data, where you have categories instead of numbers.

**clustering**, *307* Behavior that occurs in bivariate data when there are separate, distinct groups in the scatterplot; you may wish to analyze each group separately.

**coefficient of determination, $R^2$ (bivariate data)**, *317* The square of the correlation, an indication of what percentage of the variability of $Y$ is explained by $X$.

**coefficient of determination, $R^2$ (multiple regression)**, *349* A measure that indicates the percentage of the variation in $Y$ that is explained by or attributed to the $X$ variables.

**coefficient of variation**, *108* The standard deviation divided by the average, summarizing the relative variability in the data as a percentage of the average.

**complement (*not*)**, *133* An alternative event that happens only when an event does *not* happen.

**conclusion and summary**, *420* The section of a report that moves back to the "big picture" to give closure, pulling together all of the important thoughts you would like your readers to remember.

**conditional population percentages for two qualitative variables**, *512* The probabilities of occurrence for one variable when you restrict attention to just one category of the other variable.

**conditional probability**, *136* The probability of an event, revised to reflect information that another event has occurred (the probability of event A *given* event B).

**confidence interval**, *220* An interval computed from the data that has a known probability of including the (unknown) population parameter of interest.

**confidence level**, *220* The probability of including the population parameter within the confidence interval, set by tradition at 95%, although levels of 90%, 99%, and 99.9% are also used.

**constant term, *a* (bivariate data)**, *313* The intercept of the least-squares line.

**continuous quantitative variable**, *21* Any quantitative variable that is not discrete, that is, not restricted to a simple list of possible values.

**continuous random variable**, *156* A random variable for which any number in a range is a possible value.

**control chart**, *526* A display of successive measurements of a process together with a center line and control limits, which are computed to help you decide whether or not the process is in control.

**correlation coefficient, *r***, *295* A pure number between –1 and 1 summarizing the strength of the linear relationship.

**correlation matrix**, *372* A table giving the correlation between every pair of variables in your multivariate data set.

**covariance of *X* and *Y***, *297* The numerator in the formula for the correlation coefficient.

**critical *t* value**, *257* A *t* value found in the *t* table and used for the *t* test.

**critical value**, *257* The appropriate value from a standard statistical table against which the test statistic is compared.

**cross-sectional data**, *23* A data set for which the order of recording is not relevant.

**cumulative distribution function**, *80* A plot of the data specifically designed to display the percentiles by plotting the percentages against the data values.

**cyclic component**, *436* The medium-term component of a time series, consisting of the gradual ups and downs that do not repeat each year.

## D

**data mining**, *8* A collection of methods for obtaining useful knowledge by analyzing large amounts of data, often by searching for hidden patterns.

**data set**, *19* A set consisting of some basic measurement or measurements for each item, with the same piece or pieces of information recorded for each one.

**degrees of freedom**, *222* The number of independent pieces of information in the standard error.

**dependent events**, *138* Events for which information about one of them changes your assessment of the probability of the other.

**designing the study**, *5* The phase that involves planning the details of data gathering, perhaps using a random sample from a larger population.

**detailed box plot**, *77* A display of the box plot with outliers, which are labeled separately, along with the most extreme observations that are not outliers.

**deviation**, *96* The distance from a data value to the average.

**diagnostic plot (multiple regression)**, *378* A scatterplot of the prediction errors (residuals) against the predicted values, used to decide whether you have any problems in your data that need to be fixed.

**discrete quantitative variable**, *21* A variable that can take on values only from a list of possible numbers (such as 0 or 1, or the list 0, 1, 2, 3, …).

**discrete random variable**, *156* A random variable for which you can list all possible outcomes.

**dispersion**, *95* Variability, or the extent to which data values differ from one another.

**diversity**, *95* Variability, or the extent to which data values differ from one another.

## E

**efficient test**, *492* A test that makes better use of the information in the data compared to some other test.

**elasticity of *Y* with respect to $X_i$**, *386* The expected percentage change in $Y$ associated with a 1% increase in $X_i$, holding the other $X$ variables fixed; estimated using the regression coefficient from a regression analysis using the logarithms of both $Y$ and $X_i$.

**elementary units**, *19* The individual items or things (such as people, households, firms, cities, TV sets) whose measurements make up a data set.

**error of estimation**, *191* The estimator (or estimate) minus the population parameter; it is usually unknown.

**estimate**, *191* The actual number computed from the data.

**estimating an unknown quantity**, *6* The best educated guess possible based on the available data.

**estimator**, *191* A sample statistic used as a guess for the value of a population parameter.

**event**, *128* Any collection of outcomes specified in advance, before the random experiment is run; it either happens or does not happen for each run of the experiment.

**executive summary**, *418* A paragraph at the very beginning of a report that describes the most important facts and conclusions from your work.

**expected value or mean of a random variable**, *156* The typical or average value for a random variable.

**exploring the data**, *5* Looking at your data set from many angles, describing it, and summarizing it.

**exponential distribution**, *177* A very skewed continuous distribution useful for understanding waiting times and the duration of telephone calls, for example.

**extrapolation**, *327* Predicting beyond the range of the data; it is especially risky because you cannot protect yourself by exploring the data.

**extremes**, *76* The smallest and largest values, which are often of interest.

# F

**F statistic**, *468* The basis of the *F* test in the analysis of variance, it is a ratio of two variance measures used to perform each hypothesis test.

**F table**, *472* A list of critical values for the distribution of the *F* statistic when the null hypothesis is true, so that the *F* statistic exceeds the critical value a controlled percentage of the time (5%, for example) when the null hypothesis is true.

**F test (ANOVA)**, *468* A test based on the *F* statistic and used in the analysis of variance to perform each hypothesis test.

**F test (multiple regression)**, *349* An overall test to see whether or not the *X* variables explain a significant amount of the variation in *Y*.

**false alarm rate, for quality control**, *527* How often you decide to fix the process when there really isn't any problem.

**finite-population correction factor**, *203* The factor used to reduce the standard error formula when the population is small, so that the sample is an important fraction of the population adjusted standard error is:

(Finite-population correction factor) × (Standard error)

$$= \sqrt{\frac{N-n}{N}} \times S_{\overline{X}}$$

$$= \sqrt{\frac{N-n}{N}} \times \frac{S}{\sqrt{n}}$$

**five-number summary**, *76* A list of special, landmark summaries of a data set: the smallest, lower quartile, median, upper quartile, and largest.

**forecast, for time series**, *430* The expected (i.e., mean) value of the future behavior of the estimated model.

**forecast limits**, *430* The confidence limits for your forecast (if the model can produce them); if the model is correct for your data, then the future observation has a 95% probability, for example, of being within these limits.

**frame**, *191* A scheme that tells you how to gain access to the population units by number from 1 to the population size, *N*.

**frequentist (non-Bayesian) analysis**, *146* An analysis that does not use subjective probabilities in its computations, although it is not totally objective since opinions will have some effect on the choice of data and model (the mathematical framework).

# G

**grand average, for ANOVA**, *471* The average of all of the data values from all of the samples combined.

# H

**histogram**, *37* A display of frequencies as a bar chart rising above the number line, indicating how often the various values occur in the data set.

**hypothesis**, *250* A statement about the population that may be either right or wrong; the data will help you decide which one (of two hypotheses) to accept as true.

**hypothesis testing**, *7* The use of data to decide between two (or more) different possibilities in order to resolve an issue in an ambiguous situation; it is often used to distinguish structure from mere randomness and should be viewed as a helpful input to executive decision making.

# I

**idealized population**, *204* The much larger, sometimes imaginary, population that your sample represents.

**independence of two qualitative variables**, *512* A lack of relationship between two qualitative variables where knowledge about the value of one does not help you predict or explain the other; that is, the probabilities for one variable are the same as the conditional probabilities *given* the other variable.

**independent events**, *138* Two events for which information about one does not change your assessment of the probability of the other.

**indicator variable**, *392* Also called a *dummy variable*, a quantitative variable that takes on only the values 0 and 1 and is used to represent qualitative categorical data as an explanatory *X* variable.

**interaction, for multiple regression**, *389* A relationship between two *X* variables and *Y* in which a change in both of them causes an expected shift in *Y* that is different from the sum of the shifts in *Y* obtained by changing each *X* individually.

**intercept, *a***, *312* The predicted value for *Y* when *X* is 0.

**intercept or constant term, for multiple regression**, *348* The predicted value for *Y* when all *X* variables are 0.

**intersection (*and*)**, *134* An event that happens whenever one event *and* another event both happen as a result of a single run of the random experiment.

**introduction**, *419* Several paragraphs near the beginning of a report in which you describe the background, the questions of interest, and the data set you have worked with.

**irregular component**, *436* The short-term, random component of a time series representing the leftover, residual variation that can't be explained.

# J

**joint probability table**, *145* A table listing the probabilities for two events, their complements, and combinations using *and*.

# L

**law of large numbers**, *130* A rule stating that the relative frequency (a random number) will be close to the probability (an exact, fixed number) if the experiment is run many times.

**least-significant-difference test, for one-way ANOVA**, *479* Used only if the *F* test finds significance, a test to compare each pair of samples to see which ones are significantly different from each other.

**least-squares line, *Y = a + bX***, *312* The line with the smallest sum of squared vertical prediction errors of all possible lines, used as the best predictive line based on the data.

**linear model**, *318* A model specifying that the observed value for *Y* is equal to the population relationship plus a random error that has a normal distribution.

**linear regression analysis, for bivariate data**, *310* An analysis that predicts or explains one variable from the other using a straight line.

**linear relationship, in bivariate data**, *297* A scatterplot showing points bunched randomly around a straight line with constant scatter.

**list of numbers**, *35* The simplest kind of data set, representing some kind of information (a single statistical variable) measured on each item of interest (each elementary unit).

**logarithm**, *45* A transformation that is often used to change skewness into symmetry because it stretches the scale near zero, spreading out all of the small values that had been bunched together.

# M

**Mann-Whitney *U* test**, *497* A nonparametric test for two unpaired samples.

**margin of error**, *222* The distance, generally *t* times the standard error, that the confidence interval extends in each direction, indicating how far away from the estimate we could reasonably find the population parameter.

**mean**, *66* The most common method for finding a typical value for a list of numbers, found by adding up all the values and then dividing by the number of items; also called the *average*.

**mean or expected value of a random variable**, *156* The typical or average value for a random variable.

**median**, *70* The middle value, with half of the data items larger and half smaller.

**mode**, *73* The most common category; the value listed most often in the data set.

**model**, *6* A system of assumptions and equations that can generate artificial data similar to the data you are interested in, so that you can work with a single number (called a "parameter") for each important aspect of the data.

**model, mathematical model, or process (time series)**, *430* A system of equations that can produce an assortment of different artificial time-series data sets.

**model misspecification**, *371* The many different potential incompatibilities between your application and the chosen model, such as the multiple regression linear model. By exploring the data, you can be alerted to some of the potential problems with nonlinearity, unequal variability, or outliers. However, you may or may not have a problem: Even though the histograms of some variables may be skewed and even though some scatterplots may be nonlinear, the multiple regression linear model might still hold. The diagnostic plot can help you decide when the problem is serious enough to need fixing.

**modified sample size, for the sign test**, *492* The number $m$ of data values that are different from the reference value, $\theta_0$.

**moving average**, *437* A new time series created by averaging nearby observations.

**moving-average (MA) process**, *450* A process in which each observation consists of a constant, $\mu$ (the long-term mean of the process), plus independent random noise, minus a fraction of the previous random noise.

**multicollinearity**, *371* A problem that arises in multiple regression when some of your explanatory ($X$) variables are too similar. The individual regression coefficients are poorly estimated because there is not enough information to decide which one (or more) of the variables is doing the explaining.

**multiple regression**, *347* Predicting or explaining a single $Y$ variable from two or more $X$ variables.

**multiple regression linear model**, *357* A model specifying that the observed value for $Y$ is equal to the population relationship plus independent random errors that have a normal distribution.

**multivariate data**, *20* Data sets that have three or more pieces of information recorded for each item.

**mutually exclusive events**, *134* Two events that cannot both happen at once.

## N

**nominal data**, *23* Categories of a qualitative variable that do not have a natural, meaningful order.

**nonlinear relationship, in bivariate data**, *301* A relationship revealed by a plot in which the points bunch around a curved rather than a straight line.

**nonparametric methods**, *491* Statistical procedures for hypothesis testing that do not require a normal distribution (or any other particular shape of distribution) because they are based on counts or ranks instead of the actual data values.

**nonstationary process**, *453* A process that, over time, tends to move farther and farther away from where it was.

**no relationship, in bivariate data**, *299* Lack of relationship, indicated by a scatterplot that is just random, tilting neither upward nor downward as you move from left to right.

**normal distribution (data)**, *40* A particular, idealized, smooth bell-shaped histogram with all of the randomness removed.

**normal distribution (random variable)**, *165* A continuous distribution represented by the familiar bell-shaped curve.

**null hypothesis**, *250* Denoted $H_0$, the default hypothesis, often indicating a very specific case, such as pure randomness.

**number line**, *36* A straight line, usually horizontal, with the scale indicated by numbers below it.

## O

**observation of a random variable**, *155* The actual value assumed by a random variable.

**one-sided confidence interval**, *235* Specification of an interval with known confidence such that the population mean is either no less than or no larger than some computed number.

**one-sided $t$ test**, *262* A $t$ test set up with the null hypothesis claiming that $\mu$ is on one side of $\mu_0$ and the research hypothesis claiming that it is on the other side.

**one-way analysis of variance**, *468* A method used to test whether or not the averages from several different situations are significantly different from one another.

**ordinal data**, *22* Categories of a qualitative variable that have a natural, meaningful order.

**outcome**, *128* The result of a run of a random experiment, describing and summarizing the observable consequences.

**outlier**, *49* A data value that doesn't seem to belong with the others because it is either far too big or far too small.

## P

**$p$-value**, *261* A value that tells you how surprised you would be to learn that the null hypothesis had produced the data, with smaller $p$-values indicating more surprise and leading to rejection of $H_0$. By convention, we reject $H_0$ whenever the $p$-value is less than 0.05.

**paired $t$ test**, *268* A test of whether or not two samples have the same population mean value when there is a natural pairing between the two samples (for example, "before" and "after" measurements on the same people).

**parameter**, *191* A population parameter is any number computed for the entire population.

**parametric methods**, *491* Statistical procedures that require a completely specified model.

**Pareto diagram**, *525* A display of the causes of the various defects, in order from most to least frequent, so that you can focus attention on the most important problems.

**parsimonious model, for time series**, *448* A model that uses just a few estimated parameters to describe the complex behavior of a time series.

**percentage chart, for quality control**, *534* A display of the percent defective, together with the appropriate center line and control limits, so that you can monitor the rate at which the process produces defective items.

**percentile**, *76* Summary measures expressing ranks as percentages from 0 to 100, rather than from 1 to $n$, so that the 0th percentile is the smallest number, the 100th percentile is the largest, the 50th percentile is the median, and so on.

**pilot study**, *196* A small-scale version of a study, designed to help you identify problems and fix them before the real study is run.

**Poisson distribution**, *176* The distribution of a discrete random variable for which occurrences happen independently and randomly over time and the average rate of occurrence is constant over time.

**polynomial regression**, *388* A way to deal with nonlinearity in which $Y$ is predicted or explained using a single $X$ variable together with some of its powers ($X^2$, $X^3$, etc.).

**population**, *190* The collection of units (people, objects, or whatever) that you are interested in knowing about.

**population parameter**, *191* Any number computed for the entire population.

**population standard deviation**, *106* Denoted by $\sigma$, a variability measure for the entire population.

**predicted value, for bivariate data**, *313* The prediction (or explanation) for $Y$ given a value of $X$, found by substituting the value of $X$ into the least-squares line.

**prediction equation or regression equation, for multiple regression**, *348* Predicted $Y = a + b_1X_1 + b_2X_2 + \cdots + b_kX_k$, which may be used for prediction or control.

**prediction errors or residuals, for multiple regression**, *349* The differences between actual and predicted $Y$, given by $Y - $ (Predicted $Y$).

**prediction interval**, *237* An interval that allows you to use data from a sample to predict a new observation, with known probability, provided you obtain this additional observation in the same way as you obtained your data.

**primary data**, *24* Data obtained when you control the design of the data-collection plan (even if the work is done by others).

**probability**, *14* The likelihood of each of the various potential future events, based on a set of assumptions about how the world works.

**probability distribution**, *155* The pattern of probabilities for a random variable.

**probability of an event**, *129* A number between 0 and 1 that expresses how likely it is that the event will happen each time the random experiment is run.

**probability tree**, *139* A picture indicating probabilities and some conditional probabilities for combinations of two or more events.

**process, for quality control**, *524* Any business activity that takes inputs and transforms them into outputs.

**pure integrated (I) process**, *453* A time-series process in which each additional observation consists of a random step away from the current observation; also called a *random walk*.

## Q

**qualitative variable**, *22* A variable that indicates which of several nonnumerical categories an item falls into.

**quantitative variable**, *21* A variable whose data are recorded as meaningful numbers.

**quartiles**, *76* The 25th and 75th percentiles.

## R

**$R^2$**, *317* See *coefficient of determination*.

**R chart, for quality control**, *528* A display of the range (the largest value minus the smallest) for each sample, together with center line and control limits, so that you can monitor the variability of the process.

**random causes of variation**, *525* All causes of variation that would not be worth the effort to identify.

**random experiment**, *127* Any well-defined procedure that produces an observable outcome that could not be perfectly predicted in advance.

**random noise process**, *448* A random sample (independent observations) from a normal distribution with constant mean and standard deviation.

**random sample or simple random sample**, *192* A sample selected so that (1) each population unit has an equal probability of being chosen, and (2) units are chosen independently, without regard to one another.

**random variable**, *155* A specification or description of a numerical result from a random experiment.

**random walk**, *453* An observation of a pure integrated (I) time-series process that consists of a random step away from the previous observation.

**range**, *107* The largest data value minus the smallest data value, representing the size or extent of the entire data set.

**rank**, *491* Used extensively in nonparametric statistical methods, an indication of a data value's position after you have ordered the data set. Each of the numbers 1, 2, 3, ... , $n$ is associated with the data values so that, for example, the smallest has rank 1, the next smallest has rank 2, and so forth up to the largest, which has rank $n$.

**ratio-to-moving-average**, *436* A method that divides a series by a smooth moving average for the purpose of performing trend-seasonal analysis on a time series.

**reference**, *420* A note in a report indicating the kind of material you have taken from an outside source and giving enough information so that your reader can obtain a copy.

**reference value**, *252* Denoted by $\mu_0$, a known, fixed number that does not come from the sample data that the mean is tested against.

**regression analysis**, *310, 347* A type of analysis that explains one $Y$ variable from one or more other $X$ variables.

**regression coefficient, b, of Y on X, for bivariate data**, *313* The slope of the least-squares line.

**regression coefficient, for multiple regression**, *348* The coefficient $b_j$, for the $j$th $X$ variable, indicating the effect of $X_j$ on $Y$ after adjusting for the other $X$ variables; $b_j$ indicates how much larger you expect $Y$ to be for a case that is identical to another except for being one unit larger in $X_j$.

**relative frequency**, *130* For a random experiment run many times, the (random) proportion of times the event occurs out of the number of times the experiment is run.

**representative sample**, *190* A sample in which each characteristic (and combination of characteristics) arises the same percent of the time as in the population.

**research hypothesis or alternative hypothesis**, *274* Denoted by $H_1$, the hypothesis that has the burden of proof, requiring convincing evidence against $H_0$ before you will accept it.

**residual, for bivariate data**, *313* The prediction error for each of the data points that tells you how far the point is above or below the line.

## S

**sample**, *190* A smaller collection of units selected from the population.

**sample space**, *127* A list of all possible outcomes of a random experiment, prepared in advance without knowing what will happen when the experiment is run.

**sample standard deviation**, *106* A variability measure used whenever you wish to generalize beyond the immediate data set to some larger population (either real or hypothetical).

**sample statistic**, *191* Any number computed from your sample data.

**sampling distribution**, *196* The probability distribution of anything you measure, based on a random sample of data.

**sampling with replacement**, *191* A scheme in which a population unit can appear more than once in the sample.

**sampling without replacement**, *191* A scheme in which a unit cannot be selected more than once to be in the sample.

**scatterplot**, *292* A display for exploring bivariate data, $Y$ against $X$, giving a visual picture of the relationship in the data.

**seasonal adjustment**, *439* Elimination of the expected seasonal component from an observation (by dividing the series by the seasonal index for that period) so that one quarter or month may be directly compared to another to reveal the underlying trends.

**seasonal component**, *436* The exactly repeating component of a time series that indicates the effects of the time of year.

**seasonal index**, *439* An index for each time of the year, indicating how much larger or smaller this particular time period is as compared to a typical period during the year.

**secondary data**, *24* Data previously collected by others for their own purposes.

**sign test**, *492* A test used to decide whether the population median is equal to a given reference value based on the number of sample values that fall below that reference value.

**sign test for the differences**, *495* A test applied to the differences or changes when you have paired observations (before/after measurements, for example); a nonparametric procedure for testing whether the two columns are significantly different.

**skewed distribution**, *43* A distribution that is neither symmetric nor normal because the data values trail off more suddenly on one side and more gradually on the other.

**slope, *b***, *312* A measurement in units of $Y$ per unit $X$ that indicates how steeply the line rises (or falls, if $b$ is negative).

**Spread**, *95* Variability, or the extent to which data values differ from one another.

**spurious correlation**, *309* A high correlation that is actually due to some third factor.

**standard deviation (data)**, *96* The traditional choice for measuring variability, summarizing the typical distance from the average to each data value.

**standard deviation (random variable)**, *157* An indication of the risk in terms of how far from the mean you can typically expect the random variable to be.

**standard error for prediction**, *237* The uncertainty measure to use for prediction, $S\sqrt{1 + 1/n}$, a measure of variability of the distance between the sample average and the new observation.

**standard error of estimate, $S_e$**, *315* A measure of approximately how large the prediction errors (residuals) are for your data set in the same units as $Y$.

**standard error of the average**, *201* A number that indicates approximately how far the (random, observed) sample average, $\overline{X}$, is from the (fixed, unknown) population mean, $\mu$: Standard error = $S_{\overline{X}} = S/\sqrt{n}$.

**standard error of the difference**, *271* A measure that is needed to construct confidence intervals for the mean difference and to perform the hypothesis test, it gives the estimated standard deviation of the sample average difference.

**standard error of the intercept term, $S_a$, for bivariate data**, *318* A number that indicates approximately how far the estimate $a$ is from $\alpha$, the true population intercept term.

**standard error of the slope coefficient, $S_b$, for bivariate data**, *318* A number that indicates approximately how far the estimated slope, $b$ (the regression coefficient computed from the sample), is from the population slope, $\beta$, due to the randomness of sampling.

**standard error of the statistic**, *200* An estimate of the standard deviation of a statistic's sampling distribution, indicating approximately how far from its mean value (a population parameter) the statistic is.

**standard normal distribution**, *166* A normal distribution with mean $\mu = 0$ and standard deviation $\sigma = 1$.

**standard normal probability table**, *166* A table giving the probability that a standard normal random variable is less than any given number.

**standardized number**, *166* The number of standard deviations above the mean (or below the mean, if the standardized number is negative), found by subtracting the mean and dividing by the standard deviation.

**standardized regression coefficient**, *370* The coefficient $b_i S_{X_i}/S_Y$, representing the expected change in $Y$ due to a change in $X_i$, measured in units of standard deviations of $Y$ per standard deviation of $X_i$, holding all other $X$ variables constant.

**state of statistical control (in control)**, *525* The state of a process after all of the assignable causes of variation have been identified and eliminated, so that only random causes remain.

**stationary process**, *453* A process such as the autoregressive, moving-average, and ARMA models, which tend to behave similarly over long time periods, staying relatively close to their long-run mean.

**statistic**, *191* A sample statistic is any number computed from your sample data.

**statistical inference**, *220* The process of generalizing from sample data to make probability-based statements about the population.

**statistical process control**, *525* The use of statistical methods to monitor the functioning of a process so that you can adjust or fix it when necessary and leave it alone when it's working properly.

**statistical quality control**, *523* The use of statistical methods for evaluating and improving the results of any activity.

**statistically significant**, *260* A result that is significant at the 5% level ($p < 0.05$). Other terms used are *highly significant* ($p < 0.01$), *very highly significant* ($p < 0.001$), and *not significant* ($p > 0.05$).

**statistics**, *4* The art and science of collecting and understanding data.

**stem-and-leaf histogram**, *54* A histogram in which the bars are constructed by stacking numbers one on top of another.

**stratified random sample**, *205* A sample obtained by choosing a random sample separately from each of the strata (segments or groups) of the population.

**subjective probability**, *132* Anyone's opinion (use an expert, if possible) of what the probability is for an event.

**summarization**, *65* The use of one or more selected or computed values to represent the data set.

**systematic sample**, *208* A sample obtained by selecting a single random starting place in the frame and then taking units separated by a fixed, regular interval. Although the sample average from a systematic sample is an unbiased estimator of the population mean (that is, it is not regularly too high or too low), there are some serious problems with this technique.

## T

**$t$ statistic**, *257* One way to perform the $t$ test: $t = (\overline{X} - \mu_0)/S_{\overline{X}}$.

**$t$ table**, *222* The table used to adjust for the added uncertainty due to the fact that an estimator (the standard error) is being used in place of the unknown exact variability for the population.

**$t$ test or Student's $t$ test**, *274* The hypothesis test for a mean.

**$t$ tests for individual regression coefficients, for multiple regression**, *349* If the regression is significant, a method for proceeding with statistical inference for regression coefficients.

**table of contents**, *418* The section of a report that follows the executive summary, showing an outline of the report together with page numbers.

**table of random digits**, *192* A list in which the digits 0 through 9 each occur with probability 1/10, independently of each other.

**test level or significance level**, *260* The probability of wrongly accepting the research hypothesis when, in fact, the null hypothesis is true (i.e., committing a type I error). By convention, this level is set at 5%, but it may reasonably be set at 1% or 0.1% (or even 10% for some fields of study) by using the appropriate column in the $t$ table.

**test statistic**, *257* The most helpful number that can be computed from your data for the purpose of deciding between two given hypotheses.

**theoretical probability**, *131* A number computed using an exact formula based on a mathematical theory or model, such as the *equally likely* rule.

**time-series data**, *30* Data values that are recorded in a meaningful sequence.

**title page**, *418* The first page of a report, including the title of the report, the name and title of the person you prepared it for, your name and title (as the preparer), and the date.

**transformation**, *45* Replacing each data value by a different number (such as its logarithm) to facilitate statistical analysis.

**trend, for time series**, *436* The *very* long-term behavior of the time series, typically displayed as a straight line or an exponential curve.

**trend-seasonal analysis**, *436* A direct, intuitive approach to estimating the four basic components of a monthly or quarterly time series: the long-term trend, the seasonal patterns, the cyclic variation, and the irregular component.

**two-sided test**, *252* A test for which the research hypothesis allows possible population values on either side of the reference value.

**type I error**, *259* The error committed when the null hypothesis is true, but you reject it and declare that your result is statistically significant.

**type II error**, *259* The error committed when the research hypothesis is true, but you accept the null hypothesis instead and declare the result *not* to be significant.

## U

**unbiased estimator**, *191* An estimator that is correct on the average, so that it is systematically neither too high nor too low compared to the corresponding population parameter.

**Uncertainty**, *95* Variability, or the extent to which data values differ from one another.

**unconditional probability**, *136* Ordinary (unrevised) probability.

**unequal variability, in bivariate data**, *304* A problem that occurs when the variability in the vertical direction changes dramatically as you move horizontally across the scatterplot; this causes correlation and regression analysis to be unreliable. These problems may be fixed by using either transformations or a so-called weighted regression.

**union (*or*)**, *135* An event that happens whenever one event *or* an alternative event happens (or both events happen) as a result of a single run of the random experiment.

**univariate data**, *19* Data sets that have just one piece of information recorded for each item.

**unpaired $t$ test**, *270* A test to determine whether or not two samples have the same population mean value, when there is no natural pairing between the two samples (i.e., each is an independent sample from a different population).

## V

**variability, diversity, uncertainty, dispersion, or spread**, *95* The extent to which data values differ from one another.

**variable**, *19* A piece of information recorded for every item (its cost, for example).

**variable selection**, *371* The problem that arises when you have a long list of potentially useful explanatory $X$ variables and would like to decide which ones to include in the regression equation. With too many $X$ variables, the quality of your results will decline because information is being wasted in estimating unnecessary parameters. If one or more important $X$ variables are omitted, your predictions will lose quality due to missing information.

**variance**, *97* The square of the standard deviation, it provides the same information as the standard deviation but is more difficult to interpret since its units of measurement are the squared values of the original data units (such as dollars squared, squared miles per squared gallon, or squared kilograms, whatever these things are).

**Venn diagram**, *133* A picture that represents the universe of all possible outcomes (the sample space) as an outer rectangle with events indicated inside, often as circles or ovals.

## W

**weighted average**, *68* A measure similar to the average, except that it allows you to give a different importance, or *weight*, to each data item.

**Wilcoxon rank-sum test**, *497* A way to compute the result of a non-parametric test for two unpaired samples.

**within-sample variability, for ANOVA**, *472* An overall measure of how variable each sample is.

## X

**$\overline{X}$ chart, for quality control**, *528* A display of the average of each sample, together with the appropriate center line and control limits, so that you can monitor the level of the process.